HPLC FOR PHARMACEUTICAL SCIENTISTS

THE WILEY BICENTENNIAL–KNOWLEDGE FOR GENERATIONS

*E*ach generation has its unique needs and aspirations. When Charles Wiley first opened his small printing shop in lower Manhattan in 1807, it was a generation of boundless potential searching for an identity. And we were there, helping to define a new American literary tradition. Over half a century later, in the midst of the Second Industrial Revolution, it was a generation focused on building the future. Once again, we were there, supplying the critical scientific, technical, and engineering knowledge that helped frame the world. Throughout the 20th Century, and into the new millennium, nations began to reach out beyond their own borders and a new international community was born. Wiley was there, expanding its operations around the world to enable a global exchange of ideas, opinions, and know-how.

For 200 years, Wiley has been an integral part of each generation's journey, enabling the flow of information and understanding necessary to meet their needs and fulfill their aspirations. Today, bold new technologies are changing the way we live and learn. Wiley will be there, providing you the must-have knowledge you need to imagine new worlds, new possibilities, and new opportunities.

Generations come and go, but you can always count on Wiley to provide you the knowledge you need, when and where you need it!

WILLIAM J. PESCE
PRESIDENT AND CHIEF EXECUTIVE OFFICER

PETER BOOTH WILEY
CHAIRMAN OF THE BOARD

HPLC FOR PHARMACEUTICAL SCIENTISTS

Edited by

YURI KAZAKEVICH
Seton Hall University
South Orange, New Jersey

ROSARIO LoBRUTTO
Novartis Pharmaceuticals
East Hanover, New Jersey

WILEY-INTERSCIENCE
A JOHN WILEY & SONS, INC., PUBLICATION

Library of Congress Cataloging-in-Publication Data:

HPLC for pharmaceutical scientists / edited by Yuri Kazakevich, Rosario LoBrutto.
    p. cm.
  Includes bibliographical references and index.
  ISBN-13: 978-0-471-68162-5 (cloth)
  ISBN-10: 0-471-68162-8 (cloth)
  1. High performance liquid chromatography. 2. Drugs–Analysis. 3. Clinical chemistry.
  I. Kazekevich, Yuri. II. LoBrutto, Rosario.

  RS189.5.H53H75 2007
  615′.1901–dc22

                                                                      2006046395

CONTENTS

5 Normal-Phase HPLC   241

Yong Liu and Anant Vailaya

6 Size-Exclusion Chromatography   263

Yuri Kazakevich and Rosario LoBrutto

PREFACE

In the modern pharmaceutical industry, HPLC is a major analytical tool applied at all stages of drug discovery, development and production. Fast and effective development of rugged analytical HPLC methods is more efficiently undertaken with a thorough understanding of HPLC principles, theory and instrumentation. The main focus of this book is reversed-phase analysis of small molecules, although there is some attention given to LC-MS of proteins, LC-NMR, ion-exchange, size exclusion, and normal phase chromatography.

The drug discovery and development process has undergone dramatic changes particularly in the last decade. The process continues to evolve in response to new discoveries, new technologies, and increasing demand to get more drugs to the market more efficiently. Progress in drug discovery has been fueled by improvements in methodologies and technologies including automated high performance liquid chromatography (HPLC), fast HPLC, automated method development, HPLC-MS (mass spectrometry), HPLC-NMR (nuclear magnetic resonance) and high-throughput purification methods.

This book is unique in the sense that it elucidates the role of HPLC throughout the entire drug development process from drug candidate inception to marketed drug product. It is written in a manner that scientists at all levels of experience with HPLC will be able to find utility while maintaining a reasonable and manageable volume. The book covers the main theoretical and practical aspects of modern HPLC at a level that is suitable for graduate students and chromatography practitioners in industry. In addition, for the more seasoned chromatographer, a description of the specifics of HPLC applications at different stages of drug development and the latest advancements in fast, preparative, chiral and other modern LC techniques are included. The information and discussions in this book are meant to increase the chromatographer's awareness of trends in HPLC technology with emphasis on their utilization in the various aspects of drug development. Researchers are provided with the opportunity to better understand the use of HPLC not only in their respective "development silos," but also throughout their organization. Theoretical background as well as practical and pragmatic approaches and actual examples of effective development of selective and rugged HPLC methods from a physico-chemical point of view are provided throughout this book.

The contents, format and organization herein were inspired by the HPLC short courses and graduate classes we have taught on separation science to a diverse population of pharmaceutical chemist drawn from all areas of drug development. We have observed a desire for a better understanding of workflows in the various areas of drug development and how HPLC is integrated and embedded into these processes. The book is formatted to address the major functions and tasks in which HPLC is applied.

Even though, there is no "cookbook" for HPLC method development this book provides several strategies that the reader could use when presented with a particular situation. These strategies could be stored as tools in the scientists' "method development arsenal," and drawn from when needed to tackle a particular separation. Moreover, some novel approaches for implementing HPLC, fast HPLC, and hyphenated HPLC techniques towards pharmaceutical analysis are discussed. This book has the potential to serve as a useful resource for the chromatographic community. It can be used as a handbook for the novice as well as the more experienced pharmaceutical chemist who utilizes HPLC as an analytical tool to solve challenging problems regularly in the pharmaceutical industry.

The completion of this book could not have been possible without the help, inspiration and encouragement from many people. We are very grateful to our families for their understanding and support throughout the entire process of writing and editing. Also, we would like to thank our colleagues, students, friends, and peers for their helpful discussions and contributions to this work. Of special note, we would like to thank Alan Jones, Alexey Makarov, Evan O'Neill, Li Pan, Rajinder Singh, Fred Chan and Richard Vivilecchia. We express our special gratitude to Dr. Harold McNair for his kind support, guidance and mentorship over the years and his continued inspiration in our on-going endeavors. We would also like to give special thanks to our wives, Ginevra LoBrutto and Irina Kazakevich, for all their support, patience, understanding and encouragement. They have been an integral factor in allowing us to accomplish this contribution to the chromatographic and pharmaceutical community. Also, we would like to acknowledge all contributing authors who have done an excellent job in writing their respective chapters, thus allowing for facile integration of their topics into the framework of this book. We really enjoyed the many fruitful discussions with the contributors and duly acknowledge their dedication, efforts and commitment to this work. This has been truly a team effort and we believe the chromatographic community will appreciate the contents and discussions provided within this book.

CONTRIBUTORS

Ray Bakhtiar, Department of Drug Metabolism, Merck Research Laboratories, Rahway, NJ 07065

Thomas Burakowski, R & D Chemical Development, Process Development Laboratory, Boehringer Ingelheim, Ridgefield, CT 06877

Guodong Chen, Schering-Plough Research Institute, Kenilworth, NJ 07033

Maria Victoria Silva Elipe, Department of Analytical Sciences, Amgen, Inc., Thousand Oaks, CA 91320

Joseph Etse, Pharmaceutical and Analytical Research and Development, Novartis Pharmaceuticals Corporation, East Hanover, NJ 07936

Nelu Grinberg, Chemical Development, Boehringer Ingelheim Pharmaceutical, Inc., Ridgefield, CT 06877

Anton D. Jerkovich, Novartis Pharmaceuticals Corporation, East Hanover, NJ 07936

Daniel B. Kassel, Takeda Inc., San Diego, CA 92121

Irina Kazakevich, Schering-Plough Research Institute, Kenilworth, NJ 07033

Yuri Kazakevich, Department of Chemistry and Biochemistry, Seton Hall University, South Orange, NJ 07079

Ernst Kuesters, Chemical and Analytical Development, Novartis Pharma AG, CH-4002 Basel, Switzerland

Yan-Hui Liu, Schering-Plough Research Institute, Kenilworth, NJ 07033

Yong Liu, Analytical Research, Merck Research Laboratories, Rahway, NJ 07065

Rosario LoBrutto, Group Head, Pharmaceutical and Analytical Development, Novartis Pharmaceuticals, East Hanover, NJ 07936

Tapan K. Majumdar, Novartis Institute for Biomedical Research, One Health Plaza, East Hanover, NJ 07936

Michael McBrien, Advanced Chemistry Development, Inc. (ACD/Labs), Toronto, Ontario M5H 3V9 Canada

Tarun S. Patel, Pharmaceutical and Analytical Development, Novartis Pharmaceuticals, East Hanover, NJ 07936

Birendra N. Pramanik, Schering-Plough Research Institute, Kenilworth, NJ 07033

Roger M. Smith, Department of Chemistry, Loughborough University, Loughborough, Leics LE11 3TU, UK

Apryll M. Stalcup, Department of Chemistry, University of Cincinnati, Cincinnati, OH 45221

Richard Thompson, Analytical Research, Merck Research Laboratories, Rahway, NJ 07065

Francis L. S. Tse, Bioanalytics and Pharmacokinetics, Novartis Pharmaceuticals Corporation, East Hanover, NJ 07936

Anant Vailaya, Analytical Research, Merck Research Laboratories, Rahway, NJ 07065

Richard V. Vivilecchia, Novartis Pharmaceuticals Corporation, East Hanover, NJ 07936

Li-Kang Zhang, Schering-Plough Research Institute, Kenilworth, NJ 07033

PART I

HPLC THEORY AND PRACTICE

1

INTRODUCTION

Yuri Kazakevich and Rosario LoBrutto

1.1 CHROMATOGRAPHY IN THE PHARMACEUTICAL WORLD

In the modern pharmaceutical industry, high-performance liquid chromatography (HPLC) is the major and integral analytical tool applied in all stages of drug discovery, development, and production. The development of new chemical entities (NCEs) is comprised of two major activities: drug discovery and drug development. The goal of the drug discovery program is to investigate a plethora of compounds employing fast screening approaches, leading to generation of lead compounds and then narrowing the selection through targeted synthesis and selective screening (lead optimization). This lead to the final selection of the most potentially viable therapeutic candidates that are taken forward to drug development. The main functions of drug development are to completely characterize candidate compounds by performing drug metabolism, preclinical and clinical screening, and clinical trials. Concomitantly with the drug development process, the optimization of drug synthesis and formulation are performed which eventually lead to a sound and robust manufacturing process for the active pharmaceutical ingredient and drug product. Throughout this drug discovery and drug development paradigm, rugged analytical HPLC separation methods are developed and are tailored by each development group (i.e., early drug discovery, drug metabolism, pharmokinetics, process research, preformulation, and formulation). At each phase of development the analyses of a myriad of samples are performed to adequately control and monitor the quality of the prospective drug candidates, excipients, and final products. Effective and fast method development is of

HPLC for Pharmaceutical Scientists, Edited by Yuri Kazakevich and Rosario LoBrutto
Copyright © 2007 by John Wiley & Sons, Inc.

paramount importance throughout this drug development life cycle. This requires a thorough understanding of HPLC principles and theory which lay a solid foundation for appreciating the many variables that are optimized during fast and effective HPLC method development and optimization.

1.2 CHROMATOGRAPHIC PROCESS

Chromatographic separations are based on a forced transport of the liquid (mobile phase) carrying the analyte mixture through the porous media and the differences in the interactions at analytes with the surface of this porous media resulting in different migration times for a mixture components.

In the above definition the presence of two different phases is stated and consequently there is an interface between them. One of these phases provides the analyte transport and is usually referred to as the mobile phase, and the other phase is immobile and is typically referred to as the stationary phase. A mixture of components, usually called analytes, are dispersed in the mobile phase at the molecular level allowing for their uniform transport and interactions with the mobile and stationary phases.

High surface area of the interface between mobile and stationary phases is essential for space discrimination of different components in the mixture. Analyte molecules undergo multiple phase transitions between mobile phase and adsorbent surface. Average residence time of the molecule on the stationary phase surface is dependent on the interaction energy. For different molecules with very small interaction energy difference the presence of significant surface is critical since the higher the number of phase transitions that analyte molecules undergo while moving through the chromatographic column, the higher the difference in their retention.

The nature of the stationary and the mobile phases, together with the mode of the transport through the column, is the basis for the classification of chromatographic methods.

1.3 CLASSIFICATION

The mobile phase could be either a liquid or a gas, and accordingly we can subdivide chromatography into *liquid chromatography (LC)* or *gas chromatography (GC)*. Apart from these methods, there are two other modes that use a liquid mobile phase, but the nature of its transport through the porous stationary phase is in the form of either (a) capillary forces, as in *planar chromatography* (also called *thin-layer chromatography, TLC*), or (b) electroosmotic flow, as in the case of *capillary electrochromatography (CEC)*.

The next classification step is based on the nature of the stationary phase. In gas chromatography it could be either a liquid or a solid; accordingly, we distinguish gas–liquid chromatography (long capillary coated with a thin film of relatively viscous liquid or liquid-like polymer; in older systems, liquid-coated porous particles were used) and gas–solid chromatography (capillary with thin porous layer on the walls or packed columns with porous particles).

In liquid chromatography a similar distinction historically existed, since to a significant extent the development of liquid chromatography reflected the path that was taken by gas chromatography development. Liquid–liquid chromatography existed in the early 1970s, but was mainly substituted with liquid chromatography with chemically bonded stationary phases. Recently, liquid–liquid chromatography resurfaced in the form of countercurrent chromatography with two immiscible liquid phases of different densities [1]. The other form of LC is liquid–solid chromatography.

Liquid chromatography was further diversified according to the type of the interactions of the analyte with the stationary phase surface and according to their relative polarity of the stationary and mobile phases.

Since the invention of the technique, adsorbents with highly polar surface were used ($CaCO_3$—Tswett, porous silica—most of the modern packing materials) together with relatively non-polar mobile phase. In 1964, Horvath introduced a chemically modified surface where polar groups were shielded and covered with graphitized carbon black and later with chemically bonded alkyl chains. The introduction of chemically modified hydrophobic surfaces replaces the main analyte—surface interactions from polar to the hydrophobic ones, while mobile phase as an analyte carrier became polar. The relative polarity of the mobile and stationary phases appears to be "reversed" compared to the historically original polar stationary phase and non-polar mobile phase used by M. S. Tswet. This new mode of liquid chromatography became coined as *reversed-phase liquid chromatography (RP)*, where "reversed-phase" referred to the reversing of the relative polarity of the mobile and stationary phases. In order to distinguish this mode from the old form of liquid chromatography, the old became known as *normal-phase (NP)*.

The third mode of liquid chromatography, which is based on ionic interactions of the analyte with the stationary phase, is called *ion-exchange (IEX)*. The separation in this mode is based on the different affinity of the ionic analytes for the counterions on the stationary phase surface.

Specific and essentially stand-alone mode of liquid chromatography is associated with the absence or suppression of any analyte interactions with the stationary phase, which is called *size-exclusion chromatography (SEC)*. In SEC the eluent is selected in such a manner that it will suppress any possible analyte interactions with the surface, and the separation of the analyte molecules in this mode is primarily based on their physical dimensions (size). The larger the analyte molecules, the lower the possibility for them to penetrate into the porous space of the column packing material, and consequently the faster they will move through the column. The schematic of this classification is shown in Figure 1-1.

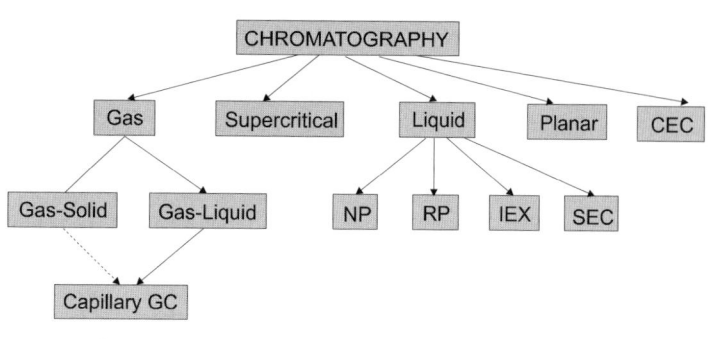

Figure 1-1. Classification of chromatographic modes.

1.4 HISTORY OF DISCOVERY AND EARLY DEVELOPMENT (1903–1933)

Chromatography as a physicochemical method for separation of complex mixtures was discovered at the very beginning of the twentieth century by Russian–Italian botanist M. S. Tswet. [2]. In his paper "On the new form of adsorption phenomena and its application in biochemical analysis" presented on March 21, 1903 at the regular meeting of the biology section of the Warsaw Society of Natural Sciences, Tswet gave a very detailed description of the newly discovered phenomena of adsorption-based separation of complex mixtures, which he later called "chromatography" as a transliteration from Greek "color writing" [3]. Serendipitously, the meaning of the Russian word "tswet" actually means color. Although in all his publications Tswet mentioned that the origin of the name for his new method was based on the colorful picture of his first separation of plant pigments (Figure 1-2), he involuntarily incorporated his own name in the name of the method he invented.

The chromatographic method was not appreciated among the scientists at the time of the discovery, as well as after almost 10 years when L. S. Palmer [4] in the United States and C. Dhere in Europe independently published the description of a similar separation processes. More information on history of early discovery and development of chromatography could be found in reference 5.

Twenty-five years later in 1931, Lederer read the book of L. S. Palmer and later found an original publications of M. S. Tswett, and in 1931 he (together with Kuhn and Winterstein) published a paper [6] on purification of xantophylls on $CaCO_3$ adsorption column following the procedure described by M. S. Tswet.

In 1941 A. J. P. Martin and R. L. M. Synge at Cambridge University, in UK discovered partition chromatography [7] for which they were awarded the Noble Prize in 1952. In the same year, Martin and Synge published a seminal paper [8] which, together with the paper of A. T. James and A. J. P. Martin [9], laid a solid foundation for the fast growth of chromatographic techniques that soon followed.

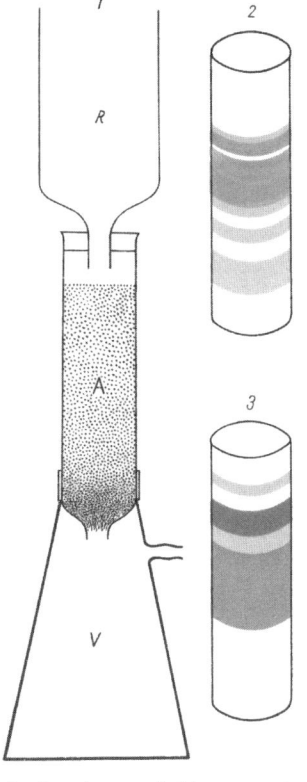

Figure 1-2. Tswet's original drawings of his experiments. From M. S. Tswet, "Chromophils in the plant and animal world" [10]. See color plate.

Chromatography was discovered by Tswet in the form of liquid–solid chromatography (LSC), but its development continued for over 50 years primarily in the form of gas chromatography and partially as thin-layer and liquid–liquid chromatography. Rebirth of liquid chromatography in its modern form and its enormously fast growth had driven this to be the dominant analytical technique in the twenty-first century which can be attributed in the most part to the pioneering work of Prof. C. Horvath at Yale University. In the mid-1960s Prof. Horvath, who previously worked on the development of a porous layer open-tubular columns for gas chromatography, had decided to use for liquid chromatography small glass beads with porous layer on their surface to facilitate the mass transfer between the liquid phase and the surface. Columns packed with those beads developed a significant resistance to the liquid flow, and Prof. Horvath was forced to build an instrument that allowed development of a continuous flow of the liquid through the column [11]. This was the origin of high-performance liquid chromatography (HPLC), and the actual name for this separation method was introduced by Prof. Horvath in 1970 at the Twenty-first Pittsburgh Conference in Cleveland, where he gave this title

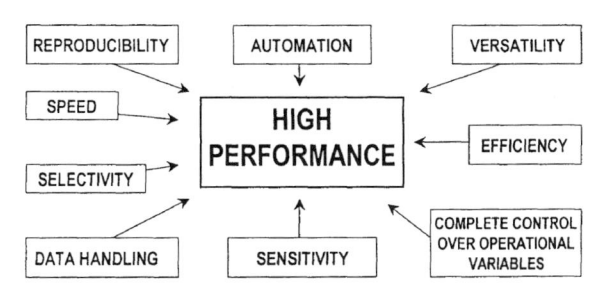

Figure 1-3. Components of performance as defined by C. Horvath. (Reprinted from reference 12, with permission.)

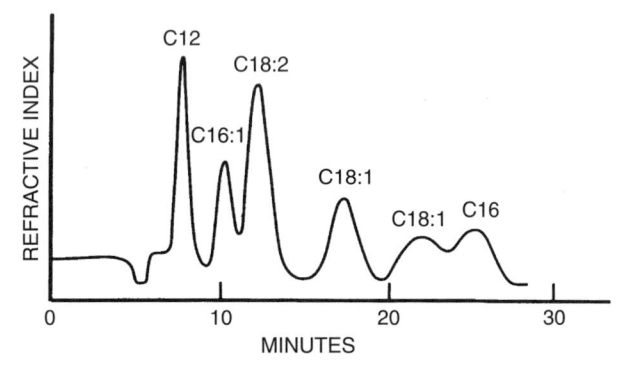

Figure 1-4. Separation of fatty acids on pellicular graphitized carbon black from the mixture of ethanol and 10^{-4} M aqueous NaOH. Refractive index detection. (Reprinted from reference 13, with permission.)

to his invited talk. Later in 2001, he further defined the meaning of the word "performance" as "an aggregate of the efficiency parameters" shown in Figure 1-3.

The first separation on a chemically modified surface with an aqueous eluent, which later got the name "reversed-phase," was also invented by Horvath Figure 1-4, he demonstrated the first reversed-phase separation of fatty acids on pellicular glass beads covered with graphitized carbon black.

1.5 GENERAL SEPARATION PROCESS

M. S. Tswet defined the fractional adsorption process, with the explanation that molecules of different analytes have different affinity (interactions) with the adsorbent surface, and analytes with weaker interactions are less retained [3].

In modern high-performance liquid chromatography the separation of the analytes is still based on the differences in the analyte affinity for the stationary phase surface, and the original definition of the separation process given at its inception almost 100 years ago still holds true.

Liquid chromatography has come a long way with regard to the practical development of HPLC instrumentation and the theoretical understanding of different mechanisms involved in the analyte retention as well as the development of adsorbents with different geometries and surface chemistry.

1.5.1 Modern HPLC Column

The separation of analyte mixtures in modern HPLC is performed in the device called the "column." Current HPLC columns in most cases are a stainless steel tube packed with very small (1–5 µm) particles of rigid porous material. Packing material is retained inside the column with special end-fittings equipped with porous frits allowing for liquid line connection (to deliver mobile phase to the column). Stainless steel or titanium frits have a pore size on the level of 0.2–0.5 µm, which allows for the mobile phase to pass through while small particles of packing material are retained inside the column.

The column is the "heart" of the chromatographic system; and it is the only device where actual separation of the analyte mixture takes place. Detailed discussion of HPLC columns and stationary phases is given in chapter 3.

1.5.2 HPLC System

Typical HPLC system consists of the following main components:

Solvent Reservoirs. Storage of sufficient amount of HPLC solvents for continuous operation of the system. Could be equipped with an online degassing system and special filters to isolate the solvent from the influence of the environment.

Pump. This provides the constant and continuous flow of the mobile phase through the system; most modern pumps allow controlled mixing of different solvents from different reservoirs.

Injector. This allows an introduction (injection) of the analytes mixture into the stream of the mobile phase before it enters the column; most modern injectors are autosamplers, which allow programmed injections of different volumes of samples that are withdrawn from the vials in the autosampler tray.

Column. This is the heart of HPLC system; it actually produces a separation of the analytes in the mixture. A column is the place where the mobile phase is in contact with the stationary phase, forming an interface with enormous surface. Most of the chromatography development in recent years went toward the design of many different ways to enhance this interfacial contact (a detailed discussion is given in Chapter 3).

Detector. This is a device for continuous registration of specific physical (sometimes chemical) properties of the column effluent. The most common detector used in pharmaceutical analysis is UV (ultraviolet), which allows monitoring and continuous registration of the UV absorbance at a selected wavelength or over a span of wavelengths (diode array detection). Appearance of the analyte in the detector flow-cell causes the change of the absorbance. If the analyte absorbs greater than the background (mobile phase), a positive signal is obtained.

Data Acquisition and Control System. Computer-based system that controls all parameters of HPLC instrument (eluent composition (mixing of different solvents); temperature, injection sequence, etc.) and acquires data from the detector and monitors system performance (continuous monitoring of the mobile-phase composition, temperature, backpressure, etc.).

1.6   TYPES OF HPLC

The four main types of HPLC techniques are NP, RP, IEX, and SEC (Section 1.2). The principal characteristic defining the identity of each technique is the dominant type of molecular interactions employed. There are three basic types of molecular forces: *ionic forces*, *polar forces*, and *dispersive forces*. Each specific technique capitalizes on each of these specific forces:

1. Polar forces are the dominant type of molecular interactions employed in normal-phase HPLC (see Chapter 5).
2. Dispersive forces are employed in reversed-phase HPLC (see Chapter 4).
3. Ionic forces are employed in ion-exchange HPLC (see Chapter 4, Section 4.10).

The fourth type of HPLC technique, size-exclusion HPLC (see Chapter 6), is based on the absence of any specific analyte interactions with the stationary phase (no force employed in this technique).

An introduction to the basic principles and typical application areas of each of the above-mentioned HPLC modes is given below.

1.6.1   Normal-Phase Chromatography (NP HPLC)

Normal-phase HPLC explores the differences in the strength of the polar interactions of the analytes in the mixture with the stationary phase. The stronger the analyte–stationary phase interaction, the longer the analyte retention. As with any liquid chromatography technique, NP HPLC separation is a competitive process. Analyte molecules compete with the mobile-phase molecules for the adsorption sites on the surface of the stationary phase. The stronger the mobile-phase interactions with the stationary phase, the lower the difference between the stationary-phase interactions and the analyte interactions, and thus the lower the analyte retention.

Mobile phases in NP HPLC are based on nonpolar solvents (such as hexane, heptane, etc.) with the small addition of polar modifier (i.e., methanol, ethanol). Variation of the polar modifier concentration in the mobile phase allows for the control of the analyte retention in the column. Typical polar additives are alcohols (methanol, ethanol, or isopropanol) added to the mobile phase in relatively small amounts. Since polar forces are the dominant type of interactions employed and these forces are relatively strong, even only 1 v/v% variation of the polar modifier in the mobile phase usually results in a significant shift in the analyte retention.

Packing materials traditionally used in normal-phase HPLC are usually porous oxides such as silica (SiO_2) or alumina (Al_2O_3). Surface of these stationary phases is covered with the dense population of OH groups, which makes these surfaces highly polar. Analyte retention on these surfaces is very sensitive to the variations of the mobile-phase composition. Chemically modified stationary phases can also be used in normal-phase HPLC. Silica modified with trimethoxy glycidoxypropyl silanes (common name: diol-phase) is typical packing material with decreased surface polarity. Surface density of OH groups on diol phase is on the level of 3–4 $\mu mol/m^2$, while on bare silica silanols surface density is on the level of 8 $\mu mol/m^2$. The use of diol-type stationary-phase and low-polarity eluent modifiers [esters (ethyl acetate) instead of alcohols] allow for increase in separation ruggedness and reproducibility, compared to bare silica.

Selection of using normal-phase HPLC as the chromatographic method of choice is usually related to the sample solubility in specific mobile phases. Since NP uses mainly nonpolar solvents, it is the method of choice for highly hydrophobic compounds (which may show very stronger interaction in reversed-phase HPLC), which are insoluble in polar or aqueous solvents. Figure 1-5 demonstrates the application of normal-phase HPLC for the separation of a mixture of different lipids.

Detailed discussion of normal-phase chromatography process, mechanism, and retention theories, as well as types and properties of used stationary phases, is given in Chapter 5.

1.6.2   Reversed-Phase HPLC (RP HPLC or RPLC)

As opposed to normal-phase HPLC, reversed-phase chromatography employs mainly dispersive forces (hydrophobic or van der Waals interactions). The polarities of mobile and stationary phases are reversed, such that the surface of the stationary phase in RP HPLC is hydrophobic and mobile phase is polar, where mainly water-based solutions are employed.

Reversed-phase HPLC is by far the most popular mode of chromatography. Almost 90% of all analyses of low-molecular-weight samples are carried out using RP HPLC. One of the main drivers for its enormous popularity is the ability to discriminate very closely related compounds and the ease of variation of retention and selectivity. The origin of these advantages could be

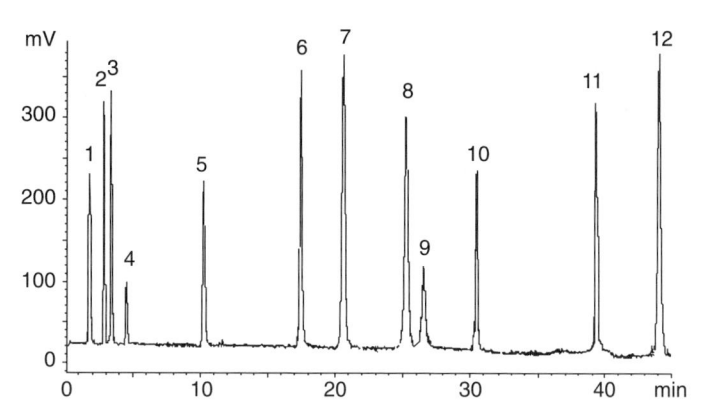

Figure 1-5. Separation of selected representatives of different lipid classes. (1) Paraffin, (2) *n*-hexadecyl palmitate; (3) cholesterol palmitate; (4) stearic acid methyl ester; (5) glycerol tripalmitate; (6) hexadecyl alchohol; (7) stearic acid; (8) cholesterol; (9) glycerol-1,3-dipalmitate; (10) glycerol-1,2-dipalmitate; (11) glycerol monopalmitate; (12) erucylamide. Column LiChrosphere® Diol (125 × 3 mm) 5-μm particles. Gradient from isooctane (A) to 60% methyl tritbutyl ether (MTBE) in 34 min + 10 min isocratic hold. (Reprinted from reference 14, with permission.)

explained from an energetic point of view: Dispersive forces employed in this separation mode are the weakest intermolecular forces, thereby making the overall background interaction energy in the chromatographic system very low compared to other separation techniques. This low background energy allows for distinguishing very small differences in molecular interactions of closely related analytes. As an analogy, it is possible to compare two spectroscopic techniques: UV and fluorescence spectroscopy. In fluorescence spectroscopy, emission registers essentially against zero background light energy, which makes its sensitivity several orders of magnitude higher than in UV spectroscopy, where background energy is very high. A similar situation is in RP HPLC, where its sensitivity to the minor energetic differences in analyte–surface interactions is very high attributed to the low background interaction energy.

Adsorbents employed in this mode of chromatography are porous rigid materials with hydrophobic surfaces. In all modes of HPLC with positive analyte surface interactions (NP, RP, IEX) the higher the adsorbent surface area, the longer the analyte retention and in most cases the better separation. The majority of packing materials used in RP HPLC are chemically modified porous silica. The properties of silica have been studied for many years [15, 16], and the technology of manufacturing porous spherical particles of controlled size and porosity is well-developed.

Chemical modification of the silica surface was also intensively studied in the last 30 years, mainly as a direct result of growing popularity of reversed-phase HPLC [16, 17]. Despite the intensive research and enormous growth of commercially available packing materials and columns, there is still no consensus on which properties the optimum RP stationary phase should have for the selective analysis of diverse sets of compounds such as pharmaceutical compounds that have a plethora of various ionizable functionalities, varying hydrophobicities, and different structural components (linear alkyl chains, aromatic rings, heterocycles, etc.). Detailed discussion of vast varieties of RP stationary phases is given in Chapter 3.

1.6.3 Ion-Exchange Chromatography (IEX)

Ion-exchange chromatography, as indicated by its name, is based on the different affinities of the analyte ions for the oppositely charged ionic centers in the resin or adsorbed counterions in the hydrophobic stationary phase. Consider the exchange of two ions A^+ and B^+ between the solution and exchange resin E^-:

$$A \cdot E + B^+ \Leftrightarrow B \cdot E + A^+$$

The equilibrium constant for this process is shown in Eq. (1-1):

$$K = \frac{[A^+][BE]}{[AE][B^+]} \tag{1-1}$$

which essentially determines the relative affinity of both cations to the exchange centers on the surface. If the constant is equal to 1, no discriminating ability is expected for this system. The higher the equilibrium constant (provided that it is greater than 1), the greater the ability of cation B^+ to substitute A on the resin surface.

Depending on the charge of the exchange centers on the surface, the resin could be either anion-exchanger (positive ionic centers on the surface) or cation-exchanger (negative centers on the surface).

Crosslinked styrene-divinylbenzene is the typical base material for ion-exchange resin. Exchange groups are attached to the phenyl rings in the structure and the degree of crosslinkage is between 5% and 20%. The higher the crosslinkage, the harder the material and the less susceptible it is to swelling, but the material usually shows lower ion-exchange capacity.

Four major types of ion-exchange centers are usually employed:

1. SO_3^-—strong cation-exchanger
2. CO_2^-—weak cation-exchanger
3. Quaternary amine—strong anion-exchanger
4. Tertiary amine—weak anion-exchanger

Analyte retention and selectivity in ion-exchange chromatography are strongly dependent on the pH and ionic strength of the mobile phase. Basic principles of the ion-exchange HPLC are discussed in Chapter 4, Section 4.10.

1.6.4 Size-Exclusion Chromatography (SEC)

SEC is the method for dynamic separation of molecules according to their size; as indicated by its name, the separation is based on the exclusion of the molecules from the porous space of packing material due to their steric hindrance. Hydrodynamic radius of the analyte molecule is the main factor determining its retention. In general, the higher the hydrodynamic radius, the shorter the retention. Historically, two different names are used for this method. In 1959 the molecular sieving principle was applied for the separation of biochemical polymers on dextran gels, and it was called *gel-filtration chromatography* (GFC) (uses aqueous-based eluents with salts). In 1961 the same principle was applied for the molecular weight determination of synthetic polymers, and the name gel-permeation chromatography (GPC) (uses primarily organic solvents such as THF) came into popular use among polymer chemists.

This is the only chromatographic separation method where any positive interaction of the analyte with the stationary phase should be avoided. In size-exclusion chromatography, the higher the molecular weight of the molecule, the greater its hydrodynamic radius, which results in faster elution. At the same time, if an analyte molecule interacts (undesired) with the stationary phase, thus increasing the retention of larger molecules, which may confound separation of molecules based solely on their hydrodynamic radius. Obviously, these two processes produce opposite effects, and analysis of the polymer molecular weight and molecular weight distribution would be impossible. This brings specific requirements to the selection of the column packing material and the mobile phase, where the mobile-phase molecules should interact with the surface of the stationary phase stronger than the polymer, thus preventing its interaction with the surface.

Polymer molecular weight determination is based on the relationship of the molecular hydrodynamic radius with the molecular weight. The radius is roughly proportional to the cubic root of the molecular weight, thus giving the impression that cubic root of the molecular weight should be proportional to the analyte retention volume. This is only observed in the regions of total exclusion and total permeation of the polymer molecules in the adsorbent porous space. A practically useful region for molecular weight determination is where partial permeation of the analyte molecules in the adsorbent porous space is observed. In this region the adsorbent pore size distribution plays the dominant role in the adsorbent ability to discriminate molecules according to their molecular weight. It was found that the logarithm of analyte molecular weight has a linear relationship with the retention volume in this region.

Hydrodynamic radius of the polymer is also dependent on the analyte interaction with the solvent. Polymer conformation and degree of the solvation varies with the variation of the solvent properties. Detailed discussion of all aspects of size exclusion chromatography is given in Chapter 6.

1.7 HPLC DESCRIPTORS (Vr, k, N, etc.)

1.7.1 Retention Volume

Modern HPLC is a routine tool in any analytical laboratory. Standard HPLC system represents a separation output in the form of chromatogram (typical modern chromatogram is shown in Figure 1-6). Each specific analyte in the chromatogram is represented by a peak. In the absence of the strong specific analyte interactions with the stationary phase and at relatively low analyte concentration, peaks are symmetrical and resemble a typical Gaussian (normal distribution) curve.

The distance of the peak maxima from the injection point expressed in time units is called retention time (t_R), and it serves as an identifier for the given analyte on that particular system. Retention time is probably the most widely used descriptor of the analyte behavior, and it is the most easily measurable parameter. However, even though it is easily measurable, it is the least universal parameter.

Analyte retention time is dependent on the mobile phase flow rate; the faster the flow rate, the smaller the analyte retention time. It is also dependent on the flow rate stability. The product of the analyte retention time and the mobile-phase flow rate is the retention volume (V_R). Analyte retention volume

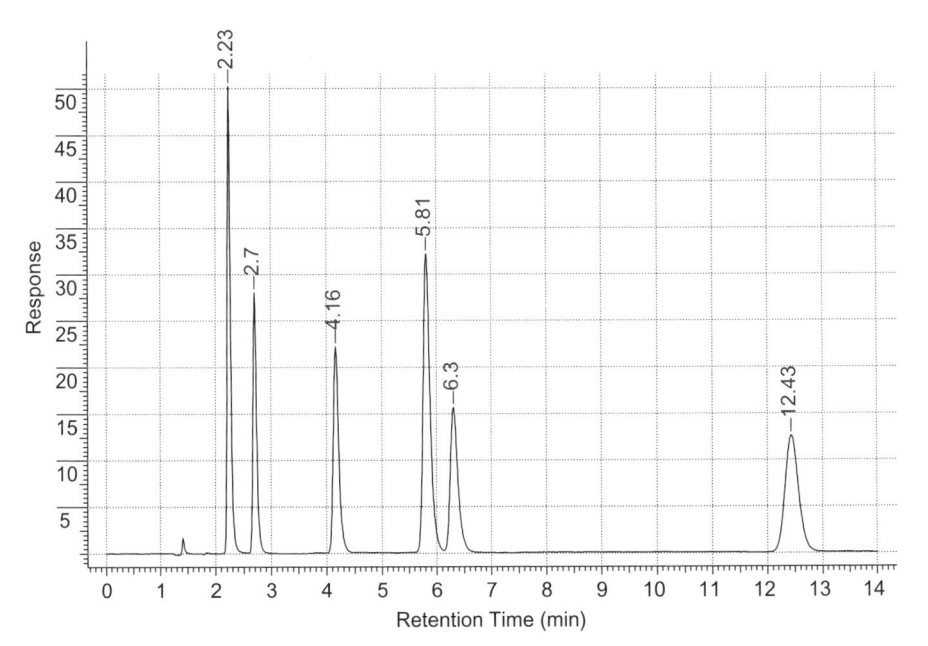

Figure 1-6. Typical modern chromatogram of a mixture of β-blockers on Zorbax Eclipse-XDB C18 column from acetonitrile/water mixture at pH = 3 (components in the sequence of increasing retention: pindolol, metoprolol, labetalol, propanolol, alprenolol, o-chloroaniline).

is more universal descriptor of the analyte behavior in the chromatographic system, since it is less dependent upon the instrumental parameters.

1.7.2 Void Volume

Even if the analyte does not interact with the stationary phase, it will not appear in the detector immediately after injection. An HPLC column is filled with small particles of porous material which have a significant volume of the liquid phase between the particles and inside their porous space, so the noninteracting analyte still has to travel through this volume before it enters the detector.

The volume of the liquid phase in the column is called "*void volume*" (V_0). Several other names are also used in the chromatographic literature: "dead volume," "hold-up volume," and sometimes "retention volume of nonretained component." In this book we will be using term "void volume."

If a particular HPLC system provides constant and stable mobile-phase flow (F), one can convert retention volume (V_R) and void volume (V_0) into the retention time (t_R) and a void time (t_0).

$$t_R = \frac{V_R}{F}, \qquad t_0 = \frac{V_0}{F} \qquad (1\text{-}2)$$

Void time can be interpreted as part of the total analyte retention time that the analyte actually spends in the mobile phase moving through the column, and for the rest of the retention time the analyte sits on the stationary phase surface.

The void volume or void time is a very important parameter, and its correct determination could be critical for the interpretation of the experimental results.

Proper experimental determination and even the definition of the void volume actually is very controversial and is a subject for continuous debates in the scientific literature (see details in Chapter 2, Section 2.13). Let us illustrate this using the definition given in the first paragraph of this section: "retention of noninteracting analyte," which is equivalent to "nonretained" component. The first valid questions are How can we find a so-called "nonretained" component? and What are the criteria of the absence of retentivity? If we assume that we find the component which does not interact with the stationary phase at all, this means that eluent molecules interact with the stationary-phase surface stronger than analyte molecules and they get adsorbed on the stationary-phase surface and consequently this "nonretained" analyte will not interact. Preferential adsorption of the eluent molecules will result in slight exclusion of the analyte from the adsorbent surface. Even the monomolecular layer of adsorbed eluent molecules provides a significant exclusion volume due to the high overall surface of the stationary phase in the column.

Ideally, to determine the total volume of the liquid phase in the column, one would need to find a component that will have the same behavior as the eluent. The best approach is to inject a sample of a deuterated eluent. Usually the elution of deuterated acetonitrile injected in the flow of pure acetonitrile in the column works the best [18]. This approach requires the use of the refractive index detector.

More detailed theoretical discussion of the importance and determination of the void volume is given in Chapter 2.

1.7.3 Retention Factor

As was mentioned in the previous section, the analyte retention consists of two parts: (1) the time the component resides in the mobile phase actually moving through the column and (2) the time the analyte is retained on the stationary phase (Figure 1-7). The difference between the total retention time (t_R) and the hold-up time is called the reduced retention time (t_R'), and corresponding difference between the analyte retention volume and the void volume is called the reduced retention volume, V_R'.

The ratio of the reduced retention volume to the void volume is a widely used dimensionless parameter called *retention factor, k.*

$$k = \frac{V_R - V_0}{V_0} = \frac{V_R'}{V_0} = \frac{t_R - t_0}{t_0} \tag{1-3}$$

Retention factor (sometimes called capacity factor) is a very convenient chromatographic descriptor since it is dimensionless and independent on the

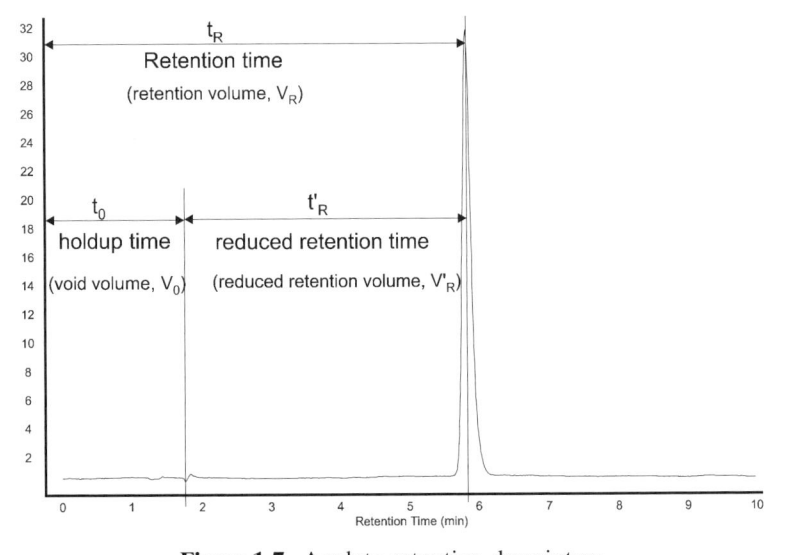

Figure 1-7. Analyte retention descriptors.

mobile phase flow rate and column dimensions. Ideally if the retention of the same analyte was measured on two instruments equipped with columns of different dimensions with the same type of a stationary phase and the same mobile phase, theoretically the retention factors of that analyte on both systems should be identical. This allows for valid comparison of the results obtained on different systems and in different laboratories. Note that accurate and correct determination of the void volume is critical for any cross-experimental comparisons.

Components positively retained on the column have positive values of the retention factors. "Nonretained" components or analytes eluted with the void volume show the retention factor equal to zero, and analytes excluded from the stationary phase surface (eluent molecules interacts with the stationary-phase surface stronger than the analyte molecules) show negative retention factors.

As a very rough first approximation the chromatographic retention process could be described on the basis of simple single equilibria of the analyte distribution between the mobile and stationary phases. The equilibrium constant of this process is proportional to the analyte retention factor

$$K \propto k \qquad (1\text{-}4)$$

The equilibrium constant, K, thermodynamically could be described as the exponent of the Gibbs free energy of the analyte's competitive interactions with the stationary phase. In liquid chromatography the analyte competes with the eluent for the place on the stationary phase, and resulting energy responsible for the analyte retention is actually the difference between the analyte interaction with the stationary phase and the eluent interactions for the stationary phase as shown in equation (1-5)

$$K = \exp\left(\frac{\Delta G_{analyte} - \Delta G_{eluent}}{RT}\right) \qquad (1\text{-}5)$$

The Gibbs free energy of the eluent interactions with the stationary phase is dependent on the eluent composition. The higher the concentration of the organic compound (usually methanol or acetonitrile) in the mobile phase, the stronger the eluent interactions with the stationary phase (ΔG_{eluent}) and correspondingly the lower the difference [in equation (1-5)], thus leading to the lower equilibrium constant and lower analyte retention.

1.7.4   Selectivity

The ability of the chromatographic system to discriminate different analytes is called selectivity (α). Selectivity is determined as the ratio of the retention factors of two analytes, or the ratio of the reduced retention times

$$\alpha = \frac{k_2}{k_1} = \frac{t_{R_2} - t_0}{t_{R_1} - t_0} \quad (1\text{-}6)$$

The increase of the selectivity in the development of the separation of a complex mixture is the primary goal of any chromatographer, because if the selectivity for the pair of analytes is equal to 1, then it does not matter how narrow your peaks or how fast your separation—you will not be able to separate these components until you increase the selectivity.

What governs the selectivity? To answer this question, let us use the phenomenological thermodynamic expressions for the retention factors (1-4) and (1-5) and apply them to the expression for selectivity (1-6):

$$\alpha = \frac{\exp\left(\dfrac{\Delta G_{analyte,2} - \Delta G_{eluent}}{RT}\right)}{\exp\left(\dfrac{\Delta G_{analyte,1} - \Delta G_{eluent}}{RT}\right)} \quad \Rightarrow \quad \exp\left(\frac{\Delta G_{analyte,2} - \Delta G_{analyte,1}}{RT}\right) \quad (1\text{-}7)$$

Equation (1-7) shows that in an ideal case the selectivity of the system is only dependent on the difference in the analytes interaction with the stationary phase. It is important to note that the energetic term responsible for the eluent interactions was canceled out, and this means that the eluent type and the eluent composition in an ideal case does not have any influence on the separation selectivity. In a real situation, eluent type and composition may influence the analyte ionization, solvation, and other secondary equilibria effects that will have effect on the selectivity, but this is only secondary effect.

The selectivity is primarily dependent on the nature of the analytes and their interaction with the stationary phase surface. If a dramatic change of the selectivity is needed for a particular separation, the best solution is the replacement of the type of the stationary phase.

Selectivity is generally not affected by the eluent composition or temperature unless these parameters modify the analyte nature (solvation, ionization, tautoermization, etc.). However, the type of solvent methanol versus acetonitrile, for example, may affect the selectivity between critical pair of components (i.e., isomers).

1.7.5 Efficiency

An analyte is injected into the column in the form of very small zone with even distribution of the analyte within this zone. While this zone is moving through the column, it gets broadened. The degree of this band-broadening is called the efficiency. There are several different theories (or mathematical approaches) used for the description of the band broadening. Martin and Synge [8] introduced the plate theory for the evaluation of the column efficiency. Plate theory assumes that the analyte is in the instant equilibrium with the stationary phase and the column is considered to be divided into a number of hypothetical plates. Each plate has a finite height (height of effective theoretical plate, HETP), and an analyte spends a finite time in this plate. This time is considered to be sufficient to achieve equilibrium. The smaller the plate height or the greater the number of plates, the more efficient the analyte exchange is between two phases, and the better the separation. That is why column efficiency is measured in number of theoretical plates. The detailed discussion of the efficiency theory is in Chapter 2.

The efficiency is the measure of the chromatographic band broadening and the number of the theoretical plates (N) in the column and is usually calculated using the following equation:

$$N = 16\left(\frac{t_R}{w}\right)^2 \tag{1-8}$$

where t_R is the analyte retention time and w is the peak width measured in time units as the distance between the intersections of the tangents to the peak inflection points with the baseline, as shown in Figure 1-8.

Column efficiency is mainly dependent on the kinetic factors of the chromatographic system such as molecular diffusion, mass-flow dynamics, properties of the column packing bed, flow rate, and so on. The smaller the particles and the more uniform their packing in the column, the higher the efficiency. The faster the flow rate, the less time analyte molecules have for diffusive band-broadening. At the same time, the faster the flow rate, the further analyte molecules are from the thermodynamic equilibrium with the stationary phase. This shows that there should be an optimum flow rate that allows achievement of an optimum efficiency for a given column. Detailed discussions of the

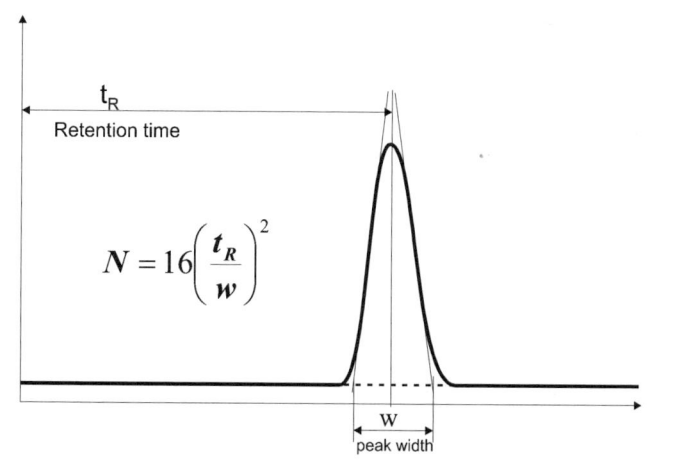

Figure 1-8. Schematic of the efficiency measurements (number of the theoretical plates in the column).

column efficiency and its optimization at different experimental conditions are discussed in Chapter 2 and Chapter 17.

Efficiency is the property of the column. Ideally, all analytes separated on the same column should demonstrate the same number of theoretical plates. Secondary equilibria of the analyte in the column or extracolumn band broadening can alter this.

Efficiency and selectivity are complementary chromatographic descriptors. A column with high efficiency can generate narrow chromatographic zones and allows the separation of analytes with low selectivity. On the other hand, if the chromatographic system has high selectivity for two analytes, they could be separated on the column with low efficiency, as shown in Figure 1-9.

Figure 1-9 demonstrates that satisfactory separation could be obtained by optimization of either efficiency or selectivity or both at the same time. Efficiency is essentially the property of the column, but selectivity is the reflection at the nature of analytes and the surface chemistry of the packing material. Combination of these descriptors would allow the characterization of the overall separation power of a particular chromatographic system.

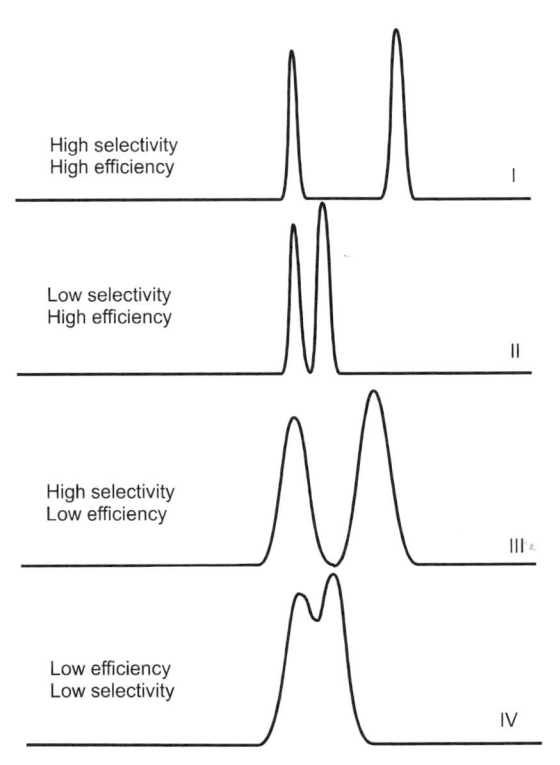

Figure 1-9. I: Peaks are narrow and far from each other, simple decrease of the column length or flow rate can significantly shorten the runtime without the loss of separation quality. II: Acceptable separation, method may not be rugged. III: Acceptable separation, quantitation reproducibility could be low. IV: Bad separation.

1.7.6 Resolution

The distance between the peak maxima reflects the selectivity of the system. The greater the distance, the higher the selectivity. The width of the chromatographic peak reflects the system band broadening and thus efficiency. Resolution, R, is defined [equation (1-9)] as the ratio of the distance between two peaks to the average width of these peaks (at baseline), and this descriptor encompasses both the efficiency and selectivity.

$$R = 2\frac{t_{R,2} - t_{R,1}}{w_2 + w_1} \tag{1-9}$$

For the resolution of a so-called "critical pair" of analytes (two analytes in the mixture that have minimal distance between them compared to all other analytes in the mixture), if they have relatively high retention factors ($k' > 5$) that their peak widths can be assumed as equal, then equation (1-9) reduces to

$$R = \frac{t_{R,2} - t_{R,1}}{w} \tag{1-10}$$

Peak width could be expressed from equation (1-8) as

$$w = \frac{4t_R}{\sqrt{N}} \tag{1-11}$$

If we select the retention of the second analyte for the calculation of the peak width, then applying equation (1-11) into expression (1-10) we get

$$R = \frac{t_{R,2} - t_{R,1}}{t_{R,2}} \cdot \frac{\sqrt{N}}{4} \tag{1-12}$$

Relatively simple algebraic conversion will bring us to so-called Master Resolution Equation:

$$R = \frac{\alpha - 1}{\alpha} \cdot \frac{k_2}{1 + k_2} \cdot \frac{\sqrt{N}}{4} \tag{1-13}$$

As we discussed above, efficiency and selectivity are complementary descriptors dependent on the different sets of chromatographic parameters. Efficiency is more dependent on the quality of the column packing, particle size, flow rate, and instrumental optimization, while selectivity is more dependent on the stationary phase properties and the nature of the analytes themselves. However, efficiency is sometimes affected by nonideal interactions of the analyte with the stationary phase (i.e., peak tailing).

TABLE 1-1. Comparison of the Variation of Selectivity and Efficiency Necessary to Increase Resolution from 1 to 1.5

Resolution	Efficiency	Selectivity
1	10,000	1.04
1.5	22,500	1.04
1.5	10,000	1.06

Improvement of the resolution of poorly resolved analytes then could be pursued in two different ways: either by increasing the efficiency or by improving the selectivity. The resolution value equal to 1.5 is usually regarded as sufficient for the baseline separation of closely eluted peaks; and if we consider that typical average efficiency of modern HPLC column is equal to 10,000 theoretical plates, then we can calculate the selectivity necessary for this separation to get a resolution of 1.5. It will be also useful to compare what would be required in terms of efficiency and selectivity to improve the resolution from 1 to 1.5. Corresponding calculations are shown in the Table 1-1.

It is obvious from the Table 1-1 that to improve the resolution from 1 to 1.5, only a slight increase of the selectivity is needed from 1.04 to 1.06 while keeping the efficiency on the level of 10,000. However, in order to obtain the same increase of the resolution while keeping the selectivity (1.04) constant, a dramatic increase of the efficiency of more than two times will be needed.

This example emphasizes that the main efforts in developing good rugged separation should be directed toward the achievement of highest possible selectivity first, which essentially means that the selection of the proper stationary-phase and mobile-phase conditions is very crucial.

REFERENCES

1. W. D. Conway and R. J. Petroski (eds.), *Modern Countercurrent Chromatography*, ACS Symposium Series 593, ACS, Washington, D.C., 1995.

2. M. S. Tswet, On the new form of adsorption phenomena and its application in biochemical analysis, *Proc.* Warsaw Natural *Biol. Soc.* **14** (1903), 20–39 (М.С. Цвет, О новой категории адсорбционных явлений и о применении их к биохимическому анализу, Труды Варшавского обещства естествоиспытателей, отделение биологии **14** (1993), 20–39.

3. M. S. Tswet, *Chromophils in Animal and Plant World*, Doctor of Science dissertation, Warsaw, 1910, p. 379.

4. L. S Palmer, *Carotinoids and Related Pigments—The Chromophlipids*, ACS Monograph Series, American Chemical Society, New York, 1922, 316 pp.

5. E. M. Senchenkova, *Michael Tswet—the Creator of Chromatography*, RAS, Moscow, 2003.

6. R. Kuhn, A. Winterstein, and E. Lederer, The xanthophylls, *Hoppe-Seyler's Z. Phyisiol. Chem.* **197** (1931), 141–160.

7. A. J. P. Martin and R. L. M. Synge, Separation of higher monoaminoacids by counter-current liquid–liquid extraction. The aminoacid composition of wool, *J. Biochem.* **35** (1941), 91–121.

8. A. J. P. Martin and R. L. M. Synge, A new form of chromatogram employing two liquids phases. 1. A theory of chromatography. 2. Application to the micro-determination of the higher monoamino-acids in proteins, *J. Biochem.* **35** (1941), 1358–1368.

9. A. T. James and A. J. P. Martin, Gas–solid partition chromatography. The separation and micro-estimation of volatile fatty acids from formic acid to dodecanoic acid, *J. Biochem.* **50** (1952), 679–690.

10. "Хромофиллы в растительном и животном мире" (Chromophils in plant and animal world) Doctor of Science dissertation, Warsaw, 1910, 380 pp. Reprinted from *Chromatographic adsorption analysis*, selected works of M. S. Tswet by Academy of Sciences of the USSR, 1946.

11. C. G. Horvath, B. A. Preiss, and S. R. Lipsky, Fast liquid chromatography. Investigation of operating parameters and the separation of nucleotides on pellicular ion exchangers, *Anal. Chem.* **39** (1967), 1422–1428.

12. C. Horvath, *My Focus on Chromatography over 40 Years*, Journal of Chromatography Library, Vol. 64, Elsevier, Amsterdam, 2001, pp. 238–247.

13. L. S. Ettre and A. Zlatkis (eds.), *75 Years of Chromatography—a Historical Dialogue*, Journal of Chromatography Library, Vol. 17, Elsevier, Amsterdam, 1979, p. 155.

14. A. Schaefer, T. Kuchler, T. J. Simat, and H. Steinhart, Migration of lubricants from food packagings, *J. Chromatogr.* A, **1017** (2003), 107–116.

15. R. K. Iler, *Chemistry of Silica*, Wiley-Interscience, New York, 1979.

16. K. K. Unger, *Porous Silica, Its Properties and Use as Support in Column Liquid Chromatography*, Journal of Chromatography Library, Vol. 16, Amsterdam, Elsevier, 1979.

17. G. V. Lisichkin (ed.), *Chemistry of Modified Surface Compounds*, PhysMathLit, Moscow, 2003.

18. J. H. Knox, R. Kaliszan, Theory of solvent disturbance peaks and experimental determination of thermodynamic dead-volume in column liquid chromatography, *J. Chromatogr.* A, **349** (1985), 211–234.

2

HPLC THEORY

Yuri Kazakevich

2.1 INTRODUCTION

The process of analyte retention in high-performance liquid chromatography (HPLC) involves many different aspects of molecular behavior and interactions in condensed media in a dynamic interfacial system. Molecular diffusion in the eluent flow with complex flow dynamics in a bimodal porous space is only one of many complex processes responsible for broadening of the chromatographic zone. Dynamic transfer of the analyte molecules between mobile phase and adsorbent surface in the presence of secondary equilibria effects is also only part of the processes responsible for the analyte retention on the column. These processes just outline a complex picture that chromatographic theory should be able to describe.

HPLC theory could be subdivided in two distinct aspects: kinetic and thermodynamic. Kinetic aspect of chromatographic zone migration is responsible for the band broadening, and the thermodynamic aspect is responsible for the analyte retention in the column. From the analytical point of view, kinetic factors determine the width of chromatographic peak whereas the thermodynamic factors determine peak position on the chromatogram. Both aspects are equally important, and successful separation could be achieved either by optimization of band broadening (efficiency) or by variation of the peak positions on the chromatogram (selectivity). From the practical point of view, separation efficiency in HPLC is more related to instrument optimization, column

---

HPLC for Pharmaceutical Scientists, Edited by Yuri Kazakevich and Rosario LoBrutto

dimensions, and particle geometry—factors that could not have continuous variation during method development except for the small influence from variation of the mobile phase flow rate. On the other hand, analyte retention or selectivity is mainly dependent on the competitive intermolecular interactions and are influenced by eluent type, composition, temperature, and other variables which allow functional variation.

2.2   BASIC CHROMATOGRAPHIC DESCRIPTORS

Four major descriptors are commonly used to report characteristics of the chromatographic column, system, and particular separation:

1. Retention factor (k)
2. Efficiency (N)
3. Selectivity (α)
4. Resolution (R)

Retention factor (k) is the unitless measure of the retention of a particular compound on a particular chromatographic system at given conditions defined as

$$k = \frac{V_R - V_0}{V_0} = \frac{t_R - t_0}{t_0} \tag{2-1}$$

where V_R is the analyte retention volume, V_0 is the volume of the liquid phase in the chromatographic system, t_R is the analyte retention time, and t_0 is sometimes defined as the retention time of nonretained analyte. Retention factor is convenient because it is independent on the column dimensions and mobile-phase flow rate. Note that all other chromatographic conditions significantly affect retention factor.

Efficiency is the measure of the degree of peak dispersion in a particular column; as such, it is essentially the characteristic of the column. Efficiency is expressed in the number of theoretical plates (N) calculated as

$$N = 16 \left(\frac{t_R}{w} \right)^2 \tag{2-2}$$

where t_R is the analyte retention time and w is the peak width at the baseline.

Selectivity (α) is the ability of chromatographic system to discriminate two different analytes. It is defined as the ratio of corresponding retention factors:

$$\alpha = \frac{k_2}{k_1} \tag{2-3}$$

Resolution (R) is a combined measure of the separation of two compounds which include peak dispersion and selectivity. Resolution is defined as

$$R = 2 \frac{t_2 - t_1}{w_2 + w_1} \tag{2-4}$$

In the following sections the chromatographic descriptors introduced above [equations (2-1)–(2-4)] will be discussed in terms of their functional dependencies, specifics, and relationships with different chromatographic and thermodynamic parameters.

2.3   EFFICIENCY

The most rigorous discussion of the formation of chromatographic zone and the mathematical description of zone-broadening is given in reference 1. Here only practically important and useful equations will be discussed.

If column properties could be considered isotropic, then we would expect symmetrical peaks of a Gaussian shape (Figure 2-1), and the variance of this peak is proportional to the diffusion coefficient (D)

$$\sigma^2 = 2Dt \tag{2-5}$$

At given linear velocity (v) the component moves through the column with length (L) during the time (t), or

$$L = vt \tag{2-6}$$

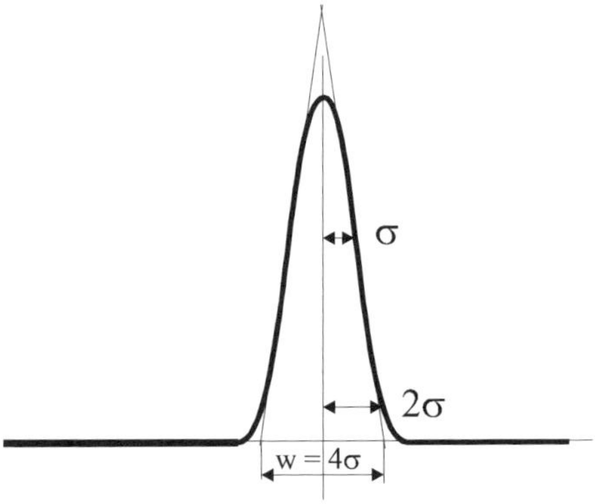

Figure 2-1. Gaussian band broadening.

Substituting t from equation (2-6) in equation (2-5), we get

$$\sigma^2 = \left(\frac{2D}{v}\right)L \qquad (2\text{-}7)$$

Expression $2D/v$ has units of length and is essentially the measure of band spreading at a given velocity on the distance L of the column. This parameter has essentially the sense of the *height equivalent to the theoretical plate* and could be denoted as H, so we get

$$H = \frac{\sigma^2}{L} \qquad (2\text{-}8)$$

Several different processes lead to the band-spreading phenomena in the column which include: multipath effect; molecular diffusion; displacement in the porous beds; secondary equilibria; and others. Each of these processes introduces its own degree of variance toward the overall band-spreading process. Usually these processes are assumed to be independent; and based on the fundamental statistical law, overall band-spreading (variance) is equal to the sum of the variances for each independent process:

$$\sigma^2_{\text{tot}} = \sum \sigma^2_i \qquad (2\text{-}9)$$

In the further discussion we assume the total variance in all cases.

In the form of equation (2-8) the definition of H is exactly identical to the plate height as it evolved from the distillation theory and was brought to chromatography by Martin and Synge [2]. If H is the theoretical plate height, we can determine the total number of the theoretical plates in the column as

$$N = \frac{L}{H} \quad \Rightarrow \quad N = \left(\frac{L}{\sigma}\right)^2 \qquad (2\text{-}10)$$

In linear chromatography, each analyte travels through the column with constant velocity (u_c). Using this velocity, we can express the analyte retention time as

$$t_R = \frac{L}{u_c} \qquad (2\text{-}11)$$

Similarly, the time necessary to move analyte zone in the column on the distance of one σ (Figure 2-1) can be defined as τ

$$\tau = \frac{\sigma}{u_c} \qquad (2\text{-}12)$$

Substituting both equations (2-11) and (2-12) into (2-10), we get

$$N = \left(\frac{t_R}{\tau}\right)^2 \tag{2-13}$$

Parameter τ in equation (2-13) is the standard deviation and expressed in the same units as retention time. Since we considered symmetrical band-broadening of a Gaussian shape, we can use Gaussian function to relate its standard deviation to more easily measurable quantities. The most commonly used points are the so-called peak width at the baseline, which is actually the distance between the points of intersections of the tangents to the peaks inflection points with the baseline (shown in Figure 2-1). This distance is equal to four standard deviations, and the final equation for efficiency will be

$$N = 16\left(\frac{t_r}{w_b}\right)^2 \tag{2-14}$$

Another convenient determination for N is by using the peak width at the half-height. From the same Gaussian function the peak width on the half-height is 2.355 times longer than the standard deviation of the same peak, and the resulting formula for the number of the theoretical plates will be

$$N = 5.545\left(\frac{t_R}{w_{1/2 h}}\right)^2 \tag{2-15}$$

Efficiency is mainly a column-specific parameter. In a gas chromatography column, efficiency is highly dependent on the flow rate. In HPLC, because of much higher viscosity, the applicable flow rate region is not so broad; within this region, variations of the flow rate do not affect column efficiency significantly.

On the other hand, geometry of the packing material and uniformity and density of the column packing are the main factors defining the efficiency of particular column. There is no clear fundamental relationship between the particle diameter and the expected column efficiency, but phenomenologically an increase of the efficiency can be expected with the decrease of the particle diameter, since the difference between the average size of the pores in the particles of the packing material and the effective size of interparticle pores decreases, which leads to the more uniform flow inside and around the particles. From Figure 2-2 it is obvious that the smaller the particles, the lower the theoretical plate height and the higher the efficiency. The general form of the shown dependence is known as Van Deemter function (2-16), which has the following mathematical form:

$$H = A + \frac{B}{v} + Cv \tag{2-16}$$

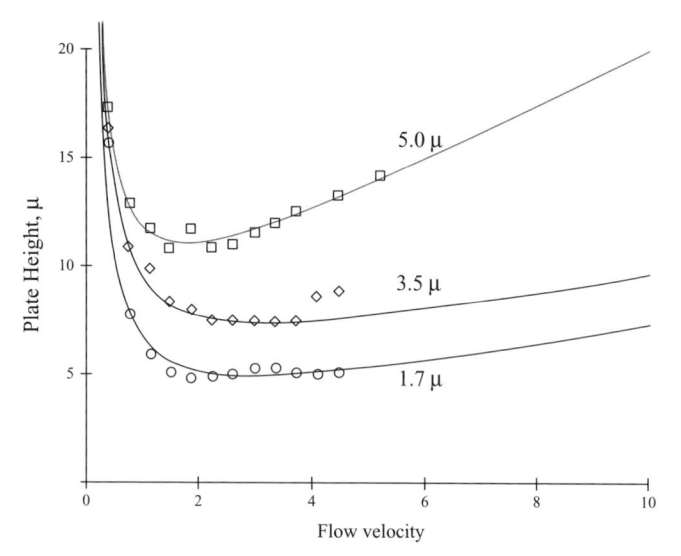

Figure 2-2. The experimental dependence of the theoretical plate height on the flow velocity for columns packed with same type of particles of different average diameter.

where v is the linear flow velocity, and A, B, and C are constants for given column and mobile phase.

Three terms of the above equation essentially represent three different processes that contribute to the overall chromatographic band-broadening.

A—represents multipath effect or eddy diffusion

B—represents molecular diffusion

C—represents mass transfer

The multipath effect is a flow-independent term, which defines the ability of different molecules to travel through the porous media with paths of different length.

The molecular diffusion term is inversely proportional to the flow rate, which means that the slower the flow rate, the longer component stays in the column and the molecular diffusion process has more time to broaden the peak.

The mass-transfer term is proportional to the flow rate, which means that the faster the flow, the greater the band-broadening. Superposition of all three processes is shown schematically in Figure 2-3.

As it could be seen from the comparison of Figure 2-2 and Figure 2-3, all dependencies of the column efficiency on the flow rate follow the theoretical Van Deemter curve. In theory there is an optimum flow rate that allows obtaining the highest efficiency (the lower theoretical plate height).

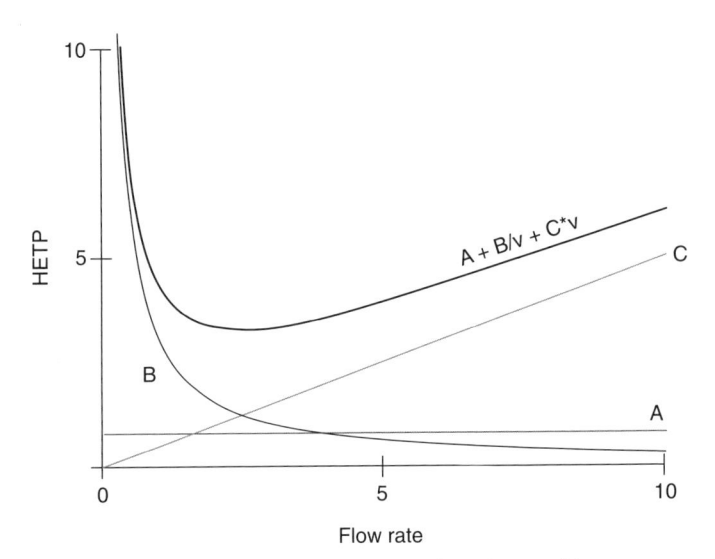

Figure 2-3. Schematic of the Van Deemter function and its components.

As follows from Figure 2-2, the lower the particle diameter, the wider the range of the flow rates where the highest column efficiency is achieved. For columns packed with smaller particles, efficiency is not as adversely affected at faster flow rates, because the mass-transfer term is lower for these columns. Essentially, this means that retention equilibrium is achieved much faster in these columns.

Faster flow rates mean higher flow resistance and higher backpressure. It is a modern trend to work with the smaller particles at high linear velocity. However, the overall efficiency of the columns packed with smaller particles (<2 μm) is not much higher compared to conventional columns with 3- to 5-μ particles. The comparison of a conventional 15-cm column with 4.6-mm internal diameter packed with 5-μm particles to a column of 15-cm length and 2-mm I.D. packed with 1.7-μm particles shows that the average efficiency of the first column is between 12,000 and 15,000 theoretical plates, and for the second column the efficiency is not much higher: It is on the level of 15,000 to 18,000 theoretical plates. This small increase of the efficiency may only slightly improve the separation; however, the comparison of the run times at the same volumetric flow rates on both columns shows that the separation on the second column can be achieved five times faster.

Of course, the ability to increase u depends on the pressure capabilities of the instrument, since pressure is directly proportional to velocity:

$$\Delta P = \frac{uL\eta\phi}{d_p^2} \qquad (2\text{-}17)$$

where ΔP is the pressure drop across the column, η is viscosity, and ϕ is the flow resistance factor [3]. Therefore, the fastest possible separation requires that the maximum pressure allowed by the instrument be used, assuming that the resolution requirement can still be met. This also means that the speed of analysis is limited by that maximum pressure. As a result, one wants to make the most of the pressure available by reducing the pressure drop across the column as much as possible. This can be achieved by working at higher temperatures, using MeCN/water mobile phases instead of methanol/water mobile phases on the same length of column or by using shorter columns.

To limit analysis time, the shortest column length possible should be used. Shorter columns have lower pressure requirements, allowing to gain an advantage in speed. It must be kept in mind, however, that N will decrease as u increases (for particles $\geq 3\,\mu m$), meaning that at faster velocities longer columns are necessary to give the required theoretical plates, thus generating greater operating pressures.

2.4 RESOLUTION

In the introduction section we define the term *resolution* as the ability of the column to resolve two analyte in two separate peaks (or chromatographic zones). In more general form than it was given before, the resolution can be defined as the half of the distance between the centers of gravity of two chromatographic zones related to the sum of their standard deviations:

$$R = \frac{X_2 - X_1}{2(\sigma_1 + \sigma_2)} \tag{2-18}$$

In case of symmetrical peaks, centers of peak gravity could be substituted with the peak maxima; and using the relationship of the peak width with its standard deviation (shown in Figure 2-1), a common expression for the resolution could be obtained:

$$R = \frac{t_{R2} - t_{R1}}{\frac{1}{2}(w_2 + w_1)} \tag{2-19}$$

The peak width in equation (2-19) could be substituted using expression (2-14), and the resulting equation is

$$R = \frac{t_{R2} - t_{R1}}{t_{R2} + t_{R1}} \cdot \frac{\sqrt{N}}{2} \tag{2-20}$$

Expression (2-20) demonstrates that resolution is proportional to the square root of the efficiency.

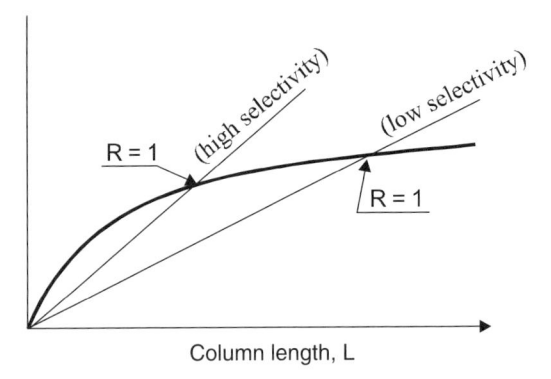

Figure 2-4. Relationship between resolution, selectivity, and column length.

From the practical point of view, in case of the lack of resolution for some specific separation there are generally two ways to improve it: Increase the efficiency, or increase the selectivity. The efficiency is proportional to the column length: The longer the column, the higher the efficiency, but equation (2-20) shows that the increase of the efficiency increases the resolution only as a square root function (as illustrated in Figure 2-4). At the same time, the increase of the column length leads to the increase of the flow resistance and backpressure, which limits the ability to further increase the column length.

If we assume that the peak widths of two adjacent peaks are approximately equal, we can rewrite expression (2-18) in the form

$$R = \frac{X_2 - X_1}{4\sigma} \qquad (2\text{-}21)$$

For symmetrical chromatographic bands, this is the ratio of the distance between peaks maxima to the peak width. The distance between peak maxima is proportional to the distance of the chromatographic zone migration, and the peak width is proportional to the square root of this distance. Figure 2-4 illustrates this relationship.

At low selectivity to achieve the same resolution, one has to use a longer column to increase efficiency and consequently operate under higher-pressure conditions. The relationship between the column length, mobile-phase viscosity, and the backpressure is given by equation (2-17), which is the variation of the Kozeny–Carman equation. Expression (2-17) predicts a linear increase of the backpressure with the increase of the flow rate, column length, and mobile phase viscosity. The decrease of the particle diameter, on the other hand, leads to the quadratic increase of the column backpressure.

Achievement of good resolution between analytes in complex chromatogram is the main goal in HPLC method development. Optimal resolution could be achieved by optimization of system efficiency, or selectivity (or

both). Relationships of the retention, selectivity, and efficiency with the resolution has been for long a subject of extensive theoretical studies [4–6], with
the goal to express resolution as a function of k, α, and N.

Unfortunately, the direct algebraic transformation of expression (2-19) into
some form of functional dependence of R on k, α, and N is impossible. Knox
and Thijssen were the first to independently propose the transformation based
on the assumption of equal peak width ($w_2 = w_1$) and consideration of the
retention of the first peak of the pair (k_1). The resulting expression is

$$R = \left(\frac{k_1}{k_1 + 1}\right)(\alpha - 1)\frac{\sqrt{N}}{4}$$
(2-22)

For closely eluted peaks and relatively high efficiency of the system, these
assumptions do not lead to the significant deviations of equation (2-22) from
true resolution given by equation (2-19).

Purnell [7] suggested to center attention on the second peak of the pair,
thus using the peak width of the second component as a base width (meaning
that the width of the first peak is equal to the width of the second peak). This
assumption leads to the following equation:

$$R = \left(\frac{k_2}{k_2 + 1}\right)\left(\frac{\alpha - 1}{\alpha}\right)\frac{\sqrt{N}}{4}$$
(2-23)

Both equations do not give a real resolution value; also, the greater the distance between peaks, the higher the error.

Said [6] suggests the use of average values instead of selection of the first
or second primary peaks, which leads to the following expression for resolution:

$$R = \left(\frac{\alpha - 1}{\alpha + 1}\right)\left(\frac{\bar{k}}{1 + \bar{k}}\right)\frac{\sqrt{N}}{2}$$
(2-24)

All these expressions give approximate values of resolution; also, the smaller
the distance between target peaks in the chromatogram, the closer the values
to the true resolution. Detailed analysis of available master resolution equations is given in the B. Karger article [4].

2.5 HPLC RETENTION

In Section 2.1 the main chromatographic descriptors generally used in routine
HPLC work were briefly discussed. Retention factor and selectivity are the
parameters related to the analyte interaction with the stationary phase and
reflect the thermodynamic properties of chromatographic system. Retention
factor is calculated using expression (2-1) from the analyte retention time or
retention volume and the total volume of the liquid in the column. Retention

time (t_R) is essentially equivalent to the retention volume ($t_R \times$ *flow rate*), provided that the mobile-phase flow rate was constant throughout the whole separation process.

Part of the total retention volume of the analyte is the void volume. Even if the analyte molecules do not interact with the column packing material, the analyte needs some time to pass through the column. This time is usually called hold-up time, dead time, or void time. The corresponding volume is either the void volume, the volume of the liquid phase in the column, or the dead volume. Analyte retention volume that exceeds the column void volume is essentially the volume of the mobile phase which had passed through the column while analyte molecules were retained by the packing material.

To derive the relationship of the analyte retention with the thermodynamic properties of chromatographic system, the mechanism of the analyte behavior in the column should be determined. The mechanism and the theoretical description of the analyte retention in HPLC has been the subject of many publications, and different research groups are still in disagreement on what is the most realistic retention mechanism and what is the best theory to describe the analyte retention and if possible predict its behavior [8, 9].

2.6 RETENTION MECHANISM

Almost 30 years ago, Colin and Guiochon mentioned in an excellent review [10] that there are essentially three possible ways to model separation mechanism. The first one is analyte partitioning between mobile and stationary phases, the second one is the adsorption of the analyte on the surface of nonpolar adsorbent, and the third one has been suggested by Knox and Pryde [11], where they assume the preferential adsorption of the organic mobile-phase modifier on the adsorbent surface followed by the analyte partitioning into this adsorbed layer.

Partitioning is the first and probably the simplest model of the retention mechanism. It assumes the existence of two different phases (mobile and stationary) and instant equilibrium of the analyte partitioning between these phases. Simple phenomenological interpretation of the dynamic partitioning process was also introduced at about the same time. Probably, the most consistent and understandable description of this theory is given by C. Cramers, A. Keulemans, and H. McNair in 1961 in their chapter "Techniques of Gas Chromatography" [12]. The analyte partition coefficient is defined as

$$K = \frac{c_S}{c_M} \tag{2-25}$$

and the analyte retention factor is defined as the ratio of the total amount of analyte in the stationary phase to the total amount of analyte in the mobile phase

$$k = \frac{q_S}{q_M} = \frac{V_S c_S}{V_M c_M} = \frac{V_S}{V_M} K \tag{2-26}$$

The fraction of the analyte in the mobile phase could be written in the form

$$R_f = \frac{q_M}{q_S + q_M} = \frac{1}{1+k} \tag{2-27}$$

and is regarded as the retardation factor. It is then assumed that R_f could be considered as the fraction of time which a component spends in the mobile phase and by multiplying mobile-phase velocity on R_f the average component velocity in the column is obtained.

$$u_c = u \frac{1}{1+k} \tag{2-28}$$

where u it the mobile-phase linear velocity and u_c is the velocity of the analyte. The retention time is the ratio of the column length and analyte velocity, thus

$$t_R = \frac{L}{u}(1+k) = t_0(1+k) \tag{2-29}$$

Converting expression (2-29) into retention volumes using $V_R = F t_R$ and $V_m = F t_0$, where F is the mobile-phase flow rate, and combining with (2-26) we get

$$V_R = V_m + V_s K \tag{2-30}$$

Later these mnemonic derivations appear in all textbooks dealing with gas and liquid chromatography. Equation (2-30) describes the retention of the analyte, which undergoes only one process of ideal partitioning between well-defined mobile and stationary phases.

In gas chromatography the analyte partitioning between mobile gas phase and stationary liquid phase is a real retention mechanism; also, phase parameters, such as volume, thickness, internal diameter, and so on, are well known and easily determined. In liquid chromatography, however, the correct definition of the mobile-phase volume has been a subject of continuous debate in the last 30 years [13–16]. The assumption that the retardation factor, R_f, which is a quantitative ratio, could be considered as the fraction of time that components spend in the mobile phase is not obvious either.

The phenomenological description of the retention mechanism discussed above is only applicable for the system with single partitioning process and well-defined stationary and mobile phases. A more general method for the derivation of retention function is based on the solution of column mass balance [17].

2.7   GENERAL COLUMN MASS BALANCE

Analyte transport through the HPLC column is considered to be one-dimensional along the axis x of the column, as shown in Figure 2-5. The analyte in the column slice dx is considered to be in instantaneous thermodynamic equilibrium. Various processes in which analyte could be involved (e.g., ionization, solvation, tatutomerism) are also considered to be at the equilibrium. To simplify the discussion and allow for the analytical solution of mass balance equation, the absence of the axial analyte dispersion is assumed.

The following assumptions on the behavior of the chromatographic systems are made:

1. Molar volumes of the analyte and mobile-phase components are constant, and compressibility of the liquid phase is negligible.
2. Adsorbent is rigid material impermeable for the analyte and mobile-phase components.
3. Adsorbent is characterized by its specific surface area and pore volume, which are evenly distributed axially and radially in the column. (This assumption is equivalent to the assumption of column homogeneity.)
4. Thermal effects are negligible.
5. System is at instant thermodynamic equilibrium.

The column void volume, V_0, is defined as the total volume of the liquid phase in the column and could be measured independently [18]. Total adsorbent surface area in the column, S, is determined as the product of the adsorbent mass and specific surface area.

Considering the classical picture of the column slice (Figure 2-5) with the mobile-phase flow rate F and length L packed with small porous particles, the analyte injected at $x = 0$ will be carried through the column with the mobile phase and its concentration in any part of the column is a function of the distance from the inlet x and the time t. The amount of the analyte entering the

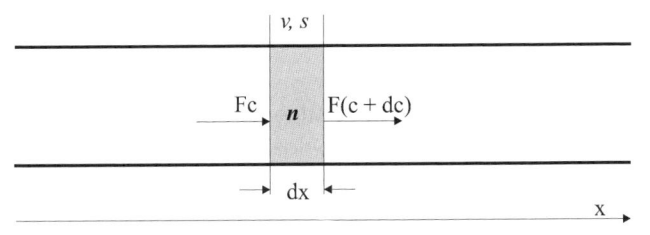

Figure 2-5. Illustration of the column slice for construction of mass balance. Mobile-phase flow F in mL/min; analyte concentration c in mol/L; n is the analyte accumulation in the slice dx in mol; v is the mobile-phase volume in the slice dx expressed as V_0/L, where L is the column length; s is the adsorbent surface area in the slice dx, expressed as S/L, where S is the total adsorbent area in the column.

cross-sectional zone dx during the time dt is equal to $Fcdt$. During the same time dt the amount of analyte leaving the zone dx could be expressed as $F(c + dc)dt$. The difference in the analyte amount entering zone dx and exiting it at the same time dt will be $Fdcdt$. Note that dc could be positive or negative. The analyte accumulation in the zone dx could be expressed in the form of the gradient of the concentration along the column axis (x)

$$Fdcdc = -F\left(\frac{\partial c}{\partial x}\right)_t dxdt \qquad (2\text{-}31)$$

Equation (2-31) represents the analyte amount accumulated in the zone dx during the time dt. This amount undergoes some distribution processes in the selected zone. These processes are the actual reason for the analyte accumulation. The analyte distribution function in the selected zone is the second half of the mass balance equation; the amount of analyte accumulated in zone dx should be equal to the amount distributed inside this zone. In general form, it could be written as

$$-F\left(\frac{\partial c}{\partial x}\right)_t dxdt = \left(\frac{\partial \psi(c)}{\partial t}\right)_x dxdt \qquad (2\text{-}32)$$

where the analyte concentration is a function of both the distance and the time and $\psi(c)$ is a distribution function that has units of the analyte amount per unit length of the column. This distribution function is the key for the solution of equation (2-32), and the definition of this function essentially determines how chromatographic retention will be described.

Equation (2-32) is simplified to

$$-F\left(\frac{\partial c}{\partial x}\right)_t = \left(\frac{\partial \psi(c)}{\partial t}\right)_x \qquad (2\text{-}33)$$

This equation states that the formation of the analyte concentrational gradient in a fixed moment of time in any place of the column should be equal to the time-dependent variation of the analyte amount.

The general solution of equation (2-33) could be obtained using two classical relationships for partial derivatives [for simple forms of the distribution function $\psi(c)$]

$$\left(\frac{\partial y}{\partial t}\right)_x = \left(\frac{\partial y}{\partial c}\right)_x \cdot \left(\frac{\partial c}{\partial t}\right)_x \qquad (2\text{-}34)$$

$$\left(\frac{\partial c}{\partial t}\right)_x = \left(\frac{\partial c}{\partial x}\right)_t \cdot \left(\frac{\partial x}{\partial t}\right)_c \qquad (2\text{-}35)$$

Substitution of equation (2-34) into equation (2-33) gives

$$-F\left(\frac{\partial c}{\partial x}\right)_t = \left(\frac{d\psi(c)}{dc}\right)_x \left(\frac{\partial c}{\partial t}\right)_x \qquad (2\text{-}36)$$

The partial derivative of the distribution function by the concentration is actually a full derivative since it is independent on the time and position of the analyte in the column given that the chromatographic system is at the equilibrium. Applying relationship (2-35), we get

$$F\left(\frac{\partial c}{\partial x}\right)_t = \frac{d\psi(c)}{dc}\left(\frac{\partial c}{\partial x}\right)_t\left(\frac{\partial x}{\partial t}\right)_c \Rightarrow F = \frac{d\psi(c)}{dc}\left(\frac{\partial x}{\partial t}\right)_c \qquad (2\text{-}37)$$

The term $(\partial x/\partial t)_c$ is linear velocity of the analyte at concentration c and can be denoted as u_c.

The final form of the simplified mass balance solution is

$$u_c = \frac{F}{\dfrac{d\psi(c)}{dc}} \qquad (2\text{-}38)$$

Since $V_R = FL/u_c$, where L is the column length, equation (2-38) could be written in its final form

$$V_R(c) = L \cdot \frac{d\psi(c)}{dc} \qquad (2\text{-}39)$$

The retention volume is essentially proportional to the derivative of the analyte distribution function defined per unit of the column length.

Further development of the mathematical description of the chromatographic process requires the definition of the analyte distribution function $\psi(c)$, or essentially the introduction of the retention model (or mechanism).

2.8 PARTITIONING MODEL

As a first example of an applicable model traditional partitioning mechanism will be considered. In this mechanism the analyte is distributed between the mobile and stationary phases, and phenomenological description of this process is given in Section 2.1. The V_m and V_s are the volume of the mobile and the volume of the stationary phases in the column, respectively. Instant equilibrium of the analyte distribution between mobile and stationary phases is assumed.

The total amount of the analyte in the column cross section dx is distributed between the mobile and stationary phases and could be written in the following form:

$$\psi(c) = v_s c_s + v_m c_m \qquad (2\text{-}40)$$

where v_s and c_s are the volumes of the stationary and mobile phases per unit of the column length ($v_s = V_s/L$, $v_m = V_m/L$) and c_s and c_m are the equilibrium analyte concentrations in the stationary and mobile phases, respectively. Since the analyte concentration in the stationary phase is an isothermal function of its concentration in the mobile phase ($c_s = f(c_m)$), using equation (2-39) we can write

$$V_R(c_m) = L \cdot \frac{d[v_s f(c_m) + v_m c_m]}{dc_m} \tag{2-41}$$

If we recall that v_m and v_s are the volumes of the mobile and stationary phase per unit of the column length, the expression (2-41) could be converted into

$$V_R(c) = V_m + V_s \frac{df(c)}{dc} \tag{2-42}$$

where $df(c)/dc$ is the derivative of the partitioning distribution function. For low analyte concentration the distribution function is assumed to be linear and its slope (derivative) is equal to the analyte distribution constant K. Equation (2-42) then could be written in the well-known form

$$V_R = V_m + V_s K \tag{2-43}$$

This equation has been derived for the model of the analyte distribution between mobile and stationary phases and is the same as expression (2-30) in Section 2.6. To be able to use this equation, we need to define (or independently determine) the volumes of these phases. The question of the determination or definition of the volume of stationary phase is the subject of significant controversy in scientific literature, especially as it is related to the reversed-phase HPLC process [19].

2.9   ADSORPTION MODEL

An alternative (or just different) description of HPLC retention is based on consideration of the adsorption process instead of partitioning. Adsorption is a process of the analyte concentrational variation (positive or negative) at the interface as a result of the influence of the surface forces. Physical interface between contacting phases (solid adsorbent and liquid mobile phase) is not the same as its mathematical interpretation. The physical interface has certain thickness because the variation of the chemical potential can have very sharp change, but it could not have a break in its derivative at the transition point through the interface. The interface could be considered to have a thickness of one or two monomolecular layers, and in RPLC with chemically modified adsorbents the bonded layer is a monomolecular layer that is more correctly considered as an interface than the separate stationary phase, as it is usually done in partitioning theory. In the adsorption model the column packing material is composed of solid porous particles with high surface area and is impermeable for the analyte and the eluent molecules.

Adsorbent nonpermeability is an important condition, since it essentially states that all processes occurs in the liquid phase. Since adsorption is related to the adsorbent surface, it is possible to consider the analyte distribution between the whole liquid phase and the surface. Using surface concentrations and the Gibbs concept of excess adsorption [20], it is possible to describe the adsorption from binary mixtures without the definition of adsorbed phase volume.

2.10   TOTAL AND EXCESS ADSORPTION

Adsorption is the process of analyte accumulation on the surface under the influence of the surface forces. Determination of the total amount of the analyte adsorbed on the surface requires the definition of the volume where this accumulation is observed, usually called the adsorbed layer volume (V_a). In chromatographic systems, adsorbents have large surface area, and even very small variation in the adsorbed layer thickness lead to a significant variation on the adsorbed layer volume. There is no uniform approach to the definition of this volume or adsorbed layer thickness in the literature [14, 21, 22].

Another approach to the expression of the analyte adsorbed on the surface is based on the consideration of the surface specific quantity which has been accumulated on the surface in excess to the equilibrium concentration of the same analyte in bulk solution. This allows avoiding an introduction of any model of adsorbed layer; as shown later, it is a fruitful approach for the description of HPLC retention.

In a liquid binary solution, this accumulation is accompanied by the corresponding displacement of another component (solvent) from the surface region into the bulk solution. At equilibrium a certain amount of the solute will be accumulated on the surface in excess of its equilibrium concentration in the bulk solution, as shown in Figure 2-6. Excess adsorption Γ of a component in binary mixture is defined from a comparison of two static systems with the same liquid volume V_0 and adsorbent surface area S. In the first system the adsorbent surface considered to be inert (does not exert any surface forces in the solution) and the total amount of analyte (component 2) will be $n_0 = V_0 c_0$. In the second system the adsorbent surface is active and component 2 is preferentially adsorbed; thus its amount in the bulk solution is decreased. The analyte equilibrium concentration c_e can only be measured in the bulk solution, so the amount $V_0 c_e$ is thereby smaller than the original quantity n_0 due to its accumulation on the surface, but it also includes the portion of the analyte in the close proximity of the surface (the portion $V_a c_e$, as shown in Figure 2-6; note that we did not define V_a yet and we do not need to define

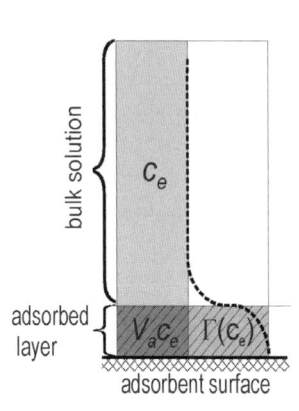

Figure 2-6. Schematic of the excess adsorption description.

it). The difference $V_0c_0 - V_0c_e$ is the excess amount accumulated on the surface on the top of what was present from the equilibrium solution. This excess amount (note that this is an amount and not a concentration) is usually related to the unit of the adsorbent surface (surface concentration) and denoted as Γ:

$$\Gamma(c_e) = \frac{V_0}{S}(c_0 - c_e) \qquad (2\text{-}44)$$

Equation (2-44) allows for the calculation of the surface-specific excessively adsorbed amount from the original analyte concentration (before adsorption) and the equilibrium analyte concentration (after adsorption equilibrium is established).

2.11   MASS BALANCE IN ADSORPTION MODEL

In any instant in the cross-section zone dx (Figure 2-5) of the column, the analyte distribution function, $\psi(c)$, could be expressed as

$$\psi(c) = v_0c_e + s\Gamma(c_e) \qquad (2\text{-}45)$$

where v_0 is the total volume of the liquid phase in that cross section (the liquid phase volume per unit of the column length), c_e is the equilibrium concentration of the analyte in the bulk liquid, s is the adsorbent surface area per unit of the column length, and $\Gamma(c_e)$ is the amount of analyte excessively adsorbed on the unit of adsorbent surface.

Expression (2-45) is essentially the analyte distribution function that could be used in the mass-balance equation (2-33). The process of mathematical solution of equation (2-33) with distribution function (2-45) is similar to the one shown above and the resulting expression is

$$V_R = V_0 + S \frac{d\Gamma(c)}{dc} \qquad (2\text{-}46)$$

This expression describes the analyte retention in binary system using only the total volume of the liquid phase in the column, V_0, and total adsorbent surface area S as parameters and the derivative of the excess adsorption by the analyte equilibrium concentration. It is important to note that the position of Gibbs dividing plane in the system has not been defined yet.

In case of injection of a very small amount of analyte, its concentration is in the linear region of adsorption isotherm (Henry region of linear variation of adsorption with the equilibrium concentration of the analyte) and the derivative could be substituted with the slope of excess adsorption isotherm, also known as Henry constant, K_H, to get

$$V_R = V_0 + SK_H \qquad (2\text{-}47)$$

This equation is very similar to the expression obtained for partitioning retention model (2-43).

It is essential that while setting the conditions for the differential mass-balance equation we did not define the function of the excess adsorption isotherm. We can now use the expression (2-46) for measurement of the model independent excess adsorption values. It is convenient to use it for the study of the adsorption behavior of binary eluents [22].

Equation (2-46) is only applicable for binary systems (analyte—single component mobile phase). Similar expression could be derived if we assume that the adsorption of the analyte does not disturb the equilibrium of the binary eluent system.

2.12   ADSORPTION OF THE ELUENT COMPONENTS

In the derivations of the retention functions so far, the adsorbed phase volume or thickness of the adsorbed layer was not introduced. The adsorbent and column parameters (surface area and void volume) independently measured are not dependent on the eluent type composition. Measurement of the void volume and adsorbent surface area is discussed in the following references 18 and 23.

In reversed-phase HPLC the binary mixture of water and organic components is the common eluent. It is obvious that organic modifier would be preferentially adsorbed on the surface of hydrophobic stationary phase, and this adsorption has been studied for more than 30 years [24–26].

Equation (2-46) is applicable for the description of adsorption behavior of binary eluent in the column. In the isocratic mode, the column is equilibrated at given composition of the eluent. Any small volume of the mixture of eluent

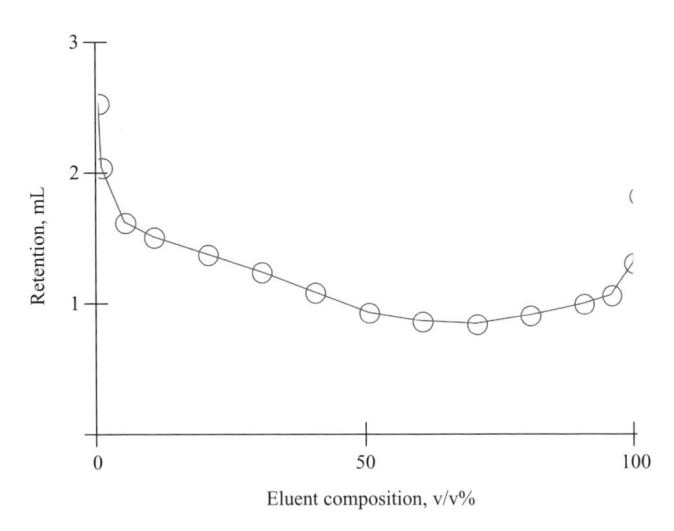

Figure 2-7. Minor disturbance peak retention dependence as the function of the acetonitrile concentration in water.

components with a small difference in the concentration compared to the eluent that is injected in the column introduces a so-called minor disturbance in the system. A minor disturbance peak will have the retention volume defined by equation (2-46). Typical experimental dependence of the minor disturbance peak retention is shown in Figure 2-7.

Integration of function (2-46) allows the calculation of the excess adsorption values [22]:

$$\Gamma(c) = \frac{1}{S}\int_0^c (V_R - V_0)\, dc \qquad (2\text{-}48)$$

Integration of this dependence through the whole concentration range actually allows the calculation of the column void volume, or the total volume of the liquid phase in the column. Since excess adsorption of pure component is equal to 0 ($\Gamma(0) = \Gamma(100) = 0$), then

$$V_0 = \frac{\displaystyle\int_{c_0=0}^{c_{max}=100} V_R(c)\, dc}{c_{max} - c_0} \qquad (2\text{-}49)$$

Since the measurement of the excess adsorption isotherm of a component in the binary system does not require a priori introduction of any model, it is possible to consider the excess adsorption isotherm as being model-independent (within a framework of adsorption process) and it is possible to derive the properties of the adsorbed layer on the basis of consideration of

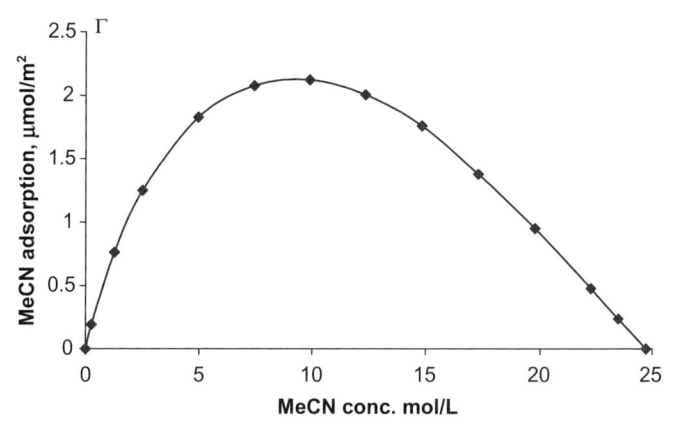

Figure 2-8. Typical excess adsorption isotherm of acetonitrile from water on the surface of reversed-phase silica.

a concentration-dependent adsorption process. A typical excess adsorption isotherm is shown in the Figure 2-8.

This plot represents the variation of an excessively adsorbed amount of acetonitrile with the variation of the equilibrium concentration of acetonitrile in the bulk solution. In the adsorption system the influence of adsorption forces exerted by the adsorbent surface are limited in their distance; consequently, we should have limited volume where adsorbed analyte accumulates. It is also assumed that liquid is uncompressible and that molar volumes of both components do not change under the influence of adsorption forces. This leads to the displacement adsorption mechanism.

At each point on the isotherm (except the origin and the endpoint) the amount shown in Figure 2-8 represents the acetonitrile quantity adsorbed on one square meter of the surface in excess to the amount of acetonitrile which is on the surface originally from equilibrium concentration, which could be calculated only if the volume of the adsorbed phase (or Gibbs dividing plane) will be introduced. Instead of introduction of the volume of the adsorbed phase, let us consider the variation of the excess adsorption value at the end of the isotherm (in the region between 15 and 25 mol/L of acetonitrile in Figure 2-8).

An increase of the acetonitrile equilibrium concentration in this region leads to the linear decrease of the excessively adsorbed amount. The adsorbed layer has a finite volume (or finite thickness), and in this region of very high acetonitrile concentration it is possible to assume that the adsorbed phase is completely filled with acetonitrile, and therefore the following expression for only this region on the isotherm could be written:

$$(c_e V_a + \Gamma(c_e))v_m = V_a \qquad (2\text{-}50)$$

where c_e is the acetonitrile equilibrium concentration, V_a is the hypothetical adsorbed layer volume, $\Gamma(c_e)$ is the excess adsorption, v_m is acetonitrile molar volume. If this layer is not completely filled with acetonitrile molecules within this concentration region, then there would be an additional degree of freedom in the excess adsorption variation with equilibrium concentration and the isotherm would not show a linear decrease at zero excess with 100% of acetonitrile in the bulk solution:

If we calculate the derivative of equation (2-50) by the equilibrium concentration, we get

$$V_a = -\frac{d\Gamma(c_e)}{dc_e} \qquad (2\text{-}51)$$

In another words, the negative slope of the excess adsorption isotherm in the linear region is equal to the volume of adsorbed layer, which was derived from the consideration of the adsorption process and not from a prior introduction of the model. A similar expression was derived by Everett [27].

The analysis of experimental excess adsorption isotherms using equation (2-50) had shown unusual results [22]. The adsorbed layer thickness of acetonitrile adsorbed from water on different types of reversed-phase adsorbents calculated as the ratio of adsorbed layer volume and adsorbent surface area appears to be on average equal to 14 Å, which is equivalent to approximately five monolayers of acetonitrile molecules adsorbed on the hydrophobic surface. At the same time, the adsorbed layer thickness of methanol adsorbed from water on the same adsorbents is equal to only 2.5 Å, which is equivalent to the monolayer-type adsorption.

From the above discussion it is logical to assume that adsorbed layer thickness τ, calculated as the ratio of V_a/S, is the maximum distance of the influence of the surface forces. The sum of the excess adsorption value and the product of the equilibrium concentration and adsorbed layer volume represent the total amount of the adsorbate in that layer for any given equilibrium concentration

$$a(c_e) = \Gamma(c_e) + V_a c_e \qquad (2\text{-}52)$$

Equation (2-52) is the total adsorption isotherm derived from experimentally measurable excess isotherm using the model of adsorption process obtained from the analysis of the experimental isotherm profile.

The adsorption isotherms of the eluent components have been studied extensively [22, 25, 26, 28, 29]. However, significant controversy still exists with regard to the interpretation of where the accumulation of the molecules of organic eluent modifier actually occurs. Horvats and further development by Dill [30, 31] stated that organic molecules penetrate between the bonded ligands of the stationary phase, and thus this monomolecular layer of bonded ligands can be considered as a stationary phase. Significant drawback of this

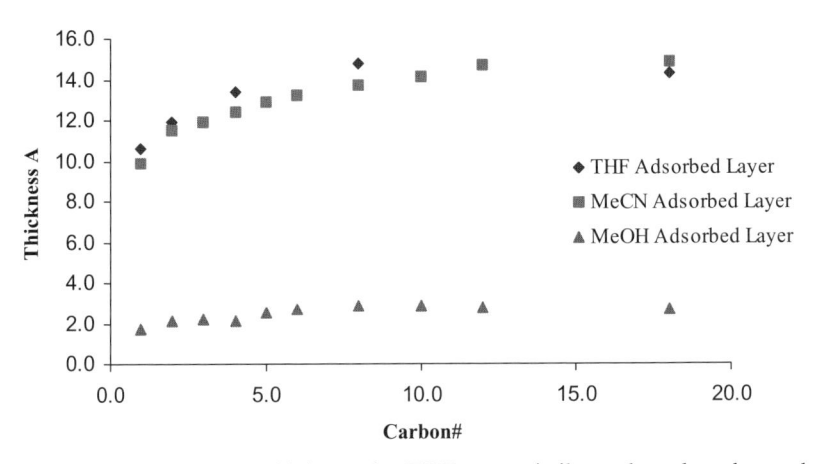

Figure 2-9. Adsorbed layer thickness for THF, acetonitrile, and methanol on adsorbents of different surface chemistry. (Reprinted from reference 22, with permission.)

theory is that stationary-phase volume is dependent on the length of the bonded chains: The shorter the chain, the less penetration is possible. The C1-type bonded phase is a limiting case: It is definitely a reversed-phase adsorbent, but there is no possible penetration between bonded trimethylsilane ligands (these ligands do not have conformational freedom), and the accumulation on the surface of this stationary phase should be considered as adsorption. Moreover, in several studies it was shown that the amount of the organic eluent component accumulated on the surface of the HPLC packing material is almost independent on the length of the bonded chains (Figure 2-9)[22].

This suggests that there is no actual penetration of the eluent and also the analyte molecules between the bonded ligands, but rather the accumulation (adsorption) of them on the surface. This was also confirmed by the study of the viscosity of bonded chains [32], which shows that the viscosity of bonded C18-type alkyl chains is at least two orders of magnitude higher than the viscosity of corresponding free alkanes (note that free octadecane is solid at room temperature, and it is almost insoluble in acetonitrile).

2.13   VOID VOLUME CONSIDERATIONS

Void volume is a critical parameter in HPLC, since most theoretical relationships in chromatographic theory deal with the retention factor, and the accuracy of the void volume determination plays an important role in all calculations.

The concept of the void volume and mobile-phase volume was briefly discussed in Chapter 1, Section 1.7.2. In the partitioning model (Section 2.8) the total volume of the liquid phase in the column is equal to the sum of the volumes of the mobile and stationary phase in the same column.

Adsorption model of HPLC retention mechanism allows clear definition of the column void volume as the total volume of the liquid phase in the column, but this model requires the use of the surface-specific retention and the correlation of the HPLC retention with the thermodynamic (and thus energetic) parameters, which is not well-developed. This model requires the selection of the standard state of given chromatographic system and relation of all parameters to that state.

In most analytical applications of HPLC, all these discrepancies are quietly and conveniently forgotten, and selection of some so-called "nonretained" component as a void volume marker is a common way for void volume measurement. In the majority of recent analytical publications, either thiourea or uracil were used as the void volume markers. As a disclaimer, we have to say here that for the purposes of analytical method development, qualitative or quantitative separation of complex mixtures which involves the use of a "nonretained" component as a void volume marker is acceptable insofar as there are no physicochemical generalization, thermodynamic development, or futher theoretical development performed upon the basis of these "pseudo" void volume determinations.

As a simple illustration of the inconsistency of the use of any nonretained components for the void volume determination, we remind the reader that in the basic assumptions and in the derivation of expressions (2-43) and (2-46), void volume of the column was considered to be constant in any eluent type and composition. Figure 2-10 illustrate the dependence of the thiourea and uracil retention in acetonitrile/water and in methanol/water as a function of organic content in the eluent. The horizontal line in Figure 2-10 shows the

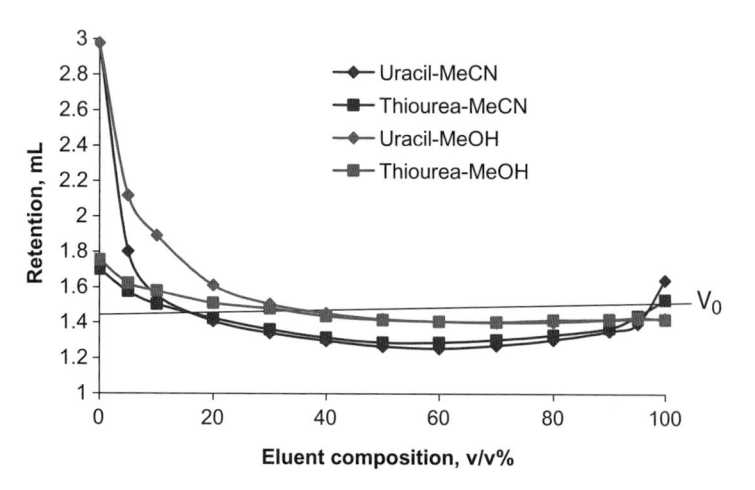

Figure 2-10. Dependence of the retention of uracil (diamonds) and thiourea (squares) on the eluent composition on Restek Allure-C18 column ($150 \times 4.6\,\mathrm{mm}$, $5\,\mu\mathrm{m}$). Acetonitrile/water (solid lines) and methanol/water (dashed lines) were used as the mobile-phase components.

volume of the liquid phase in the column measured according to equation (2-49). Note that the retention of both markers for most used eluent compositions is lower than the volume of the liquid phase. This indicates the exclusion of the marker molecules from the adsorbent surface and preferential adsorption of the organic eluent components.

The use of the true volume of the liquid phase in the column as the void volume can lead to the principal difficulties in the interpretation of the retention of polar analytes that are also excluded from the contact with the adsorbent surface. The retention volume of these analytes will be lower than the column void volume, and thus their retention factors will be negative. A logarithm of negative retention factors does not exist that shows the applicability limit of the approximate theory described above. In a general sense the void volume should not change as a function of the type and organic composition. Table 2-1 demonstrates the compatibility of the void volume measured using different thermodynamically consistent methods.

2.14 THERMODYNAMIC RELATIONSHIPS

If we disengage ourselves from the consideration of mobile and stationary phases and assume (as a very rough approximation) that column void volume is the volume of the mobile phase ($V_m = V_0$) and leave stationary-phase volume undefined (as V_s), we can use basic expression for the retention factor (from Chapter 1) and expression (2-43) to get

TABLE 2-1. Void Volume Values Measured with MeCN/Water, MeOH/Water, and THF/Water and a Deuterated Eluent Component for a Set of Columns Packed with Adsorbents Modified with Mono-alkylsilanes of Different Chain Length

Chain Length	$V_{labeled}$[a] from MeCN	V_{md}[b]			RSD%
		MeCN/Water	MeOH/Water	THF/Water	
1	1.914	1.893	1.954	2.015	2.76
2	1.895	1.909	1.951	1.974	1.90
3	1.850	1.838	1.86		0.6
4	1.885	1.876	1.898	1.910	0.79
5	1.860	1.846	1.880		0.92
6	1.833	1.818	1.847		0.79
8	1.827	1.814	1.869	1.865	1.49
10	1.811	1.797	1.825		0.77
12	1.770	1.765	1.807		1.29
18	1.723	1.713	1.751	1.718	0.98

[a]Retention volume of deuterated acetonitrile eluted using pure acetonitrile flow.
[b]Integral average of the minor disturbance peak retention dependence on the eluent concentration.

Source: Reprinted from reference 18, with permission.

$$k = \frac{V_s}{V_0} K \qquad (2\text{-}53)$$

which means that the chromatographic retention factor is proportional to the partitioning equilibrium constant (assuming partitioning retention mechanism with only partitioning equilibrium in the column). Thermodynamic equilibrium constant is the exponent of the Gibbs free energy of the system, so we can write

$$\ln(k) = \frac{\Delta G}{RT} + \ln\left(\frac{V_s}{V_0}\right) \qquad (2\text{-}54)$$

Most chromatographic systems with absence of the secondary equilibria effects (such as analyte ionization, specific interactions with active adsorption sites, etc.) show linear dependencies of the logarithm of the retention factors on the inverse temperature, as shown in Figure 2-11.

Figure 2-12 represents the temperature dependencies of a homologous series of alkylbenzenes retention at 60% MeCN/40% water on a Phenoemenex Luna-C18 column. The intercepts for each analyte is different from others, which essentially means that each analyte requires the determination of its own stationary phase volume.

Application of the partitioning mechanism for the description of the retention process leads to another theoretical consequence applicable to ideal chromatographic systems; that is, only one retention mechanism is present and no secondary equilibria effects are observed. Liquid chromatography is a competitive process, where analyte molecules compete with the eluent molecules for the retention on the stationary phase; based on that, the standard state of

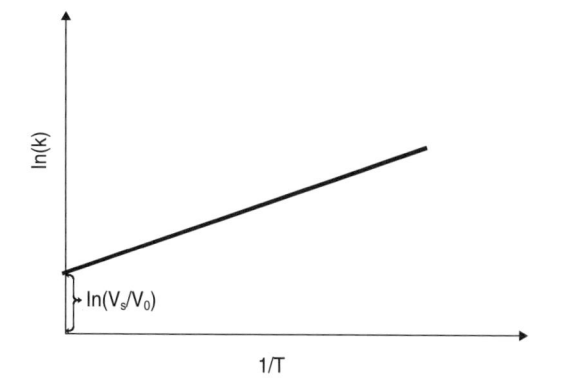

Figure 2-11. Ideal dependence of the HPLC retention on the temperature.

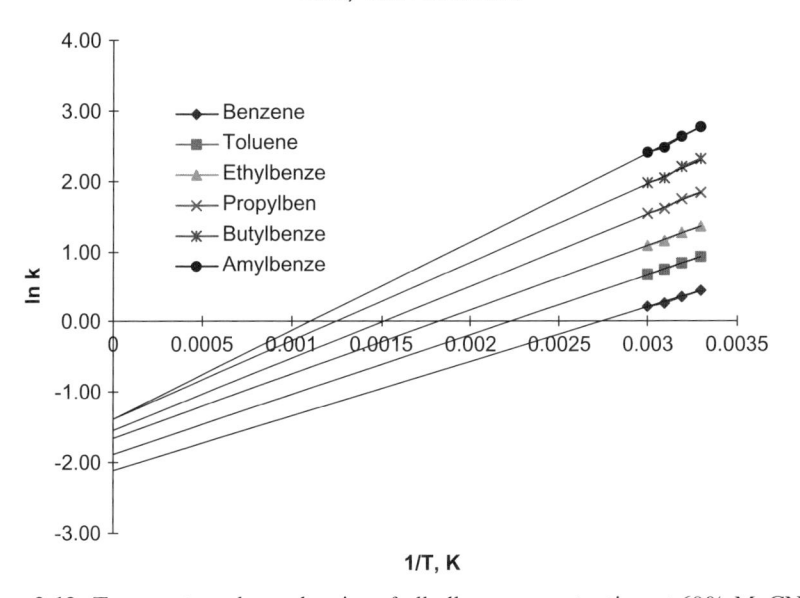

Luna, 60% Acetonitrile

Figure 2-12. Temperature dependencies of alkylbenzenes retention at 60% MeCN/40 water on Luna-C18 column.

a chromatographic system as the system without the eluent molecules can be defined. Equation (2-54) could be rewritten in the form

$$\ln(k) = \frac{\Delta G_{an} - \Delta G_{el}}{RT} + \ln\left(\frac{V_s}{V_0}\right) \qquad (2\text{-}55)$$

where ΔG_{an} is the Gibbs free energy of the analyte molecules in the chromatographic system, and ΔG_{el} is the Gibbs free energy of the chromatographic system with only the eluent. If we inject the small amount of labeled eluent molecules, then $\Delta G_{an} = \Delta G_{el}$ and from the expression (2-55) we get $k = V_s/V_0$, or $V_R - V_0 = V_s$. This result is consistent with the partitioning process, since we only consider the analyte partitioning between the mobile and stationary phases and assume that the eluent is the mobile phase only.

Let us consider the selectivity expression introduced in Chapter 1:

$$\alpha = \frac{k_2}{k_1} \qquad (2\text{-}56)$$

Expression (2-56) could be modified (assuming that none of the parameters are equal to 0) in the following form:

$$\alpha = \exp(\ln(k_2) - \ln(k_1)) \qquad (2\text{-}57)$$

Substituting expression (2-57) for each analyte into equation (1-5), we get

$$\alpha = \exp\left[\frac{\Delta G_{an.2} - \Delta G_{an.1}}{RT}\right] \tag{2-58}$$

Expression (2-58) contains only the Gibbs free energies of the analyte interactions in the column and no eluent-related terms. This means that in ideal
systems (in the absence of secondary equilibria effects) the eluent type or the
eluent composition should not significantly influence the chromatographic
selectivity. This effect could be illustrated from the retention dependencies of
alkylbenzenes on a Phenoemenex Luna-C18 column analyzed at various ace-
tonitrile/water eluent compositions (Figure 2-13, Table 2-2).

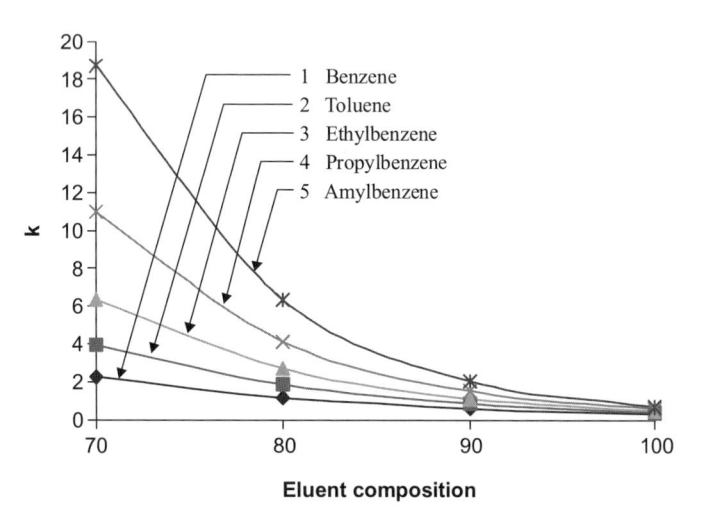

Figure 2-13. Retention of alkylbenzenes on Luna-C18 column from acetonitrile/water
eluent.

**TABLE 2-2. Selectivity Between Alkylbenzenes at
Different Eluent Compositions on Reversed-Phase
Column**[a]

MeCN Composition	Selectivity		
	Peak Numbers		
	2, 1	3, 2	4, 3
70	1.75	1.6	1.45
80	1.72	1.58	1.40
90	1.70	1.65	1.43
100	1.71	1.62	1.42

[a]Retention dependencies are shown in Figure 2-13.

2.14.1 Effect of the Eluent Composition

Effect of the eluent composition could be discussed on the basis of equation (2-55). In the simple case of binary eluent (organic/water mixture), we can consider in the first approximation that Gibbs free energy of the eluent interaction with the packing material surface is a linear function of the eluent composition

$$\Delta G_{el} = \Delta G_{H_2O} + x\Delta G_{org} \qquad (2\text{-}59)$$

where ΔG_{el} is an overall Gibbs free energy of the eluent in the chromatographic system, ΔG_{H_2O} is the free energy of water, ΔG_{org} is the Gibbs free energy of organic eluent component, and x is the concentration of the organic component in the eluent. This consideration assumes an ideal behavior of the system and does not account for adsorption of the eluent components on the surface of the stationary phase, despite the fact that a fairly reasonable description of the effect of the eluent composition on the analyte retention is obtained. Substituting equation (2-59) into expression (2-55), we get

$$\ln(k) = \frac{\Delta G_{an} - \Delta G_{H_2O} + x\Delta G_{org}}{RT} + \ln\left(\frac{V_s}{V_0}\right) \qquad (2\text{-}60)$$

Equation (2-60) represents linear dependence of the logarithm of the retention factor on the eluent composition:

$$\ln(k) = a + x \cdot b \qquad (2\text{-}61)$$

where a and b are constants. Physical sense of these constants is complex and very approximate. Expression (2-61) predicts the exponential dependence of the analyte retention on the eluent composition, in most ideal or close to ideal chromatographic systems, as shown in Figures 2-14 and 2-15.

   This simple dependence of the retention factors on the eluent composition could be used for practical rough estimation of the retention variation with the variation of the concentration of organic modifier in the mobile phase. If the retention of particular analyte was measured at two different eluent compositions, then the retention of that same analyte at any other composition could be roughly evaluated from the following expression:

$$k(x) = \exp\left[\frac{\ln(k_1) - \ln(k_2)}{x_1 - x_2}(x - x_1) + \ln(k_1)\right] \qquad (2\text{-}62)$$

where $k(x)$ is the analyte retention at eluent composition x (concentration is expressed in the mole fractions or volume factions), and k_1 and k_2 are the experimental analyte retention factors measured at the eluent composition of x_1 and x_2 accordingly.

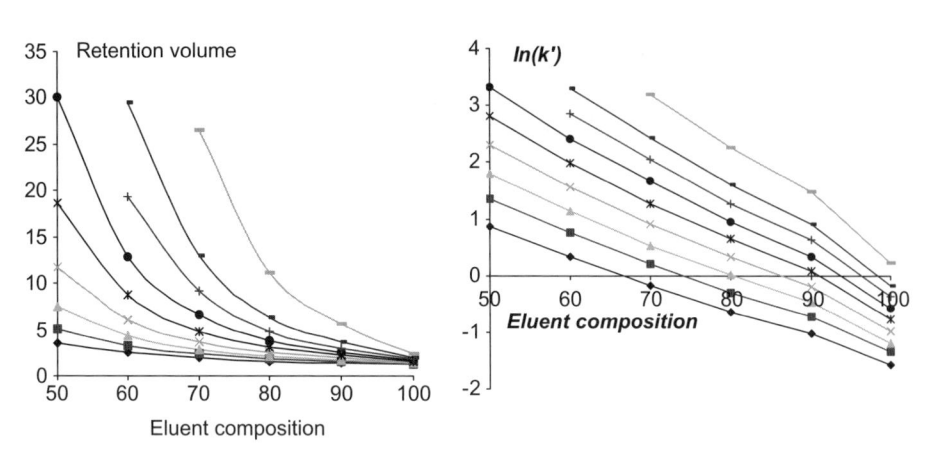

Figure 2-14. Retention of alkylbenzenes (benzene–octylbenzene) on Agilent Zorbax-Eclipse XDB C18 column from different composition of acetonitrile/water eluent. Left pane: Retention volumes; right pane: Retention factors.

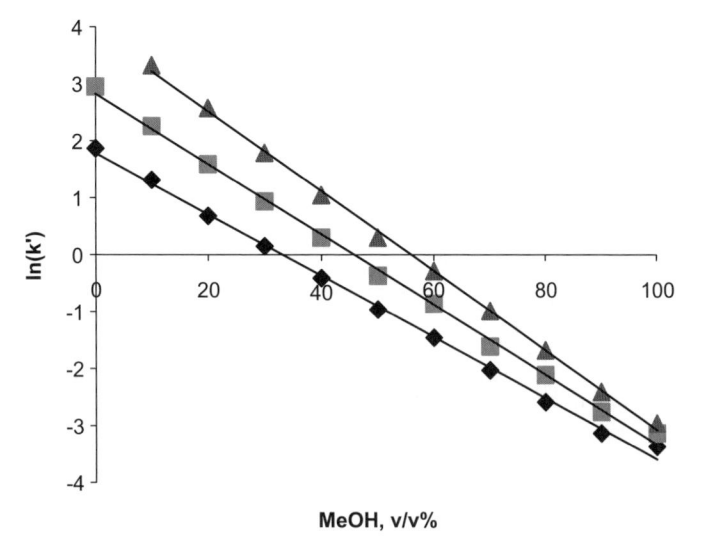

Figure 2-15. Retention of alkylpyridines on Phenomenex Luna-C18 column from methanol/water eluent. (diamonds, pyridine; squares, 4-methylpyridine; triangles, 4-ethylpyridine).

2.15 ADSORPTION-PARTITIONING RETENTION MECHANISM

Formation of a thick adsorbed layer of acetonitrile on the surface of reversed-phase adsorbent allows the introduction of a two-stage model of the analyte retention process. The first process is the partitioning of the analyte molecules from the bulk eluent into the adsorbed acetonitrile layer, and the second process is the analyte adsorption on the surface of the packing material.

The binary eluent adsorption equilibrium is considered to be not disturbed by the injection of a small amount of the analyte (essentially the third component in the system). In an isocratic mode at a fixed eluent composition, the organic adsorbed layer is a stationary phase for the analyte to partition into. The analyte can partition into the adsorbed layer followed by consequent adsorption on the surface of the reversed-phase adsorbent. The overall retention is a superposition of two consecutive processes. Since the eluent component adsorption could be measured independently and adsorbed layer volume could be represented as a function of the mobile phase composition, the analyte retention also could be expressed as a function of the eluent composition.

This model assumes the absence of any disturbance of the acetonitrile adsorption equilibrium by the analyte. This assumption limits the application of this model toward very low analyte concentrations—typical conditions for linear elution chromatography.

The analyte distribution function in the column cross section dx could be written in the following form:

$$\psi(c_e) = v_m c_e + v_s c_s + s\Gamma(c_s)$$ (2-63)

where v_m is the volume of the equilibrium liquid phase in the selected cross section of the column with thickness dx, c_e is the analyte equilibrium concentration in the mobile phase, v_s is the volume of the adsorbed acetonitrile layer, c_s is the analyte concentration in that phase at the equilibrium with the mobile phase and with the surface, s is the adsorbent surface area, and Γ is the excess adsorption of the analyte on the surface from the stationary phase (where the analyte concentration is c_s).

Because of the assumption made above, the solution is limited to the linear region of analyte partitioning and adsorption isotherms. The analyte distribution between two liquid phases (eluent and adsorbed phase) at equilibrium could be described as follows:

$$c_s = K_p c_e$$ (2-64)

where K_p is the analyte distribution constant between the eluent and the adsorbed organic layer. Since the analyte is injected in very low amount, its adsorption on the surface of bonded phase is also assumed to be in the Henry region. The adsorption process could be described as

$$\Gamma(c_s) = K_H c_s$$ (2-65)

or substituting from equation (2-64)

$$\Gamma(c_s) = K_H K_p c_e$$ (2-66)

where K_H is a Henry adsorption constant.

The final form of the distribution function, accounting for $v_0 = v_m + v_s$, therefore will be

$$\psi(c_e) = [v_0 + (K_p - 1)v_s + sK_H K_p]c_e \qquad (2\text{-}67)$$

Applying this function into the mass-balance equation (2-33) and performing the same conversions [Eqs. (2-34)–(2-39)], the final equation for the analyte retention in binary eluent is obtained. In expression (2-67) the analyte distribution coefficient (K_p) is dependent on the eluent composition. The volume of the acetonitrile adsorbed phase is dependent on the acetonitrile adsorption isotherm, which could be measured separately. The actual volume of the acetonitrile adsorbed layer at any concentration of acetonitrile in the mobile phase could be calculated from equation (2-52) by multiplication of the total adsorbed amount of acetonitrile on its molar volume. Thus, the volume of the adsorbed acetonitrile phase (v_s) can be expressed as a function of the acetonitrile concentration in the mobile phase (V_s (c_{el})). Substituting these in equation (2-67) and using it as an analyte distribution function for the solution of mass balance equation, we obtain

$$V_R(c_{el}) = V_0 - V_s(c_{el}) + K_p(c_{el})[V_s(c_{el}) + SK_H] \qquad (2\text{-}68)$$

$V_R(c_{el})$ is the analyte retention as a function of the eluent concentration, V_0 is the total volume of the liquid phase in the column, $V_s(c_{el})$ is the volume of adsorbed layer as a function of eluent composition, $K_p(c_{el})$ is the distribution coefficient of the analyte between the eluent and adsorbed phase, S is the adsorbent surface area, and K_H is the analyte Henry constant for its adsorption from pure organic eluent component (adsorbed layer) on the surface of the bonded phase.

Practical applicability of equation (2-68) has been demonstrated for the retention of neutral and polar compounds in the region of the completed acetonitrile layer [22]. This concept has also been applied for the explanation of unusual chromatographic behavior of PF_6^- ions on reversed-phase columns with acetonitrile and methanol as organic eluent modifiers [33]. PF_6^- ions are essentially precursors or typical anions of now popular ionic liquids, which show unusual retention dependencies with increasing acetonitrile concentration. Their retention increases with the increase of the acetonitrile content from 0 to approximately 20 v/v%, and then the retention starts to decrease at higher acetonitrile concentrations in the mobile phase. PF_6^- is a liophilic ion soluble in pure acetonitrile up to almost 0.5 M, and thus it can be retained in the acetonitrile adsorbed layer. In a high aqueous mobile phase an increase of the acetonitrile concentration leads to the increase of the amount of acetonitrile adsorbed on the surface, thus increasing the adsorption capacity for the retention of PF_6^- ions. Upon further increase of acetonitrile concentration, this leads to an increase of the elution strength of the eluent, and consequently the retention for PF_6^- ion starts to decrease. Methanol, on the other hand, does not show any significant effect on the PF_6^- chromatographic retention behavior. All these effects were successfully described mathematically using equation (2-68) [33].

2.16   SECONDARY EQUILIBRIA

Analyte ionization, tautomerization, or solvation equilibrium in the chromatographic column has a profound effect on the retention and efficiency. These effects are known as secondary equilibria effects [34, 35]. The effect of the analyte ionization on the retention has been extensively studied [36, 37]. Fundamental work by C. Horvath and co-workers created a solid foundation in this field [30, 38] for ionic equilibria of

$$[AH] \Leftrightarrow [A^-] + [H^+] \tag{2-69}$$

with the equilibrium constant expressed as

$$K_a = \frac{[A^-] \cdot [H^+]}{[AH]} \tag{2-70}$$

On the basis of the assumption that a retention factor is a measure of the stoichiometric mass distribution of the analyte between stationary and mobile phases, they derive the following expression for the overall retention factor:

$$k = \frac{k_0 + k_1 \dfrac{[H^+]}{K_a}}{1 + \dfrac{[H^+]}{K_a}} \tag{2-71}$$

where k_0 is the hypothetical retention factor of a neutral base and k_1 is the retention factor for an ionized base. This equation represents sigmoidal dependence of the basic analyte retention on the mobile-phase pH with the inflection point corresponding to pH = pK_a (Figure 2-16).

As it was discussed in Section 2.7, if the model of the analyte behavior in the column is defined, then this model can be applied in the mass-balance equation. The resulting solution of the mass-balance equation is the expression for the analyte retention behavior, and it is only valid in the frame of the selected model.

So far the solution of the mass-balance equation for models with a single dominating process (partitioning or adsorption) was discussed in Sections 2.8 and 2.9. In both cases the solutions have similar form, with the difference in the definition of the parameters (volumes of the mobile and stationary phases in the case of partitioning; total volume of the liquid phase and adsorbent surface area in the case of adsorption model).

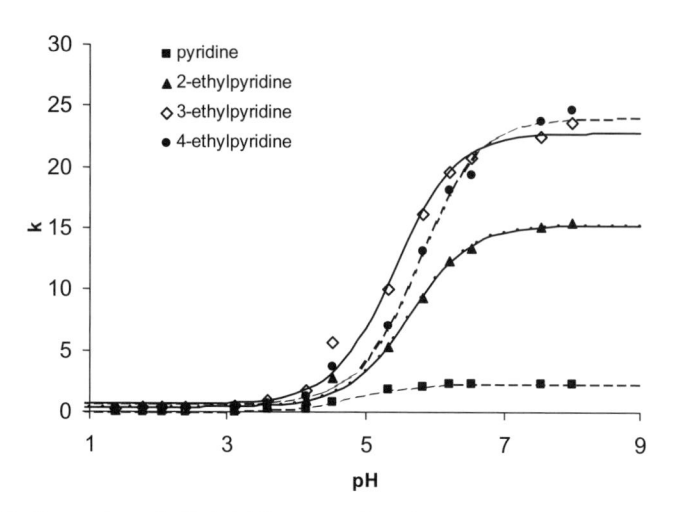

Figure 2-16. Retention of alkylpiridines as a function of mobile phase pH. (Reprinted from reference 39, with permission.)

2.16.1 Inclusion of Secondary Equilibria in the Mass Balance

HPLC analytes involved in adsorption equilibrium processes could also participate in other types of processes, such as ionization, tautomerization, solvation, and so on. The analyte behavior in adsorption equilibrium is certainly affected by the presence of secondary equilibrium, and if it could be considered as an independent process based on the time and position of the analyte within the column, this equilibrium could be incorporated in the solution of mass-balance equation.

Let us assume the presence of ionic equilibrium for an ionizible basic analyte:

$$B + H^+ = BH^+ \tag{2-72}$$

Above equilibrium is dependent on the mobile-phase pH and the relationship between ionic and nonionic form of the analyte is described by Henderson–Hasselbalch equation

$$\frac{c_{BH^+}}{c_B} = 10^{pK_a - pH} \quad \text{or} \quad \frac{c_{BH^+}}{c_B} = \frac{[H^+]}{K_i} \tag{2-73}$$

where K_i is the analyte ionization constant. Both ionic and nonionic forms of the analyte can be adsorbed on the surface, and the absence of the ionic equilibrium in the adsorbed state is assumed; this means that if the analyte gets adsorbed in the ionic form, it stays in this form on the surface.

The left-hand side of equation (2-33) expresses the presence of the gradient of the analyte in the column cross section. In the case of an ionizable analyte, there are two forms of the analyte present, and using expression (2-73) the left-hand side of equation (2-33) should be written in the form

$$-F\left(\frac{\partial c}{\partial x}\right)_t dxdt = -F\left(\frac{\partial(c_B + c_{BH^+})}{\partial x}\right)_t dxdt = -F\left(1 + \frac{[H^+]}{K_i}\right)\left(\frac{\partial c_B}{\partial x}\right)_t dxdt \quad (2\text{-}74)$$

The analyte distribution function in the column cross section dx (Figure 2-5) includes both forms of the analyte in the bulk solution and on the absorbent surface:

$$\psi(c) = v_0 c_B + v_0 c_{BH^+} + s\Gamma_B(c_B) + s\Gamma_{BH}(c_{BH^+}) \quad (2\text{-}75)$$

where v_0 is the volume of the liquid phase in the column per unit of the column length, c_B is the equilibrium concentration of the neutral form of basic analyte, c_{BH^+} is the equilibrium concentration of the protonated form of basic analyte, s is the surface area of the adsorbent in the column per unit of the column length; $\Gamma_B(c_B)$ is excess adsorption of the neutral form of the analyte as the function of the equilibrium concentration of corresponding form; $\Gamma_{BH}(c_{BH^+})$ is the excess adsorption of the ionic form as the function of the equilibrium concentration of this form of the analyte.

Using this distribution function in equation (2-33), we get

$$-F\left(\frac{\partial c}{\partial x}\right)_t = \left(\frac{\partial[v_0 c_B + v_0 c_{BH^+} + s\Gamma_B(c_B) + s\Gamma_{BH}(c_{BH^+})]}{\partial t}\right)_x \quad (2\text{-}76)$$

The equilibrium concentration of the ionic analyte form (c_{BH^+}) in the second term could be substituted with c_B from expression (2-73):

$$-F\left(1 + \frac{[H^+]}{K_i}\right)\left(\frac{\partial c_B}{\partial x}\right)_t = \left(v_0\frac{\partial c_B}{\partial t}\right)_x + \left(v_0\frac{[H^+]}{K_i}\frac{\partial c_B}{\partial t}\right)_x$$
$$+ \left(s\frac{\partial\Gamma_B(c_B)}{\partial t}\right)_x + \left(s\frac{\partial\Gamma_{BH}(c_{BH^+})}{\partial t}\right)_x \quad (2\text{-}77)$$

The same transformations of variables as in equations (2-34) and (2-35) could be used to get

$$F\left(1 + \frac{[H^+]}{K_i}\right)\left(\frac{\partial c_B}{\partial x}\right)_t = \left[v_0\left(1 + \frac{[H^+]}{K_i}\right) + s\left(\frac{d\Gamma_B(c_B)}{dc_B} + \frac{[H^+]}{K_i}\frac{d\Gamma_{BH}(c_{BH^+})}{dc_B}\right)\right]$$
$$\left(\frac{\partial c_B}{\partial x}\right)_t\left(\frac{dx}{dt}\right)_c \quad (2\text{-}78)$$

The expression $(dx/dt)_c$ is the linear velocity of the analyte in the column u_c, and we get the final equation for the retention volume of ionizable analyte using the same conversions as in equations (2-33)–(2-39).

$$V_R = V_0 + S \frac{\left(\dfrac{d\Gamma_B(c_B)}{dc_B} + \dfrac{[H^+]}{K_i} \dfrac{d\Gamma_{BH}(c_{BH^+})}{dc_B} \right)}{\left(1 + \dfrac{[H^+]}{K_i} \right)} \qquad (2\text{-}79)$$

Equation (2-79) is the general form describing retention of ionizable analytes. Since it was derived with the assumption that injected analyte does not noticeably disturb the eluent adsorption equilibrium in the column, it is only applicable for very low analyte concentrations. At these low analyte concentrations, the slope of the excess adsorption isotherm is assumed to be constant and we can substitute the derivatives of the excess adsorption functions for both forms of the analyte with corresponding Henry constants (K_B and K_{BH}):

$$V_R = V_0 + S \frac{\left(K_B + \dfrac{[H^+]}{K_i} \right) K_{BH^+}}{\left(1 + \dfrac{[H^+]}{K_i} \right)} \qquad (2\text{-}80)$$

At proton concentration at least hundred times higher than the analyte ionization constant for a basic analyte (pH is two units lower than pK_a), expression (2-80) reduces to $V_R = V_0 + SK_{BH^+}$. It essentially represents the retention volume of only ionic form of the analyte. At a pH at least two units higher than the basic analyte pK_a (suppressed ionization conditions), expression (2-80) reduces to $V_R = V_0 + SK_B$ and it represents the retention of only nonionic form of the analyte at conditions where protonation is completely suppressed. Corresponding capacity factors for neutral and protonated forms of basic analyte could be written in the form

$$k_B = \frac{S}{V_0} K_B \quad \text{(neutral)} \qquad (2\text{-}81)$$

$$k_{BH} = \frac{S}{V_0} K_{BH^+} \quad \text{(protonated)} \qquad (2\text{-}82)$$

Substituting K_B and K_{BH^+} from expressions (2-81) and (2-82) into equation (2-80) and expressing the overall analyte retention in the form of retention factor, we get

$$k = \frac{k_B + k_{BH^+} \dfrac{[H^+]}{K_a}}{1 + \dfrac{[H^+]}{K_a}} \tag{2-83}$$

This is exactly the same expression for the retention of ionizable basic analyte as a function of mobile-phase pH, which was first derived by Horvath and Melander in 1977 [38].

Equation (2-80) expresses the retention of an ionizable basic analyte as a function of pH and three different constants: ionization constant (K_a); adsorption constant of ionic form of the analyte (K_{BH^+}); and adsorption constant of the neutral form of the analyte (K_B). These three constants describe three different equilibrium processes, and they have their own relationships with the system temperature and Gibbs free energy with respect to the particular analyte form.

Modeling of the retention of aniline ($pK_a = 4.6$) with derived equation (2-83) had shown that with the increase of the temperature at mobile-phase pH = 4, this will lead to the increase of the model basic analyte retention, whereas at pH above 6 with the increase of the column temperature, this will lead to a corresponding decrease in the model basic analyte retention (Figure 2-17).

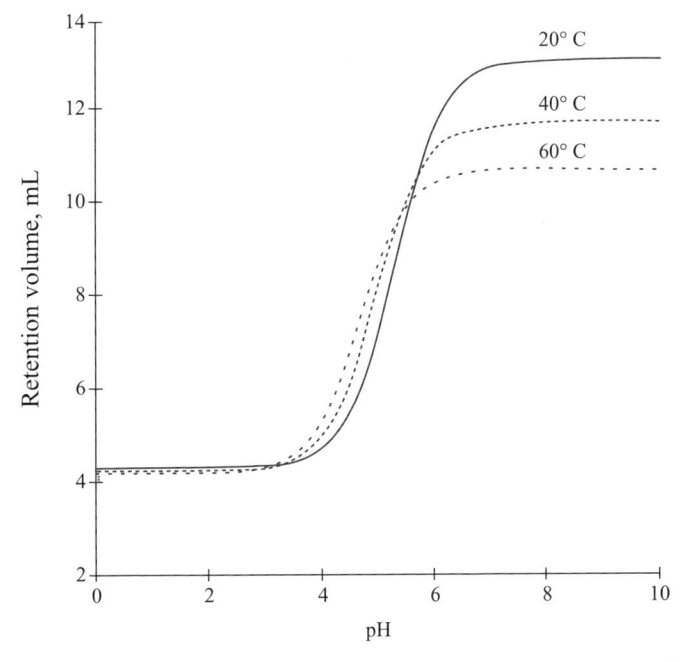

Figure 2-17. Predicted retention dependence on the mobile-phase pH at three different temperatures.

Corresponding dependencies in van't Hoff coordinates, although almost linear, show different directions in their slope, which is dependent on the particular pH chosen for this particular model (Figure 2-18). Note that in this model the effect of temperature on the change of the dissociation constants of the buffer species and model basic analyte species were taken into account on the basis of standard relationship of the equilibrium constant with the temperature ($K_i = \exp[\Delta G/RT]$). It has been shown that the dissociation constants of particular acidic species and basic species show some specific variations of their ionization constants with temperature in methanol/water and acetonitrile/water mobile phases [40, 41].

2.16.2 Salt Effect

Salt modifiers and chromatographic application of ionic liquids in the last few years has generated significant interest [42, 43]. In HPLC these compounds are used as mobile-phase additives for the enhancement of basic analyte reten-

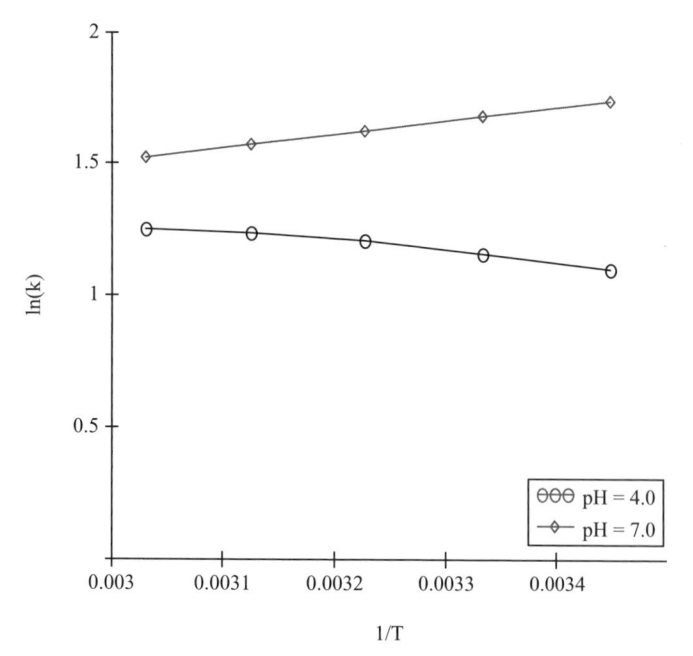

Figure 2-18. So-called van't Hoff plot for the retention of the same analyte at two different pH values of the mobile phase; all other conditions are identical. This is to illustrate that any classical thermodynamic treatments are not applicable to the chromatographic system with multiple different processes involved.

tion and apparent efficiency. Being added to the mobile phase at relatively low concentrations (1–100 mM), they could be considered as liophilic (oil-loving) ions. Chromatographic behavior of these liophilic ions (chaotropic ions) should be similar to ion-pairing agents (amphiphilic ions) with the lack of their adverse effects, such as permanent adsorption on the surface of the stationary phase and slow system equilibration.

A chromatographic system consisting of the binary eluent (e.g., acetonitrile–water) with liophilic salt added in low concentration (not more than 100 mM), along with basic analyte, is considered here. Adsorption behavior of acetonitrile in the column has been discussed in Section 2.12, and we assume that low concentrations of liophilic salt additives and injection of small amount of the analyte does not noticeably disturb its adsorption equilibrium.

If the excess adsorption of acetonitrile on the given column was measured before, the adsorbed layer volume could be described as

$$V_{al}(c_{MeCN}) = (\Gamma(c_{MeCN}) + V_a c_{MeCN})v_{MeCN} \qquad (2\text{-}84)$$

where V_{al} is the volume of the acetonitrile adsorbed layer at any given concentration of acetonitrile (c_{MeCN}), while V_a is the maximum adsorbed layer volume determined from the linear portion of the excess adsorption isotherm (as discussed in Section 2.12); v_{MeCN} is the molar volume of acetonitrile.

The liophilic ions present in the mobile phase at given concentration c_{lip} will be distributed between the acetonitrile adsorbed layer and mobile phase. This distribution could be represented by the distribution constant, $K_{lip}(c_{MeCN})$, which is the function of the acetonitrile content.

The nature of the analyte interactions with liophilic ions could be electrostatic attraction, ion association, or dispersive-type interactions. Most probably all mentioned types are present. Ion association is essentially the same as an ion-pairing used in a general form of time-dependent interionic formation with the average lifetime on the level of 10^{-6} sec in water–organic solution with dielectric constant between 30 and 40. With increase of the water content in the mobile phase, the dielectric constant increases and approaches 80 (water); this decrease the lifetime of ion-associated complexes to approximately 10^{-8} sec, which is still about four orders of magnitude longer than average molecular vibration time.

We only assume the existence of specific interactions of positively charged basic analyte and liophilic ions. In our system the following equilibria coexist:

1. Analyte ionization (pH-dependent)
2. Analyte ion association (pH and salt concentration-dependent process)
3. Partitioning of all three analyte forms (neutral, ionized, and ion-associated) into the adsorbed acetonitrile layer
4. Adsorption of all three analyte forms on the surface of a stationary phase

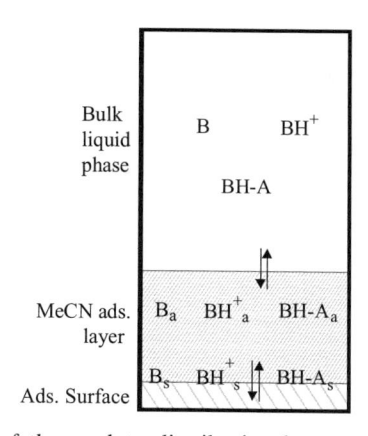

Figure 2-19. Schematic of the analyte distribution between mobile phase, adsorbed organic layer, and adsorbent surface in the presence of ionic and ion-pairing equilibria. B, neutral form of basic analyte; BH^+, protonated basic analyte; BH-A, ion-paired protonated basic analyte. Subscript a denotes adsorbed acetonitrile layer; subscript s denotes adsorption on the surface of adsorbent.

Coexisting species in different phases of the column are schematically shown in Figure 2-19.

The analyte molecules are distributed between the mobile phase, the acetonitrile adsorbed layer, and the adsorbent surface. The analyte could be in neutral, ionic, and ion-associated form, assuming that only neutral and ion-paired analyte could partition into the organic adsorbed layer and subsequently be adsorbed on the surface. This discussion is limited to the hypothetical energetically homogeneous surface of the reversed-phase adsorbent where residual silanols are effectively shielded by the alkyl bonded layer with high bonding density. The effect of accessible residual silanols, although much discussed in the literature, has never been estimated quantitatively in direct experiments and thus could not be included in any theoretical considerations. The total amount of analyte in the bulk solution ($n_{m.p.}$) is represented as a sum of the concentrations of each form of the analyte multiplied by the mobile-phase volume:

$$n_{m.p.} = (v_0 - v_a)(c_B + c_{BH^+} + c_{BH-A}) \tag{2-85}$$

The analyte amount in acetonitrile adsorbed layer (n_a) and the excess adsorbed on the surface (Γ_s) could be also expressed as

$$n_a = v_a(c_{B_a} + c_{BH_a^+} + c_{BH-A_a}) \tag{2-86}$$

$$\Gamma_s = s(\Gamma_B + \Gamma_{BH^+} + \Gamma_{BH-A}) \tag{2-87}$$

Ionic analyte equilibrium is described by equation (2-73). The ion-pairing equilibrium also could be described as a function of counterion concentration from the following equilibrium:

$$BH^+ + A^- \xleftrightarrow{K_p} BH \cdot A \qquad (2\text{-}88)$$

where K_p is an ion-pairing constant. The concentration of ion pairs could be expressed using equation (2-73) as

$$c_{BH-A} = K_p \frac{[H^+]}{K_i} c_p c_B \qquad (2\text{-}89)$$

where c_p is the counterion concentration. Total amount of the analyte in the mobile phase [expression (2-85)] could be now written as a function of only one concentration (c_B):

$$n_{m.p.}(c_B) = (v_0 - v_a)\left(1 + \frac{[H^+]}{K_i} + K_p \frac{[H^+]}{K_i} c_p\right) c_B \qquad (2\text{-}90)$$

Assuming that only neutral and ion-paired analytes can partition into the acetonitrile adsorbed layer, corresponding partitioning constants should be introduced. The K_a is the constant of nonionic analyte partitioning in the acetonitrile adsorbed layer; K_p^a is the constant for ion-pair partitioning into the acetonitrile adsorbed layer. Similar adsorption constants should be introduced for the description of the surface adsorption processes. The K^H is the constant of the analyte adsorption on the surface of the adsorbent; K_p^H is the constant of the ion-pair adsorption on the surface of the adsorbent.

Analyte distribution function for the mass balance is the sum of the analyte quantities in all phases:

$$\psi(c) = (v_0 - v_a)\left(1 + \frac{[H^+]}{K_i} + K_p \frac{[H^+]}{K_i} c_p\right) c_B + v_a\left(K_a + \frac{K_p^a K_p}{K_i} c_p [H^+]\right) c_B$$

$$+ S\left(K^H K_a + \frac{K_p^H K_p^a K_p}{K_i} c_p [H^+]\right) c_B \qquad (2\text{-}91)$$

where K_i is the analyte ionization constant; K_p is the ion-pairing or association constant for protonated analyte; c_p is the concentration of counterions; $[H^+]$ is the concentration of protons. To express the analyte retention volume, equation (2-91) should be substituted into equation (2-39), and the final function is essentially the derivative of the analyte distribution by total analyte concentration in the mobile phase. Since in the mobile phase the analyte is in three different forms, dc_{tot} could be expressed as

$$dc_{tot} = \left(1 + \frac{[H^+]}{K_i} + \frac{K_p[H^+]}{K_i} c_p\right) dc \qquad (2\text{-}92)$$

The retention volume of the analyte as a function of the proton concentration and concentration of counteranions is shown below:

$$V_R(c_p,[H^+]) = V_0 - V_a + V_a \frac{K_a + \dfrac{K_p^a K_p}{K_i} c_p[H^+]}{1 + \dfrac{H}{K_i} + \dfrac{K_P}{K_i} c_p[H^+]} + S \frac{K^H K_a + \dfrac{K_p^H K_p^a K_p}{K_i} c_p[H^+]}{1 + \dfrac{H}{K_i} + \dfrac{K_P}{K_i} c_p[H^+]}$$

$$(2\text{-}93)$$

Each constant in the equation above represents single equilibrium process, which is assumed to be independent on other equilibria in the column. Equation (2-93) describes the retention of basic ionizable analytes in reversed-phase chromatographic system with binary eluents and liophilic counteranions added. Similar expression could be derived for the behavior of anionic analytes in the presence of liophilic countercation.

The theoretical retention dependencies for basic analyte calculated using equation (2-93) with variation of the mobile-phase pH and counterion concentration are shown in Figure 2-20.

The pH dependencies of the basic analyte retention in Figure 2-20 are not classic sigmoidal, which is supposed to plateau at low pH, and no change in retention should occur for ionized bases with a further decrease in mobile-phase pH. However, as shown in Figure 2-20, a slight increase of the retention is observed with decrease of mobile-phase pH. This effect has been observed

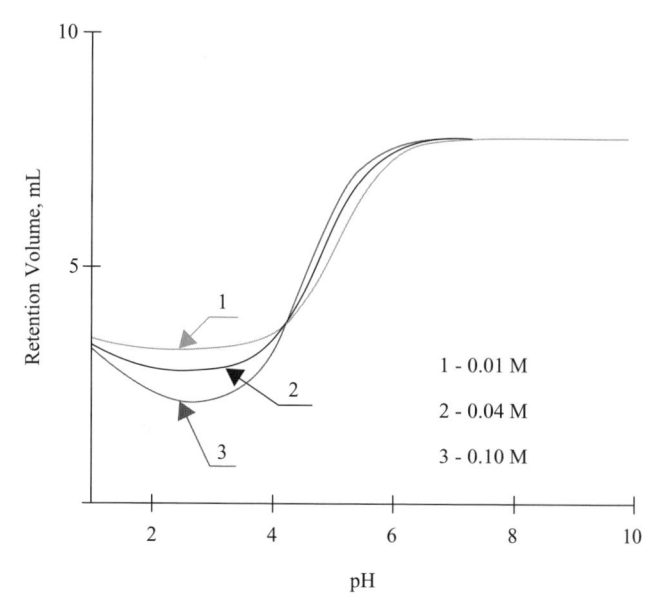

Figure 2-20. Retention of basic analyte as a function of mobile phase pH at three different counterion concentrations.

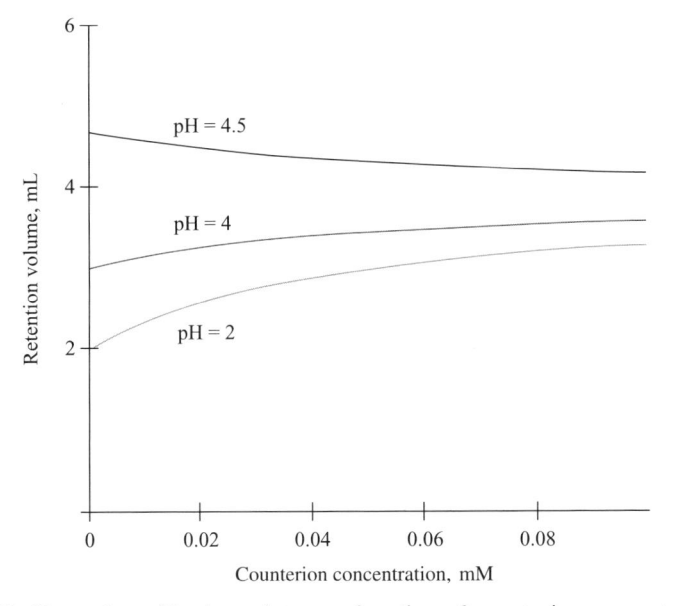

Figure 2-21. Retention of basic analyte as a function of counterion concentrations for three different mobile-phase pH values.

experimentally [39] and was attributed to the increase of the acidic modifier counteranion concentration (acidic modifier was used to decrease of mobile-phase pH) (Figure 2-21). Further description of this effect is discussed in Chapter 4.

2.17   GRADIENT ELUTION PRINCIPLES

Significant number of modern HPLC separation methods are based on gradient elution or continuous variation of the eluent composition during the analytical experiment. Theoretical description of the gradient separation is very complex and, to large extent, nonexistent. It is possible to use basic HPLC theory for phenomenological explanation of gradient retention dependencies.

Linear velocity of chromatographic zone as a function of the eluent composition $u(c)$ [as derived in equation (2-38)] in the most simple model of the retention process (partitioning) is

$$u(c) = \frac{FL}{V_m + V_s K(c)} \tag{2-94}$$

where F is the mobile-phase flow rate, L is the column length, V_m is the volume of the mobile phase, V_s is the volume of the stationary phase, and $K(c)$ is the analyte partitioning constant, which is a function of the mobile-phase composition c. In the simple form the function of partitioning coefficient dependence on the eluent composition (in volume fraction units, x) could be expressed following the discussion in Section 2.14.1 as

$$K(x) = \exp\left(\frac{\Delta G_{an} - x \cdot \Delta G_{el}}{RT}\right) \qquad (2\text{-}95)$$

where ΔG_{an} is the free Gibbs energy of the analyte interaction with the stationary phase, and ΔG_{el} is the free Gibbs energy of the organic eluent component interaction with the stationary phase.

For a linear gradient, eluent composition is a linear function of time:

$$x(t) = A + Bt \qquad (2\text{-}96)$$

where A is the starting eluent composition and B is the gradient ramp. Substituting equations (2-96) and (2-95) into equation (2-94), the function of the analyte velocity as a function of time could be obtained:

$$u(t) = \frac{FL}{V_m + V_s \exp\left(\dfrac{\Delta G_{an} - (A + Bt) \cdot \Delta G_{el}}{RT}\right)} \qquad (2\text{-}97)$$

Analyte retention time is the time that an analyte travels through the column, so the integral of the analyte linear velocity by time should be equal to the column length:

$$L = \int_0^{t_R} u(t)\,dt \qquad (2\text{-}98)$$

Analyte retention time is the upper integration limit. In general, equation (2-97) could be integrated in an analytical form. Figure 2-22 shows the

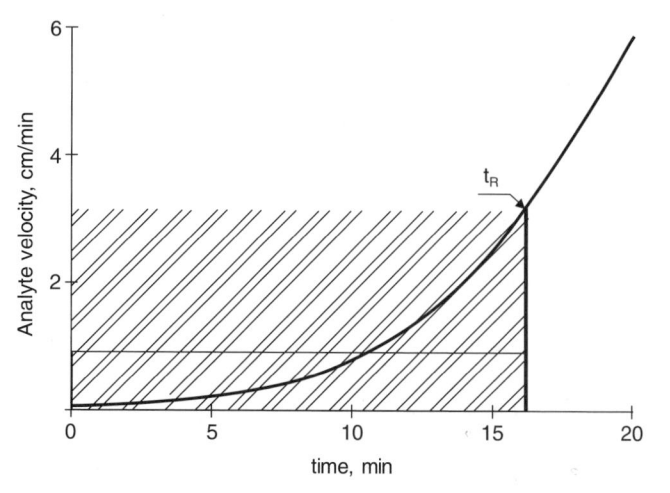

Figure 2-22. Analyte linear velocity in the column as function of time in linear gradient conditions.

dependence of the analyte velocity on the gradient time with linear gradient from 0% to 100% of acetonitrile in 20 minutes on a 15-cm reversed-phase column. Area under the analyte velocity function is equal to 15 cm [as follows from equation (2-98), which is equal to the column length].

Optimization of the gradient separation could be expressed in the form of a mathematical algorithm. Usually from a couple of gradient runs of the same analyte mixture, this is sufficient to calculate empirical constants for the equation similar to equation (2-97). These algorithms are implemented in most of the optimization software, such as DryLab®, ACDLabs®, and ChromSword®.

2.18   TYPES OF ANALYTE INTERACTIONS WITH THE STATIONARY PHASE

Assumption of the presence of single partitioning mechanism of analyte chromatographic retention has been the basis for the development of various methods for the evaluation of specific analyte interaction energies from retention data [44–46]. All these methods are only applicable in ideal chromatographic systems with proven absence of secondary equilibria effects, and all require specific assumptions regarding the volume of the stationary phase. Equation (2-43) is the main basis for these theories.

Carr et al. [47–52] used the solvophobic theory to elucidate the retention of solutes in a reversed-phased HPLC system on nonpolar stationary phases. It is assumed that the free energy of transfer of an analyte molecule from the mobile phase to the stationary phase, ΔG, can be regarded as a linear combination of the energies, ΔG_i, from the interaction of various molecular subunits according to the following equation:

$$\Delta G = \sum \Delta G_i \qquad (2\text{-}99)$$

Abraham and Roses [53] expressed the chromatographic relation in the general form called linear solvation energy relationships (LSER):

$$\ln k = c + s\pi_2^H + rR_2 + a\sum \alpha_2^H + b\sum \beta_2^H + vV_x \qquad (2\text{-}100)$$

where k is the retention factor; c is the intercept of the linear function of the LSER equation; π_2^H is the index of solvent polarizability; R_2 is the excess molar refraction; a is the correction term for the dipolar polarizability term π_2^H of the solute; $\Sigma\alpha_2^H$ and $\Sigma\beta_2^H$ are the overall hydrogen-bond donor (HBD) acidity and overall hydrogen-bond acceptor (HBA) basicity parameters, respectively; and V_x is the molecular volume of the solute computed according to Abraham and McGowan [54]. The solvatochromic parameters of the analyte of interest may be obtained from the literature [55, 56]. The coefficients r, x, a, b, and v of the solvatochromic parameters are determined using multiple linear regression analysis of ln k (obtained experimentally) and the solvatochromic parameters. The signs and magnitudes of the coefficients depict the direction and relative strength of different kinds of solute/stationary and solute/mobile-phase interactions. The most influential interactions governing RP-HPLC retention on alkyl and phenyl-type bonded phases were determined to be hydrogen-bond acceptor (HBA) basicity (b coefficient) and the solute molecular volume (v coefficient) [47, 57–58]. However, there is a noticeable difference between the aromatic and aliphatic phases such that solutes in general are less retained on the aromatic phases as compared to the aliphatic phases. The lower retention observed on the aromatic phases could be explained by the v coefficients. The v coefficients of the aromatic phases are substantially smaller than the aliphatic phases. The comparatively small v coefficient for aromatic phases was attributed to the stronger adsorption of eluent component onto the aromatic bonded phase than adsorption of mobile-phase component onto the aliphatic bonded phase. The adsorption of methanol was suggested to alter the stationary-phase environment more like a pseudo-liquid stationary phase, thus reducing the corresponding LSER coefficients [52]. While all solvatochromic parameters coefficients for aromatic phases were distinctively smaller than that of aliphatic phases, with the exception of coefficient r, the rR_2 term reflects the tendency of the system (mobile/stationary) phases to interact with solutes through π- and n-electron pairs [58]. The r coefficient and the ensuing polarizability interaction were identified as one of the potential differences in selectivity between aromatic and aliphatic phases. However, as shown by Carr et al. [50, 52], the retention selectivity predicted using LSER were not in complete agreement with experimental results. They concluded that the data shown using the LSER model does not properly account for all molecular interactions involved in RP-HPLC. Interactions such as ion exchange, Lewis acid–base interactions, and shape recognition were excluded.

The LSER theory allows the association of the analyte retention behavior with the energetic components of the analyte interactions with the stationary phase. On the other hand, this theory is based on the assumption of the applicability of equation (2-43) to the analyte retention process, and this equation only allows the existence of one single partitioning retention mechanism. Coexistence of several different retention mechanisms, or the presence of secondary equilibria in a chromatographic system, effectively does not allow the applicability of this theory to the description of the analyte retention [59].

2.19 CONCLUSION

HPLC theory is too broad of a topic to be comprehensively covered by one chapter or even by one book. This chapter was intended to create a general understanding of the relationships between different aspects of chromatographic theory, such as analyte retention behavior and retention mechanism.

Participation of the analyte in different processes in the column (ionization, solvation, and other secondary equilibria processes) introduces a complex retention mechanism, which could not be described by a single equilibrium constant and, consequently, does not allow facile extraction of representative thermodynamic parameters of intermolecular interactions in the HPLC system.

Different models of the retention in these complex systems do not allow for accurate prediction of the analyte retention, but they assist in the understanding of the processes governing analyte migration through the column and also help in the selection of starting conditions and intelligent optimization of a particular separation.

Specific topics discussed in this chapter have not been fully developed yet, and we hope that this material may inspire research activity in this direction, which will benefit the entire chromatographic community.

REFERENCES

1. C. Giddings, *Unified Separation Science*, Wiley-Interscience, New York, 1991.

2. A. J. P. Martim and R. L. M. Synge, A new form of chromatogram employing two liquids phases. 1. A theory of chromatography. 2. Application to the microdetermination of the higher monoamino-acids in proteins, *J. Biochem.* **35** (1941), 1358–1368.

3. A. E. Scheidegger, *The Physics of Flow Through Porous Media*, University of Toronto Press, 1960.

4. B. Karger, A critical examination of resolution equations for gas–liquid chromatography, in A Zlatkis (ed.), *Advances in Gas Chromatography* Preston Technical Abstracts, Evanston IL, 1967, pp. 1–9.

5. H. A. C. Thijssen, Gas–liquid chromatography a contribution to the theory of separation in open hole tubes, *J. Chromatogr.* **11** (1963), 141–150.

6. A. S. Said, Resolution, speparability, and conditions for fraction collection in preparative gas chromatography, *J. Gas Chromatogr.* **2** (1964), 60–71.

7. J. H. Purnell, Correlation of separating power and efficiency of gas-chromatographic columns, *J. Chem. Soc.* (1960), 1268–1274.

8. A. Vailaya and C. Horvath, Retention in reversed-phase chromatography: Partition or adsorption? *J. Chromatogr. A* **829** (1998), 1–27.

9. F. Gritti and G. Guiochon, Critical contribution of nonlinear chromatography to the understanding of retention mechanism in reversed-phase liquid chromatography, *J. Chromatogr. A* **1099** (2005), 1–42.

10. H. Colin and G. Guiochon, Introduction to reversed-phase high-performance liquid chromatography, *J. Chromatogr.* **141** (1977), 289–312.

11. J. H. Knox and A. Pryde, Performance and selected applications of a new range of chemically bonded packing materials in high-performance liquid chromatography, *J. Chromatogr.* **112** (1975), 171–188.

12. C. Cramers, A. Keulemans, and H. McNair, Techniques of gas chromatography, in E. Heftmann (ed.), *Chromatography*, 1961, Reinhold, New York.

13. E. H. Slaats, W. Markovski, J. Fekete, and H. Poppe, Distribution equilibria of solvent components in reversed-phase liquid chromatographic columns and relationship with the mobile phase volume, *J. Chromatogr.* **207** (1981), 299–323.

14. H. Poppe, Distribution isotherms in reversed-phase systems, *J. Chromatogr. A*, **656** (1993), 19–36.

15. K. S. Yun, C. Zhu, and J. F. Parcher, Theoretical relationships between the void volume, mobile phase volume, retention volume, adsorption, and Gibbs free energy in chromatographic processes, *Anal. Chem.* **67** (1995), 613–619.

16. C. A. Rimmer, C. R. Simmons, and J. G. Dorsey, The measurement and meaning of void volumes in reversed-phase liquid chromatography, *J. Chromatogr. A* **965** (2002), 219–223.

17. G. Guiochon, S. G. Shirazi, and A. M. Katti, *Fundamentals of Preparative and Nonlinear Chromatography*, Academic Press, New York, 1994.

18. I. Rustamov, T. Farkas, A. Ahmed, F. Chan, R. LoBrutto, H. M. McNair, and Y. V. Kazakevich, Geometry of chemically modified silica, *J. Chromatogr. A* **913** (2001), 49–63.

19. T. L. Chester and J. W. Coym, Effect of phase ratio on van't Hoff analysis in reversed-phase liquid chromatography, and phase-ratio-independent estimation of transfer enthalpy, *J. Chromatogr. A* **1003** (2003), 101–111.

20. J. W. Gibbs, *On the Equilibrium of Heterogeneous Substances. Collected Works.* Longmans, New York, 1928, pp. 55–353.

21. F. Riedo and E. Kovats, Adsorption from liquid mixtures and liquid chromatography, *J. Chromatogr.* **239** (1982), 1–28.

22. Y. V. Kazakevich, R. LoBrutto, F. Chan, and T. Patel, Interpretation of the excess adsorption isotherms of organic eluent components on the surface of reversed-phase adsorbents: Effect on the analyte retention *J. Chromatogr. A* **913** (2001), 75–87.

23. F. Chan, L. S. Yeung, R. LoBrutto, and Y. V. Kazakevich, Characterization of phenyl-type HPLC adsorbents, *J. Chromatogr. A* **1069** (2005), 217–224.

24. N. L. Ha, J. Ungvaral, and E. Kovats, Adsorption isotherm at the liquid–solid interface and the interpretation of chromatographic data, *Anal. Chem.* **54** (1982), 2410–2421.

25. G. Foty, C. de Reyff, and E. S. Kovats, Method of chromatographic determination of excess adsorption from binary liquid mixtures, *Langmuir.* **6** (1990), 759–766.

26. R. M. McCormick and B. Karger, Distribution phenomena of mobile-phase components and determination of dead volume in reversed-phase liquid chromatography, *Anal. Chem.* **52** (1980), 2249.

27. D. H. Everett, Thermodynamics of adsorption from solutions I, *J. Chem. Soc., Faraday Trans I*, **60** (1964), 1803–1813.

28. Yu. A. Eltekov and Y. V. Kazakevich, Investigation of adsorption equilibrium in chromatographic columns by the frontal method, *J. Chromatogr.* **365** (1976), 213–219.

29. H. L. Wang, U. Duda, and C. J. Radke, Solution adsorption from liquid chromatography, *J. Colloid Interface Sci.* **66** (1978), 152–165.

30. W. R. Melander and C. Horvath, Reversed-phase chromatography, in C. Horvath (ed.), *HPLC, Advances and Perspectives*, Vol. 2, Academic Press, New York, 1980, pp. 114–303.

31. K. A. Dill, The mechanism of solute retention in reversed-phase liquid chromatography, *J. Phys. Chem.* **91** (1987), 1980–1988.

32. A. Yu. Fadeev, G. V. Lisichkin, V. K. Runov, and S. M. Staroverov, Diffusion of sorbed pyrene in the bonded layer of reversed phase silicas: Effect of alkyl chain length and pore diameter, *J. Chromatogr.* **558** (1991), 31–42.

33. Y. V. Kazakevich, R. LoBrutto, and R. Vivilecchia, Reversed-phase high-performance liquid chromatography behavior of chaotropic counteranions, *J. Chromatogr. A* **1064** (2005), 9–18.

34. H. Poppe, Secondary equilibria and their interaction with chromatographic transport, *J. Chromatogr. A* **506** (1990), 45–60.

35. J. Barbosa, I. Toro, and V. Sanz-Nebot, Secondary chemical equilibria in high-performance liquid chromatography: influence of ionic strength and pH on retention of peptides, *J. Chromatogr. A* **823** (1998), 497–509.

36. M. Roses and E. Bosch, Influence of mobile phase acid–base equilibria on the chromatographic behavior of protolytic compounds, *J. Chromatogr. A* **982** (2002), 1–30.

37. R. LoBrutto, A. Jones, and Y. V. Kazakevich, Effect of Counteranion Concentration on HPLC Retention of Protonated Basic Analytes, *J. Chromatogr. A* **913** (2001), 191–198.

38. C. Horvath, W. Melander, and I. Molnar, Liquid chromatography of ionogenic substances with nonpolar stationary phases, *Anal. Chem.* **49** (1977), 142–154.

39. R. LoBrutto, A. Jones, Y. V. Kazakevich, and H. M. McNair, Effect of the eluent pH and acidic modifiers in high-performance liquid chromatography retention of basic analytes, *J. Chromatogr. A* **913** (2001), 173–187.

40. A. Albert and E. P. Serjeant, *The Determination of Ionisation constants*, 3 ed., Chapman & Hall, London, 1984; R. G. Bates, *Determination of pH, Theory and Practice*, John Wiley & Sons, New York, 1964.

41. C. B. Castells, L. G. Gagliardi, C. Ràfols, M. Rosés, and E. Bosch, Effect of temperature on the chromatographic retention of ionizable compounds: I. Methanol–water mobile phases, *J. Chromatogr. A* **1042** (2004), 23–36.

42. D. W. Armstrong, L. He, and Y.-S. Liu, Examination of ionic liquids and their interaction with molecules, when used as stationary phases in gas chromatography, *Anal. Chem.* **71** (1999), 3873.

43. C. F. Poole, Chromatographic and spectroscopic methods for the determination of solvent properties of room temperature ionic liquids, *J. Chromatogr. A* **1037** (2004), 49–82.

44. Á. Sándi and L. Szepesy, Characterization of various reversed-phase columns using the linear free energy relationship: I. Evaluation based on retention factors, *J. Chromatogr. A* **818** (1998), 1–17.

45. L. Choo Tan and P. W. Carr, Study of retention in reversed-phase liquid chromatography using linear solvation energy relationships: II. The mobile phase, *J. Chromatogr. A* **799** (1998), 1–19.

46. R. Kaliszan, M. A. van Straten, M. Markuszewski, C. A. Cramers and H. A. Claessens, Molecular mechanism of retention in reversed-phase high-performance liquid chromatography and classification of modern stationary phases by using quantitative structure–retention relationships, *J. Chromatogr. A* **855** (1999), 455–486.

47. O. Sinanoglu and B. Pulman (eds.), *Molecular Associations in Biology*, Academic Press, New York, 1968, pp. 427–445.

48. M. J. Kamlet, J. L. M. Abboud, M. H. Abraham, and R. W. Taft, Linear Solvation energy relationships. 23. A comprehensive collection of the solvatochromic para-menters, pi, alfa, and bets, and some methods for simplifying the generalized solvatochromic equation, *J. Org. Chem.* **48** (1983), 2877–2887.

49. P. W. Carr and J. Zhao, Comparison of the retention characteristics of aromatic and aliphatic reversed-phases for HPLC using linear solvation energy relationships, *Anal. Chem.* **70**, (1998), 3619–3628.

50. P. W. Carr, M. Reta, P. C. Sadek, S. C. Rutan, Comparative study of hydrocarbon, fluorocarbon and aromatic bonded RP-HPLC stationary phases by linear solvation energy relationships, *Anal. Chem.* **71** (1999), 3484–3496.

51. P. W. Carr, L. C. Tan, and M. H. Abraham, Study of retention in reversed-phase liquid chromatography using linear solvation energy relationships, *J. Chromatogr. A* **752** (1996), 1–18.

52. P. W. Carr and J. Zhao, Approach to the concept of resolution optimization through changes in the effective chromatographic selectivity, *Anal. Chem.* **71** (1999), 2623–2632.

53. M. H. Abraham and M. Roses, Hydrogen bonding. 38. Effect of solute structure and mobile phase composition on reversed-phase high-performance liquid chromatographic capacity factors, *J. Phys. Org. Chem.* **7** (1994), 672–683.

54. M. H. Abraham and J. C. McGowan, The use of characteristic volumes to measure cavity terms in reversed phase liquid chromatography, *Chromatographia* **23** (1987), 243–246.

55. J. L. E. Rebusaet and R. Vieskar, Characterization of π–π interactions which determine retention of aromatic compounds in reversed-phase liquid chromatography, *J. Chromatogr. A* **841** (1999), 147–154.

56. J. Horak, N. M. Maie, and W. Lindner, Investigation of chromatographic behavior of hybrid reversed-phase materials containing electron donor–acceptor systems: II Contribution of π–π aromatic interactions, *J. Chromatogr. A* **1045** (2004), 43–58.

57. M. J. Kalmet, J. L. M. Abboud, M. H. Abraham, and R. W. Taft, Linear solvation energy relationships. 23. A comprehensive collection of the solvatochromic para-meters, π, α, and β, and some methods for simplifying the generalizaed solvatochromic equation, *J. Org. Chem.* **48** (1983), 2877–2887.

58. M. H. Abraham, Scales of solute hydrogen bonding: Their construction and application to physicochemical and biochemical processes, *Chem. Soc. Rev.* **22** (1993), 73–83.

59. Y. V. Kazakevich, High performance liquid chromatography retention mechanisms and their mathematical descriptions, *J. Chromatogr. A* **1126** (2006), 232–243.

3

STATIONARY PHASES

Yuri Kazakevich and Rosario LoBrutto

3.1 INTRODUCTION

The column is the only device in the high-performance liquid chromatography (HPLC) system which actually separates an injected mixture. Column packing materials are the "media" producing the separation, and properties of this media are of primary importance for successful separations.

Several thousands of different columns are commercially available, and when selecting a column for a particular separation the chromatographer should be able to decide whether a packed, capillary, or monolithic column is needed and what the desired characteristics of the base material, bonded phase, and bonding density of selected column is needed. Commercial columns of the same general type (e.g., C18) could differ widely in their separation power among different suppliers. Basic information regarding the specific column provided by the manufacturer, such as surface area, % carbon, and type of bonded phase, usually does not allow prediction of the separation or for the proper selection of columns with similar separation patterns.

Great varieties of different columns are currently available on the market. Four distinct characteristics could be used for column classification:

1. Type (monolithic; porous; nonporous)
2. Geometry (surface area; pore volume; pore diameter; particle size and shape; etc.)

HPLC for Pharmaceutical Scientists, Edited by Yuri Kazakevich and Rosario LoBrutto
Copyright © 2007 by John Wiley & Sons, Inc.

3. Surface chemistry (type of bonded ligands; bonding density; etc.)
4. Type of base material (silica; polymeric; zirconia; etc)

All these characteristics are interrelated. Variations of porosity which include pore diameter can affect both the adsorbent surface area and the bonding density. The type of base material affects adsorbent surface chemistry. Therefore, in our discussion we combine these characteristics in two major classes: geometry and surface chemistry.

Most geometry-related properties of packing materials are related to the column efficiency and flow resistance: particle size, particle shape, particle size distribution, packing density, and packing uniformity. Surface-chemistry-related properties are mainly responsible for the analyte retention and separation selectivity.

Adsorbent surface area, pore volume, and pore diameter are the properties of significant importance. HPLC retention is generally proportional to the surface area accessible for a given analyte (Chapter 2). Surface area accessibility is dependent on the analyte molecular size, adsorbent pore diameter, and pore size distribution.

The chemical nature of the ligands bonded on the surface of support material defines the main type of chemical interactions of the surface with eluent and analyte molecules. In essence, all C18-type columns should be similar with regard to their main interaction type, namely, hydrophobic interactions: Methylene selectivity of all C18-type columns are virtually identical [1]. Bonded phases of the same type differ in their ability to suppress (or shield) other types of interactions (ionic; dipole) exerted by the base material (e.g., silica). Energy of these unwanted interactions is about 10 times greater than the energy of dispersive interactions [2]. Due to the exponential nature of the relationship between retention and interaction energy even the presence of 1% or less of these active centers in the packing material surface can significantly affect the analyte retention.

Bonding density is the primary parameter in evaluation of the quality of the bonded material. Usually the higher the bonding density, the better the shielding effect, although care should be taken in cross-evaluation of similar columns on the basis of their bonding density. Surface geometry can also significantly affect bonding density. Base material with smaller pores has higher surface area; however, bonding density is usually lower due to the smaller pores.

All parameters of the packing material are interrelated in their influence on the chromatographic performance of the column. The quality of an HPLC column is a subjective factor, which is dependent on the types of analytes and even on the chromatographic conditions used for the evaluation of the overall quality.

Long-term column stability (pH and temperature) and batch-to-batch reproducibility are probably the most important quality characteristics to be considered in column selection in the pharmaceutical industry. Nevertheless, these criteria should be evaluated with caution when selecting the column evaluation parameters. Long-term stability of retention and efficiency characteristics are usually different, depending on the testing conditions (mobile phase, temperature, and analyte probes). Efficiency is usually fairly stable at low mobile-phase pH, while retention of the probe analytes may show a drift in retention. However, the retention is generally stable at high pH while efficiency could be deteriorated.

3.2 TYPE OF PACKING MATERIAL (POROUS, NONPOROUS, MONOLITHIC)

Majority of packing materials used in HPLC are porous particles with average diameters between 3 and 10 µm. For most pharmaceutical applications, 3-µm particle sizes are recommended. Porosity provides the surface area necessary for the analyte retention (usually between 100 and 400 m^2/g). Interparticle space is large enough to allow up to 1–3 mL/min flow within acceptable pressure range (however, the pressure drop across the column depends on the particle size, length of column, temperature of separation, and type of mobile-phase composition).

Introduction of small nonporous spherical particles in the mid-1990s [3, 4] was an attempt to increase efficiency by eliminating dual column porosity. In the column packed with porous particles, interparticle space is about 100-fold larger than pores inside the particles, and liquid flow around the particles is also faster; this leads to the significant band broadening. Unfortunately, elimination of particle porosity dramatically decreases adsorbent surface area, thereby decreasing the column loading capacity. Columns packed with small (1.5 µm) nonporous particles also require ultra-microinjection volumes and a corresponding increase of detector sensitivity.

The introduction of monolithic columns in the 1990s was another and more successful attempt to increase column permeability while decreasing the gap in column dual porosity. Macropores in the monolith are between 4000 and 6000 Å in diameter, and they occupy almost 80% of the column volume. Compared to the conventional packed column with 5- or even 3-µm particles, the silica skeleton in monolith is only approximately 1 µm thick, which facilitates accessibility of the adsorbent surface inside the mesopores of the skeleton (pores between 20 and 500 Å in diameter are usually called mesopores). Comparison of the spherical packing material and monolithic silica is shown in Figure 3-1.

3.3 BASE MATERIAL (SILICA, ZIRCONIA, ALUMINA, POLYMERS)

In modern liquid chromatography, almost all reversed-phase separations are performed on chemically modified adsorbents. Analyte interactions with the

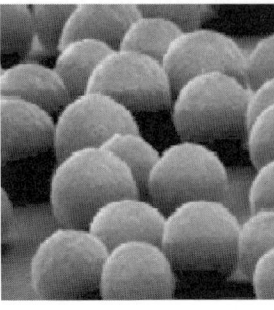

Figure 3-1. SEM pictures of HPLC silica particles (5 μm) and silica monolith. (Reprinted with permission from reference 5.)

stationary phase surface are the primary factor for successful separations. Most commercial adsorbents reflect their surface chemistry in their names (e.g., C18, C8, Phenyl, etc.) while the base material used usually is not specified, although its properties are very important.

Specific parameters of the base of packing material are:

- Surface area
- Pore size
- Pore volume
- Pore size distribution
- Particle shape
- Particle size
- Particle size distribution
- Structural rigidity
- Chemical stability
- Surface reactivity
- Density and distribution of the surface reactive centers

Surface area is directly related to the analyte retention [Equation (2-47) in Chapter 2]. Generally, the higher the surface area, the greater the retention.

Pore size is a critical parameter for the surface accessibility. Molecules of different size could have different accessible surface area due to the steric hindrance effect (bigger molecules might not be able to penetrate into all pores). Pore size is also related to the surface area. Assuming that all pores of the base material are cylindrical and neglecting the networked porous structure (assuming straight and not interconnected pores), it is possible to write the following expressions for the surface area and pore volume [6]:

$$S = 2\pi RL, \qquad V = \pi R^2 L \tag{3-1}$$

where S is a surface area of one gram of porous adsorbent; R is the average pore radius; V is a pore volume of one gram of adsorbent; and L is a total length of all pores in the same one gram of the adsorbent.

It may be interesting for the reader to estimate an approximate length of all pores in 1 g of average adsorbent. Surface area of average HPLC adsorbent is on the level of 300 m^2/g and average pore diameter is 100 Å. One gram of silica is an approximate amount that is usually packed into the standard 15-cm-long HPLC column (4.6-mm I.D.). If you calculate the length of all pores in this column [using equation (3-1) and express it in meters or kilometers], you will get a feeling of what you are dealing with when you are using HPLC.

If we take a ratio of the above expressions, we get simple relationship between these parameters:

$$\frac{S}{V} = \frac{2}{R} \tag{3-2}$$

K. K. Unger [6] found that in most cases, expression (3-2) shows a 15% discrepancy between measured and estimated adsorbent surface, which is very good when we take into account the above assumptions made in its derivation.

The most commonly used base material is silica (SiO_2), the most common substance on the Earth and thoroughly studied in the last two centuries. An excellent monograph on the properties of silica was published by Iler [7]. Development of modern HPLC techniques promoted advancement in porous silica technology. Almost all silica-based HPLC packings manufactured in the twenty-first century are very uniform spherical porous particles with narrow particle and pore size distribution.

Silica has one significant drawback: It is soluble at high pH, although chemical modification with high bonding density of attached alkylsilanes extends its stability range to over pH 10.

Another porous base material suggested in the last decade as an alternative to silica is zirconia. Zirconia is stable in a very wide pH range (pH 1–14), but zirconia surface has relatively low reactivity (more difficult to bond different functional groups to the surface), which significantly limits a selection of available stationary phases.

Polymer-based materials have been on the market for more than 30 years. Crosslinked styrene-divinylbenzene and methylmethacrylate copolymers are the most widely used. These materials show high pH stability and chemical inertness. Their rigidity and resistance to the swelling in different mobile phases is dependent on the degree of crosslinkage.

Practical application of these materials for the separation of small molecules are somewhat limited due to the presence of microporosity. Gaps between cross-linked polymer chains are on the level of molecular size of low-molecular-weight analytes. These analytes could diffuse inside the body of a polymer-based packing material, which produce drastically different retention of a small portion of injected sample than the rest of it. At the same time, polymers are the main packing material for size-exclusion chromatography.

3.4 GEOMETRY

3.4.1 Shape (Spherical/Irregular)

Recent technological advancements made spherical particles widely available and relatively inexpensive. Columns packed with spherical particles exhibit significantly higher efficiency, and columns packed with irregular particles are seldom used and are becoming nonexistent for analytical scale separations.

3.4.2 Particle Size Distribution

Packing materials are characterized by the average diameter of their particles and the distribution of the particle size around the average value.

The particle size distributions shown in Figure 3-2 are for spherical packing materials with nominal particle size of 10 µm. Distributions of different batches are symmetrical, with average width of approximately 50% of nominal diameter. Most critical for HPLC application is the presence of very small particles (fines) less than 0.5 µm. These small particles are usually fragments of crushed particle (porous silica is a fragile material). These fine particles will steadily migrate in the column toward the exit frit and clog it. These particles will eventually dramatically decrease the column efficiency, and peak distortion is usually observed for all peaks in the chromatogram.

Particle size distribution itself does not affect chemical behavior of HPLC adsorbent, although it is known to influence the efficiency of packed column. Packings with wide particle size distribution contain a significant amount of

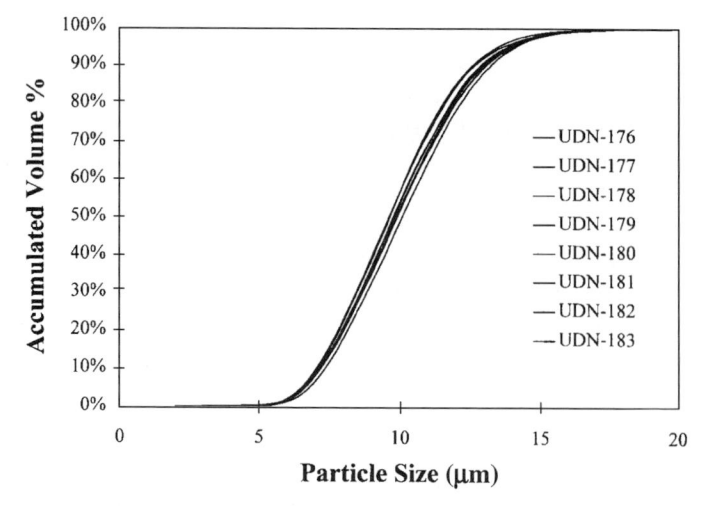

Figure 3-2. Example of cumulative particle size distribution of HPLC packing materials (Waters µ-Bondapack). Average particle size is 10 µm (inflection point). (Reprinted from reference 8, with permission.)

fine particles, which increases column backpressure; a big size difference between particles in the column decreases overall column efficiency.

Halasz and Naefe [9] and Majors [10] suggested that if the distribution is not wider than 40% of the mean, then acceptable flow resistance and column efficiency can be obtained. The narrower the particle size distribution, the better and the more reproducibly the columns could be packed. Generally accepted criteria is that 95% of all particles should be within 25% region around the mean particle diameter [11, 12].

3.4.3 Surface Area

Surface area of HPLC adsorbents is probably the most important parameter, although it is almost never used or accounted for in everyday practical chromatographic work. As shown in the theory chapter (see Chapter 2), HPLC retention is proportional to the adsorbent surface area. The higher the surface area, the greater the analyte retention, although as we discuss later, depending on the surface geometry, analytes of a different molecular size could effectively see different surface areas on the same adsorbent.

The experimental methods for the measurement of the surface area of porous silica is fairly well established. Nitrogen or argon adsorption isotherms at the temperature of liquid nitrogen (77 K) are used in accordance with BET (Brunauer, Emmet, Teller) theory [13] for the calculation of the total surface area per unit of adsorbent weight. There are different variations of BET theory available as well as different instrumental approaches [14] for the measurement of nitrogen isotherms. For proper characterization of mesoporous (pore diameter is greater than 20 Å) adsorbents, the static measurement of adsorption isotherm with proper equilibration at each measured point is preferable. Detailed discussion of all aspects of nitrogen adsorption isotherms and related theories could be found in the classic book *Adsorption, Surface Area and Porosity* by Gregg and Sing [15]. Full nitrogen adsorption isotherm (adsorption and desorption branches) is shown in Figure 3-3.

The region between 0.05 and 0.25 relative pressures is called the BET region, and it is used for the determination of the so-called monolayer capacity—the amount of nitrogen molecules adsorbed on the sample surface in a compact monolayer fashion. The BET equation represents the dependence of amount of adsorbed nitrogen as a function of the relative equilibrium pressure (p/p_0):

$$\frac{\dfrac{p}{p_0}}{n\left(1-\dfrac{p}{p_0}\right)} = \frac{1}{n_m C} + \frac{(C-1)}{n_m C}\frac{p}{p_0} \tag{3-3}$$

where n_m is the monolayer capacity, C is energetic constant of nitrogen interaction with the surface; p/p_0 is the relative equilibrium pressure, and n is the

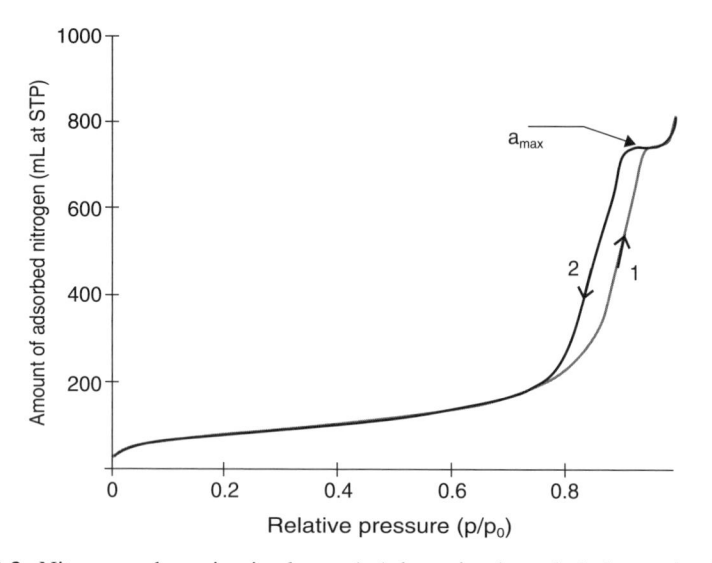

Figure 3-3. Nitrogen adsorption isotherm. 1, Adsorption branch; 2, desorption branch.

amount adsorbed. Figure 3-3 shows the experimental dependence of the amount of nitrogen adsorbed on the surface versus the relative pressure (pressure of nitrogen at the equilibrium in the gas phase over the adsorbent related to the saturation pressure at the temperature of the experiment). Equation (3-3) is the linear form of the function shown in Figure 3-3, but only in the relatively low pressure region between 0.05 and 0.25 where the formation of adsorbed monolayer is complete and BET theory is valid. The plot of the experimental points in $p/p_0/(1 - p/p_0)$ versus p/p_0 allows for linear minimization and calculation of C and n_m values [15].

It is generally assumed that a nitrogen molecule occupies 16.4 $\mathring{A}^2$ on the polar silica surface. The adsorbent surface area is then calculated as a product of the total amount of nitrogen in the monolayer (n_m) and the nitrogen molecular area (16.4 $\mathring{A}^2$).

3.4.4 Pore Volume

At higher relative pressures, above 0.7 in Figure 3-3 a fast increase of the adsorbed amount of nitrogen is observed. This region is attributed to the process of capillary condensation of nitrogen inside the adsorbent pores. This increase is observed until the whole pore volume is filled with liquid nitrogen. When relative equilibrium pressure approaches the saturation pressure and all pores are already filled with liquid nitrogen, a small flat section on the adsorption isotherm is usually observed (a_{max}). This section indicates the completion of the pore filling with condensed nitrogen, and it could be used for accurate determination of the adsorbent pore volume:

$$V_{\text{pore}} = V_L \frac{a_{\max} P}{RT} \tag{3-4}$$

where V_L is the molar volume of liquid nitrogen (34.7 mL/mol); $a_{\max}$ is the maximum amount of nitrogen in the pores expressed in milliliters at 1 atm and 25°C; P is the pressure (1 atm); R is the gas constant ($0.082 \frac{L \cdot atm}{K \cdot mol}$); and T is the temperature, 298 K.

The desorption branch of nitrogen isotherm is typically used for the determination of the pore size distribution. The only important factor that should be carefully verified for each adsorbent is the presence of microporous structure. If the micropores (pores with diameter less than 20 Å) are present in base material, the actual surface to which HPLC analyte might be exposed will be different from the surface measured with nitrogen adsorption. This is due to the size difference of nitrogen molecule and practically any HPLC analyte molecule. Bigger molecules will have steric hindrance in micropores, and any interpretation of the HPLC retention related to the surface area will be erroneous. In addition, proper chemical modification of adsorbents with micropores is essentially not possible. Minimum pore diameter acceptable in HPLC adsorbents is approximately 50 Å. Adsorbent pore size provided by the manufacturer is the diameter corresponding to the maximum of the pore size distribution curve, obtained from the adsorption branch of nitrogen isotherm. The distribution of the pores could vary significantly as it is shown in Figure 3-4.

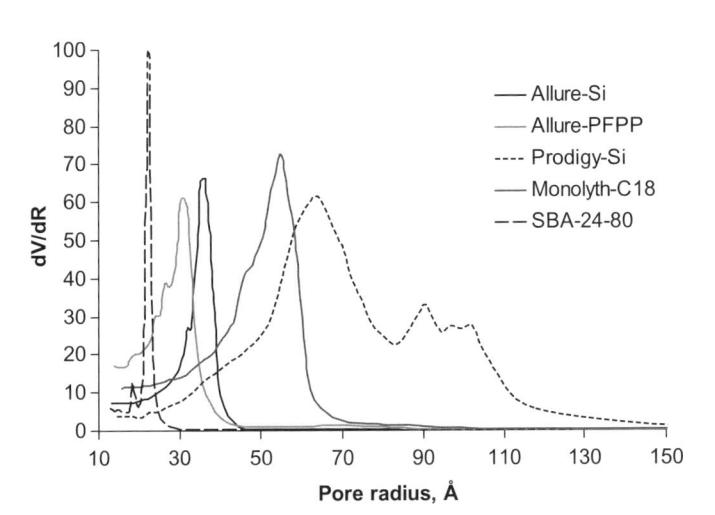

Figure 3-4. Pore size distribution of different HPLC materials. Allure Silica (Restek); Allure-PFPP (Restek), Prodigy-Silica (Phenomenex); Chromolith C18 (Merck KgaA, Germany); and research-type ordered silica with highly uniform pores of 50-Å pore diameter.

3.4.5 Surface Geometry

The roughness of the silica surface could introduce the steric hindrance of the surface accessibility similar to the effect of the micropores. In the discussion above, we assume the ideal tubular geometry of the silica surface. The use of different probe molecules for the BET measurement of silica surface area (such as N_2, Ar, Kr, benzene, etc.) leads to significant difference in the surface area values for the same silica sample. It was suggested that silica surfaces possess the property of fractals [16]; this essentially means that molecules of different size will see a different surface area.

The surface constructed of ridges and valleys could be considered as an example of a fractal surface. The slopes of these ridges are also constructed by smaller ridges, and with the higher magnification even smaller ridges are visible. As an example, if a big ball is used to roll over this surface, it will see only the big ridges and thus a relatively small total surface. If, on the other hand, a smaller ball is rolling over the same surface, it will see much more of smaller ridges and a lot of them, resulting in a much higher effective surface area. The smaller the probing ball, the finer the surface roughness it will see and correspondingly higher surface will be detected.

Molecular nitrogen will see a significant surface area due to its small size comparable to the dimensions of the surface roughness, while bigger molecules such as pyrene will not be able to see all ridges and valleys and will see a significantly lower effective surface area. These factors have been studied extensively [17] for silica, and authors have found fractal factors to vary between 2 and 3, depending on the silica synthesis, treatment, and so on.

Adsorbent surface area (S) is measured as a product of molecular area (σ) of a probe substance and the number of the molecules (N) in complete adsorbed monolayer. On the fractal surface the total number of molecules in the monolayer is dependent on its roughness and could be expressed as

$$N \sim \sigma^{-\frac{D}{2}} \tag{3-5}$$

where D is a fractal number. Since $S = N\sigma$, the adsorbent surface could be expressed as follows:

$$S \sim \sigma \cdot N \sim \sigma^{(2-D)/2} \tag{3-6}$$

On the flat surface, D is equal to 2 and only in this case the surface area is not dependent on the size of probe molecule.

The higher the fractal number, the less accessible the surface (quasi-three-dimensional or rough surface). For silica with a pore diameter of 10 nm and higher, the fractal factor has a tendency to be between 2.05 and 2.3, which is close to the flat surface. Figure 3-5 illustrates the apparent decrease of accessible silica surface in the form of a fraction of the total surface with the increase of the fractal number of this surface (roughness). For these types of

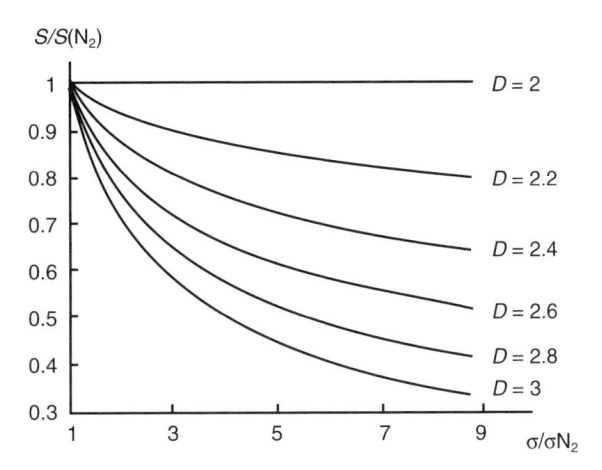

Figure 3-5. Dependence of the surface accessible for the probe molecule on the surface fractal number. The probe molecule size is given as the ratio to the nitrogen (S/S_{N_2}), and the accessible surface is shown as the fraction of the surface measured by nitrogen. (Reprinted from reference 18, with permission.)

silica samples, all molecules with MW <5000 should see practically the same surface [18].

3.5 ADSORBENT SURFACE CHEMISTRY

Interaction of the analyte molecules with the adsorbent surface is the driving force of HPLC separations. The surface of HPLC packing material should have specific interactions with different analytes, and at the same time the packing material itself should be mechanically and chemically stable. Variation of the adsorbent surface chemistry is achieved via chemical modification of the base material surface—that is, chemical bonding of the specific ligands. Chemical modification of the surface has two purposes: (a) shielding of the surface of base material and (b) introduction of the specific surface interactions.

3.5.1 Surface Chemistry of the Base Material

Base material provides mechanically stable rigid porous particles (mostly spherical) for reversed-phase HPLC adsorbents. Particle porosity on the mesoporous level (30 to 500-Å diameter) is necessary to provide high specific surface area for the analyte retention. Surface of the base material should have specific chemical reactivity for further modification with selected ligands to form the reversed-phase bonded layer. Base material determines the mechanical and chemical stability—the most important parameters of future (modified) reversed-phase adsorbent.

3.5.1.1 Mechanical Stability (Rigidity).

Particles of packing material are subject to significant mechanical stress under the column packing procedure and sometimes during column operation (pressure shock, or fast release of excessive pressure). The rigidity of material is, to a large extent, dependent on its surface tension (or surface energy), which is a function of material surface chemistry. Chemical modification of base material significantly alters this parameter, and the rigidity of modified material usually is not the same as for original base silica [18].

3.5.1.2 Chemical Stability.

Hydrolytic stability of base material is the most important parameter because most reversed-phase HPLC separations are performed in water/organic eluents with controlled pH. Selection of the mobile phase pH is mainly dictated by the properties of the ionizable analytes to ensure that they are in one predominate ionization state.

Chemical modification of the adsorbent surface significantly alters practically all properties of the base material. Dense coverage of the adsorbent surface with inert ligands usually expands chemical stability of the packing material.

3.5.2 Silica

Silica (SiO_2) is the most widely used base material for HPLC adsorbents. The majority of HPLC packings are silica-based. The chemistry of silica, the methods of silica's controlled synthesis, the surface structure of silica, and the properties of silica have been studied for over two centuries. It is possible to control a synthesis of ideally spherical particles with predefined pore size and pore size distribution, as well as the synthesis of monolithic silica rods. Porous silica provides the high surface area necessary for successful separation; at the same time, silica particles are very hard (mechanically strong), which allow them to withstand harsh packing conditions and flow of viscous liquids. Silica is not shrinking or swelling when exposed to the different solvents. Even though silica has an array of advantageous properties, it has some drawbacks. The main one is its solubility in water at high pH. The other is an extreme polarity of its surface.

3.5.2.1 Synthesis of Silica.

Silica used in HPLC is an amorphous, porous solid, which could be obtained by different synthetic procedures.

Colloidal Sol–Gel Procedure. Silica sol is passed through nonaqueous media, where it forms spherical droplets, which rapidly solidify into the hydrogel beads [19]. Solid beads are dried and calcinated at around 600–1000°C. This synthetic procedure usually gives spherical particles of silica with significant amount of impurities (Na, Fe, B, etc.) in the body and on the surface of material. These impurities can increase the acidity of the surface silanols, thus lowering the pK_a of the respective silanol groups.

Polycondensation Procedure. This method is essentially the controlled poly-merization of tetraethoxysilane (TES). In the first step, TES is partially hydrolyzed in a viscous liquid, which then is emulsified in the mixture of ethanol and water and then undergoes further hydrolytic polycondensation when catalyzed. The formed solid beads of hydrogel are washed and dried into porous silica. This procedure allows synthesis of highly pure silica particles, which is essential for HPLC.

Some recent modifications of the polycondensation process allow synthe-sis of silica with organic moieties embedded into the bulk material [20, 21], which manufacturers claim to give advantages over conventional silica in terms of higher pH and temperature stability.

3.5.2.2 Surface Silanoles. The surface of amorphous silica is constructed of several different terminal groups. The major portion of the silica surface is covered with single (isolated or free) silanols. Free silanols contain a silicon atom that has three bonds in the bulk structure, and the fourth bond is attached to a single hydroxyl group.

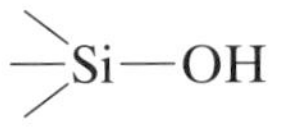

$$\begin{array}{c}\diagdown\\\diagup\end{array}\!\!Si\!-\!OH$$

The calcination process at high temperatures (over 800°C) often removes water molecules from adjacent silanols, leading to a formation of a siloxane bond. This process is known as *dehydroxylation* [18]. Dehydroxylated silica is very inert, but it can slowly absorb water and rehydroxylate.

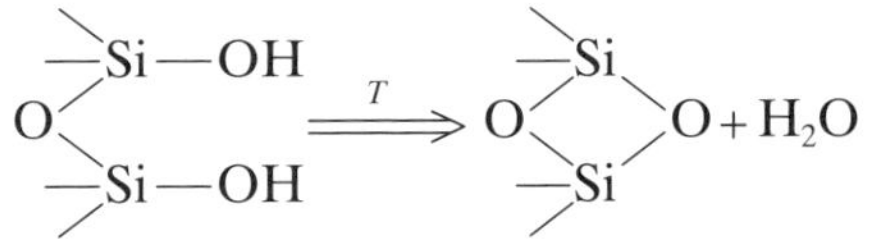

$$O\!\!\begin{array}{c}\diagup Si\!-\!OH\\\diagdown Si\!-\!OH\end{array}\xrightarrow{T} O\!\!\begin{array}{c}\diagup Si\\\diagdown Si\end{array}\!\!O + H_2O$$

Some adjacent silanols can hydrogen-bond to each other, which requires rel-atively close position of silanols usually observed in the β-kristobalite form of silica. Typical chromatographic packing material is estimated to have not more than 15% of its surface with β-kristobalite-type silanols arrangement.

In the chromatographic literature an additional type of surface silanols is often mentioned [8]; the geminal silanol shown below contains two hydroxyl groups attached to one silicon atom.

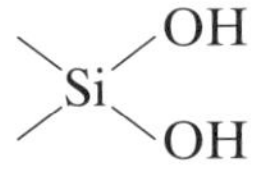

$$\begin{array}{c}\diagdown\\\diagup\end{array}\!\!Si\!\!\begin{array}{c}\diagup OH\\\diagdown OH\end{array}$$

Peri and Hensley [22] proposed the existence of these groups on silica surface, although their existence on the surface has not been confirmed. These groups have only been experimentally observed on monomeric organosilicon com-pounds in solution.

The density of the silanole groups on the silica surface is the most important parameter defining surface reactivity and polarity and is also claimed to have negative effects on chromatographic properties of modified adsorbents.

Many attempts have been made to measure silanol surface density (α_{OH}). Iler [7] estimated α_{OH} to be equal to 8 groups/nm^2 on the basis of the [100] face of β-cristobalite. However, most porous amorphous silicas show surface silanol concentration on the level of 4.6 to 5 groups/nm^2 [6].

3.5.3 Silica Hybrid

In the last decade, several composite base packing materials were successfully introduced into the market [20, 21]. The primary driving force in developing these materials was the attempt to use all benefits of well-known porous silica and suppress its drawbacks. The two main drawbacks are solubility at high pH and high surface density of silanols.

Modification of the base silica synthesis by addition of methylthrietoxysilane allows the introduction of methyl groups on the silica surface. A schematic representation of the expected surface is shown in Figure 3-6. Surface concentration of methyl groups is dependent on the reagent ratio.

The position of methyl groups within the silica body is essentially random, and their appearance on the surface may not be favorable during the polycondensation process. The presence of terminating methyl groups within silica body may decrease the mechanical stability of base silica, and this is also dependent on the reagent ratio. Authors claim mechanical stability comparable to that of regular HPLC silica [23] while the amount of surface silanols is significantly reduced [24, 25].

Another approach to manufacturing hybrid silica (Gemini) was introduced by Phenomenex [26]. They synthesize layered hybrid silica where the core of the particle is regular silica and the surface is covered by the layer of organic-embedded silica. This allows better control of the porous structure because

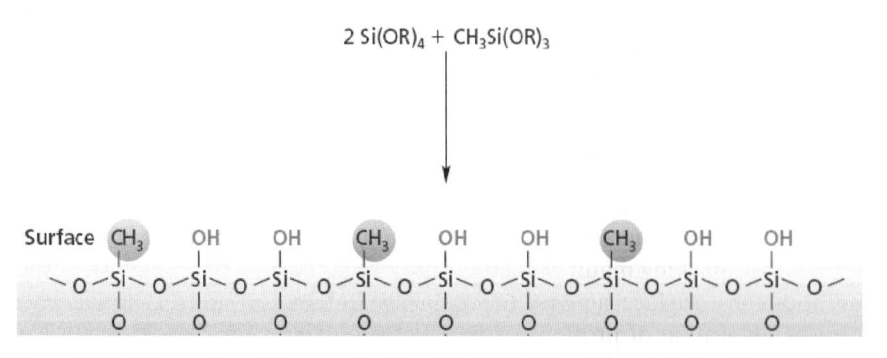

Figure 3-6. Schematic of the synthesis of hybrid silica. (Reprinted with permission from reference 23.)

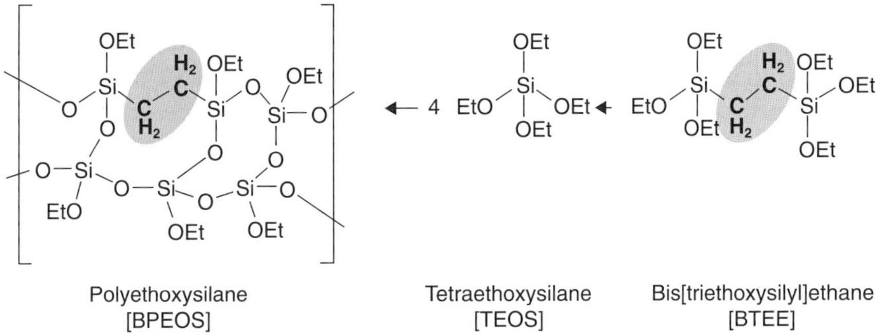

Polyethoxysilane [BPEOS]　　　Tetraethoxysilane [TEOS]　　　Bis[triethoxysilyl]ethane [BTEE]

Figure 3-7. Schematic of the formation of bridged hybrid silica. (Reprinted with permission from reference 28.)

traditional synthesis of base silica is used and surface silanole concentration is decreased while maintaining the mechanical strength of silica.

Recently, further development of hybrid materials allows the introduction of organic bridged silica (Figure 3-7). Embedded organic groups do not have terminating function any more but actually are forming an organic bridge between silicon atoms. According to authors, this hybrid particle, having an empirical formula $SiO_2(O_{1.5}SiCH_2CH_2SiO_{1.5})_{0.25}$, is synthesized by the co-condensation of 1,2-bis(triethoxysilyl)ethane (BTEE, 1 equiv.) with TEOS (4 equiv.) [20]. The resulting hybrid material shows better pH stability because Si–C covalent bonds are much less prone to hydrolysis than Si–O–Si bonds. Surface energy is also significantly reduced, as estimated by comparison of C-constants of the BET equation (BET C-constant represents the energy of nitrogen interaction with the surface [15]). Usual values for regular silica are between 100 and 150, while on hybrid silica a nitrogen interaction with the surface is significantly weaker and the C-constant value drops to 49, which is comparable to that of phenyl-modified silicas [27].

A decrease in the surface energy is associated with the decrease in the surface silanols concentration, but this decrease is not reflected on the ability of this material to accept surface modification. Chemical modification with octadecilsilane ligands resulted in 3.2- to 3.3-µmol/m^2 surface density, which is typical for most modern regular modified silicas [20].

3.5.4 Polymeric Packings

Variation of the mobile-phase pH is one of the most powerful tools in controlling the separation for ionizable analytes. The main drawback of silica-based HPLC packing materials is their narrow applicable pH range. The other limitations are surface activity (or polarity), which for specific applications (such as separation of proteins or biologically active compounds) could play a major role. All these factors are the driving force for the search in alternative base materials for HPLC packings.

The majority of polymer-based packing materials are polystyrene-divinylbenzene crosslinked copolymers. While PS–DVB packings have the advantages of chemical stability at wide pH range, they suffer from the disadvantage of yielding lower chromatographic efficiencies than silica-based bonded phase packings of the same particle size. Even a high degree of crosslinkage in three-dimensional polymer structure leave sufficient space between polymer chains for small analyte molecules to diffuse into the body of the polymer. These "micropores" cause noticeable increase in the broadening of chromatographic zone. In another words, column efficiency is lower due to a slow intraparticle sorption rate and due to slow diffusion of solute molecules within the polymer matrix [29–36].

3.5.5 Zirconia (Metal Oxides)

Introduction of all these materials on the market is driven primarily by their superior stability at high mobile-phase pH and temperature range.

There is a very limited selection of commercially available materials due to higher inertness of the metal oxide surface, and there are almost no reproducible methods of chemical surface modification [37]. Most of the surface chemistry alteration is achieved by coating and not bonding. Control of the surface area and porosity is also limited.

The commercial availability of zirconia-based HPLC packings are mainly related to the enormous extensive research of P. Carr and other workers [37, 38]. They applied zirconia as the starting material for a number of different polymer-coated RP phases.

Carr and others have described the preparation and properties of polybutadiene (PBD) and polyethyleneimine (PEI), as well as aromatic polymer-coated and carbon-clad zirconia-based RP phases. The preparation of PBD-coated zirconia and the chromatographic evaluation of these phases have been described extensively by Carr, McNeff, and others [39–41]. From these studies, the authors conclude that at least for neutral analytes PBD zirconia-coated phases behave quite similar with respect to retention and efficiency compared to silica-based RP phases [42]. For polar and ionic analytes, however, substantial differences with respect to retention, selectivity, and efficiency have been reported [43].

3.5.6 Porous Carbon (or Carbon-Coated Phases)

For many years, different research groups attempted to create porous material with ideal graphite surface and strong enough to be used in HPLC. The advantage of this material would be that two main interactions can occur with analytes: hydrophobic and π–π interactions on an essentially planar surface of graphite (Figure 3-8). The use of graphitized carbon black in gas chromatography had shown significant predictability of retention and specificity for the separation of conformational isomers, and similar advantages are expected for these adsorbents in HPLC.

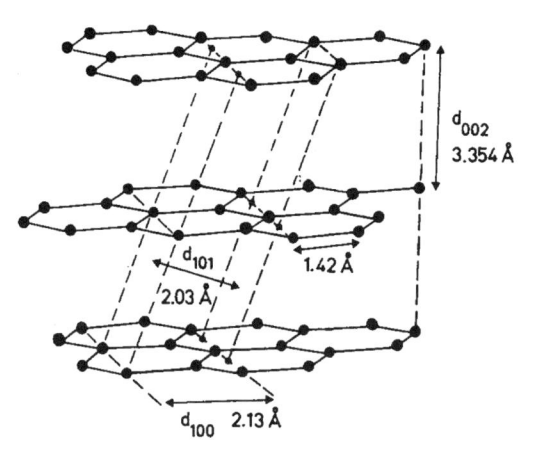

Figure 3-8. Atomic structure of porous graphitic carbon. (Reprinted from reference 48, with permission.)

The first commercial HPLC packing material with graphitized carbon surface was made commercially available at the end of the 1980s under the name Hypercarb [44–46]. On the atomic level, porous graphitic carbon (PGC) is composed of flat sheets of hexagonally arranged carbon atoms (about 10^5 atoms per sheet) [47]. Edges of graphite sheets are expected to be partially oxidized with the formation of hydroxyl, carbonyl, and carboxylate groups [49].

PGC-type packing material shows unique chromatographic properties, since it is more hydrophobic than conventional C18 phases and it shows significantly higher methylene selectivity [50]. This material has high chemical stability in the wide pH range, and it has unique selectivity for the separation of polar compounds because of its high polarizability [51, 52]. This adsorbent is the primary choice if the separation of conformational isomers is required, because the planar nature of the main part of its surface provides the basis for isomeric separation (Figure 3-9).

3.6 SURFACE OF CHEMICALLY MODIFIED MATERIAL

In the preparation of reversed-phase packing material the main purpose of chemical modification is to convert polar surface of base material into the hydrophobic surface which will exert only dispersive interactions with any analyte.

In porous packing materials with 10-nm average pore diameter, 99% of the available surface area is inside the pores. Conversion of highly polar silica with high silanol density (4.8 groups/nm^2) [7] into the hydrophobic surface requires dense bonding of relatively thick organic layer which can effectively shield the surface of base silica material.

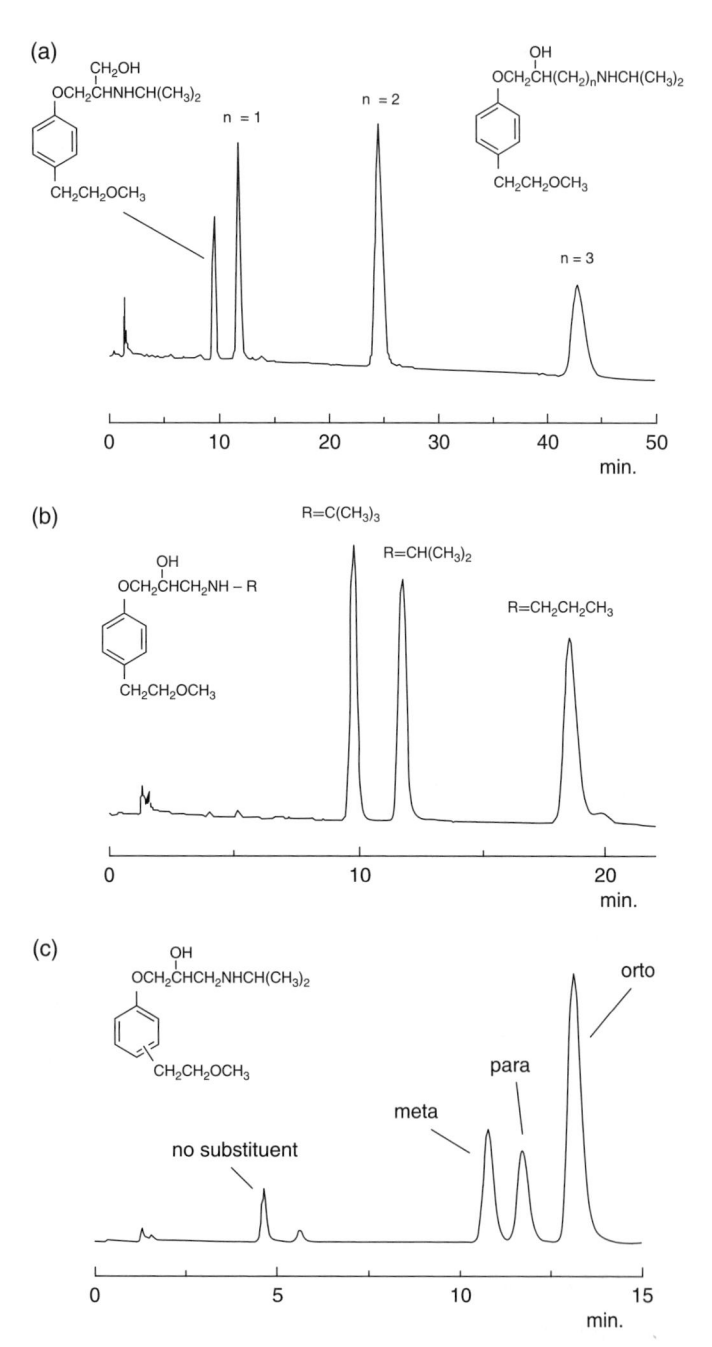

Figure 3-9. Selectivity of porous graphitic carbon for positional and conformational isomers under reversed-phase conditions. (Reprinted from reference 53, with permission.)

3.6.1 Limits of Surface Modification

A wide variety of different ligands have been bonded on silica surface [18] and used as HPLC packing materials. The most traditional and widely used are shown in Table 3-1.

Molecular volumes and maximum ligand length shown in Table 3-1 allows for the estimation of theoretical maximum bonding density. The maximum possible thickness of bonded layer could not exceed the length of ligand in all-trans conformation. If we divide the molecular volume of a ligand ($Å^3$/molecule) by its length ($Å$), we get the minimum possible molecular area ($Å^2$) that the ligand can occupy on the surface with the densest bonding. Reciprocal value will be the number of ligands per unit of surface area, and this is only valid for a flat surface. The majority of the surface of porous material is concave. Assuming a cylindrical pore model (standard assumption for silica surface), the maximum bonding density (d_b) corrected for the surface curvature is shown in equation (3-10) (and illustrated in Figure 3-10) as the function of bonding density on the pore radius (R) and ligand length (l) and molecular volume (V_l) is

$$d_b = \frac{1}{V_t}\left(l - \frac{l^2}{2R}\right) \tag{3-7}$$

In the literature the bonding density is often expressed in either number of moles (or micromoles) per square meter (μmole/m^2) or in number of molecules (bonded ligands) per square nanometer. The relationship between these two units is the simple ratio

$$d_b\left(\text{molecule}/\text{nm}^2\right) = \frac{6.022 \times 10^{23}}{10^{24}} d_b\left(\mu\text{mol}/\text{m}^2\right) \tag{3-8}$$

Calculated values are only a theoretical maximum; in reality, the average bonding density for C1–C18 alkyl ligands varies between 3 and 4 μmol/m^2 or between 1.8 and 2.4 groups/nm^2. Generally, less than half of available silanols (4.8 groups/nm^2 or 8 μmol/m^2) are reacted with bonded ligands; the other half is left on the surface. Because it is impossible to modify all available silanols, it is important to shield them, in order to make them inaccessible for analytes.

3.6.2 Chemical Modification

Surface silanols could react with many different functional groups to form the so-called bonded phase. The majority of the bonding agents used are chlorosilanes, although ethoxysilanes and sometimes alcohols are also used. Practically all commercially available chromatographic phases are made using silanization modification process.

TABLE 3-1. Structures and Geometry Parameters of the Most Common Bonded Ligands

#	Structure	Name	Structural Formula (Å³)	Volume[a] (Å³/molec)	Length[b] (Å)
1		C1		150	3.7
2		C4		259	7.5
3		C8		365	13
4		C18		630	24
5		Phenyl		303	9.2

6	Phenyl-hexyl	408	14
7	Perfluoro-phenyl	363	10.6
8	Nitrile (cyano)	248	7.8
9	Amino-proyl	238	7
10	Diol	314	11.5

[a]Molar volume evaluated using ACD software (version 9.3).
[b]Ligand length calculated using ACD software for the distance between anchoring oxygen and foremost atom in all-trans conformation of corresponding ligand.

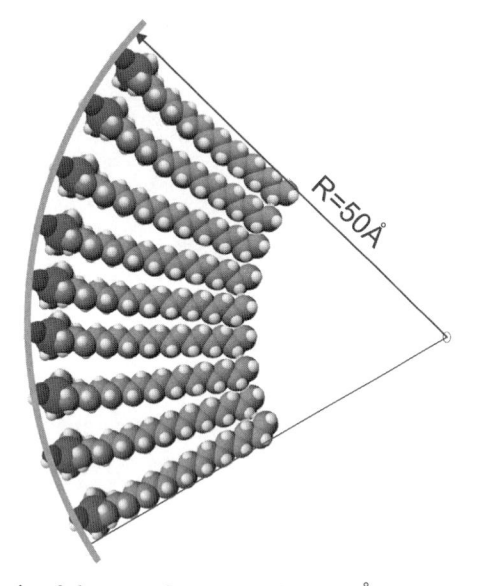

Figure 3-10. Schematic of the pore fragment with 100-Å diameter modified with C18 ligands in all-trans conformation illustrating the steric hindrance effect.

3.6.2.1 Types Modification Reactions. Chemical modification of hydroxylated surface of base material in more than 90% of commercially available HPLC stationary phase is performed by its reaction with alkylchlorosilane-type modifiers, reaction (3-9). HCl is generated as a byproduct, and typically a scavenger for HCl is employed in the synthesis. For most common bonded phases the two side groups are methyl; however, other side groups such as isopropyl and *tert*-butyl have been used.

$$|Si\text{–}OH + Cl\text{–}Si(Me)_2\text{–}n\text{-alkyl} = |Si\text{–}O\text{–}Si(Me)_2\text{–}n\text{-alkyl} + HCl \qquad (3\text{-}9)$$

$$|Si\text{–}OH + HO\text{–}alkyl = |Si\text{–}O\text{–}alkyl + H_2O \qquad (3\text{-}10)$$

On the silica surface, silanols can react with alcohols forming Si–O–R bonds, reaction (3-10). This type of reaction allows for a very high bonding density, but hydrolytic cleavage of Si–O–R bonds is relatively high and this material is not practically applicable for HPLC support.

Chemical modification of the silica surface with hydridosilane-type modifiers as shown in schematic (3-11) offers some advantages:

$$Si\text{–}OH + H_3Si\text{–}R = Si\text{–}O\text{–}Si(H_2)\text{–}R + H_2 \qquad (3\text{-}11)$$

A byproduct of this reaction is H_2, which easily leaves the reaction zone and leads to higher bonding density [18].

3.6.2.2 Surface Modification with Chlorosilanes. Chlorosilanes are volatile and very sensitive to the presence of trace amounts of water. Vapor-phase silanization reactions with rigorous control of the absence of water show that Cl_nSiMe_{4-n}-type chlorosilanes react with aerosil (nonporous microparticulate silica) as

$$|Si-OH + Cl_nSiMe_{4-n} \rightarrow |Si-O + SiMe_{4-n}Cl_{n-1} + HCl \qquad (3\text{-}12)$$

that is, only one surface silanol participates in this reaction [54]. Unger [6], on the other hand, advocates the possibility of the formation of bridging bonds:

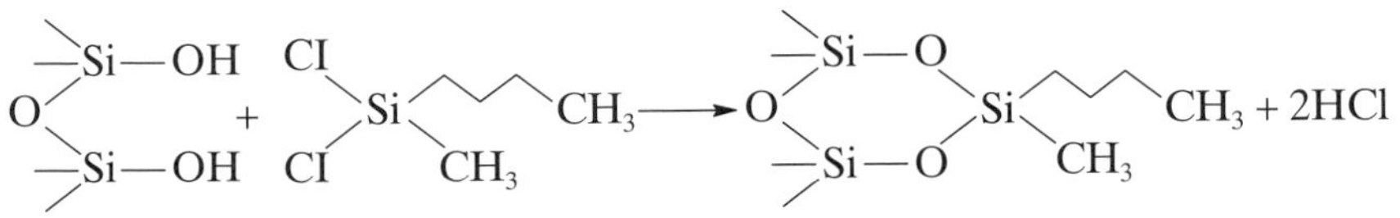

Analysis of interatomic distances for these types of bonds shows that adjacent silanols on the surface should be at a distance of approximately 2.5 Å, which corresponds to that distance in β-crystobalit and leads to the surface silanol density of 8 groups/nm^2. The majority of porous silica have a silanol density of 4.8 groups/nm^2, and there is low probability of forming bridging bonds. Lisichkin and co-workers [18] discussed the interpretation of silica surface as a mixture of different microcrystalline surfaces of quartz, β-tridimite, and β-crystobalite. They estimate that approximately 15% of silica surfaces have a β-crystobalite structure. This means that this bridging reaction shown above is possible but only 15% of available silanols could react in this manner, while trifunctional bonding is virtually impossible.

The silanization reaction in liquid phase (usually in dry toluene) is technologically more convenient, although it is practically impossible to ensure the complete absence of water in a reaction mixture that leads to hydrolysis of chlorosilanes. Using deuterium exchange, Roumeliotis and Unger [55] analyzed surface silanol concentration before and after the modification with different reagents (Table 3-2).

All monofunctional modifiers decrease the surface concentration of silanols by 50% (depending on the alkyl ligand) compared to original silica (7.6 μmol/m^2), while difunctional modifiers only slightly reduces the surface concentration of silanols by 12% and trifunctional modifiers apparently increases the amount of silanols. This reflects actual hydrolysis of nonreacted chlorine groups of the alkyl di- and trifunctional modifiers and their conversion into silanols.

When a difunctional modifier is used, it mainly reacts with only one silanole, and the second chlorine is converted into hydroxyl by the reaction with either residual water in the solvent media or during the adsorbent washing process after the main reaction is completed. As the result, each surface silanole is effectively substituted with another hydroxyl on the attached ligand and the total amount of silanols almost does not change.

TABLE 3-2. Surface Silanol Concentration Change after Chemical Modification [18]

Modifier	Carbon Content	Surface Concentration (umol/m^2)		
		Ligands	OH Groups	(Lig + OH)-Silica
$Cl_3SiC_8H_{17}$	7.36	2.35	11.70	6.42
$Cl_2Si(Me)C_8H_{17}$	8.43	2.40	6.74	1.51
$ClSi(Me_2)C_8H_{17}$	9.24	2.35	3.73	−1.55
$ClSiMe_3$	4.18	3.37	2.98	−1.28
$ClSi(Me_2)C_4H_9$	7.13	2.97	3.68	−0.98
$ClSi(Me_2)C_8H_{17}$	10.43	2.71	3.73	−1.19
$ClSi(Me_2)C_{12}H_{25}$	11.73	2.20	3.81	−1.62
$ClSi(Me_2)C_{16}H_{33}$	16.67	2.36	4.22	−1.05
Original silica			7.63	

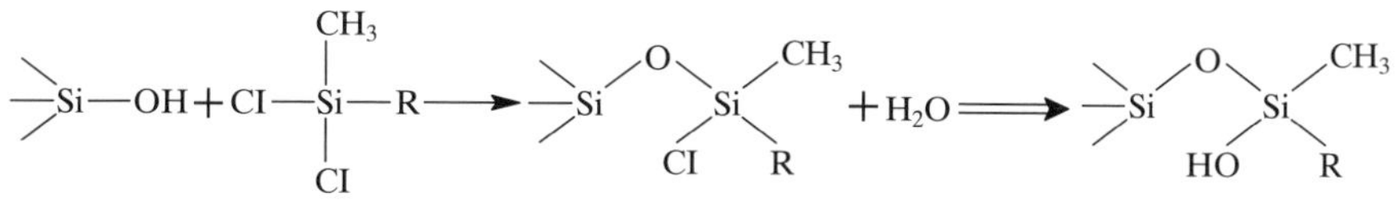

When a trifunctional modifier is used each surface silanole is substituted with two hydroxyls, as shown in the following reaction:

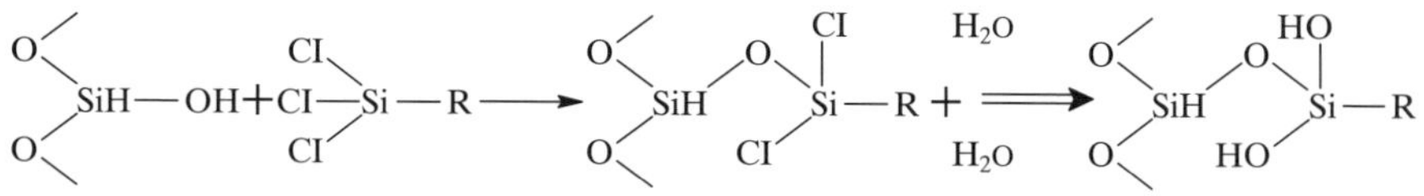

Theoretically, the sum of surface concentrations of bonded ligands and residual silanols should be equal to the original silanols concentration for the reaction with monofunctional reagents. In reality, there is an approximately 1-μmol/m^2 difference; and the shorter the bonded alkyl chain, the smaller the difference, at least in the range between C4 and C12 ligands. The difference itself could only be explained by the effectiveness of the surface shielding for the deuteron exchange; and the longer the bonded chain, the more effective the shielding for comparable bonding density.

3.6.2.3 Monomeric and Polymeric Bonding. Difunctional and trifunctional reagents are often used for the formation of so-called polymeric bonding. In the previous section we discussed the controversy of the surface reactions of mono-, di-, and trifunctional reagents. At present, there is no definite conclusion regarding the possibility of the formation of bridging bonding, but most agree that it could not exceed 15% of the total amount bonded. For the formation of the bond shown in Figure 3-11 the distance between surface silanols should be similar to that on β-kristobalite (approximately

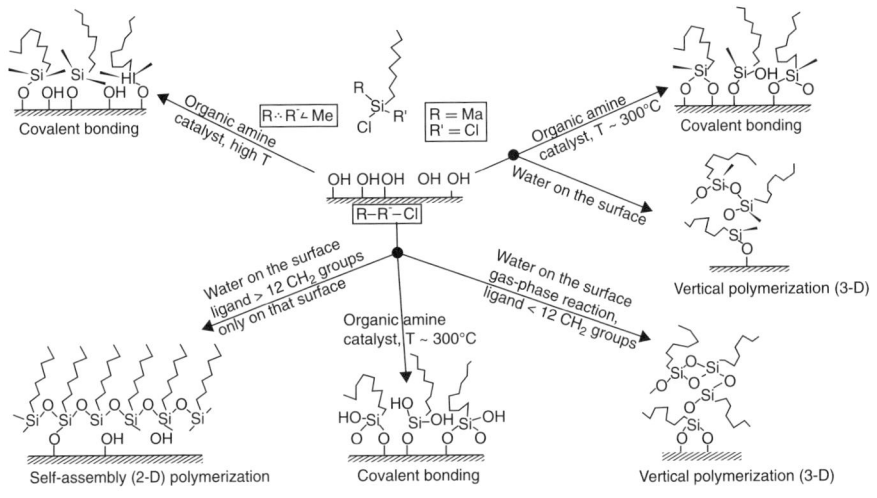

Figure 3-11. Schematic of different types of bonding. (Reprinted from reference 18, with permission.)

2 Å), which corresponds to a very high surface concentration of silanols (4.8 groups/nm^2). According to some estimations, thermally calcinated porous silica may have small sections of its surface in the form of microcrystalline β-kristobalite (approximately 10–15%), and the formation of bridged (difunctional) bonding could be observed only on this type of surface.

The probability of the formation of trifunctional bonding is absolutely negligible.

In most cases, di- and trifunctional reagents will polymerize in the presence of water forming linear (for difunctional) and branched alkoxysilanes (for trifunctional modifiers). Trace amounts of water are always present on the surface of polar silica.

Figure 3-11 illustrates possible pathways of silica surface modification with mono-, di-, and trifunctional reagents.

3.6.2.4 Mixed Bonding.
In some cases, bonding of different ligands on the silica surface is desirable. On porous materials with small pore diameter (below 80 Å), steric hindrance restricts the achievement of high bonding density. In this case, bonding of a mixture of randomly distributed short alkyl ligands and long chains allow for better shielding of polar silica surface.

The practical synthesis of the uniformly distributed mixed bonded phases is discussed in reference 18, where the authors indicate that reproducible synthesis can be very challenging.

The authors in reference 18 also indicate that in most practically applicable synthetic procedures for bonding of C18-type modifiers on a silica surface, the formation of the bonded layer proceeds according to the spot-formation process (in the presence of amine catalyst). In other words, octadecylsilanes

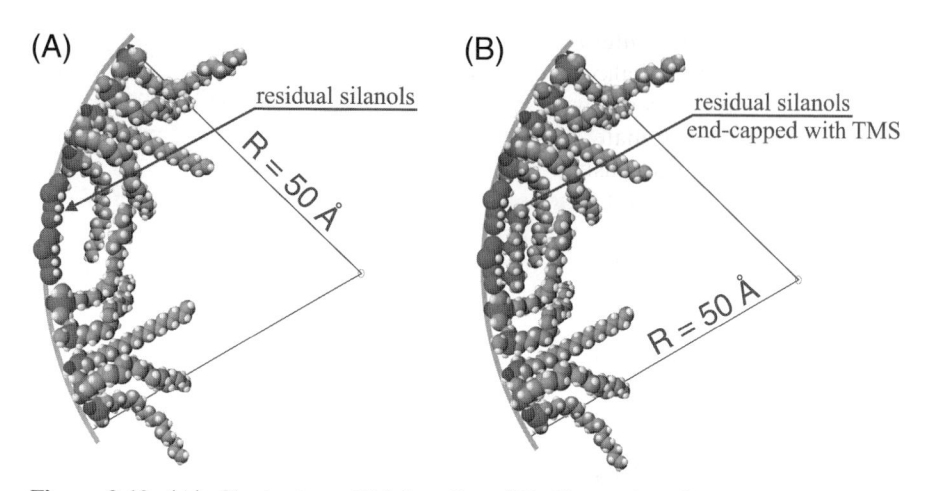

Figure 3-12. (A) Cluster-type C18 bonding. (B) Cluster bonding with end-capped silanols.

are bonded close to each other, thereby forming clusters, and there is significant space between those clusters (schematically shown in Figure 3-12).

3.6.2.5 End-Capping.

Secondary silanization with TMS is usually performed with the intention of covering accessible residual silanoles left after the main modification step. Possible cleavage of the main ligands (substitution) may result as a consequence [18].

Figure 3-12 shows the hypothetical structure of a bonded layer. Cluster-type bonding of alkyl ligands leaves significant areas that are not covered due to the concave nature of the inner silica surface and bulkiness of alkyl chains. We discussed before that only about 50% of all available silanols actually react with a surface-modifying agent while the other half are left unreacted. Some of these so-called "residual" silanols are left underneath the closely bonded alkylsilanes, and others are between these bonded clusters. The former are essentially inaccessible, whereas the latter are "accessible residual silanols." These accessible residual silanols are believed to produce a lot of unwanted effects in reversed-phase HPLC (peak tailing, bonded phase instability, fronting, etc.).

The remedy of these unwanted effects is the process called "end-capping," which is essentially second-stage surface modification with small ligands that can squeeze between patches of C18 ligands and react with accessible residual silanols, thus potentially deactivating them.

Despite the widely accepted opinion that endcapping really covers accessible residual silanols with relatively inert "mushrooms" of trimethylsilanes, we can refer the reader to the old work of Berendsen et al. [56], where they did a rigorous analysis of alkyl-modified silica before and after end-capping. In all their samples, carbon content decreases after end-capping. This can only mean that long-chain bonded alkylsilanes are substituted with short trimetylsilanes that actually have higher bonding energy. If this process is indeed occurring, it is impossible to estimate actual coverage with TMS and original ligands. Chromatographic testing usually shows the decrease of peak tailing and other effects usually attributed to the influence of a reduced amount of a residual silanols.

It is possible that patches of accessible residual silanols are indeed covered with TMS and at the same time a fraction of bonded long-chain ligands (mainly at the borders of these patches) are substituted with TMS also. Overall packing material becomes more hydrophobic, since the amount of accessible silanols decreases with end-capping, but carbon content is decreased also.

3.6.3 Types of Bonded Phases

3.6.3.1 Alkyl-Type Phases (C1–C18, C30). Probably 90% of all reversed-phase columns are alkyl-type bonded phases. An enormous amount of publications are devoted to the classification, standardization, and comparison of these phases. In their book *Practical HPLC Method Development*, Snyder and Kirkland [57] indicate that reversed-phase retention for nonpolar and non-ionic compounds generally follows the retention pattern C1 < C4 < C8 ≈ C18. At the same time, they refer to the comparison of C18-type columns from different manufacturers and find dramatic variation in the retention of both polar and nonpolar compounds at the same conditions on different columns.

HPLC retention expressed either in absolute values (V_R) or in relative terms (retention factor) is a complex function of a multitude of parameters. Type of bonded phase is only one of them. Surface area of base material, pore size, bonding density, end-capping, and even the column history all can significantly affect analyte retention.

The only valuable comparison of different columns found so far is so-called methylene selectivity. Methylene selectivity is the slope of the dependence of the logarithm of retention factor of the members of homologous series on the number of carbon atoms in their alkyl chain (Figure 3-13). This parameter reflects the surface interaction energy of one CH_2 group and is found to be very similar on all C18-type columns (and even C8).

3.6.3.2 Phenyl-Type Phases. Phenyl-type phases have been studied for a long time [58, 59]. The presence of a phenyl ring on the surface of a bonded phase introduces so-called π–π interactions with some analytes that are capable of these types of interactions. This introduces an additional specificity for HPLC separations on these stationary phases. Compared to common alkyl-type phases, phenyl columns show lower methylene selectivity; in other words, the separation of members of homologous series will be less selective on phenyl columns than on alkyl-modified phases.

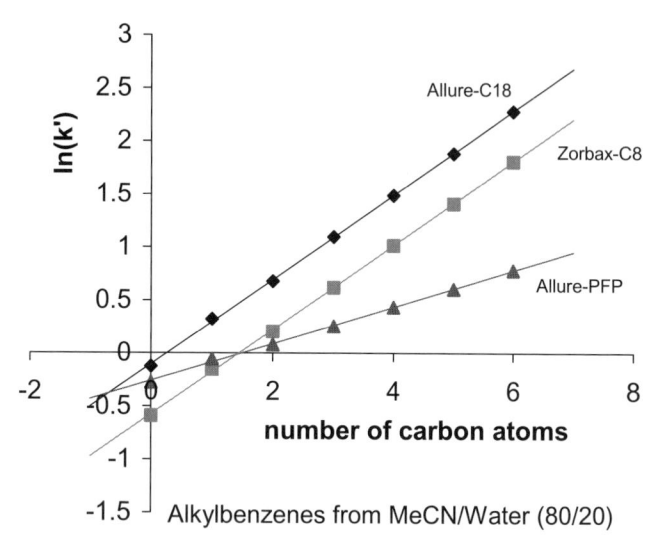

Figure 3-13. Slope of the logarithm of homologous series retention.

3.6.3.3 Polar Embedded Stationary Phases. The introduction of polar embedded stationary phases was inspired by the need to develop reversed-phase methods in high aqueous mobile phases for separation of either highly polar or ionic compounds. At these conditions, the hydrophobic surface of chemically bonded material has limited wettability. In pure aqueous mobile phase the "phase collapse" effect was observed. The consequence of this "phase collapse" is that after some period of time the retention of any analytes on the column dramatically decreases, although it does not become equal to the void volume, but very close to it. The first proposed explanation was that bonded alkyl ligands are "collapsed" or do not allow analyte penetration between bonded chains. Recently, it was shown that alkyl chains are always "collapsed" [60]. They tend to minimize their energy by assuming such conformations that allow the maximum number of contacts between these chains. The study of apparent viscosity of alkyl-bonded layers showed that it is approximately three orders of magnitude higher than the viscosity of corresponding free *n*-alkanes. Recall that *n*-octadecyl is solid at room temperature; this means that octadecyl chains immobilized on the surface form a rock solid material.

Recent studies of the "phase collapse" effect show that after steady removal of a wetting agent (any previously used organic mobile-phase modifier), water does not wet the hydrophobic inner surface of porous material, and the flow of the mobile phase through the porous space and corresponding transport of the analyte molecules to the surface inside the pores is suppressed. The effect is essentially equivalent to an approximately 100-fold decrease of the adsorbent surface area (the majority of the surface area is inside the pores of packing material). This effect is reversible, and pumping of an organic eluent component restores the flow through the porous space, although, depending on the adsorbent porosity, it may require up to several hours of organic flow. The smaller the pores, the longer it takes.

Polar embedded stationary phases include a nonionizible polar group embedded into the bonded chain. Symmetry-Shield RP18 (Waters) and Supelcosil ABZ (Supelco) are examples of this type of bonding. Another example is Synergi Fusion, (Phenomenex) which uses a polar embedded ligand and a hydrophobic ligand. The polar embedded ligands do not contain nitrogen functionality. Other types of commercially available polar embedded phases include XBridge Shield RP18 (Waters), Synergy Polar RP, Synergy Hydro-RP (Phenomenex), Zorbax SB-AQ (Agilent), YMC-Pack ODS-AQ (YMC), Dionex Acclaim PA (Dionex), and HiChrom ACE 3 AQ (ACE).

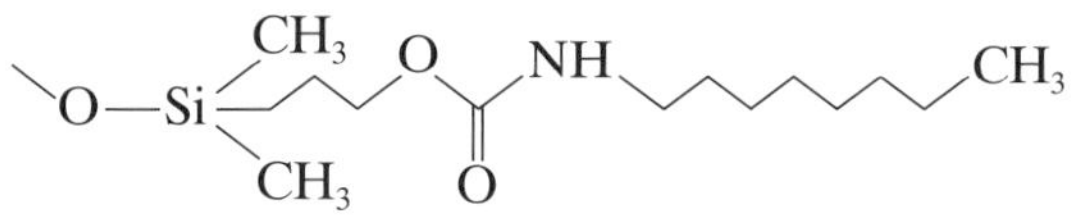

Symmetry Shield

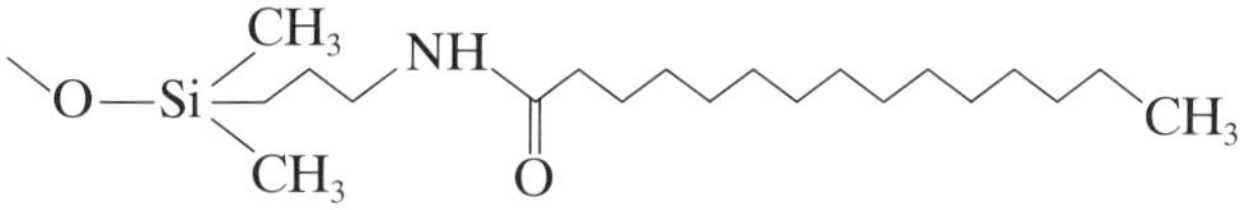

Supelcosil ABZ

The second reason in the introduction of polar groups in the bonded ligands is that these groups interact with residual silanols, which make the silanols effectively inactive for the interaction with polar or basic analytes. Sometimes these phases are also end-capped with polar end-capping groups. These phases also show a significant difference in selectivity compared to conventional C18-type phases. In some cases they show improvement in the peak shape for basic components.

3.6.4 Structure of the Bonded Layer

Bonded alkyl ligands are flexible, although their flexibility is significantly restricted by immobilization on the surface; also, the higher the ligand bonding density, the less flexibility these chains have.

The main goal of chemical modification of the surface is to create a preferably uniform surface with a selected type of interactions. Energetic uniformity of the surface is dependent on the ligand's bonding density, distribution, and conformations. An additional desirable feature is hydrolytic stability, which in most cases is achieved by proper shielding of anchoring bonds.

Shielding of the base material, leading to the elimination of the influence of residual silanols, is essentially the main focus in the development of most recent packing materials.

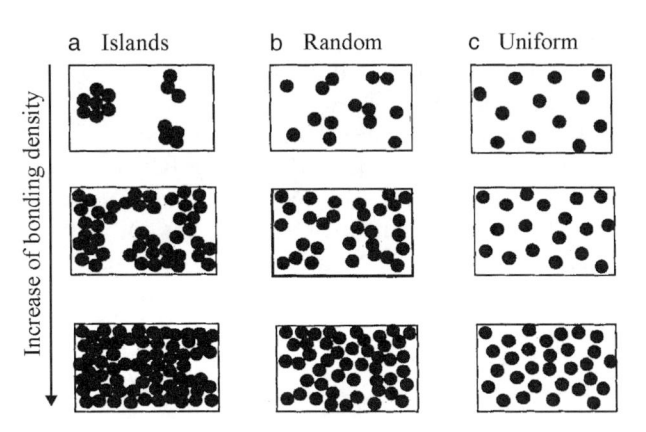

Figure 3-14. Distribution of bonded ligands at low, average, and high bonding density for island, random, and uniform bonding. (Reprinted from reference 18, with permission.)

The efficiency of the substrate shielding is dependent on the local density of bonded ligands and their overall distribution on the surface. Three different types of surface ligand distribution could be distinguished: random, uniform, and island-like. These distributions are illustrated in Figure 3-14.

Type of bonding distribution is significantly dependent on the conditions under which the bonding reaction is performed. Type of catalyst (usually amine), reagent concentration, temperature, and presence of residual adsorbed water on the substrate surface affect this process. Detailed discussion of this could be found in reference 61.

Arrangement of the bonded ligands on the surface, along with conformation of these ligands and their occupied volume, is also important for understanding of the shielding effect. Berendsen et al. [62] studied the variation of the specific pore volume with surface modification. Recently, this obvious decrease of adsorbent pore volume was correlated with molecular volume of bonded ligands [63]. It was shown that molecular volume of bonded ligands calculated from the volume of bonded layer and bonding density corresponds to that for liquid normal alkanes of similar chain length (Figure 3-15).

A similar study was performed on packed columns at HPLC conditions, and it was found that at any composition of the mobile phase, regardless of the type of organic modifier (THF, methanol or acetonitrile), the volume of the bonded layer corresponds to the densest arrangement of bonded ligands [60].

Molecular conformation of bonded ligands and their degree of freedom is dependent on bonding density. The higher the bonding density, the lower the number of possible conformations and thus the less mobility the bonded chains have. Immobilization of ligands on the surface already restrict their mobility, so if we compare the state of free C18 molecules (*n*-octadecyl) with immobilized octadecyl, we can expect more rigid (or solid-like) behavior of immobilized chain. Indeed, the study of the viscosity of bonded layers [64]

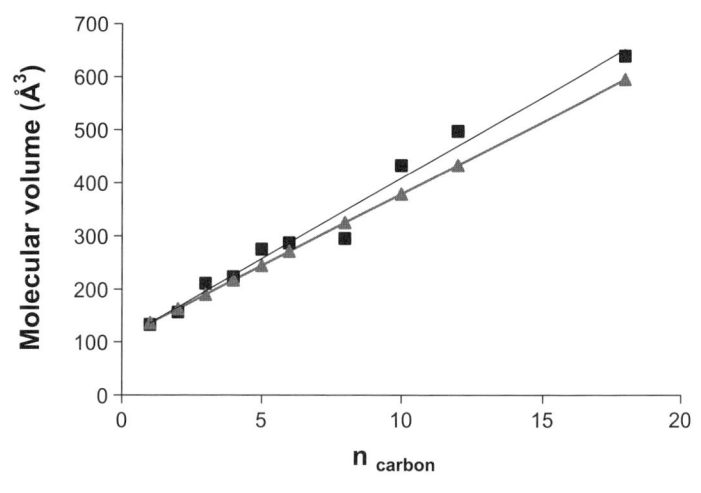

Figure 3-15. Correlation of molecular volume of bonded n-alkyl ligands with that for liquid n-alkanes [63].

showed that the apparent viscosity of alkyl chains in the bonded layer increased by approximately two orders of magnitude compared to the viscosity of corresponding n-alkanes. Conformation of bonded ligands could be interpreted as shown in Figure 3-16.

3.6.5 Density of Bonded Ligands

Bonding density is a surface concentration of bonded ligands, and it is expressed in either number of moles per square meter (μmol/m^2) or in number of groups per square nanometer (groups/nm^2). Unit conversion is shown in equation (3-7).

The term *bonding density* is constantly used throughout this book. This is probably the most important characteristic of the bonded phase. The higher the bonding density, the more hydrophobic the adsorbent surface, the better the shielding of residual silanols, and the higher the hydrolytic stability.

Above we briefly discussed the limitations on possible values of bonding density (Section 3.6.1). Molecular volume of bonded ligands is one of these limitations. If the most stretched conformation of a ligand has length l and its molar volume is v, then the minimum area, ω, it occupies on the surface will be $\omega = v/l$ and the maximum bonding density will be $d_{max} = 1/\omega$. This maximum density is calculated for the flat surface, while on the concave internal surface of the pore the maximum density is lower and could be expressed as

$$d_{b \cdot max} = \frac{1}{v}\left(l - \frac{l^2}{2R}\right) \tag{3-13}$$

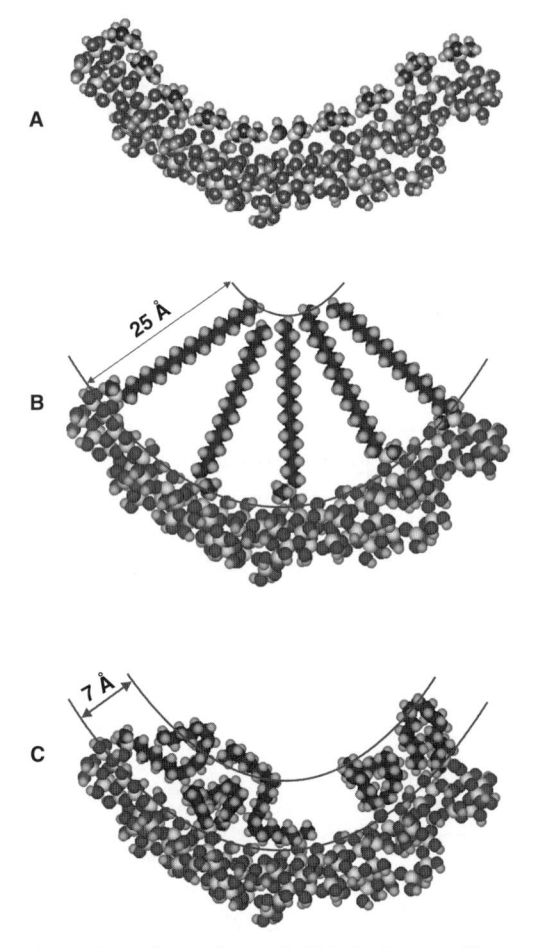

Figure 3-16. Hypothetical conformations of C18 chains on silica surface. (Reprinted from reference 63, with permission.)

where v is molar volume of the ligand, l is maximum length of the ligand, and R is pore radius. The molecular volume for a C18 ligand is estimated as 630 Å^3 with a maximum length in all-trans conformation is 24 Å.

Maximum density means that the whole volume of a surface layer of thickness l is actually filled with bonded ligands and is calculated for a uniform cylindrical surface. In reality, pores are interconnected and have a certain pore size distribution as well, and there are often gaps between bonded molecules (accessible silanols), so real values are usually lower as it could be seen in Figure 3-17.

Bonding density cannot be measured directly; the measured value is either (a) carbon content on modified material, (b) weight loss after ashing in the oven, or (c) thermogravimetric weight loss between 200°C and 700°C. Carbon

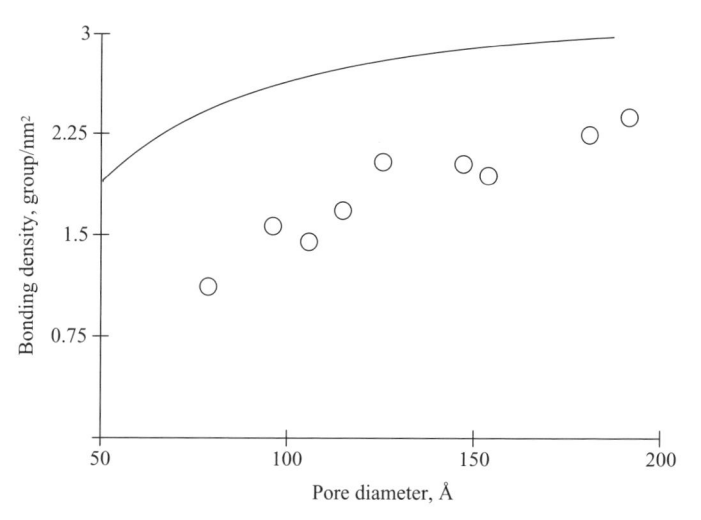

Figure 3-17. Dependence of maximum bonding density on the adsorbent pore diameter. Circles are the experimental points from references 63, and 65–67.

TABLE 3-3. Comparison of the Carbon Load and Surface Density Values

Adsorbent	Carbon Load (%w/w)	Surface Area (m^2/g)	Bonding Density ($\mu Mol/m^2$)
1	0.24	3.6	4.0
2	12	320	2.3

content is usually expressed as a percent of carbon atoms in a modified sample by weight. Sometimes it is regarded as an indirect measure of the surface hydrophobicity. We have to emphasize that this is highly misleading because it is not related to the available surface area. As an illustration we can compare two adsorbents modified with C18 chains (Table 3-3).

The first adsorbent in Table 3-3 has only 0.24% of carbon, while the second adsorbent has almost a 50-fold higher content (12%); this indicates that the second adsorbent has 50 times more bonded ligands than the first adsorbent. On the other hand, the second adsorbent has 80 times higher surface area and all these bonded ligands are spread on this surface, yielding lower bonding density than the first adsorbent.

Carbon content is obtained from CHN analysis of modified adsorbent sample. This analysis is usually based on the determination of the amount of CO_2 gas produced after combustion of the sample of known weight in the oxygen atmosphere—the standard procedure for the determination of the bulk formula of organic compounds. In contrast to the organic compounds, modified HPLC adsorbents contain a noncombustible silica matrix. This matrix can melt at the combustion conditions set on CHN analyzer, and this melting can sometimes prevent complete combustion of ligands bonded inside the small pores of the sample (occlusion). It is advisable to compare the data obtained at the different combustion conditions (amount of oxygen in the oven, speed of heating of inductive oven, etc.).

Bonding density is calculated using the value of carbon content (from CHN analysis) in carbon %w/w and the specific surface area, obtained from the BET analysis of base silica. Calculation is performed using the equation originally introduced by Berendsen and de Galan [68]:

$$d_b[\mu mol/m^2] = \frac{10^6 C\%}{1200 n_c - C\%(MW - 1)} \cdot \frac{1}{S} \qquad (3\text{-}14)$$

where $C\%$ is the carbon percent by weight, n_c is the number of carbons in bonded ligand, MW is the molecular weight of bonded ligand, and S is the silica surface area. This expression is only applicable for monomeric bonding. In the case of polymeric bonding, the effect of crosslinkage should be taken into account, although it is usually insignificant.

Calculation of the bonding density for polymeric bonding is more accurate if $C\%$ from CHN analysis is used, because weight loss during ashing or TGA analyses is dependent on the decomposition pattern. In the case of a vertically polymerized bonded phase, decomposition may involve a siloxane backbone, and association of the weight loss with the ligands molecular weight is essentially impossible. Carbon content, on the other hand, is only dependent on the original reagent structure.

Carbon content determination is often substituted by the measurement of the weight loss of the modified sample heated in the oven from 200°C to 900°C. It is assumed that all physically adsorbed molecules are removed before reaching 200°C and that above 200°C, only the decomposition of organic bonded ligands is occurring. Detailed study of thermogravimetric analyses of chemically modified adsorbents shows some significant deviations. Figure 3-18 shows thermograms of silica modified with alkyl ligands of different length measured from 40°C to 800°C.

One can note that the decomposition of bonded ligands starts slightly before 200°C and that the decomposition/combustion of methyl groups attached to silica (C1) occurred above 400°C. At 600°C the decomposition of bonded organic layer is completed. Above 800°C, additional weight loss is usually observed. This weight loss is the major source of errors in bonding density determination, and it is not related to the decomposition of any organic but rather dehydroxylation of base silica and loss of water molecules occluded in the silica matrix.

Calculation of the bonding density on the basis of weight loss data is only approximate since it requires the assumption of the chemical decomposition pattern—for example, the following conversion:

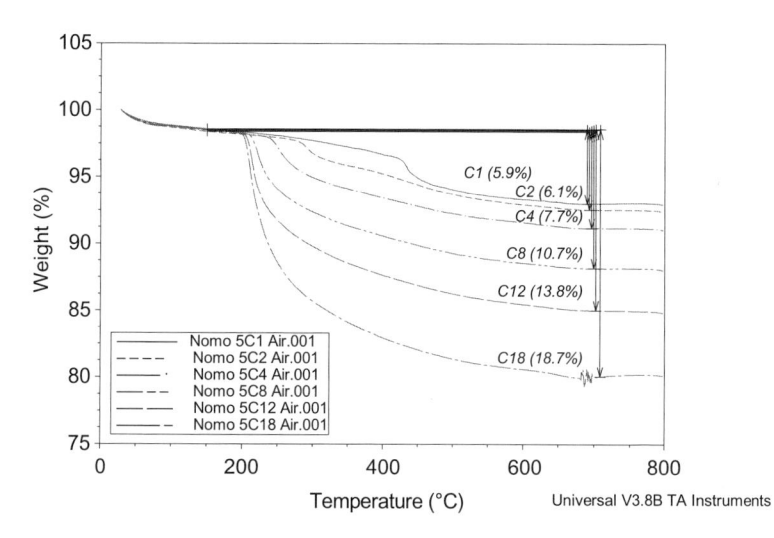

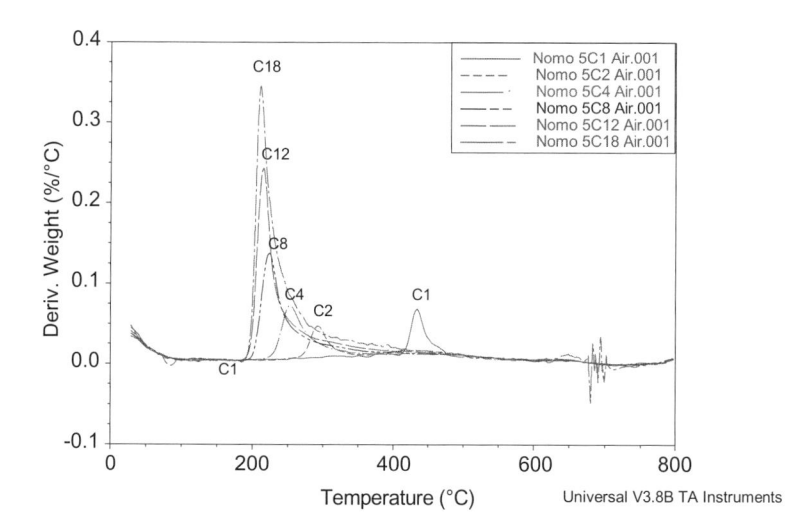

Figure 3-18. Series of thermograms and their derivatives for adsorbents modified with alkysilanes of different chain length.

The weight loss will be only 232 Da per one C18 bonded ligand (MW_{loss}). A weight loss of only 232 is observed because both CH_3 and C18 groups at silica atom will be substituted with OH groups, so the actual loss of all carbons and hydrogens (283 Da total) will be alleviated with 3 hydroxyls (51 Da) leaving the total loss to be equal to 232 Da. Each bonded ligand increases the weight of base silica during modification process by 310 Da (MW_{lig}). The percent of the weight loss (P_{loss}) during thermal decomposition could be expressed as

$$P_{loss} = \frac{d_b S MW_{loss}}{d_b MW_{lig} S + 1} 100 \qquad (3\text{-}15)$$

Solving this expression for bonding density (d_b), we obtain

$$d_b = \frac{P_{loss}}{100 \, MW_{loss} - P_{loss} MW_{lig}} \cdot \frac{1}{S} \qquad (3\text{-}16)$$

The actual decomposition of bonded ligands can be different from what we assume in the above calculation (weight loss), and corresponding bonding density values will be very approximate if based on the data from thermo-gravimetry or "ashing" techniques. The formation of trihydroxysilane is highly doubtful, and active silica dehydration could be observed at temperatures above 600°C [69].

3.6.6 Residual Silanoles

As we mentioned above, chemically modified silica always retains a significant amount of residual silanols. In fact, at least half of the original silanols on the silica surface remain unreacted after chemical modification. Silanols have acidic nature, but their ionization constants are highly dependent on the surface environment and on silica purity. Silanol pK_a is estimated to be between 5 and 7 [7].

Proton donor ability of surface silanols is believed to be the source of peak tailing for analytes with proton acceptor functionality (usually basic analytes). The presence of impurities such as iron, boron, and aluminum [70] in bulk silica decreases the silanol pK_a and decrease the hydrolytic stability of bonded phases.

All these effects are strongly dependent on the surface concentration of bonded ligands (bonding density), and despite the presence of a significant amount of unreacted silanols on the silica surface, only the silanols accessible for analytes should be considered.

This shielding effect and variable conformation of bonded ligands makes the estimate of the amount of accessible residual silanols virtually impossible.

In general, the more uniform the silica surface and the higher the bonding density, the less the residual silanols effect one should expect on that HPLC adsorbent.

3.6.7 Surface Area of Modified Adsorbent

The measurement of the silica surface area is well established; however, the assessment of the surface area of chemically modified porous silica brings a host of problems. It is obvious that chemical modification should decrease the surface area of porous material, but the degree of this decrease will be dependent on the density of bonding of organic ligands on the surface. If the bonding density is not very high, which leaves significant conformational freedom to the bonded chains, this may cause a significant roughness of the new surface. Also, the assessment of the surface area of the modified adsorbent is complicated due to the uncertainty of the molecular area of the adsorbate used for the surface area measurement. The most common adsorbate is nitrogen, and its molecular area on the silica surface is assumed to be equal to $16.2\,\text{Å}^2$. In the scientific literature [15, 71] there are number of indications that this area is significantly higher on the hydrophobic surface. Essentially this is due to the increased lateral freedom of nitrogen molecule on the nonpolar surface.

As an example of the variation of the surface area of modified adsorbents, we discuss the series of silicas modified with alkylsilanes of different length [63]. In Figure 3-19 the dependence of the surface area on the number of carbon atoms of attached ligands are shown.

The calculation of the adsorbent surface area using BET theory involves the estimation of the molecular cross-sectional area of nitrogen molecule [15]. In general it is assumed to be equal to $16.2\,\text{Å}^2$ per nitrogen molecule on the surface, but this value is the subject of intense criticism during the past 30 years [72].

It has been shown that nitrogen occupies a larger area on hydrophobic surfaces than on polar surfaces. The nitrogen molecular area on hydrophobic surfaces is estimated to be between 19 and $22\,\text{Å}^2$ [71].

Adsorbent surface area is calculated as the product of the monolayer capacity estimated from BET equation and nitrogen molecular area, ω_{N_2}. If the

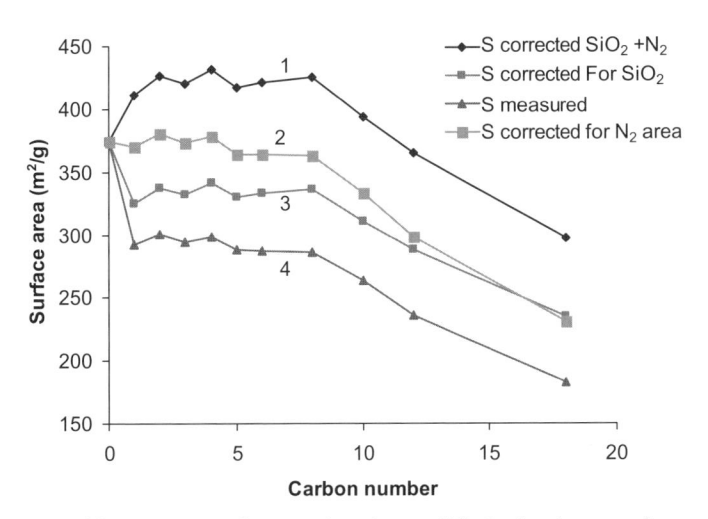

Figure 3-19. Different ways of measuring the modified adsorbent surface area.

specific surface area of a hydrophobic adsorbent was determined using the value of $\omega_{N_2} = 16.2\,\text{Å}^2$, then the total adsorbent area would be underestimated. Correct comparison of the surface area of different modified adsorbents should also be done relative to 1 g of bare silica adsorbent.

The surface area dependencies on the number of carbons of the modified adsorbents measured with LTNA corrected for the different factors described above are shown in Figure 3-19. A plot of the directly measured surface area values shows a significant drop of the surface area between bare silica and C1-modified adsorbent (Figure 3-19, curve 4). Bonding a monolayer of relatively small trimethylsilane molecules could not cause this (almost $70\,\text{m}^2/\text{g}$) drop in a surface area value. However, the increase of the effective molecular area of nitrogen from 16.2 to $20.5\,\text{Å}^2$ will cause exactly this error in the adsorbent surface measurement. The measured surface area values were corrected using a nitrogen molecular area of $20.5\,\text{Å}^2$. (Figure 3-19, curve 2)

On the other hand, a correct comparison of the surface areas of modified materials should be done by relating measured values to the same original silica weight (Figure 3-19, curve 3). As could be seen from Figure 3-19, an addition of the correction for the mass of silica to the surface already corrected for the increased molecular area of the nitrogen leads to the significant increase of the surface areas for adsorbents modified with relatively short alkyl chains (Figure 3.19, curve 1).

Although it looks strange, a restricted mobility of anchored short alkanes may actually cause an appearance of significant surface heterogeneity (roughness) which leads to the increase of the "monolayer" capacity and results in an increase of the total surface area. This is probably not the true surface area value that will play any role in the HPLC process, and it is doubtful that any of these values may be relevant for the description of HPLC retention.

The assessment of the surface area of modified adsorbent is extremely uncertain. Accounting for different types of corrections may lead to the significant differences in the calculated surface areas of the same material. Probably the dominating factor is the roughness of the surface of modified silica. Differences in the alkyl chain conformations may lead to the completely different values for the same material. Possible occlusion of nitrogen molecules between alkyl chains adds additional uncertainty [73].

One definite surface parameter, which is also common to all related modified adsorbents, is the surface area of original bare silica. It is probably the most appropriate to relate all processes to the surface of bare silica (making corresponding corrections for modified adsorbents on the fraction of the silica in 1 g of actual material).

Total surface area of silica in the HPLC column could be calculated using the following equation, provided that V_0, V_{ex}, V_{pore}, S_{SiO_2}, and bonding density (d_b) are known.

$$S_{SiO_2}^{total} = \frac{S_{SiO_2}(V_0 - V_{ex})}{V_{pore}\left[1 + S_{SiO_2}d_b(58 + 14n_c)\right]} \qquad (3\text{-}17)$$

where n_c is the number of carbons in the main alkyl chain of the bonded phase.

All these complications could be largely avoided if we use the surface of base silica instead. This is more beneficial for chromatographic purposes, because this will simplify the comparison of different adsorbents made on the basis of the same silica, and it also simplifies the assessment of the total surface area in the column.

Indeed, if we have two columns packed with two different reversed-phase adsorbents made from the same silica, and if, for example, the total surface area in both columns is the same, then the difference in the retention of the same analyte on both columns could be only associated with the effect of the analyte–bonded-phase interactions. If, on the other hand, we would estimate the actual surface area of modified adsorbents and find that it is significantly different, then the difference in the retention cannot be associated with the interactions alone since the available area is different, and we would need to consider surface specific interactions and surface specific retention.

3.7 POLYMER-BASED ADSORBENTS

The polymer gels are usually synthesized by suspending the monomers, a crosslinking reagent, and a polymerization initiator in a suitable porogenic diluent and rapidly stirring the mixture in the presence of water to produce organic-phase droplets. The size of these droplets defines the size of the resulting polymer particles. The crosslinking reagent forms a rigid three-dimensional structure the polymer. The porogenic diluent provides the macroporous structure to the gel. The proper selection of the porogen allows the control of the pore size, pore size distribution, and pore geometry.

Polymer-based adsorbents are not widely used in the HPLC analysis of small molecules, mainly because of the presence of micropores in the structure of polymer resins [29]. These micropores may trap small molecules, and their relatively slow diffusion leads to significant band broadening and overall loss of separation efficiency.

Ion-exchange resins are used to separate molecular ions (cations, anions, amino acids, proteins, peptides) based on their interaction with the ionic resin. Ion-exchange groups are bonded to inert polymeric particles that are generally 3–30 μm in diameter to make either anionic or cationic polymers. Anionic polymers are known as cation exchange resins, and these resins can be strong or weak cation exchange resins that are dependent upon the anionic group that is bonded to the polymer. Strong cation-exchange resins usually contain strong acidic groups such as aromatic of aliphatic sulfonic acid or phosphonic acid, and weak cation-exchange resins contain aliphatic (pK_a 4–5) or aromatic carboxylic (pK_a 3–4) or phenolic (pK_a 8–10) groups bonded to the polymer. Cationic polymers, on the other hand, are known as anion-exchange resins, and these resins can also be weak or strong anion exchanges. Strong anion-exchange resins usually contain strong basic groups such as quaternary ammonium groups (always charged regardless of pH of the mobile phase), and weak anion-exchange resins usually contain primary (pK_a 7–8), secondary (pK_a 7–9), or tertiary amine groups (pK_a 9–10) (with the weaker anion-exchange resins containing primary amine groups, since they have a lower pK_a).

Most ion-exchange resins based on organic polymers are made by the process of suspension polymerization. The monomers can be neutral as in the case of styrene, divinylbenze, methyl acrylate, and acryonitirle, and the resulting polymer beads are then chemically modified to introduce the acidic or basic functionality. Styrene-divinylbenzene-based ion exchangers are usually more hydrophobic than their more hydrophilic counterparts. The methacrylate matrix offers a more intermediate polarity and a less hydrophobic surface than styrenic-based materials.

The number of groups that are bonded to the polymeric matrix is known as ion-exchange capacity. Ion-exchange capacity is a measure of the quanity of functional groups in the polymer and is usually denoted in the literature as milliequivalents of functionality per gram of polymer. The greater the ion-exchange capacity, the greater the analyte loading and/or retention of desired ions. Typical values are from 0.5–5 mequiv./g. For example, a sodium polystyrene-based sulfonate strong cation-exchange resin is prepared by suspension polymerization of a mixture of styrene and divinylbenzene to make small polymeric beads. The beads are then sulfonated using either concentrated sulfuric acid or chlorosulfonic acid and then neutralized with sodium hydroxide to give the functionalized product [74]. The ion-exchange capacity can be controlled by using highly crosslinked packings, which results in a lower degree of sulfonation; alternatively, a lesser crosslinked polymeric matrix could be used, and the degree of the sulfonation can be controlled to the desired level [8].

For styrene-based strong or weak anion exchangers the styrene-divynylbenzene polymeric beads are usually chloromethylated using alkyl (methyl or ethyl) methylchloroether to form the chloromethyl–polystyrene intermediate. Then the intermediate is further reacted with a tertiary, secondary, or primary amine to make the desired ion-exchange resin (strong or weak). Also, the synthesis of hydrophilic ion exchangers based on hydrophilic methacrylate resins have gained popularity for the separation of proteins where contributions from ionic and hydrophobic interactions (these would be suppressed compared to styrene-based phases) would allow for suitable retention in pharmaceutical analysis. In order to synthesize these phases, the hydroxyl groups of the hydrophilic metharcylate resin (e.g., 2-hydroxyethyl methacrylate) are reacted with various groups to make the desired resin [75].

Some advantages of ion-exchange polymeric resins are that they are generally compatible with 0–100% organic or aqueous solvents, and also different mobile phases do not cause the supports to shrink or swell. The higher crosslinking allows for a macroporous resin that is less prone to shrinking or swelling in aqueous media. Also, these ion-exchange polymeric resins are relatively stable from pH 1–13 and can operate up to temperature of 80°C. However, for some polymeric ion-exchange phases the extended stability of the phase at high pH > 8 is often compromised when working at temperatures greater than 50°C, and it is recommended that the chromatographer should seek the specific stability limits (max temperature at certain pH ranges) directly from the column manufacturer. For example, although methacrylate-based packings are stable at pH 2–12, they may be more prone to hydrolytic attack due to ester linkage in the polymeric structure.

3.8 STATIONARY PHASES FOR CHIRAL SEPARATIONS

Separation of optically active isomers is one of the most important areas of HPLC application in the pharmaceutical industry. Since most of biological systems are predominantly homochiral, different enantiomers of the same drug could have different effect and potency, and the development of enantioselective analytical (and preparative) separation methods is very important. Detailed description of chiral HPLC separation is given in Chapter 22 of this book; here we only briefly review the specifics of distinctive types of chiral stationary phases (CSP).

3.8.1 Polysaccharide-Coated Phases

The inherent chiral nature and availability of natural polymers, such as cellulose and amylose, were the primary reasons of their use in chiral separations. The ability of cellulose to separate racemic mixtures was first observed in paper chromatography [76, 77]. The breakthrough point in the use of cellulose and amylose in modern HPLC was achieved with the development of CSPs where saccharides were adsorbed on the surface of aminopropyl-modified macroporous silica, [78, 79].

Silica-based material provides mechanical stability and high efficiency, while adsorbed polysaccharide helices offer chiral selectivity to the wide range of enantiomeric analytes.

Polysaccharide-coated CSP are prepared by coating of their benzoate or carbamate derivatives on substrate from a solution with the following evaporation of the solvent. These phases can be used under RP conditions with aqueous eluents or in normal-phase conditions with limited number of solvents to avoid dissolution of the coated layer [80, 81].

The nature and the position of the substituents introduced into the benzene ring of the carbamate derivatives essentially define the chiral recognition ability of these CSPs [82]. Higher long-term stability of polysaccharide phases was achieved with covalent bonding of polysaccharide to the surface of the support [83].

3.8.2 Pirkle-Type Phases

In 1980 W. H. Pirkle introduced packing material with covalently bonded optically active (R)-N-(2-naphthyl) alanine for use as a chiral stationary phase in HPLC [84]. The general principle of enantioselectivity of these phases is the difference in the overall interaction energy between two enantiomers. Space orientation and position of active centers of Pirkle-type phase allows one enantiomer to achieve stronger interaction because the geometry of the molecule allows its active groups to position close to the active centers of a CSP, while other enantiomers, due to their geometry, are unable to achieve similar interaction, which causes the first enantiomer to be retained longer on the surface of this CSP. Pirkle-type phases are very compound-specific, and the successful separation achieved for one type of molecule may not be obtained for another, similar analyte.

3.8.3 Protein Phases

Proteins with their inherent chiral nature are used in CSPs in coated and immobilized form. Bovine serum albumin was the first used in SCP in immobilized form [85] for the separation of acidic and neutral compounds.

The α_1-acid glycoprotein (AGP) was introduced as chiral selector in 1983 by Hermannson [86]. This phase found wide application in the enantioseparation of basic drugs.

Separations on AGP-type phases show unusual temperature and pH dependences. An example of the separation on chiral AGP column is shown in Figure 3-20, together with the pH dependence of the enantiomers' retention.

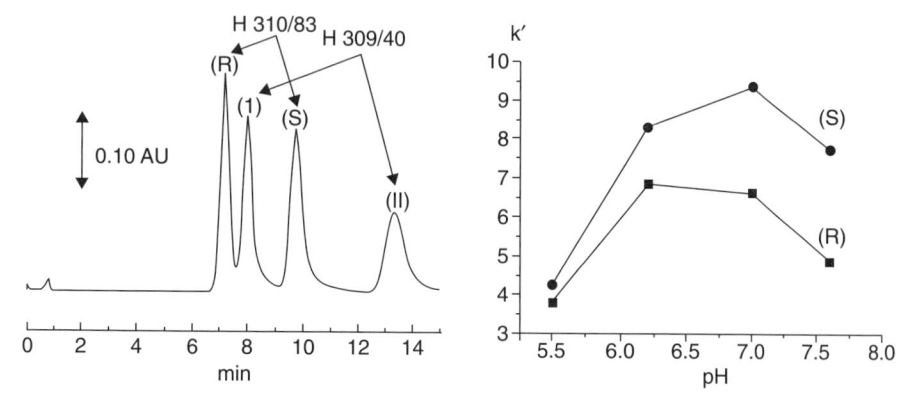

Figure 3-20. Enantioselective separation of (R,S)-3-hydroxymethyl-2-methyl-9-phenyl-7H-8,9-dihydropyrano[2,3-c]-imidazo[1,2-a]pyridine (solute 1) and (R,S)-3-methylethylether-2-methyl-9-phenyell-7H-8,9-dihydropyrano[2,3-c]-imidazo[1,2-a]pyridine on chiral-AGP column and pH dependence of the retention of (R) and (S) forms of solute 1. Mobile phase: phosphate buffer (pH 7)–acetonitrile (9:1). Column temperature 40°C. (Reprinted from reference 87, with permission.)

3.8.4   Molecular Imprinted Polymers for Chiral Separations

Molecular imprinted polymers (MIPs) attracted significant attention in the last decade as chiral stationary phases targeted for the separation of selected compounds used as templates during the synthesis of packing material [88].

The stationary phase in MIP is essentially synthesized around the molecule of interest, and the nature of adsorptive active centers on the surface could be selected, synthetic ingredients with selected centers are included in the reaction mixture together with emulgating agent and pore-forming compounds. After polymerization the template molecules are extracted and material is grinded, and sieved particles are packed into HPLC column. Because the active centers were oriented around enantio-pure template molecules during the polymerization step, they retain their position in the matrix after template removal, and these cavities possess enormous selectivity toward the selected form of enantiomers. Enantioselectivity of MIP is on the level between 10 and 30, which could never be achieved on any other types of CSPs. The other side of the coin is that due to that enormous selectivity, the kinetics of the mass transfer of selected molecule into the cavity is slow and these columns show enormous tailing (Figure 3-21). Most of the successful applications of MIP are for preparative separation in the form of solid-phase extraction where efficiency is not a critical issue.

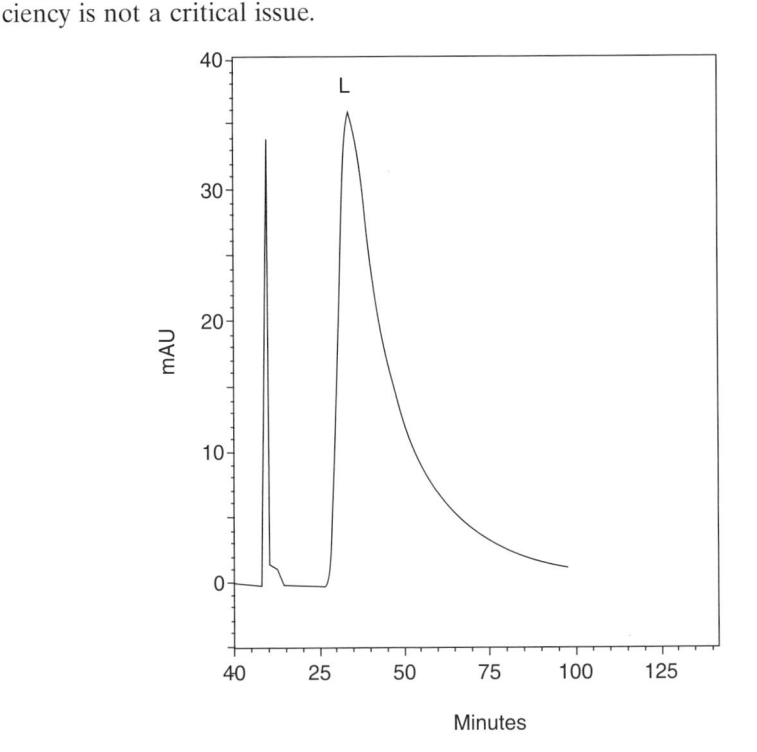

Figure 3-21. Example of chiral separation of *N*-Boc-l-Trp on MIP. (Reprinted with permission from reference 89.)

3.9   COLUMNS

In the previous sections of this chapter we mainly discussed the properties of HPLC packing materials (adsorbents) and the influence of their physical and chemical properties on HPLC retention. In the following sections we will concentrate primarily on the properties of HPLC column itself.

A modern HPLC column is either stainless steel or plastic tube filled with packing material (adsorbent) and arranged with end-fittings designed to provide sealed connection with the eluent inlet and outlet lines and to retain packing material inside while allowing liquid to pass through.

A column is essentially a device that holds a stationary phase in place, allowing the mobile phase to carry an injected sample through and allowing analytes to interact with available surface. As we discussed in Section 3.2, the efficiency is mainly dependent on the column type (packed, monolithic, or capillary) and particle size for packed columns, or through-pore diameter for monolithic columns.

3.9.1   Capillary / Monolithic / Packed Columns

Comparison of the porous structure of different columns was discussed in Section 3.2; here we emphasize that with a packed column the ratio of particle size to the average interparticle pores (space) is on the level of 3–3.5 while with monolithic columns trough-pores are on the level of 6000 Å and silica material is only about 1 u thick, which makes this ratio 0.5–0.2 or about 10 times smaller, thus significantly decreasing the time needed for analyte molecules to diffuse into the mesoporous space for the interaction with main surface. This allows for much faster flow rates without the loss of the dynamic equilibrium conditions (otherwise known as the slow mass transfer term (C) in the Van Deemter equation).

Cumulative pore volume distribution in the packed column is shown as solid line in Figure 3-22. Note that bimodal type of porosity distribution creates the difference in the liquid flow through the column, thus limiting column efficiency. Silica monolith pore size distribution is also shown in the Figure 3-22. The main difference with packed columns is the volume of major channels. The size of main channels in the monolith bed is comparable to the average size of interparticle space in the packed bed, but its volume is significantly higher. This reduces the column flow resistance, and monolith columns can operate at three to four times higher flow rates compared to similar packed columns. Faster flow rates up to 5–10 mL/min are normal for these types of columns.

A monolithic column can be defined as a column consisting of one piece of solid with a defined pore structure that possesses interconnected skeletons and interconnected flow paths (through-pores) and mesopores. A monolithic silica column consists of micron-sized skeletons (1–2 μm), through-pores (up to 8 μm), and mesopores (10–12 nm). The mesopores provide the needed

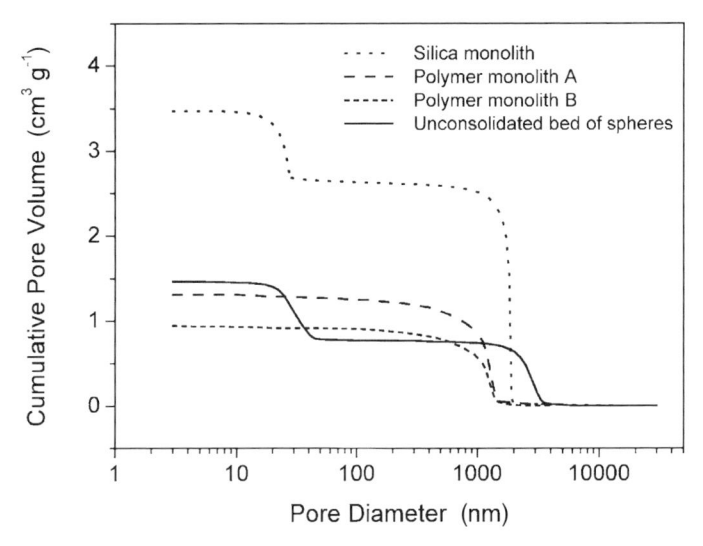

Figure 3-22. Cumulative pore volume distribution of different HPLC columns indicating monomodal pore size distribution for polymer monolith and bimodal distributions for both packed particulate silica and silica monolith columns. (Reprinted from reference 90, with permission.)

surface area for retention. SEM photographs of monolithic silica columns have shown that both the silica skeletons and the through-pores are co-continuous. A monolithic column with small-sized skeletons and large through-pores have a large (through-pore size)/(skeleton size) ratio, commonly 1–4, and can simultaneously provide high permeability and high column efficiency [90–93].

When monolithic silica columns are prepared in a fused silica capillary, the silica network structure can be bonded to the tube wall. They can be used as a column directly after preparation, or as a reversed-phase adsorbent after alkyl or some other type of modification. The porosity of monolithic silica columns is much greater than that of a particle-packed column. A major difference is seen in interstitial porosities: 65–70% for monolithic silica prepared in a mold, and higher than 80% for those prepared in a capillary, compared to 40% for a particle-packed column. A comparison of the separations of cytochrome triptic digest on packed and monolithic colums is shown in Figure 3-23. The separations are nearly identical except that on monolithic column it is ten times faster. Figure 3-24 shows the dependence of the backpressure generated on the system as a function of the flow rate for packed column and a set of different monolithic columns. The slope on all monolithic columns is the same, and it is approximately five times lower than that on a packed column. Additional information on fast HPLC on monolithic columns is given in Chapter 17.

Columns packed with nonporous particles have also been available on the market for more than 10 years. These columns do not have bimodal overall

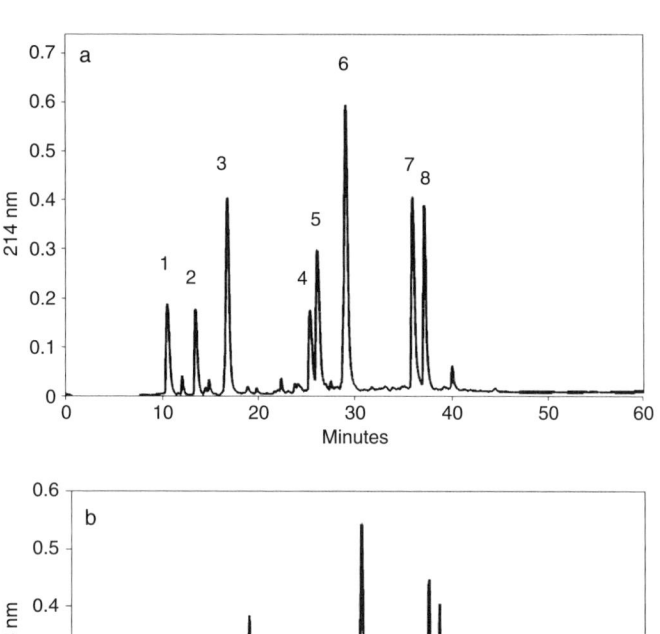

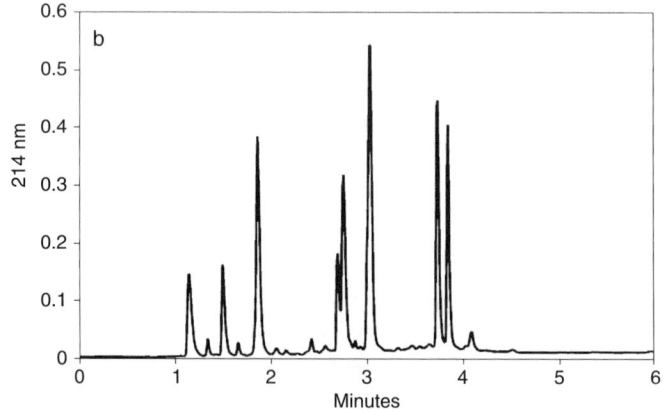

Figure 3-23. RP separation of bovine cytochrome c tryptic digest at flow rates (A) 1.0 mL/min with 60-min gradient and (B) 10.0 mL/min with 6-min gradient. (Reprinted from reference 95, with permission.)

pore size distribution and thus demonstrate exceptional efficiency. The main drawback of these columns is their available surface area and consequently their loading capacity.

If 1 g of conventional porous packing material has an average of 300 m²/g surface area, the same 1 g of nonporous packing material with particle diameter of just 1 μm has only 2.7 m²/g or about 100 times lower surface area. This means that these columns require approximately 100 times lower injection volume and higher detector sensitivity.

Waters recently introduced columns packed with 1.7- to 1.8-μm porous particles and a specially designed HPLC system, which can operate at pressures up to 1000 atm. These columns do not show significant increase of the efficiency compared to conventional columns, but higher flow rates increase the speed of analysis and an increase of resolution per unit time may be obtained.

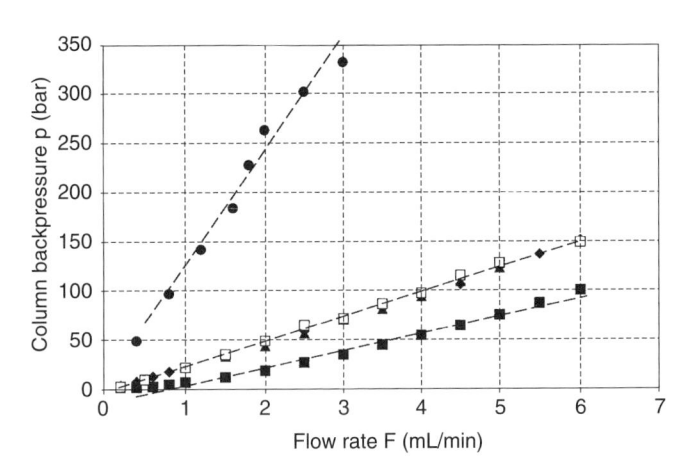

Figure 3-24. Column backpressure as a function of the flow rate. Backpressure comparison on conventional and monolithic column (silica) Circles denote Purospher-C18 packed column; closed squares denote equipment pressure drop without the column; other symbols denote different monolithic columns, all in one line. (Reprinted from reference 96, with permission.)

3.9.1.1 Chromatographic Properties of Monolithic Columns.

The major chromatographic features of monolithic silica columns are (a) high permeability based on large through-pores and (b) a large number of theoretical plates per column due to small-sized skeletons (Figure 3-25). There are two main types of monolithic silica columns: (1) prepared in a mold, only C8 and C18 available (4.6 mm i.d. × 5 or 10 cm, through-pore size = 2 μm, skeleton size = 1 μm) and (2) the capillary type, none commercially available; however, some experimental research-type phases exist (conventionally, 50- to 100-μm i.d. × 25–130 cm, through-pore size = 8 μm, skeleton size = 2 μm).

The main benefit of monolithic columns is the separation speed. As an example of the practical use of ultrafast separation, the two-dimensional HPLC separation of tryptic digest BSA is shown. The fractions of the column effluent from the first dimension (IEX) separation was collected for 2 min each and injected into the second-dimension monolithic column (RP). The run time for the second dimension could not be more than 2 min (more than fraction collection time), and the monolithic column provided fast and efficient separation (Figure 3-26).

Vervoort et al. [98]. did an important theoretical modeling study of the influence of monolith structure heterogeneity on the efficiency. They confirm that in silica monolith the plate height to a large extent is determined by pore heterogeneity, and improvement in the uniformity of the through-pores in a monolith structure may result in 2- to 2.5-fold efficiency increase.

Currently the majority of commercially available columns are packed with porous particles. Monolithic columns are rapidly gaining popularity, although

(a)   (b)   (c)

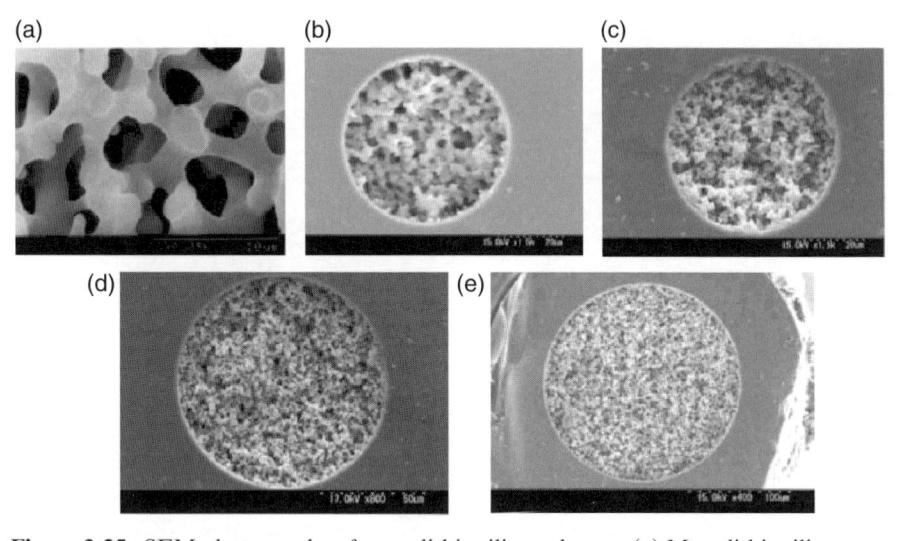

(d)   (e)

Figure 3-25. SEM photographs of monolithic silica columns: (a) Monolithic silica prepared from TMOS in a test tube; (b–e) prepared from a 3:1 mixture of TMOS and MTMS in fused silica capillary; (b) 50-mm i.d., silica skeleton size 2 mm, through-pore size 4.5 mm; (c) 50 mm i.d.; (d) 100 mm i.d., and (e) 200-mm i.d. (c)–(e) Silica skeleton size 1.5 mm, through-pore size 2.0 mm. (Reprinted from reference 97, with permission.)

there is still a long road ahead before they get a comparable market share. The third type of column is capillaries, which are also commercially available but used exclusively for research purposes and for applications with low sample amounts. Capillary columns could be either packed or monolith or even just simple open-tubular. All of them are mainly used in capillary electrochromatography (packed and monolith types) or in capillary electrophoresis (open tubular type).

3.9.1.2 System Requirements. Column dimensions bring some limitations to the HPLC system. Generally, the smaller the column, the smaller the optimum injection volume and detector flow-cell. Volume of the injected sample essentially determines the starting width of chromatographic zone. All theoretical descriptions of column efficiency are based on the infinitely small original zone width. If the estimated theoretical column efficiency is expected to be 10,000 theoretical plates on a 4.6 × 150-mm column, then the injection of 100-μL sample on this column will decrease the efficiency for early eluting compounds by about five times. A simple example can illustrate this. Considering the same column (4.6 × 150 mm) with 10,000 theoretical plates efficiency, one can easily estimate the expected peak width using well-known expression

$$N = 16\left(\frac{V_R}{w_b}\right)^2 \Rightarrow w_b = \frac{4V_R}{\sqrt{N}}\qquad(3\text{-}19)$$

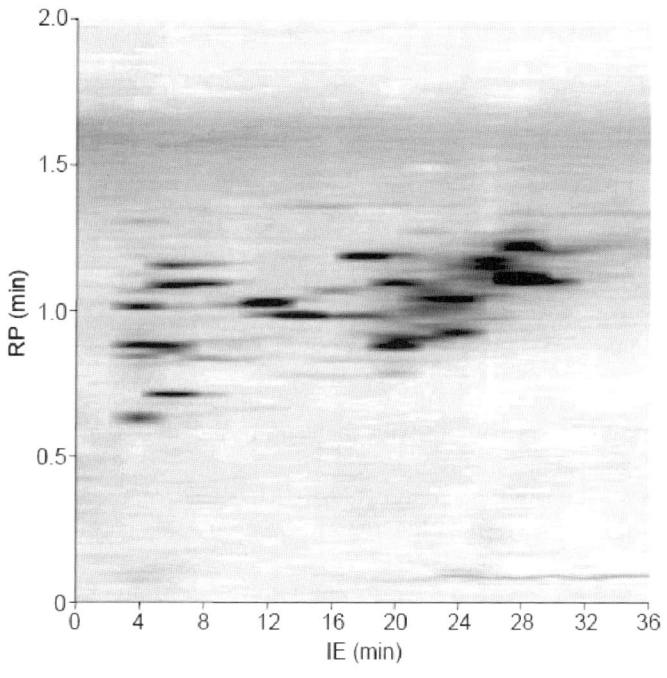

Figure 3-26. Two-dimensional separation of tryptic digest of BSA by simple 2D-HPLC. First-D cation exchange separation (horizontal axis), MCI CQK-31S 5-mm (50 mm, 2.1-mm i.d.) column, 2nd-D reversed-phase separation (vertical axis), monolithic silica-C18 Chromolith Flash (25 mm, 4.6 mm i.d.) column. The fractionation at the 1st-D was done every 2 min. (Reprinted from reference 98, with permission.)

For the mixture of benzene and phenanthrene eluted on a C18-type column with 70/30 acetonitrile/water eluent (1 mL/min flow rate), benzene retention will be 2.2 mL and phenanthrene will be eluted at 14.6 mL. Column void volume is estimated to be approximately 1.7 mL. Using above expression, one can calculate the expected peak width for benzene (88 μL) and phenanthene (584 μL). If the injection volume is 100 μL, then the injected original peak width will be higher than expected peak width of eluted first peak (early eluting component).

On the other hand, for highly retained analyte (phenanthrene) the final peak width will not be significantly greater than theoretically estimated. This is because this component shows strong interaction with the stationary phase; and while it is being loaded on the column, it will tend to adsorb on the surface immediately after it comes into contact with the stationary phase; as a result, even though it takes up to 6 sec to transfer all 100 μL to the column, phenanthrene will be concentrated on the top portion of the column, the so-called adsorption compression effect (Figure 3-27).

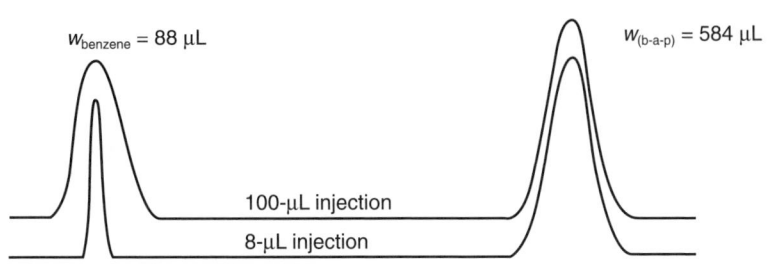

Figure 3-27. Effect of the injection volume on the peak width of early and late eluting analytes.

This example emphasizes the importance of balancing the column dimensions and the injection volume, with the required analysis time and sensitivity. If the main goal is the speed of analysis, then a smaller the injection volume is better; if, on the other hand, the sensitivity is critical, then it is possible to use a large injection volume, but the chromatographic conditions should be adjusted in such a way that target analytes will have strong interactions with the stationary phase, and an adsorption compression effect will compensate the loss of efficiency.

As a general rule, the injection volume should be at least 10 times smaller than the theoretical peak width of a component eluting at the void volume. For the example shown above, this volume should be less than 8 μL.

Another useful rule is that the volume of the detector flow cell should be less than the estimated injection volume. Flow cell volume could be considered as the hypothetical peak slice used for digital quantitation. The amount of analyte in the flow cell at the specific moment of time is equal to the average concentration in the cell multiplied by its volume. The smaller the peak slice, the more accurate the quantitation. At least 15–20 points needed for the proper description of symmetrical Gaussian peak. In the example above, we estimated the optimum injection volume as one-tenth of the earliest peak, and the flow cell volume should be even smaller, 5 μL for the above example.

3.9.1.3 Loadability.

Column loadability is a controversial subject because, as it is commonly accepted, the appearance of asymmetrical peaks (peak tailing) is regarded as an overloading phenomenon. On the other hand, asymmetrical peaks are often observed for only one or a few components of separated mixture, while others show symmetrical peaks in a wide concentration range as shown in Figure 3-28. The overloading effect is independent of the wavelength of the detection.

The other overloading effect that is rarely observed in HPLC separations is on the columns with very low surface area, like columns packed with nonporous particles. In figure 3-29 the retention of 0.1% solution of benzene is shown. When only 1 μL is injected, the resulting peak shows only slight tailing; however, when the injection volume is increased to 5 μL, a severe shoulder in

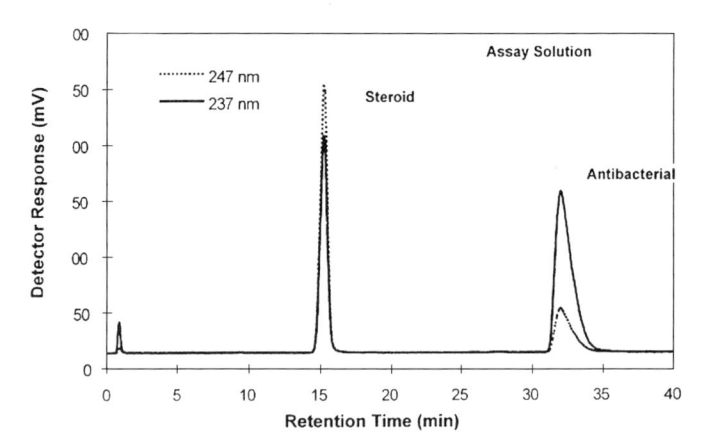

Figure 3-28. Elution profile of two components. (Reprinted from reference 99, with permission.)

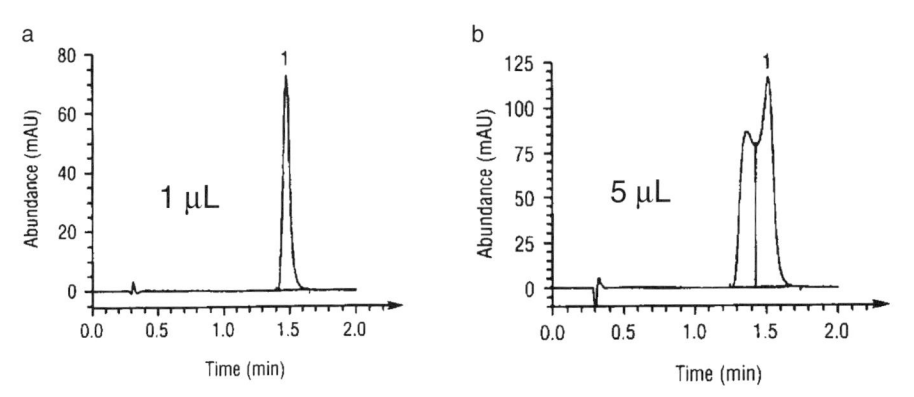

Figure 3-29. Retention of benzene on Kovasil-RP18 column (50×4.6 mm, nonporous particles 1.5μm).

the front of the peak is observed. This is because in 5μL of 0.1 v/v% solution, the amount of benzene molecules is too large for the available ($\sim 4 \text{m}^2$) total adsorbent surface in the column. While a portion of benzene molecules is adsorbed on the surface, the rest is carried by the mobile phase and eluted before the retained peak (this is an oversimplified description of the actual phenomenon and represents an extreme case of overloading).

Peak tailing is the most commonly observed effect of sample overloading. In essence, in most cases this effect is associated with nonlinear adsorption isotherms. In Chapter 2 the relationship of the retention volume and the derivative of the excess adsorption isotherm of the analyte on given stationary phase surface was derived. If the isotherm is linear within the injected concentration region, all components of the chromatographic zone are moving with the constant velocity. If, on the other hand, the isotherm is convex to the y-axis, then the higher the concentration, the lower the derivative of the adsorption isotherm and, consequentially, the faster the velocity of this particular concentration. In other words, a chromatographic zone with higher concentration will move faster, causing the sharpening of the peak front and formation of the noticeable tail. It is believed that the adsorption isotherm has a linear region at very low concentrations; at these concentrations, peaks are symmetrical. If the analyte concentration exceeds this linear region, its peak will show tailing.

Any types of adsorption isotherm functions do not have any region with constant derivative, and natural existence of this region would mean the break in the continuous variation of system properties, which could not exist. This quasi-linear region essentially means the region where the dispersion processes (diffusion, band broadening) significantly exceeded the effects of isotherm nonlinearity, and chromatographic peaks appear almost symmetrical (within the accuracy of our detection and data acquisition system).

The third overloading effect is usually associated with the presence of accessible residual silanols or other strong adsorption sites on the surface of stationary phase. If the surface concentration of these sites is very low, then a small portion of components sensitive to these sites (usually protonated basic analytes) will get adsorbed on them, thus increasing the formation of the peak tail. Neutral and nonionizible analytes are usually not sensitive for these effects.

3.9.2   Column Cleaning

An HPLC column could be considered as an accumulation medium, such that any type of compound dissolved in the mobile phase—whether it is an injected sample or simply an impurity from the mobile phase or packing solvent—will be somehow retained on the column. Chemically modified reversed-phase columns have immobilized flexible ligands attached to the base material surface. In Section 3.6 we discussed the behavior of these chains and demonstrated that from a chromatographic point of view they essentially behave as a solid surface. The penetration of the analyte molecules in between the bonded ligands, although possible, has very low probability. The diffusion of the probe within the bonded layer is estimated to be about 1000 times slower than the diffusion of the analyte in the liquid phase. This means that some molecules could be trapped in the stationary phase and be retained there for a long time. These trapped molecules could significantly alter the surface properties of the packing material and gradually change the retention pattern for analytes analyzed on that column and could eventually lead to gradual elution of ghost peaks in samples that are analyzed in a sequence. The main source of that contamination of the bonded phase is the HPLC mobile phase for used column and the solvent(s) used (for packing a brand new column). To calculate approximate column contamination we assume an average analysis time of 30 min with 1-mL/min flow with 50% MeCN: 50% water mobile phase that is about 30 mL of eluent that passes through the column per injection. Assuming that the total amount of impurities in most HPLC grade solvents is estimated on the level of 3 ppm (3 mg per liter of solution). Assuming 100 injections on that adsorbent surface with that mobile phase, this equates to about 9 mg of potential impurities absorbed on the stationary phase. How much of a typical stationary phase surface would these impurities cover over 100 injections? Assume that potential impurities have a molecular weight of 100 g/mol and a molecular area of 100 Å^2/molecule and assume that the typical surface area of an adsorbent is 300 m^2/g.

$$\%S_{covered} = \frac{A \cdot N_a \cdot mg_A}{\text{Total surface area} \cdot MW_A}$$

$$= \frac{100 \text{Å}^2/\text{molecule} \cdot 6 \times 10^{23} \text{ molecules}/\text{mol} \cdot 9 \text{ mg}}{300 \text{ m}^2 \cdot 100 \text{ g}/\text{mol}} \cdot 100 = 18\%$$

Therefore about 20% of the stationary phase surface could potentially be covered with these impurities.

Gradual trapping of some impurities in the bonded phase result in the gradual decrease of the column performance, appearance of strong adsorption centers, slow variation of the retention of some analyte, loss of the efficiency, and so on. But this trapping is not an irreversible chemisorption process, and in most cases it is not associated with the degradation of the stationary phase due to the hydrolytic cleavage of bonded ligands or silica dissolution; however, these processes could take place, sometimes to a much higher degree than stationary-phase contamination. If column performance degradation is observed and this column has not been used at the extreme conditions [too high or too low mobile-phase pH and high temperatures (above 50°C)], then performance decay is most probably associated with contamination.

The question is, Could these reversed-phase columns be cleaned in a fast and efficient manner? The goal is to force the stationary phase to release trapped molecules as fast as possible. However, usual cleaning procedures with washing with high concentration of organic usually require very long washing times. When a column is flushed for any extended period of time, even with very strong solvent, the stationary phase is essentially in the equilibrium with used solvent, which is not really that effective in releasing trapped molecules and takes a very long time to desorb off contaminants that came from the mobile phase or impurities from samples that did not elute. The equilibrium in the column has to be greatly disturbed to force the release of trapped molecules, and this disturbance could be induced by a sharp change of the eluent composition or solvent type. For a reversed-phase column a very efficient removal of these contaminants could be achieved by a simple sharp change from the pure water to pure acetonitrile, and this process is repeated several times as shown in Figure 3-30 for a 150- × 4.6-mm-i.d. column.

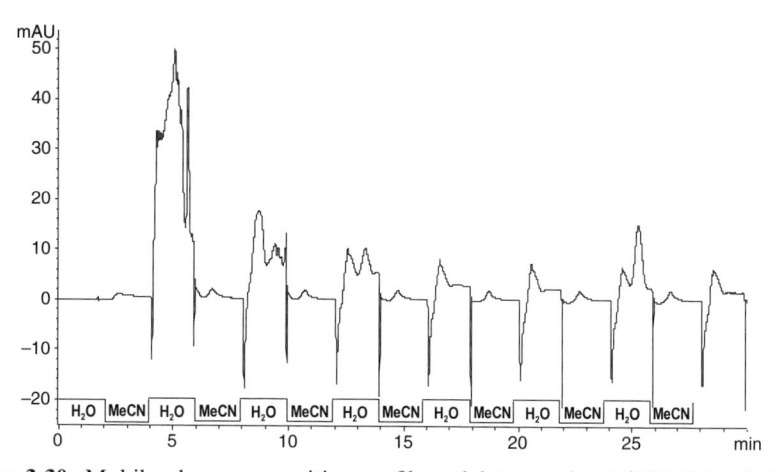

Figure 3-30. Mobile-phase composition profile and detector signal (UV 205 nm) during the column cleaning procedure.

It is advisable to start cleaning process with pure water to wash out possible salt and avoid their precipitation in pure acetonitrile. As could be seen in Figure 3-30, after each acetonitrile front the burst of the detector signal occurs, which is related to the release of the contaminants from the column. These bursts are gradually decreased with each cycle. The solvent fronts of different polarity disturb the phase equilibrium and leads to the release of trapped impurities. Usually 6 to 10 repeated cycles of the alternating solvent fronts is more than sufficient for thorough column cleaning. Each front should be about one column volume to avoid equilibration with the mobile phase. Therefore, for a 150- × 4.6-mm column (1.7-mL void volume) at 1-mL/min eluent flow, each step could be approximately 1.7 min; for a 150- × 3.0-mm-i.d. column (0.7-mL void volume) at 0.7 mL/min, each step could be approximately 1 min. Also, a more stringent procedure would entail solvents of different polarity following the sequence of water, acetonitrile, and THF and repeating the cycle as described above.

3.9.3 Column Void Volume

The void volume is the volume of the liquid phase inside the column. The importance of this parameter has been discussed in Chapter 2. Despite the very long debates, this is still a subject of significant controversy. Essentially, anyone who intends to measure the column void volume has to answer the question if he/she wants correct or estimated (convenient) measurements.

3.9.3.1 Measurement of Void Volume. It is generally believed that the injection of so-called "unretained" component can be used for the measurement of the column void volume. The biggest challenge is to find a compound that is really unretained. In the last 40 years, many different analytes have been suggested as "void volume markers," and most of them, if not all, did not pass a simple test: Their retention changes with the variation of the eluent composition. In theory, if we accept the above definition of the void volume, then this value should not change with the variation of practically any chromatographic parameters (type and concentration of organic modifier), except probably temperature.

A void volume marker should travel through the column in the same way as do eluent molecules. When binary (or tertiary) eluents are used, molecules of different eluent components actually travel differently through the column. In the theory chapter we discuss the adsorption of the organic eluent components on the surface of the reversed-phase column; thus, in general, water molecules will move through the column faster than organic molecules. Knox and Kaliszan [100] suggested that the actual void volume will be the weighted average of the retention of each component:

$$V_0 = V_{H_2O} x_{H_2O} + V_{org} x_{org} \qquad (3\text{-}20)$$

where x_{H_2O} and $x_{org.}$ are the mole fractions of the eluent components in the mixture.

The simplest possible solution is to use a single-component eluent; for example, inject deuterated acetonitrile in the flow of neat acetonitrile. The isotopic effect (the difference in the retention of molecules with different isotopes), although it exists, could be assumed negligible.

All these methods require the use of the refractive index of mass spectrometry (MS) detectors, since the detection of the isotopes is not possible with conventional ultraviolet (UV) detectors.

These difficulties are inspiring the search for other alternative methods for the void volume measurements. The main question anyone should answer is, What should the required accuracy be? For any prediction or thermodynamic-type work, the void volume values should be very accurate. For pure analytical purposes or for the comparison of different columns, the use of some markers is justifiable insofar as all experimental conditions are kept the same all the time.

In everyday method development practice, it is important to ensure the separation of target compounds, matrix components, and other impurities. The elution of the analyte at the void volume means that it did not interact with the stationary phase and thus could not be separated from other components that do not interact with the surface either. To ensure the analyte interaction with the stationary phase, it is usually recommended to choose chromatographic conditions when any component of interest elutes with at least 1.5 void volume values or even greater. The error of the void volume determination for these purposes could be 20% or even greater (insofar as these void volume values are not used for any calculations but just to estimate where the least retained analye elutes). The use of uracil, thiourea, or allantoin as analytical void volume markers is most common in practical analytical work.

3.9.3.2   *Other Experimental Methods.* It is probably suitable to discuss here
column porous structure. Porous space of a conventional packed column consists of the interparticle volume (V_{ip}—space around particles of packing) and
pore volume (V_p—space inside porous particles). The sum of those two constitutes the column void volume. The void volume marker ("unretained")
should be able to evenly distribute itself in these volumes while moving
through the column. Only in this case the statistical center mass of its peak
will represent the true volume of the liquid phase in the column. In other
words, its chromatographic behavior should be similar to that of the eluent
molecules in a monocomponent eluent. If a chosen void volume marker compound has some preferential interaction with the stationary phase compared
to that of the eluent molecules, it will show positive retention and could not
be used as void marker. If on the other hand it has weaker interaction, it will
be excluded from the adsorbent surface and will elute faster than the real void
time, meaning that it also could not be used. For any analytical applications
(when no thermodynamic dependences are not extracted from experimental
data), 10% or 15% error in the determination of the void volume are acceptable. It is generally recommended to avoid elution of the component of interest with a retention factor lower than 1.5. Accurate methods for the
determination of the column void volume are discussed in Chapter 2.

3.9.4   Mass of Adsorbent in the Column

The mass of the packing material in the column is a largely underestimated
parameter. As it was discussed in the theory chapter, the retention is proportional to the total surface area of the adsorbent in the column. External
methods of the surface area measurement determine the specific surface area
(in square meters per gram of the material), and the total surface in the column
is a product of the specific surface area of base material and the mass of the
adsorbent in the column.

The retention factor, according expression (2-1) from Chapter 2, is

$$k = \frac{V_R - V_0}{V_0}, \qquad \text{where } V_R = V_0 + SK \Rightarrow k = \frac{S}{V_0} K \qquad (3\text{-}21)$$

This shows that the retention factor is proportional to the total surface area
of the adsorbent. On the other hand, the selectivity, α, is the ratio of the capacity factors and is independent on the total surface area:

$$\alpha = \frac{k_2}{k_1} = \frac{K_2}{K_1} \qquad (3\text{-}22)$$

Below we compare the retention in the form of traditional retention factors
(k') and surface-specific retention factors (k_s) of six different analytes on four
C18-type columns from four different vendors (Figure 3-31).

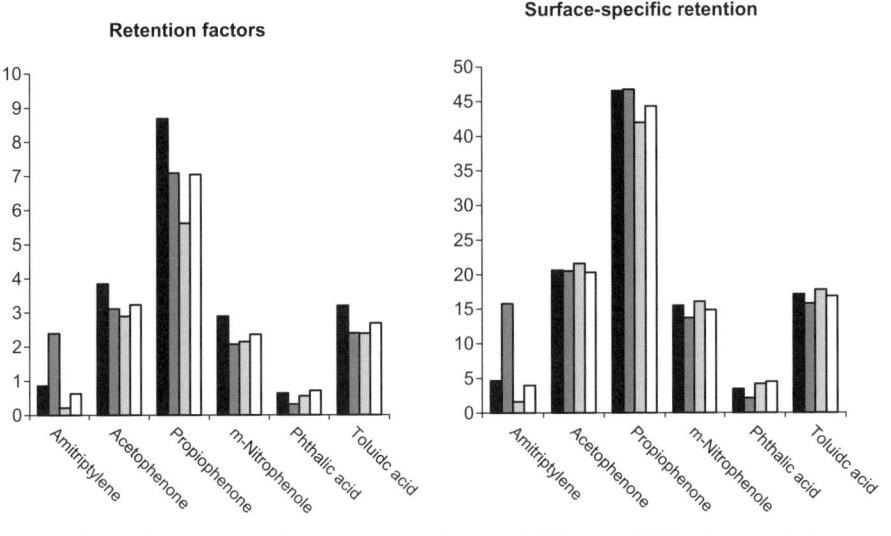

Figure 3-31. Comparison of the retention factors of different ODS columns. Columns: Luna-C18(2) (Phenomenex), Zorbax XDB C18 (Agilent), SunFire C18 (Waters), YMC Pro-Pac C18 (YMC). Conditions: MeCN/water, 0.1% H_3PO_4 pH 2.1.

TABLE 3-4. Relative Standard Deviation of Conventional and Surface-Specific Retention Factors on Different Columns

	RSD %	
Analyte	k'	k_s
Amitriptylene	92.5	98.7
Acetophenone	12.6	2.8
Propiophenone	17.6	5.7
m-Nitrophenol	15.6	5.8
Phthalic acid	30.8	29.1
Toluidc acid	14.5	4.1

Table 3-4 shows the relative standard deviation in the retention of the same analyte on all columns. For ionized analytes eluted very close to the void volume, the retention difference between all columns is significant, but for analytes retained on the column, the RSD of k_s is much smaller than k' and essentially within the experimental error range.

Unfortunately, the measurement of the total surface area of the adsorbent in the column is a tedious process that involves several different experimental techniques, and unless column manufacturers choose to provide this information, it is not feasible that pharmaceutical scientist would be willing to measure it.

TABLE 3-5. Comparison of the Calculated and Measured Mass of Adsorbent in the Column (for Five Different Columns)

Column	V_o (mL)	V_{ip} (mL)	V_p (mL/g)	m_{calc} (g)	m_{meas} (g)	Error (%)
1	1.956	1.035	0.804	1.146	1.117	2.553
2	1.955	1.041	0.804	1.137	1.142	−0.454
3	1.912	1.030	0.778	1.134	1.125	0.771
4	1.845	1.026	0.727	1.127	1.150	−2.039
5	1.693	1.005	0.531	1.296	1.269	2.102

Here we provide a brief description of the methodology for the estimation of the mass of adsorbent in the column and total surface area.

Note. The total surface area of the reversed-phase adsorbent in the column is understood as the total surface area of the base unmodified material in the column, since the surface area of the modified material essentially could not be defined, as discussed in the Section 3.6.7.

The estimation of the amount of adsorbent in the column is based on the comparison of the total pore volume ($V_{p,tot}$ [mL]) of the adsorbent in the column with the specific pore volume (V_p [mL/g]) of the same adsorbent determined from the full nitrogen adsorption isotherm. The ratio of these two values will give the adsorbent mass. Total pore volume in the column is determined as the difference of the column void volume (V_0) and the interparticle volume (V_{ip}).

Interparticle volume could be measured using GPC as the total exclusion volume of high-molecular-weight polymers, and the void volume could be accurately measured as the elution volume of deuterated acetonitrile eluted with neat acetonitrile. The example of these measurements and comparison with the adsorbent mass determined by unpacking the column and weighing the dried adsorbent are shown in Table 3-5.

$$m = \frac{V_0 - V_{ip}}{V_p} \tag{3-23}$$

As could be seen from the table, the error does not exceed 2.5%, which is very good for the determination that involves so many different techniques.

REFERENCES

1. N. S. Wilson, J. W. Dolan, L. R. Snyder, P. W. Carr, L. C. Sander, Column selectivity in reversed-phase liquid chromatography III. The physico-chemical basis of selectivity, *J. Chromatogr. A* **961** (2002), 217–236.

2. J. Israelachvili, *Intermolecular and Surface Forces*, Academic Press, Amsterdam, 2002.

3. L. Jelinek and E. Kovats, True surface areas from nitrogen adsorption experiments, *Langmuir* **10** (1994), 4225–4231.

4. T. Issaeva, A. Kourganov, and K. Unger, Super high speed liquid chromatography of proteins and peptides on non-porous Micra NPS-RP packings, *J. Chromatogr. A* **846** (1999), 13–23.

5. F. C. Leinweber and U. Tallarek, Chromatographic performance of monolithic and particulate stationary phases, *J. Chromatogr. A* **1006** (2003), 207–228.

6. K. K. Unger, *Porous Silica*, Journal of Chromatography Library, Vol. 16, Elsevier, Amsterdam, 1979.

7. R. K. Iler, *The chemistry of Silica*, Wiley-Interscience, New York, 1979.

8. U. D. Neue, *HPLC Columns. Theory, Technology, and Practice*, Wiley-WCH, New York, 1997.

9. I. Halasz and M. Naefe, Influence of column parameters on peak broadening in high-pressure liquid chromatography, *Anal Chem.* **44** (1972), 76–84.

10. R. E. Majors, High-performance liquid chromatography on small particle silica gel, *Anal. Chem.* **44** (1972), 1722–1726.

11. C. Dewaele and M. Versele, Influence of the particle size distribution of the packing material in reversed-phase high-performance liquid chromatography, *J. Chromatogr.* **260** (1983), 13–21.

12. W. Strubert, Herstellung von Hochleistungssäulen Fü die schnelle Flüssigkeits-Chromatographie, *Chromatographia* **6** (1973), 50–52.

13. S. Brunauer, P. H. Emmet, and E. Teller, Adsorption of gases in multimolecular layers, *J. Am. Che. Soc.* **60** (1938), 309–319.

14. W. D. Harkins and G. Jura, Surfaces of solids, *J. Am. Chem. Soc.* **66** (1944), 1362–1366.

15. S. J. Gregg and K. S. W. Sing, *Adsorption, Surface Area and Porosity*, Academic Press, London, 1976.

16. B. Mandelbrot, *The Fractal Geometry of Nature*, San Francisco, Freeman, 1982.

17. A. Y. Fadeev, O. R. Borisova, and G. V. Lisichkin, Fractality of porous silica: A comparison of adsorption and porosimetry data. *J. Colloid Interface Sci.* **183** (1996), 1–19.

18. G. V. Lisichkin (ed.), *Chemistry of Bonded Surface Compounds*, Fizmatlit, Moscow, 2003.

19. M. Le Page and de Vires, French Patent No. 1475929 (1967).

20. K. D. Wyndham, J. E. O'Gara, T. H. Walter, K. H. Glose, N. L. Lawrence, B. A. Alden, G. S. Izzo, C. J. Hudalla, and P. C. Iraneta, Characterization and evaluation of C18 HPLC stationary phases based on ethyl-bridged hybrid organic/inorganic particles, *Anal. Chem.* **75** (2003), 6781–6788.

21. Y.-F. Cheng, T. H. Walter, Z. Lu, P. Iraneta, B. A. Alden, C. Gendreau, U. D. Neue, J. M. Grassi, J. L. Carmaody, J. E. O'Gara, and R. P. Fisk, Hybrid organic–inorganic particle technology: Breaking through traditional barriers of HPLC separations, *LC-GC* **18** (2000), 1162–1172.

22. J. B. Peri and A. L. Hensley, The surface structure of silica gel, *J. Phys. Chem.* **72** (1968), 2926–2933.

23. Y.-F. Cheng., T. H. Walter, Z. Lu, P.Iraneta, B. A. Alden, C.Gendreau, U. D. Neue, J. M. Grassi, J. L. Carmody, J. E. O'Gara, and R. P. Fisk, Hybrid organic–inorganic

particle technology: Breaking through traditional barriers of HPLC separations, *LC-GC* **18** (2000), 1162–1172.

24. A. Méndez, E. Bosch, M. Rosés, and U. D. Neue, Comparison of the acidity of residual silanol groups in several liquid chromatography columns, *J. Chromatogr. A* **986** (2003), 33–44.

25. U. D. Neue, A. Méndez, K. Tran, and P. W. Carr, The combined effect of silanols and the reversed-phase ligand on the retention of positively charged analytes, *J. Chromatogr. A* **1063** (2005), 35–45.

26. Personal communications.

27. F. Chan, L. S. Yeung, R. LoBrutto, and Y. V. Kazakevich, Characterization of phenyl-type HPLC adsorbents, *J. Chromatogr. A* **1069**, (2005), 217–224.

28. XBridge Columns brochure, Waters, Inc., 2006.

29. N. Tanaka and M. Araki, Polymer-based packing materials for reversed-phase liquid chromatography, in J. C. Giddings, E. Grushka, and P. R. Brown (eds.), *Advances in Chromatography*, Vol. 30, Marcel Dekker, New York, 1989, pp. 81–122.

30. D. Gowanlock, R. Bailey, and F. F. Cantwell, Intra-particle sorption rate and liquid chromatographic bandbroadening in porous polymer packings I. Methodology and validation of the model, *J. Chromatogr. A* **726** (1996), 1–23.

31. F. Nevejans and M. Verzele, On the structure and chromatographic behavior of polystyrene phases, *J. Chromatogr.* **406** (1987), 325–342.

32. L. D. Bowers and S. Pedigo , Solvent strength studies on polystyrene-divinylbenzene columns *J. Chromatogr.* **371** (1986), 243–251.

33. N. Tanaka, T. Ebata, K. Hashizume, K. Hosoya, and M. Araki, Polymer-based parking materials with backbones for reversed-phase liquid chromatography. Performance and retention selectivity, *J. Chromatogr.* **475** (1989), 195–208.

34. J. V. Dawkins, L. L. Lloyd, and F. P. Warner, Chromatographic characteristics of polymer-based high-performance liquid chromatography packings, *J. Chromatogr.* **352** (1986), 157–167.

35. J. Li, L. M. Litwinson, and F. F. Cantwell, Intra-particle sorption rate and liquid chromatographic bandbroadening in porous polymer packings II, *J. Chromatogr. A* **726** (1996), 25–31.

36. J. Li and F. F. Cantwell, Intra-particle sorption rate and liquid chromatographic bandbroadening in porous polymer packings III, *J. Chromatogr. A* **726** (1996), 37–44

37. J. Nawrocki, M. Rigney, A. V. McCormick, and P. W. Carr, Chemistry of zirconia and its use in chromatography, *J. Chromatogr.* **657** (1993), 229–282.

38. P. T. Jackson and P. W. Carr, Study of polar and Nonpolar substituted benzenes and aromatic isomers on carbon-coated zirconia and alkyl bonded phases, *J. Chromatogr. A* **958** (2002), 121–129.

39. M. P. Rigney, E. F. Funkenbusch, and P. W. Carr, Physical and chemical characterization of microporous zirconia, *J. Chromatogr.* **499** (1990), 291–304.

40. J. Yu and Z. El Rassi, Reversed-phase liquid chromatography with microspherical octadecyl-zirconia bonded phases, *J. Chromatogr.* **631** (1993), 91–106.

41. J. Li and P. W. Carr, A study of the efficiency of polybutadiene-coated zirconia as a reversed-phase chromatographic support, *Anal. Chem.* **69** (1997), 2193–2201.

42. J. Dai, X. Yang, and P. W. Carr, Comparison of chromatography of octadecyl silane bonded silica and polybutadiene-coated zirconia phases based on diverse set of cationic drugs, *J. Chromatogr. A* **1005** (2003), 63–82.

43. C. J. Dunlap, C. V. McNeff, D. Stoll, and P. W. Carr, Zirconia stationary phases for extreme separations, *Anal. Chem.* **73** (2001), 598A–607A.

44. M. T. Gilbert, J. H. Knox, and B. Kaur, Porous glassy carbon, a new columns packing material for gas chromatography and high performance liquid chromatography, *Chromatographia* **16** (1983),138–145.

45. J. H. Knox and M. T. Gilbert, UK Patent 7 939 449.

46. J. H. Knox and M. T. Gilbert, US Patent 4 263 268.

47. T. Hanai, Separation of polar compounds using carbon columns, *J. Chromatogr. A* **989** (2003), 183–196.

48. J. H. Knox, B. Kaur, and G. R. Millward, Structure and performance of porous graphitic carbon in liquid chromatography, *J. Chromatogr.* **352** (1986), 3–25.

49. P. Ross and J. H. Knox, Carbon-based packing materials for liquid chromatography. Structure, performance, and retention mechanisms, in J. C. Giddings, E. Grushka, and P. R. Brown (eds.), *Advances in Chromatography*, Vol. 37, Marcel Dekker, New York, 1997, pp. 73–119.

50. J. Kříž, E. Adamcová, J. H. Knox, and J. Hora, Characterization of adsorbents by high-performance liquid chromatography using aromatic hydrocarbons porous graphite and its comparison with silica gel, alumina, octadecylsilica and phenyl-silica, *J. Chromatogr. A* **663** (1994), 151–161.

51. N. Tanaka, K. Kimata, K. Hosoya, H. Miyanishi, and T. Araki, Stationary phase effects in reversed-phase liquid chromatography, *J. Chromatogr. A* **656** (1993), 265–287.

52. M.-C. Hennion, V. Coquart, S. Guenu, and C. Sella, Retention behaviour of polar compounds using porous graphitic carbon with water-rich mobile phases, *J. Chromatogr. A* **712** (1995), 287–301.

53. A. Karlsson, M. Berglin, and C. Charron, Robustness of the chromatographic separation of alprenolol and related substances using a silica-based stationary phase and selective retention of metoprolol and related substances on a porous graphitic carbon stationary phase, *J. Chromatogr. A* **797** (1998), 75–82.

54. V. A. Tretikh, V. V. Pavlov, K. I. Tkachenko, and A. A. Chuiko, Distribution of structural hydroxyl groups on an Aerosil surface, *Theor. Eksp. Khim.* **11** (1975), 415–417.

55. P. Roumeliotis and K. K. Unger, Structure and properties of *n*-alkyldimethylsilyell-bonded silica reversed-phase packings, *J. Chromatogr.* **149** (1978), 211–224.

56. G. E. Berendsen, K. A. Pikaart, and L. de Galan Preparation of various bonded phases for HPLC using monochlorosilanes, *J. Liquid Chromatogr.* **3** (1980), 1437–1464.

57. L. R. Snyder, J. J. Kirkland, and J. L. Glajch, *Practical HPLC Method Development*, 2nd ed., John Wiley & Sons, New York, 1997.

58. T. Hanai and J. Hubert, Selectivity of phenyl-bonded silica gel, *J. Chromatogr.* **291** (1984), 81–89.

59. S. Heron and S. Tchapla, Retention mechanism in RPLC: Specific effects of propiophenyl and multifunctional bonded silicas, *J. Chromatogr. A*, **725** (1996), 205–218.

60. Y. V. Kazakevich, R. LoBrutto, F. Chan, and T. Patel, Interpretation of the exces adsorption isotherms, *J. Chromatogr. A* **913** (2001), 73–87.

61. A. Y. Fadeev and G. V. Lisichkin, Adsorption of new and modified inorganic sorbents, in A. Dabrovski and V. A. Tyertykh (eds.), *Series Studies in Surface Science and Catalysis*, Elsevier, Amsterdam, 1995, p. 191.

62. G. E. Berendsen, K. A. Pikaart, and L. De Galan, Preparation of various bonded phases for HPLC using monochlorosilanes, *J. Liq. Chromatogr.* **3** (1980), 1437–1464.

63. I. Rustamov, T. Farcas, F. Ahmed, F. Chan, R. LoBrutto, H. M. McNair, and Y. V. Kazakevich, Geometry of chemically modified silica, *J. Chromatogr.* **913** (2001), 49–63.

64. A. Y. Fadeev, S. M. Staroverov, B. K. Runov, and G. V. Lisichkin, Diffusion of sorbed pyrene in the bonded layer of reversed-phase silicas: Effect of alkyl chain length and pore diameter, *J. Chromatogr.* **558** (1991), 31–42.

65. A. Y. Fadeev, and S. M. Staroverov, Calculation of the characteristics of the bonded layers of reversed stationary phases, *J. Chromatogr.* **465** (1989), 233–240.

66. W. Cheng and M. McCown, Effect of alkyl chain length on surface silanization of silica, *J. Chromatogr*, **318** (1985), 173–185.

67. C. Cao, A. Y. Fadeev, and T. J. McCarthy, Reaction of organosilanes with silica in carbon dioxide, *Langmuir* **17** (2001), 757–762.

68. G. E. Berendsen and L. de Galan, Preparation and chromatographic properties of some chemically bonded phases for reversed-phase liquid chromatography, *J. Liq. Chromatogr.* **1** (1978), 561–586.

69. A. V. Kiselev, Y. S. Nikitin, I. I. Frolov, and Y. I. Yashin, Problems of selectivity and efficiency in liquid–solid chromatography, *J. Chromatogr.* **91** (1974), 201–206.

70. J. Nawrocki, The silanols group and its role in liquid chromatography, *J. Chromatogr. A* **779** (1997), 29–71.

71. N. E. Buyanova, R. V. Zagrafskaya, A. P. Karnaukhov, and A. S. Shepelina, Determination of the area occupied by argon, nitrogen and krypton molecules in a monolayer, *Kineti. Catal.* **24** (1983), 1011–1023.

72. K. S. W. Sing, D. H. Everett, R. A. W. Haul, et al., Reporting physisorption data for gas/solid systems, *Pure Appl. Chem.* **57** (1985), 603–656.

73. E. H. Slaats, J. C. Kraak, W. J. T. Burgman, and H. Poppe, Study of the influence of competition and solvent interaction on retention in liquid–solid chromatography by measurement if activity coefficients in the mobile phase, *J. Chromatogr.* **149** (1978), 255–270.

74. L. Hughes, Ion exchange resins—The technology behind the mystery, *Pharmaceut. Technol. Eur.* **17** (2005), 38–42.

75. P. R. Haddad and P. E. Jackson, *Ion Chromatography*, Journal of Chromatography Library, Vol. 46, Elsevier, Amsterdam, 1990.

76. C. E. Dent, Behavior of sixty amino acids and other ninhydrin-reacting substances on phenol-collidine filter paper chromatograms, with notes as to the occurrence of some of them in biological fluids, *Biochem. J.* **43** (1948), 169–180.

77. M. Kotake, T. Sakan, N. Nakamura, and S. Senoh, Resolution into optical isomers of some amino acids by paper chromatography, *J. Am. Chem. Soc.* **73** (1951), 2973–2974.

78. Y. Okamoto, M. Kawashima, and K. Hatada, Chromatographic resolution. 7. Useful chiral packing materials for high-performance liquid chromatographic resolution of enantiomers: Phenylcarbamates of polysaccharides coated on silica gel, *J. Am. Chem. Soc.* **106** (1984), 5357–5359.

79. A. Ichida, T. Shibata, I. Okamoto, Y. Yuki, N. Namikoshi, and Y. Toga, Resolution of enantiomers by HPLC on cellulose derivatives, *Chromatographia* **19** (1984), 280–287.

80. Y. Okamoto and E. Yashima, Polysaccharide derivatives for chromatographic separation of enantiomers, *Chem. Int. Ed.* **37** (1998), 1021–1043.

81. E. Yashima, C. Yamamoto, and Y. Okamoto, Polysaccharide-based chiral LC columns, *Synlett* (1998), 344–360.

82. Y. Okamota and Y. Kaida, Polysaccharide derivatives as chiral stationary phases in HPLC, *J. High Resolution Chromatogr.* **13** (1990), 708–712.

83. P. Franco, A. Senso, L. Oliveros, and C. Minguillon, Covalently bonded polysaccharide derivatives as chiral stationary phases in high-performance liquid chromatography, *J. Chromatogr.* **906** (2001), 155–170.

84. W. H. Pirkle, D. W. House, and J. M. Finn, Broad spectrum resolution of optical isomers using chiral high-performance liquid chromatographic boned phases, *J. Chromatogr.* **192** (1980), 143–158.

85. S. Allenmark, B. Bomgren, and H. Boren, Direct liquid chromatographic separation of enantiomers on immobilized protein stationary phases: III. Optical resolution of a series of N-aroyl d, l-amino acids by high-performance liquid chromatography onbovine serum albumin covalently bound to silica, *J. Chromatogr.* **264** (1983), 63–68.

86. J. Hermansson, Direct liquid chromatographic resolution of racemic drugs using α_1-acid glycoprotein as the chiral stationary phase, *J. Chromatogr.* **269** (1983), 71–80.

87. A. Nystrfm and A. Karlsson, Enantiomeric resolution on chiral-AGP with the aid of experimental design: Unusual effects of mobile phase pH and column temperature, *J. Chromatogr.* **763** (1997), 105–113.

88. G. Wulff, Molecular recognition in polymers prepared by imprinting with templates, *Am. Chem. Soc. Symp. Ser.* **308** (1986), 186–230.

89. R. Sun, H. Yu, H. Luo, and Z. Shen, Construction and application of a stoichiometric displacement model for retention in chiral recognition of molecular imprinting *J. Chromatogr. A* **1055** (2004), 1–9.

90. H. Minakuchi, K. Nakanishi, N. Soga, N. Ishizuka, and N. Tanaka, *Anal. Chem.*, **68** (1996), 3498–3501.

91. M. S. Fields, Silica Xerogel as a continuous column support for high-performance liquid chromatography, *Anal. Chem.* **68** (1996), 2709–2712.

92. N. Ishizuka, H. Minakuchi, K. Nakanishi, N. Soga, H. Nagayama, K. Hosoya, and N. Tanaka, *Anal. Chem.* **72** (2000), 1275–1280.

93. M. Motokawa, H. Kobayashi, N. Ishizuka, H. Minakuchi, K. Nakanishi, H. Jinnai, K. Hosoya, T. Ikegami, and N. Tanaka, *J. Chromatogr. A* **961** (2002), 53–63.

94. L. Xiong, R. Zhang, and F. E. Reginer, Potential of silica monolithic columns in peptide separations, *J. Chromatogr.* **1030** (2004), 187–194.

95. B. Bidlingmeyer, K. K. Unger, and N. van Doehren, Comparative study on the column performance in microparticulate 5-μm C18-bonded and monolithic C18-bonded reversed-phase columns in high-performance liquid chromatography, *J. Chromatogr. A* **832** (1999), 11–16.

96. T. Ikegami and N. Tanaka, Monolithic columns for high-efficiency HPLC separations, *Curr. Opin. Chem. Biol.* **8** (2004), 527–533.

97. N. Tanaka, H. Kimura, D. Tokuda, K. Hosoya, T. Ikegami, N. Ishizuka, H. Minakuchi, K. Nakanishi, Y. Shintani, M. Furuno, K. Cabrera, Simple and comprehensive two-dimensional reversed-phase HPLC using monolithic columns, *Anal. Chem.* **76** (2004), 1273–1281.

98. N. Vervoort , P. Gzil , G. V. Baron, and G. Desmet, Model column structure for the analysis of the flow and band-broadening characteristics of silica monoliths, *J. Chromatogr.* **1030** (2004) 177–186.

99. J. Zhu and C. Coscolluella, Chromatographic assay of pharmaceutical compounds under column overloading *J. Chromatogr. B: Biomed. Sci. Appl.* **741** (2000), 55–65.

100. J. Knox and R. Kaliszan, Theory of solvent disturbance peaks and experimental determination of thermodynamic dead-volume in column liquid chromatography, *J. Chromatogr.* **349** (1985), 211–234.

4

REVERSED-PHASE HPLC

Rosario LoBrutto and Yuri Kazakevich

4.1 INTRODUCTION

Over 25 years ago, Horvath and Melander, in their fundamental work [1], discussed the reason behind the explosive popularity of reversed-phase liquid chromatography (RPLC) for analytical separations. It was estimated that about 80–90% of all analytical separations were performed in RPLC mode, and the authors noted that "the variation of eluent composition alone extends both retention and selectivity in HPLC [high-performance liquid chromatography] over an extremely broad range." They compared gas chromatography with HPLC, citing "in gas chromatography a plurality of stationary phases has found practical application whereas HPLC tends toward the use of very limited number of columns and optimization of the separation by manipulating the composition of the mobile phase." To some extent the statement is true even today, except that with introduction of capillary columns in GC today, only a very limited number of stationary phases are used, while in HPLC during the last 25 years of development, thousands of different stationary phases have been introduced. Practically all reversed-phase separations are carried out on stationary phases with chemically modified hydrophobic surfaces. Minor variations in the surface chemistry and geometry can lead to noticeable differences in surface interactions and, as a result, to differences in chromatographic selectivity. Specific stationary-phase properties and their influence on the chromatographic retention, selectivity, and efficiency are discussed in detail in Chapter 3.

HPLC for Pharmaceutical Scientists, Edited by Yuri Kazakevich and Rosario LoBrutto
Copyright © 2007 by John Wiley & Sons, Inc.

Mobile phase (eluent) is by far the major "tool" for the control of analyte retention in RPLC. Variations of the eluent composition, type of organic modifier, pH, and buffer concentration provide the chromatographer with a valuable set of variables for successful development of a separation method.

Mobile-phase pH affects the analyte ionization and thus its apparent hydrophobicity and retention. Most pharmaceutical analytes, API (active pharmaceutical ingredient), in-process intermediates, reaction samples, drug substances, raw materials, drug products, and other types of samples generated during the drug development life cycle are ionizable, and their retention is affected by the mobile-phase pH. At the same time, the pHs of aqueous–organic mixtures are different from the pH of the aqueous component itself. The relationship between measured pH of the aqueous phase and the actual pH of the eluent will be discussed, and approaches on how to correlate the HPLC retention to actual eluent pH will be elaborated. The influence of temperature and type and concentration of organic on analyte and pH modifier ionization and its relation to HPLC retention will also be described.

All the choices the chromatographer has in terms of bonded phase, aqueous phase modifier, and organic modifier can have synergistic effects on the analyte retention and selectivity in reversed-phase chromatography. These parameters will be discussed in this chapter, with specific examples illustrating the power of the selection of the most suitable parameters for control of the analyte retention and selectivity.

4.2   RETENTION IN REVERSED-PHASE HPLC

The basis for the analyte retention in reversed-phase chromatography is the competitive interactions of the analyte and eluent components with the adsorbent surface. The stronger the interactions of the analyte with the surface, the longer its retention. Selectivity or the ability of chromatographic system to discriminate between different analytes is also dependent on differences in the surface interactions of the analytes.

Historically, reversed-phase chromatography could be traced back to the work of Howard and Martin [2], who treated an adsorbent surface (of Kiselgure) with dimethylchlorosilane followed by coating of this nonpolar surface with paraffin oil employing methanol–acetone mixtures as the mobile phase. They treated the retention process as partitioning of the analyte between the mobile phase and paraffin oil, which served as a stationary phase (alkylchlorosilane treatment of the polar surface serves only the purpose of increasing wettability by paraffin oil). For many years the advancement in the developments in HPLC essentially followed the development of phases used for gas chromatography. In the middle of the 1960s, modification of the silica gel surface with hexadecyltrichlorosilane was introduced for GC [3]. Following this, Stewart and Perry [4] suggested that this material would be the best possibility for the advancement of "liquid–liquid" chromatography (the term

RPLC was coined). Later, Majors [5] introduced porous silica microparticles modified with alkylsilanes, a packing material that is almost exclusively used in reversed-phase HPLC today.

This brief historical overview of RPLC development is far from the full description of all significant achievements made in the past; however, the primary goal is to show the path of the development, which was, to a larger extent, in the tail of GC development. Consequently, the models and the descriptions of the retention mechanism were essentially transferred from gas–liquid partition chromatography.

Partitioning describes the transfer of the analyte molecules from one phase into another, where the phase is an isotropic macroscopic object with definite physicochemical characteristics. A monomolecular layer of bonded ligands could not be considered as a phase, although following the terminology widely accepted in the literature the term stationary phase is used to essentially denote a solid surface of immobile packing material in the column.

The retention mechanism in modern RPLC is a superposition of different types of dynamic surface equilibria. Main equilibria governing the analyte retention is the adsorption of the analyte molecule on the surface of packing material. The description of the analyte retention on the basis of this main adsorption equilibrium could be expressed as

$$V_R = V_0 + SK \tag{4-1}$$

where V_0 is the total volume of the liquid phase in the column (void volume), S is the adsorbent surface area, and K is the adsorption equilibrium constant. This expression assumes ideal analyte behavior in the chromatographic system at very low analyte concentration. As follows from equation (4-1), the equilibrium constant, K, has units of length (i.e., volume/m^2) and, as such, could not be used as a general thermodynamic equilibrium constant (unitless), but rather as a coefficient representing the analyte retention volume per unit of the adsorbent surface (e.g., μL/m^2). More general expressions and detailed adsorption-based description of the analyte retention in reversed-phase HPLC is given in Chapter 2 of this book.

While dynamic distribution of the analyte between the mobile phase and adsorbent surface is a primary process, there are many secondary processes in the chromatographic system that significantly alter the overall analyte retention and selectivity. Detailed theoretical discussion of the influence of secondary equilibria on the chromatographic retention is also given in Chapter 2.

The analyte nature and its appearance (e.g., ionization state) in the mobile phase are also factors that affect the retention mechanism. Eluent pH influences the analyte ionization equilibrium. Eluent type, composition, and presence of counterions affect the analyte solvation. These equilibria are also secondary processes that influence the analyte retention and selectivity and are of primary concern in the development of the separation methods for most pharmaceutical compounds.

This brief descriptive overview of the reversed-phase process emphasizes the complexity of the retention mechanism and the necessity to consider the influence of different and independent processes on the analyte retention. Since the governing process in the analyte retention is the adsorption equilibrium, the influence of the surface packing material (stationary phase) on the analyte retention in RPLC is described in Section 4.3.

4.3 STATIONARY PHASES FOR RPLC

The introduction of chemically modified stationary phases has had a remarkable impact in the field of liquid chromatography. Successful development and improvement in the technology of manufacturing reproducible bonded layers has revolutionized many chromatographic techniques. Porous silica stationary phases have been modified with ligands of various chemistry and size. The composition and the structure of the bonded organic layer is varied by changing the size of the modifier, specific surface area of the adsorbent, and the bonding density. Chemical bonding of organic ligands with high bonding density on the inner surface of silica pores alters the adsorbent geometry. The effect of surface modification on adsorbent geometric parameters (surface area, pore volume, pore size) has been investigated on several different silica gels [6–8]. It was shown that a decrease in mean pore diameter and in pore volume are associated with the molecular volume of bonded ligands and bonding density. Similar effects were also observed by other researchers [9, 10] Clearly, surface modification has a significant impact on the adsorbent geometry of reversed-phase columns, which will also influence the separation mechanism itself [11]. These effects are discussed in detail in Chapter 3.

Silica-based packing materials dominate in applications for RP separations in the pharmaceutical industry. Hydrophobic surface of these packings typically are made by covalent bonding of organosilanes on the silica surface. This modification involves the reaction of monofunctional alkyldimethylchlorosilanes with the surface silanol groups. Octadecylsilane was the first commercially available silica-based bonded phase and is still the most commonly utilized [12]. Also, alkyl-type ligands of different number of carbon atoms (C1, C4, C8, C12) are often used as well as phases with phenyl functionality; also, polar end-capped, polar embedded phases have been introduced [13–15]. Polar embedded phases provide an additional avenue for potential modification of the chromatographic selectivity, and some of these phases offer an enhancement of retention of polar analytes [16]. These phases can be used with high aqueous mobile phases, even 100% aqueous, without loss of analyte retention that sometimes could be observed for more hydrophobic phases.

Screening several different types of stationary phases during method development for a particular separation is often useful because different columns usually have different selectivity for components in a sample, as can be seen for a forced degradation sample analyzed on three different types of reversed-phase columns using 0.1 v/v% TFA (Figure 4-1) and phosphate buffer, pH 7 (Figure 4-2) mobile phases. Mobile-phase pH can also provide an alternate means of varying the separation selectivity as well.

Other silica-based phases that are available include phenyl and fluorinated alkyl and phenyl-bonded phases. The phenyl and fluorinated phases offer the potential for π–π interactions and show different selectivity in comparison to

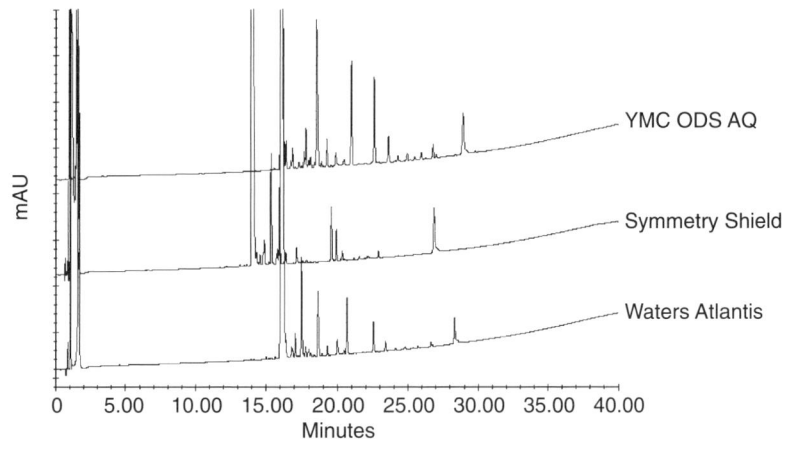

Figure 4-1. Effect of column type on selectivity. Mobile phase: Low pH. (A) 0.1 v/v % TFA. (B) 0.1 v/v% TFA in MeCN. Linear gradient from 5% B to 80% B in 40 min, 220 nm. Temperature, 40°C; flow rate, 1.0 mL/min; column dimensions, 150 × 3.0 mm; particle sizes, 3.5 μm for Symmetry Shield and Atlantis and 3.0 μm for YMC ODS AQ. (Courtesy of Markus Krummen, Novartis Pharmaceuticals.)

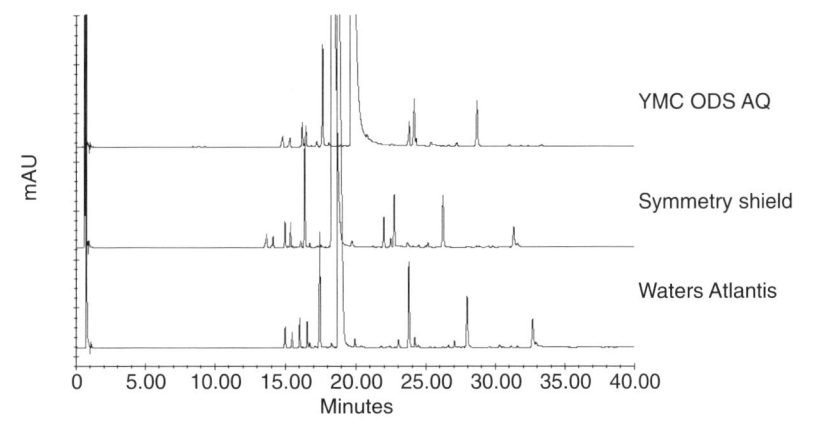

Figure 4-2. Effect of column type on selectivity. Mobile phase: High pH. (A) 10 mM K_2HPO_4, pH 7.0. (B) In MeCN, linear gradient from 5% B–80% B in 40 min, 220 nm. Temperature, 40°C; flow rate, 1.0 mL/min; column dimensions, 150 × 3.0 mm; particle sizes, 3.5 μm for Symmetry Shield and Atlantis and 3.0 μm for YMC ODS AQ. (Courtesy of Markus Krummen, Novartis Pharmaceuticals.)

the alkylsilane phases [17–21]. The fluorinated phases have shown some size and shape selectivity, particularly for aromatic molecules [22, 23]. Moreover, with phenyl-type phases, selectivity/separation differences could be obtained when methanol or acetonitrile is employed. Acetonitrile is an electron-rich organic modifier, which could modify the π–π interactions between the solute and the aromatic moiety of the stationary phase. Methanol, on the other hand, is a proton donor and does not contain π electrons, and therefore its influence on the analyte retention would be principally different [24–26]. It is generally recognized that the type of organic eluent modifier employed plays a dominant role in separation selectivity, although the mechanism of its influence on the analyte retention still remains a subject of intense investigation.

Most silica-based reversed-phase packing materials have a relatively narrow applicable pH range. Below pH 2, the linkage of the bonded phase to the silica substrate is prone to hydrolytic cleavage. Above pH 7, the silica substrate is prone to dissolution, particularly in aqueous-rich mobile phases. In addition, basic compounds may exhibit peak asymmetry above pH 3 due to secondary interactions between the ionized form of the solute and accessible residual silanols. Some new developments in column chemistry have been adopted to address the issues of limited pH working range and reduction of surface density of silanols. The use of hybrid materials allowed for the introduction of organic bridged silica in which an organic bridge is formed between silicon atoms. Resulting hybrid material have been claimed by vendors to show better pH stability at pHs >7 since Si–C covalent bond is much less prone to hydrolysis than Si–O–Si bonds. However, the stability of phases depends on many factors such as the operating pH, type, and concentration of organic modifier and salt concentration, operating temperature, and operating back-pressure. Another approach to manufacturing hybrid silica (Gemini) was introduced by Phenomenex. A layered hybrid silica is synthesized such that the core of the particle is regular silica and the surface is covered by a layer of organic-embedded silica also lending itself to greater pH stability. These stationary phases are further discussed in Chapter 3.

The narrow pH stability range of silica-based packing materials leads to the continuous search for alternative packings that may provide greater pH stability. The options include polymer-based, zirconia-based, and carbon-based phases. The polymer-based columns include poly(styrene-divinyl benzene) and divinylbenzenemethacrylate. These polymer-based columns tend to be stable in the pH 0–14 range. However, lower efficiencies on these polymeric columns relative to silica-based columns are usually obtained due to slower mass transfer kinetics. These phases are also prone to swelling/shrinking as a function of the mobile-phase composition. Retention and selectivity is based on a combination of hydrophobic and π–π interactions [27]. Zirconia is nearly insoluble at pH 1–14 and is stable at temperatures greater than 150°C. The zirconia surface is positively charged up to pH ~ 8, after which it becomes negatively charged [28]. Surface charge, however, is also influenced by adsorption of mobile-phase anions that are hard Lewis bases. The adsorption of hard Lewis bases such as phosphate ion results in ion-exchange sites offering different selectivities than silica [29, 30]. A comparison of polybutadiene (PBD)-coated zirconia and octadecylsilane (ODS) phases indicated that ion exchange is the dominant interaction for basic solutes on the PBD phases while hydrophobic interactions dominate on the ODS phases when phosphate is in the mobile phase [31]. Carbon-based columns are chemically stable over pH range 1–14. These phases are very hydrophobic compared to alkylsilane phases and thus are useful for the separation of polar compounds. However, they strongly, sometimes irreversibly, retain very hydrophobic solutes. Graphitized carbon phases are very suited for the separation of positional and conformational isomers, since the majority of their surface is an ideal graphite plane. Porous graphitized carbon consists of multiple graphite microcrystals and thus offers significant difference in the planar interactions for conformational isomers. Intercrystalline dislocations (irregularities in the crystalline structure), on the other hand, are places of higher surface energy and because the whole material is a conductor, they can be chemically active, which reduce column lifetime and should be taken into account if chemically labile compounds should be separated.

4.4   MOBILE PHASES FOR RPLC

Mobile phases commonly used in reversed-phase HPLC are hydro-organic mixtures. The most common reversed-phase organic modifiers include methanol and acetonitrile and/or combinations of these two modifiers. Other mobile-phase modifiers such as tetrahydrofuran, IPA, and DMSO [32] have been also used for minor selectivity adjustment; however, they are not common due to their high backpressure limitations and/or high background UV absorbance.

The concentration of organic modifier in the eluent is the predominant factor that governs the retention of analytes in RPLC. Highly purified solvents (HPLC grade) are recommended in order to minimize contamination of the stationary phase with impurities of the solvents and reduction of the background absorbance if they contain impurities that have UV chromophores >190 nm.

Considerations for choice of mobile-phase solvents include compatibility between solvents, solubility of the sample in the eluent, polarity, light transmission, viscosity, stability, and pH. The mobile-phase solvents should be miscible and should not trigger precipitation when they are mixed together. For example, dichloromethane and water are immiscible at most compositions and should not be used as mobile-phase components. Similarly, high concentrations of phosphate buffer should not be used with high levels of acetonitrile because the phosphate will eventually precipitate out, resulting in damage in the pump head and blockage of the column frit. The sample should also be soluble in the mobile phase to avoid precipitation in the column. Light

TABLE 4-1. Lower Wavelength Limit of UV Transparency[a] for the Most Typical Solvents Used in HPLC

Solvent	UV Cutoff
Acetonitrile	190
Isopropyl alcohol[b]	205
Methanol	205
Ethanol[b]	205
Uninhibited THF	215
Ethyl acetate[b]	256
DMSO[b]	268

[a]Usually determined as the wavelength at which the absorbance of the neat solvent in a 1-cm cell is equal to 1 AU (absorbance unit) with water used as reference.
[b]Uncommon reversed-phase solvent, may be used in small quantities to adjust selectivity.

transmission is an important parameter when using UV detection; see Table 4-1 for UV cutoffs of common reversed-phase organic modifiers.

Solvents with high UV cutoffs such as acetone (UV cutoff 330 nm) and ethyl acetate (UV cutoff 256 nm) cannot be used for analyses at low wavelengths such as 210 nm. Acetonitrile has a very low UV cutoff (<190 nm) and is one of the contributing factors toward its common use as a solvent for reversed-phase separations. Methanol, ethanol, and isopropanol have a UV cutoff of <205 nm, and at higher organic concentrations the mobile phase transmits less light. It is generally recommended to work at wavelengths >210 nm with these solvents. Also the viscosity of the mobile phase plays an important role in the backpressure generated in the HPLC column (pressure drop). The viscosity is not a linear function and is dependent upon the type and concentration of the organic solvent as well as the operation temperature (Figures 4-3 and 4-4) [33–35]. Also, highly viscous solvents such as methanol and isopropanol can lead to reduced diffusion rates, resulting in peak broadening as well as creating excessively high backpressures in the column. Solvents such as tetrahydrofuran (THF) and other ethers are prone to oxidation to form peroxides. These peroxides can react with the solute or with other mobile-phase components, causing the appearance of spurious peaks.

4.4.1 Eluent Composition and Solvent Strength of the Mobile Phase

HPLC retention is sometimes explained as the result of competitive interactions of the analyte and eluent molecules with the stationary phase. From this point of view the stronger the eluent interactions with the adsorbent surface, the lower the analyte retention, which leads to the term "eluent strength."

In the development of reversed-phase separation methods the organic part of the eluent is considered the strong solvent. Increasing the fraction of the

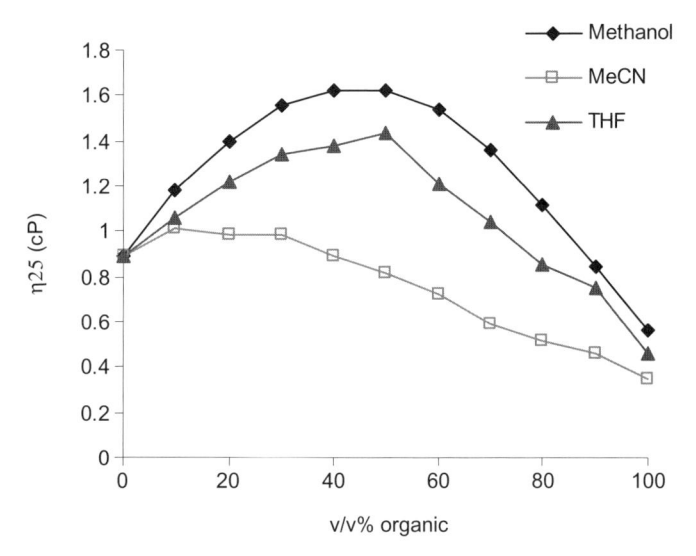

Figure 4-3. Viscosity as a function of organic/water composition values obtained from references 33–35.

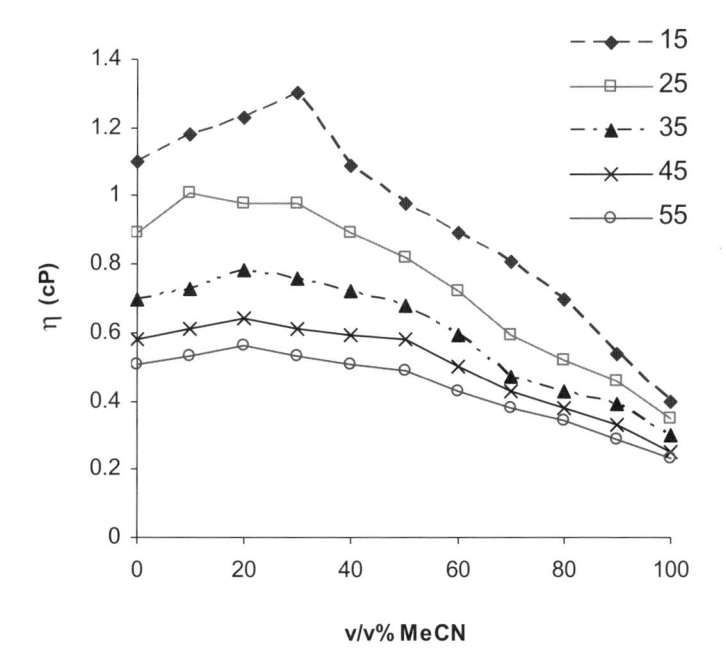

Figure 4-4. Viscosity as a function of acetonitrile/water composition from 15°C–55°C. Values obtained from references 33–35.

organic solvent increases the solvent strength and allows for elution of the species in a mixture, resulting in smaller analyte retention factors or retention volumes.

Analyte HPLC retention is a competitive process, and in an ideal form assuming only analyte–eluent competition for the stationary phase surface and in the absence of any secondary equilibria, one can write

$$k = \frac{V_R - V_0}{V_0} = \frac{S}{V_0} K \tag{4-2}$$

where K is a thermodynamic equilibrium constant, which can be expressed as

$$K = \exp\left(\frac{\Delta G_{\text{analyte}} - \Delta G_{\text{eluent}}}{RT}\right) \tag{4-3}$$

where $\Delta G_{\text{analyte}}$ is the free Gibbs energy of the analyte interaction with adsorbent surface and ΔG_{eluent} is the corresponding free Gibbs energy for eluent. Assuming that the aqueous portion of the reversed-phase eluent is inert and does not interact with the reversed-phase surface, along with using the principle of energetic additivity, one can assume that the free Gibbs energy of the eluent interaction with the stationary phase is proportional to the concentration of organic modifier in the mobile phase.

$$\Delta G_{\text{eluent}} = \frac{c_{\text{org.}}}{c_{\text{max}}} \cdot \Delta G_{el} \tag{4-4}$$

where ΔG_{el} is the free Gibbs energy of the interaction of neat organic phase with the surface, $c_{\text{org.}}$ is the current concentration of organic modifier in the mobile phase, and c_{max} is molar concentration of neat organic phase. Substituting equations (4-4) and (4-3) into equation (4-2) and taking the logarithm leads to equation (4-5):

$$\ln(k) = \frac{\Delta G_{\text{analyte}}}{RT} + \ln\left(\frac{S}{V_0}\right) - c_{\text{org.}} \frac{\Delta G_{el}}{c_{\text{max}} RT} = A + B c_{\text{org.}} \tag{4-5}$$

where $A = \Delta G_{\text{analyte}} / RT + \ln(S/V_0)$ and $B = \Delta G_{el} / c_{\text{max}} RT$ are constants and the logarithm of retention factor is a linear function of the eluent composition. Note that this is only applicable in the absence of secondary equilibria effects, which will be discussed later in this chapter.

Therefore, increasing the concentration of the organic modifier generally leads to an exponential decrease in the analyte retention volume. The general rule of thumb is that for every 10 v/v% increase in organic modifier there is a two- to threefold decrease in the analyte retention factors for analytes with molecular weights of less than 1000 Da. Figure 4-5.

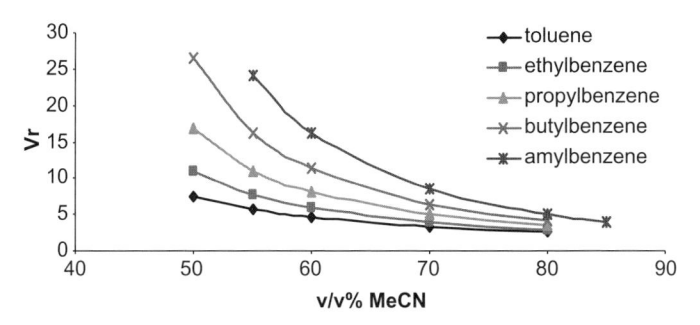

Figure 4-5. Retention volume versus % acetontirle composition. Chromatographic conditions: column, 15 cm × 0.46 cm Zorbax Eclipse XDB-C8; eluent, 0.1% H_3PO_4 50–80% MeCN; flow rate, 1 mL/min; temperature: 25°C.

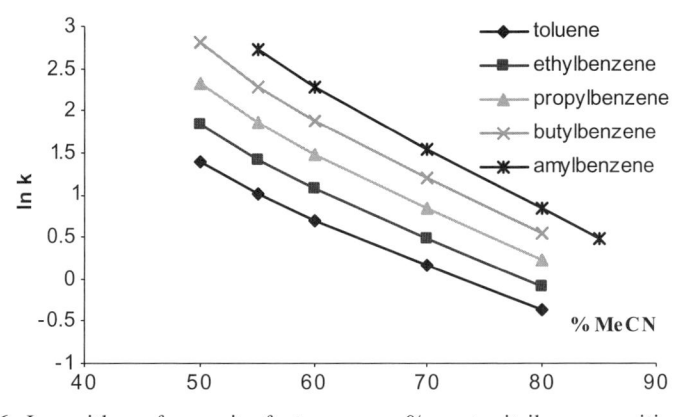

Figure 4-6. Logarithm of capacity factor versus % acetonitrile composition. (Conditions same as in Figure 4-5.)

The logarithm of the retention factor (k) versus the organic composition is usually taken to be almost linear over a limited range in reversed-phase systems (see Figure 4-6). In the eluent concentration region between 50% and 80% of organic component, the slopes of the retention of homologous compounds were the same for all homologs. However, if a wider organic eluent concentration region is studied as in Figure 4-7, a nonlinear dependence of logarithm of retention factor versus the organic composition is observed.

The dependence of k (retention factor) on the volume percentage of the modifier is a subject of great controversy. One school of thought claims a linear dependence [36, 37], whereas another advocates a quadratic relationship [38, 39] and indicates that deviation from linearity will be more pronounced at high concentrations of the modifier.

Several different theories have been proposed for the description of the influence of the eluent composition on the analyte retention in reversed-phase

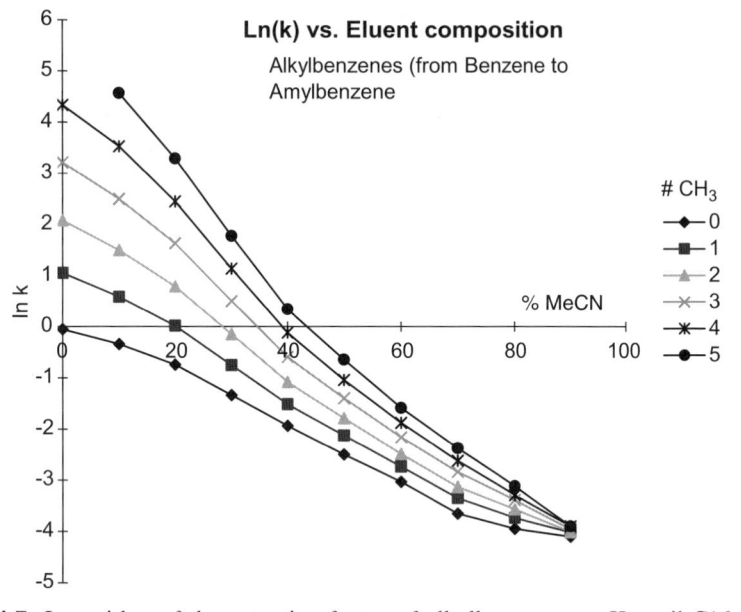

Figure 4-7. Logarithm of the retention factor of alkylbenzenes on Kovasil-C16 (non-porous silica) as a function of the acetonitrile content.

HPLC. Probably the very first consistent theory was introduced by Soczewin-ski [40, 41] for normal-phase separations. He suggested the equation which in the simplified form reads

$$\ln(k) = \ln(k_2) - S\ln(x) \tag{4-6}$$

where k_2 is the hypothetical extrapolated retention factor for the analyte eluted with pure solvent 2 (strongest solvent), S is an adsorbent surface area, and x is a molar fraction of the second eluent component. Almost at the same time, Snyder [36] introduced the concept of eluotropic strength essentially on the basis of the correlation with Hildebrand solubility parameter. The influence of eluotropic strength, ε, of an eluent retention factor was suggested in the form

$$\ln(k) = \ln(k_w) + A(\varepsilon_w - \varepsilon^0) + \ln C \tag{4-7}$$

where C is a complex function of molecular volumes and molecular areas of the eluent components.

Later, Snyder et al. [42] introduced simplified semiempirical equation

$$\log(k) = \log(k_w) - S\phi \tag{4-8}$$

where k_w is the extrapolated analyte retention factor in pure water, ϕ is the volume fraction of the organic eluent modifier, and S is the slope of this linear function specific for a particular organic modifier used and the nature of the solute (most important is the molecular weight). For small molecules the S values for methanol and acetonitrile are generally in the range of 3–4 [43], and for biomolecules they are more than 50. It later appears that S values are not exactly solvent-specific but rather dependent on the type of column used. Horvath and Melander indicated very strong dependence of the S parameter on the type of the bonded phase [44].

In a simplified form, it is generally accepted that the logarithm of the retention factor shows linear variation with the volume fraction of the eluent composition [45] similar to expression (4-5) above. This statement has to be taken only as a first and very rough approximation, since many deviations from this rule have been reported [46, 47] especially for acetonitrile/water mixtures as shown for n-hexanol and n-octanol in Figure 4-8 [47] and phenol and toluene in Figure 4-9.

From a practical point of view, the concept of linearity of the logarithm of the analyte retention factor could be used only for the rough estimation of the eluent composition variation. Also, the curvature of this dependence can show further deviations from linearity if the analyte is changing its ionization state at varying organic composition.

4.4.2 Type of Organic Modifier

Mobile-phase strength depends not only on the concentration of the organic modifier, but also on the type of organic modifier used. There were many attempts to create some type of mathematical correlation between the acetonitrile and methanol and THF concentrations which is supposed to result in similar retention of the analytes. More comprehensive estimates of sliding scales of solvent strength of different organic modifiers have been given by Schoenmakers et al. [48, 49] This is known as the same elutropic strength. The solvent strength of the most common organic eluents used at the same volume percentage (v/v%) in reversed-phase chromatography would be: methanol < acetonitrile < tetrahydrofuran. For example, if a similar retention of a neutral compound or an ionizable compound in its fully ionized or neutral state is to be achieved with methanol/water eluent compared to acetonitrile/water eluent on a C18 adsorbent, then an increased concentration of methanol is needed in the mobile phase (about 10 v/v% more of methanol for every 1 v/v% of acetonitrile would be needed for similar elution). If similar retention is to be achieved with acetonitrile/water mobile phase versus THF/water mobile phase on a C18 adsorbent, approximately 10 v/v% more of acetonitrile is needed for every 1 v/v% of THF. Note that these general rules serve only as an approximation because the retention of an analyte in methanol/water versus acetonitrile/water system may be dependent on many parameters lending to different interactions of the analyte with the solvent and/or with the bonded phase.

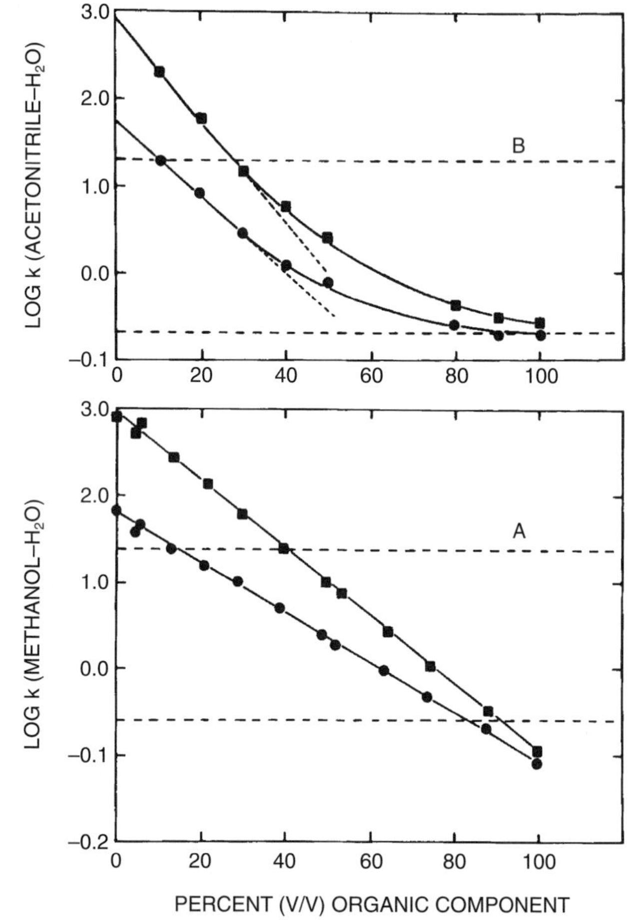

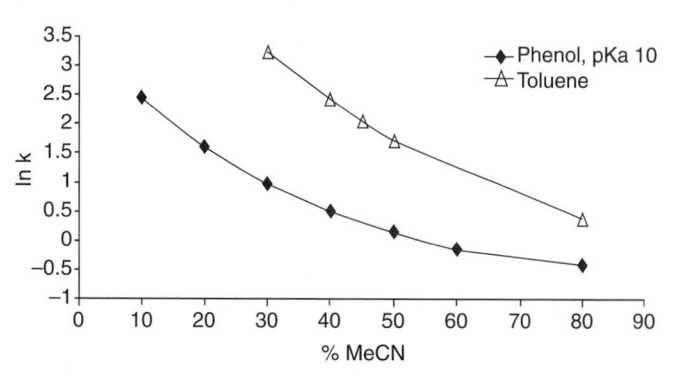

Figure 4-8. Logarithm of retention factors of *n*-hexanol and *n*-octanol on octadecyl-silica at different water/organic compositions. (Reprinted from reference 47, with permission.)

Figure 4-9. Chromatographic conditions column: 15-cm × 0.46-cm Chromegabond WR-EX C18. Eluent: Buffer/10–80% MeCN. Buffer: 15 mM sodium acetate, pH 4; flow rate, 1 mL/min.

The principal difference in the behavior of acetonitrile and methanol, the most common eluent modifiers, was recently shown [50] where acetonitrile and THF forms a thick multimolecular adsorbed layer on the surface of reversed-phase adsorbent (C1–C18 and phenyl phases), while methanol is adsorbed only in monomolecular fashion. This brings a principal difference in the analyte retention mechanism in these two hydro-organic systems. Different retention mechanisms and their theoretical description are discussed in the Chapter 2.

In a binary eluent system (acetonitrile-water), an adsorbed organic phase with finite thickness and composition different from the bulk mobile phase is preferentially accumulated near the surface of the bonded phase. The organic layer accumulated near the bonded ligands could behave as a liquid stationary phase in reversed-phase HPLC, and it contributes to the overall analyte retention process.

In this scenario, an adsorbed organic layer with a different composition than the bulk mobile phase is formed. An analyte may distribute itself from the bulk eluent into the adsorbed organic layer. This adsorbed organic layer possesses a certain thickness which depends upon the concentration of the organic component in the eluent composition and its adsorption isotherm on the surface of the packing material used. The general retention process is comprised of two processes: (1) the analyte partitions between the bulk eluent and the adsorbed organic layer; (2) the portion of the analyte partitioned into the organic layer is then distributed between this layer and the surface of the modified adsorbent.

Overall analyte retention in acetonitrile/water eluent is the superposition of different processes: partitioning and adsorption. The volume of acetonitrile adsorbed layer is also dependent on the eluent composition (v/v% acetonitrile). This essentially may provide the explanation for the nonlinear behavior of the logarithm of the retention factors as a function of the eluent composition for acetonitrile as opposed to methanol, which forms only monomolecular layer and analyte retention factors generally show linear logarithmic dependence on the eluent composition (v/v% methanol).

4.4.3 Selectivity as a Function of Type and Concentration of Organic Composition

Ideally the eluent composition should not affect the selectivity between two species if their ionization state is not changing with an increase in the organic composition see Section 2.14 for details).

Generally the selectivity of neutral components is not affected by changing the organic composition. However, for ionizable components, changing the organic composition may affect changes in the analyte ionization state and lead to changes in selectivity. For example, in all experiments shown in Figure 4-10 the aqueous portion of the mobile phase had a pH of 7, and noticeable

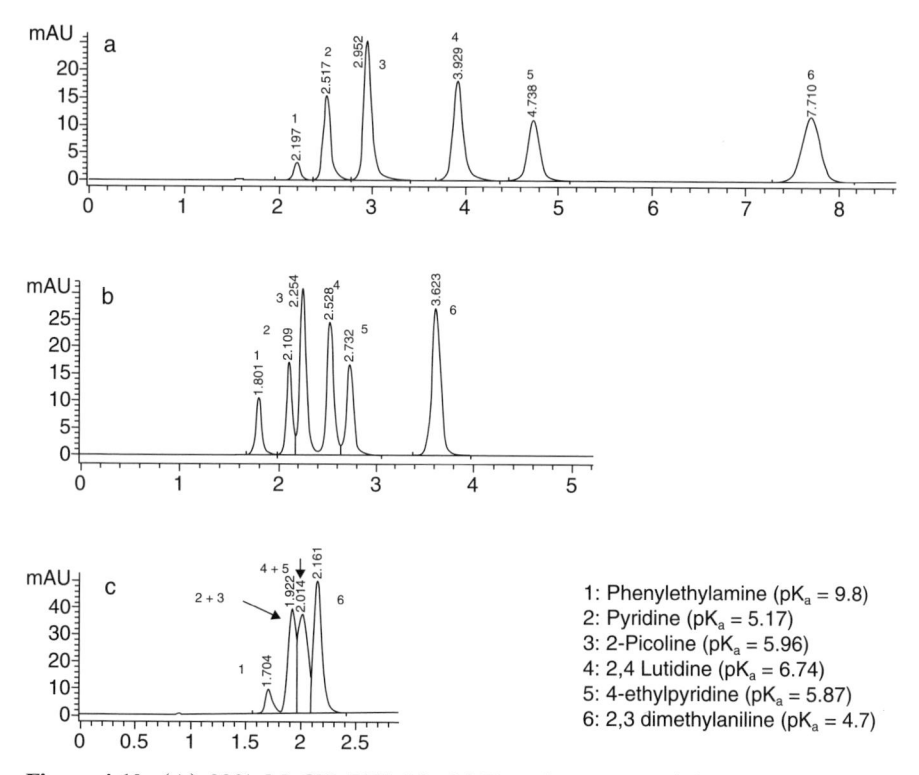

Figure 4-10. (A) 30% MeCN: 70% 20mM Phosphate, pH 7. (B) 50% MeCN: 50% 20mM Phosphate, pH 7. (C) 80% MeCN: 20% 20mM Phosphate, pH 7.

TABLE 4-2. Effect of Organic Composition on the Chromatographic Selectivity

v/v% MeCN	$\alpha = k_2/k_1$				
	1, 2	2, 3	3, 4	4, 5	5, 6
30%	1.36	1.36	1.59	1.31	1.86
50%	1.61	1.18	1.29	1.17	1.62
80%	1.29	1.00	1.37	1.00	1.21

changes in selectivity were observed for critical pairs (i.e., 1,2 and 2,3 and 3,4 …) as the organic composition is increased from 30 v/v% acetonitrile to 80 v/v% acetonitrile (Table 4-2). This may be attributed to a change in the pH of the aqueous portion of the mobile phase as well as variation in the analyte ionization state upon the addition of organic component. This change in selectivity for these ionizable species indicates that their degree of ionization varies with the change in organic composition which contribute to the variation in

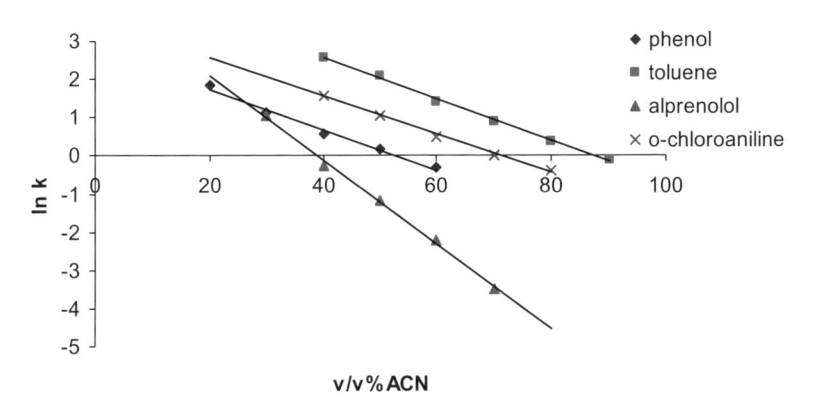

Figure 4-11. Chromatographic conditions column: 15-cm × 0.46-cm Phenomenex Luna C18(2), 5 μm. Aqeous: 15 mM sodium acetate buffer adjusted to pH 4.5 with acetic acid. Organic: MeCN, composition: 20–90 v/v% Acetonitrile; flow rate, 1 mL/min; temperature, 30°C.

selectivity. The effect of organic content on changes in mobile-phase pH and analyte ionization will be discussed in Section 4.5.

Variations in the selectivity are sometimes observed with the change in the type of organic modifier due to the specifics of the analyte–solvent interactions (solvation) and the specific adsorption behavior of the organic modifier. In the following example the effect of type and concentration of methanol and acetonitrile modifiers on the retention of acidic, basic, and neutral analytes is discussed.

The separation of four analytes [neutral analyte (toluene), strong basic compound (alprenolol, $pK_a = 9$), weakly basic compound (o-choloroaniline, $pK_a = 2.5$), and weak acidic compound (phenol, $pK_a = 10$)] was performed on a conventional C18 phase using a sodium acetate mobile phase (pH 4.5) with either acetonitrile (Figure 4-11) or methanol (Figure 4-12) as the organic portion of the eluent. The pH of the mobile phase was chosen to be at least two units away from the analytes pK_a's in the mixture, such that each of the analytes were analyzed in their fully ionized or neutral states across the range of the organic compositions studied.

In Figures 4-11 and 4-12, linear dependence of the analyte retention is observed in both hydro-organic systems for all analytes in the limited region of the eluent compositions studied. Note that in order to obtain a similar retention factor for each of the analytes using methanol, a higher amount of methanol compared to acetonitrile was necessary (roughly a 10 v/v% increase in methanol content was applied to elute components at similar retention using acetonitrile). However, with the alprenolol (analyzed in its protonated state at pH 4.5) a 20–25 v/v% greater concentration of methanol was required compared to acetonitrile, to elute alprenolol at a similar retention. Also,

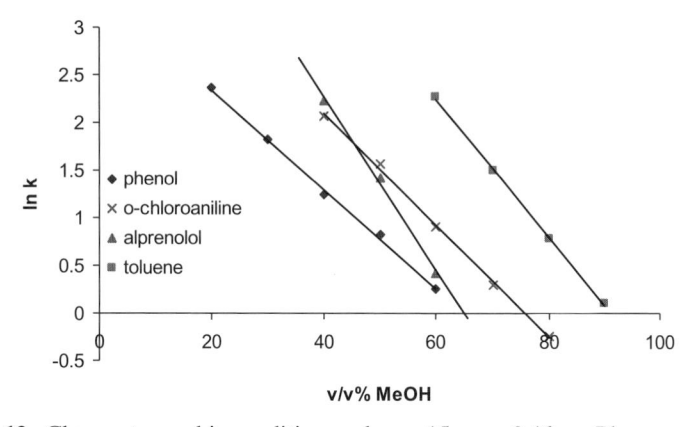

Figure 4-12. Chromatographic conditions column: 15-cm × 0.46-cm Phenomenex Luna C18(2), 5 μm. Aqeous: 15 mM sodium acetate buffer adjusted to pH 4.5 with acetic acid. Organic: MeOH, composition: 20–90 v/v% methanol; flow rate: 1 mL/min; temperature, 30°C.

differences in selectivity for the compounds in this mixture were observed when the two different organic eluents were used. This demonstrates the "power" of using different types of organic eluent to assist in optimizing the separation selectivity. It is important to note that the slope factor (*S*) given in equation (4-8) must be different for each studied analyte if a selectivity change with change in organic concentration is to be observed between two analytes. Thus, it is possible to optimize both the retention and selectivity by varying the mobile-phase composition in isocratic mode or varying the gradient slope in gradient mode. Varying the gradient slope (change in v/v% organic per unit time) is a very useful approach for adjustment of separation selectivity during reversed-phase HPLC method development.

Under reversed-phase conditions, due to differences in the hydrogen bonding capabilities, polarizability, and different absorption characteristics of acetonitrile, methanol, and THF, the use of different modifiers offer substantial differences in selectivity. As an example, an LC-MS method was required to monitor labeled cortisol/cortisone in the presence of unlabeled cortisone and other possible interfering metabolites (Figure 4-13). Separations using YMC-ODS-AQ column exhibited differences in selectivity that could be further enhanced by using different organic modifiers to separate the eight components of interest (Figure 4-14) [51]. Selectivity differences can also be obtained by using small percentages (<10%) of a third solvent component for more challenging separations. For example, in a water/acetonitrile system the addition of less than 10% THF or methanol in acetonitrile may offer desired selectivity differences.

Changes in selectivity can be obtained by changing the type of stationary phase and the type and concentration of organic for both neutral and

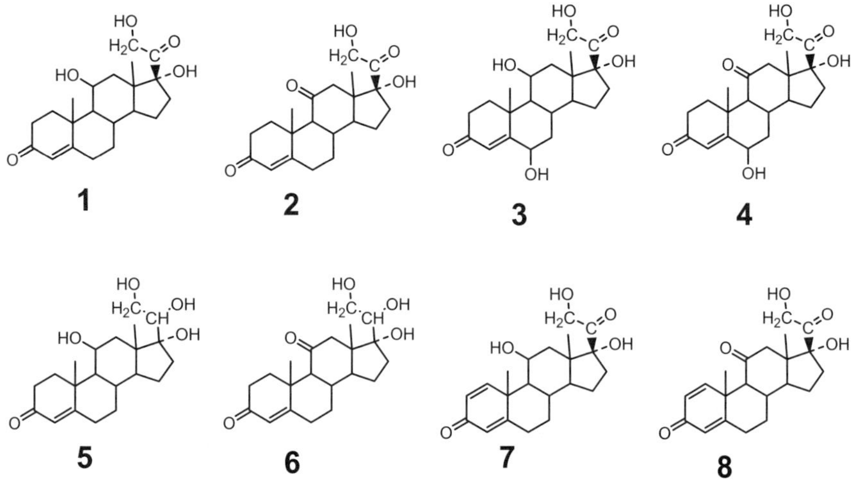

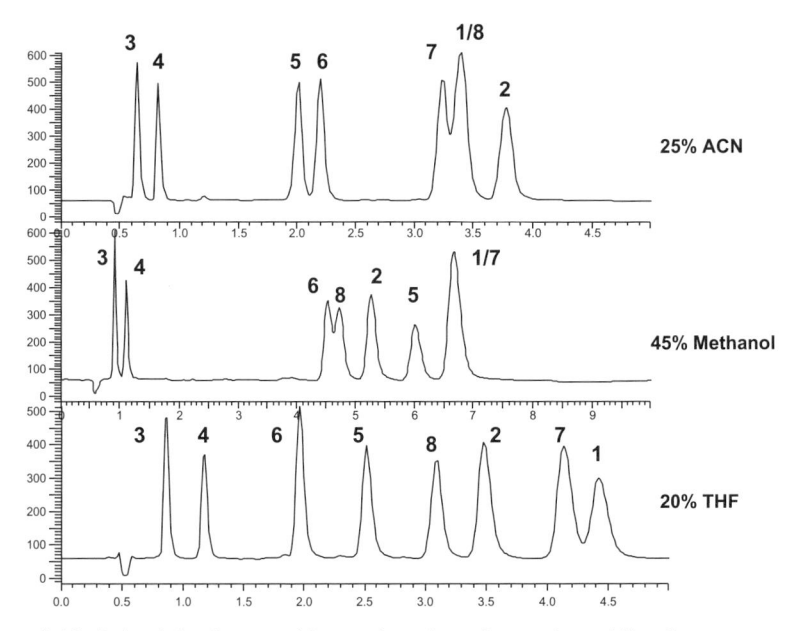

Figure 4-13. Steroids. 1, Cortisol; 2, cortisone; 3, 6β-OHF; 4, 6β-OHE; 5, 20β-DHF; 6, 20β-DHE; 7, prednisolone; 8, prednisone.

Figure 4-14. Selectivity for steroids as a function of organic mobile-phase component. Chromatograms showing the elution order of all eight congeners as a function of organic modifier with 0.1% formic acid as the buffer phase on YMC ODS-AQ column at ambient temperature. (Top panel) 25% acetonitrile, 1.5-mL/min flow rate; (middle panel) 45% methanol, 1.2-mL/min flow rate; (bottom panel) 20% tetrahydrofuran, 1.5-mL/min flow rate. (Reprinted from reference 51, with permission.)

ionizable compounds. However, the pH can have a dramatic effect on the change of the separation selectivity for ionizable compounds. In Section 4.5 the importance for judicious choice and control of pH for the separation of ionizable compounds is discussed.

4.5   pH EFFECT ON HPLC SEPARATIONS

Most pharmaceutical compounds contain ionizable functionalities such as amino, pyridinal, or carboxylic groups. Mobile-phase pH and composition are among the main parameters used to control HPLC retention of most pharmaceutical compounds and to optimize separations. The introduction of new packings that are stable over a wider pH range up to pH 12 allows for a broader applicability of mobile-phase pH as a retention/selectivity adjustment parameter [52, 53]. The pH of the mobile phase has a strong influence on the retention of protolytic solutes and should be controlled in reversed-phase HPLC. Buffers are recommended to control the pH stability of the mobile phase. Common buffers are shown in Table 4-3. Note that the common volatile buffers trifluoroacetate (pK_a 0.5), acetate (pK_a 4.8), and formate (pK_a 3.8) can be used for mass-spectrometric detection; however, they have significant background absorption, depending on their concentration at wavelengths below 220 nm. This usually leads to descending baselines when running gradient separations (since the aqueous portion of the mobile phase is being diluted with the organic and there is a consequent decrease in the background absorbance). It is generally recommended to add the same concentration of acid modifier or salt buffer that is in the aqueous phase to the organic phase to suppress the descending baseline effect (*Note*: Check the solubility of buffering agents in organic phase). However, even though this leads to a flatter baseline, it still reduces the detection sensitivity because the mobile phase is absorbing at the wavelength of interest (<220 nm).

4.5.1   Mobile-Phase pH. Practical Considerations

The pH specified in analytical methods for pharmaceutical analysis should be that of the aqueous solvent. Note that the addition of organic modifier to aqueous buffer generally results in a shift in mobile-phase pH. The pK_a of the solute is also subject to variation and is dependent on the type and concentration of the organic modifier in the eluent.

Practical Recommendations

The pH of the aqueous portion of the mobile phase must be accurately controlled with a calibrated pH meter. That is, the desired pH of the aqueous portion of the mobile phase must be within the calibrated pH range. The calibration standards should be close in ionic strength and temperature of the buffer solutions that will

TABLE 4-3. Commonly Used Buffers for Reversed-Phase HPLC

Buffer	pK_a	Buffer Range	UV Cutoff (nm)
TFA[a]	0.5	Up to 1.5	210 nm (0.05 v/v%)
Phosphoric acid	2.1	1.1–3.1	<200 nm (10 mM)
Mono/dihydrogen phosphate			
	7.2	6.2–8.2	
	12.3	11.3–13.3	
Formic acid[a]	3.8	2.8–4.8	210 nm (10 mM.)
Acetic acid[a]	4.8	3.8–5.8	210 nm (10 mM)
Citrate	3.1	2.1–4.1	230 nm (10 mM)
	4.7	3.7–5.7	
	5.4	4.4–6.4	
Trisamino methane (TRIS)	8.1	7.1–9.1	205 nm (10 mM)
Borate	9.2	8.2–9.2	<200 nm (10 mM)
Dimethylamine	10.7	9.7–11.7	<200 nm (10 mM)
Triethylamine	10.8	9.8–11.8	<200 nm (10 mM)
Ammonium hydroxide[a]	9.3	8.3–10.3	<200 nm (10 mM)
Pyrrolidine[b]	11.3	10.3–12.3	<205 nm (10 mM)
N-Methyl pyrrolidine[c]	10.3	9.3–11.3	<205 nm (10 mM)
Piperidine	11.1	10.1–11.1	<205 nm (10 mM)
Ammonium acetate[a]		3.8–5.8 (acetate)	210 nm (10 mM)
		8.3–10.3 (ammonium)	
Ammonium bicarbonate[a]	pK_1 6.4	5.4–7.4, (carbonate,1)	<200 nm (10 mM)
	pK_2 10.3	9.3–11.3 (carbonate,2)	
	9.3	8.3–10.3 (ammonium)	

[a]Volatile buffers; these can be used for LC-MS.
[b]Cannot be used in the state of California.
[c]Sometimes purity of this reagent is poor and UV cutoff could be greater.

be measured. Typical standards include pH 1, pH 2, pH 4, pH 7, and pH 10. Some pH meters allow a three-point calibration curve, and others allow for a two point calibration curve. In any event, the appropriate criteria for the slope of the calibration curve should be met and recorded in a log book and/or notebook and should be calibrated each time prior to use. The pH of the buffer should be controlled within ±0.1 pH units. Additionally, buffers can be made by weighing a known amount of salt of the acid or base to make the buffer and adding, if necessary, a known volume of acid or base to generate the desired pH. Once this is performed, the pH of the aqueous portion of the mobile phase can be measured as an additional precautionary measure.

The impact of the pH in hydro-organic mixtures on the analyte ionization and retention will be thoroughly discussed. The impact of pH on analyte UV absorbance will be discussed in the method development chapter, Chapter 8 (Section 8-6).

4.5.2   Analyte Ionization (Acids, Bases, Zwitterions)

A simple rule for retention in reversed-phase HPLC is that the more hydrophobic the component, the more it is retained. By simply following this rule, one can conclude that any organic ionizable component will have longer retention in its neutral form than in the ionized form. Analyte ionization is a pH-dependent process, so significant effect of the mobile-phase pH on the separation of complex organic mixtures containing basic or acidic components can be expected.

Ionization of the analyte could be expressed by one of the following equilibria:

$$HA \Leftrightarrow A^- + H^+ \qquad \text{for acidic components} \qquad (4\text{-}9a)$$

$$B + H_2O \Leftrightarrow BH^+ + OH^- \qquad \text{for basic components} \qquad (4\text{-}9b)$$

For HA a weak acid the products of the dissociation are hydrogen ion (H^+) and an anion (A^-), which is the conjugate base. Equilibrium constants for acids can be written in the following form:

$$K_a = \frac{[A^-] \cdot [H^+]}{[AH]} \qquad (4\text{-}10)$$

Using the definition for the pH (Hendersen–Hasselbalch form), one can rewrite

$$pK_a = pH + \log\left(\frac{[AH]}{[A^-]}\right) \qquad (4\text{-}11)$$

Similar expressions could be written for bases.

$$K_b = \frac{[BH^+] \cdot [OH^-]}{[B]} \qquad (4\text{-}12)$$

$$pK_b = pOH + \log\left(\frac{[B]}{[BH^+]}\right) \qquad (4\text{-}13)$$

Weak acids and bases exist in equilibrium with their ions, and their equilibrium constants (K_a) are small. The position of equilibrium is measured by the equilibrium constant K_a [equation (4-10)] and K_b [equation (4-12)]. The dissociation constant of water is 10^{-14} ($K_w = 10^{-14}$). Since the neutralization reaction of the equimolar quantities of weak acid and base result in the formation of water and salt, the following relationship holds:

$$K_a K_b = K_w \qquad (4\text{-}14)$$

The usefulness of relationship (4-14) is that for any base the K_b value can be found from the K_a of its conjugate acid. Conversely, the K_a value of any acid can be found from K_b of its conjugate base.

4.5.3   pK_a and pK_b Relationship

Using the pX notation, we commonly speak of the pK_a of an acid and the pK_b of a base, where $pX = -\log X$. Therefore,

$$pK_a = -\log K_a, \qquad (4\text{-}15)$$

$$pK_b = -\log K_b, \qquad (4\text{-}16)$$

Because $K_a K_b = K_w$, it follows that $pK_a + pK_b = pK_w$. Furthermore, the value of K_w at 25°C is 1.0×10^{-14} and so $pK_w = 14.00$ at 25°C.

For any acidic or basic compound the pK values can be expressed as pK_a or pK_b; however, in the literature the pK_a values are most often used.

4.5.4   Retention of Ionizible Analytes in Reversed-Phase HPLC

Primary equilibrium in the chromatographic system is the analyte distribution between mobile phase and the surface of packing material. If the analyte could be present in the mobile or stationary phase in two or more different forms and there is an equilibria between these forms, this equilibria is usually called "secondary."

Because different forms of analyte usually show different affinity to the stationary phase, secondary equilibria in HPLC column (ionization, solvation, etc.) can have a significant effect on the analyte retention and the peak symmetry. HPLC is a dynamic process, and the kinetics of the secondary equilibria may have an impact on apparent peak efficiency if its kinetics is comparable with the speed of the chromatographic analyte distribution process (kinetics of primary equilibria). The effect of pH of the mobile phase can drive the analyte equilibrium to either extreme (neutral or ionized) for a specific analyte. Concentration and the type of organic modifier affect the overall mobile phase pH and also influence the ionization constants of all ionogenic species dissolved in the mobile phase.

It is generally preferable to use a mobile-phase pH that is at least one unit away from the pK_a values of the analytes in the mixture, so each analyte will be in a predominant single ionization state (>90%) during the chromato-

graphic run. This essentially suppresses the effect of secondary ionization equilibria on the analyte retention.

Neutral and ionic forms of any analyte have significant differences in their
apparent hydrophobicity and thus tend to migrate though the column with different velocity. This causes an instantaneous disturbance of ionic equilibrium
in the chromatographic zone microenvironment. The equilibration kinetics of
analyte ionic equilibrium has a profound effect on the analyte peak shape. If
the kinetics of ionization is slow, the ionic species will tend to move faster than
the neutral species, causing a significant broadening of the composite chromatographic zone (kinetics of ionization is in almost every single case faster
than the chromatographic kinetics by orders of magnitude).

The kinetics of the ionic equilibration is also dependent on the analyte solvation. The greater the analyte solvation, the slower the equilibration kinetics. Solvation shell restricts the protonation or deprotonation of the analyte.
Solvation is also influenced by the eluent ionic strength. With an increase of
the concentration of ions in the analyte microenvironment, there is a corresponding decrease in the analyte solvation, thus increasing the ionic equilibration kinetics. The increase of the eluent ionic strength usually improves the
analyte peak shape even if the mobile-phase pH is close to the analyte pK_a.

The development of HPLC methods where the mobile-phase pH is close
to the analyte pK_a is not recommended because of potential distorted peaks
(which can be amended by the increase of the salt or buffer concentration);
these methods may not be rugged and will not be easily transferable to other
laboratories (manufacturing facility). Any minor variations in the mobile-
phase pH in this case can lead to the significant variations in the analyte retention and separation selectivity.

Since the pK_a is a characteristic constant of the specific analyte, from equation (4-11), one can conclude that relative amounts of neutral and ionic forms
of the analyte could be easily adjusted by varying the mobile-phase pH. Moreover, if the eluent pH is at least two units away from the component pK_a, more
then 99% of the analyte will be in either ionic or neutral form. If the eluent
pH is at least one unit away from the component pK_a, more than 90% of the
analyte will be in either ionic or neutral form.

In Figure 4-15, a titration curve for aniline is shown such that if the analyte
was placed in an acidic medium such as HCl, it would be >99% ionized (0%
neutral, bottom plateau), and as the analyte was titrated with a strong base
such as NaOH the analyte would become progressively more neutral (0% ionization, top plateau). In Figure 4-16 it can be seen that the analyte in its ionized
form shows the lowest retention (low plateau); and as the analyte becomes
progressively more neutral, this leads to increased retention. Eventually when
the pH of the medium exceeds the 2 units greater than the analyte pK_a, the
retention is essentially unchanged (Figure 4-16, top plateau).

4.5.4.1 Basic Compounds. The primary retention dependence of the ionizable analyte versus the mobile-phase pH for basic components will have the
form shown in Figure 4-16, assuming that no predominate secondary interac-

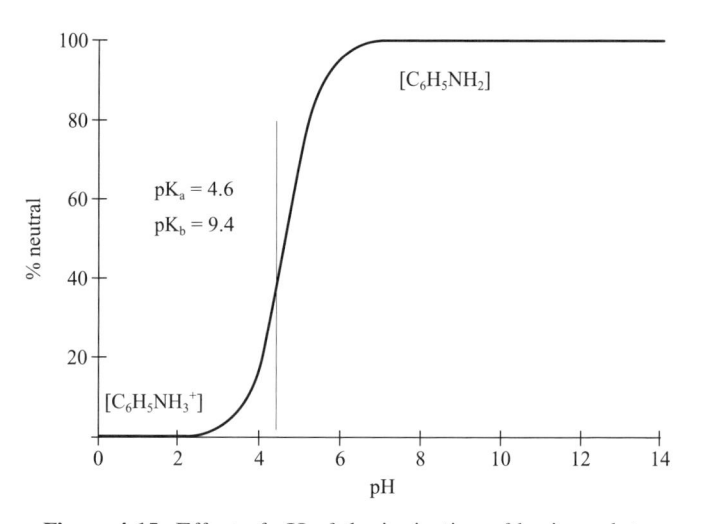

Figure 4-15. Effect of pH of the ionization of basic analyte.

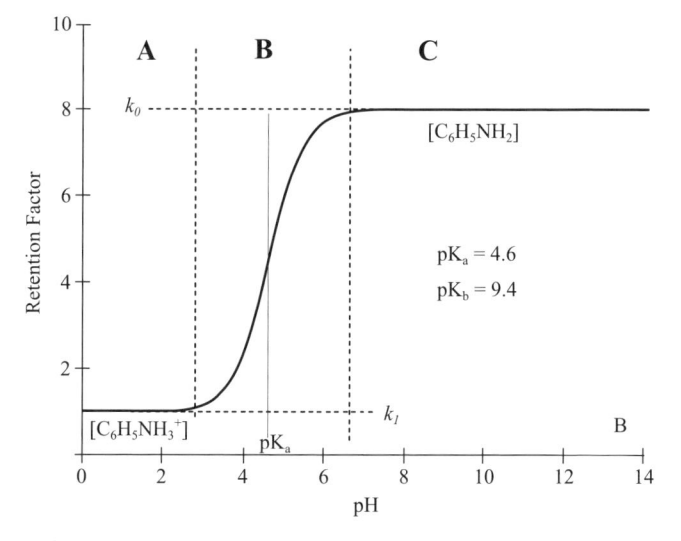

Figure 4-16. Effect of pH of the retention of basic analyte.

tions are occurring with the stationary phase (i.e., interactions with residual silanols).

The retention dependence of basic components on the pH of mobile phase could be subdivided into three regions (Figure 4-16).

A. Fully protonated analyte (cationic form), which shows the lowest retention. The analyte is in the most hydrophilic form. Its interactions with the hydrophobic stationary phase are suppressed. A compound in its ionic form is more hydrophilic, so it tends to have less interaction with hydrophobic stationary phase and also tends to be more solvated with protic solvents. This may cause the significant decrease in the retention of ionic components. If methanol, a protic solvent, is used as an organic modifier, it can participate in the analyte solvation. The inclusion of the methanol molecules in the analyte solvation shell adds some hydrophobicity to the solvated molecular cluster, and this may lead to the significant distortion of the peak shape. This will be further discussed in Chapter 8.

B. Partial protonation region. Coexistence of two analyte forms (protonated and deprotonated) in the mobile phase in equilibrium may cause poor peak shape and unstable retention. Since analyte in the neutral form has much stronger retention, its molecules tend to interact with the stationary phase stronger and reside there longer. This causes a shift of the ionization equilibrium in the mobile phase with a greater proportion of protonated molecules (more hydrophilic) at the front of chromatographic band. The overall process depends on the superposition of the ionization and adsorption processes and their relative kinetics. Usually, a slight change of the mobile-phase pH greatly shifts the analyte retention in this region.

C. Analyte in its neutral form (the most hydrophobic), which shows the longest retention.

4.5.4.2   Acidic Compounds. Similar retention curves can be obtained for acidic components, but obviously their retention dependence will be the mirror image of that for basic analytes (Figure 4-17).

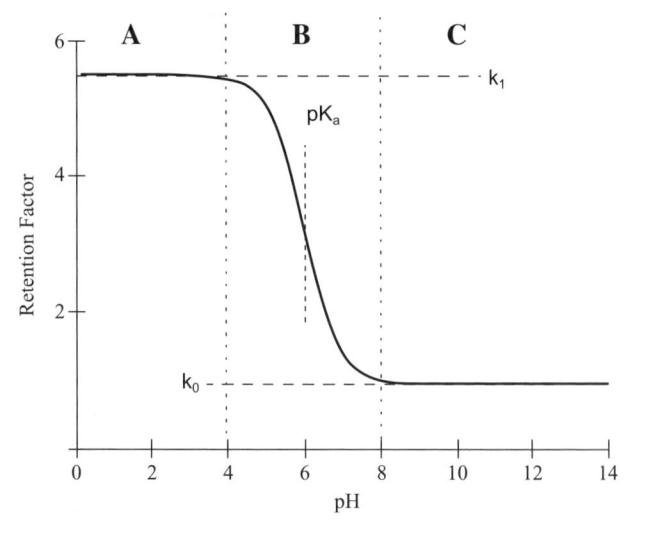

Figure 4-17. General retention dependence of acidic analyte on the mobile-phase pH. The inflection point of the curve corresponds to the component pK_a.

These retention profiles could be described by the following equation as a function of eluent pH and analyte pK_a [54].

$$k = \frac{k_0 + k_1 \dfrac{[H^+]}{K_{a(B^+)}}}{1 + \dfrac{[H^+]}{K_{a(B^+)}}}$$

(4-17)

or

$$k = \frac{k_0 + k_1 10^{(pK_a - pH)}}{1 + 10^{(pK_a - pH)}}$$

(4-18)

where for bases k_1 is the limiting retention factor of the protonated form and is represented by the lower plateau in Figure 4-16, and k_0 is the limiting retention factor of the neutral form and is represented by the higher plateau in Figure 4-16. However, for acids k_0 is the retention factor of the anionic form represented by the lower plateau in Figure 4-17, and k_1 is the retention factor of the neutral form represented by the higher plateau in Figure 4-17. For both acids and bases, k is the retention factor at a given pH and k_a is the analyte ionization constant.

If compounds are analyzed in region A or C retention variation with mobile-phase pH will be minimal. Methods employing a mobile-phase pH which corresponds to these regions are generally more rugged. On the other hand, each region has its own drawback. Therefore, during the selection of the starting HPLC conditions, one has to account for all possible effects. Some of the drawbacks are discussed below.

4.5.4.3 Effects in Region A. Basic analytes show relatively low retention (analyte in its ionic form) and may even elute in the void. The employment of chaotropic additives may be needed to enhance the retention of the protonated basic analytes (see Section 4.10) However, acidic analytes show longer retention times because the acidic analyte would be analyzed in its neutral form.

4.5.4.4 Effects in Region B. In this region there is coexistence of appreciable quantities of both ionic and nonionic forms of the analyte. Significant loss of apparent efficiency for both acidic and basic analytes may be present in this region. Peaks broaden and sometimes have a weird shape but are mostly tailing or fronting, depending on the ionic strength of the mobile phase. Very unstable retention is observed, and minor changes in pH or composition of the mobile phase will significantly shift retention. Minor changes of the eluent composition can cause change in selectivity.

4.5.4.5 Effects in Region C. Very long retention for basic analytes thus requires working with high organic concentration of the mobile phase. Acidic components will be in their anionic form at high pH. Organic analytes in their anionic form usually are strongly solvated and may be completely excluded from the pore space of the packing material. This may cause very early elution of the analytes which is usually not adjustable by the eluent composition. Also, silica is soluble at high pH. If the column has some accessible silanols, prolonged operation at high pH may cause steady degradation of the packing material. This brings a loss of the efficiency due to the formation of voids in the column, or steady change of component retention.

4.5.5 Case Studies: Effects of pH on Ionizable Analyte Retention

Knowledge of the pK_a of the analytes in the mixture is very important. Significant changes in retention and even reversals in elution order can be observed. Take, for example, two analytes: one basic and one acidic, both with pK_a values of approximately 4. If a mixture of these two analytes were analyzed at pH 2.3 and pH 6.0, the base would show lower retention at pH 2.3 and higher retention at pH 6.0, and the acidic component would show higher retention at pH 2.3 and lower retention at pH 6.0 (Figure 4-18). Depending on the analytes' relative hydrophobicity in their ionized form, they may even elute before the void volume, and the component(s) eluting prior to the void volume can and should not be quantified.

Acidic components may be weak acids such as phenolic compounds (8–10) or stronger acids such as carboxylic acids (pK_a 3–4), and the specific analyte pK_a would be dependent on the substitution on the aromatic ring and/or the

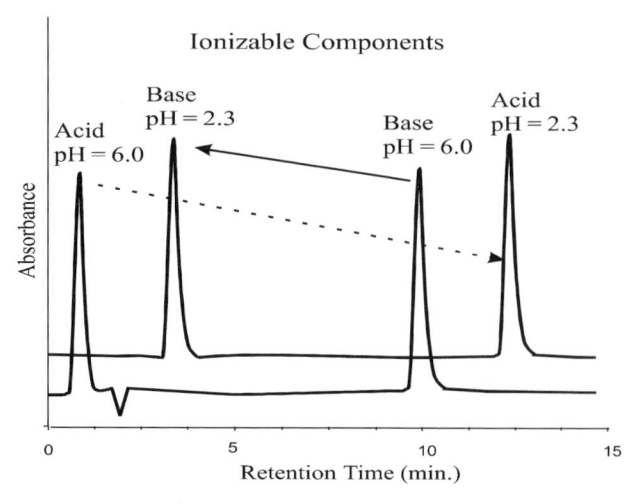

Figure 4-18. Effect of mobile-phase pH on the retention of ionizable compounds. (Reprinted from reference 55, with permission.)

neighboring groups on the alkyl chain if the ionizable species is not on the aromatic ring. Basic components may also be weak bases such as aniline (pK_a 4.6) or stronger bases such as secondary or tertiary amines (pK_as 7–10). Figure 4-19 shows theoretical retention factor versus pH curves for these weakly/strongly acidic and weakly/strongly basic compounds. For unknown species in a mixture, the retention can be determined as a function of pH run under isocratic conditions to ascertain if the analytes of interest are acidic or basic in nature.

For a mixture of acidic components (carboxylic acid), decreasing the pH of the aqueous portion of the mobile phase from 9 to 2 led to the enhancement of the retention of the acidic analytes (Figure 4-20). At aqueous mobile-phase

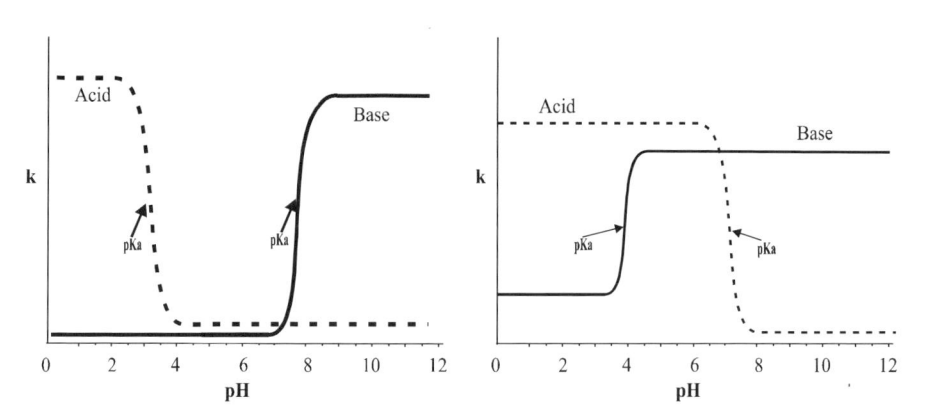

Figure 4-19. Theoretical retention versus pH profiles for acidic and basic components. (A) Strong acid and strong base. (B) Weak acid and weak base. (Reprinted from reference 55, with permission.)

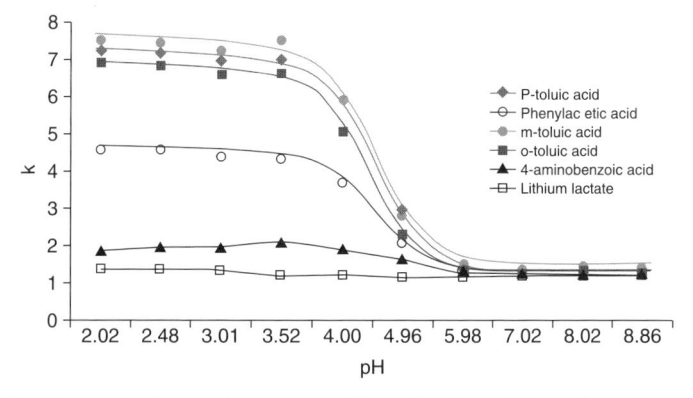

Figure 4-20. Theoretical retention versus pH profiles for acidic and a zwitterionic component. Chromatographic conditions: Column: 15-cm × 0.46-cm Phenoemenex Luna C18(2), 5 µm; 70% 15 mM K_2HPO_4 adjusted pH 2–9 with phosphoric acid; 30% MeCN; flow rate, 1 mL/min; temperature, 25°C.

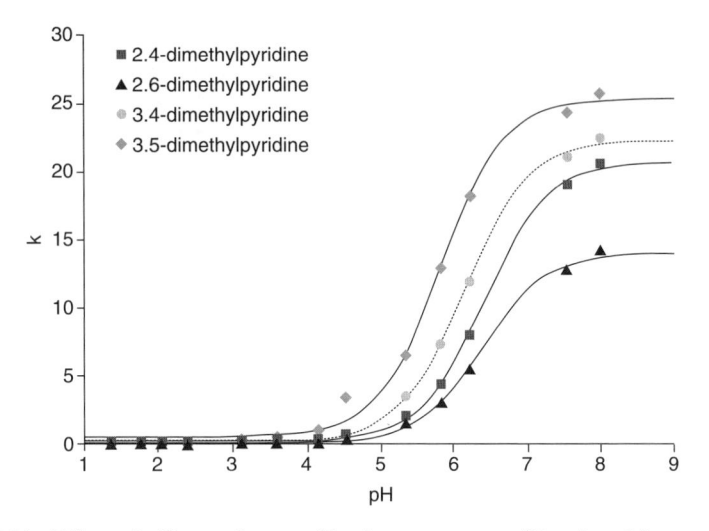

Figure 4-21. Effect of pH on mixture of basic components. (Reprinted from reference 56, with permission.)

pH values greater than and equal to 6, all the acidic components eluted close to the void volume and resolution of all the components was not obtained.

In the pH region from 3.5 to 5, the aqueous pH of the mobile phase was close to that of the analyte pK_a values, which led to a dramatic change in retention and peak skewing. Working in this pH region would not be recommended because small variations in the mobile-phase pH would lead to undesired changes in retention and selectivity. However, as the aqueous pH was changed from 3 to 2, only minor variations in retention were observed for all the components in the mixture. This also led to the optimal selectivity and resolution of all components in the mixture.

The opposite scenario for retention dependence versus pH is observed for basic compounds (Figure 4-21) [56]. In the mixture of basic components shown in Figure 4-21 (pyridinal species), increasing the pH of the aqueous portion of the mobile phase from 1.5 to 9 led to the enhancement of the retention of the basic analytes. At aqueous mobile-phase pH values of 7 and greater, the components exhibited a high retention. However, as the aqueous pH was changed to below 4, the compounds eluted close to the void volume. Generally, it is recommended that very polar bases are analyzed at pHs where the analyte is in its neutral form.

Variation of the mobile-phase pH is a powerful parameter to enhance the chromatographic selectivity and retention for mixtures of basic, acidic, and neutral compounds. Figure 4-22 shows that for neutral analytes (benzamide and flavone) the mobile-phase pH has no effect on the chromatographic retention [57]. However, for the organic acids (hydroxyisophthalic acid and fenoprofen, pK_a 4.5) at pH 2, which is at least 2 pH units below the acid analyte pK_a values, maximum retention is obtained, while at pH 7 and 12 there is a

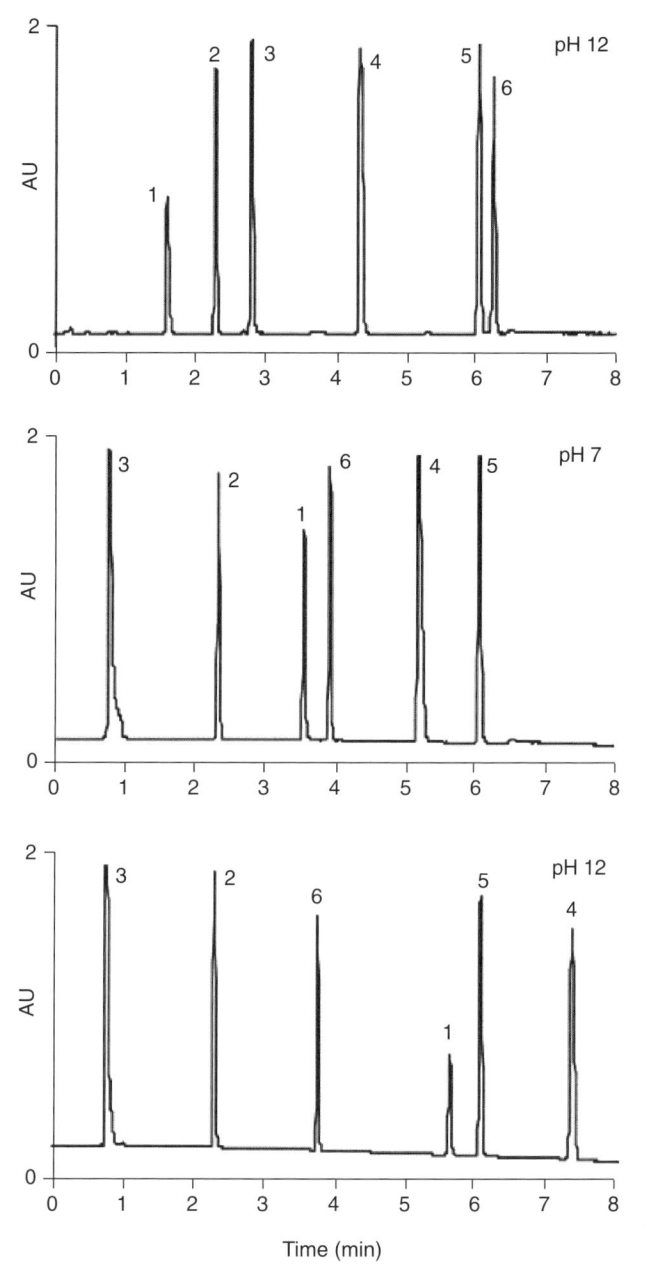

Figure 4-22. Selectivity differences at pH 2, 7, and 12 on Waters XBridge C18 column. Analytes: 1, doxylamine (base); 2, benzamide (neutral); 3, hydroxyisophthalic acid (acid); 4, doxepine (base); 5, flavone (neutral); 6, 5 fenoprofen (acid). (Reprinted from reference 57, with permission.)

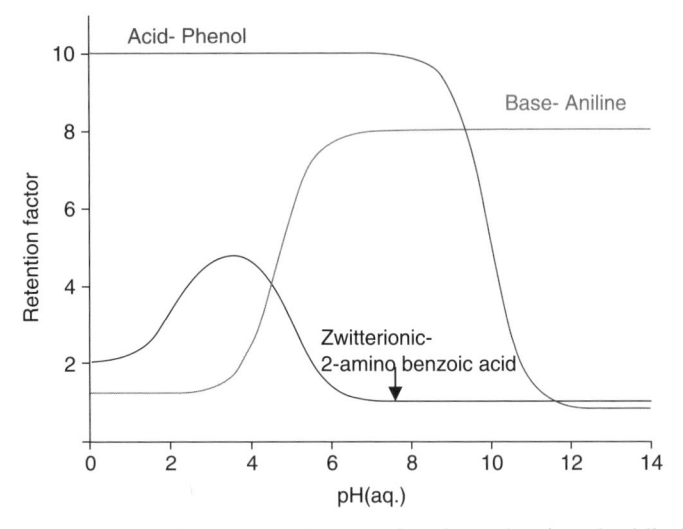

Figure 4-23. Effect of analyte pH on the retention dependencies of acidic, basic, and zwitterionic analytes.

significant decrease in retention for both these acids and hydroxyisophthalic acid is nonretained and elutes in the void. For the bases (doxepin, pK_a 9.3; doxylamine, pK_a 9.2) at pH 12, which is at least 2 pH units above their pK_a values, maximum retention is obtained, and at pH 2 the lowest retention for both basic compounds is obtained.

Zwitterionic components contain both acidic and basic functionalities. Depending on the distance between their pK_a values, two distinct sigmodial dependencies may be observed, one for the acid and one for the base, and the overall retention dependence is usually in the shape of a bell curve (inverted or upright). These are usually observed when the pK_a values are greater than 2 units apart. In Figure 4-23 a theoretical curve (bell-shaped) for a zwitterionic compound, 2-amino benzoic acid, is shown. The pK_a of the basic functionality is 2.1, and that of the acidic functionality is 5.0. On the other hand, if a zwitterionic compound contained an acidic functionality with pK_a 2 and a basic functionality with pK_a 5, the bell-shaped curve would be inverted.

The HPLC retention of zwitterionic analytes essentially follows the dependence shown in Figure 4-23. A practical example is demonstrated with Benazepril HCl (SS and SR isomers). Analyte retention was monitored as a function of pH using 10mM ammonium phosphate buffer at pH 2.1–7.1/acetonitrile (70/30) on a phenyl-hexyl column (Figure 4-24). The pK_a values for benazepril in purely aqueous media are 3.7 for the acidic moiety and 4.6 for the basic moiety. However, since Benazepril is being analyzed in a hydro-organic media the analyte ionization can be shifted, depending on the type and percent of organic modifier employed. Bell-shaped retention dependence of both Benazepril diastereomers is clearly visible in Figure 4-24.

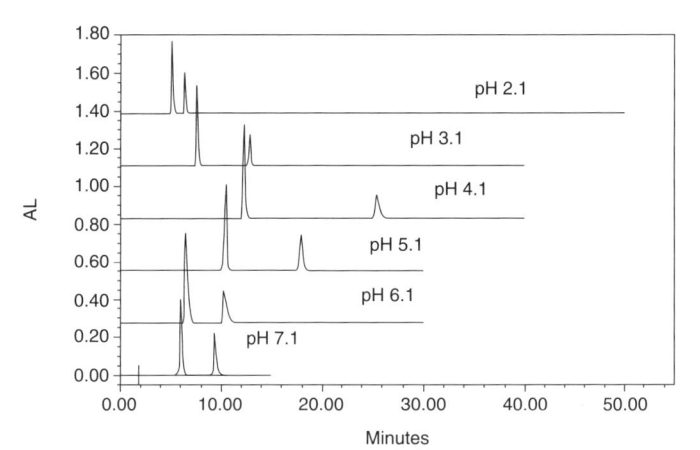

Figure 4-24. Benazepril hydrochloride SS and SR isomers on Phenomenex phenyl-hexyl column, 150 × 4.6 mm, 70% ammonium phosphate buffer ($_w^w$pH 2.1–7.1), 30 v/v% MeCN, 1 mL/min, 25°C. (Courtesy of Rajinder Singh.)

There were selectivity differences as a function of pH from 2.1 to 7.1 (pH of the aqueous phase), with concomitant changes in retention. This demonstrates that separations of ionizable compounds in RPLC should be performed where the molecules are in one predominate ionization state(s) to avoid significant changes in retention and selectivity with minor changes in pH.

4.5.6 Mobile-Phase pH

For the separation of ionogenic (ionizable) solutes, the variations of mobile-phase pH can lead to extreme changes in selectivity. The mobile-phase (eluent) pH affects the ionization of ionogenic species and consequently their HPLC retention. However, the pH of the aqueous phase is not equivalent to the pH of the aqueous/organic eluent, and consequently the variation of the mobile-phase composition leads to the variation in pH under both isocratic and gradient conditions [58–60]. Therefore the pH shift of the mobile phase upon the addition of the organic modifier is imperative for a proper description of the ionogenic analyte retention process and should be accounted in the development of the HPLC separations of such compounds. This will lead to more robust and rugged methods.

4.5.6.1 *pH Scales in Water–Organic Mixtures.*

There have been many studies about the determination of pK values of acids [61–65] in acetonitrile/water mixtures. Several models have been used to relate the pK with the solvent composition [66, 67]. Typically, the pH of acidic buffers upon the addition of acetonitrile or methanol increases to higher values, and the pH of basic buffers upon the addition of acetonitrile or methanol decreases to lower values. On the basis of IUPAC rules and recommendations [68, 69], Bosch and co-workers have studied the different pH scales that are employed in pH measurement of hydro-organic mixtures [64]. The three different pH scales are usually considered the ^w_wpH, ^s_wpH, and ^s_spHs. The ^w_wpH scale is one in which the electrode system is calibrated with aqueous buffers, and the pH of the aqueous portion of the mobile phase is measured prior to the addition of the organic modifier. The ^s_wpH scale is where the electrode system is calibrated with aqueous buffers, and the pH of the hydro-organic mobile phase is measured after the addition of the organic modifier. The ^s_wpH has also been described in the literature as the apparent pH or pH^{app}. [70, 71] The ^s_spH scale is where the electrode system is calibrated with buffer-organic mixtures of the same composition as the mobile phase, and the pH of the hydro-organic mobile phase is measured after the addition of the organic modifier [68, 69] This sometimes has been referred to in the literature as pH* [72]. The s/w and s/s scales could be interconverted by equation (4-19). This equation includes the difference of the liquid junction potentials (E_j), together with the primary medium effect $-\log(^s_w\gamma_H^0)$ [73]. Espinosa et al. [73] have stated that residual liquid junction potential can be assumed to be negligible if the junction potential of the electrode system in the measurement solution in solvent s (^sE_j) is close to the liquid junction potential in the calibration solution in water (^wE_j). Therefore $\delta \simeq -\log(^s_w\gamma_H^0)$ and δ is the difference of ^s_spH and ^s_wpH scales.

$$\delta = E_j - \log\left(^s_w\gamma_H^0\right) = {}^s_w\mathrm{pH} - {}^s_s\mathrm{pH} \qquad (4\text{-}19)$$

The δ is a constant for each mobile-phase composition and type of buffer system employed. The δ term determined for various acetonitrile/water and methanol/water mixtures ranging from 10 to 60 v/v% by Roses, Espinosa, and co-workers [73, 74] are shown in Table 4-4 and Table 4-5, respectively.

TABLE 4-4. Delta Values for Various Acetonitrile/Water Compositions [73, 74]

Volume fraction MeCN (ϕ):	0	0.1	0.2	0.3	0.4	0.5	0.6
Mole fraction:	0	0.04	0.079	0.13	0.186	0.26	0.339
δ:	0	−0.01	−0.02	−0.06	−0.13	−0.26	−0.44

TABLE 4-5. Delta Values for Various Methanol/Water Compositions [73, 74]

Volume fraction MeOH (ϕ):	0	0.1	0.2	0.3	0.4	0.5	0.6
Mole fraction:	0	0.047	0.1	0.16	0.229	0.308	0.4
δ:	0	0.01	0.03	0.05	0.09	0.13	0.18

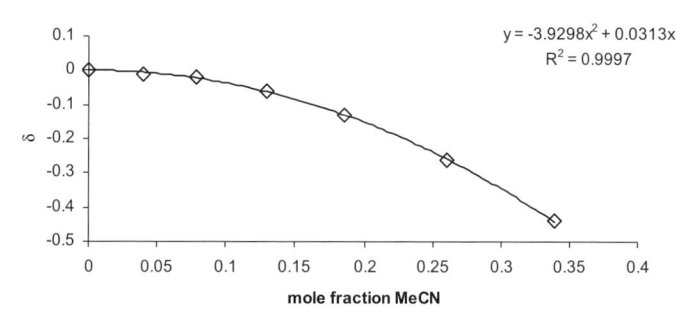

Figure 4-25. Variation in the δ quantity with mole fraction of acetonitrile.

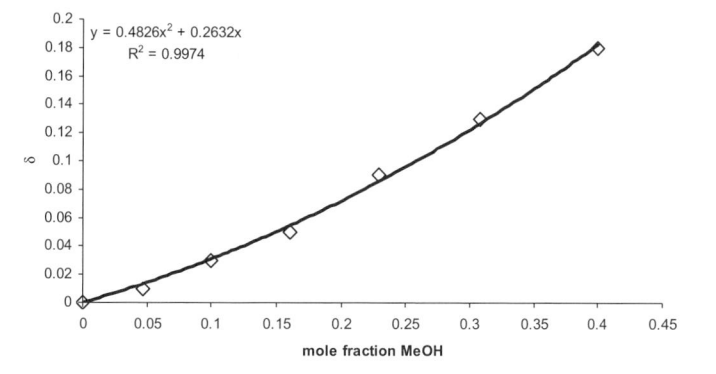

Figure 4-26. Variation in the δ quantity with mole fraction of methanol.

An empirical formula representing the variation of the δ quantity with mole fraction of acetonitrile (χ) from the values in Table 4-4 could be determined using equation (4-20). The dependence of δ versus the mole fraction of acetonitrile is shown in Figure 4-25.

$$\delta = (-3.93)\chi_{MeCN}^2 + 0.03\chi \qquad (4\text{-}20)$$

An empirical formula representing the variation of the δ quantity with mole fraction of methanol (χ) from the values in Table 4-5 could be determined using equation (4-21). The dependence of delta versus the mole fraction of methanol is shown in Figure 4-26.

$$\delta = (0.4826)\chi_{MeOH}^2 + 0.2632\chi \qquad (4\text{-}21)$$

Similarly, the delta values as a function of any volume composition up to 60 v/v% acetonitrile [i.e., is equivalent to 0.6 volume fraction (ϕ)] and methanol can be determined using equations (4-22a) and (4-22b). [74]

$$\delta = \frac{-0.446\phi_{MeCN}^2}{1 - 1.316\phi_{MeCN} + 0.433\phi_{MeCN}^2} \tag{4-22a}$$

$$\delta = \frac{0.09\phi_{MeOH} - 0.11\phi_{MeOH}^2}{1 - 3.15\phi_{MeOH} + 3.51\phi_{MeOH}^2 - 1.35\phi_{MeOH}^3} \tag{4-22b}$$

Note, however, that the difference between $_w^s\text{pH}$ and $_s^s\text{pH}$ is a constant value for each mobile-phase composition, and the difference between $_w^s\text{pH}$ and $_s^s\text{pH}$ depends not only on the type and concentration of mobile-phase composition, but also on the particular solution being measured [74–76]. However, these values can serve as estimates for converting from $_w^s\text{pH}$ to $_s^s\text{pH}$ or $_w^s pK_a$ to $_s^s pK_a$.

The authors claim that the δ values could be directly used with other electrode systems or by other laboratories, given that the residual liquid junction potential of the respective system is negligible [74–76]. This can be a convenient way to convert from the $_w^s\text{pH}$ scale to $_s^s\text{pH}$ scale as Espinosa et al. have described [73].

4.5.6.2 *Effect of Organic on Modifier Ionization–pH Shift.* Typically, most reversed-phase HPLC methods use monoprotic or polyprotic acidic buffers. The determination of pK values of acids in acetonitrile/water mixtures and methanol/water mixtures have been reviewed in the literature [61–65, 67, 77] Several excellent reviews have been published on this topic by Roses and Bosch. [74, 75] The $_s^s\text{pH}$ can be determined directly from $_w^s\text{pH}$ by the following relationship as shown in equation (4-19).

For example, seven aqueous solutions of 10 mM dipotassium monohydrogen phosphate (adjusted with phosphoric acid) with initial $_w^w\text{pH}$ (pH 2–9) were prepared in five acetonitrile/water compositions ranging from 10 to 50 v/v% of acetonitrile, and the $_w^s\text{pH}$ was determined. $_s^s\text{pH}$ was calculated using equation (4-19), and the final values are shown in Table 4-6. In Figure 4-27 the $_s^s\text{pH}$ values were plotted versus the acetonitrile concentration ranging from 10 to 50 v/v%. It was shown that the $_s^s\text{pH}$ of the eluent increases with an increase of acetonitrile content. For the buffers that had initial $_w^w\text{pH}$ values between 2 and 9, the slopes of the plots of $_s^s\text{pH}$ versus v/v% acetonitrile concentration are essentially independent of the initial aqueous pH with $R^2 > 0.98$. There is an increase (or upward shift of the pH) of approximately 0.22 pH units for every 10 v/v% of acetonitrile added, indicating a change in the acidic modifier's dissociation constant (change in the modifier's pK_a).

The change in the mobile-phase pH of a particular buffer as a function of the organic compositions will be referred to as the *pH shift* in the following sections in this book. For acidic buffers/modifiers, the relative increase in the pH will be dependent upon the type and concentration of acidic modifier and

TABLE 4-6. ^s_s**pH Values of 10 mM Monohydrogen Phosphate Buffer Adjusted with Phosphoric Acid in Various MeCN Compositions**

v/v% MeCN				^s_spH^a			
0	2.09	3.11	4	5.12	6.11	7.01	8.9
10	2.3	3.28	4.34	5.47	6.48	7.24	9.06
20	2.48	3.46	4.56	5.69	6.69	7.46	9.26
30	2.65	3.64	4.74	5.87	6.87	7.64	9.48
40	2.96	3.91	4.92	6.12	7.12	7.89	9.75
50	3.24	4.18	5.32	6.34	7.33	8.14	
Slope	0.023	0.021	0.024	0.024	0.023	0.022	0.021
R^2	0.986	0.988	0.982	0.991	0.988	0.998	0.991

^{a}Corrected for delta at each organic composition using δ_{avg} values from reference 73.

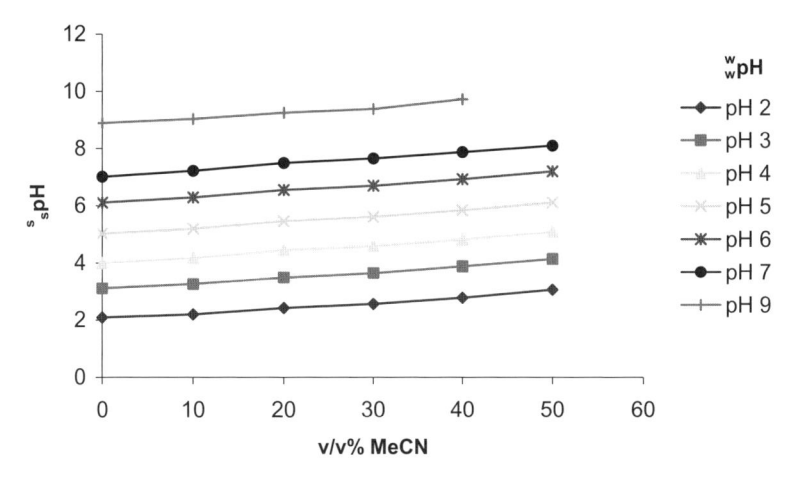

Figure 4-27. Effect of concentration of acetonitrile on the pH shift for a 10 mM mono-hydrogen phosphate buffer.

organic eluent. However, several other typically used acidic buffers such as acetate, dihydrogen phosphate, dihydrogen citrate, hydrogen citrate, and citrate and boric acid show a similar pH shift with an increase of acetonitrile organic modifier. These acids bear a similar trend in increase of the ^s_spH with increasing amounts of v/v% acetonitrile. The ^s_spH values determined by Espinosa et al. and Subirats et al. in the acetonitrile concentration range from 10 to 60 v/v% are shown in Table 4-7 and correspond to approximately 0.2–0.3 pH units increase per 10 v/v% acetonitrile [64, 78]. A conservative value of 0.2 pH units per 10 v/v% increase in acetonitrile will be used throughout the text to denote the acidic modifier pH shift of the aqueous portion of the mobile phase with the addition of acetonitrile.

The variation of the pK_a of acidic modifiers with the addition of methanol to the aqueous portion of the mobile phase bears a similar upward trend.

TABLE 4-7. $_s^s$pH Values of the Acids Studied as Buffer Components in Acetonitrile/Water Mixtures[a]

10 mM Buffer[b]	$_s^s$pH in % acetonitrile by volume							Slope per 10 v/v% MeCN	R^2
	0	10	20	30	40	50	60		
Acetic/acetate	4.74	4.94	5.17	5.44	5.76	6.15	6.62	0.31	0.978
Phosphoric/ dihydrogen phosphate	2.21	2.39	2.62	2.8	3.11	3.42	3.75	0.26	0.986
Dihydrogen phosphate/ hydrogen phosphate	7.23	7.4	7.6	7.82	8.08	8.38	8.73	0.25	0.985
Citric/dihydrogen citrate	3.16	3.31	3.49	3.68	3.9	4.16	4.45	0.21	0.987
Dihydrogen citrate/ hydrogen citrate	4.79	4.95	5.14	5.35	5.6	5.91	6.28	0.24	0.979
Hydrogen citrate/citrate	6.42	6.62	6.85	7.11	7.4	7.74	8.13	0.28	0.987

[a]Values in the table are from references 64 and 78.
[b]Adjusted pH with either concentrated HCl or NaOH.

However, the variation in the positive slope for $_s^s$pK_a values in methanol/water mixtures is smaller than for acetonitrile/water mixtures because methanol is more similar to water. The typical increase in $_s^s$pH values of acidic modifiers in methanol/water mixtures is about 0.15 pH units per 10 v/v% methanol.

4.5.6.3 Acidic Modifiers: pH Shift and Correlation with Dielectric Constant.

The $_s^s$pK variation of acids is related to changes in the electrostatic interactions upon addition of organic media. pH is the negative log of the concentration of protons that are the result of the acid dissociation (for acidic buffers). With the increase of the content of organic molecules in the solution, the dissociation is decreasing (with the decrease of dielectric constant the stabilization of dissociated ions is decreased), thus increasing the solution pH. As was discussed by Espinosa et al. [79], the pH shift occurs because an increase in organic leads to a change of the dielectric constant of the hydro-organic solution. As the organic content increases, the dielectric constant of the mobile phase decreases. In our studies with a decrease in the dielectric constant of the eluent composition (increasing acetonitrile composition) the $_s^s$pK_a of the dipotassium monohydrogen buffer was observed to increase in a linear fashion at all pHs (Figure 4-28). As the organic content increases, the dielectric con-

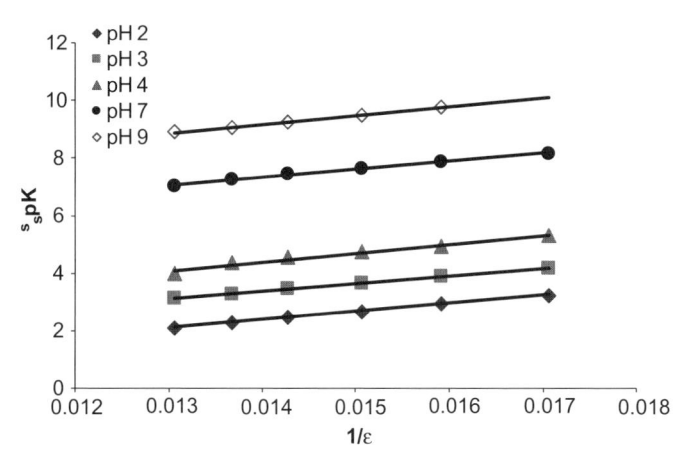

Figure 4-28. Influence of the dielectric constant on the ${}_s^s pK_a$ of acidic buffer from pH 2 to 9.

stant of the mobile phases decreases. The dielectric constant is expected to influence the position of the equilibrium in ionic secondary chemical equilibria of acidic compounds [80–83]. The solvent has the ability to disperse electrostatic charges via ion–dipole interactions, which is inversely proportional to the dielectric constant of the solvent composition. The lower the dielectric constant, the lower the ionization constant of the acid, K_a, and consequently greater pK_a values are obtained.

4.5.6.4 Basic Modifiers: pH Shift. Basic mobile-phase modifiers such as NH_4^+/NH_3 (${}_w^w pH$ 9) and $BuNH_3^+/BuNH_2$ (${}_w^w pH$ 10) show a decrease in their pK_a values with increasing organic content [74]. These basic modifiers have an average pH decrease on the order of –0.05 to –0.1 pH units per 10 v/v% acetonitrile. The minimum of the ${}_s^s pH$ values as a function of acetonitrile composition for basic modifiers is reached at approximately 30–50 v/v% MeCN. Upon further increase in MeCN concentration the ${}_s^s pH$ of the basic modifier will increase. For example, ammonium/ammonia basic modifier ${}_s^s pH$ values in acetonitrile/water mixtures are: 0% MeCN: *9.29*, 10% MeCN: *9.27*, 20% MeCN: *9.21*, 30% MeCN: *9.17*, 40% MeCN: *9.19*, 50% MeCN: *9.21*, 60% MeCN: *9.34* [64]. For $BuNH_3^+/BuNH_2$ (${}_w^w pH$ 10), basic modifier ${}_s^s pH$ values in acetonitrile/water mixtures are: 0% MeCN: *10.00*, 20% MeCN: *9.78*, 40% MeCN: *9.63*, 60% MeCN: *9.79* [64]. For basic modifiers a decrease in pH is also observed with increase of methanol content on the order of 0.1 pH units per 10 v/v% methanol.

4.5.6.5 Amphoteric Buffers: pH Shift. When buffers that contain both ionizable cations and anions such as ammonium acetate or ammonium phosphate are used, the change in the buffer pH (pH shift) is dependent on the pH of the starting buffer. For example, with an ammonium acetate buffer with the addition of organic modifier, there is an upward pH shift up to $_w^wpH$ 6 (due to acetate counterion) and a downward pH shift when $_w^wpH > 7$ (due to ammonium counterion). These effects are prevalent in both acetonitrile/water and methanol/water systems, as shown in Tables 4-8 and 4-9, respectively. The changes in pH slopes are (a) approximately constant and positive for $_w^wpH <$

TABLE 4-8. Calculated $_s^spH$ Values of 50 mM Ammonium Acetate at Different Acetonitrile/Water Compositions[a]

Buffer	$_s^spH$ in% MeCN by volume							Slope per 10 v/v% MeCN	R^2
	0	10	20	30	40	50	60		
50 mM Acetic acid	4.67	4.86	5.08	5.34	5.68	6.04	6.46	0.30	0.981
50 mM Amm. acetate	2.67	2.8	2.98	3.16	3.5	3.84	4.23	0.26	0.964
50 mM Amm. acetate	3.01	3.15	3.33	3.54	3.86	4.19	4.6	0.26	0.968
50 mM Amm. acetate	4.06	4.21	4.43	4.66	5.01	5.33	5.75	0.28	0.977
50 mM Amm. acetate	5.07	5.23	5.49	5.74	6.11	6.43	6.88	0.30	0.981
50 mM Amm. acetate	6.07	6.24	6.48	6.71	7.05	7.33	7.69	0.27	0.988
50 mM Amm. acetate	6.96	7.06	7.16	7.29	7.5	7.67	7.94	0.16	0.969
50 mM Amm. acetate	7.94	7.9	7.85	7.81	7.9	7.97	8.15	−0.04[a]	0.998[b]
50 mM Amm. acetate	8.94	8.88	8.84	8.76	8.8	8.8	8.87	−0.06[a]	0.984[b]
50 mM Amm. acetate	9.95	9.88	9.85	9.76	9.8	9.8	9.88	−0.06[a]	0.968[b]

[a]All $_w^spH$ data were obtained from reference [84], and $_s^spH$ values were calculated using δ values from reference 73. The pHs were adjusted with formic acid and ammonium hydroxide.
[b]The slope and R^2 were determined from 0–30 v/v% acetonitrile.

TABLE 4-9. Calculated $_s^spH$ Values of 50 mM Ammonium Acetate at Different Methanol/Water Compositions

Buffer	$_s^spH$ in% MeOH by Volume							Slope per 10 v/v% MeOH	R^2
	0	10	20	30	40	50	60		
10 mM Acetic acid	4.76	4.96	5.15	5.36	5.57	5.8	6.03	0.21	0.999
50 mM Amm. acetate	2.67	2.8	2.94	3.06	3.22	3.37	3.55	0.15	0.997
50 mM Amm. acetate	3.01	3.15	3.24	3.36	3.5	3.65	3.86	0.14	0.986
50 mM Amm. acetate	4.06	4.17	4.26	4.38	4.52	4.71	4.92	0.14	0.976
50 mM Amm. acetate	5.07	5.16	5.28	5.42	5.6	5.8	6.03	0.16	0.977
50 mM Amm. acetate	6.07	6.15	6.26	6.4	6.57	6.75	6.93	0.15	0.983
50 mM Amm. acetate	6.96	7.0	7.05	7.05	7.11	7.16	7.25	0.04	0.950
50 mM Amm. acetate	7.94	7.9	7.8	7.69	7.63	7.56	7.53	−0.07	0.979
50 mM Amm. acetate	8.94	8.89	8.79	8.66	8.56	8.44	8.34	−0.10	0.992
50 mM Amm. acetate	9.95	9.92	9.79	9.68	9.59	9.47	9.35	−0.10	0.989

[a]All $_w^spH$ data were obtained from reference 84, and $_s^spH$ values were calculated using δ values from Table 4-5. The pHs were adjusted with formic acid and ammonium hydroxide.

6 where the solution is buffered by the acetic/acetate pair in the solution and (b) constant and negative for $_w^w$pH > 7 where the solution is buffered by the ammonium/ammonia pair.

Also, the organic content is expected to influence the dissociation constant of acidic analytes, resulting in an increase in the acidic analyte pK_a and this could be described as the *acidic analyte pK_a shift*, which is discussed in Section 4.6. On the other hand, the organic eluent will affect the dissociation of basic analytes in the opposite direction, resulting in a decrease in the basic analyte pK_a, and is discussed in the Section 4.6 as the *basic analyte pK_a shift*.

4.5.7 Analyte Dissociation Constants

The pK_a is an important physicochemical parameter. The analyte pK_a values are especially important in regard to pharmacokinetics (ADME—absorption, distribution, metabolism, excretion) of xenobiotics since the pK_a affects the apparent drug lipophilicity [59]. Potentiometric titrations and spectrophometric analysis can be used for pK_a determination; however, if the compound is not pure, is poorly soluble in water, and/or does not have a significant UV chromophore and is in limited quantity, its determination may prove to be challenging.

Dissociation constants of ionizable components can be determined using various methods such as potentiometric titrations [85] CE, NMR, [86] and UV spectrophotometric methods [87]. Potentiometric methods have been used in aqueous and hydro-organic systems; however, these methods usually require a large quantity of pure compound and solubility could be a problem. Potentiometric methods are not selective because if the ionizable impurities in an impure sample of the analyte have a pK_a similar to that of the analyte, this could interfere with determining the titration endpoint. If the titration endpoint is confounded, then these may lead to erroneous values for the target analyte pK_a.

Liquid chromatography has also been widely used for the determination of dissociation constants [88–92] since it only requires small quantity of compounds, compounds do not need to be pure, and solubility is not a serious concern. However, the effect of an organic eluent modifier on the analyte ionization needs to also be considered. It has been shown that increase of the organic content in hydro-organic mixture leads to suppression of the basic analyte pK_a and leads to an increase in the acidic analyte pK_a compared to their potentiometric pK_a values determined in pure water [74].

Knowledge of pK_a for the target analyte and related impurities is particularly useful for commencement of method development of HPLC methods for key raw materials, reaction monitoring, and active pharmaceutical ingredients. This practice leads to faster method development, rugged methods, and an accurate description of the analyte retention as a function of pH at varying organic compositions. Relationship of the analyte retention as function of mobile-phase pH ($_s^s$pH) is very useful to determine the pK_a of the particular analyte in the hydroorganic mixture and can be extrapolated to predict the $_w^w pK_a$ of the analyte. Reversed-phase HPLC in isocratic mode can be used for the pK_a determination of new drug compounds.

4.5.8   Determination of Chromatographic pK_a

The general procedure for the chromatographic determination of the pK_a is to run at least 5 pH experiments isocratically to construct a pH (on the x-axis) versus retention factor (or retention, on the y-axis) plot. The concentration of organic in the mobile phase should be selected to elute the most hydrophilic species (ionized form) with a $k' > 1$. If the compound is acidic, the elution of the fully ionized species will be obtained at 2 pH units greater than the analyte pK_a. If the compound is basic, the elution of the fully ionized species will be obtained at 2 pH units less than the analyte pK_a. The organic composition chosen must also be able to elute the neutral species within a reasonable retention time (i.e., <30 min). A short column with narrow internal diameter (i.e., 5.0×3.0 mm, using flow rate of 1.5 mL/min) that is stable from $_w^w pH$ 2–11 should be used for these studies. The mobile phase could be made from 15 mM potassium phosphate, and the pH can be adjusted with either HCl or NaOH from 2 to 11.

   If the target analyte is a basic compound, then the lowest pH mobile phase could be run first, to obtain the retention of the ionized species. At least 25 column volumes (1 column volume = $\pi \times$ radius of column$^2 \times$ length of column $\times 0.7$) should pass through the column in order to obtain stable retention at each pH used. There is no need to run blank injections. Multiple injections of the analyte should be made; and once a stable retention is obtained at a particular pH, the next pH can be evaluated. This is repeated throughout the whole pH range from low pH to high pH. A representative chromatogram overlay at the various pH values is shown in Figure 4-29 for a basic compound (compound M). The retention factor (or retention) is then plotted versus the $_s^s pH$ of the mobile phase. A representative plot of the retention dependencies versus the $_s^s pH$ of the mobile phase at 30 v/v% acetonitrile compositions is shown in Figure 4-30. Using nonlinear regression analysis software, the $_s^s pK_a$ of the analyte can be determined. For the example given in Figure 4-29 the $_s^s pK_a$ of compound M at 30 v/v% acetonitrile was determined to be 3.9 (Figure 4-30). Knowing the $_s^s pK_a$ of the analyte and the type and concentration of organic modifier used, the $_w^w pK_a$ of the analyte can be calculated. For acetonitrile/water systems the $_w^w pK_a$ can be calculated by the following empirical formula for basic and acidic compounds:

$$_w^w pK_a = {_s^s pK_a} + (x\% \text{ organic}) * B \qquad \text{(basic compounds)} \qquad (4\text{-}23)$$

$$_w^w pK_a = {_s^s pK_a} - (x\% \text{ organic}) * A \qquad \text{(acidic compounds)} \qquad (4\text{-}24)$$

where $B = 0.02$ (corresponds to basic analyte pK_a shift per 10 v/v% MeCN) and $A = 0.03$ (corresponds to acidic analyte pK_a shift per 10 v/v% MeCN).

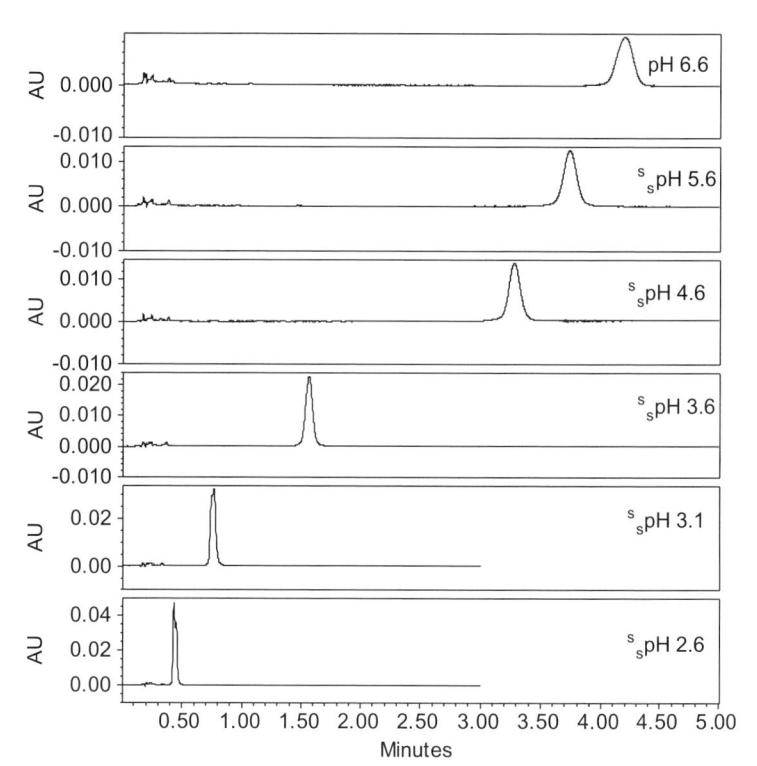

Figure 4-29. Column: Acquity BEH C18 1.7 μm, 2.1∗50 mm, flow rate, 0.8 mL/min, temperature, 35°C, injection 2-μL full loop, run time 3–5 min, detection 215 nm. Strong wash: 0.1% NH$_4$OH 50/50 MeCN/H$_2$O. Weak wash: 90/10 H$_2$O/MeCN. Mobile phase A: 15 mM K$_2$HPO$_4$ adjusted with HCl. Mobile phase B: MeCN. Starting pressure: ~9000 psi, isocratic 30 v/v% MeCN.

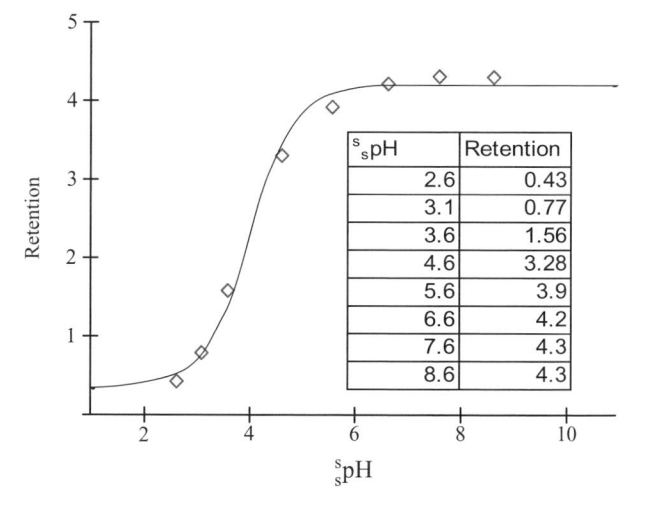

$_s^s$pH	Retention
2.6	0.43
3.1	0.77
3.6	1.56
4.6	3.28
5.6	3.9
6.6	4.2
7.6	4.3
8.6	4.3

Figure 4-30. Retention versus $_s^s$pH for compound M at 30 v/v% acetonitrile.

TABLE 4-10. pK Values for Compound M at Various Organic Compositions

	pK_a		
pK_a	30 v/v%	40 v/v%	50 v/v%
$_s^s pK_a$	3.9	3.65	3.5
Estimated $_w^w pK_a$	4.5	4.45	4.5

The basic and acidic analyte pK_a shift values will be discussed in Section 4.6. Using equation (4-23), the $_w^w pK_a$ at 30 v/v% acetonitrile was estimated to be 4.5. $_w^w pK_a = 3.9 + (30$ v/v% MeCN)*0.02 = 4.5. Similar pH studies were conducted with 40 and 50 v/v% MeCN compositions, and the respective $_s^s pK_a$ (experimental) and $_w^w pK_a$ (predicted) values are shown in Table 4-10. These results agree well with the potentiometric value of 4.4 for this compound M.

4.6 EFFECT OF ORGANIC ELUENT COMPOSITION ON ANALYTE IONIZATION

As discussed in Section 4.5.6, the increase of the organic content in hydro-organic mixture leads to suppression of the basic analyte pK_a and to an increase in the acidic analyte pK_a. Accounting for the pH shift of the mobile phase and analyte pK_a shift upon the addition of organic modifier is necessary for the chromatographer to analyze the ionogenic samples at their optimal pH values.

In order to avoid any secondary equilibrium effects on the retention of ionogenic analytes, it is preferable to use the mobile-phase pH either two units greater or less than the analyte pK_a in the particular hydro-organic media that is employed. Therefore, one must account for the pH shift of the mobile phase upon the addition of the organic modifier for a proper description of the ionogenic analyte retention process. However, the effect of organic eluent modifier on the analyte ionization needs to also be considered. It has been shown that increase of the organic content in hydro-organic mixture leads to suppression of the basic analyte pK_a and an increase in the acidic analyte pK_a compared to their potentiometric pK_a values determined in pure water [74, 79]. Accounting for the pH shift of modifier in the mobile phase and analyte pK_a shift upon the addition of organic modifier, this will allow the chromatographer to analyze the ionogenic samples at their optimal pH values.

4.6.1 Effect of Organic Modifier on Basic Analyte pK_a Shift

In order for proper description of the basic analyte retention versus the mobile-phase $_s^s pH$, the pH shift of the aqueous portion of the mobile phase must be

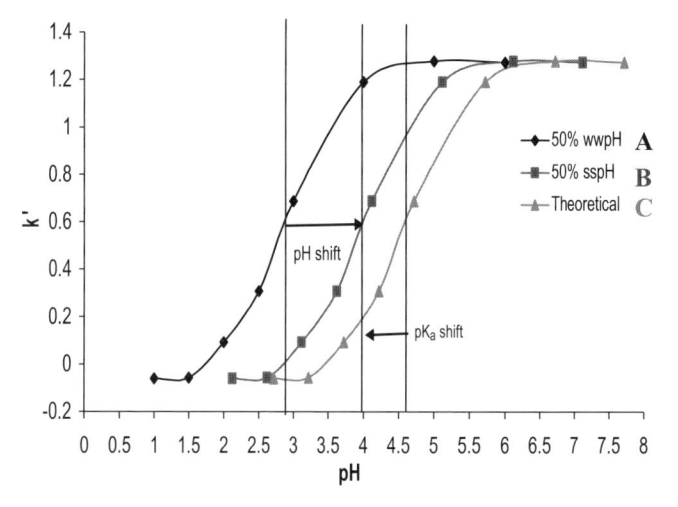

Figure 4-31. Retention versus $_w^w$pH and $_s^s$pH for aniline at 50 v/v% MeCN. (15 mM phosphate buffer adjusted with phosphoric acid.) See color plate.

taken into account. Figure 4-31 is a plot of the retention factor of aniline plotted versus two different pH scales: $_w^w$pH (Figure 4-31, line A) and $_s^s$pH (Figure 4-31, line B). Moreover, a theoretical curve of the retention dependence versus pH of the mobile phase was constructed for aniline, based on its potentiometric pK_a of 4.6 in a purely aqueous system (Figure 4-31, line C). The inflection point of the dependence of k' versus pH corresponds to the analyte pK_a at a particular hydro-organic composition. As can be seen, the plot of retention factor versus. $_w^w$pH (Figure 4-31, line A) does not correspond to pK_a from the theoretical curve (Figure 4-31, line C). The pK_a difference between these two curves is actually the combination of two individual shifts occurring in opposing directions: acidic mobile-phase upward pH shift and the basic analyte downward pK_a shift. The difference between the $_w^w$pH and $_s^s$pH curve is due to the pH shift of the aqueous portion of the acidic mobile phase which is caused by a change in the dissociation in the acidic buffer in the particular hydro-organic eluent. After the retention factor is plotted versus $_s^s$pH (Figure 4-31, line B), the pK_a determined still does not correspond to the pK_a from the theoretical curve (Figure 4-31, line C). The difference between the $_s^s$pH curve and the theoretical curve could be attributed to a change of the basic analyte ionization state at a particular hydro-organic composition upon addition of acetonitrile in the mobile phase, and this is denoted as the basic analyte pK_a shift.

Figure 4-32 is a plot of the retention factor of aniline versus the $_s^s$pH of the hydro-organic mixture (pH shift of the aqueous portion of the mobile phase is accounted for) from 10 to 50 v/v% MeCN using the values from Table 4-11. In the graph for all organic compositions a sigmoidal dependence of retention factor versus $_s^s$pH is obtained and the plateau regions are the limiting factors for the fully ionized and neutral forms of the analyte. The inflection point of

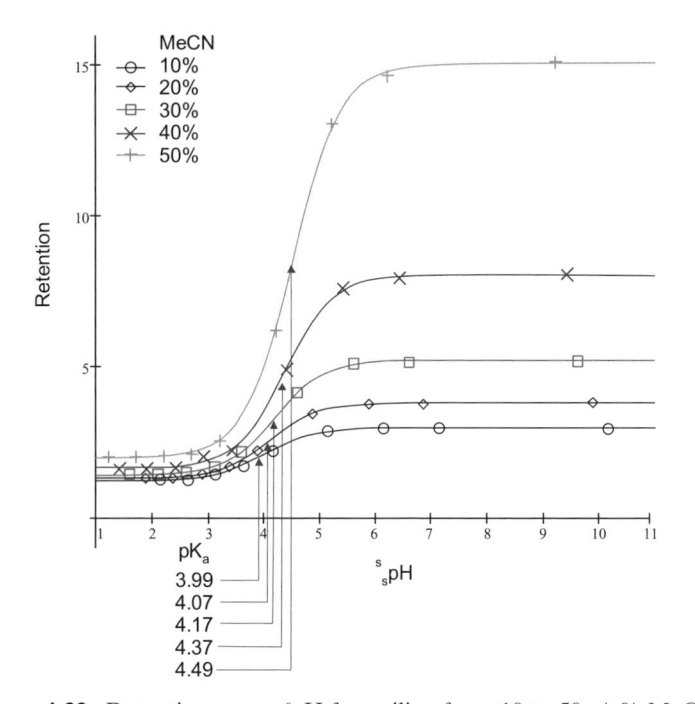

Figure 4-32. Retention versus $_s^s$pH for aniline from 10 to 50 v/v% MeCN.

TABLE 4-11. Retention Volume of Aniline as a Function of $_s^s$pH (10–50 v/v% Acetonitrile)

$_s^s$pH	50	$_s^s$pH	40	$_s^s$pH	30	$_s^s$pH	20	$_s^s$pH	10
2.62	1.225	**2.36**	1.294	**2.08**	1.406	**1.89**	1.587	**1.69**	2.002
3.12	1.419	**2.86**	1.393	**2.58**	1.461	**2.39**	1.624	**2.19**	2.043
3.62	1.701	**3.36**	1.658	**3.08**	1.645	**2.89**	1.987	**2.69**	2.069
4.12	2.193	**3.86**	2.21	**3.58**	2.145	**3.39**	2.182	**3.19**	2.549
5.12	2.848	**4.86**	3.42	**4.58**	4.11	**4.39**	4.885	**4.19**	6.172
6.12	2.961	**5.86**	3.749	**5.58**	5.081	**5.39**	7.572	**5.19**	13.04
7.12	2.954	**6.86**	3.76	**6.58**	5.136	**6.39**	7.925	**6.19**	14.64
10.12	2.961	**9.86**	3.774	**9.58**	5.18	**9.39**	8.043	**9.19**	15.115

the dependence of k versus $_s^s$pH corresponds to the analyte $_s^s$pK_a at a particular hydro-organic composition.

In Figure 4-33 the analyte $_w^w$pK_a and $_s^s$pK_a is plotted versus 0–50 v/v% MeCN. It is shown that even after correcting for the pH shift of the mobile phase upon addition of organic at each organic composition, the chromatographic $_s^s$pK_a at

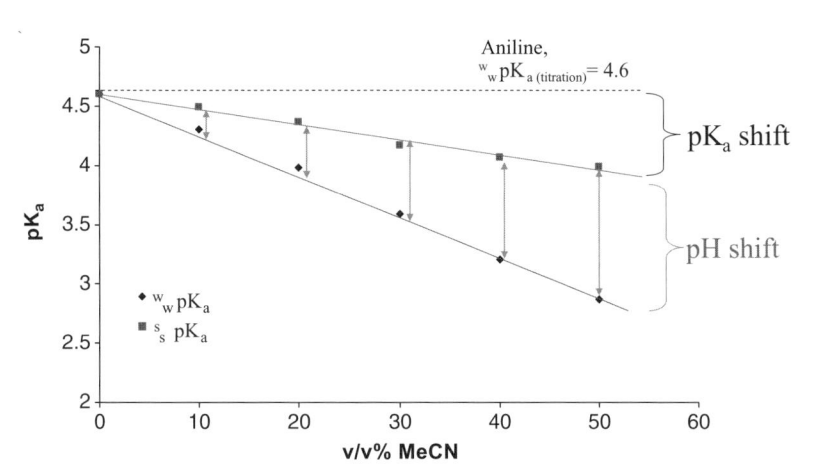

Figure 4-33. Effect of organic composition on aniline pK_a shift.

TABLE 4-12. $_s^spK_a$ **Values of Basic Compounds as a Function of Acetonitrile Composition**

	v/v% MeCN						Slope	R^2
	$pK_a{}^*$	$_s^spK_a$	$_s^spK_a$	$_s^spK_a$	$_s^spK_a$	$_s^spK_a$		
	0	10	20	30	40	50		
Aniline, pK_a 4.6	**4.6**	4.49	4.37	4.17	4.07	3.99	−0.013	0.985
4-Fluoro aniline pK_a, **4.65**	**4.65**	4.49	4.35	4.13	4.01	3.92	−0.015	0.988
3-Bromoaniline pK_a, **3.53**	**3.53**	3.35	3.14	2.88	2.70	2.37	−0.023	0.992
3-Chloroaniline pK_a, **3.52**	**3.52**	3.34	3.18	2.92	2.73	2.43	−0.021	0.991
2-Fluoro aniline pK_a, **3.2**	**3.2**	3.05	2.84	2.59	2.37		−0.019	0.993
4-Chloroaniline pK_a, **3.99**	**3.99**	3.88	3.66	3.43	3.29	3.15	−0.018	0.989
3-Fluoroaniline pK_a, **3.58**	**3.58**	3.43	3.25	3.02	2.86	2.64	−0.019	0.996
4-Bromoaniline pK_a, **3.88**	**3.88**	3.78	3.55	3.32	3.16	3.01	−0.018	0.989

* potentiometric

all of the organic compositions do not correlate to analytes potentiometric pK_a value determined in the aqueous solvent (pK_a 4.6). A decrease of 0.13 pK_a units per 10% v/v MeCN for aniline was determined (basic analyte pK_a shift). Similar negative slopes for other monosubstituted aromatic amines were determined (~ 0.13–0.23 pK_a units per 10% MeCN) were obtained. (Table 4-12). Linear relationships for $_s^spK_a$ values in acetonitrile/water mixtures up to

50 v/v% acetonitrile were obtained ($R^2 > 0.98$) (Table 4-12). The downward change in pK_a as a function of the v/v% MeCN between 0 and 50 v/v% MeCN agreed well with the $_s^s pK_a$ values determined by Espinosa et al. [93] for aniline, 0.14 pK_a unit decrease per 10 v/v% acetonitrile, and 4-chloro aniline 0.18 pK_a unit decrease per 10 v/v% acetonitrile.

The analyte pK_a shift upon addition of acetonitrile can be estimated by using the slope of this dependence (0.2 pK_a units decrease per 10% MeCN). This will be denoted as the basic analyte pK_a shift for further discussions in the book. The decrease in the analyte pK_a for basic compounds in acetonitrile/water has been attributed to the breaking of the water structure by addition of organic solvent which consequently changes its ionization equilibria [79, 76, 94]. Therefore, specific solvation effects for certain classes of compounds could lead to different slopes of the change in the pK_a as a function of the type and concentration of organic composition. Roses et al. has published parameters for prediction of the slopes and intercepts of the linear correlations between the $_s^s pK_a$ values in acetonitrile/water mixtures and the $_w^w pK_a$ values in pure water for aliphatic carboxylic acids, aromatic carboxylic acids, phenols, amines, and pyridines [93]. Similar parameters have been determined for this family of compounds for methanol/water mixtures [80]. Using these parameters for each family of compounds for a particular type of organic, the a_s and b_s terms could be determined and the following empirical equation was determined:

$$_s^s pK_a = a_s \cdot {}_w^w pK_a + b_s \qquad (4\text{-}25)$$

This empirical equation could be used to estimate the analyte $_s^s pK_a$ values for different classes of acidic and basic compounds in particular acetonitrile/water or methanol/water compositions.

4.6.2   Effect of Organic Modifier on Acidic Analyte pK_a Shift

In order for proper description of the acidic analyte retention versus the mobile-phase pH, the pH shift of the aqueous portion of the mobile phase must be taken into account. Plot of the retention factor of 2-4dihydroxybenzoic acid versus two different pH scales [$_w^w$pH (Figure 4-34, line A) and $_s^s$pH (Figure 4-34, line B)] is shown in Figure 4-34. A theoretical curve (Figure 4-34, line C) of the retention dependence versus pH of the mobile phase was constructed for 2,4-dihydroxy benzoic acid, based on its potentiometric pK_a of 3.2 in a purely aqueous system. The inflection point of the dependence of k' versus pH corresponds to the analyte $_s^s pK_a$ at a particular hydro-organic composition. The difference between the $_w^w$pH (Figure 4-34, line A) and the $_s^s$pH curve (Figure 4-34, line B) for the acidic analyte is due to the difference between the pH of aqueous portion of the mobile phase ($_w^w$pH) and the actual mobile phase pH ($_s^s$pH).

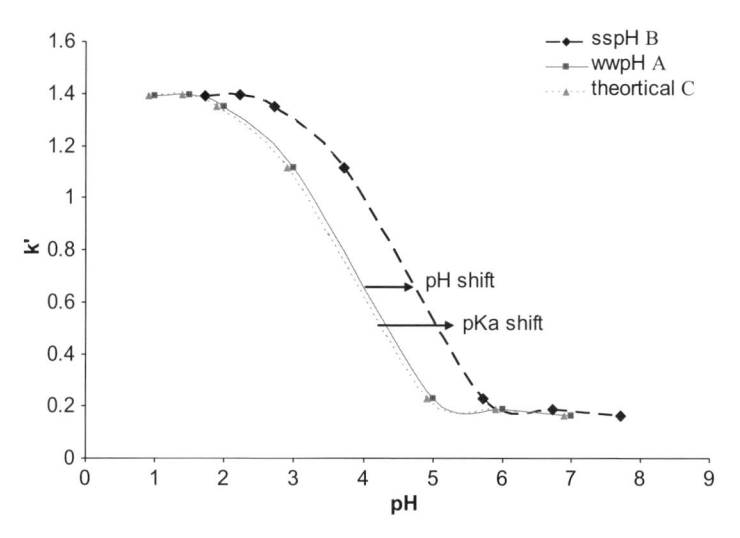

Figure 4-34. Retention versus $_w^w$pH and $_s^s$pH for 2,4-dihydroxybenzoic acid at 35 v/v% MeCN. (15 mM phosphate buffer adjusted with phosphoric acid.)

However, the $_s^s$pK_a obtained after correction for the pH shift of the mobile phase (Figure 4-34, line B) does not correspond to $_w^w$pK_a from the theoretical curve (Figure 4-34, line C). The overall difference between the theoretical curve and $_s^s$pH curve is due to the pK_a shift of the acidic analyte. The difference between the $_s^s$pH curve and the theoretical curve is due to a change of the acidic analyte ionization state at a particular hydro-organic composition upon addition of acetonitrile in the mobile phase. In essence, there is a larger pK_a shift for the 2,4-dihydroxybenzoic acid than for the phosphoric acid (used as a buffer).

Dependencies of 2,4-dihydroxybenzoic acid retention factors versus the $_s^s$pH of the hydro-organic mixture (pH shift of the aqueous portion of the mobile phase is accounted for) at different organic compositions (from 10 to 35 v/v% MeCN) are shown in Figure 4-35. In this graph a sigmoidal dependence of retention factor versus pH is obtained and the plateau regions are the limiting factors for the fully ionized and neutral forms of the analyte. The inflection point of the dependence of k' versus $_s^s$pH corresponds to the acidic analyte pK_a at a particular hydro-organic composition.

In Figure 4-36 the acidic analyte $_w^w$pK_a and $_s^s$pK_a values determined as a function of acetonitrile composition from 10 to 35 v/v% MeCN are shown. It is shown that even after correcting for the pH shift of the mobile phase upon addition of organic, the chromatographic $_s^s$pK_a values does not correlate to the pK_a that was determined by titration in aqueous solvents, $_w^w$pK_a. An increase of 0.27 pK_a units per 10% v/v MeCN for 2,4-dihydroxybenzoic acid was determined. A similar slope for other mono- and disubstituted aromatic benzoic

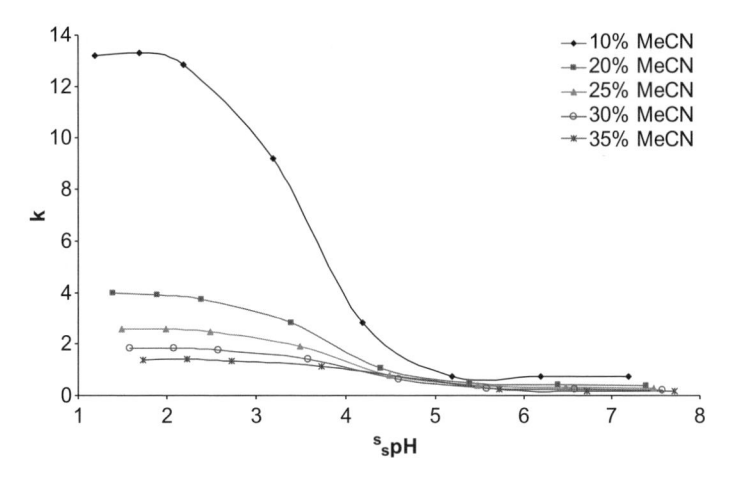

Figure 4-35. Retention versus $_w^w$pH and $_s^s$pH for 2,4-dihydroxybenzoic acid from 10 to 35 v/v% MeCN.

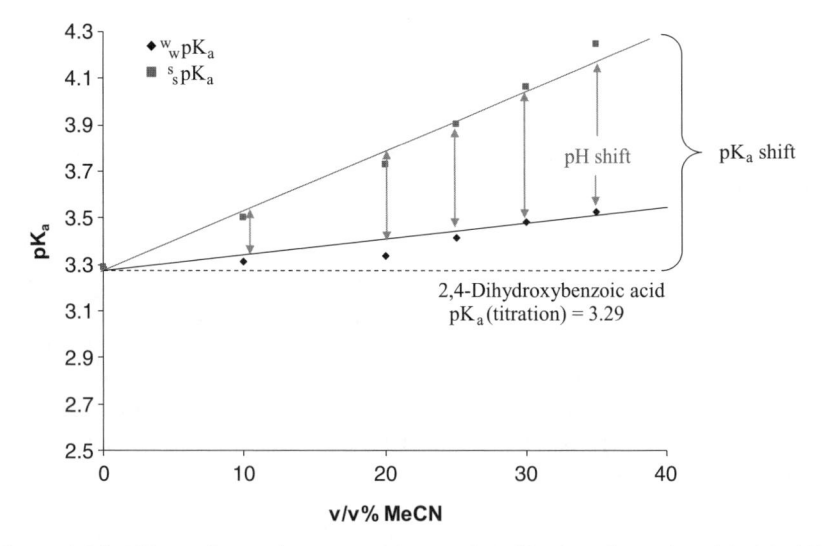

Figure 4-36. Effect of organic composition on 2,4-dihydroxybenzoic acid pK_a shift.

acids was determined ($\sim$ 0.27–0.42 pK_a units per 10% MeCN, Table 4-13). The average upward slope of 0.3 pK_a units upon 10 v/v% addition of MeCN will be denoted as the acidic analyte pK_a shift further in the book.

Also for weakly acidic analytes such as mono- and disubstituted phenols [74, 76], increases of 0.2–0.3 pK_a units per 10 v/v% acetonitrile were obtained: phenol, 0.33 pK_a units; 3,5-dicholorphenol, 0.21 pK_a units; 3-bromophenol, 0.32 pK_a units; 4-chlorophenol, 0.30 pK_a units per 10% acetonitrile.

TABLE 4-13. $^s_s pK_a$ **Values of Acidic Compounds as a Function of Acetonitrile Composition**

Compound	pK_a*	$^s_s pK_a$	$^s_s pK_a$	$^s_s pK_a$	$^s_s pK_a$	$^s_s pK_a$	Slope	R^2
				v/v% MeCN				
	0	10	20	25	30	35		
2,4-Dihydroxybenzoic	3.29	3.50	3.73	3.90	4.06	4.25	0.027	0.986
Benzoic acid	4.20	4.41	4.70	4.87	5.04	5.23	0.030	0.990
Salicylic	3.00	3.18	3.39	3.53	3.67	3.84	0.024	0.985
2,4,5-Trimethoxybenzoic	4.24	4.89	5.22	5.40	5.56	5.77	0.042	0.979
2,3,4-Trimethoxybenzoic	4.24	4.40	4.75	4.94	5.12	5.30	0.031	0.982
2,3,4-Trihydroxybenzoic	3.30	3.43	3.78	3.95	4.17	4.28	0.030	0.974
2,5-Dihydroxybenzoic	3.01	3.22	3.35	3.47	3.66	3.84	0.022	0.960
3,5-Dihydroxybenzoic	3.96	4.27	4.64	4.70	4.85	5.12	0.032	0.987

* potentiometric pK_a

4.7 SYNERGISTIC EFFECT OF pH, ORGANIC ELUENT, AND TEMPERATURE ON IONIZABLE ANALYTE RETENTION AND SELECTIVITY

Ideally, increasing the concentration of organic in the mobile phase will lead to a decrease in the retention of components in reversed-phase HPLC, along with to a decrease in resolution, while the selectivity should remain constant. The eluent composition should not affect the selectivity between two species if their ionization state is not changing with an increase in the organic composition. However, since the organic eluent does lead to changes in the mobile-phase pH and analyte pK_as, changes in selectivity may be observed at certain pH values. This may lead to only small changes or no changes in ionizable analyte's retention with an increase of organic concentration. In an ideal case, the plot of the logarithm of the retention factor versus the acetonitrile composition should give a linear dependence.

In Figure 4-37 the natural logarithm of the retention factor of aniline at different $^w_w pH$ values is plotted versus the acetonitrile/buffer eluent composition. Different slopes of retention dependence are obtained at a certain eluent pH versus eluent composition. Comparison of aniline analyzed at $^w_w pH \geq 6$, (analyzed in its predominately neutral form) or at $^w_w pH$ 2, (analyzed in its fully ionized form) at all acetonitrile compositions shows that the logarithmic retention of the neutral and fully ionized species varies linearly with the acetonitrile concentration (Figure 4-37). At both these pH regions the analyte does not change its ionization state with an increase of the acetonitrile composition. However, at $^w_w pH$ 3 no significant change in retention is observed from 20 to 50 v/v% MeCN. Due to the upward pH shift of the acidic modifier, the $^s_s pH$ of the eluent at 20% and 50%, respectively, is 3.4 and 4.0. On the other hand, for the basic analyte due to downward pK_a shift upon increase of the organic concentration from 20% to 50%, the $^s_s pK_a$ decreases from 4.4 to 4 (values from

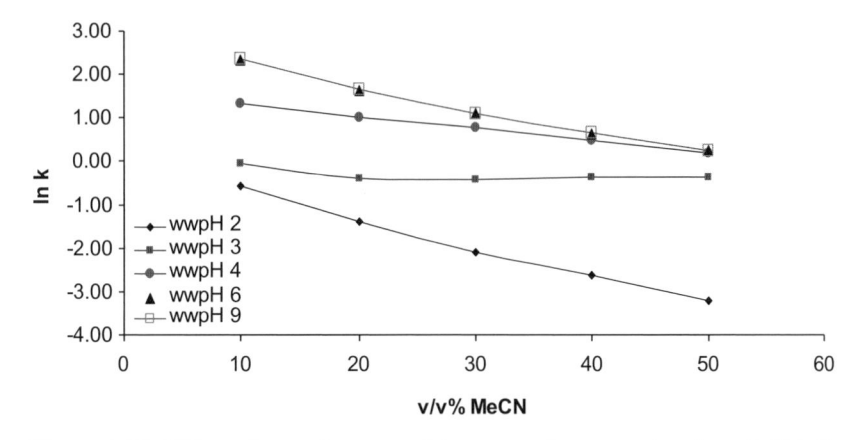

Figure 4-37. Effect of organic composition on analyte retention from $_w^w$pH 2–9.

TABLE 4-14. $_s^s$pK_a for Aniline at 10–50% MeCN

	pot pKa	$_s^s$pKa	$_s^s$pKa	$_s^s$pKa	$_s^s$pKa	$_s^s$pKa
	0	10	20	30	40	50
Aniline, pK_a 4.6	**4.6**	4.49	4.37	4.17	4.07	3.99

Table 4-14). Therefore, with an increase of organic concentration at $_w^w$pH 3 the analyte is being analyzed more progressively in its neutral state. The $_s^s$pK_a of aniline at acetonitrile compositions from 10 to 50 v/v% is shown in Table 4-14.

An increase of the acetonitrile concentration in general leads to an exponential decrease of the analyte retention. However, aniline is becoming less ionized upon increase of organic content in the eluent. Therefore, increasing organic content at a certain aqueous pH has a supposition of two opposite effects on the overall analyte retention: (1) an increase in analyte retention due to a decrease in an analyte ionization since analyte pK_a decreases with increase of organic content, which leads to analysis of analyte in a more neutral state, and (2) a decrease in analyte retention due to decreased analyte interaction with the stationary phase, which decreases hydrophobic interaction. This is clearly observed at $_w^w$pH 3, where no significant change in retention is observed from 20 to 50 v/v% MeCN and the two effects are in essence compensating each other.

The separation selectivity can be significantly affected as a result of different pH shift of different buffers even at the same organic composition. For example, if two buffers are prepared at the same pH, one using an acidic buffer such as phosphate and another using a basic buffer such as ammonia, both at $_w^w$pH 8, the separation of a mixture of ionizable components could be different. This could be attributed to the different mobile-phase $_s^s$pH after the aqueous is mixed with the organic. Espinosa et al. [64] analyzed N,N-dimethyl-

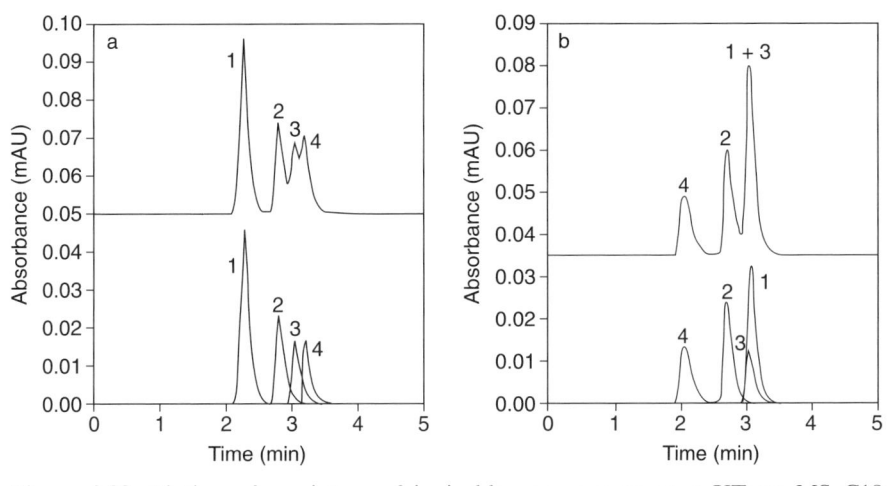

Figure 4-38. Elution of a mixture of ionizable components on a XTerra MS C18 (Waters) column with a 60% ACN mobile phase prepared from $^w_wpH = 8.0$. (A) Phosphate buffer. (B) NH_4^+/NH_3 buffer. Compounds are (1) 2-nitrophenol, (2) 2,4,6-trimethylpyridine, (3) 3-bromphenol, and (4) NN dimethylbenzylamine. (From reference 64, with permission.)

benzylamine ($^w_wpK_a$ 8.8), 2-nitrophenol ($^w_wpK_a$ 7.14), 2,4,6-trimethylpyridine ($pK_a = 7.33$), and 3-bromophenol ($pK_a = 9.00$) on a Waters XTerra MS C18 column using a ^w_wpH 8 ammonia buffer and a phosphate buffer both containing 60 v/v% acetonitrile. At this acetonitrile composition the ^s_wpH of the ammonia buffer is estimated to be about 7.7 [84], and the ^s_wpH of the phosphate buffer is estimated to be about 9.1. Using the phosphate buffer, the basic compound N,N-dimethylbenzylamine (compound 4 in Figure 4-38) was more retained (analyzed predominately in its neutral state) than with the ammonia buffer since in the ammonia buffer the analyte was predominately in a more ionized state. On the other hand, the 2-nitrophenol (compound 1 in Figure 4-38) in the phosphate buffer was less retained (analyzed predominately in its ionized state) while in the ammonia buffer it was more retained, since it was being analyzed in a lessionized state.

4.8   EXAMPLES OF APPLYING pH SHIFT AND ANALYTE pK_a SHIFT RULES

When developing separation methods for analytes with known pK_a values, determination of the starting mobile-phase pH is highly advisable. This estimation may help to avoid strange analyte retention behavior during further method optimization and variation of the mobile-phase composition. Below we include several examples where the methodology of the combined pH and pK_a shift evaluation is outlined.

Example 1. Putting it All Together: Analyzing a Base in Its Ionized Form. For example, 2,4-dimethylpyridine (base), your target analyte, has a $^w_w pK_a$ of 6.7 and the eluent conditions are 50% MeCN and 50% phosphate buffer. What should the pH of the phosphate buffer be in order to obtain the basic analyte in its fully ionized form?

Step 1. First account for the downward pK_a shift for the basic analyte upon addition of organic. For every 10 v/v% increase in acetonitrile, the $^s_s pK_a$ of the analyte decreases by 0.2 pK_a units.

Step 2. Once this $^s_s pK_a$ is determined, the $^s_s pH$ at which the analyte would be fully ionized needs to be determined. This corresponds to 2 pH units less than the $^s_s pK_a$ of 5.7 as shown in step 2 below.

Step 3. Then account for the pH shift of the acidic portion of the mobile phase upon addition of acetonitrile. For every 10 v/v% increase in acetonitrile, the pH of the acidic buffer increases by approximately 0.2 pH units. This would correspond to a 1.0 pH unit increase as shown in step 3 below.

Step 4. Then determine what the maximum $^w_w pH$ of the aqueous portion of the buffer should be prepared at, taking into account the pH shift of the aqueous portion of the mobile phase upon addition of organic as shown in step 4 below. Therefore the optimal pH to analyze this compound would be at aqueous mobile-phase $^w_w pH$ of <2.7 using isocratic mode. This pH is also applicable for gradient mode separations given that the analyte of interest will elute at 50 v/v% of acetonitrile or less.

1:   $6.7 - (5 \times 0.2) = 5.7$ Downward analyte pK_a shift
2:             $5.7 - 2 = 3.7$ $^s_s pH$ at which basic analyte would be fully ionized
3:             $5 \times 0.2 = 1.0$ Upward pH shift of the aqueous acidic buffer
                                                 upon addition of organic (50 v/v% MeCN)
4:             $3.7 - 1.0 = 2.7$ Max $^w_w pH$ of the aqueous portion of the mobile
                                                 phase in order to have analyte in fully ionized
                                                 form at 50 v/v% MeCN

For aromatic amines with pK_a values in the range less than $^w_w pK_a < 5$, this presents a problem. Most reversed-phase columns are not stable at the $^w_w pH$ below 1.5. This limits the chromatographer to work at either lower organic composition (low upward pH shift of the mobile phase and concurrent lower downward basic analyte pK_a shift) or analysis of basic compounds in their neutral form.

Example 2. Analyzing a Base in its Neutral Form. In this example we will use the same compound as in the example above [2,4-dimethylpyridine (base) with pK_a of 6.7] along with the same eluent conditions (50% MeCN and 50% phosphate buffer). The goal is to calculate pH of the buffer, in order to obtain the basic analyte in its fully neutral form.

Step 1. First, account for the downward pK_a shift for the basic analyte upon addition of organic. For every 10 v/v% increase in acetonitrile, the $_s^spK_a$ of the analyte decreases by 0.2 pK_a units.

Step 2. Once this $_s^spK_a$ is determined, the $_s^spH$ at which the analyte would be a fully neutral form needs to be determined. This corresponds to 2 pH units greater than the $_s^spK_a$ of 5.7, as shown in step 2 below.

Step 3. Then account for the pH shift of the acidic portion of the mobile phase upon addition of acetonitrile. For every 10 v/v% increase in acetonitrile, the pH of the acidic buffer increases by approximately 0.2 pH units. This would correspond to a 1.0 pH unit increase, as shown in step 3 below.

Step 4. Then determine what the minimum $_w^wpH$ of the aqueous portion of the buffer should be prepared at taking into account the pH shift of the aqueous portion of the mobile phase upon addition of organic as shown in step 4 below. Therefore the optimal pH to analyze this compound would be at an aqueous mobile phase pH of >6.7.

1:   $6.7 - (5 \times 0.2) = 5.7$ Downward analyte pK_a shift
2:         $5.7 + 2 = 7.7$ $_s^spH$ at which basic analyte would be fully neutral
3:         $5 \times 0.2 = 1.0$ Upward pH shift of the aqueous acidic buffer
                          upon addition of organic (50 v/v% MeCN)
4:         $7.7 - 1.0 = 6.7$ Min. $_w^wpH$ of the aqueous portion of the mobile
                          phase in order to have analyte in fully neutral
                          form at 50% v/v% MeCN

In this scenario as the acetonitrile concentration is raised, as would occur in gradient mode, there is no risk in analyzing this molecule near the pK_a. As the organic content increases above 50 v/v%, there is a greater gap between the $_s^spK_a$ of the basic analyte and the $_s^spH$ of the mobile phase (only if an acidic buffer is used).

Example 3. Analyzing an Acid in Its Neutral Form. For example, if your target is an acidic analyte with a pK_a of 4.0 and the eluent conditions are 40% MeCN and 60% aqueous, what should be the pH of the phosphate buffer in order to obtain the acid analyte in its fully neutral form?

Step 1. First account for the upward pK_a shift for the acidic analyte upon addition of organic. For every 10 v/v% increase in acetonitrile the $_s^spK_a$ of the analyte increases by 0.3 pK_a units.

Step 2. Once this $_s^spK_a$ is determined, the $_s^spH$ at which the analyte would be fully neutral form needs to be determined. This corresponds to 2 pH units lower than the $_s^spK_a$ of 5.2, as shown in step 2 below.

Step 3. Then account for the pH shift of the acidic portion of the mobile phase upon addition of acetonitrile. For every 10 v/v% increase in acetonitrile the pH of the acidic buffer increases by approximately 0.2 pH units. This would correspond to a 0.8 pH unit increase, as shown in step 3 below.

Step 4. Then determine what the maximum $_w^w$pH of the aqueous portion of the buffer should be prepared at taking into account the pH shift of the aqueous portion of the mobile phase upon addition of organic, as shown in step 4 below. Therefore the optimal pH to analyze this compound would be at aqueous mobile phase pH of <2.4.

1:    $4.0 + (4 \times 0.3) = 5.2$ Upward analyte pK_a shift
2:                $5.2 - 2 = 3.2$ $_s^s$pH at which acidic analyte would be fully neutral
3:            $4 \times 0.2 = 0.8$ Upward pH shift of the aqueous acidic buffer
                                              upon addition of organic (40% MeCN)
4:            $3.2 - 0.8 = 2.4$ Max. $_w^w$pH of the aqueous portion of the mobile
                                              phase in order to have analyte in fully neutral
                                              form at 40% MeCN

Example 4. Zwitterionic Components. Let us go back to Figure 4-24, for the separation of the Benazepril diastereomers on a phenyl-hexyl column, and see if we could have predicted the appropriate $_w^w$pH to perform the separation just by taking into consideration the analyte pK_a values and applying the pH shift and pK_a shift rules. Generally, for every 10 v/v% of acetonitrile there is an upward pK_a shift of 0.3 for the acidic analyte pK_a and a downward shift of 0.2 for the basic analyte pK_a. In 30 v/v% acetonitrile the apparent pK_a for the acidic portion of Benazepril HCl will shift to about $3.7 + 0.9 = 4.6$ (pK_a upward shift of 0.9), and for the basic site of Benazepril HCl the apparent pK_a will shift to about $4.6 - 0.6 = 4$ (pK_a downward shift of 0.6). Generally, at ±2 pH units from each pK_a, the analyte (at each respective ionization center) remains in one predominate ionization state. Therefore in 30 v/v% MeCN, at a mobile phase $_s^s$pH below $_s^s$pH 2.6 the acidic functionality is expected to be neutral, and at a mobile phase $_s^s$pH greater than 6.6 the acidic functionality is expected to be fully ionized. Similarly, in 30 v/v% MeCN, at $_s^s$pH below 2.0 the basic functionality is expected to be fully protonated and at a mobile phase $_s^s$pH greater than 6.0 the basic functionality is expected to be predominately neutral. Therefore, taking both ionizable functionalities into consideration at mobile phase $_s^s$pH values between 2.0 and 6.6, it is expected to see changes in retention due to the change in the ionization states of both the acidic and basic functionalities. Indeed, in the experiment when Benazepril was analyzed at $_w^w$pH 2.1–6.1 or respectively $_s^s$pH 2.7–6.7 (ammonium phosphate buffer), significant changes in retention were observed. There were selectivity differences as a function of $_s^s$pH from 2.7 to 6.7 with concomitant changes in retention. Separations for ionizable compounds in RPLC are to be performed where the molecules are in one predominate ionization state(s) to avoid significant changes in retention with minor changes in pH. For Benazepril, in order to have both ionizable functionalities in one predominate ionization state, at 30 v/v% MeCN the optimal $_s^s$pH of the mobile phases to run the separation is at $\leq$ $_s^s$pH 2.0 or $_s^s$pH $\geq$ 6.6. By applying the pK_a shift selection rules we would have been able to predict the optimal pH to perform the separation at. These effects of analyte retention changing as a function of pH should be independent on the type of column employed.

4.9   EFFECT OF TEMPERATURE ON ANALYTE IONIZATION

The use of elevated temperatures for the reversed-phase HPLC separation of mixtures has been used primarily for increasing column efficiency or shortening run time [95, 96] and enhancing separation selectivity [97]. Elevated temperatures also increase solute solubility and diffusivity. Column efficiency is also expected to increase with temperature as diffusion rate increases. However, temperature can also affect the dissociation constants of the ionizable components, and this can lead to anomalous retention behavior of these compounds as a function of temperature (i.e., increases in retention with increase in temperature) [98–102]. The pH of a phosphate buffer and acetate buffer are not significantly affected by change in temperature [103–108]. For acidic analytes, depending on the type of acid and its intrinsic properties, the analyte pK_a may not vary as a function of temperature. Phenolic and carboxylic acids pK_a's do not vary significantly with a change of temperature. However, basic analytes may experience greater changes in their ionization constants with increase of the temperature [109]. The weaker an acid, the greater the change in the analyte pK_a (mainly seen for basic compounds) with a change in temperature. Essentially for basic compounds the analyte is being analyzed in its more neutral form with an increase in the temperature and may experience increases in retention at higher analysis temperatures. For example, McCalley found that the retention of nortriptyline at pH 7 and quinine at pH 7 increased as the temperature was increased from 20°C to 60°C. [110] Also, Buckenmaier et al. [111] (Figure 4-39) found that for the bases amitriptyline, benzylamine, nortriptyline, and quinine a continuous retention increase for these compounds from 30°C to 60°C with increasing temperature could be attributed to temperature-dependent pK_a shifts (on the order of − 0.03 pK_a units/K). However, the retention of the quaternary amine compounds remained constant and/or experienced slight decrease in their retention.

The $_s^s pK_a$ of five weak electrolytes of different chemical nature (butylamine, N,N-dimethylaniline, phenol, and benzoic acid) in 50% methanol/water at 20–50°C were determined by Castells et al. [108], and the values are shown in Table 4-15. The effect of temperature was the greatest for the basic compound butylamine, and a lesser effect was observed for the weaker bases pyridine and N,N-dimethylaniline and the weakly acidic phenol.

Therefore the temperature of the separation should also be taken into consideration when performing method development, especially for basic compounds. Basic compounds that have pK_a values >6 usually experience the greatest changes in retention with increase in temperature. The pK_a values of these basic compounds decrease with an increase in temperature, thereby making them more neutral when analyzed at higher temperatures.

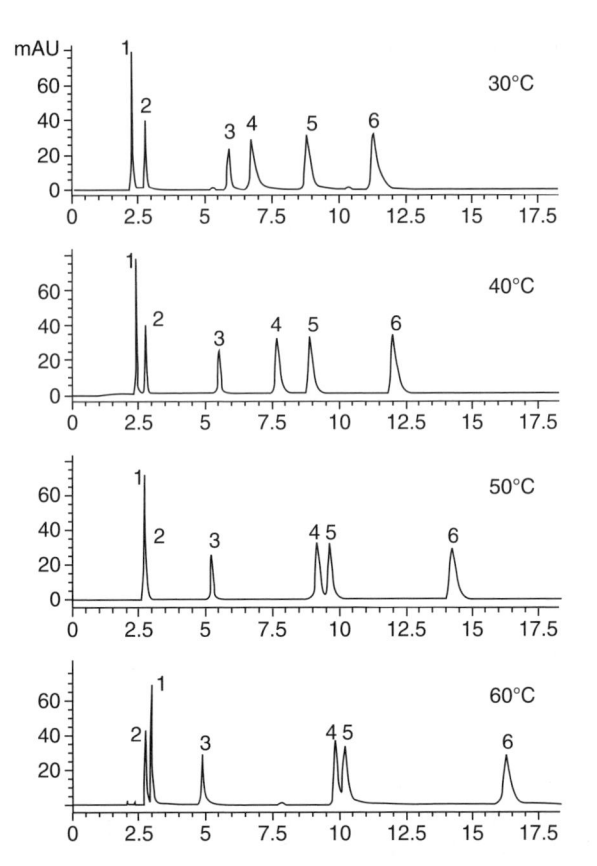

Figure 4-39. Effect of temperature on ionizable analyte retention at $_w^s$pH 7.8 using a ODS3V column in temperature range 30–60°C. (1) Benzylamine ($_w^s$pK_a 8.96, 30°C). (2) BteN (quaternary amine), (3) berberine chloride, (4) quinine ($_w^s$pK_a, 8.3, 30°C) (5) protriptyline ($_w^s$pK_a 10.0, 30°C) (6) nortriptyline ($_w^s$pK_a 9.7, 30°C). Flow rate: 1 mL/min. Mobile phase: $_w^s$pH 7.8, acetonitrile-phosphate buffer (40:60, v/v), with ionic strength (I) maintained at 0.05 M. Buffer concentration adjusted to maintain I constant at different temperatures. (Reprinted from reference 111, with permission.)

TABLE 4-15. $_s^s$pK_a Values for Acidic and Basic Compounds as a Function of Temperature

50 v/v% Methanol–Water Temperature (K)	Phenol $_s^s$pK_a	Butylamine $_s^s$pK_a	Pyridine $_s^s$pK_a	N,N-Dimethylaniline $_s^s$pK_a
293	11.05	10.05	4.14	4.36
298	10.97	9.89	4.08	4.28
303	10.9	9.68	4.02	4.21
308	10.82	9.53	3.96	4.11
313	10.73	9.36	3.91	4.03
318	10.67	9.23	3.85	3.97
323	10.63	9.1	3.81	3.89

Values from reference 108

4.10   ION-INTERACTION CHROMATOGRAPHY

4.10.1   Introduction

In reversed-phase HPLC with water/organic eluents, ionic interactions always play an important role in regard to analyte retention, solvation, ionic equilibria, and other processes. To some extent, chromatographic effects and practical use of ionic interactions have been discussed in the previous sections of this chapter. In this section the influence of the ionic additives in the mobile phase on the retention of ionic or ionizable analytes will be discussed.

Ion-interaction chromatography is an intermediate between reversed-phase and ion-exchange chromatography. Introduction of amphiphilic and liophilic ions into the mobile phase causes their adsorption on the hydrophobic surface of packing material with subsequent transformation into a pseudo ion-exchange surface. Ionic interactions with charged analytes can occur in the mobile phase and with counterions that may be adsorbed on the stationary-phase surface.

Amphiphilic ions are usually molecules with relatively long alkyl chain and have a charged group at one end. These substances are surfactants (reason for the name "soap chromatography") and possess highly localized charge. In the chromatographic system these molecules are accumulated at the interface between the hydrophobic stationary phase and water/organic eluent. They are oriented at the interface so that the charged part of the molecule remains in the eluent and the hydrophobic part (alkyl chain) is adsorbed on the stationary phase surface. This forms the charged surface, and excessive surface charge should be compensated by the accumulation of the counterions in the mobile phase at the close proximity of the surface (Figure 4-40) forming the corresponding electrical double layer.

Specific effect of amphiphilic ions on the retention of charged analytes was observed by many researchers more than three decades ago [112–115].

Remarkable number of different names was introduced to these methods. The technique has been called "soap chromatography" [113], "solvent-generated ion-exchange" [114], "ion-interaction" [115], and "ion-pair" [116]. Researchers introduced a similar number of different theories for the description of the effect of ionic mobile-phase additives on the retention of charged analytes; essentially, each specific name for this technique corresponds to its own distinct retention theory. Melander and Horvath [116] divided existing theories into two main groups: stoichiometric [113, 114, 117–119] and nonstoichiometric [120–133].

Phenomenologically, two different mechanisms could be envisioned: (a) the formation of the ion pair between the analyte and amphiphilic counterion with subsequent adsorption of this complex on the stationary phase and (b) adsorption of the amphiphilic counterion itself on the stationary phase surface and subsequent retention of charged analyte in essentially an ion-exchange mode. Melander and Horvath [117] concluded that, in reality, probably both mechanisms coexist in the chromatographic system.

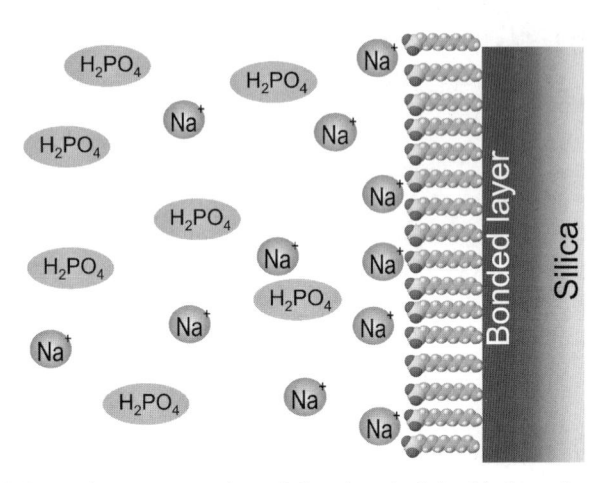

Figure 4-40. Schematic representation of the electrical double layer in reversed-phase ion-pair chromatography. See color plate.

4.10.2 Double Layer Theory

The influence of the formation of the electrostatic potential on the adsorbent surface on the retention of a charged analyte could be introduced in an over-simplified form, assuming an ideal partitioning model on a flat surface, where the retention factor is related to the free Gibbs energy as

$$k_0 = \varphi \exp\left(\frac{\Delta G^0}{RT}\right) \qquad (4\text{-}26)$$

In the presence of the double layer the free Gibbs energy of the analyte interaction with the surface will be

$$\Delta G_t^0 = \Delta G^0 + zF\Psi_0 \qquad (4\text{-}27)$$

where z is the analyte charge and Ψ_0 is the electrostatic surface potential created by the adsorbed ion-pairing agent.

Stáhlberg [134, 135] introduced the application of the Gouy–Chapman double layer theory for the retention of small ionic analytes in ion-exchange and ion-pair chromatography (Figure 4-41). The resulting equation for the retention factor is

$$k = \frac{A}{V_0}\left(\frac{\sigma K_{ch} \exp\left(-\dfrac{zF\Psi_0}{RT}\right)}{-zF} + \int_0^1 \left(\exp\left(-\frac{zF\Psi(x)}{RT}\right) - 1 \right) dx \right) \qquad (4\text{-}28)$$

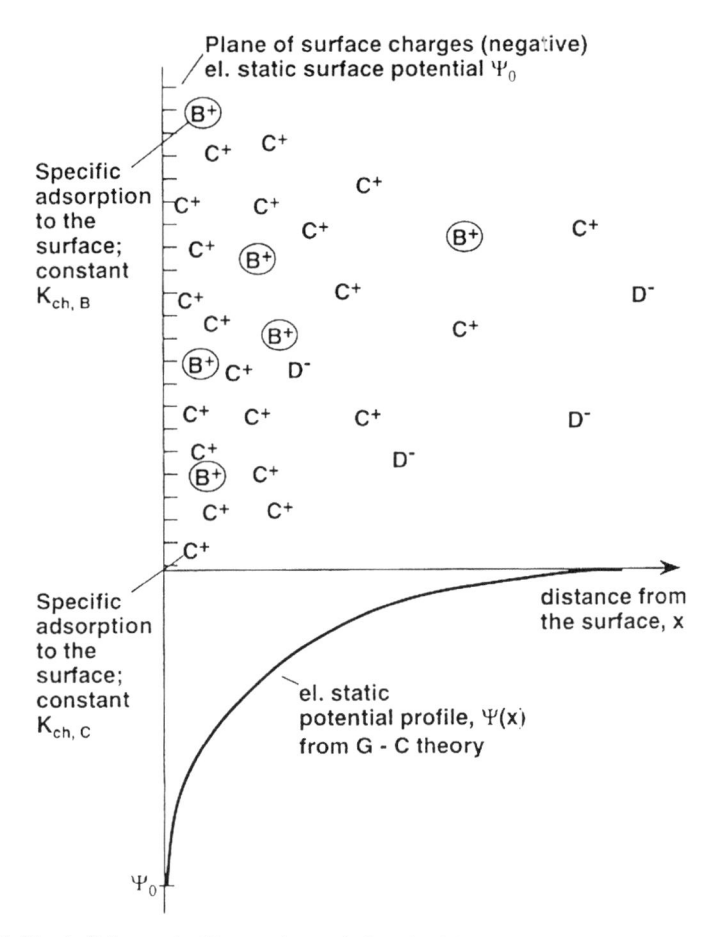

Figure 4-41. A Schematic illustration of the double layer model for ion-exchange chromatography of small ions according to Stählberg. Reprinted with permission from [136].

where K_{ch} is the analyte association constant and $\Psi(x)$ is the double layer profile.

Most of these theories have one significant drawback: They are all derived for the flat open adsorbent surface, but HPLC adsorbents are porous materials with average pore size on the level of 100 Å for bare material. After chemical modification of the original silica surface, the effective pore diameter decreases and the properties of electric double layer in the confined space of small pores are significantly different from that on the flat surface.

Detailed and comprehensive discussion of double-layer-based theories of ion-pair and ion-exchange chromatography is given in good review by Stählberg [136].

4.10.3 Ion Pairs

The double layer theory describes the process of ion-exchange and ion-interaction chromatography from the point view of distributed electrostatic field effect on the charged analyte retention. Other approaches have a more stoichiometric character and describes the analyte retention on the basis of the formation of ionic pairs and their subsequent retention on the adsorbent surface.

In 1926, Bjerrum [137] used Debye–Hückel theory to describe ion association and took into account the interaction of ions within a short range. He introduced an ion-pair concept, gave a definition of ion pairs as neutral species formed by electrostatic attraction between oppositely charged ions in solution, and showed how ion-pair formation was dependent on the ions size (radius of ions), solvent (dielectric constant), and temperature.

Later Bjerrum's theory was supported by the work of Kraus [138], who showed importance of the dielectric constant, and Atherton [139], who demonstrated the existence of ion pairs using electron spin resonance spectroscopy. The formation of ion pairs may be studied by various methods: conductance studies, UV–visible spectrometry, IR spectrophotometry, partition, distribution, or solvent extraction. The lifetime of ion pairs was determined to be at least 10^{-5} sec, which is equivalent to about 10^8 molecular vibrations, demonstrating that ion pairs can be considered as independent species [140]. Today, the ion-pair formation as independent species is widely accepted.

Ion-pair formation is due only to outer-sphere interactions and no chemical bond of any kind is formed. The work of Sadek and Fuoss [141], and later Roberts and Szwarc [142], showed that an ion pair can exist in two forms: as a tight or intimate ion pair, or as a loose or solvent separated ion pair, depending on the nature of solvent–ion interactions. The behavior of the solvent and its affinity for the ion pair can be explained by the solvation theory proposed by Higuchi et al. [143]. They classified ion pairs based on charge accessibility as shown in Figure 4-42. For the first type of ion pair, the cation is large and liophilic except for the positively charged center and the anion would contain an exposed anionic surface (high negative charge per unit area). The anionic portion of this ion pair can be solvated by liophilic molecules having an exposed partial positively charged surface, (dipolar molecules with acidic protons such as phenol; also alcohols). For the second type of ion pair the anion is large and liophilic except for the negatively charged center, and the cation would contain an exposed cationic surface (high positive charge per unit area). The cationic portion of this ion pair can be easily solvated by liophilic molecules having an exposed partial negatively charged surface (these include solvating species that have nucleophilic sites such as ethers, ketones, and amides). The third type of ion pair represents the ion pair with deeply buried charges (no exposed electrically unbalanced surface). In this third case, no solvation would be needed to extract this ion pair by a nonpo-

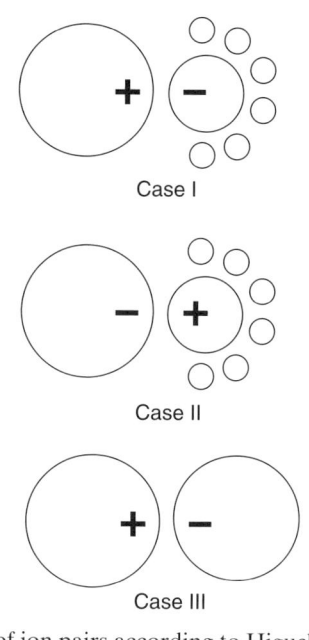

Case I

Case II

Case III

Figure 4-42. Classification of ion pairs according to Higuchi et al. (Reprinted from reference 143, with permission).

lar solvent. Higuchi examined the extraction of dextromethorphan hydrombromide pairs into various organic solvent–cyclohexane mixtures and showed that for a chloroform–cyclohexane mixture, the extraction ability exceeded that of either a 1-pentanol–cyclohexane mixture, where the dielectric constants are higher. This observation is very important to note in regard to chromatographic systems. If coulombic forces participate in the formation of ion pairs, then ion pairs are formed only if the ions approach each other and are separated by a critical separation distance (D) given by the Bjerrum equation:

$$D = z^+ z^- \frac{e^2}{2\varepsilon kT} \tag{4-29}$$

In equation (4-29) z^+ and z^- are the ionic charges, e is electron charge, ε is dielectric constant, k is Boltzmann's constant, and T is the absolute temperature. When the ion separation distance is less than D, ion pairing is regarded as taking place. The dielectric constant of a solvent plays a significant role. At the same critical distance (D) needed for ion-pair formation, a solvent with a high dielectric constant such as water (~80) will be less favorable for ion-pair formation compared to a solvent that has a lower dielectric constant (<40, such as acetonitrile and methanol).

On the other hand, Higuchi showed the importance of solvent mediated effects on the formation of ion pairs that include noncoulombic contributions such as hydrogen bonding and hydrophobic properties of the ions. Therefore, not only does the dielectric constant of the solvent, play a role but other molecular properties of the solvent play an important role in the formation of ion pairs.

Although ion pairing was initially investigated in the field of physical chemistry, the concept was rapidly adopted in colloidal chemistry, analytical chemistry, and the pharmaceutical sciences. Higuchi et al. [144] have reported numerous methods for performing extraction of ionized solutes into organic phases in which ions of opposite charge are added to the aqueous phases, resulting in ion pairing between the solute ion and pairing ion. The resulting complex is neutral and poorly hydrated and can be easily transferred to the organic phases. The ion-pair extraction method is widely used in the pharmaceutical and analytical sciences.

In chromatographic practice, the use of classical ion-pairing reagents is usually recommended as a last resort for the separation of very hydrophilic ions that could not be shifted from the void volume by any other means (i.e., change in mobile-phase pH, type of column, change in organic concentration). The reason for careful use of such a powerful method of specific selectivity and retention adjustment is the consequent irreversible adsorption of classical ion-pairing agents on the surface of the reversed-phase adsorbent. The degree of retention of the ionized solutes is directly proportional to the surface charge density generated by the adsorption of the counterion. The amount of ion-pairing reagent adsorbed on the surface at a particular concentration in the mobile phase is dependent on the hydrophobicity of the alkyl portion of the ionpair reagent. The amount of pairing agent adsorbed by the stationary phase from the eluent has been determined using the breakthrough method described by Knox and Hartwick [119]. Once that amount and surface area of stationary phase are determined, the surface concentration ($\mu mol/m^2$) allows for the calculation of the absorption isotherms. Figure 4-43 shows the adsorption isotherms of a series of alkylsulfates on the surface of reversed-phase adsorbent and shows the column uptake dependence on type and concentration of the reagents. It is observed that for the reagents with small alkyl chains (octyl) the uptake increases but then levels off, indicating that the column is becoming saturated with the reagent.

The retention of the ionized compound is predominately dictated by the uptake of the reagent and the resulting charge on the surface of the column. Figure 4-44 [119] shows the variation of the retention factor of tyrosinamide as a function of the alkylsulfates in the mobile phase.

It has been shown that similar separations may be obtained by reagents of varying hydrophobicity when the reagent concentration in the mobile phase is adjusted to give the same molar uptake by the column (see Figure 4-45). However, this is only true when the solvent ionic strength is kept constant while the mobile phase concentration of the ion pairing ion is increased.

It is logical to expect that with the increase of the concentration of an ion-pairing agent, the retention of oppositely charged analytes increases while

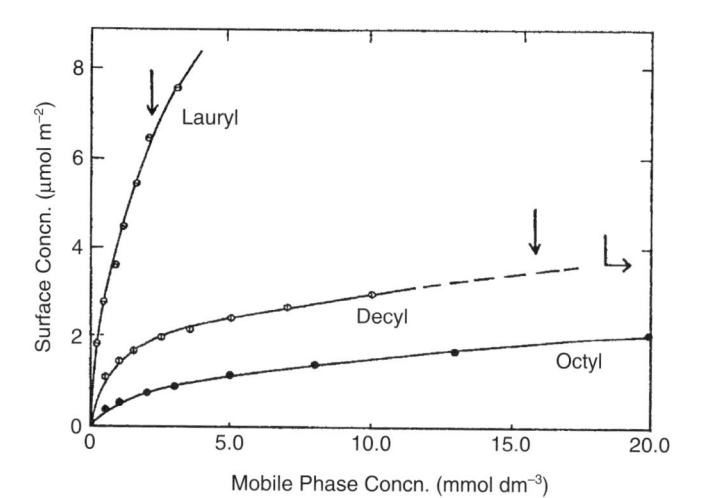

Figure 4-43. Adsorption isotherms of alkylsulfates on Hypersil-ODS from methanol/water (20/80) with 0.02 M phosphate buffer at pH 6.0. (Reprinted from reference 119, with permission.)

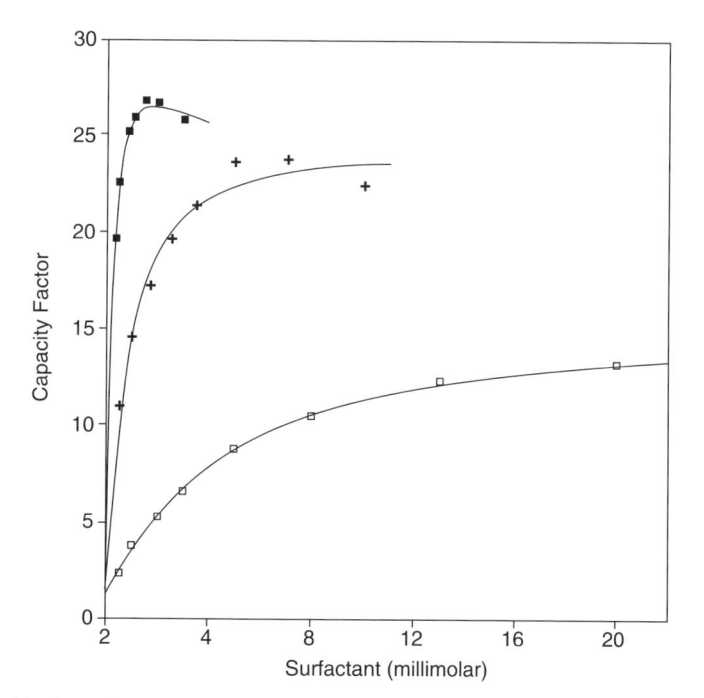

Figure 4-44. Capacity factor of tyrosinamide versus concentrations of dodecyl sulfate (upper curve), decyl sulfate (middle curve), and octyl sulfate (lower curve). (Reprinted from reference 119, with permission.)

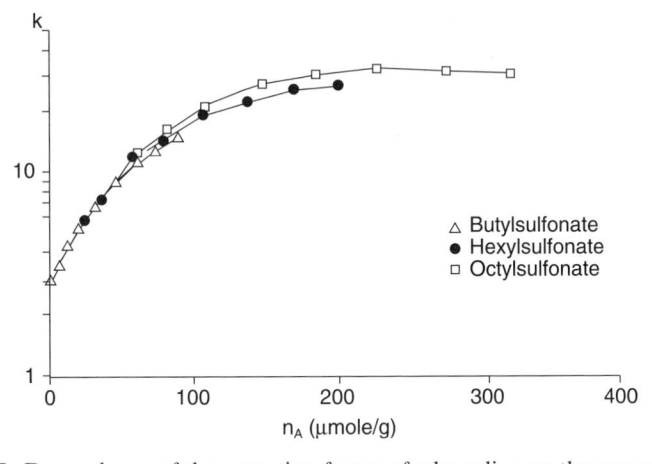

Figure 4-45. Dependence of the retention factor of adrenaline on the concentration of amphiphilic ions on the stationary phase surface. Retention factor shown in logarithmic scale. (Reprinted from reference 136, with permission.)

similarly charged analytes as the ion pairing reagent will elute faster. This indeed has been observed experimentally (Figure 4-46). Figure 4-45 shows the similar retention dependencies of adrenaline retention for different amphiphilic ions adsorbed on the surface of the reversed-phase material, indicating that at the same surface concentration of any amphiphilic ion adsorbed, the retention of basic analyte is the same; thus the retention is dependent on the surface charge density of adsorbed ions. Comparison of Figures 4-46 and 4-45 indicates that the retention of a charged analyte in ion-pairing mode is dependent on the adsorption of ion-pairing ions on the surface of the stationary phase and not on its concentration in the mobile phase.

Same were also observed by Knox in a salt-controlled methanol-aqueous eluent for the analysis of normetadrenaline as a function of octyl, decyl, and lauryl sulfates [119].

In the contrast to the irreversible adsorption of amphiphilic ions on the reversed-phase surface, the liophilic ions shows relatively weak interactions with the alkyl chains of the bonded phase. Liophilic means oil-loving. These liophilic ions are usually small inorganic ions and they possess an important ability for dispersive type interactions. They are (a) characterized by significant delocalization of the charge, (b) primarily symmetrical, (c) usually spherical in shape, and (d) absence in surfactant properties.

The presence of these ions in aqueous solution was found to disrupt the water structure [146]; in other words, they introduce chaos into structured ionic solution that hence are given the name "chaotropic" ions [147]. The effect of chaotropic ions on the disruption of the solvation shell was mainly studied in

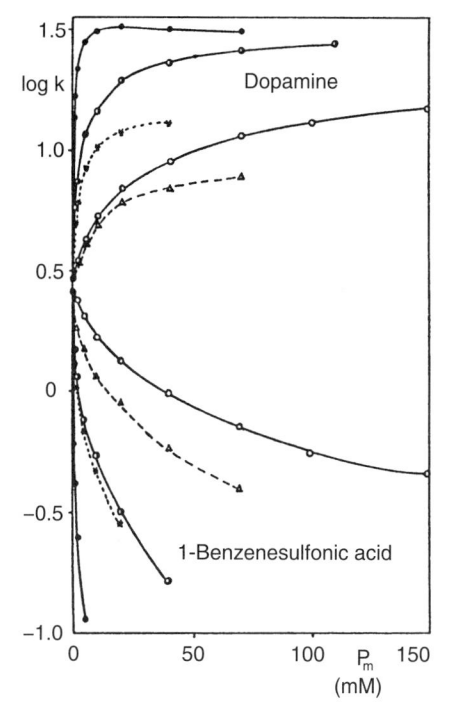

Figure 4-46. Logarithm of the retention of dopamine and 1-benzenesulfonic acid on reversed-phase column as a function of the mobile-phase concentration of ion-pairing additives (pH 2.1). Column: Hypersil-ODS, $T = 25°C$; constant ionic strength was maintained by addition of NaH_2PO_4; open circles, butylsulfate; triangles, cyclohexylsulfamic acid; ×, d-camphor-10-sulfonic acid; half-closed circles, 1-hexanesulfonate; black circles, octansulfonate. (Reprinted from reference 145, with permission.)

the field of biochemistry, where it was shown that they can impact the conformational and the solvation behavior of proteins and peptides [146, 147]. Inorganic ions were arranged according to their ability to disrupt a water solvation shell in the so-called Hofmeiser series [148]. An increase of chaotropicity [149] has a relatively vague phenomenological description, which is essentially related to the increase in hydrophobicity as a result of charge delocalization and significant polarizability. In the sequence

$$H_2PO_4^- < CF_3COO^- < BF_4^- < ClO_4^- < PF_6^-$$

a greater possibility for charge delocalization and higher overall electron density is seen from left to right, with a simultaneous increase in the symmetry. This leads to an increasing ability of these ions to participate in dispersive interactions.

4.10.4 Chaotropic Effect

Study of the effect of liophilic ions on the retention of ionic analytes in reversed-phase HPLC has led to the development of yet another possible theory of their influence on the chromatographic retention of basic compounds [150–152]. Ionic analytes in water/organic mixtures are solvated. The solvation shell suppresses the analyte's ability for hydrophobic interactions with the stationary phase, thus effectively decreasing the analyte's retention. Controlled disruption of the solvation shell allows for control of the analyte retention. Presence of the counterions in the close proximity to the ionic solvated analyte leads to the disruption of the analyte solvation shell. This effect is known as chaotropic control for the retention of ionic compounds in reversed-phase chromatography. Counteranions that have a less localized charge, high polarizability, and lower degree of hydration show a significant effect on the retention of protonated basic analytes and are known as chaotropic ions. Chaotropic ions change the structure of water in the direction of greater disorder. Therefore, the solvation shell of the basic analytes may be disrupted due to ion interaction with the chaotropic anions.

With the increase of the counteranion concentration, the solvation of the protonated basic analyte decreases. The primary sheath of water molecules around the basic analytes is disrupted, and this decreases the solvation of the basic analyte. The decrease in the analyte solvation increases the analyte hydrophobicity and leads to increased interaction with the hydrophobic stationary phase and increased retention for the basic analytes.

The chaotropic effect is dependent on the concentration of the free counteranion and not the concentration of the protons in solution at pH < basic analyte pK_a. This suggests that change in retention of the protonated basic analyte may be observed with the increase in concentration of the counteranion by the addition of a salt at a constant pH as shown in Figure 4-47 for a pharmaceutical compound containing an aromatic amine with a pK_a of 5.

In the example in Figure 4-47, the retention of pharmaceutical analyte X was first altered by decrease of mobile-phase pH (Figure 4-47A), and in the second case (Figure 4-47B) the pH was maintained constant and the concentration of counteranion was increased via addition of its sodium salt. The resulting effect on the retention of basic analyte is strikingly similar if both dependencies are plotted against the concentration of free counteranions of ClO_4^-, as shown in Figure 4-48.

Disruption of the basic analyte solvation shell should be possible with practically any counteranion employed, and the degree of this disruption will be dependent on the "chaotropic nature" of the anion. Chaotropic activity of counteranions has been established according to their ability to destabilize or bring disorder (bring chaos) to the structure of water [148, 149].

Even a very low counteranion concentration in the mobile phase will cause significant initial disruption of the solvation shell, thus leading to the significant increase of the analyte retention, while in the high concentration region

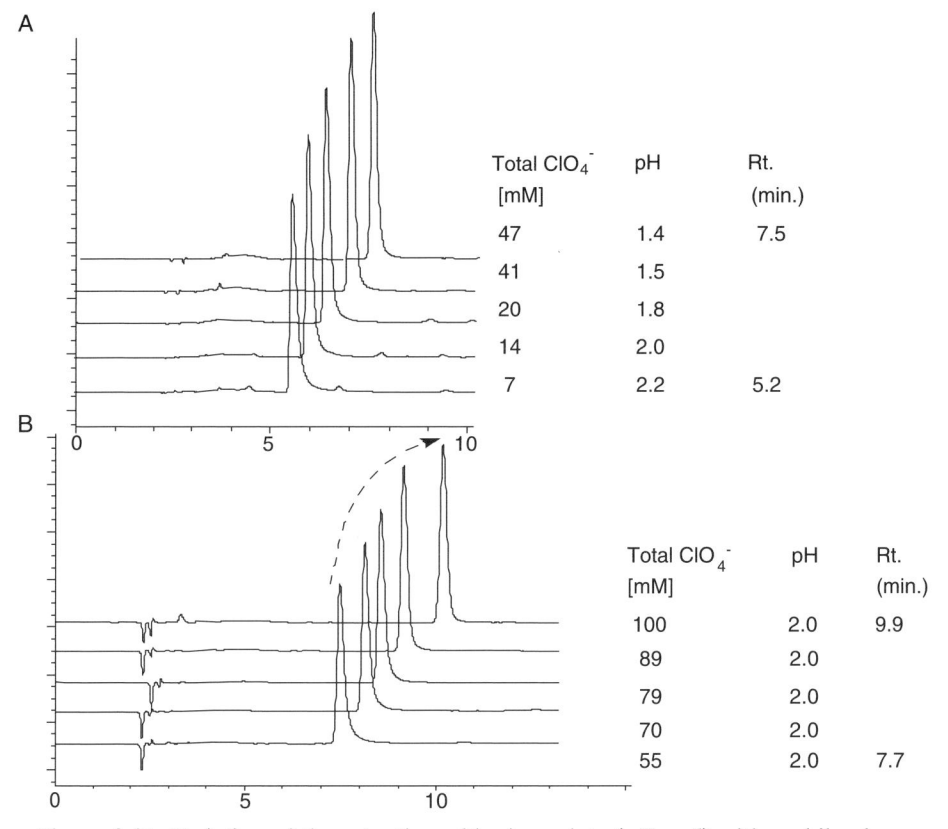

A

Total ClO$_4^-$ [mM]	pH	Rt. (min.)
47	1.4	7.5
41	1.5	
20	1.8	
14	2.0	
7	2.2	5.2

B

Total ClO$_4^-$ [mM]	pH	Rt. (min.)
100	2.0	9.9
89	2.0	
79	2.0	
70	2.0	
55	2.0	7.7

Figure 4-47. Variation of the retention of basic analyte ($pK_a > 5$) with mobile-phase pH (A) and counteranion concentration (B). (Reprinted from reference 185, with permission.)

a type of a saturation effect is observed (Figure 4-49). Logically, at high counteranion concentration when all solvation shells are fully disrupted, any further increase of the counteranion concentration should not cause any additional retention increase.

As was shown above, the chaotropic effect is related to the influence of the counteranion of the acidic modifier on the analyte solvation and is independent on the mobile-phase pH, provided that complete protonation of the basic analyte is achieved. Analyte interaction with a counteranion causes a disruption of the analyte solvation shell, thus affecting its hydrophobicity. Increase of the analyte hydrophobicity results in a corresponding increase of retention. This process shows a "saturation" limit, when counteranion concentration is high enough to effectively disrupt the solvation of all analyte molecules. A further increase of counteranion concentration does not produce any noticeable effect on the analyte retention.

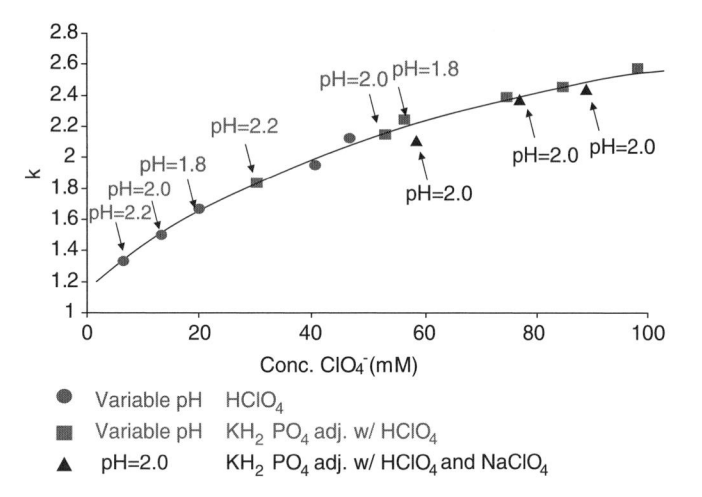

Figure 4-48. Retention of basic analyte ($pK_a > 5$) as a function of ClO_4^- counteranion concentration with variable pH (circles), fixed pH (triangles), and variable pH with phosphate buffer (squares). (Reprinted from reference 185, with permission.)

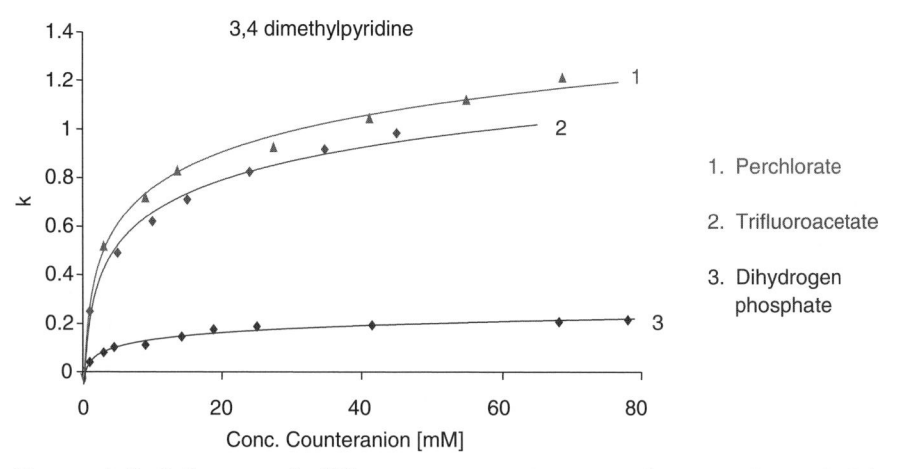

Figure 4-49. Influence of different counteranions on the retention of 3,4-dimethylpyridine. (Reprinted from reference 185, with permission.)

4.10.4.1 Chaotropic Model. If the counteranion concentration is low, some analyte molecules have a disrupted solvation shell, and some do not due to the limited amount of counteranions present at any instant within the mobile phase. If we assume an existence of the equilibrium between solvated and desolvated analyte molecules and counteranions, this mechanism could be described mathematically [151].

The assumptions for this model are:

1. Analyte concentration in the system is low enough that analyte–analyte interactions could be considered nonexistent.
2. The chromatographic system is in thermodynamic equilibrium.

The analyte solvation–desolvation equilibrium inside the column could be written in the following form:

$$B_s^+ + A^- \Leftrightarrow B^+ \ldots A^- \qquad (4\text{-}30)$$

where B_s^+ is a solvated basic analyte, A^- is a counteranion, and $B^+ \cdots A^-$ is the desolvated ion-associated complex. The total amount of analyte injected is [B], analyte in its solvated form is $[B_s^+]$, and analyte in its desolvated form is denoted as $[B^+ \cdots A^-]$, indicating its interaction with counteranions.

The equilibrium constant of reaction (4-30) is

$$K = \frac{[B^+ \ldots A^-]}{[B_s^+][A^-]} \qquad (4\text{-}31)$$

Total analyte amount is equal to the sum of the solvated and desolvated forms of analyte

$$[B] = [B_s^+] + [B^+ \ldots A^-] \qquad (4\text{-}32)$$

The fraction of solvated analyte could be expressed as

$$\theta = \frac{[B_s^+]}{[B]} \qquad (4\text{-}33)$$

The fraction of the desolvated analyte in the mobile phase could be expressed as

$$1 - \theta = \frac{[B^+ \ldots A^-]}{[B]} \qquad (4\text{-}34)$$

Substituting expressions (4-33) and (4-34) into expression (4-31), we can write an expression for the equilibrium constant:

$$K = \frac{1-\theta}{\theta \cdot [A^-]} \qquad (4\text{-}35)$$

Solving equation (4-35) for θ (solvated fraction), we get

$$\theta = \frac{1}{K[A^-]+1} \tag{4-36}$$

Expression (4-36) shows that the solvated fraction of the analyte is dependent on the counteranion concentration and desolvation equilibrium parameter.

Completely solvated analyte has a low retention factor (even if it is equal to 0), which we denote as k_s, while the corresponding retention factor for desolvated form is denoted as k_{us}.

Assuming that solvation–desolvation equilibrium is fast, we can express the overall retention factor of injected analyte as a sum of the retention factor of solvated form multiplied by the solvated fraction (θ) and the retention factor of the desolvated form multiplied by the desolvated fraction ($1 - \theta$), or

$$k = k_s \cdot \theta + k_{us} \cdot (1-\theta) \tag{4-37}$$

Substituting θ in equation (4-37) from (4-36), we get

$$k = k_s \left(\frac{1}{K[A^-]+1} \right) - k_{us} \left(\frac{1}{K[A^-]+1} \right) + k_{us} \tag{4-38}$$

and the final form can be rewritten as

$$k = \frac{k_s - k_{us}}{K \cdot [A^-]+1} + k_{us} \tag{4-39}$$

This equation has three parameters: k_s is a "limiting" retention factor for solvated analyte, k_{us} is a "limiting" retention factor for desolvated analyte, and K is a desolvation parameter [151]. The description of the experimental results with function (4-39) is shown in Figure 4-50. Expression (4-39) in principle allows for the calculation of the solvation equilibrium constant from experimental chromatographic data.

4.10.4.2 *Effect of Different Counteranions.* The chaotropic theory was shown to be applicable in many cases where small inorganic ions were used for the alteration of the retention of basic pharmaceutical compounds [153–157]. Equation (4-39) essentially attributes the upper retention limit for completely desolvated analyte to the hydrophobic properties of the analyte alone. In other words, there may be a significantly different concentration needed when different counterions are employed in the eluent for complete desolvation of the analyte. Therefore, the resulting analyte hydrophobicity and thus retention characteristics of analyte in completely desolvated form should be essentially independent on the type of counteranion employed. Experimental results, on the other hand, show that the use of different counterions

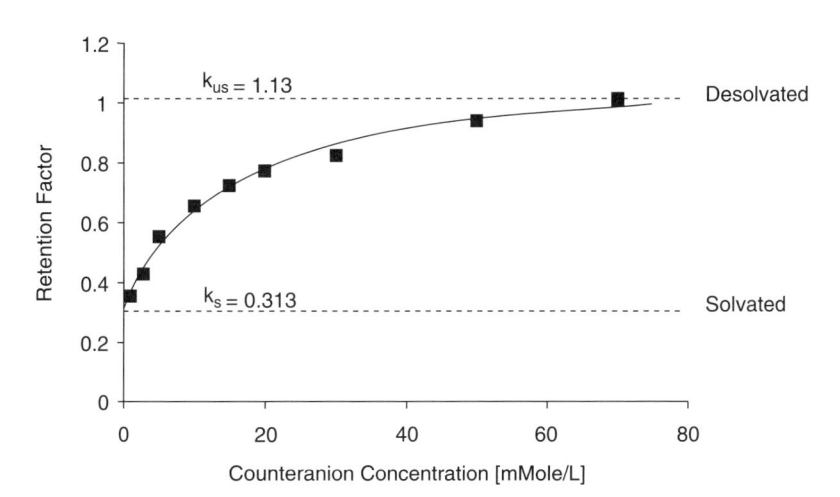

Figure 4-50. Experimental dependence of the retention of basic analyte on the counteranion concentration (points), along with corresponding theoretical curve for this effect calculated using equation (4-39). (Reprinted from reference 185, with permission.)

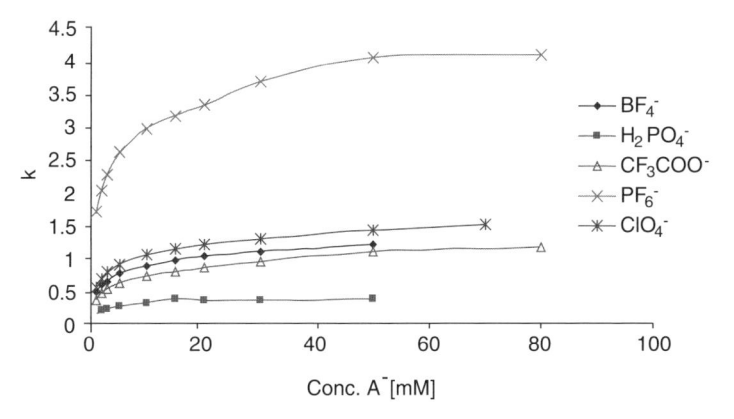

Figure 4-51. Retention factor variations for acebutolol analyzed with different chaotropic agents. (Reprinted from reference 156, with permission.)

leads to the different retention limits of completely desolvated analyte. Figure 4-51 clearly illustrates this effect. This discrepancy could be explained by the presence of two simultaneous processes: the desolvation and ion association (ion pairing). The effect of the counterion concentration on the analyte retention in both processes (desolvation and ion pairing) have Langmurian shape [156], and overall retention is a superposition of both effects.

4.10.4.3 Retention of the Counteranions. Three distinct processes could be envisioned in the effect of chaotropic ions on the retention of basic analytes:

1. Classic ion pairing involves the formation of essentially neutral ion pairs and their retention according to the reversed-phase mechanism.
2. In the chaotropic model, counteranions disrupt the analyte solvation shell, thus increasing its apparent hydrophobicity and retention.
3. Liophilic counteranions are adsorbed on the surface of the stationary phase, thus introducing an electrostatic component into the general hydrophobic analyte retention mechanism.

In their recent papers, Guiochon and co-workers are essentially advocating for the domination of the first process [158–160]. They are explaining the counteranion effect on the basis of the formation of a neutral ionic complex, followed by its adsorption on the hydrophobic stationary phase. Similarity in adsorption behavior of anionic and cationic species is interpreted as a confirmation of their adsorption in the form of neutral complexes.

The retention of ionic components on reversed-phase columns is essentially regarded as ion-pair chromatography, which has been extensively developed by Horvath [161] and Sokolovski [162, 163] in the form of stochiometric adsorption of ionic species and by Stählberg in the form of adsorption of ions and formation of an electrical double layer [164].

The adsorption of amphiphilic ions was experimentally confirmed about 30 years ago, while the actual interaction of the small liophilic ions with hydrophobic stationary phase in reversed-phase conditions was found only recently [165].

Most probably all three mechanisms exist while one of them is dominating, depending upon the eluent type, composition, and adsorbent surface properties.

For acetonitrile/water systems it was found that acetonitrile forms thick adsorbed layer on the surface of hydrophobic bonded phase, while methanol adsorption from water formed a classical monomolecular adsorbed layer [166]. The thick adsorbed layer of acetonitrile provides a suitable media for the adsorption of liophilic ions on the stationary phase adding an electrostatic component to the retention mechanism, while monomolecular adsorption of methanol should not significantly affect adsorption of ions.

The study of the retention of chaotropic anions (BF_4^-, perchlorate, and PF_6^-) was performed using acetonitrile/water eluents on alkyl- and phenyl-type phases with LC–MS detection (electrospray, negative ion mode) [165]. At all mobile-phase conditions with acetonitrile/water PF_6^- ion exhibits the greatest retention, and this is the most liophilic ion in the Hoffmeister series. This ion has the highest degree of charge delocalization and highest polarizability, which facilitates its possible dispersive (or van der Waals) interactions. These properties allow this ion to interact with acetonitrile. Other anions have similar properties, but their ability for dispersive interactions is lower then PF_6^-. At acetonitrile concentrations up to 20 v/v% acetonitrile, all ions exhibit a maximum retention.

General dependence of the analyte retention on the eluent composition in reversed-phase HPLC shows an exponential decay with the increase of the organic modifier concentration. This is usually described in the following form:

$$\ln(k) = a + xb \tag{4-40}$$

where k is a retention factor, x is the eluent composition, and a and b are constants. This relationship has a thermodynamic background because in the partitioning retention model the retention factor is proportional to the distribution equilibrium constant, which in turn is an exponent of the excessive free Gibbs energy of the analyte in the chromatographic system. Excessive free Gibbs energy is the difference of the analyte potential in the stationary phase and its potential in the eluent. This is only true if retention is a result of a single process on the adsorbed surface (e.g., partitioning, or adsorption). If, on the other hand, the retention mechanism is complex, retention dependencies will not adhere to equation (4-40).

The thick acetonitrile layer adsorbed on the bonded phase surface acts as a pseudo-stationary phase, thus making retention in acetonitrile/water systems a superposition of two processes: partitioning into the acetonitrile layer and adsorption on the surface of the bonded phase. Based on the model described in reference 166, analyte retention could be represented in the following form:

$$V_R(c_{el}) = V_0 + (K_p(c_{el}) - 1)V_{ads} + SK_H K_p(c_{el}) \tag{4-41}$$

where $V_R(c_{el})$ is the retention volume of analyte ions as a function of the eluent composition, V_0 is the void volume, $K_p(c_{el})$ is the equilibrium constant for the distribution of the analyte ions between the eluent and adsorbed layer, V_{ads}, S is the adsorbent surface area, and K_H is the adsorption equilibrium constant for analyte ions adsorption from neat acetonitrile on the corresponding stationary phase.

Semiempirical expression was derived for the description of the retention of chaotropic counteranions in reversed-phase conditions [165]. Overall expression for the description of the retention dependencies of analyte ions versus eluent composition will have only four unknowns and allow numerical approximation of experimental retention data (shown as a function of the mole fraction of organic eluent component).

$$V_R(x) = V_0 - V_{ads.}(x) + A \cdot \exp\left(\frac{\Delta G_{MeCN} - x\Delta G_{el.}}{RT}\right) \cdot (V_{ads.}(x) + SK_H) \tag{4-42}$$

Essentially equation (4-42) describes the retention volume of the analyte as a sum of the mobile-phase volume ($V_0 - V_{ads}$, assuming that adsorbed

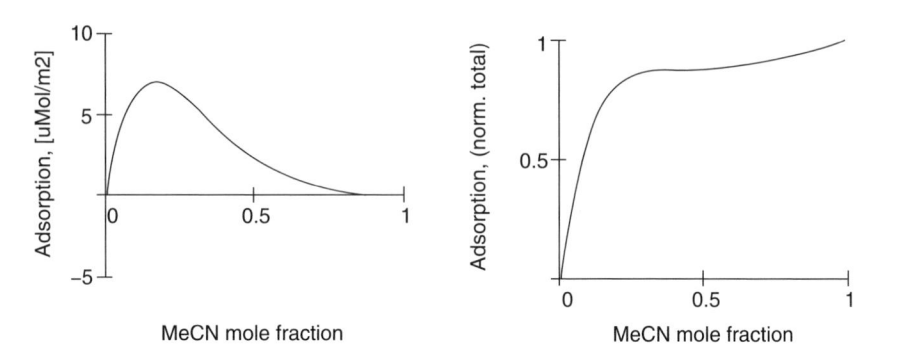

Figure 4-52. Acetonitrile excess adsorption isotherm from water on Zorbax Eclipse XDB-C8 adsorbent (left); normalized filling of adsorbed layer (right). (Reprinted from reference 165, with permission.)

acetonitrile layer is stagnant) and an energetic term that describes analyte (in this case, chaotropic anion) partitioning into the adsorbed layer and its adsorption on the stationary phase surface. Volume of the adsorbed layer on top of the bonded phase is also a function of the acetonitrile concentration in the mobile phase (Figure 4-52).

Coefficient ΔG_{el} in equation (4-42) has a meaning of energetic span of partitioning constant in the whole concentration region, and it reflects (a) the excessive interactions of studied ions with water and acetonitrile and (b) structural organization of molecules.

The suggested phenomenological model describes the retention of PF_6^- ions on different reversed-phase columns very well. Average deviation of calculated values from experimentally measured values is on the level of 1%, which confirms that indeed a superposition of several processes govern the retention of liophilic ions in acetonitrile/water systems. Experimental values along with the theoretical curves are shown in Figure 4-53.

The multilayered character of acetonitrile adsorption creates a pseudo-stationary phase of significant volume on the surface, which acts as a suitable phase for the ion accumulation. In the low organic concentration region (from 0 to 20 v/v% of acetonitrile), studied ions show significant deviation from the ideal retention behavior (decrease in ion retention with increase in acetonitrile composition) due to the formation of the acetonitrile layer, and significant adsorption of the chaotropic anions was observed. This creates an electrostatic potential on the surface in which there is an adsorbed acetonitrile layer, which provides an additional retentive force for the enhancement of the retention of protonated basic analytes. When the dielectric constant is lower than 42 [167], this favors the probability of ion pair formation in this organic enriched layer on top of the bonded phase.

However, at high concentration of organic (>25 v/v%) in the mobile phase the retention of counteranions start to decrease, and this is attributed to the

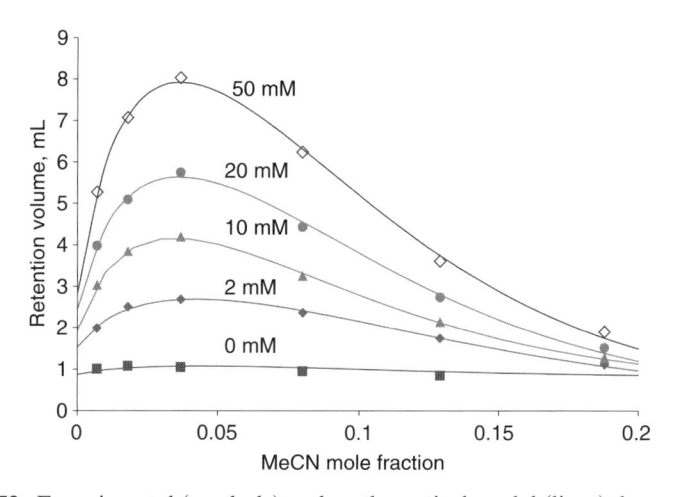

Figure 4-53. Experimental (symbols) and mathematical model (lines) dependencies of PF$_6$ retention on Allure-PFP (perfluorinated propyl-phenyl phase) column versus the acetonitrile composition (shown in molar fractions) at different ionic strength (0, 2, 10, 20, and 50 mM adjusted with NH$_4$Cl). (Reprinted from reference 165, with permission.)

normal effect of the increase of the organic composition in the mobile phase on the retention of the analyte, which shows an exponential decay. The schematic of the retention mechanism of basic analytes in the presence of liophilic ions in acetonitrile/water mobile phase is depicted in Figure 4-54. Acetonitrile forms an adsorbed layer where liophilic ions are soluble due to their ability for dispersive interactions with π-electrons of acetonitrile. The presence of counterions in that layer create additional electrostatic retentive factor for positively charged analyte. The complex form of the liophilic ions adsorption on the stationary phase as a function of organic concentration should be also reflected on the retention of basic analytes, and this was experimentally observed (Figure 4-55 [168]). Note that analyte relative retention increase is only observed in acetonitrile/water systems, where a thick adsorbed organic layer is formed, whereas in methanol/water systems, methanol only forms a monomolecular adsorbed layer that does not provide additional capacity for the retention of liophilic ions. Also, methanol does not have π-electrons, thereby significantly decreasing its ability for dispersive interactions with liophilic ions.

Hexafluorophosphate retention dependencies similar to the one shown in Figure 4-56 [169] were observed on different stationary phases, but only when acetonitrile was used as an organic eluent component. If acetonitrile was substituted with methanol, the effect of the increase of PF$_6$ retention with the increase of organic concentration disappears. This indicates that liophilic ions show strong dispersive interactions with acetonitrile and have little affinity to the hydrophobic adsorbent surface—as opposed to the amphiphilic ions, which

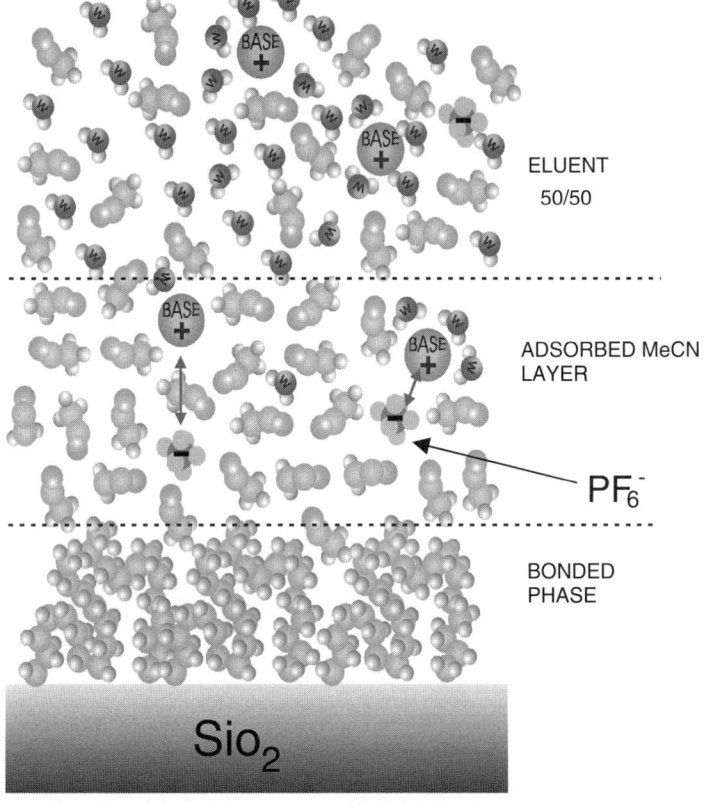

Figure 4-54. Schematic of the retention mechanism of basic analyte on reversed-phase material in water/acetonitrile eluent in the presence of liophilic ions (PF_6^-). See color plate.

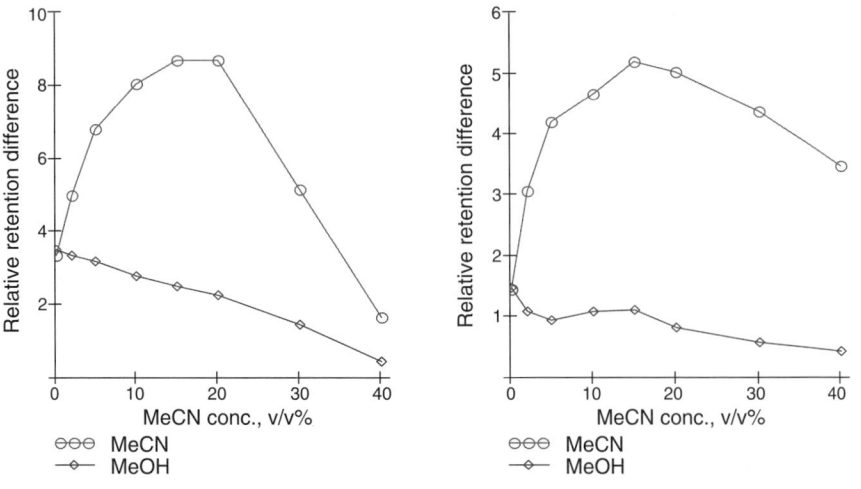

Figure 4-55. Relative adjusted retention of aniline (PF_6/no-PF_6 ratio) on Allure-PFPP (left) and Zorbax-C18 (right) columns from acetonitrile (circles) and from methanol (diamonds). (Reprinted from reference 168, with permission.)

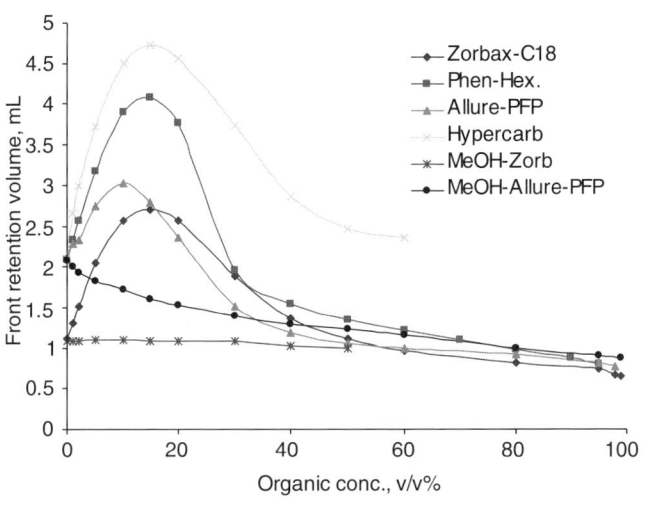

Figure 4-56. Overlay of the retention volumes of PF_6^- front (0.05 mM concentration of NH_4PF_6 in the solution) on all four columns measured from acetonitrile/water and methanol/water mixtures. (Reprinted from reference 169, with permission.)

show significant and often irreversible adsorption on the surface of the reversed-phase adsorbents regardless of type of organic modifier.

Overall, liophilic ions (usually small ions capable for dispersive interactions) provide a useful means for selective alteration of the retention of basic analytes. Influence of these ions on the column properties is fully reversible, and equilibration requires minimal time (usually less than an hour, or about 10 to 20 column volumes). On the other hand, the mechanism of their effect is very complex and is dependent on the type of organic modifier used and on the concentration applied. Theoretical description and mathematical modeling of this process is a subject for further studies.

4.10.4.4   Effect of the Counteranion Type and Concentration on Peak Efficiency and Asymmetry.   Theoretically, a column can generate a certain maximum number of theoretical plates at the optimum flow rate. This number should be independent of the type of the analyte and mobile phase. In reality, any secondary processes, energetic surface heterogeneity, or restrictions in sorption–desorption kinetics in the column will result in the specific decrease of the efficiency for a particular compound.

Increasing the chaotropic counteranion concentration of perchlorate, hexafluorophosphate, and tetrafluoroborate in the mobile phase for basic compounds studied led to an increase in the apparent efficiency of the system until the maximum plate number for the column is achieved [153]. In Figure 4-57A the efficiency for three basic ophthalmic drug compounds increases relatively fast when the concentration of counteranion BF_4^- was increased from 1 mM

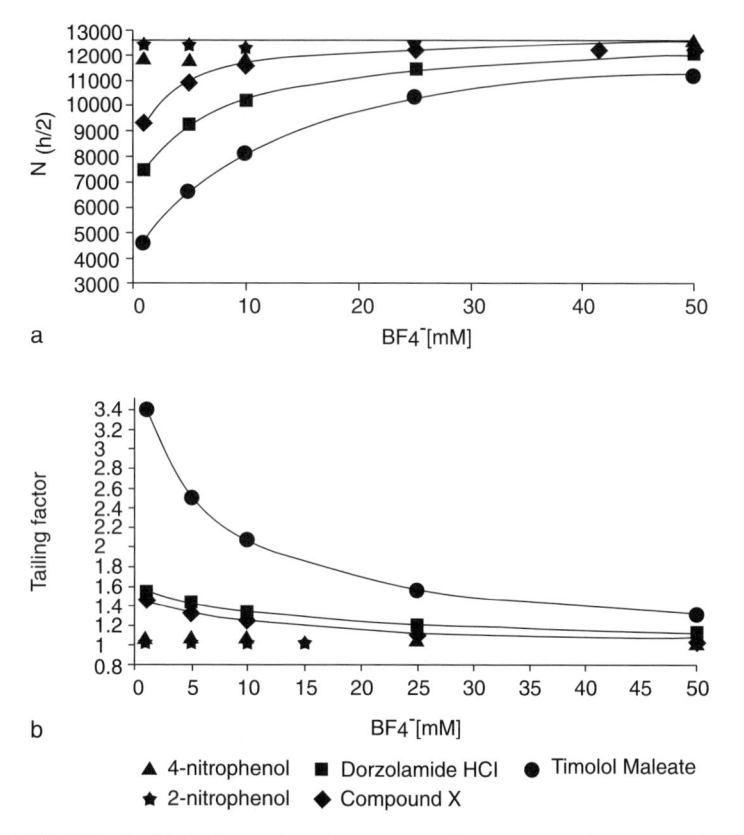

Figure 4-57. Effect of tetrafluoroborate concentration on analyte apparent efficiency and tailing factor. Column: Zorbax Eclipse XDB-C8. Mobile phase: 0.1 v/v% phosphoric acid + xBF$_4$ [1–50 mM]; acetonitrile, ophthalmic compounds (10% acetonitrile), phenols (25% acetonitrile). (A) N$_{(h/2)}$ versus tetrafluoroborate concentration. (B) Tailing factor versus tetrafluoroborate concentration. (Reprinted from reference 153, with permission.)

to 10 mM. Then upon further increase of the counteranion concentration, the efficiency of the basic compounds increases slowly until it achieves the maximum column efficiency (phenols, neutral markers). Also with an increase of BF$_4^-$ counteranion concentration, the tailing factor of basic compounds decreases and approaches the tailing factor of the neutral analytes, phenolic compounds (Figure 4-57B).

It has been shown that the PF$_6^-$ counteranion has had the greatest effect on the improvement of the peak asymmetry at low concentrations compared to other chaotropic additives. At the highest concentration of counteranions (PF$_6^-$, ClO$_4^-$, BF$_4^-$), the number of plates for most of the basic compounds studied was similar to that of the neutral markers. In contrast, the neutral markers, phenols, showed no significant changes in retention and efficiency with increased counteranion concentration.

One of the origins of peak tailing in chromatography can be attributed to energetic surface heterogeneity with overloading of highly energetic adsorption sites [170–175]. Moreover, possible ion-exchange types of interactions with these sites could lead to slow sorption–desorption of solute molecules from the strong sites compared to the weak sites, leading to a further increase in band tailing [176, 177]. It also has been shown by McCalley and others that basic analyte sample loading may also have an effect on peak efficiency [170, 178, 179]. Thus a decrease in sample load has led to the improvement in the efficiency of basic compounds. However, it is sometimes necessary to inject large sample sizes to enable the detection of small impurities with consequent increase in basic analyte tailing factor and decrease in peak symmetry.

However, chaotropic additives can be added to the mobile phase to suppress secondary interactions with the stationary phase. The adsorption of chaotropic counteranions in the adsorbed organic phase on top of the bonded phase can add an electrostatic component to the retention as well as suppressing some undesired secondary interactions leading to peak tailing of protonated basic compounds. The following trend in increase of basic analyte retention factor and decrease of tailing factor was found: $PF_6^- > ClO_4^- \sim BF_4^- > H_2PO_4^-$ [153].

Figure 4-58 shows an overlay of chromatograms for labetalol with different analyte loads from 1 to 50μg using a 10mM dihydrogen phosphate mobile

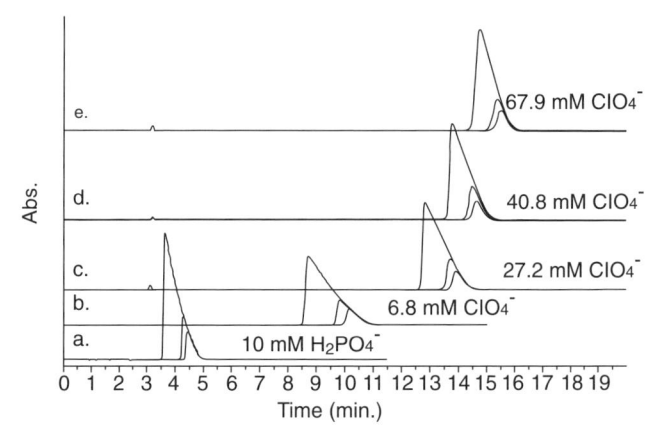

Figure 4-58. Chromatographic overlays of Labetalol analyzed at different analyte concentrations using increasing mobile phase concentration of perchlorate anion. Chromatographic conditions: Column: Zorbax Eclipse XDB-C8. Analyte load: 3.3, 6.5, 31.2μg. (a) 75% 0.1 v/v% H_3PO_4: 25% acetonitrile, (b) 75% 0.05 v/v% $HClO_4$, 25% acetonitrile, (c) 75% 0.2 v/v% $HClO_4$, 25% acetonitrile, (d) 75% 0.4 v/v% $HClO_4$, 25% acetonitrile, (e) 75% 0.5 v/v% $HClO_4$, 25% acetonitrile. (Reprinted from reference 153, with permission.)

phase, at increasing perchlorate anion concentrations. These overlays reveal a typical pattern where the peak tails for different analyte loads coincide, indicating a so-called "thermodynamic overload" that occurs when analyte concentration exceeds the linear region on the adsorption isotherm, and this isotherm curvature inevitably leads to right-angled peaks [180–182].

The greater the chaotropic counteranion concentration, the higher the adsorption capacity and the straighter the analyte isotherm, which results in a shorter tail. Excessive electrostatic interactions are relatively weak in the presence of significant amount of counteranions in the mobile phase, and this would lead to the relatively low initial isotherm slope. Electrostatic interactions are relatively long-distance, which would explain relatively high adsorption capacity and the nonexponential shape of the peak tail. With an increase in counteranion concentration at all analyte loadings, an increase in peak efficiency and decrease in peak tailing can be achieved [153].

Increasing the load of basic analytes in order to increase analyte sensitivity can lead to a decrease in apparent peak efficiency and increase in peak tailing. However, if an analysis must be performed at a relatively high sample load, the addition of a chaotropic additive may be employed to increase the apparent peak efficiency and symmetry. Much higher loading capacities could be obtained by operating columns with these mobile-phase additives without substantial deterioration in efficiency.

4.10.4.5 Applications in the Pharmaceutical Industry.

Since a great majority of drugs include basic functional groups, HPLC behavior of basic compounds has attracted significant interest [183]. Therefore, reversed-phase HPLC separation of organic bases of different pK_a values is of particular importance in the pharmaceutical industry. It is generally recommended that the chromatographic analysis of basic compounds to be carried at 2 pH units less than the analyte pK_a. However, at these conditions the elution of protonated basic compounds may be close to the void volume. Another option might be to analyze these compounds in their neutral form (mobile-phase pH 2 units above the analyte pK_a). Note that going to higher pH values might not be feasible due to the pH stability limit of the packing material, or long analysis times might be obtained for the basic analyte in its neutral form. The advantages of employing chaotropic mobile-phase additives at a pH where the basic analyte is in its fully protonated form provides the chromatographer an additional approach to adjust basic analyte retention and chromatographic selectivity without the need of changing type of column, pH, or organic modifier. The retention behavior of basic compounds containing primary, secondary, tertiary, and quaternary amines can be enhanced as a function of the concentration of chaotropic mobile-phase additives (ClO_4^-, PF_6^-, BF_4^-, $CF_3CO_2^-$) at a low pH. However, it has also been observed that different inorganic counteranions at equimolar concentrations lead to a concomitant increase in retention as well as peak symmetry and increased loading capacity. This was first observed when the chaotropic approach was implemented for the analysis of substituted

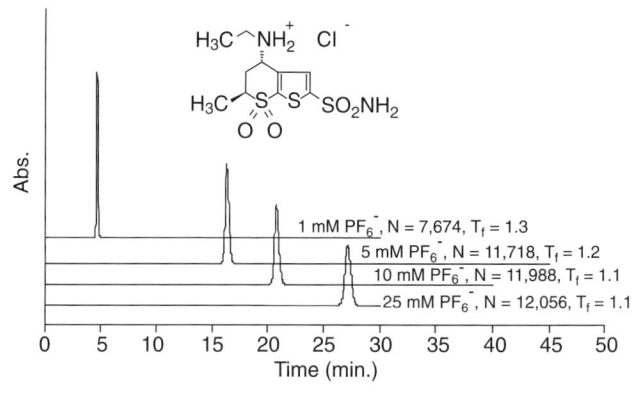

Figure 4-59. Effect of hexafluorophosphate concentration on analyte retention, peak efficiency, $N_{(h/2)}$, and tailing factor. Chromatographic conditions: Column: Zorbax Eclipse XDB-C8. Mobile phase: 90% 0.1 v/v% phosphoric acid + xPF$_6$ [1–25 mM]; 10% acetonitrile; flow rate, 1.0 mL/min; temperature, 25°C; analyte load, 1 µg; wavelength, 254 nm. (Reprinted from reference 153, with permission.)

pyridines and aromatic amines and ophthalmic pharmaceutical compounds [184]. Later Roberts et al. [155] also observed similar effects during the analysis of primary, secondary, and tertiary benzyl amines and antidepressants.

The analysis of Dorzolamide HCl at pH 2 with phosphoric acid shows early elution. The addition of hexafluorophosphate to the mobile phase leads to an enhancement of the retention. Figure 4-59 is an overlay of Dorzolamide HCl chromatograms at four increasing PF$_6^-$ concentrations. As the concentration increased, peak tailing decreased, and peak efficiency and analyte retention increased. Figure 4-60 shows the effect of different counteranions on basic analyte retention and peak efficiency. Depending upon the desired selectivity between a neutral component and a charged basic analyte, a particular chaotropic counteranion could be employed.

Moreover, if a method is to be developed with a chaotropic additive that does not have a buffering capacity, a buffer such as phosphate, maybe employed and the increase in retention can be modulated by the addition of the salt of the chaotropic additive as was shown in Figure 4-48. This approach is particularly useful, especially if other ionogenic species are present in the pharmaceutical mixture. The retention of only the protonated basic compounds can be selectively altered by judicious choice of type and concentration of chaotropic mobile-phase additive without any further mobile-phase pH adjustment. A chaotropic approach could be used for the separation of very polar basic compounds as a fast screening method for the resolution of closely eluting basic species without resorting to changing pH, mobile-phase composition, or type of column (Figure 4-61). These methods are especially useful in reaction monitoring where only a few species are present and reaction conversion needs to be determined.

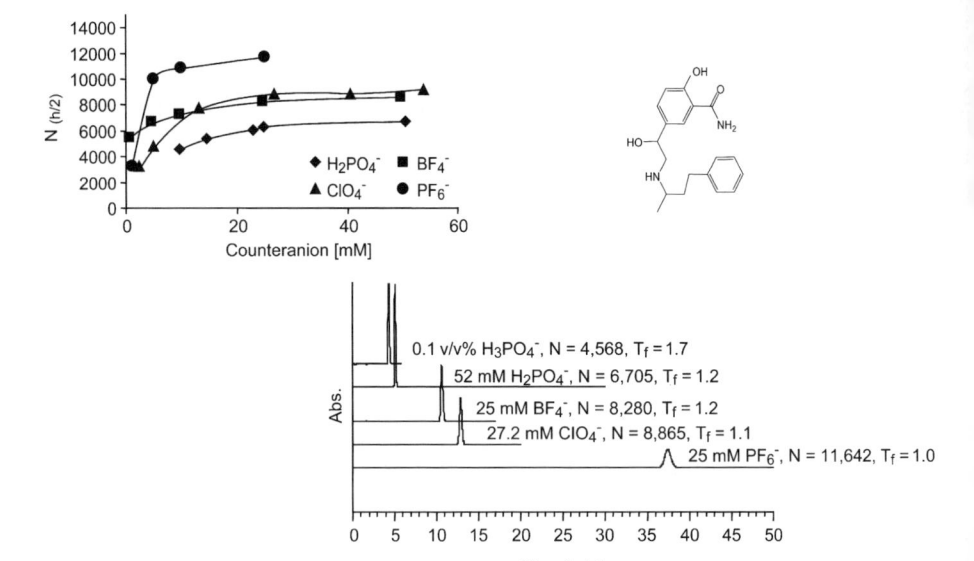

Figure 4-60. Effect of counteranion type and concentration on analyte retention, peak efficiency, $N_{(h/2)}$, and tailing factor, T_f. Chromatographic conditions: Column: Zorbax Eclipse XDB-C8. Mobile phase: 75% aqueous; 25% acetonitrile; flow rate, 1.0 mL/ min; temperature, 25°C; wavelength, 225 nm. (Reprinted from reference 153, with permission.)

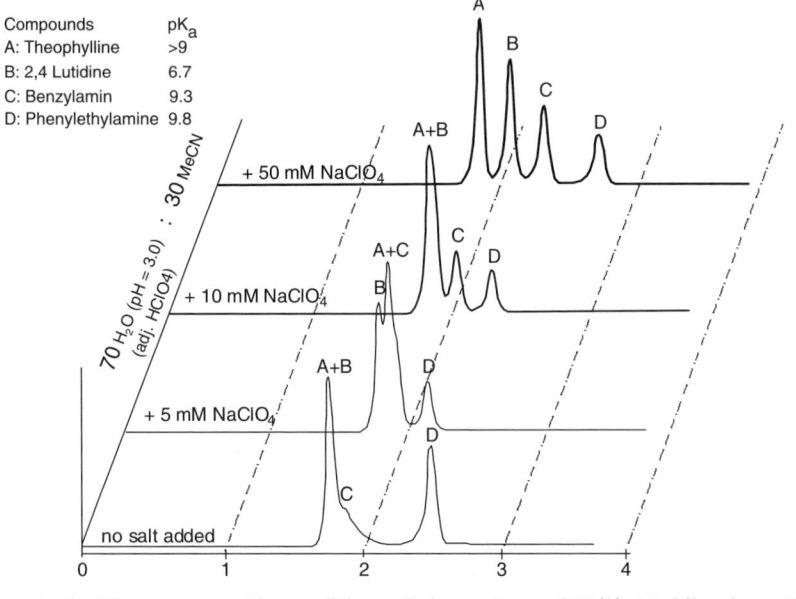

Figure 4-61. Chromatographic conditions: Column: Luna C18(2). Mobile phase: 70% aqueous, 30% acetonitrile. pH adjusted to pH 3 with perchloric acid + x mM ClO_4 adjusted with $NaClO_4$; flow rate, 1.0 mL/min; temperature, 25°C.

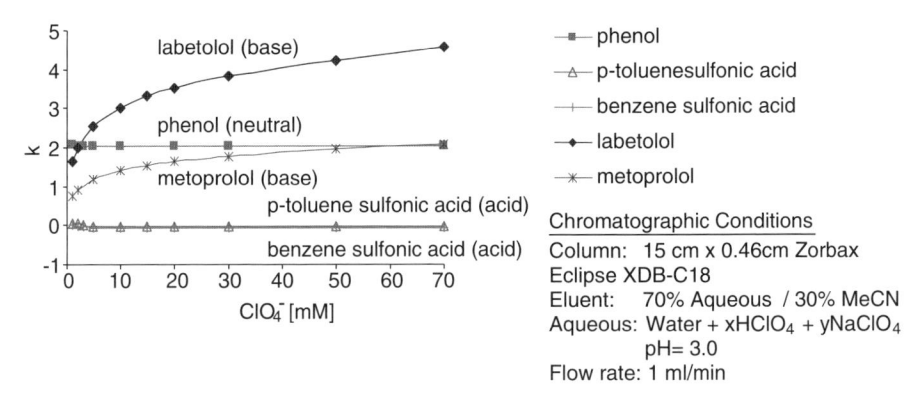

Figure 4-62. Retention factor of acidic, neutral, and basic analytes versus perchlorate concentration. (Reprinted from reference 150, with permission.)

The separation selectivity of a mixture of acidic, basic, and neutral compounds can be altered with the addition of chaotropic mobile-phase additives (Figure 4-62). The retention of the basic compounds can be increased by addition of chaotropic counterions in the mobile phase, while the retention of neutral and acidic compounds is generally unaffected. This is particularly useful during the development of impurity profile methods in the pharmaceutical industry where the retention of a polar protonated basic impurity may be adjusted such that adequate separation selectivity is obtained when unionizable, acidic, or basic (in neutral form) impurities in the drug substance are present. In Figure 4-62 the retention of protonated basic compounds, metoprolol and labetalol, increase while the retention of phenol (in its neutral state) remains constant.

The effect of different chaotropic mobile-phase additives can also assist the chromatographer in achieving adequate retention and resolution of critical pairs in complex mixtures [156].

As shown in Figure 4-63 at an equimolar concentration of chaotropic mobile-phase additives, the greatest increases in retention and resolution between critical pairs of components was achieved employing a 30 mM PF_6^- mobile-phase additive. An increase in retention and an increase in the peak efficiency were obtained, leading to an increase in the resolution of critical pairs of components (A/B and C/D).

The enhancement of retention of very polar basic species is also very important for pharmaceutical applications. For example, Voglibose (VGB) is very hydrophilic; therefore in RPLC even under high aqueous conditions using a mobile phase with 0.1 v/v% of phosphoric acid, it will elute before the void volume. Moreover, VGB has a weak UV chromophore and a very low absorbance at 190 nm. The dose strength for a particular formulation is very low at 0.2 mg, which presents a problem for dissolution in 900-mL vessels, which necessitates the need to either (a) derivatize VGB to produce a suitable

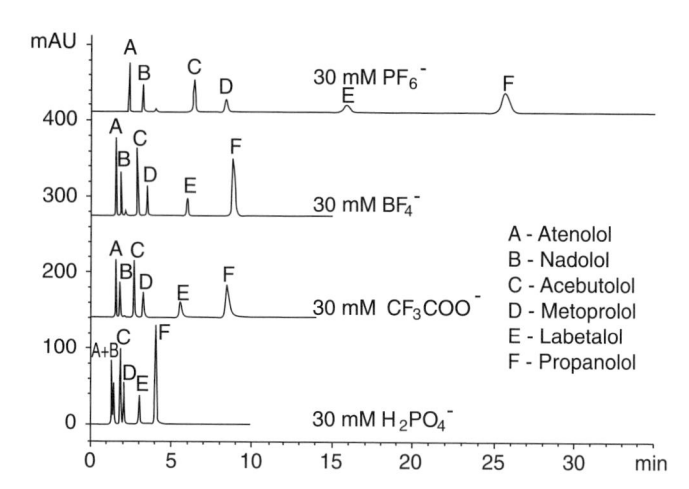

Figure 4-63. Chromatograms of a mixture of beta blockers with different inorganic anions in the mobile phase. Chromatographic conditions: Column: Zorbax Eclipse XDB-C18 (150 × 4.6mm). Mobile phase: Aqueous (pH 3.0)–acetonitrile (70:30); flow rate, 1mL/min. Detection: UV at 225nm. (Reprinted from reference 156, with permission.)

chromophore, (b) use fluorescent or mass spectrometric detection, or (c) use UV detection at 190nm with a suitable chaotropic modifier in the mobile phase. Due to the weak chromophore of VGB, the use of trifluoroacetic acid as a chaotropic agent is prohibitive since it will have a high background absorbance and will significantly reduce the method sensitivity. However, hexaflurophosphate, the most chaotropic ion in the series, is UV transparent at 190nm. In Figure 4-64, 50mM of hexafluorophosphate was added to a 90% 0.1 v/v% phosphoric acid–10% acetonitrile mobile phase, and significant enhancement of the analyte retention was obtained. The VGB peak was eluted and adequately retained and could be properly quantitated.

4.10.4.6 Mulitply Charged Species.
Liophilic anions were also shown to have an effect on the retention of multiply charged peptides. Mono-, di-, tri-, and tetralysine were analyzed with and without the addition of chaotropic anion in the mobile phase. Mono-, di-, tri-, and tetralysine have two, three, four, and five positively charged residues, respectively.

Lysine and lysine analogs (Figure 4-65) are very polar molecules and typically show early elution on traditional C18 reversed phases. Also, lysine has a weak chromophore, and UV detection below 215nm is necessary. Detection of lysine may be challenging when TFA is employed since TFA has a UV cutoff of 210nm. Therefore experiments were performed with the strongest chaotropic agent, hexafluorophosphate, that does not absorb in this UV region. Mobile phases were prepared by keeping the pH constant at 2.1 using 0.1 v/v% H_3PO_4, and the concentration of PF_6 was increased by the addition of KPF_6.

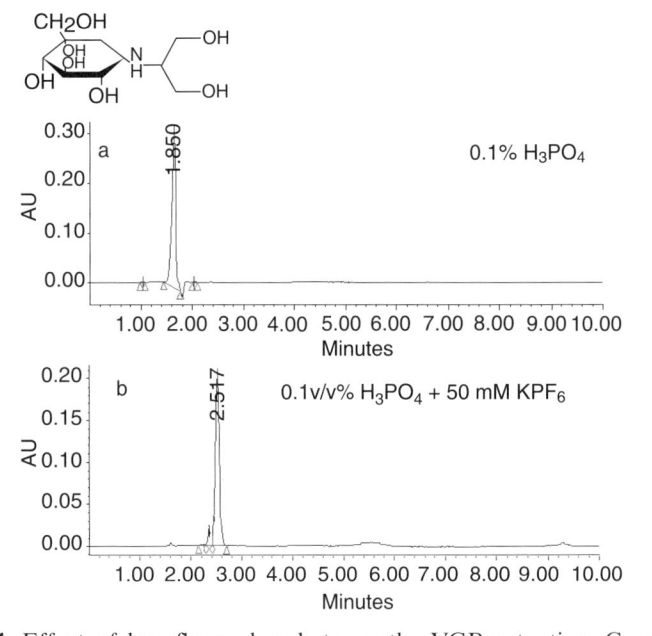

Figure 4-64. Effect of hexafluorophosphate on the VGB retention. Conditions are shown in Figure 4-63.

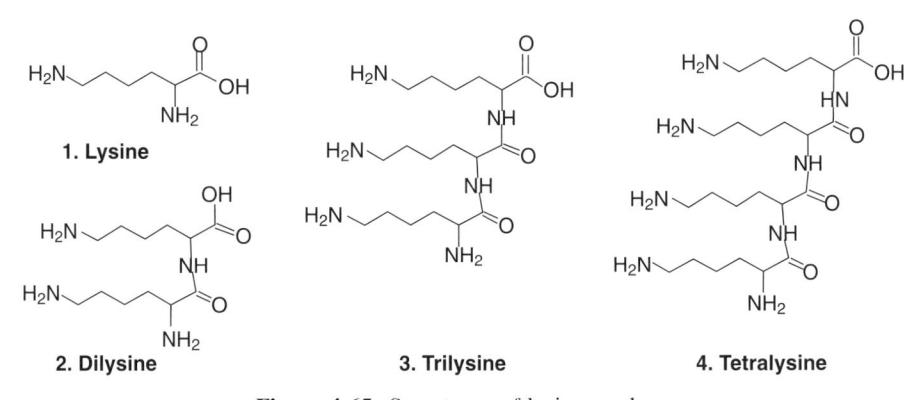

Figure 4-65. Structures of lysine analogs.

When dihydrogen phosphate up to 20 mM was used as the mobile-phase additive, no significant change in the retention was observed as the concentration of the dihydrogen phosphate ion was increased (Figure 4-66). The more polar peptide, tetralysine, elutes first whereas the more hydrophobic peptide, lysine, elutes last.

Keeping the pH constant at pH 2.1 with phosphoric acid, the concentration of hexafluorophosphate was increased from 0 to 50 mM. With the increase of

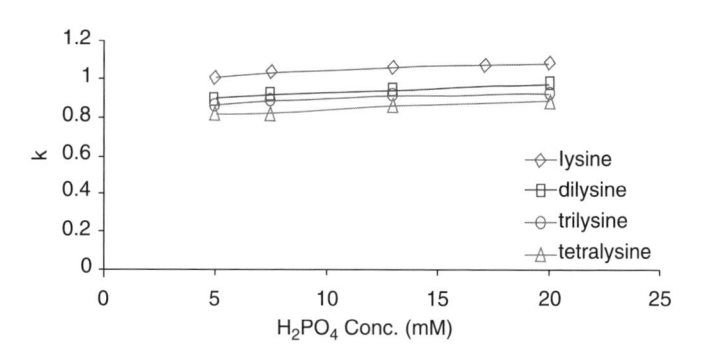

Figure 4-66. Retention of lysine analogs as a function of KH_2PO_4 concentration. Mobile phase: 30% acetonitrile; 70% 0–20 mM KH_2PO_4 adjusted to pH 2.1 with H_3PO_4. Column: Waters Symmetry C18, 150 × 3.0 mm, 3.5 µm; flow rate, 0.4 mL/min; column temperature, 35°C; detection, 210 nm.

the hexafluorophosphate concentration, a significant increase in the retention of the lysine analogs was observed (Figure 4-67). Selectivity changes were also observed upon increasing the hexafluorophosphate concentration in the mobile phase. Using 0–2 mM hexafluorophosphate anion, the peptides elute in the order according to their polarity: When employing hexafluorophosphate concentrations greater than 5 mM, the elution order reverses because of enhanced electrostatic interaction between peptides with greater charge density and PF_6 anion. This will lead to decreased analyte solvation and consequently increased retention. The employment of chaotropic additives for very polar amino acids can be used as an effective approach to enhance the retention and selectivity of the separation.

4.10.4.7 Chaotropic Mobile-Phase Additives (Concluding Remarks). The effect of small inorganic ions of liophilic nature (chaotropic agents) on the retention of basic analytes in reversed-phase HPLC is, to some extent, similar to the effect of amphiphilic mobile-phase additives (ion-pairing agents). Chaotropic counteranions essentially introduce reversible secondary equilibria in the chromatographic system without irreversible modification of the surface and significant alteration of the retention of neutral analytes. They facilitate mass transfer by disruption of the solvation of charged analytes and introduce weak electrostatic component in the retention process, which allows for flexible alteration of the separation selectivity and enhancement of apparent efficiency.

The use of chaotropic counteranions for a chromatographic separation is beneficial as a method development strategy. These modifiers may replace the need for changing column type and/or addition of hydrophobic "ion-pairing" reagents for the more challenging separations. Further studies are needed to fully elucidate the detailed mechanism of chaotropic mobile-phase additives.

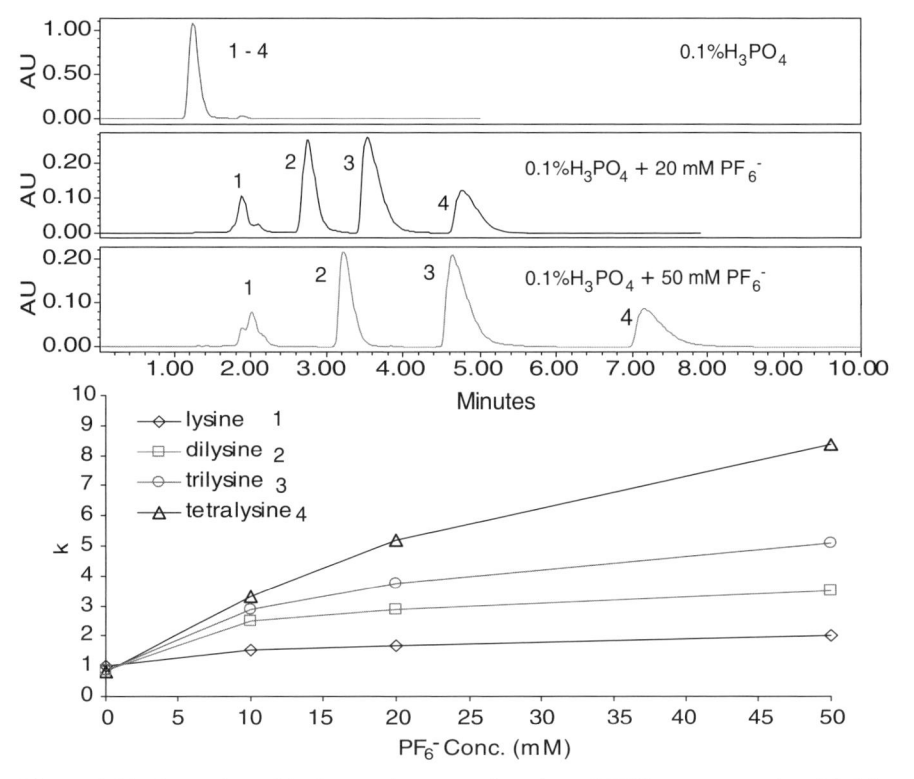

Figure 4-67. Retention of lysine analogs as a function of KPF$_6$ concentration. Mobile phase: 30% acetonitrile; 70% 0.1 v/v% H$_3$PO$_4$ (pH 2.1) concentration of PF$_6^-$ ions adjusted with KPF$_6$ (0–50 mM). Column: Waters Symmetry C18, 150 × 3.0 mm, 3.5 μm; flow rate, 0.4 mL/min; column temperature, 35°C; detection, 210 nm.

4.11 CONCLUDING REMARKS

Analyte retention in reversed-phase chromatography is a complex process based on superposition of many different and relatively weak interactions. The exposure of the analyte molecules to the significant surface of packing material during their migration through the column enhances the differences in these interactions and thus allows for unique selectivity between otherwise similar analytes.

The main process determining analyte retention is the hydrophobic inter-actions with the stationary phase and its competition with organic mobile-phase additive. This simplistic description allows for a rough estimation of the analyte retention and, in principle, is applicable only for very ideal systems, like the separation of alkylbenzenes in methanol/water mixtures.

Analyte ionization, solvation, tautomerization, and other processes alter-ing the analyte appearance in the solution adds significant complexity to the overall retention process. Adsorption of the eluent components and additives on the surface of packing material also influence the analyte migration through the column.

In this chapter we discussed the influence of most known secondary equilibria effects as well as the utilization of the organic eluent component absorption on the surface to describe the analyte retention in reversed-phase HPLC.

RP HPLC is an area of intensive research. Over 5000 papers are published yearly on the theory, development, and practical applications of reversed-phase chromatography. In this chapter we present our vision of the current state of the RP HPLC. We hope that it will be useful for practical chromatographers in their efforts to develop efficient and selective separation methods, and we also hope that it will be encouraging for researchers studying different aspects of HPLC separations.

REFERENCES

1. C. Horvath and W. Melander, Reversed-phase chromatography, in C. Horvath (ed.), *High Performance Liquid Chromatography Advances and Perspectives*, Vol. 2, Academic Press, New York, 1980, pp. 113–303.

2. G. A. Howard and A. J. P. Martin, The separation of the C12–C18 fatty acids by reversed-phase partition chromatography, *Biochem J.* **46** (1950), 532–538.

3. E. W. Abel, F. H. Polland, P. C. Uden, and G. Nickless, A new gas–liquid chromatographic phase, *J. Chromatogr.* **22** (1966), 23–28.

4. H. N. M. Stewart and S. G. Perry, A new approach to liquid partition chromatography, *J. Chromatogr.* **37** (1968), 97–98.

5. R. E. Majors, High-performance liquid chromatography on small particle silica gel, *Anal. Chem.* **44** (1972), 1722–1726.

6. J. L. Bass, B. W. Sands, and P. W. Bratt, *Proceedings of the Silanes Surfaces and Interfaces, Symposium*, Snowmass, Colorado, June 19–21, 1985, pp. 267–279.

7. I. Rustamov, T. Farcas, F. Ahmed, F. Chan, R. LoBrutto, H. M. McNair, and Y. V. Kazakevich, Geometry of chemically modified silica, *J. Chromatogr. A* **913** (2001), 49–63.

8. Y. V. Kazakevich, R. LoBrutto, F. Chan, and T. Patel, Interpretation of the excess adsorption isotherms of organic eluent components on the surface of reversed-phase adsorbents. Effect on the analyte retention, *J. Chromatogr. A* **913** (2001), 75–87.

9. Y. Bereznitski, M. Jaroniec, and M. E. Gangoda, Characterization of silica-based octyl phases of different bonding density: Part II. Studies of surface properties and chromatographic selectivity, *J. Chromatogr. A* **828** (1998), 59–73.

10. L. C. Sander, C. J. Glinka, and S. W. Wise, Silanes, Surfaces, and Interfaces, *Proceedings of the Silanes Surfaces and Interfaces, Symposium*, Snowmass, Colorado, June 19–21, 1985, 431–447.

11. S. Kitahara, K. Tanaka, T. Sakata, and H. Muraishi, Porosity change of silica gels by the alkoxylation of their surfaces, *J. Colloid Interface Sci.* **84** (1981), 519–525.

12. J. J. Kirkland, Development of some stationary phases for reversed-phase high-performance liquid chromatography, *J. Chromatography A* **1060** (2004), 9–21.

13. J. E. O'Gara, B. A. Alden, T. H. Walter, J. S. Peterson, C. L. Niederlaender, and U. D. Neue, Simple preparation of a C8 HPLC stationary phase with an internal polar functional group, *Anal. Chem.* **67** (1995), 3908–3813.

14. T. L. Ascah and B. Feibush, Novel, highly deactivated reversed-phase for basic compounds, *J. Chromatogr.* **506** (1990), 357–369.

15. T. Czajkowska and M. Jaroniec, Selectivity of alkylamide bonded-phase with respect to organic acids under reversed-phase conditions, *J. Chromatogr. A* **762** (1997), 147–158.

16. J. Layne, Characterization and comparison of the chromatographic performance of conventional, polar-embedded, and polar-endcapped reversed-phase liquid chromatography stationary phases, *J. Chromatogr. A* **957**, (2002), 149–164.

17. W. Ecknig, B. Trung, R. Radeglia, and U. Gross, Group separation of alkyl-substituted aromatic Hydrocarbons by high-performance liquid chromatography using perfluorocarbon modified silica gel, *Chromatographia* **16** (1982), 178–182.

18. S. L. Richheimer, M. C. Kent, and M. W. Bernart, Reversed-phase high-performance liquid chromatographic method using a pentafluorophenyl bonded phase for analysis of tocopherols, *J. Chromatogr. A* **677** (1994), 75–80.

19. L. K. Shao and D. C. Locke, Determination of Paclitaxel and related taxanes in bulk drug and injectable dosage forms by reversed phase liquid chromatography, *Anal. Chem.* **69** (1997), 2008–2016.

20. M. Euerby, A. McKeown, and P. Petersson, Chromatographic classification and comparison of commercially available perfluorinated stationary phases for reversed-phase liquid chromatography using principal component analysis, *J. Sep. Sci.* **26** (2003), 295–306.

21. U. Neue, K. Van Tran, P. Iraneta, and B. Alden, Characterization of HPLC packings, *J. Sep. Sci.* **26** (2003), 174–186.

22. F. M. Yamamoto and S. Rokushika, Retention properties of the fluorinated bonded phase on liquid chromatography of aromatic hydrocarbons, *J. Chromatogr. A* **898** (2000), 141–151.

23. K. Jinno, H. Nakamura, and F. M. Yamamoto, Retention characteristics of fluorinated bonded silica phase in reversed-phase liquid chromatography, *Chromatographia* **39** (1994), 285–293.

24. G. Thevenon-Emeric, A. Tchapla, and M. Martin, Role of π–π interactions in reversed-phase liquid chromatography, *J. Chromatogr.* **550** (1991), 267–283.

25. C. Grosse-Rhode, H. G. Kicinski, and A. Kettrup, Comparison of two anthryl-modified silica stationary phases for HPLC separation of PAHs and Nitro-PAHs, *Chromatographia* **29** (1990), 489–494.

26. M. Salo, H. Vuorela, and J. Halmekoski, Effect of organic modifier on the retention of retinoids in reversed-phase liquid chromatography, *Chromatographia* **36** (1993), 147–151.

27. J. Reubsaet and R. Vieskar, Characterization of π–π interactions which determine retention of aromatic compounds in reversed-phase liquid chromatography, *J. Chromatogr. A* **841** (1999), 147–154.

28. J. Nawrocki, M. P. Rigney, A. McCormick, and P. W. Carr, Chemistry of zirconia and its use in chromatography, *J. Chromatogr. A* **657** (1993), 229–282.

29. J. Nawrocki, C. Dunlap, A. McCormick, and P. Carr, Part I. Chromatography using ultra-stable metal oxide-based stationary phases for HPLC, *J. Chromatogr. A* **1028** (2004), 1–30.

30. J. Nawrocki, C. Dunlap, J. Li, J. Zhao, C. McNeff, A. McCormick, and P. Carr, Part II. Chromatography using ultra-stable metal oxide-based stationary phases for HPLC, *J. Chromatogr. A* **1028** (2004), 31–62.

31. J. Dai, X. Yang, and P. Carr, Comparison of the chromatography of octadecyl silane bonded silica and polybutadiene-coated zirconia phases based on a diverse set of cationic drugs, *J. Chromatogr. A* **1005** (2003), 63–82.

32. S. D. West, The prediction of reversed-phase HPLC retention indices and resolution as a function of solvent strength and selectivity, *J. Chromatogr. Sci.* **25** (1987), 122–129; and S. D. West, Correlation of retention indices with resolution and selectivity in reversed-phase HPLC and GC, *J. Chromatogr. Sci.* **27** (1989), 2–12.

33. K. Valko, L. R. Snyder, and J. Glajch, Retention in reversed-phase liquid chromatography as a function of mobile phase composition, *J. Chromatogr.* **656** (1993), 501–520.

34. T. Hamoir, D. L. Massart, W. King, S. Kokot, and K. Douglas, Prediction of initial chromatographic conditions in reversed-phase high-performance liquid chromatography, *J. Chromatogr. Sci.* **31** (1993), 393–400.

35. K. Valko and P. Siegel, New chromatographic hydrophobicity index (φ_0) based on the slope and the intercept of the $\log k'$ *versus* organic phase concentration plot, *J. Chromatogr.* **631** (1993), 49–61.

36. L. R. Synder, *Principles of Adsorption Chromoatography*, Marcel Dekker, New York, 1968.

37. E. Soczewinski and M. Waksmundzka-Hajons, Effect of type and concentration of organic modifier in aqueous eluents on retention in reversed-phase systems, *J. Liq. Chromatogr.* **3** (1980), 1625–1636.

38. P. J. Schoenmakers, H. A. H. Billiet, R. Tijssen, and L. de Galan, Gradient selection in reversed-phase liquid chromatography, *J. Chromatogr.* **149** (1978), 519–537.

39. P. Jandera, H. Colin, and G. Guiochon, Interaction indexes for prediction of retention in reversed-phase liquid chromatography, *Anal. Chem.* **54** (1982), 435–441.

40. E. Soczewniski and G. Matysik, Two types of R_M-composition relationships in liquid–liquid partition chromatography, *J. Chromatogr.* **32** (1968), 458–471.

41. E. Soczewinski, Solvent composition effects in thin-layer chromatography systems of the type silica gel–electron donor solvent, *Anal. Chem.* **41** (1969), 179–182.

42. L. R. Snyder, J. W. Dolan, and J. R. Gant, Gradient elution in high-performance liquid chromatography: I. Theoretical basis for reversed-phase systems, *J. Chromatogr.* **165** (1979), 3–30.

43. U. D. Neue, *HPLC Columns*, Wiley-VCH, New York, 1997.

44. C. Horvath, W. Melander, and I. Molnar, Solvophobic interactions in liquid chromatography with nonpolar stationary phases, *J. Chromatogr.* **125** (1976), 129–156.

45. A. Vailaya and C. Horvath, Retention in reversed-phase chromatography: Partition or adsorption? *J. Chromatogr. A* **829** (1998), 1–27.

46. P. W. Carr, J. Li, A. J. Dallas, D. I. Eikens, and L. C. Tan, Revisionst look at solvophobic driving forces in versed-phased liquid chromatography, *J. Chromatogr. A* **656** (1993), 113–133.

47. B. L. Karger, J. R. Grant, A. Hartkopf, and P. H. Weiner, Hydrophobic effects in reversed-phase liquid chromatography, *J. Chromatogr.* **128** (1976), 65–78.

48. P. J. Schoenmakers, H. A. H. Billet, and L. de Galan, Influence of organic modifiers on the retention behaviour in reversed-phase liquid chromatography and its consequences for gradient elution, *J. Chromatogr.* **185** (1979), 179–195.

49. P. J. Schoenmakers, H. A. H. Billet, and L. de. Galan, Systematic study of ternary solvent behaviour in reversed-phase liquid chromatography, *J. Chromatogr.* **218** (1981), 261–284.

50. Y. V. Kazakevich, R. LoBrutto, F. Chan, and T. Patel, Interpretation of the excess adsorption isotherms of organic eluent components on the surface of reversedphase adsorbents. Effect on the analyte retention, *J. Chromatogr.* **913** (2001), 75–87.

51. P. Zhuang, R. Thompson, and T. O'Brien, A retention model for polar selectivity in reversed phase chromatography as a function of mobile phase organic modifier type, *J. Liq. Chrom. Rel. Technol.* **28** (2005), 1345–1356.

52. U. D. Neue, C. H. J. Phoebe, K. Tran, Y. Cheng, and Z. Lu, Dependence of reversedphase retention of ionizable analytes on pH, concentration of organic solvent and silanol activity, *J. Chromatogr. A* **925** (2001), 49–67.

53. U. D. Neue, E. S. Grumback, J. R. Mazeo, K. Tran, and D. M. Diehl, Chapter 6 (Method development in reversed phase chromatography), in I. D. Wilson (ed.), *Bioanalytical Separations, Handbook of Analytical Separations*, Vol. 4, 2003, pp. 185–214. Elsevier, Amsterdam.

54. C. Horvath, W. Melander, and I. Molnar, Liquid chromatography of ionogenic substances with nonpolar stationary phases, *Anal. Chem.* **49** (1977), 142–154.

55. R. LoBrutto and Y. V. Kazakevich, Retention of ionizible components in reversedphase HPLC, in S. Kromidas (ed.), *Practical Problem Solving in HPLC*, Wiley-VCH (2000), New York, pp. 122–158.

56. R. LoBrutto, A. Jones, Y. V. Kazakevich, and H. M. McNair, Effect of the eluent pH and acidic modifiers on the HPLC retention of basic analytes, *J. Chromatogr. A* **913** (2001), 173–187.

57. K. Jenkins, D. Diehl, D. Morrison, and J. Mazzeo, Utilizing XBridge™ HPLC columns for method development at pH extremes, *LC-GC Europe*, September 2, 2005, pp. 92–93.

58. P. Nikitas and A. Pappa-Louisi, New equations describing the combined effect of pH and organic modifier concentration on the retention in reversed-phase liquid chromatography, *J. Chromatogr. A* **971** (2001), 47–60.

59. R. Kaliszan, P. Wiczling, and M. J. Markuszewski, pH gradient high-performance liquid chromatography: Theory and applications, *J. Chromatogr. A* **1060** (2004), 165–175.

60. S. Heinisch and J. L. Rocca, Effect of mobile phase composition, pH and buffer type on the retention of ionizable compounds in reversed-phase liquid chromatography: Application to method development, *J. Chromatogr. A* **1048** (2004), 183–193.

61. J. Barbosa, J. L. Beltran, and V. Sanz-Nebot, Ionization constants of pH reference materials in acetonitrile–water mixtures up to 70% (w/w), *Anal. Chim. Acta* **288** (1994), 271–278.

62. S. Rodinini and A. Nese, Standard pH values for potassium hydrogen phthalate reference buffer solutions in acetonitrile–water mixtures up to 70 wt.% at various temperatures, *Electrochim. Acta* **32** (1987), 1499–1505.

63. J. Barbosa and V. Sanz-Nebot, Standard pH values for phosphate buffer reference solutions in acetonitrile–water mixtures up to 50% (m/m), *Mikrochim. Acta* **116** (1994), 131–141.

64. S. Espinosa, E. Bosch, and M. Roses, Retention of ionizable compounds on HPLC. 12. The properties of liquid chromatography buffers in acetonitrile–water mobile phases that influence HPLC retention, *Anal. Chem.* **74** (2002), 3809–3818.

65. E. Bosch, S. Espinosa, and M. Roses, Retention of ionizable compounds on high-performance liquid chromatography: III. Variation of pK values of acids and pH values of buffers in acetonitrile–water mobile phases, *J. Chromatogr. A* **824** (1998), 137–146.

66. E. Bosch, P. Bou, H. Allemann, and M. Roses, Retention of ionizable compounds on HPLC: pH scale in methanol–water and the pK and pH values of buffers, *Anal. Chem.* **68** (1996), 3651–3657.

67. E. Bosch, C. Rafols, and M. Roses, Variation of acidity constants and pH values of some organic acids in water–2-propanol mixtures with solvent composition. Effect of preferential solvation, *Anal. Chim. Acta* **302** (1995), 109–119.

68. T. Mussini, A. K. Covingon, P. Longhi, and S. Rondinini, Criteria for standartization of pH measurements in organic solvents and water + organic solvent mixtures of moderate to high permittivities, *Pure Appl. Chem.* **57** (1985), 865–876.

69. P. R. Mussini, T. Mussini, and S. Rondinini, Reference value standards and primary standards for pH measurements in D_2O and aqueous–organic solvent mixtures: New accessions and assessments. *Pure Appl. Chem.* **69** (1997), 1007–1014.

70. J. L. M. Van de Venne, J. L. H. M. Hendrikx, and R. S. Deelder, Retention behaviour of carboxylic acids in reversed-phase column liquid chromatography, *J. Chromatogr.* **167** (1978), 1–16.

71. D. Skyora, E. Tesarova, and M. Popl, Interactions of basic compounds in reversed-phase high-performance liquid chromatography influence of sorbent character, mobile phase composition, and pH on retention of basic compounds, *J. chromatogr. A* **758** (1997), 37–51.

72. R. G. Bates, *Determination of pH: Theory and Practice*, 2nd ed., Wiley, New York, 1964.

73. S. Espinosa, E. Bosch, and M. Roses, Retention of ionizable compounds on HPLC. 5: pH scales and the retention of acids and bases with acetonitrile–water mobile phases, *Anal. Chem.* **72** (2000), 5193–5200.

74. M. Roses and E. Bosch, Influence of mobile phase acid–base equilibria on the chromatographic behavior of protolytic compounds, *J. Chromatogr. A* **982** (2002), 1–30.

75. M. Roses, Determination of the pH of binary mobile phases for reversed phase liquid chromatography, *J. Chromatogr. A* **1037** (2004), 283–298.

76. S. Espinosa, E. Bosch, and M. Roses, Retention of ionizable compounds in high-performance liquid chromatography. IX: Modeling retention in reversed-phase liquid chromatography as a function of pH and solvent composition with acetonitrile–water mobile phases, *J. Chromatogr. A* **947** (2002), 47–58.

77. E. Bosch, P. Bou, H. Allemann, and M. Roses, Retention of ionizable compounds on HPLC: pH scale in Methanol–Water and the pK and pH values of buffers, *Anal. Chem.* **68** (1996), 3651.

78. X. Subirats, E. Bosch, and M. Roses, Retention of ionisable compounds on high performance liquid chromatography. XV: Estimation of the pH variation of aqueous buffers with the change of the acetonitrile fraction of the mobile phase, *J. Chromatogr. A* **1059** (2004), 33–42.

79. S. Espinosa, E. Bosch, and M. Roses, Acid–base constants of neutral bases in acetonitrile–water mixtures, *Anal. Chim. Acta* **454** (2002), 157–166.

80. F. Rived, I. Canals, E. Bosch, and M. Roses, Acidity in methanol–water, *Anal. Chim. Acta* **439** (2001), 315–333.

81. N. A. Izmailov, *Electrokhimiya Rastvorov (Electrochemistry of Solutions)*, State University, Kharkov, 1959.

82. N. A. Izmailov and V. N. Izmailova, The effects of solvation of ions and molecules on acid dissociation in solution, *Zh. Fiz. Khim.* **29** (1955), 1050–1063.

83. L. Sucha and S. Kotrly, *Solution Equilibria in Analytical Chemistry*, Van Nostrand Reinhold, London, 1972.

84. I. Canals, K. Valko, E. Bosch, A. P. Hill, and M. Roses, Retention of ionizable compounds on HPLC. 8. Influence of mobile-phase pH change on the chromatographic retention of acids and bases during gradient elution, *Anal. Chem.* **73** (2001), 4937–4945.

85. R. Ruiz, C. Rafolis, M. Roses, and E. Bosch, A potentially simpler approach to measure aqueous pK_a of insoluble basic drugs containing amino groups, *J. Pharmaceut. Sci.* **92** (2003), 1473–1481.

86. C. Geradin, M. In, L. Allouche, M. Haouas, and F. Taluelle, *In situ* pH probing of hydropthermal solutions by NMR, *Chem. Mater.* **11** (1999), 1285–1292.

87. E. Chufan, F. Suvire, R. Enriz, and J. C. Pedregosa, A potentiometric and spectrophotometric study on acid–base equilibria in ethanol–aqueous solution of acetazolamide and related compounds, *Talanta* **49** (1999), 859–868.

88. R. LoBrutto, A. Jones, A., Y. V. Kazakevich, and H. M. McNair, Effect of the eluent pH and acidic modifiers in high-performance liquid chromatography retention of basic analytes, *J. Chromatogr. A* **913** (2001), 173–187.

89. J. L. Beltran, N. Sanli, G. Fonrodona, D. Baron, G. Ozkan, and J. Barbosa, Spectrophotometric, potentiometric and chromatographic pK_a values of polyphenolic acids in water and acetonitrile–water media, *Anal. Chim. Acta* **484** (2003), 253–264.

90. J. Barbosa, D. Barron, E. Jimenez-Lozano, and V. Sanz-Nebot, Comparison between capillary electrophoresis, liquid chromatography, potentiometric and spectrophotometric techniques for evaluation of pK_a values of zwitterionic drugs in acetonitrile–water mixtures, *Anal. Chim. Acta* **437** (2001), 309–321.

91. M. Bartolini, C. Bertucci, R. Gotti, V. Tumiatti, A. Cavalli, M. Recanatini, and V. Andrisano, Determination of the dissociation constants (pK_a) of basic acetyl-

cholinesterase inhibitors by reversed-phase liquid chromatography, *J. Chromatogr. A* **958** (2002), 59–67.

92. F. Oumada, C. Rafols, M. Roses, and E. Bosch, Chromatographic determination of aqueous dissociation constants of some water-insoluble nonsteroidal antiinflammatory drugs, *J. Pharm. Sci.* **91** (2002), 991–999.

93. S. Espinosa, E. Bosch, and M. Roses, Retention of ionizable compounds in high performance liquid chromatography 14. Acid–base pK values in acetonitrile–water mobile phases, *J. Chromatogr. A* **964** (2002), 55–66.

94, Z. Pawlak, Solvent effects on acid–base behavior. Acidity constants of eight protonated substituted pyridines in (acetonitrile + water), *J. Chem. Thermodynam.* **19** (1987), 443–447.

95. W. S. Hancock, R. C. Chloupek, J. J. Kirkland, and L. R. Synder, Temperature as a variable in reversed-phase high-performance liquid chromatographic separations of peptide and protein samples: I. Optimizing the separation of a growth hormone tryptic digest, *J. Chromatogr. A* **686** (1994), 31–43.

96. F. D. Antia and C. Horvath, High-performance liquid chromatography at elevated temperatures: Examination of conditions for the rapid separation of large molecules, *J. Chromatogr.* **435** (1988), 1–15.

97. N. M. McNern, H. K. Edskes, and D. D. Shukla, Purification of hydrophilic and hydrophobic peptide fragments on a single reversed phase high performance liquid chromatographic column, *Biomed. Chromatogr.* **7** (1993), 15–19.

98. D. V. McCalley, Effect of temperature and flow-rate on analysis of basic compounds in high-performance liquid chromatography using a reversed-phase column, *J. Chromatogr. A* **902** (2000), 311–321.

99. P. L. Zhu, L. R. Synder, J. W. Dolan, N. M. Djordjevic, D. W. Hill, L. Van Heukelem, and T. J. Waeghe, Combined use of temperature and solvent strength in reversed-phase gradient elution III. Selectivity for ionizable samples as a function of sample type and pH, *J. Chromatogr. A* **756** (1996), 51–62.

100. Y. Mao and P. W. Carr, Separation of selected basic pharmaceuticals by reversed-phase and ion-exchange chromatography using thermally tuned tandem columns, *Anal. Chem.* **73** (2001), 4478.

101. R. J. M. Vervoort, E. Ruyter, A. J. J. Debets, H. A. Claessens, C. A. Cramers, and G. J. de Jone, Characterization of reversed-phase stationary phases for the liquid chromatographic analysis of basic pharmaceuticals by thermodynamic data, *J. Chromatogr. A* **964** (2002), 67–76.

102. A. Clausen, T. Dowling, and G. Bicker, Description of the retention behavior and chromatographic measurement of the change in pK_a with temperature of a diastereomeric pair of isoleucine derivatives, *J. Liq. Chromatogr. Relat. Technol.* **25** (2002), 705.

103. K. L. Rowlen and J. M. Harris, Raman spectroscopic study of solvation structure in acetonitrile/water mixtures, *Anal. Chem.* **63** (1991), 964–969.

104. R. G. Bates, M. Paabo, and R. A. Robinson, Interpretation of pH measurements in alcohol–water solvents, *J. Phys. Chem.* **67** (1963), 1833–1838.

105. M. Woodhead, M. Paabo, R. A. Robinson, and R. G. Bates, Buffer solutions of tris(hydroxymethyl)aminomethane for pH control in 50 weight per cent methanol 10°C to 40°C, *Anal. Chem.* **37** (1965), 1291.

106. S. Rondinini, P. R. Mussini, and T. Mussini, Reference values standards and primary standards for pH measurements in organic solvents and water + organic solvant mixturea of moderate to high permittivities, *Pure Appl. Chem.* **59** (1987), 1549–1560.

107. M. Pababo, R. A. Robinson, and R. G. Bates, Dissociation of 4-aminopyridinium ion in 50 weight per cent methanol–water and related acidity functions from 10°C to 40°C, *Anal. Chem.* **38** (1966), 1573.

108. C. B. Castells, C. Rafols, M. Roses, and E. Bosch, Effect of temperature on pH measurements and acid–base equilibria in methanol–water mixtures, *J. Chromatogr. A* **1002** (2003), 41–53.

109. A. Albert and E. P. Serjeant, *The Determination of Ionization Constants*, 2nd ed., Chapman and Hall, Edinburgh, 1971.

110. D. V. McCalley, Effect of temperature and flow-rate on analysis of basic compounds in high-performance liquid chromatography using a reversed-phase column, *J. Chromatogr. A*. **902** (2000), 311–321.

111. S. M. C. Buckenmaier, D. V. McCalley, and M. R. Euerby, Rationalization of unusal changes in efficiency and retention with temperature shown for bases in reversed phase HPLC at intermediate pH, *J. Chromatogr. A* **1060** (2004), 117–126.

112. C. Horvath, W. Melander, and I. Molnar, Liquid chromatography of ionogenic substances with nonpolar stationary phases, *Anal. Chem.* **49** (1977), 142–154.

113. J. H. Knox and G. R. Laird, Soap chromatography—a new high-performance liquid chromatographic technique for separation of ionizable materials: Dyestuff intermediates, *J. Chromatogr.* **122** (1976), 17–34.

114. J. C. Kraak, K. M. Jonker, and J. F. K. Huber, Solvent-generated ion-exchange systems with anionic surfactants for rapid separations of amino acids, *J. Chromatogr.* **142** (1977), 671–688.

115. B. A. Bidlingmeyer, S. N. Deming, W. P. Price, Jr., B. Sachok, and M. J. Petrusek, Retention mechanism for reversed-phase ion-pair liquid chromatography, *J. Chromatogr.* **186** (1979), 419–434.

116. W. Melander and C. Horvath, Mechanistic study on ion-pair reversed-phase chromatography, *J. Chromatogr.* **201** (1980), 211.

117. W. R. Melander and C. Horvath, in M. T. W. Hearn (ed.), *Ion Pair Chromatography: Theory and Biological and Pharmaceutical Applications*, Vol. 31, Marcel Dekker, New York, pp. 27–275.

118. A. Tilly Melin, Y. Askemark, K.-G. Wahlund, and G. Schill, Retention behaior of carboxylic acids and their quaternary ammonium ion pairs in reversed phase chromatography with acetonitrile as organic modifier in the mobile phase, *Anal. Chem.* **51** (1979), 976–983.

119. J. H. Knox and R. A. Hartwick, Mechanism of ion-pair liquid chromatography of amines, neutrals, zwitterions and acids using anionic heteroatoms, *J. Chromatogr.* **204** (1981), 3–21.

120. F. F. Cantwell and S. Puon, Mechanism of chromatographic retention of organic ions on a nonionic adsorbent, *Anal. Chem.* **51** (1979), 623–632.

121. S. Afrashtehfar and F. F. Cantwell, Chromatographic retention mechanism of organic ions on a low-capacity ion exchange adsorbent, *Anal. Chem.* **54** (1982), 2422–2427.

122. R. A. Hux and F. F. Cantwell, Surface adsorption and ion exchange in chromatographic retention of ions on low-capacity cation exchangers, *Anal. Chem.* **56** (1984), 1258–1263.

123. F. F. Cantwell, in J. A. Marinsky and Y. Marcus (eds.), *Ion-Exchange and Solvent Extraction*, Vol. 9, Marcel Dekker, New York, 1985.

124. F. F. Cantwell, Retention model for ion-pair chromatography based on double-layer ionic adsorption and exchange, *J. Pharm. Biomed. Anal.* **2** (1985), 153–164.

125. H.-J. Liu and F. F. Cantwell, Electrical double-layer model for ion-pair chromatographic retention on octadecylsilyl bonded phases, *Anal. Chem.* **63** (1991), 2032–2039.

126. H.-J. Liu and F. F. Cantwell, Electrical double-layer model for sorption of ions on octadecylsilyl bonded phases including the role of residual silanol groups, *Anal. Chem.* **63** (1991), 993–1000.

127. L. L. M. Glavina and F. F. Cantwell, Origin of indirect detection in the liquid chromatography of a neutral sample with an ionic probe using an ODS bonded phase and aqueous mobile phase, *Anal. Chem.* **65** (1993), 268–276.

128. S. G. Weber and J. D. Orr, Establishment and determination of interfacial potentials and stationary phase dielectric constant in reversed-phase liquid chromatography, *J. Chromatogr.* **322** (1985), 433–441.

129. S. G. Weber, Theoretical and experimental studies of electrostatic effects in reversed-phase liquid chromatography, *Talanta* **36** (1989), 99–106.

130. J. Ståhlberg, The Gouy–Chapman theory in combination with a modified langmuir isotherm as a theoretical model for ion-pair chromatography, *J. Chromatogr.* **356** (1986), 231–245.

131. J. Ståhlberg and A. Furängen, An experimental verification of the electrostatic theory for ion pair chromatography, *Chromatographia* **24** (1987), 783–789.

132. J. Ståhlberg and I. Haegglund, Adsorption isotherm of tetrabutylammonium ion and its relation to the mechanism of ion pair chromatography, *Anal. Chem.* **60** (1988), 1958–1964.

133. J. Ståhlberg and A. Bartha, Extension of the electrostatic theory of reversed-phase ion-pair chromatography for high surface concentrations of the adsorbing amphiphilic ion, *J. Chromatogr.* **456** (1988), 253–265.

134. A. Bartha, J. Ståhlberg, and F. Szokoli, Extension of the electrostatic model of reversed-phase ion-pair high-performance liquid chromatography to include the effect of the eluent pH, *J. Chromatogr.* **552** (1991), 13–22.

135. J. Ståhlberg, Electrostatic retention model for ion-exchange chromatography, *Anal. Chem.* **66** (1994), 440–449.

136. J. Ståhlberg, Retention models for ions in chromatography, *J. Chromatogr. A* **855** (1999), 3–55.

137. N. Bjerrum, Ionic association. I. Influence of ionic association on the activity of ions at moderate degrees of association, *Kgl. Danske Videnskab Selskab* **7** (1926), 1–48.

138. C. A. Kraus, The ion-pair concept: Its evolution and some applications, *J. Phys. Chem.* **60** (1956), 129–141.

139. B. M. Atherton and S. Y. Weissman, Association between sodium and naphthalenide ions, *J. Am. Chem. Soc.* **83** (1961), 1330–1334.

140. D. J. W. Grant and T. Higuchi, *Solubility Behavior of Organic Compounds*, John Wiley & Sons, New York, 1990, pp. 399–433.

141. H. Sadek and R. M. Fuoss, Electrolyte–solvent interaction. IV. Tetrabutylammonium bromide in methanol–carbon tetrachloride and methanol–heptane mixtures, *J. Am. Chem. Soc.* **76** (1954), 5897–5901.

142. R. C. Roberts and M. Szwarc, The chemistry of the radical anion of tetraphenylethylene, *J. Am. Chem. Soc.* **87** (1965), 5542–5548.

143. T. Higuchi, A. Michaelis, and A. Hurwitz, Ion pair extraction of pharmaceutical amines. Role of dipolar solvating agents in extraction of dextromethorphan, *Anal. Chem.* **39** (1967), 974–979.

144. T. Higuchi, A. Micaelis, and J. H. Rytting, Role of solvating agents in promoting ion pair extraction, *Anal. Chem.* **43** (1971), 287–294.

145. A. Bartha, G. Vigh, H. Billet, and L. de Galan, Effect of the type of ion-pairing reagent in reversed-phase Ion-pair chromatography, *Chromatographia* **20** (1985), 587–590.

146. J. R. Rydall and P. M. Macdonald, Investigation of anion binding to neutral lipid membranes using 2H NMR, *Biochemistry* **31** (1992), 1092–1099.

147. Y. Hatefi and W. G. Hanstein, *Proc. Natl. Acad. Sci. USA* **62** (1969), 1129.

148. K. D. Collins and M. W. Washabaugh, The Hofmeister effect and the behavior of water at interfaces, *Q. Rev. Biophys.* **8** (1985), 323–422.

149. M. G. Cacace, E. M. Landay, and J. J. Ramsden, The Hofmeistrer series: Salt and solvent effects on interfacial phenomena, *Q. Rev. Biophys.* **30** (1997), 241–277.

150. R. LoBrutto, A. Jones, Y. Kazakevich, and H. M. McNair, Effect of the eluent pH and acidic modifiers on the HPLC retention of basic analytes, *J. Chromatogr. A* **913** (2001), 173–187.

151. R. LoBrutto, A. Jones, and Y. Kazakevich, Effect of counteranion concentration on HPLC retention of protonated basic analytes, *J. Chromatogr. A* **913** (2001), 189–196.

152. R. Thompson, Z. Ge, N. Grinberg, D. Ellison, and P. Tway, Mechainistic aspects of the stereospecific interaction for aminoindanol with a crown ether column, *Anal. Chem.* **67** (1995), 1580–1587.

153. L. Pan, R. LoBrutto, Y. V. Kazakevich, and R. Thompson, Influence of inorganic mobile phase additives on the retention, efficiency and peak symmetry of protonated basic compounds in reversed-phase liquid chromatography, *J. Chromatogr. A* **1049** (2004), 63–73.

154. K. Pilorz and I. Choma, Isocratic reversed-phase high-performance liquid chromatographic separation of tetracyclines and flumequine controlled by a chaotropic effect, *J. Chromatogr. A* **1031** (2004), 303–311.

155. J. M. Roberts, A. R. Diaz, D. T. Fortin, J. M. Friedle, and S. D. Piper, Influence of the Hofmeister series on the retention of amines in reversed-phase liquid chromatography, *Anal. Chem.* **74** (2002), 4927–4933.

156. A. Jones, R. LoBrutto, and Y. V. Kazakevich, Effect of the counter-anion type and concentration on the liquid chromatography retention of β-blockers, *J. Chromatogr. A* **964** (2002), 179–187.

157. J. Flieger, The effect of chaotropic mobile phase additives on the separation of selected alkaloids in reverse phase high performance liquid chromatography, *J. Chrom. A* **1113** (2006), 37–44.

158. F. Gritti and G. Guiochon, Role of the buffer in retention and adsorption mechanism of ionic species in reversed-phase liquid chromatography: I. Analytical and overloaded band profiles on Kromasil-C$_{18}$, *J. Chromatogr. A* **1038** (2004), 53–66.

159. F. Gritti and G. Guiochon, Effect of the ionic strength of salts on retention and overloading behavior of ionizable compounds in reversed-phase liquid chromatography: I. XTerra-C$_{18}$, *J. Chromatogr. A* **1033** (2004), 43–55.

160. F. Gritti and G. Guiochon, Effect of the ionic strength of salts on retention and overloading behavior of ionizable compounds in reversed-phase liquid chromatography: II. Symmetry-C$_{18}$, *J. Chromatogr. A* **1033** (2004), 57–69.

161. C. Horvath, W. Melander, I. Molnar, and P. Molnar, Enhancement of retention by ion-pair formation in liquid chromatography with nonpolar stationary phases, *Anal. Chem.* **49** (1977), 2295.

162. A. Sokolowski, Zone formation in ion-pair HPLC. I. Effects of adsorption of organic ions on established column equilibria, *Chromatographia* **22** (1986), 168–173.

163. A. Sokolowski, Zone formation in ion-pair HPLC. II. System peak retention and effects of desorption of organic ions on established column equilibria, *Chromatographia* **22** (1986), 177–183.

164. I. Hagglund and J. Ståhlberg, Ideal model of chromatography applied to charged solutes in reversed-phase liquid chromatography, *J. Chromatogr. A* **761** (1997), 3–47.

165. Y. V. Kazakevich, R. LoBrutto, and R. Vivilecchia, Reversed-phase HPLC behavior of chaotropic counteranions, *J. Chromatogr. A* **1064** (2005), 9–18.

166. Y. V. Kazakevich, R. LoBrutto, F. Chan, and T. Patel, Interpretation of the excess adsorption isotherms of organic eluent components on the surface of reversed-phase adsorbents: Effect on the analyte retention, *J. Chromatogr. A* **913** (2001), 75–87.

167. D. R. Crow, *Principles and Applications of Electrochemistry*, Chapman and Hall Chemistry Textbook Series, New York, 1974.

168. I. L. Kazakevich, Ph. D. thesis, Seton Hall, NJ, 2005.

169. I. L. Kazakevich, N. H. Snow, Adsorption behavior of hexafluorophosphate on selected bonded phases, *J. Chromatogr. A* **119** (2006), 43–50.

170. S. M. C. Buckenmaier, D. V. McCalley, and M. R. Euerby, Overloading study of bases using polymeric RP-HPLC columns as an aid to rationalization of overloading on silica-ODS phases, *Anal. Chem.* **74** (2002), 4672–4681.

171. M. J. Wirth, D. J. Swinton, and M. D. Ludes, Adsorption and diffusion of single molecules at chromatographic interfaces, *J. Phys. Chem. B.* **107** (2003), 6258–6268.

172. F. Gritti, G. Gotmar, B. J. Stanley, and G. Guiochon, Determination of single component isotherms and affinity energy distribution by chromatography, *J. Chromatogr. A* **988** (2003), 185–203.

173. T. Fornstedt, G. Zhong, and G. Guiochon, Peak tailing and slow mass transfer kinetics in nonlinear chromatography, *J. Chromatogr. A* **742** (1996), 55–68.

174. T. Fornstedt, G. Zhong, and G. Guiochon, Peak tailing and mass transfer kinetics in linear chromatography, *J. Chromatogr. A* **741** (1996), 1–12.

175. G. Gotmar, T. Fornstedt, and G. Guiochon, Peak tailing and mass transfer kinetics in linear chromatography dependence on the column length and the linear velocity of the mobile phase, *J. Chromatogr. A* **831** (1999), 17–35.

176. J. C. Giddings, *Unified Separation Science*, Wiley Interscience, New York, 1991.

177. J. W. Dolan and L. R. Snyder, *Troubleshooting LC Systems*, Humana Press, Clifton, NJ, 1989.

178. D. V. McCalley, Influence of sample mass on the performance of reversed phase columns in the analysis of strongly basic compounds by high performance liquid chromatography, *Journal of Chromatogr. A* **793** (1998), 31–46.

179. D. V. McCalley, Selection of suitable stationary phases and optimum conditions for their application in the separation of basic compounds by reversed-phase HPLC, *J. Sep. Sci.* **26** (2003), 187–195.

180. F. Riedo and E. S. Kovats, Adsorption from liquid mixtures and liquid chromatography, *J. Chromatogr. A* **239** (1982), 1–32.

181. J. F. K. Huber and R. G. Gerritse, Evaluation of dynamic gas chromatographic methods for the determination of adsorption and solution isotherms, *J. Chromatogr.* **58** (1971), 137–158.

182. H. L. Wang, J. L. Duda, and C. J. Radke, Solution adsorption from liquid chromatography, *J. Colloid Interface Sci.* **66** (1978), 152–159.

183. R. S. DeWitte, M. McBrien, and E. Kolovanov, Intelligent optimization of HPLC separations based on chemical structures benefiting from unified knowledge base, presentation by Advanced Chemistry Development, Toronto, Canada, March 5–8, 2001, New Orleans, Pittcon 2001.

184. R. LoBrutto and Y. V. Kazakevich, Effect of chaotropic anions on analyte retention, in *Proceedings, 22nd International Symposium on High Performance Liquid Phase Separations and Related Techniques*, St. Louis, May 2–8, 1998.

185. R. LoBrutto and Y. V. Kazakevich, Chaotropic effects in RP-HPLC, in E. Grushka and N. Grinberg (eds.), *Advances in Chromatography*, Vol. 44, Taylor & Francis, Boca Raton, FL, (2006), pp. 291–314.

5

NORMAL-PHASE HPLC

Yong Liu and Anant Vailaya

5.1 INTRODUCTION

High-performance liquid chromatography (HPLC) is a separation tool *par excellence* for the analysis of compounds of wide polarity. Since its inception approximately four decades ago, HPLC has revolutionized numerous disciplines of science and technology. Among the various modes of HPLC, reversed-phase and normal-phase chromatography (NPC) are employed most commonly in separation. Normal-phase chromatography was the first liquid chromatography mode, discovered by M. S. Tswett in 1903, and it is well established as evidenced by a plethora of books and articles that have been published in recent years. In this chapter we describe a simplified overview of the theory and practice of normal-phase chromatography.

5.2 THEORY OF RETENTION IN NORMAL-PHASE CHROMATOGRAPHY

Unlike the more popular reversed-phase chromatographic mode, normal-phase chromatography employs polar stationary phases, and retention is modulated mainly with nonpolar eluents. The stationary phase is either (a) an inorganic adsorbent like silica or alumina or (b) a polar bonded phase containing cyano, diol, or amino functional groups on a silica support. The mobile phase is usually a nonaqueous mixture of organic solvents. As the polarity of the mobile phase decreases, retention in normal-phase chromatography

HPLC for Pharmaceutical Scientists, Edited by Yuri Kazakevich and Rosario LoBrutto
Copyright © 2007 by John Wiley & Sons, Inc.

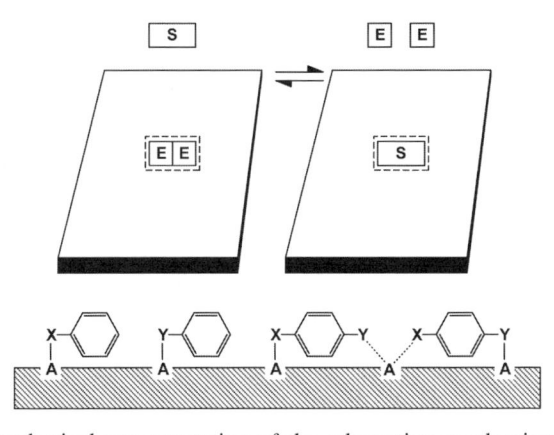

Figure 5-1. Hypothetical representation of the adsorption mechanism of retention in normal-phase chromatography. S denotes sample molecule, E denotes molecule of strong polar solvent, and X and Y are polar functional groups of the stationary phase. Prior to retention, the surface of stationary phase is covered with a monolayer of solvent molecules E. Retention in normal-phase chromatography is driven by the adsorption of S molecules upon the displacement of E molecules. The solvent molecules that cover the surface of the adsorbent may or may not interact with the adsorption sites, depending on the properties of the solvent. (Reprinted from reference 1, with permission.)

increases. Figure 5-1 illustrates the mechanism of retention in NPC [1]. Retention is governed by the extent to which the analyte molecules displace the adsorbed solvent molecules on the surface of the stationary phase. This retention model based on adsorption was first proposed by Snyder [2–5] to describe retention on silica and alumina adsorbents and later extended to explain retention on polar bonded phases, such as diol-, cyano-, and amino-bonded silica. Snyder assumed a homogeneous surface so that adsorption energies for solute and solvent molecules are constant. The stoichiometry of solute–solvent competition can be given by

$$S_m + nE_a \leftrightarrow S_a + nE_m \tag{5-1}$$

m and a refer to solute (S) and solvent (E) molecules in the mobile and adsorbed phases, respectively. n is the coefficient that takes into account different adsorption cross sections for solute and solvents; that is, adsorption of a solute molecule displaces n solvent molecules in the adsorbed monolayer. For a binary mobile-phase system consisting of a weak nonpolar solvent and a strong polar solvent, adsorption of the weak solvent can be ignored. Therefore, solute retention can be expressed by

$$\ln(k_2) = \ln(k_1) - \frac{A_S}{A_E} \ln(N_E) \tag{5-2}$$

Here, A_S is the solute cross-sectional area, A_E is the molecular area of the strong solvent, N_E is the mole fraction of the strong solvent in the mobile phase, k_2 is retention factor of the solute in the binary mobile-phase mixture, and k_1 is the retention factor in the strong solvent alone.

Yet another adsorption-based retention model similar to that of Snyder was proposed by Soczewinski [6] to describe the retention in NPC. It assumes that retention in NPC is the product of competitive adsorption between solute and solvent molecules for active sites on the stationary phase surface. The stationary-phase surface consists of a layer of solute and/or solvent molecules, but, unlike the former, the latter model assumes an energetically heterogeneous surface where adsorption occurs entirely at the high-energy active sites, leading to discrete, one-to-one complexes of the form

$$S_m + qE_a\text{-}A^* \leftrightarrow S\text{-}A^* + qE_m \tag{5-3}$$

A^* is an active surface site and q refers to the number of substituents on a solute molecule that are capable of simultaneously interacting with the active site. This equation takes into account the possibility of an analyte molecule's interaction with multiple sites. Based on this model, the solute retention factor can be expressed by the following equation, which is similar to Snyder's:

$$\log k_2 = d - q \log N_E \tag{5-4}$$

where d is a constant. Comparison of the two models reveals that both predict a linear $\log k_2$ versus $\log N_E$ plot. Snyder's model predicts that the slope of this line should be the ratio of the molecular areas of solute and solvent, whereas Soczewinski's model predicts that the slope is the number of strongly adsorbing substituent groups (number of adsorption sites on the analyte) on the solute.

In practice, it was found that equations (5-1) and (5-2) are most reliable for less polar solvents and solute molecules on alumina or silica stationary phases only. Neither of the models is entirely satisfactory in the forms presented, particularly for predicting retention behavior on bonded stationary phases. These phases contain strongly adsorbing active sites as assumed in Soczewinski's model, but the solute molecular area and not just polar substituents are known to play an important role in competitive adsorption as assumed by Snyder. Furthermore, secondary solvent effects resulting from solute–solvent interactions in both the mobile and adsorbed phases are not taken into consideration in either model. These effects, such as hydrogen bonding, give rise to some of the most useful changes in retention and often are an important source of chromatographic selectivity [7, 8].

Another experimental deviation from equations (5-1) or (5-2) was determined to be due to the localization of solvent molecules onto the adsorption sites of stationary phase resulting from silanophilic interactions. When the polar substitution groups of a solvent molecule interact strongly with the polar groups on the surface of the column packing, they become attached or localized onto the stationary-phase surface. An important consequence of solvent localization is the apparent change in the solvent strength value of a polar solvent. (Solvent strength is presented by ε^0, which is determined empirically by using polyaromatic hydrocarbons that do not localize but lie flat on a surface. Solvent with larger value of ε has stronger elution power [1].) Consequently, the solvent strength does not vary linearly with the concentration of the stronger solvent for a binary mixture where one solvent is stronger than the other [7]. There is competition between the two solvents for the active sites of the adsorbent and the stronger solvent will preferentially adsorb, resulting in a more concentrated adsorbed layer of the stronger solvent. For instance, the dependence of solvent strength for several binary mixtures on alumina as adsorbent shows a large increase in solvent strength due to a small increase in the concentration of a polar solvent at low concentrations. But at the other extreme, a relatively large change in the concentration of the polar solvent affects the solvent strength of the mobile phase to a lesser extent. In the case of low concentration of polar solvent before the localization on the surface of stationary phase reaches saturation, a small change of the polar solvent concentration can greatly affect the number of polar active sites on the column packing. As a consequence, significant variations of analytes retention are observed. Once the polar active sites of the stationary phase are localized completely, change of polar solvent concentration will have a smaller impact on analyte retention.

These deficiencies were addressed by revising Snyder's model as follows [8]. To account for the preferential adsorption of solute and solvent onto the strong sites, empirical A_S and N_E values larger than those calculated from molecular dimensions are used based on experimental observation. The revised model acknowledges the tendency of polar molecules to localize on the strongly adsorbing active site and expresses solute retention in terms of the solvent strength as follows:

$$\log k_2 = \log k_1 - a' A_S \left(\varepsilon_1^0 - \varepsilon_2^0 \right) \tag{5-5}$$

where a' is an adsorbent activity factor, ε_1^0 and ε_2^0 are solvent strengths for solvent 1 and 2, and A_S is the analyte cross-sectional area on the adsorbent surface. The "analyte" cross-sectional area can be predicted from molecular dimensions. Secondary solute–solvent interactions are incorporated into the revised model by adding extra terms denoted by Δ for each of the solvents as follows:

$$\log k_2 = \log k_1 - a' A_S \left(\varepsilon_1^0 - \varepsilon_2^0 \right) + \left(\Delta_2 - \Delta_1 \right) \tag{5-6}$$

When a nonlocalizing, nonpolar solvent such as hexane is employed as a weak solvent, the equation can be further simplified so that

$$\log k_2 = \log k_h - a' A_s \varepsilon_2^0 + \Delta_2 \qquad (5\text{-}7)$$

assuming hexane does not induce any secondary solvent effects and its solvent strength is zero. Here k_h is the analyte retention factor in pure hexane. Equation (5-7) has been found useful to understand the fundamental principles governing the retention behavior as far as solute, solvent, and bonded-phase properties are concerned. For instance, by fitting equation (5-7) to the experimental NPC data, the extent of solute localization can be determined by comparing the slopes of a $\log k_2$ versus ε_2^0 plot, provided that the molecular cross section can be estimated accurately.

5.3 EFFECT OF MOBILE PHASE ON RETENTION

Selection of suitable mobile-phase system is critical in NPC to achieve the desired separation [4]. In general, a suitable solvent should have the following properties: low viscosity, compatibility with detection system (for instance, solvent should be transparent at wavelength of detection if UV is used as detector), available in pure state, low flammability and toxicity, highly inert, and adequate solubility for solutes. Unlike RPLC, analytes become less retained as solvent strength (solvent polarity) increases. Solvent strength in NPC can be represented by ε^0, and values of ε^0 for some commonly used NPC solvents are listed in Table 5-1 for silica as column packing [1]. Relative solvent strength for other NPC column packings such as alumina and polar bonded phases follow the same trend as in the table; that is, larger values of ε^0 are obtained for more polar solvents. Ideally, the mobile-phase strength should be chosen to maintain analyte retention factor within the optimum range of $1 \leq k' \leq 5$ with selectivity values sufficient to reach a satisfactory resolution.

In general, binary mobile phases, such as a mixture of a nonselective solvent hexane with a polar solvent, are used for NPC separations. If separation cannot be achieved by adjusting mobile phase strength (change the concentration of one of the components in a binary mixture), then variation of polar solvent nature has to be pursued. Snyder has developed a useful scheme to classify solvents (nonelectrolytic solvents) nature based on their interactions with solutes and the stationary phase [9]. This approach should not be taken as concrete rules but rather as a phenomological approach. The property of a solvent is characterized by the three most important parameters, which are its proton-acceptor (Xe), proton-donor (Xd), and dipole-donor (Xn) affinity. Each of these contributes to the overall polarity of the solvent, which in turn is related to its chromatographic strength. Rohrschneider determined the values of these parameters from distribution coefficients of test solutes such as ethanol, dioxane, and nitrobenzene [10]. A medium polar solvent—such as chloroform, which has a polarity of 4.31—involves 31% proton acceptor, 35% proton donor, and 34% dipole interactions. If the parameter values of the solvents are plotted on a triple coordinate system, various solvents can be grouped into

TABLE 5-1. NPC Solvent Strength (ε^0) and Selectivity[a] of Various Solvents Employed in HPLC

Solvent	ε^0	Localization	Basic?	UV[b]
Hexane, heptane, octane	0.00	No	c	201
1,1,2-Triflurotrichloroethane (Freon FC-113)	0.02	No	c	235
Chloroform	0.26	No	c	247
1- or 2-Chloropropane	0.28	No	c	225
Methylene chloride	0.30	No	c	234
2-Propyl ether	0.32	Minor	c	217
1,2-Dichloroethane	0.34	No	c	234
Ethyl ether	0.38	Yes	Yes	219
MTBE[d]	0.48	Yes	Yes	225
Ethyl acetate	0.48	Yes	No	256
Dioxane	0.51	Yes	Yes	215
Acetonitrile	0.52	Yes	No	192
THF	0.53	Yes	Yes	230
1- or 2-Propanol	0.60	Yes	e	214
Methanol	0.70	Yes	e	210

[a]Silica used as absorbent.
[b]Minimum UV wavelength; assumes that maximum baseline absorbance (100% B) is 0.5 AU.
[c]Solvent basicity is irrelevant for nonlocalizing solvents.
[d]Methyl *t*-butyl ether.
[e]Different selectivity due to presence of proton donor group.
Source: Reprinted from Ref. 1, with permission.

eight classes (Figure 5-2) [9]. Solvents within each class should show similar selectivity for a set of components, while the nature of solvents from different classes are quite different and may impart differences in selectivity for the same set of components. In NPC method development, replacing solvents belonging to the same selectivity class cannot offer substantial variation in chromatographic separation. Therefore, it is recommended to select solvents that are placed close to the apices of the triangle for maximum selectivity. Common solvents in group I are isopropyl ether and MTBE, group VII solvents include dichloromethane and 1,2-dichloroethane, and chloroform and fluoro-alcohols constitute group VIII solvents. Solvent mixtures having the same elution strength but different selectivities are called isoelutropic mobile phases.

Binary mixtures, however, have only limited abilities for controlling mobile-phase selectivity. Therefore, ternary and even quaternary mobile phases that contain two or more different polar solvents along with a nonpolar solvent are often used to achieve the required selectivity. If the ratio of the concentration of two polar solvents is constant but the sum of the their concentration is being changed with respect to that of the nonpolar solvent, the effect on retention is much the same as when the concentration of the single strong solvent

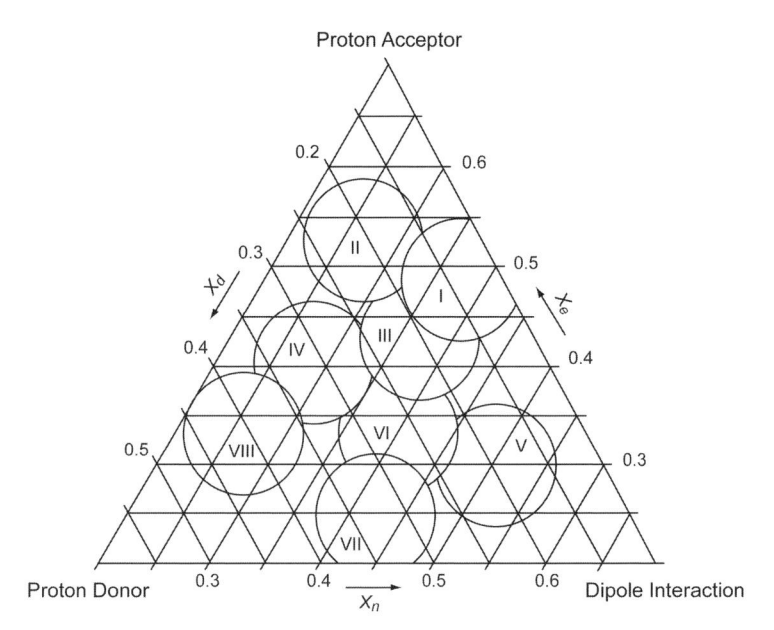

Figure 5-2. Snyder's selectivity triangle for solvents. (Reprinted from reference 9, with permission.)

changed in a binary mobile phases. On the other hand, if the sum of the two polar solvents stays constant but the ratio is variable, larger effects on the selectivity of separation are observed than in the system where the ratio is constant. This is attributable to changes in dipole–dipole and proton–donor–acceptor interactions between polar solvents and the analytes. Such selectivity tuning is the main purpose of using ternary mobile phases in NPC. A phenomenological approach for the appropriate selection of ternary mobile mixture based on Snyder's solvent selectivity triangle concept combined with a statistical approach can be applied [11–15]. As can be seen in Figure 5-4, a seven-run design is used. A primary binary solvent mixture such as hexane-MTBE with the solvent strength that is convenient for the separation is first selected. This binary mixture represents one corner of the selectivity triangle. Two other binary mixtures, namely, hexane-dichloromethane and hexane-chloroform, having the composition with the same solvent strength, are then tested. As shown in Figure 5-3, the area bound by the sides of the triangle formed by MTBE, dichloromethane, and chloroform defines the selectivity domain in which the optimum mobile-phase composition will be found. Next, separations are performed with three different ternary mobile-phase systems produced by mixing an equal volume of each of the binary solvents. Thus, the three experiments are set in the middle of triangle. Finally, the analysis is carried out by mixing in the three binary mixtures in equal ratio. By comparing the seven chromatograms obtained in the above experiment, optimum

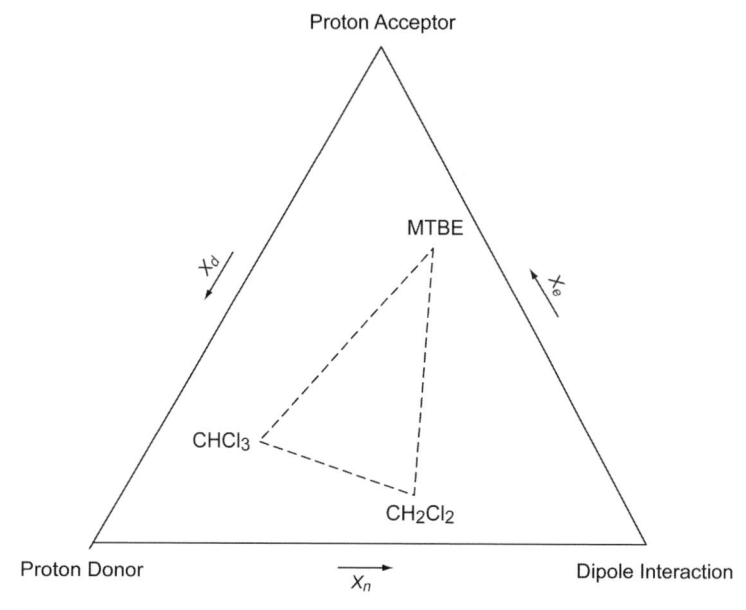

Figure 5-3. Selected solvents for mobile-phase optimization in NPC. (Reprinted from reference 11, with permission.)

solvent composition for the separation can be easily identified. Figure 5-4 demonstrates the triangle reduction method whereby the same procedure is repeated, starting from a smaller triangle—for instance, as defined by apices 2, 4, and 5, which corresponds to an area where the resolution is the highest—until an optimum mobile-phase mixture is determined for adequate resolution of the separated mixtures. Furthermore, optimum solvent composition can also be obtained by regression analysis with data obtained from the seven runs experiment [14].

Separation of acidic or basic analytes on NPC generally results in significant peak tailing due to the strong hydrogen-bonding interactions with silanol group on the stationary phase. Therefore, acidic or basic additive such as TFA (trifluoroacetic acid) or DEA (diethylamine) are often included in the mobile-phase system to minimize the hydrogen-bonding interactions.

5.4 SELECTIVITY

5.4.1 Effect of Analyte Structure

In NPC, analytes retentions generally increase in the following sequence: alkane < alkenes < aromatic hydrocarbons ≈ chloroalkanes < sulfides < ethers < ketones ≈ aldehyde ≈ esters < alcohols < amides ≪ phenols, amines, and carboxylic acids [16]. The retention also depends to some extent on the

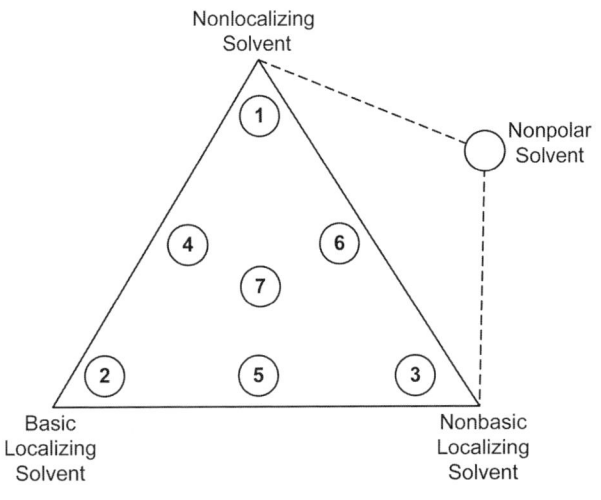

Figure 5-4. Procedure for selectivity optimization in NPC based on mixtures with hexane of nonlocalizing solvent (CH_2Cl_2), a basic-localizing solvent (MTBE), and a nonbasic localizing solvent (ACN or ethyl acetate). All mobile phases are of equal strength. (Reprinted from reference 1, with permission.)

hydrocarbon part of the solutes. Unlike RPLC, however, analytes become less retained as the size of alkyl chains increases. Furthermore, the separation in homologous series is less satisfactory than in RPLC. According to Soczewinski' model, analyte can have multiple interaction sites simultaneously when the adsorption sites interacts with a specific steric position of functional groups in the solute molecules with multiple functional groups. On the other hand, molecules with other positions of functional groups may have weaker or absent multiple sites interaction with the stationary phase (e.g., ortho versus meta versus para positions on an aromatic solute). This feature makes the use of NPC very suitable for the separation of positional isomers. In addition, difference in the retention and selectivities of molecules of similar polarities, but different shapes, such as rigid planar, rod-like, or of a flexible chain structure, are often observed in NPC.

5.4.2 Types of Stationary Phases

In order to accomplish the desired separation, the selection of appropriate stationary phase and eluent system is imperative. The most commonly used stationary phases in normal-phase chromatography are either (a) inorganic adsorbents such as silica and alumina or (b) moderately polar chemically bonded phases having functional groups such as aminopropyl, cyanopropyl, nitrophenyl, and diol that are chemically bonded on the silica gel support [16]. Other phases that are designed for particular types of analytes have also proved to be successful. These include modified alumina [17], titania [18], and zirconia [19–21].

Since the stationary phase in normal phase chromatography is more polar than the mobile phase, analyte retention is enhanced as the relative polarity of the stationary phase increases and the polarity of the mobile phase decreases. Retention also increases with increasing polarity and number of adsorption sites in the column. This means that retention is stronger on adsorbents with larger specific surface areas (surface area divided by the mass of adsorbents). Generally, the strength of interaction with analytes increases in the following order: cyanopropyl < diol < aminopropyl ≪ silica ≈ alumina stationary phases. However, strong selective interactions may change this order. The use of silica columns is less convenient for analytical applications. However, isomer and preparative separation favors the use of unmodified silica. Basic analytes are generally very strongly retained by the silanol groups in silica gel, and acidic compounds show increased affinities to aminopropyl silica columns. Aminopropyl and diol-bonded stationary phases prefer compounds with proton–acceptor or proton–donor functional groups as in alcohols, esters, ethers, and ketones, whereas dipolar compounds are usually more strongly retained on cyanopropyl silica than on aminopropyl or diol silica. Alumina phase has unique application in the separation of compounds with different numbers or spacing of unsaturated bonds. This is because alumina favors interaction with π electrons and often yields better selectivity than silica [16].

Despite the many desirable properties of silica, its limited pH stability (between 2 and 7.5) is also a major issue in NPC when strong acidic or basic mobile-phase additives are used to minimize interactions. Hence, other inorganic materials such as alumina, titania, and zirconia, which not only have the desired physical properties of silica but also are stable over a wide pH range, have been studied. Recently, Unger and co-workers [22] have chosen a completely new approach where they use mesoporous particles based not only on silica but also on titania, alumina, zirconia, and alumosilicates. These materials have been used by the authors to analyze and separate different classes of aromatic amines, phenols, and PAHs (polyaromatic hydrocarbons).

Bonded stationary phases for NPC are becoming increasingly popular in recent years owing to their virtues of faster column equilibration and being less prone to contamination by water. The use of iso-hydric (same water concentration) solvents is not needed to obtain reproducible results. However, predicting solute retention on bonded stationary phases is more difficult than when silica is used. This is largely because of the complexity of associations possible between solvent molecules and the chemically and physically heterogeneous bonded phase surface. Several models of retention on bonded phases have been advocated, but their validity, particularly when mixed solvent systems are used as mobile phase, can be questioned. The most commonly accepted retention mechanism is Snyder's model, which assumes the competitive adsorption between solutes and solvent molecules on active sites of the silica surface. Several studies have shown that this model is applicable for diol- [23, 24], cyano- [23, 25], and aminopropyl-bonded silica [26, 27].

5.5 APPLICATIONS

5.5.1 Analytes Prone to Hydrolysis

NPC is ideally suited for the analysis of compounds prone to hydrolysis because it employs nonaqueous solvents for the modulation of retention. An example of the use of NPC in the analysis of a hydrolysable analyte was demonstrated by Chevalier et al. [28] for quality control of the production of benorylate, an ester of aspirin. A major issue in benorylate production is the potential formation of impurities suspected of causing allergic side effects; therefore monitoring of this step is critical to quality control. The presence of acetylsalicylic anhydride prohibited the use of RPLC since it can be easily hydrolyzed in the water-containing mobile phase. However, an analytical method based on the use of normal-phase chromatography with alkylnitrile-bonded silica as the stationary phase provided an ideal solution to the analysis. Optimal selectivity was achieved with a ternary solvent system: hexane–dichloromethane–methanol, containing 0.2 v/v% of acetic acid to prevent the ionization of acidic function and to deactivate the residual silanols. The method was validated and determined to be reproducible based on precision, selectivity, and repeatability.

Another application that demonstrates the advantages of using NPC for the separation of analytes prone to hydrolysis is the reaction monitoring for the formation of 9,10-anthraquinone [29]. Anthraquinone is an important intermediate in the manufacturing of various dye products but also is used as a catalyst in the isomerization of vegetable oils. It is produced in large amount by Friedel–Crafts reaction of phthalic anhydride with benzene in the presence of $AlCl_3$ catalyst.

The development of a normal-phase HPLC method was warranted due to the presence of phthalic anhydride, which is unstable in water. Analysis in organoaqueous solvent systems that are used in RPLC would lead to an on-column reaction forming the respective carboxylic acid degradation product. Figure 5-5 shows the chromatogram obtained for the separation of 9,10-anthraquinone from the reactants and impurities on a silica column. The method was successfully applied to monitor the reaction conversion and also to determine the stability of 9,10-anthraquinone at the specified storage conditions.

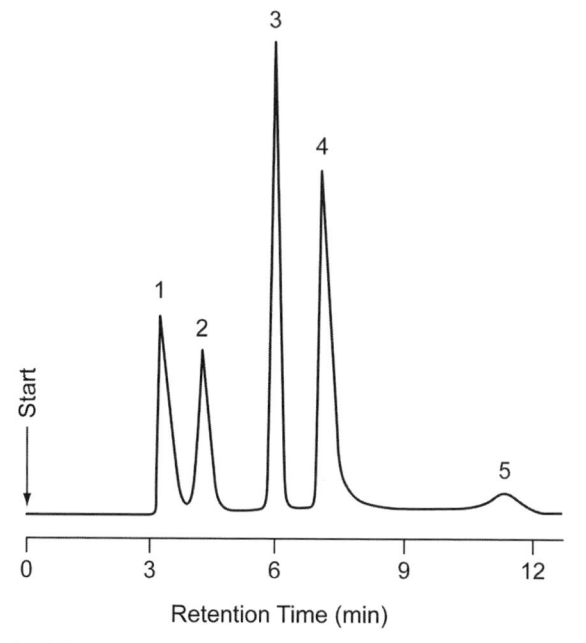

Figure 5-5. Typical chromatogram of a reaction mixture collected during the course of reaction of phthalic anhydride with benzene in the presence of $AlCl_3$, as catalyst. Peaks: 1, benzene; 2, anthraquinone; 3, phthalic anhydride; 4, maleic anhydride; 5, unknown. Chromatographic conditions: Column: Spherisob silica, $250 \times 4.6\,mm$, $10\,\mu m$; mobile phase, n-heptane–ethanol–chloroform–acetic acid $(89:5:5:1, v/v/v/v)$; flow rate, $1\,mL/min$; detection, UV at 254 nm; temperature, 27°C. (Reprinted from reference 29, with permission.)

In addition, sometimes a normal-phase HPLC method at subambient temperature must be applied for analytes that are extremely prone to hydrolysis. In the synthesis of leukotriene D_4 antagonist, accurate quantitation of mesylate intermediate is essential for process optimization. Owing to its inherent instability, analysis of mesylate intermediate must be carried out under normal-phase conditions with nonprotic solvents; however, significant cyclization of mesylation was still observed in such condition at room temperature. The authors concluded that the on-column reaction of the mesylate was silica-catalyzed cyclization. By conducting the normal-phase HPLC analysis at −30°C, it was demonstrated that on-column cyclization was adequately inhibited [30].

5.5.2 Extremely Hydrophobic Compounds

NPC has been used in the analysis hydrophobic compounds such as polyaromatic hydrocarbons [31–33]. An interesting example of an application of NPC involving extremely hydrophobic compounds was recently offered by Liu and

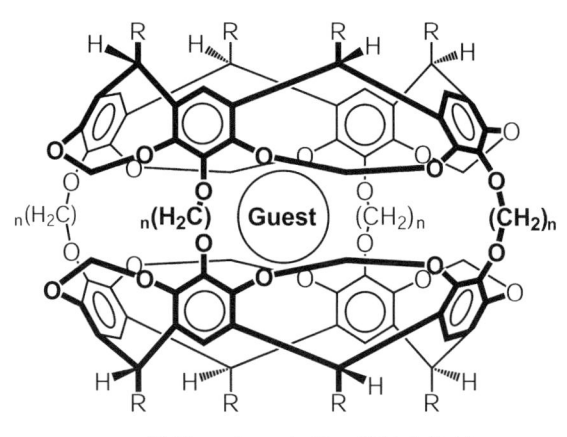

1⊙Guest: n=4; R = (CH₂)₂C₆H₅

Figure 5-6. A schematic of hemicarcerplex. (Reprinted from reference 34, with permission.)

Warmuth [34] when they adopted NPC for the analysis of supermolecules, such as hemicarcerplexes. Hemicarcerplexes are complexes formed with hemicarcerand host and guest molecules. As shown in Figure 5-6, hemicarcerands possess a very hydrophobic structure with molecular weight over 2000 and is insoluble in protic solvents. A normal-phase HPLC method was developed using a silica column with dichloromethane and diethylether as the mobile-phase system. The authors demonstrate that the chromatographic retention of hemicarcerplexes is mainly dominated by its size. Furthermore, a linear relationship between the logarithmic retention factor and the size of the hemicarcerplexes was observed for linear guest molecules independent of their polarity.

5.5.3 Separation of Isomers

With more and more complex molecules being investigated as drug candidates, isomer separation has become increasingly challenging. Despite being a workhorse analytical tool, reversed-phase chromatography is limited in its ability to distinguish between isomers [35–39]. On the other hand, NPC has established itself as the technique of choice for the separation of positional isomers as well as stereoisomers due to the specific nature of interactions. The separation of positional isomers of alkyl-substituted polyaromatic hydrocarbons (PAHs) in the petroleum industry is an example where NPC has been employed successfully. Alkyl substitution significantly increases PAH retention in RPC so that alkyl-substituted PAHs with low ring number have retention times close to some of the nonsubstituted higher-ring-numbered PAHs, resulting in co-elution. Wise et al. [40] reported that aminopropyl-bonded silica phase yields a separation sequence of PAHs solely based on the number of conjugated rings independent of the type of alkyl substitution.

Separations involving *cis/trans* isomers can also be accomplished by employing NPC. An example of this application is the separation of tricyclic antidepressant doxepin, which is marketed as a mixture of geometric isomers in a *cis/trans* ratio of 15:85 [41]. When a spherisorb silica column is used with a hexane–methanol–nonylamine mobile-phase system, the *cis* isomer of doxepin elutes first. The structures of the two isomers and the chromatographic separation are shown in Figure 5-7. NPC has also been successfully employed in the separation of *cis/trans* isomers of steroids. Four diastereomers

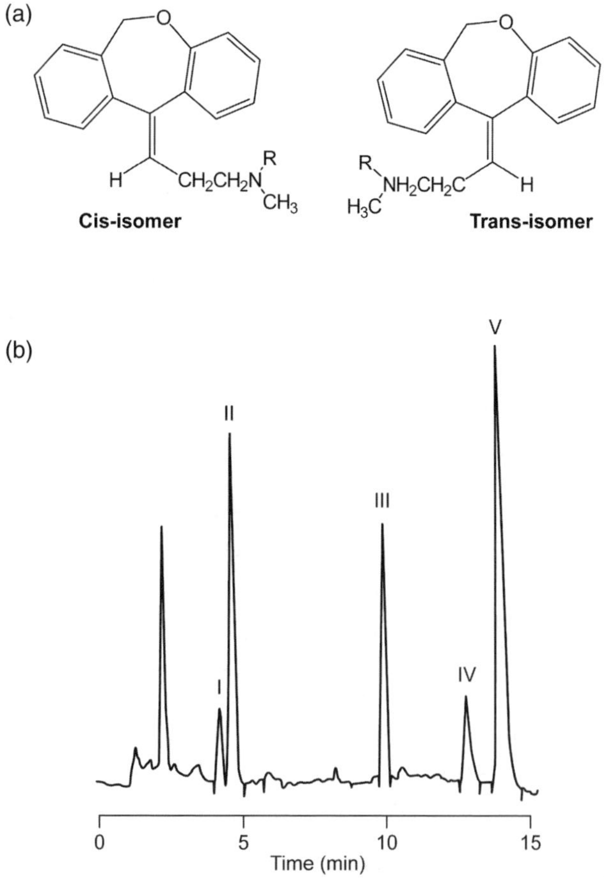

Figure 5-7. (a) Structures of *cis* and *trans* isomers of Doxepin and (b) normal-phase chromatographic separation of isomers of doxepin. I, *cis*-doxepin; II, *trans*-doxepin; III, nortriptyline; IV, *cis-N*-desmethyldoxepin; V, *trans-N*-desmethyldoxepin. Chromatographic conditions: Column: Spherisob silica, 150 × 4.5 mm, 3 μm. Hexane:methanol: nonylamine, 95:5:0.3 (v/v/v); flow rate, 1.0 mL/min; detection, 254 nm; temperature, 23°C. (Reprinted from reference 41, with permission.)

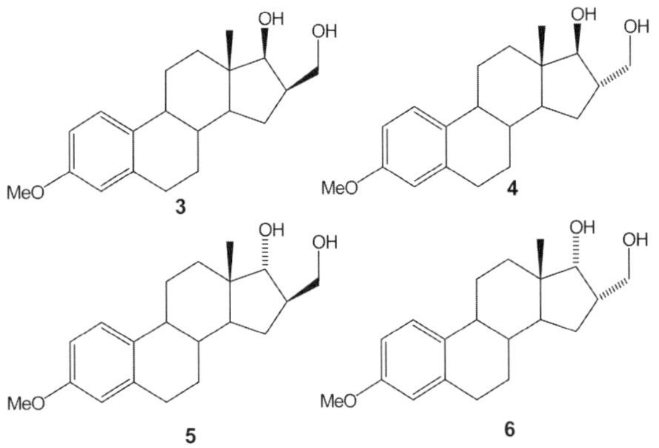

Figure 5-8. Structures of *cis* and *trans* isomers of steroids. (Reprinted from reference 42, with permission.)

consisting of two pairs of *cis/trans* isomers (see Figure 5-8 for structures) were separated using a silica column and hexane-dichloromethane-2-propanol mobile-phase system as shown in Figure 5-9 [42].

Other interesting examples of positional isomer separation involving NPC are (a) the separation of dihydrodipyridopyridopyrazines, a new family of antitumor agents, on a silica Nucleosil 50 Å-10μm column [43] and (b) the separation of celecoxib isomers by Chiralpak AD column [44]. NPC was also employed successfully for the resolution of (a) four configurational isomers of a steroidal calyx pyrrole [45], (b) regio- and stereoisomers of eicosanoids [46], (c) retinal and retinol isomers [47], and (d) several *E/Z* isomers pairs of vitamin A [48].

In certain cases, such as the separation of PAHs obtained from a coal liquefaction process, using reversed-phase HPLC is complicated as sample preparation is elaborate. This is due in large part to the fact that most complex fuel-related materials contain compounds that are not usually soluble in acetonitrile, the solvent of choice in reversed-phase HPLC. Here, NPC, which employs a variety of solvents, offers an alternative to the analysis of such samples. Separation of five well-studied coal liquefaction process stream samples was achieved and 19 isomers were resolved when NPC was used [33]. The method employed a tetrachlorophthalimidopropyl-modified silica column (TCPP) with a charge-transfer mechanism.

One of the most challenging tasks in isolating secondary metabolites from fermentation broths is the removal of numerous structural analogs of the desired product formed by the host organism. Pneumocandin B_0 is a potent antifungal agent produced as a recently discovered secondary metabolite by the fermentation of *Zalerion arboricola* [49]. Pneumocandin B_0 is the product of interest, with a molecular weight of 1069 Da. Pneumocandin C_0, which differs from B_0 only by a single carbon shift of hydroxyl group, is a key impurity co-produced by the fermentation. This impurity is proved to be intractable by reversed-phase chromatography or crystallization. The isomer was successfully

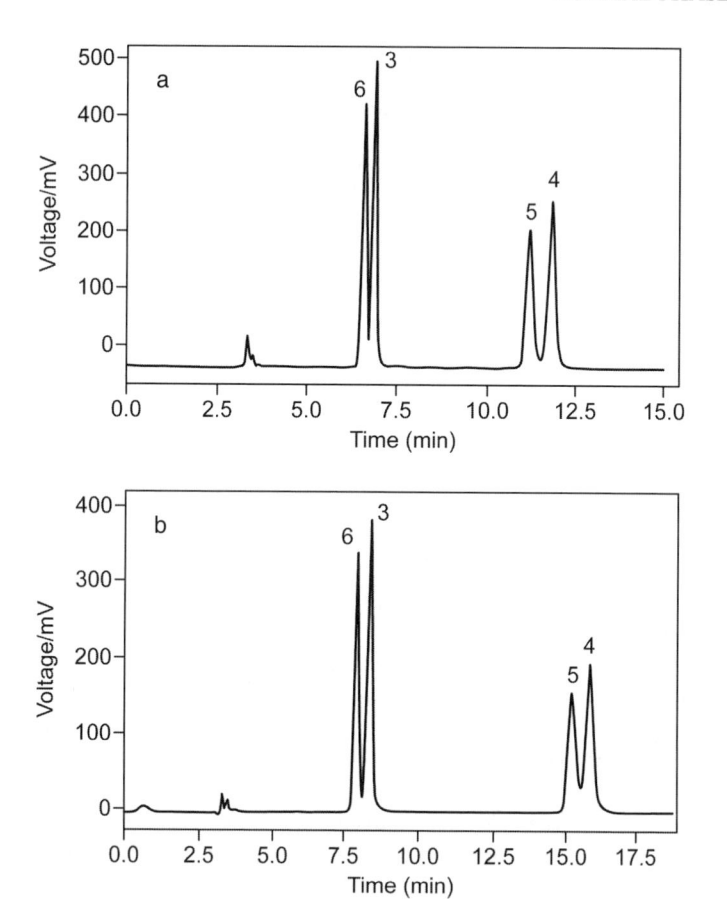

Figure 5-9. Chromatograms of isomers 3–6. Chromatographic conditions: Column: APEX silica, 250 × 4.6 mm, 5 μm (Jones chromatography); mobile phase, hexane-dichloromethane-2-propanol. (a) 82:10:8 (v/v/v), (b) 84:10:6 (v/v/v); flow rate, 1 mL/min; detection, 280 nm; temperature, ambient. (Reprinted from reference 42, with permission.)

separated in NPC mode by employing LiChrospher Silica stationary phase and an ethyl acetate/methanol/water mixture (86/7/7) mobile-phase system.

5.5.4   Carbohydrates

NPC has also found some applications in the field of carbohydrate analysis. Typical stationary phases used for this application are alkyl amine-, diol-, or polyol-bonded silicas [50–53]. Alkyl amino-bonded silicas are commonly used for the separation of saccharides and oligosaccharides in various matrixs, such as food or biological fluids. Although water is used as part of mobile phase, the retention behavior of carbohydrate follows the NPC retention behavior.

Carbohydrates are eluted in the order of increasing polarity, and retention decreases when water content increases. With an aminopropyl silica column, Koizumi et al. [54] showed the resolution of d-glycooligosaccharides up to a degree of polymerization of 30–35 under isocratic conditions, with a binary mixture of acetonitrile/water. Similarly, with the use of a polyamine polymer resin-bonded silica or an amino-bonded silica, separation of maltooligosaccharides up to a degree of polymerization of 28 was achieved [55].

5.5.5 Separation of Saturated/Unsaturated Compounds

The surface of the silica may be dynamically coated with transition metals, and the selectivities observed can be attributed to the complexes between the metal ions and the analyte species [56]. The use of silver-impregnated silica (adsorption of salts of transition metals on the silica surface) has been used for the analysis of saturated and unsaturated fatty acid methyl esters (FAME) and triacylglycerols (TAG) [57]. The retention of the unsaturated FAME and TAG can be attributed to the stability of the complex that is formed between the π electrons of the carbon–carbon double bonds and the silver ions. The predominant interaction for saturated analytes is with the polar silanol groups. The secondary interactions are those of the silver ions with the unpaired electrons of the carbonyl oxygens of the analytes. The amount of silver adsorbed onto the silica and the pH (employment of acidic or basic modifiers) have been determined to have an effect on the retention and resolution of certain acidic and basic compounds and fatty acids [58].

5.6 CONCLUSIONS

In normal-phase chromatography, polar stationary phases are employed and solutes become less retained as the polarity of the mobile-phase system increases. Retention in normal-phase chromatography is predominately based upon an adsorption mechanism. Planar surface interactions determine successful use of NPC in separation of isomers. The nonaqueous mobile-phase system used in NPC has found numerous applications for extremely hydrophobic molecules, analytes prone to hydrolysis, carbohydrates, and saturated/unsaturated compounds. In the future, with the advent of new stationary phases being developed, one should expect to see increasingly more interesting applications in the pharmaceutical industry.

REFERENCES

1. L. R. Snyder, J. J. Kirkland, and J. L. Glajch, Non-ionic samples: Reversed- and normal phase HPLC, in *Practical HPLC Method Development*, 2nd ed., Wiley, New York, 1997, pp. 266–289.

2. L. R. Snyder, *Principles of Adsorption Chromatography; the Separation of Non-ionic Organic Compounds*, Marcel Dekker, New York, 1968.

3. L. R. Snyder and T. C. Schunk, Retention mechanism and the role of the mobile phase in normal-phase separation on amino-bonded-phase columns, *Anal. Chem.* **54** (1982), 1764–1772.

4. L. R. Snyder, J. L. Glajch, and J. J. Kirkland, Theoretical basis for systematic optimization of mobile phase selectivity in liquid–solid chromatography: solvent–solute localization effects, *J. Chromatogr.* **218** (1981), 299–326.

5. L. R. Snyder, Liquid–solid chromatography. New Insights into retention on bonded-phase packings, *LC Magazine* **1** (1983), 478–482.

6. E. Soczewinski, Solvent composition effects in thin-layer chromatography systems of the type silica gel-electron donor solvent, *Anal. Chem.* **41** (1969), 179–182.

7. C. Marcel, J. Alain, Normal-Phase Liquid Chromatography, in E. Katz, R. Eksteen, P. Schoenmakers, and N. Miller (ed.), *Handbook of HPLC*, Marcel Dekker, New York, 1998, pp. 325–363.

8. J. G. Dorsey and W. T. Cooper, Retention mechanisms of bonded-phase liquid chromatography, *Anal. Chem.* **66** (1994), 857A–867A.

9. L. R. Snyder, Classification of the solvent properties of common liquids, *J. Chromatogr. Sci.* **16** (1978), 223–234.

10. L. Rohrschneider, Solvent characterization by gas–liquid partition coefficients of selected solutes, *Anal. Chem.* **45** (1973), 1241–1247.

11. J. L. Glajch, J. J. Kirkland, K. M. Squire, and J. M. Minor, Optimization of solvent strength and selectivity for reversed-phase liquid chromatography using an interactive mixture-design statistical technique, *J. Chromatogr.* **199** (1980), 57–79.

12. J. L. Glajch and J. J. Kirkland, Optimization of selectivity in liquid chromatography, *Anal. Chem.* **55** (1983), 319A–336A.

13. R. Lehrer, The practice of high-performance LC with four solvents, *Int. Lab.* **11** (1981), 76–88.

14. R. D. Snee, Experimenting with mixtures, *ChemTech* **9** (1979), 702–710.

15. J. J. Kirkland, J. L. Glajch, and L. R. Snyder, Practical optimization of solvent selectivity in liquid–solid chromatography using a mixture-design statistical technique, *J. Chromatogr.* **238** (1982), 269–280.

16. P. Jandera, Comparison of various modes and phase systems for analytical HPLC, in K. Valkó (ed.), *Handbook of Analytical Separations*, Separation Methods in Drug Synthesis and Purification, Vol. 1, Elsevier, New York, 2000, pp. 1–71.

17. C. Laurent, H. Billiet, and L. De Galan, On the use of alumina in HPLC with aqueous mobile phases at extreme pH, *Chromatographia* **17** (1983), 253–258.

18. K. Murayama, H. Nakamura, T. Nakajima, K. Takahashi, and A. Yoshida, Preparation and evaluation of octadecyl titania as column-packing material for high-performance liquid chromatography, *Microchem. J.* **49** (1994), 362–367.

19. J. Nawrocki, M. Rigney, A. McCormick, and P. W. Carr, Chemistry of zirconia and its use in chromatography, *J. Chromatogr. A* **657** (1993), 229–282.

20. H. J. Wirth and M. T. W. Hearn, High-performance liquid chromatography of amino acids, peptides and proteins CXXX. Modified porous zirconia as sorbents in affinity chromatography, *J. Chromatogr. A* **646** (1993), 143–151.

21. D. A. Whitman, T. P. Weber, and J. A. Blackwell, Chemometric characterization of Lewis base-modified zirconia for normal phase chromatography, *J. Chromatogr. A* **691** (1995), 205–212.

22. U. Trüdinger, G. Müller, and K. K. Unger, Porous zirconia and titania as packing materials for high-performance liquid chromatography, *J. Chromatogr. A* **535** (1990), 111–125.

23. P. L. Smith and W. T. Cooper, Retention and selectivity in amino, cyano and diol normal bonded phase high-performance liquid chromatographic columns, *J. Chromatogr.* **410** (1987), 249–265.

24. A. W. Salotto, E. L. Weiser, K. P. Caffey, R. L. Carty, S. C. Racine, and R. L. Snyder, Relative Retention and column selectivity for the common polar bonded-phase columns: The diol-silica column in normal-phase high-performance liquid chromatography, *J. Chromatogr.* **498** (1990), 55–65.

25. E. L. Weiser, A. W. Salotto, S. M. Flach, and R. L. Snyder, Basis of retention in normal-phase high-performance liquid chromatography with cyano-propyl columns, *J. Chromatogr.* **303** (1984), 1–12.

26. L. D. Olsen and R. J. Hurtubise, Mobile phase effects on aromatic hydroxyl compounds with an aminopropyl column and interpretation by the Snyder model, *J. Chromatogr.* **479** (1989), 5–16.

27. L. R. Snyder and T. C. Schunk, Retention mechanism and the role of the mobile phase in normal-phase separation on amino-bonded-phase columns, *Anal. Chem.* **54** (1982), 1764–1772.

28. G. Chevalier, P. Rohrbath, C. Bollet, and M. Caude, Identification and quantitation of impurities from Benorilate (Salipran) by high-performance liquid chromatography, *J. Chromatogr.* **138** (1977), 193–201.

29. S. Husain, R. Narsimha, S. Khalid, and R. R. Nageswara, Application of normal- and reversed-phase high-performance liquid chromatography for monitoring the progress of reactions of anthraquinone manufacturing processes, *J. Chromatogr. A* **679** (1994), 375–380.

30. J. O. Egekeze, M. C. Danieiski, N. Grinberg, G. B. Smith, D. B. Sidler, H. J. Perpal, G. R. Bicker, and P. C. Tway, Kinetic analysis and subambient temperature chromatography of an active ester, *Anal. Chem.* **67** (1995), 2292–2295.

31. C. H. Marvin, S. Mehta, D. Lin, B. E. McCarry, and D. W. Bryant, Relative genotoxicities of PAH of molecular weight 252 amu in coal tar-contaminated sediment, *Polycyclic Aromat. Compd.* **20** (2000), 305–318.

32. O. Ferroukhi, N. Atik, S. Guermouche, M. H. Guermouche, P. Berdague, P. Judenstein, and J. P. Bayle, High performance liquid chromatography of aromatic and polyaromatic hydrocarbons on a new chemically bonded liquid crystal phase, *Chromatographia* **52** (2000), 564–568.

33. D. E. McKinney, D. J. Clifford, L. Hou, M. R. Bogdan, and P. G. Hatcher, High performance liquid chromatography (HPLC) of coal liquefaction process streams using normal-phase separation with diode array detections, *Energy Fuels* **9** (1995), 90–96.

34. Y. Liu and R. Warmuth, A "through-shell" binding isotope effect, *Angew. Chem. Int. Ed.* **44** (2005), 7107–7110.

35. M. M. Mendes-Pinto, A. C. Ferreira, M. P. Oliveira, and P. Guedes de Pinho, Evaluation of some carotenoids in grapes by reversed- and normal-phase

liquid chromatography: A qualitative analysis, *J. Agric. Food. Chem.* **52** (2004), 3182–3188.

36. H. Shan, J. Pang, S. Li, T. B. Chiang, W. K. Wilson, and G. J. Schroepfer, Chromatographic behavior of oxygenated derivatives of cholesterol, *Steriods* **68** (2003), 221–233.

37. L. Cossignani, M. S. Simonetti, and P. Damiani, Structural Changes of Triacylglycerol and diacylglycerol fractions During Olive Drupe Ripening, *Eur. Food Res. Technol.* **212** (2001), 160–164.

38. O. Froescheis, S. Moalli, H. Liechti, and J. Bausch, Determination of lycopene in tissues and plasma of rats by normal-phase high-performance liquid chromatography with photometric detection, *J. Chromatogr. B* **739** (2000), 291–299.

39. L. Zhou, Y. Wu, B. D. Johnson, R. Thompson, and J. M. Wyvratt, Chromatographic separation of 3,4-difluorophenylacetic acid and its positional isomers using five different techniques, *J. Chromatogr. A* **866** (2000), 281–292.

40. S. A. Wise, S. N. Chesler, H. S. Hertz, L. R. Hilpert, and W. E. May, Chemically-bonded aminosilane stationary phase for the high-performance liquid chromatographic separation of polynuclear aromatic compounds, *Anal. Chem.* **49** (1977), 2306–2310.

41. J. Yan, J. W. Hubbard, G. McKay, and K. K. Midha, New micro-method for the determination of lamotrigine in human plasma by high-performance liquid chromatography, *J. Chromatogr. B* **691** (1997), 131–138.

42. J. Wölfling, G. Schneider, and A. Péter, High-performance liquid chromatographic methods for monitoring of isomers of 17-Hydroxy-16-hydroxymethyl-3-methoxyestra-1,3,5(10)-triene, *J. Chromatogr. A* **852** (1999), 433–440.

43. F. Himbert, R. Pennanec, G. Guillaumet, and M. Lafosse, Preparative liquid chromatography and centrifugal partition chromatography for purification of new anticancer precursors, *Chromatographia* **60** (2004), 269–274.

44. D. Screenivas Rao, M. K. Srinivasu, C. Lakshmi Narayana, and O. G. Reddy, LC separation of ortho and meta isomers of Celecoxib in bulk and formulations using a chiral column, *J. Pharm. Biomed. Anal.* **25** (2001), 21–30.

45. M. Dukh, P. Drašar, I. Černy, V. Pouzar, J. A. Shriver, V. Kral, and J. L. Sessler, Novel deep cavity calix[4]pyrroles derived from steroidal ketones, *Supramol. Chem.* **14** (2002), 237–244.

46. P. Demin, D. Reynaud, and C. R. Pace-Asciak, High-performance liquid chromatographic separation of fluorescent esters of Hepoxilin enantiomers on a chiral stationary phase, *J. Chromatogr. B* **672** (1995), 282–289.

47. G. N. Nöll and C. Becker, High-performance liquid chromatography of non-polar retinoid isomers, *J. Chromatogr. A* **881** (2000), 183–188.

48. R. M. Duarte Fávara, M. H. Iha, and M. L. P. Bianchi, Liquid chromatographic determination of geometrical retinol isomers and carotene in enteral feeding formulas, *J. Chromatogr. A* **1021** (2003), 125–132.

49. A. E. Osawa, R. Sitrin, and S. S. Lee, Purification of pneumocandins by preparative silica-gel high-performance liquid chromatography, *J. Chromatogr. A* **831** (1999), 217–225.

50. M. Verzele, G. Simoens, and F. Van Damme, A critical review of some liquid chromatography systems for the separation of sugars, *Chromatographia* **23** (1987), 292.

51. S. C. Churms, Carbohydrates as a tool for oriented immobilization of antigens and antibodies, *J. Chromatogr.* **500** (1990), 555–583.

52. B. Herbreteau, Review and state of sugar analysis by high performance liquid chromatography, *Analysis* **20** (1992), 355–374.

53. M. Wuhrer, C. M. Koeleman, A. M. Deelder, and C. H. Hokke, Normal-phase nanoscale liquid chromatography–mass spectrometry of underivatized oligosaccharides at low-femtomole sensitivity, *Anal. Chem.* **76** (2004), 833–838.

54. K. Koizumi, T. Utamura, and Y. Okada, Analyses of homogeneous D-gluco-oligosaccharides and -Polysaccharides (degree of polymerization up to about 35) by high-performance Liquid chromatography and thin-layer chromatography, *J. Chromatogr.* **321** (1985), 145–157.

55. K. Koizumi, T. Utamura, and S. Hizukuri, Two high-performance liquid chromatographic columns for analyses of malto-oligosaccharides, *J. Chromatogr.* **409** (1987), 396–403.

56. R. LoBrutto, in J. Cazes (ed.), *Encyclopedia of Chromatography*, Marcel Dekker, New York, 2001.

57. O. R. Adlof, Normal-phase separation effects with lipids on a silver ion high-performance liquid chromatography column, *J. Chromatogr. A* **764** (1997), 337–340.

58. H. Okamoto, K. Kakamu, D. Nobuhara, and J. Ishii, Effect of silver-modified silica on retention and selectivity in normal-phase liquid chromatography, *J. Chromatogr. A* **722** (1996), 81–85.

6

SIZE-EXCLUSION CHROMATOGRAPHY

Yuri Kazakevich and Rosario LoBrutto

6.1 SEPARATION OF THE ANALYTE MOLECULES BY THEIR SIZE

Size-exclusion chromatography (SEC) separates polymer molecules and bio-molecules based on differences in their molecular size. The separation process in simplified form is based on the ability of sample molecules to penetrate inside the pores of packing material and is dependent on the relative size of analyte molecules and the respective pore size of the absorbent. The process also relies on the absence of any interactions with the packing material surface. Two types of SEC are usually distinguished:

1. Gel permeation chromatography (GPC)—separation of synthetic (organic-soluble) polymers. GPC is a powerful technique for polymer characterization using primarily organic solvents.
2. Gel filtration chromatography (GFC)—separation of water-soluble biopolymers. GFC uses primarily aqueous solvents (typically for aqueous soluble polymers, proteins, etc.).

Physical and chemical properties of polymers are dependent on their molecular weight and molecular weight distribution. The separation principle in SEC is based on the forced transport of the polymer molecules through the porous stationary-phase media under the conditions of suppressed interactions of the

HPLC for Pharmaceutical Scientists, Edited by Yuri Kazakevich and Rosario LoBrutto
Copyright © 2007 by John Wiley & Sons, Inc.

polymer analyte with the surface. The mobile-phase eluent is selected in such way that it interacts with the surface of packing material stronger than the polymer. Under these conditions, the smaller the size of the molecule, the more it is able to penetrate inside the pore space and the movement through the column is retarded. On the other hand, the bigger the molecular size, the higher the probability the molecule will travel around the particles of the packing material and, thus, is eluted earlier. The molecules are separated in order of decreasing molecular weight, with the largest molecules eluting from the column first and smaller molecules eluting last (Figure 6-1).

Molecules larger than the pore size do not enter the pores and elute together as the first peak in the chromatogram and this is called total exclusion volume which defines the exclusion limit for a particular column. Molecules that can enter the pores diffuse into the internal pore structure of the gel to an extent depending on their size and the pore size distribution of the

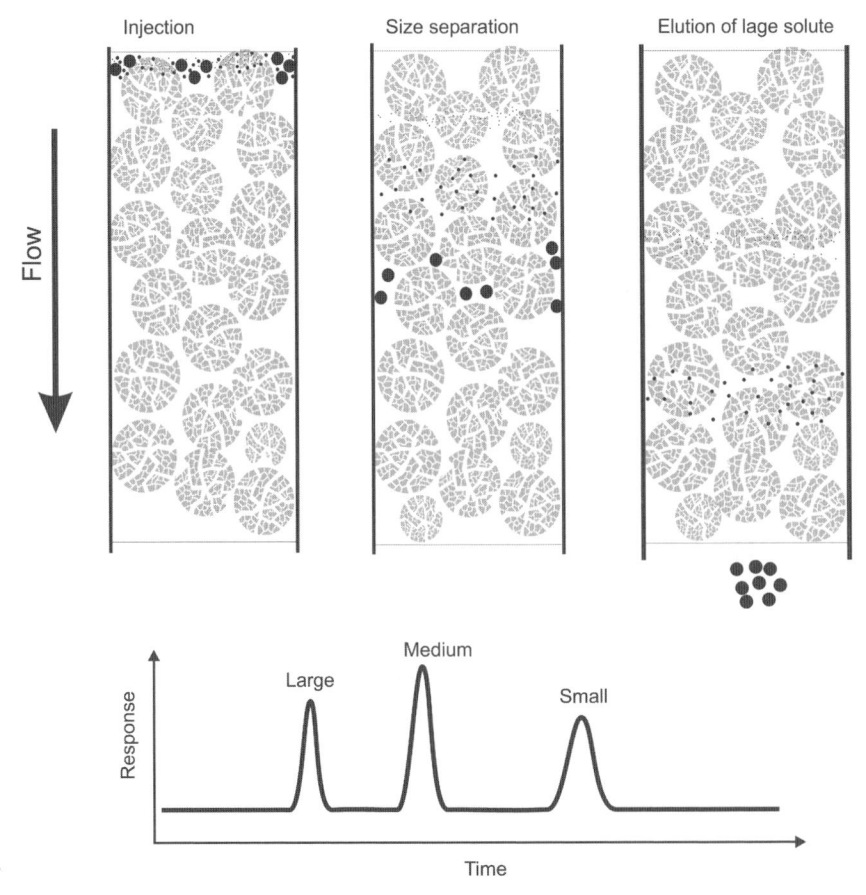

Figure 6-1. Illustrative description of separation in SEC.

gel. The molecules will have an average residence time in the particles that depends on the molecules size and shape in the particular mobile phase. Therefore, different molecules have different total residence times in the column. This portion of a chromatogram is called the selective permeation region (the effective volume in which separation can occur). Molecules that are smaller than the pore size can enter all pores, have the longest residence time on the column, and will elute all together as the last peak in the chromatogram. This last peak in the chromatogram determines the total permeation limit for a particular column. The largest elution volume (retention volume) in any given SEC column is equal to the total mobile-phase volume in the column (known as the void volume, V_0). The exclusion range indicates the molecular weight of solutes above which all solutes having a molecular weight greater than the exclusion limit. These analytes will elute at the same retention time as a single peak. A specific column can be used for separation of solutes with molecular weights that are within the molecular weight window between the exclusion and permeation limits (Figure 6-2).

Separation process in SEC is based on the actual size of the molecules, which in turn reflects the molecular weight of the polymer. The resulting SEC

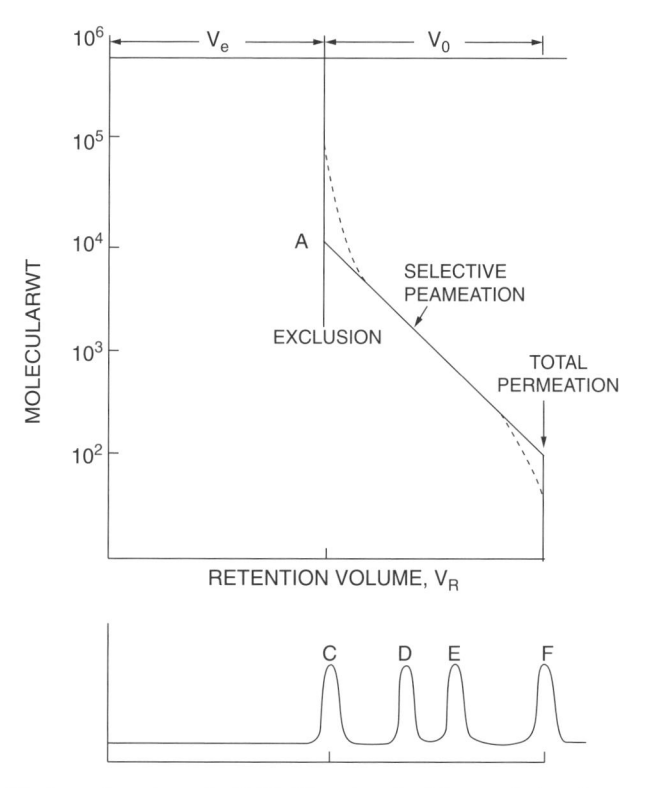

Figure 6-2. Elution of analytes in SEC. (Reprinted with permission from reference 1.)

chromatogram reflects the size distribution of the polymer sample, and its relationship with the molecular weight distribution which lays the foundation of the SEC theory.

6.2 MOLECULAR SIZE AND MOLECULAR WEIGHT

A polymer molecule in solution has a certain shape that strongly depends on the type of polymer, type of solvent, temperature, and other conditions. Usually a polymer forms some kind of globular species whose size is dependent on the degree of solvation by solvent molecules.

This globe could be described by its volume (v) and hydrodynamic radius (R).

Hydrodynamic radius (radius of gyration) of the polymer in the solution could be expressed in the form

$$R = \left(\frac{3}{4} \pi M[\eta] \right)^{\frac{1}{3}} \tag{6-1}$$

where $[\eta]$ is intrinsic viscosity and M is molecular weight.

For some polymers that are not flexible, the effective R is used to represent the radius of the sphere. The parameter R is equivalent to the mechanical behavior of the polymer in solution. Viscosity is the simplest parameter of the polymer solution. From the Stokes and Einstein equations, the volume of the equivalent sphere is proportional to the product of intrinsic viscosity and polymer molecular weight:

$$[\eta]M = 2.5 N_A v \tag{6-2}$$

where $[\eta]$ is intrinsic viscosity, M is molecular weight, N_A is Avogadro's number, and v is the volume of the equivalent sphere. As one could see, the intrinsic viscosity is an important parameter related to the molecular weight of the polymer and its molecular volume. By definition, intrinsic viscosity is a limit of the ratio of the specific viscosity of the polymer solution to its concentration at $c \to 0$, or it is the y-intercept of the dependence of η_{sp} versus concentration.

Polymer molecules of a different nature but with the same molecular weight usually have different hydrodynamic radii. This is due to the differences in coil flexibility, intramolecular interactions, and, most importantly, the differences in their interactions with the solvent. This essentially means that if two different polymers analyzed at identical SEC conditions show similar peaks with identical elution volume, it does *not* confirm that the molecular weights of these polymers are identical. It only indicates that at the given conditions the gyration radii of the molecules are the same, causing similar elution.

The nature of the solvent also has a significant effect on the polymer conformation and thus on its gyration radius and molecular volume. If the solvent–polymer interactions are favorable or essentially prevailing over the interactions between different segments of the same polymer, then we can expect a high degree of solvation and the polymer globe will swell. For instance, if polystyrene is dissolved in toluene, due to the similar nature of the solvent and polystyrene monomer, toluene will solvate polymer molecules and their gyration radius increases. On the contrary, if the same polystyrene is dissolved in tetrahydrofuran (THF), then interactions between polystyrene segments prevail over the interactions with THF. As a consequence, the size of the polymer globe in THF is relatively small, especially in comparison to that in toluene.

From equation (6-2), one can conclude that intrinsic viscosity is proportional to the polymer molecular volume. On the other hand, the effective molecular volume is also the function of the molecular weight and the type of used solvent (or the nature of the solvent–polymer and polymer–polymer interactions). The intrinsic viscosity is an exponential function of the molecular weight with fixed coefficients for any specific polymer and solvent.

$$[\eta] = K \cdot M^{\alpha} \tag{6-3}$$

This expression is known as the Mark–Houwink equation, and K and α are constants for any given pair of polymer and solvent. These constants are tabulated and could be found for most known polymers in reference 2.

6.3 SEPARATION MECHANISM

Eluent flow through the chromatographic column packed with porous packing material has a velocity distribution depending on the pathway. Flow around the adsorbent particles is the fastest. Flow through the pore space is much slower. Since the smallest molecules can penetrate all of the pores, they can be distributed in the whole liquid volume of the column and their average migration speed is therefore the slowest. Molecules of intermediate size may penetrate into the pore space but may not come close to the pore walls, so their center of mass will be allocated closer to the center of the pores where flow velocity is higher. Their average migration speed is higher. The biggest molecules experience steric hindrance in permeation inside the packing pore space and move through the column primarily around the particles with the fastest possible speed. As a result, the biggest molecules come out of the column first, and the smallest ones come out last.

Obviously, all molecules that are not able to penetrate into the pore space, move with the same velocity. Retention volume of all these molecules is the same and is called exclusion volume, also known as total exclusion. The total exclusion volume is a characteristic of a particular column which determines its upper separation limit.

6.4   CALIBRATION

SEC calibration establishes the relationship of a particular elution volume with specific molecular weight of the polymer (Figure 6-3) [3]. For calibration the elution volume of the solutions of polymer standards with known narrow molecular weight distributions are measured. An example of a separation is shown in Figure 6-4 [4]. In SEC, *hydrodynamic volume* of the polymer molecules is being measured rather than the actual mass of a particular species. The hydrodynamic volume is the space a particular polymer molecule occupies when it is in solution. The molecular weight can be approximated from SEC data from the relationship between molecular weight and hydrodynamic volume for particular known standards. However, the relationship between hydrodynamic volume and molecular weight is not the same for all polymers, so only an approximate measurement can be obtained.

A series of commercially available polystyrene standards can be used for calibration. The elution volume (elution time multiplied by flow rate) corresponding to a particular peak in the chromatogram is related to the molecular weight of a particular polystyrene. After assignment of the molecular weight for each component to its elution volume, the logarithms of the molecular weight of the standards are plotted against their elution volumes in order to construct a calibration curve (Figure 6-3). Each combination of column, polymer, and solvent has its own calibration curve.

The same polymer molecules could have different sizes in different solvents, and two molecules of different polymers might have the same size despite their

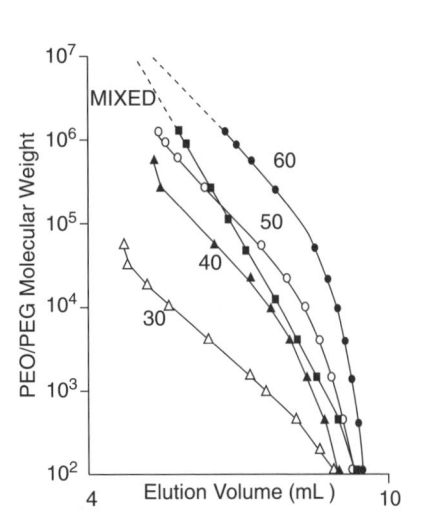

Figure 6-3. Calibration curves for a set of AquaGel (Polymer Laboratories) columns designed for the separation of water soluble polymers. Calibration using PEO and PEG standards. (Reprinted from reference 3, with permission from Polymer Laboratories Inc.)

Polystyrenes (Wide MW Range)

Column:	Phenogel 10 μm 10⁵, 10⁴, 10³Å	Sample:	1. 1,560,000	MW
Dimensions:	300 × 7.8 mm		2. 260,000	MW
Mobile Phase:	THF		3. 94,000	MW
Flow Rate:	1.0 mL/min		4. 30,000	MW
Detection:	Differential Refractometer		5. 6,100	MW
Injection Volume:	100 μm 0.25% w/v		6. 845	MW
Temperature:	Ambient		7. 146	MW

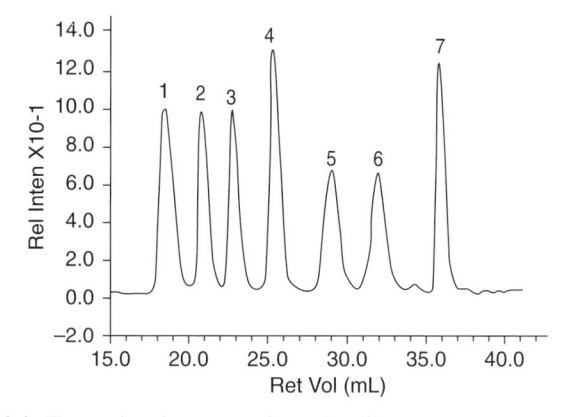

Figure 6-4. Example of a separation of calibration mixture of polystyrene standards. (Reprinted from reference 4, with permission from Phenomenex.)

different molecular weight. So the calibration curve for the certain polymer is valid only if the standards used are of the same nature and used eluent was of the same type.

If two different polymers in the same solvent have the same intrinsic viscosity, then their molecular weights are related as

$$K_1 \cdot M_1^{\alpha_1} = K_2 \cdot M_2^{\alpha_2} \qquad (6\text{-}4)$$

This makes it possible to use standards of one polymer for characterization of another if the corresponding Mark–Houwink constants are known. For most known polymers, Mark–Houwink constants are tabulated. For example, a polystyrene (PS)-based calibration could be used for characterization of polymethylmethacrylate (PMMA).

Retention volume in SEC is proportional to the size of the polymer molecules in solution. In addition, as discussed in Section 6.2, equation (6-2), the product of the intrinsic viscosity of the polymer and its molecular weight is proportional to the hydrodynamic molecular volume. These relationships allowed Benoit et al. [5, 6] to introduce a universal molecular weight calibration.

In conventional molecular weight calibration, the dependence of the retention volumes of a series of narrow molecular weight distribution standards with known average masses is plotted against the logarithms of their molecular weights. In universal calibration, the logarithm of the product of the intrinsic viscosity [η] and molecular weight M (essentially hydrodynamic volume) is plotted against retention volume. Hydrodynamic molecular volume is directly related to the retention volume if only the steric separation mechanism is involved. Benoit et al. [5, 6] found that plots of the logarithm of hydrodynamic volumes versus corresponding retention volumes for a series of narrow standards of different polymers in different solvents resulted in a single calibration curve, as shown in Figure 6-5.

The combination of the differential refractive index (RI) detector and on-line viscometer allows the direct use of the universal calibration and thus true molecular weight determination. The RI detector is concentration-sensitive, and the viscometer records specific viscosity. The ratio of the specific viscosity to the concentration is equivalent to intrinsic viscosity (as discussed in Section 6.1), and the continuous dependence of this ratio versus the retention volume could be related to the universal calibration curve, thus allowing the correlation of each point on the chromatogram with the true molecular weight.

More detailed information on this type of GPC analysis can be found in the book by Yau et al., *Modern Size-Exclusion Liquid Chromatography* [7].

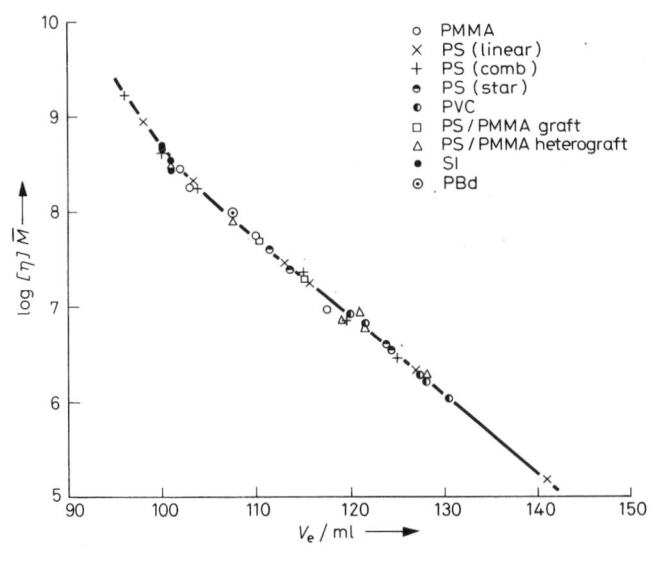

Figure 6-5. Universal Benoit calibration. (Reprinted from reference 5, with permission.)

6.5 COLUMNS

Polymer-based packing materials are the main type of adsorbent used in size-exclusion HPLC. Most SEC analyses of synthetic organic polymers are performed on rigid or semirigid crosslinked polystyrene gel materials (styrene–divinylbenzene copolymers with different degree of crosslinkage). These materials require careful selection of separation conditions and solvent, since they can shrink and swell in different solvents and temperatures, thus changing their separation power. A major requirement of size-exclusion separation is the complete absence of any interactions between analyzed components and the surface of packing material in the column. Polymeric packings do not have any active surface groups and could be synthesized with controlled porosity. For the separation of biopolymers in aqueous media, gel-filtration chromatography is used with stationary phases that have a mildly hydrophilic surface, which is usually required to avoid noticeable hydrophobic interactions with the surface. These types of stationary phases for GFC include (a) dextrans, agaroses, polyacrylamide, or mixtures of these components which are suitable for low- or medium-pressure chromatography and (b) porous silica-based media which are more suitable for higher-pressure applications.

The separation range in size-exclusion chromatography for a particular column is relatively narrow, and it lies between the total volume of the liquid phase in the column (void volume) and the exclusion volume, V_e. The difference between these two volumes is the total pore volume of the packing material in the column. Indeed, if some molecules of studied polymers are small enough to penetrate inside all pores of the packing material, they will elute with the column void volume. On the other hand, polymers with significant molecular size that cannot penetrate inside the particles will all travel together around the particles and elute early with exclusion volume.

To obtain wider separation range, longer and wider column dimensions are used. The greater the amount of packing material in the column, the higher the total pore volume and the higher the difference between void and exclusion volumes.

Another important parameter is the ability of the column to discriminate different ranges of molecular masses. This range is dependent on the size of the pores of packing material and pore size distribution. Column manufacturers usually provide the standard calibration curves of polystyrene standards in THF for each column, as shown in Figure 6-6 [8].

One of the most important requirements for the GPC column is the absence of the specific interactions with the studied polymer. The more inert the surface of the packing material, the better. Early applications of GPC separation were sometimes performed on porous glass particles with controlled porosity [9]. The ease of the manufacturing of the controlled porosity particles had determined this choice, but it was not always possible to find an

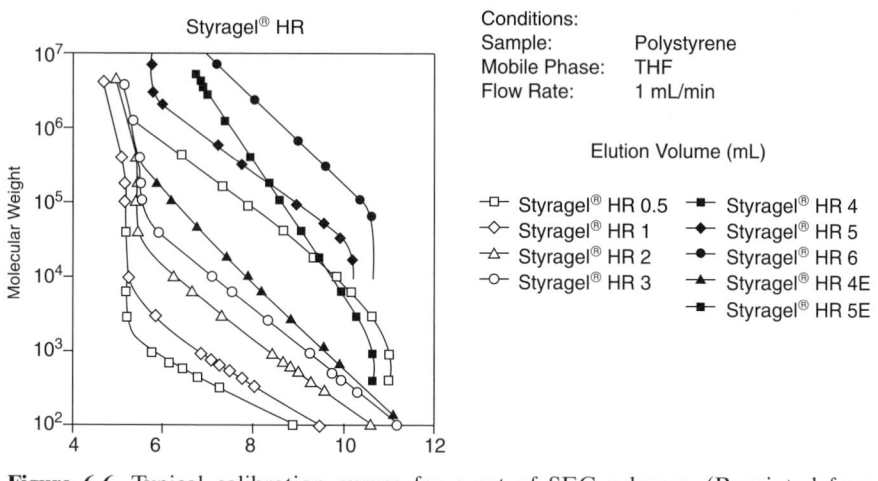

Figure 6-6. Typical calibration curves for a set of SEC columns. (Reprinted from reference 8, with permission.)

eluent that would suppress interactions of the polymer with the glass surface. Modern GPC columns are primarily made of styrene–divinylbenzene copolymer. Increased technological advances allow preparation of the rigid porous particles with relatively narrow pore size distribution from this copolymer.

The presence of the aromatic rings in the body of the packing material is prone to π–π-type interactions, which could be a problem for the separation of polymers with significant aromaticity, like polyimids.

Traditional reversed-phase columns appear to offer the most inert surface since the alkyl-type bonded ligands at high bonding density can only participate in weak dispersive interactions, which could be suppressed by practically any organic solvent. An example of the calibration curve made on a commercial reversed-phase column is shown in Figure 6-7.

For the analysis of water-soluble polymers (such as surfactants, oligosaccharides, PEGs, lignosulfonates, polyacrylates, polysaccharides, PVA, cellulose derivatives, PEO, polyacrylic acids, polyacrylamides, hyaluronic acids, CMC, starches, gums) and for separations of oligomers and small molecules, columns that are comprised of macroporous material with hydrophilic functionalities may be used. The requirement for these columns in SEC mode is to eliminate or minimize ionic and hydrophobic effects that make aqueous SEC (otherwise known as GFC) very demanding. The interaction of analytes with neutral, ionic, and hydrophobic moieties must be suppressed. It is often necessary to modify the eluent (addition of salt) in order to avoid sample-to-sample and sample-to-column interactions that can result in poor aqueous SEC separations and low recoveries.

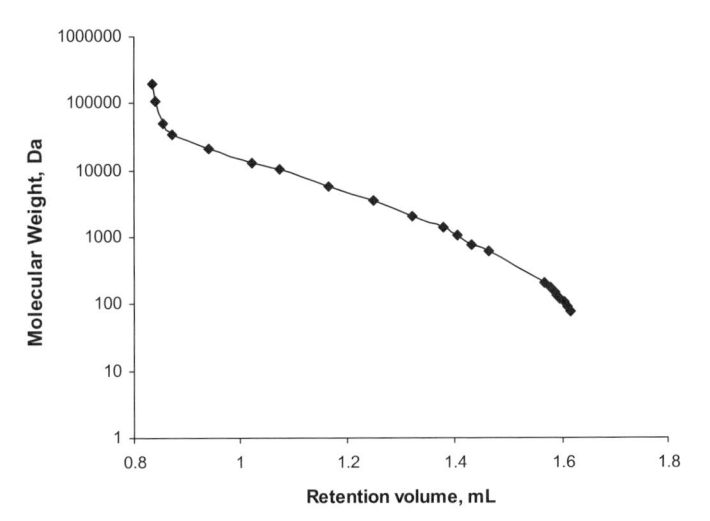

Figure 6-7. Calibration curve for low-molecular-weight region. Column: Prodigy-ODS2 (pore size 150 Å). Effective separation region for polystyrenes in THF from 100 to 50,000 Da.

6.6   MOLECULAR WEIGHT DISTRIBUTION

It is impossible to find a sample of a synthetic polymer in which all the chains have the same molecular weight. Instead, a distribution of molecular weights is reported. Some of the polymer chains will be much larger than the others, and some will be significantly smaller. The largest number of similar chains will be populated around a central point of the distribution, the highest point on the curve called the "molecular weight distribution."

All synthetic polymers consist of the mixture of the molecules of the same nature but different size (length, or number of repeat units), so we have to deal with average molecular weights when we work with polymers. The average molecular weight can be calculated in different ways, and each approach has its own value. In the Mark–Houwink equation (6-3) the intrinsic viscosity is related to the molecular weight; this is the viscous average weight.

Two other important average molecular weights are number average (M_N) and weight average (M_W). The first one is the weight calculated by average number of the molecules

$$M_n = \frac{\sum N_i M_i}{\sum N_i} = \frac{\sum \varphi_i}{\sum \dfrac{\varphi_i}{M_i}} \tag{6-5}$$

where N_i is the number of the molecules in fraction i (or slice of GPC chromatogram), M_i is the molecular weight in that fraction, and ϕ_i is the area fraction on the chromatogram from a concentration sensitive detector (RI). This molecular weight is usually measured by osmometry.

Weight average molecular weight of the polymer is usually measured by light scattering and is defined as

$$M_w = \frac{\sum M_i^2 \cdot N_i}{\sum M_i N_i} = \frac{\sum M_i \varphi_i}{\sum \varphi_i} \qquad (6\text{-}6)$$

The ratio of weight average to number average molecular weight is called the polydispersity, r_d. The wider is the molecular weight distribution the higher the polydispersity value. The r_d for unimolecular polymer is equal to 1. All natural polymers, such as peptides, DNA, and saccharides, have polydispersity equal to 1.

6.7   EFFECT OF ELUENT

For the method development of the SEC separation, the main attention should be focused on the suppression of analyte interactions with the surface of packing material. This usually requires a careful selection of the SEC column. In GPC, commercially available synthetic organic polymer columns are usually packed with styrene–divinylbenzene copolymer particles, which are only capable of weak dispersive interactions. Any possible analyte interactions with the surface could be suppressed by using a strong solvent, which will be preferentially adsorbed on the packing material surface. Selection of such a solvent is limited since the polymer solubility in that particular solvent needs to be considered. Tetrahydrofuran is the most common solvent used for most GPC separations, although for polyimids and other high-temperature polymers the use of special solvents such as n-methylpyrrolidone may be necessary.

The development of GFC methods generally has similar considerations, with much higher requirements for the packing material. GFC deals with water-soluble polymers; thus the main solvent is water, with some additives. The suppression of possible surface interactions has to be done with these additives and with careful selection of the packing material.

6.8   EFFECT OF TEMPERATURE

Temperature of the system is also an important parameter since the viscosity of the polymer solutions has significant dependence on the temperature. Refractive index and viscometric detectors are the most common detectors used in GPC, and their responses are both dependent on the temperature. Both of these detectors have relatively low sensitivity, resulting in the necessity to use concentrated polymer solutions, which in turn increases the sample viscosity. It is highly recommended to stabilize the temperature of the whole system and work at elevated temperatures. At higher temperature the solution viscosity decreases, which allows working with lower backpressures. The lower backpressure also leads to a more stable liquid pump operation and a more stable flow of the mobile phase. Even a slight variation in the mobile phase flow rate results in the significant error in the molecular weight determination. Typical requirements for the flow stability: GPC system should possess a flow variation of less than 0.2%.

6.9 DETECTORS

A refractive index (RI) detector is probably the most widely used in GPC. The main advantage of an RI detector is that any polymer solution will generate a response. This detector has several disadvantages:

- Low sensitivity (2–3 orders of magnitude lower than UV)
- Very sensitive to pressure, flow, and temperature fluctuations

Two specialized detectors have been developed with specific applications for SEC. The first is the laser light-scattering detector, which has been around for almost 20 years but now has new electronics and computer data acquisition capabilities. Substitution of bulky He–Ne gas lasers with small, inexpensive diode lasers has greatly reduced the size and cost of laser light-scattering detectors, and the development of reference flow viscometers has provided similar size and cost advantages for viscometer detectors.

Each detector measures a different and complementary variable. The light-scattering detector gives a response that is proportional to molecular weight and concentration. Likewise, the viscometer detector response is proportional to the intrinsic viscosity and concentration.

6.10 SOLVING MASS BALANCE ISSUES

SEC can be used to solve mass balance issues that may be encountered in reversed-phase chromatography applications—for example, if the area% does not agree with the assay% value for pharmaceutical analysis of a particular active pharmaceutical ingredient or intermediate. The discrepancy in mass balance could be due to species that are not observed at the wavelength of detection using RPLC; the sample could contain salt, water, residual solvents, and even polymers. These polymer species may not have been eluted using the RPLC conditions. GPC may be employed to elucidate if high-molecular-weight species are present in the sample (leading to discrepancy in mass balance). The higher-molecular-weight species may or may not have a UV chromophore, so detection with UV and/or RI should be employed. This technique is most suitable for quality control and polymer screening, as well as for troubleshooting during the analysis of pharmaceutical samples.

6.11   AQUEOUS SEC APPLICATIONS

For aqueous SEC, ionic interactions need to be suppressed and the eluent usually modified by the addition of salt and/or the adjustment of pH. For water-soluble polymers with hydrophobic character, the addition of a weak organic solvent (methanol) is sometimes required to inhibit hydrophobic interactions. A general approach for the separation of the aqueous polymeric samples is shown in Figure 6-8.

In biological separations, the choice of mobile phase is critical and must consider both column performance and the maintenance of biological function of the sample and the aim of separation. Figure 6-9 represents a preparative separation of a mixture of proteins at pH 7 in phosphate buffer. Typical buffers used for classical protein handling are acceptable for gel filtration. These include denaturing buffers (6 M guanidine hydrochloride or 8 M urea) and those containing nonionic detergents (Tween-20, Brij-35 at concentrations from 0.01 to 0.1% v/v). All silica-based gel filtration columns possess a slight negative charge. This charge is likely due to unreacted

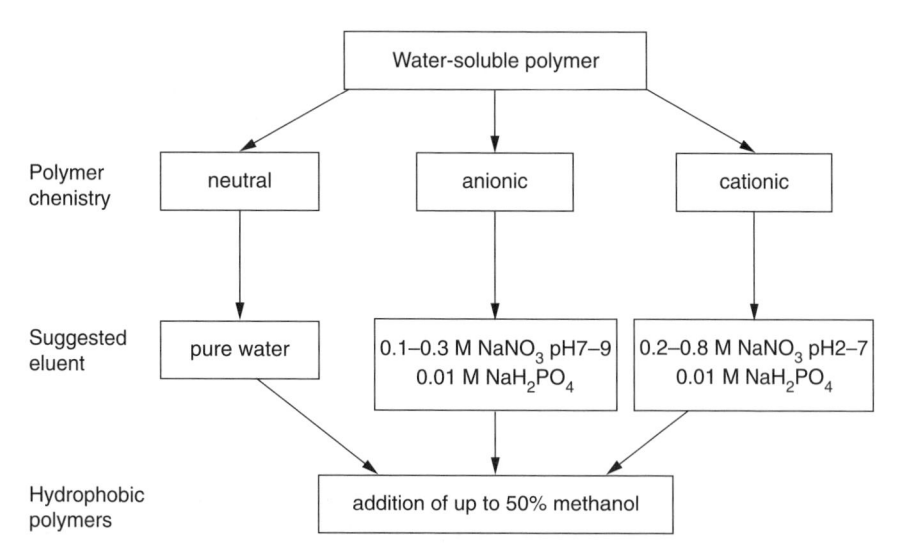

Figure 6-8. Principles of the selection of acceptable column type and mobile-phase type for the GFC separation of water-soluble polymers. (Reprinted from reference 3, with permission.)

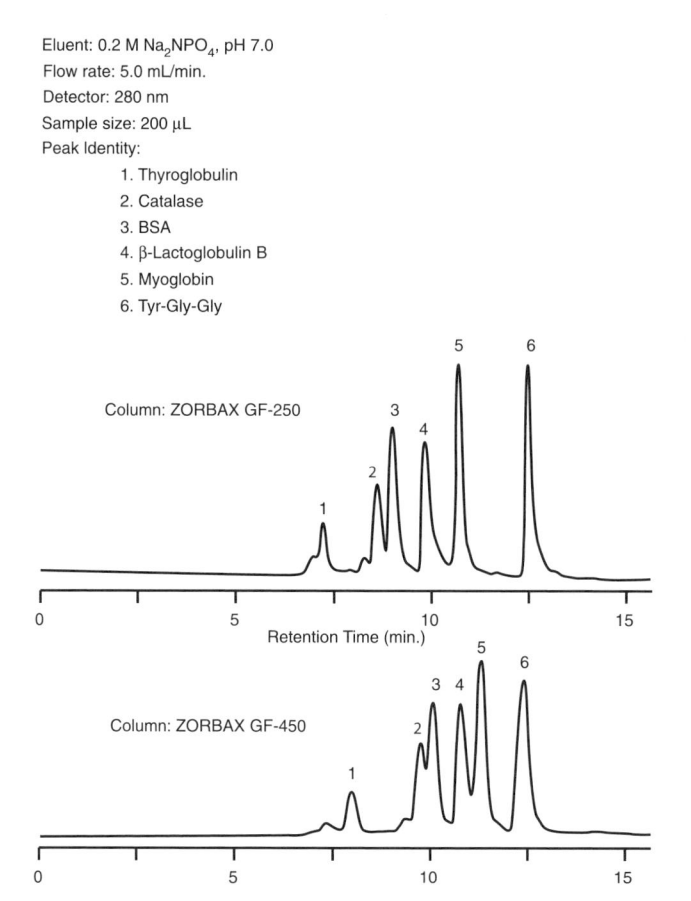

Eluent: 0.2 M Na_2NPO_4, pH 7.0
Flow rate: 5.0 mL/min.
Detector: 280 nm
Sample size: 200 µL
Peak Identity:
    1. Thyroglobulin
    2. Catalase
    3. BSA
    4. β-Lactoglobulin B
    5. Myoglobin
    6. Tyr-Gly-Gly

Column: ZORBAX GF-250

Retention Time (min.)

Column: ZORBAX GF-450

Figure 6-9. Separation of proteins on Zorbax GF-250 and GF-450 preparative columns. (Reprinted from reference 10, with permission.)

silanol groups on the silica surface. The separation of biomacromolecules at typical buffer concentrations of 0.1–0.5 M is recommended. Typical buffers include sodium phosphate, Tris-HCl, and sodium acetate. The addition of 50–100 mM NaCl may also be added to suppress any undesired ionic interactions between the charged solute and the stationary phase. Also, the pH of the mobile phase is very important to suppress nonideal ion-exchange interactions of the biomacromolecules with the stationary phase. If ion-exchange effects are observed, they can potentially be suppressed by either changing the pH or increasing the ionic strength of the buffer. Using buffers with pH values above 8.5 does cause a slow base-catalyzed dissolution of the silica packing, but can be tolerated for short periods at the expense of reduced column life.

Other types of GFC columns are chemically modified crosslinked polystyrene gels with sulfo groups on the surface. They are suitable for the

Column:	Rezex RPM-Monosaccharide
Dimensions:	300 × 7.8 mm
Order No:	00H-0135-K0
Mobile Phase:	Water
Flow Rate:	0.6 mL/min
Detection:	RI
Temperature:	75°C
Sample:	1. Stachyose
	2. Maltose
	3. Glucose
	4. Xylose
	5. Galactose
	6. Fructose
	7. Meso-Erythritol
	8. Mannitol
	9. Salicin
	10. Xylitol
	11. Sorbitol

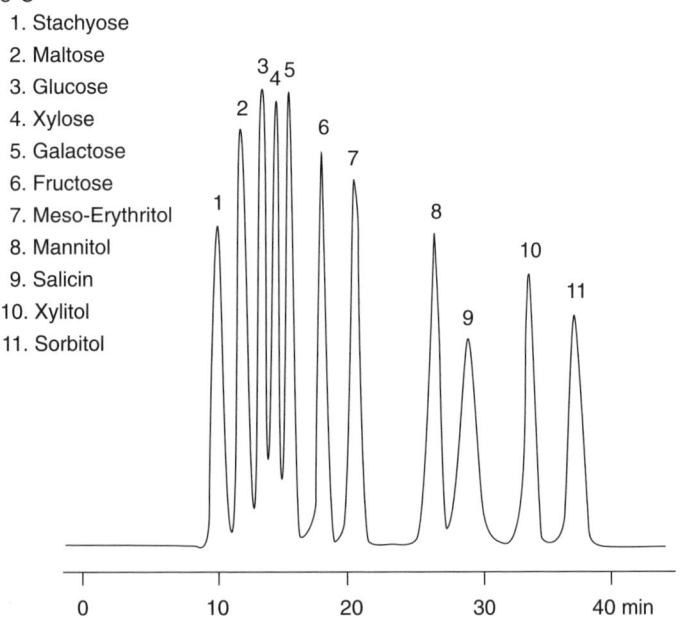

Figure 6-10. Example of a separation of dietary fiber mix on GPC column from Shodex. (Reprinted from reference 4, with permission from Phenomenex.)

separation of water-soluble polymers like polysaccharides, starches, and celluloses. An example of such a separation is shown in Figure 6-10. Some other references that review the details of size-exclusion chromatography are available [11–13].

REFERENCES

1. L. Snyder and J. J. Kirkland, (eds.), *Introduction to Modern Liquid Chromatography*, 2nd ed. John Wiley & Sons, New York, 1979.
2. J. Brandrup, E. H. Immergut, E. A. Grulke, A. Abe, and D. R. Bloch, *Polymer Handbook*, 4th ed., John Wiley & Sons, New York, 2003.
3. www.polymerlabs.com
4. www.phenomenex.com

5. H. Benoit, Z. Grubisic, P. Rempp, D. Decker, and J.-G. Zillox, Liquid-phase chromatographic study of branched and linear polystyrenes of known structure, *J. Chim. Phys.* **63** (1966), 1507–1514.

6. Z. Grubisic, P. Rempp, and H. J. Benoit, Universal calibration for gel permeation chromatography, *Polymer Sci. B, Polymer Lett.* **5** (1967), 753–759.

7. W. W. Yau, J. J. Kirkland, and D. D. Bly, *Modern Size-Exclusion Liquid Chromatography*, Wiley-Interscience, New York, 1979.

8. *Waters Catalog*, Waters Inc., Milford, MA, 2006.

9. B. G. Belenkii and L. Z. Vilenchik, *Modern Liquid Chromatography of Macromolecules*, Journal of Chromatography Library, vol. 25, Elsevier Amsterdam, 1983.

10. Agilent Zorbax PrepHT GF-250 datasheet, Agilent Technologies, 2004.

11. Chi-San Wu, *Handbook of Size Exclusion Chromatography and Related Techniques*, Marcel Dekker, 2004.

12. J. Wen, T. Arakawa, and J. S. Philo, Review: Size exclusion chromatography with on-line light scattering, absorbance, and refractive index detectors for studying proteins and their interactions, *Anal. Biochem.* **240** (1996), 155.

13. M.-I. Aguilar, *HPLC of Peptides and Proteins*, Vol. 251, Humana Press, Totowa, NJ, 2004.

7

LC/MS: THEORY, INSTRUMENTATION, AND APPLICATIONS TO SMALL MOLECULES

Guodong Chen, Li-Kang Zhang, and Birendra N. Pramanik

7.1 INTRODUCTION

The discovery of a new drug is a challenging task that includes (a) identification of a biochemical target for certain diseases and (b) screening of a large number of compounds from libraries of compounds arising from synthetic chemistry, combinatorial chemistry, and natural product isolation for lead generation. The lead compound is then optimized based on biological activity, selectivity, pharmacokinetic property, and metabolism. This process produces a large volume of samples requiring rapid and accurate analysis, with the speed of analysis contributing directly to the drug discovery cycle time.

As one of primer analytical techniques, mass spectrometry (MS) developed from nineteenth-century physics, starting with the pioneering work of J. J. Thomson on the electrical discharges in evacuated tubes. In 1913, Thomson wrote "I feel sure that there are many problems in Chemistry which could be tackled with far greater ease by this than any other method. The method is surprisingly sensitive—more so than that of Spectrum Analysis—requires infinitesimal amount of material, and does not require this to be specially purified. . . ." Indeed, MS offers speed, high sensitivity and isotopic specificity. This technique separates mixtures of ions on the basis of mass-to-charge ratios,

HPLC for Pharmaceutical Scientists, Edited by Yuri Kazakevich and Rosario LoBrutto
Copyright © 2007 by John Wiley & Sons, Inc.

providing the molecular weight of a compound and its structural information from fragment ions. It is widely used for identification and quantification of known/unknown organic compounds. Rapid development of MS in recent decades has further expanded its role in the structural characterization of small molecules in the drug discovery process [1].

In combination with chromatographic separation techniques, principally in the form of high-performance liquid chromatography (HPLC) / MS (LC/MS), mass spectrometry has become the principal method of mixture analysis in pharmaceutical research and development [2]. Early discovery research often involves library compounds analysis using high-throughput LC/MS methods. Identification and quantification of drug metabolites is essential in drug metabolism and pharmacokinetic studies. Structural characterization of impurities and decomposition products in bulk drug substances is an integral part of pharmaceutical product development. LC/MS combines the high-resolution separation capability of HPLC with MS detection and characterization ability, playing important roles in all these aspects of drug discovery process.

7.2 IONIZATION METHODS AND LC/MS INTERFACES

Different physical principles can be used to separate and measure ions (charged particles) with different mass-to-charge ratios under high vacuum conditions, and this has resulted in a variety of mass spectrometers [3]. In principle, the functioning of all mass spectrometers in generating mass spectra involves four steps: (1) introduction of the sample; (2) ionization of the sample molecule to convert the neutral molecules to ions in the gas phase (ionization method); (3) sorting of the resulting gas-phase ions by mass-to-charge ratios (mass analyzer); and (4) detection of separated ions. Critical components of a mass spectrometer include ion source and mass analyzer. Depending on desired applications and sample types, different mass spectrometers can be utilized to perform specific analytical tasks.

7.2.1 Ionization Methods

An ideal ionization source for a mass spectrometer should provide a high ionization efficiency and a high stability of ions for subsequent mass analysis by mass analyzers. In addition, control of internal energy deposited on ions during ionization should be achievable in the ionization source to control the degree of fragmentation. It is also desirable to couple ionization source with chromatographic separation techniques, especially with HPLC. Various ionization methods have been developed over the years, including electron impact (EI), chemical ionization (CI), desorption ionization (DI), matrix-assisted laser desorption/ionization (MALDI), desorption electrospray ionization (DESI), electrospray ionization (ESI), and atmospheric pressure chemical ionization (APCI). Note that ESI and APCI are part of LC/MS interfaces that will be

TABLE 7-1. Summary of Ionization Methods

Ionization Method	Ionization Agent	Strengths	Limitations
EI	Electrons (~70 eV)	Extensive fragmentation, reproducible spectra, searchable large reference compound EI libraries	Limited to volatile/ nonpolar molecules
CI	Gaseous ions	Abundant molecular ions with controllable fragmentation	Limited to nonpolar and moderately polar molecules, limited fragmentation
APCI	Corona discharge/ gaseous ions	Operative at atmospheric pressure, easy interface to HPLC, abundant molecular ions	Limited to low to moderately polar molecules
DI	Energetic particles (atoms, ions), photons	Abundant molecular ions from high-mass compounds	Difficult to interface to HPLC
ESI	Electrical/thermal/ pneumatic energy	Operative at atmospheric pressure, easy interface to HPLC, multiple-charged ions for large biomolecules	Poor results for nonpolar molecules, limited fragmentation

discussed in separate sections. Table 7-1 summarizes some characteristics of different ionization methods.

EI and CI are two early developed ionization methods. They are extremely useful for ionizing volatile compounds. In EI process, molecules are ionized by collisions with energetic electrons (typically 70 eV) produced from a heated filament. It produces highly reproducibly mass spectra with extensive fragmentation of molecular ions. Thus, library searching with existing EI mass spectra is possible for unknown identifications. The nature of fragmentation in EI often leads to lower abundances or absence of molecular ions. On the other hand, CI is a soft ionization method that generates mainly molecular ions by ion/molecule reactions of regent ions with analyte molecules [4].

Commonly used reagent gases include methane, ammonia, and isobutane. The internal energy deposition or fragmentation of ions determines the appearance of mass spectra. It can be controlled by selecting appropriate reagent gases. In terms of proton transfer reactions in CI-MS, the relative proton affinity between the analyte and the reagent gas determines whether the analyte will be ionized. The analyte with a higher proton affinity than that of the reagent gas will be ionized, while the analyte with a lower proton affinity than that of the reagent gas will not be ionized. Furthermore, the difference in proton affinities of reagent gas and analyte is largely responsible for the extent of fragmentation if ionization of the analyte occurs.

EI and CI methods are complementary to each other, providing molecular weight and structural information. As an illustration, Figure 7-1 shows an EI-MS spectrum of mometasone furoate, an anti-inflammatory steroid drug. A very low abundant molecular ion at m/z 520 is visible. However, the base peak in the spectrum is the fragment ion at m/z 295 corresponding to the loss of furoate ring, HCl, and a moiety of [COCH$_2$Cl]. Other fragment ions in the spectrum yield structurally characteristic fragmentations for this molecule. In contrast, a CI-MS spectrum of the same compound exhibits the protonated molecular ion as the most abundant peak at m/z 521 along with some fragment ions (Figure 7-2). The appearance of these two spectra clearly demonstrates the utility of EI-MS and CI-MS methods.

An inherent limitation for EI and CI methods is the requirement that the sample analyzed must be volatile. Both methods do not produce MS data for

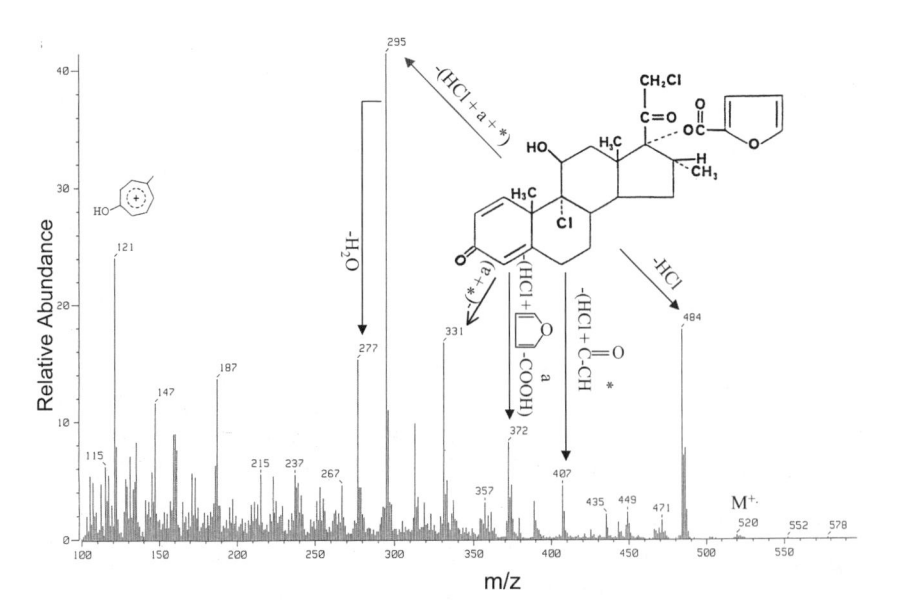

Figure 7-1. EI-MS spectrum of mometasone furoate.

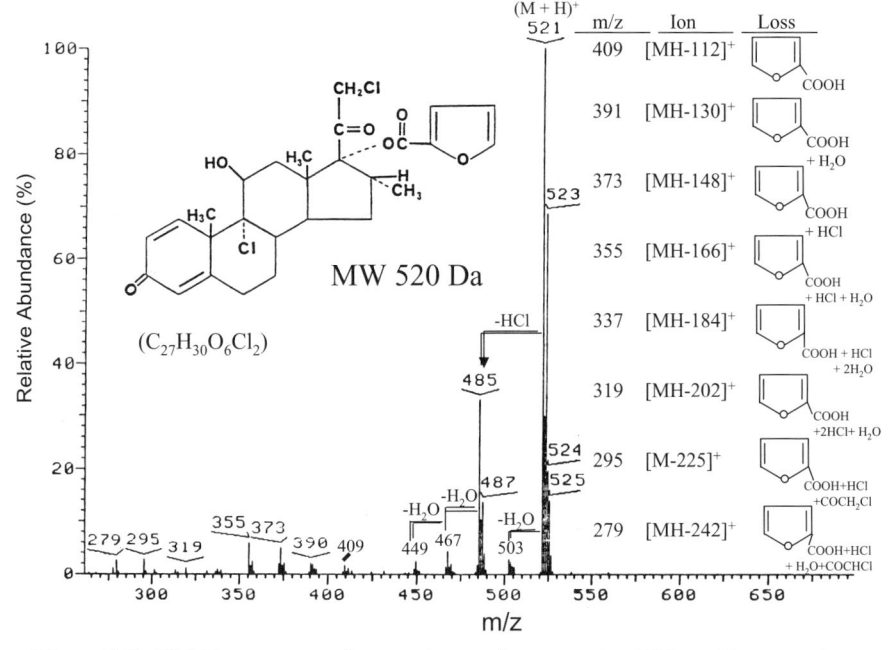

Figure 7-2. CI-MS spectrum of mometasone furoate using NH_3 as CI reagent gas.

polar compounds. One solution to this limitation is to employ DI methods to ionize nonvolatile samples with high molecular weights [5]. In the DI process, energetic particles or photons impact onto samples on a surface and result in the liberation of intact molecular ions via selvedge region without direct transfer of the energy to the sample molecules. The particle bombardment includes keV atoms (e.g., Ar, fast atom bombardment [6]), keV ions (e.g., Cs^+, liquid secondary ion MS [7]), and MeV ions (e.g., plasma desorption [8]). Both fast atom bombardment (FAB) and liquid secondary ion MS (LSI) utilize large excess matrix for absorption, excitation, and relaxation of energetic particles, producing mainly molecular ions of interest. They are well-suited for studies of natural products and small polar compounds with molecular weights of a few thousand daltons (Da). Another important DI method is MALDI, which employs a UV or IR absorbing matrix in large excess with samples (5000:1) to absorb the photon energy from laser irradiation [9]. This method generates mostly singly charged molecular ions with molecular weight as high as 500 kDa. MALDI has a high ionization efficiency for large biomolecules with supersensitivity in the range of low-femtomole level. It has become one of most widely used ionization methods in biological mass spectrometry.

The latest addition to DI methods is DESI [10]. It directs an aqueous spray from an electrospray apparatus onto the sample on a surface. In the process, the fast nebulizing gas jet transports the charged droplets and impacts the surface in the absence of matrix, carrying away analyte molecules. This approach has been successfully applied to analysis of small molecules and proteins. The unique characteristic of DESI is that it operates under ambient conditions. All other DI methods as described above normally require vacuum operation conditions, and sample manipulation during experiments is not feasible. DESI lifts the restriction on the vacuum constraints and can be very flexible in carrying out novel experiments. Potential applications include forensic analysis, explosive detection, and biological imaging experiments in tissues.

7.2.2 Historical View of Interfaces

A critical component of the LC/MS system is the interface that connects an HPLC system to a mass spectrometer. The basic requirements for a successful interface include maintaining chromatographic performance (minimum additional peak broadening), high transfer efficiency from LC to MS, and no degradation in mass spectrometric performance. Historically, a main challenge in LC/MS interfaces was that high liquid flows from HPLC make it very difficult to maintain the high vacuum required for the function of a mass spectrometer. A number of different LC/MS interfaces have been developed over the years to address this issue and overcome the difficulty [3].

7.2.2.1 Direct Liquid Introduction. One of the first attempted experiments to introduce liquids into a mass spectrometer was to minimize the amount of liquid into an MS, removing solvent by the vacuum system and ionizing the analyte in the gas phase. The pioneering work carried out by Tal'roze et al. [11] described the simplest direct liquid introduction interface. In their experiments, solvent was introduced into the mass spectrometer through a capillary at a flow rate below 1 μL/min. The ionization of analytes occurred by EI. The low flow rate used in the experiments was a limitation and would not give good sensitivities for analytes. In 1970s, McLafferty's group employed a direct liquid introduction (DLI) interface to directly introduce a small fraction (<1%) of the liquid from HPLC into the ion chamber of a CI mass spectrometer [12]. The solvent acted as the ionizing reagent. The maintenance of the vacuum was assisted by using large pump systems and differential pumping. Micro-and nanobore chromatography (<1-mm-i.d. column) were suitable for DLI. A detection sensitivity of picogram level was achieved for full-scan analysis.

7.2.2.2 Moving Belt System. Initially developed by McFadden et al. [13], the moving belt system was based on the physical method of evaporation of the mobile phase through heat and vacuum that leave analytes as a thin coating on a continuously cycling polyimide belt. The analytes were transported from atmospheric pressure region to the vacuum of the ion source through differentially pumped vacuum locks. Ionization methods used included EI and CI for volatile analytes. The system has excellent enrichments and efficiencies, although it is often limited to the analysis of compounds which could have been analyzed by gas chromatography (GC)/MS.

7.2.2.3 Thermospray.

The thermospray interface was introduced and developed by Blakley and Vestal [14]. In their approach, a liquid flow from HPLC was directed through a resistively heated capillary connecting to the MS ion source. The heat and vacuum would evaporate the solvent from a supersonic beam of mobile phase produced in the spray, creating charged small microdroplets. These small liquid droplets were further vaporized in the heated ion source. Ions present in the ion source were then transferred to the mass analyzer, and residual vapors were pumped away.

Ionization process in thermospray involves ion desolvation/evaporation from charged liquid droplets and gas-phase ion/molecule reactions. Both positively and negatively charged ions are formed in the process. Volatile buffers such as ammonium acetate are often used as part of HPLC effluent. These buffer ions act as CI reagent ions to form either protonated or deprotonated ions. The gas-phase proton affinity (for positive ion) or acidity (for negative ion) of the analyte relative to the buffers will determine whether the analyte will be ionized. If the proton affinity of the analyte is lower than that of the reagent ion, the analyte will not be ionized. This CI-like aspect of the ionization process results in thermospray mass spectra containing mostly molecular ions. When buffer is not used in thermospray experiments, an external ionization method is often applied, including EI filament and discharge ionization. These supplemental modes produce solvent-related CI mass spectra.

A main advantage of thermospray is that it can handle commonly used HPLC eluents at higher flow rates (up to 2 mL/min) and generate good results for polar, nonvolatile, and thermolabile compounds. However, the sensitivity of the method is highly compound-dependent and not particularly attractive to high-molecular-weight compounds.

7.2.2.4 Continuous-Flow FAB.

Continuous-flow FAB is a modified form of FAB method [15, 16]. In this modified method, the HPLC effluent with added FAB matrix (usually 5% aqueous glycerol) is continuously transported through a fused-silica capillary to the tip of a FAB probe residing inside of the ion source. The HPLC liquid with matrix deposited on the tip of the FAB probe is subjected to atom bombardment for ionization of analytes. The matrix addition can be done either pre-column or post-column, although post-column addition is preferred. The acceptable liquid flow rate in continuous-flow FAB is less than 10 μL/min. Flow splitting or the use of capillary chromatography is often required in the experiments. A major advantage in this method is the reduced chemical noise since much less matrix is used in continuous-flow FAB than in standard FAB experiments. This has led to improved detection limits to subpicomole range. Significantly, this interface allows the LC/MS analysis of biomolecules that are traditionally analyzed by DI methods.

7.2.3 Common Interfaces

The early developed LC/MS interfaces as described above have played important roles in the evolution of LC/MS interfaces. However, their applicability,
sensitivity, and robustness are very limited. The overwhelming popularity of
LC/MS today is largely due to the development of atmospheric pressure ionization (API) interfaces, including ESI and APCI.

7.2.3.1 Electrospray. The first description of ESI was made by Zeleny in
1917 [17]. He described how a high electrical potential applied to a capillary
caused the solvent to break into small droplets. In late 1960s and early 1970s,
Dole and co-workers attempted to generate gas phase ions from macromolecules in solution using an atmospheric pressure electrostatic sprayer by ion
mobility spectrometry [18, 19]. In the late 1970s, Thomson and Iribarne successfully demonstrated the production of macro-ions from electrically charged
droplets using MS [20, 21]. The very first applications of ESI were reported
independently by Yamashita and Fenn [22] and Aleksandrov et al. [23] in
the mid-1980s. Now ESI has become one of the most successful ionization
methods / interfaces used in mass spectrometry [24].

The basic ESI apparatus consists of a spray needle at high electrical potential (4–5 kV), a thermal/pneumatic desolvation chamber, and the vacuum interface (Figure 7-3). The ESI process is electrophoretic in nature. It may involve
the generation of charged micro-droplets under a high electrical field and the
subsequent evaporation of droplets using either a drying gas (N_2) or thermal
desolvation. The first step of ion formation is the droplet formation at the
needle tip when the high electrical field causes ions of the same polarity to

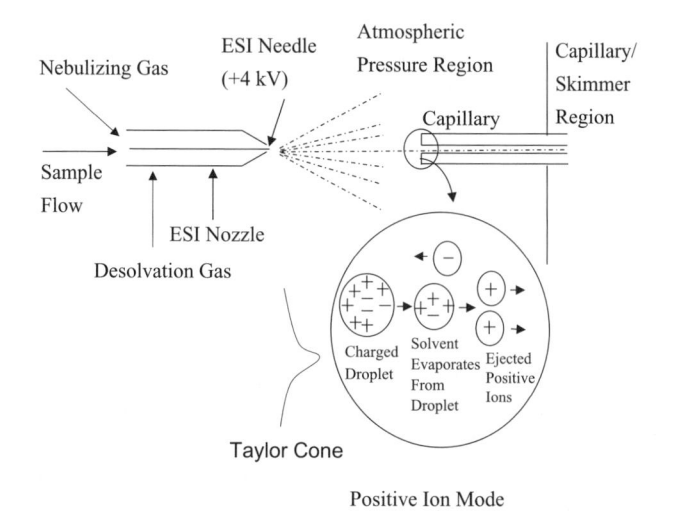

Figure 7-3. ESI source schematic diagram. The ion formation is illustrated in the positive ion mode.

form "Taylor cone" on the solution surface and emit charged droplets. The second step of ion formation is solvent removal, and its process is somewhat debatable. One theory is Dole/Fenn's coulombic explosion. When the initially formed droplets become smaller droplets due to evaporation of solvents, the surface charge density increases and the coulombic forces exceed the surface tension (Rayleigh stability limit), with the droplets breaking into smaller droplets. Further evaporation process with Rayleigh droplet fragmentation produces analyte ions [22, 25].

One of the most important features in ESI is the formation of multiply charged ions for proteins/peptides [26]. Since a mass spectrometer measures mass-to-charge ratios of a compound, the multiply charged ions will appear in the mass spectrum at m/z values that are fractions of the mass (MW) of the ion. This allows the detection of high molecular weights of proteins/peptides using a standard quadrupole mass analyzer (3000-Da mass range). In addition, the detection of multiply charged ions provides precise measurements of molecular weights of proteins/peptides via the deconvolution method. A mass accuracy of better than 0.01% can be achieved for proteins with masses up to 100 kDa [27]. Another important characteristic of ESI is the softness of the ionization. It is a very mild process and can generate mainly molecular ions with little fragmentation. For small molecules, the singly charged molecular ions usually dominate the mass spectrum. The third characteristic of ESI is the simplicity of the source design and its operation at atmospheric pressure, allowing ESI to be readily coupled to HPLC. It is important to note that a low flow rate (~200 μL/min) of the sample solution is required in order to maintain a stable spray in ESI. Thus, flow splitters are often utilized in ESI-LC/MS applications. This does not reduce the concentration sensitivity of ESI since ESI responses are directly related to the concentration of the analyte entering the ion source. However, the mass sensitivity can be substantially increased with a lower flow rate if the same concentration sensitivity is maintained ($c = m/v$). This has led to the wide use of nano-spray (~nL/min) LC/MS for analysis of proteins and peptides, achieving femtomole sensitivity [24].

7.2.3.2 *Atmospheric Pressure Chemical Ionization.* APCI is closely related to ESI. It was developed by Horning et al. [28] in the early 1970s. Figure 7-4 illustrates a typical APCI source. The sample solution is introduced into a nozzle spray device similar to that used in ESI, but without the high electrical potential applied to the nozzle. The nebulizing gas (usually N_2) is often added to assist the desolvation/ionization process. Although a heater at a temperature of 400–500°C is used to vaporize solvents, minimal degradation of the sample occurs. A corona discharge needle at a high voltage (3–5 kV) is responsible for producing a discharge current and inducing solvent ionizations. The generated solvent reagent ions react with analyte molecules via gas-phase ion/molecule reactions and produce analyte ions. Clearly, the ion formation process is separated from solvent evaporation process in APCI (in contrast to ESI), allowing the use of solvents unfavorable for ion formation. For example,

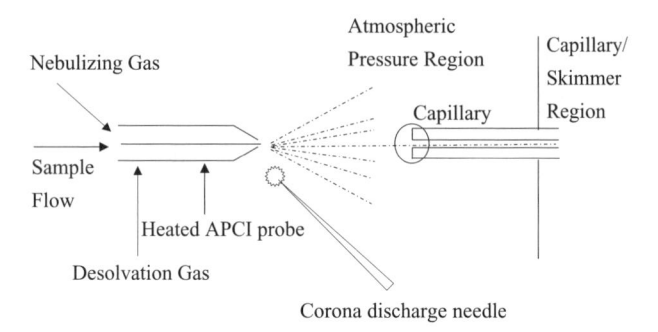

Figure 7-4. APCI source schematic diagram.

low-polarity solvents generally used in normal-phase chromatography can be evaporated for APCI ionization. Unlike ESI, APCI does not form multiple-charge ions for high mass compounds, and its response is more directly related to the absolute amount of analytes. APCI achieves optimal performance at high flow rates (1–2 mL/min), making it ideal as the LC/MS interface for connecting to conventional HPLC without flow splitters. The other features of APCI include the appearance of CI-like mass spectra and suitability for analysis of volatile or semivolatile compounds.

7.2.4 Special Interfaces

There are other LC/MS interfaces that are less commonly used than ESI and APCI, but are often employed by researchers for analysis of nonpolar or neutral compounds, including particle beam and atmospheric pressure photoionization (APPI).

7.2.4.1 Particle Beam. Particle beam interface, also known as MAGIC (monodisperse aerosol generation interface for chromatography), was developed by Browner and co-workers in the early 1980s [29]. It uses a momentum separator to eliminate volatile solvents and to transport analyte in the form of micro-aggregates particles to the EI/CI source of a mass spectrometer. Typically, the LC effluent is forced into a small nebulizer using a helium gas flow to form aerosol droplets. These uniform droplets go through a desolvation chamber and evaporate into particles that are further separated from solvent vapors by a multistage momentum separator. The flow rate of liquid samples in particle beam interface ranges from 0.1 mL/min to 0.5 mL/min. The interface is limited to relatively volatile compounds. One of advantages in using a particle beam interface is the database searching capability with the EI-MS library for structural identifications.

7.2.4.2 Atmospheric Pressure Photoionization. Photoionization has been used in mass spectrometry to ionize a variety of compounds [3]. The formation of ions involves the absorption of a photon by the molecule and ejection of an electron from the molecule to form the radical cation. The necessary condition for the ionization to occur is that the photon energy has to exceed the ionization potential of the molecule of interest. Like ESI and APCI, APPI uses nebulizer and vaporizer for desolation. The ionization occurs at atmospheric pressure with UV light source [30]. The standard UV lamp has a photon energy of about 10 eV that is sufficiently high to ionize most organic molecules. Common HPLC solvents and permanent gases usually have higher ionization potentials that will not be ionized. This results in relatively noise-free mass spectra, as opposed to ESI or APCI. In some cases, analyte molecules may exhibit higher ionization potentials (>10 eV) and the direct photoionization may not produce ions. Then, addition of a large excess of a dopant such as toluene and acetone is necessary to yield charge carriers for ionizing analytes of interest. The undesired consequence of this dopant-assisted APPI is the increase of background ions.

The potential application of APPI includes analysis of compounds (nonpolar and neutral analytes) that are not effectively ionized by ESI or APCI. APPI appears to be less influenced by matrix suppression as seen in ESI or APCI. It can serve as a complementary ionization source to ESI/APCI.

7.3   MASS ANALYZERS

The basic function of a mass analyzer is to measure the mass-to-charge ratios of ions (charged particles) and provide a means of separating the ions. The operating principles of mass analyzers depend on interactions of charged particles with electrical or magnetic fields. Commonly used mass analyzers include magnetic sector, quadrupole, ion trap, time-of-flight (TOF), and Fourier transform ion cyclotron resonance (FT-ICR). The combination of different mass analyzers can provide additional capabilities of performing mass spectrometry/mass spectrometry (MS/MS) or tandem MS experiments for structural characterization. Table 7-2 lists some characteristics of various mass analyzers.

7.3.1   Magnetic Sector

When an ion passes through a magnetic filed perpendicularly, it moves along a circle with the radius where the centrifugal force is balanced by the magnetic force. Essentially, the magnetic sector acts as a momentum-to-charge analyzer. It brings a divergent beam of ions to focus (direction focusing). If all ions have the same kinetic energy, the magnetic sector can behave as a mass analyzer. The mass-to-charge ratios of ions are directly related to the magnetic field strength and the radius of the curvature, but inversely related to the accelerating voltage in the ion source.

In a realistic situation, the ions produced in the ion source always have a certain distribution of ion kinetic energy that will impact the mass resolution

TABLE 7-2. Characteristics of Mass Analyzers

Mass Analyzer	Quantity Being Measured	Mass Range (Da)	Mass Resolution[a]	Dynamic range[b]
Magnetic sector	Momentum per charge	$>10^4$	$>10^4$	$>10^6$
Quadrupole	Path stability	$>10^3$	Unit resolution	$>10^4$
Ion trap	Path stability	$>10^3$	Unit resolution, $>10^3$ at slow scan speed	$>10^3$
Time-of-flight	Time	$>10^5$	$>10^4$	$>10^3$
FT-ICR	Frequency	$>10^5$	$>10^5$	$>10^4$

[a]Mass resolution is defined as $m/\Delta m$, where Δm is defined as mass difference at full width at half-maximum (FWHM).
[b]Dynamic range is defined as the range of either ion counts or sample concentration over which a linear response is obtained.

$R (= m/\Delta m)$. In order to compensate for the distribution in ion kinetic energy, an electrostatic analyzer capable of separating ions according to kinetic energy-to-charge ratios can be used in combination with magnetic sector. Each of these two devices independently focuses ions for direction, and together they give velocity focusing—that is, equal and opposite dispersions for velocity. This effect is termed *double focusing* (direction focusing and velocity focusing), and the concept was realized in the early work of Aston [31]. The double focusing instrument minimizes ion kinetic energy distribution and gives a high mass resolution independent of mass, providing accurate mass measurement capability. Other important features include large dynamic range and high-energy collision activation capability for structural elucidation studies. Although the magnetic sector instrument has played an important role in the structural characterization of small molecules, its role in LC/MS is much less significant because of sensitivity issue and the difficulty in interfacing with LC.

7.3.2 Quadrupole

The quadrupole is a device in which electrical potentials of RF and DC are applied to opposite pairs of a linear array of four parallel rods with hyperbolic cross sections (Figure 7-5A). The ion motion under the electrical field can be described by the Mathieu equation [32]. In general, an ion moving through the rod assembly only experiences the force in the plane normal to the direction of ion motion (z-direction). Only ions that are stable in this plane will remain in the rod assembly and eventually reach the ion detector. For a given mass-to-charge ratio of ions, the stability relies on the size of the rod assembly, oscillation frequency, RF voltage, and DC voltage. The mass analysis is performed by sweeping DC and RF voltages, while maintaining their ratio and oscillation

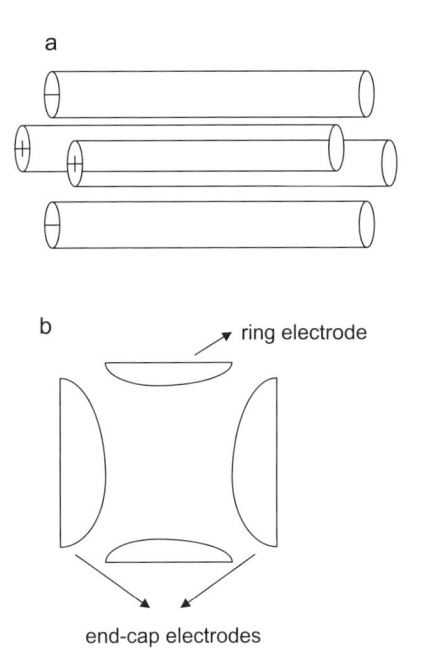

a

b ➤ ring electrode

end-cap electrodes

Figure 7-5. Schematic diagram of (A) quadrupole and (B) 3-D quadrupole ion trap.

frequency constant. This mass-selective stability scan mode allows ions of different mass-to-charge ratios to be stable and pass through the device. Those ions of higher or lower mass than the desired ones are ejected from the rod assembly without passing through. The mass resolution of quadrupole depends on the ratio of DC-to-RF voltages. Typically, a quadrupole is operated as a mass analyzer with unit mass resolution.

A quadrupole is small and relatively inexpensive. It serves as an excellent collision cell for collision activations of ions and ion/molecule reactions. It can also be used as a broadband ion transmission device. A quadrupole is readily coupled with other mass analyzers for MS/MS experiments. One of the most popular configurations is a triple quadrupole mass spectrometer that has found wide applications in LC/MS and LC/MS/MS (see section on Tandem MS).

7.3.3 Ion Trap

The quadrupole ion trap is a three-dimensional (3-D) analog of the linear quadrupole [33]. It consists of two end-cap electrodes with hyperbolic cross sections and one ring electrode located between the end caps (Figure 7-5B). The RF voltage is applied to the ring electrode and the ground potential is normally operated on the end caps. A rotationally symmetric electric quadrupole field is generated in the device to trap ions with stable motions described by Mathieu equation. The mass spectra are obtained by raising RF voltage so as to cause the ions to become unstable and be ejected from the trap through a hole in the end-cap electrode. In this mass-selective instability scan mode, ions of increasing mass-to-charge ratios are ejected and detected as the RF voltage is raised.

The trapped ions possess characteristic oscillation frequencies. The stable motion of ions in the trap is assisted by the presence of a helium buffer gas (1 mtorr) to remove kinetic energies from ions by collisions. When a supplementary AC potential, corresponding to the frequency of a certain m/z ion, is applied to the end-cap electrode, ions are resonantly ejected from the trap. This method of resonance ejection is used to effectively extend the mass-to-charge ratio of the ion trap. Some other characteristic features of a 3-D ion trap include high sensitivity, high resolution with slow scan rate, and multiple-stage MS capability (see the section on tandem MS). In addition, it is inexpensive and small in size. As a result, a 3-D ion trap is widely used in LC/MS and LC/MS/MS applications.

One of the inherent limitations to a 3-D ion trap is the ion storage capacity because of the relative small volume inside the trap. The space-charge effect can be significant when the ion population reaches above 10^6 ions, impacting mass resolution, mass accuracy, sensitivity, and dynamic range. In order to overcome this limitation, a 2-D linear ion trap has been designed to further improve the performance of 3-D ion trap [34]. Its quadrupole structure has a hyperbolic rod profile, similar to the conventional quadrupole rod. In one of the designs in a 2-D linear ion trap, the quadrupole rod is cut into three axial sections (front section, center section, back section). Appropriate DC and RF potentials are applied to the three sections to contain ions along the axis in the central section of the device. The ion detection is achieved by ejecting ions out of the trap through a hole in the center section. Dual ion detectors along the center section have been used to improve the sensitivity. The advantages of a 2-D linear ion trap over a 3-D ion trap are the increased ion storage capacity (at least 10 times more than 3-D ion trap) and higher trapping efficiencies, leading to a better sensitivity and a larger dynamic range. The use of 2-D linear ion trap is gaining popularity in LC/MS and LC/MS/MS applications, either as a stand-alone instrument or in combination with other mass analyzers (see the section on tandem MS).

7.3.4 Time-of-Flight

In a time-of-flight (TOF) mass spectrometer [35], the ions generated in the ion source are accelerated through a known potential and travel through a flight tube to reach the ion detector (Figure 7-6). The ion arrival time at the detector is measured, and it is directly related to the m/z values of ions. It takes a longer time for heavy ions to reach the detector, while light ions arrive at the detector earlier.

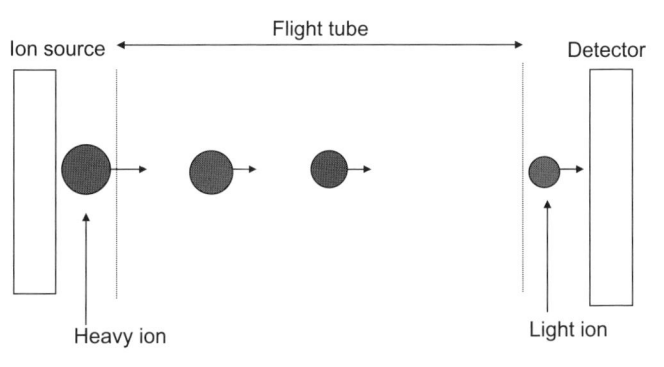

Figure 7-6. Schematic diagram of a linear time-of-flight mass spectrometer.

In spite of the simplicity of the method, the high performance of TOF depends on the corrections regarding flight time differences of the ions with the same m/z values that are caused by initial kinetic energy spread, initial angular spread, and initial position of ion formation spread. A well-established method for velocity differences is to use the reflectron to compensate for the kinetic energy differences [36]. In this method, more energetic ions penetrate deeper into the reflectron field so that these ions will have a longer flight path and arrive at the detector at the same time as ions with less kinetic energy. Other instrument improvements have also been made to achieve high performance capability. The main data acquisition method in TOF is time-to-digital converter. It is a time counting device and provides high time resolution. In contrast to scanning mass analyzers (magnetic sector, quadrupole and ion trap), all ions in TOF are almost simultaneously detected, further improving sensitivity.

One of the unique aspects of TOF is the requirement of pulsing ions for measurements. Thus TOF is best coupled with ion sources producing ions in pulses. Naturally, MALDI is ideal for the combination with TOF. In fact, MALDI-TOF is one of the most widely used systems for analysis of large biomolecules. For a continuous ion source, the ions can be stored for a short period of time and pulsed out for analysis. In the case of ESI, orthogonal injection provides efficient injection of ions from ESI source to TOF. Other important characteristics of TOF also include high mass range (over 100,000 Da), high resolution (over 10,000), and good mass accuracy for accurate mass measurements (less than 5 ppm). These features make TOF attractive for high-resolution LC/MS and LC/MS/MS applications, especially with the combination of quadrupole and TOF system (see the section on tandem MS).

7.3.5 FT-ICR

FT-ICR is derived from earlier work of E. O. Lawrence on ion cyclotron resonance in the 1930s [37]. Its further development using image current detection and Fourier transform was carried out by Comisarow and Marshall in 1970s [38]. With its unsurpassed high performance, FT-ICR has become one of the most powerful mass spectrometers.

The operating principle of FT-ICR is different from other mass analyzers. The central component of the system is a cylindrical analyzer cell, located in the homogeneous region of a very high magnetic field (3–9 tesla), capable of trapping and storing gas-phase ions in an ultrahigh vacuum chamber (10^{-9} to 10^{-10} torr). Ions in a strong magnetic field exhibit cyclic motion with an angular ion-cyclotron frequency in a plane perpendicular to the direction of the magnetic field. This frequency depends on the ion's mass-to-charge ratio and the magnetic field strength, independent of the ion kinetic energies. Upon excitation by an RF voltage, trapped ions in the cell incur the cyclotron motion and induce an image current that is recorded as a time-domain signal containing multiple cyclotron frequencies. The resulting time-domain signal can be Fourier transformed into the frequency-domain signal where each frequency is related to a particular m/z value. The frequency spectrum is readily converted into a mass spectrum.

One of the characteristics in FT-ICR is the super-high mass resolution and mass accuracy. This is largely due to the fact that FT-ICR measures frequencies of ions and such measurements can be made with high precision and accuracy. The ultrahigh vacuum condition employed in FT-ICR also contributes to a longer coherent cyclotron motion of ions and, hence, longer image decay time or transient time directly related to mass resolution. Like an 3-D ion trap, the dynamic range of FT-ICR is limited to the number of ions present in the cell, typically about 10^6, before Coulombic repulsion starts to have serious effects on ion motion and mass resolution. The ion detection in FT-ICR is non-destructive, and reexcitation/remeasurement is possible. All ions with a broad mass range are detected simultaneously in FT-ICR, resulting in excellent sensitivity. As a trapping device, FT-ICR provides the capability of multiple-stage MS analysis for complex molecules (see the section on tandem MS). Recent instrument development also allows facile integration of HPLC with FT-ICR for LC/MS and LC/MS^n applications with unmatched high mass resolution and mass accuracy.

7.3.6 Tandem MS

Mass spectrometry/mass spectrometry (MS/MS) or tandem MS allows the examination of individual ions in a mixture of ions [39]. The ions of interest serve as surrogates for neutral molecules. They are mass-selected and are dissociated upon collisions, and the resulting product ions are analyzed by a second mass measurement. Such experiments generate structurally informative fragment ions for structural characterization of complex molecules.

In a typical MS/MS experiment, two mass analyzers (or two mass-analysis events) are utilized to measure the parent and product ions. The first mass analyzer is usually set to select the ion of interest (i.e., the parent ion), which then undergoes dissociation and generates product ions. The product ions are analyzed by the second mass analyzer. The dissociation of ions in MS/MS experiments can be accomplished by collisions with a target gas introduced into the mass spectrometer (collision-induced dissociation, CID [40]), collisions with a surface (surface-induced dissociation, SID [41]), or photodissociation. The commonly used dissociation method is CID because of its simplicity in implementation. The relative large cross section for the efficient excitation/dissociation process in CID also contributes to the wide use of CID in MS/MS experiments. Typical collision gas used includes argon and nitrogen.

Historically, MS/MS began with metastable ion studies in a sector instrument [42]. A nonintegral and broad mass peak from the metastable ion usually appears in a normal mass spectrum, relating the parent ions to the product ions. It can provide valuable information for interpretation of fragmentation patterns. Modern MS/MS experiments often rely on the scanning of the second mass analyzer to record the product ions. This type of scan mode is a product ion scan that is widely used to obtain structural information of the parent ion (and its neutral form of the parent molecule). The other two forms of MS/MS scan modes include parent ion scan and neutral loss scan. In the parent ion scan, a product ion is mass-selected and a mass scan is performed to obtain all parent ions that fragment to generate this particular product ion. It is often used to identify a class of compounds defined by their characteristic fragment ions. In the case of detection of phosphorylated peptides, a parent ion scan of the fragment ion at m/z 79 (PO_3^-) in the negative ion MS/MS mode often indicates the presence of phosphopeptides. A neutral loss scan involves the identification of all parent ions that fragment by the loss of a neutral molecule. It is very useful in identifying the presence of a functional group for a group of compounds. For example, a neutral loss scan of 98 Da (H_3PO_4 or HPO_3/H_2O) from the phosphopeptide in a positive-ion MS/MS experiment can also be used to confirm the existence of a phosphopeptide.

MS/MS experiments can be performed with a variety of instruments that are either tandem-in-space or tandem-in-time. The tandem-in-space instruments are comprised of multiple mass analyzers, including multiple sector instrument, triple quadrupole (QQQ), and quadrupole-TOF (Q-TOF). In a triple quadrupole, all three MS/MS scan modes can be performed [43]. As an illustration, Figure 7-7 shows a schematic diagram of the triple quadrupole in the product ion scan mode. The first quadrupole (Q1) is set to pass the desired parent ion, and the resulted product ions from CID are obtained by scanning the third quadrupole (Q3). The CID occurs in the second quadrupole (Q2), which is operated in the RF-only mode to allow all ions passing through. A collision gas (Ar or N_2) is introduced into the Q2 to facilitate the CID. Triple quadrupole is commonly used for both qualitative and quantitative studies. The latest addition in tandem-in space instruments is Q-TOF [44]. Its configuration is similar to that of triple quadrupole except that the third quadrupole in QQQ is replaced by a TOF analyzer (Figure 7-8). With the presence of a TOF analyzer, Q-TOF offers high-resolution LC/MS capability and accurate

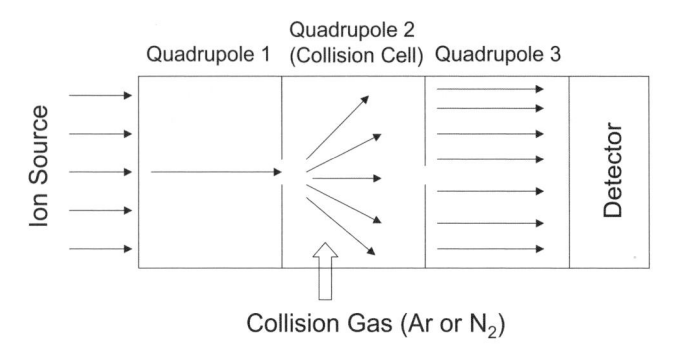

Figure 7-7. Operation of a triple quadrupole mass spectrometer in the product ion mode.

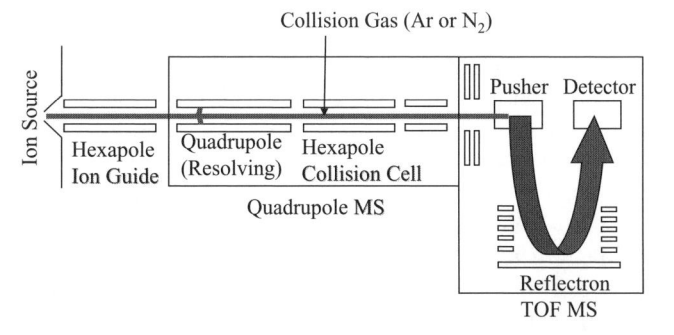

Figure 7-8. Schematic diagram of a Q-TOF mass spectrometer.

mass measurement for trace-level components in mixtures. It has increased sensitivity because of the fast acquisition of near simultaneous detection of all ions in TOF-MS. Additional gain in sensitivity can be achieved by storing and pulsing ions out of the second quadrupole (collision cell) into the TOF analyzer. The most commonly used MS/MS mode in Q-TOF is the product ion scan in which the first quadrupole is used to select the parent ion and the TOF analyzer is set to acquire the product ions resulting from the collisions taking place in the second quadrupole. The main application of Q-TOF is qualitative analysis for structural elucidations. The detection scheme (ion counting) in Q-TOF limits its dynamic range for quantitative work.

The tandem-in-time instruments are mostly ion-trapping devices, including ion trap and FT-ICR. They operate in a time sequence in the scan function to yield MS/MS data, mostly product ion spectra. No additional mass analyzer is required. In the case of an ion trap, the scan function begins with the isolation of ions of interest with ejection of all other ions from the ion trap, followed by (a) translational excitation of ions by applying a supplementary RF voltage to the trap and (b) mass analysis of the product ions using resonant ejection.

The entire process can be repeated to multiple-stage mass analysis (MS^n). Similar to an ion trap, an FT-ICR is capable of performing MS^n experiments in a time sequence with high sensitivity (simultaneous detection of all the ions) and super-high resolution. Both of these two instruments are excellent tools for structural characterization of complex molecules.

A third type of MS/MS instruments is a hybrid of tandem-in-space and tandem-in-time devices, including the Q-trap (QQ-2D-linear trap) [45] and the ion trap-FT-ICR (2D-linear ion trap-FT-ICR) [46]. The Q-trap takes the configuration of triple quadrupole, with the third quadrupole replaced by a 2D-linear ion trap. The uniqueness of this design is that the 2D-linear ion trap component can be used to perform either (a) a normal quadrupole scan function in the RF/DC mode or (b) a trap scan function by applying the RF potential to the quadrupole. It is well-suited for both qualitative and quantitative studies. In the case of ion Trap-FT-ICR, it combines ion accumulation and MS^n features of a 2D-linear ion trap with excellent mass analysis capability (mass resolution, mass accuracy, and sensitivity) of FT-ICR.

The inherent analytical advantages of MS/MS experiments are enhanced specificity and improved signal-to-noise ratio. It is useful in isomer differentiation, protein sequencing, and mixture analysis. MS/MS experiments are often performed in combination with HPLC for analysis of complex mixtures.

7.4 ROLE OF INSTRUMENTAL PARAMETERS ON IONIZATION EFFICIENCY IN LC/MS

Ionization efficiency of a compound defines the conversion of a neutral molecule to a charged particle (ion). It directly affects the detection limit and sensitivity of a mass spectrometer in LC/MS system. There are several instrumental parameters that play important roles in the ionization efficiency of a compound. These parameters (and their names) are usually vendor-specific.

7.4.1 Optimization of Ionization Settings

A typical schematic diagram of an LC/MS system is shown in Figure 7-9. Typically, a liquid sample is injected into the HPLC system by auto-sampler. After column separation, the sample is detected by both a UV detector and a mass spectrometer. The optimization of ion signals depends on a number of factors.

7.4.1.1 Ionization Mode. The first selection of ionization settings is the choice of ionization mode, i.e., positive ion mode or negative ion mode. This depends on the structures of analytes. For basic compounds (i.e., amines), a positive ion mode can be used to form a protonated or cationized molecule. For acidic compounds, a deprotonated molecule is formed in the negative ion mode. The negative ion MS offers selectivity and sensitivity since only limited

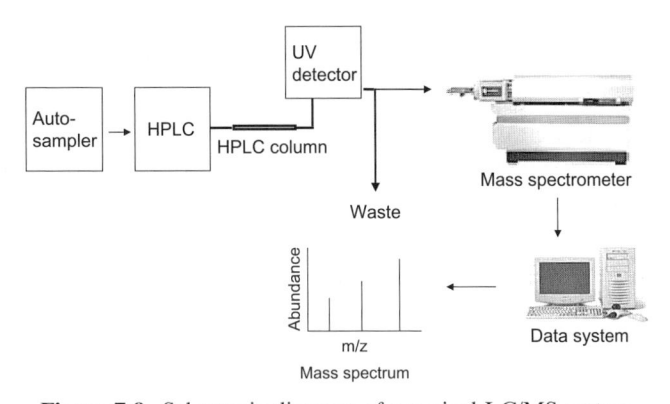

Figure 7-9. Schematic diagram of a typical LC/MS system.

compounds can be ionized in this mode. If a compound exists in the form of salts (i.e., quaternary ammonium salts), it may be necessary to perform both positive ion and negative ion LC/MS experiments to obtain structural information. It is common to perform alternating positive ion and negative ion LC/MS experiments for initial assessment of unknown compounds. Modern instruments have the capability to switch rapidly between positive ion and negative ion modes. Its limitation is the reduced analysis time on a specific mode for a compound, which might be ionizable only in a specific mode. Optimum mobile-phase conditions are different for positive ion and negative ion modes (please see sections on the effect of mobile-phase composition for detailed discussions). Thus, alternating between positive and negative ion modes may not provide the best sensitivity for a specific mode.

Another selection of ionization mode is the choice of ionization method—that is, ESI or APCI. This also depends on the structures of compounds. ESI works best for most polar molecules, while APCI performs better with low to moderately polar molecules. Alternating between ESI and APCI modes is suitable for detection of unknown compounds so that the best ionization method can be selected for signal optimization. As in the case of polarity mode selection, HPLC conditions also contribute to the performance of ESI/APCI (see later sections for discussions).

7.4.1.2 Ionization Voltage. One of important instrumental parameters is a high voltage used in both ESI and APCI, including capillary voltage and corona discharge voltage. In LC/ESI-MS, a strong electrical field is generated by supplying a capillary voltage, typically 4 to 5 kV. In LC/APCI-MS, a lower voltage is often applied to a corona pin, usually 2.5 to 3 kV. These high voltages can be adjusted to maximize the sensitivity. In addition, the position of ESI capillary or APCI probe may have impact on ion signals and should be optimized.

7.4.1.3 Cone Voltage (Orifice Voltage). A third instrumental parameter that can be used to induce some in-source fragmentation in structural determination is cone (orifice) voltage. It is applied to a sampling cone or orifice to extract ions from the atmospheric pressure region of the ion source into the vacuum region of the mass spectrometer. As cone voltage is increased, the ions are accelerated quickly through this region, undergoing collisions with solvent vapor and desolvation gas and resulting in some fragmentation. Although cone voltage fragmentation is not selective (no mass selection of parent ions), it can be effective in generating some fragment ions. In practice, cone voltage can be adjusted to either detect molecular ions of a compound (low cone voltage) or obtain fragmentation information (high cone voltage). An example is illustrated in Figure 7-10. In the case of 25-V cone voltage, the abundant peak in the spectrum corresponds to the molecule ion—that is, $[M + H^+]$ at m/z 325 and $[M + CH_3CN + H^+]$ at m/z 366 (Figure 7-10A). When the cone voltage is increased to 50 V, the dominant ions in the spectrum are fragment ions—that is, m/z 74, 115, 252 (Figure 7-10B). Clearly, the settings of cone voltage directly affect the appearance of the spectrum.

7.4.1.4 Desolvation Gas Flow and Temperature Settings. One of the key instrumental parameters affecting ionization efficiency of a compound is desolvation gas flow and temperature setting. For LC/ESI-MS, a nebulizing gas (N_2) passing through the ESI probe is normally set between 70 and 90 L/hr to assist in the generation of aerosol of liquid droplets from the sample solution. The ion source block temperature is commonly set from 100°C to

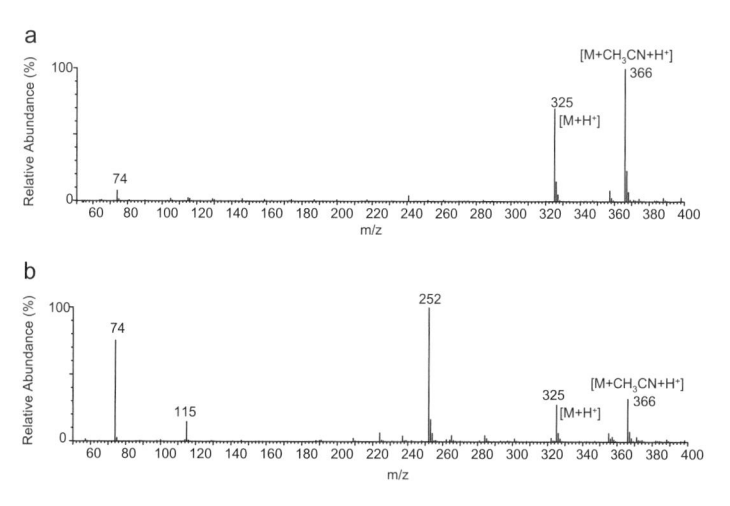

Figure 7-10. ESI mass spectra of a compound at (A) 25-V cone voltage and (B) 50-V cone voltage.

150°C, depending on the solvent flow rate and water contents of solvent. A heated desolvation gas (N_2) is delivered through the ion source to help the evaporation and removal of the solvent. A higher flow rate of desolvation gas is often needed when a higher solvent flow rate is used. For example, a typical flow rate of 400 L/hr is maintained for a solvent flow rate at 50 μL/min. The desolvation temperature varies from 100°C for less than 10 μL/min of solvent flow to about 400°C for over 50 μL/min of solvent flow. For LC/APCI-MS, the APCI probe temperature is normally set at 400°C, which can be higher for involatile samples or lower for volatile samples. It is primarily used to evaporate the solvent and heat the nebulizing gas (N_2). The typical ion source block temperature setting is at 150°C. The desolvation gas (N_2) at about 150 L/hr and nebulizing gas at full scale are common in APCI-MS.

7.4.2   Effect of Flow Rate

The flow rate in LC/MS affects both the separation and the performance of a mass spectrometer. Depending on the column used in separation, different flow rates are applied. For example, optimum flow rate of 1.0 mL/min may be employed for a 4.6-mm-i.d. column. A 2.1-mm-i.d. column has optimum flow rate of 0.2 mL/min. For LC/ESI-MS, ESI-MS is compatible with LC capillary columns and conventional analytical columns. The optimum performance of ESI may require a flow splitting for flow rates of 0.5 to 1.0 mL/min. One of unique aspects of ESI is the concentration dependence. The ESI response depends on the concentration of the sample in solution, not the total amount of sample into the source. Thus, a flow splitting does not reduce the ESI signals. On the contrary, it may increase sensitivity due to the optimum flow rate going into the ESI source with the flow splitting since the desired flow rate for stable ESI spray is about 200 μL/min or less. In addition, the flow splitting increases the periods between ion source cleaning since less liquid is going into the ion source. The common approach in flow splitting is a post-column split with a zero dead volume T-piece. The split ratio can be measured by collecting two liquid flow volumes from the split (one going to the ion source and the other going to the waste container) at a fixed time interval. The adjustment of ratio can be made by changing the length (or internal diameter) of tubing connected to the split. For example, shorter (or larger internal diameter) tubing directing the flow to the ion source will lead to the increase of the liquid flow to the ESI source. For LC/APCI-MS, APCI is compatible with 4.6-mm- or 2.1-mm-i.d. columns. Flow rates of up to 2 mL/min can be directly entered into the ion source to achieve optimum signals. No splitting of flow is required.

As mentioned previously, the temperature settings and gas flow rates in the ion source will have to be adjusted to optimize ionization efficiency of a compound, depending on the solvent flow rates.

7.5 EFFECT OF MOBILE-PHASE COMPOSITION ON IONIZATION EFFICIENCY IN LC/MS

7.5.1 Choice of Solvents

Commonly used reversed-phase LC solvents, including water, acetonitrile, and methanol, are ideal for LC/MS. All reversed-phase solvents need to be degassed prior to LC/MS analysis to maintain the stability of ion signals. This can be achieved by sonification, helium sparging, or vacuum membrane degassing. When solvents of high aqueous content are to be used, the source and probe temperatures should be raised to assist desolvation in the ion source. Normal-phase solvents such as dichloromethane, hexane, toluene, and other hydrocarbons are not suitable for ESI-MS because a polar mobile phase for ionization is needed in ESI. These normal-phase solvents and their typical solutes are sufficiently volatile to be analyzed by APCI and work well with APCI-MS.

7.5.2 Choice of Mobile-Phase Additives

Mobile-phase additives are used in HPLC to control the pH and ensure efficient and reliable separations. They also have to be compatible with ESI or APCI conditions. If the pH of the mobile phase needs to be reduced for better LC separations, the most suitable additives in LC/MS are acetic acid and formic acid with typical concentrations ranging from 0.1% to 1%. Note that addition of acids will suppress ionization in negative ion mode. Weakly acidic compounds may not form deprotonated ions under acidic conditions. If the pH of the mobile phase needs to be increased to enhance LC separations, ammonium hydroxide (0.1% to 1%) is suitable. Weakly acidic compounds can be ionized effectively in negative ion mode. Triethylamine is another additive that may be useful to enhance ionization of other compounds in negative ion mode because it is basic. It should be cautioned that the presence of triethylamine might suppress ionization of other compounds in the positive ion mode. A commonly used volatile salt in LC/MS to buffer mobile phases is ammonium acetate (<0.1 M). It is used to replace nonvolatile salts such as phosphates because these nonvolatile salts tend to crystallize in the ion source and block the source, suppressing ionization of analytes.

Some other unsuitable additives in LC/MS include surfactants/detergents and inorganic acids (sulfuric acid or phosphoric acid). The former additives tend to suppress the ionization of other compounds, while the latter can cause corrosion of metal parts in the source. As an ion-pairing agent, trifluoroacetic acid (TFA) reduces ESI signals in the positive ion mode and may completely suppress ionization in the negative ion mode. TFA is often used for peptides and protein analysis with a level of less than 0.1%.

When desired mobile-phase additives are used in LC/ESI-MS, unexpected protonated or deprotnated ions can sometimes be formed while sampling strongly basic or acidic solutions, respectively. This is often referred to as "wrong-way-round ionization" where such ions are not expected to exist in appreciable concentrations in solution [47]. The source of ionizing protons can come from the surface enrichment of electrolytically produced protons in the droplet surface layer for near-neutral pH solutions of low ionic strength. For strongly basic solutions with ammonia, gas-phase proton transfer from ammonium ions can lead to protonated species. For neutral or high pH solutions of high ionic strength, discharge-induced ionization might be responsible for the generation of protonated ions [48]. Similarly, deprotonated ions from acidic solutions in negative ion ESI-MS may be produced by basic gas-phase species in negative ion discharges. Thus, when analyzing a compound that is neutral, the ionization can occur as "wrong-way-round ionization" or post-column ionization with addition of acidic eluent (i.e., acetic acid). The existing HPLC method may not need to be modified to promote ionization of analytes.

7.5.3 Adduct Formation

Mobile phase additives and impurities in the sample solution can add to the complexity of mass spectrum in LC/MS. Noncovalent complexes such as protonated dimeric ions $[2M + H^+]$ are common. Other common adduct ions include complex cations ($[M + NH_4^+]$, $[M + Na^+]$, $[M + K^+]$, etc.), solvent adduct ions ($[M + H_2O + H^+]$, $[M + CH_3OH + H^+]$, $[M + CH_3CN + H^+]$, $[M + Cl^-]$, $[M + CH_3COO^-]$, etc.), and multimeric ions ($[3M + H^+]$, $[2M + Na^+]$, etc.). The formation of these adduct ions can provide additional information on the assignment of molecular weight for unknowns. However, excess alkali metal cations can suppress ionization of other compounds and may need to be removed prior to analysis. In some cases, it might be necessary to add a trace level of cation source to the sample to promote complex cations if the protonation does not readily occur. Discussions on this subject are detailed in the section on MS interpretation.

7.5.4 Effect of Analyte Concentration

As envisioned previously, ESI response is concentration-dependent. A linear response versus concentration is up to the maximum concentration of about 10^{-5} M. When analyte concentration exceeds this limit, the ESI response levels off. This is because ESI intensity is proportional to the surface concentration of an ion. At about 10^{-5} M, the droplet surface is completely saturated and higher concentration will not increase the total number of surface charges available for ion formation [25]. This will impact the high concentration end of quantitation using LC/ESI-MS. For the low concentration end, the detection limit depends on the sensitivity of LC/MS system, including efficient ion transfer/detection and removal of chemical noise in the system.

7.5.5 Selected Ion Monitoring and Multiple Reaction Monitoring

Two approaches are often used to improve the detection limit, including selected ion monitoring (SIM) and multiple reaction monitoring (MRM). In LC/MS studies, it is often desirable to increase detection sensitivity by limiting the mass analyzer scan to just one ion—that is, SIM. In this mode, a single ion of interest is monitored continuously by a mass spectrometer and no other ions are detected. This results in significant improvement of signal-to-noise ratio. SIM trades specificity for sensitivity. In general, the sensitivity in SIM is increased by a factor of 100 to 1000 over full-scan mass spectra. This can be quite useful in detection and quantification of specific compounds at low levels.

Another related technique, MRM, involves a MS/MS experiment that is often performed in a triple quadrupole mass spectrometer. Typically, Q1 is set to pass the parent ion and Q2 is used as a collision cell to fragment the parent ion. The third quadrupole is to transmit only diagnostic product ion from the parent ion. The advantage is the increased specificity from MS/MS experiment. MRM provides the highest-duty cycle scan in a triple quadrupole and is widely used in quantitating multiple analytes in a complex matrix (examples are given in the section on pharmacokinetic studies of drugs).

7.6 MS INTERPRETATION

Mass spectrum interpretation is essential to solve one or more of the following problems: establishment of molecular weight and of empirical formula; detection of functional groups and other substituents; determination of overall structural skeleton; elucidation of precise structure and possibly of certain stereochemical features. As detailed in the previous sections, ESI and APCI are two of the most effective interfaces for LC/MS that have been developed. Thus, the focus of the discussion will be on interpretation of mass spectra obtained by ESI or APCI.

7.6.1 Molecular Weight and Empirical Formula Determination

The qualitative applications of mass spectrometry are based on the determination of the mass-to-charge (m/z) ratio of an analyte. ESI and APCI MS are well known as a "soft" ionization technique. This means that relatively small amount of energy are needed to ionize the analyte. The efficiency of ion formation depends on a molecule's ability to associate and carry a charge. In a positive ion experiment, this ionization process can be defined by the following simple protonation reaction:

$$M + H^+ \rightarrow MH^+$$

7.6.1.1 Pseudomolecular Ions. In contrast to the traditional MS, the highest mass peaks in ESI/APCI spectra are not always the molecular ion of interest. Instead, pseudomolecular ions, or noncovalent complex ions, are commonly observed. The pseudomolecular ions are generally formed by the analyte–adduct interaction in the solution system that is preserved as a result of the soft ionization of the ESI/APCI process. These ions are also formed by analyte-adduct gas-phase collisions in the spray chamber [49]. The exact mechanisms of how the analyte adducts are formed in ESI/APCI still remain unresolved at this point. More often than not, the adduct ion formation is a major cause for the low detection limit for ESI/APCI MS. However, these associative processes have also created interest in the study of drug–protein/drug–oligonucleotide gas-phase complexes that benefit from the ability of ESI/APCI MS analysis.

The binding of Na^+, NH_4^+, and other background species to an analyte is often seen in the ESI-MS analysis (Table 7-3). The formation of pseudomolecular ions depends explicitly on the structure and functionality of the species involved in the formed complex, as well as the instrument conditions [50–52]. One of the first studies on pseudo-molecular ion formation in ESI-MS was reported by Fenn and co-workers [22, 53] in 1984. They showed the effect of flow rate, source temperature, and probe voltage on methanol, water, and acetonitrile solvent systems containing various additives such as LiCl, NaCl, $(CH_3)_4NI$, and HCl. The formation of pseudomolecular ions of 30 organic compounds with NH_4^+, Na^+, K^+, Cs^+, and Ca^{2+} was later illustrated by Kebarle and co-workers [49, 54–56]. They demonstrated that the analyte ion sensitivities decrease with the presence of salts in the solution. The same group found the detection limits of those organic bases to be in the sub-femtomole to attomole range when the salt concentration was below 10^{-4} M. In 1996, Leize

TABLE 7-3. Common Pseudomolecular Ions

Pseudomolecular Ions	Mass *(m/z)*
$[M + Na]^+$	M + 23
$[M + K]^+$	M + 39
$[M + Li]^+$	M + 7
$[M + Na + K - H]^+$	M + 61
$[M + H + NH_3]^+$	M + 18
$[M + H + ACN]^+$	M + 42
$[M + H + MeOH]^+$	M + 33
$[M + Na + ACN]^+$	M + 64
$[M + K + ACN]^+$	M + 80
$[M + H + CH_3CH_2NH_2]^+$	M + 46
$[M + Cl]^-$	M + 35
$[M + CH_3COO]^-$	M + 59
$[M + CF_3COO]^-$	M + 113

et al. [57] demonstrated that Cs⁺ showed the greatest ionization efficiency relative to Li⁺, Na⁺, K⁺, and Rb⁺ in equimolar mixtures of LiCl, NaCl, KCl, RbCl, and CsCl solution. They found that the solvation energy played an important role in the determination of the ionization efficiency [57].

To illustrate the pseudomolecular ion formation, we analyzed a mixture of Verapamil (Scheme 1) and a small peptide A (ALILTLVS) by ESI-MS in the positive ion mode. ESI-MS of the mixture produced the molecular ions of Verapamil ($[M_1 + H^+]$) and peptide A ($[M_2 + H^+]$) at m/z 455 and 829, respectively (Figure 7-11). The spectrum also shows two peaks at m/z 477 and m/z 851 for the sodium adduct of Verapamil ($[M_1 + Na^+]$) and peptide A ($[M_2 + Na^+]$, respectively. In some cases, sodiated dimeric ions ($[2M + Na^+]$) and/or protonated dimeric ions ($[2M + H^+]$) can be observed as well in the ESI mass spectrum.

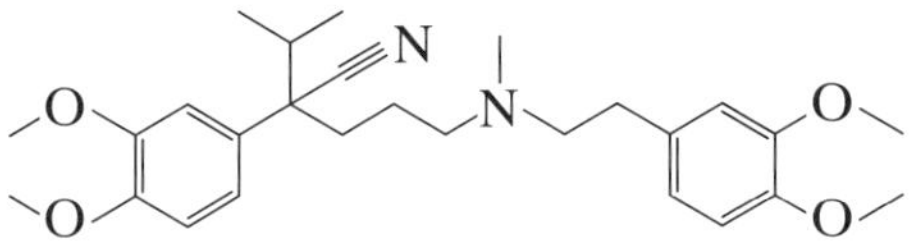

Scheme 1. Structure of Verapamil.

Depending on sample types and ionization conditions, $[M + Na^+]$, $[M + NH_4^+]$, and other pseudomolecular ions (Table 7-3) could be prominent ions in the mass spectra. An example was reported by Zhou and Hamburger [58] for the characterization of secondary metabolites of plant and microbial origin

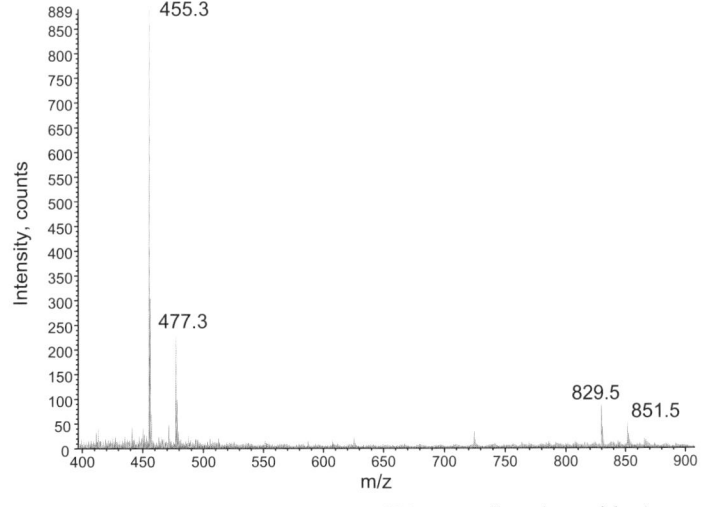

Figure 7-11. ESI mass spectra of Verapamil and peptide A.

by ESI-MS. A number of different compounds, including macrolides, peptides, aminoglycosides, polyethers, polyenes, alkaloids, terpenoids, purines, amindes, phenolics, and glycosides, were characterized based on the dominant pseudo-molecular ion observed. They found that with an APCI interface, most of the compounds produced $[M + H^+]$ and/or $[M - H]^-$ ions instead of the pseudo-molecular ions [58], except some poorly functionalized and very thermolabile compounds. More recently, Schug and McNair [59, 60] investigated the pseudomolecular ion formation by ESI-MS operated in the negative ion mode. They performed extensive ESI-MS experiments designed to determine the underlying principles in the formation of proton-bound dimer ions and sodium-bridged dimer ions from the halide-substituted benzoic acid deriva-tives [60] and six acidic pharmaceuticals [59]. The relative gas-phase basicity and proton affinity had significant effects on the formation of the sodium-bridged dimer ion [60]. For those compounds lacking highly acidic sites and less prone to undergo deprotonation, chloride ion attachment in the presence of chlorinated solvents such as chloroform can promote the formation of $[M + Cl^-]$ in the negative ion mode [24]. This is very useful when compounds do not respond well in the negative ion ESI mode.

Another complicating factor in molecular ion determination is the forma-tion of solvent–analyte noncovalently bound complexes (i.g., $[M + CH_3CN + H^+]$, $[M + MeOH + H^+]$, etc.). The relative intensities of these solvent cluster ions depend on the components in the solution phase, ionization mode, spray voltage, capillary temperature, sheath gas pressure, and auxiliary gas flow. Zhao et al. [61] reported that acetonitrile could be reduced to ethyl amine under ESI conditions (Scheme 1). They demonstrated that the "M + 46" ion in the mass spectrum represented the ethyl amine adduct ion ($[M + CH_3CH_2NH_2 + H^+]$) when the ESI-MS was performed by infusion of the com-pound in acetonitrile and water (1% $HCOOH$ + 1% NH_4OH) (1:1; v:v). Moreover, they showed that the same analyte produced a moderate $[M + CD_3CH_2NH_2 + H^+]$ (M + 49) signal when acetonitrile-d_3 was used as the organic solvent (Scheme 2).

$$CH_3CN \xrightarrow[\text{ESI}]{+ 4H} CH_3CH_2NH_2$$

$$CD_3CN \xrightarrow[\text{ESI}]{+ 4H} CD_3CH_2NH_2$$

Scheme 2.

Pseudomolecular ion formation is an artifact common to most ESI/APCI-MS analysis. Although these ions cause the complicating of the mass spectra, they were also found to be useful for the confirmation of the molecular ion of an analyte of interest.

7.6.1.2   Isotopic Abundance. The isotopic pattern of a molecular ion peak provides valuable chemical information and can be used to determine the formula of the molecule. Most of the common elements encountered in organic molecules have more than one isotope, except fluorine, iodine, and phosphorus (Table 7-4). This results in "isotopic clusters" produced in the mass spectrum. One example is for chlorine, which has two isotopes of 35 Da and 37 Da with a characteristic isotopic ratio of 3:1. On ionization of an organic compound containing one atom of chlorine (e.g., 4-chloro-phenlyamine), approximately 75.77% of the molecular ions will have a mass of 127 Da, and 24.23% will have a mass of 129 Da. For the MS analysis of a small molecule, the molecular ion is generally assigned to the peak representing the most abundant isotope.

The isotopic pattern may complicate the molecular weight assignment; on the other hand, it will also provide valuable reference for recognizing the type and number of element in a molecule. The characteristic patterns resulting from multiple isotopic contributions of the chlorine, bromine, and sulfur isotopes are shown in Table 7-5. One example is illustrated in the CI-MS spectrum of mometasone furoate (Figure 7-2), displaying prominent

TABLE 7-4. Natural Isotope Abundance and Exact Masses of Common Elements

Element	Symbol	Nominal Mass (Da)	Exact Mass	Abundance
Hydrogen	H	1	1.00783	99.99
	D or ^{2}H	2	2.01410	0.01
Carbon	^{12}C	12	12.0000	98.91
	^{13}C	13	13.0034	1.09
Nitrogen	^{14}N	14	14.0031	99.6
	^{15}N	15	15.0001	0.37
Oxygen	^{16}O	16	15.9949	99.76
	^{17}O	17	16.9991	0.037
	^{18}O	18	17.9992	0.20
Fluorine	F	19	18.9984	100
Silicon	^{28}Si	28	27.9769	92.28
	^{29}Si	29	28.9765	4.70
	^{30}Si	30	29.9738	3.02
Phosphorus	P	31	30.9738	100
Sulfur	^{32}S	32	31.9721	95.02
	^{33}S	33	32.9715	0.74
	^{34}S	34	33.9679	4.22
Chlorine	^{35}Cl	35	34.9689	75.77
	^{37}Cl	37	36.9659	24.23
Bromine	^{79}Br	79	78.9183	50.5
	^{81}Br	81	80.9163	49.5
Iodine	I	127	126.9045	100

TABLE 7-5. Isotopic Abundances for Ions Containing Different Numbers of Sulfur, Chlorine, and Bromine Atoms

Number of Cl Atoms	Mass	Number of Br Atoms (0)	Number of Br Atoms (1)	Number of Br Atoms (2)	Number of Br Atoms (3)	Number of Br Atoms (4)	Number of Sulfur Atoms (1)	Number of Sulfur Atoms (2)
0	A	100	100	51.5	34.3	17.6		
	A+2		97.3	100	100	68.5		
	A+4			48.7	97.3	100		
	A+6				31.6	64.9		
	A+8					15.8		
1	A	100	77.3	44.2	26.5	14.4		
	A+2	32	100	100	85.8	60.8		
	A+4		24.1	69.3	100	100		
	A+6			13.4	48.5	79.4		
	A+8				7.8	30		
	A+10							
2	A	100	62	38.7	20.8	12.1		
	A+2	64	100	100	74	54.8		
	A+4	10.2	45	88.8	100	100		
	A+6		6.2	31.2	63.1	93.3		
	A+8			3.8	18.3	46.4		
	A+10				2	11.5		
	A+12					1.1		
	A						100	100
	A+1						0.8	1.6
	A+2						4.4	8.8
	A+3							0.07
	A+4							0.19

molecular ions at m/z 521 ([M + H$^+$]), 523, and 525, of relative abundances 100:70:14. The isotopic intensity pattern is in good agreement with that of two chlorine atoms (Table 7-5), due to the ions $C_{27}H_{31}O_6{}^{35}Cl_2$, $C_{27}H_{31}O_6{}^{35}Cl^{37}Cl$, and $C_{27}H_{31}O_6{}^{37}Cl_2$. The characteristic isotopic patterns resulting from combinations of the isotope peaks can be used to ascertain elemental composition of the corresponding ion.

7.6.1.3 Nitrogen Rule. If the nominal molecular weight of an analyte appears to be an even mass number, the compound contains an even number of nitrogen atoms (or no nitrogen atoms). On the other hand, if the nominal molecular weight of an analyte appears to be an odd mass number, the compound contains an odd number of nitrogen atoms. This so-called "Nitrogen Rule" is very useful for determining the nitrogen content of an unknown compound. In the case of Verapamil (Scheme 1), the molecular weight of the compound is 420 Da, indicating an even number of nitrogen atoms in the molecule.

7.6.1.4 Hydrogen/Deuterium (H/D) Exchange. The exchange of hydrogen for deuterium in organic molecules has been used in mass spectrometry for structural studies in both solution phase and gas phase. It also has wide applications in the structural studies of proteins [24]. This method measures the difference in molecular weight of a compound before and after the deuterium exchange to determine the exchangeable hydrogens in a molecule for structural elucidations. For example, one can determine the number of labile hydrogen atoms from the mass shift X of [M + H$^+$] in H$_2$O to [M + D$^+$] in D$_2$O as X − 1. The exchangeable hydrogens are usually bound to N, O, or S atoms in functional groups such as OH, NH, NH$_2$, or COOH. In the case of mometasone furoate (structure shown in Figure 7-1), there is only one exchangeable hydrogen atom from the OH group in the molecule.

There are two general approaches to set up H/D exchange LC/MS experiments. One is to use deuterium oxide as sheath liquid to introduce it to the ESI/APCI source as a post-column addition. The actual exchange takes place in the ion source. This method can enable some degree of H/D exchange without change of chromatographic separation (i.e., retention time). However, back-exchange can occur and contribute to incomplete exchange due to the presence of H$_2$O in solvents or inadequate amounts of D$_2$O. The other approach is to couple ESI/APCI-MS with deuterated mobile phases such as D$_2$O or CH$_3$OD (less commonly used) for on-line LC/MS analysis of mixtures. The change of chromatographic retention time due to the use of deuterated mobile phases should not be an issue because of the use of mass identifications. This approach provides accurate measurements of exchangeable hydrogens in a molecule to assist structural elucidation (examples will be presented in later section on identification of drug metabolites).

7.6.1.5 Accurate Mass Measurement. Mass accuracy measurement, typically reported as parts per million (ppm), is essential for elemental-composition assignment. Traditional accurate mass measurements are carried out in a magnetic sector mass spectrometer, requiring relatively large quantities of materials. The development of modern MS instrumentation (i.e., QTOF, FT-ICR, etc.) has allowed accurate mass determinations of small molecules as well as biomolecules that are present at very low levels.

An internal mass calibration is generally needed to achieve mass measurement accuracy of 5 to 10 ppm with a Q-TOF MS analysis [62–64]. Internal calibration is based on mixing one or several internal standards or calibrants of known molecular weight with the analyte and then using the known masses to calibrate the mass measurements of unknowns that coexist in the sample mixture.

Currently, FT-ICR MS provides the highest mass resolving power and mass accuracy among all the mass spectrometric methods. Using external calibration, FT-ICR MS is capable of achieving mass measurement accuracies of 1 ppm or better. Internal calibration can provide an order-of-magnitude greater mass accuracy than external calibration for the FT-MS analysis.

One example is the high-resolution ESI-FT-ICR MS analysis of a mixture of Verapamil and peptide A. The following MS data were obtained for the mono-isotopic molecular ion and the isotopic molecular ions of peptide A using an external calibration (calculated mass; mass error in ppm): (829.5393, 0.1 ppm), (830.5423, 1.3 ppm), (831.5450, 1.0 ppm), and (832.5476, 1.2 ppm). Similar results were obtained for the Verapamil (data not shown).

Accurate mass measurement is important in establishing compound identity. For example, an unknown with a mass of m/z 122.0606 ([M + H$^+$]) can be C_7H_8NO; it cannot be $C_5H_{13}NCl$ (m/z 122.0731), $C_3H_8NO_4$ (m/z 122.0447), $C_4H_{12}NO_3$ (m/z 122.0811), C_4H_9NOCl (m/z 122.0367), or $C_8H_{12}N$ (m/z 122.0964). A mass measurement accuracy of 102 ppm is required to distinguish these elemental compositions. Thus, an unequivocal elemental composition of a compound can be obtained with sufficient high mass measurement accuracy (i.e., <5 ppm) along with other information.

7.6.1.6   Double Bond Equivalency (DBE).

To evaluate whether a formula, $C_xH_yN_zO_n$, is a reasonable elemental composition for a certain mass, one can calculate the DBE (numbers of rings and double bonds) of the formula. The calculation is based on the valences of elements involved, as shown in equation (7-1). For example, pyridine (C_5H_5N) has a calculated DBE of 4 (= 5 − 5/2 + 1/2 + 1), indicating the ring and three double bonds in this molecule. For benzene (C_6H_6), its calculated DBE is 4 (= 6 − 6/2 + 1), suggesting the ring and three double bonds in the molecule as well.

A more general case, $I_yII_nIII_zIV_x$, was suggested by McLafferty [65], where I = H, F, Cl, Br, I; II = O, S; III = N, P; and IV = C, Si.

$$\mathrm{DBE}(R + db\ ) = x - y/2 + z/2 + 1 \qquad (7\text{-}1)$$

The calculated value 12 found for mometasone furoate ($C_{27}H_{30}O_6Cl_2$) represents five rings and seven double bonds of this molecule [equation (7-2)]. Note that although an oxygen atom is present in the formula of mometasone furoate, it does not contribute to the calculation of DBE. The intensity of the molecular ion usually parallels the chemical stability of the molecule, and compounds with high numbers of rings and double bonds (DBE) often show higher molecular-ion abundance than those with low DBE. This is consistent with the abundant molecular ion peak observed in the CI-MS spectrum of mometasone furoate (Figure 7-2).

$$\text{DBE}(R + db) = 27 - 32/2 + 1 = 12 \qquad (7\text{-}2)$$

7.6.2 Fragmentation Pattern

Fragmentation pattern of a molecule, after ionization in the mass spectrometer, can be used to obtain structural information. The fragment ions from singly charged parent ions are the ions observed at low-mass range of a MS/MS spectrum. The fragmentation chemistry and mechanisms are reasonably understood. The prominent abundant fragment ions are the most stable fragments that tend to be formed. The fragmentation processes also depend on the stability of the transition states by which the ions are produced.

Many of the fragment ions observed in the product-ion spectra are formed by collision-induced heterolytic cleavage. For example, the formation of a product ion, [M + H⁺ − HX], can be explained by a 1,4 hydrogen rearrangement mechanism (Scheme 3). The product ion is formed by the neutral loss of HX, where X can be a heteroatom or a more electronegative group. A less common fragmentation mechanism by homolytic cleavage is also observed in tandem MS experiments. In this case, the driving force for the fragmentation of an ion is dependent on the stabilities of the resulting ion and the radical species relative to the energy of the initial ionic species. For instance, the formation of stable product ions, including acylium ion, benzylic ion, and allylic carbonium ion, are able to promote homolytic cleavage.

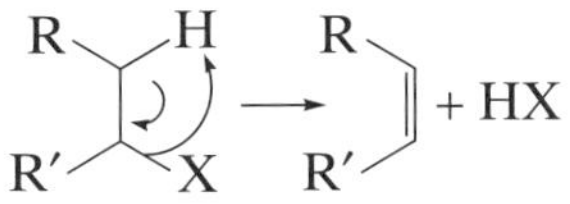

Scheme 3.

Charge-remote fragmentation is defined as a class of gas-phase decompositions that occur physically remote from the charge site [66–70]. Although the mechanism of charge-remote fragmentation is still debatable (Scheme 4) [67], it has been proven useful in the structural determination of long-chain or poly-ring molecules, including fatty acids, phospholipids, glycolipids, triacylglycerols, steroids, peptides, ceramides, and so on.

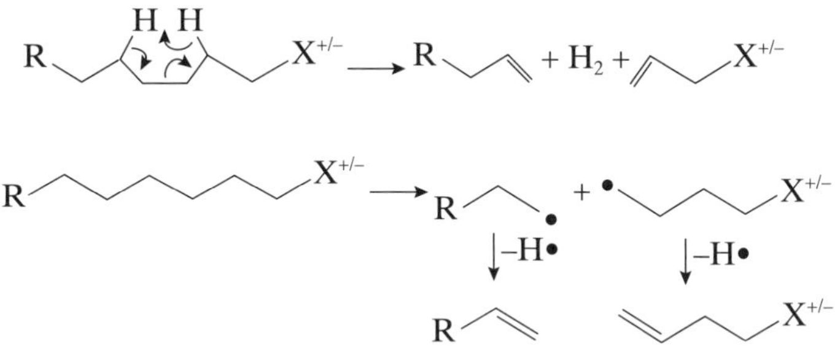

Scheme 4.

It is possible to derive structural information from the fragmentation pattern in a spectrum. The appearance of prominent peaks at certain mass numbers is empirically correlated with certain structural features. For example, the mass spectrum of an aromatic compound is usually dominated by a peak at m/z 91, corresponding to the tropylium ion. Structural information can also be obtained from the differences between the masses of two peaks in a spectrum. For instance, a fragment ion occurring 20 mass numbers below the molecular ion strongly suggests a loss of a HF moiety. Thus, a fluorine atom is likely to be present in the substance analyzed.

In addition, the knowledge of the principles governing the mode of fragmentation of ions makes it possible to confirm the structure assigned to a compound. This information is often used to determine the juxtaposition of structural fragments and thus to distinguish between isomeric substances. Reasonable guesses can be made as to which fragment ions to be expected in a mass spectrum if the isomeric substances are known.

The molecular ions formed in ESI/APCI and their fragment ions are usually even electron ions, that is, $[M + H^+]$. In some cases, radical cations $M^{+\bullet}$ can be formed in ESI-MS [24], depending on their structures and ESI conditions. In the case of Florfenicol (SCH 25298, Scheme 5, **I**), an antibacterial agent in Animal Health, its product ion mass spectrum only displays a very low abundant molecular ion peak at m/z 358 (data not shown). An abundant peak corresponding to the loss of water is formed by heterolytic fragmentation as illustrated in Scheme 5 (**II**, m/z 340). The **II** likely decomposes to **III** (m/z 320) by a neutral loss of HF, and this fragmentation is promoted by the formation of a substituted tropylium ion (Scheme 5). The fragmentation pattern of Florfenicol is characterized by an unusual feature of a most abundant peak occurring at odd mass, that is, m/z 241. The **IV,** a radical cation, is likely formed by homolytic cleavage of the sulfur–carbon bond in **III**, with loss of the methanesulfinic radical (Scheme 5).

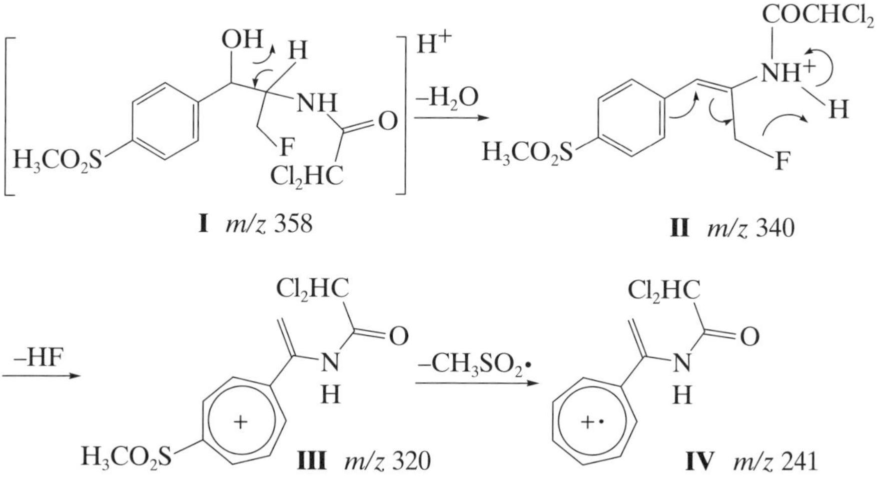

Scheme 5.

7.7 PRACTICAL APPLICATIONS

Mass spectrometry is a powerful and effective technology in drug discovery and development. This section will concentrate on the practical applications of LC/MS in problem solving, including high-throughput LC/MS analysis for combinatorial chemistry, structural characterization of impurities and decomposition products in bulk drug substances, and identification and quantification of drug metabolites.

7.7.1 High-Throughput LC/MS for Combinatorial Chemistry

The application of combinatorial chemistry to the synthesis of potential therapeutic agents has received increasing attention such that combinatorial chemistry is now an important tool in modern drug discovery [71]. Automated approaches capable of screening large libraries of small molecules have resulted in the successful application of LC/MS in combinatorial chemistry. Current trends for further integration of LC/MS techniques with new instrumental development have generated structure-based assays for drug discovery.

To assess the quality of a combinatorial chemistry library, it is essential to determine the purity and quantity of the expected products. Commercial software, developed by instrument manufacturers, has made possible the unattended and rapid analysis of tens of thousands of individual components of a specific library. The application of LC/MS in high-throughput screening of combinatorial libraries has been reviewed by several authors [72–78].

An important application of LC/MS in relation to combinatorial synthesis is the introduction of open-access LC/MS instrumentation. The dedicated open-access software packages are available from most instrument manufacturers. Using an open-access LC/MS system, organic chemists can readily obtain the molecular weight information of reaction products and monitor the progress of chemical synthesis. The chemist just needs to log-in to the computer system, assign an identification code to the sample, and select the type of LC/MS experiments to be performed. The automatic sample analyses can be performed rapidly using short HPLC columns and fast gradients, with run times typically of 5 min or less.

Multichannel ESI inlets have been developed to enhance sample throughput for large combinatorial library analysis. A multiplexed electrospray interface (MUX), which enables four- and eight-channel parallel introduction from four or eight LC systems into a multiplexed ESI source, was introduced in 1999. These systems, when coupled with TOF-MS, can provide not only the high-throughput capacity needed for library analysis, but also accurate mass determination of drug candidates and their synthetic by-products. Fang et al. [79] reported that they had coupled a nine-channel MUX-TOF MS system to conduct eight parallel high-throughput accurate mass LC/UV/TOF MS analysis. They used one of the nine channels to introduce reference standard as the lock mass to calibrate the instrument. The mass accuracies were found to be better than 5 and 10 ppm for 50% and 80% of the samples, respectively, from a single batch analysis of 960 samples [79]. The average root mean square (RMS) errors of the accurate mass measurements of the molecular weight of the combinatorial library samples were determined to be 10 ppm [79].

The use of FTMS has been growing rapidly in drug discovery and pharmaceutical development [80]. One opportunity for utilizing the ultrahigh mass resolving power of FTMS is in characterization of complex mixtures, such as combinatorial libraries. Burton et al. [81] conducted multiple accurate mass measurements for 41 compounds, using three different approaches. The absolute mean errors were 5.2 ($\sigma = 7.4$ ppm), 0.7 ($\sigma = 0.9$ ppm), and 0.8 ($\sigma = 1.0$ ppm) ppm for the external, conventional internal, and dual ionization internal calibrations, respectively. In another application, Nawrocki et al. [82] employed a 4.7-T external-source ESI FT-ICR mass spectrometer to analyze small-peptide libraries, demonstrating the feasibility of analyzing several combinatorial libraries containing 100 to 10,000 small peptides. Furthermore, by comparing the FTMS data with computer-simulated combinatorial library mass spectra, the authors were able to monitor the diversity and degeneracy of the library syntheses.

A well-established method for drug discovery is the utilization of a biological assay to screen a large library of small organic molecules for their ability to bind target biopolymers (i.e., protein) in a specific assay [83, 84]. In general, the MS-based technologies have the advantage that only small amounts of protein reagent are required. Recently, Annis et al. [85, 86] reported a high-throughput affinity selection–mass spectrometry assay to screen mass-encoded 2500-member combinatorial libraries. A schematic representation of the method is shown in Figure 7-12. Combination of the protein and a small

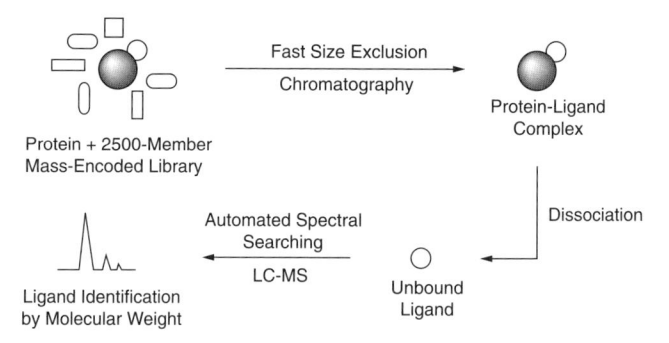

Figure 7-12. Diagram of the automated ligand identification system (ALIS) affinity selection–mass spectrometry method. (Reprinted from reference 85, with permission of Elsevier Science.)

molecule library leads to the formation of a complex of the protein with any suitable library member. Size-exclusion chromatography (SEC) is then employed to rapidly separate the protein target, along with any small molecules bound to the target, from any unbound small molecules. The SEC band containing the complex is immediately transferred to a reversed-phase chromatography column (60°C and pH 2). This step serves to denature the target, thereby dissociating the previously bound small molecules from the complex. The unbound small molecules are directly introduced into a high-resolution mass spectrometer for analysis. By using the affinity selection–mass spectrometry method, Annis and co-workers discovered a bioactive ligand for the anti-infective target *Escherichia coli* dihydrofolate reductase (DHFR) [85, 86].

7.7.2 Characterization of Impurities and Decomposition Products in Bulk Drug Substances

One of the major applications of LC/MS in pharmaceutical analysis is the identification of impurities and degradation products in pharmaceuticals. Often, impurities are synthetic by-products, starting materials, or degradation products. Drug regulatory agencies require the purity of a pharmaceutical to be fully defined. Impurities that are present at a level of 0.1% or greater relative to the active ingredient need to be characterized to comply with the regulatory requirements. This is important to ensure that the pharmacological and toxicological effects are truly those of the drug substances and not due to the impurities.

7.7.2.1 Characterization of Impurities by LC/MS. The impurities in pharmaceuticals are mainly formed during the synthetic process from starting materials, intermediates, and by-products. Generally, the impurities in staring materials and intermediates are not required to be characterized by the regulatory agencies. These impurities, however, are likely to contain components that affect the purity of the final manufactured pharmaceutical. By-products are often generated during synthesis and are one of the major sources of pharmaceutical impurities. The identification of the by-products often allows the Development Operations to refine the manufacturing process to minimize impurities and, thus, to maximize yield.

Traditionally, the impurities are isolated and purified by off-line HPLC and then characterized by using FT-IR, NMR, MS, and X-ray crystallography, among others. The main limitation associated with this approach is that relatively large sample quantities are needed for analysis, and the process can be very labor-intense. In contrast, LC/MS and LC/MS/MS are highly sensitive techniques requiring typically less than 1 µg of material for analysis. In certain cases, if the impurities are found at very low levels in the drug substance, extraction procedures are used to concentrate them to detectable levels.

The capabilities of separating a mixture containing highly varied concentrations of analyte and structural characterization of impurities have led to the increased use of LC/MS. Nicolas and Scholz [87] illustrated the characterization of a number of DuP 941 (Scheme 6) impurities by LC/MS(/MS). The five unknown impurities were labeled A, D, E, F, and G along with the known impurities B and C in the total ion chromatogram of DuP 941 obtained from LC/MS analysis (Figure 7-13B). Because the UV-visible absorption and response factor of related compounds tend to be similar, while their MS ionization efficiencies can be significantly different, it is always useful to record the UV chromatogram (Figure 7-13A) as well as the mass spectra for the identification of impurities. The protonated molecular ions ([M + H$^+$]) of the impurities A, D, E, F, G were found to be at m/z 392, 339, 324, 482, and 558, respectively. The molecular ions of the impurities were selected for tandem MS analysis in order to identify the unknown structures (Figure 7-13C). The product-ion spectra for the five unknown impurities were shown in Figure 7-14. The impurity A was found to be a by-product (Figure 7-15A) during the synthetic process. Major fragment ions were observed at m/z 319, 305, and 261

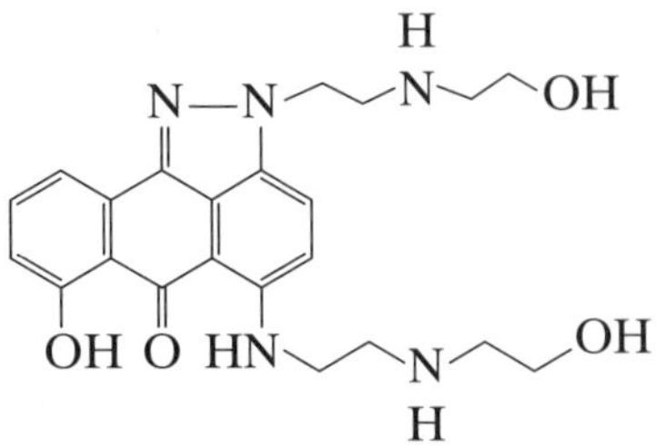

Scheme 6. Structure of DuP 941. (Reprinted from reference 87, with permission of Elsevier Science.)

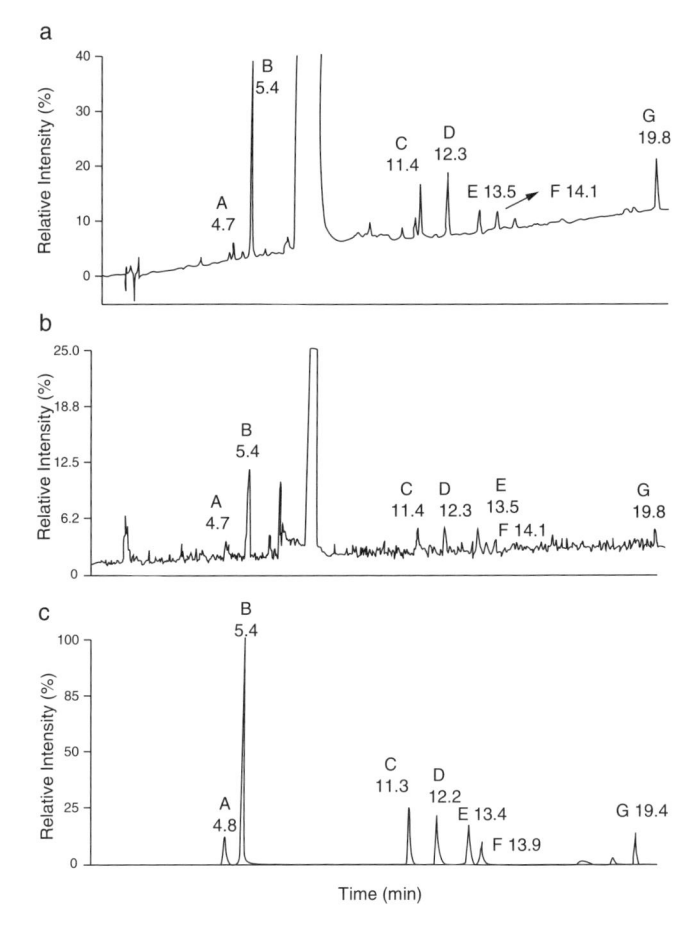

Figure 7-13. DuP 941 lot 3 chromatograms: (A) LC/UV; (B) LC/MS; (C) LC/MS/MS. A and B were acquired from a single injection using an HP 1090. UV and MS detectors were in series. C was acquired from a subsequent injection. (Reprinted from reference 87, with permission of Elsevier Science.)

(Figure 7-14A). The base peak at m/z 305 might arise from the neutral loss of a 2-vinylamino-ethanol. The product ions formed by the neutral loss of 73 (2-methyleneamino-ethanol) and 44 (Ethenol) Da from the precursor ion of m/z 392 were also consistent with the proposed structure (Figure 7-15A). The production spectra of impurities D and E were similar in two ways. Both have a base peak at m/z 88 which was produced when an N-(2-hydroxyethyl) aminoethyl group was cleaved from the molecule. Both produce a less intense product ion that was 61 Da less than the precursor ion (Figure 7-15; m/z 278 in D, and m/z 263 in E). This 61-Da loss was found to be the loss of a 2-aminoethanol neutral. Further studies suggested that impurities D and E are photo-decomposition products of DuP 941[87]. Based on the LC/MS/MS data,

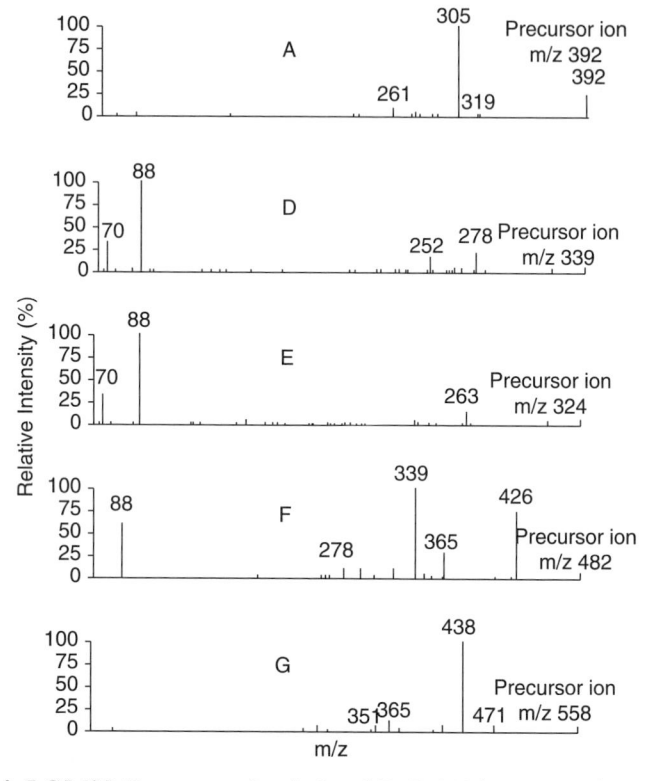

Figure 7-14. LC/MS/MS spectra of unisolated DuP 941 impurities. (Reprinted from reference 87, with permission of Elsevier Science.)

possible structures were also assigned to the other impurities (Figure 7-15F,G). The fragmentation patterns of the product-ion spectra (Figure 7-14F,G) were found to be consistent with the proposed structures. In addition, the same group found that the MS/MS fingerprint reproducibility was very good for all the unknowns examined and could be applied as a useful tool for the on-line characterization of impurities in bulk drug substances [87].

7.7.2.2 Characterization of Decomposition Products by LC/MS. The use of LC/MS and LC/MS/MS also has a significant impact on degradant characterization. The degradation profiles are critical to the safety and potency assessment of the drug candidate for clinical trials. Drug degradation in formulations is highly complex and often unpredictable. The degradation products usually arise from the ingredients used in dosage formulation and/or in the process of formulation where temperature, humidity, and light may all play a role. The degradants can be generated from hydrolysis, oxidation, adduct formation, dimerization, rearrangement, and often the combination of these processes.

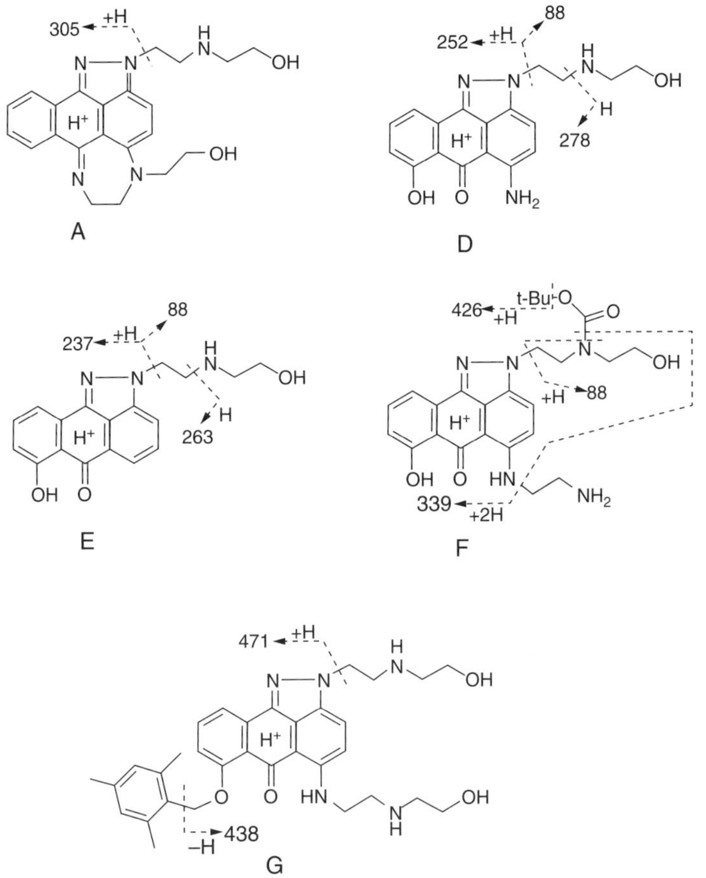

Figure 7-15. Possible structures of unisolated DuP 941 impurities showing proposed CID fragmentations. (Reprinted from reference 87, with permission of Elsevier Science.)

To accelerate drug development, various stress-testing protocols had been designed to emulate stresses the compound might experience during manufacturing processes and storage. These methods exposed drug candidates to forced degradation conditions such as acid, base, heat, oxidation, and exposure to light. A successful identification of the degradation products can help formulation scientists to understand the degradation mechanism of drug candidate and improve the clinical formulation development.

There had been numerous reports in the literature that involved LC/MS and LC/MS/MS for characterization of degradation products [1, 88–97]. An early example of the rapid structure elucidation of drug degradants induced by acid, base and heat by LC/MS was reported by Rourick et al. in 1996 [90]. In general, the LC/UV/MS provided the UV and molecular weight data, whereas the tandem LC/MS provided substructural information. The same group demonstrated that the similar procedure could be applied to obtain the structural information of the degradation products of paclitaxel (Taxol) [89]. Recently, Feng et al. [92] investigated the oxidative degradation products of an antifungal agent, SCH 56592 (Scheme 7), by both LC/MS and LC/NMR analysis. Four major oxidative degradation products of SCH 56592 were characterized, and the oxidation was found to be occurred at the piperazine ring in the center of the drug molecule.

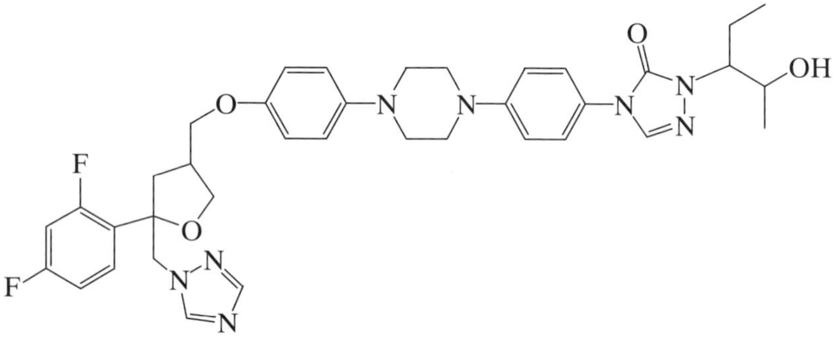

Scheme 7. Structure of SCH 56592.

Shipkova et al. [94] demonstrated the use of on-line high-resolution LC/ESI-MS using a magnetic sector mass spectrometer for analysis of minor components in complex mixtures. Everninomicin (SCH 27899; Figure 7-16) belongs to an important group of oligosaccharide antibiotics isolated from the fermentation broth of *Micromonospora carbonaceae*. The compound was degraded in ammonium hydroxide solution at pH 10, and the total ion chromatogram of the reaction mixture was obtained by LC/ESI-MS analysis (Figure 7-17). The ESI-MS analysis was performed in the negative ion mode. The deprotonated molecular ion of SCH 27899 was recorded at m/z 1628 ([M − H]$^-$), which indicated that the molecular weight of SCH 27899 was 1629 Da. Based on the LC/MS results of the base stressed sample, nine degradants eluted at 6.1 min (MW 712), 7.5 min (MW 951), 10.2 min (MW 1496), 12.1 min

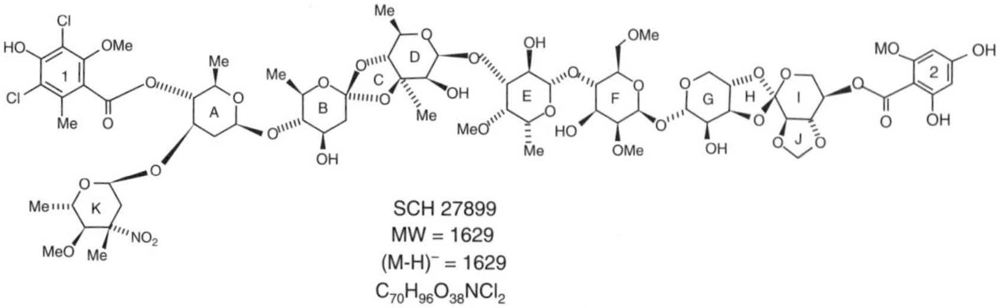

Figure 7-16. Structure of SCH 27899.

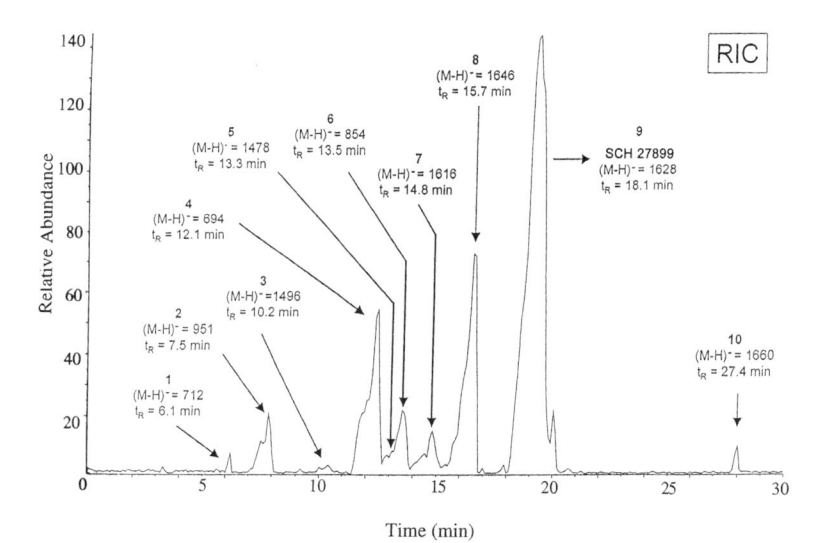

Figure 7-17. Reconstructed ion chromatogram (RIC) of a bulk drug substance (SCH 27899) degraded in ammonium hydroxide solution at pH 10, displaying all the identified mixture components 1–10. (Reprinted from reference 94, with permission of John Wiley & Sons.)

(MW 694), 13.3 min (MW 1478), 13.5 min (MW 854), 14.8 min (MW 1616), 15.7 min (MW 1646), and 27.4 min (MW 1660) in the LC/MS chromatogram were characterized as degradant **1** to **10**, respectively (Figure 7-17). A mass measurement accuracy of 0.4 to 1.9 ppm was achieved for the exact mass measurements, using PEG sulfates as internal calibration substance. Based on the high-resolution MS information, empirical formulae were obtained for all components (data not shown). The MS/MS spectra of source-produced fragment ions were used to characterize the unknown structures due to the limit sample quantities. The structures of the degradants were proposed based on the fragmentation studies, as well as high-resolution data (Figure 7-18). The authors found that degradants **1** to **5** were the hydrolysis products of SCH 27899, whereas degradant **6** was hydrolysis and oxidation products of SCH 27899. For example, degradant **2** was formed via ring opening of ortho-ester C, whereas degradant **5** was formed by loss of a terminal aromatic group (2) from SCH 27899.

The unique ability of FTMS to provide an exact-mass measurement for each of the ions produced in multiple-stage tandem MS (MS^n) assists greatly in the structure determination of degradation products [97]. Winger et al. [96] carried out LC FTMS and tandem MS experiments for the detection and identification of various degradants from drug candidates. This approach drastically reduced the time required for isolation and purification of substantial quantities of material and expedited the identification process.

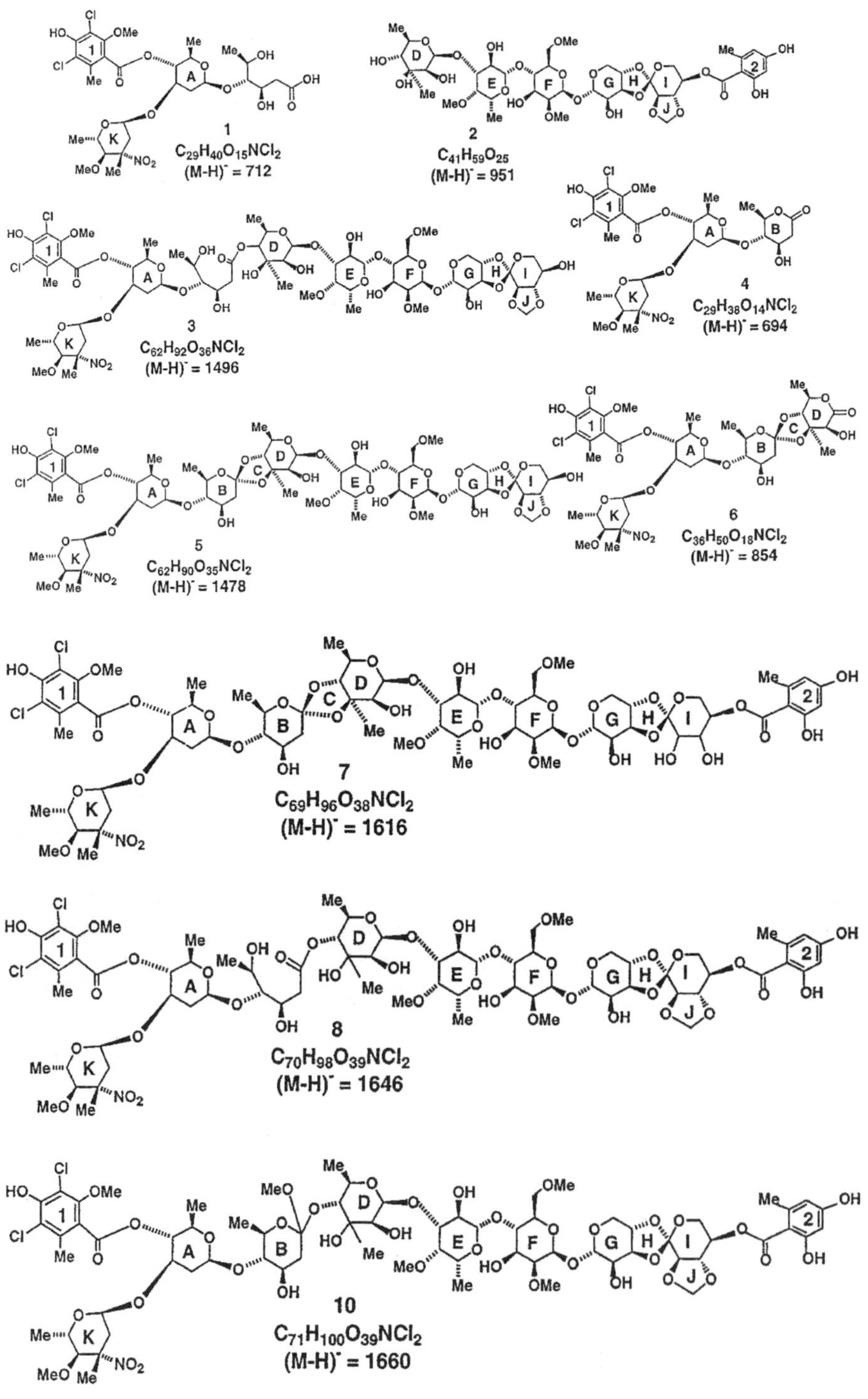

Figure 7-18. Proposed structures for the detected mixture components 1–10. (Reprinted from reference 94, with permission of John Wiley & Sons.)

7.7.3 Pharmacokinetic Studies of Drugs

Pharmacokinetic (PK) properties have been recognized as one of the most important attrition factors in the drug development. There has been substantial increase in both the variety and the number of newly synthesized drug candidates in the discovery stage. Therefore, significant resources were invested in developing high-throughput analytical methods to support *in vivo* studies. Owing to the unique combination of selectivity and sensitivity, LC/MS and LC/MS/MS have become one of the most widely used techniques to determine drug concentrations in biological matrices.

7.7.3.1   SIM and MRM in Quantitation Studies. The quantitative analysis of drug candidates are usually performed using HPLC in combination with a UV detector or a radioactivity detector after these compounds have been extracted from a physiological fluids. The main limitations associated with the UV method, however, are lack of analyte specificity and insufficient sensitivity. The radioactivity method is also limited by the requirement of expensive radiolabeled compounds, though it generally provides sufficient sensitivity for the determinations. Alali et al. [98] reported the use of LC/SIM-MS for the quantitative determination of the ketotifen in human plasma by using pizotifen as a reference compound (Scheme 8). The reference compound was a structural analog of ketotifen and was used to increase the precision of the assay. The calibration plasma samples, fortified with the reference compound, were prepared at seven concentrations, ranging from 0.5 to 20 ng/mL. The plasma samples were then extracted with *tert*-butyl methyl ether and injected into an AQA single quadruple LC/MS system. The LC/MS analysis involved the use of SIM to monitor the molecular ions (MH$^+$) that correspond to the ketotifen (*m/z* 296) and the reference compound (pizotifen, *m/z* 310). The retention times for ketotifen and pizotifen were found to be 14.3 min and 17.2 min, respectively. The standard calibration curve was found to be linear in the range from 0.5 to 20.0 ng/mL for ketotifen in human plasma. The accuracy and precision of this assay were highlighted in Table 7-6.

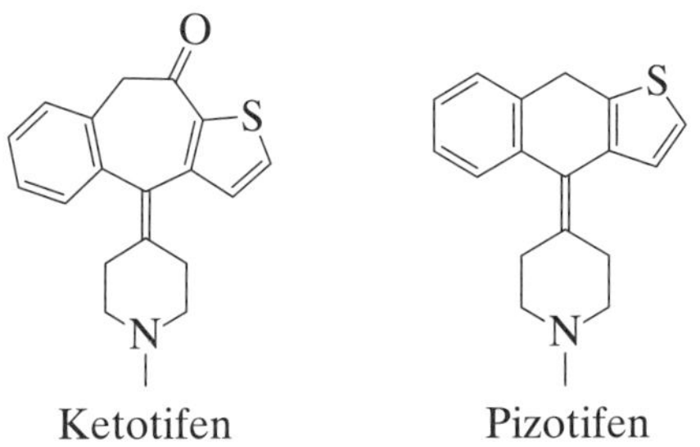

Ketotifen   Pizotifen

Scheme 8. Structures of ketotifen and pizotifen. (Reprinted from reference 98, with permission of Elsevier Science.)

TABLE 7-6. Statistical Analysis of Back-Calculated Normalized Ketotifen Concentrations of the Calibration Standards in Human Plasma

Concentration (ng/mL)	Mean	S.D.[a]	Precision as CV[b]	Accuracy	RE[c] (%)
0.5	1.0100	0.1311	12.98	101.00	1.00
1.0	1.0553	0.0590	5.59	105.53	5.53
2.0	1.0212	0.0942	9.22	102.12	2.12
5.0	1.0460	0.0731	6.99	104.60	4.60
10.0	1.0446	0.0602	5.76	104.46	4.46
15.0	1.0066	0.0441	4.38	100.66	0.66
20.0	0.9824	0.0211	2.15	98.24	−1.76

[a]Standard deviation.
[b]Coefficient of variation.
[c]Relative error.
Source: Reprinted from reference 98, with permission of Elsevier Science.

LC/SIM-MS analysis permitted an initial assessment of the level of drug substance present in the physiological fluids. However, in order to accelerate the drug discovery cycle, it is critical to develop rapid and efficient analytical method to support the needs for high-throughput screening. Nowadays, the high-resolving power HPLC coupled with a tandem mass spectrometer is a widely used quantitation technique due to its inherent accuracy, sensitivity, and selectivity [91, 99, 100]. Triple quadrupole mass spectrometers equipped with either an ESI or APCI ionization source are most commonly used for the quantitative assays. When a triple quadrupole mass spectrometer is operated in the MRM mode, the analyte is identified and detected not only by means of its molecular ion but also by means of a selected fragment ion. Thus, the LC/MRM-MS enables quantitation of ng/mL levels of drug candidates in physiological fluids [101, 102].

The use of MRM methods for quantitative bioanalysis often reduces sample preparation and analysis time. The MRM method that used LC/ESI-MS/MS for the quantitative analysis of an anticancer drug, Yondelis™ (Ecteinascidin 743, ET-743, trabectedin, Scheme 9), in human plasma was demonstrated by Rosing et al. [103]. The full-scan mass spectrum of ET-743 (MW 762) contained an abundant $[MH^+ - H_2O]$ ion at m/z 744 as a result of loss of water molecules from in-source CID (spectrum not shown). The internal standard, ET-729 (Scheme 9, MW 747), exhibited similar performance in the full-scan mass spectrum; an abundant $[MH^+ - H_2O]$ ion at m/z 730 was produced. The product ion spectra of ET-743 and ET-729 exhibited the most abundant fragment ions at m/z 495 and m/z 479, respectively (spectra not shown). The product ion at m/z 495 ($C_{27}H_{31}N_2O_7$) was formed in the collision cell after cleavage of the sulfur bond and ester binding at C-11′ [103].

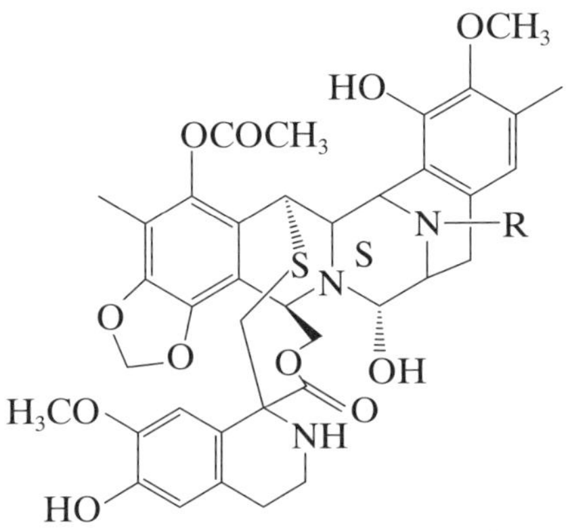

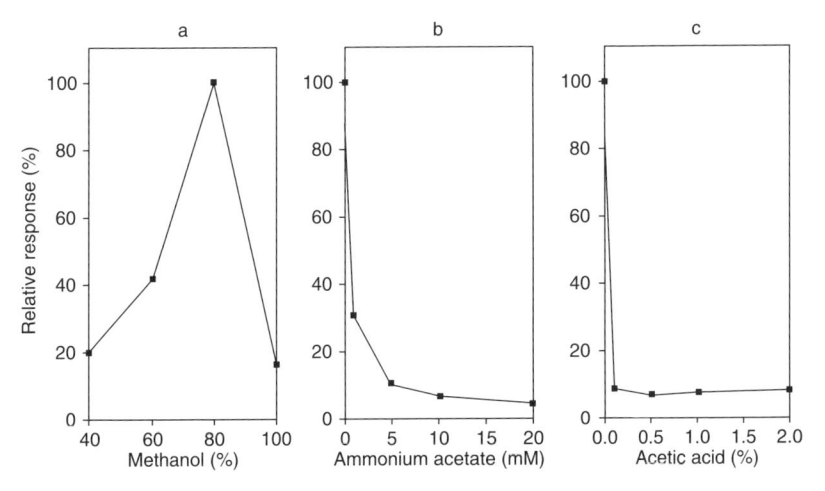

Scheme 9. Structures of ET-743 (R = CH3) and ET-729 (R = H). (Reprinted from reference 103, with permission of John Wiley & Sons.)

In order to optimize the LC/MS/MS system, the authors investigated the effects of methanol content, ammonium acetate concentration, and the percentage of acetic/formic acids in the mobile phase on the ESI response (m/z 744 to 495 transition) (Figure 7-19). They found that the best ESI response was obtained at ~80% (v/v) of methanol (Figure 7-20A). They also discovered that the ammonium acetate (5 mM) could be applied as a buffer in the mobile phase to achieve better reproducible separation between ET-743 and the internal standard. It is well known that the addition of acetic acid or formic acid in the mobile phase can suppresse the ionization of residual silanols on silica-based reversed-phase columns for LC/MS analysis (Figure 7-19C). In positive ion mode, however, the acids can form an ion pair with the MH⁺

Figure 7-19. Effect of (A) methanol content, (B) ammonium acetate concentration, and (C) percentage of acetic acid in the mobile phase on the relative ion response in ESI–MS/MS. (Reprinted from reference 103, by permission of John Wiley & Sons.)

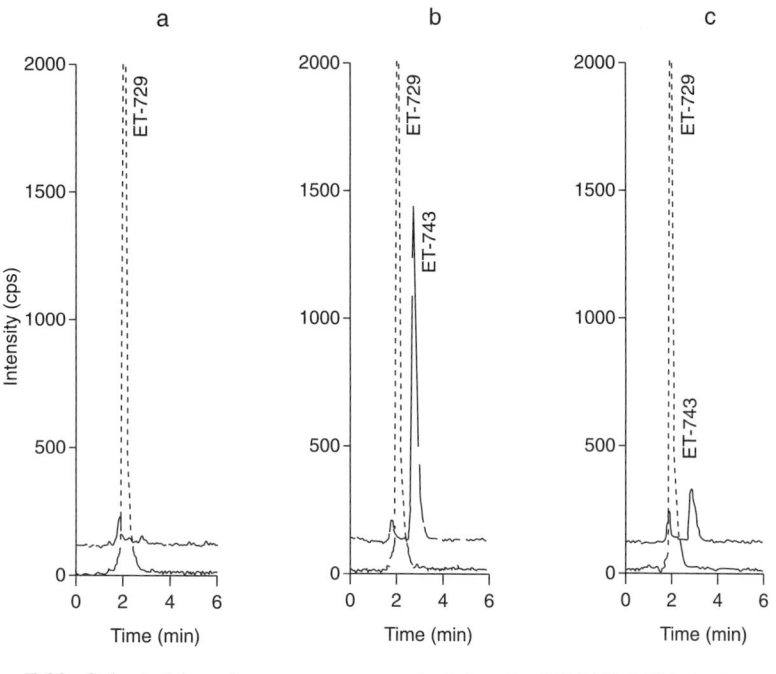

Figure 7-20. Selected ion chromatograms typical for the LC/ESI–MS/MS determination of ET-743 in human plasma: blank plasma (A) and plasma samples collected at the end (B, 0.1088 ng/mL) and 48 hr after cessation of a 24-hr infusion at a dose level of 100 μg/m² (C, 0.0183 ng/mL). The solid line represents the transition of m/z 744.4 to 495.0 (ET-743), and the dashed line represents the transition of m/z 730.6 to 478.8 (internal standard ET-729). (Reprinted from reference 103, with permission of John Wiley & Sons.)

pseudomolecular ion of the analyte in solution. The ion-paired molecules will not eject from the electrospray droplet into the gas phase, and thus the ESI sensitivities were reduced (Figure 7-19C). The addition of a concentration of 0.4% (v/v) formic acid in the mobile phase gave the best ESI response. Representative MRM chromatograms of ET-743 and ET-729 were shown in Figure 7-20. To validate the LC/MRM-MS method, the calibration standards were analyzed in three analytical runs. Concentrations were back-calculated from the ratio of the peak areas of ET-743 and the internal standard (Table 7-7). Deviations from the mean calculated concentrations over three runs were between −5.0 and 3.4, and the precision ranged from 1.0 to 14.9 for all concentrations. The assay performance data suggested the mean overall extraction recovery of ET-743 from human control plasma was between 98.8 ± 10.5% ($n = 3$) [103].

7.7.3.2 On-Line Sample Preparation for Biological Fluids.

There is a continuing demand for high-throughput bioanalytical method based on

TABLE 7-7. Calibration Concentrations (n = 2) Back-Calculated from the Ratio of the Peak Areas of ET-743 and ET-729

	Concentration (ng/mL)					
	0.01	0.05	0.25	0.80	1.80	2.50
1	0.0101	0.0463	0.259	0.865	1.960	2.137
2	0.0102	0.0467	0.232	0.759	1.834	2.886
3	0.0100	0.0494	0.242	0.789	1.789	2.647
Mean	0.0101	0.0475	0.244	0.804	1.861	2.557
RSD (%)[a]	1.0	3.6	5.6	6.8	4.8	14.9
Deviation (%)[b]	1.0	−5.0	−2.4	0.5	3.4	2.3

[a]Relative standard deviation.
[b]Deviation of the nominal concentration.
Source: Reprinted from reference 103, with permission of John Wiley & Sons.

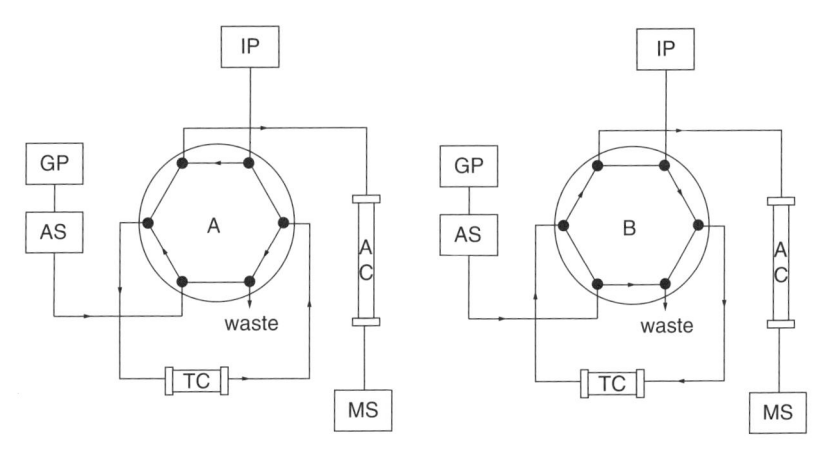

Figure 7-21. Analytical setups for the assay. IP, isocratic pump; GP, gradient pump; AS, autosampler; TC, trapping column; AC, analytical column; MS, mass spectrometer. (Reprinted from reference 113, with permission of John Wiley & Sons.)

LC/MS/MS to support the clinical development [104–108]. The conventional sample preparation procedures require labor-intensive solid-phase extraction (SPE) sample pretreatment steps and extensive method development time for the LC/MS drug analysis. One of the popular alternatives to SPE is the resurgence of the on-line SPE or column switching techniques [109–111]. More recently, this strategy has been further developed to determine the drug candidates in human plasma by LC/MS/MS analysis [104, 109, 112].

The utility of the LC/MS/MS method involving on-line sample clean-up for direct analysis of ET-743 in plasma has been demonstrated by Stokvis et al. [113]. Basically, the LC/MS/MS analysis system consists of two HPLC pumps connected by a six-port switching valve (Figure 7-21). In the first step (Figure 7-21, position A), the analyte was retained on the trapping column, whereas

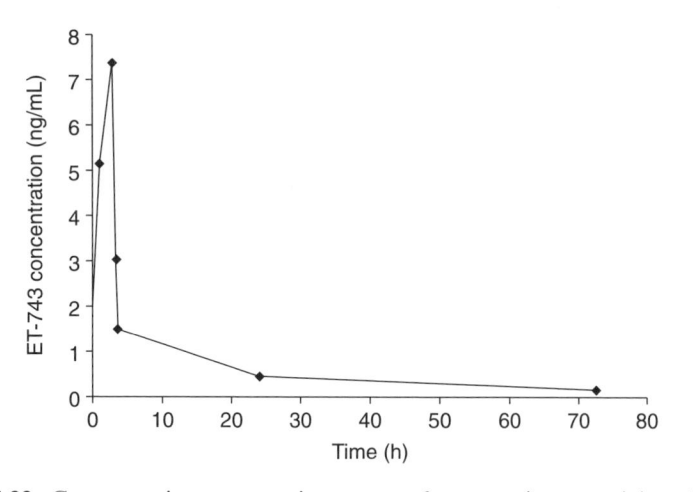

Figure 7-22. Concentration versus time curve for a patient receiving $1.2 \, mg/m^2$ Yondelis intravenously for 3 hr. (Reprinted from reference 113, with permission of John Wiley & Sons.)

the potentially interfering endogenous components were washed out. In a second step (Figure 7-21, position B), the trapping column was switched in-line with an analytical column, and the analyte along with the internal standard were back-flushed into the analytical column. With this on-line automated quantitation method, the calibration curves were found to be linear with a correlation coefficient better than 0.9939. The resulting lower limit of quantitation (LLOQ) was 0.05 ng/mL using 100 μL of plasma samples. They also demonstrated the application of the assay for the analysis of ET-743 in phase II clinical studies (Figure 7-22). The results indicated that the ET-743 concentrations were above the LLOQ level 72 hr after IV administration, therefore, suggested the applicability of the assay in phase II pharmacokinetic studies with ET-743 [113]. The on-line sample preparation assay provided a simple LC/MS/MS method that significantly reduced the off-line sample preparation time and increased higher throughput biological sample determination capability.

7.7.3.3 Matrix Effects. The matrix effects (ion suppression or ion enhancement) can compromise the selectivity and sensitivity of LC/MS/MS methods for the determination of drug concentrations in biological matrices. One common matrix effect is ion suppression due to co-eluting components that can affect the ionization efficiency of the analyte of interest. The matrix effects are major causes for errors in precision, accuracy, linearity, and reproducibility of the quantitation methods based on LC/MS/MS [104, 108, 114–118]. It is critical to overcome such effects in quantitative bioanalysis by LC/MS/MS.

Many researchers have investigated a variety of methods for the control of the matrix effects. The use of a stable isotopically labeled internal standard for the quantitative LC/MS/MS assay for Kahalalide F in human plasma was demonstrated by Stokvis et al. [115]. They found that the switching from an analogous to the isotopically labeled internal standard significantly improved the accuracy and precision of the assay by reducing the matrix effect [115]. This was probably due to the assumption that the isotopically labeled internal standard generally co-elutes with the analyte of interest, and it would experience the same extent of matrix effect as the analyte.

A number of reports in the literature indicate that matrix effect is dependent on ionization type, sample preparation, and bio-fluid type [104, 108, 114, 116–119]. In 2003, Dams et al. [119] demonstrated that the APCI was less susceptible to matrix effect, thereby allowing for simplification of sample preparation prior to LC/MS/MS analysis without jeopardizing the quality of quantitative data, as shown in Table 7-8. More recently, Souverain et al. [117] investigated the matrix effect in bio-analysis of illicit drugs with both LC/ESI–MS/MS and LC/APCI–MS/MS. Four procedures—liquid–liquid extraction (LLE), SPE, protein precipitation (PP) with acetonitrile (ACN), and protein precipitation with perchloric acid (PA)—were tested to evaluate

TABLE 7-8. Results of Post-column Infusion Experiments, Showing Time (Minutes) and Extent (−%: suppression; +%: enhancement) of the Observed Matrix Effects

| | Electrospray Ionization | | |
	Urine	Oral Fluid	Plasma
Direct injection	1–2 min: −90% 8–9.5 min: −20%	1–2 min: −70%	
Dilution	1–2 min: −60–95%	ND	
Protein precipitation	1–2 min: −85% 2–8 min: −15% 18–18.5 min: −20%	1–2 min: −70% 17.5–22 min: −50%	1–2.5 min: −75% 20–22 min: −65% 22.5–23 min: +150%
Solid-phase extraction	1–2 min: −40% 8 min: −30%	1–2 min: −10% 12–20 min: −10–15%	1–2 min: −35% 20–22 min: −50%
	Atmospheric Pressure Chemical Ionization		
	Urine	Oral Fluid	Plasma
Direct injection	1–2.5 min: −97.5% 6–7 min: −50%	NDa	
Dilution	1–2 min: −70%	NDa	
Protein precipitation	1–2 min: −99% 5.5 min: −40% 6–6.5 min: −50%	NDa	1–2.5 min: −60%
Solid-phase extraction	1–3 min: −20%	1–3 min: −20%	NDa

Source: Reprinted from reference 119, with permission of Elsevier.

the efficiency of the separation of matrices and analytes by sample preparation. The LC/APCI–MS/MS was found to be less vulnerable to the matrix effect than LC/ESI–MS/MS, and the LLE appeared to be the most efficient extraction procedure with an APCI source [117]. Alternatively, turbulent flow chromatography coupled with tandem MS could be applied with protein precipitation to reduce matrix effects, as long as the "ion suppression window" was small compared to the assay window [108].

Korfmacher and co-workers [108, 116] illustrated that exogenous (outside) material could be the cause of matrix suppression. Generally, the nonpolar exogenous material leaching out from the plastic tubes used to store the plasma samples eluted late in the chromatographic run. If the analytes and internal standards co-eluted with this suppression agent, their responses would be significantly reduced. In order to avoid "exogenous matrix effects" in LC/MS analyses, one should use the same brand of tubes for the processing and storing of both standard and sample plasma [108, 116]. Matrix effects from anticoagulant (i.e., Li-heparin) may also affect the performance of an assay significantly. Thus, it needs to be evaluated during a validation or prior to the analysis of human plasma samples.

7.7.4 Identification of Drug Metabolites

The metabolite characterization of a new chemical entity (NCE) in various drug discovery stages is crucial in assessment of the safety of a drug for human use [108, 120–125]. The identification of metabolites may reveal the metabolically labile portions of a molecule in a particular drug series. This information can be used by the synthetic chemists to synthesize compounds that are less susceptible to metabolism and, consequently, have a lower elimination rate and a longer half-life.

In general, drugs are metabolized to more polar, hydrophilic entities, thereby facilitating their elimination from the body. There are two major pathways for metabolism. Biotransformation reactions catalyzed by enzymes (i.e., cytochrome P450), including oxidation, reduction, and hydrolysis, are usually referred to as phase I metabolism—for example, oxidation of aliphatic or aromatic carbon, N-oxidation, and so on. On the other hand, the reactions that involve the addition of bulky and polar groups through conjugation to a nucleophilic site on the drug molecule are referred to as phase II metabolism [108, 121, 126]—for example, glucuronidation, sulfation, and so on. Both phase I and phase II metabolism may occur in parallel for particular compounds [108, 121, 126]. The high sensitivity, selectivity, and mass accuracy of the LC/MS technique has allowed itself to be used as a routine analytical tool for drug metabolism studies [127–130].

Recently, Hop et al. [131] used a combination of LC/MS, LC/MS/MS, and NMR techniques to identify metabolites of a substance P (Neurokinin 1 receptor) antagonist, compound **A** (Scheme 10), in rat hepatocytes and rat plasma. In both *in vitro* and *in vivo* studies, the samples were prepared for analysis by

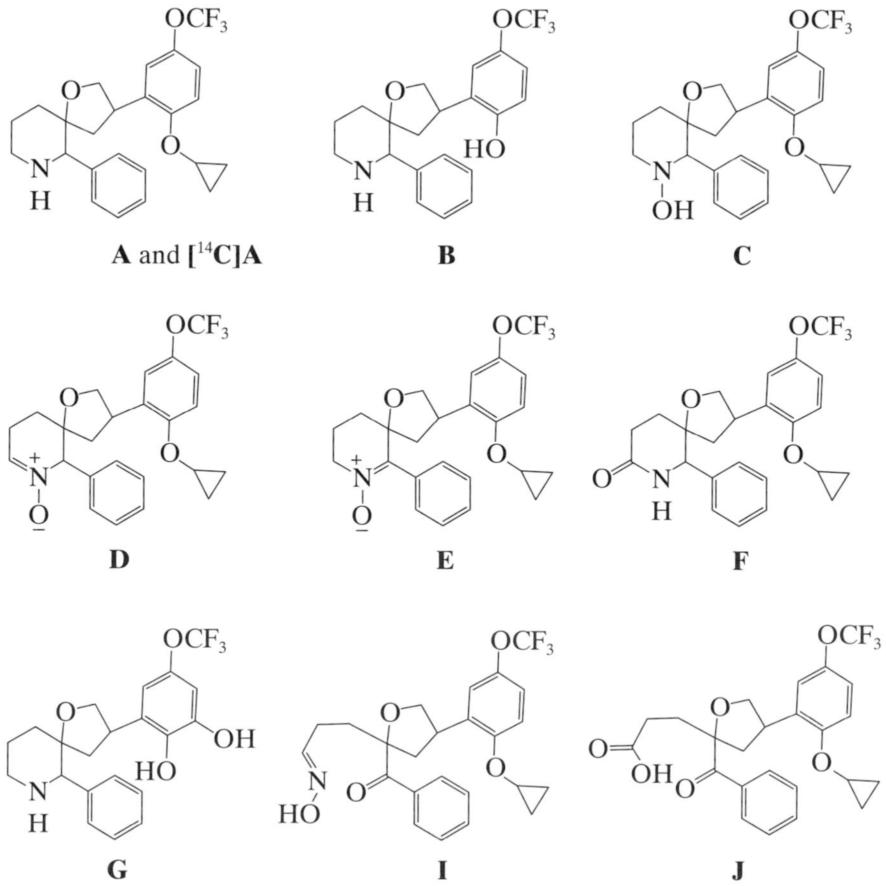

Scheme 10. Structures of compound A and its metabolites. (Reprinted from reference 131, with permission of the American Society for Pharmacology and Experimental Therapeutics.)

protein precipitation using acetonitrile followed by centrifugation. The supernatant was profiled with LC/MS and LC/MS/MS. The analysis involved the use of the product-ion spectra of compound **A** and ^{14}C-labeled compound **A** as the structural templates for the identification of metabolite structures. A comparison of the product-ion spectrum of compound A with that of the ^{14}C-labeled compound **A** suggested that the fragments at m/z 231, 215, 203, 191, and 175 were associated with the trifluoromethoxy phenyl moiety, whereas the fragments at m/z 184, 172, 159, 131, 91, and 56 were associated with the phenyl piperidine moiety. The assignment of the fragment ions was also confirmed by high-resolution LC/MS/MS analysis (Table 7-9). These product ions served as diagnostic markers for structural modification. Based on their product-ion spectra, nine major metabolites—the *O*-dealkylated metabolite (Scheme 10,

TABLE 7-9. Assignment of the Fragment Ions Observed in the Product Ion Spectrum of Compound A, C24H26NO3F3, Using Q-TOF II

Fragment Ion (Da)	Observed Mass (Da)	Formula Assignment (Da)	Theoretical Mass (Da)	Error (ppm)
416	416.1839	$C_{24}H_{25}NO_2F_3$	416.1837	0.5
231	231.0628	$C_{11}H_{10}O_2F_3$	231.0633	2.2
215	215.0325	$C_{10}H_6O_2F_3$	215.0320	2.4
203[a]	203.0677	$C_{10}H_{10}O_2F_3$	203.0684	−3.1
203[a]	203.0304	$C_9H_6O_2F_3$	203.0320	−7.8
191	191.0325	$C_8H_6O_2F_3$	191.0320	2.4
184	184.1137	$C_{13}H_{14}N$	184.1126	5.8
175	175.0378	$C_8H_6OF_3$	175.0371	4.4
172	172.1136	$C_{12}H_{14}N$	172.1126	5.4
159	159.1053	$C_{11}H_{13}N$	159.1048	3.4
131	131.0873	$C_{10}H_{11}$	131.0861	9.1
91	91.0526	C_7H_7	91.0548	−23.9[b]

[a]The signal at m/z 203 was composite.
[b]The signal at m/z 91 was outside the calibrated mass range resulting in a larger error.

Source: Reprinted from reference 131, with permission of the American Society for Pharmacology and Experimental Therapeutics.

TABLE 7-10. Most Abundant and Structure Characteristic Fragment Ions Observed in the Product Ion Mass Spectra of Compound A and Its Metabolites B-J

Compound	MW (Da)	Most Abundant and Structure Characteristic Fragment Ions									
A	433	172	131	184	191	175	159	215	231	91	56
[14C]A	435	172	131	184	193	177	159	217	233	9.1	56
B	393	172	184	191	159	131	174	203	217	91	56
C	449	172	175	131	146	191	188	170	215	231	91
D	447	129	162	144	117	191	175	215	131	91	231
E	447	105	196	188	162	170	86	69	191	175	215
F	447	162	145	117	186	175	19.1	103	215	91	203
O	409	172	184	131	159	207	174	174	215	91	56
11	585	410									
I	463	105	175	215	191	159	117	91			
J	464	105	215	175	191	203	159				

Source: Reprinted from reference 131, with permission of the American Society for Pharmacology and Experimental Therapeutics.

B), the hydroxylamine metabolite (Scheme 10, **C**), the nitrones (**D** and **E**), the lactam metabolite (Scheme 10, **F**), the hydroxylated *O*-dealkylated metabolite (**G**), the glucuronide of metabolite **G** (Scheme 10, **H**), the oxime metabolite (Scheme 10, **I**), and the keto acid metabolite (Scheme 10, **J**)—were characterized by high-resolution LC/MS and LC/MS/MS (Table 7-10). The authors found that the major circulating metabolite observed *in vivo* was generated by oxidative deamination of the piperidine ring yielding a keto acid metabolite, **J** (Scheme 11) [131]. Other metabolites, which might be the intermediates for formation of the keto acid, were also observed in the radiochromatogram of rat plasma (spectrum not shown) [131].

Scheme 11. Mechanism for the formation of metabolite J via oxidative deamination. (Reprinted from reference 131, with permission of the American Society for Pharmacology and Experimental Therapeutics.)

One of the goals of metabolite characterization is to identify the metabolic pathways and to determine whether or not any potentially reactive or toxic metabolites are formed [108, 121, 127, 132]. It is generally accepted that toxicities can stem from drug bioactivation *in vivo*, thus identifying the potential toxic metabolites is crucial in the lead optimization process [108, 121, 127, 132]. The generation of acyl glucuronide and glutathione (GSH) metabolites are important biotransformation pathways for many drugs and xenobiotics [127, 132, 133]. For example, the formation of acyl glucuronide conjugate forced the withdrawal of four marketed drugs due to hepatotoxicity [134].

The hydroxylation or oxidation of heterocyclic atoms (i.e., nitrogen) is a common phase I oxidative reaction. The metabolites generated by the oxidation at the N-atom are known as *N*-oxides. Several *N*-oxides are reported to be carcinogenic and/or to exhibit toxicological effects [135–137]. Chowdhury and co-workers demonstrated the application of LC/APCI-MS(/MS) to distinguish *N*-oxide metabolites from hydroxylated metabolites [138, 139]. In LC/APCI-MS, the molecular ions of *N*-oxides are found to undergo thermal deoxygenation at elevated temperatures [138, 139]. The resulted product ion ([MH$^+$ – 16]) had been attributed to the loss of elemental oxygen from the protonated *N*-oxide. On the other hand, the [MH$^+$ – 16] ions were not produced in the LC/APCI-MS spectra of hydroxylated metabolites [138, 139]. Since this thermal deoxygenation is unique to *N*-oxide metabolites, it can also be used to differentiate *N*-oxides from other hydroxylated metabolites.

Other strategies, such as on-line H/D exchange LC/-MS/MS and chemical derivatization, were well-established approaches for characterization of small molecules, including drug metabolites [121, 140–142]. Historically, the H/D exchange for structural elucidation has been used for a number of years [143], including determination of the affinity constants for protein–ligand interactions and for quantifying the conformational changes associated with ligand binding to proteins [144]. The determination of the number of exchangeable hydrogen atoms in a structure can provide additional information for structural characterization, such as the differentiation between *N*- or *S*-oxide formation and hydroxylation in drug metabolism studies. Ohashi et al. [142] demonstrated the on-line H/D exchange for characterization of metabolites of promethazine (MW 284 Da) (Scheme 12). M1 and M2, metabolites of promethazine, gave rise to pseudomolecular ions at *m*/*z* 301 (Figures 7-23a and 7-24a), which are 16 Da higher than the parent compounds. This suggested that M1 and M2 might be oxidation metabolites of promethazine—that is, addition of oxygen or a hydroxyl group to the phenothiazine. However, the discrimination between the oxidation types could not be achieved by analysis of the product-ion spectra (Figures 7-23b and 7-24b).

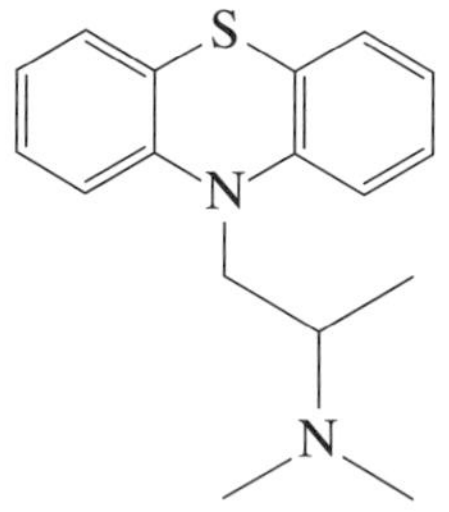

Scheme 12. Structure of promethazine. (Reprinted from reference 142, with permission of Elsevier Science.)

Therefore, the H/D exchange method was evaluated by the authors to differentiate between these possibilities. The on-line H/D exchange LC/MS experiment using D_2O in the mobile-phase-generated $[M_D + D^+]$ ions at *m*/*z* 302 and 303 for M1 (Figure 7-23c) and M2 (Figure 7-24c), respectively. This revealed that M1 had no exchangeable hydrogen atom in its structure, which ruled out the possibility of the hydroxyl structure. The authors assumed M1 to be an *S*-oxidated metabolite of promethazine. On the other hand, M2 had one exchangeable hydrogen atom, and thus it was assumed to be a hydroxylated metabolite. The fragmentation patterns of the product-ion spectra (Figures 7-23d and 7-24d) were found to be consistent with the proposed structures.

7.8 CONCLUSIONS

LC/MS has become one of the most powerful analytical techniques in the drug discovery and development process, as illustrated in this chapter. The unique

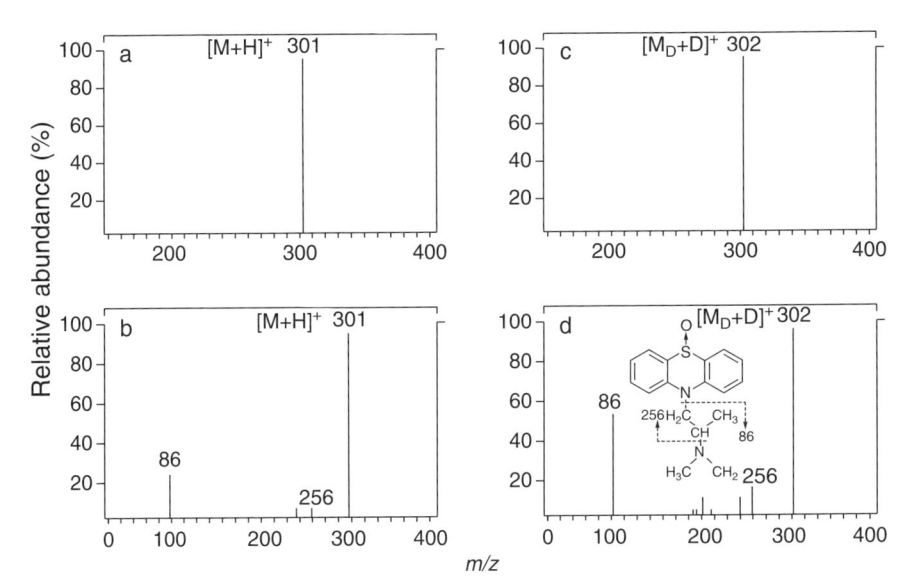

Figure 7-23. (a) ESI mass spectrum of M1, a metabolite of promethazine. (b) Product-ion spectrum of M1. (c) ESI mass spectrum of M1 in D₂O. (d) Product-ion spectrum of M1 in D₂O. (Reprinted from reference 142, with permission of Elsevier Science.)

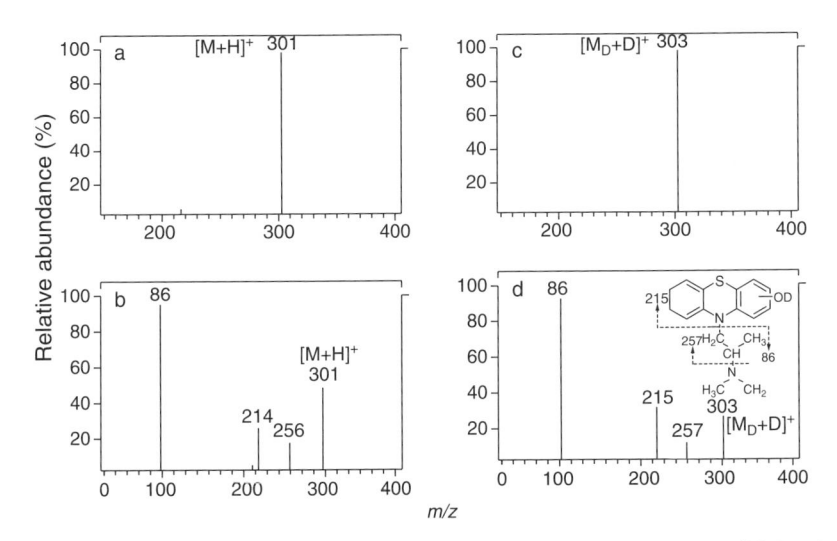

Figure 7-24. (a) ESI mass spectrum of M2, a metabolite of promethazine. (b) Product-ion spectrum of M2. (c) ESI mass spectrum of M2 in D₂O. (d) Product-ion spectrum of M2 in D₂O. (Reprinted from reference 142, with permission of Elsevier Science.)

analytical merits of this hyphenated technique include sensitivity, accuracy, speed, and applicability. LC/MS is widely used for both qualitative and quantitative analysis in various applications, such as high-throughput analysis, structural characterization of impurities and degradants in bulk drug substances, and identification and quantification of drug metabolites. With the ever-evolving technological advances in mass spectrometry and separation science, LC/MS will continue to play important roles in drug discovery in the future.

ACKNOWLEDGMENT

The authors would like to thank Dr. John J. Piwinski for his support on the projects.

REFERENCES

1. B. N. Pramanik, P. L. Bartner, and G. Chen, The role of mass spectrometry in the drug discovery process, *Curr. Opin. Drug Discovery Dev.* **2** (1999), 401–417.

2. M. S. Lee, *LC/MS Applications in Drug Development*, John Wiley & Sons, New York, 2002.

3. R. G. Cooks, G. Chen, and P. Wong, Mass spectrometers, in G. L. Trigg (ed), *Encyclopedia of Applied Physics*, VCH Publishers, New York, 1997, pp. 289–330.

4. M. S. B. Munson and F. H. Field, Chemical ionization mass spectrometry. I. General introduction, *J. Am. Chem. Soc.* **88** (1966), 2621–2630.

5. K. L. Busch and R. G. Cooks, Mass spectrometry of large, fragile, and involatile molecules, *Science (Washington, DC, United States)* **218** (1982), 247–254.

6. M. Barber, R. S. Bordoli, R. D. Sedgwick, A. N. Tyler, and B. V. Bycroft, Fast atom bombardment mass spectrometry of bleomycin A2 and B2 and their metal complexes, *Biochemi. Biophysi. Res. Communi.* **101** (1981), 632–638.

7. A. Benninghoven, H. W. Werner, and F. G. Rudenauer, *Secondary Ion Mass Spectrometry: Basic Concepts, Instrumental Aspects, Applications and Trends*, John Wiley & Sons, New York, 1986.

8. R. D. Macfarlane and D. F. Torgerson, Californium-252 plasma desorption mass spectroscopy, *Science* **191** (1976), 920–925.

9. M. Karas, D. Bachmann, U. Bahr, and F. Hillenkamp, Matrix-assisted ultraviolet laser desorption of non-volatile compounds, *Int. J. Mass Spectrom. Ion Processes* **78** (1987), 53–68.

10. Z. Takats, J. M. Wiseman, B. Gologan, and R. G. Cooks, Mass spectrometry sampling under ambient conditions with desorption electrospray ionization, *Science* **306** (2004), 471–473.

11. V. L. Tal'roze, G. V. Karpov, I. G. Gorodetskii, and V. E. Skurat, Capillary system for introducing liquid mixtures into an analytical mass spectrometer, *Z. Fizi. Khim.* **42** (1968), 3104–3112.

12. P. J. Arpino, B. G. Dawkins, and F. W. McLafferty, Liquid chromatography/mass spectrometry system providing continuous monitoring with nanogram sensitivity, *J. Chromatogr. Sci.* **12** (1974), 574–578.

13. W. H. McFadden, H. L. Schwartz, and S. Evans, Direct analysis of liquid chromatographic effluents, *J. Chromatogr.* **122** (1974), 389–396.

14. C. R. Blakley and M. L. Vestal, Thermospray interface for liquid chromatography/mass spectrometry, *Anal. Chem.* **55** (1983), 750–754.

15. R. M. Caprioli, T. Fan, and J. S. Cottrell, A continuous-flow sample probe for fast atom bombardment mass spectrometry, *Anal. Chem.* **58** (1986), 2949–2954.

16. Y. Ito, T. Takeuchi, D. Ishii, and M. Goto, Direct coupling of micro high-performance liquid chromatography with fast atom bombardment mass spectrometry, *J. Chromatogr.* **346** (1985), 161–166.

17. J. Zeleny, Instability of electrified liquid surfaces, *Phys. Rev.* **10** (1917), 1–6.

18. L. L. Mack, P. Kralik, A. Rheude, and M. Dole, Molecular beams of macroions. II. *J. Chem. Phys.* **52** (1970), 4977–4986.

19. M. Dole, L. L. Mack, R. L. Hines, R. C. Mobley, L. D. Ferguson, and M. B. Alice, Molecular beams of macroions, *J. Chem. Phys.* **49** (1968), 2240–2249.

20. J. V. Iribarne, P. J. Dziedzic, and B. A. Thomson, Atmospheric pressure ion evaporation-mass spectrometry, *Int. J. Mass Spectrom. Ion Phys.* **50** (1983), 331–347.

21. B. A. Thomson and J. V. Iribarne, Field-induced ion evaporation from liquid surfaces at atmospheric pressure, *J. Chem. Phys.* **71** (1979), 4451–4463.

22. M. Yamashita and J. B. Fenn, Electrospray ion source. Another variation on the free-jet theme, *J. Phys. Chem.* **88** (1984), 4451–4459.

23. M. L. Aleksandrov, L. N. Gall, N. V. Krasnov, V. I. Nikolaev, V. A. Pavlenko, and V. A. Shkurov, Ion extraction from solutions at atmospheric pressure—a method for mass-spectrometric analysis of bioorganic substances, *Dokl. Akad. Nauk SSSR* **277** (1984), 379–383.

24. B. N. Pramanik, A. K. Ganguly, and M. L. Gross, *Applied Electrospray Mass Spectrometry*, Marcel Dekker, New York, 2002.

25. P. Kebarle, A brief overview of the present status of the mechanisms involved in electrospray mass spectrometry, *J. Mass Spectrom.* **35** (2000), 804–817.

26. J. B. Fenn, M. Mann, C. K. Meng, S. F. Wong, and C. M. Whitehouse, Electrospray ionization for mass spectrometry of large biomolecules, *Science* **246** (1989), 64–71.

27. R. D. Smith, J. A. Loo, C. G. Edmonds, C. J. Barinaga, and H. R. Udseth, New developments in biochemical mass spectrometry: Electrospray ionization, *Anal. Chem.* **62** (1990), 882–899.

28. E. C. Horning, M. G. Horning, D. I. Carroll, I. Dzidic, and R. N. Stillwell, New picogram detection system based on a mass spectrometer with an external ionization source at atmospheric pressure, *Anal. Chem.* **45** (1973), 936–943.

29. R. C. Willoughby and R. F. Browner, Monodisperse aerosol generation interface for combining liquid chromatography with mass spectroscopy, *Anal. Chem.* **56** (1984), 2625–2631.

30. D. B. Robb, T. R. Covey, and A. P. Bruins, Atmospheric pressure photoionization: An ionization method for liquid chromatography–mass spectrometry, *Anal. Chem.* **72** (2000), 3653–3659.

31. F. W. Aston, *Isotopes*, Edward Arnold & Co., London, UK.

32. P. H. Dawson (ed.), *Quadrupole Mass Spectrometry and its Applications*, Elsevier, Amsterdam, The Netherlands, 1976.

33. R. G. Cooks, G. Chen, and C. Weil, Quadrupole mass filters and quadrupole Ion traps, in R. M. Caprioli, A. Malorni, and G. Sindona (eds.), *Selected Topics in Mass Spectrometry in the Biomolecular Sciences*, Series C, Vol. 504 Kluwer Academic Publishers, Dordrecht, 1997, pp. 213–238.

34. J. C. Schwartz, M. W. Senko, and J. E. P. Syka, A two-dimensional quadrupole ion trap mass spectrometer, *J. Am. Soc. Mass Spectrom.* **13** (2002), 659–669.

35. M. Guilhaus, Principles and instrumentation in time-of-flight mass spectrometry. Physical and instrumental concepts, *J. Mass Spectrom.* **30** (1995), 1519–1532.

36. B. A. Mamyrin, V. I. Karataev, D. V. Shmikk, and V. A. Zagulin, Mass reflectron. New nonmagnetic time-of-flight high-resolution mass spectrometer, *Z. Eksp. Teoreticheskoi Fiz.* **64** (1973), 82–89.

37. E. O. Lawrence and M. S. Livingston, The production of high speed light ions without the use of high voltages, *Phys. Rev.* **40** (1932), 19–35.

38. M. B. Comisarow and A. G. Marshall, Fourier transform ion cyclotron resonance spectroscopy, *Chem. Phys. Lett.* **25** (1974), 282–283.

39. F. W. McLafferty, Tandem mass spectrometry, *Science* **214** (1981), 280–287.

40. K. R. Jennings, Collision-induced decompositions of aromatic molecular ions, *Int. J. Mass Spectrom. Ion Phys.* **1** (1968), 227–235.

41. M. A. Mabud, M. J. Dekrey, and R. G. Cooks, Surface-induced dissociation of molecular ions, *Int. J. Mass Spectrom. Ion Processes* **67** (1985), 285–294.

42. R. G. Cooks, J. H. Beynon, R. M. Caprioli, and G. R. Lester, *Metastable Ions*, Elsevier, Amsterdam, The Netherlands, 1973.

43. R. A. Yost and C. G. Enke, Selected ion fragmentation with a tandem quadrupole mass spectrometer, *J. Am. Chem. Soc.* **100** (1978), 2274–2275.

44. I. V. Chernushevich, A. V. Loboda, and B. A. Thomson, An introduction to quadrupole-time-of-flight mass spectrometry, *J. Mass Spectrom.* **36** (2001), 849–865.

45. G. Hopfgartner, E. Varesio, V. Tschaeppaet, C. Grivet, E. Bourgogne, and L. A. Leuthold, Triple quadrupole linear ion trap mass spectrometer for the analysis of small molecules and macromolecules, *J. Mass Spectrom.* **39** (2004), 845–855.

46. J. E. P. Syka, J. A. Marto, D. L. Bai, S. Horning, and M. W. Senko, Novel linear quadrupole ion trap/FT mass spectrometer: Performance characterization and use in the comparative analysis of histone H3 post-translational modifications. *J. Proteome Res.* **3** (2004), 621–626.

47. B. A. Mansoori, D. A. Volmer, and R. K. Boyd, Wrong-way-round electrospray ionization of amino acids, *Rapid Commun. Mass Spectrom.* **11** (1997), 1120–1130.

48. S. Zhou and K. D. Cook, Protonation in electrospray mass spectrometry: Wrong-way-round or right-way-round? *J. Am. Soc. Mass Spectrom.* **11** (2000), 961–966.

49. M. G. Ikonomou, A. T. Blades, and P. Kebarle, Electrospray-ion spray: A comparison of mechanisms and performance, *Anal. Chem.* **63** (1991), 1989–1998.

50. J. M. Daniel, S. D. Friess, S. Rajagopalan, S. Wendt, and R. Zenobi, Quantitative determination of noncovalent binding interactions using soft ionization mass spectrometry, *Int. J. Mass Spectrom.* **216** (2002), 1–27.

51. J. S. Brodbelt, Probing molecular recognition by mass spectrometry, *Int. J. Mass Spectrom.* **200** (2000), 57–69.

52. J. S. Brodbelt, Analytical applications of ion-molecule reactions, *Mass Spectrom. Rev.* **16** (1997), 91–110.

53. M. Yamashita and J. B. Fenn, Negative ion production with the electrospray ion source, *J. Phys. Chem.* **88** (1984), 4671–4675.

54. P. Kebarle, Ion-molecule equilibria, how and why, *J. Am. Soc. Mass Spectrom.* **3** (1992), 1–9.

55. L. Tang and P. Kebarle, Effect of the conductivity of the electrosprayed solution on the electrospray current. Factors determining analyte sensitivity in electrospray mass spectrometry, *Anal. Chem.* **63** (1991), 2709–2715.

56. M. G. Ikonomou, A. T. Blades, and P. Kebarle, Investigations of the electrospray interface for liquid chromatography/mass spectrometry, *Anal. Chem.* **62** (1990), 957–967.

57. E. Leize, A. Jaffrezic, and A. Van Dorsselaer, Correlation between solvation energies and electrospray mass spectrometric response factors. Study by electrospray mass spectrometry of supramolecular complexes in thermodynamic equilibrium in solution, *J. Mass Spectrom.* **31** (1996), 537–544.

58. S. Zhou and M. Hamburger, Application of liquid chromatography–atmospheric pressure ionization mass spectrometry in natural product analysis. Evaluation and optimization of electrospray and heated nebulizer interfaces, *J. Chromatogr. A* **755** (1996), 189–204.

59. K. Schug and H. M. McNair, Adduct formation in electrospray ionization. Part 1: Common acidic pharmaceuticals, *J. Sep. Sci.* **25** (2002), 760–766.

60. K. Schug and H. M. McNair, Adduct formation in electrospray ionization mass spectrometry II. Benzoic acid derivatives, *J. Chromatogr. A* **985** (2003), 531–539.

61. X.-G. Zhao, J. Ma, H. Feng, J. Wu, and Z.-M. Gu, Loss of hydroxyl radical in CAD MS/MS for identification of oxidized drug metabolites, *Proceedings of the 52nd ASMS Conference on Mass Spectrometry and Allied Topics* May 23–27, 2004, Nashville, Tennessee.

62. K. Clauwaert, S. Vande Casteele, B. Sinnaeve, D. Deforce, and W. Lambert, Exact mass measurement of product ions for the structural confirmation and identification of unknown compounds using a quadrupole time-of-flight spectrometer: A simplified approach using combined tandem mass spectrometric functions, *Rapid Commun. Mass Spectrom.* **17** (2003), 1443–1448.

63. J.-C. Wolff, C. Eckers, A. B. Sage, K. Giles, and R. Bateman, Accurate mass liquid chromatography/mass spectrometry on quadrupole orthogonal acceleration time-of-flight mass analyzers using switching between separate sample and reference sprays. 2. Applications using the dual-electrospray ion source, *Anal. Chem.* **73** (2001), 2605–2612.

64. S. Ilard, F. Caradec, P. Jackson, and W. Luijten, Identification of an *N*-(hydroxysulfonyl)oxy metabolite using *in vitro* microorganism screening, high-resolution and tandem electrospray ionization mass spectrometry, *Rapid Commun. Mass Spectrom.* **14** (2000), 2362–2366.

65. F. W. McLafferty, *Interpretation of Mass Spectra,* 3rd ed., University Science Books, Mill Valley, CA, 1980.

66. M. L. Gross, Charge-remote fragmentation: An account of research on mechanisms and applications, *Int. J. Mass Spectrom.* **200** (2000), 611–624.

67. C. Cheng and M. L. Gross, Applications and mechanisms of charge-remote fragmentation, *Mass Spectrom. Rev.* **19** (2000), 398–420.

68. J. Adams and M. J. Songer, Charge-remote fragmentations for structural determination of lipids, *TrAC, Trends Anal. Chem.* **12** (1993), 28–36.

69. M. L. Gross, Charge-remote fragmentations: Method, mechanism and applications, *Int. J. Mass Spectrom. Ion Processes* **118–119** (1992), 137–165.

70. J. Adams, Charge-remote fragmentations: analytical applications and fundamental studies, *Mass Spectrom. Rev.* **9** (1990), 141–186.

71. K. C. Nicolaou, R. Hanko, and W. Hartwig, *Handbook of Combinatorial Chemistry, Volume 1: Drugs, Catalysts, Materials*, John Wiley & Sons, New York, 2002.

72. W. M. A. Niessen, Progress in liquid chromatography–mass spectrometry instrumentation and its impact on high-throughput screening, *J. Chromatogr. A* **1000** (2003), 413–436.

73. A. Triolo, M. Altamura, F. Cardinali, A. Sisto, and C. A. Maggi, Mass spectrometry and combinatorial chemistry: A short outline, *J. Mass Spectrom.* **36** (2001), 1249–1259.

74. R. D. Sussmuth and G. Jung, Impact of mass spectrometry on combinatorial chemistry, *J. Chromatogr. B: Biomed. Sci. Appl.* **725** (1999), 49–65.

75. J. N. Kyranos and J. C. Hogan, High-throughput characterization of combinatorial libraries generated by parallel synthesis, *Anal. Chem.* **70** (1998), 389A–395A.

76. J. A. Loo, Mass spectrometry in the combinatorial chemistry revolution, *Eur. Mass Spectrom.* **3** (1997), 93–104.

77. Y. G. Shin and R. B. van Breemen, Analysis and screening of combinatorial libraries using mass spectrometry, *Biopharm. Drug Dispos.* **22** (2001), 353–372.

78. C. Enjalbal, J. Martinez, and J. L. Aubagnac, Mass spectrometry in combinatorial chemistry, *Mass Spectrom. Rev.* **19** (2000), 139–161.

79. L. Fang, M. Demee, J. Cournoyer, T. Sierra, C. Young, and B. Yan, Parallel high-throughput accurate mass measurement using a nine-channel multiplexed electrospray liquid chromatography ultraviolet time-of-flight mass spectrometry system, *Rapid Commun. Mass Spectrom.* **17** (2003), 1425–1432.

80. D. Chakravarti, P. C. Mailander, K.-M. Li, S. Higginbotham, and H. L. Zhang, Evidence that a burst of DNA depurination in SENCAR mouse skin induces error-prone repair and forms mutations in the H-ras gene, *Oncogene* **20** (2001), 7945–7953.

81. R. D. Burton, K. P. Matuszak, C. H. Watson, and J. R. Eyler, Exact mass measurements using a 7 tesla fourier transform ion cyclotron resonance mass spectrometer in a good laboratory practices-regulated environment, *J. Am. Soc. Mass Spectrom.* **10** (1999), 1291–1297.

82. J. P. Nawrocki, M. Wigger, C. H. Watson, T. W. Hayes, and M. W. Senko, Analysis of combinatorial libraries using electrospray Fourier transform ion cyclotron resonance mass spectrometry, *Rapid Commun. Mass Spectrom.* **10** (1996), 1860–1864.

83. J. W. Armstrong, A review of high-throughput screening approaches for drug discovery, *Am. Biotechnol. Lab.* **17** (1999), 26–28.

84. K. R. Oldenburg, Current and future trends in high throughput screening for drug discovery, *Annu. Rep. Med. Chem.* **33** (1998), 301–311.

85. D. A. Annis, J. Athanasopoulos, P. J. Curran, J. S. Felsch, and K. Kalghatgi, An affinity selection–mass spectrometry method for the identification of small molecule ligands from self-encoded combinatorial libraries. Discovery of a novel antagonist of *E. coli* dihydrofolate reductase, *Int. J. Mass Spectrom.* **238** (2004), 77–83.

86. D. A. Annis, N. Nazef, C.-C. Chuang, M. P. Scott, and H. M. Nash, A general technique to rank protein–ligand binding affinities and determine allosteric versus direct binding site competition in compound mixtures, *J. Am. Chem. Soc.* **126** (2004), 15495–15503.

87. E. C. Nicolas and T. H. Scholz, Active drug substance impurity profiling. Part II. LC/MS/MS fingerprinting, *J. Pharm. Biomed. Anal.* **16** (1998), 825–836.

88. F. Erni, Liquid chromatography–mass spectrometry in the pharmaceutical industry: Objectives and needs, *J. Chromatogr.* **251** (1982), 141–151.

89. K. J. Volk, S. E. Hill, E. H. Kerns, and M. S. Lee, Profiling degradants of paclitaxel using liquid chromatography–mass spectrometry and liquid chromatography–tandem mass spectrometry substructural techniques, *J. Chromatogr. B: Biomed. Sci. Appl.* **696** (1997), 99–115.

90. R. A. Rourick, K. J. Volk, S. E. Klohr, T. Spears, E. H. Kerns, and M. S. Lee, Predictive strategy for the rapid structure elucidation of drug degradants, *J. Pharm. Biomed. Anal.* **14** (1996), 1743–1752.

91. M. S. Lee and E. H. Kerns, LC/MS applications in drug development, *Mass Spectrom. Rev.* **18** (1999), 187–279.

92. W. Feng, H. Liu, G. Chen, R. Malchow, and F. Bennett, Structural characterization of the oxidative degradation products of an antifungal agent SCH 56592 by LC-NMR and LC-MS, *J. Pharm. Biomed. Anal.* **25** (2001), 545–557.

93. Y. Wu, The use of liquid chromatography–mass spectrometry for the identification of drug degradation products in pharmaceutical formulations, *Biomed. Chromatogr.* **14** (2000), 384–396.

94. P. A. Shipkova, L. Heimark, P. L. Bartner, G. Chen, and B. N. Pramanik, High-resolution LC/MS for analysis of minor components in complex mixtures: Negative ion ESI for identification of impurities and degradation products of a novel oligosaccharide antibiotic, *J. Mass Spectrom.* **35** (2000), 1252–1258.

95. Z. Zhao, Q. Wang, E. W. Tsai, X. Z. Qin, and D. Ip, Identification of losartan degradates in stressed tablets by LC-MS and LC-MS/MS, *J. Pharm. Biomed. Anal.* **20** (1999), 129–136.

96. B. E. Winger and C. A. J. Kemp, Characterization of pharmaceutical compounds and related substances by using HPLC FTICR-MS and tandem mass spectrometry. *Am. Pharm. Rev.* **4** (2001), 55–63.

97. L.-K. Zhang, D. Rempel, B. N. Pramanik, and M. L. Gross, Accurate mass measurements by Fourier transfor mass spectrometry, *Mass Spectrom. Rev.* **24** (2005), 286–309.

98. F. Q. Alali, B. M. Tashtoush, and N. M. Najib, Determination of ketotifen in human plasma by LC-MS, *J. Pharm. Biomed. Anal.* **34** (2004), 87–94.

99. Y. Hsieh, J.-M. Brisson, and G. Wang, Fast HPLC-MS/MS analyses for small molecules, *Am. Pharm. Rev.* **6** (2003), 14–20.

100. E. Brewer and J. Henion, Atmospheric pressure ionization LC/MS/MS techniques for drug disposition studies, *J. Pharm. Sci.* **87** (1998), 395–402.

101. M. Jemal, High-throughput quantitative bioanalysis by LC/MS/MS, *Biomed. Chromatogr.* **14** (2000), 422–429.

102. D. O'Connor, Automated sample preparation and LC-MS for high-throughput ADME quantification, *Curr. Opin. Drug Discovery Dev.* **5** (2002), 52–58.

103. H. Rosing, M. J. X. Hillebrand, J. M. Jimeno, A. Gomez, and P. Floriano, Quantitative determination of Eteinascidin 743 in human plasma by miniaturized high-performance liquid chromatography coupled with electrospray ionization tandem mass spectrometry, *J. Mass Spectrom.* **33** (1998), 1134–1140.

104. G. Hopfgartner and E. Bourgogne, Quantitative high-throughput analysis of drugs in biological matrices by mass spectrometry, *Mass Spectrom. Rev.* **22** (2003), 195–214.

105. Y. Hsieh, K. Ng, and W. A. Korfmacher, Development and application of single-column direct plasma injection procedures for drug candidate assays using HPLC-MS/MS, *Am. Pharm. Rev.* **5** (2002), 88–90.

106. K. A. Cox, R. E. White, and W. A. Korfmacher, Rapid determination of pharmacokinetic properties of new chemical entities: *In vivo* approaches, *Comb. Chem. High Throughput Screening* **5** (2002), 29–37.

107. K. A. Cox, K. Dunn-Meynell, W. A. Korfmacher, L. Broske, and A. A. Nomeir, Novel *in vivo* procedure for rapid pharmacokinetic screening of discovery compounds in rats, *Drug Discovery Today* **4** (1999), 232–237.

108. W. A. Korfmacher (ed.), *Using Mass Spectrometry for Drug Metabolism Studies*, Taylor & Francis CRC Press, Boca Raton, FL, 2005.

109. B. L. Ackermann, A. T. Murphy, and M. J. Berna, The resurgence of column switching techniques to facilitate rapid LC/MS/MS based bioanalysis in drug discovery, *Am. Pharm. Rev.* **5** (2002), 54–63.

110. W. Roth and K. Beschke, Fully-automated assay by liquid chromatography for routine drug monitoring in body fluids—method development with biological samples, *J. Pharm. Biomed. Anal.* **2** (1984), 289–296.

111. W. Roth, K. Beschke, R. Jauch, A. Zimmer, and F. W. Koss, Fully automated high-performance liquid chromatography. A new chromatograph for pharmacokinetic drug monitoring by direct injection of body fluids, *J. Chromatogr.* **222** (1981), 13–22.

112. B. L. Ackermann, M. J. Berna, and A. T. Murphy, Recent advances in use of LC/MS/MS for quantitative high-throughput bioanalytical support of drug discovery, *Curr. Top. Med. Chem. B* (2002), 53–66.

113. E. Stokvis, H. Rosing, L. Lopez-Lazaro, and J. H. Beijnen, Simple and sensitive liquid chromatographic quantitative analysis of the novel marine anticancer drug Yondelis (ET-743, trabectedin) in human plasma using column switching and tandem mass spectrometric detection, *J. Mass Spectrom.* **39** (2004), 431–436.

114. C. Chin, Z. P. Zhang, and H. T. Karnes, A study of matrix effects on an LC/MS/MS assay for olanzapine and desmethyl olanzapine, *J. Pharm. Biomed. Anal.* **35** (2004), 1149–1167.

115. E. Stokvis, H. Rosing, L. Lopez-Lazaro, J. H. M. Schellens, and J. H. Beijnen, Switching from an analogous to a stable isotopically labeled internal standard for

the LC-MS/MS quantitation of the novel anticancer drug Kahalalide F significantly improves assay performance, *Biomed. Chromatogr.* **18** (2004), 400–402.

116. H. Mei, Y. Hsieh, C. Nardo, X. Xu, and S. Wang, Investigation of matrix effects in bioanalytical high-performance liquid chromatography/tandem mass spectrometric assays: Application to drug discovery, *Rapid Commun. Mass Spectrom.* **17** (2003), 97–103.

117. S. Souverain, S. Rudaz, and J.-L. Veuthey, Matrix effect in LC-ESI-MS and LC-APCI-MS with off-line and on-line extraction procedures, *J. Chromatogr. A* **1058** (2004), 61–66.

118. J.-P. Antignac, K. de Wasch, F. Monteau, H. De Brabander, F. Andre, and B. Le Bizec, The ion suppression phenomenon in liquid chromatography–mass spectrometry and its consequences in the field of residue analysis, *Anal. Chim. Acta* **529** (2005), 129–136.

119. R. Dams, M. A. Huestis, W. E. Lambert, and C. M. Murphy, Matrix effect in bioanalysis of illicit drugs with LC-MS/MS: Influence of ionization type, sample preparation, and biofluid, *J. Am. Soc. Mass Spectrom.* **14** (2003), 1290–1294.

120. R. T. Borchardt, R. M. Freidinger, T. K. Sawyer, and P. L. Smith (eds.), *Integration of Pharmaceutical Discovery and Development: Case Histories,* Kluwer Academic Publishers, Dordrecht, The Netherlands, 1998.

121. D. Q. Liu and C. E. C. A. Hop, Strategies for characterization of drug metabolites using liquid chromatography–tandem mass spectrometry in conjunction with chemical derivatization and on-line H/D exchange approaches, *J. Pharm. Biomed. Anal.* **37** (2005), 1–18.

122. F. Naganeo and K. Iwasaki, Application of high throughput LC/MS on drug metabolism and pharmacokinetics in drug discovery, *J. Mass Spectrom. Soc. Jpn* **52** (2004), 137–141.

123. T. A. Baillie, Drug discovery and development in the post-genome era. Can we rationally design safer drugs? *Adv. Mass Spectrom.* **16** (2004), 1–18.

124. C. E. C. A. Hop, Applications of quadrupole-time-of-flight mass spectrometry to facilitate metabolite identification, *Am. Pharm. Rev.* **7** (2004), 76–79.

125. W. A. Korfmacher, Lead optimization strategies as part of a drug metabolism environment, *Curr. Opin. Drug Discuss Dev.* **6** (2003), 481–485.

126. R. Kostiainen, T. Kotiaho, T. Kuuranne, and S. Auriola, Liquid chromatography/atmospheric pressure ionization-mass spectrometry in drug metabolism studies, *J. Mass Spectrom.* **38** (2003), 357–372.

127. D. C. Evans, A. P. Watt, D. A. Nicoll-Griffith, and T. A. Baillie, Drug-protein adducts: An industry perspective on minimizing the potential for drug bioactivation in drug discovery and development, *Chem. Res. Toxicol.* **17** (2004), 3–16.

128. E. J. Oliveira and D. G. Watson, Liquid chromatography–mass spectrometry in the study of the metabolism of drugs and other xenobiotics, *Biomed. Chromatogr.* **14** (2000), 351–372.

129. N. Kobayashi, Quantitative analysis of pharmaceuticals in clinical samples by LC/MS/MS method, *Gendai Kagaku Zokan* **31** (1997), 284–289.

130. S. Ekins, B. J. Ring, J. Grace, D. J. McRobie-Belle, and S. A. Wrighton, Present and future *in vitro* approaches for drug metabolism, *J. Pharm. Toxicol. Methods* **44** (2000), 313–324.

131. C. E. C. A. Hop, Y. Wang, S. Kumar, M. V. S. Elipe, and C. E. Raab, Identification of metabolites of a substance P (neurokinin 1 receptor) antagonist in rat hepatocytes and rat plasma, *Drug Metab. Dispos.* **30** (2002), 937–943.

132. D. C. Evans and T. A. Baillie, Minimizing the potential for metabolic activation as an integral part of drug design, *Curr. Opin. Drug Discovery Dev.* **8** (2005), 44–50.

133. A. Ghosal, N. Hapangama, Y. Yuan, J. Achanfuo-Yeboah, R. Iannucci, S. Chowdhury, K. Alton, J. E. Patrick, and S. Ziaida, Identification of human UDP-glucuronosyltransferase enzyme(s) responsible for the glucuronidation of ezetimibe (ZETIA), *Drug Metab. Dispos.* **32** (2004), 314–320.

134. E. L. Karen, D. A. Paul, J. W. Steffie, D. U. Himmelstein, S. M. Wolfe, and D. H. Bor, Timing of new black box warnings and withdrawals for prescription medications, *J. Am. Med. Assoc.* **287** (2002), 2215–2220.

135. T. Sugimura, K. Okabe, and M. Nagao, The metabolism of 4-nitroquinoline-1-oxide, a carcinogen. 3. An enzyme catalyzing the conversion of 4-nitroquinoline-1-oxide to 4-hydroxyaminoquinoline-1-oxide in rat liver and hepatomas, *Cancer Res.* **26** (1966), 1717–1721.

136. T. R. Bosin and R. P. Maickel, Mass spectra of carcinogenic 4-hydroxylamino-quinoline-*N*-oxides, *Res. Commun. Chem. Pathol. Pharmacol.* **6** (1973), 813–820.

137. M. Kiese, The biochemical production of ferrihemoglobin-forming derivatives from aromatic amines, and mechanisms of ferrihemoglobin formation, *Pharmacol. Rev.* **18** (1966), 1091–1161.

138. R. Ramanathan, A. D. Su, N. Alvarez, N. Blumenkrantz, and S. K. Chowdhury, Liquid chromatography/mass spectrometry methods for distinguishing *N*-oxides from hydroxylated compounds, *Anal. Chem.* **72** (2000), 1352–1359.

139. W. Tong, S. K. Chowdhury, J.-C. Chen, R. Zhong, K. B. Alton, and J. E. Patrick, Fragmentation of *N*-oxides (deoxygenation) in atmospheric pressure ionization: Investigation of the activation process, *Rapid Commun. Mass Spectrom.* **15** (2001), 2085–2090.

140. D. Q. Liu, C. E. C. A. Hop, M. G. Beconi, A. Mao, and S.-H. L. Chiu, Use of on-line hydrogen/deuterium exchange to facilitate metabolite identification, *Rapid Commun. Mass Spectrom.* **15** (2001), 1832–1839.

141. W. Lam and R. Ramanathan, In electrospray ionization source hydrogen/deuterium exchange LC-MS and LC-MS/MS for characterization of metabolites, *J. Am. Soc. Mass Spectrom.* **13** (2002), 345–353.

142. N. Ohashi, S. Furuuchi, and M. Yoshikawa, Usefulness of the hydrogen–deuterium exchange method in the study of drug metabolism using liquid chromatography–tandem mass spectrometry, *J. Pharm. Biomed. Anal.* **18** (1998), 325–334.

143. K. Biemann, *Mass Spectrometry: Organic Chemical Applications*, McGraw-Hill, New York, 1962.

144. M. M. Zhu, D. L. Rempel, Z. Du, and M. L. Gross, Quantification of protein–ligand interactions by mass spectrometry, titration, and H/D exchange: PLIMSTEX, *J. Am. Chem. Soc.* **125** (2003), 5252–5253.

8

METHOD DEVELOPMENT

Rosario LoBrutto

8.1 INTRODUCTION

The primary focus of this chapter in on general approaches and considerations toward development of high-performance liquid chromatography (HPLC) methods for separation of pharmaceutical compounds, which may be applied within the various functions in the drug development continuum. It is very important to understand the aim of analysis and the requirements for a particular method to be developed. The aim of analysis of each HPLC method may vary for each developmental area in the drug development process and specific examples are given in Section 8.2.

General method development considerations that apply to all reversed-phase methods are discussed in Section 8.3. These include properties of the analyte, detector, mobile phase, stationary phase, and gradient considerations. Building upon this knowledge, strategies for method development for target analytes in which the structure is known and not known are given as general guidelines. This material is reinforced with several method development case studies emphasizing the approaches used and the shortcomings that were encountered during the method development continuum. Also, a method development flow chart for gradient separations is provided in Section 8.5.6 (Figure 8-37) which can be used as an excellent starting point for the development of HPLC methods.

Also, throughout this chapter we focus on analytical challenges a pharmaceutical scientist encounters during method development; these include speed

HPLC for Pharmaceutical Scientists, Edited by Yuri Kazakevich and Rosario LoBrutto

of separation, sample preparation, extraction issues, solution stability, detection sensitivity (effect of pH), and separations for structurally similar species.

The final part of the chapter provides a refresher on "pK_a from an analytical chemist's perspective," the drivers for choosing normal phase versus reversed phase as a separation mode for a particular analysis, and instrument/system consideration, and it concludes with a very interesting section on column testing within the framework of bonded phase stability (effect of pH and type of buffer) and probing column selectivity.

8.2 TYPES OF METHODS

There are many HPLC methods that are developed during the process of drug development. The chromatographer needs to understand the aim of analysis in order to make judicious choices prior to the commencement of method development and the implications it may have on the final method that is developed. The following should be considered: method development time, the maximum run time for analysis, the number of samples expected per week, the complexity of the mixture, the structure of the main analyte (physicochemical properties), possible degradation pathways (i.e., hydrolysis, oxidative, photolysis, dehydration, thermolytic, racemization), and whether the analyte or analytes are ionizable.

In Section 8.2, the aim of analysis is emphasized especially for the API (active pharmaceutical ingredient) and the drug product. The workflows and the rationale at major decision points during synthetic processing steps where HPLC can be applied in process development are elaborated upon. For example, a fast method is needed to monitor reaction conversion of two components. However, a more complex method would be needed for stability-indicating purposes where multiple degradation products, synthetic by-products, and excipient peaks need to be resolved from the active pharmaceutical ingredient.

8.2.1 Key Raw Materials

The purity of the starting materials needs to be determined in order to ensure the purity of the final compound and to avoid any undesired secondary reactions that may carry forward in the downstream chemistry. Starting materials can be classified as raw materials or key raw materials. The latter are raw materials in which some part of the raw material structure is incorporated into the final structure of the drug substance. For the raw materials, usually an identification test and concentration of the reagent suffices although the purity of these materials is sometimes also deemed as a necessity. However, for key raw materials the purity of this reagent must be known. The analysis of the key raw materials depends on the nature of the substance (i.e., volatility) and a variety of methods could be used such as GC and HPLC. To ensure the quality of the key raw materials, the level and type of impurities present in these materials need to be determined and appropriate specifications need to be set. These can be determined by running a small-scale synthesis (i.e., front run) with the key raw material; and if an intermediate with an acceptable purity is obtained in the next step of the synthesis, the level and type of impurities present in the key raw materials can be considered acceptable. Note that the impurities may carry forward in the downstream chemistry and/or may react to form new synthetic by-products. If these synthetic by-products are carried forward to the final API, these impurities must be qualified in the appropriate toxicological studies if they are above a certain level (also, any potential genetoxic impurities must be identified and well-controlled with sensitive analytical methods). Different lots from the same manufacturer and/or lots of key raw materials from different manufacturers are usually tested.

In the following example, the importance of determining the quality of the key raw material is highlighted. Trimethoprim (TMP) (2-4-diamino-5-(3,4,5-trimethoxybenzyl)pyrimidine) is an antibacterial, folic acid antagonist that is usually used with sulfamethoxazole as a treatment for urinary tract infections. Shown in Figure 8-1 is one synthetic scheme of TMP [1]. The starting material that is used is 3,4,5-trimethoxy benzaldehyde (a key raw material). Depending on the quality of this key raw material, the impurities from this starting material may further react in the subsequent synthetic steps (downstream chemistry) at the benzaldehyde functionality to produce undesired synthetic by-products. TMP manufactured by five different manufacturers from three countries of origin were analyzed by the same HPLC chromatographic method; two impurities, Impurity 1 and Impurity 2, were found in multiple lots of TMP, and the levels depended upon the country of origin (Figure 8-2). [2].

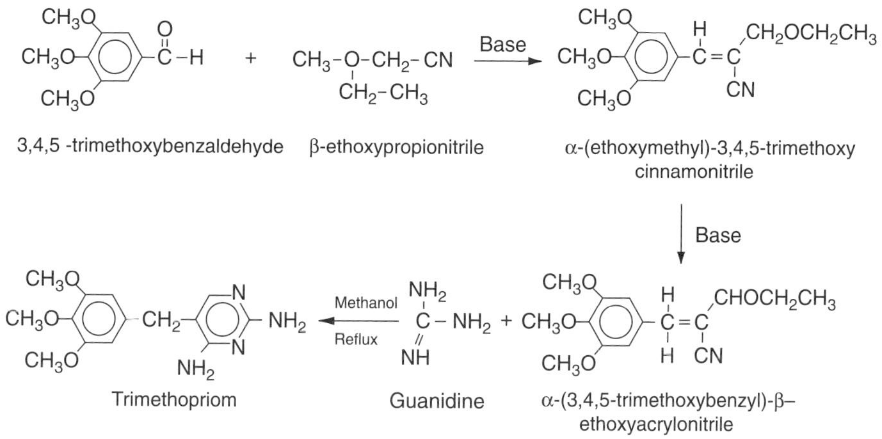

Figure 8-1. Synthesis of TMP. (Reprinted from reference 1, with permission.)

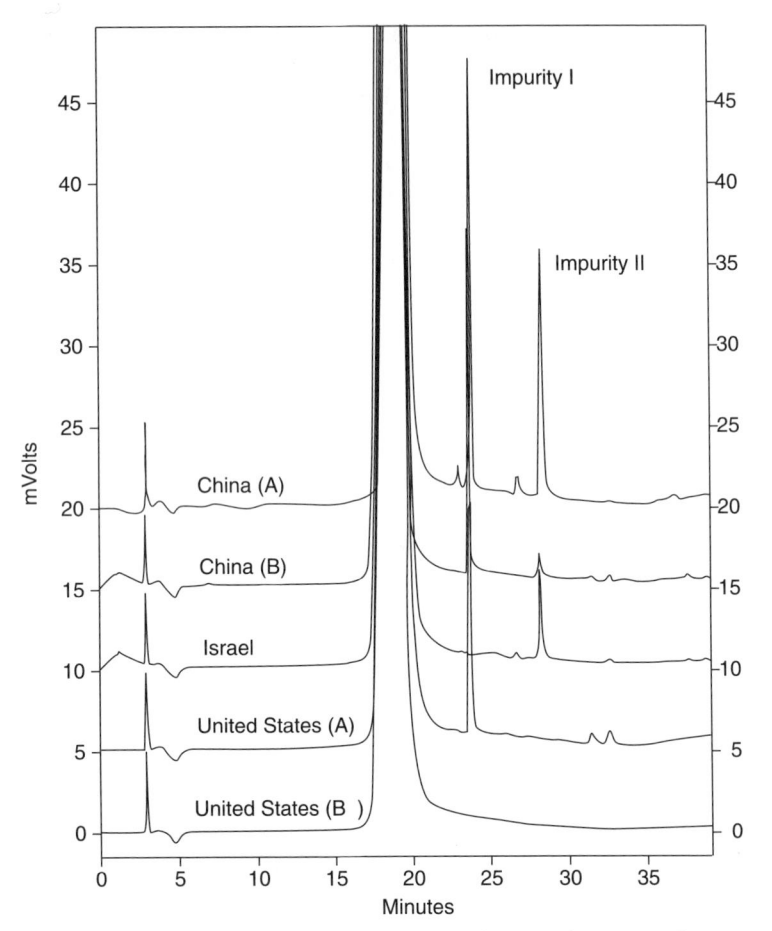

Figure 8-2. HPLC profiles of TMP drug substance from various manufacturers using the same chromatographic method. (Reprinted from reference 2, with permission.)

Impurity 1 and Impurity 2, were identified as 2,4-diamino-5-(-4-ethoxy-3,5-dimethoxybenzyl) pyrimidine and 2,4-diamino-5-(3-bromo-4,5-dimethoxybenzyl) pyrimidine, respectively (structures shown in Figure 8-3), using LC/MS, MS/MS, and NMR. Both these synthetic by-products were presumed to come from the starting key raw material 3,4,5-trimethoxybenzaldehyde.

This emphasizes that appropriate control of the starting material is needed. If the purity of the starting material is not controlled adequately, then recrystallization steps may be needed in later steps to remove synthetic by-products in the API that are above a qualified level which may result in decreased yield of the synthesis. Chromatographic analysis of the key raw material should be employed to determine the limits of the impurities in 3,4,5-trimethoxyben-

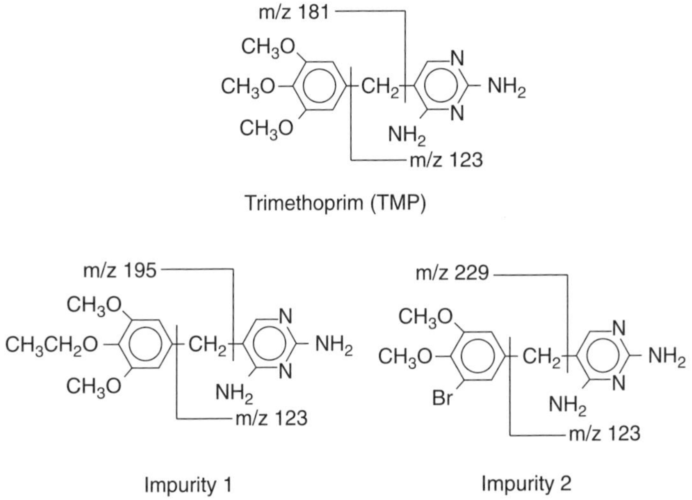

Trimethoprim (TMP)

Impurity 1   Impurity 2

Figure 8-3. Trimethoprim (API) and two potential synthetic by-products. (Reprinted from reference 2, with permission.)

TABLE 8-1. Reporting Thresholds of Impurities [3]

Maximum Daily Dose[a]	Reporting Threshold[b,c]	Identification Threshold[c]	Qualification Threshold[c]
≤2 g/day	0.05%	0.10% or 1.0 mg/day intake (whichever is lower)	0.15% or 1.0 mg/day intake (whichever is lower)
>2 g/day	0.03%	0.05%	0.05%

[a]The amount of drug substance administered per day.
[b]Higher reporting thresholds should be scientifically justified.
[c]Lower thresholds can be appropriate if the impurity is unusually toxic.

zaldehyde (individual and total) that would give acceptable levels of the resulting downstream impurities in the API. This would include determining the purity of the key raw material with a defined method and relating the purity of the key raw material from different lots/manufacturers to the quality of the final drug substance.

In general, the detection and identification of impurities present in API and formulations play an integral role in the drug development process and methods need to be developed to adequately resolve these species and quantitate them. The International Conference on Harmonization (ICH) [3] has published a guidelines on impurities in new drug substances and new drug products (see Table 8-1). The acceptable limit of the impurities in drug substances is dependent upon the maximum daily dose and the qualification threshold; however, lower thresholds are sometimes deemed necessary if the impurity is known to be unusually potent or to produce toxic or unexpected pharmacological effects through genetoxicity studies, general toxicity studies, and/or in-silico assessment.

8.2.2   Drug Substance (Active Pharmaceutical Ingredient)

During process development of a new compound, many samples are generated which include reaction mixtures from process monitoring, batch and waste layers from extractions (aqueous and/or organic), batch concentrates, mother liquors and supernatants, and isolated solid intermediates; in the event that PrepHPLC is used, column fractions are analyzed to determine the yield and purity of each fraction so they may be pooled. Impurities may be formed during the manufacture of the API, and these impurities may be related to the starting materials, process-related impurities, synthetic intermediates, or degradation products, and suitable methods need to be developed to control these processes.

8.2.2.1   Reaction Conversion. Methods that monitor reaction conversion should ensure the resolution of all solvents and in-process impurities from the reactants and the desired intermediate. The goal here is to monitor product A going to product B—in essence, measuring the disappearance of A or completion of the reaction if reagent A is used in excess. Also, the concentration of product B may be needed as well as the purity of product B to control any undesired by-products.

In Figure 8-4 the starting material is reacted in a one pot reaction to form the intermediate. At 30 minutes the reaction is only 30% complete. By further monitoring, it was determined that the reaction must be allowed to continue

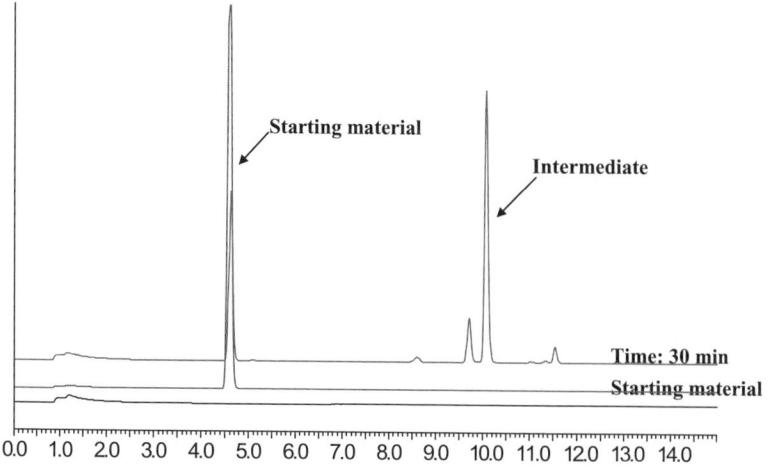

Figure 8-4. Reaction conversion of starting material to intermediate.

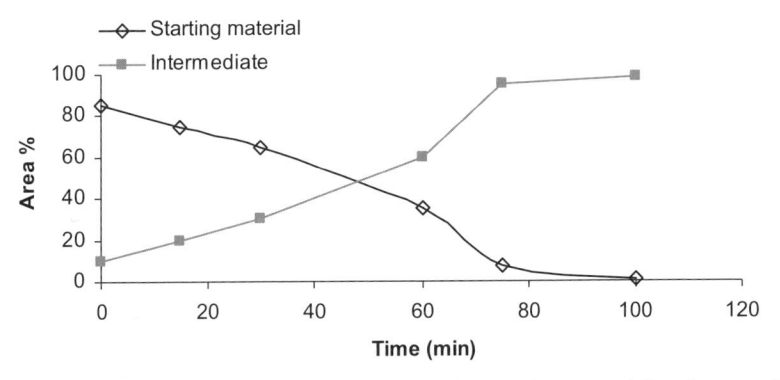

Figure 8-5. Reaction monitoring of converting the starting material to intermediate.

for at least 100 minutes in order to reach completion (Figure 8-5). During this process optimization stage, the concentration of the reagents, the temperature of the reaction, mixing conditions, and other processing conditions are optimized and HPLC is used as a tool to monitor the reaction. It must be determined how long the reaction should proceed in order to form the desired intermediate in good yield and for how long the intermediate is stable in solution prior to going to the next step (hold point stability).

These reaction monitoring analyses should be fast because the reaction time scales may be in the order of minutes to hours. In-line flow injection analysis or spectroscopic methods are sometimes used to monitor reactions (reaction conversion) that are on the minute time scale and for reactions that involve hazardous materials because by the time the samples are analyzed by an off-line chromatographic method, the reaction has gone to completion and/or undesired by-products may have been formed. However, off-line HPLC may still be needed to determine the purity of the desired intermediate present in the reaction solvent before proceeding to the next step of the reaction, and fast HPLC methods can be employed. If the reaction conversion is along the time scale of an hour or greater, fast HPLC methods are used implementing nonporous stationary-phase materials, monolithic columns, and columns packed with sub-2 μm particles (uPLC, xPLC, fPLC) (more information on fast HPLC methods is provided in Chapter 17). It is advantageous to have short methods to analyze these reaction conversion samples. The in-process samples may have to be diluted with an appropriate solvent to quench the reaction and to have the desired components within the linear range of the chromatographic method. The diluent must be chosen such that it does not react with the components in the mixture. Also, all blank system suitability and standards samples should be run on the chromatographic system prior to injection of the reaction sample to ensure that the HPLC system is working properly and that no downtime in the reporting of the results to the process chemist and/process engineer. For the more conservative chromatographer, two HPLC systems can be set up in the event that an instrument breaks down to ensure no downtime in the reporting of the results.

8.2.2.2   Concentration Determination of In-Process Samples. The concentration of the unisolated desired product in solution at a particular intermediate step may also need to be determined by HPLC. A data calculation sheet such as Excel with the response factors of the standards and the dilution factor of the sample could be incorporated in the data calculation sheet prior to injection of reaction sample to facilitate the results reporting for the concentration of the intermediate in solution. Hence, only the area of the desired intermediate in solution needs to be populated in the spreadsheet, and the concentration result then can be determined. The determined concentration of the intermediate in solution ensures adequate charging of the raw materials used in the further steps of the synthesis. Also, this intermediate in solution is sometimes further concentrated and the concentration is monitored until the desired concentration is obtained. A solvent switch step is sometimes performed, and the HPLC method must be able to selectively separate the reaction solvents (if they are UV active) from the desired intermediate and potential impurities that may be formed. These reaction solvents may include toluene, inhibited THF with cresol or BHT (if inhibited with BHT, this is very hydrophobic, so proper elution of this additive may be necessary), ethyl acetate, and so on. Sample preparation here is also important, and the appropriate diluent must be determined to ensure solubility of all components and no reactivity with the sample analyte.

Sometimes when extractions are performed to remove undesired by-products, the concentration of the desired product in the organic and aqueous layers are also determined. The concentration in the organic layers are deemed the most important (contains the desired product), and the concentrations in the aqueous layers are determined later to ensure mass balance and overall yield of the reaction. The aqueous layers are usually enriched with the undesired by-products and are good samples to use during the development of the HPLC method in the early stages of the synthetic development. The pH of these layers is usually checked as well to ensure that the proper amount of acid or base has been added to the reaction mixture either to quench the reaction or to drive the desired product into the organic layer.

8.2.2.3   Purity of Intermediates. Determining the purity of the desired product in the organic layers is important to ensure that an adequate number of aqueous washes removed the unwanted by-products. This organic layer may be carried forward to the next step without any further isolation. However, if the intermediate will be isolated, then the purity and weight percent of the isolated intermediate needs to be determined to ensure mass balance and determine overall yield of the reaction. The purity of the intermediates needs to be evaluated in order to determine if synthetic by-products generated in a prior step will react in the downstream chemistry or, if they do not react in further steps of the synthesis, to determine what their rejection will be in subsequent steps of the synthesis. Eventually, these methods may also have to be transferred to a manufacturing facility and will be used as critical process controls before proceeding to the next step of the synthesis. Methods should be able to separate the API from the synthetic process impurities including the penultimate intermediate and potential degradation products and any potential solvents that may be observed, depending on the type of detection employed. The final method should be able to resolve all impurities from the active as well as from each other. The optimal goal is to have one method to resolve all isolated intermediates, penultimate products, synthetic by-products, and potential degradation products from the API.

Eventually, a stability-indicating method needs to be validated and included in the IND (investigational drug substance), IMPD (investigational medicinal product dossier), and NDA (new drug application). Stability-indicating methods must be able to resolve potential degradation products from the active substance that may increase or form during the storage (certain temperature/certain humidity).

There are generally two types of stability studies that are performed: long-term and accelerated. The purpose of the accelerated stability studies is to help set the shelf life at the recommended storage conditions and predict the amount of degradation that could be anticipated under long-term storage conditions. However, at the predicted shelf life timepoint (number of months or years) the sample stored at the recommended storage condition must be run to confirm the shelf-life prediction from the accelerated stability studies. Therefore the stability-indicating methods should be rugged and robust and meet all validation requirements that will allow for determining the purity of the active pharmaceutical ingredient throughout the duration of the supportive stability studies. Method validation is discussed in detail in Chapter 9.

8.2.3 Drug Product

Methods that are used during the development of a drug product formulation should be able to assess reacting or catalyzing excipients and any undesired reactions leading to degradation products. Methods should be able to separate the active pharmaceutical ingredient (API) from the drug product degradation products, excipients, excipient degradation products, and any synthetic impurities that are present in the API. These are usually performed during early phase development and are known as excipient compatibility studies. Both binary (API + excipient) and excipient mixtures + API are used to assess which excipient or formulation blend will lead to the most stable drug formulation. Moreover, other preformulation studies that are also performed include solubility studies and solution stability studies at various pH values (Figure 8-6). HPLC with UV detection is the most applicable technique to use

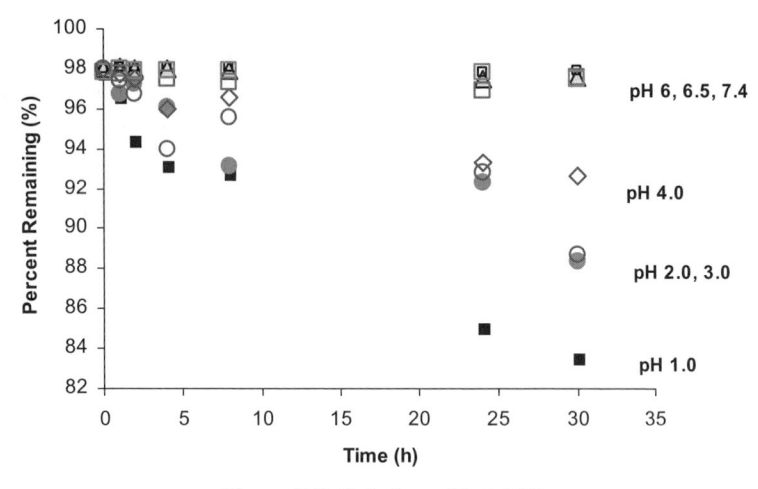

Figure 8-6. Solution pH stability.

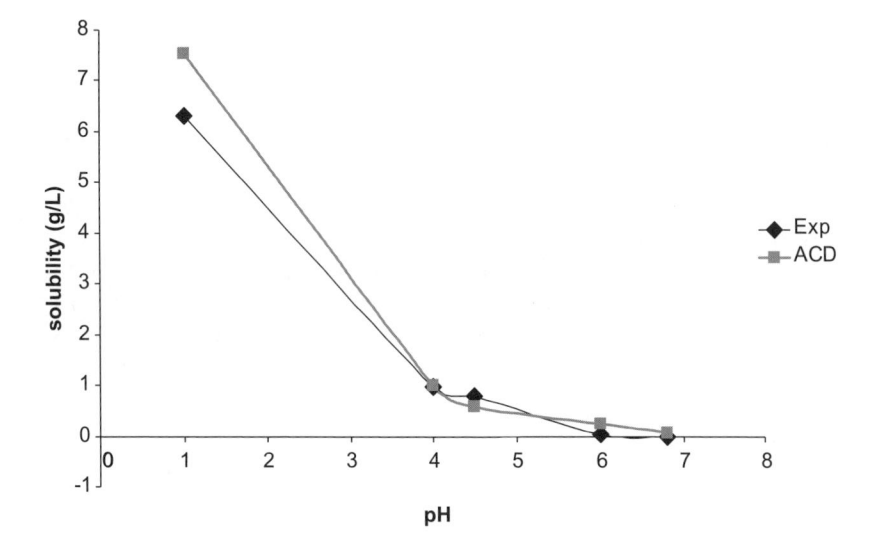

Figure 8-7. ACD prediction versus HPLC experimental results for the solubility of a basic compound.

for these studies. Fast gradient methods on short columns are usually used in most cases for solubility determination. Some software programs such as ACD (advanced chemistry development software) can be used to estimate the solubility as a function of pH and can be used as a starting point to estimate the appropriate dilution of the different solutions prepared at the different pH values. For example, in Figure 8-7 if the target concentration for the HPLC assay is 1 mg/mL for this basic compound, then the sample at pH 1 (8 mg/mL

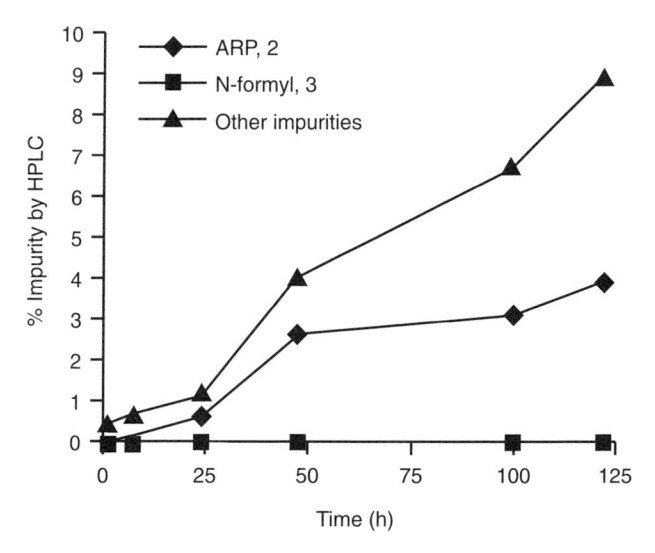

Figure 8-8. Temperature stress study of a amine hydrochloride salt at 85°C. (Reprinted from reference 4, with permission.)

estimated by ACD) should be diluted by eight times, and all the other samples at pH values from pH 4 to pH 7 should be injected neat. It is a very convenient way to run solubility samples from the different pHs, and this avoids a lot of trial in error in trying to find the optimal dilution scheme for each sample at a particular pH. Moreover, the predicted ACD values and the experimental values from pH 1–8 in this example were in very good agreement. This is used only for estimation since sometimes the ACD prediction of the solubility for the ionized form of bases and acids shows greater deviation than for the solubility predicted for the neutral form of the molecules.

Reaction of the excipients with the API must be carefully controlled and well understood. For example, it is known that primary amines + reducing sugars result in brown pigments (Malliard reaction). It was shown that the amine of a hydrochloride salt (fluoxetine) reacts with a reducing sugar, lactose (a common excipient), to form a glycosyl amine [4]. Then the amine can rearrange via an Amadori rearrangement to form a 1-amino-2-keto sugar and is colored. This indicates that both free bases and hydrochloride salts could react with reducing sugars, and this process can be accelerated at higher temperatures. The degradation of this process can be monitored by HPLC, as shown in Figure 8-8. For hydrochloride salts, it is speculated that reaction is preceded by loss of HCl gas from the hydrochloride salt. It is generally recommended to determine the chloride content of the formulation under different stability stress conditions to assess if there is a loss of HCl, since this may change the solubility of the compound (i.e., free base may be less soluble than the HCl salt). If there is loss of HCl this can change the inherent stability of the compound in the formulation (i.e., free base versus the salt). The chloride content can be determined using ion chromatography and/or silver nitrate titration.

Stability-indicating methods (assay/purity) are required to monitor the possible degradation products that may arise due to API-excipient interactions, production process, poor or inappropriate packaging materials, or improper storage. In the drug product, impurities that are related to the synthetic by-products of the API are not specified in the drug product because the levels are quantified in the API. This holds true only if it has been determined that the impurity levels will not increase upon storage at specified long-term storage conditions in the drug product. Therefore, for the drug product formulations, as per the current ICH guidelines on impurities in new drug products [5], the reporting thresholds are 0.1% for drugs with maximum daily dose of ≤1 g/day and 0.05% for drugs with maximum daily dose of >1 g/day. For more description on the reporting, identification, and qualification of impurities in drug products, the reader is referred to attachment 1 from the ICH, Guidance for the Industry: Q3B(R): Impurities in New Drug Products (shown in Table 8-2).

Eventually, a stability-indicating method needs to be validated and to be included in the IND, IMPD, and NDA. These methods should be rugged and robust and meet all validation requirements at each particular stage of drug development. This method may be used also for content uniformity, and

TABLE 8-2. Thresholds for Degradation Products in New Drug Products [5]

Maximum Daily Dose[a]	Threshold[b,c]
Reporting Thresholds	
≤1 g	0.1%
>1 g	0.05%
Identification Thresholds	
<1 mg	1.0% or 5 μg TDI, whichever is lower
1–10 mg	0.5% or 20 μg TDI, whichever is lower
>10 mg to 2 g	0.2% or 2 mg TDI, whichever is lower
>2 g	0.10%
Qualification Thresholds	
<10 mg	1.0% or 50 μg TDI, whichever is lower
10–100 mg	0.5% or 200 μg TDI, whichever is lower
>100 mg to 2 g	0.2% or 3 mg TDI, whichever is lower
>2 g	0.15%

[a]The amount of drug substance administered per day.
[b]Thresholds for degradation products are expressed either as a percentage of the drug substance or as total daily intake (TDI) of the degradation product. Lower thresholds can be appropriate if the degradation product is unusually toxic.
[c]Higher thresholds should be scientifically justified.

dissolution assuming the length of analysis is short. The methods that are used for content uniformity and dissolution may not have to separate the active compound from all the impurities and/or degradation products assuming a separate stability-indicating method has been established. The methods for dissolution and content uniformity may offer the chromatographer some advantages in terms of high-throughput analysis by allowing chromatographer to use shorter methods employing shorter columns, and other fast HPLC approaches because the resolution of the impurities from each other may not be deemed necessary.

8.2.4 Achiral Versus Chiral Methods

Stereoisomers are classified by symmetry as either enantiomers or diastereomers. Enantiomers have identical physical properties except for the direction of optical rotation. Diastereomers are basically stereoisomers that are not enantiomers of each other. A pair of enantiomers exists for all molecules containing a single chiral center and have the opposite configuration at each of the stereo centers. The maximum number of stereoisomers for a compound with n stereo centers is 2^n. Diastereomers, on the other hand, have the same configuration at one of the two centers and have the opposite configuration at the other.

The assessment of the isomeric purity of substances is imperative because isomeric impurities may have unwanted toxicologic, pharmacologic, or other effects. These impurities may be carried through a synthesis and preferentially react at subsequent steps to yield an undesirable level of another impurity. One isomer of a series may produce the desired effect, while the other isomer may be inactive or even produce some undesired effect. Large differences in activity between stereoisomers illustrate the need to accurately assess isomeric purity of pharmaceutical, agricultural, or other chemical entities. Often these differences exist between enantiomers, the most difficult stereoisomers to separate. The development of analytical methods used to control the stereochemical purity of intermediates, drug substances, and drug products is imperative because these methods will ensure a robust chemical process that can ultimately be transferred to the manufacturing facility.

The separation of chiral compounds will be discussed in Chapter 22. However, the separation of diastereomers can be accomplished using achiral stationary phases. Another alternative is the use of chiral columns for the separation of diastereomers in either the reversed-phase or normal-phase mode. The use of achiral bonded phases without chiral additives, such as phenyl and alkyl bonded phases for the separation of diastereomeric pharmaceutical compounds, is acceptable. Different selectivities can be obtained by employing stationary phases containing varying functionalities (phenyl, polar embedded moieties). The effect of aqueous mobile-phase pH, temperature, and type of organic eluent (acetonitrile versus methanol) can also play a dramatic role on the separation selectivity of diastereomeric compounds.

8.3 DEFINING THE METHOD

Prior to the initiation of method development, all the known information about the analyte such as its structure, physical and chemical properties, toxicity, purity, hygroscopitiy, solubility, and stability should be determined. These data may be available from preformulation reports, early drug discovery sample screening reports, from the literature on similar compounds, or from past experience with similar compounds. However, many times this information is not available, so preventive measures must be taken in order to ensure that the analyte does not degrade or change during the method development scouting experiments. Implementation of a tray cooler, preparation of fresh samples, storage of samples in a refrigerator, and protection of the solids and sample solutions from light represent some common preventive measures.

Throughout the course of development, usually there are multiple components to be analyzed in sample matrices. The best-case scenario would be to obtain isolated standards of the degradation products, in-process impurities, and synthetic precursors to determine their elution and to ensure that the components are well-resolved from each other. However, in early development, isolated standards or even pure intermediates may not be available, so samples such as mother liquors or supernatant or crude samples (before recrystallization) enriched with potential impurities or degradation products can also be used to challenge the method. Depending upon the structure of the compound, potential degradation products can be predicated and forced degradation samples can be generated; these samples can also be used during method development.

The goals or requirements of the HPLC method that needs to be developed should be known as well as the analytical figures of merit, which include the required detection limits, selectivity, linearity, range, and accuracy and precision.

The potential use of this method needs to be considered: if any regulatory requirements are to be met, if the method is used to analyze multiple samples, or if the method will be eventually transferred to the production site. Some other additional requirements may include sample throughput, analysis time, and instrument limitations.

Also, mass balance should always be a consideration during method development. Generally, for a drug substance method it should be established if the area percent method (peak area normalization) and the weight percent method (on a dry basis) are giving similar results, whereas for a drug product method it should be established if peak area normalization and the assay method (based on label claim) are giving similar results. If a bias is obtained between HPLC area percent (peak area normalization) and HPLC weight percent (assay), this may indicate the presence of co-eluting impurities, impurities with different response factors, and/or inadequate elution of all impurities present in the sample. Note that HPLC area percent and weight percent terminology is usually used by analytical chemists working with the API and peak area normalization, and assay is usually used by analytical chemists working with the drug product.

8.4   METHOD DEVELOPMENT CONSIDERATIONS

There are many factors to consider when developing methods. The initial steps include collecting as much information about the analyte in regard to the physicochemical properties (pK_a, log P, solubility) and determining which mode of detection would be suitable for analysis (i.e., suitable wavelength in case of UV detection). Sample preparation, which includes centrifugation, filtration, and/or sonication and type of diluent, plays an integral role in method development because this may affect the chromatography and the recovery of the analytes. Determination of the solution stability in the diluent is also important during early method development. If the solution is not stable, it will become increasingly more challenging to compare subsequent method development analysis. Choice of the mobile-phase and gradient conditions is dependent on the ionogenic nature of the analyte and the hydrophobicity of the analytes in the mixture respectively. This is a crucial step in the method development process because these two factors will probably have the most impact on the change in the analyte selectivity, especially for ionizable compounds. Also, the type of stationary phase is very important mainly in regard to bonded phase stability at the operational mobile-phase pH. Different stationary phases can and do provide differences in selectivity; however, the change in selectivity is much less predictable compared to varying the pH of the mobile phase to obtain the desired selectivity.

8.4.1   Sample Properties

8.4.1.1   Analyte Structure and pK_a. In this preliminary step, the ionogenic nature of the compound of interest should be determined. If the target analyte is neutral, the eluent pH will not affect its retention. However, the structure of this neutral molecule must be assessed, to postulate if a potential ionogenic degradation product may be formed during stress testing and stability testing. If this is the case, the HPLC method must be capable of adequately retaining and separating this "potential ionogenic" species from the active and other degradation products or impurities. For example, if you have a compound that has an amide bond, one of the potential degradation products may be carboxylic acid as a result of acid/base hydrolysis or degradation due to micro-enviromental pH of the formulation. Therefore, in an eluent that has a high pH, the potential acidic impuritiy may be in its ionized form which may result in the elution of the potential degradation product with or even before the void volume.

If the target analyte is ionizable, the pK_a of the analyte should be determined or obtained. Software packages such as ACD (Advanced Chemistry

TABLE 8-3. pK_a of Some Common Functional Groups

Group on Aromatic	pK_a^a	Acid/Base
Linear alcohol	>12	Acid
Carboxylic acid, –COOH	4–5	Acid
Thiol, –SH (aromatic)	6–7	Acid
Phenol, –OH	10–12	Acid
Alkyl amine (pri, sec, tert)	>9	Base
Aromatic amine, –NH$_2$, –NR$_2^b$	4–6	Base
Pyridinal	5–7	Base
Morpholine	8–9	Base
Piperidine	10–11	Base
Imidazole	6–8	Base

^{a}Addition of R(methyl, ethyl, etc.) group on aromatic ring or on NR$_2$ will cause an increase of compound pK_a due to electron-donating effects from methyl groups.
^{b}Substitution in general of halogens on aromatic ring will decrease compound pK_a. *Example*: o-chloroaniline pK_a = 2.6, aniline pK_a = 4.6.

Development), Pallas (CompuDrug Chemistry Ltd, Budapest, Hungary), and so on, may be used to get an estimated pK_a value for the ionizable functionalities on the molecule. In Section 8.7, discussion on analyte pK_a from an analytical chemist's perspective is given.

Also, discussions with the early-phase drug discovery and/or preformulation groups may provide this information because they may have already ascertained the pK_a of the molecules by titration or other experimental methods (πION, etc). Table 8-3 shows some of the more common ionizable functionalities present in pharmaceutical compounds.

As was discussed in Chapter 4, the optimal pH to commence method development is at a pH that is at least 1–2 units from the analyte pK_a in the particular hydro-organic mixture that is employed. For isocratic experiments, this is easily determined by varying the pH of the aqueous phase and monitoring the retention versus the pH, which generally results in a sigmodial type of dependence assuming that only one type of ionization center is present. In the event that there are two ionization sites that are acidic and basic, there are competing effects on the retention because multiple ionization equilibria exist and the overall effect on the retention is dependent on the relative hydrophobicities of the species present at a particular pH. Knowledge of the log P for the drug of interest and potential degradation products, metabolites, and synthetic impurities is usually helpful to give insight into the types of stationary phases and organic content needed to elute and/or retain all the components in the mixture.

8.4.1.2 Solubility of Components and Diluent Effects (Matrix Effects).
Solubility of the analyte is also very important. Solubility of a particular durg

compound is a prerequisite for any salt selection program. Salt formation during a salt selection program provides a means of altering the physicochemical and resultant biological characteristics of a drug substance without modifying its chemical structure, and most compounds with a suitable acidic or basic functionality can potentially be transformed into its salt form. The free acid/free base and their corresponding salts will all have different solubilities in the diluent. Generally, salt formation is associated with an increase in the compound's solubility. For example, the free base and phosphate-salt hydrate of codeine have aqueous solubilities of 8.3 and 435 mg/mL, respectively [6].

The analyte must be soluble in the diluent and must not react with any of the diluent components. It must be determined if the impurities in the drug substance observed are actual impurities from the synthesis or if they are formed *in situ* in the diluent. The diluent should match to the starting eluent composition of the assay to ensure that no peak distortion will occur, especially for early eluting components. Usually, this type of peak distortion occurs for compounds and/or impurities that elute at $k' < 2$. If the analyte is more soluble in the diluent than the starting eluent composition, the compound will tend to reside in the "solvent plug" being injected onto the column and a peak fronting or skewing may occur (see Figure 8-9). In Figure 8-9, peak skewing is occurring with the increase of the concentration of methanol in the diluent (chromatograms 2–5). However, the solvation of the analyte by the diluent and mobile phase components may also play a role, and peak distortion may occur. In Figure 8-10A [7], benzoic acid (diluent: 50% MeOH:50% water) analyzed in 50% MeOH:50% water eluent shows significant peak distortion. In this eluent, this acidic analyte is ionized. The analyte in its ionized form is expected to show early elution on a C18 column; however, solvation of the ionized analyte with methanol in the mobile phase forms a partially hydrophobic shell that could be retained on the reversed-phase adsorbent. The

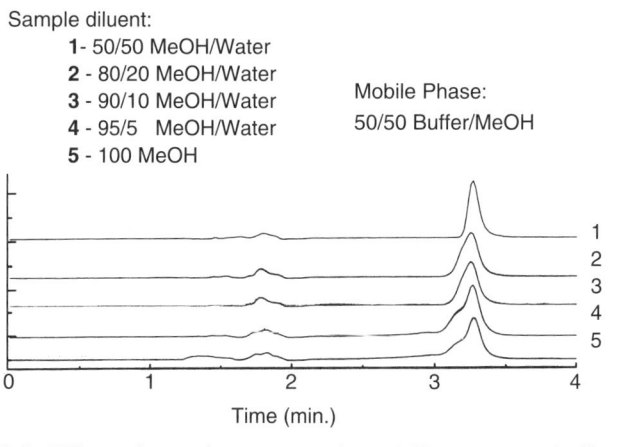

Figure 8-9. Effect of organic concentration of diluent on peak distortion.

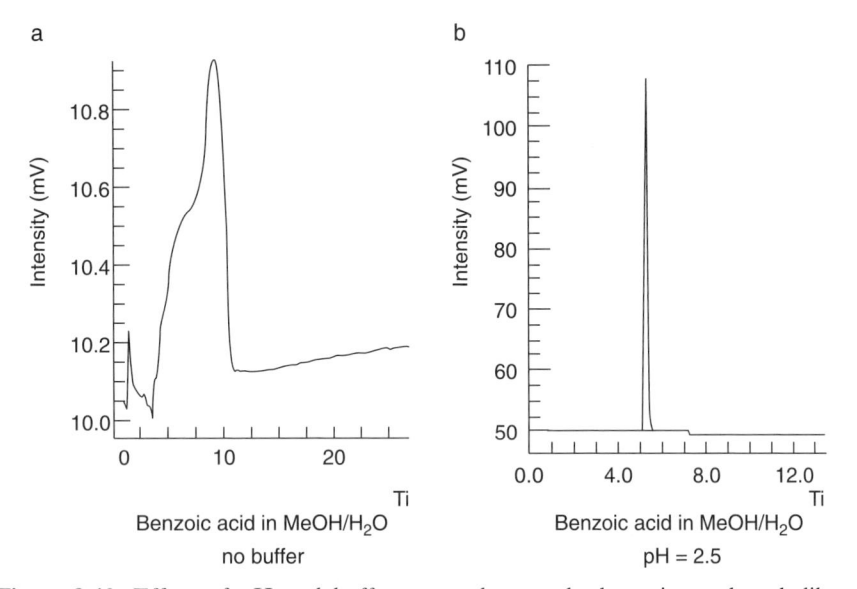

Figure 8-10. Effect of pH and buffer on analyte peak shape in methanol diluent (150- × 4.6-mm C18 column, 1 mL/min, void volume 1.3 mL). (Reprinted from reference 7, with permission.)

peak distortion could be attributed to secondary equilibria in the mobile phase (solvation/desolvation equilibria). However, in Figure 8-10B [6], when the acidic analyte (diluent: 50% MeOH:50% water) is analyzed in its predominantly neutral form in the 50% MeOH:50% (0.05 v/v% phosphoric acid, pH 2.5), the analyte is not prone to this solvation/desolvation equilibria and an adequate peak shape and retention is obtained. On the other hand, in Figure 8-11A, when the acidic analyte (diluent: 50% MeCN:50% water) is analyzed in 50% MeCN:50% water mobile phase, the ionized analyte elutes prior to the void volume and no peak distortion occurs because acetonitrile is not able to solvate the analyte since it cannot form hydrogen bonds. Also, in Figure 8-11B when the acidic analyte is analyzed in its predominately neutral form in 50% MeCN:50% (0.05 v/v% phosphoric acid, pH 2.5) mobile phase, the analyte is well-retained and no peak distortion occurs. The acidic analyte is less retained with acetonitrile than with methanol used as a mobile-phase organic modifier due to the greater elution strength of acetonitrile.

Also, peak distortion may occur for early eluting eluting analyte when its retention is close to the retention of the organic modifier in the diluent. The organic component of the diluent may not have a UV chromophore, and the detection of this organic component may be only possible using refractive index detection. Generally, acetonitrile and methanol elute close to the void volume; however, THF and isopropanol elute later. Therefore, co-solvent mixtures with THF/water or isopropanol/water may be an effective way to prevent

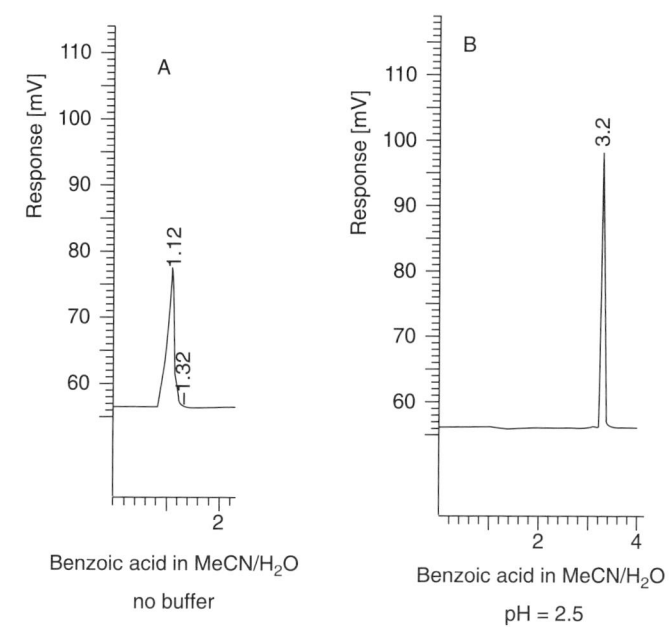

Figure 8-11. Effect of pH and buffer on analyte peak shape in MeCN diluent (150- × 4.6-mm C18 column, 1 mL/min, void volume 1.3 mL).

peak distortion of early eluting peaks such as acids (maleic, tartaric, lactic, fumaric, citric) [8].

There are also other scenarios that should be considered for peak distortion of early eluting components.

Scenario 1. If a compound has a greater solubility in acetonitrile and the diluent is 100% acetonitrile, with the starting eluent composition being 95% aqueous, pH 2:5% acetonitrile, the early eluting compound may show a peak skewing. Diode array spectra should be obtained to elucidate that this is not an on-column degradation product and/or a coeluting species. If the diode array spectra is the same across this distorted peak, it can be deemed as spectrally homogenous and may be possibly due to a diluent effect. However, if the diluent was changed to 95% aqueous–pH 2/5% acetonitrile, and the peak did not show skewing then the proper diluent has been determined.

Scenario 2. If the diluent was changed to 95% aqueous, pH 2/5% acetonitrile, the sample may not be soluble in this diluent. Therefore serial dilutions may be made such that the compound is first diluted in a diluent in which it is most soluble in and then further diluted to target concentration with a diluent that would be as similar as possible to the staring mobile-phase composition.

Scenario 3. If the diluent was 95% aqueous, pH 2/5% acetonitrile and the sample does show a good peak shape but reinjection of the same solution over time shows an additional impurity that is increasing, this may suggest reaction with the diluent. Multiple steps may be taken here; use an autosampler with a tray cooler to decrease the rate of reaction, adjust the pH of the diluent if pH is catalyzing the reaction in the protic solvent, and/or increase the amount of organic in the diluent. If increasing the organic concentration in the diluent does help in suppressing the formation of the additional impurities but as a consequence peak splitting is observed, then it would be recommended to try another column in order to enhance the retention of the early eluting component such that the compound/impurity would be more retained at a higher initial organic composition. Also, precolumn derivatization may be required to ensure that the desired product is not reacting with diluent prior to analysis. If derivatization is to be employed, then the type, concentration, and derivatization time all need to be explored. This would be considered as a last resort. Other approaches may include the use of aprotic solvents as a diluent and the use of normal-phase chromatography.

Generally, reaction with diluent and mobile phase is sometimes observed for compounds that contain keto functionalities (gem diol, oxycontin [9], active aldehyde [10], active esters such as mesyl sulfonates [11, 12], and enolate intermediates [13]), so protic solvents such as aqueous/methanol should be avoided or derivatizatoin may be required either precolumn or *in situ*.

Scenario 4. Buffered eluents must be used when analyzing ionizable species. Ionizable species are prone to solvation by the mobile-phase components and the solvation equilbira may lead to poor peak shapes. In Figures 8-12A and 8-12B, two acidic compounds, benzoic acid (pK_a 4.2) and sorbic acid (pK_a 4.8), are analyzed at pH 3.5 (a pH lower than the analyte pK_a) and at pH 7.0 (a pH greater than the analyte pK_a). Acceptable peak shapes are obtained at both

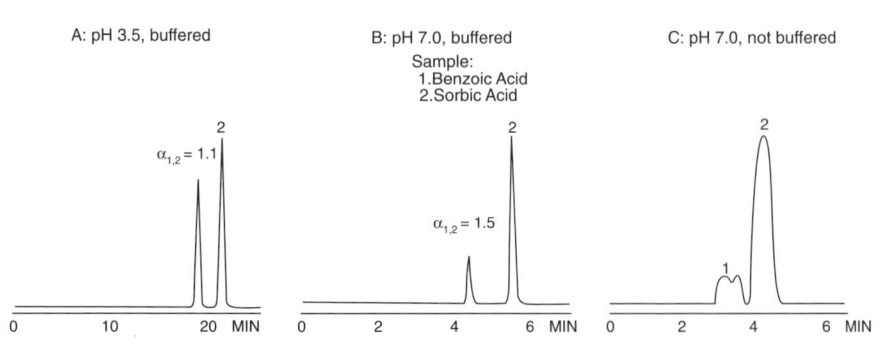

Figure 8-12. Effect of pH and buffer on the peak shapes of ionizable analytes. (Reprinted from reference 14, with permission.)

these pH values with buffered eluents on this BioBasic C18column. [14] However, when these compounds are analyzed at pH 7.0 without a buffer (only water and methanol), the peak shapes are distorted (Figure 8-12C). This could be related to the solvation of ionizable species by both the methanol and water which, due to the different secondary equilibria processes, leads to peak distortion.

8.4.2 Detector Considerations

Choice of the proper detection scheme is dependent on the properties of the analyte. Different types of detectors are available such as ultraviolet (UV), fluorescence, electrochemical, light scattering, refractive index (RI), flame ionization detection (FID), evaporative light scattering detection (ELSD), corona aerosol detection (CAD), mass spectrometric (MS), NMR, and others. However, the majority of reversed-phase and normal-phase HPLC method development in the pharmaceutical industry is carried out with UV detection. In this section the practical use of UV detection will be discussed.

A wavelength for UV detection must be chosen so that an accurate mass balance may be determined. Therefore, if area% normalization is to be used, then all the impurities and the active pharmaceutical ingredient must have similar relative response factors (area response/weight). This is sometimes difficult because the impurities may have different electron-donating or electron-withdrawing functional groups, attached to the aromatic ring and/or the impurities may have more complex conjugated systems and the absorption spectra have been shifted to longer or shorter wavelengths compared to the parent compound. Therefore the UV spectra of target analyte and impurities must be taken and overlaid with each other, and the spectra should be normalized due to different amounts present in the mixture. A wavelength must be chosen such that adequate response is obtained for the active and that at least a 0.05 v/v% solution of the active at target concentration could be quantified (S/N greater than 10). The wavelength chosen should not be on a distinct slope of the spectrum, and the relative difference in the absorbance at a certain wavelength is not significantly different from the impurities/degradation products present. Figure 8-13 shows the diode array overlay for an API and its related impurities. The optimal wavelength for detection is 280 nm because the impurities and API have similar absorbance at this particular wavelength. Although at 250 nm all the compounds have similar absorbance and even higher absorbance compared to 280 nm for some of the components. However, at the 250 nm wavelength greater variability in the response factors may be obtained if an analysis is run on different systems with different detectors (Figure 8-13). Most detectors are calibrated at ±2 nm. If an analysis were to be carried out at 250 nm the spectral bandwidth becomes very important. The spectral bandwidth is dependent on the slit width. The linearity of the detector is inversely proportional to the spectral bandwidth (as the spectral bandwidth gets narrower, the linearity gets better).

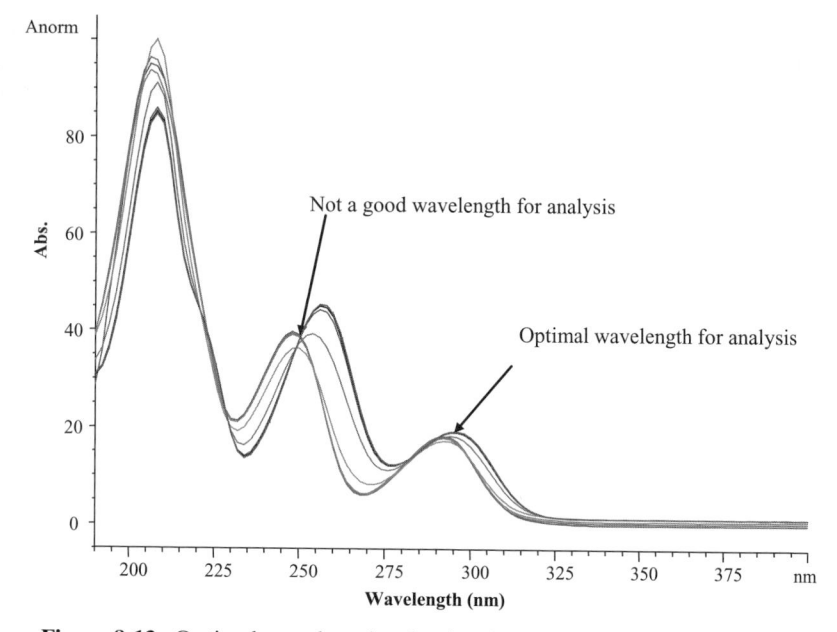

Figure 8-13. Optimal wavelength selection for API and related impurities.

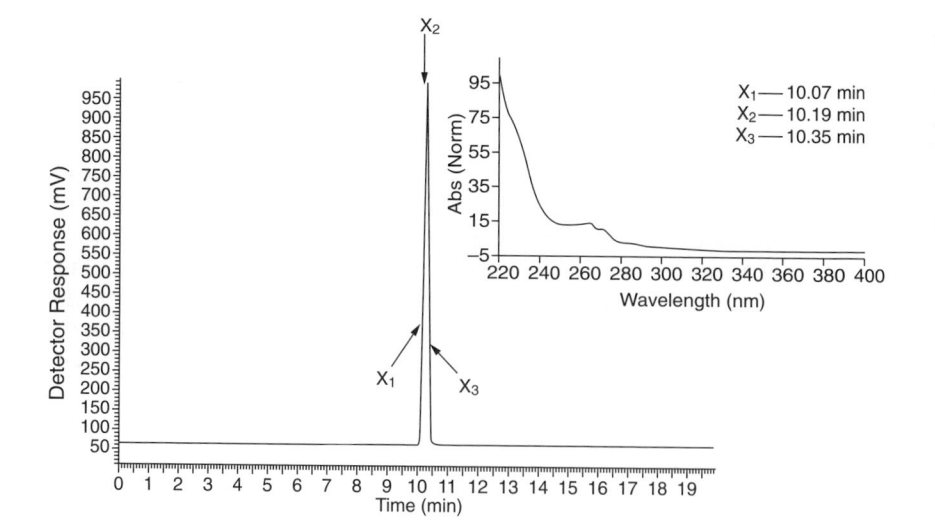

Figure 8-14. Determination of peak homogeneity: Diode array detection (DAD). (Reprinted from reference 10, with permission.)

Also the spectral homogeneity of the peak of interest must be taken into consideration. Diode array spectra at least three points across the peak should be taken to ensure the peak is spectrally homogenous see Figure 8-14. If the peak is not spectrally homogenous, the overlay of the spectra will show

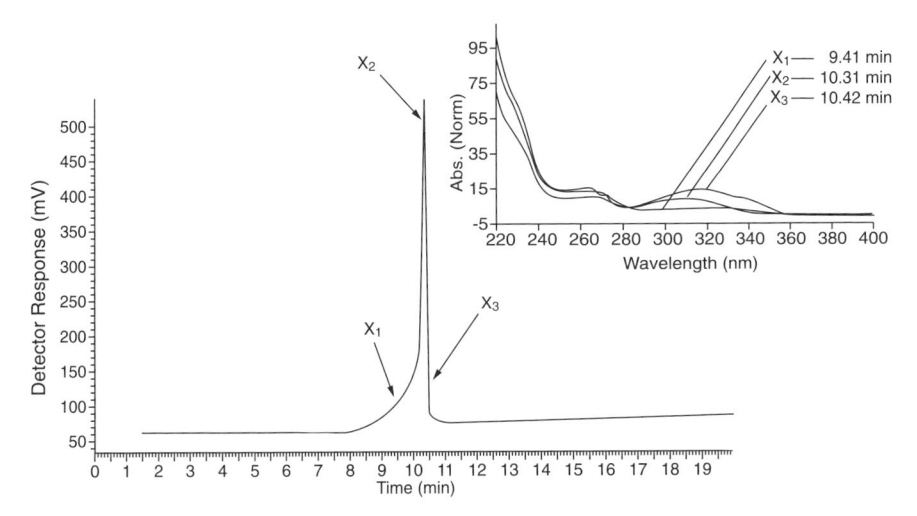

Figure 8-15. Diode array detection for elucidation of coeluting species. (Reprinted from reference 10, with permission.)

distinct differences (see Figure 8-15). However, even if the diode array spectra do overlay, this does not absolutely ensure that the peak does not contain any coeluting impurities, because the impurities could have similar diode array spectra and/or if there is a low level of a coeluting species with a different diode array spectrum, it may not be determined by this approach. In these cases, MS detection needs to be employed to ensure MS spectral homogeneity. MS spectra are taken across the peak and the MS spectra across the peak should not show the presence of any other coeluting species of different masses. This does not absolutely ascertain that the peak is homogeneous since isomers of the same compound will have the same $[M + H]$ and is indistinguishable from the parent compound. Also, the impurity that may be coeluting may not have an appreciable ionization efficiency at the particular mobile-phase and mass spectrometric conditions.

An example of where using diode array detection may not be helpful is shown in Figure 8-16. Note that for this reaction mixture (convergent synthesis) the desired product 1 has the same diode array spectra as synthetic precursors 2 and 3. If these two synthetic precursors had coeluted with 1, they would not have been able to be deconvoluted. This stresses the importance of running LC-MS in a parallel to diode array studies during method development.

8.4.3 Solution Stability and Sample Preparation

It should be determined if the drug substance being analyzed is stable in solution (diluent). During initial method development an autosampler tray cooler

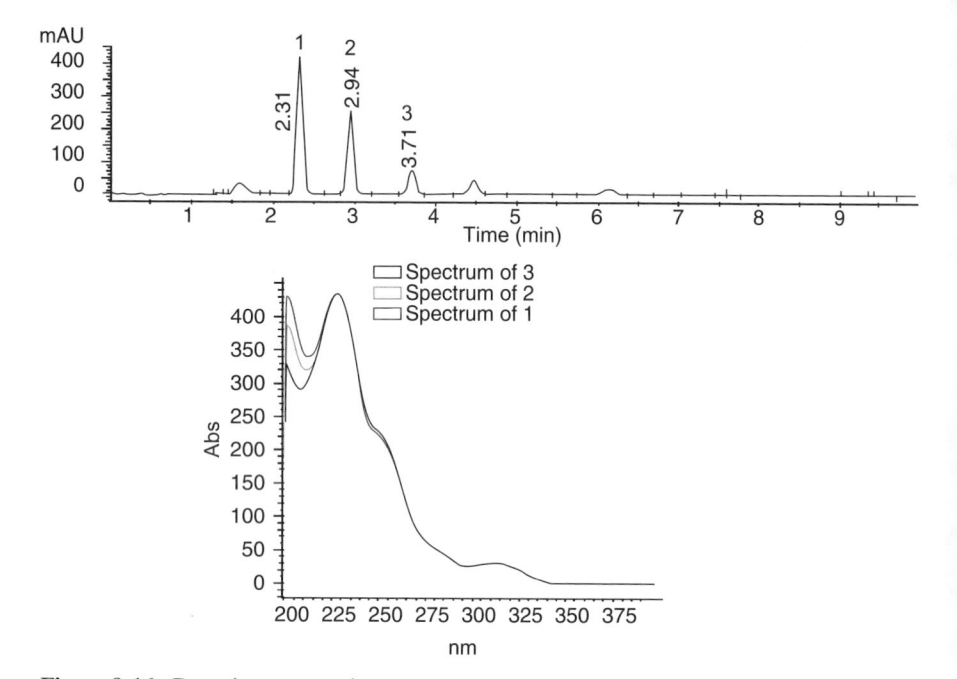

Figure 8-16. Reaction conversion of a convergent synthesis (2 + 3) to 1 and overlay of diode array profiles.

and preparations of the solutions in amber flasks should be performed until it is determined that the active component is stable at room temperature and does not degrade under normal laboratory conditions. Also, since it is not known if dimeric species or more hydrophobic compounds are present in the sample solutions during the initial method development or are formed in stability studies, gradient elution should always be performed with a hold at higher organic conditions (or up to the buffer stability limit).

The reduction of downtime of the instrument (i.e., operations of pump components, injectors, and detectors) can be controlled to some degree if sample solutions are filtered and/or centrifuged; the use of a 0.2- or 0.45-µm-pore-size filter is generally recommend for removal of particulates [15]. Filtration as a preventive maintenance tool for HPLC analyses is well-documented in the literature [16–18].

Sample preparation is a critical step of method development that the analyst must investigate. For example, the analyst should investigate if centrifugation (determining the optimal rpm and time) shaking and/or filtration of the sample is needed, especially if there are insoluble components in the sample. This is usually more prevalent with excipient/DS mixtures and with slurry solutions obtained during the synthesis steps of the API. Syringe filters are routinely used to remove particulate contamination/insoluble components from samples prior to chromatographic analysis.

The objective is to demonstrate that the sample filtration does not affect the analytical result due to adsorption and/or extraction of leachables. A word of caution here is that filter studies should be performed to ensure that no adsorption of the compound on the filter is observed. This is particularly the case with protein and peptide samples. Note that for proteins and peptides the impact of centrifugation (speed and time) must be investigated because this may lead to increased aggregate formation. Also, for protein and peptides the initial concentration of the sample could also have an impact on the concentration gradient of the sample in the centrifuge tube, and the concentration of the top, middle, and bottom portions should be assessed.

The effectiveness of the syringe filters is largely determined by their ability to remove contaminants/insoluble components without leaching undesirable artifacts (i.e., extractables) into the filtrate. Extractables are often the result of inappropriate material construction and improper handling of the device during the manufacturing process. Particular attention should be paid to potential extractables from the membrane and housing material. The sample preparation procedure should be adequately described in the respective analytical method that is applied to a real in-process sample or a dosage form for subsequent HPLC analysis. The analytical procedure must specify the manufacturer, type of filter, and pore size of the filter media. Also, it must be known if the particular filter type is compatible with the type of analyte, organic solvents, and pH of the solution to be filtered.

The following procedure may be used to determine if there is any absorption on the filter. A stock solution is prepared at the target concentration. One aliquot of the stock solution is centrifuged, and other aliquots from the centrifuged stock solution are filtered through the desired filters (pre-wet with 5 mL of diluent) and the results compared. If any additional peaks are observed in the filtered samples, then the diluent must be filtered to determine if a leachable component is coming from the syringe filter housing/filter. In Figure 8-17 a solid oral dosage form was prepared at 1 mg/mL concentration. The initial stock solution was centrifuged (no filter) and two additional samples from the centrifuged solution were filtered with a nylon filter and a cellulose filter. The area counts (Table 8-4) of all three solutions were compared, and it was shown that significant absorption was observed on the nylon 66 filter. Further optimization of the sample preparation would include removing the centrifugation step and just filtering the supernatant (solution above the undissolved excipients) with the cellulose acetate filter.

Another example includes the recovery (mass %) of API and degradation products of API from two 100-mg tablet (5 tablets) sample solution clarified by filtration and clarified by means of centrifugation. The data in the Table 8-5 demonstrates that the two methods of sample clarification are equivalent and that the filtration procedure (0.2-μm Nylon filter, with 5 mL pre-wet) is

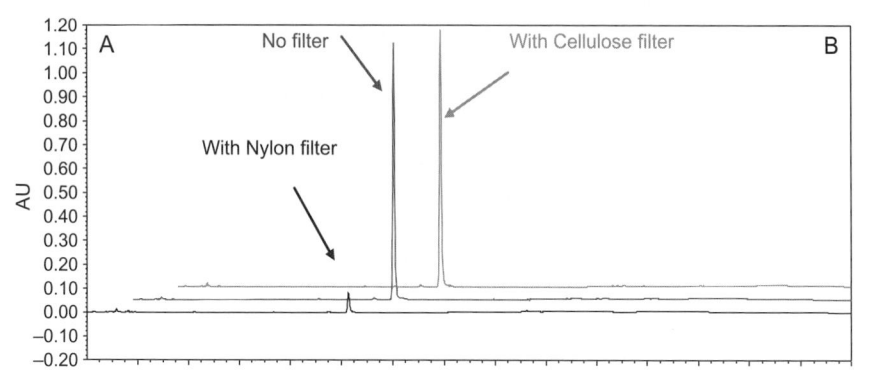

Figure 8-17. Comparison of filtered (nylon filter versus cellulose filter) versus no filter (centrifuged). Column: Luna C18 (2). Mobile phase: (A) 10 mM ammonium bicarbonate, pH 7.5; (B) MeCN, linear gradient from 0 to 15 minutes, 20—70% of B. Sample concentration: 1 mg/mL.

TABLE 8-4. Area Counts for Centrifuged/Filtered Solutions

Type of Sample Preparation	Area Counts
30-mm Nylon filter, 0.2 μm	484,155
No filter (supernatant solution)	5,612,755
13-mm Cellulose acetate filter, 0.45 μm	5,633,064

TABLE 8-5. Filter Evaluation Results for API Assay-Related Substance Samples

Sample Name	API (%)	Impurity 1 (%)	Impurity 2 (%)	Impurity 3 (%)	Impurity 4 (%)
Filtered samples					
1	99.0	0.079	0.082	0.025	0.042
2	99.4	0.079	0.082	0.025	0.042
3	99.5	0.079	0.084	0.025	0.043
Mean	99.3	0.079	0.083	0.025	0.042
% S_{rel}	0.3%	0.0%	1.4%	0.0%	1.4%
Centrifuged samples					
1	99.1	0.080	0.080	0.025	0.042
2	98.9	0.081	0.080	0.025	0.042
3	100.0	0.078	0.078	0.025	0.042
Mean	99.3	0.080	0.079	0.025	0.042
% S_{rel}	0.6%	1.9%	1.5%	0.0%	0.0%

adequate and does not cause any specific absorption of the active and/or impurities.

Other considerations for sample preparation include incorporation of methanol in the sample preparation scheme, especially if a second dilution is used (check for sample reactivity). The impact on peak shape (diluent/mobile phase mismatch for components with $k < 2$) should also be considered. Sample preparation usually constitutes approximately 70% of solvent usage, and incorporating methanol for routine sample preparation can lead to reduction in solvent costs.

8.4.4 Choice of Stationary Phase

Ideally for a reversed-phase separations, the retention factors (k) for all components in a sample should lie between 1 and 10 to achieve separation in a reasonable time. For a given stationary phase the k of a particular component can be controlled by changing the solvent composition of the mobile phase. However, the impact of eluent composition will depend on the type of stationary phase and the nature of the components in the mixture. In reversed-phase HPLC the most common solvent mixtures are: water and acetonitrile, water and methanol, and water and THF. The elution strength increases as the organic portion of the modifier increases. Thus, to optimize a chromatographic separation, the concentration of the organic modifier is adjusted so that the k of the components in the sample are in the range of 1 to 10. However, sometimes due to the hydrophobic nature of the compound, even high concentrations of organic modifier will not allow elution of all components in a single run and the chromatographer can try one or a combination of the following approaches: (1) Use a stronger modifier; (2) apply a steeper gradient; (3) use a less hydrophobic stationary phase. Detailed discussion of the reversed-phase separation principles and separation optimization is given in Chapter 4.

The type of column chosen for a particular separation depends on the compound and the aim of analysis. Pharmaceutical companies may have a preferred list of columns that have good demonstrated performance in regard to pH/temperature stability. These columns that have been selected by a specific laboratory are known to be stable within predefined pH and temperature regions in which method development/column screening are employed. A good understanding of the chemical stability of the stationary phases is needed, and some examples are shown in Section 8.10.

Screening columns from each of the following various column classes should provide for the desired chromatographic selectivity, even for the most challenging separations: (1–3) C8 or C18 stable at pH < 2, pH 2–8, and pH > 8–11; (4) phenyl; (5) pentafluorphenyl; (6) polar embedded and stationary phases that could be run in 100% aqueous. A certain number of columns in each of the six column classes and subclasses could be chosen as standard columns that the chromatographers choose as a first choice for performing method development. These standard columns could be chosen based on some set of internal criteria (ie., chromatographic selectivity for a set of compounds, bonded phase stability, and lot-to-lot reproducibility). The criteria for selection may include that the column is stable for a certain number of column volumes (efficiency, tailing factor, retention time criteria for predefined probe analytes) at the recommended max and min pH at a particular maximum temperature. By tracking the column usage (number of column volumes run at a particular pH/temperature), this will reduce the number of system suitability failures and decrease the cost of the consumables for a particular laboratory. Moreover, this information should be shared among the analytical chemists in the different line functions (DMPK, Drug Substance, Drug Product, Preformulation, TechOps, or PharmOps) to ensure that these columns are readily available and that the practical experience can be shared for the selected columns within a particular company. Also, it is generally recommended not to use the same column for multiple projects, especially when performing release and stability testing.

For more hydrophobic compounds, a stationary phase that has a lower surface area should be used. For very polar compounds that cannot be retained on traditional C18 phases, less hydrophobic columns such as C4 and polar embedded stationary phases could be used. However, all this is also dependent on the pH of the analysis since some columns are not stable at low pH (<2) and higher pH (>7) for extended periods of time. This should be taken into careful consideration when defining a column(s) during the development of a method.

Moreover, the effect of pH on a particular compound's retention needs to be determined first before exploring the retentivity and selectivity of different columns. The strategy and choice of the optimal pH for analysis was discussed in Chapter 4 and is further reinforced in the case studies within this chapter. After the optimal pH is chosen for the separation and the gradient has been optimized on a particular column and the optimal selectivity still has not been achieved between critical pairs, then a column screening can be performed. For method column screening, generally columns with 10-cm or 5-cm × 3.0-mm i.d. could be used that are packed with 3-μm particles. Implementation of a column switcher that can use six different types of stationary phases such as two types of C18 from different vendors, phenyl, two polar embedded, and pentafluorphenyl is suggested.

In Figure 8-18, a mixture of acids and bases was analyzed on three types of columns: phenyl, polar embedded, and C18 column. Significant differences in selectivity were obtained. The separation could be further optimized by modifying the gradient slope and employing off-line method development tools such as Drylab for further optimization and resolution of the critical pairs.

Moreover, once a particular column or columns that have provided the best selectivity are chosen, an automated method optimization may be performed. This would include employment of an integrated HPLC method development system such as AMDS/Drylab such that the gradient slope/temperature

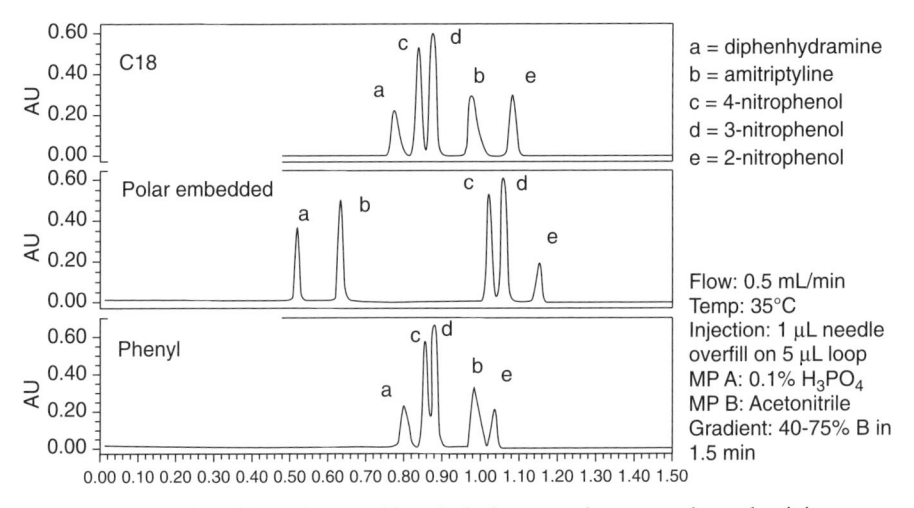

Figure 8-18. Effect of type of bonded phase on the separation selectivity.

can be further optimized on multiple columns that had shown the best selectivity.

8.4.5 Mobile-Phase Considerations

8.4.5.1 Choice of pH. If analytes are ionizable, the proper mobile-phase pH must be chosen based on the analyte pK_a so the target analyte is in one predominate ionization state ionized or neutral. If possible, method development at both of these defined mobile-phase pH values is encouraged to maximize the potential gains that may be obtained in regard to selectivity (for the neutral and ionized forms of the target analyte and related substances).

Alteration of the mobile-phase pH is one of the greatest tools in the "chromatographers toolbox" allowing simultaneous change in retention and selectivity between critical pair of components. Analytes may be analyzed in their ionic form or neutral form. This may be dependent on the type of analysis that is required. If fast analysis is required, then analysis of the component in its ionized form may be acceptable if the desired resolution from the matrix components is achieved. However, if adequate resolution of the active from its process-related impurities/degradation products/excipients are not obtained, then mobile-phase additives may be added to the mobile phase or the mobile-phase pH may be adjusted so the analyte may be analyzed in its neutral form in order to potentially enhance the selectivity/resolution between critical pairs of components. Increasing flow rate, increasing temperature (up to column stability limit at a particular pH), increasing the concentration of the organic eluent, and using shorter columns with narrower dimensions may be used to obtain more desirable run times. However, speed does not come without a price, and the influence of the aforementioned parameters on the resolution of the critical pairs in a mixture/sample needs to be evaluated.

8.4.5.2 Buffers. In order to develop rugged HPLC methods, knowledge of choosing the right buffer is very important. Buffers that are selected should have a good buffering capacity at the specified mobile-phase pH. Also, the concentration of the buffer should be at least 10 mM to provide the needed ionic strength to suppress any undesired analyte solvation effects that may lead to poor peak shapes. Methods that specify a phosphate buffer in the pH range of 4 to 6, or an acetate buffer in the range of 6 to 7, are, unfortunately, not good buffers. These buffers are not just useless in these pH ranges, they complicate the preparation of mobile phase unnecessarily and give the analyst a false sense of controlling the reproducibility of the separation.

Optimum buffering capacity occurs at a pH equal to the pK_a of the buffer. In general, you can expect most buffers to provide adequate buffering capacity for controlling mobile-phase pH only within ±1 unit of their respective pK_a values. Beyond that, buffering capacity may be inadequate.

Also, buffers are great media for growing bacteria. It is recommended to have at least 10 v/v% of organic in the aqueous phase to prevent bacterial growth.

Table 4-3 in Chapter 4 lists some commonly used buffers for reversed-phase HPLC. In this table the buffers and their respective pK_a values, and UV cutoffs are listed. Since it is becoming more common to find HPLC interfaced to mass spectrometers, volatile buffers for LC/MS applications are also indicated.

8.4.5.3 General Considerations for Buffers. The type of buffer that is chosen will depend on the wavelength of the method and the concentration of organic in the mobile phase. A judicious choice of type and concentration of buffer must be made to ensure mobile-phase compatibility.

- Phosphate is more soluble in methanol/water than in acetonitrile/water or THF/water.
- Some salt buffers are hygroscopic. If an analyst makes a 20 mM buffer and the original buffer salt contains 20 w/w% water, then the buffer concentration would be 16 mM. This may lead to changes in the chromatography (increased tailing of basic compounds, and possibly selectivity differences).
- Ammonium salts are generally more soluble in organic/water mobile phases than potassium salts, and potassium salts are more soluble than sodium salts.
- TFA can degrade with time, is volatile, absorbs at low UV wavelengths, and is not a buffer at pH > 1.5.

- Citrate buffers can attack stainless steel. When using these buffers, be sure to flush them out of the system as soon as the analysis is completed, but this is a recommendation for any buffer system.
- Microbial growth can quickly occur in buffered mobile phases that contain little or no organic modifier. This growth will accumulate on column inlets and can damage chromatographic performance.
- At pH greater than 7, phosphate buffers accelerates the dissolution of silica and severely shortens the lifetime of silica-based HPLC columns. If possible, organic buffers should be used at pH greater than 7.
- Ammonium bicarbonate buffers usually are prone to pH changes and are usually stable for only 24 to 48 hours. The pH of this mobile phase tends to become more basic due to the release of carbon dioxide.
- After buffers are prepared, they should be filtered through a 0.2-μm filter.
- A "test tube test" should be conducted to determine if the buffer at the concentration it is prepared will precipitate in the column/system when it is exposed to the highest organic concentration in the gradient. The temperature should also be considered as well. Buffers generally will have a higher solubility at higher temperatures. The test tube test can be performed by preparing the mobile phase in a 10-mL test tube and then putting the test tube in the refrigerator and/or water bath (to mimic higher temperatures) to determine if any precipitation occurs. The results of "test tube tests" of phosphate buffers (10 and 25 mM) in various acetonitrile/water compositions at room temperature and 5°C are given in Table 8-6.
- Mobile phases should be degassed if an on-line degasser is not available on the HPLC system.

Also, the purity of the buffer should be taken into consideration. Small amounts of trace impurities can absorb in the UV wavelength of interest and cause a high background absorbance, thus suppressing the limit of detection for a particular analysis. One such case is with N-methyl pyrrolidine, although it does not absorb above 210 nm; sometimes the use of this reagent is not feasible unless the wavelength of detection is greater than 225 or 254 nm due to the presence impurities with chromophores that absorb in that region. If mobile phase A had 50 mM N-methyl pyrrolidine that was contaminated with some low-level impurities and mobile phase B had MeCN and a linear gradient was run from 5% MeCN to 95% MeCN and the wavelength that was being monitored was 210 nm, a decrease in the baseline would be observed due to a dilution effect of the buffer impurity background absorption. The same behavior is usually observed when TFA, acetic acid, and/or formic acid are used in the aqueous portion of the mobile phases and a wavelength of <220 nm is used. Note this is also dependent on the concentration of the acidic/basic modifier employed. It is generally recommended to use the same concentration of TFA or other acid (UV absorbing) in both the aqueous and organic portions of the

TABLE 8-6. Solubility of Some Typical Phosphate Buffers at Room Temperature and 5°C

Buffer Salt	Buffer Concentration (mM)	Initial pH Value	Adjusted pH Value	Appearance of Solution with Following Ratio of ACN/Buffer								
				70:30		75:25		80:20		85:15	90:10	95:5
				RT^a	5°C	RT^a	5°C	RT^a	5°C	RT^a	RT^a	RT^a
Ammoniumphosphate monobasic	25	4.5	3	Soluble	Soluble	Soluble	Soluble	Soluble	Soluble	Precip.	Precip.	Precip.
Di-ammonium hydrogenphosphate	25	8	7	Soluble	Soluble	Soluble	Soluble	Precip.	Soluble	Precip.	Precip.	Precip.
Di-ammonium hydrogenphosphate	10	8	7	Soluble	Soluble	Soluble	Soluble	Soluble	Soluble	Soluble	Soluble	Precip.
Di-potassium hydrogenphosphate	10	9.1	7	Soluble	Soluble	Soluble	Soluble	Soluble	Soluble	Precip.	Precip.	Precip.

^{a}RT, room temperature.
Source: Courtesy of Alfons Roth, Novartis Pharmaceuticals.

mobile phase in order to suppress any type of baseline shift during a gradient run. Therefore, allowing the concentration of the buffer additive to be kept constant throughout the entire gradient run. If the analyte has another wavelength maxima that is greater than 220 nm and has a high molar extinction coefficient at that alternate wavelength, then it could be used for monitoring the analysis. Depending on the molar extinction coefficient, at this higher wavelength (even though not the primary maxima) a higher S/N ratio may be obtained for the LOQ solution since the mobile-phase additives will not absorb in this higher-wavelength UV region. Therefore there is less interference from the background buffer absorbance at this particular wavelength.

8.4.5.4   *Concentration of Buffers.*   A buffer concentration in the range of 10 to 50 mM is adequate for most reversed-phase applications. However, sometimes the concentration of the buffer does lead to improvement of peak shape, presumably because the cation of the buffer suppresses silanophilic interactions of the protonated base with accessible ionized residual silanols.

This concentration should also be low enough to avoid problems with precipitation when significant amounts of organic modifiers are used in the mobile phase and, in the case of phosphate buffers, low enough to minimize the abrasive effect on pump seals. It is seldom advisable to use a buffer concentration more than 100 mM and less than 10 mM. It is mainly ionic strength and not a buffer capacity that is governing the peak distortion in less than 10 mM concentration range—at these conditions the solvation equilibrium is slow, causing peak distortion.

There are cases where solute retention in reversed-phase HPLC is affected by buffer concentration. These cases are usually confined to situations where there are ion exchange interactions taking place between basic solutes and acidic silanols on the surface of silica stationary phase supports. Sometimes increasing the ionic strength and/or maintaining constant ionic strength throughout the gradient separation does help with the improvement of the peak shape, and this can be associated with a suppression of ion-exchange processes of the analyte and the accessible ionized residual silanols. The buffer cation, sodium, ammonium, or potassium competes with the analyte for these undesired sites. Increasing the cation concentration can sometimes lead to an improvement of the peak asymmetry (reduction of tailing) for protonated basic compounds.

Generally, this will occur when separating protonated basic compounds using reversed-phase columns packed with stationary phases that have significant silanol activity, and it is most often observed when using mobile phases with pH values greater than 3. Increasing the concentration of the mobile phase buffer, and thereby increasing the ionic strength of the mobile phase, will sometimes suppress this ion-exchange interaction and reduce this "secondary retention" effect.

8.4.5.5   *Practical Aspects of Preparing a Buffered Mobile Phase.*   The following steps are suggested for preparing a buffered mobile phase:

- Define the appropriate $_w^w pH$ for the separation and then select an appropriate buffer. Refer to Table 4-5 to determine the appropriate buffer for your application.
- Prepare an aqueous buffer solution of the desired concentration and $_w^w pH$.
- Measure the pH of the solution and adjust, if necessary, to the desired $_w^w pH$ with dilute acid or dilute basic solution. When adjusting the $_w^w pH$ of a buffer solution, make sure to wait until the solution reaches equilibrium after adding additional acid or base before measuring the pH.

If you are performing an isocratic separation, combine the aqueous buffer solution with the appropriate organic modifier (e.g., methanol or acetonitrile) to produce the desired mobile phase and let the solution come to equilibrium. You may choose to measure and adjust the $_w^s pH$ of the final mobile phase (hydro-organic mixture) rather than the $_w^w pH$ of the aqueous buffer solution. However, measuring the $_w^s pH$ of an aqueous/organic solution is more tedious (especially with MeCN >50 v/v%, long equilibration times) compared to measuring the pH of a purely aqueous solution, so it is highly recommended that you measure and adjust the $_w^w pH$ before adding the organic modifier.

8.4.5.6   Choice of Organic Modifier. Selection of the organic modifier type could be viewed as relatively simple: The usual choice is between acetonitrile and methanol (rarely THF). In Chapters 2 and 4 the principal difference in the behavior of methanol and acetonitrile in the column is discussed. In short, methanol shows more predictable influence on the analyte elution, and the logarithm of the retention factor shows linear variation with the concentration of methanol in the mobile phase. Often for the effective separation of complex mixtures of related compounds, this ideal behavior is not a benefit and greater effect of the type and organic concentration on the separation efficiency is required. Acetonitrile as an organic modifier may offer these variations due to the introduction of a dual retention mechanism. The dual retention mechanism was discussed in Chapter 2.

The viscosity of water/organic mixtures should be considered as an additional parameter in the selection of organic modifier. Acetonitrile/water mixtures show roughly 2.5 times lower viscosity than equivalent methanol/water eluents; this means that one can use 2.5 faster flow rates with acetonitrile as organic modifier and develop faster separation methods.

Acetonitrile is not ionogenic and is not a hydrogen bonding agent, but its four π-electrons offer strong dispersive interactions that should also be taken into account in the solvent selection. Changing the type of organic eluent may have an effect on the resulting selectivity of the two species in a mixture. There is no definite way to predict if changing the type and concentration of the organic eluent will impart a difference in selectivity of the closely eluting species. However, once the method pH and column are chosen, the organic portion of the mobile phase could be changed to another solvent to probe if any changes in selectivity could be obtained. The types of solvent that are recommended are pure acetonitrile, pure methanol, and a mixture of acetonitrile/methanol. Sometimes, a small addition of THF (up to 5 v/v%) or isopropanol to either acetonitrile or methanol may lead to changes in the selectivity. The adsorption of the organic eluent component on the stationary phase and the interactions of the eluent with the analyte molecules play a significant role in determining the resultant selectivity of the separation. However, changing the type of organic employed may also lead to increased selectivity, but other considerations such as method sensitivity should be explored. Methanol and IPA both absorb below 220 nm, making the determination of low-level impurities difficult especially if they have a low molar extinction coefficient. Also, the use of more viscous solvents precludes the use of higher flow rates due to the higher column backpressure. The stability of the mobile phase should also be considered. THF is known to form peroxides. However, most analytical chemists use THF that is not inhibited, since the inhibitor (BHT or cresol) absorbs <254 nm and leads to high background absorbance. This could be a safety concern. Most laboratories monitor the shelf life of solvents that can form peroxides, and generally it is limited to a 6-month shelf life after it has been opened.

8.4.6 Gradient Separations

8.4.6.1 Isocratic Versus Gradient Separations. Traditionally, isocratic separations are deemed as more reproducible than gradient separations. Indeed, constant eluent composition means equilibrium conditions in the column and the actual velocity of compounds moving through the column are constant; analyte–eluent and analyte–stationary-phase interactions are also constant throughout the whole run. This makes isocratic separations more predictable, although the separation power (the number of compounds which could be resolved) is not very high. The peak capacity is low; and the longer the component is retained on the column, the wider is the resultant peak.

Gradient separation significantly increases the separation power of a system mainly because of the dramatic increase of the apparent efficiency (decrease of the peak width). The condition where the tail of a chromatographic zone is always under the influence of a stronger eluent composition leads to the decrease of the peak width. Peak width varies depending on the rate of the eluent composition variation (gradient slope). Using a shallow gradient (slow composition variation), there is still some increase of the peak width with an increase of retention, while using a very sharp gradient the peak width of late eluting analytes could actually decrease. This effect becomes smaller with the decrease of the sample injection volume and optimization of the instrumental setup to minimize extra-column volume.

On the other hand, the steeper the gradient, the further the separation conditions are from the equilibrium, and reproducibility of these methods could be challenging. Most of the modern HPLC equipment can reproduce gradient profile with high accuracy, but permanent variation of the eluent composition and usually pH also stresses the column, which may decrease its lifetime. Some recently developed columns can withstand repetitive stress for 10,000 to 20,000 column volumes. Further description of the development of fast HPLC methods using a gradient mode is found in Chapter 17.

8.4.6.2 Changing Gradient Slope. Gradient elution is usually employed with complex multicomponent samples since it may not be possible to get all components eluted between k 1 and 10 using a single solvent strength under isocratic conditions. This leads to the "general elution problem" where no one set of conditions is effective in eluting all components from a column in a reasonable time period while still attaining resolution of each component. This necessitates the implementation of a gradient. Employing gradients shallow or steep allows for obtaining differences in the chromatographic selectivity. This would be attributed to the different slopes of the retention versus organic composition for each analyte in the mixture. Therefore, changing the gradient slope or steepness of the gradient is an important variable that should be considered in reversed-phase method development for controlling the retention of components and adjusting the selectivity for components. Take, for example, the two gradient runs in Figure 8-19, where a shallow and steep gradient were

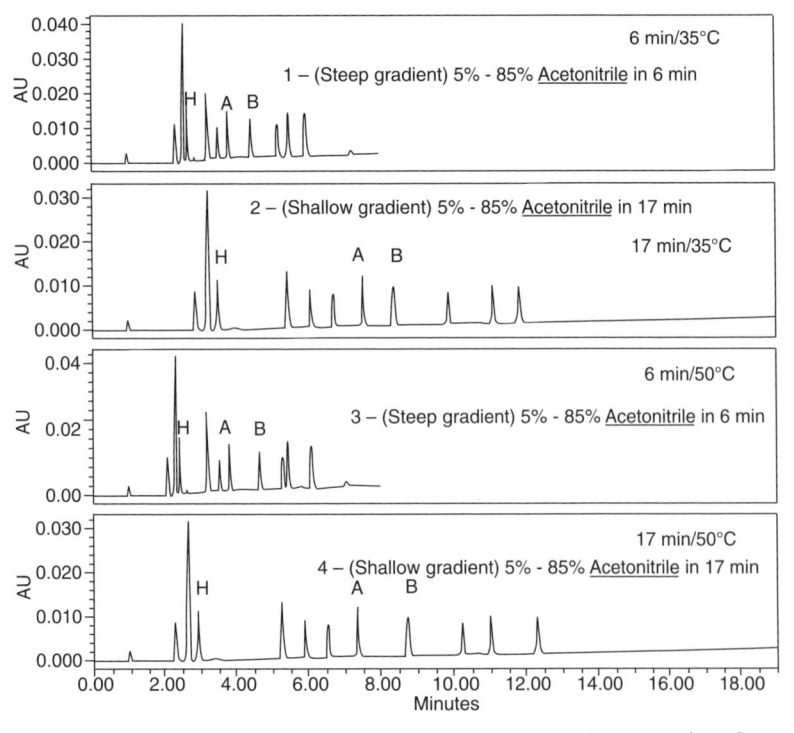

Figure 8-19. Effect of gradient slope on the chromatographic separation. Symmetry C18, 4.6 × 100 mm, 3.5 μm.

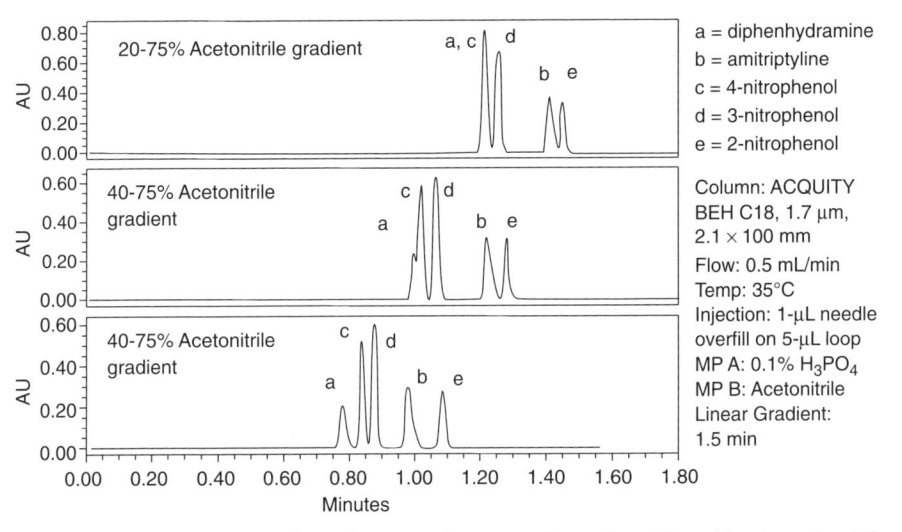

Figure 8-20. Effect of gradient slope on the separation selectivity of basic and acidic analytes on Waters BEH C18 column, sub-2-μm particles.

run at both 35°C and 50°C. Increase of separation selectivity is observed for the impurities eluting between peak H and peak A using the shallow gradient for both temperatures (35°C and 50°C). Using the steep gradient at both temperatures, a coelution of two of the impurities eluting between H and A was obtained. In Figure 8-20, the effect of three gradients (from shallow to steep) on the retention of basic and acidic analytes is shown. In this case the steeper gradient provided for greater separation selectivity than the shallow gradient, even though all components eluted faster. This demonstrates that the resolution, the selectivity, and the peak capacity for a particular separation needs to be challenged during the gradient optimization process.

Once the desired separation selectivity is obtained and if a faster separation is required, the gradient slope should be kept constant so the selectivity would not be affected. This can be accomplished by a proportional change of gradient time and flow rate. In order to decrease analysis time by a factor of $2X$ while keeping retention factor constant, the flow rate should be doubled while the gradient time is reduced by $2X$. The proportional change in flow rate and gradient time should not affect the selectivity and result in the same retention factors. However, increasing flow rate is limited by the pressure limitation of the HPLC equipment. For more information on gradient separation and developing fast HPLC methods, see Chapter 17 and *Practical HPLC Method Development* from Synder et al. [19].

8.4.6.3 Linear Versus Multi Step Gradients. If the gradient separation mode is selected, it is generally recommended to use a simple linear gradient because of better reproducibility, less column stress, and better resolution of

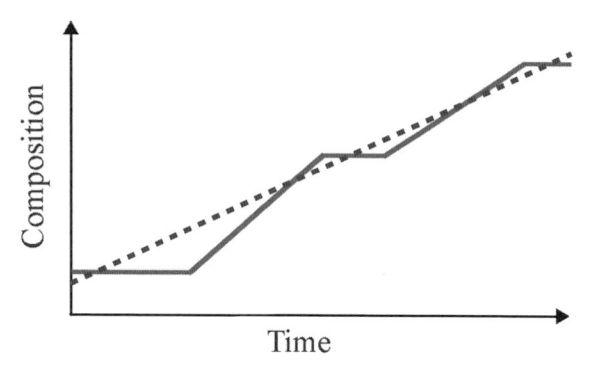

Figure 8-21. Equivalent linear and multistep gradient profiles.

critical pairs. If any isocratic step is included in the run, the peak width of all analytes is broadened during this step and the whole method loses its main advantage, namely, high apparent efficiency (which results in higher resolution). The substitution of multistep gradient shown in Figure 8-21 with equivalent linear gradient (dashed line) will definitely affect the selectivity between analytes, but it will also significantly decrease the peak width (especially for the late eluting analytes), thus increasing the resolution. This resolution increase obtained with a linear gradient often outweights the small variation gained in selectivity from multi step gradients.

8.4.6.4 Equilibration. When a gradient method is used, the column must be allowed to equilibrate at the starting mobile-phase conditions prior to the next sample injection and the start of the next gradient run. Generally, if a gradient is used from 5% organic to 95% organic on bonded phases that are not polar embedded or that should not be run at 100% aqueous conditions, at least three to five column volumes should pass through the column before starting the next injection. The gradient method can be programmed to immediately change (after completion of the separation) from 95% organic to 5% organic in <0.1 min (this depends on instrument limitations). This sharp change of the eluent composition will actually help to clean the column (see detailed discussion in Chapter 2).

How is a column volume (V_{column}) calculated? A good rule of thumb is that the volume of the column can be approximated from the volume of a cylinder. Seventy percent of the column volume is approximately the volume inside the porous space and around the particles:

$$V_{column} = 0.7\pi R^2 L \qquad (8\text{-}1)$$

The time needed to equilibrate the column is determined by the flow rate. The number of column volumes (three to five column volumes) can be divided by

TABLE 8-7. Equilibration Times for Columns of Different Dimensions

Length (cm)	Diameter (cm)	Column Volume (CV) (mL)	3 CV (mL)	5 CV (mL)
15	0.46	1.74	5.2	8.7
15	0.3	0.74	2.2	3.7
10	0.46	1.16	3.5	5.8
10	0.46	1.16	3.5	5.8
10	0.21	0.24	0.7	1.2
5	0.46	0.58	1.7	2.9
5	0.3	0.25	0.7	1.2

the flow rate (mL/min) to determine the time (in minutes) needed. Therefore, the lower the flow rate, the longer the equilibration time.

Some typical equilibration times for various column dimensions are shown in Table 8-7; however, these should only be used as a guide. If complete equilibration is not achieved, early eluting components may show differences in retention from run to run. An experiment could be run such that three different methods could be run with different equilibration times. For example, if a 15-cm × 4.6-mm i.d. column and a flow rate of 1 mL/min was used, then the equilibration times for the three methods would be 5 min (3 CV), 9 min (5 CV), and 11 min (6 CV) equilibration times, respectively. If the retention of the early eluting components are consistent (less than 1% variation in retention time) in all three methods, then the lowest equilibration time could be used. However, if the early eluting components show greater variation in their retention time with the 5-min equilibration time compared to the methods with the 9- and 11-min equilibration time, then an equilibration time of greater than 5 min is warranted. Optimization of the optimal equilibration time is required for reproducible methods.

Other considerations include differences in dwell volumes from the different HPLC systems. The dwell volume should be determined for all the systems in the laboratory and based on these determinations, this should be factored into the calculation of the equilibration time. For example, if the maximum dwell volume of all the systems in a particular laboratory to which the method is transferred to is 2 mL and you are running on an instrument at 1 mL/min that has a dwell volume of 1 mL, then you should add an extra minute of equilibration time. This becomes extremely important during method transfers where the instruments in the receiving laboratory may be different.

8.5   METHOD DEVELOPMENT APPROACHES

8.5.1   If Analyte Structure Is Known

Determine if analytes are acidic, basic, or neutral. This will allow the chromatographer to choose a pH such that the analyte is being analyzed predominately in one ionization state. Use the rules for pH shift and pK_a shift to ensure that the analyte is one predominant ionization state and choose the appropriate mobile-phase pH (see Chapter 4).

Some general guidelines are as follows: If your target analyte is an acidic analyte ($pK_a \geq 3$), use a 0.2 v/v% phosphoric acid mobile phase. If target analyte is a basic analyte ($pK_a \geq 7-9$), use an ammonium acetate buffer (pH 5.8) to analyze in its ionized form or use a 25 mM ammonium hydroxide buffer (pH 10.5) or 25 mM N-methyl pyrrolidine buffer (pH 10.5) to analyze in its neutral form. Use a 10-cm × 3.0-mm column packed with 3-μm particles and intermediate polarity phase such as a C8 column that is stable for the pH at which you may be running the probe separation. Run in the gradient mode using an acidic buffer or a basic buffer from 5% to 95% of organic component, or up to the buffer solubility limit over 10 min, and use an isocratic hold for 10 min to ensure elution of hydrophobic components. Use a flow rate of 0.8–1 mL/min flow rate and 40°C temperature. Injection volume should be on the order of 5–20 μL and the concentration of the analyte should be 0.5–1 mg/mL. This corresponds to approximately 5–20 μg injected on column. On the other hand, for neutral analytes higher analyte loading such as 50–100 μg maybe used since nonideal interactions with the stationary phase are less prevalent. Note that for ionizable compounds, especially basic compounds when analyzed in their ionized state, higher mass on column may lead to mass overload of "hot spots" on the bonded phase and poor peak efficiencies may be observed. Try not to load more than 10 μg on column for basic compounds. Usually greater loading capacity is obtained for basic compounds when they are analyzed in their neutral state. Note that for columns with larger inner diameters such as 4.6 mm, larger sample loads may be acceptable.

Once the probe gradient is run, check the diode array purity; and if LC-MS is available, run as well to check for peak homogeneity. If you have any known precursors or impurities, run them as well to ensure resolution from the main component and to make sure they are adequately retained. The main analyte should elute between k 2–5. If the main component elutes at $k = 2-5$ and is spectrally pure and the impurities all elute k > 1, the method is complete. If the retention factor of the impurities is below 1, then an isocratic hold at the initial organic composition should be implemented until the minor component (impurity) elutes k > 1 and then a linear gradient can be implemented. The method could be further optimized by increasing the flow rate as long as the backpressure limitation of the system has not been reached. A general rule of thumb is that the backpressure should not exceed 85% of the maximum back-pressure for a particular HPLC system.

If resolution is not achieved between a critical pair, the use of a shallower gradient can be investigated. If that does not increase the resolution, then a longer column (15-cm column, packed with 3-μm particles of the same stationary phase type) should be used with a reduced flow rate of 0.7 mL/min (due to backpressure limitations).

If separation is still not achieved, consider using a different organic modifier such as mixture of MeCN and methanol to possibly induce changes in the selectivity. Also, the wavelength of detection must be considered, especially if MeOH is used due to its UV cutoff (absorbs <210 nm). If methanol does give you desired selectivity, then an analyst needs to determine if sufficient response (S/N > 10:1) is obtained at desired LOQ (i.e., 0.05% solution of target), especially if the wavelength for detection is <210 nm.

If changing the organic modifier does not work, consider changing the mobile-phase pH (analyze the molecule in a different ionization state). For example, if a basic compound was originally analyzed under basic conditions (pH $\gg$ pK_a), try to use acidic conditions (pH $\ll$ pK_a) with the acetonitrile in the initial gradient. If that still does not work, then consider using a different stationary phase (phenyl or polar embedded) employing the initial gradient, with initial aqueous mobile phase and acetonitrile organic modifier, and repeat the process that was performed on the original column used for initial method development. The final method optimization may include varying the gradient slope, column temperature, and flow rate.

Note that multiple pH values and columns can be screened in gradient mode at the same time as well. This will increase the efficiency/probability of obtaining the best column/conditions and the best demonstrated chromatographic selectivity. Note that the aqueous phase pH values that would be chosen for these pH/column screening studies should be based on knowledge of the physicochemical properties of the molecule, taking into consideration the mobile-phase pH and analyte pK_a shifts in the hydro-organic media.

8.5.2 If Method Is Being Developed for Separation of Active and Unknown Component

Define the criteria for the method such as the LOQ, maximum run time, wavelength detection, and so on. Look at the structure of the target analyte (estimate pK_a) or use ACD (advanced chemistry development) and determine the best pH to run the method. Try to use shorter columns for gradient scouting experiments (5 cm × 4.6 mm) packed with 3-μm columns or use a high-pressure system (max pressure 15,000 psi) with 10-cm × 2.1-mm, 1.7-μm particles. Use 35–45°C as starting temperature. If pH scouting studies are needed, run a probe linear gradient using 0.2 v/v% phosphoric acid on a short column (5-cm × 4.6-cm column) to determine the isocratic conditions for the pH studies. Run pH studies isocratically to determine the desired pH region to understand the behavior of the impurities in the analyte mixture. The desired pH region of the aqueous phase is the pH region where the retention of the components in the mixture do not significantly change their retention as a function of the pH of the aqueous phase. Track impurities using diode array if possible. Run a linear gradient at a pH within the desired pH region and hold at high organic concentration on 5-cm × 4.6-mm column. If you obtain sufficient resolution, then you are finished. If you need more resolution, then use a 15-cm × 3-mm i.d. column. If resolution is obtained, then you are finished. If desired resolution/selectivity is not obtained, then screen different organic modifiers/different stationary phase types. Note that the separation of the critical pair may be obtained on an alternate stationary phase that offers additional selectivity. In addition to the weak dispersive types of interaction that are available on a C8 or C18 phase, phenyl phases may provide additional interactions such as π–π-type interactions and may assist in providing additional selectivity. If the impurities/active are very polar, the use of polar embedded phases may provide additional selectivity by introduction of a secondary type of interaction such as hydrogen bonding close to the surface in the organic-enriched layer. Alternatively, for basic compounds, different counteranions could be introduced in the mobile phase in order to increase retention of protonated basic amines. These are known as chaotropic reagents and were discussed in Section 4.10. This may lead to increased retention and increased selectivity between critical pairs of components.

Moreover, if the laboratory has an automated method development system, then this could be used to determine the best set of gradient conditions to give the best resolution between the critical pair or pairs on multiple columns. When using an automated method development system such as AMDS (Waters, MA) for gradient optimization, generally two types of organic modifiers are used at two different temperatures employing a steep/shallow gradient on two to six columns. Based on these scouting runs and the users' acceptance criteria for the method, a resolution map is generated by input of the data into Drylab, Chromsword, ACD, or another program. From this resolution map the best conditions are chosen and optimized (change in flow rate, multistep gradient ramps, etc.), and these conditions are run to confirm that the method that was predicted is indeed representative of the actual separation. This is typically called a verification run. AMDS relies on constraints of the DryLab model. Note that Drylab is not suitable for the following types of compounds:

- Chiral compounds
- Achiral isomers or diastereomers
- Inorganic ions
- Carbohydrates
- Proteins and peptides

The DryLab model utilized in Waters AMDS has additional requirements: The number of sample components should not exceed 12; peak area% should be greater than 1%. These requirements are necessary to achieve greater prediction accuracy only. Any discrepancies could be corrected manually in DryLab using the data entry screen by manually entering the retention of the components from the scouting runs (to assign the peaks with a certain number). DryLab has been used for the method development of model drug candidates and their degradation products, by optimization of temperature and gradient slope, and the historical review on the milestones and concepts in the development of DryLab software is given in references 20–23.

8.5.3 Defining System Suitability

System suitability parameters with their respective acceptance criteria should be a requirement for any method. This will provide an added level of confidence that the correct mobile phase, temperature, flow rate, and column were used and will ensure the system performance (pump and detector). This usually includes (at a minimum) a requirement for injection precision, sensitivity, standard accuracy (if for an assay method), and retention time of the target analyte. Sometimes, a resolution requirement is added for a critical pair, along with criteria for efficiency and tailing factor (especially if a known impurity elutes on the tail of the target analyte). This is added to ensure that the column performance is adequate to achieve the desired separation.

System suitability requirements for retention time, efficiency, resolution, and tailing factor are set based on prior method challenging experiments and prior method development experience. This is a dynamic process; and as the user gains more experience with the method, the breadth of the acceptance criteria is further expanded until the method is finally validated for the intended purpose.

Two examples are given for setting system suitability requirements for challenging separations. In the first example, if a separation is to be carried out where the retention of the target analyte may have a greater propensity to vary with slight changes in pH, tighter controls for the pH requirement should be implemented, where the pH of the aqueous phase should be controlled to ±0.05 units. Moreover, some preliminary experiments should be performed using an aqueous mobile-phase pH ±0.2 units from the desired pH to determine if this will have an effect on the critical pairs in the separation and what the desirable retention time window is. This information is useful to define the system suitability criteria for the method. Also, it is recommended to run the separation on different lots of columns to see if there is any lot-to-lot variability. Preferably, running the separation on columns that were made from different batches of base silicas is desirable. Also, obtaining columns from different synthetic bonding batches made on the same batch of silica is also desirable. In the example shown in Figure 8-22 for a drug product that contains two actives, three different columns from three different lots of base silica were used and the pH of the aqueous mobile phase was varied from 5.7 to 6.1, with the target pH being 5.9. Some of the specific system suitability parameters and acceptance criteria that were set included tailing factors (5% peak height), retention time windows for peaks A and B, and sensitivity requirement. Some of the selected system suitability parameters were set to the following:

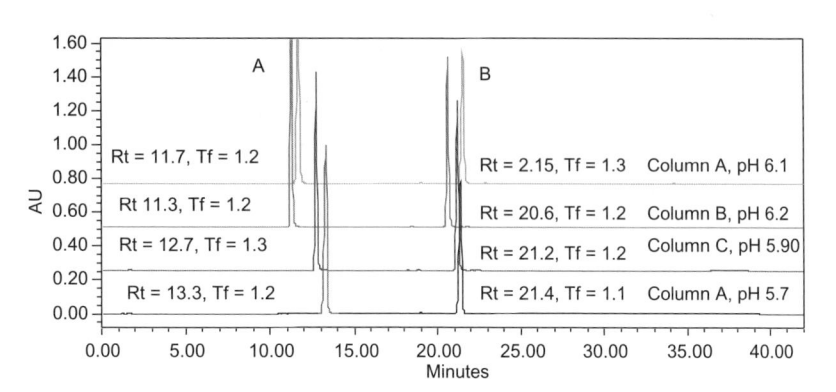

Figure 8-22. Waters XBridge 150- × 3.0-mm, 3.5-μm C18 column. Column temperature 40°C [(A) 90%: 20 mM ammonium phosphate buffer: 10% MeCN, (B) 100% MeCN]. Gradient: 10% A to 85% B over 38 min. Flow: 0.6 mL/min.

System Suitability Parameters

- Tailing factor (5% peak height) for peak B ≤ 1.5
- Tailing factor (5% peak height) for peak A ≤ 1.5
- Rt for peak A must be 12.0 ± 1.3 min
- Rt for peak B must be 21 ± 1.0 min
- The S/N of the LOQ solution (0.05%) for both actives A and B must be ≥10:1

In the second example, if it is known that a potential degradation product can occur and will elute close to the active, a resolution requirement should be set for this critical pair. When trying to set a resolution requirement between critical pairs of impurities, standard samples containing the critical pair should be readily available. However, standard samples may not be available with all critical impurities so the standard may be spiked with authentic impurities. If authentic impurities are not available or are in limited quantity, then the drug substance may be degraded in solution using mild stress conditions to produce a decomposition product or products that can be used to define a resolution requirement for a critical pair. The mild stress conditions should produce decomposition products *in situ* in a fast time scale. In the following example in Figure 8-23, the drug substance was stressed with 3% hydrogen peroxide for 1 hr at 25°C and 80°C to generate impurity A. At 80°C, suitable degradation was obtained to determine the resolution requirement between impurity A and the active B (target analyte). This requirement was set because it was postulated that this drug substance could be readily oxidized. Indeed in solid state stability studies, minor amounts of the impurity A (oxidized impurity) were observed under accelerated conditions (40°C/75% RH, 3 months).

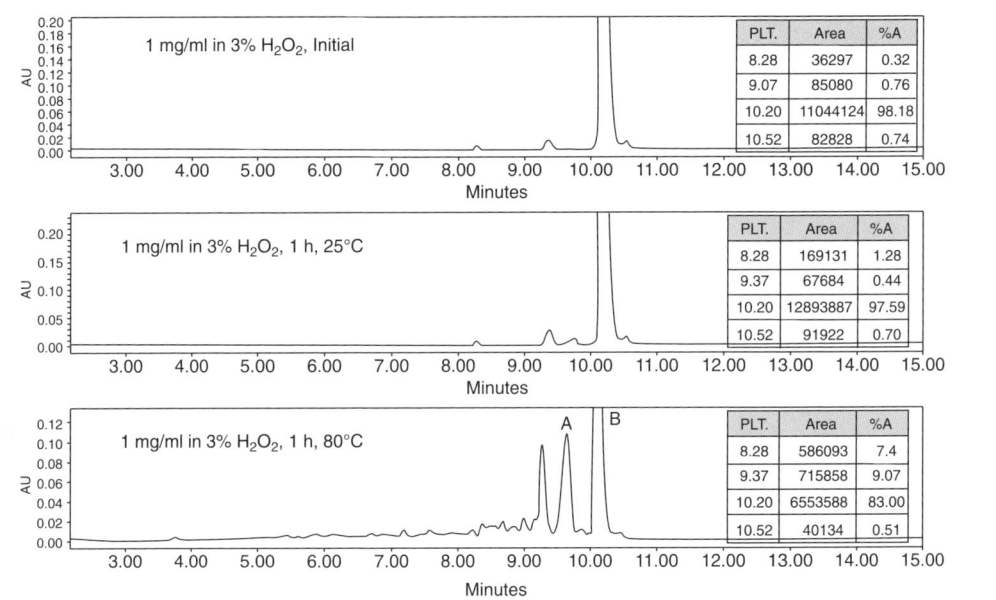

Figure 8-23. *In situ* degradation for generation of system suitability solution.

Another example in regard to *in situ* degradation for generation of a system suitability sample is given in the literature [24].

8.5.4 Case Study 1: Method Development for a Zwitterionic Compound

Method development for the analysis of a zwitterionic drug substance by reversed-phase HPLC was undertaken. The zwitterionic compound A contains an acidic functionality, ($_w^w$pK_a 4.0) and a basic functionality ($_w^w$pK_a 3.0). Both of these pK_a values were determined using ACD Labs (Advanced Chemistry Development, Toronto, Canada) software. Given this information, the chromatographer could apply the pH and pK_a selection rules (including pH and pK_a shifts) outlined in Sections 4.5 and 4.6 in Chapter 4 to select the optimal pH to work at in order to avoid working near the pK_a values of either of the ionizable functionalities. The following case study will illustrate (a) why working at pH values at or near the pK_a values of the API will lead to separations that may not be robust and (b) what influence the pH has on the inherent retention of intermediate compound A and related synthetic by-products. These experiments could be conducted as an exercise to further understand the effect of pH on the retention of the species in the sample of interest since the synthetic by-products may have different ionizable functionalities then the parent compound (intermediate).

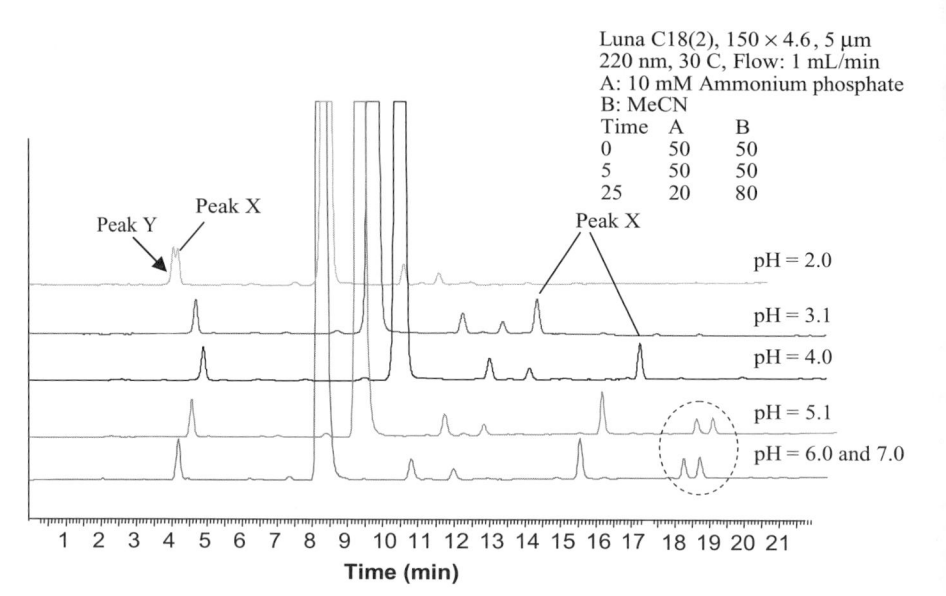

Figure 8-24. $_w^w$pH study on zwitterionic compound A on a Phenomenex Luna C18 (2) column. Method conditions are indicated in the figure.

8.5.4.1 Gradient Screening.

An initial method development was performed using a Phenomenex Luna C18 (2) column with acetonitrile as the organic mobile-phase component, and the aqueous portion was a 10 mM ammonium monohydrogen phosphate buffer adjusted to pH 2 with phosphoric acid. Initially, a linear gradient was used from 60% to 80% MeCN with a hold at 80% MeCN for 10 minutes. An early eluting component was observed close to the void volume using this probe gradient. Also, no peaks were seen to elute during the 80% MeCN isocratic hold. Therefore, a new gradient method (shown in Figure 8-24) with an initial isocratic hold to retain the more polar species and removal of the latter isocratic hold at 80% MeCN was used. The new method employed an isocratic hold at 50% MeCN for 5 min, and then a linear gradient was run from 50% MeCN to 80% MeCN from 5 to 25 minutes. Note that a 150- × 4.6-mm column was used, but a 150- × 3.0-mm could have been easily used with proper adjustment of the flow rate.

8.5.4.2 pH Screening Study.

Once the probe gradient method is selected, a pH study can be conducted. The pH study in gradient mode was carried out using 10 mM ammonium monohydrogenphosphate as a buffer. The $_w^w$pH of the aqueous portion of the eluent was adjusted to 2, 3, 4, 5, 6, and 7 with phosphoric acid. Phosphate is not a buffer at pH 4 and 5, but this is only used for the pH screening experiment. In the event that pH 4 or 5 was deemed acceptable for the separation, a suitable buffer that has buffering capacity in that region would be chosen. Note that upon changing the $^{W}_{w}$pH of the mobile phase, at least 25 column volumes of the new mobile phase were passed through the column prior to sample analysis. This step should be a general requirement when performing $^{W}_{w}$pH scouting studies. An alternative approach is to perform repeat injections of the intermediate after changing the mobile phase $^{W}_{w}$pH until consistent retention times are obtained for all components in the mixture, which would deem that the column is adequately equilibrated.

As can be seen from Figure 8-24, the retention of all of the impurities and the API are dependent on the pH of the mobile phase. The intermediate and the related impurities exhibited lower retention at low pH. Retention increased initially with increasing pH, where it reached a maximum and then decreased as the pH was further increased. The optimal pH range to carry out further method development was determined to be $^{W}_{w}$pH 6–7, where the retention of the API and related impurities did not change as a function of the $^{W}_{w}$pH.

The intermediate and its related impurities are zwitterionic in nature and contain both acidic and basic functionalities; this is confirmed by its bell-shaped dependence on the mobile-phase pH (Section 4.5 in Chapter 4). Peak X elutes before the main component at low $^{W}_{w}$pH (2.0), but it elutes after the main component at pH values higher than 2.0 ($^{W}_{w}$pH 3–7).

8.5.4.3 *Peak Tracking.* During method development, if it is observed that critical pairs are changing elution order, the use of diode array and/or LC-MS should be employed to assist in peak tracking. In this particular example comparing the elution of impurity X and the intermediate, it was believed that the elution order had switched at $^{W}_{w}$pH 2 and $^{W}_{w}$pH > 3.1. Therefore the reversal of elution order was determined by comparing the diode array spectrum of impurity X and was further confirmed by LC-MS (note that ammonium bicarbonate mobile phase was used for pH 7 LC-MS analysis and TFA was used for pH 2 LC-MS analysis, with both using ESI in the positive ion mode). Note that the diode array profiles of impurity X did not directly overlay at pH 2 and pH 7, and an isobestic point (where two substances absorb at a certain wavelength of light to the same extent) was observed that can be attributed to changes in conjugation of the aromatic ring when analyzed at different pH values (see Section 8-6 for more information on the effect of pH on changes in UV absorbance). Peak tracking at different pH values by diode array sometimes is a challenging task, especially if the analyst wants to compare the UV spectrum of the impurity present at different ionization states. This was the driver to perform LC-MS analysis in order to confirm the $[M + H]^+$ ion of this impurity species. Indeed, when LC-MS analysis was performed, it was confirmed that this impurity had shifted elution order when the pH of the mobile phase was changed from 2 to 7. An extracted ion spectrum of the $[M + H]^+$ ion of impurity X at pH 7 was performed for facile identification of the impurity.

8.5.4.4 *Anomalies During Method Development.* Further evaluation of the chromatograms in Figure 8-24 revealed that some late eluting peaks were observed with the pH 5, 6, 7 mobile phases, and those peaks were not observed when the lower pH mobile phases were used. In order to troubleshoot if these peaks are indeed present in the sample or artifacts, it should be determined if the late eluting peaks are (1) synthetic process impurities with different ionizable functionalities or (2) impurities formed in the sample solvent (indicating lack of solution stability). In order to make this assessment, the stability of the intermediate in the diluent was challenged. In this case study the solution was stored at room temperature under normal light conditions and the diluent was acetonitrile. The experiments for ^w_wpH 2–4 were performed on day 1, and those for ^w_wpH 5–7 were performed on day 2 ($\approx$36 hr after initial preparation). A further investigation was performed by preparing a fresh stock solution and storing one-half of the solution in the refrigerator (4°C) for 36 hr while the other half of the solution was stored in a clear volumetric flask on the bench (ambient conditions), and it was determined that these impurities are actually formed in the diluent at room temperature under normal light conditions (see Section 14.8.1 for further details). The solutions were determined to be light-sensitive. The case study message is that fresh solutions should be prepared daily in amber volumetric flasks and a tray cooler should be used when possible when the stability of the sample in solution has not yet been determined.

8.5.4.5 *Method Selectivity and Choice of Column.*

Generally during method development, multiple columns at various pH values can be screened in isocratic or gradient using a column switcher or commercially available method development systems that have the ability of running five or more columns. The reason is that different stationary-phase types may provide a different selectivity and give the chromatographer additional confidence in resolving potential co-eluting species. In this case study, the separation performed on a Luna C18(2) (Phenomenex, Torrance, CA) was compared to the separation performed on a polar end-capped column, Synergy-Hydro-RP (Phenomenex, Torrance, CA). Similar trends in the retention dependence relative to ^w_wpH were observed for all impurities and intermediate on both types of columns, since the effect of pH on the analyte retention is a function of the analyte ionization state (Figures 8-24 and Figure 8-25). However, differences in selectivity and differences in the magnitude of the retention can be related to stationary-phase type and surface area of the column, respectively. Differences in selectivity were observed between peak X and impurity Y at ^w_wpH 2 when comparing these two columns. The Hydro-RP column showed greater selectivity at ^w_wpH 2 between impurity Y and impurity X. Note that the late eluting degradation products present in Figure 8-24 at ^w_wpH 5–7 were not observed using this column, since the samples were stored protected from light. It was determined that the storage conditions were important to minimize the degradation product formation. Although similar retention profiles were obtained at ^w_wpH 6 and 7 (desired pH range for the separation) on both columns, the pH stability of the Phenomenex Luna C18 (2) (up to ^w_wpH 10) is

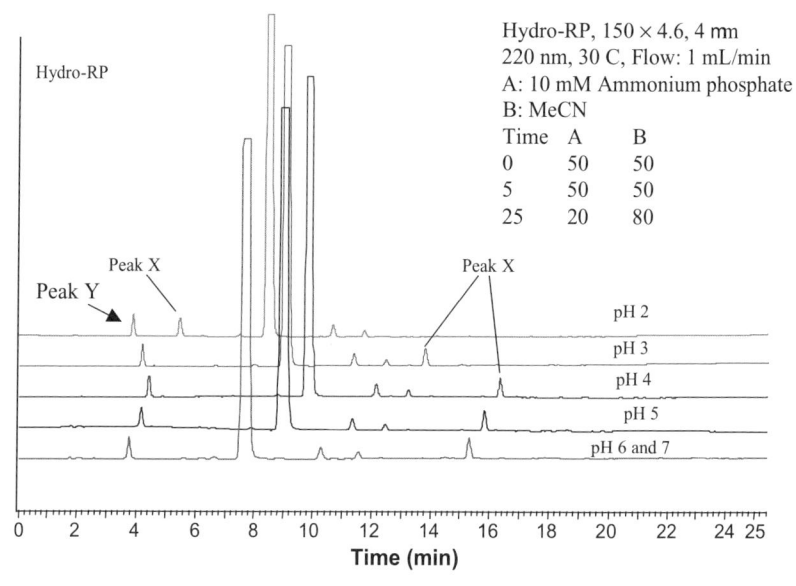

Figure 8-25. pH study of zwitterionic compound A on a Phenomenex Synergy-Hydro-RP column. Method conditions are indicated in the figure.

greater than that of the Phenomenex Hydro-RP (up to $_w^w$pH 7.5). The suitable pH and temperature range of a particular column should always be a consideration when performing method development and choosing the column to perform further method optimization experiments, especially when working at higher pH values and/or higher temperatures.

8.5.4.6 Method Optimization. Having established that a monohydrogen diammonium phosphate buffer with $_w^w$pH adjusted to 7.0 (with phosphoric acid) was the best mobile-phase pH to use (no variation in retention of peaks for $_w^w$pH 6–7) at the optimized gradient conditions (indicated in Figure 8-24), further optimization was carried out by varying the temperature of the separation. Temperature can lead to increase in the apparent efficiency of the separation and can lead to improved mass transfer.

A temperature study was conducted from 15°C to 40°C on the Luna C18 (2) column using a solution of intermediate diluted in MeCN that was stored for 4 days at room temperature under normal light conditions (Figure 8-26). As the temperature was increased from 15°C to 40°C, the resolution between an impurity eluting on the tail of the intermediate and the intermediate increased, as did the resolution for the potential degradation products eluting later in the separation. Also, the tailing factor for the intermediate decreased upon increasing the temperature to 40°C.

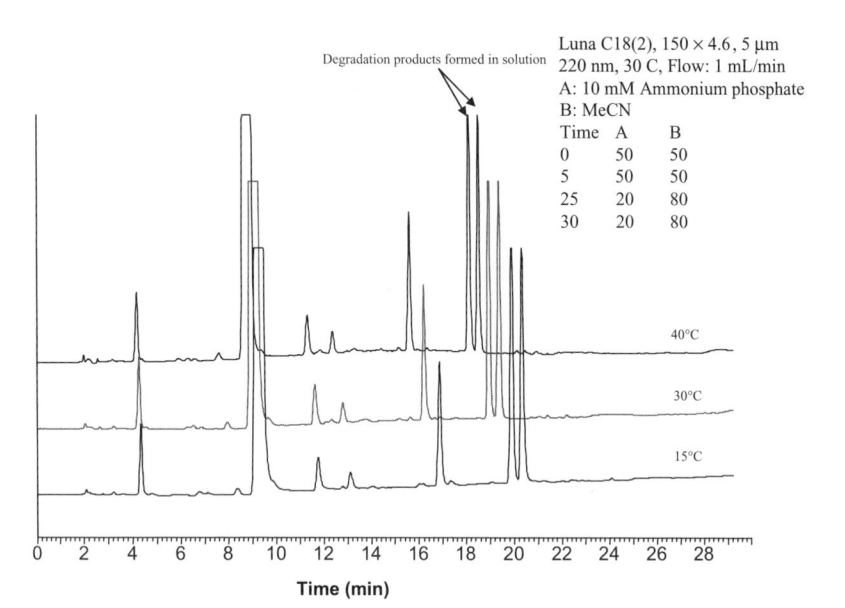

Figure 8-26. Temperature study on Luna C18 (2). Method conditions are indicated in the figure.

8.5.4.7 Final Optimization. Having established that a monohydrogen diammonium phosphate buffer with ^w_wpH adjusted to 7.0 (with phosphoric acid) and ACN was the most effective mobile phase, the gradient conditions were finalized to the following conditions: 50/50 ACN/buffer, held for 5 min and ramped up to 80% ACN over 20 min with a final hold of 5 min at 80 v/v% acetonitrile. The hold at higher organic concentration is usually employed for compounds during early development in the event that more hydrophobic species are formed either during the processing (i.e., change in synthetic routes, hold point stability, etc.) or during solid-state stability studies of the active pharmaceutical ingredient or drug product.

8.5.5 Case Study 2: Influence of pH, Temperature, and Type and Concentration of Solvent on the Retention and Selectivity of Acidic (Phenolic) Compounds

The HPLC retention behavior of a para bromo-monosubstituted phenol intermediate and its ortho isomer were investigated. The ortho isomer is a common impurity generated during the synthesis of the desired para isomer intermediate. It was critical to control this impurity because it would react at the hydroxyl functionality in the downstream chemistry to produce unwanted synthetic impurities in the API (active pharmaceutical ingredient). Implementing a recrystallization procedure to remove these downstream synthetic impurities, although efficient, would reduce the overall yield of the API. Therefore, control of the starting material, para bromo phenol, was deemed necessary. The retention of these two isomeric species was found to be highly dependent on the eluent pH, type of organic modifier, and temperature. It was determined that a reversal of elution order could be obtained where the minor isomer elutes prior to the major isomer by optimization of the eluent pH, temperature, and change of type of organic modifier from acetonitrile to methanol.

8.5.5.1 *Effect of pH on the Retention/Selectivity of the Isomers.* The first step in method development is to understand the effect of pH on the separation characteristics of the method. The pK_a values of the ortho and the para isomers was estimated by ACD (Advanced Chemistry Development software) to be 9.0 and 9.5, respectively. Obviously the best pH to carry out the separation would be at pH that is less than 2 units lower than the analyte that has the lowest pK_a. This would be at ^w_wpH values less than 7.0. However, to illustrate the effect of pH on the separation selectivity of the isomers, a controlled pH study at isocratic conditions was conducted.

Figure 8-27 (k versus ^w_wpH) and Figure 8-28 (selectivity versus ^w_wpH) show the effect of pH on the retention of the para and ortho isomers at a constant mobile-phase composition of 50:50 15 mM KH_2PO_4:acetonitrile, at 25°C over the aqueous ^w_wpH range 2.0–10.7 analyzed on a Luna C18(2) (Phenomenex, Torrance, CA) column. Both of these isomeric compounds are acidic, and it is expected that an increase in the mobile-phase pH will cause a decrease in the analyte retention because these compounds are becoming progressively more ionized. At 25°C for these isomers analyzed at $^w_wpH < 8$ the undesired isomer, ortho isomer, is eluting after the para isomer and at $^w_wpH > 9$ the ortho isomer elutes before the para isomer (desired elution order).

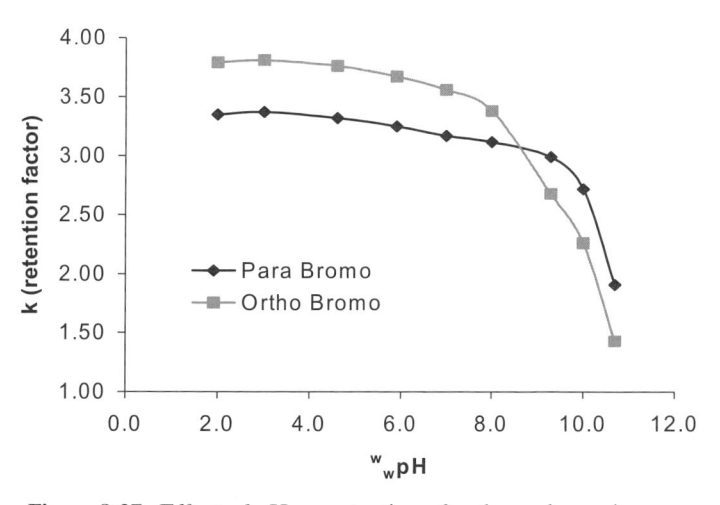

Figure 8-27. Effect of pH on retention of ortho and para isomers.

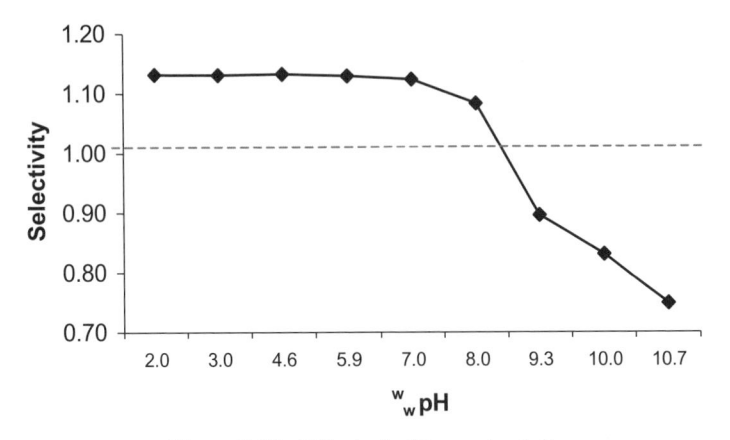

Figure 8-28. Effect of pH on selectivity.

8.5.5.2   Effect of Organic Concentration on the Retention/Selectivity of the Isomers.   The effect of organic concentration on the selectivity and retention (Figure 8-29) of para and ortho isomers at three different pH values (^w_wpH 2, ^w_wpH 8, and ^w_wpH 8.6) were determined. The optimal pH for the separation is at ^w_wpH 2. However, studies were performed at ^w_wpH 8, and ^w_wpH 8.6 to illustrate why working at these higher pH values would not be ideal for the separation from a robustness point of view. Although favorable changes in selectivity may occur at a pH near the pK_a values of the components in the mixture, the method may not be robust due to minor changes in pH and organic concentration.

% ACN Study ^w_wpH 2.0.   The amount of acetonitrile in the mobile phase was varied over the following range: 42–55% MeCN. Plots of ln (k) versus % organic for both isomers were linear in this v/v% acetonitrile region (R^2 = 0.999).

Increasing the v/v% acetonitrile led to a decrease in the retention of both isomers (Figure 8-29A), and a decrease in resolution between both isomers however the selectivity and efficiency for each isomer remained constant. Ideally, the eluent composition should not affect the selectivity between two species if their ionization state is not changing with an increase in the organic composition (see Section 2.14 in Chapter 2 for details).

Also, the efficiency did not change as a function of the organic composition because the capacity factor of the analytes was such that extra-column band broadening (column and extra-column effects) leading to peak dispersion were avoided. The selectivity also was constant within this studied organic composition range at pH 2. The decrease in resolution from 4.1 to 3.2 was obtained upon increasing the organic composition from 42% to 55% acetonitrile.

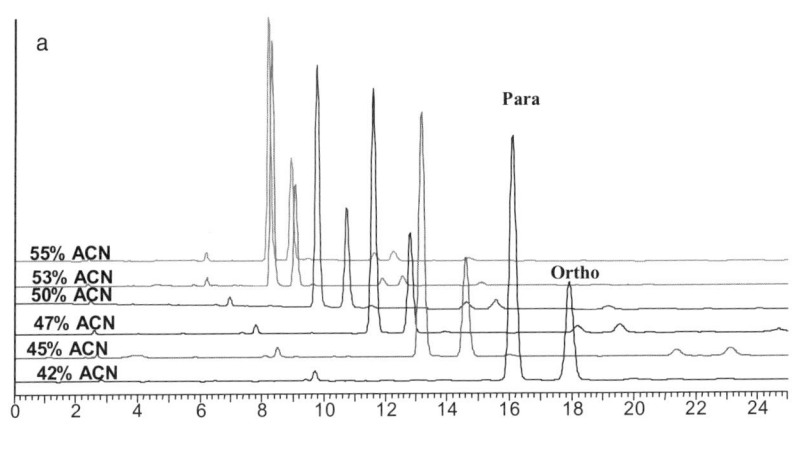

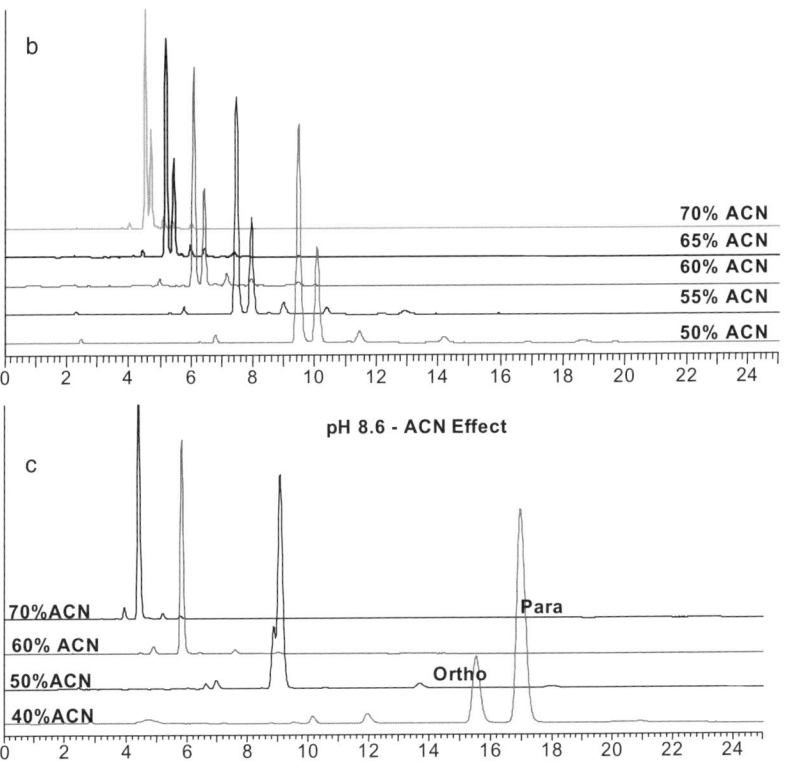

Figure 8-29. Effect of organic content on the analtye retention. (A) $_w^w$pH 2, (B) $_w^w$pH 8.0, (C) $_w^w$pH 8.6.

% ACN Study $_w^w pH$ 8.0. A similar study was carried out at $_w^w pH$ 8 (Figure 8-29B). The range of v/v% acetonitrile studied was from 50% to 70%. Increasing the v/v% acetonitrile over the range studied caused a decrease in resolution to the point where baseline resolution between the isomers was lost. Also at this pH and these acetonitrile compositions studied, the selectivity between the isomers was independent of % organic. Plots of ln k versus % organic for both isomers were linear in this v/v% acetonitrile region ($R^2 = 0.997$).

% ACN Study pH 8.6. A similar study was carried out at $_w^w pH$ 8.6 (Figure 8-29C), where the pH of the mobile phase is approaching the analyte pK_a values. The pK_a of the para isomer is greater than pK_a of the ortho isomer as observed by the inflections points in Figure 8-27. At $_w^w pH$ 8.6 the ortho isomer elutes prior to the para isomer. The range of % acetonitrile studied was 40–70 v/v%. Plots of ln (k') versus % organic for both isomers showed slight curvature at high organic concentrations, and a change in the selectivity between the two isomers was observed. The addition of organic component to an aqueous mobile phase shifts the pH of the acidic aqueous eluent (phosphate buffer) upward (0.2 pH units per 10 v/v% acetonitrile; see discussion in Section 4.5) and shifts the acidic analyte ionization state of the isomers to higher pK_a values (~0.3 pK_a units per 10 v/v% acetonitrile, determined in an independent study). Since $_w^w pH$ 8.6 is approaching the analyte pK_a values, small changes in the eluent pH (hydro-organic mixture pH) could have an effect on the overall analyte retention and selectivity. In essence, these analytes are being analyzed in their more neutral form with increasing amount of organic component in the mobile phase, making them more hydrophobic. However, the increase in the organic leads to a decrease in the analyte retention for both isomers due to a decrease in the analyte hydrophobicity. Two effects that are acting upon the retention of the isomers in opposite directions could provide an explanation for the curvature in the plot of these ionizable species. Also, this could account for the change in selectivity with an increase in % organic component at $_w^w pH$ 8.6 (Figure 8-30). This should not be a pH for further method development, since the method would not be robust in regard to slight changes of $_w^w pH$. However, it was shown that although suitable selectivity at $_w^w pH$ 8.6 (40% MeCN) was obtained, the method may not be reproducible from day to day, due to slight changes in mobile-phase $_w^w pH$, which could particularly cause issues during method transfer when the method is transferred to another site or facility in later stages of development.

8.5.5.3 Effect of Temperature on the Retention/Selectivity of the Isomers.
The next variable investigated was the effect of temperature on the analyte retention. The effect of temperature on the retention and selectivity of the para and ortho isomers at $_w^w pHs$ 2, 8, and 8.6 was studied (Figures 8-31, 8-32, and 8-33). The effect of temperature could be used to optimize the run time and the apparent efficiency of the separation. At a buffer pH of $_w^w 2.0$, the effect

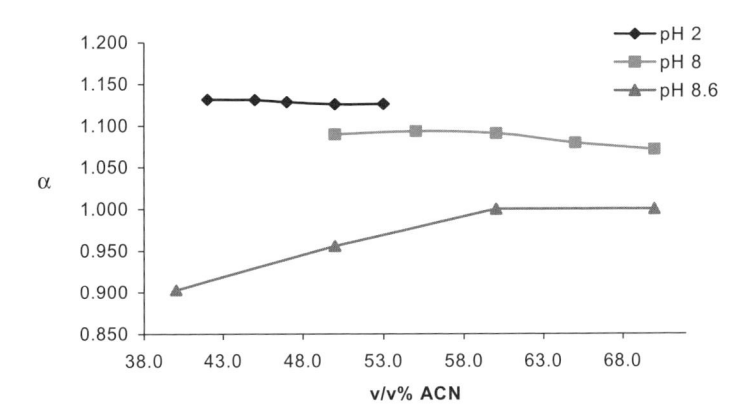

Figure 8-30. Selectivity versus organic composition at ^w_wpH 2, 8, and 8.6.

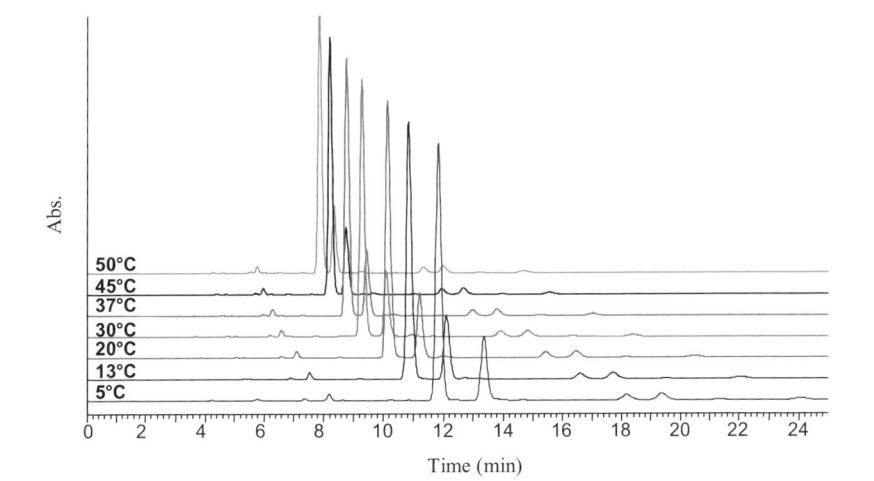

Figure 8-31. Effect of temperature on the analyte retention, ^w_wpH 2.

of temperature on retention/resolution was studied over the temperature range 5°C–50°C with 50/50 acetonitrile/buffer (Figure 8-31). The retention and resolution of both isomers decreased with increasing temperature, with the major isomer (para isomer) eluting first at all temperatures studied. This study was also conducted at a buffer ^w_wpH of 8.0 over the range 5–70°C at 50/50 acetonitrile/buffer (Figure 8-32). The retention and resolution of both isomers again decreased with increasing temperature; however, a reversal in elution order was observed when the temperature was increased above 50°C, where the ortho isomer eluted prior to the para isomer. The temperature study was also performed at ^w_wpH 8.6. At 5°C, the para isomer eluted first, at 20°C they

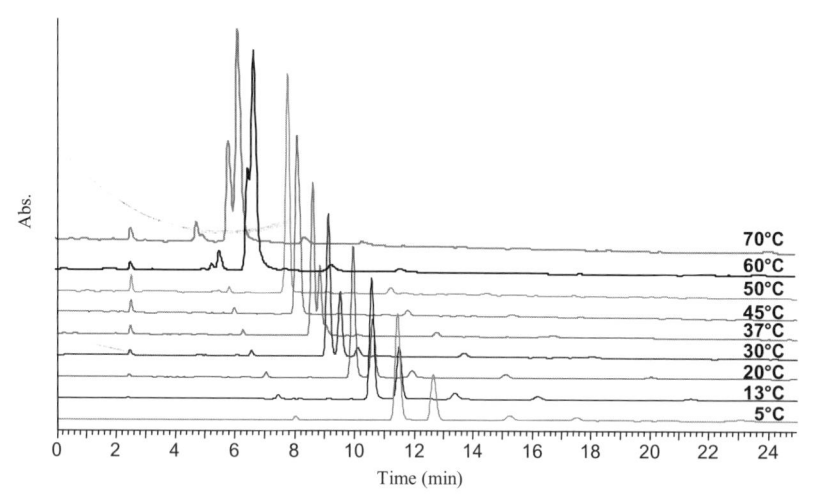

Figure 8-32. Effect of temperature on the analyte retention, $_w^w$pH 8.

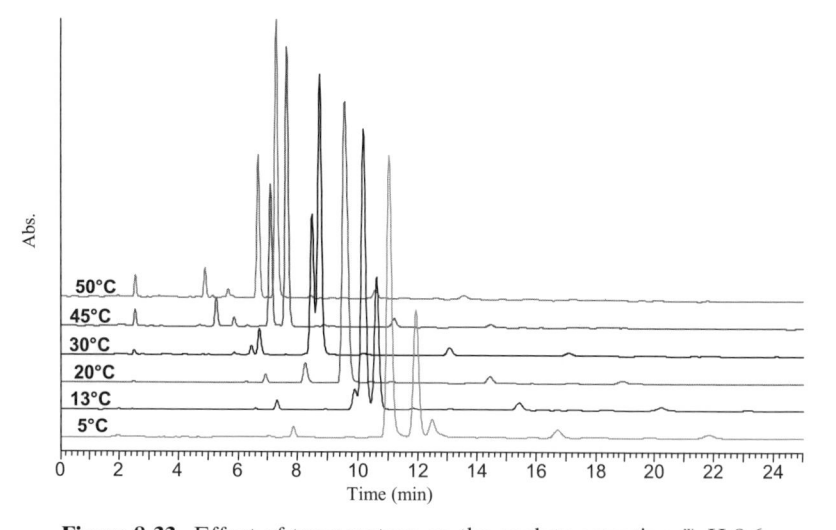

Figure 8-33. Effect of temperature on the analyte retention, $_w^w$pH 8.6.

co-eluted, and at temperatures greater than or equal to 30°C a reversal of elution order was obtained where the ortho isomer eluted prior to the para isomer (Figure 8-33). Figure 8-34 shows the selectivity changes as a function of temperature at $_w^w$pH 2, $_w^w$pH 8, and $_w^w$pH 8.6. Significant changes in the selectivity are observed at higher temperatures at $_w^w$pH 8 and $_w^w$pH 8.6, further indicating that the analyte ionization state is indeed changing as a function of temperature. Therefore it is not recommended to work at a pH near the analyte pK_a, since changes in temperature of the chromatographic system or

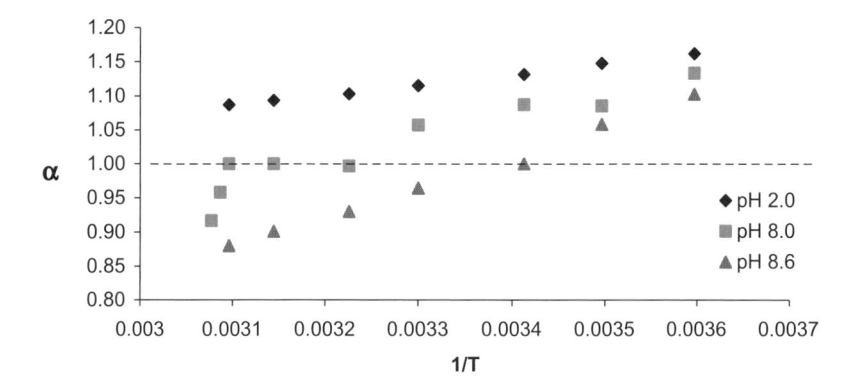

Figure 8-34. Selectivity as a function of inverse temperature at $_w^w$pH 2, 8, and 8.6.

laboratory conditions could cause significant changes in the retention and selectivity due to potential changes in the analyte ionization state (see Section 4.9).

Using the acetonitrile/buffer compositions, interesting changes in the selectivity were observed by working at a pH near the pK_a of the compounds, and the desired elution order was obtained where the minor ortho isomer eluted prior to the para isomer. However, this occurred at $_w^w$pH 8.6, and this was not an optimal pH for the final method because the method would not be robust (pH is close to analyte pK_a values). Therefore, the effect of changing type of organic eluent on the retention and selectivity of the para and ortho isomers was conducted at $_w^w$pH 2, where the method would be more robust and provide a more suitable alternative. Using gradient mode, at $_w^w$pH 2 in an acetonitrile eluent at 15°C the minor isomer, elutes after the major isomer (Figure 8-35). However, when a methanol eluent was employed at the same temperature a reversal in elution order was obtained (Figure 8-36). This can be attributed to the differences of the analytes interaction with the stationary phase and the adsorbed organic layer on top of the collapsed bonded phase. It has been shown that acetonitrile forms multilayer adsorption on top of the bonded phase, while methanol only shows monomolecular adsorption. In binary eluents with MeCN/water the retention mechanism involves a combination of analyte distribution between the eluent and an acetonitrile adsorbed layer, followed by analyte adsorption on the surface of the bonded phase (see Chapters 2 and 3 for further description). However, with methanol/water eluents the retention mechanism is predominately driven by an adsorption type mechanism because the methanol adsorbs in a form of a monomolecular layer. Hence, due to the difference in the analyte partition coefficients and their adsorption on the stationary phases, changes in selectivities could be obtained for two components when using either methanol/water or acetonitrile/water eluents. Varying the type of organic modifier in RPLC separations is recommended for the separation of isomers.

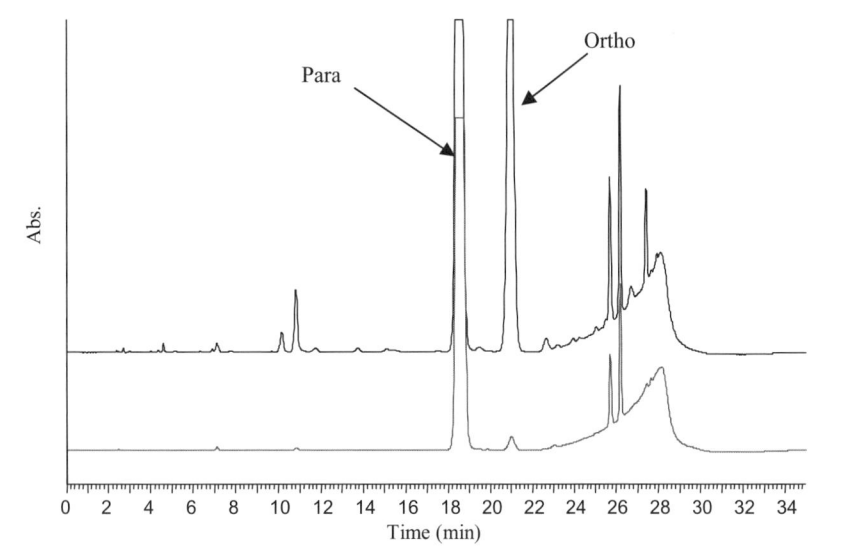

Figure 8-35. Luna C18 (2), 1 mL/min, 220 nm. (A) 0.1 v/v% H₃PO₄, (B) MeCN, 0–18 min isocratic at 42% MeCN and then linear gradient from 42% B to 75% B over 10 min, 15°C.

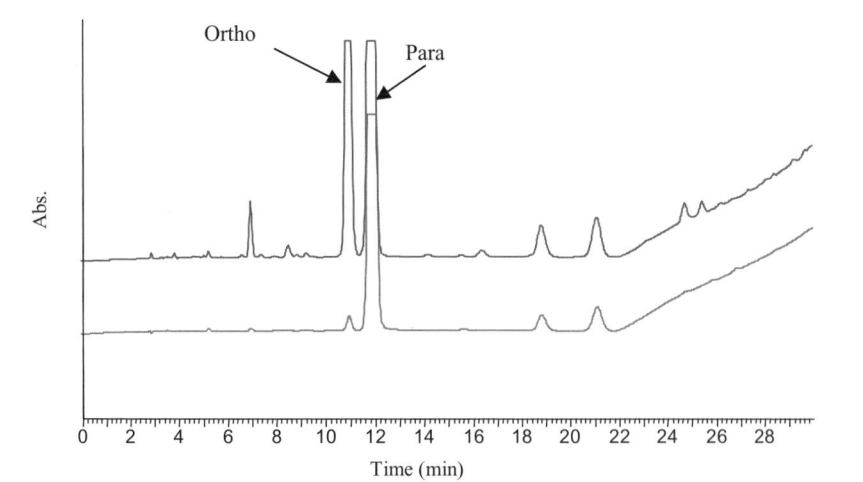

Figure 8-36. Luna C18 (2), 1 mL/min, 220 nm. (A) 0.1 v/v% H₃PO₄, (B) MeOH, 0–18 min isocratic at 65% MeOH and then linear gradient from 65% B to 90% B over 10 min, 15°C.

8.5.5.5 Case Study 2: Concluding Remarks. Extreme changes in selectivity and reversal of elution order of phenolic isomeric compounds were obtained after changing either the pH of the mobile phase, the temperature of the system, or the type of organic eluent employed. Changes in the analyte ionization state were observed upon increasing the acetonitrile composition as well as the temperature. Method development for ionizable analytes requires a judicious choice of the mobile-phase conditions and system parameters in order to perform the analysis of the compounds in their desired ionization state. Choosing the optimal parameters in the "chromatographer's toolbox" allows for the development of rugged and reproducible methods.

8.5.6 Case Study 3: Method Development for a Diprotic Basic Compound

A case study is presented for the method development of a diprotic base compound. The first step in method development is to look at the chemical structure of the analyte and to determine if there are any ionizable sites on the molecule. If there are ionizable sites then their respective pK_a values should be determined. The pK_a values may have been already determined by the preformulation group and close communication with that group would avoid duplication of work. However, commercially available programs such as ACD Labs (Advanced Chemistry Development, Toronto, Canada) are also available to allow for in-silico prediction of the analyte pK_a. Also, using selected fragments of the molecule can also be helpful for pK_a determination of the desired molecule because the pK_a values for each of these fragments of the molecule may be readily available from the literature. This is only an estimate at best, but can guide the chromatographer down the right path for initial mobile-phase pH selection.

In the following case study for this pharmaceutical compound M, the method development scenario and rationale for each iteration in the method development process is highlighted. Also, a method development flow chart for gradient separations is included which can be used as a general strategy for method development (Figure 8-37). References are made in this case study to the flow chart in Figure 8-37.

8.5.6.1 Step 1: Analyze the Molecule. In Step 1 of the flow chart (Figure 8-37) it is recommended to analyze the molecule from a physicochemical point of view (knowledge of the pK_a, log P, log D). The structure of the compound must be analyzed to determine the ionizable functionalities, and the pK_a values of each ionizable group should be determined. In this case, Product M is a diprotic base with two pK_a values 3.3 and 5.3 estimated by ACD. This compound contains an aromatic basic functional group (pyridinal nitrogen) with an electron-withdrawing group, chlorine, in the meta position and also contains electron-donating groups on the same aromatic ring. Electron-withdrawing groups such as chlorine tend to intensify the positive charge of the anilinium/pyridinum ion; this destabilizes the ion relative to the free

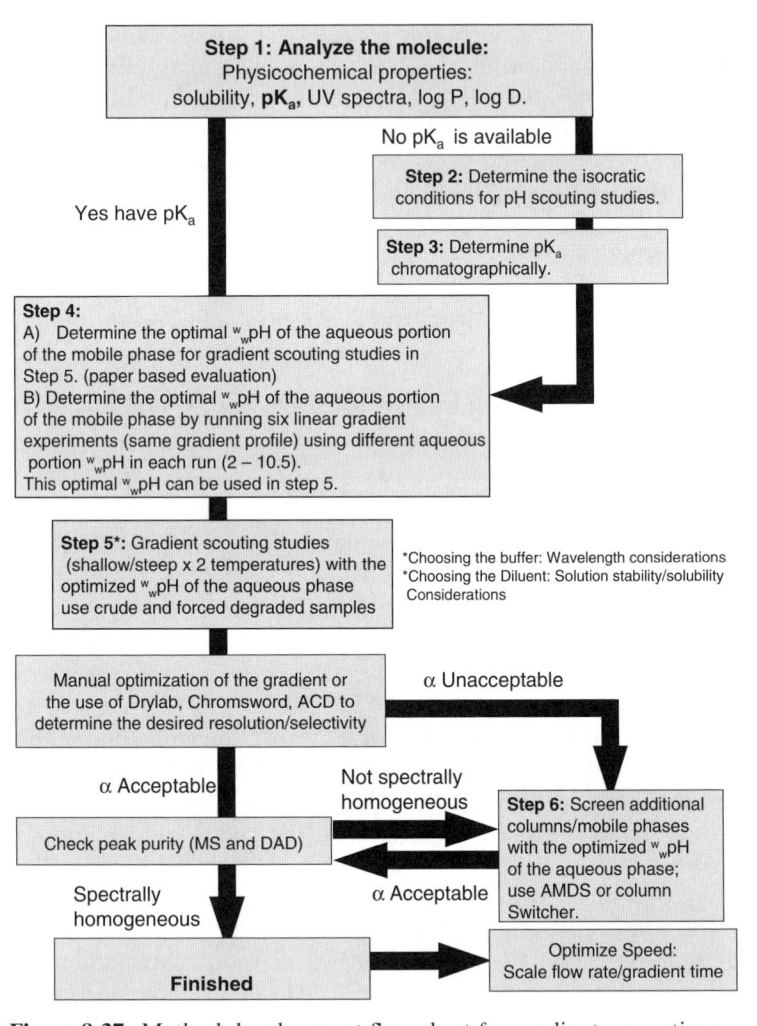

Figure 8-37. Method development flow chart for gradient separations.

amine/pyridine, therefore decreasing the basicity compared to pyridine (pK_a 5.2). If one does not have a program to predict pK_a, the pK_a of this analyte could be estimated to be close to that of meta chloro pyridine, $^w_w pK_a$ 2.95. The other basic functionality contains a phenyl group attached to a morpholine group. The pK_a of morpholine is 8.8; but because the phenyl ring is attached to the nitrogen group, this leads to resonance stabilization and consequently leads to a reduction of the analyte pK_a.

Note that because this compound does have multiple basic functionalities, two ionization equilibria could be written for this amphoteric species. At mobile $^w_w pH$ values between 3 and 5 the existence of multiple species is expected. Since the two pK_a values are close to one another ($\sim 2\,pK_a$ units apart), the inflection points overlap, making titration and/or chromatographic

pK_a prediction for each ionizable functionality difficult. At ^W_WpH 4.3, the basic site (pyrdinal nitrogen) is predominately neutral (90%) and the other basic site (morpholine nitrogen) is predominately ionized (90%). In turn, it would be expected to observe only one inflection point from both the potentiometric titration and chromatographic determination of the pK_a.

Alternatively, if the pK_a of the analyte is not known and software is not available for in-silico prediction of the pK_a, the chromatographer can go directly to step 2 (determine the isocratic conditions for pH scouting studies) and step 3 (determine pK_a chromatographically) in the flow chart in Figure 8-37. The analyte could be analyzed at six different pH values at a particular organic composition (isocratic mode) for an estimate of the analyte ^S_SpH and for determination of the suitable chromatographic conditions to analyze the analyte.

If the pK_a is known, the chromatographer can go directly to step 4.

8.5.6.2 Step 2: Determine the Isocratic Conditions for pH Scouting Experiments. If the chromatographer intends to determine the chromatographic pK_a and understand the influence of mobile-phase pH on the target analyte retention, pH scouting studies need to be performed in isocratic mode. In order to begin this process, the appropriate set of isocratic conditions to adequately retain the analyte in its fully ionized form and to elute the analyte in its fully neutral state needs to be determined. Usually a steep gradient run is used to estimate the initial isocratic elution conditions. In an example shown in Figure 8-38 a probe linear gradient from 5 to 95 v/v% acetonitrile

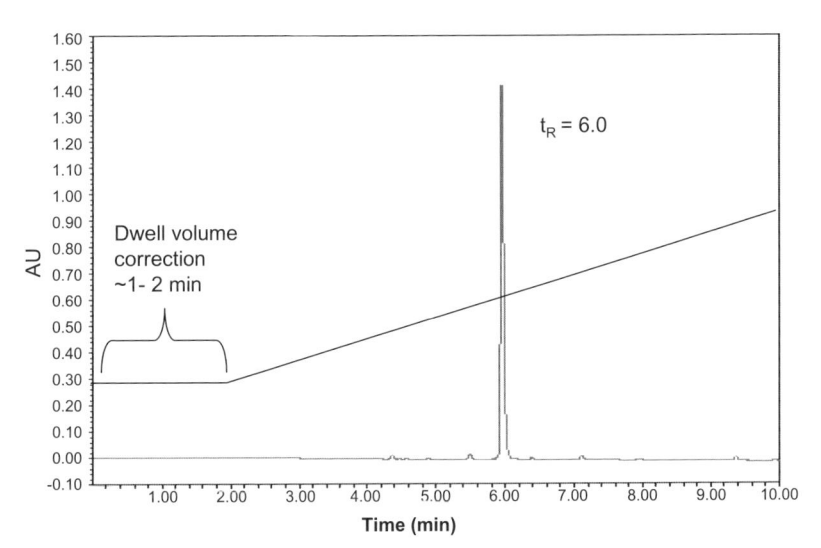

Figure 8-38. Mobile phase A: 0.2 v/v% H_3PO_4. Mobile phase B: Acetonitrile, linear gradient from 5% B to 95% B over 10 min. Column: Luna C8(2) 150 × 4.6 mm. Injection volume, 10 μL; flow, 1.0 mL/min; wavelength, 300 nm, column temperature, 35°C.

from 0–10 min was run and the target basic analyte (in its predominately ionized form) eluted at 6.0 min. The ^w_wpH of the mobile phase was 1.9 and the flow rate was 1.0 mL/min. Note that the selection of the UV wavelength of detection should have already been performed using off-line UV, or alternatively a diode array detector can be used and the proper wavelength for detection would consequently be extracted during the data processing. The concentration of organic in isocratic mode, which is necessary for the adequate retention of the analyte in its ionized form, now can be estimated. The ^w_wpH of the aqueous portion of the mobile phase was chosen to be 1.9 (0.2 v/v% H_3PO_4) and the flow rate was 1.0 mL/min. Using the linear gradient from 5 to 95 v/v% acetonitrile over 10 min, the v/v% organic per minute was calculated as 9 v/v% ACN/min (Scheme 1). However, the dwell volume (V_D) of the instrument must be accounted for because the actual gradient does commence within the column until about 1–2 min (depending on the instrument and instrumental setup). The dwell volume of most common HPLC systems is 1–2 mL and can be easily determined (see reference 19 for details). However, at 1 mL/min flow rate (F), an estimate of a 1.5-min dwell time (t_D) was used for this purpose ($t_D = V_D/F$). Note that the velocity of the analyte moving through the column under gradient conditions is not constant and follows a pseudo-exponential profile (see Chapter 2, Section 2.17). The estimation given in Scheme 1 serves as an approximation to determine the starting isocratic elution conditions from the probe gradient run. By taking into account the gradient slope of 9% ACN/min and accounting for the dwell time (1.5 min) and elution time (6 min from Figure 8-38) of the analyte from the probe gradient run, the estimated isocratic composition to elute the analyte at the same retention as in the gradient probe run can be calculated as shown. The estimated isocratic composition in which the analyte would elute at 6 min ($k \sim 3.5$) is estimated as 41 v/v% acetonitrile ±10% acetonitrile using 0.2 v/v% H_3PO_4 (^w_wpH 1.9) mobile phase. The isocratic conditions chosen to perform the pH scouting study was 30 v/v% acetonitrile.

Gradient slope × (Elution time from probe gradient run – Dwell time)
= Isocratic % organic composition ± 10% organic composition

$$\text{Gradient Slope} = \frac{90\% \text{ ACN}}{10 \text{ min}} = 9\% \frac{\text{ACN}}{\text{min}}$$

$$\text{Dwell Time} = \frac{1.5 \text{ mL}}{1.0 \frac{\text{mL}}{\text{min}}} = 1.5 \text{ min}$$

$$9\% \text{ ACN/min} \times (6 \text{ min} - 1.5 \text{ min}) = 41\% \text{ ACN } \pm 10\% \text{ ACN}$$

Scheme 1. Estimation of isocratic conditions from gradient probe run.

8.5.6.3 *Step 3: Determine pK$_a$ Chromatographically.*

The retention of the analyte can be determined at different eluent pH values and can be used to determine its pK$_a$ in a particular hydro-organic mixture and assist the chromatographer in proper pH selection for the aqueous portion of the mobile phase. The goal is to avoid a pH region where minor changes in pH can adversely affect the retention of the target analyte. If the pK$_a$ values are not known, it is suggested to perform the pH scouting experiments from pH 1.5 to 10 (at least five $_w^w$pH values should be investigated). In this example, the pK$_a$ values of the two ionization centers were predicted by ACD, and the $_s^s$pH values chosen were at least 1 unit less than the lowest $_s^s$pK$_a$ in the molecule and at least 1 unit greater than the highest $_s^s$pK$_a$ of the molecule. Generally, the pK$_a$ of a basic compound decreases by about 0.2 pK$_a$ units per 10 v/v% acetonitrile (see Chapter 4, Sections 4.5 and 4.6) and if 30 v/v% acetonitrile is used, it is expected to lead to a reduction of 0.6 pK$_a$ units (0.2*3 = 0.6 pK$_a$ unit basic analyte pK$_a$ shift) for both basic ionization centers. Therefore this correlates with a $_s^s$pK$_a$ of 2.7 for pyridinal nitrogen (i.e., 3.3 − 0.6 = 2.7) and with a $_s^s$pK$_a$ of 4.7 for the morpholinal nitrogen (i.e., 5.3 − 0.6 = 4.7). The expected pK$_a$ determined by chromatography is to be midpoint of these two $_s^s$pK$_a$ values (i.e., 3.7).

In step 3, for this study the upper $_s^s$pH for the mobile phase to be prepared was determined to be $_s^s$pH 6.7 (at least two units greater than the highest $_s^s$pK$_a$ of the molecule). The lower $_s^s$pH for the mobile phase (containing 30 v/v% MeCN) that should be prepared for this study should be 1.7, but this would mean that an aqueous mobile-phase $_w^w$pH of 1.1 would have to be prepared to obtain a $_s^s$pH of 1.7 (see Chapter 4, Section 4.5 for pH shift). Remember that the pH shift of the mobile phase for a phosphate buffer is approximately 0.2 pH units in the upward direction for every 10 v/v% acetonitrile. In this case, not to compromise the stability of the packing material (column chosen has recommended a lower pH limit of $_w^w$pH 1.5), a pH of $_w^w$pH 1.6 was chosen to be prepared which correlates to a $_s^s$pH of 2.2 ($_w^w$pH 1.6 + 0.6 units upward pH shift upon addition of 30 v/v% acetonitrile). Most definitely the final method will not be set at this low pH, since the analyte would exist in multiple ionization states; however, the experiment was performed at this low pH to elucidate the effect of the pH on the analyte retention in this low-pH region.

Five to six $_s^s$pH values would then be chosen between $_s^s$pH 2.2 and 6.7 to run the pH study at isocratic conditions (30 v/v% MeCN). In Figure 8-39 the chromatographic retention as a function of the $_s^s$pH are shown. Note that blanks do not need to be run, just multiple injections of the same analyte. Once the retention of multiple injections of the target analyte is achieved, the column is deemed to be equilibrated with the mobile phase. Usually after eluting 25 column volumes through the column, the column is assumed to be equilibrated (this may not be the case if an ion-pairing reagent is used). In this study, three injections at each pH were run. Only the last injection is shown in Figure 8-39 and Table 8-8. One recommendation is to perform the pH study either

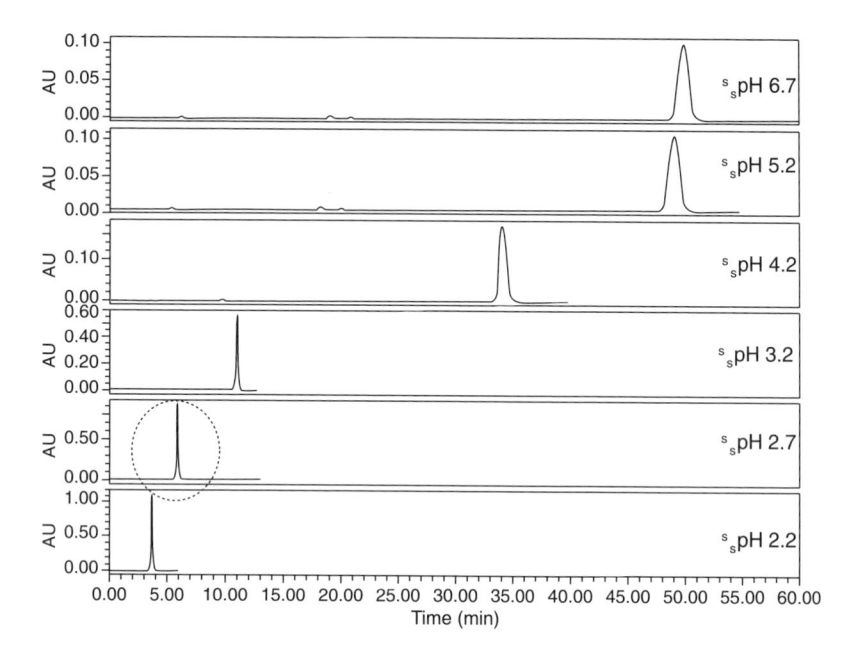

Figure 8-39. Phenomenex Luna 3u C8(2) Column. [150 × 4.6 mm, 3 μm]. Mobile phase: 10 mM K$_2$HPO$_4$, acetonitrile (70 : 30, v/v), pH adjusted with w/ H$_3$PO$_4$; flow rate, 1.0 mL/min; injection volume, 10 μL; wavelength, 247 nm., column temperature, 35°C.

TABLE 8-8. Retention of Diprotic Compound at $_w^w$pH and $_s^s$pH

$_w^w$pH	$_s^s$pH	t_R
6.1	6.7	49.8
4.6	5.2	49.1
3.6	4.2	34.3
2.6	3.2	11.2
2.1	2.7	6.0
1.6	2.2	3.8

from low to high pH or from high to low pH for faster equilibration between each successive pH experiments.

The k (retention factor) values are then plotted versus the $_s^s$pH values, and the inflection point of this sigmoidal relationship could be taken as the $_s^s$pK_a of that particular compound at particular hydro-organic mixture. The $_s^s$pK_a determined at 30 v/v% MeCN was determined to be 3.9 (using nonlinear regression analysis program MathCad 8). This corresponds well to our original estimation of $_s^s$pK_a 3.7.

The retention of this analyte leveled off between pH values $_w^w$pH 4.6 ($_s^s$pH ~5.2) and $_w^w$pH 6.1 ($_s^s$pH ~6.7), where the analyte is in its neutral form. No lower limit plateau is observed where the analyte would be in its fully ionized form. Therefore, further analysis for this compound should be carried out at pH values > ^s_spH 6.1 using no less than 30 v/v% acetonitrile in the mobile phase. Note that at even higher organic compositions the basic analyte pK_a is further reduced (up to 60 v/v% acetonitrile) and the pH of the aqueous portion (adjusted with acidic modifier) of the mobile phase is further being shifted upwards. This is a favorable situation because the analyte is being analyzed in a more neutral state with an increase in the organic concentration. Hence this is the situation when gradient elution is used and higher organic content is employed.

All these analyses were conducted on a 15- × 4.6-mm, 3-μm column. However, the pH scouting analysis could have been performed faster. A 5-cm × 4.6-mm column on a conventional low-pressure HPLC system could have been used, further reducing the analysis times of each run by at least three times. Also, a 5-cm or 10-cm × 2.1-mm column, packed with 1.7-μm particles, could have been used in conjunction with an ultrahigh-pressure HPLC system, allowing for further reduction of the analysis time for the pH scouting experiments (see Figure 4-29 in Chapter 4).

8.5.6.4   Step 4A: Determine the ^w_wpH of the Aqueous Portion of the Mobile Phase for Gradient Screening Studies (Paper based evaluation). After gaining confidence with the approaches described in Step 2 and 3 in Figure 8-37 or if the analyte pK_a, is known, then the chromatographer can go directly from Step 1 to Step 4A in Figure 8-37 to determine what the starting mobile-phase ^w_wpH should be in order to perform the gradient screening studies.

There are three items that need to be considered:

1. **In what form will the molecule be analyzed (neutral or ionized)?** For this particular molecule we want to analyze the molecule in its neutral form.
2. **The pK_a shift of the ionizable analyte.** For this example, since the analyte is basic, the downward pK_a shift for basic analytes must be accounted for. The working pH should be at least 2 pH units above the basic analyte pK_a to be fully neutral. One pH unit could also be used (analyte is approximately 90% in neutral form).
3. **The pH shift of the mobile phase.** In this example, an acetate buffer was chosen. The upward pH shift (acidic) of the acetate buffer upon addition of the organic must be accounted for. If the buffer contains both acidic and basic functionalities (i.e., ammonium acetate), the pH shift is dependent on the pH that is chosen. This is based on the respective pK_a of the counteranion and countercation of the buffer species employed. For example, at pH values below 7 the acidic pH shift rule would apply for the acetate counteranion, and at pH values greater than 7 the basic pH shift rule would apply for the ammonium countercation (see Section 4.5).

If Product M, a diprotic base, is to be analyzed in its neutral form, the higher $^w_wpK_a$ of Product M (which is 5.3) needs to be considered because the other $^w_wpK_a$ of 3.3 is less basic. Let us use to try to determine at what ^w_wpH the analyte would be in its neutral form at eluent conditions of 30 v/v% MeCN and 70 v/v% acidic buffer.

The goal is to calculate ^w_wpH of the buffer in order to obtain the basic analyte in its fully neutral form.

Step A. First, account for the downward pK_a shift for the basic analyte upon addition of organic. For every 10 v/v% increase in acetonitrile, the $^s_spK_a$ of the analyte decreases by $0.2 pK_a$ units. (Highest analyte pK_a is 5.3 and considering 30 v/v% of acetonitrile in the mobile phase the pK_a shift is $(0.2 * 3 = 0.6)$

$$5.3 - (3 * 0.2) = 4.7 \ (^s_spK_a)$$

Step B. Once $^s_spK_a$ is determined, the ^s_spH at which the analyte would be in its fully neutral form (>99%) needs to be determined. This corresponds to ^s_spH that is 2 pH units greater than the $^s_spK_a$ of 4.7 (calculated above). *Note that if one wanted to determine the ^s_spH in which the analyte would be ≥90% of its neutral form, this would correspond to working ^s_spH 1 unit greater than the $^s_spK_a$ of 4.7. (This is also acceptable from a method robustness point of view.)*

$$4.7 + 2 = 6.7 \ (^s_spH)$$

Step C. Then account for the pH shift of the acetate buffer (acidic buffer) upon addition of acetonitrile. For every 10 v/v% increase in acetonitrile, the pH of the acidic buffer increases by approximately 0.2 pH units. This would correspond to a 0.6 pH unit increase:

$$3 * 0.2 = 0.6$$

Step D. Then determine what the minimum ^w_wpH of the aqueous portion of the buffer should be by taking into account the upward pH shift of the aqueous portion of the mobile phase upon addition of organic. Therefore the optimal pH to analyze this compound would be at an aqueous mobile phase pH of ≥6.1.

6.7 − 0.6 = 6.1   Max pH of the aqueous portion of the mobile phase in order to have analyte in fully neutral form (>99%) at 30 v/v% MeCN.

The prediction of ^w_wpH 6.1 for the aqueous portion of the mobile phase using Steps A–D agrees well with the actual experiments that were performed in Step 3, where the retention of Product M was independent of the pH when the ^w_wpH is greater than 4.6. ^w_wpH, greater than 4.6 were used for further studies in Step 5.

8.5.6.4.1   Step 4B: Determination of the Optimal $_w^wpH$ of the Aqueous Portion of the Mobile Phase in Gradient Mode. Alternatively, the chromatographer may proceed directly from Step 1 to Step 4B, to determine the optimal $_w^wpH$ of the aqueous portion of the mobile phase under *gradient* conditions. This can be accomplished by running six linear gradient experiments (same gradient profile, e.g. from 5% to 85% of organic over 20 minutes) using different aqueous portion $_w^wpH$ in each run ($_w^wpH$: 2—10.5). This would allow for the determination of the desired $_w^wpH$ region under gradient conditions: a pH region in which the retention of the components in the sample does not change significantly as a function of the $_w^wpH$. An optimal $_w^wpH$ (within the desired pH region) of the aqueous portion of the mobile phase can then be selected and used in Step 5.

8.5.6.5   Step 5: Gradient Scouting Studies with the Optimized pH of the Aqueous Phase. Once the optimal $_w^wpH$ is known, the gradient conditions can be optimized for obtaining the best selectivity and resolution of all critical pairs. Multiple samples could be run which include a crude sample (better to use a sample that has an elevated amount of impurities) and forced degraded samples. Two gradients can be run at two temperature (35°C and 50°C)—one with a shallow slope (i.e., 5% MeCN to 95% MeCN over 20 min) and one with a steep slope (i.e., 5% MeCN to 95% MeCN over 8 min)—and then the gradient can be modified accordingly if needed. If the optimal selectivity and resolution of all critical pairs cannot be obtained and/or the target analyte is not spectrally homogeneous, go to Step 6, Figure 8-37. (Screen different columns/ mobile phases with the optimized $_w^wpH$ of the aqueous phase using an automated method development system or column switcher.)

Alternatively, the results from the gradient runs for each sample can be inputted into Drylab, ACD, or Chromsword for further optimization (see Sections 8.5.6.11). For the predicted experimental conditions (i.e., gradient slope, temperature, flow rate), if desired selectivity and resolution can be obtained, an experiment can be run for verification. The peak purity for the main analyte (MS and DAD detection) should be checked in the verification run. If the desired selectivity and/or the target analyte are not spectrally homogeneous, go to Step 6, Figure 8-37.

In this case study, further method development was carried out on a crude sample using a 10 mM ammonium acetate buffer that has a $_w^wpH$ 5.8 (note acetate has suitable buffering capacity from pH 3.8 to 5.8). Two gradient runs (shallow/steep gradient slope) were performed. The best chromatography was obtained in gradient mode with a linear gradient from 5% acetonitrile to 95% acetonitrile over 20 min (shallow slope), with a 3-min hold at 95% acetonitrile. A hold at higher organic is usually recommended in the early stages of development to ensure the elution of very hydrophobic components. It is also recommended to employ a high organic hold for stability-indicating methods for the API and drug product in the event that higher molecular weight species (i.e., hetero or homo dimers of the API, API containing hydroxyl/amino group which could react with stearic acid to form a more hydrophobic degradation product) are formed in the solid state upon storage.

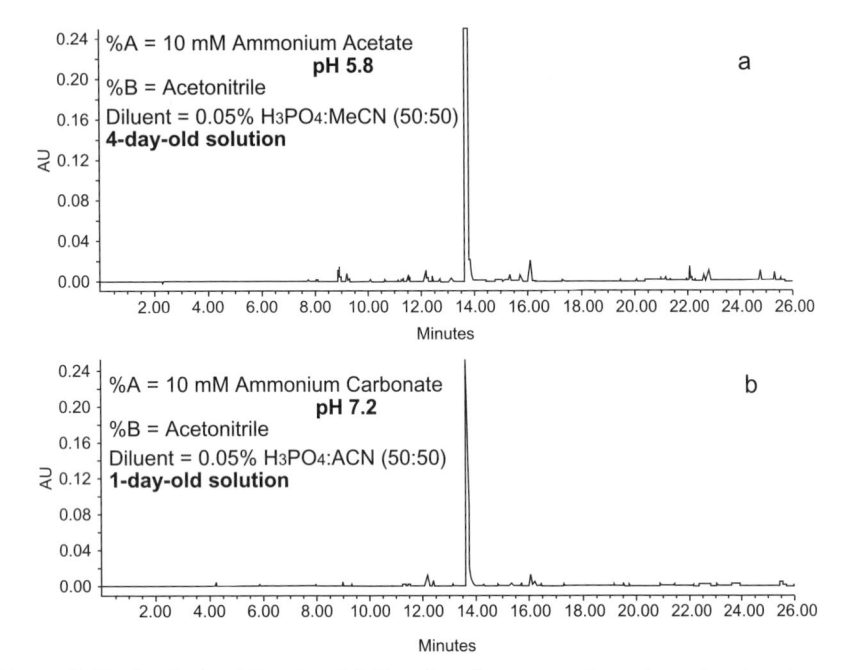

Figure 8-40. Analysis of Product M (free base) as neutral species using two types of volatile buffers. Chromatographic conditions: Column: Luna C8(2) 150 × 4.6 mm. Mobile phase: Aqueous (see A and B for exact conditions), acetonitrile. Wavelength, 247 nm; column temperature, 40°C; flow, 1 mL/min; injection volume, 10 μL. Linear gradient from 5% acetonitrile to 95% acetonitrile over 20 min, with 3-min hold at 95% acetonitrile.

Also, another driver for choosing the ammonium acetate buffer was that it is also LC-MS compatible. Also, this acetate buffer could be used since Product M has maximum wavelength for absorbance at 247 nm, and there is no background absorbance from the buffer at this wavelength. With this linear gradient employed, the target analyte elutes at approximately 14 min (Figure 8-40). To determine the approximate concentration of acetonitrile in which the analyte elutes, the following calculation can be performed. Using the gradient slope of 4.5% MeCN/min and the dwell time of 1.5 min (dwell volume of system is 1.5 mL and flow rate is 1.0 mL/min), the analyte elutes at approximately 50% acetontrile: [14 min − 1.5 min] × 4.5%/min = 56% MeCN. Note that the velocity of the analyte movement through the column using gradient conditions is not constant and follows a pseudoexponential profile and that this estimation just serves as an approximation to determine approximate elution conditions to ensure that the analyte is being analyzed in one predominate ionization state accounting for pH and analyte pK_a shift. At this organic eluent composition range (50% MeCN ± 10% MeCN) the analyte is still predominately in its neutral state (since the $^s_s pK_a$ is further lowered upon addition of organic). Note that as the acetonitrile content is increased up to 60 v/v% acetonitrile, the pK_a of basic compounds generally continues to decrease while the pH of the mobile phase (acidic buffers) continues to increase, which allows for the analyte to remain further in its neutral state.

8.5.6.6 *Choosing the Buffer: Wavelength Considerations.*

In this step (Step 5, Figure 8-37) the choice of buffer should be optimized, taking into consideration the buffer capacity at a particular pH and the background absorbance of the buffer and the absorbance of the analyte species.

For this particular example, two different volatile buffers were used: 10 mM ammonium acetate buffer ($_w^w$pH 5.8) and 10 mM ammonium bicarbonate ($_w^w$pH 7.2). Figure 8-40 shows the analysis carried out with these two LC-MS-compatible mobile phases: 10 mM ammonium acetate solution (Figure 8-40A) and 10 mM ammonium bicarbonate (Figure 8-40B) with the same gradient conditions and no changes in the target analyte retention. The chromatography in each system was independent of the different buffer species or the pH employed (since analyte is in its fully neutral form in this pH region). The 10 mM ammonium acetate buffer has a UV cutoff <220 nm due to acetate ion.

For this example the wavelength of detection was at 247 nm because the compound has a maximum absorbance at 247 nm. The acetate buffer is transparent at this wavelength. However, if the desired wavelength for analysis is 210 nm, the ammonium carbonate buffer should be used due to its lower UV cutoff (<200 nm at 10 mM concentration). Ammonium carbonate buffers should be prepared fresh daily (max 48 h) because it releases CO_2 and this may alter the mobile-phase pH to higher values.

8.5.6.7 *Choosing the Diluent: Solution Stability/Solubility Considerations.*

The solution stability in the diluent should be investigated at this stage (Step 5, Figure 8-37) to determine if the target analyte is degrading in the diluent and if any special considerations such as protecting the solution from light or storing solution at cooler temperatures need to be applied during further method development experiments.

In both examples shown in Figure 8-40, an acidic diluent containing acetonitrile was used as the sample preparation solvent. In Figure 8-40A for the sample stored in the acidic diluent for 4 days at room temperature shows some degradation compared to Figure 8-40B for the sample stored in the acidic diluent for one day at room temperature. The free base had limited solubility in water: acetonitrile diluents, so the addition of dilute acid to the diluent was necessary. Free bases should be more soluble in acidic diluents because they would be in their ionized forms, and free acids should be more soluble in basic diluents. The solubility of the target analyte should be challenged, and the pH of the diluent and/or increase in the concentration of organic in the diluent may have to be adjusted to enhance the compounds' solubility. Another factor that needs to be considered is the stability of the analyte in the diluent. Product M was found not to be very stable in acidic diluent. So going forward a sample tray cooler, 4°C, was used to enhance the stability of the sample in the diluent, and it was determined to remain stable for at least 72 h at 4°C and for no more than 24 h at room temperature.

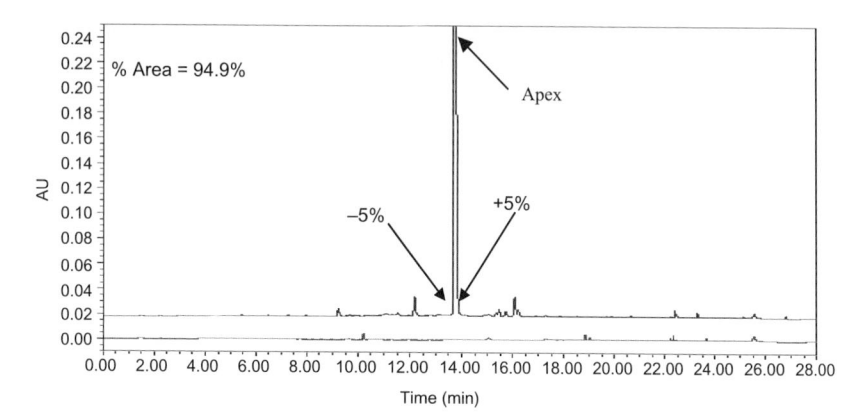

Figure 8-41. Determination of peak homogeneity. Chromatographic conditions: Column: Luna C8(2) 150 × 4.6 mm. Mobile phase: 10 mM NH₄OAc, pH 5.8, acetonitrile. Column temperature, 40°C; flow, 1.0 mL/min; injection volume, 10 μL. Linear gradient from 5% acetonitrile to 95% acetonitrile over 20 min with 3-min hold at 95% acetonitrile.

8.5.6.8 *Checking for Peak Purity.*

If the selectivity of all the components in the separation is acceptable, then the next step is to check for the peak purity of the target analyte. Spectral purity of the active peak was determined using PDA (photodiode array) detector. Diode array detection is used to determine if the peak is spectrally homogeneous. This is performed by overlaying spectra from at least three points across the peak. This is done by extracting spectra at the apex, rise and fall of the peak (Figure 8-41). The background needs to be subtracted from before and after the peak. The diode array spectra is shown in Figure 8-42, and it was determined that the peak was not spectrally homogeneous because the diode spectra do not overlay; this indicates that an impurity is co-eluting with the main compound on the right-hand side of the peak. Also, LC-MS studies can be done in parallel to determine the molecular weight of the unknown co-eluting impurity.

If a peak is determined to be spectrally homogeneous, this does not guarantee that there are no co-eluting species. For example, a spectrally homogeneous peak could contain a co-eluting impurity that is present in a small quantity and may not be able to be differentiated from the target analyte spectra and/or an impurity may have the same diode array spectra as the main component. In both these instances the overlay of the diode array would not allow for determination of co-eluting species. In this example, there was co-elution noted, and it was found that changing the gradient slope on this C8 column did not provide for any increased selectivity between the co-eluting species and the target analyte. Therefore, further method optimization experiments were performed to resolve the impurity from active. This warranted proceeding to Step 6, Figure 8-37: Screen different columns/mobile phases using an automated method development system or column switcher.

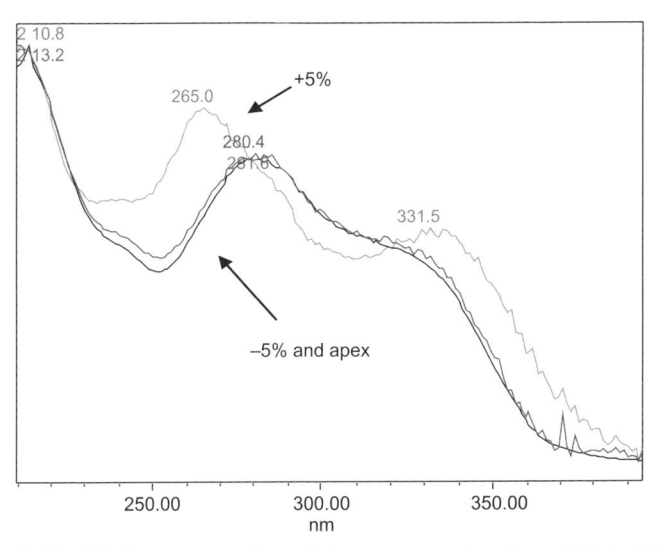

Figure 8-42. Diode array profiles of the target analyte from 200 to 400 nm.

8.5.6.9 *Using of Mass Spectrometric Detection and Diode Array Detection in Parallel.*
The use of mass spectrometry in a parallel is recommended during method development, especially if a co-eluting species is determined by diode array analysis.

A single quadrupole mass spectrophotometer which may be available in analytical departments of most pharmaceutical companies, can be used for this purpose. Mass spectra should be taken across the peak (–5%, apex, and +5%) as with the PDA. This was performed for Product M, and it was determined that the peak is not spectrally homogeneous; also, an impurity with an odd number of nitrogens was determined to be co-eluting. Furthermore, due to the isotope peak present at a 1:3 ratio compared to the [M + H] ion, this suggested that the impurity also contained a chlorine atom as the parent compound. The parent molecule, Product M, also shows the same isotopic pattern, indicating that this part of the molecule remains intact in the impurity. The LC-MS studies further confirmed the co-elution of a species with the target analyte, and the next step was to proceed to Step 6, Figure 8-37 (screen different columns/mobile phases using an automated method development system or column switcher).

8.5.6.10 *Step 6: Screen Different Columns/Mobile Phases Using an Automated Method Development System or Column Switcher.*
If the desired selectivity is not obtained or the peak is not spectrally homogeneous from **Step 5**, then different columns (two to six) and mobile phases can be screened using a column switcher. The typical mobile phases that could be screened include acetonitrile, methanol, and acetonitrile/methanol mixtures using the same $_w^w$pH

aqueous phase. However, other approaches include using an automated method development tool such as the Waters AMDS, which may facilitate method development and optimization. Using this system method, optimization can be performed by varying several chromatographic parameters in order to satisfy a user's set criteria.

AMDS allows the user to vary the following chromatographic parameters:

- Organic solvent
- Gradient
- Column temperature
- Type of column

When using such a system, the analytical goals for the separation must be defined. The following conditions were chosen:

- Organic solvents: methanol, acetonitrile
- Aqueous: 10 mM NH_4OAc, pH 5.8, adjusted with acetic acid
- Columns: Luna C18 (2) 3 μm (150 × 4.6 mm), Sunfire C18 3.5 μm (150 × 4.6 mm)
- Resolution more important than run time
- Run time < 10 min
- Minimum resolution > 3.0 (between identified peaks)
- Pressure < 3000 psi

In this case study, two different C18 columns from different manufacturers were used. Alternatively, other stationary phase types could also be used such as a polar embedded phase and a C18 phase. Some systems come also equipped with a six-column switcher and in that case, two different types of polar embedded phases, phenyl phase, pentafluorophenyl phases, two different C18 phases (of different bonding density) and an alternate C8 phase could be used.

In Figure 8-43, the following chromatograms were run at two different temperatures, using two different gradients (shallow and steep) on two different columns using methanol as the organic modifier. Due to low solvent strength of methanol, the main compound did not elute with the steep short gradient or eluted late with the shallow longer gradient. Therefore it is recommended that if the chromatographer uses pure methanol as the organic modifier, experiments should be performed by starting at higher methanol content (i.e., 25% methanol) using the same gradient time frames, or a mixture of methanol/acetonitrile could be used (30:70) employing a gradient from 10% to 95% organic. The addition of methanol may provide additional selectivity compared to pure acetonitrile. Note that the buffer solubility limit of the particular buffer an analyst is working with must be known; otherwise, precipitation

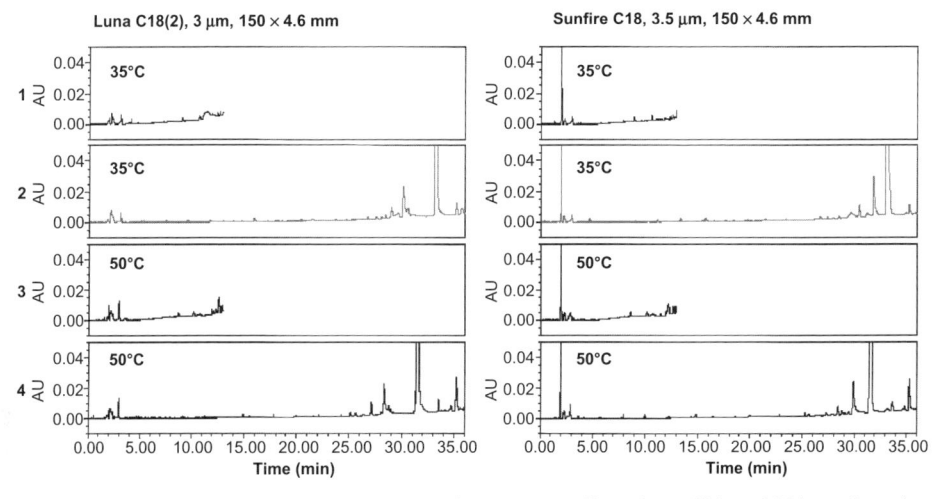

Figure 8-43. For both columns, run 1 used a steep gradient from 5% to 90% methanol in 13 min, run 2 used a shallow gradient from 5% to 90% methanol in 36 min, run 3 used a steep gradient from 5% to 90% methanol in 13 min, and run 4 used a shallow gradient from 5% to 90% methanol in 36 min. Runs 1 and 2 were at 35°C, and runs 3 and 4 were at 50°C. Flow rate for all four runs was 0.8 mL/min.

of the buffer salts will occur in the system/column. Ammonium acetate is fully soluble using 95% methanol and/or 95% acetonitrile. Generally, with a 10 mM phosphate buffer, it is soluble at 45°C up to 90% methanol and 80% acetonitrile.

The same experiments were performed with acetonitrile as the organic modifier, Figure 8-44. However, in all runs (Figure 8-44) with both columns with the acetonitrile mobile phase, all components eluted regardless of the steepness of the gradient employed. In all runs the lower temperature (35°C) seemed to give the best resolution of all the impurities from each other and the main component. Acetonitrile is the preferred organic solvent for this separation. Greater resolution between the critical pair (main component and impurity eluting after main component) is observed on the Sunfire column (Figure 8-45B) versus the Luna column (Figure 8-45A). The target analyte peak was also determined to be spectrally homogeneous using diode array detection, and the impurity that eluted after the main component (RRT 1.04) had a similar spectrum to the impurity that was co-eluting in the original separation, on the C8 column in Figure 8-42. However, further optimization was necessary to shorten run time.

8.5.6.11 Drylab for Gradient Optimization. Drylab was used for further optimization of the run time. All the chromatographic results from the four gradients run on the Waters Sunfire C18 were entered into DryLab to

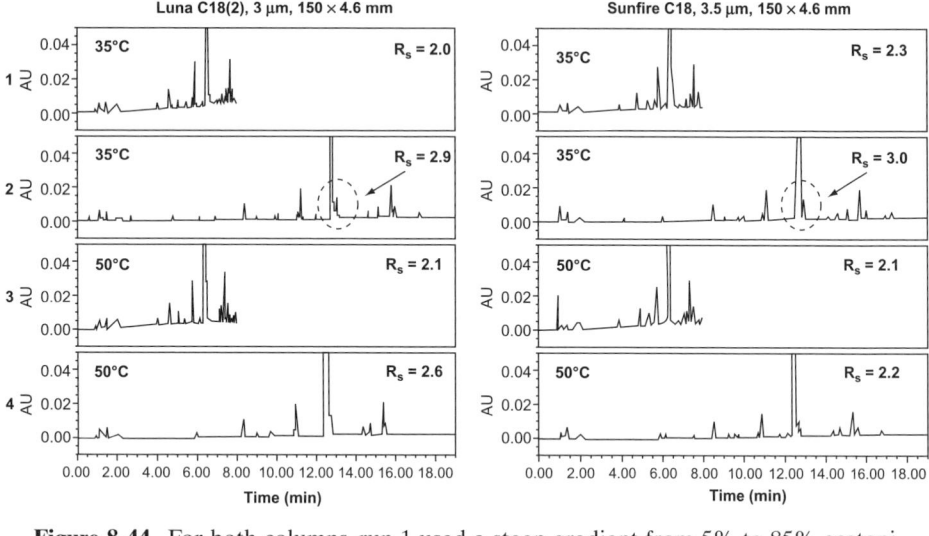

Figure 8-44. For both columns, run 1 used a steep gradient from 5% to 85% acetonitrile in 8 min, run 2 used a shallow gradient from 5% to 85% acetonitrile in 19 min, run 3 used a steep gradient from 5% to 85% acetonitrile in 8 min, and run 4 used a shallow gradient from 5% to 85% acetonitrile in 19 min. Runs 1 and 2 were at 35°C, and runs 3 and 4 were at 50°C. Flow rate for all four runs was 1.5 mL/min.

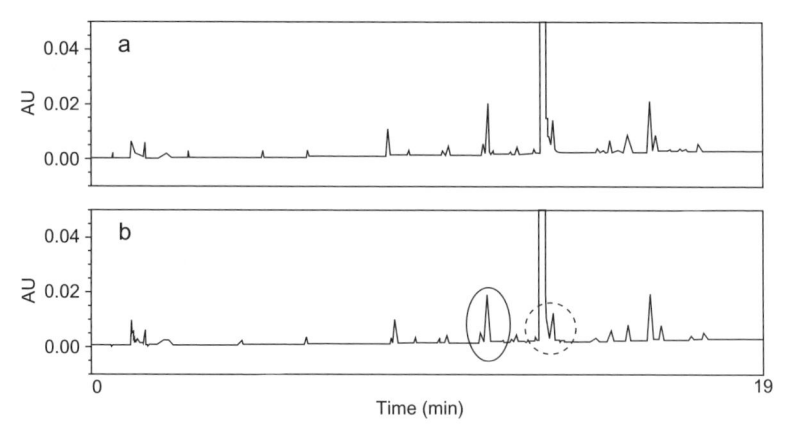

Figure 8-45. Automated method development. (**A**) Luna C18(2), 3 mm, 150 × 4.6 mm, 10 mM NH₄OAc, pH 5.8. Acetonitrile: Shallow gradient (5–85% acetonitrile in 19 min); flow, 1.5 mL/min; temperature, 35°C. (**B**) Sunfire C18, 3.5 µ, 150 × 4.6 mm, 10 mM NH₄OAc, pH 5.8. Acetonitrile gradient (5–85% acetonitrile in 19 min): Flow, 1.5 mL/min; temperature, 35°C.

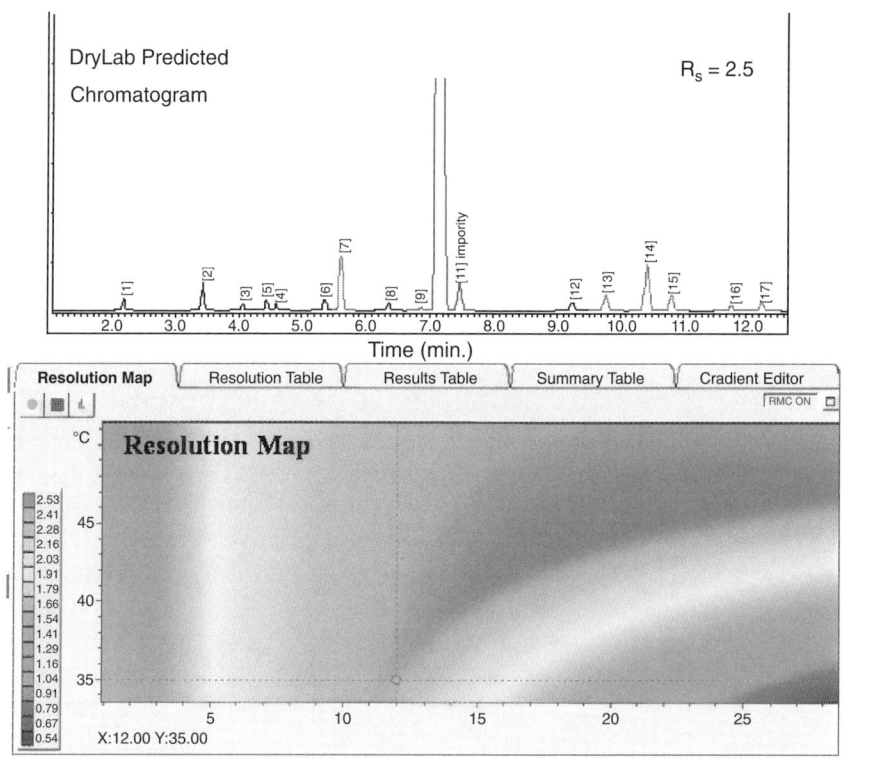

Figure 8-46. Resolution between active (compound M) and Impurity 11. See color plate.

determine the resolution map for the active and *impurity 11* eluting close to the main component (Figure 8-46). In a similar fashion, a resolution map could be generated for all components in the mixture (Figure 8-47). Using the resolution map shown in Figure 8-46, it was determined that the optimal run time was 12 min (35% acetonitrile to 75% acetonitrile over 10 min and 2-min hold at 75% acetonitrile) at a temperature of 35°C. The DryLab-predicted chromatogram is shown in Figure 8-46. Using these conditions, an actual verification run was performed using the predicted conditions, and excellent agreement was obtained with the Drylab-predicted result (Figure 8-48). The run time was decreased without compromising the resolution between the active and the impurity eluting after the main component.

Note that Drylab gives accurate predictions for the main analyte and impurities only if the analyte ionization state is not changing with increasing organic. This was accounted for in Steps 2, 3, and 4. However, for other impurities in the sample Drylab may not give an accurate prediction if the analyte (impurities/degradation products) are changing ionization state with increasing organic.

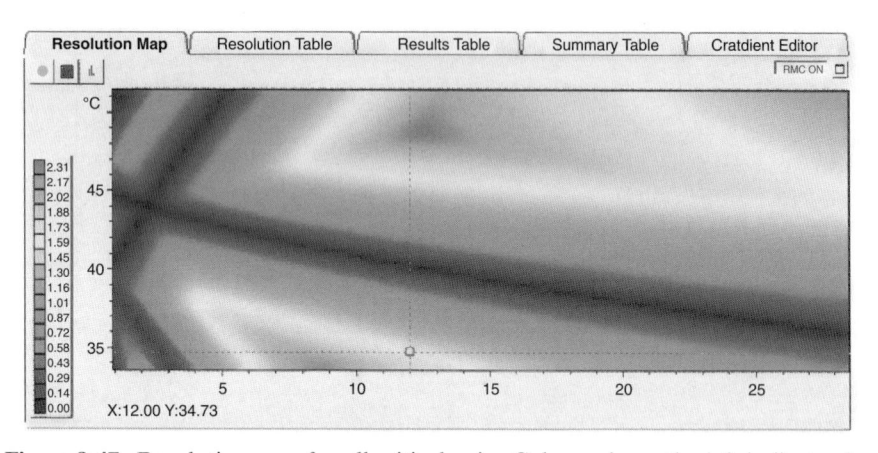

Figure 8-47. Resolution map for all critical pairs. Color scale on the left indicates the minimum resolution that is predicted for a particular color in the resolution map. The *x* axis is the gradient time and the *y* axis is the temperature. The crosshair can be moved to obtained the predicted conditions for optimal resolution of all critical pairs. See color plate.

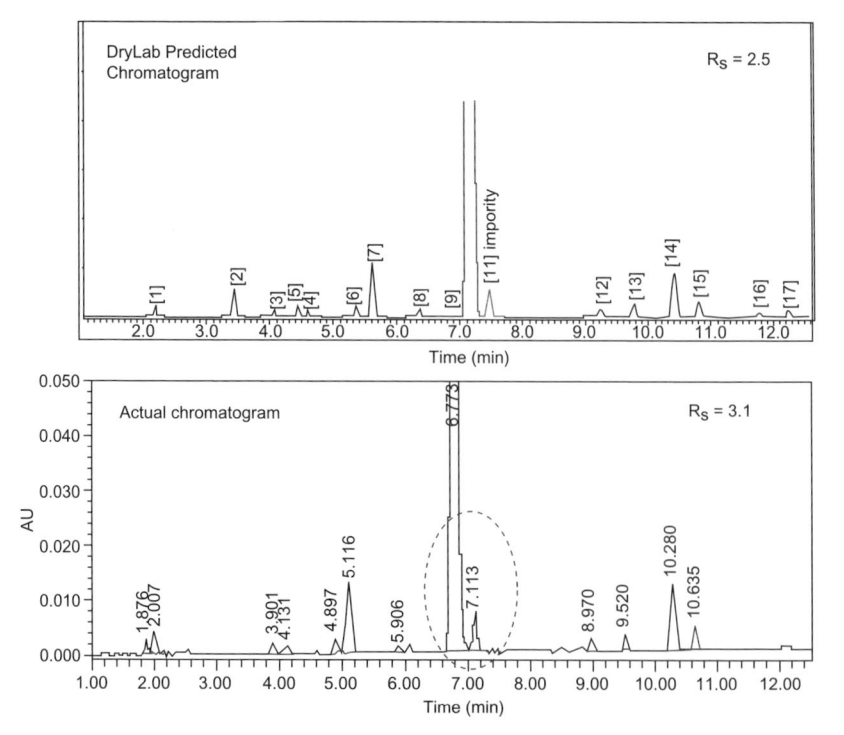

Figure 8-48. (Top) Drylab-predicted chromatogram versus (bottom) verification run.

8.5.6.12 Running Forced Degradation Samples. Prior to initiating method development as was mentioned in Section 8.4, the structure must be analyzed to predict most probable degradation products. For this compound in this case study, it has an amide functionality, so most probable impurity is the carboxylic acid impurity. It was indeed determined by a forced degradation study (acidic conditions) that it was the major acid hydrolysis degradation product. Using the method that was optimized by Drylab, a forced degradation sample was run that was stressed at 50°C for 1 week in pH 1 diluent. It was noted that a major impurity was formed (Figure 8-49) and was determined to be the carboxylic acid impurity by MS analysis. This compound contains an amide and is prone to hydrolysis. Scheme 2 shows the potential degradation pathway. Therefore to enhance the retention of this potential degradation product the initial organic composition of the gradient was reduced from 35% to 25 v/v% acetonitrile (Figure 8-49). The carboxylic acid impurity has an enhanced retention at lower organic composition and had adequate resolution between active and impurity eluting after main component was still obtained.

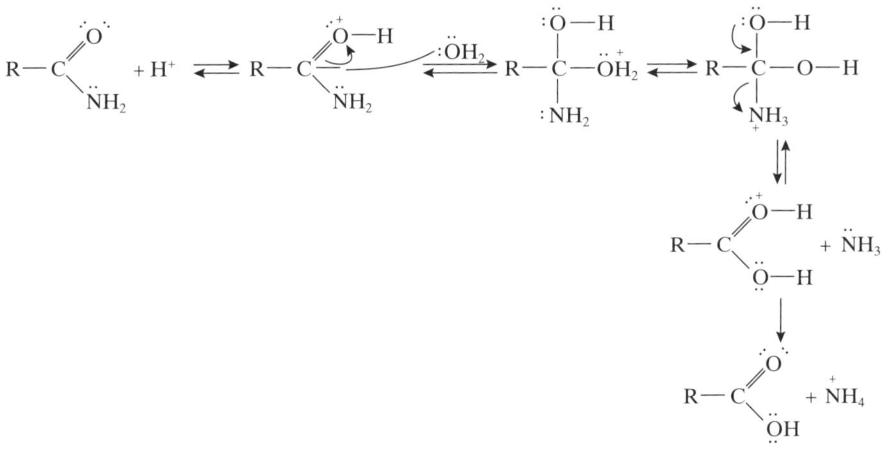

Scheme 2.

Using this optimized method shown in Figure 8-49 that starts at 25 v/v% acetonitrile, LC-MS studies were performed to determine the $[M + H]^+$ ion of the impurity that has been resolved from the main peak. The mass spectrum of Product M was taken and was shown to be spectrally homogeneous. The mass spectrum of the impurity (RRT 1.04) that has now been resolved from the main peak was also taken. The UV and the total ion chromatograms are shown in Figure 8-50. This impurity, RRT 1.04, has the same $[M + H]^+$ ion that was co-eluting with the main component in the initial separation on the C8

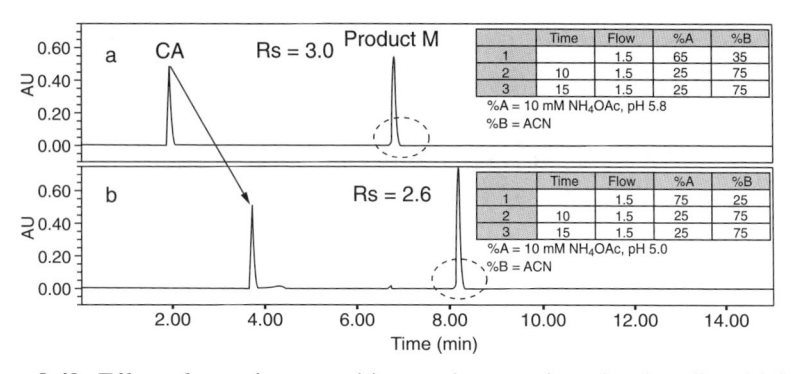

Figure 8-49. Effect of organic composition on the retention of carboxylic acid degradation product. Column: Waters Sunfire C18, 3.5 μm, 150 × 4.6 mm. (A) 10 mM NH₄OAc, pH 5.8. (B) Acetonitrile: Flow, 1.5 mL/min; temperature, 35°C.

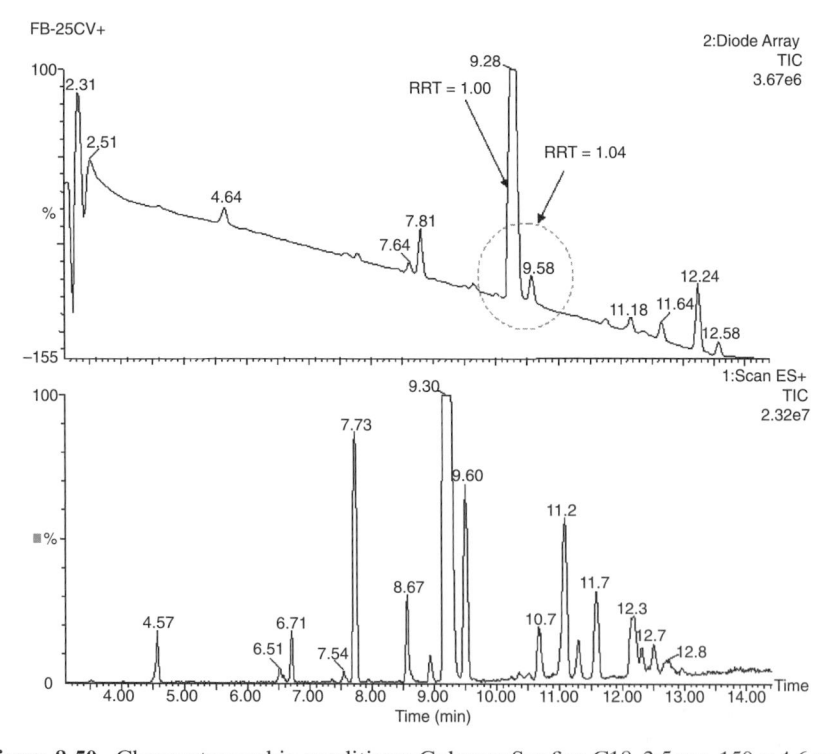

Figure 8-50. Chromatographic conditions: Column: Sunfire C18, 3.5 μm, 150 × 4.6 mm. Mobile phase: 10 mM NH₄OAc, pH 5.8; acetonitrile, 25% acetonitrile to 75% acetonitrile over 10 min and 5-min hold at 75% acetonitrile. Wavelength: 247 nm, column temperature, 35°C, flow, 1.3 mL/min; injection volume, 10 μL; MS conditions flow split 10: 1. ESI: + ion mode, single quadrupole, Z-Q. Capillary, + 3.5 kV; cone, 25 V; source temperature, 150°C; desolvation temperature, 400°C; cone gas flow, 113 L/hr; desolvation gas flow, 419 L/hr.

column (Figure 8-41). This confirms that the optimized method was able to resolve the impurity from the main component.

8.5.6.13 Final Optimization. The method could then be further optimized for speed by increasing the flow rate and decreasing gradient time (t_g) proportionally or by decreasing gradient range ($\Delta\%$) and decreasing t_g proportionally. These are further discussed in Chapter 17, Section 17.3. The temperature could be increased as well, but the chromatographer should be aware that although increases in speed may be realized, the selectivity may change with modifications in the temperature.

Other potential improvements in the method could include using a smaller (e.g., 3 mm) i.d. column while using the same length column and particle size of the packing material. A 3.0-mm-i.d. column can be used to reduce solvent waste, since columns with smaller diameters have reduced column volume and require use of lower flow rates, and therefore they can decrease solvent waste by at least 60%. A simple calculation to achieve equivalent retention on a smaller-i.d. column at the same linear velocity is shown in equation (8-2).

For example, if the original method used a 15-cm × 4.6-mm column with a 1.5 mL/min flow rate, what would be the equivalent flow rate for a 15-cm × 3-mm column? Using the following equation, this can be calculated.

$$\frac{\Pi r^2 L_{3.0\text{-i.d.column}}}{\Pi r^2 L_{4.6\text{-i.d.column}}} = \frac{r^2_{3.0\text{-i.d.column}}}{r^2_{4.6\text{-i.d.column}}} = 0.425 \tag{8-2}$$

If you divide the 1.5-mL/min flow rate by 0.425, then a flow rate of 0.64 mL/min is obtained. Moreover, use of lower flow rates can lead to enhanced ionization efficiency using ESI (no flow splitting). Increasing the flow rate increases droplet size, which decreases the yield of gas-phase ions from the charged droplets.

8.5.6.14 Case Study 3: Concluding Remarks. There is no cookbook for method development. The strategies presented an approach that could be taken for effective method development and optimization. In summary, a steep gradient was used initially to predict the suitable isocratic conditions for determining the most suitable pH for the method. The retention behavior of active as a function of pH (isocratic) was determined. The best mobile-phase pH for further gradient experiments was determined. Also, ACD was shown to be able to estimate the pK_a of the molecule; and by applying the rules based on pH shift of the mobile phase and pK_a shift of the analyte upon addition of organic component, the optimal pH for analysis was predicted. In order to elucidate if there was co-elution of impurities with the target analyte spectral homogeneity was assessed using both PDA and LC-MS. If possible, use AMDS/Dry Lab for method optimization and then use MS to confirm the separation of active species from possible co-eluting species. MS/MS analysis can be performed for further structural elucidation of the impurities. Deuterated experiments can be performed to support structural assignments.

8.5.7   Case Study 4: Structural Elucidation Employing a Deuterated Eluent

The fine structural details of analytes could be further defined by a deuterium-exchange experiments that measures the number of exchangeable protons in each molecule. The number of exchangeable protons in a molecule can be determined based on the mass shift. This technique allows an understanding of which protons are susceptible to exchange, but also can be used to differentiate compounds of the same molecular weight that have a different number of exchangeable protons. Deuterium exchange provides strong evidence to support degradation product and synthetic by-product elucidation. Take, for example, the two compounds 5-aminoindazole and 1-aminoindan, both of which have an $[M + H]^+$ of 133.9. An HPLC method was developed to separate these two compounds (Figure 8-51). The mass spectra for each compound are shown in Figures 8-52A and 8-52B. Both of these compounds show $[M + H]^+$ ions of 133.9.

If an analytical chemist were to discern between these two compounds, MS/MS analysis could be performed, but if only a single quadrupole instrument was available, what additional experiments could the analytical chemist perform on the single quadrupole instrument? The use of deuterated mobile phases could be used. Since 5-aminoindazole has three exchangeable protons and 1-aminoindan has only two exchangeable protons, by using a deuterated mobile phase the $[M + D]^+$ species would be different. Both these compounds were run by the same HPLC method as in Figure 8-51, but with a deuterated mobile phase (70% D_2O:30% MeCN). The mass spectra of the two components were taken from this chromatographic run. In Figure, 8-53,

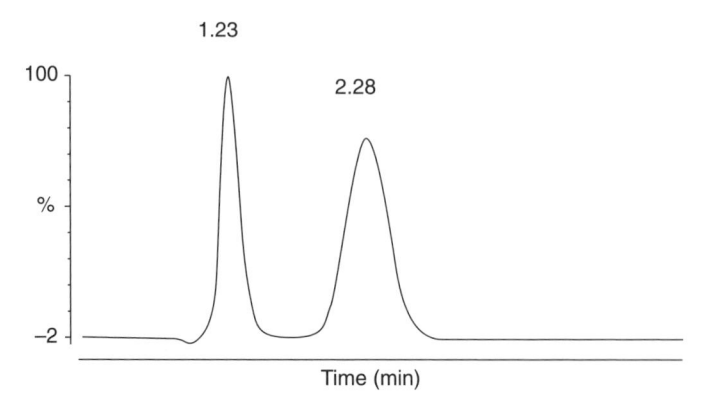

Figure 8-51. HPLC separation of 5-aminoindazole (RT at 1.23 min) and 1-aminoindan (RT at 2.28 min), 70% H_2O:30% MeCN. Flow, 0.5 mL/min; Waters symmetry shield, 50×4.6 mm, 5 μm.

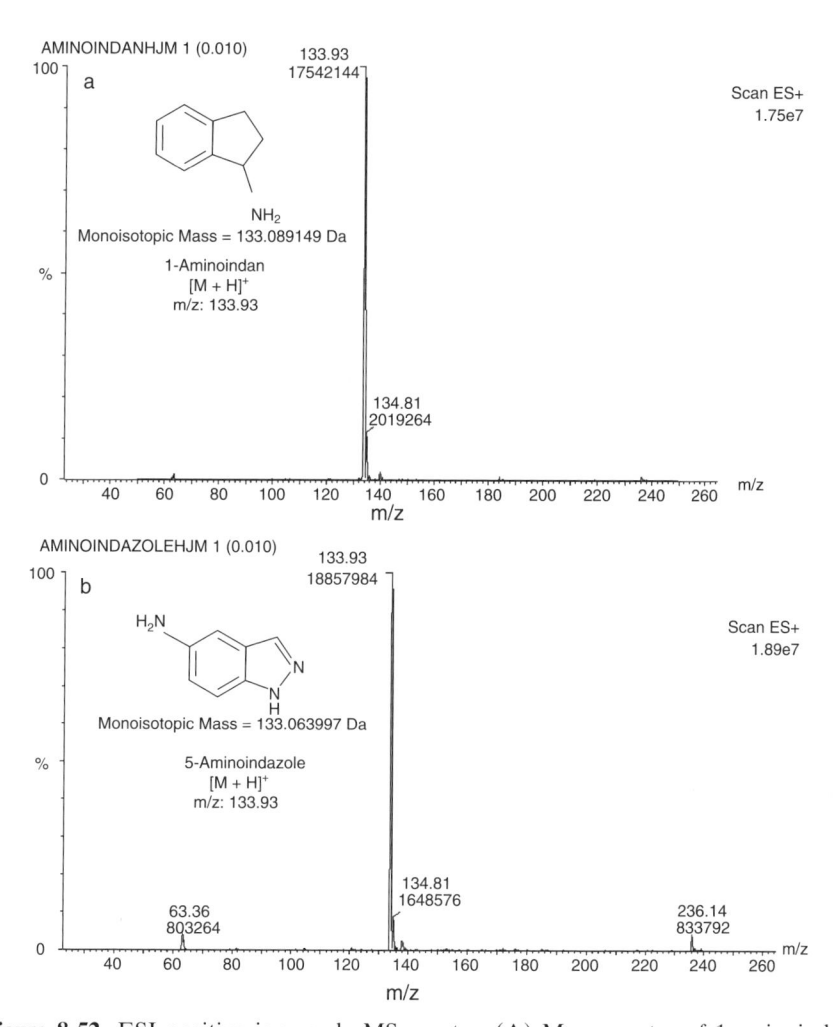

Figure 8-52. ESI positive ion mode MS spectra. (A) Mass spectra of 1-aminoindan (ESI positive ion mode). (B) Mass spectra of 5-aminoindazole (ESI positive ion mode).

the mass of the first component on the chromatogram increases by 5, indicating the presence of three exchangeable protons (one for each hydrogen atom replaced with D and two from D^+), which suggests that this analyte is 5-aminoindazole. In Figure 8-54 the mass of the second compound increases by 4; therefore only two exchangeable protons (one for each hydrogen atom replaced with D and two from D^+) are present, suggesting that the analyte is 1-aminoindan. The assignments of each species were confirmed by running individual standards of each compound. Knowledge of the number of labile H atoms in a molecule is useful for assisting in the elucidation of proposed impurity structures.

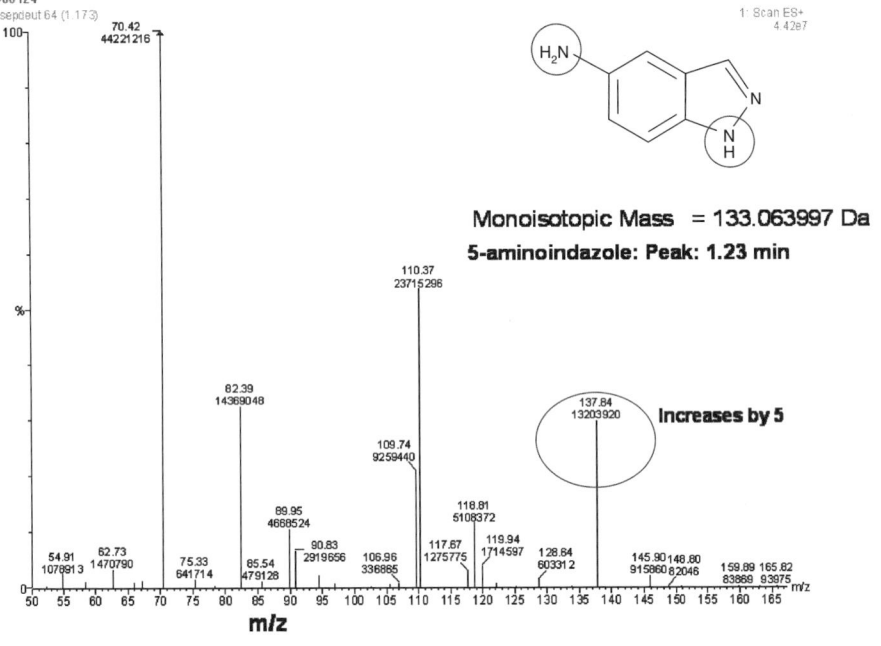

Figure 8-53. Mass spectra of 5-aminoindazole in deuterated mobile phase.

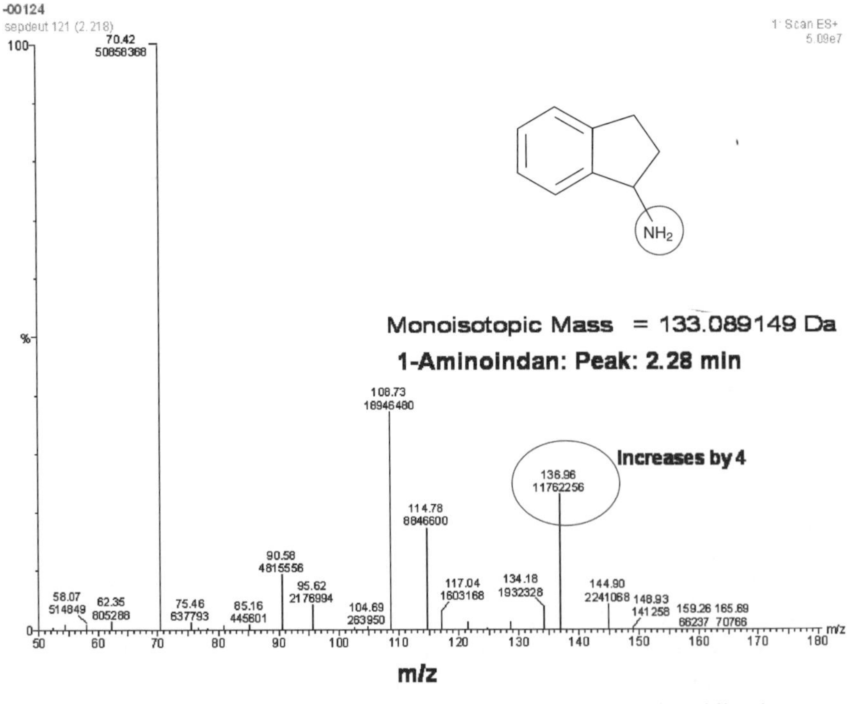

Figure 8-54. Mass spectra of 1-aminoindan in deuterated mobile phase.

8.6 EFFECT OF pH ON UV ABSORBANCE

The extent to which an organic molecule absorbs electromagnetic radiation in the ultraviolet region (UV light) depends on the structure of the molecule. Generally, molecules that contain one single double bond absorb weakly in the UV region. However, if multiple double bonds are present in a molecule and they are conjugated, the molecule absorbs more strongly and the absorbance is shifted to longer wavelengths.

The pH of the mobile-phase effects the ionization of ionogenic solutes and also the analyte UV response. The change in pH can change the electronic structure of the molecule and result in changes in the molar absorptivity and the absorption maximum of the molecule [25]. Ionization of aromatic compounds containing a pyridinal nitrogen, amino, carboxylic acid, and phenolic group can lead to significant changes of their UV response. Understanding the effects of charge delocalization and conjugation on the UV response and detection wavelength will allow the chromatographer to choose the proper pH and wavelength of detection to obtain a method with higher sensitivity. Silverstein et al. [26] and Shenk [27] provide a good overview for predicting how the structure of a molecule and its environment will affect its molar absorptivity and the wavelength of the absorption maximum.

Most applications of absorption spectroscopy to organic compounds are based upon transitions for n or π electrons to the π^* excited state. Energies required for these processes bring absorption peaks into the spectral region (200–700 nm). π-electrons are further delocalized by conjugation. The effect of this delocalization is to lower the energy level of the π^* orbital and give it less antibonding character and as a result absorption maxima are shifted to longer wavelengths [28].

UV spectra of aromatic hydrocarbons are characterized by three sets of bands (E1, E2, and B bands) that originate from $\pi \rightarrow \pi^*$ transitions. Generally the E2 and B bands are of most interest to chromatographers, since the solvent cutoff for most mobile phases is <200 nm.

For example, benzene has strong absorption peaks at

E1:	184 nm,	$\varepsilon_{max} \sim 60{,}000$
E2:	204 nm,	$\varepsilon_{max} = 7{,}900$
B:	256 nm,	$\varepsilon_{max} = 200$

Table 8-9 shows E2 and B bands for some organic molecules. Auxochromes are a functional group that does not itself absorb in the UV region but have the effect of shifting chromophore peaks to longer wavelengths and increasing their intensity. The —OH and —NH$_2$ groups have an auxochromic effect on benzene chromophore. These substituents have at least one pair of n electrons capable of interacting with π electrons of the ring. This stabilizes the π^* state and lowers its energy. The phenolate anion auxochromic effect is more pronounced than for phenol because the anion has an additional pair of

TABLE 8-9. Molar Absorptivity Values for Neutral, Acid, and Basic Species[a]

Compound	Molecular Formula	E_2 Band		B Band	
		λ_{max} (nm)	ε_{max}	λ_{max} (nm)	ε_{max}
Benzene	C_6H_6	204	7,900	256	200
Naphthalene	$C_{10}H_8$	286	9,300	312	289
Toluene	$C_6H_5CH_3$	207	7,000	261	300
Chlorobenzene	C_6H_5Cl	210	7,600	265	240
Phenol	C_6H_5OH	211	6,200	270	1,450
Phenolate ion	$C_6H_5O^-$	235	9,400	287	2,600
Thiophenol	C_6H_5SH	236	10,000	269	700
Aniline	$C_6H_5NH_2$	230	8,600	280	1,430
Anilinium ion	$C_6H_5NH_3^+$	203	7,500	254	160

[a]Values from reference 28.

unshared electrons. Aniline has a pair of n electrons capable of interacting with the π electrons of ring. This stabilizes the π^* state by the relationship shown in Equation (8-3), thereby lowering its energy [28]. With a decrease in protonation, the absorption maxima would be shifted to longer wavelengths and increasing intensities and a red shift occurs. However, upon protonation the nonbonding electrons are lost by formation of the anilinium cation, and the auxochromic effect disappears as a consequence.

$$E = hv = \frac{hc}{\lambda} \qquad (8\text{-}3)$$

The change in the mobile-phase $_s^s$pH at a constant organic composition may have an effect on an ionizable analyte's UV response (Figure 8-55A). Also, at constant $_w^w$pH as the organic concentration is increased, this may also lead to a change in the analytes absorbance at a particular wavelength. Increasing concentration of the organic shifts the pH of the mobile phase upward (for an acidic modifier), and changes in UV absorbance may be observed (Figure 8-55B).

At 232 nm there is a decrease in aniline's absorbance as this analyte becomes progressively more ionized. A plot of the UV absorbance at a particular wavelength versus the $_w^w$pH of the aqueous phase will lead to a sigmoidal dependence (Figure 8-56). The inflection point corresponds to the analyte pK_a (not corrected for pH shift of the mobile phase). When performing method development experiments a judicious choice for the wavelength of the detection should be carefully considered because this can lead to desired/undesired effects (change in sensitivity at particular wavelength as a function of pH) on the resulting chromatography. Figure 8-57 demonstrates that a greater response for aniline is observed at $_w^w$pHs where the analyte is in

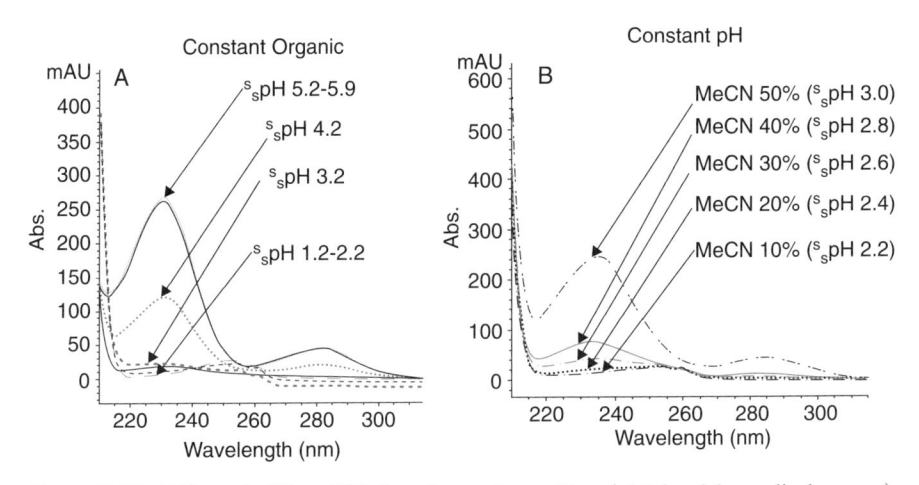

Figure 8-55. Effect of pH on UV absorbance for aniline (obtained from diode array). (A) 10 v/v% acetonitrile and pH of 15 mM $K_2HPO_4 \cdot 7H_2O$ adjusted to ${}^w_w$pH 1–9 with H_3PO_4. (B) ${}^w_w$pH2.0 and acetonitrile concentration changed from 10 to 50 v/v%.

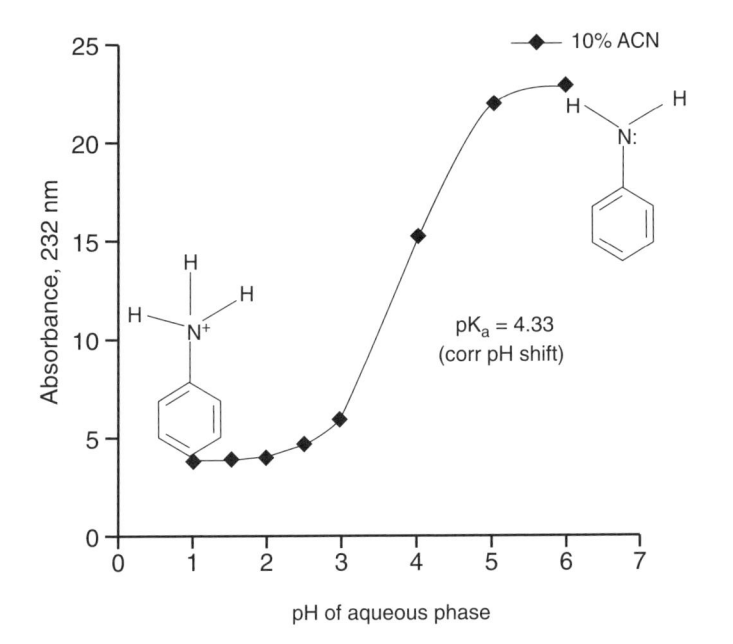

Figure 8-56. Absorbance at 232 nm versus the ${}^w_w$pH of the aqueous phase. Mobile phase contains 10 v/v% acetonitrile.

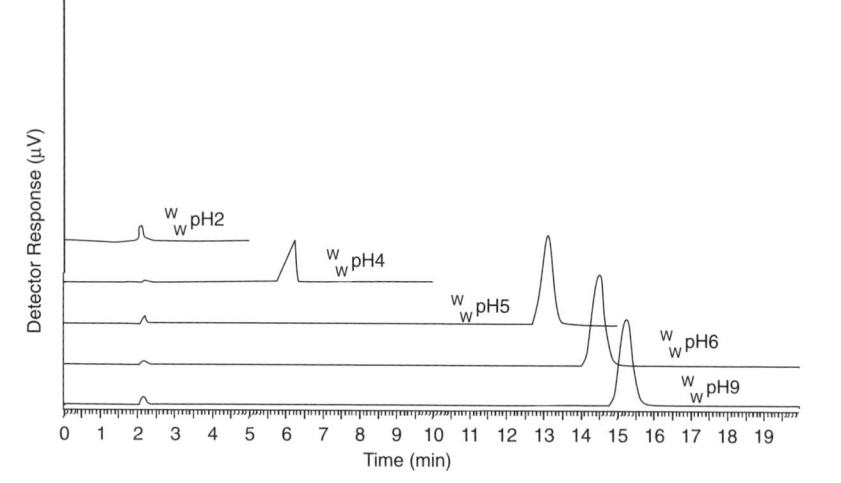

Figure 8-57. Effect of pH on UV absorbance for aniline. Conditions: Column: 15-cm × 0.46-cm Luna C18(2). Eluent: 90% aqueous:10% MeCN. Aqueous: 15 mM $K_2HPO_4 \cdot 7H_2O$ adjusted to $_w^w pH$ 1.5–9 with H_3PO_4, Flow rate, 1 mL/min; temperature, 25°C; detection, PDA.

its neutral state. As the analyte ionization state varies with pH so does the conjugation. In some cases, the wavelength in a specific region does not vary with the pH and the sensitivity of the analysis will not change as a function of pH as seen for 2,4-dihydroxybenzoic acid, at 280 nm in Figure 8-58.

8.7   ANALYTE pK_a—FROM AN ANALYTICAL CHEMIST'S PERSPECTIVE

In order to avoid any secondary equilibrium effects on the retention of ionogenic analytes, it is preferable to use a mobile-phase pH either two units greater or less than the analyte pK_a. Therefore knowledge of the analyte pK_a is very important. A basic understanding of how functional group substitution on a molecule affects the pK_a of the ionizable group on the substrate is given. An exhaustive description of all the nuances of analyte substitution on analyte pK_a is not included in this section. However, further details can be found in the references 29–31.

8.7.1   Aromatic Acids

Effect of Analyte Substitution on Analyte pK_a. The acidity of substituted phenols or carboxylic acids depend upon the substituent attached to the cor-

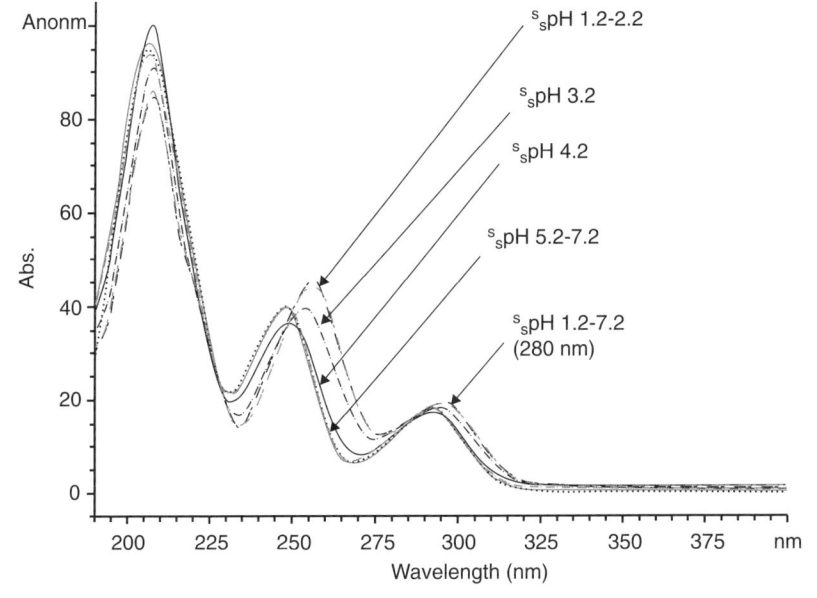

Figure 8-58. Effect of pH on UV absorbance for 2,4-dihydroxybenzoic acid. **Conditions**: Column: 15-cm × 0.46-cm Luna C18(2). Eluent: 90% aqueous:10% MeCN. Aqueous: 15 mM $K_2HPO_4 \cdot 7H_2O$ adjusted to ${}^w_w$pH 1–7 with H_3PO_4. Flow rate, 1 mL/min; temperature, 25°C; detection, PDA.

responding substrate, phenol, or carboxylic acid Aromatic acids with an electron-withdrawing substituent are more acidic because these substituents stabilize the ion by delocalizing the negative charge. Aromatic acids with electron-donating groups are less acidic because the substituents destabilize the ion by localizing the charge.

8.7.1.1   Electron-Withdrawing Effects—Aromatic Acids. Electron-withdrawing groups in the nucleus of the substrate increases the acidity. Inductive effect usually falls off with distance: ortho (*o*) > meta (*m*) > para (*p*). However, electron-withdrawing mesomeric effects also play a role when the electron-withdrawing substituent is in the *o*- or *p*-position (see Figure 8-59). This promotes ionization by stabilization (through delocalization) of resultant anion.

8.7.1.2   Electron-Donating Groups—Aromatic Acids. The effect of electron-donating groups such as alkyl groups attached to the benzene nucleus are small. These substituents destablize the phenoxide anion and disturb the interaction of the negative charge with delocalized *p* orbitals of the aromatic nucleus, shown in the following table.

Compound	pK_a
phenol	9.95
o-MePhenol	10.28
m-MePhenol	10.08
p-MePhenol	10.19

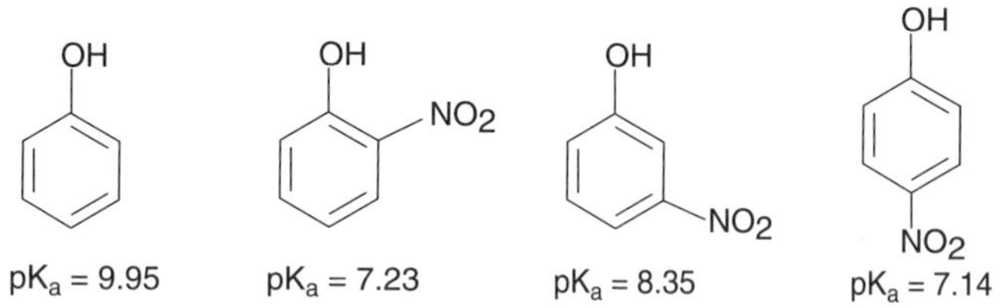
OH
OH
NO2
OH
NO2
OH
NO2
pKa = 9.95
pKa = 7.23
pKa = 8.35
pKa = 7.14

Figure 8-59. pK_a values of aromatic acid (phenols) species with electron-withdrawing groups.

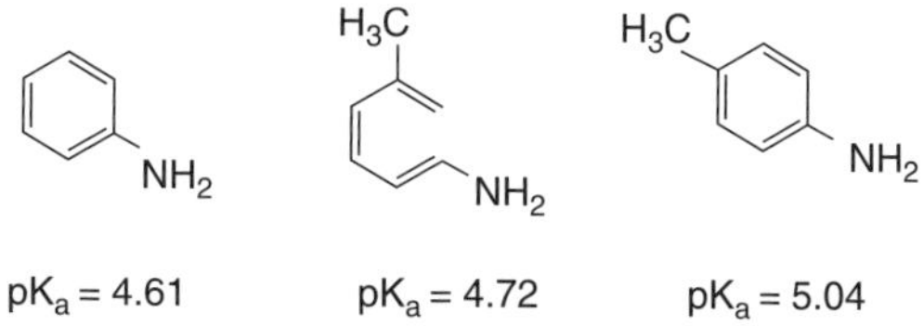
NH2
H3C
NH2
H3C
NH2
pKa = 4.61
pKa = 4.72
pKa = 5.04

Figure 8-60A. pK_a values of aromatic amine species with electron-donating groups.

8.7.2 Amines

8.7.2.1 Arylamines. Arylamines like aliphatic amines are basic. A lone pair of nonbonding electrons on nitrogen can bond to acids, yielding an arylammonium salt. Base strength of arylamines are lower than aliphatic amines. A stronger base corresponds to a less acidic ammonium ion (higher pK_a). A weaker base corresponds to a more acidic ammonium ion (lower pK_a).

8.7.2.2 Aromatic Amines—Electron-Donating Groups. Electron-donating groups tend to disperse the positive charge of the anilinium ion, and this stabilizes the ion relative to the amine. Electron-donating groups increase the basicity. Electrons are being pushed toward nitrogen and makes the fourth pair more available for sharing with acid. These activating substituents make the aromatic ring electron-rich. Some examples are shown in Figure 8-60A. Electron donors ($-CH_3$, $-NH_2$, $-OCH_3$) increase the basicity of arylamines.

8.7.2.3 Aromatic Amines—Electron-Withdrawing Groups. Electron-withdrawing groups tend to intensify the positive charge of the anilinium ion, and this destabilizes the ion relative to the amine. This increase the reactivity of an aromatic

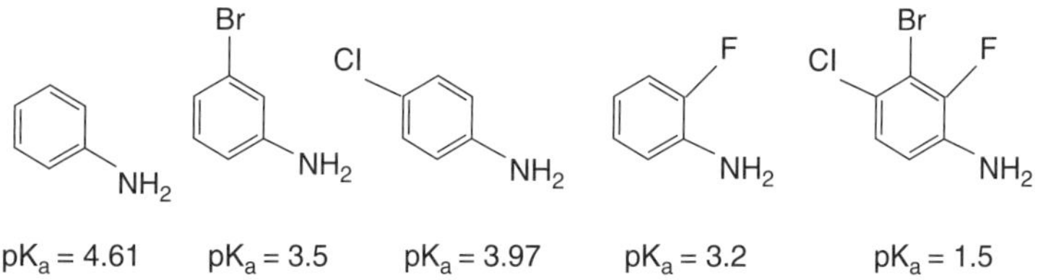
Br
Cl
F
Cl
Br
F
NH2
NH2
NH2
NH2
NH2
pKa = 4.61
pKa = 3.5
pKa = 3.97
pKa = 3.2
pKa = 1.5

Figure 8-60B. pK_a values of aromatic amine species with electron-withdrawing groups.

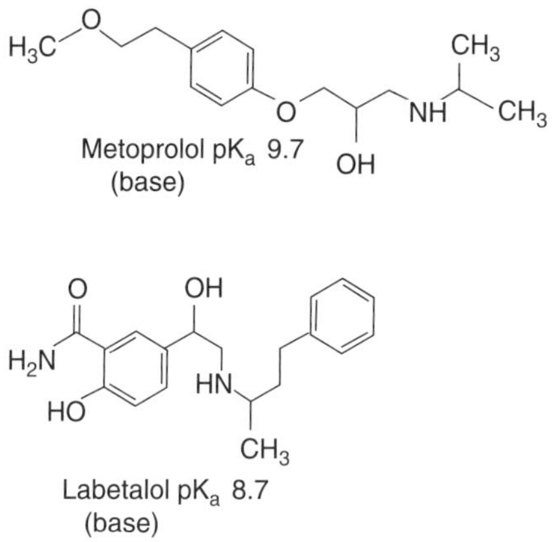
H3C
O
CH3
O
NH
CH3
OH
Metoprolol pKa 9.7
(base)
O
OH
H2N
HN
HO
CH3
Labetalol pKa 8.7
(base)

Figure 8-61. pK_a values of secondary amine species of two pharmaceutical compounds.

ring toward electrophilic substitution. Electron-withdrawing groups decrease the basicity and pull electrons away from nitrogen and make the fourth pair less available for sharing with acid. These deactivating substituents make the aromatic ring electron-poor. Some examples are shown in Figure 8-60B. Electron-withdrawing groups (–C1, $-NO_2$, –CN) decrease arylamine basicity.

8.7.2.4 Alkyl Amines and Amides. Nonaromatic secondary and tertiary amines have pK_a values greater than 8, as shown for two beta blockers in Figure 8-61. Amides are nonbasic, poor nucleophiles and do not protonate in aqueous acids. As with carboxylic acids, the resonance stabilization of the negative charge of the CH_3CONH- rationalizes the higher acidity of the amide. pK_a values of amides are typically greater than 15.

8.8 REVERSED-PHASE VERSUS NORMAL-PHASE SEPARATIONS

Reversed-phase HPLC is the dominant method used for the majority of pharmaceutical applications (>95%). Normal-phase chromatography may be required for separations that are not compatible with reversed-phase mode.

Solutes that are labile (i.e., reacts with protic solvents) or exhibit poor solubility in aqueous media are prime candidates for normal-phase chromatography. Normal phase is well-suited for the separation of isomers and diastereomers, as well as for separating compounds with saturated and unsaturated side chains. Generally, the greater is the amount of unsaturation the greater the retention due to increased polarizability of double bond.

Diol phases are a good starting point for normal-phase application. Silica, amino, and cyano are alternative phases. Silica tends to strongly retain solutes that can interact with its highly active sites. Hexane or heptane modified with a polar organic solvent is generally utilized as the mobile phase. The polar organic solvent can be chosen based on it physicochemical properties (dipole, hydrogen bond acceptor/donor). Generally, small changes of the polar organic solvent can cause large changes in retention, and this should be investigated during method development. Common solvents include ethanol, isopropanol, tetrahydrofuran, ethyl acetate, and dichloromethane. The level of water in the solvents needs to be controlled as well, since differences in retention may be observed. Additives such as trifluoroacetic acid or triethylamine can be used to reduce interactions with the highly active sites of silica, allowing for reduced retention and improved peak shape. A further description of normal-phase chromatography can be found in Chapter 5.

Normal-Phase Chromatography Example. Vitamin E, an antioxidant, is a complex made up of tocopherols and tocotrienols (Figure 8-62), which are sometimes used to stabilize formulations. Tocopherols are a series of related benzopyranols with a C16 saturated side chain. Tocotrienols contain three double bonds on the C16 side chain [32].

Could you predict the elution order of the alpha, beta, gamma, and delta isomers in the normal-phase mode? Note that in the normal phase, the less hydrophobic the compound and the more substituents that could potentially hydrogen bond to the stationary phase, the greater the affinity for the stationary phase and the longer the retention. The order of elution for the alpha, beta, gamma, and delta isomers for both the tocopherols and the tocotrienols series is the same (Figure 8-63) [32]. The order of elution for beta and gamma would be hard to predict because they have very similar hydrophobicity and same number of potential hydrogen bonding moieties. Their differences in elution order depend on the planarity of the molecule and its interaction with the stationary phase. In normal-phase chromatography, the more unsaturated molecules, tocotrienols, elute later compared to the tocopherols, which have a saturated side chain, and this could be attributed to the increased polarizability of the double bond [33]. Comparing the separation to reversed mode, the elution order is reversed, where the retention is as follows: delta tocopherol < gamma tocopherol < beta tocopherol < alpha tocopherol < alpha tocopheryl acetate (Figure 8-64).

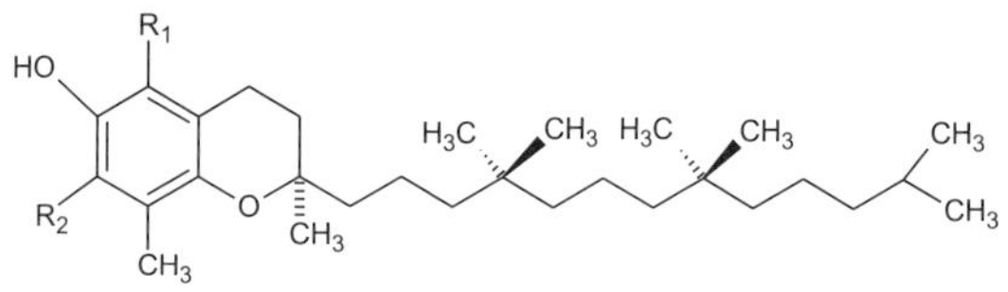

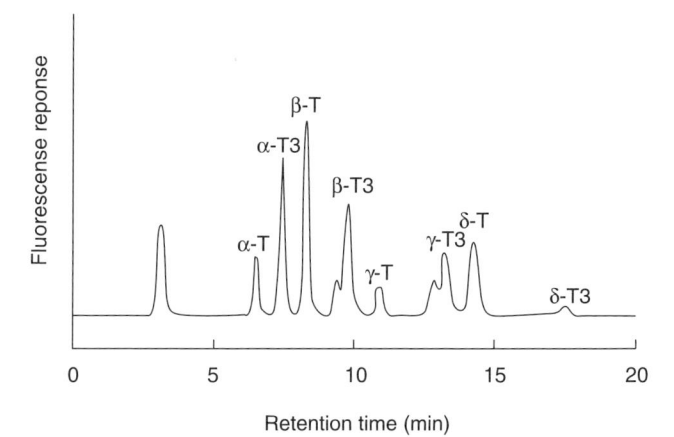

Tocopherols

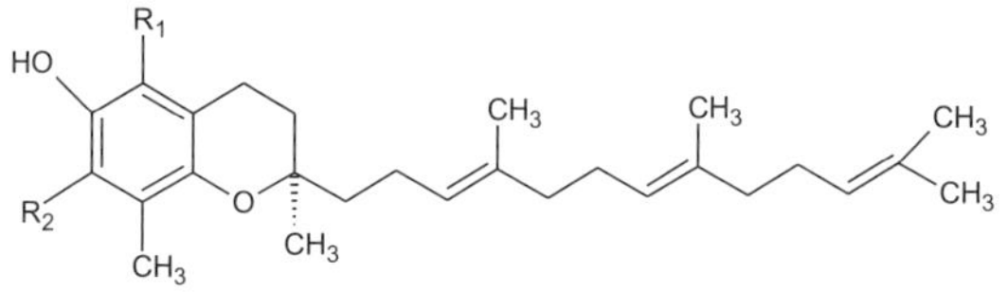

Tocotrienols

Tocopherols	Tocotrienols	R_1	R_2
α-Tocopherol (α-T)	α-Tocotrienol (α-3)	Me	Me
β-Tocopherol (β-T)	β-Tocotrienol (β-3)	Me	H
γ-Tocopherol (γ-T)	γ-Tocotrienol (γ-3)	H	Me
δ-Tocopherol (δ-T)	δ-Tocotrienol (δ-3)	H	H

Figure 8-62. Structures of substituted tocopherol and tocotrienols.

Figure 8-63. HPLC conditions: Genesis silica column (250 × 4.6 mm, 4 μm). Flow rate, 1.5 mL/min. Mobile phase: Hexane-1,4-dioxane (96:4). Fluorimetric detection: Fluor LC 304 (excitation @ 294 nm and emission @ 326 nm). (Reprinted from reference 32, with permission.)

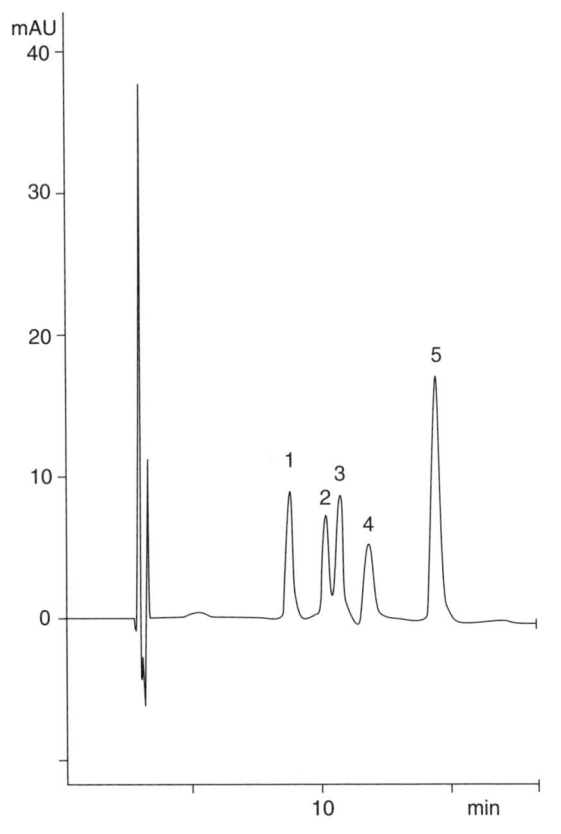

Figure 8-64. Separation of synthetic tocopherols by reversed-phase HPLC (280 nm). (1) δ-tocopherol, (2) γ-tocopherol, (3) β-tocopherol, (4) α-tocopherol, (5) α-tocopheryl acetate. (Reprinted from reference 33, with permission.)

8.9   INSTRUMENT/SYSTEM CONSIDERATIONS

The four common causes for high-performance liquid chromatography (HPLC) column failure include column clogging at the inlet frit (from samples/mobile phase), voids generated in the column, strongly adsorbed impurities from solvent/sample, and chemical attack of the stationary phase from the mobile phase or analytes. Procedure for removal of strongly adsorbed impurities from sample/mobile phase was discussed in Chapter 3, Section 3.9.2.

8.9.1   Column/System Backpressure

Column backpressure gives a good indication of how the column and/or system are operating. The initial backpressure of the column should be checked prior to running a method. The backpressure with column attached at initial run conditions should be stated in the method. If high column backpressure is observed, the column should be taken off and a ZDV (zero dead volume) should be installed and system backpressure recorded and compared to know system backpressure for that particular system under a certain set of mobile-phase conditions. Note that the system backpressure will be different from instrument to instrument due to the different types of tubings that are employed. This depends on the tubing diameters and total tubing length. The system backpressure is also dependent on the mobile-phase composition, the type of organic modifier, the temperature, and the flow rate. If the system backpressure value is known on that particular system with a certain set of eluent and column conditions and the backpressure value is higher than normal, then the cause of this increased backpressure needs to be investigated. A recommendation is to start removing tubing, starting at the outlet and working your way back to the injector and then the pump, until you see a drop of pressure; then replace that piece of tubing that is leading to the problem with higher backpressure. Also, it is possible that an in-line filter may be clogged. The in-line filter is usually made of a 0.5-μm porosity frit and is located either between the pump and injector or between the injector and the column. It is recommended to change this in-line filter on a monthly basis (given continuous flow through the system during that month), especially if buffered eluents are used. These filters should be readily available, and an analyst should know how to replace them to avoid waiting for a metrologist or vendor engineer or contract engineer to change them. Also, a proper maintenance log for each HPLC must be maintained in a regulated environment, and any type of maintenance should be properly recorded according to the pharmaceutical departments standard operating procedures.

8.9.2 Column Inlet and Outlet Frits

One of the most common symptoms of column failure is high column backpressure. Plugging is the most frequently encountered problem by analytical chemists or analysts. Injection of samples containing particulates, along with wear of pump piston seals and injector valve rotor seals, will eventually block the column inlet, causing high column backpressure, and shorten the normal lifetime of the column. Backpressure also generally increases as particulate matter accumulates on the inlet frit of the column and may lead to band distortion of the peaks in the chromatogram. Columns packed with 5-μm particles typically use 2-μm porosity inlet frits to contain the packing; for the 3-μm particles, 0.5- or 2-μm porosity inlet frits are used, depending on the manufacturer. However, the outlet frits for the 5- and 3-μm packings are typically 2 μm and 0.5 μm, respectively, but this is also dependent on the column manufacturer. Columns packed with sub-2-μm particles may have porosity inlet and outlet frits on the order of ≤0.5 μm. While the smaller particle sizes offer several advantages, including higher resolution, higher separation capability, and lower volumes of mobile phase, they are also more susceptible to premature plugging by particulates.

To a certain extent, particulate matter buildup on the frit is unavoidable, but there are some several simple practices that can slow down this process such as filtering the mobile phase and the sample prior to injection. Also, centrifuging the sample is recommended for at least 2 min at 10,000 rpm. If it is determined that a column inlet frit is blocked, then backflushing the column might help. For backflushing, remove the column from the system and connect the outlet end (normal direction flow) to the pump and put the inlet end into a beaker and pump at low flow rate 0.5 mL/min using 80% MeCN : 20% water for at least 20 column volumes, (column volume can be estimated by the volume of the cylinder, $\pi R^2 L$ for a 150- × 3.0-mm column the column volume is ~1.1 mL) to displace any particles from the frit. Make sure to pump directly from the column to waste and not through the detector in order to keep contaminants from going into the flow cell/detector. Before backflushing any column, it is recommended to discuss with the column manufacturer if this may adversely affect the integrity of the column.

8.9.3 Seals

Maintenance and care of the pump piston seals is recommended. Buffers and other types of salt additives (i.e., ion pairing reagents, chaotropic additives) that are not soluble in organic solvents should not be allowed to reside in the LC system when there is no flow through the HPLC system. Also, the buffer solubility limit in a particular solvent system should be known to prevent precipitation of the buffer salts in the HPLC system. A test tube precipitation test can be used to determine if the concentration of the organic will trigger precipitation (see Section 8.4.5.3 for details). If the buffer solution is left in a "dry" LC system, the buffer salts can evaporate on the piston surface behind the pump seal, thereby creating an abrasive coating of salt crystals that will damage the seal over time. Therefore it is recommended to wash the HPLC system with acetonitrile/water (20:80) for at least 30 min before the system is shut down to remove any potential buffer residues. A shut-down method with this wash method is recommended at the end of the sequence. Pump piston seals usually last for at least 6 months (if system is continually used throughout that duration) and are usually replaced during the preventive maintenance on the HPLC system. Injector rotor valve seals can last for greater than 10,000–20,000 cycles, and these should changed on an annual basis during the yearly preventive maintenance/calibration of the instrument.

8.9.4 Mobile-Phase Preparation

The operations of pump components, injectors, and detectors can be expected to be less troublesome when mobile phases are filtered. For HPLC applications, the 0.2-μm-pore-size filter is typically selected for removal of particulates that may arise from physical contaminants such as fibers. Generally, it is recommended to filter both the aqueous and organic portions of the mobile phase independently. (Note that some filter materials are soluble in organic solvents; check the specification of the filter before use.) Note that premixed mobile phases with organic and aqueous should not be filtered (using vacuum filtration) since this may change the final composition of the organic in the filtered mobile phase. Also the mobile phases must be covered to avoid evaporation of the buffer components (TFA, acetate, bicarbonate, ammonium hydroxide) and the solvent (especially if premixed mobile phases are used). Also this prevents dust and other particulates from contaminating the mobile phase. The mobile phase should be covered with suitable caps, and the use of aluminum foil is also encouraged. One of the most common mistakes that analysts make is the use of polymeric products to cover the mobile phase (parafilm), since this is not compatible with organic solvents and may lead to potential contamination of the mobile phase with polymeric components. It is also recommended to add at least 5–10 v/v% of organic solvent to the aqueous phase to prevent microbial growth.

The frit on the tubing connecting the mobile-phase reservoir to the pump also keeps particulate matter from entering the system. Usually a 5-μm porosity frit is used, without generating significant resistance to flow. A simple test for blockage of this frit is to determine if mobile-phase siphons freely through the frit and tubing. This can be performed by disconnecting the inlet tubing at the proportioning manifold (low-pressure-mixing system) or pump inlet (high-pressure mixing). The mobile phase will then siphon through this tubing. A delivery of 10 times the pump requirement is generally recommended. For example, if you normally operate the pump at 1 mL/min, no less than 10 mL/min should flow through the siphon [34]. If the flow is less than desired, the frit should be replaced or the frit could be sonicated in a solution of IPA/dilute nitric acid (1 v/v%).

8.9.5 Guard Columns

A guard column is a small column that contains packing material similar to that in the analytical column. The pore size and particle size of the material can be the same as the packing material in the analytical column. The guard column has a frit at each end, and the frits trap particulate matter. According to the column vendor claims, these guard columns are supposed to help to prevent contamination of the analytical column, but they also may lead to decrease in the efficiency of the peaks of interest due to the addition of extra-column volume. It may prevent contamination in the short term, especially for biological samples; however, on a long-term basis, since contaminants are also molecules and they are moving along the column, usually moving slower than analytes, they may elute from the guard column into the main column (i.e., leading to the generation of ghost peaks). The guard column is an excellent mixing chamber and can lead to band dispersion. However, guard columns are often used and recommended for analysis of proteins, lipids, and other high-molecular-weight species, if the sample is biological in origin (urine, plasma, etc.), to prevent contamination of the column with matrix components. The need for replacing the guard column depends on the matrix, the number of injections, and whether decreased performance is observed (change in efficiency for isocratic separations by more than 10% and gradient method change in apparent efficiency by more than 20%).

8.9.6 Instrument/System Considerations (Concluding Remarks)

There are quite a few techniques that can be used to extend the useful life of an HPLC column; these include employing "column shock" method after several hundred injections, running at lower operating temperatures, not running at pH extremes, and using moderate concentrations of buffers, to name a few. However, the column should be considered as a disposable item. This might be difficult to accept, considering that the typical column costs approximately $500. However, the cost of the column should be viewed in terms of overall analysis costs. Assume that each analysis an analyst performs is on the order of $5 per sample. If the column lasts for 500 injections, this means that the column contributes to 20% of the cost. Many analysts may get 1000–2000 samples analyzed before a column dies, so the cost per sample would be even less: 10% of the cost and 5% of the cost, respectively.

8.10 COLUMN TESTING (STABILITY AND SELECTIVITY)

There are numerous tests in the literature in regard to the quest for defining universal tests to probe the silanol activity and to define a set of mixture of compounds for selectivity assessment and comparison of reversed phase stationary phases. The tests include different probe molecules run under different conditions (pH and organic composition, isocratic and gradient test procedures). Even the different tests on the same column may lead to different results (in terms of selectivity and silanol activity ranking). The reader is referred to a review article by Rogers and Dorsey [35], which captures some of the more commonly used tests procedures for assessing silanol activity (i.e., Goldberg, Verzele and Dewaele, Engelhardt, Mutton, Engelhardt/Lobert test), and to the following papers on column selectivity [36–39].

8.10.1 Column Selectivity Testing

An area of intense investigation is choosing the column with the best selectivity. The selectivity is dependent on the bonded phase (i.e., bonded ligand, silanol activity), the probe analytes, the pH of the mobile phase, the type and concentration of the buffer, and the type and concentration of the organic modifier, especially if the probe analytes are ionizable. For a given set of probes the selectivity may be high on a specific column; however, for another set of analytes ("your pharmaceutical compounds") the selectivity may be poor. There is no universal selectivity test that can ensure that a particular column will give the desired selectivity for a set of compounds. However, if enough knowledge is gained on a set of columns in regard to selectivity with certain probe compounds and these have been good predictors of columns demonstrating adequate selectivity for pharmaceutical compounds in a particular company, then these may be used to screen new columns that come on the market that have demonstrated good bonded-phase stability within a particular pH range (see Section 8.10.2).

To ascertain if the selectivity between a set of probe compounds is reproducible from column to column with the same stationary phase, analysis on different columns should be performed. For example, three different columns from different lots of base silica should be used as well as three columns from the same lot of base silica. This will test the batch-to-batch reproducibility and the intra-lot reproducibility. At least 10 injections should be made on each column with the different probe mixes, and the selectivity across the columns from injection to injection should be compared.

One particular set of probes for testing selectivity differences at low pH (0.1 v/v% phosphoric acid/MeCN) that seems to work well is using the set of the following compounds: amitriptyline, diphenhydramine, o-, m-, and p-nitrophenols (0.1 mg/mL, diluent 80/20, water/MeCN, injected volume 2μL, wavelength 220 nm) (MIX 1). The mobile phase consists of (A) 0.1 v/v% phosphoric acid, pH 2 and (B) acetonitrile. For a 4.6- × 50-mm column packed with 3-μm particles the gradient would be 40–75% B in 1.5 min, hold at 75% for 0.5 min, flow rate 2.0 mL/min, temperature 40°C. Another set of probe analytes could include a mixture of some of the following primary, secondary, tertiary, and quaternary amines (i.e., pyridine, aniline, methyl benzylamine, diphenhydramine, amitriptyline, berberine-Mix 2) using a gradient of 5–80% B in 5 min, hold at 80% for 0.5 min, flow rate 1.5 mL/min. It is also recommended to run an isocratic test using pyridine and diphenhydramine to determine the tailing for these particular compounds. This test should be run at pH 2 (0.1% phosphoric acid) and pH 7 (10 mM ammonium phosphate) using 10–30% of acetonitrile.

The following are some examples using the MIX-1 as a probe mixture for selectivity challenging. In Figure 8-65, this test was used to observe if there were any selectivity differences for a C18 column. Two columns from three different lots of base silica were used for this study (total six columns). It was observed that selectivity differences were prevalent for the ionized basic species in this particular mobile-phase system (only one column from each of two lots is shown). The same test was run at higher concentrations of phosphoric acid (0.2 v/v%), and similar results were obtained.

In Figure 8-66, the same selectivity test mix (MIX 1) and conditions were used for a polar embedded column. Multiple columns from three different lots

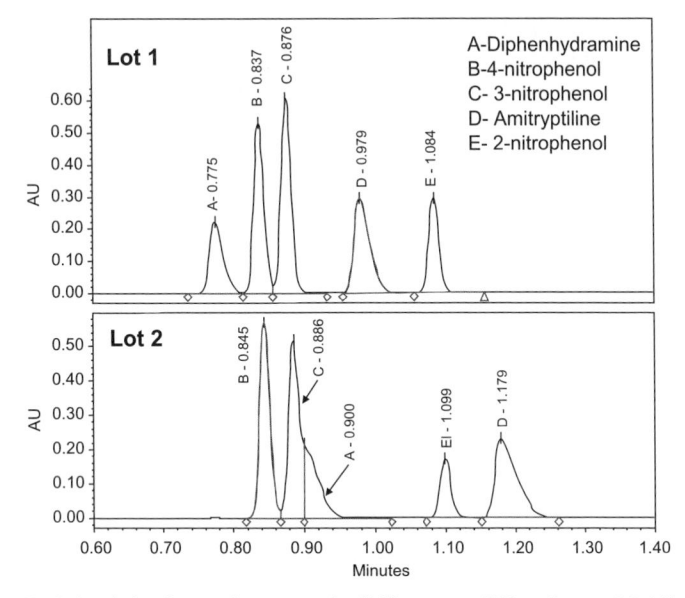

Figure 8-65. Selectivity lot to lot reproducibility test. C18 column. Mobile phase A: 0.1% H_3PO_4 in H_2O. Mobile phase B: MeCN. Gradient: 40–75% B in 1.5 min, hold 75% for 0.5 min. Flow, 0.5 mL/min; temperature, 35°C, injection, 1 µL; sample concentration, 0.1 mg/mL; diluent, 20/80, acetonitrile/water.

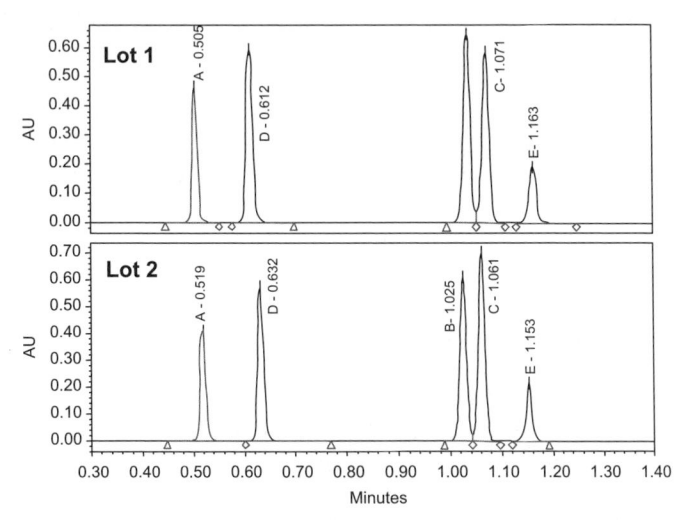

Figure 8-66. Selectivity lot-to-lot reproducibility test. Polar embedded C18 column. Mobile phase A: 0.1% H_3PO_4 in H_2O. Mobile phase B: MeCN. Gradient: 40–75% B in 1.5 min, hold 75% for 0.5 min. Flow, 0.5 mL/min; temperature, 35°C; injection, 1 µL; sample concentration, 0.1 mg/mL. Diluent: 20/80, ACN/water.

of base silica were used for this study. No selectivity differences were observed (only one column from each of the two lots shown in Figure 8-66).

Therefore, a system suitability test with a known set of probes such as MIX 1 could be used as an internal test to provide further confidence in regard to the batch-to-batch reproducibility of the packing material and/or to observe if the bonded phase has been compromised. This could also be used to probe the lot-to-lot reproducibility of new types of stationary phases that are available on the market. Once this simple and fast system suitability test is performed with MIX 1 and acceptable results are obtained using a set of defined acceptance criteria, the analysts may commence with his/her analytical method and run the specific system suitability test stated in the method for their particular target pharmaceutical analyte.

8.10.2 Column Stability Testing

Due to extended usage of the column under certain pH and temperature conditions, the bonded phase could become compromised, leading the poor chromatographic peak shapes, loss in efficiency, and loss and even sometimes increase in analyte retention. Sometimes, peak shapes can become dramatically distorted upon increased usage of the column without any change in the retention time, and this may also be dependent on the type of analyte in the mixture (base versus neutral). However, if there is a loss of bonded phase, then losses in retention may be observed for all components (neutral and basic). However, if loss of bonded phase and end-capping reagent and/or change to stationary-phase surface occurs and greater exposure of the residual silanols is prevalent, then the increase in retention and peak tailing for protonated basic components may be observed while the retention for neutral compounds may decrease.

Columns for a particular laboratory can be chosen based on some set of internal criteria. One of the criteria to select a column should be such that the column is stable for a certain number of column volumes (efficiency, tailing factor, and retention time criteria for predefined probe analytes) at the recommended maximum and minimum pH at a particular maximum temperature. This would allow the chromatographer to employ such phases with a significant degree of confidence and ensure the robustness of the stationary phase during method development and for release and stability testing.

Another example of evaluating column stability is shown in Figures 8-67, 8-68, 8-69, and 8-70. The efficiency and retention time of several acidic, basic, and neutral components and the column backpressure were monitored as a function of the number of column volumes in isocratic mode [40]. In all the studies the same isocratic conditions, flow rate, and temperature were used and only the pH of the aqueous portion of the mobile phase was changed. In Figure 8-67 an acidic mobile phase employing 0.5 v/v% TFA was used, in Figure 8-68 a 10 mM ammonium acetate (pH 5.8) was used, and in Figure 8-69 a 4 mM N-methyl pyrrolidine buffer (pH 11.5) was used. In all three pH column

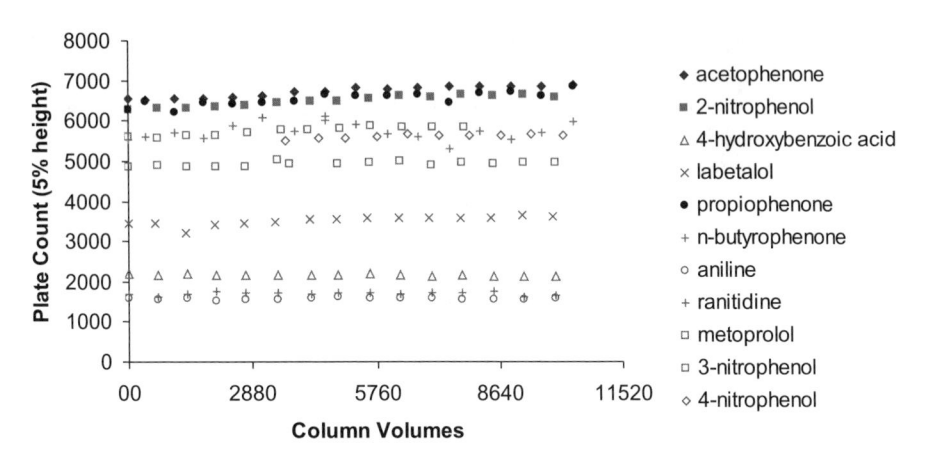

Figure 8-67. C18 sub-2-μm column. Column temperature, 30°C; injection volume, 1 μL; flow rate, 0.7 mL/min; wavelength, 225 nm. Mobile phase: 80% A/20% B (isocratic). A, 0.5 v/v% TFA; B, acetonitrile.

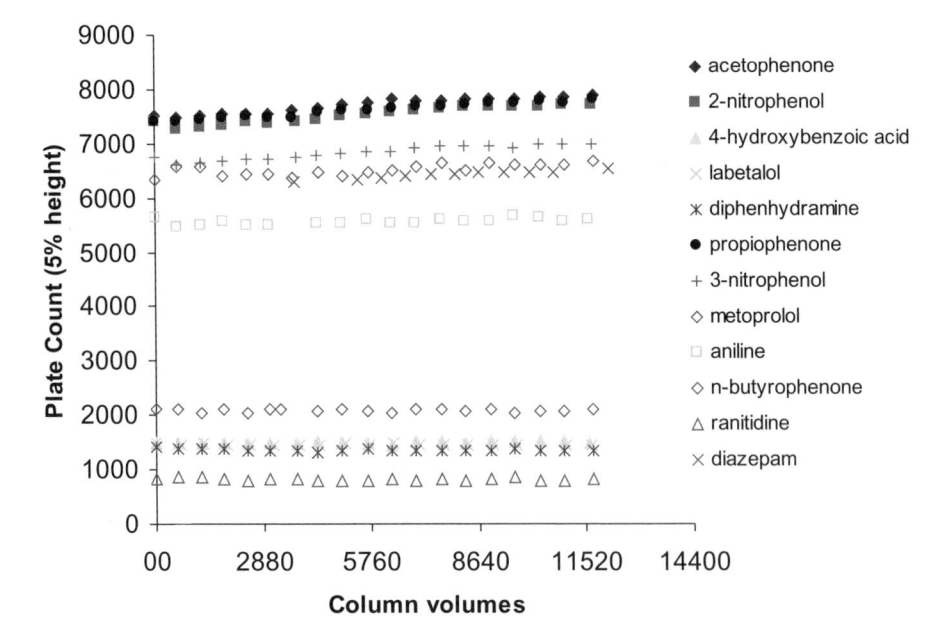

Figure 8-68. C18 sub-2-μm column. Column temperature, 30°C; injection volume, 1 μL; flow rate, 0.7 mL/min; wavelength, 225 nm. Mobile phase: 80% A/20% B (isocratic). A, pH 5.8 [10 mM ammonium acetate, pH 5.8, adjusted with acetic acid]; B, acetonitrile.

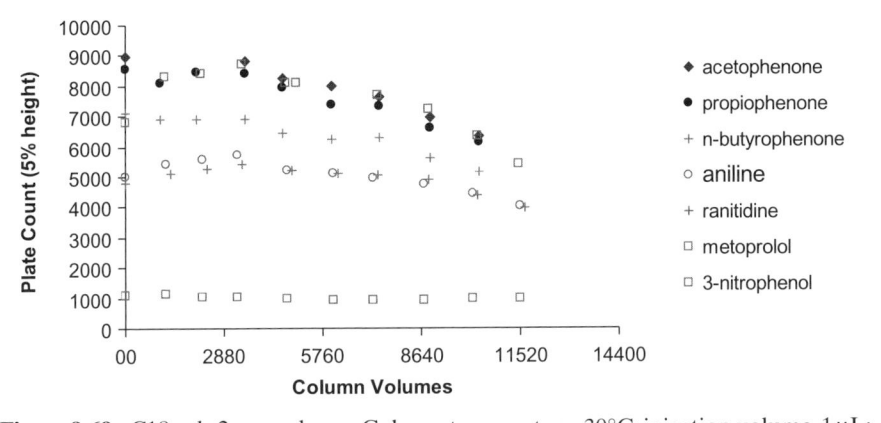

Figure 8-69. C18 sub-2-μm column. Column temperature, 30°C; injection volume, 1 μL; flow rate, 0.7 mL/min; wavelength, 225 nm. Mobile phase: 80% A/20% B (isocratic). A, pH 11.4 [4 mM *N*-methyl pyrrolidine]; B, acetonitrile.

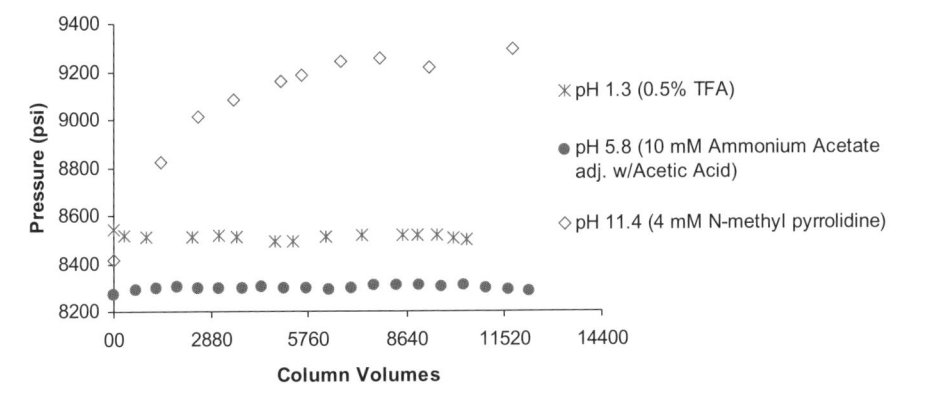

Figure 8-70. Backpressure at ^w_wpH 1, 3, 5.8, and 11.4 as a function of column volumes.

stability studies the retention for all the components showed less than a 1% change in retention over the duration of the study (>10,000 column volumes). Also, for the 0.5% TFA mobile phase and the 10 mM ammonium acetate (pH 5.8) mobile phase, no significant changes in efficiency were observed for all components (<2%) over the duration of the study (>10,000 column volumes). However, with the *N*-methyl pyrrolidine (^w_wpH 11.4) mobile phase there was a significant loss in the efficiency of all the components (~20%) with increasing number of column volumes of the mobile phase. The column backpressure was also monitored Figure 8-70, for all three mobile phases, and no significant change in the backpressure was observed for the TFA and ammonium acetate mobile phase; however, about a 10% increase in the column backpressure was observed for the *N*-methyl pyrrolidine (pH 11.4) mobile phase. This could suggest the potential dissolution of the silica/restructuring of the packing material, which led to the restricted flow through the column, giving a rise in the overall column backpressure. Also, the peak shapes of all components started to show increased distortion from approximately 8000 column volumes to 11,500 column volumes.

The pH/temp stability limits of the columns usually provided by the vendors are general recommendations. However, if determination of the "representative" column stability is required, the analyst should use column stability testing conditions similar to those used in their laboratories for testing their samples. The overall column stability depends on many factors such as whether the study is performed in either isocratic or gradient modes, whether an organic or inorganic buffer is used, type and concentration of organic modifier, temperature, pH of the mobile phase, type of counteranion and countercation of buffer, the flow rate, and backpressure (mechanical stability). Also, the results of the analysis could potentially be confounded if the aqueous component of the mobile phase does not contain any organic due to increased probability of microbial growth, which may lead to the clogging of the inlet frit and cause peak distortion for all peaks in the chromatogram. Therefore, if one is going to use only aqueous solely in line A and organic solely in line B, the aqueous portion A should be replaced every 48 hours.

8.10.3 Choice of Buffer Related to Bonded-Phase Stability

The type of pH modifier to make a desired mobile phase pH also has an effect on the column stability, and this is indirectly related to the peak efficiency and the retention of the analyte. As an increasing number of column volumes of the mobile phase are traversed through the column, the stability of the packing material could be comprised. Rearrangement of the packing bead leads to the loss of efficiency, dissolution of silica leads to loss in efficiency and retention, and hydrolytic decomposition of the bonded phase could impact the peak shape and retention. Different compounds, such as neutral compounds, acidic compounds, and basic compounds, could show different behaviors.

Different types of buffers at the same ionic strength and $_w^w$pH can have a significant impact on the dissolution of silica. The dissolution of silica is usually measured by the silicomolybdate colorimetric method [41]. When determining the bonded-phase stability using different run buffers (effect of buffer counteranion or countercation), the same $_s^s$pH must be used. The $_s^s$pH values (pH of the mobile phase: aqueous + organic) may be different from the aqueous portion of the mobile phase and may obscure if the dissolution of the silica is directly related to the type of anion/cation and/or the pH. Generally, with the addition of organic solvents the pH of the mobile phase decreases for basic buffers and increases for acidic buffers (see Section 4.5 for more details).

Generally, the lower the concentration of the buffer, the slower the dissolution rate of the silica. The rate of the dissolution of silica or the increase of

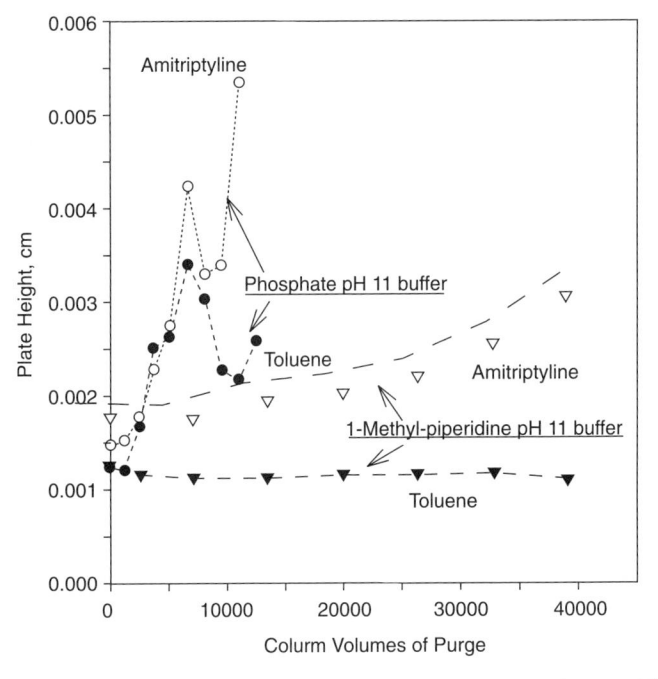

Figure 8-71. Effect of $_w^w$pH 11 buffer type on bidentate-C18P columns, 15 × 0.46 cm. Mobile phase: 55% acetonitrile:45% 0.05 M buffer (1-methyl-piperidine buffer was adjusted to $_w^w$pH 11 with HCl and dipotassium monohydrogen phosphate buffer was adjusted to $_w^w$pH 11 with NaOH). Flow rate, 1.0 mL/min, temperature, 23°C; UV, 215 nm. (Reprinted from reference 43, with permission.)

the silica solubility can be dramatically increased at elevated temperatures. An excellent publication on the subject of effect of buffers on silica-based column stability in reversed-phase HPLC is given by Claessens et al. [42].

An example is shown in Figure 8-71, where the column stability was evaluated with two different buffers, phosphate buffer ($_w^w$pH 11) and 1-methyl piperidine ($_w^w$pH 11) [43]. Plots of H (plate height) or N versus number of column volumes can be generated. Remember that H is inversely proportional to the efficiency:

$$H = \frac{L}{N} \tag{8-4}$$

With the phosphate buffer both components, basic amitriptyline and neutral toluene show poorer efficiency as the number of column volumes of mobile phase are passed through the column, indicating that the stationary-phase structure is becoming compromised (potential voids in the column). Moreover, if column voids are formed, it will effect the efficiency of all components in the mixture (neutral and ionizable). However, with 1-methyl-piperidine buffer, only the basic compound, amitriptyline, shows a decrease in efficiency with increased column purges of mobile phase, indicating that the surface is becoming more active toward this compound and secondary interactions are becoming more prevalent (slight modification of the surface can lead to these effects). Toluene is inert to these changes on the surface. In this study one buffer contains acidic functionality (phosphate) and the other buffer contains a basic functionality. Once 55% MeCN is added, the mobile phase pH ($_s^s$pH) made with the phosphate buffer (an acidic buffer) is approximately $_s^s$pH 12 [44] and the mobile phase pH ($_s^s$pH) made with the N-methyl-piperidine buffer (basic buffer) is approximately $_s^s$pH 10. This is due to the pH shift of the mobile phase upon addition of organic. Hence, the comparison of these two buffers may be confounded because the column is being exposed to two different mobile-phase $_s^s$pHs. In a more vigorous study the phosphate buffer would be prepared at $_w^w$pH 9 and the resulting pH of the hydro-organic mixture would be $_s^s$pH10. Then the two buffers could be adequately compared, since the column would be exposed to the same apparent pH ($_s^s$pH). These types of comparisons should be considered when comparing the column stability data provided by different vendors.

Also, mechanical stress or dissolution of the silica rearrangement of the packing bed may occur, which will lead to the loss of efficiency. This may lead to the formation of column voids at the head of the column and channeling within the packing bed, which will lead to a decrease in the peak efficiency for all peaks in the chromatogram. In Figure 8-72, a gradient was employed

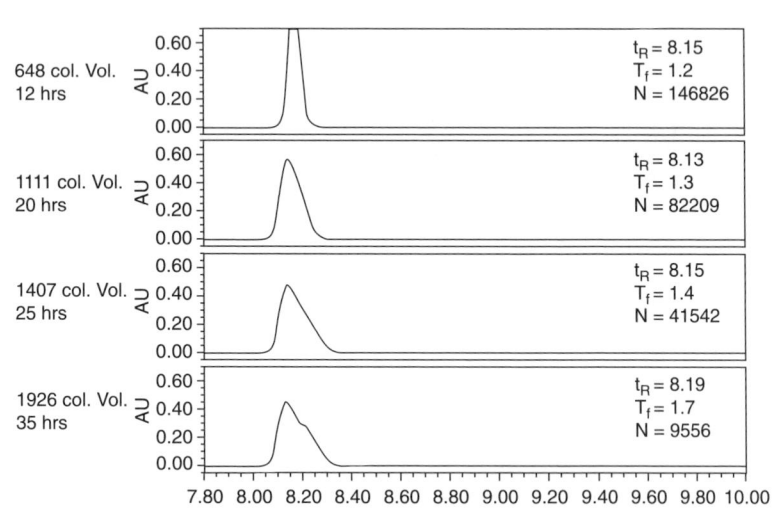

Figure 8-72. Chromatographic conditions: C18 3.5-μm column. 150 × 4.6mm; wavelength, 247 nm; column temperature, 55°C; flow, 1.5 mL/min; injection volume, 10 μL; %A = 10 mM NH$_4$OAc, pH 5.8; %B = acetonitrile, 25% B to 75% B over 10 min, hold at 75% B for 5 min.

and the apparent efficiency decreased with increased number of column volumes of the mobile phase. Note that while the peak shape was dramatically distorted, there was no change in the retention time. This occurred for all peaks in the chromatogram (other peaks not shown). However, if there is loss of bonded phase, then losses in retention may be observed for all components (neutral, acidic, and basic compounds). However, if loss of bonded phase, end-capping reagent, and/or change to stationary-phase surface occurs and greater exposure of the residual silanols are prevalent, then an increase in retention and peak tailing for protonated basic components may be observed (secondary interactions with residual silanols) while the retention for neutral compounds may decrease (due to increased hydrophilicity of the surface).

8.11 CONCLUDING REMARKS

A well-defined method development plan with clear aim of analysis is critical to the success for fast and effective method development. The general approach for the method development for the separation of pharmaceutical compounds was discussed, emphasizing that modifications in the mobile phase (organic and pH) play a dramatic role on the separation selectivity. The knowledge of the pK_a of the primary compound is of utmost importance prior to the commencement of HPLC method development. Moreover, pH screening experiments can help to discern the ionizable nature of the other impurities (i.e., synthetic by-products, metabolites, degradation products, etc.) in the mixture.

The separation of a complex mixture of different ionizible and nonionizible organic components can be challenging, and the development of a rugged separation method can be an adventure. A multitude of approaches can be applied; however, the one that seems to work the best is to screen a limited set of columns at a certain predefined pH range (1–2 pH units below or above the target analyte pK_a in a particular hydro-organic mixture) and determine the best permutation and combination of column/mobile phase in order to obtain the specific selectivity that is desired for critical pair of components in the mixture (API, biological fluid, drug product, etc.). There are thousands of different columns on the market, and the selection of one that will allow a simple separation of your mixture is more a "black magic" than a science, but usually the most common columns to carry out initial method development include those comprised of C18, polar embedded/end-capped and/or phenyl bonded ligands. Final optimization can be performed by changing the temperature, gradient slope, and flow rate as well as the type and concentration of mobile-phase modifiers.

REFERENCES

1. J. E. Hoover, in *Remingtons's Pharmaceutical Sciences*, 15th ed., Mack Printing, Easton PA, 1974, p. 1158.

2. G. J. Lehr, T. L. Barry, G. Petzinger, G. M . Hanna, S. W. Zito, Isolation and identification of process impurities in trimethoprim drug substance by high-performance liquid chromatography, atmospheric pressure chemical ionization liquid chromatography/mass spectrometry and nuclear magnetic resonance spectroscopy, *J. Pharm. Biomed. Anal.* **19** (1999), 373–389.

3. ICH, Guidance for the Industry: Q3A: Impurities in new drug substances, 2003.

4. D. D. Wirth, S. W. Baertschi, R. A. Johnson, S. R. Maple, M. S. Miller, and K. Diana, Maillard reaction of lactose and fluoxetine hydrochloride, a secondary amine, *J. Pharm. Sci.* **87** (1998), 31–39.

5. ICH, Guidance for the Industry: Q3B(R): Impurities in new drug products, November 2003.

6. J. D. Higgins II and W. L. Rocco, Pharma preformulation, *Today's Chemist* (2003), 22–26.

7. R. LoBrutto and Y. V. Kazakevich, Retention of ionizible components in reversed-phase HPLC, in S. Kromidas (ed.), *Practical Problem Solving in HPLC*, Wiley-VCH, New York, 2000, pp. 122–158.

8. E. Loeser and P. Drumm, Using strong injection solvents with 100% aqueous mobile phases in RP-LC, *J. Sep. Sci.* (2006), accepted for publication.

9. A. Skogolf, R. M. Orrnaf, D. Wu, and P. J. Palermo, Peak fronting in reversed-phase high-performance liquid chromatography: A study of the chromatographic behavior of oxycodone hydrochloride, *J. Pharm. Biomed. Anal.* **19** (1999), 669–678.

10. R. LoBrutto, Y. Bereznitski, T. J. Novak, L. DiMichele, L. Pan, M. Journet, J. Kowal, and N. Grinberg, Kinetic Analysis and subambient temperature online on-column derivatization of an active aldehyde, *J. Chromatogr. A* **995** (2003), 67–78.

11. V. Antonucci and L. Wright, Development of practical chromatographic methods for the analysis of active esters, *J. Liq. Chrom. Relat. Technol.* **24** (2001), 2145–2159.

12. J. O. Egegkeze, M. C. Danielski, N. Grinberg, G. B. Smith, D. R. Sidler, H. J. Perpall, G. R. Bicke, and P. C. Tway, Kinetic analysis and subambient temperature chromatography of an active ester, *Anal. Chem.* **67** (1995), 2292–2295.

13. A. Abrahim, R. Hartman, Z. Ge, B. Mao, and J. Marcoux, Development of a derivatization method, coupled with reverse phase HPLC, for monitoring the formation of an enolate intermediate, *J. Liq. Chrom. Relat. Technol.* **25** (2002), 1049–1062.

14. A. N. Heyman and R. Henry, Importance of Controlling Mobile Phase pH, Keystone Technical bulletin, 99-06 (page 2, Figure 2), http://www.hplcsupply.com/pdf/App_9.pdf#search='Biobasic%20and%209906%20and%20henry'.

15. M. L. Mayer, Selecting filters for chromatographic applications, *LC-GC* **14** (10) (1996), 902–905.

16. M. L. Mayer, Filtration: Preventive maintenance for HPLC, *Am. Lab.* **29** (1997), 34–37.

17. J. A. Dean, *Analytical Chemistry Handbook*, McGraw-Hill, New York, 1995.

18. J. C. Merrill, Avoiding problems in HPLC through filtration, *Am. Lab.* **19** (1987), 74–81.

19. L. R. Synder, J. J. Kirkland, and J. L. Glach, *Practical HPLC Method Development*, 2nd ed., John Wiley & Sons, New York, 1997.

20. T. H. Hoang, D. Cuerrier, S. McClintock, and M. Di Maso, Computer-assisted method development and optimization in high-performance liquid chromatography, *J. Chromatogr. A* **991** (2003), 281–287.

21. I. Molnar, Computerized design of separation strategies by reversed-phase liquid chromatography: Development of DryLab software, *J. Chromatogr. A* **965** (2002), 175–194.

22. R. Kaliszan, P. Haber, T. Baczek, D. Siluk, and K. Valko, Lipophilicity and pKa estimates from gradient high-performance liquid chromatography, *J. Chromatogr. A* **965** (2002), 117–127.

23. J. W. Dolan, Temperature selectivity in reversed-phase high performance liquid chromatography, *J. Chromatogr. A* **965** (2002), 195–205.

24. U. Rose, *In situ* degradation: A new concept for system suitability in monographs of the European Pharmacopeia. *J. Pharm. Biomed. Anal.* **18** (1998), 1–14.

25. D. Parriott, *A Practical Guide to HPLC Detection*, 1st ed., Academic Press, Harcourt Brace Jovanovich, San Diego, 1993.

26. R. M. Silverstein, C. G. Bassler, and T. C. Morrill, *Spectrometric Identification of Organic Compounds*, 3rd ed., John Wiley & Sons, New York, 1974, Chapter 5.

27. G. H. Shenk, *Absorption of Light and Ultraviolet Radiation: Fluorescence and Phosphorescence*, Chapters 1 and 2, Allyn and Bacon, Boston, 1973.

28. D. A. Skoog, F. J. Holler, and T. A. Nieman, *Principles of Instrumental Analysis*, 5th ed., Harcourt Brace & Company, Philadelphia, 1998.

29. J. A. Dean, *Langes Handbook of Chemistry*, 14th ed., McGraw-Hill Education—Europe, 1992.

30. D. H. Ripin and D. A. Evans pK_a table, http://daecr1.harvard.edu/pdf/evans_pKa_table.pdf.

31. S. Ege, *Organic Chemistry*, 3rd ed., D. C. Heath and Company, Lexington, MA, 1994.

32. A. Kamal-Eldin, S. Gorgen, J. Petterson, and A. Maya Lampi, Normal phase high performance liquid chromatography of tocopherols and tocotrienols: Comparison of different chromatographic columns, *J. Chromatogr. A* **881** (2000), 217–227.

33. A. Schieber, M. Marx, and R. Carle, Simultaneous determination of carotenes and tocopherols in ATBC drinks by high performance liquid chromatography, *Food Chem.* **76** (2002), 357–362.

34. J. W. Dolan, Extending column life, *LC-GC Europe*, LC Troubleshooting 2005 yearbook, www.lcgceurope.com.

35. S. D. Rogers and J. G. Dorsey, Review: Chromatographic silanol activity test procedures: The quest for a universal test, *J. Chromatogr. A* **892** (2000), 57–65.

36. D. H. Marchand, K. Croes, J. W. Dolan, L. R. Snyder, R. A. Henry, K. M. R. Kallury, S. Waite, and P. W. Carr, Column Selectivity in reversed phase liquid chromatography, VIII. Phenylalkyl and fluoro-substituted columns, *J. Chromatogr. A* **1062** (2005), 65–78.

37. T. Baczek, R. Kaliszan, K. Novotan, and P. Jandera, Comarpative characteristics of HPLC Columns based on quantitative structure–retention relationships (QSRR) and hydrophobic-subtraction model, *J. Chromatogr. A* **1075** (2005), 109–115.

38. L. R. Synder, J. W. Dolan, and P. W. Carr, Review: The hydrophobic-subtraction model of reversed phase column selectivity, *J. Chromatogr. A* **1060** (2004), 77–116.

39. J. J. Gilroy, J. W. Dolan, and L. R. Synder, Column selectivity in reversed-phase liquid chromatography IV. Type-B alkyl-silica columns, *J. Chromatogr. A* **1000** (2003), 757–778.

40. R. LoBrutto, A. Jerkovich, A. Jones, T. Prowse, and R. Vivilecchia, UPLC™—A critical look at system/column performance and method transfer considerations for pharmaceutical analytes, HPLC 2005 Conference, Stockholm, Sweden, 2005.

41. *NIOSH Manual of Analytical Methods*, 4th ed., Issue 2, Method 7601: Silica, crystalline, August 15, 1994, pp. 1–5.

42. H. A. Claessens, M. A. van Straten, and J. J. Kirkland, Effect of buffers on silica based column stability in reversed phase high performance liquid chromatography, *J. Chromatogr. A* **728** (1996), 259–270.

43. J. J. Kirkland, J. B. Adams, Jr., M. A. van Straten, and H. A. Claessens, Bidentate silane stationary phases for reversed-phase high-performance liquid chromatography, *Anal. Chem.* **70** (1998), 4344–4352.

44. M. Roses, Determination of the pH of binary mobile phases for reversed phase chromatography, *J. Chromatogr. A* **1037** (2004), 283–298.

9

METHOD VALIDATION

Rosario LoBrutto and Tarun Patel

9.1 INTRODUCTION

The method validation process is to confirm that the method is suited for its intended purpose. Although the requirements of validation have been clearly documented by regulatory authorities [ICH, USP, and FDA], the approach to validation is varied and open to interpretation. Validation requirements differ during the development process of pharmaceuticals. The method validation methodologies in this chapter will focus on the method requirements for preliminary and full validation for both drug substance and drug product. Preliminary method validation is generally performed in the earlier phases of development up to Phase IIa because at this time ICH Q2A and Q2B [1] are not yet binding. A more extensive validation (full validation) is performed for methods used in later stages of drug development (after Phase IIa) and for methods that will be used to evaluate marketed products. Specific requirements or methodologies for validation depending on the life cycle of the potential drug candidate in each specific area in the drug development process will be addressed in the corresponding chapter.

An analytical method is a laboratory procedure that measures an attribute of a raw material, drug substance, or a drug product. Analytical method validation is the process of demonstrating that an analytical method is reliable and adequate for its intended purpose. Any method that is utilized to determine results during drug substance and formulation development will have to be validated. Reliable data for release of clinical supplies, stability, and setting shelf life can only be generated with appropriate validated methods.

HPLC for Pharmaceutical Scientists, Edited by Yuri Kazakevich and Rosario LoBrutto
Copyright © 2007 by John Wiley & Sons, Inc.

Validation of high-performance liquid chromatography (HPLC) methods focus mainly on the following:

- Identification tests
- Quantitative measurements of the content of related substances*
- Semiquantitative and limit tests for the control of related substances*
- Quantitative tests for the assay of major components (e.g., drug substance and preservatives) in samples of drug substance or drug product (assay, content uniformity, dissolution rate, etc.)

Moreover, HPLC methods that are described in pharmacopeias may not have to be validated but should be verified, if necessary. Well-characterized reference materials with documented purity should be used throughout the validation study, especially during full development. Validation experiments and analyses must be carried out on fully qualified and calibrated instrumentation, and some references have been published on this subject [2–6].

Analytical method validation is established through documented evidence demonstrating the accuracy, precision, linearity, selectivity, ruggedness, and/or robustness of that particular test method which will be utilized to generate test results for a drug substance or drug product. Different test methods require different validation parameters. All analytical procedures require some form of method validation, regardless of whether the test method is utilized for the testing of Good Laboratory Practice (GLP) toxicology, shelf-life determination (stability indicating), in-process controls [7], clinical release, or release of products for open market [8]. As development of the project progresses and as more analytical and product-specific information is acquired, the analytical methods evolve and are gradually updated. The extent of validation increases and the documentation is completed.

During the early development phase, depending on the analytical laboratory, generic validation protocols may be used because project-specific protocols are not required. Sometimes an internal Standard Operating Procedure (SOP) suffices and a generic validation protocol does not need to be used. Usually, for Phase I, validation experiments may be carried out concurrently with the analysis of the first batch of clinical supplies or the first delivery of drug substance to be used for clinical supplies. However, depending on the pharmaceutical organization method validation may need to be performed prior to the analysis of material that will be used for clinical supplies.

For analytical method validation during full development (after final synthesis has been set for drug substance and after final market formulation has been set for drug product) corresponding to the definitive control procedure for new drug application (NDA), a specific validation protocol has to be written. Before start of the experimental work, the protocols must be written

*Related substances described in this chapter encompass degradation products, and synthetic by-products.

by an analytical chemist and approved by a quality assurance department. Some of the items that are necessary to be specified in the validation protocol are listed below:

- The analytical method for a given product or drug substance
- The test to be validated
- The test parameters for each test, including type and number of solutions and number of injections
- The acceptance criteria for each parameter based on an internal SOP (product or method-specific adaptations may be necessary and are acceptable, if justified)
- List of batches of drug substance and/or drug products
- For a drug product the grade/quality of the excipients used in the formulation
- List of reference materials to be used in the validation experiments
- Information on the instruments and apparatus to be used
- Responsibilities [author, chemists, analytical research project leader, quality assurance (QA), etc.]

Depending upon a company's culture, a method validation protocol could be simple (listed items above) or exhaustive (in addition to the listed items above, each parameter to be validated is described in detail): How solutions are going to be made, the experimental design, how the calculations are going to be performed, any software to be utilized (e.g., Excel). If a full-length protocol is required within a particular company, then the writing of this protocol and approval of the protocol would need to be completed prior to the commencement of the validation work. Otherwise, there may be many deviations to the protocol which will be needed to be referenced to in the final method validation report. Some companies also have templates for the validation reports, thereby allowing for facile population of the results. Once populated, the file is reviewed to determine if all validation parameters and acceptance criteria were met. If they were not, then a deviation is added and the proper justification must be given. If it is deemed that the justification is not appropriate, then an action plan for the specific figure of merit in question is determined (i.e., repeat analysis, change of the analytical procedure, and revalidation). Also, if the analytical method has not yet been approved at the time of writing the validation protocol, it is recommended to attach a final draft of the method to the protocol. The final HPLC method must also be approved with the validation report submission.

9.2 VALIDATION REPORT

A validation report is written during early and full development, and approval by QA is required. Existing method validation data from earlier stages of development may be used for full development if the HPLC method has not changed. Minor changes such as change in equilibration time may be acceptable, and the preliminary validation performed for early phase may be used. These data can be referred to in the validation report, and reference to the original data must be given.

The validation report should contain reference to the analytical methods (specific code number used as identifier within the pharmaceutical organization) and the corresponding drug substance or product name. Note that for early-phase method validation reports the results maybe filled in a predefined table and compared against the acceptance criteria. However, for late-phase validation, more explicit reports are generated explaining each and every experiment, with detailed steps of sample and standard preparation.

The list of reference materials (reference standards with the appropriate certificate of analysis) as well as the list of calibrated and qualified instruments used in the validation experiments should be documented in the report. For drug substances the list of the batches of drug substances, notebook number/reference number for any individual impurities, or solutions or intermediates used should be listed. For drug products the list of the batches of drug substances, drug product, and the grade/quality of excipients should be listed. The test parameters and acceptance criteria should be listed together with the results for each test, and the results passed or failed should be indicated. The validation report should also contain whether the method validation was successful and if any changes had to be applied to the analytical method, and then the final analytical method must be resubmitted for QA approval.

9.3 REVALIDATION

After any major changes in the HPLC method (solution preparation, experimental conditions, etc.) or after change in route of synthesis of the drug substance or drug product manufacture (change of process, change of equipment, change of analytical procedure), it must be assessed whether a new validation or a partial validation is required, addressing all the validation parameters that may be affected by the methodological change. If revalidation is not deemed necessary, then the reasons behind the decision must be documented in the revision history of the test method and the proper change control initiated. The revision of the test method and any documents that refer to the original method, such as the analytical specifications, will then be approved by QA. When revalidation is deemed necessary, the reason for change must be documented and any new validation activities must be performed according to an approved, updated HPLC validation protocol. The results would then be documented in an update of the validation report or a supplement to the original validation report.

9.4 ASSIGNMENT OF VALIDATION PARAMETERS

The type and degree of validation depends on the nature of the test. In particular, methods described in pharmacopeias may not have to be validated but should be verified, if needed. Different test methods require different validation parameters. As development of the project progresses and as more analytical and product-specific information is acquired, the analytical methods evolve and are gradually updated. The extent of validation increases and the documentation is completed. Table 9-1 outlines the validation parameters that are usually required for the early development stage, and Table 9-2 outlines the validation parameters that are usually required for the full development stage.

The proposed acceptance criteria in Table 9-3 should be included in the validation protocol, especially for the full development stage.

There are numerous method validation examples in the literature [9–18]. Each company has their own approach and own set of acceptance criteria for different analytical assays, but these approaches must be within the confines of their line unit QA department and be in accordance with any regulatory provisions. In the next section a description for each of the parameters to be validated (figures of merit) are described in detail and examples are given for each.

TABLE 9-1. Early Development

Validation Parameters	Type of Tests to Be Validated		
	Identity	Weight Percent/Assay/Content Uniformity/Dissolution	Impurity Testing: Quantitative Test[a]
Specificity	Yes	Yes	Yes
Linearity	No	Yes[b]	Yes[b]
Accuracy	No	Yes[c]	Yes[d]
Precision (repeatability)	No	Yes	Yes[e]
Limit of detection	No	No	Yes[f]
Limit of quantitation	No[g]	No	Yes[d]
Stability of the solutions	No	Yes	Yes

[a]If impurities not available, with drug substance.
[b]Four points may be adequate.
[c]For drug product only (assay/CU/dissolution).
[d]A spiking experiment carried out is adequate at this stage (only possible if impurity/impurities are available).
[e]At least triplicate analysis.
[f]Not required, but recommended.
[g]For the identity test of a 0-mg formulation (placebo), it may be necessary to document the absence of drug substance, and an LOQ determination will then be required.

TABLE 9-2. Full Development

Validation Parameters	Identity	Weight Percent/Assay/Content Uniformity/Dissolution	Impurity Testing: Quantitative Test
		Type of Tests to Be Validated	
Specificity[a]	Yes	Yes	Yes
Linearity	No	Yes	Yes
Accuracy	No	Yes	Yes
Precision (repeatability)	No	Yes	Yes
Precision (intermediate precision)[b]	No	Yes	Yes
Precision (reproducibility)	No	[c]	[c]
Range	No	Yes	Yes
Limit of detection	No	No	Yes[e]
Limit of quantitation	No[d]	No	Yes
Stability of the solutions	No	Yes	Yes
Robustness	[f]	Yes	Yes

[a]Lack of specificity of one analytical procedure may be compensated for by other supporting analytical procedures.
[b]In cases where reproducibility has been performed, intermediate precision not needed.
[c]In exceptional cases.
[d]For the identity test of a 0-mg formulation (placebo) it may be necessary to document the absence of drug substance and an LOQ determination will then be required.
[e]Not required by ICH, but recommended.
[f]May be required, depending on the nature of the test.

9.4.1 Accuracy

The test for accuracy is intended to demonstrate the closeness of agreement between the value found and the value that is accepted either as a conventional true value or as an accepted reference value [19]. Therefore, accuracy can be defined as the agreement between the result obtained with method being validated and an accepted reference value. The accuracy can be inferred from precision, linearity, and specificity. The results for the method being validated can be compared to the results with those of a well-characterized, independent method. These results may be compared to an alternate reversed-phase HPLC method (phenyl versus C18 or separation run at different pH using the same column) using the same detection scheme. In some, cases an orthogonal method is used to demonstrate accuracy. The methods should differ with respect to separation mode and therefore provide orthogonal information concerning related substances and degradation products. For example, one method would use reversed-phase (RP) separation mode on a C18 column, and the second method would use a strong cation exchange (SCX) column [20]. The orthogonal methods may show different selectivities toward the degradation products, thereby demonstrating the orthogonal nature of the two separation techniques. The accuracy would be demonstrated

TABLE 9-3. Proposed Acceptance Criteria for Drug Product (DP) and Drug Substance (DS)

Quality Characteristics	Parameter to be Validated	Acceptance Criteria
Identity	Selectivity/specificity	All known peaks are separated. Major (API) peak is "pure" [Peak purity angle ≥ peak threshold angle]. {DS and DP} For the identity test of a 0-mg formulation (placebo), it may be necessary to document the absence of drug substance, and an LOQ determination will then be required. {only DP}
Dissolution (drug product)	Accuracy (mean) • Recovery • S_{rel} for recovery Precision • Repeatability • Intermediate precision Linearity • Correlation coefficient • y-intercept (absolute value) • Residual standard deviation Stability of solutions • Sample • Reference standard Specificity • HPLC Range (basket/paddle)	 95–105% ≤2.5% $S_{rel} \leq 2.0\%$, $n \geq 6$ {at Q time} Project specific. $n \geq 6$ $r \geq 0.990$ ≤5% ≤2.5% ≤2.0% change over specified time ≤2.0% change over specified time No interference from placebo solution at the retention time of API. IR: ±30% of specified range MR,SR: From 50% of Q-value to 130% of label claim.
Content uniformity (CU)	Precision Accuracy Stability of solutions	As defined in assay
Drug product	Specificity Linearity • Correlation coefficient • y-intercept • Residual standard deviation Range	Chromatographic peaks are separated. No indication of interference from placebo solution at the retention time of API. $n \geq 6$ $r \geq 0.990$ ≤5.0% ≤2.0% At least 70–130% of declared content
Assay—drug product	Accuracy (mean)—DP • Recovery • S_{rel} for recovery	 98.0–102.0% ≤2.0%, $n \geq 9$ (at least three concentrations)

TABLE 9-3. *Continued*

Quality Characteristics	Parameter to be Validated	Acceptance Criteria
Weight percent— drug substance	Accuracy—DS • Comparison of methods (i.e., titration, DSC, PSA) Precision	% difference of the mean of two methods $\leq$2.0%
	• Repeatability	$S_{rel} \leq 2.0\%, n \geq 6$, DP $S_{rel} \leq 1.0\%, n \geq 6$, DS
	• Intermediate precision	$S_{rel} \leq 2.0\%, n \geq 4$ [when combined from two analysts]
	Linearity	$n \geq 6$
	• Correlation coefficient	$r \geq 0.998$
	• y-intercept	$\leq$2.0%
	• Residual standard deviation	$\leq$2.0%
	Stability of solutions	
	• Sample	$\leq$2.0% change over specified time (DP)
	• Reference standard	$\leq$2.0% change over specified time (DP)
	• Sample	$\leq$1.0% change over specified time (DS)
	• Reference standard	$\leq$1.0% change over specified time (DS)
	Specificity	
	• HPLC	Chromatographic peaks are separated. No indication of interference from placebo solution at the retention time of API. No indication of another peak under the API peak.
	Range	At least 80–120% of declared content (100% = concentration X of final sample stock solution)
	Ruggedness/robustness	$\leq$1.0% difference for a defined range of intentionally altering sensitive parameters (pH of mobile phase, column, temperature, flow rate, wavelength, etc.)
Drug product-Related substances (degradation products)	Precision • Repeatability	Level < 0.1%, $S_{rel} \leq 30\%, n \geq 6$ Level 0.1–<0.2%, $S_{rel} \leq 20\%, n \geq 6$ Level 0.2–<0.5%, $S_{rel} \leq 10\%, n \geq 6$ Level 0.5–<5%, $S_{rel} \leq 5\%, n \geq 6$ Level $\geq$ 5%, $S_{rel} \leq 2.5\%, n \geq 6$
Drug substance (synthetic by-products and degradation products)	• Intermediate precision [all replicates combined from two analysts]	Level < 0.1%, $S_{rel} \leq 40\%, n \geq 4$ Level 0.1–<0.2%, $S_{rel} \leq 30\%, n \geq 4$ Level 0.2–<0.5%, $S_{rel} \leq 15\%, n \geq 4$ Level 0.5–<5%, $S_{rel} \leq 7.5\%, n \geq 4$ Level $\geq$ 5%, $S_{rel} \leq 4.0\%, n \geq 4$
	Specificity • HPLC	Known peaks are separated. No indication of interference from placebo solution at the retention time of API. No indication of another peak under the API peak.

TABLE 9-3. *Continued*

Quality Characteristics	Parameter to be Validated	Acceptance Criteria
	Linearity	$n \geq 6$
	• Correlation coefficient	$r \geq 0.990$, DP and $r \geq 0.998$, DS
	• y-intercept	Level < 0.5%: $\leq 25\%$
		Level 0.5–<1%: $\leq 10\%$
		Level $\geq 1\%$: $\leq 5.0\%$
	• Residual standard deviation	Level < 0.2%: $\leq 20\%$
		Level 0.2–<0.5%: $\leq 10\%$
		Level 0.5–<5%: $\leq 5.0\%$
		Level $\geq 5\%$: $\leq 2.5\%$
	Range	LOQ to 120% of specification limit of largest impurity or related substance
	LOD	Peak signal/noise ratio $\geq 3 : 1$
	LOQ	Peak signal/noise ratio $\geq 10 : 1$ and $S_{rel} \leq 10\%$, $n \geq 5$
	Accuracy (mean)	Level < 0.2%: 70–130%
	• Recovery	Level 0.2–<0.5%: 80–120%
		Level 0.5–<5%: 90–110%
		Level $\geq 5\%$: 95–105%
	• S_{rel} for recovery	Level < 0.5%: $\leq 10\%$,
		Level 0.5–<5%: $\leq 5\%$
		Level $\geq 5\%$: $\leq 2.5\%$
		For all, $n = 9$ (at least three concentrations), a weighted average maybe used based on the level and the S_{rel}.
	Stability of solutions [report two decimal places]	
	• Reference standard	Level < 5% of theoretical 100% concentration — Change $\leq 10\%$ over specified time
		Level $\geq 5\%$ of theoretical 100% concentration — Change $\leq 2.0\%$ over specified time
	• Sample	Related substances (impurities)
		Level < 0.5% — Change $\leq 20\%$ over specified time
		Level 0.5–<5% — Change $\leq 10\%$ over specified time
		Level $\geq 5\%$ — Change $\leq 5\%$ over specified time
		No new peak $\geq$ reporting level
	Ruggedness/robustness	Defined based on an experimental design and data (sensitive parameters and a range for each parameter in the final test method)

if the overall purity in both of the methods would still be the same according to a predefined set of acceptance criteria. Different types of separation methods could also be used to show accuracy. For example, if normal-phase chromatography was used as the parent method, this could be compared to a separation obtained using supercritical chromatography. In another example, an electrophoretic method using capillary electrochromatography or capillary electrophoresis could be compared to an HPLC separation.

Also, the HPLC weight percent (assay) method of the drug substance can be compared to nonchromatographic methods such as nuclear magnetic resonance (NMR) [21], Phase Solubility Analysis (PSA), and DSC [22] and to nonspecific titration and spectrophotometric assay methods that may have been used in early development before the qualification of a reference standard. Potentiometric titration methods using nonaqueous or aqueous titrations are only amenable to ionizable compounds and are nonspecific because the impurities may contain the same ionizable functionality as the parent compound being titrated. Titration is a nonspecific method because synthetic by-products in drug substance may have a pK_a similar to that of the main component (the endpoints for the by-products and the drug substance may overlap in this case) and results may be biased, leading to a higher weight percent of the material.

However, these titration methods can be used in early development when a reference standard is not available. Also, the spectrometric-based assay methods such as ultraviolet (UV) may be nonspecific because most of the drug substance impurities contain a similar chromophore as the parent molecule. If UV is used, UV absorption is measured at one or more wavelengths and the absorbance value is recorded for a particular concentration. Sandor Gorog has critically evaluated the difference between specific and nonspecific assay methods in the European and US Pharmacopoeias [23]. The difference between the mean and the accepted true value with a defined confidence interval should be reported in the acceptance criteria.

The accuracy can also be demonstrated by recovery of drug substance spiked into a placebo for a drug product. The accuracy can also be demonstrated by recovery of the impurity spiked to a drug substance or into a placebo with drug substance. The percentage recovery with the certain acceptance criteria at each defined level is reported.

Accuracy should be assessed using a minimum of nine determinations at a minimum of three concentration levels covering the specified range (e.g., 3 concentrations/3 replicate preparations of each in the total analytical procedure) within the ranges shown in Table 9-4.

Accuracy is performed to determine recovery of an active or degradation products from a drug product or recovery of related substances from a drug substance. The experiment is designed to recover the total amount of active or degradation product from a drug product or a specific impurity or impurities from a drug substance. For recovery of the active for assay and CU, a known amount of drug substance in solution is spiked into the placebo blend. The influence of sample preparation steps for tablets must be taken into con-

TABLE 9-4. Minimal Concentration Ranges for Accuracy Test (Wider Range May Be Used)

Type of Analytical Procedure	Range, at Minimum, to Be Covered
Assay (content)	80–120% of declared content
Assay (CU)	70–130% of declared content
Assay (dissolution)	±30% of specified range (for immediate release dosage form).
	If the specification for a controlled release product (modified release or sustained release) covers a region from 20% (after 1 hr) to 90% (after 24 hr), the validated range would cover 50% of 1-hr limit (20% × 50% = 10%) to 130% of the label claim (label claim × 1.3).
Degradation products/impurities	Reporting level to at least 120% of specification limit.

sideration such as grinding, sonication, and extraction. For assay determination, the experiment setup is straightforward. A minimum of three concentrations (centered around the target concentration) and three replicates are prepared at each concentration (one injection each) to make a total of nine determinations. The minimum three concentrations should be 70%, 100%, and 130% of the target concentration. If this is used, then these accuracy solutions could be used for both content uniformity and assay method if they are indeed the same method. Usually in early development the same method is used. In later development if a new fast CU method is developed (<5 min), it would have to be revalidated for recovery. In the following theoretical example, for a lyophilized drug substance, a placebo solution is made with the excipients and is diluted with the drug substance X stock solution (2.67 mg/mL) and further diluted to the desired concentration with the proper diluent. The nominal 100% level solution without placebo is used as an external calibration standard. The range explored is shown in Table 9-5. An example of the preparation is shown in Table 9-6. The percent recovery of drug substance X from the nominal 70%, 85%, 100%, 115%, and 130% sample solutions was then determined by an external calibration standard (shown in Table 9-7). The average percent recovery is 99.3% ($n = 15$), and at each level the recovery is found to be within the acceptance criteria of 98.0–102.0%. For validation of HPLC analytical methods during the early phase, an acceptance criteria of 95–105% for recovery may be acceptable if the specifications are outside the range of 95–105% (i.e., 90–110%). The specifications may never be tighter than the acceptance criteria. Moreover, the S_{rel} [%RSD (relative standard deviation)] determined from the overall percent recovery is 0.18% ($n = 15$) and passes the acceptance criteria of $S_{rel} \leq 2.0\%$.

For degradation products, the same procedure as the assay can be followed, except the amount should be from reporting level to at least 120% of the

TABLE 9-5. Range for Recovery Experiment for CU and Assay Method to Be Defined in Method Validation Protocol

Target Concentration of Solutions (%)	Target Concentration of Solutions (mg/mL)	Amount Injected (µg)	Number of Preparations/Number of Injections
130	1.30	13.0	3/1
115	1.15	11.5	3/1
100	1.00	10.0	3/1
85	0.85	8.5	3/1
70	0.70	7.0	3/1

TABLE 9-6. Example of Actual Sample Preparation in Method Validation Protocol

Target Concentration of Solutions (%)	Milliliters of Stock Solution (2.67 mg/mL) used (mL)	Placebo Added (mg)	Volumetric Flask Used (mL)	Actual Final Concentration of Solution (µg/mL)
130	12	85.1	25	1281.6
115	11	85.1	25	1174.8
100[a]	10	85.1	25	1068
85	8	85.1	25	854.4
70	13	170.3	50	694.2

[a]Nominal 100% level without placebo is used as a calibration standard.

TABLE 9-7. Recovery Results for Assay and Content Uniformity Method

Sample Name	Actual Concentration (µg/mL)	Average Actual Concentration (µg/mL)	% Found	Average % Found	% Recovery
70%-Recovery–1	694.2	694.2	69.4808	69.5036	99.29
70%-Recovery–2	694.2		69.5307		
70%-Recovery–3	694.2		69.4992		
85%-Recovery–1	854.4	854.4	84.3520	84.5353	99.45
85%-Recovery–2	854.4		84.5062		
85%-Recovery–3	854.4		84.7478		
100%-Recovery–1	1068	1068	99.2699	99.1599	99.16
100%-Recovery–2	1068		99.0610		
100%-Recovery–3	1068		99.1488		
115%-Recovery–1	1174.8	1174.8	114.566	114.4453	99.52
115%-Recovery–2	1174.8		114.3211		
115%-Recovery–3	1174.8		114.4488		
130%-Recovery–1	1281.6	1281.6	128.6508	128.8411	99.11
130%-Recovery–2	1281.6		128.8216		
130%-Recovery–3	1281.6		129.0509		
			% Average recovery =		99.31
			SD		0.18
			%RSD		0.18

specification limit. Note that the reporting level can never be lower than the limit of quantitation (LOQ) of the method. However, during early-phase validation for drug products, if authentic degradation products are not available, then low amounts of API are added (LOQ to 120% specification limit of largest impurity) to the placebo and the recovery experiment is performed.

Once degradation products of known purity become available (isolated or synthesized), a spiked recovery experiment should be performed. For drug substances these may be directly spiked into the drug substance (DS), and for the drug product (DP) these may be spiked into the DS + placebo. This spiked experiment is conducted to determine whether a sample preparation procedure is able to completely extract active and degradation products from the sample matrix. For drug substances, a known amount of spiked impurities (authentic samples) is added to the active pharmaceutical ingredient and the recovery experiment is performed. The purity (A% – water – residual solvents – inorganic impurities) of the impurity that will be spiked must be known as well in order to calculate the actual amount added to the respective DS so that the theoretical amount of the impurity that would be in the solution can be determined. This is because the API may have some amounts of the same known degradation products may already present in the API drug substance. This must be accounted for in the calculation. Therefore, the amount of the impurity that is present in the matrix (DS) must be known. The total of the spiked amount of impurity and the amount of impurity that is present in the drug substance must be used to determine the overall amount of impurity. Also, for drug products when authentic degradation products are added to placebo in presence of API, the purity factor of the isolated degradation product that is spiked needs to be taken into account.

In the example shown in Table 9-8 and Table 9-9, a recovery experiment is performed for a drug substance that has 0.2% (area percent normalization) of

TABLE 9-8. Spiked Recovery Experiment

Target concentration of DS (mg/mL)	1	100 mg in 100 mL diluent		
Stock solution of impurity A (mg/mL)	0.1001	11 mg in 100 mL diluent		
Purity of impurity A	91%			

Amount DS (mg)	mL Stock A	mL Diluent	Concentration of Spiked Impurity (mg/mL)	Spiked % of DS target (1 mg/mL)
100	1	99	0.001001	0.1001
100	2	98	0.002002	0.2002
100	4	96	0.004004	0.4004

TABLE 9-9. Spiked Recovery Results

Impurity A % in Matrix	Impurity A % Spiked	Theoretical Overall	Actual Overall	Recovery Overall
0.2	0.1	0.3	0.31	103.3%
0.2	0.2	0.4	0.42	105.0%
0.2	0.4	0.6	0.59	98.3%

impurity A. The specification limit for this impurity is 0.3%, and 120% of the specification limit is 0.36%. Therefore a recovery experiment will be preformed where 0.1%, 0.2%, and 0.4% of impurity A will be spiked into the DS in solution. A stock solution of impurity A can be made. Depending on the desired percent of the impurity to be spiked in the DS, an aliquot of stock solution A is added to the 100 mg of the DS and then diluted to volume with the diluent. The purity factor of impurity A must be taken into account. In the following example, 11 mg was multiplied by the 0.91 (purity factor of the impurity) to give a total of 10 mg of impurity A. This spiked amount is added to the total (known as the theoretical overall total) as shown in Table 9-9. The actual overall is the percent of the impurity determined by HPLC analysis. Then the percent recovery can be determined (actual overall/theoretical overall) × 100.

9.4.1.1   Filter Check. If for the drug product the sample preparation procedure (recovery procedure) requires filtering the sample solution prior to the solution being injected into an HPLC system, then a check for adsorption of the components onto the filter membrane must be performed. The experiment should be set up to conduct the filter step and centrifuge on the same solution. So, for the same solution (reference standard solution as well as a sample solution), an aliquot of solution is passed through a membrane filter and collected after 2, 4, 6, 8, and 10 mL. In addition, the same solution (not filtered) is centrifuged and supernatant is collected. All solutions are then injected on a HPLC system. Since the identical solution has gone through different paths, the peak areas from the chromatogram should be identical (with some acceptable variability due to injection precision of the analytical instrumentation). If there is no change in peak areas between centrifuged and filtered solutions, then it can be stated that the membrane for that particular filter does not cause adsorption of the analyte(s). If the peak areas between centrifuged and filtered solution are different (filtered solution shows smaller peak areas than the centrifuged solution), then it can be stated that the membrane in that particular filter is adsorbing the analyte(s). However, sometimes an increase in peak areas is observed as greater volumes are passed through the filter (e.g., 2–6 mL). Therefore, the minimum volume that needs to be passed through the filter to get constant peak areas that are comparable to the centrifuged peak areas must be determined. If those areas are within (2.0%), then the designated amount of volume that is needed to pass through the filter before the solution can be collected for HPLC analysis must be noted. These types of filter experiments must be performed every time the drug product formulation has changed (change in excipients and/or ratios of excipients/DS). Also, even the same membrane filter type from different vendors can give different results due to changes in the housing of that particular membrane filter, and these should also be investigated.

9.4.1.2 Completeness of Extraction.

For drug products containing constituents that are insoluble in the extraction medium used in the analytical procedure, it may be deemed adequate to perform a separate test for completeness of extraction (in addition to recovery experiments as described above). The completeness of extraction can be evaluated two ways: kinetically (over some elapsed time t) and thermodynamically (change in volume).

For time experiments, when the initial recovery experiment is completed (when using a real drug product, not from spiked experiments), the solution that is left over is set aside for time t (usually 24 hours at a temperature condition known not to affect the inherent stability of the solution). After time t, the solution is re-shaken by hand and then re-injected to determine assay value.

For the volume experiment, the initial recovery experiment is repeated using the actual drug product using an increased volume of sample solvent. For example, if the procedure is stated to extract the content of a drug product with 50 mL of solvent, then further experiments would dictate the use of 75 mL or 100 mL of solvent for this experiment.

A generalized procedure to evaluate both kinetic and thermodynamic factors is provided in Table 9-10. This would require that an actual drug product sample is extracted and analyzed as per procedure described in the analytical method with the extraction time of t_0 and extraction volume of V_0 stated in the method. Six more experiments would be conducted such that longer extraction times t_1, t_2 and t_3 are used and higher extraction volumes V_1, V_2, and V_3 are used. The extraction volumes employed should use the same extraction time specified in the method (t_0).

TABLE 9-10. Generalized Procedure to Evaluate Both Kinetic and Thermodynamic Factors

Experiment Number	Extraction Time	Extraction Volume
1	t_0	V_0
2	t_1	V_0
3	t_2	V_0
4	t_3	V_0
5	t_0	V_1
6	t_0	V_2
7	t_0	V_3

By performing these two experiments (time and volume), it will show whether the initial extraction procedure is sufficient or has any shortcomings. If any shortcomings are observed, then a new extraction procedure must be included in the method (i.e., longer time and/or higher amount of extraction solvent). Most likely, for modified release drug products, time is essential (higher recovery is observed over time). Change in volume will usually have an impact in cases where the solubility of an API is on the edge for that particular sample preparation solvent with the excipients present in the matrix. If this is the case, then the procedure should be modified to extract with higher volume of sample preparation solvent.

Lastly, to really prove that the sample preparation procedure or the recovery procedure is completely extracting the API and degradation products, utilization of a homogenizer must be considered.

In general, homogenizers are utilized for automated sample preparation procedures in workstations such as TPWII (Caliper Life Sciences, 68 Elm Street, Hopkinton, MA 01748 or www.caliperls.com). Homogenizers are made up of a stainless steel blade that rotates up to 20,000 rpm. The following example illustrates why homogenizers play a vital role in sample preparation. A modified release (MR) product under development gives a drug release profile for at least 8 hours. This corresponds to the release rate of drug substance or the API from the dosage form within 8 hours at 37°C in the dissolution media (apparatus I as described in the United States Pharmacopoeia (USP) ⟨711⟩ at 100 rpm). For example, when a sample preparation procedure was developed for assay and degradation products for a modified release drug product, it was determined that a sonication time of about 30 minutes with about 4 hours of mechanical shaking provided adequate extraction efficiency of the drug from the dosage form. In contrast, if a stand-alone homogenizer was utilized for the same procedure for this dosage form, then the total sample extraction time was found to be about 5 minutes with intermittent stops required by the system software (e.g., TPWII takes about 2 to 5 seconds from end of one pulse to the start of another pulse). The homogenizer provides the energy needed to break the dosage form and to extract the API very efficiently when compared to conventional sonication and mechanical shaking. The final sample preparation procedure was finalized as follows:

Two pulses at 8000 rpm for 10 seconds each

Six pulses at 15,000 rpm for 15 seconds each

Soak/settle time for 2 minutes (to allow all particles to settle to the bottom of the vessel, which allowed for a facile filtration step)

9.4.2 Precision

Precision provides an indication of random errors and can be broken down into repeatability and intermediate precision. This procedure should only be performed when the entire analytical method procedure is finalized.

Repeatability represents the simplest situation and involves analysis of replicates by the same analyst, generally one injection after the other. Repeatability tests are mandatory for all tests delivering numerical data. Repeatability is divided into two parts: injection repeatability and analysis repeatability (multiple preparations) [24].

Validation of the precision of an HPLC method occurs at three stages. The first stage is injection precision (injection repeatability) based on multiple injections of a single preparation of a sample on a particular sample on a given day. The set of criteria is given for area (% area normalization) methods (DS and DP) based on %RSD of peak area. The second stage is analysis repeatability where multiple preparations and multiple injections of a sample are analyzed by the same chemist on the same day. The third stage is intermediate precision and is usually performed by different analysts, on a different system, on a different day on the same DP or DS batch to determine the variability of the analytical test. The intermediate precision test may give indications to potential issues that may arise during method transfer. Relative standard deviation or coefficient of variation (S_{rel} or %RSD) is used to assess if the adequate precision has been obtained.

If automation is utilized, then an intermediate precision test is required to compare results obtained through manual testing versus automated testing (if all solvent composition and analyte concentrations of all actives are identical in both methods).

9.4.3 Linearity

The purpose of the test for linearity is to demonstrate that the entire analytical system (including detector and data acquisition) exhibits a linear response and is directly proportional over the relevant concentration range for the target concentration of the analyte. It is recommended to perform the linearity of the API and related substances independently; and once linearity has been demonstrated, another linearity could be performed containing both API and specific related substance if necessary. For this reason, a stock solution of each substance (API, degradation products, synthetic by-product) must be prepared separately (one per solution), and a serial dilution from this stock solution must be injected into an HPLC system (constant injection volume). There are two major reasons to perform a linearity test on each solution independently. First, each substance may not be pure and the linearity test for each component may become confounded. This is especially true if the active drug substance contains the impurity that linearity is being performed on and/or if the impurity contains the active drug substance as an impurity. Second, when each substance is studied independently, the calculation of relative response factor (RRF) is much easier to determine. The ranges that should be covered for the linearity test are described in Table 9-11.

At least five concentrations within the range specified above for the linearity test should be used. When a linearity test is needed for an assay and

TABLE 9-11. Recommended Ranges for Linearity Tests

Type of Analytical Procedure	Range to be Covered
Drug Substance	
Weight percent	80–120% of target concentration
Impurities	LOQ or reporting level to at least 120% of specification
Drug Product	
Assay (content)	80–120% declared content
Assay (CU)	70–130% declared content
Assay (dissolution rate)	±30% of specified range
Impurities/degradation products	LOQ or reporting level to at least 120% of specification
Assay and degradation products	LOQ or reporting level to 120% of assay content [only if 100% reference standard is utilized to calculate low level of degradation products]

degradation products test, it is generally recommended that five or more concentrations should be utilized to cover the entire range. The focus should be on both (a) the lower limit (LOQ to 1.0%) for the degradation products and (b) the higher limit (80–120%) for the assay of the active. If the assay method is also used for content uniformity, the range 80–120% should be expanded accordingly to 70–130%.

As stated in the recovery section, if authentic degradation products or impurities are not available, then an API may be utilized to perform the linearity test at the lower concentration range (reporting level to 1.0%). In addition for drug products, if assay and degradation products are calculated from a single 100% reference standard solution (mass percent) or area percent normalization, then two independent linearity tests must be generated to demonstrate linearity at the degradation product level as well as at the assay level. Hence, if the 100% standard is used to quantitate the levels of degradation products, the slopes of the low-level linearity and the high-level linearity curves must be compared. If the criteria for agreement between the two slopes are not met, then for quantitation of the degradation products a lower concentration of standard (usually between 0.5% and 5.0%) is used to calculate the degradation products (related substances). This will be further discussed in the next section.

Acceptability of linearity data is often judged by examining the correlation coefficient and y-intercept of the linear regression line for the area response versus concentration plot and residual standard deviation (standard error compared to the calculated y-value at a certain target % level). Correlation coefficients of >0.990 (DP) or 0.998 (DS) are generally considered as evidence of acceptable fit of the data to the regression line. The y-intercept and %RSD acceptance criteria for DP and DS depend on the linearity range being tested, and the proposed criteria are shown in Table 9-3. Although these are very practical ways of evaluating linearity data, they are not true measures of linearity [25, 26]. The coefficient of correlation can be subject to misinterpretation and may give a misrepresentation of linearity, since different datasets can yield identical regression statistics [27, 28]. The parameters, correlation coefficients, y-intercept, and %RSD by themselves can be misleading and should not be used without a visual examination of the response versus concentration plot [29]. A more statistically sound approach to examine linearity would include examining the residuals from a linear regression. The residuals are the distances of the experimental points from the fitted regression line, measured in a direction parallel to the response axis. Analysis of the residuals provides further support that the calibration curve would be deemed linear if the residual response shows a normal distribution with a zero mean [30]. Although correlation coefficients of the linear regression can be >0.99, the plots of the response factor versus the concentration can shed light if there are any apparent deviations from linearity. A slope close to zero (response factor versus concentration) would indicate that a linear response is obtained over the specified concentration range. An additional acceptance criterion that could be considered is that the response factor will show %RSD of ≤2.0% across all concentration levels between 80% and 120% of the target concentration (assay). Also, this %RSD acceptance criterion could be applied to the low-level linearity regions such that the response factor will show %RSD of ≤10% across all concentration levels between LOQ to 120% of impurity specification level. Additionally, this comparison of the response factor can be used to help justify the LOQ above and beyond the typical S/N >10, injection precision requirements, and low-level linearity requirements. A simple test would be to compare the response factor difference between the proposed LOQ and the 5× LOQ value concentration and also to compare the response factor difference between the proposed LOQ and the maximum concentration tested in the low-level linearity experiment. Both of these percent difference values should show ≤10% difference to provide additional support for qualifying the proposed LOQ as the official LOQ for the method.

9.4.3.1 Linearity Example (Assay and Content Uniformity).

An example for linearity for Assay and Content Uniformity is given. The target concentration is 1.0 mg/mL for this particular drug substance D. Table 9-12 shows the table that could be included in a method validation protocol stating the concentrations that will be tested from 50–130% of the target concentration. The sample preparation procedure is indicated in Table 9-13. The linearity results and the relative response factors at each concentration are shown in Table 9-14. The response factor is calculated by peak area divided by concentration at each concentration level. A typical graph for linearity is obtained such that the concentration is plotted on the x axis and the area counts are plotted on the y axis Figure 9-1. The %RSD of 0.51% for the calculated response factors at all concentrations is reported in Table 9-14. In general, %RSD should be

TABLE 9-12. Linearity for Assay for Drug Product in Method Validation Protocol

Prepare and analyze solutions at the following levels:

Target Concentration of Solutions (%)	Concentration (mg/mL)	Amount Injected (µg)	Number of Injections
130	1.30	13.0	2
115	1.15	11.5	2
100	1.00	10.0	2
85	0.85	8.5	2
70	0.70	7.0	2
50	0.50	5.0	2

Criteria:
Linearity
Correlation coefficient: r ≥ 0.998.
y-intercept: ≤2.0% when compared to the calculated y-value at the 100% level.
Residual standard deviation: ≤2.0% (standard error compared to the calculated y-value at 100% level).

TABLE 9-13. Example of Actual Sample Preparation for Compound D in Method Validation Protocol

Solution Number	Target Concentration of Solutions (%)	Stock Solution[a] Used (mL)	Added Sample Solvent (mL)	Final Concentration of Solution (µg/mL)
1	130	12	25	1281.6
2	115	11	25	1174.8
3	100[b]	4	10	1068
4	85	8	25	854.4
5	70	13	50	694.2
6	50	9	50	480.6

[a]Stock solution concentration of active X is 2.67 mg/mL.
[b]100% level without placebo is used as a calibration standard.

less than 2.0% for assay methods (80–120% of target). A plot of response factor versus concentration is shown in Figure 9-2. The near-zero slope (0.03) of the response factor plot indicates that a linear response is obtained over this concentration range. Table 9-15 shows the regression analysis performed by Excel using the available Add-In functionality ToolPak. In Table 9-15, the y value using the 100% standard is calculated ($y = mx + b$), where x is the concentration of the 100% standard. The %RSD (0.51%) is calculated as well and is defined as the standard error/y. In order to identify if there are significant deviations from the assumed linearity, an investigation of the residuals should

TABLE 9-14. Linearity Results (Assay and Content Uniformity)

Concentration (μg/mL)	Peak Area	Average	Response Factors (RF)	
480.6	5,040,447		10,487.8	
480.6	5,044,555	5,042,501	10,496.4	
694.2	7,282,317		10,490.2	
694.2	7,269,300	7,275,808.5	10,471.5	Avg RF: 10495.5
854.4	8,910,381		10,428.8	
854.4	8,926,904	8,918,642.5	10,448.2	SD (RF): 53.6
1,068	11,323,083		10,602.1	
1,068	11,319,630	11,321,356.5	10,598.9	%RSD (RF): 0.51
1,174.8	12,345,022		10,508.2	
1,174.8	12,309,278	12,327,150	10,477.8	Slope: 0.03
1,281.6	13,427,476		10,477.1	
1,281.6	13,404,485	13,415,980.5	10,459.2	

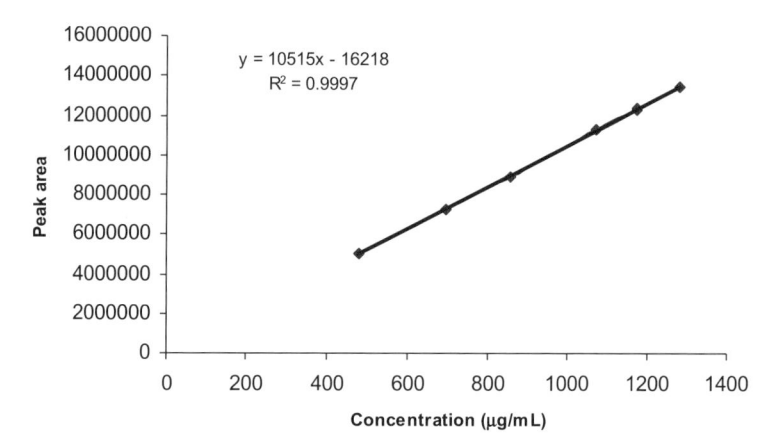

Figure 9-1. Assay and CU linearity from 50% to 130% of the target.

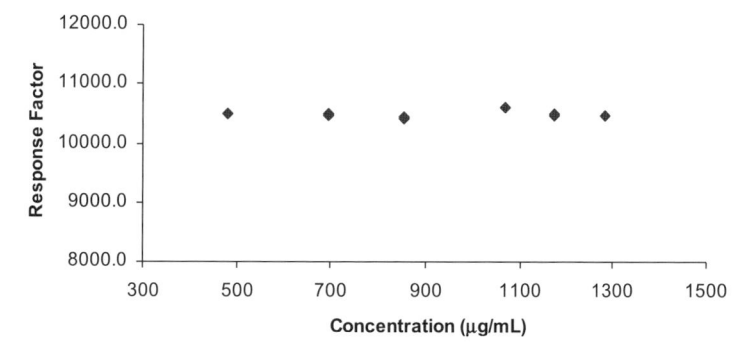

Figure 9-2. Response factor versus concentration (μg/mL) at the assay level.

TABLE 9-15. Regression Analysis Using Analysis ToolPak Add-In Functionality from Excel

Summary Output
Regression Statistics

Multiple R:	0.999838398
R square:	0.999676821
Adjusted R square:	0.999644503
Standard error:	57705.10759
Observations:	12

ANOVA

	df	SS
Regression:	1	1.03002E+14
Residual:	10	33298794422
Total:	11	1.03035E+14

	Coefficients	Standard Error
Intercept = b	−16218.49184	57793.36424
X Variable 1 = m	10515.47644	59.78890334
$x = 1068\,\mu g/mL$		
$Y =$	11214310.34	$y = mx + b$ at 100% level
Y-intercept (abs. value) =	0.14	(Intercept/y)*100
%RSD =	0.51	(Std error/y)*100

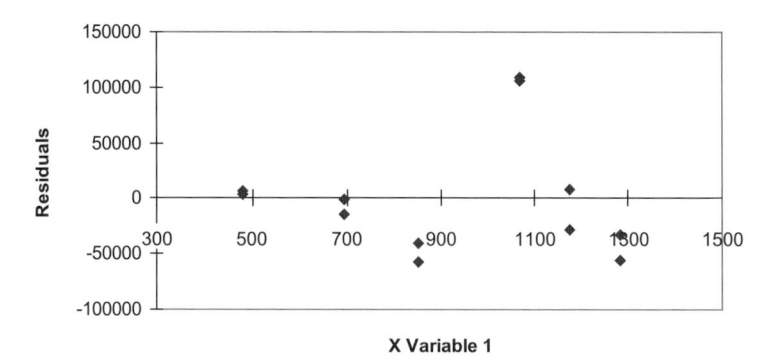

Figure 9-3. Residual plot where X Variable 1 is the concentration of the API in solution.

be performed; this is demonstrated in Figure 9-3. The residuals should be randomly distributed around a true mean of zero. The plot of residuals is usually plotted versus the x values (concentration in this example). If a linear, unweighted regression model is used, the residual plot should show random behavior in a constant range, and it should not be biased in one direction or the other or should not show a systematic or curved pattern of the residuals

TABLE 9-16. Comparison of Results to Acceptance Criteria

Parameter	Acceptance Criteria	Results
Linearity		
• Correlation coefficient	$r \geq 0.9998$	0.999 (Table 9-15)
• y-intercept	$\leq 2.0\%$	0.14% (Table 9-15)
• Residual standard deviation	$\leq 2.0\%$	0.51% (Table 9-14)
Range	At least 70–130%[a]	50–130% (100% is 1.0 mg/mL)

[a]This range is used because this linearity is for both assay and content uniformity.

TABLE 9-17. Linearity for Related Substances for Compound D

Target % of Solutions	Target Concentration (mg/mL)	Target Amount Injected (μg)	Number of Injections
9.0	0.09	0.9	2
7.0	0.07	0.7	2
5.0	0.05	0.5	2
3.0	0.03	0.3	2
2.0	0.02	0.2	2
1.0	0.01	0.1	2
0.4	0.004	0.04	6

[a]Note that in this theoretical example the specification limit for compound D is set to 7.0% and since no isolated impurity exists the active pharmaceutical ingredient is used for the low level linearity.

Criteria: Linearity

        Correlation coefficient: $r \geq 0.998$

        y-intercept: $\leq 5.0\%$ when compared to the y-value at the 1.0% level.

        Residual standard deviation:

        (0.4–9.0%): $\leq 4.6\%$[a]; calculation based on 1.0% level standard

        LOQ: S/N $\geq 10{:}1$, $S_{rel} \leq 10\%$, $n \geq 5$

[a]Justification for acceptance criteria is set based on the weighted value from 7 levels:

        Criteria set at %RSD $\leq 2.5\%$ for 5.0, 7.0 and 9.0%

        Criteria set at %RSD $\leq 5\%$ for 1.0, 2.0 and 3.0%

        Criteria set at %RSD $\leq 10\%$ for 0.4%

[31–33]. The following acceptance criteria have been met as shown in Table 9-16 for the linearity part of this validation.

9.4.3.2 Linearity Example (Related Substances, Low-Level Linearity). An example for linearity of related substances is given. The target concentration is 1.0 mg/mL for compound D (drug substance). The maximum specification limit of a particular impurity is 7.0%. Therefore the linearity range should be from LOQ to at least 120% of maximum specification limit, which would be from 0.4% to 9.0% of target concentration. Table 9-17 is the table that could be included in a method validation protocol stating the concentrations that will be tested are from 0.4% to 9.0% of the target concentration with the

TABLE 9-18. Linearity Results for Related Substances

% of Target (1.0 mg/mL)	Concentration (µg/mL)	Peak Area	Response Factors (RF)	
0.3738	3.738	35,860	9,593.365436	
0.3738	3.738	35,941	9,615.034778	
0.3738	3.738	35,749	9,563.670412	
0.3738	3.738	35,210	9,419.475655	
0.3738	3.738	35,183	9,412.252541	Avg. RF (0.4%)—
0.3738	3.738	35,237	9,426.698769	LOQ: 9,505.1
1.068	10.68	107,181	10,035.67416	
1.068	10.68	107,194	10,036.89139	
2.136	21.36	214,144	10,025.46816	Avg. RF (2.1%)—
2.136	21.36	213,809	10,009.78464	5XLOQ: 10,017.6
3.204	32.04	328,167	10,242.41573	
3.204	32.04	328,013	10,237.60924	%Difference (Avg. RF)
5.1264	51.264	530,457	10,347.55384	(0.4%:2.1%): 5.4
5.1264	51.264	532,297	10,383.44647	
6.408	64.08	662,721	10,342.08801	
6.408	64.08	661,735	10,326.701	
8.544	85.44	890,138	10,418.28184	Avg. RF (8.5%):
8.544	85.44	889,655	10,412.62875	10,415.5
	Overall Avg. RF: 9,991.6	SD (RF): 381.1	%RSD (RF): 3.8	%Difference (Avg. RF) (0.4%:8.5%): 9.6

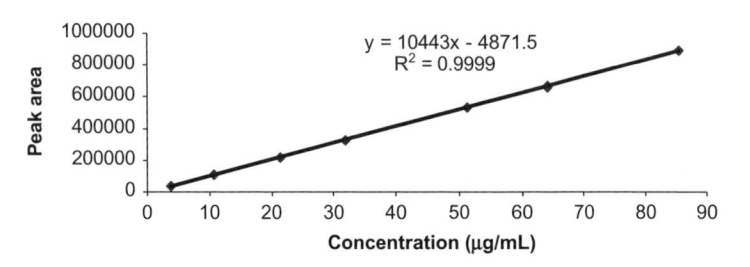

Figure 9-4. Low-level linearity from 0.4% to 9% of the target.

proposed acceptance criteria. The linearity results and the relative response factors at each concentration are shown in Table 9-18. The graph for linearity where area response was plotted versus analyte concentration is shown in Figure 9-4. The response factors for all concentrations in the low-level region is calculated and reported in Table 9-18 (%RSD = 3.8%). In general, %RSD should be less than 10.0% for related substance methods in the low-level linearity region. A response factor versus concentration plot is provided in Figure 9-5. The slope (~10) of the response factor plot indicates that there is some deviation of linear response toward the lower end of the concentration range studied. An additional test was conducted where the response factor differ-

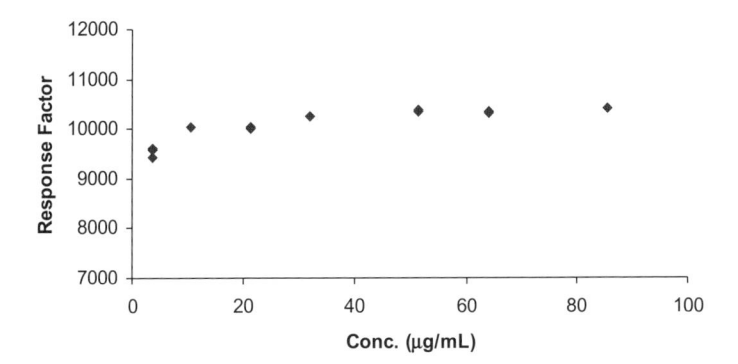

Figure 9-5. Response factor versus concentration (μg/mL) at the related substance level.

ence between the proposed LOQ (0.4%) and 5× the proposed LOQ value concentration (2.1%) was determined, and a value of 5.4% was obtained which was deemed acceptable (≤10% is acceptable). The response factor difference between the proposed LOQ (0.4%) and the maximum concentration tested (8.5% of target) in the low-level linearity experiment was determined, and a value of 9.6% was obtained which was deemed acceptable. Both percent difference values of 5.4% and 9.6%, respectively, showed less than a 10% difference and could be considered to be acceptable (Table 9-18). However, it would be recommended to use a low-level standard of 1.0% to quantitate impurities in the region of 0.4% to 9.0% because dilute standards of greater than 1.0% would lead to underestimation of impurities below 1.0%.

Table 9-19 shows the regression analysis performed by Excel using the available Add-In functionality ToolPak. In Table 9-19, the y value of 4.6% was calculated using the 1.0% standard ($y = mx + b$), where x is the concentration of the 1.0% standard and acceptance criteria given in Table 9-17 of ≤5.0% was met. In Table 9-19, the %RSD (standard error/y) was calculated as 2.1%, and this had met the acceptance criterion of ≤4.6% given in Table 9-17. In order to identify if there are significant deviations from the assumed linearity, an investigation of the residuals should also be performed; this is demonstrated in Figure 9-6. The residuals were randomly distributed around a true mean of zero.

The acceptance criterion for the % RSD is calculated as follows:

$$\frac{(3 \times 2.5\%) + (3 \times 5\%) + (1 \times 10\%)}{7} = \frac{7.5 + 15 + 10}{7} = \frac{32.5}{7} = 4.6\%$$

9.4.3.3 Low-Level Linearity Versus High-Level Linearity: When Do You Use a Dilute Standard to Calculate Impurity Levels? There are two ways to determine the related substances: one in terms of mass percent using a reference standard (at 100% level of API or dilute reference standard) and the

TABLE 9-19. Regression Analysis Using Analysis ToolPak Add-In Functionality from Excel

Summary Output

Regression Statistics

Multiple *R*:	0.99997499
R Square:	0.999949981
Adjusted *R* square:	0.999946855
Standard error:	2216.504059
Observations:	18

ANOVA

	df	*SS*
Regression:	1	1.57145E + 12
Residual:	16	78,606,243.91
Total:	17	1.57153E + 12

		Coefficients	*Standard Error*
	Intercept = *b*	−4,871.508798	770.5743113
	X Variable 1 = *m*	10,443.38585	18.46542508
	x = 10.68 for the 1% level		
1%	*Y* =	106,663.8521	$y = mx + b$ at 1% level
Pass	*Y*-intercept =	−4.57	(Intercept/*y*)*100
Pass	RSD =	2.08	(Std. error)*100

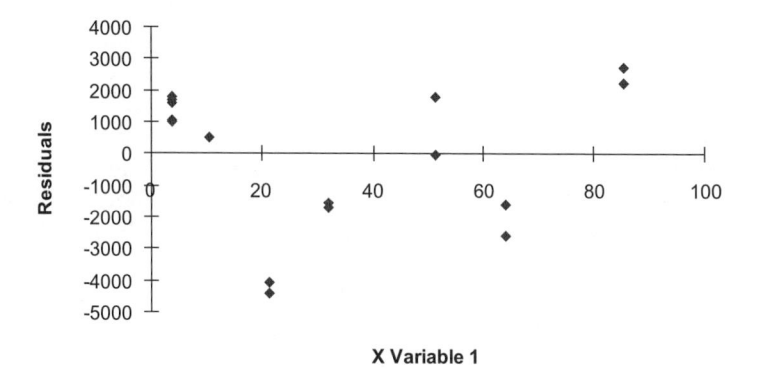

Figure 9-6. Residual plot where X Variable 1 is the concentration of the API in solution.

second using area percentages (area normalization). The percent of each impurity using either approach should be the same, assuming that the impurities have the same response factor as the API and there is no deviation from linearity throughout the whole concentration range from LOQ to 130% of API. The impurities can be calculated against a 100% standard of the API in terms of mass percent only if there is no deviation from linearity throughout the whole concentration range; otherwise, impurities can be overestimated or underestimated. Two separate linear regression calculations for the two different concentration ranges should be performed: (1) lower range from reporting level to 120% of impurity specification and (2) upper range from 70% to 130% of assay specification.

If impurities are calculated with respect to a 100% reference standard solution or based on area percentages, an additional criterion should be set such that the difference in slopes of upper range and slopes of lower range of the linearity curves is ≤4.0%. Also, the response factors from the lower-range linearity curve and the upper-range linearity curve should be compared and the overall %RSD should be ≤5.0% (this can be dependent on weighted average of the acceptance criteria for %RSD of the response factor at the low- and high-level linearity regions). In the event that this criterion is not met and there is deviation from linearity in the low-level region the impurities should be calculated against an appropriate dilute reference solution of the drug substance that is comparable to the level of impurities that are typically present.

Consider the two examples previously described for linearity at the assay and related substance levels shown in Tables 9-12 to 9-19 and in Figures 9-1 to 9-6. Comparison of the slopes shows a percent difference of less than 1%. The %RSD of all the response factors from 0.4% to 130% of the target concentration is 3.8% (Table 9-20), which is less than the 5.0% proposed acceptance criteria for %RSD of the response factors across the whole concentration range. Also, Figure 9-7 shows that when the response factor was plotted against all the analyte concentrations used in both linearity curves from 0.4% (3.738 μg/mL) to 130% (1282 μg/mL) of target, there is a deviation from linearity in the low-level analyte concentration region. Even though all acceptance criteria were met for linearity in both regions, it would be recommended not to use the 100% standard to quantitate impurities and not to use area normalization but instead to use a dilute standard to calculate the mass % of impurities. It would be recommended to use a low-level standard of 1.0% to quantitate impurities because standards of greater than 1.0% would lead to underestimation of impurities below 1.0%. The use of a low-level standard is generally recommended because it is an excellent way to determine if there is a mass balance discrepancy. This is especially important if the method will be used to release new batches of clinical supplies or when the method is used for supportive stability studies. If a discrepancy in mass balance is observed, this would trigger an investigation to elucidate the root cause for the mass balance issue (i.e., new impurity with different response factor, new impurity with no UV chromophore).

9.4.4 LOD/LOQ

Limit of detection (LOD) and limit of quantitation (LOQ) are determined for all impurity tests (including residual analysis during cleaning verification).

TABLE 9-20. Response Factors from 0.4–130% of Target Concentration

Concentration (μg/mL)	Peak Area	Response Factors (RF)
3.738	35,860	9,593.4
3.738	35,941	9,615.0
3.738	35,749	9,563.7
3.738	35,210	9,419.5
3.738	35,183	9,412.3
3.738	35,237	9,426.7
10.68	107,181	10,035.7
10.68	107,194	10,036.9
21.36	214,144	10,025.5
21.36	213,809	10,009.8
32.04	328,167	10,242.4
32.04	328,013	10,237.6
51.264	530,457	10,347.6
51.264	532,297	10,383.4
64.08	662,721	10,342.1
64.08	661,735	10,326.7
85.44	890,138	10,418.3
85.44	889,655	10,412.6
480.6	5,040,447	10,487.8
480.6	5,044,555	10,496.4
694.2	7,282,317	10,490.2
694.2	7,269,300	10,471.5
854.4	8,910,381	10,428.8
854.4	8,926,904	10,448.2
1,068	11,323,083	10,602.1
1,068	11,319,630	10,598.9
1,174.8	12,345,022	10,508.2
1,174.8	12,309,278	10,477.8
1,281.6	13,427,476	10,477.1
1,281.6	13,404,485	10,459.2
Avg. overall RF: 10,193.2	SD overall RF: 386.4	%RSD: 3.8

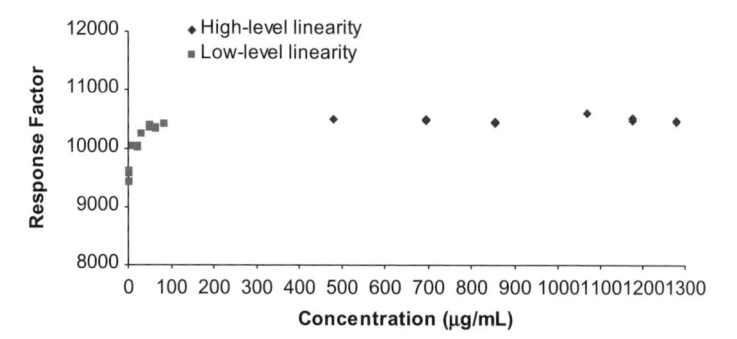

Figure 9-7. Response factor versus concentration (μg/mL) from the related substance level to 130% assay level.

The definition of LOD is defined by the USP as "a parameter of limit tests. It is the lowest concentration of analyte that can be detected, but necessarily not quantititated, under the stated experimental conditions" [USP ⟨1225⟩ "Validation of Compendial Methods"]. In contrast, LOQ is defined as "a parameter of quantitative assays for low levels of compounds in sample matrices, such as impurities in bulk drug substances and degradation products in finished pharmaceuticals. The LOQ is the lowest concentration in a sample that may be measured with an acceptable level of accuracy and precision, under the stated experimental conditions" [USP ⟨1225⟩ "Validation of Compendial Methods"]. The LOQ must be lower than or equal to the reporting level or reporting threshold (RTH), as defined in ICH Q3B (R1) guidelines, which is based on maximum daily dose (MDD) for any drug product. For example, if MDD for any drug product is >1 g, then RTH = 0.1%. If MDD is ≤1 g, then RTH = 0.05%. If the test procedure is not able to attain this limit, the reasons should be discussed on a pure scientific basis with the appropriate health authorities.

Each separation method, including HPLC, has a minimal concentration at which it will be able to detect and quantitate with some level of confidence. Statistically speaking, limit of detection (LOD) is defined for a peak that gives a signal-to-noise ratio of about 3:1, and limit of quantitation (LOQ) is defined for a peak that gives a signal-to-noise ratio of about 10:1. Some analytical chemists may prefer to inject five or six consecutive injections of these solutions and then calculate %RSD of peak area for the peak of interest. LOD giving %RSD of ≤20% and LOQ giving %RSD of ≤10% are acceptable values throughout the industry and are accepted by health authorities as well.

To determine LOD and LOQ, progressive dilutions of the analyte are prepared and analyzed. For HPLC, the normal range for LOD/LOQ is between 0.01% and 0.2% for non-peptide/protein-related products; however, for protein/peptides the LOD/LOQ range is typically between 0.1% and 0.5%.

From the injected series, a peak is selected whose height hs is about 3 to 10 times larger than the signal noise hn, as defined below:

hn = largest deviation (positive or negative) of the detector signal from the average baseline level measured over a span of at least 10 peak widths from the retention time of the target analyte. hn must be measured in the same units as hs. hn may be obtained from the same chromatogram as hs or from a blank injection (Figure 9-8A) (if other peaks elute within ±10 peak widths of target analyte).

hs = peak height of the analyte, measured from the average baseline level of the top of the peak, measured in the same units as hn as shown in Figure 9-8B.

Wh = peak width at half-height.

For HPLC, the LOD and LOQ can be defined (in mass or concentration unit):

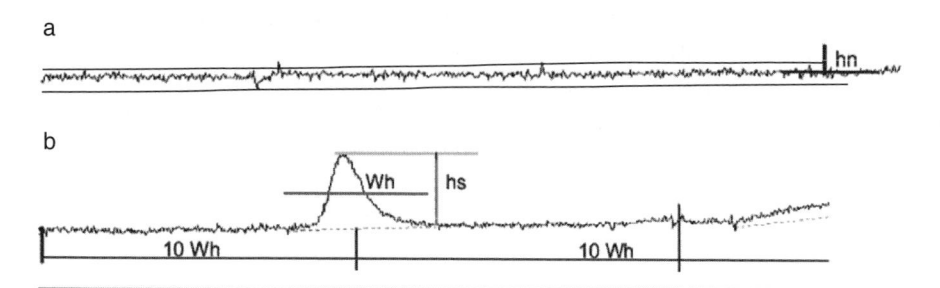

Figure 9-8. (A) Signal noise from blank chromatogram. (B) Peak height measurement for calculation of LOQ.

$$LOD = C_S \frac{3}{S/N}$$

$$LOQ = C_S \frac{10}{S/N}$$

where $S/N = h_s/2h_n$ and Cs is the amount or concentration of injected analyte.

It is also recommended to express the LOD and LOQ in percent of the declared amount of drug substance. Additional criteria for LOQ should be that this concentration be included in the low-level linearity curve and that all criteria for linearity must be met.

9.4.5 Relative Response Factors

The relative response factor (RRF) of 1 defines that an impurity and active at identical concentrations have the same analytical response. Generally, no corrections for RRF need to be performed if the RRF of particular impurities are between 0.8 and 1.2. If the relative response factors are outside this region, the impurities can be overestimated or underestimated. For example, if impurity X has an RRF of 0.5 compared to the active, then it would be underestimated; and if impurity Y has an RRF of 1.5 compared to the active, then it can be overestimated. In these cases the RRF must be taken into account when determining the percent of related substances during evaluation of the purity of the DS or the DP. The following equation could be used.

$$C = \frac{PAA_i \times 100}{PAA + \sum PAA_i} \times \frac{1}{RRF} \tag{9-1}$$

where PAA_i is the peak area of the individual related substance in the test solution, PAA is the peak area of active in the drug product test solution, and RRF is the relative response factor of impurity to active.

Also, if the there are two actives in the drug product and the impurities in the drug product from one active are being quantitated versus the other active the variant ratio (DS1 versus DS2) must be taken into account. Assume DS1 is the active that the DS2 impurities are being quantitatied against. Therefore, if the DS1:DS2 ratio was 1:2 the variant ratio (VR) of 0.5 would be used. If it was a 1:4 variant, then a factor of 0.25 would be used. The following equation can be used to take into consideration the RRF and the VR:

$$C = \frac{PAA_i \times 100}{PAA + \sum PAA_i} \times VR \times \frac{1}{RRF} \qquad (9\text{-}2)$$

where PAA_i is the peak area of the individual related substance in the test solution, PAA is the peak area of active in the drug product test solution, VR is the variant ratio of DS1 to DS2 (e.g., 1 for the 1:0 variant, 0.5 for the 1:2 variant, 0.25 for the 1:4 variant), and RRF is the relative response factor of DS2 to DS1.

9.4.6 Stability of Solution

The objective of this experiment is to demonstrate that the sample and reference standard solutions prepared according to the respective method are stable at least during normal duration of an analytical sequence (it is recommended usually to do solution stability at 24, 48, and 72 hours). The acceptance criteria are given in Table 9-21. It should have already been determined during the method development stage if the diluent is suitable for the sample preparation and the diluent does not react with the active and/or excipients in the matrix. It is recommended to analyze samples and a reference standard solution at intervals between initial timepoint and time t (in hours or days) against freshly prepared reference standard solutions. A sample solution and a reference standard solution are prepared and analyzed immediately. The same solutions are then stored at normal ambient laboratory conditions and at 5°C. After time t (hours or days), the solutions initially prepared (bring the 5°C solution to room temperature or whatever temperature set in the method prior to injection) are reanalyzed against freshly prepared reference standard solution. The solutions (sample or reference standard) are said to be stable

TABLE 9-21. Acceptance Criteria for Stability of Solution

Test Type		Acceptance Criteria
Assay, CU or dissolution	±2.0%	
Weight percent (DS)	±1.0%	
Related substances/area%	±1.0%	
Degradation product/	Level <0.5%	Change ≤20% over specified time
impurities	Level 0.5–<5%	Change ≤10% over specified time
	level ≥5%	Change ≤5% over specified time
	No new peak	≥ reporting level

for time t when stored at certain conditions when the acceptance criteria are met.

For drug substances it is acceptable to compare the peak area response at initial timepoint and at the defined time t (generally 24, 48, and 72 hours), assuming that the HPLC system is running, system suitability is run daily, and original mobile phase is used for all analysis (i.e., to avoid any potential change in the analyte ionization due to slight variation of mobile-phase pH). The change in response is reported and compared to initial. For early-phase validation for drug product a similar approach can be used such that the reference solution (standard) and the drug product are re-injected at initial timepoint and then re-injected at time t. The change in response is reported and compared to initial.

The driver for determining solution stability is to determine the length of time the samples are stable in the respective diluents. For example, if a run sequence is for 24 hours and the solutions are only stable for 12 hours, the later samples analyzed after 12 hours will lead to inaccurate results. At times the solution stability in the respective diluent will be greater at 5°C than at room temperature; therefore, solution stability at room temperature and refrigerator conditions must be evaluated. This may be beneficial in the case when there are instrument problems and the entire sequence needs to be rerun. With proper determination of the solution stability, the original solutions prepared could be run on a suitable HPLC without having to re-prepare the solutions. Also, for some compounds the solutions may be stable for x number of days regardless of temperature in which they are stored; however, they may be prone to photocatalyzation in the diluent, so proper storage conditions may need to be defined. In this case, three solutions may be used for the solution stability experiment: control (room temperature/protected from light, amber volumetric flask), room temperature (clear flask), refrigerator (clear flask).

Depending upon the outcome, a comment should be added to the analytical method stating the duration within which the solutions remain stable (e.g., the time within which the acceptance criteria are met).

9.4.7 Ruggedness/Robustness

The definition in regard to ruggedness given by USP (USP ⟨1225⟩ "Validation of Compendial Methods") is as follows: "The ruggedness of an analytical method is the degree of reproducibility of test results obtained by the analysis of the same samples under a variety of normal test conditions, such as different laboratories, different analysis different instruments, different days, etc. Ruggedness is normally expressed as the lack of influence on test results from the operational and environmental variables of the analytical method. Ruggedness is a measure of reproducibility of test results under normal expected operational conditions from laboratory-to-laboratory and from analyst-to-analyst." Practically speaking, ruggedness is another name for intermediate precision, where two analysts, from two different laboratories, on two different days, utilizing different instrumentations, lot numbers of columns, reagents, solvents, and chemicals, follow the identical test method to test the identical sample. Generally, it is not considered necessary to study these effects individually, and the use of an experimental design (matrix) is encouraged.

The obtained results from both analysts are grouped together to determine whether this additive precision is acceptable or not. For example, if each analyst prepared two sample preparations API at target concentration for intermediate precision, then a total of four values are pooled together (additive precision) as stated in Table 9-3 of Assay, Precision; $S_{rel} \leq 2.0\%$, $n \geq 4$. In addition to an additive precision requirement, some laboratories also include an acceptance criterion (for example, absolute mean difference $\leq 2\%$) for mean value. For example, if analysts 1 and 2 prepare three sample preparations each, then additive precision is calculated from a total of six values (three from each analyst). In addition, the mean value obtained by analyst 1 ($n = 3$) is compared against the mean value obtained from analyst 2 ($n = 3$), in which it must pass an absolute difference (between the two means) of $\leq 2.0\%$.

Robustness tests examine the effect that operational parameters have on the analysis results. The following definition of Robustness is taken from the ICH glossary [34]:

> The evaluation of robustness should be considered during the development phase and depends on the type of procedure under study. It should show the reliability of an analysis with respect to deliberate variations in method parameters. If measurements are susceptible to variations in analytical conditions, the analytical conditions should be suitably controlled or a precautionary statement should be included in the procedure.

Robustness is performed on a given test method, but each parameter is deliberately modified one at a time to determine its effect on the final result (e.g., total % degradation products, assay level, etc.). Prior to evaluation of robustness testing, a set of system suitability parameters must be set. This system suitability is a set of acceptance criteria for a particular method. Typical system suitability criteria include ranges for retention time (R_t), RRT (relative retention time), apparent efficiency (N, determined from the peak width at certain peak height; i.e., 5%, 10%, or 50% peak height are the most typical), resolution (R_s), tailing factor (T_f, at a certain peak height), asymmetry factor, and so on. Typically, these system suitability criteria and their respective ranges are defined during method development. During the evaluation of robustness testing, the series of system suitability requirements must be met to ensure that the validity of the analytical procedure is maintained whenever used.

For the determination of a method's robustness, chromatographic parameters are varied within a realistic range and the quantitative influence of the variables is determined. If the influence of the parameter is within a previously specified tolerance, the parameter is said to be within the method's robustness range. Obtaining data on these effects will allow to judge whether a method needs to be revalidated when one or more of the parameters are changed [35].

For HPLC methods, the parameters that can be deliberately modified are mobile-phase composition, pH of the mobile phase (if applicable), ionic strength of the aqueous portion of the mobile phase, concentration of mobile-phase additive (chaotropic, ion-pairing), gradient slope (if applicable), initial hold time for gradient (if applicable), flow rate, column (different lots and suppliers), column temperature, injection volume, autosampler temperature (if applicable), and wavelength. After changing these parameters, it must be assessed whether the system suitability requirements can be met for the particular HPLC method. Consider the following two examples shown below.

9.4.7.1 Effect of Modifier Concentration.
For example, during the method validation for a particular drug product the concentration of the acidic modifier (phosphoric acid) was varied. The target concentration of the acidic modifier was 0.6 v/v% phosphoric acid. The concentration of the acidic modifier was varied from 0.4 v/v% to 0.8 v/v%. At all concentrations studied except for 0.4 v.v% and 0.8 v/v% phosphoric acid, the recommended system suitability requirements were met (Table 9-22). In Figure 9-9, significant changes in retention were obtained for impurity C, which is the penultimate intermediate of the drug substance. It is basic in nature, and this impurity is greatly influenced by the phosphoric acid concentration. This may be attributed to the chaotropic effect described in Chapter 4, Section 4-10. Since the target phosphoric acid concentration is 0.6 v/v%, minor changes in the acid concentration of ±0.1 v/v% from the target have no impact on any of the system suitability criteria. However, larger changes than 0.1 v/v% of phosphoric acid could lead to system suitability failures.

9.4.7.2 Effect of Variation of Temperature.
Another variable that should be studied is the effect of temperature on the separation, especially if the method is sensitive to temperature and/or will be transferred to a manufacturing facility (receiving laboratory) that resides in an environment that leads

TABLE 9-22. System Suitability Parameters Determined as a Function of Concentration of H_3PO_4 (v/v%)

System Suitability	0.4%	0.5%	0.6%	0.7%	0.8%
Rs > 4 between compound A and compound B	6.6	6.5	6.2	6.3	6.2
Rt for compound B 8.2–9.2 min	8.9	8.9	8.8	8.8	8.8
Tailing factor of compound B (5% peak height) ≤2.3	2.3	2.2	2.1	2.0	1.9
RRT compound C = 2.1–2.3	1.9	2.1	2.2	2.3	2.4

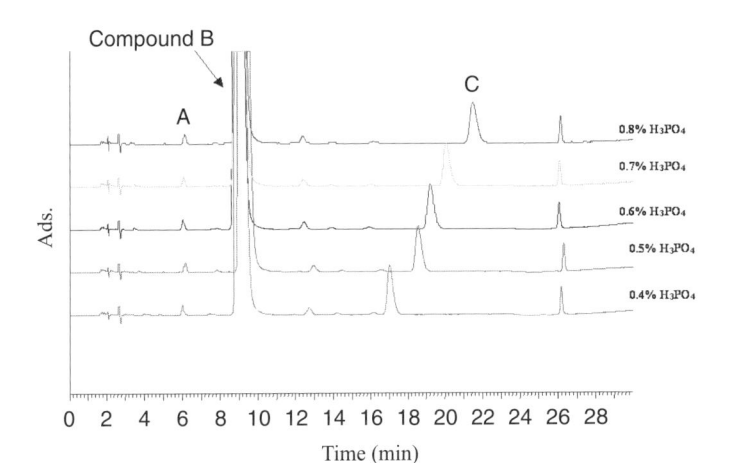

Figure 9-9. Effect of acidic modifier concentration on the analyte retention.

TABLE 9-23. Retention Times of Compound B, Synthetic By-products and Degradation Products

Peak Name	17°C	18°C	19°C	20°C	21°C	22°C	23°C	24°C	25°C
A	5.5	5.6	5.8	5.8	5.9	6.0	6.0	6.1	6.2
B	8.0	8.1	8.2	8.4	8.5	8.6	8.7	8.8	9.0
D	10.7	10.8	11.0	11.1	11.3	11.4	11.6	11.7	11.9
E	11.2	11.4	11.6	11.7	11.9	12.1	12.3	12.5	12.7
F	12.4	12.6	12.9	13.0	13.2	13.5	13.7	14.0	14.6[a]
G	14.3	14.3	14.4	14.4	14.4	14.5	14.6	14.6	14.6[a]
H	16.2	16.3	16.5	16.5	16.5	16.7	16.9	16.9	17.0
C	18.7	18.7	18.7	18.6	18.5	18.4	18.4	18.4	18.3
I	25.3	25.4	25.6	25.7	25.8	25.9	26.0	26.1	26.2

[a]Denotes co-elution of components.

to a higher ambient temperature than the transfer laboratory. The target temperature of this example method was 21°C. It was deemed necessary to choose this temperature in order to obtain the desired resolution between the critical pairs of synthetic by-products/degradation products and meet the other system suitability criteria. A temperature study was conducted from 17°C to 25°C at 1°C temperature intervals. Table 9-23 shows the retention times of a sample spiked with various synthetic by-products and potential forced degradation products analyzed at various temperatures from 17°C to 25°C. Co-elution of impurities F and G were observed at 25°C. At 17–18°C the retention of compound B was outside of the system suitability requirements. At 19–23°C, minor changes in selectivity were observed, but all system suitability criteria were met. Also, within the temperature range of 19°C to 23°C,

TABLE 9-24. System Suitability Parameters Determined as a Function of Temperature

System Suitability	17°C	18°C	19°C	20°C	21°C	22°C	23°C	24°C	25°C
RS > 4 between compound A and compound B	6.4	6.6	6.3	6.4	6.4	6.4	6.7	6.6	6.5
Tailing factor of compound B (5% peak height) ≤2.3	2.1	2.1	2.2	2.1	2.1	2.2	2.1	2.1	2.1
RRT compound C = 2.1–2.3	2.35	2.31	2.28	2.22	2.18	2.14	2.11	2.04	2.03
RT for compound B 8.2–9.2 min	8.0	8.1	8.2	8.4	8.5	8.6	8.7	8.8	9.0

acceptable resolution of synthetic by-products/degradation products from main compound, as well as resolution of critical pairs of impurities, was obtained. When dealing with methods that are temperature-sensitive, additional criteria should be set for resolution between critical pairs of impurities, and the selectivities/retention of the active and related substances should be closely examined. Only within the temperature range of 19°C to 23°C could all the system suitability criteria shown in Table 9-24 be met. Additional system suitability requirements for resolution between critical pairs of impurities could also be added, especially for impurities F and G (if the authentic impurities are available).

The degree of ruggedness and robustness of a particular test method depends on the stage of the drug development process. For drug substance (active pharmaceutical ingredient) and formulation development, both ruggedness and robustness experiments to study a new molecular entity (NME) in very early stages (Phase I clinical trials) would be much more limited compared to later stages in development when the final API synthesis is set and the final market image has been defined. For the first case (phase I clinical trials), very minimal ruggedness/robustness testing may be required. However, this really depends on a company's culture, since it is not a requirement from any health authorities. However, for the second case (after Phase IIa), an exhaustive study varying many critical parameters and multi-laboratory studies would be required.

9.4.8 Specificity

An investigation of specificity should be conducted during the validation of identification test, the determination of impurities, and the assay. The procedures used to demonstrate specificity will depend on the intended objective of the analytical procedure. It is not always possible to demonstrate that an analytical procedure is specific for a particular analyte (complete discrimination). In this case a combination of two or more analytical procedures is recommended to achieve the necessary level of discrimination (e.g., for optically active substances, in addition to an achiral HPLC method, a chiral HPLC method may be added).

Identity is a general requirement for dosage forms. When determining specificity for identity, the assay and related substances or the content uniformity methods can be used. Assay and content uniformity methods are quantitated by external reference standard. This identity test confirms that the correct active ingredient (s) is present and is present in correct ratio if multiple variants are available. The method could also be used for post-packaging analysis. The general requirements are that the sample and standard chromatograms should correspond in retention time and normalized peak area within ±10%.

The easiest way to perform specificity for any HPLC method is to perform this test in conjunction with a forced decomposition study. The utilization of mass spectrometry (MS) detector (in series) after a Photo Diode Array (PDA) detector to obtain more information is encouraged (in terms of mass-to-charge ratio of parent ions, initial fragmentation pattern, and peak purity).

Specificity is confirmed when an API peak is "pure" (confirmed by PDA and/or MS) and there is no interference from placebo solution (placebo dissolved in sample preparation solvent) at the retention time of an API peak.

9.4.9 Forced Degradation Studies (Solid State and Solution)—Drug Substance and Drug Product

Forced degradation studies are usually performed during the salt selection process for the drug substance (DS). In drug product (DP) development, the forced degradation studies of DS in the presence of excipients are first performed during the pre-formulation stage to assist in the selection of the most formidable compounds and excipients. This may lead to the development of more suitable formulations, packaging, and change in storage and manufacturing conditions as the optimal formulation is defined to be used in clinical studies. Forced degradation testing is often repeated when the final drug substance route and market formulation is defined and when the compound enters phase 3 clinical trials. A good overview of forced degradation testing according to the regulatory guidance documents, with emphasis on what should be considered for late clinical phases and for registration application dossiers (i.e., marketing authorization applications or new drug applications), is provided by the Impurity Profiling Group [36]. Forced degradation studies (sometimes referred to as stress testing) are also performed in order to demonstrate specificity during the development and validation of stability-indicating methods. These studies are usually performed at conditions exceeding that of accelerated storage conditions.

Forced degradation studies may provide information in regard to degradation pathways and degradation products that could form during storage of the drug substance or the drug product. The main goal of forced degradation studies is to effectively produce samples containing representative and realistic degradation products. These degradation products should be assessed if they are (a) related to the drug substance or the excipients or (b) due to drug substance–excipient interactions under certain forced degradation conditions. A delicate balance of efficiency and severity/duration of stress conditions is needed. Overstressing can destroy relevant compounds or generate irrelevant compounds. Understressing may fail to generate important degradation products. The extent of degradation targeted should be approximately anywhere from 5% to 10%. The other goal is that the potential degradation products that are generated should be resolved from the active component during development of a stability-indicating HPLC method. The assessment of peak purity using diode array and LC-MS detection are usually employed. These degradation products that are generated during the forced degradation studies can be identified, and the determination of degradation pathways and mechanisms for the drug substance and drug product can be elucidated.

Forced degradation studies are carried out either in the solution state and/or in the solid state. Usually the forced degradation testing is carried out on one batch of drug substance and/or one formulation blend (capsules and tablets). This forced degradation testing should not be part of a formal stability program.

The temperature/humidity conditions used may be more severe than the typical accelerated stability testing conditions in order to generate potential degradation products in a reasonable time. The typical forced degradation conditions include thermolytic, hydrolytic, oxidative, photolytic (in excess of ICH conditions), high pH (alkaline conditions), and low pH (acidic conditions). Outlined in Table 9-25 and Table 9-26 are some solid-state and solution forced degradation studies, respectively, that could be conducted. In the following

TABLE 9-25. Solid-State Forced Degradation Studies

Stress	Test Conditions	Duration
Thermal (closed container)	50°C and 80°C (ambient RH)	1 wk and 2 wks
Thermal/oxidative (open container)	50°C and 80°C (ambient RH)	1 wk and 2 wks
Thermal/humidity (open container)	40°C/75% RH	1 wk and 2 wks
Light (closed container)	Ambient	Maximum 1.2 million lux hours and 200 watt hours/square meter
Light/oxidative (open container)	Ambient	Maximum 1.2 million lux hours and 200 watt hours/square meter

TABLE 9-26. Solution Forced Degradation Studies

Test Factor	Test Conditions	Duration
pH	10 mg in 2 mL water 10 mg in 2 mL of 0.1 N HCl 10 mg in 2 mL of 0.1 N NaOH All in amber volumetric flasks and at room temperature.	1 day and 3 days[a]
Oxidation (H_2O_2)	10 mg/2 mL 3% H_2O_2 At 5°C and room temperature in amber volumetric flasks. If DS is not soluble, then pH modification may be necessary.	1, 2, and 3 days
Oxidation (metal ion catalyzed)	10 mg/2 mL water containing 100 ppm Fe_3^+, Ni_2^+, Cu_2^+ saturated with bubbled oxygen in amber volumetric flasks.	1, 2, and 3 days
Oxidation (saturated with oxygen)	10 mg/2 mL saturated with bubbled oxygen in amber volumetric flasks.	1, 2, and 3 days
Oxidation (free radical initiator: AIBN)	Ratio of DS to AIBN (1:10) in MeCN and/or MeCN/water (50:50) at 40°C, in amber flask.	4 hr, 1 day, 2 day
Oxidation (pressurized oxygen)	10 mg/2 mL with pressurized oxygen in closed headspace vial (80–300 psi), protect from light.	1, 2, and 3 days
Oxidation + light	Depending on which oxidative stress causes most oxidation, a new solution is prepared and is put in light chamber.	6 hr, 24 hr
Light	50 mg/10 mL water Ambient	Maximum 1.2 million lux hours and 200 watt hours/square meter, 6 hr, 1 day, 2 days
Heat	10 mg in 2 mL water at 50°C	6 hours, 1 day, 2 days

[a]If no degradation occurs, consider storing solution in GC oven at 50°C for 4 hr.

section we will elaborate on oxidative, hydrolytic, and photolytic forced degradation (stress testing).

For oxidation, different stressing schemes can be used, and this depends generally on the structure of the drug substance (active component): auto-oxidation, metals, peroxide-mediated, peroxy-mediated, bubbled oxygen, and pressurized oxygen. Auto-oxidation involves a free radical initiator such as AIBN (2,2'-azobisisobutyronitrile) or AMVN (2,2'-azobis(2,4-dimethylvaleronitrile) to initiate oxidation [37] and has been used to mimic long-term room temperature degradation related to oxidation. The concentration of AIBN that could be explored includes 1:1, 1:10, and 1:100 ratio of DS to AIBN. The typical solvents used include 50% acetonitrile: 50% water and 100% acetonitrile, and the stress study is usually conducted at 40°C. Some compounds are very susceptible to transition-metal-catalyzed oxidation, and the API solution can be stressed with various amounts of different transition metals. For peroxide-mediated forced degradation, dilute hydrogen peroxide (0.3–3%), can be used, but hydroxyl radicals are very reactive but unselective and may overstress the sample. For peroxy-radical-mediated forced degradation, peroxy radicals react with low-energy C—H bonds or C=C bonds and have been deemed as a more "realistic" oxidative system. Amines, sulfides, and carbon–carbon double bonds are susceptible to oxidation using hydrogen peroxide to give corresponding N-oxides, sulfoxides (or sulfones), and epoxides, respectively. These functional groups are prone to electrophlilic attack, and the reactions are ionic and do not involve free radicals [38]. Another, more predictive oxidative system would include forced degradation by purging the headspace with oxygen. The typical pressure that could be applied is 80–300 psi. Bubbled oxygen can be used to generate $^1O_2$. Using this type of oxidation would be advisable if API possesses certain functionalities such as sulfides, dienes/polyenes, imidazoles, purines, furans, and other heterocycles, since they are susceptible to reaction with $^1O_2$ [39–41]. For formulation development, forced degradation studies with $^1O_2$ should be considered, especially for liquid-based formulations (suspensions and solutions). Also, excipients should be carefully evaluated for their ability to generate $^1O_2$ upon hv, especially flavors and dyes [42]. More detailed examples about potential degradation pathways related to structure are given in the formulation chapter and in reference 38.

Another common pathway of degradation is hydrolysis, which is sometimes called solvolysis because the reaction may involve the pharmaceutical cosolvent such as ethyl alcohol or polyethylene glycol. The solvents such as water can act as nucelophile and attack the electropositive centers on the active pharmaceutical ingredient or excipients. Most compounds that are susceptible to hydrolysis include compounds that have a labile carbonyl functionality such as esters, lactones, and lactams. Also, the substituents near the carbonyl moieties of these compounds can have a dramatic effect on the reaction rates, and the substituent groups may exert electronic (inductive and resonance), steric, and/or hydrogen bonding effects that can affect the inherent stability of the compounds in the solid and solution state [43]. In order to determine potential hydrolysis degradation products of the API or the drug product, stress testing in water, acid, and base is conducted usually at room temperature protected from light. Note that the reaction rates of hydrolysis reactions can be further catalyzed under the acidic or alkaline conditions. If degradation is not observed under these conditions, the forced degradation studies using the same solutions can be performed at accelerated temperatures (i.e., 40–60°C).

Photolysis is another common degradation pathway because light-sensitive drugs can be affected by either sunlight or artificial light sources such as fluorescent light. The different reaction types that can be initiated photo-chemically include, reduction, N-dealkylation, hydrolysis, oxidation, isomer-ization, ring alteration, polymerization, or removal of various substituents like halogens or carboxyl groups (Figure 9-10) [44]. Note the forced degradation studies should be performed both with the API and the drug product (blend) because excipients can initiate, propagate, or participate in photochemical reactions. A control sample (sample protected from light) should always be run in parallel during the photodegradation study. Further details on photo-stability stress testing can be found in reference 45.

Generally the reporting of degradation studies are not required for an investigational new drug (IND) application. These studies are usually per-formed to develop confidence that the HPLC method used to analyze the drug substance and the drug product is stability-indicating. However, forced degra-dation studies are required at the submission of a new drug application (NDA). Any significant degradation products should be isolated and/or char-acterized by the appropriate analytical techniques (LC-MS, LC-NMR, NMR, UV, etc.), and a full report of these degradation studies would need to be per-formed [46]. Procedures for the preparation and/or isolation (if performed) and the methodology for the structural determination should be reported even in the event that the impurity cannot be identified. The physical and chemical properties of the isolated degradation product (where applicable) should also be assessed. Also, the mechanism and kinetics of the formation of each degra-dation product should be assessed as well as the order of the reaction. This may be challenging in some cases if the degradation process occurs by various types of oxidation [47–49]. Also, for drug products a distinction must be made between degradation products that are related to the drug substance itself, drug substance–excipient interactions, and those related solely to excipients.

9.5 DISTINGUISHING DRUG-RELATED AND NON-DRUG-RELATED DEGRADATION PRODUCTS

It is very important to distinguish which degradation products are related to the placebo or the drug substance. Therefore, the placebo should undergo similar degradation protocol as the drug product and the drug substance. The following questions should be asked when performing the studies.

- Is the new peak that is generated an impurity from the drug substance, a degradation product, or the placebo itself?
- At what level is the impurity/degradation product present?
- Is it a process-related impurity and, if so, at what step of the process is it formed?
- Is it a degradation product? If so, under what degradation condition is it formed?

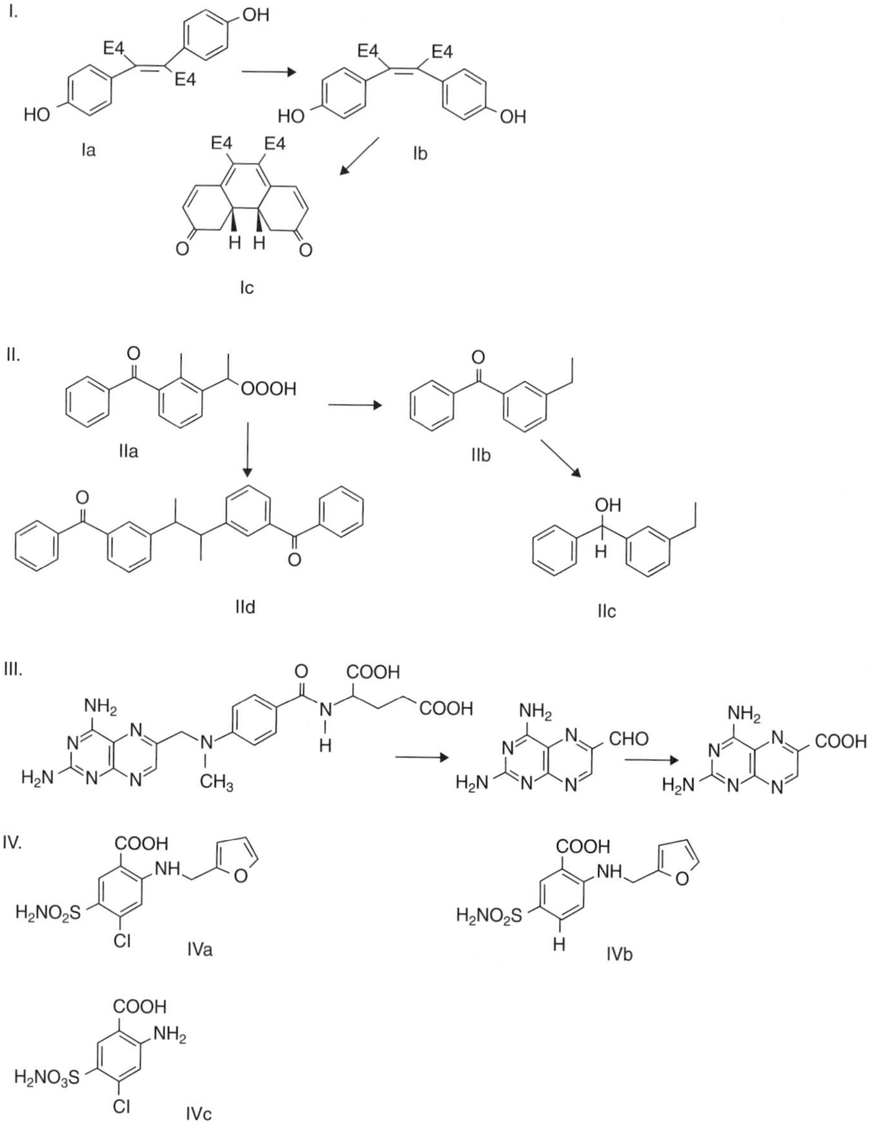

Figure 9-10. Examples of photoinduced reactions in drug molecules. (I) Photoinduced isomerization (Ib), cyclization and enol–keto isomerization (Ic) of stilboestrol (Ia). (II) Photoinduced reactions of ketoprofen (IIa) and its degradation product (IIb); decarboxylation (IIb), reduction (IIc), and dimerization (IIc) products of the drug. (III) *N*-dealkylation of methotrexate followed by oxidation. (IV) Dehalogenation (IVb) and photohydrolysis (IVc) of frusemide (IVa) [44].

9.5.1 Drug Product Stress

Stress testing should be conducted on final qualitative formulation. This should be a collaborative study with the analytical team. A gradient method should always be employed with an isocratic hold at high organic to ascertain if any hydrophobic degradation products are present.

9.5.1.1 Combination Products.
For combination products, additional questions need to be asked such as, Are there any potential chemical reactions between actives? and What is the potential impact on physical properties of the product (e.g., hygroscopicity)? Therefore, forced degradation of the combined actives in solution and the solid state should be explored to probe all potential modes of degradation. This, in essence, will lead to more stable formulations.

9.5.1.2 Experimental Approaches.
Forced degradation studies (solid state and solution state) should be conducted for defining a suitable stability indicating method. Outlined in Table 9-25 and Table 9-26 are some solid state and solution forced degradation studies respectively that could be conducted. An initial forced degradation scouting study, to help define the time frames for a particular forced degradation study, should be conducted. For example in Figure 9-11 a probe forced degradation study of a heated solution at 80°C was conducted at 1, 4, and 8 hours. As can be seen in Table 9-27, 8 hours was sufficient to generate about 10% degradation products. The results of the forced degradation scouting study can also be used if adequate degradation is observed (5–10% total degradation products). The forced degradation should be carried out on the drug substance, the formulation blends (oral drug products), and the placebo blends all in parallel to determine if the excipients or

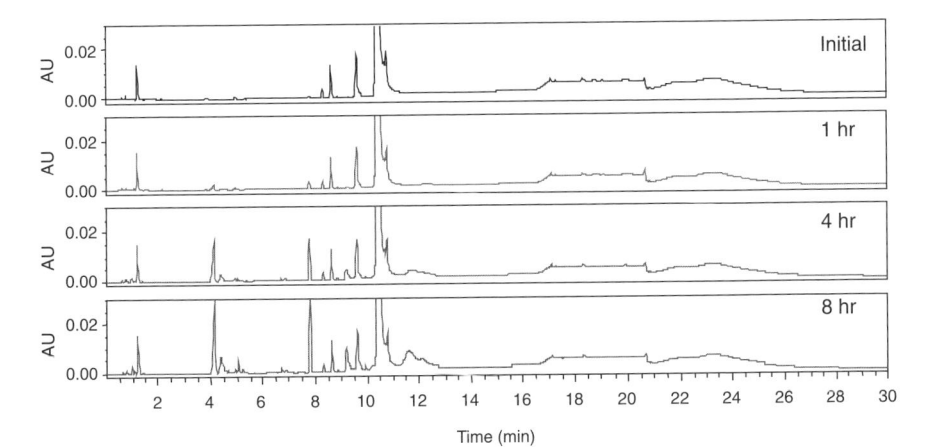

Figure 9-11. Forced degradation solution heat stress study: 1 mg/mL of API in solution. Top to bottom: initial, 1, 4, and 8 hours at 80°C. (stress conditions)

TABLE 9-27. Heat Stress Study of a 1 mg/mL Solution of API at 80°C

RT (min):	4.100	4.133	4.400	4.933	5.017	5.183	6.733	6.883	7.850	8.283	8.367
RRT:	0.394	0.397	0.422	0.474	0.482	0.498	0.646	0.661	0.754	0.795	0.803
Initial:				<0.05	<0.05				<0.05	<0.05	
1 hr:		0.06		<0.05	<0.05				0.10	0.08	0.09
4 hr:	0.29	0.45	0.19	<0.05	<0.05	<0.05	<0.05	<0.05	0.49	<0.05	0.10
8 hr:	0.62	0.88	0.53	0.08	0.15	0.07	<0.05	<0.05	0.95	<0.05	0.07

RT (min):	8.667	8.883	9.233	9.617	9.950	10.417 (Active)	10.683	10.750	11.700	12.383	Total Impurities (A%)
RRT:	0.832	0.853	0.886	0.923	0.955	1.000	1.026	1.032	1.123	1.189	
Initial:	0.38	<0.05	<0.05	0.72	<0.05	97.06	0.38	0.98	<0.05	<0.05	2.6
1 hr:	0.09	<0.05	<0.05	0.71	<0.05	97.11	0.00	0.90	0.10	0.05	2.7
4 hr:	0.38	<0.05	0.31	0.73	<0.05	94.87	0.37	0.91	0.49	0.08	4.7
8 hr:	0.36	<0.05	0.63	0.72	<0.05	91.50	0.36	0.92	0.95	0.36	8.2

impurities in the excipients are participating in generating additional forced degradation products.

9.6 CONCLUDING REMARKS

This chapter described the fundamentals and figures of merit for HPLC method validation in pharmaceutical analysis. Special considerations for addressing linearity, recovery, and setting system suitability requirements were presented. The validation process is to confirm that the method is suited for its intended purpose and to prove the capabilities of the test method. The definitions of method validation parameters are well explained by health authorities. Although the requirements of validation have been clearly documented by regulatory authorities, the approach to validation is varied and open to interpretation, and validation requirements differ during the development process of pharmaceuticals (early to late phase). However, the requirements also vary by different health authorities. This point is very well demonstrated by Shabir [50], shedding light on the differences and similarities between validation requirements of the US Food and Drug Administration, the US Pharmacopeia, and the International Conference on Harmonization.

REFERENCES

1. www.ICH.gov
2. C. C. Chan, H. Lam, Y. C. Lee, X.-M. Zhang, (eds.), *Analytical Method Validation and Instrument performance verification*, Wiley Interscience, New York, 2004.
3. D. Paroitt, Performance verification testing of high performance liquid chromatography equipment, *LC-GC* **12** (1994), 135.
4. V. Grisanti and E. Zachowski, Operation and performance qualification, *LC-GC* **20** (2002), 355.
5. J. W. Dolan, Performance qualification of LC systems, *LC-GC* **20** (2002), 842.
6. M. Zooubair El Fallah, Performance verification: a regulatory burden or an analytical tool? *LC-GC* **17** (1999), 343.
7. P. W. Wrezel, I. Chion, and M. S. Hussain, Validation and implementation of in-process control HPLC assays for active pharmaceutical ingredients, *LC-GC North Am.* **22** (2004), 1006–1009.
8. S. Ahuja and S. Scypinski (eds.), *Handbook of Modern Pharmaceutical Analysis*, Academic Press, London, 2001.
9. M. Bouma, B. Nuijen, M. T. Jansen, G. Sava, F. Picotti, A. Flaibani, A. Bult, and J. H. Beijnen, Development of a LC method for pharmaceutical quality control of the antimetastatic ruthenium complex NAMI-A, *J. Pharm. Biomed. Anal.* **31** (2003), 215–228.
10. C. Akay and S. A. Ozkan, Simultaneous LC determination of trimethoprim and sulphamethoxazole in pharmaceutical formulations *J. Pharm. Biomed. Anal.* **30** (2002), 1207–1213.

11. D. Ciutaru, I. Badea, L. Lazar, D. Nicolescu, and A. Tudose, A HPLC validated assay of paclitaxel's related impurities in pharmaceutical forms containing Cremophor® EL, *J. Pharm. Biomed. Anal.* **34** (2004), 493–499.

12. Y. Shao, R. Alluri, M. Mummert, U. Koetter, and S. Lech, A stability-indicating HPLC method for the determination of glucosamine in pharmaceutical formulations, *J. Pharm. Biomed.* Anal. **35** (2004), 625–631.

13. N. H. Zawilla, M. A. Mohammad, N. M. El kousy, and S. M. El-Moghazy Aly, Determination of meloxicam in bulk and pharmaceutical formulations, *J. Pharm. Biomed. Anal.* **32** (2003), 1135–1144.

14. K. R. Tambwekar, R. B. Kakariya, and S. Garg, A validated high performance liquid chromatographic method for analysis of nicotine in pure form and from formulations, *J. Pharm. Biomed. Anal.* **32** (2003), 441–450.

15. A. M. Y. Jaber, H. A. Al Sherife, M. M. Al Omari, and A. A. Badwan, Determination of cetirizine dihydrochloride, related impurities and preservatives in oral solution and tablet dosage forms using HPLC, *J. Pharm. Biomed. Anal.* **36** (2004), 341–350.

16. R. K. Verma, S. Garg, A validated high performance liquid chromatographic method for analysis of isosorbide mononitrate in bulk material and extended release formulations, *J. Pharm. Biomed. Anal.* **30** (2002), 583–591.

17. J. Milano, L. M. Morsch, and S. G. Cardoso, LC method for the analysis of oxiconazole in pharmaceutical formulations, *J. Pharm. Biomed. Anal.* **30** (2002), 175–180.

18. Y. Sun, K. Takaba, H. Kido, M. N. Nakashima, and K. Nakashima, Simultaneous determination of arylpropionic acidic non-steroidal anti-inflammatory drugs in pharmaceutical formulations and human plasma by HPLC with UV detection, *J. Pharm. Biomed. Anal.* **30** (2003), 1611–1619.

19. ICH Q2A: Validation of analytical methods (definitions and terminology), 1994.

20. W. Demond, R. A. Kenley, J. L. Italien, D. Lokensgard, G. Weilersbacher, and K. Herman, Orthogonal HPLC methods for quantitating related substances and degradation products of pramlintide, *AAPS PharmSciTech* **1** (2000), article 6.

21. G. Maniara, K. Rajamoorthi, S. Rajan, and G. W. Stockton, Method performance and validation for quantitative analysis by 1H and 31P NMR spectroscopy. Applications to analytical standards and agricultural chemicals, *Anal. Chem.* **70** (1998), 4921–4928.

22. D. Giron, Thermal analysis of drugs and drug products, in J. Swarbick and J. C. Boylan (eds.), *Encyclopedia of Pharmaceutical Technology*, Marcel Dekker, (2002), 2766–2793.

23. S. Gorog, The sacred cow: The questionable role of assay methods in characterizing the quality of bulk pharmaceuticals, *J. Pharm. Biomed. Anal.* **36** (2005), 931–937.

24. J. D. Orr, I. S. Krull, and M. E. Swartz, Validation of Impurity methods, Part II, *LC-GC North Am.* **21** (2003), 1146–1152.

25. R. Cassidy and M. Janoski, *LC-GC* **10** (1992), 692.

26. C. A. Dorschel, J. L. Ekmanis, J. E. Oberholtzer, F. V. Warren, and B. A. Bidlingmeyer, *Anal. Chem.* **61** (1989), 951A.

27. S. Burke, Regression and calibration, *LC-GC Europe* Online Supplement Statistics and Data Analysis (2001), 13–18.

28. M. M. Kiser and J. W. Dolan, Selecting the best curve fit, *LC-GC Europe* **March** (2004), 138–143.

29. J. M. Green, A practical guide to analytical method validation, *Anal. Chem.* **68** (1996), 305A–309A.

30. Analytical Methods Committee: Is my calibration linear? AMC technical brief, No. 3, December 2000.

31. J. Erner, Chapter 2.4, in J. Ermer and J. H. McB. Miller (eds.), *Method Validation in Pharmaceutical Analysis*, Wiley-VCH, New York, 2005.

32. EURACHEM: The fitness for purpose of analytical methods. A Laboratory Guide to Method Validation and Related Topics. http://www.eurachem.ul.pt, 1998.

33. K. Baumann, Regression and calibration for analytical separation techniques. Part II: Validation, weighted and robust regression, *Process Control and Quality* **10** (1997), 75–112.

34. ICH guidelines Q2B, Validation of Analytical Procedures: Methodology, November 1996.

35. http://www.labcompliance.com/methods/meth_val.htm, accessed May 16, 2006.

36. S. Klick, P. G. Muijselaar, J. Waterval, T. Eichinger, C. Korn, T. K. Gerding, A. J. Debets, C. Sanger-van de Griend, C. van den Beld, G. W. Somsen, and G. J. De Jong, Toward a generic approach for stress testing of drug substances and drug product, *Pharm. Technol.* **February** (2005), 48–66.

37. G. Boccardi, Autoxidation of drugs: prediction of degradation impurities from results of reactions with radical chain initiators, *Il Farmaco* **49** (1994), 431.

38 G. Boccardi, Oxidative susceptibility texting, Chapter 7, in S. W. Baertschi (ed.), *Pharmaceutical Stress Testing: Predicting Drug Degradation*, Vol. 153, Drugs and the Pharmaceutical Sciences, Taylor and Francis, 2005.

39. M. Wasserman (eds.), *Singlet Oxygen*, Academic Press, New York, 1979.

40. A. A. Frimer (ed.), *Singlet O_2*, Vols. II and III, CRC Press, Boca Raton, FL, 1985.

41. Foote, Pryor, (eds.), *Free Radicals in Biology*, Vol. II, Academic Press, New York, 1976, 85–133.

42. R. Seburg, Photosensitized Singlet Oxygen-Mediated Degradation of an Angiotensin II Antagonist, Losartan Potassium, Merck Research Laboratories, http://www.iirusa.com/forceddegradation/files/IIR_p1038_Seburg.pdf#search='AIBN%20and%20forced%20degradation%20and%20HPLC', accessed May 15, 2006.

43. J. K. Guillory and R. I. Poust, Chemical kinetics and drug stability, Chapter 6 in G. S. Banker and C. T. Rhodes, *Modern Pharmaceutics*, 3rd ed., Marcel Dekker, New York, 1996, pp. 139–165.

44. H. H. Tønnesen, Formulation and stability testing of photolabile drugs, *Int. J. Pharm.* **225** (2001), 1–14.

45. E. Fasani and A. Albini, Photostability stress testing, Chapter 10 in *Pharmaceutical Stress Testing: Predicting Drug Degradation*, edited by S. W. Baertschi, 1st edition, Vol. 153, Drugs and the Pharmaceutical Sciences Series (Taylor and Francis), 2005.

46. FDA, Center for Drug Evaluation and Research, Submitting Documentation for the Stability of Human Drugs and Biologics, Rockville, MD, February 1987.

47. S. W. Hovorka and C. Schoneich, Oxidative degradation of pharmaceuticals: Theory, mechanisms and inhibition, *J. Pharm. Sci.* **90** (2001), 253–269.

48. K. C. Waterman, R. C. Adami, K. M. Alsante, J. Hong, M. S. Landis, F. Lombardo, and C. J. Roberts, Stabilization of Pharmaceuticals to Oxidative Degradation, *Pharm. Dev. Tech.* **7** (2002), 1–32.

49. S. R. Byrn, R. R. Pfeiffer, and J. G. Stowell, *Solid State Chemistry of Drugs*, 2nd ed., SSCI Inc., West Lafayette, IN, 1999.

50. G. A. Shabir, Validation of high-performance liquid chromatography methods for pharmaceutical analysis: Understanding the differences and similarities between validation requirements of the US Food and Drug Administration, the US Pharmacopeia and the International Conference on Harmonization, *J. Chromatogr. A* **987** (2003), 57–66.

10

COMPUTER-ASSISTED HPLC AND KNOWLEDGE MANAGEMENT

Yuri Kazakevich, Michael McBrien, and Rosario LoBrutto

10.1 INTRODUCTION

In modern high-performance liquid chromatography (HPLC), computers in a broad sense are used in every instrumental module and at every stage of analysis. Computers control the flow rate, eluent composition, temperature, injection volume, and injection process. Detector output signal is converted from analog form into the digital representation to recognize the presence of peaks, and then at higher level of computer analysis a chromatogram is obtained. All these computer-based functions are performed in the background, and the chromatographer usually does not think about them.

The second level of computer utilization in HPLC is extraction of valuable analytical and physicochemical information from the chromatogram. This includes standard analytical procedures of peak integration, calibration and quantitation, and more complex correlation of the retention dependencies with variation of selected parameters.

At the third (and probably highest) level, a computer is used for the sophisticated analysis of many different experimental results stored in databases. This level is usually regarded as a knowledge management level and can have quite a variety of different goals:

- Selection of the starting conditions for method development by using information of similar separations

HPLC for Pharmaceutical Scientists, Edited by Yuri Kazakevich and Rosario LoBrutto
Copyright © 2007 by John Wiley & Sons, Inc.

- Optimization of the existing method, to speed up the analysis, increase ruggedness of the chromatographic method, and so on
- Review of a multitude of data from different experiments and their correlation with information from other physicochemical methods
- Cross-laboratory information exchange (early drug discovery, preformulation groups, drug metabolism and pharmokinetic groups, drug substance and drug product groups)

In this chapter the third level of computer-assisted HPLC—the use of expert systems (like Drylab [1], AutoChrom™ [2], and ChromSword® [3]) for effective method development—is discussed.

Computer-assisted method development has received a great deal of attention from management within the pharmaceutical industry, mainly from the perspective of cost savings associated with faster and more efficient development. Adoption and incorporation of the tools in day-to-day workflows has been relatively limited due in part to a reluctance of chromatographers to believe that computers can replace the intuition of the expert chromatographer. With the present state-of-the-art, there is little question that computers can play a role in efficient method development. However, it must be accepted that computers are a supplement to, rather than a replacement for, the knowledge of the method development chromatographer.

Two main types of software tools exist that are directly applicable to the problem of chromatographic method development.

1. Optimization or experimental design software packages for modeling the chromatographic response as a function of one or more method variables. These can also play a key role in data management of the considerable information that results from rigorous method development exercises.

2. Structure-based prediction software predicts retention times or important physicochemical processes based on chemical structures. Application databases store chromatographic methods for later retrieval and adaptation to new samples with similar structures and physicochemical parameters.

10.2 PREDICTION OF RETENTION AND SIMULATION OF PROFILES

In Chapters 2, 3, and 4, all aspects of the analyte retention on the HPLC column are discussed. There are many mathematical functions describing retention dependencies versus various parameters (organic composition, temperature, pH, etc.). Most of these dependencies rely on empirical coefficients. Analyte retention is a function of many factors: analyte interactions with the stationary and mobile phases; analyte structure and chemical properties; structure and geometry of the column packing material; and many other parameters. The theoretical functional description of the influence of the eluent composition, mobile-phase pH, salt concentration, and temperature, as well as the influence of the type of organic modifier and type of salt added to the mobile phase, are discussed in detail in Chapter 2 and 4.

Currently, eluent composition, column temperature, and eluent pH are the only continuous parameters used as the arguments in functional optimization of HPLC retention. However, other parameters such as ionic strength, buffer concentration and concentration of salts and/or ion-pairing reagents can be taken into account, and mathematical functions for these can be constructed and employed.

The simplest and the most widely used forms of retention time prediction for analytical scale HPLC are based on the empirical linear dependence of the logarithm of the retention factor on the eluent composition.

10.2.1   General Thermodynamic Basis

Association of the chromatographic retention factor with the equilibrium constant is the basis for all optimization or prediction algorithms. As was shown in Chapter 2, this association is only very approximate and should be used with caution.

In short, an approximate mathematical description of the retention factor dependences on the eluent composition and temperature is written in the form

$$k = \exp\left(\frac{\sum \Delta G_{\text{an.frag.}}}{RT} - \phi \frac{\Delta G_{\text{el.}}}{RT} \right) \qquad (10\text{-}1)$$

where ϕ is the molar fraction of the organic eluent modifier, $\Delta G_{\text{el.}}$ is the Gibbs free energy of the organic eluent modifier interaction with the stationary phase; R is the gas constant; T is the absolute temperature, and $\Delta G_{\text{an.frag.}}$ is the Gibbs free energy of the interactions of structural analyte fragments with the stationary phase.

Equation (10-1) is based on the assumption of simple additivity of all interactions and a competitive nature of analyte/eluent interactions with the stationary phase. The paradox is that these assumptions are usually acceptable only as a first approximation, and their application in HPLC sometimes allows the description and prediction of the analyte retention versus the variation in elution composition or temperature. For most demanding separations where discrimination of related components is necessary, the accuracy of such prediction is not acceptable. It is obvious from the exponential nature of equation (10-1) that any minor errors in the estimation of interaction energy, or simple underestimation of mutual influence of molecular fragments (neglected in this model), will generate significant deviation from predicted retention factors.

10.2.2 Structure–Retention Relationships

Many attempts to correlate the analyte structure with its HPLC behavior have been made in the past [4–6]. The Quantitative structure–retention relationships (QSRR) theory was introduced as a theoretical approach for the prediction of HPLC retention in combination with the Abraham and co-workers adaptation of the linear solvation energy relationship (LSER) theory to chromatographic retention [7, 8].

The basis of all these theories is the assumption of the energetic additivity of interactions of analyte structural fragments with the mobile phase and the stationary phase, and the assumption of a single-process partitioning-type HPLC retention mechanism. These assumptions allow mathematical representation of the logarithm of retention factor as a linear function of most continuous parameters (see Chapter 2). Unfortunately, these coefficients are mainly empirical, and usually proper description of the analyte retention behavior is acceptable only if the coefficients are obtained for structurally similar components on the same column and employing the same mobile phase.

To date, the shortcomings in the theoretical [22] and functional description of HPLC column properties make all these theories insufficient for practical application to HPLC method design and selection.

In the past, several theoretical models were proposed for the description of the reversed-phase retention process. Some theories based on the detailed consideration of the analyte retention mechanism give a realistic physicochemical description of the chromatographic system, but are practically inapplicable for routine computer-assisted optimization or prediction due to their complexity [9, 10]. Others allow retention optimization and prediction within a narrow range of conditions and require extensive experimental data for the retention of model compounds at specified conditions [11].

Probably the most widely studied is the solvophobic theory [12] based on the assumption of the existence of a single partitioning retention mechanism and using essentially equation (10-1) for the calculation of the analyte retention. Carr and co-workers adapted the solvophobic theory [12, 13] and LSER theory [11, 14–17] to elucidate the retention of solutes in a reversed-phase HPLC system on nonpolar stationary phases.

The free energy of transfer of a molecule from the mobile phase to the stationary phase, ΔG, can be regarded as a linear combination of the free retention energies, ΔG_i, arising from various molecular subunits (solvatochromic parameters). Many solvatochromic parameters for some analytes could be found in the literature [18–21]. The signs and magnitudes of the coefficients depict the direction and relative strength of different kinds of solute/stationary and solute/mobile phase interactions contributing to the retention in the investigated matrix [11–15]. The most influential factors governing RP-HPLC retention on alkyl and phenyl-type bonded phases were determined to be hydrogen bonding and the solute molecular volume [12, 13, 20, 23]. The hydrogen bonding is measured as the effect of complexation between hydrogen-bond acceptor (HBA) solutes and hydrogen-bond donor (HBD) bulk phases [24]. The solute molecular volume is comprised of two terms: One measures the cohesiveness of the chromatographic phases (both the mobile and stationary phases) and the other is the dispersive term that measures the ability of the chromatographic phases to interact with solutes via dispersive forces.

10.3   OPTIMIZATION OF HPLC METHODS

10.3.1   Off-Line Optimization

The most common software tools used for chromatographic method development are optimization packages. All of these tools take advantage of the fact that the retention of a given compound will change in a predictable manner as a function of virtually any continuous chromatographic variable.

The classic example (and certainly most common application) of computer-assisted chromatographic optimization is eluent composition, commonly called solvent strength optimization. The chromatographer performs at least two experiments varying the gradient slope for gradient separations or concentration of organic modifier for isocratic separations at a certain temperature. The system is then modeled for *any* gradient or concentration of organic modifier. A simplistic description of the chromatographic zone migration through the column under gradient conditions is given in Chapter 2. At isocratic conditions the linear dependence of the logarithm of retention factor on the eluent composition is used for optimization:

$$\ln(k) = A\varphi + B \tag{10-2}$$

where k is the retention factor of the compound, ϕ is the fraction of organic solvent in the mobile phase, and A and B are constants for a given compound, chromatographic column, and solvent system. Based on a few experiments, the constants in the expression can be extracted, and retention of each compound can be predicted.

This optimization approach can be used to model both retention times and selectivities due to the fact that both the A and B terms are unique for a given analyte.

The typical output from method optimization software is a resolution map, as shown in Figure 10-1. The map shows resolution of the critical pair (two closest eluting peaks) as a function of the parameter(s). The example shows resolution as a function of gradient time (slope of the gradient). The resolution map has several advantages as an experimental display tool: It forms a concise summary of experiments performed, it allows the chromatographer to select areas of interest and communicate the expected result, and it facilitates the viewing of data that would allow for a more robust separation.

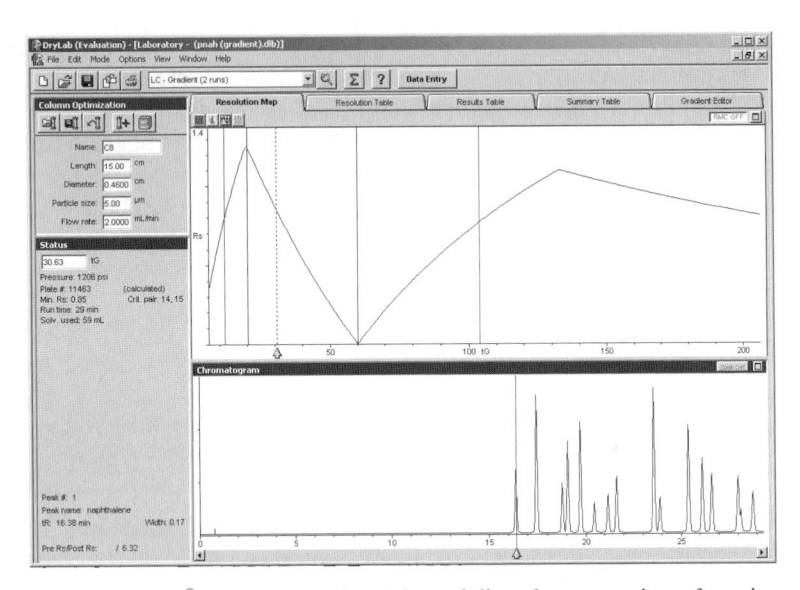

Figure 10-1. DryLab® software version 3.0 modeling the separation of a mixture of naphthalenes. Resolution of the critical pair (the two peaks that elute closest together) is denoted as a function of time of gradient. Experimental runs are shown as solid lines on the resolution map; selected prediction is a dashed line.

Optimization of the eluent composition is commonly based on the linear relationship of $\ln k$ to ϕ (10-4) and generally applicable for ideal chromatographic systems with unionizible analytes in methanol/water mixtures. It is commonly assumed that:

- A single partitioning-like equilibrium process dominates in the retention mechanism.
- Analyte ionization changes do not occur in the pertinent solvent range.
- Column property changes do not occur over the course of the experiment.

Like in any optimization tool, the chromatographer should be wary of extrapolation beyond the scope of the training experiments. Behavior of certain parameters, like temperature and solvent strength, is fairly easily modeled. Other parameters, such as buffer concentration and pH, can be much more difficult to model. In these cases, interpolation between fairly closely spaced points (actual experiments that were performed) is most appropriate. Figure 10.2 shows a resolution map for a two-dimensional system in which solvent composition and trifluoroacetic acid concentration are simultaneously optimized. The chromatographer has collected systematic experiments at TFA concentrations of 5, 9, 13, and 17 mM and acetonitrile concentrations of 30, 50, and 70 v/v% for a series of small molecules on a Primesep 100 column.

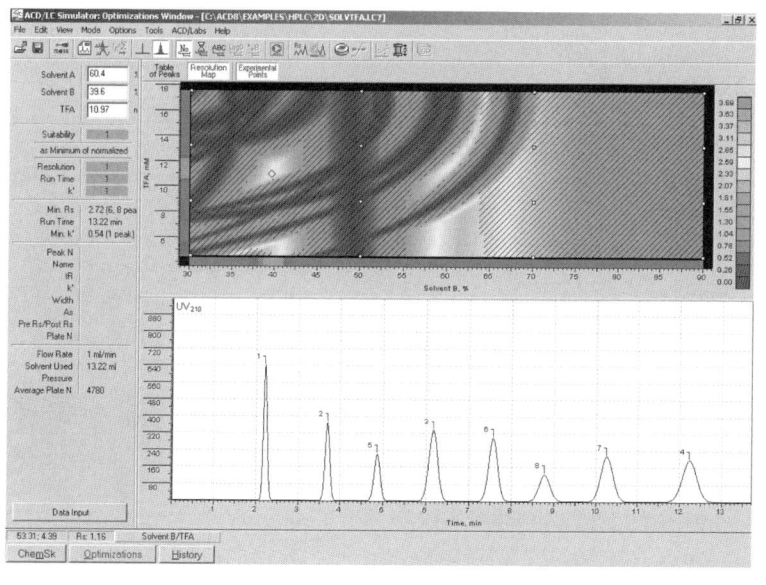

Figure 10-2. ACD/LC Simulator™ 9.0 modeling the separation of a series of compounds as a function of solvent composition and TFA concentration (mM). Experiments are shown as white dots on the resolution map with the predicted optimal method shown in yellow. See color plate.

Note. The type and concentration of the organic eluent can cause a pH shift of the aqueous portion of the mobile phase as well as change the ionization state of the analyte in a particular hydro-organic mixture. Temperature can also lead to change in the ionization constants of analytes.

Even when chromatographers are careful to keep buffer strengths constant during modification of organic solvent strengths, effective analyte pK_a changes and mobile-phase pH changes as a result of solvent strength, which can cause changes in ionization state of compounds, changes in the resultant mobile-phase pH, and/or changes in the behavior of chromatographic columns [25]. Departures from linearity can be particularly striking in acetonitrile as opposed to methanol. For systems in which the greatest possible quality of method is required in terms of resolution, run time, and robustness, the results from predictions should be verified against experimental data and, where necessary, nonlinear predictions should be used to refine the model and to locate the optimal conditions.

Computer-assisted optimization of parameters has not been universally accepted, primarily due to a lack of ease of use. All compounds must be tracked across all experiments, and all retention times must be introduced to the system for each component. This is sometimes difficult because significant variations in the retention and elution order could be observed for certain analytes. With diode array detection, even if the different analytes have distinct diode array profiles, the analytes with low concentration in the mixture may still be difficult to track. The use of MS detection can assist in the detection of the peaks in the different experiments, with the assumption that they are not isomers of each other. Software vendors have begun to address much of this with the implementation of automated peak-tracking systems (see Section 10.3.4.2) and direct transfer of experimental information from chromatography data systems.

Advantages of this technique are the efficiency of development of methods, structured development profiles, and effective reporting of what was performed during the different method development iterations. In addition, it is possible to model the effect of parameter variation on the robustness of methods in addition to general chromatographic figures of merit: apparent efficiency, tailing, resolution of critical pairs, backpressure of system, total run time.

10.3.2 On-Line Optimization

Recently there has been renewed interest in automated method development in which the optimization software directly interfaces with the instrument in order to run or suggest new experiments based on the prior results that generated the initial resolution maps. In the late 1980s, a number of approaches to this problem were attempted, but none of these tools prevailed, due in part to the challenges of tracking peaks between experiments.

The current second-generation tools offer more promise due to (a) a focus on secondary detection techniques for peak tracking and (b) better automation tools offered by instrument vendors.

The advantages of on-line automation are the achievement of time savings in relation to the chromatographic method development time. The software can make decisions at any time of the day or night and can immediately communicate this information to the instrument after the completion of the experiment. There is also a more subtle benefit to the link of optimization software to the chromatography data system. Method development "wizards" with drop-down menus/user-defined fields can simplify the process of configuring the instrument sequence/method prior to a method development session.

Disadvantages of on-line optimization lie primarily in the maturity of this technology. If manual method development is based on the experience and intuition, the automated method development in principle should follow the logic of chromatographic theory, which unfortunately is not yet developed enough to provide a logical guide for automated optimization. Software and instrument vendors are relying on the statistical optimization with minimal use of available theoretical developments and only on the level of simple partitioning mechanism and energetical additivity. The capacity of software innovators to address detection limit, peak-tracking, and artificial intelligence issues remains in question at present, but the considerable commitment by instrument and software vendors points to the future value of these tools. As spectroscopic peak-tracking algorithms mature, the effectiveness of the tools will grow considerably.

10.3.3 Method Screening

There are some chromatographic parameters that do not readily lend themselves to optimization. There have been some efforts to quantify the selectivity in chromatographic columns [26, 27], but it is often difficult to achieve targeted values for each of the parameters involved without custom preparation of materials. Experimental mobile-phase pH values must typically be very close together in order to enable subsequent pH optimization. Column and pH choice are critical to the selectivity of a given system, so it is clear that their effects should not be ignored. One solution to this problem is to screen different columns and pH values prior to commencing any kind of optimization. The screening results are reviewed, and optimization systems at a particular pH are designed accordingly.

With the advent of column switchers and more reproducible alternative column materials, it is now quite feasible to screen multiple pH values—for example, at high, medium, and low pH—using scouting gradients in order to choose the column and pH at which to perform further optimization experiments. This is a particularly tempting scenario when few or no chemical structures are available for the synthetic by-products or degradation products in the sample, or when samples are particularly complex. Recently there has been considerable development on systems for selection of optimal pH and type of column concomitantly [28].

For complex samples, it can be time-consuming and challenging to review all the results of system screens objectively. In addition, online optimization precludes the direct involvement of the chromatographer. For this reason, it is desirable to use some numerical description of the potential effectiveness of a given set of conditions so the on-line optimization software can trigger further separations on the chromatographic system.

Screening review tools cannot work solely based on the venerable "resolution of the critical pair" approach; the results of an initial screen must be able to give nonzero results even with co-elution of two components, when the resolution of the critical pair will, of course, be zero.

Additionally, a suitability approach involving criteria related to run time is unwise, since run time can be fine-tuned based on solvent strength or flow rate in final optimization. Rather, at the screening stage, the chromatographer should be focused on sufficient selectivity to form the basis of an eventual successful separation, and then finc-tuning can be performed. There are a number of different measures of the desirability of an initial screen, including average resolution, resolution of critical pair, selectivity of critical pairs, and so on. The chromatographer need not be intimately familiar with the nuances of every rating system available. The only key is to be certain that appropriate

TABLE 10-1. Numerical Approaches to Ranking Separations

Approach	Basis	Application
Minimum resolution (resolution of the critical pair)	Resolution of closest-eluting peaks (R_{CP})	Final model of separation
Method suitability	Product or minimum of various criteria: run time, resolution of critical pair, and resistance of viability to small changes in conditions	Final model of separation (customizable)
Mean resolution	Average resolution	Assessment of selectivity
Run time (RT) versus Target (t) and Maximum (M)	$N = 1$ if RT < Target; $N = 0$ if RT > maximum; $N = 1 - (RT - t)/(M - t)$	Evaluating suitability of solvent strength and column choice
Equidistance	deviation from equal peaks resolution $$N = \frac{\text{RunTime} - t_0}{n-1}$$	Comparison of starting systems
Resolution score	average value of normalized resolutions between all the peaks detected on a chromatogram $$\text{RsScore} = \frac{\sum Rs_n}{N-1}$$	Comparison of starting systems

rating systems are used at appropriate times. Table 10-1 shows some common approaches to the rating of chromatographic column screens [29].

10.3.4 Method Optimization

All approaches to method optimization based on multiple experiments have the requirement that all components be detected and that they be tracked between runs. For complex samples, this is typically the most labor-intensive aspect of method development. For unattended method development, the instrument is required to monitor the change in retention of each component automatically. The historical limitations to this technology have been a key stumbling block in the widespread adoption of automated method development.

10.3.4.1 Peak Matching in Method Optimization. An initial solution to the problem of peak tracking across multiple experiments was the isolation of each impurity on a preparative or semiprep scale, followed by injection of each component individually. The chromatographic world has essentially rejected this concept outright. Very few chromatographers have the time or willingness to isolate standards for each component. The use of crude samples and mother liquors enriched with synthetic byproducts for initial method development of drug substance is recommended. Another approach is to look at the molecule of interest and predict most probable degradation product(s) and use forced degraded samples for initial method development. For example, if a compound contains ester functionality, then acidic stress conditions can be employed to discern the retention of the carboxylic acid degradation product and other resultant degradation products. As another example, if a compound contains a pyridinal functionality, it might be subject to oxidation and catalyzed under light stress conditions; therefore a forced degradation solution in the presence of peroxide/light can be used to generate the resultant *N*-oxide degradation product.

10.3.4.2   LC/UV-Vis and LC/MS. Hyphenated detection in modern chromatography has led to a great deal of interest in automated and semiautomated peak tracking based on diode array and mass spectral data. While several algorithms have been published for the utilization of hyphenated data for peak tracking [30, 31] based on a spectral match angle approach [32], there are few commercially available tools. Multivariate "chemometric" approaches seem to have the most potential for future success. There are two main commercially available approaches to peak tracking using diode array data. In the Waters® AMDS system using DryLab, peaks are tracked based on a library search technique, using match angles for extracted spectra. Essentially, after peak-picking, spectra are extracted and searched against a library formed from the spectra from other chromatograms.

ACD/AutoChrom uses the "mutual automated peak matching" [33] or UV-MAP approach based on extraction of pure variables from diode array data. The UV-MAP algorithm applies abstract factor analysis (AFA) followed by iterative key set factor analysis to the augmented data matrix in order to extract retention times for each of the selected experiments.

No commercial system for peak-tracking based on mass spectrometry (MS) data has been published to date. Recently [34], a customized MS-based peak tracking tool was reported using algorithms connecting to the Agilent ChemStation Plus chromatography data system. This algorithm uses a logic-based approach to the extraction of molecular weights from MS data. Components are assigned based on isotope ratio confirmation, adduct assignment, and elution characteristics. Retention time extraction was reported to be approximately 80% successful, with failures primarily attributed to insufficient ionization of components. A similar approach is used in the ACD/AutoChrom product, combining MS with diode array detection in order to address some issues with low signal individual detectors.

Disadvantages. Neither the MS and ultraviolet (UV) detectors provide a complete solution alone. UV spectra simply are not unique enough to differentiate between closely related compounds. Under the conditions typically used for liquid chromatography, compounds may fail to give sufficient ionization with MS detection. In addition, the modification of conditions that is inherent to method development causes spectral and ionization changes. All of these provide a tremendous challenge to software designers, but initial results appear promising.

Advantages. This is critical technology to enable both automated and routine application of computer-assisted optimization. The manual effort required for traditional approaches to data interpretation in chromatographic method development is quite considerable.

10.3.4.3 Composite Samples and Data Management. A recent trend in pharmaceutical development has been the development of methods for the resolution and quantitation of related compounds based on a "proactive" strategy. During early drug development, a large number of different tests will be conducted on prospective drug candidates, including impurities analysis for stability indicating methods. The development of methods for this purpose is problematic because final synthetic routes and formulations are not yet established, so the resultant impurity profiles will change as the synthesis is optimized and the final market image is defined. However, in order to avoid impeding the development process, it is important to have quantitative methods readily at hand and then modify them if needed as the drug development process continues.

Many groups have chosen to approach this problem from the point of view of development of methods for all anticipated compounds such that practically any sample configuration can be treated with the same method, or with only slightly altered set of conditions.

One of the more common approaches to method development in the drug substance and drug product groups in the pharmaceutical industry is to first generate forced-decomposition samples (using mild conditions, not more than 5–10% degradation) based on treatment of the compounds with various stress conditions including, typically, UV light, heat, acid, base, and peroxide. These decomposed samples are injected separately, and then a method is designed to separate components in the forced degradation samples as if they were all present in the same sample. The development of methods for these "composite samples" is typically required to be exceedingly rigorous. Columns, solvent systems, and pH values will be screened, and multidimensional optimization performed. The software tools that have been discussed in this chapter are invaluable for this kind of project. However, there is an additional challenge with this kind of method development. The amount of raw data generated in this kind of project can be particularly daunting.

Before embarking on choosing the optimal conditions for optimization, generally a pH screen (at least five pH values) in either gradient or isocratic mode is performed to determine the most suitable pH ranges for the active pharmaceutical ingredient (at least one unit below or above the target analyte pK_a in a particular hydro-organic system). This results in at least five experiments on one column using LC/DAD detection. Once the acceptable pH ranges are determined (where analyte is predominately in its ionized state or in its neutral state), then column screening can be performed if necessary. If we consider the second step in method development for the API sample as the screening of six columns at two pH values with shallow gradient slope, we can see that 12 initial screening methods will be generated, with at least two hyphenated chromatographic traces in the form of LC/MS and LC/DAD data. This will result in managing at least 24 hyphenated data traces. If a steep gradient slope was also investigated, this would increase the number of hyphenated traces to 48. Then, in the third step, the two best columns at a particular pH with a particular gradient condition would then be chosen for analysis of the API sample, blank and the five different stressed samples. Thus for this wave, the chromatographer must review and manage 28 hyphenated data traces (14 LC/MS, 14 DAD). In all of the method development experiments described with this approach the chromatographer would have to manage a total of 43 chromatograms.

For the massive amounts of data collected with complex samples, obviously peak-matching tools as discussed in Section 10.3.4.2 become quite invaluable. In addition, it is critical to manage the complex data in an efficient manner. If the user is reduced to cut-and-paste for peak tables, or even to transfer from the raw data, any reexamination of the data can be very confusing. Typical chromatography data systems organize data by filenames or by sample/ project. However, it is critical in this case to organize data according to the experimental method, since for each method there are multiple chromatographic traces, each contributing to the overall, effective experimental result. Recently, software has been designed to manage analytical data in this manner; the data for original traces is sorted by the chromatographic method, tracing for which sample/condition set the data were collected.

In the project architecture, information is grouped according to experimental conditions, or "experiments." Multiple detector traces are arranged for each subsample, with subsamples organized by experiment. Experiments are grouped according to waves that are designed for optimization and/or screening objectives. Finally, one or more waves composes a method development project. Figure 10-3 shows the AutoChrom workspace window that shows the organization of chromatograms for individual subsamples in a forced degradation study, with the summary of the components in the composite chromatogram. Multiple detectors for each subsample have been "collapsed" in this view to enable the view of all subsamples at once. Figure 10-4 shows the overall data hierarchy.

The advantages of this kind of organization system are clear. Any issues with accuracy of transcription arc alleviated. Since peak tables are automatically extracted from the data traces, there is no need for cut-and-paste functions. However, the destination path must be set prior to the transfer, and also the proper integration thresholds must be configured. Data can be part of multiple optimization/screening waves at the same time. In addition, there are

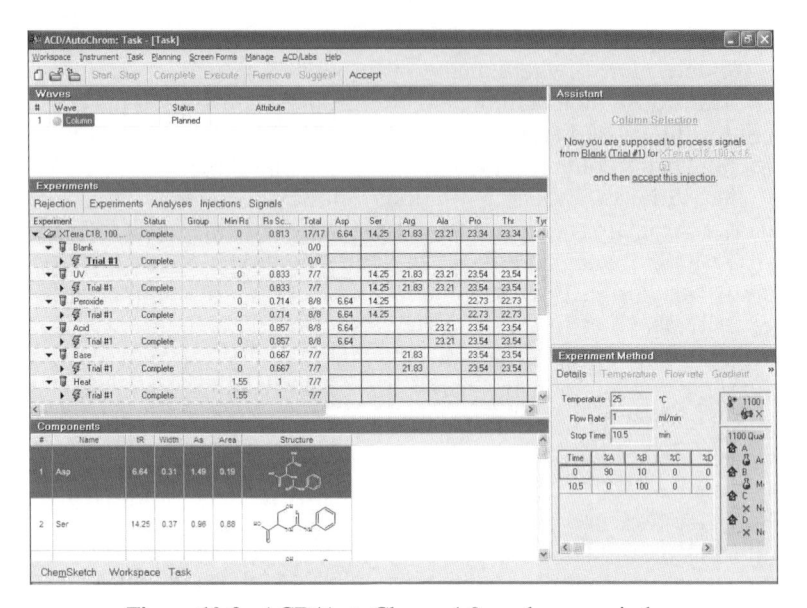

Figure 10-3. ACD/AutoChrom 1.0 workspace window.

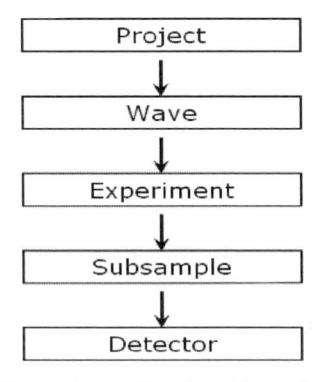

Figure 10-4. The data hierarchy.

considerable advantages with regard to speed. Since the peak tables are extracted directly from the hyphenated data and summarized in the project window, the user has access to all peak data without loading the full datasets. The raw spectral data are loaded on demand.

The primary disadvantage of this approach is in terms of setup of the system. If the approach is not combined with instrument control, then a process must be devised for efficient transfer of information to the data system.

10.4 STRUCTURE-BASED TOOLS

It is uncommon for the method development chromatographer to have absolutely zero information with regard to the chemical structures present in a given sample. Typically, at least one or more compounds are known. There are several software tools intended to enable the chromatographer to leverage on knowledge of these structures in order to enhance the method development process. These include knowledge management tools such as application databasing, prediction of physicochemical parameters, and structure-based retention time prediction.

10.4.1 Knowledge Management

Building and deploying a chromatographic and spectral database has a goal to turn disparate experiments into a global chromatographic knowledge base by archiving applications according to chemical structure. This could result in a global-wise knowledge base, searchable and retrievable with all relevant information, experimental tests, and results stored. This would allow for a more efficient workflow with a homogeneous repository for all relevant data, allowing users to process, evaluate, compare, and generate reports in one environment.

Success is no longer just about capturing better data—it's your ability to share that knowledge to help improve the organization's productivity. With improvements in instruments and personnel productivity, today's laboratories are producing significant quantities of scientific data.

How can pharmaceutical companies convert the results of this productivity into knowledge? Data need to be captured, processed, and interpreted for immediate use, as well as stored and managed to support future product development. The value of data increases when all researchers are able to access, share, and leverage each other's knowledge. Software/databases that can bridge all instruments, data sources, and information centers to meet these challenges head on, is encouraged.

The motivation toward saving methods including chromatographic and spectral data is that the information can be communicated to other groups working on the same or similar compounds in other divisional areas. Software that can incorporate the tools for creating a chromatographic/spectral knowledge base would be needed to achieve this endeavor. The database design could include the chromatography and spectral acquisition details, and these data could be correlated with structures of drug compounds and their associated impurities, degradation products, metabolites, and so on. If a good starting point could be defined, then scientists can save time in their method development journey.

Programs that allow for structures or partial structures searching can be used to assist with the selection of starting points. These data could be easily searched for. The method development work that a chromatographer plans to employ may have been performed prior in early development or in another department within the organization (data can be shared across oceans). These data could be included in a separations/spectral knowledge base. Based on chemical structures, chromatographers can build on what was done in the past and/or use the previous conditions as excellent starting points for analysis.

The main advantages include:

- Structure-based searches—internal database
- Access to commercially available applications
- Linking chromatographic methods to the structures
- Linking spectral data (MS, NMR, 2D-NMR, IR, UV) to the structures
- Finding applications based on functionality
- Finding information needed to duplicate an experiment
- Contains information/avenue to evaluate and modify an experiment prior to attempting it
- Sharing information cross-functionally (DS, drug substance; DP, drug product; DMPK, drug metabolism pharmacokinetics; EDD, early drug discovery)

However, as with any technology, a reality check needs to be performed and it has to be determined if implementing such a database will add value to the organization. An evaluation of the current workflows needs to be performed, and a critical gap analysis should be completed.

The following questions should be analyzed in the preparation of database implementation:

- Do the processing and interpretation of analytical data need to be accelerated? If so, in what ways?
- How do we share data now? How do we want to share data in the future?
- Is it that the retrieval of data needs to be faster? If so, can we quantify how much faster it needs to be?
- Is it that the creation of reports needs to be easier and faster?
- How do we currently share data across the globe, especially within multi-national pharmaceutical companies that have research and development divisions worldwide (United States, Europe, Asia, etc.)

Other pertinent questions that could arise during the paper evaluation process include.

- Need to identify if there is global interest?
- What is the speed of the data retrieval?
- Can the database be easily interfaced with the different analytical instrumentations available worldwide (Chromeleon, Empower, MassLynx, etc.)

- What linkages to research databases are needed?
- Is this technology maturing to the point where it will have a major impact to our business?
- Is the software user-friendly?
- Can it be supported by IT? What platforms are available?
- Will analysts use it?

10.4.2 Applications Databases

One of the primary questions that have plagued method development chromatographers is, "Where do I start?" This question applies equally to any school of thought, whether the chromatographer uses no optimization tools, uses computer-assisted optimization, or even uses on-line optimization. In any of these cases, the chromatographer must choose a proper starting point.

One approach to this problem is to use methods developed in the past as a knowledge base for the determination of a starting point. Stored methods are retrieved, and method development sessions can be designed based on the past work performed in different line units of the organization (early drug discovery, preformulation group, DS and DP groups). A key point here is the need for chemical structures to assist with locating similar compounds. It is not likely that researchers will find their compounds of interest unless they have been studied before (unless there are comparator products and a USP monograph has been written and method inputted into the chromatographic database), but substructure and structure similarity searches can find similar compounds that have been the focus of earlier development. It is likely that these methods can be an excellent pointer to new opportunities.

Structure-based separation databases integrated with other analytical and pharmaceutical information provides a basis for a significant increase of development efficiency.

If analytical chemists from the various areas of drug development (drug metabolism, preformulation, formulation, drug substance) enter their separations of the target compounds into the database and link the structures of the potential impurities/degradation products/metabolites identified, this provides a plethora of information to groups developing methods in later phases of the drug development continuum. This is useful as an interactive tool for sharing information across groups or functions avoiding replication of method development of difficult separations. It can provide more suitable starting points to further develop/optimize the needed separations in the different functional areas. The use and organization of this type of database will be discussed.

It is important that a distinction be made between chemical formulae and chemical structures. For databases with any type of diversity to be realized, the chemical formula cannot provide effective retrieval of compounds. Structure-based searches can take three different approaches:

- Structure
- Substructure
- Structure similarity [35]

Structure searches look for molecules that are identical in every way. Substructure searches can be used to target functionalities that the chromatographer deems to be instrumental to the separation at hand. Structure similarity searches are the primary tool in the application database; structures are ranked numerically according to similarity, with essentially all reactive groups taken into account. There are a number of different approaches to structure similarity, including Tanimoto, Dice, cosine, Hamming distance, and Euclidean distance [35]. All of these approaches rank structure similarity between 0 and 1, but will give different values. However, the overall ranking of structures tends to be very similar. To date, no structure similarity search algorithm has emerged as clearly superior for purposes of modeling chromatographic behavior.

Application databases have been particularly popular in the world of chiral method development (Figure 10-5). While it has been observed that small changes in compounds can result in loss of effectiveness (separation selectivity) for a given method, the results of searches can be used to create targeted method screens that can reduce the time and expense of development [36].

Most commercially available applications databases contain some capacity for update of user applications. This is a key capability, because the most relevant structures are likely to be found within the organization, rather than outside. When updating applications, it is extremely useful to have

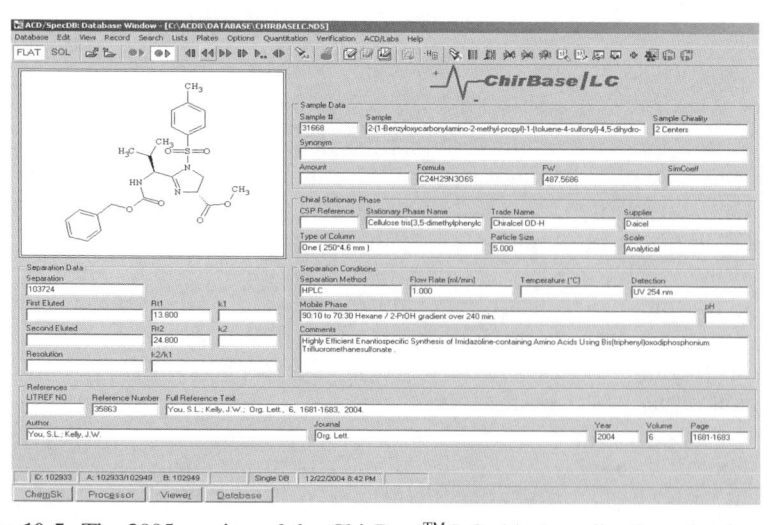

Figure 10-5. The 2005 version of the ChirBase™ LC chiral applications database contains over 100,000 entries.

compatibility with the original chromatography data system, such that methods are read directly from the original datafile, rather than input manually. However, if manual inputs are required, then form-based inputs with the most common variables should be used to maintain consistency of the information inputted. Other fields can be searchable as well; and for any searchable field to provide any meaningful hits in the future, these fields "must" be populated.

The ease of use of structure similarity search means that chromatographers can mine these tools for organizational and/or published knowledge in this area in a few seconds.

Additionally, any effort to accumulate a knowledge base should be accompanied with careful control of data consistency.

10.4.3 Structure-Based Prediction

10.4.3.1 Prediction of Physicochemical LogP, LogD, pK_a. There are three main physicochemical terms of use to the experienced chromatographer. These are LogP, LogD, and pK_a.

LogP (octanol/water partition coefficient) is the classic measure of hydrophobicity of an uncharged species. There are a number of LogP prediction systems available, including PrologP^{TM}, clogP^{TM}, ACD/Logp DB^{TM}, and others. These systems are consistent in that they estimate the hydrophobicity of the compound based on contributions from characterized fragments (Figure 10-6). The accuracy of these predictions is generally quite good, but the relevance of LogP to liquid chromatography is questionable due to ionization of

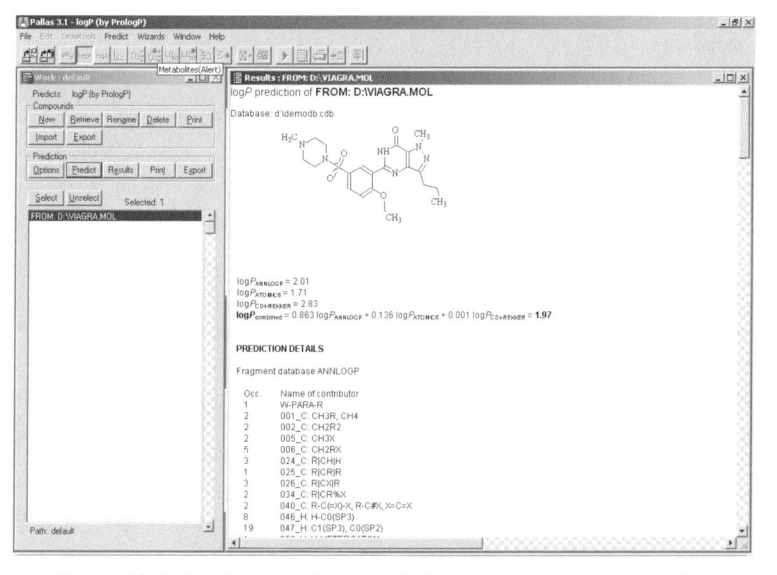

Figure 10-6. Prediction of LogP of Viagra with Pallas version 3.1.

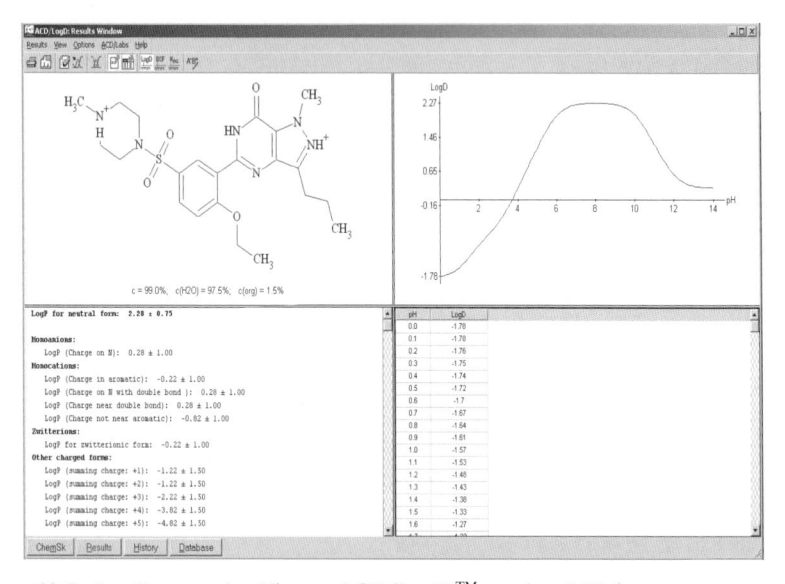

Figure 10-7. LogD curve for Viagra. ACD/LogD™ version 9.00 (note two tautomeric forms were predicted and only one shown).

many compounds of interest. However, LogP calculations can give a very fast estimation of the compounds' general nature—that is, Is my compound hydrophilic, hydrophobic, or very hydrophobic?

LogD is the measure of the hydrophobicity of a species *as it exists in solution*. The distinction between LogD and LogP is based on the pK_a for the compound, and thus while LogP is a simple numerical value, LogD is a function of pH.

LogD curves can be very useful in chromatographic method development, since they can assist with the design of robust separations. The flat areas of the LogD curve (Figure 10-7), represent pH ranges that should give stable retention times as a function of pH in that region. However, this is only true for the neutral form of the basic compound and the neutral form of the acidic compound. For basic compounds (or basic functionalities) the lower the pH, the more the ionic equilibrium is shifted toward the protonated form of the analyte, which continually increases its concentration in the aqueous phase and decreases its content in oil phase. Therefore there is no plateau region at low pH. However, for an acidic compound (or acidic functionalities), as the pH is increased, the ionic equilibrium is shifted toward the ionized form of the analyte, which will result in continually increasing the acidic analytes concentration in the aqueous phase and decreasing its content in the oil phase. A decrease in the LogD versus pH curve would be observed at these higher pHs.

The single physicochemical parameter of most importance to the liquid chromatographer is the analyte pK_a. The pK_a values of the various ionizable functionalities for Viagra are shown in Figure 10-8. Ionization of an analyte

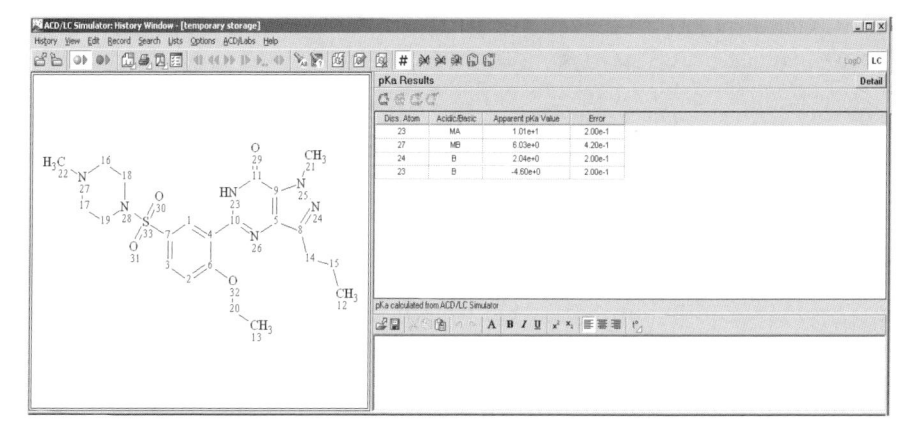

Figure 10-8. Prediction of the pK_a values of ionizable groups in Viagra. ACD/LC Simulator version 9.00.

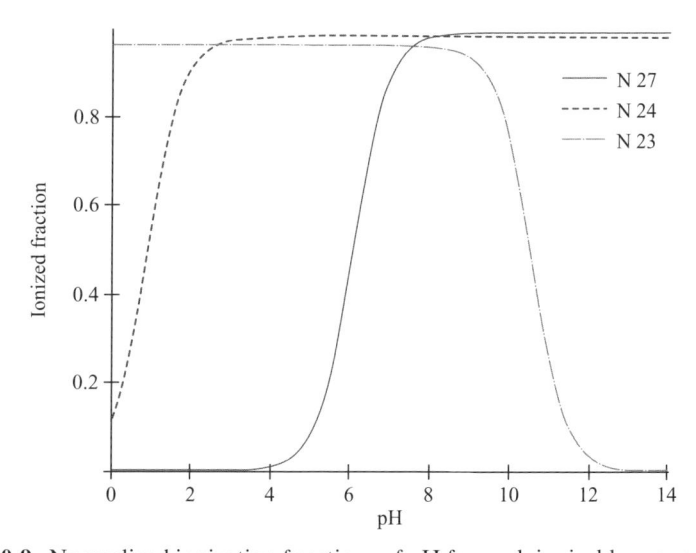

Figure 10-9. Normalized ionization functions of pH for each ionizable group of Viagra (structure shown in Figure 10-8).

can affect detection limits, peak shape, selectivity, and robustness of the method [37].

In general, the rule of thumb when using analyte pK_a values during the design of chromatographic experiments is that the pH of the mobile phase should be at least one to two units away from the pK_a values of the ionizable species. Examination of the ionization curve as a function of pH (Figure 10-9) for Viagra can help rationalize this choice, and the chromatographer should work at a pH region that is on the plateau region of this curve which corresponds to pH 8–9 of the aqueous phase. When the eluent pH is close to the pK_a values of the species, more than one form of the compound will be present in appreciable quantities in the system. Small changes in pH will alter this proportion greatly (shift the equilbria to more ionized or more neutral, depending on the direction of the change in pH), resulting in large changes in overall retention time.

The basis of the ACD/pK_a™ algorithm is classification of the compound prior to prediction using Hammett-based linear free energy calculation (sigma constants are used as descriptors of the electron withdrawal or donation characteristics of substituents connected to the ionization center). This approach is amended to account for ionic forms of polyelectrolytes and reference compounds. Transmission effects for compounds that have distal substituents are also considered. The prediction approach is based on the study of almost 16,000 compounds with over 30,000 experimental pK_a values. Also considered in the calculation are [38] tautomeric equilibria, proton migration, covalent hydration, vinylology, ring-breaking approximations, ring-size correction factors, steric effects, and variable charge effects.

The pK_a prediction has not yet reached a high level of accuracy. An error of ±0.5 pK units is to be expected, but there will of course be situations where errors will be higher, particularly with compounds that are dissimilar from the compounds that were studied to formulate the prediction system. Some pK_a prediction packages are trainable such that experimental values for related compounds (or indeed the compounds themselves) can be stored and used to increase accuracy in subsequent predictions. Besides, the pK_a itself is not a solid physical constant of a particular compound; its value is dependent on many environmental conditions, such as solution media, dielectric constant, temperature, ionic strength, and even method of measurement. The average error for the literature values obtained in different laboratories for the same compound has been on the order of ±0.5 pH units.

System training is the first step to get better pK_a prediction with a small set of compounds; but as the database of similar compounds get larger, the accuracy of pK_a prediction gets better. In general, pK_a calculation for an ionizable group on compounds that have potential for intramolecular hydrogen bonding with another moiety on the aromatic ring; when two ionization centers are close to each other, compounds that have an ortho substituent near the ionization center (especially electron-withdrawing) and compounds that have various tautomeric forms can be challenging.

The main complication with using aqueous pK_a values in chromatography lies in the profoundly nonaqueous nature of most reversed-phase systems today. The presence of organic mobile-phase modifiers affects both the pK_a of the analyte and the effective pH of the buffer (see Chapter 4).

To date, there has been no software system that addresses this problem despite the well-known trends in analyte pK_a shift and mobile-phase pH shift with increasing levels of acetonitrile and methanol up to 60 v/v% [39–42]. However, relatively simple calculations may be applied by the chromatographer to perform the necessary correction for analyte pK_a shift and mobile-phase pH shift (see Chapter 4 for details).

10.4.3.2 Prediction of Retention Times: LC Simulator, ChromSword.
Recently there has been renewed interest in the prediction of analyte retention based on chemical structures as opposed to an experiment-based optimization scheme. These tools can be of use, particularly in support of an application database. However, the accuracy of the tools, particularly for gradient experiments, can be inadequate for routine work.

There are two main approaches to the prediction of retention times based on chemical structures. Both use a training set of compounds to characterize the system prior to creation of a prediction expression. The first (used in ChromSword®) uses experimental retention times for a set of prescribed compounds to create an expression based on molar volume and energy of interaction with water [43]:

$$\ln k = a(V)^{2/3} + b(\Delta G) + c \tag{10-3}$$

where V is the molecular volume of the solute, ΔG is the energy of interaction of a solute with water, and a, b, and c are constant parameters describing the characteristics of the particular chromatographic system, including solvent and column. This unique approach takes a simplistic view of reversed-phase retention mechanisms and neglects ionization of the predicted compounds. The terms can be refined once the chromatographer collects experimental data for the system; however, this somewhat defeats the purpose of retention time prediction.

A second approach is based on physicochemical parameters, used in ACD/LC Simulator. The prediction of RP retention times based on physicochemical parameters assumes that the primary retention mechanism is hydrophobicity of the compound as a function of its ionic form at a given pH. The general approach is given as

$$\log k = a\mathrm{LogD} + bT + c \tag{10-4}$$

where LogD is the octanol/water partition coefficient of the compound in the ionic form in which it exists in this solvent system, and T is a supplemental term that could be molar volume, molecular weight, molar refractivity, or an ion exchange term.

The physicochemical approach to retention time prediction has the advantage of accounting for the pH of the system by explicitly calculating pK_a values of the species.

No structure-based system of retention time prediction to date has explicitly addressed the issue of the pH changes that result from the inclusion of organic modifiers to the aqueous portion of the mobile phase and the resulting effects on both mobile-phase pH and analyte pK_a shifts. Temperature effects can also lead to changes in analyte dissociation constants as well and are not accounted for in structure-based retention time prediction systems.

While a great deal of effort has been expended on the problem of predicting retention times for compounds based on a given chromatographic system, the results have been questionable to date. Simply put, liquid chromatographic systems are complex, and it is very challenging to create a comprehensive prediction approach that can predict retention times for all compounds based on a few characterization experiments. In addition, there is little escaping the fact that most chromatographers do not know every compound in their system and will be forced to run scouting gradients regardless of the predictions that arise. The key for these method development chromatographers is usually to know the general characteristics of the compound: Is it hydrophilic or hydrophobic? and What is the appropriate pH range at which to work?

Advantages. Retention time prediction is a fast, effective way to get an idea of the approximate retention time that can be expected for a given compound under a given set of conditions.

Disadvantages. Error levels still remain a concern, particularly with gradient systems and ionizable compounds. Systematic studies have not been published to date, but average errors in k for gradient systems can approach 30%. Also, both ChromSword and LC Simulator require a reasonable training set of compounds in order to characterize a chromatographic method for a particular compound.

10.4.3.3 Generic Method Selection: ChromGenius. Many of the limitations of structure-based prediction of retention time can be alleviated by the use of a "federation of local models" rather than one model designed to optimize the retention of any given compound. This kind of approach requires a large knowledge base of chromatographic behavior of various compounds.

ACD/ChromGenius™ is a tool designed for prediction of retention times based on the "federation of local models" approach [44]. The prediction process is shown in Figure 10-10. One or more known compounds known or suspected to be present in the sample are input into the system. For each method/compound combination, the software selects the most relevant previously studied compounds based on one of several structure similarity searches. This group of relevant compounds (typically about 20 compounds chosen by structure similarity search) is used in conjunction with multiple linear regression analysis to generate a prediction equation relating predicted physicochemical parameters (i.e., LogD) to retention time.

The large number of compounds enables satisfactory description of the retention behavior of the target analyte within the limited range of chromatographic conditions. The greater relevance of the compounds in the database compared to target analyte reduces the effects of unmodeled phenomena, since any compound that is predicted will have the most similar compounds

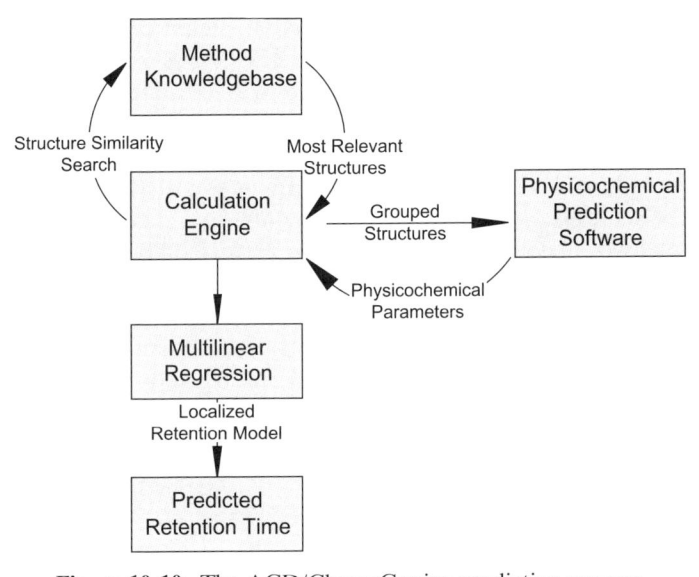

Figure 10-10. The ACD/ChromGenius prediction process.

in the knowledge base used to create a custom-designed expression relating physicochemical properties to the analyte retention time. In addition, the self-diagnosis system (a leave-one-out diagnosis system in which every compound in the knowledge base is predicted as if it is not present and then compared to the true experimental value) in ChromGenius allows for automated gradient correction. Inherently, equation (10-4) makes the assumption that chromatographic conditions are identical for all compounds that are studied under a given set of conditions. However, under gradient conditions, only compounds that elute at the same time as each other will "experience" the same chromatographic conditions. For this reason, to minimize errors, it is necessary to modify the predictions based on the expected elution time of the compound. This is accomplished by a "leave-one-out" approach that compares predicted retention times with experimental for the entire chromatographic knowledgebase. Typically, compounds (target analytes) with low retention (eluting prior to the midpoint of the gradient) will have predicted elution times that are lower than actual experimental values. Compounds that elute late in the gradient will typically have later predicted retention times than actual experimental values. With enough data points, it is possible to create an expression that applies a correction factor based on the elution time of the compound (target analyte) that this "virtual" prediction is being conducted for.

Limitations lie primarily in the size of the required knowledgebase and in the gradient effects on effective mobile phase pH and pK_a, but generally predictions are accurate enough to enable logical selection of method development starting points. The most rigorous data available using this approach are for 2006 compounds that have been predicted using a proper leave-one-out approach. Average error was on the level of 13% for prediction of these 2006 compounds [45]. The diversity of the analytes was extremely broad, indicating that this could be indicative of any new compound that would be introduced to the system. For late eluting compounds using fast gradient systems, this has produced good results. The highest relative error was attributed to prediction of the retention of the early-eluting compounds.

In manual work, method selection can be performed by manual inspection of the predicted chromatograms for resolution and run time, or according to the user-defined automated rating process. The automated selection process can be customized on a similar basis as the selection process in method optimization software, with requirements based on targeted run time, resolution, or a combination of the two; this suitability approach is described in equation (10-5):

$$S_{\text{overall}} = S_{\text{RT}} \cdot S_{\text{Res}} \cdot S_{k'} \tag{10-5}$$

where if retention time of the last eluting component (t_{exp}) < target (t), then $S_{\text{RT}} = 1$, if t_{exp} > maximum (m), then $S_{\text{RT}} = 0$, else $S_{\text{RT}} = (t_m - t_R)/(t_m - t_t)$; if resolution of the critical pair (R_{exp}) > target (t), then $S_{\text{Res}} = 1$, if R_{exp} < minimum (m), then $S_{\text{Res}} = 0$, else $S_{\text{Res}} = (R_{\text{exp}} - R_m)/(R_t - R_m)$; and if the retention index of the first eluting peak (k'_{exp}) > target (t), then $S_{k'} = 1$, if k'_{exp} < minimum (m), then $S_{k'} = 0$, else $S_{k'} = (k'_t - k'_{\text{exp}})/(k'_t - k'_m)$. S_{RT} is suitability based on run time, S_{Res} is suitability based on resolution, and $S_{k'}$ is suitability based on retention. It can be seen that the form of equation (10-5) is such that suitability terms for resolution, run time, and retention are combined to give an overall suitability value between zero and one. An example of the output of this approach (method ranking system) is shown in Figure 10-11.

Advantages. The software (ACD/ChromGenius) is simple to apply and is accurate in structure-based predictions, even for gradient experiments.

Disadvantages. The primary disadvantage of this software is that the user rarely will know all structures in the sample, and thus he or she will likely still need to do at least some optimization prior to utilizing the method.

10.5   CONCLUSION

There is no doubt that computers will play an important role in the future of chromatographic method development. Increased emphasis on streamlining of method development process dictates that added efficiencies need to be located. The optimization tools that have often been neglected seem to be undergoing a renaissance in the form of connectivity and ease-of-use improvements, such that even for routine development these optimization tools can

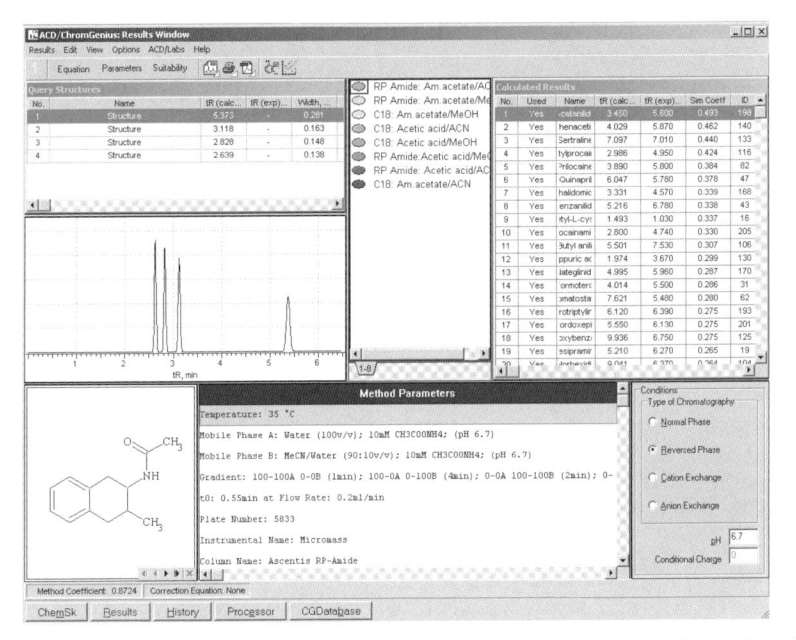

Figure 10-11. The method ranking system. Colors reflect suitability rankings based on user criteria for retention, resolution, and run time as per equation (10-5).

play a role. No single tool will suit the needs of all chromatographers for all applications. Walk-up laboratories for LC/MS/DAD evaluation of sample purity to support synthesis may need nothing more than structure-based generic method selection, while research and development laboratories for development of API (active pharmaceutical ingredient) and drug product may require rigorous simultaneous optimization of multiple chromatographic parameters.

Structure-based tools such as applications databases and physicochemical prediction will not replace chromatographers, but rather allow them to use their knowledge of chromatography, relating it to the analyte functionality and making excellent choices of initial conditions that reap faster, better separations in less time.

REFERENCES

1. www.rheodyne.com/products/chromatography/drylab/index.asp.

2. www.autochrom.com.

3. S. V. Galushko, Software for method development in HPLC, *GIT Spezial Chromatogr.* **16** (1996), 88–93.
4. R. Kaliszan, *Structure and Retention in Chromatography. A Chemometric Approach*, Harwood Academic Publishers, Amsterdam, 1997.
5. J. Jiskra, H. A. Claessens, C. A. Cramers, and R. Kaliszan, Quantitative structure–retention relationships in comparative studies of behavior of stationary phases under high-performance liquid chromatography and capillary electrochromatography conditions, *J. Chromatogr. A* **977** (2002), 193–206.
6. R. Kaliszan, Quantitative structure–retention relationships applied to reversed-phase high-performance liquid chromatography, *J. Chromatogr. A* **656** (1993), 417–435.
7. P. C. Sadek, P. W. Carr, R. M. Doherty, M. J. Kamlet, R. W. Taft, and M. H. Abraham, Study of retention process in reversed-phase high-performance liquid chromatorgraphy by use of the solvatochromic comparison method, *Anal. Chem.* **57** (1985), 2971–2978.
8. M. H. Abraham, Scales of solute hydrogen bonding: Their construction and application to physicochemical and biochemical processes, *Chem. Soc. Rev.* **22** (1993), 73–83.
9. G. Guiochon, S. G. Shirazi, and A. M. Katti, *Fundamentals of Preparative and Nonlinear Chromatography*, Academic Press, New York, 1994.
10. C. Horvath, W. R. Melander, and I. Molnar, Solvophobic interactions in liquid chromatography with nonpolar stationary phases, *J. Chromatogr.* **125** (1976), 129–156.
11. P. W. Carr, M. Reta, P. C. Sadek, and S. C. Rutan, Comparative study of hydrocarbon, fluorocarbon, and aromatic bonded RP-HPLC stationary phases by linear solvation energy relationships, *Anal. Chem.* **71** (1999), 3484–3496.
12. M. J. Kamlet, J. L. M. Abboud, M. H. Abraham, and R. W. Taft, Linear solvation energy relationships. 23. A comprehensive collection of the solvatochromic parameters, π, α, and β, and some methods for simplifying the generalized solvatochromic equation, *J. Org. Chem.* **48** (1983), 2877–2887.
13. O. Sinanoglu and B. Pulman (eds.), *Molecular Associations in Biology*, Academic Press, New York, 1968, pp. 427–445.
14. P. W. Carr and J. Zhao, Comparison of the retention characteristics of aromatic and aliphatic reversed-phases for HPLC using linear solvation energy relationships, *Anal. Chem.* **70** (1988), 3619–3628.
15. P. W. Carr, L. C. Tan, and M. H. Abraham, Study of retention in reversed-phase liquid chromatography using linear solvation energy relationships I. The stationary phase, *J. Chromatogr. A* **752** (1996), 1–18.
16. P. W. Carr and J. Zhao, Approach to the concept of resolution optimization through changes in the effective chromatographic selectivity, *Anal. Chem.* **71** (1999), 2623–2632.
17. M. H. Abraham and M. Roses, Hydrogen bonding. 38. Effect of solute structure and mobile phase composition on reversed-phase high-performance liquid chromatographic capacity factors, *J. Phys. Org. Chem.* **7** (1994), 672–684.
18. J. C. McGowan, *J. Chem. Technol. Biotechnol.* **34A** (1984), 38.
19. M. H. Abraham and J. C. McGowan, The use of characteristic volumes to measure cavity terms reversed phase liquid chromatography, *Chromatographia* **23** (1987), 243–246.

20. J. L. E. Reubsaet and R. Vieskar, Characterisation of π–π interactions which determine retention of aromatic compounds in reversed-phase liquid chromatography, *J. Chromatogr. A* **841** (1999), 147–154.

21. J. Horak, N. M. Maier, and W. Lindner, Investigations on the chromatographic behavior of hybrid reversed-phase materials containing electron donor–acceptor systems: II. Contribution of π–π aromatic interactions, *J. Chromatogr. A* **1045** (2004), 43–58.

22. D. DeVault, Theory of chromatography, *J. Am. Chem. Soc.* **65** (1943), 532.

23. M. H. Abraham, M. Roses, C. F. Poole, and S. K. Poole, Hydrogen bonding. 42. Characterization of reversed phase high-performance liquid chromatographic C18 stationary phases, *J. Phys. Org. Chem.* **10** (1997), 358–368.

24. M. Salo, H. Vuorela, and J. Halmekoski, Effect of the organic modifier on the retention of retinoids in reversed-phase liquid chromatography, *Chromatographia* **36** (1993), 147–151.

25. J. W. Dolan, L. R. Snyder, L. C. Sander, P. Haber, T. Baczek, and R. Kaliszan, Reversed-phase liquid chromatographic separation of complex samples by optimizing temperature and gradient time: III. Improving the accuracy of computer simulation, *J. Chromatogr. A* **857** (1999), 41–68.

26. M. R. Euerby and P. Petersson, Chromatographic classification and comparison of commercially available reversed-phase liquid chromatographic columns using principal component analysis, *J. Chromatogr. A* **994** (2003), 13–36.

27. J. W. Dolan, A. Maule, D. Bingley, L. Wrisley, C. C. Chan, M. Angod, C. Lunte, R. Krisko, J. M. Winston, B. A. Homeier, D. V. McCalley, and L. R. Snyder, Choosing an equivalent replacement column for a reversed-phase liquid chromatographic assay procedure, *J. Chromatogr. A* **1057** (2004), 59–74.

28. J. Pellett, P. Lukulay, Y. Mao, W. Bowen, R. Reed, M. Ma, R. C. Munger, J. W. Dolan, L. Wrisley, and K. Medwid, "Orthogonal" separations for reversed-phase liquid chromatography, *J. Chromatogr. A* **1101** (2006), 122–135.

29. L. R. Snyder, J. J. Kirkland, and J. L. Glajch, *Practical HPLC Method Development*, 2nd ed. 2004, Wiley-Interscience, Hoboken, NJ.

30. D. W. Hill, T. R. Kelley, and K. J. Langner, Computerized library search routine for comparing ultraviolet spectra of drugs separated by high-performance liquid chromatography, *Anal. Chem.* **59** (1987), 350–353.

31. E. C. Nicolas and T. H. Scholz, Active drug substance impurity profiling. Part I. LC/UV diode array spectral matching, *J. Pharm. Biomed. Anal.* **16** (1998), 813–824.

32. M. E. Swartz and P. R. Brown, Use of mathematically enhanced spectral analysis and spectral contrast techniques for the liquid chromatographic and capilally electrophoretic detection and identification of pharmaceutical compounds, *Chirality* **8** (1996), 67–76.

33. A. Bogomolov and M. McBrien, Mutual peak matching in a series of HPLC-DAD mixture analyses, *Anal. Chim. Acta* **490** (2003), 41–58.

34. G. Xue, A. D. Bendick, R. Chen, and S. S. Sekulic, Automated peak tracking for comprehensive impurity profiling in orthogonal liquid chromatographic separation using mass spectrometric detection, *J. Chromatogr. A* **1050** (2004), 159–171.

35. P. Willet, J. M. Barnard, and G. M. Downs, Chemical similarity searching. *J. Chem. Inf. Sci.* **38** (1998), 983–996.

36. M. McBrien, Scientific Computing and Instrumentation online, March 13, 2005.

37. S. Kromidas (ed.), *HPLC Made to Measure: A Practical Handbook for Optimization*, Wiley-VCH, Hoboken, NJ, 2006.

38. R. S. DeWitte and E. Kolovanov, Predicting molecular physical properties, in R. T. Borchardt, E. H. Kerns, C. A. Lipinski, D. R. Thakker, and B. Wang (eds.), *Pharmaceutical Profiling in Drug Discovery for Lead Selection*, AAPS Press, Arlington VA, 2004.

39. S. Espinosa, E. Bosch, and M. Roses, Retention of ionizable compounds on HPLC. 5. pH Scales and the retention of acids and bases with acetonitirle–water mobile phases, *Anal. Chem.* **72** (2000), 5193.

40. R. LoBrutto, A. Jones, Y. V. Kazakevich, and H. M. McNair, Retention of ionizable compounds on HPLC. 8. Influence of mobile-phase pH change on the chromatographic retention of acids and bases during gradient elution, *J. Chromatogr. A* **913** (2001), 173–187.

41. I. Canals, K. Valko, E. Bosch, A. P. Hill, and M. Roses, Retention of ionizable compounds on HPLC. 8. Influence of mobile phase pH change on the chromatographic retention of acids and bases during gradient elution, *Anal. Chem.* **73** (2001), 4937–4945.

42. M. Roses, Determination of the pH of binary mobile phases for reversed-phase liquid chromatography, *J. Chromatogr. A* **1037** (2004), 283–298.

43. S. V. Galushko, A. A. Kamenchuk, and G. L. Pit, Calculation of retention in reversed-phase liquid chromatography: IV. ChromDream software for the selection of initial conditions and for simulating chromatographic behavior, *J. Chromatogr. A* **660** (1994), 47–59.

44. www.chromgenius.com.

45. M. McBrien, E. Kolovanov, and R. Taylor, Highly accurate prediction of retention times in standard gradient chromatographic systems based on physicochemical parameters, March 9–14, 2003, Pittsburgh conference, Orlando, FL.

PART II

HPLC IN THE PHARMACEUTICAL INDUSTRY

11

THE EXPANDING ROLE OF HPLC IN DRUG DISCOVERY

Daniel B. Kassel

11.1 INTRODUCTION

Great efficiencies have been achieved in the drug discovery process as a result of technological advances in target identification, high-throughput screening, high-throughput organic synthesis, just-in-time *in vitro* ADME (absorption, distribution, metabolism, and excretion), and early pharmacokinetic screening of drug leads. These advances, spanning target selection all the way through to clinical candidate selection, have placed greater and greater demands on the analytical community to develop robust high-throughput methods. This review highlights the various roles of high-performance liquid chromatography/mass spectrometry (HPLC/MS) in drug discovery and how the field has evolved over the past several years since the introduction of myriad high-throughput drug discovery technologies. Included are significant developments in HPLC/MS to support target selection (proteomics), biological screening and assay development, high-throughput compound analysis and characterization, UV- and mass-directed fractionation for unattended, automated compound purification, and high-throughput *in vitro* ADME screening.

Focus within the pharmaceutical industry has been to increase the likelihood of successfully developing clinical candidates by optimizing the components of the discovery process (i.e., spanning target identification → chemical

HPLC for Pharmaceutical Scientists, Edited by Yuri Kazakevich and Rosario LoBrutto
Copyright © 2007 by John Wiley & Sons, Inc.

design → synthesis → compound analysis and purification → registration → biological and ADME screening. By optimizing each step in the iterative discovery process, it is expected that the compound attrition rate will be reduced dramatically as compounds advance into preclinical development. Both HPLC and LC/MS enjoy important roles throughout the discovery process, as will be highlighted in detail in this review. Once considered primarily an enabling tool for medicinal chemists, HPLC and LC/MS are now key technologies incorporated at just about every stage of the drug discovery process. Drug discovery programs typically initiate, as follows. Assuming that the relevant therapeutic area (e.g., oncology, metabolic diseases, inflammation, pain, CNS, etc.) has been selected, the next step is to identify a biological target relevant to the disease. As will be discussed shortly, numerous technological advances in the field of analytical chemistry (e.g., nanocolumn HPLC/MS/MS) that have greatly facilitated protein/target identification have been made since the human genome initiative was launched. Following on the heels of target selection is the requirement to establish tools for "just-in-time" high-throughput screening of compound repositories (so-called corporate collections) and synthetic libraries as a means for identifying initial hits/actives. In combination with structure–activity relationship (SAR) data generated from these high-throughput screens, chemists incorporate knowledge of protein three-dimensional structures and utilize computational tools (i.e., *in silico* methods that measure diversity and "drug-likeness" as well two-dimensional and three-dimensional pharmacophore models [descriptors] that predict biological activity) to support iterative compound design, synthesis, and biological testing. Once the hits or actives have been identified, the process of hit refinement and lead optimization is initiated. At this stage, a chemistry team is established and both parallel synthesis and more traditional medicinal chemistry strategies are incorporated to rapidly converge on qualified leads (so-called hit-to-lead stage). HPLC and LC/MS play an extremely important role in the hit-to-lead stage of discovery, providing key enabling analysis and purification capabilities to the medicinal chemist. Furthermore, activities that were traditionally relegated to drug metabolism and pharmacokinetics departments within development organizations are now integrated into early discovery so as to provide early measurements and predictions of *in vivo* properties. Again, LC/MS has played an extremely important role in enhancing the drug developability of these hits and leads. All of these advances have helped to streamline the discovery phase of pharmaceutical drug discovery and development and are presented within.

11.2 APPLICATIONS OF HPLC/MS FOR PROTEIN IDENTIFICATION AND CHARACTERIZATION

The human genome initiative that took a stronghold on biotechnology companies in the early 1990s through the first few years of the twenty-first century spawned a completely new field that had analytical chemistry as its cornerstone. Specifically, high-resolution capillary and nano-column HPLC coupled with tandem mass spectrometry became one of the tools of choice for characterizing proteins and identifying potential therapeutic protein targets. Although capillary HPLC/MS/MS was applied as early as 1989–1991 to the characterization of proteins and for identifying sites of post-translational modification [1–4], the field took off in earnest following the genomics boom and became known as proteomics, coined by Wilkins et al. [5]. In essence, the mandate of the proteomics field since its inception has been to identify differences at the protein level, in cells, tissues, plasma, and so on, between a disease state and control ("normal"). The basic premise is that proteins will be either up- or down-regulated (i.e., over- or underexpressed) in the disease state relative to "normal" state, and these differences can be identified and quantified by mass spectrometry. There have been several analytical advances made in the field of proteomics since its inception, far too numerous to capture in this review. One noteworthy advance in proteomics is the technique of multidimensional protein identification technology (MUDPIT), developed by Yates and co-workers, which has been used widely in place of the more laborious, less automated method of 2D-polyacrylamide electrophoresis [6]. MUDPIT is a column chromatography method whereby ion-exchange chromatography is used in the first dimension of chromatography to simplify the complexity of the complex mixture of peptides by separating them based on charge followed by reversed-phase HPLC for the higher-resolution separation based on molecular weight and hydrophobicity. An equally important development in the field of proteomics has been isotope-coded affinity tags (ICAT) technology, a method whereby isotopic labeling of peptides containing cysteine residues is performed so as to facilitate peptide quantitation and identification of putative biological targets [7]. The reader is directed to the following review in the field of proteomics for more information [8].

A wealth of preclinically validated targets has emerged as a result of mouse genetics [9] and siRNA technology [10]. For both techniques, a single gene knockout is performed, and the effect of the deletion is monitored/evaluated. Proteomics, on the other hand, generally takes a shotgun approach to identifying the targets that are relevant and specific to the disease. Unfortunately, because many diseases are polygenic in origin and because protein pathways are extremely complex (e.g., intracellular protein signaling pathways [11]), proteomics has been best at identifying a short list of "candidate" protein targets rather than a single protein target completely unique to the disease. The challenge has been to sift through all the proteins that have been identified as altered in a disease state relative to healthy state, and this has proved extremely challenging.

The focus of proteomics has turned to identifying potential biomarkers of disease. A biomarker, by definition, is (a) a molecular indicator for a specific biological property or (b) a feature or facet that can be used to measure the progress of disease or the effects of treatment. As an example, a biomarker for Type II diabetes is higher fasting blood glucose levels relative to age-matched controls. Another, more definitive biomarker of type II diabetes is elevated HbA1c levels. For many diseases, however, the relevant biomarkers are less well understood. This is especially true in the fields of oncology and inflammation research. Biomarker research is a particularly intense area of focus for many pharmaceutical companies, with new departments being formed for the purpose of identifying both preclinical and clinical biomarkers to facilitate their drug discovery and development programs. Like the field of proteomics, the field of biomarker research is far too vast to warrant its review here. A very nice review article by the late Wayne Colburn, a pioneer in diabetes biomarker research, describes this maturing field [12].

11.3 APPLICATIONS OF HPLC/MS/MS IN SUPPORT OF PROTEIN CHEMISTRY

Independent of the tool used to identify the protein target, whether it be mouse genetics, siRNA technology, or proteomics, once a protein has been identified as a suitable target for drug discovery, the next step in the drug discovery process is to express and purify the protein (carried out combining molecular biology and protein chemistry techniques) in sufficient quantities so as to support biological screening, X-ray crystallography, and any other drug discovery studies requiring purified protein material. The traditional method for assessing protein expression and purification has been to use 1D-polyacrylamide gel electrophoresis. 1D-PAGE is capable of separating proteins based on molecular weight and charge (pI). However, the technique is unable to provide more than a crude assessment of protein molecular weight. Recently, open-access or walk-up LC/MS has been incorporated into protein chemistry and molecular biology labs and has greatly facilitated confirmation of protein expression [13–15]. Generic gradient LC-MS methods are used to trap and elute expressed, purified proteins by RP-HPLC/ESI/MS. Open-access protein QC is a bit more challenging than its small-molecule counterpart in that not all proteins "fly" by electrospray ionization, identifying a "universal" HPLC method for their separation can be challenging, and instrument calibration and mass accuracy are of paramount importance. We developed a fast, 5-minute protein QC method using a Poroshell 1-mm-i.d. column and found the method to be satisfactory for the vast majority of protein separations and analyses performed in our laboratory. To achieve adequate mass accuracy for protein molecular weight determinations, an external calibration with myoglobin is performed at the beginning and end of each overnight queue of protein samples so as to ensure that the instrument calibration is maintained over the course of the batch analysis. Molecular weights of deconvoluted protein spectra are then compared to the predicted protein molecular weight, and the results are captured graphically (in the form of a microtiter plate view) as well as in tabular format, amenable to database uploading, as shown in Figure 11-1.

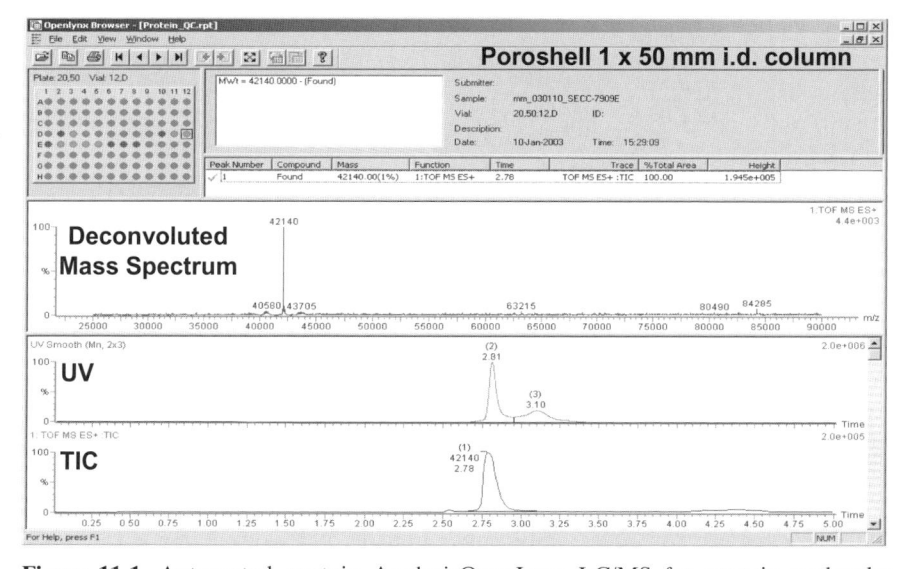

Figure 11-1. Automated protein AnalysisOpenLynx LC/MS for protein molecular weight confirmation.

11.4 APPLICATIONS OF HPLC/MS/MS IN SUPPORT OF ASSAY DEVELOPMENT AND SCREENING

The overwhelming majority of biological assays have been developed in microtiter plate format (typically 96-well, 384-well, 1536-well) and with parallel detection methods such as fluorescence polarization. The vast majority of druggable targets, including enzymes, ligand gated ion channels, and G-protein-coupled receptors, are all amenable to screening in high-throughput microtiter plate format.

In general, serial-based chromatographic methods, such as HPLC and HPLC/MS, are unable to compete with the high-throughput screening technologies. However, a small number of targets, such as those involved in mediating protein–protein interactions, are not well-suited to HTS methodologies. For this class of targets, HPLC coupled with mass spectrometry has proved to be a very reliable, albeit lower throughput, alternative. The technique that has been used most widely for directly assessing protein–small molecule and protein–protein interactions is affinity chromatography–mass spectrometry. Kassel et al. [16] presented one of the first papers coupling affinity chromatography with mass spectrometry. In their work, a two-dimensional LC/LC/MS method was developed to assess protein–ligand binding. Affinity chromatography was used in the first dimension of separation, followed by reversed-phase chromatography coupled with mass spectrometry for the identification of binders. Kaur et al. [17] showed the power of size exclusion chromatography (SEC) coupled with reversed-phase HPLC/MS for identifying ligands for a receptor derived from a 576-component combinatorial library. Today, size-exclusion columns are available in microtiter plate format, permitting higher-throughput characterization of protein–protein and protein–ligand interactions.

Berman et al. [18] pioneered one of the earliest applications of HPLC in support of assay development. They showed the power of HPLC for the determining preferred substrates of the enzyme collagenase, a metalloprotease. Complex mixtures (pools of 100 components each) of probe substrates for collagenase were prepared by combinatorial methods. Each of the pooled libraries was incubated with enzyme. Substrate disappearance (turnover) and product appearance profiles were monitored by HPLC and the optimal substrate(s) identified. Recently, Lambert et al. [19] published a two-dimensional LC/LC/MS method for the identification and optimization of substrates for TNF convertase. Scientists at Nanostream, Inc., a company dedicated to high-throughput HPLC, introduced a parallel capillary LC/fluorescence method to support screening for kinase inhibitors. Their method complements the more traditional (and higher-throughput) fluorescence-based screening approach but offers the advantage of chromatographic separation of phosphorylated and unphosphorylated products, thereby reducing background interference.

Another emerging role of HPLC/MS is in support of cell-based assays for which no direct measures of drug effect are possible and require indirect methods for detection. A recent publication by Clark et al. highlights the power of LC-MS for screening inhibitors of HMG-CoA reductase (a rate-determining enzyme in the cholesterol biosynthesis) [20]. In addition, Thibodeaux et al. [21] and Xu et al. [22] reported on methods for directly assessing the cell-based activity of inhibitors of the metabolic disease target, 11β-hydroxysteroid dehydrogenase-1 (11β-HSD-1). LC-MS was used to measure the effect of 11β-HSD-1 inhibitors on the intracelleular conversion of cortisol and cortisone using LC/MS/MS.

11.5   SOURCES OF COMPOUNDS FOR BIOLOGICAL SCREENING

Once the assay and assay format have been decided upon, the next step in the discovery process is to initiate compound screening for the purpose of identifying hits or lead compounds. The fundamental requirement is that the assay results identify a collection of actives or "hits." The definition of "hit" varies between organizations, but most accept the definition that the compound shows a confirmed structure, shows a confirmed dose response, exhibits an IC50 ≤ 10 μM potency, and is a member of a chemotype that is amenable to analoging and fast follow-on synthesis.

What is the source of these initial actives or hits? There is a wide array of compound sources. Generally, pharmaceutical and biotechnology organizations initiate screening by accessing their internal compound repositories (so-called corporate collections or compound archives). Often, the corporate collections are not particularly diverse but are biased to the therapeutic focus(es) of the organization. Consequently, the screening libraries are often augmented by addition of commercially available screening libraries that are (a) gene-family focused (e.g., GPCR-targeted libraries, kinase-targeted libraries, etc.) and/or (b) general diversity sets. Further augmentation of the initial screening activities is to include custom synthesis compound libraries (typically produced by automated high-throughput organic synthesis (HTOS) methodologies, such as those described by Nikolaou et al. [23].

One of the challenges with compound collections is that they are historical by nature. For large Pharma, it is not uncommon for corporate collections to include compounds that were synthesized more than 25 years ago. At the time of synthesis, it can be presumed that the compounds met the purity criteria for compound registration. However, it can also be presumed that a high likelihood exists that the compounds have degraded over extended storage time. Another reason for poorer quality of compound collections is attributable to the fact that most compounds are stored as DMSO stock solutions as opposed to storage as solid materials. Storage of compounds in DMSO is done primarily for the reason that (a) DMSO is considered a "universal" solvent and (b) solutions are much easier to handle in plate-based high-throughput biological screening systems. However, the drawback to DMSO is that it is a very hydroscopic solvent and unless the compounds are stored under inert conditions, they are prone to hydrolysis. Kozikowski et al. [24] evaluated the effect of freeze/thaw cycles on stability of compounds stored in DMSO.

Until very recently, with the introduction of high-throughput analytical technology, these compound sources were far too large to merit re-analysis and/or re-purification and hence were screened "as is." The result was (and has been observed frequently) that hits could not be reconfirmed during follow-on bioassay screening, and subsequent evaluation of the compounds by techniques such as HPLC/MS and NMR showed that the expected compound was not pure and, in some cases, was completely absent! The adage "garbage in, garbage out" became a mantra of many high-throughput screening laboratories and forced companies to take a much more serious look at the quality of their compound collections. Morand et al. [25] from Proctor and Gamble set out to fully assess the quality of their >500,000 compound corporate collection. They achieved this goal through incorporation of a massively parallel flow injection–mass spectrometry system, capable of analyzing a plate of samples in less than 2 minutes. The throughput of their technique was one to two orders of magnitude faster than typical flow injection–mass spectrometry systems used for reaction monitoring [26].

In addition to quality control over compound collections, the issue of purity of synthetic libraries derived using combinatorial chemistry quickly came under the microscope. In the early to mid-1990s, "combichem" became a household word throughout the pharmaceutical industry and was believed to be a key technology that would revolutionize drug discovery. The basis of combinatorial chemistry was the ability to perform split-mix synthesis on solid support and to take advantage of the combinatorial nature of the process to generate vast arrays of compounds. Combinatorial libraries were purported to be pure, owing to the fact that they were synthesized on solid support and amenable to extensive washing to remove excess reagents and, therefore, directly amenable to high-throughput screening. However, these combinatorial libraries synthesized on solid support suffered from the same problems that have long plague solution-phase synthesis—that is, the generation of unexpected and unwanted by-products. Due to the shear size of these compound libraries and the relatively small amounts available following resin cleavage, it was not possible to either characterize or purify the expected products. Conventional split-mix combinatorial methods, though still popular with some bench chemists, have been replaced largely by the technique of directed parallel solution and parallel solid-phase organic synthesis.

11.6   HPLC/MS ANALYSIS TO SUPPORT COMPOUND CHARACTERIZATION

Combinatorial chemistry paved the way for high-throughput, parallel organic synthesis techniques, now mainstream in the pharmaceutical and biotechnology industries for lead generation activities. The ability to synthesize compound libraries rapidly using automated solution-phase and solid-phase parallel synthesis has led to a dramatic increase in the number of compounds now available for high-throughput screening. The unprecedented rate by which compound libraries are now being generated has forced the analytical community to implement high-throughput methods for their analysis and characterization.

As early as 1994, groups adopted high-speed, spatially addressable automated parallel solid-phase and solution-phase synthesis of discretes [27–31]. Both solution-phase and solid-phase parallel synthesis permits the production of large numbers as well as large quantities of these discrete compounds, eliminating the need for extensive decoding of mixtures and re-synthesis following identification of "active" compounds in high-throughput screening of combinatorial libraries. Importantly, parallel synthesis is performed readily in microtiter plate format amenable to direct biological screening, as was touched upon earlier. The relative ease of automation of parallel synthesis led to a tremendous in flux of compounds for lead discovery and lead optimization.

Almost all of the analytical characterization tools (e.g., HPLC, NMR, FTIR, and LC/MS) are serial-based techniques, and parallel synthesis is inherently parallel. Consequently, this led rapidly to a new bottleneck in the discovery process (i.e., the analysis and purification of compound libraries). Parallel synthesis suffers from some of the same shortcomings of split and mix synthesis (e.g., the expected compound may not be pure, or even synthesized in sufficient quantities). The analytical community was faced with the decision of how to analyze these parallel synthesis libraries.

The traditional method for assessing compound purity has been to perform the following: Purify the desired product to homogeneity by crystallization or column chromatography (e.g., RP- or NP-HPLC), acquire a 1D-NMR and 2D-NMR spectrum on the isolated product, obtain confirmatory molecular weight information by mass spectrometry, perform a C, H, and N combustion analysis, generate an exact mass measurement (to within 5 ppm of the expected mass) by high-resolution mass spectrometry, and determine the amount of isolated product by weight—all prior to compound submission and biological screening. In the era of high-throughput compound library synthesis, however, this extensive characterization is simply not possible. Therefore, groups have focused principally on a limited number of analytical measurements for compound identity and purity—in particular, LC/MS analysis incorporating orthogonal detection methods, such as UV and evaporative light scattering detection (ELSD) and flow-probe 1D-NMR [32]. The most commonly employed technique for characterizing compound libraries is to incorporate LC/MS with electrospray or atmospheric pressure chemical ionization with UV and ELSD and, more recently, photoionization [33].

LC/MS emerged as the method of choice for the quality control assessment to support parallel synthesis because the technique, unlike flow injection mass spectrometry, provides the added measure of purity (and quantity) of the compound under investigation. In addition, "universal-like" HPLC gradients (e.g., 10% to 90% acetonitrile in water in 5 minutes) have been found to satisfy the separation requirements for the vast majority of combinatorial and parallel synthesis libraries. Fast HPLC/MS has been found to serve as good surrogate to conventional HPLC for assessing library quantity and purity [34–37]. Fast HPLC/MS is simple in concept. It involves the use of short columns (typically 4.6 mm i.d. × 30 mm in length) operated at elevated flow rates (typically 3–5 mL/min).

Typically, short columns are used for compound analysis because they allow for fast separations to be carried out at ultrahigh flow rates. Also, these columns tend to be more robust than narrow bore columns (1-mm and 2-mm i.d.) (i.e., less clogging is experienced and longer lifetimes are observed when these columns are subjected to unfiltered chemical libraries). A typical LC/MS analysis consists of injection a small aliquot (10–30 μL) of the reaction mixture (total concentration of 0.1–1.0 mg/mL) and performing the separation using a "universal" gradient of 10–90% Buffer B in 2–5 minutes. Buffer A is typically H_2O containing 0.05% trifluoroacetic acid (or formic acid), and Buffer B is typically acetonitrile containing 0.035% trifluoroacetic (or formic acid). HPLC columns are operated typically at flow rates of 3–5 mL/min (depending on their dimensions), and the cycle time between injections is 3–5 minutes.

An example of a fast LC/MS analysis of a combinatorial library component is shown in Figure 11-2. Fast LC/MS run times incorporating these short columns is typically between 3 and 5 minutes including re-equilibration.

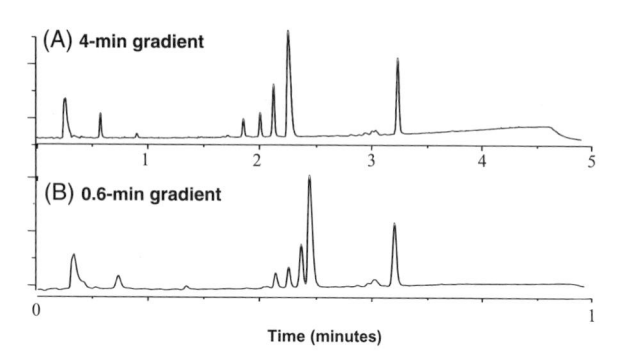

Figure 11-2. (A) A 4-minute HPLC/MS separation of a solution-phase parallel synthesis library. The gradient profile for fast HPLC/MS was 10–90% acetonitrile in H_2O in 4 minutes with a 1-minute equilibration time. (B) A 1-minute, total cycle time chromatographic separation of the same crude product. (Reprinted from reference 42, with permission.)

Recent reports by Kyranos et al. [38] suggest that "pseudo-chromatography" (in essence, step elution chromatography) provides a more rapid and reliable assessment of the quality of library synthesis than methods such as flow injection mass spectrometry.

11.6.1 Purity Assessment of Compound Libraries

The issue of compound purity has received a great deal of attention over the last several years as more and more chemists have adopted high-throughput organic synthetic protocols but are unwilling to compromise the quality of the molecules submitted for biological evaluation. The general consensus target purity of a compound library compound before it is to be archived or screened for biological activity is between 90% and 95% pure. This purity criterion is more stringent than in the past, where 85–90% (based on UV detection) was considered acceptable. This may be attributed primarily to a shift toward smaller, focused (or biased) libraries than larger, diverse collections of compounds. The majority of mass spectrometry manufacturers now offer software packages that aid in the automatic determination of purity.

UV chromatograms are typically used, rather than the total ion current chromatogram, to assess purity. This is because the total ion current chromatogram is a measure of a compound's "ionizability," which is well known to vary dramatically from one compound to the next. Orthogonal detection methods, such as chemiluminescence nitrogen detection (CLND) [39] and ELSD [40, 41], have been proposed to be more universal detection methods than UV and hence are being used with increasing frequency to assess reaction yields and purity. CLND, as indicated from its name, measures the amount of nitrogen in a sample. In this method, a compound is transferred to a high-temperature oxygen reaction chamber (set to 1000°C) whereby the compound undergoes rapid decomposition to form nitrous oxide (NO). The liberated NO reacts with ozone (O_3) to form metastable NO_2, which is selectively detected by release of a photon.

CLND has been demonstrated to be a valuable tool for quantifying low quantities of material and has been shown to be particularly well-suited to NP-HPLC and SFC-MS, for the principal reason that separations are carried out using solvents that do not contain nitrogen (i.e., CO_2 and CH_3OH). ELSD measures the mass (quantity) of the material directly, is often presented as being a molecular-weight-independent detector, and is a tool that has gained wide-scale acceptance for on-line quantification of compound libraries. An example of a separation of a four-component library incorporating UV, ELSD, CLND, and MS detection is shown in Figure 11-3. Using these various detectors, the chemist is able to obtain measures of purity of their library with greater confidence than when relying solely upon LC/UV/MS data.

An example of automated purity assessment of a compound analyzed by LC/UV/MS is shown in Figure 11-4. In this example, purity is assessed at two different wavelengths, λ_{220} and λ_{254}. Excel macros are used for automated

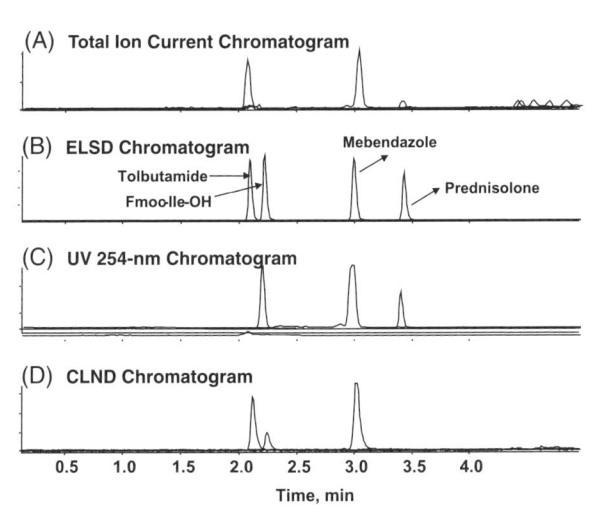

Figure 11-3. Column flow rate was 5 mL/min. A portion of the column effluent was split to each of the three detectors (CLND, 200 µL/min; ELSD, 200 µL/min; MS, 100 µL/min). A make-up flow of 50/50 MeOH/H_2O (300 µL/min) was added to the flow stream diverted to the mass spectrometer ion source. Mass spectra were acquired using electrospray ionization with no special modifications to the ion source. (A) Total ion current chromatogram showing two of the four components ionize efficiently under electrospray ionization conditions. (B) ELSD chromatogram of the four components, all showing comparable response. (C) UV chromatogram (254 nm) shows some selectivity in detection as does. (D) CLND detection.

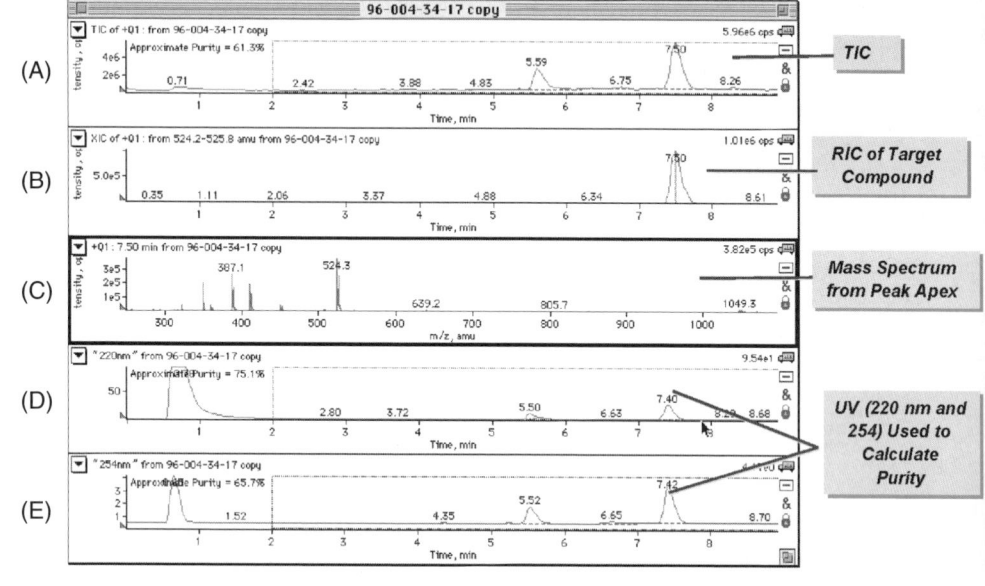

Figure 11-4. Purity assessment is a critical component in the decision process by the chemist as to whether their isolated compound is of sufficient quality to be submitted for compound registration and biological testing. To facilitate automated and rapid purity assessment of compound libraries, applescripts and visual basic scripts are used. (A) Total ion current chromatogram shows two components. (B) Extracted ion chromatogram for the expected product identifies its retention time. (C) Mass spectrum observed for the expected product. (D) UV 220-nm chromatogram indicates the expected product is approximately 75% pure. (E) UV 254-nm chromatogram indicates the expected product is approximately 66% pure.

post-data acquisition processing with associated graphical representations of data to facilitate analysis. For libraries generated in microtiter plate format, the results of each individual well may be color-coded to reflect relative degrees of purity.

More often, as described earlier, compound purity is reported taking into account the purities determined from the UV, ELSD, and CLND detectors. In some instances, purity assessment has been made based on the intensity of the expected ion in the mass spectrum relative to the sum of the intensities of all ions in the spectrum. This method, however, is only a very crude estimate of purity, because ionization efficiencies for compounds can vary widely within and between classes of compounds. Though LC/MS (with UV and/or ELSD detection) has been adopted as the method of choice for assessing the quality and quantity of material prepared by parallel synthesis techniques, a decision still needs to be made by each respective organization as to what constitutes acceptable quality before submitting a sample for biological testing.

Some groups have evaluated ultra-fast chromatography separations (so-called ballistic, pseudo-chromatography) in order to provide a snapshot of the sample purity [42, 43]. The major drawback to the ballistic chromatography technique is that column resolution is reduced when operating at these suboptimal linear velocities. Also, the pseudo-chromatography approach is best suited to applications where purity assessment is secondary to rapid compound profiling.

11.7 PURIFICATION TECHNOLOGIES FOR DRUG DISCOVERY

Historically, it was believed that solid-phase synthesis protocols eliminate the need for purification because excess reagents are removed readily by extensive washing. Unfortunately, even for solid phase peptide synthesis, final products, acid-cleaved from the resin are found to be far from pure. Furthermore, parallel solution-phase synthesis has found greater popularity, because it is readily automated and extends the "portfolio" of reactions available to the chemist for high-throughput parallel synthesis. The limitations with solid-phase synthesis and the movement toward parallel solution phase synthesis are forcing numerous groups to evaluate and implement a variety of purification strategies.

A prevailing assumption is that if the chemistry is sufficiently high-yielding during the process development phase of synthesis, then it is reasonable to expect comparably high yields during the production phase of synthesis. In process development, a subset of the total library to be synthesized is rigorously optimized to maximize reaction yield. During production, it is assumed that the vast majority of members of the library will behave similarly and that the desired product will be the major component in the well. The reality is that far too often, the biological activity cannot be tracked to a single component or, in some instances, to the expected product in the well. Groups attempting to elucidate the active component(s) of the well have expended significant effort, only to find that the activity does not correlate with a single component within the sample. Consequently, more and more groups have embraced the value of "quality in, quality out" and are now applying the same analytical rigor to parallel synthesis chemical products as they have for more classical medicinal chemistry synthesis. These activities have enhanced the quality of structure–activity relationships (SAR) and structure–inactivity relationships (SIR) that can be derived from the assaying of these compounds for biological activity.

Numerous techniques are available to the organic chemist to support library purification. Zhao et al. [44] published an extensive review on compound library purification strategies, including HPLC, liquid–liquid extraction, solid-supported liquid–liquid extraction, solid-phase extraction, ion-exchange chromatography, and countercurrent chromatography, among others. This review focuses exclusively on HPLC- and HPLC/MS-based purification methods.

11.7.1   UV-Directed Purification

Both activity and inactivity data are being used increasingly to generate SAR and direct subsequent synthetic efforts. Consequently, organizations have recognized the importance of confirming the purity of compounds prior to screening, and not only those compounds for which activity is observed. In order to minimize false positives and false negatives, it is advantageous to assay only high-quality compounds. Therefore, great effort has been devoted to the development of automated purification technology designed to keep pace with the output of high-throughput combinatorial/parallel synthesis.

Automated methods are now available to the chemist to perform high-throughput purification. Although HPLC has long been a method available to the chemist for product purification, only recently have these systems been designed for unattended and high-throughput operation. Weller et al. [45] were one of the first groups to demonstrate "walk-up" high-throughput purification of parallel synthesis libraries based on HPLC and UV detection. An open-architecture software interface enabled chemists to select the appropriate separation method from a pull-down menu and initiate an unattended automated reversed-phase UV-based fraction collection. Fractionation was achieved using a predetermined UV threshold. Multiple fraction collectors were daisy-chained in order to provide a sufficient footprint for fraction collection. Since the early work of Weller et al., a number of commercial systems have been introduced for walk-up preparative LC/UV purification (including Gilson, Hitachi, and Shimadzu, to name a few).

One of the challenges associated with UV-based purification systems is that multiple fractions are collected for every sample injected. Although user-defined adjustable triggering parameters (e.g., UV thresholds for initiating and terminating fraction collection) can be used to reduce the total number of fractions, all, to some extent, will contain impurities. The exact number of chromatographic peaks for a given sample will be hard to predict, and therefore the footprint for fraction collection will be difficult to predict. Experience has shown that it is not uncommon for 5–10 fractions to be collected per injection. When purifying only a small number of samples (<10), it is neither particularly challenging to collect the fractions nor challenging to perform post-purification analysis so as to identify the fraction(s) containing the desired product. However, when attempting to purify compound libraries (e.g., 96-well plates of samples), the number of fractions and the time it takes to identify the relevant fraction(s) fast becomes a bottleneck. Schultz et al. [46] addressed the fraction collection issue and streamlined post-fraction collection processing (including evaporation, re-constitution and post-purification analysis) by collecting fractions directly into 48-well microtiter plates. Their method was particularly well-suited to semipreparative purifications (using smaller-inner-diameter columns to support low milligram quantities).

In order to gain further efficiencies into UV-based purification of compound libraries, numerous groups have developed automated high throughput UV-based purification systems coupled with on-line mass spectrometric detection. Kibbey [47] was one of the first scientists to implement a fully automated preparative LC/MS system for combinatorial library purification. His approach was to perform a scouting analytical run prior to purification so as to optimize chromatographic method and fraction collection parameters. Fraction location and molecular weight information were captured through a custom LIMS system. The added mass spectrometric information greatly facilitated deconvoluting of collected fractions and streamlined their purification process. Hochlowski [48] describes a service-based purification factory incorporating UV and ELSD detection coupled with mass spectrometry that supports purification of over 200 compounds per day.

More recently, intelligent UV-based systems for preparative scale purification of combinatorial libraries have been introduced, utilizing knowledge of retention time of the expected product based on a pre-analytical evaluation followed by preparative HPLC with UV-based fractionation using a narrow time collection window. Yan et al. [49] coined the "accelerated retention window" method as a tool for improving high-throughput purification efficiency. In their method, a high-throughput parallel LC/MS analysis is performed prior to preparative purification to confirm that the expected product is indeed contained within the well and to identify the approximate retention time of the expected synthetic product. Only those compounds found to be $\geq 10\%$ pure based on the analytical run are candidates for final product purification. Furthermore, the information from the high-throughput parallel analysis was uploaded to a stand-alone preparative LC system for final product purification. Fraction collection was initiated using an accelerated retention time window method so as to accelerate the preparative HPLC analysis. Additional refinements of UV-based purification strategies have been made recently, allowing for further simplification of the fraction collection and post-purification analysis step. In one embodiment, Karancsi et al. [50] implemented a "main component" fraction collection method based on UV-triggering that supports the "holy grail" of high-throughput purification, that being the one compound/one fraction concept [50].

11.7.2 Mass-Directed Preparative Purification

The technique of preparative LC/MS, introduced in the late 1990s was the first technique to greatly simplify the purification process. For the first time, preparative LC/MS (PrepLCMS) methods allowed the concept of one compound/one fraction to be realized [51–55]. In the Prep LC-MS mode, the mass

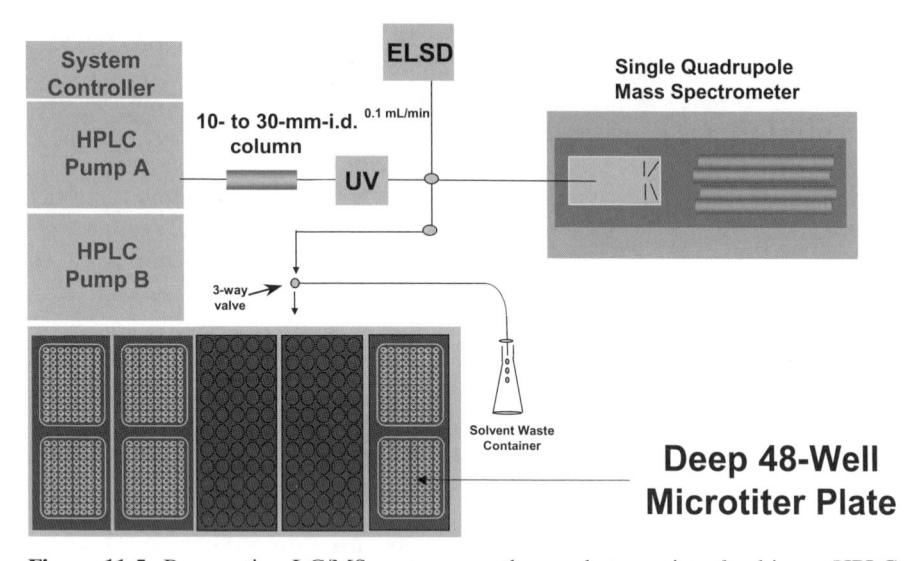

Figure 11-5. Preparative LC/MS systems on the market consist of a binary HPLC system, a combined autosampler/fraction collector (footprint of a Gilson 215 inject/collect liquid handler shown in the figure), and a single quadrupole mass spectrometer.

spectrometer is used in this mode as a highly selective detector for mass-directed fractionation and isolation. This technique provides a means for reducing dramatically the number of HPLC fractions collected per sample and virtually eliminates the need for post-purification analysis to determine the mass of the UV-fractionated compound. Preparative LC/MS is now widely incorporated in the pharmaceutical industry. Systems for preparative LC/MS are configured in numerous ways and are operated in numerous ways, including an expert user mode, walk-up or open access mode, or a project team setting, supporting small teams of chemists working on similar chemistries. All components of the system are under computer control and are hence truly automated. Components of these systems are nearly identical to stand-alone HPLC systems with the addition of a flow splitter device to divert a small portion of the column flow to the mass spectrometer for on-line detection and fraction collector triggering. Typical systems are configured in an automated analytical/preparative mode of operation. In this configuration, the chemist is able to select between a variety of column sizes for either analytical, semi-preparative, or preparative separations. The HPLC, switching valves, mass spectrometer, and fraction collector are under complete computer control, as shown in Figure 11-5. In some instances, a solvent pump is added to deliver a methanol make-up flow to the mass spectrometer. The flow splitter and extra solvent pump serve the primary purpose of reducing the potential for overloading of sample into the ion source. An advantage of the flow splitter and make-up pump is that it reduces the trifluoroacetic acid (ion pairing) in the

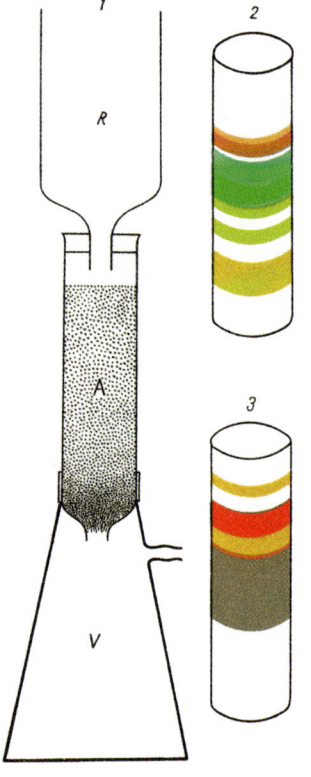

Figure 1-2. Tswet's original drawings of his experiments. From M. S. Tswet, "Chromophils in the plant and animal world" [10].

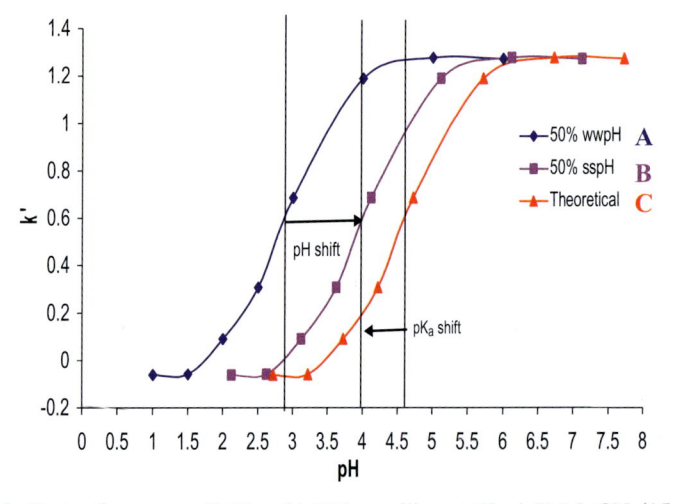

Figure 4-31. Retention versus $_w^w$pH and $_s^s$pH for aniline at 50 v/v% MeCN. (15mM phosphate buffer adjusted with phosphoric acid).

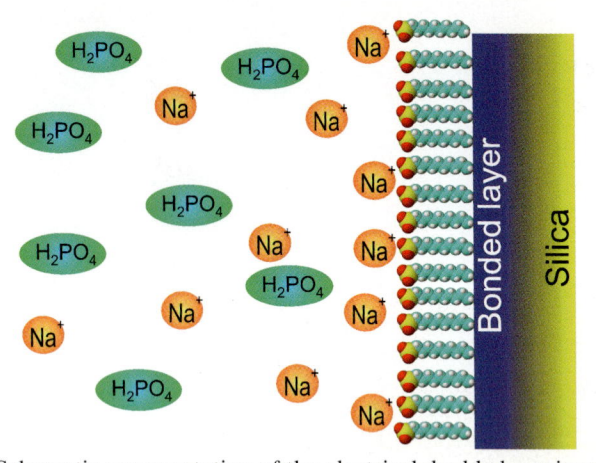

Figure 4-40. Schematic representation of the electrical double layer in reversed-phase ion-pair chromatography.

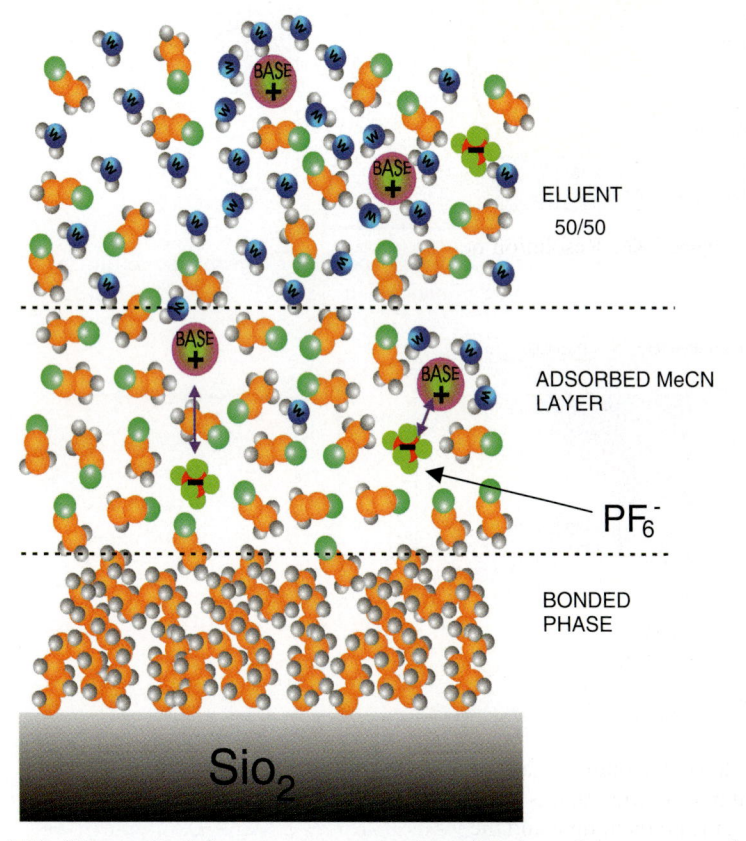

Figure 4-54. Schematic of the retention mechanism of basic analyte on reversed-phase material in water/acetonitrile eluent in the presence of liophilic ions (PF_6^-).

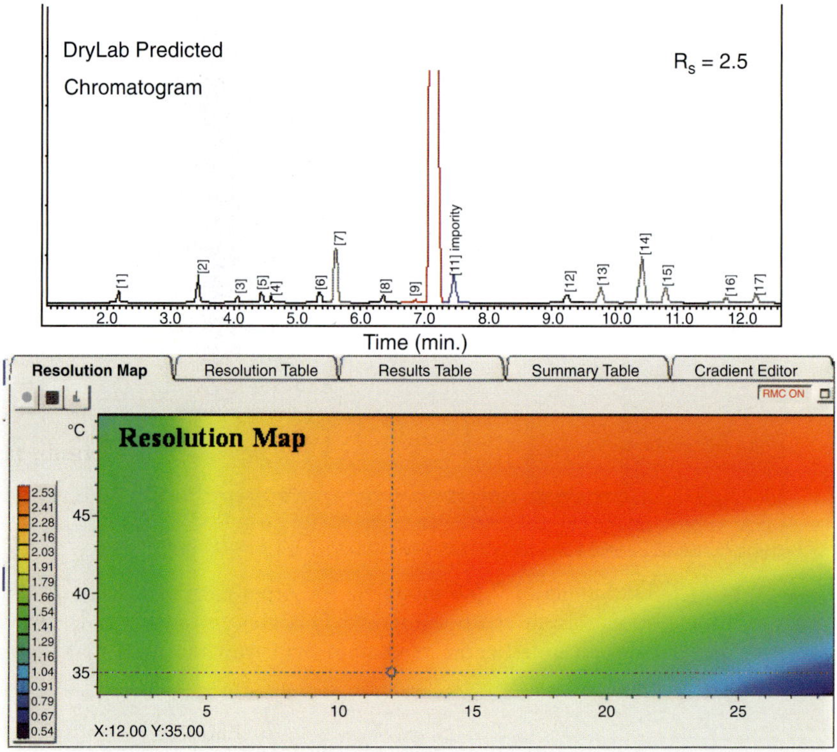

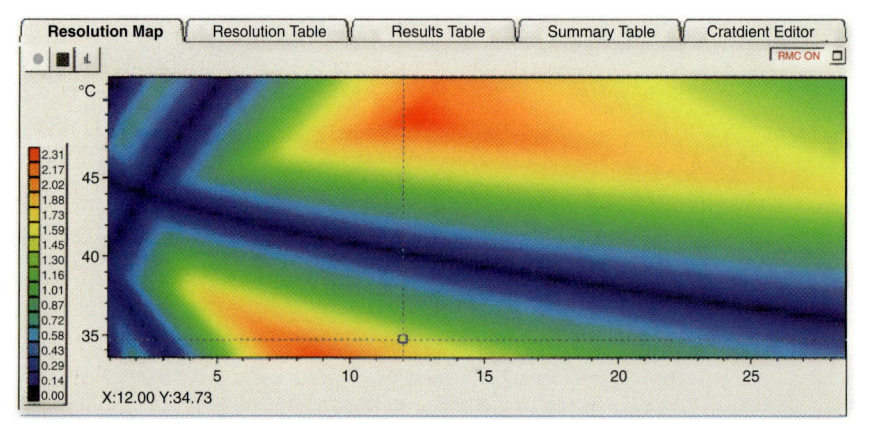

Figure 8-46. Resolution between active (compound M) and Impurity 11.

Figure 8-47. Resolution map for all critical pairs. Color scale on the left indicates the minimum resolution that is predicted for a particular color in the resolution map. The x axis is the gradient time and the y axis is the temperature. The crosshair can be moved to obtained the predicted conditions for optimal resolution of all critical pairs.

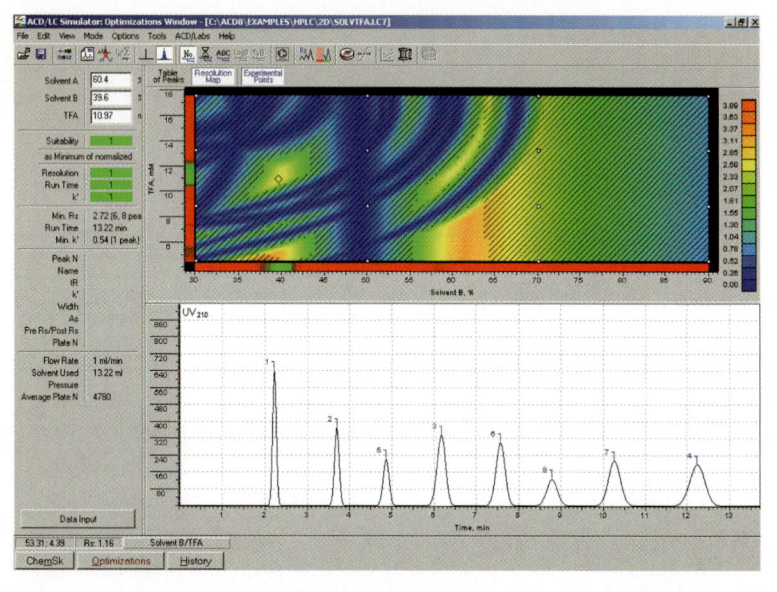

Figure 10-2. ACD/LC Simulator™ 9.0 modeling the separation of a series of compounds as a function of solvent composition and TFA concentration (mM). Experiments are shown as white dots on the resolution map with the predicted optimal method shown in yellow.

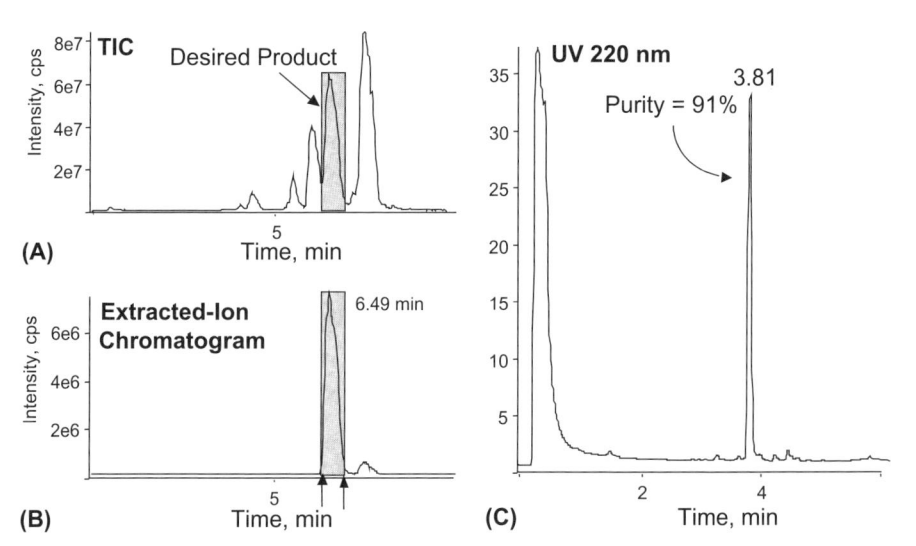

Figure 11-6. Fifty milligrams of a crude reaction product was solubilized in 1 mL of 50/50 MeOH/DMSO and injected onto a 20-mm × 50-mm-i.d. reversed-phase column. Separation was achieved using a gradient of 10–90% ACN in 7 minutes. (A) TIC chromatogram shows five components well-separated. (B) Extracted ion chromatogram (XIC) for expected product shows a single, prominent peak at 6.49 minutes. Fraction collection was initiated and terminated, as indicated by the arrows directly below the XIC peak. (C) Post-purification analysis of the isolated component shows that the compound was purified to approximately 90% level.

ion source, which can affect the sensitivity of detection for acidic library components.

An example of a mass-guided fractionation of a combinatorial library is shown in Figure 11-6. In this example the crude reaction product is only about 30% pure. The component of interest shows a prominent single chromatographic peak when monitoring specifically for its corresponding mass. Post-purification analysis of this singly isolated fraction (based on mass-directed fractionation) demonstrates that the compound of interest was purified to greater than 90%. Had a UV-based fractionation system been used in this particular example, at least five individual fractions would have been isolated. Extending this to a 96-component library synthesized in microtiter plate format (and assuming this compound was representative of the quality of the members of the library), a UV-based approach would have led to approximately 400–500 fractions requiring reanalysis to pinpoint the desired product. This not only would be a time-consuming reanalysis process but also would require significant time to transfer the appropriate fractions to a screening plate for biological assessment.

Debate exists as to whether UV-based or mass-based fraction collection is the more appropriate tool for purifying compound libraries. The choice of technique probably should be governed by the relative importance of any given sample and the purification throughput requirements at any moment in time. As a simple rule of thumb, during the earlier stages of the discovery process, where larger numbers and smaller quantities (<50 mg) of compound are needed for biological screening and early ADME/PK screening, a mass-spectrometry-based fraction collection system probably makes the most sense, since the total number of isolated fractions can be reduced to a minimum. At later stages of a discovery program (e.g., during late stage lead optimization, where a smaller number and larger (≥100 mg) quantities of compounds are being evaluated for *in vivo* efficacy), UV-based methods might take priority. Independent of the debate, it is widely agreed that mass spectrometry serves as a highly sensitive and selective detector for analysis and purification of compound libraries. Mass-triggered fraction collection enables compound libraries to be purified based solely on their expected product mass. Libraries can be purified maintaining a "one compound-one fraction" model, which facilitates sample tracking, registration, and biological testing and enables screening results to be readily correlated with synthetic structure.

11.8 HIGHER-THROUGHPUT PURIFICATION STRATEGIES

11.8.1 Fluorous Split-Mix Library Synthesis and Preparative LC/MS De-Mixing

The interest in combinatorial chemistry (split-mix technology) as a means for generating large compound collections plummeted over the past several years, due primarily for the reasons of poor quality control over synthesis and complex decoding strategies. Conventional split-mix technology was replaced by a number of parallel synthesis strategies that did not require the de-mixing or de-coding step. Recently, a relatively new technique called fluorous synthesis, developed by Luo et al. [56], offers an interesting twist on the conventional split-mix combinatorial chemistry approach. In fluorous synthesis, a mixture of substrates is paired with a series of perfluoroalkyl (Rf) phase tags and is taken through multiple synthetic reaction steps culminating in a mixture of tagged products. Demixing is achieved by tag-controlled fluorous HPLC preparative separation, the fractions are stripped of solvent, and the tag is removed to yield individually purified and structurally distinct products. A recent publication by Zhang et al. [57] describes both the synthetic and analytical HPLC strategy for a 420-component fluorous library. Members of Syrrx' analytical team, in collaboration with Zhang and co-workers, recently developed a higher throughput mass-directed purification method to support purification of their fluorous tagged libraries [58]. Figure 11-7 shows the UV and total ion current chromatogram (TIC) for a pool of five compounds from the 420-component library. Separation of the five-component mixture was achieved in less than 6 minutes. Fraction collection was initiated when the

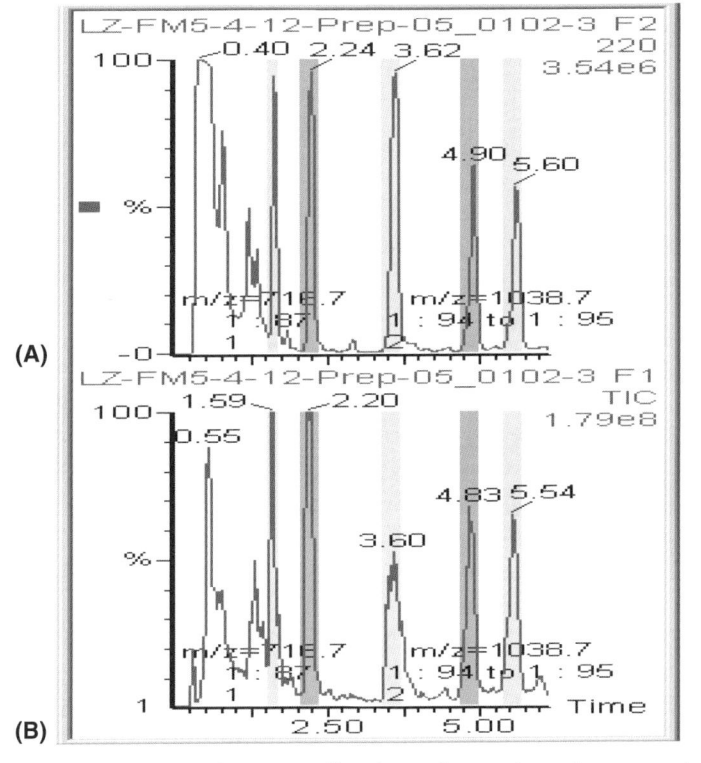

Figure 11-7. A five-component fluorous split-mix crude reaction mixture was injected onto a 20-mm × 50-mm-i.d. reversed-phase column. (A) UV chromatogram and (B) Total ion current chromatogram. Compounds were purified using mass-directed fraction collection (peaks highlighted).

expected $[M + H]^+$ ion for each of the tagged products exceeded the pre-set ion intensity threshold, and fraction collection was terminated when the ion intensity for the expected product(s) dropped below a second pre-set ion intensity threshold value. Combining split-mix fluorous synthesis with high-speed chromatography provides a means for rapidly generating large numbers and large quantities of highly purified druggable molecules.

11.8.2 Parallel Analysis and Parallel Purification

The synthetic throughput achievable by the medicinal chemist (having adopted parallel synthesis strategies) has rendered analysis and purification one of the key (and possibly rate-limiting) steps in the discovery process. Although advances in sample analysis throughput have been clearly demonstrated, there is a limit as to how fast a separation and analysis can be achieved

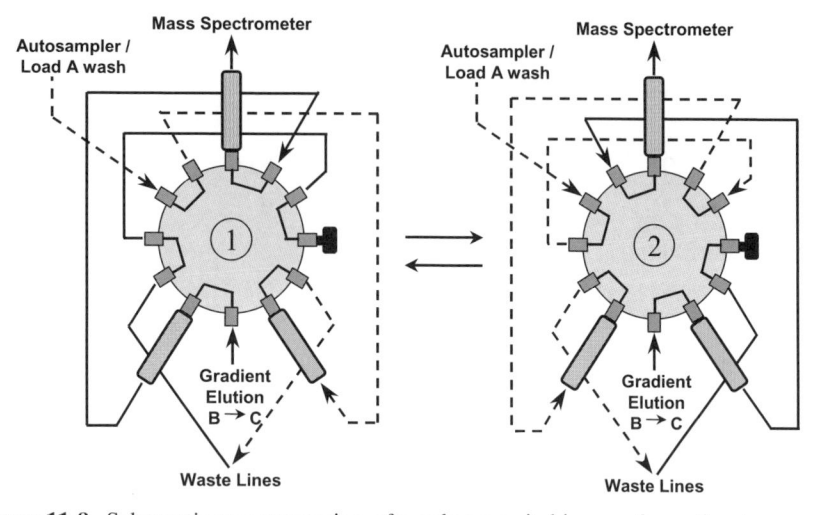

Figure 11-8. Schematic representation of a column switching configuration to support analysis from one column while the second column is equilibrating.

while maintaining good separation efficiency and quality analysis. Two techniques that have been developed to increase throughput without compromising column chromatography are (a) rapid column switching and regeneration systems for enhanced-throughput serial-based analysis and (b) parallel chromatography methods. A simple and elegant modification of the LC/MS method is to incorporate a set of switching valves and a third pump to reduce cycle time between injections, as shown in Figure 11-8. While one column is being used to perform the LC/MS analysis, the other column is being regenerated. An alternative use of 10-port switching valves is to allow for rapid serial sampling between columns. This technique works well for samples that are amenable to either isocratic or step elution. While one sample is being loaded onto one column, the contents of the other column are eluted into the ion source.

In order to increase sample throughput while maintaining high-quality analytical data, groups have begun to perform separations in parallel [59–63]. Numerous groups have independently developed parallel sample introduction techniques, although the MUX ion source from Micromass/Waters is the only one commercially available. By performing analyses in parallel, chromatographic integrity can be maintained while effectively addressing sample throughput. Di Biasi et al. [59] and Wang et al. [60] presented novel ion source interfaces enabling four to eight samples to be processed in parallel, thereby increasing the sample analysis throughput dramatically over conventional, serial-based LC/MS analyses. Commercially available parallel spray interfaces consist of a multiple spray head assembly and a blocking device (e.g., rotating plate), enabling individual sprayers to be sampled at specific and defined time intervals. Although the multiple sprays are delivered to the mass spectrometer simultaneously, they are sampled in a time-dependent manner.

Performing parallel analysis of compound libraries offers many potential advantages over serial-based LC/MS analytical methods, the most obvious of which is dramatically increased compound analysis throughput. Using single-channel HPLC-based purification systems, routine sample throughput of up to 192 reaction mixtures per 24-hour day was reported [64]. With parallel HPLC systems, it has been reported that the theoretical throughput increases to 384 samples per day for a two-channel system and to 768 samples per day for a four-channel system.

Parallel LC/MS analysis is readily achieved incorporating one or more of the following: a parallel autosampling device, a set of HPLC pumps configured to divert flow through an array of HPLC columns, a parallel UV detector, an array of analog detectors (e.g., ELSD), and a mass spectrometer configured either with or without an indexing device (i.e., MUX ion source) to facilitate independent sampling of individual sprayers. Some groups have found it possible to use a binary HPLC pumping system and split the flow equally between the array of columns using a simple valco tee. A recent report by Xu et al. [65] showed that a system configured for 8-column parallel analysis using this simple flow splitting technique was capable of analyzing well over 1000 compounds without failure.

A schematic of a commercially available system that was configured in our laboratory for parallel analysis and purification is shown in Figure 11-9. This four-channel parallel LC/MS purification system consists of a binary HPLC system, an autosampler configured with four injection, a multichannel UV detector, a quadrupole mass spectrometer equipped with an MUX ion source which monitors four flow streams simultaneously, and four independently

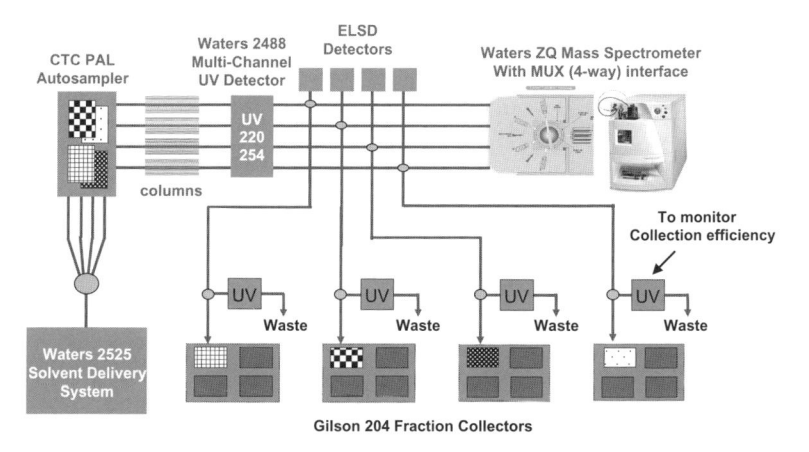

Figure 11-9. Four-channel LC/MS system to support higher throughput analysis and purification. A Waters 2525 solvent delivery system is used to deliver flow to an array of C18 analytical or preparative columns.

controlled fraction collectors. The binary HPLC pump is operated at a total flow rate of 40 mL/min, and the flow is split equivalently into four 10 mL/min flow streams by aid of a valco manifold tee.

Each of the four streams passes through its own injection valve on the parallel autosampler and into one of an array of four YMC-Pack, 5-μm, 10-mm-i.d. × 100-μm ODS-A HPLC columns. The effluent from each column passes through the multichannel UV detector where the separate chromatographic flow streams are monitored at two UV wavelengths (λ_{220} and λ_{254}). The column effluent is diverted through an in-house modified parallel flow splitter unit at the outlet of the parallel UV detector. This parallel splitter diverts the majority of flow (99.5%) from each of the column effluents toward dedicated Gilson 204 fraction collectors (FC). The remainder of flow (i.e., 50 μL/min or 0.5% of each of the total column flows) is merged with one of the four 0.18-mL/min methanol streams and detected in the MUX multiplexed ion source. Xu et al. [65] recently described recent advances in this laboratory for MUX-based parallel preparative LC/MS purification of compound libraries. In the work described, each of the flow streams is sampled in a time-dependent manner, and each of the individual sprayers and their associated ion signals are linked to unique fraction collectors to facilitate independent parallel purification.

A potential limitation to this approach is duty cycle. The duty cycle for mechanical blocking devices is approximately 0.6 sec (50-msec rotation time between each position of the spray assembly and 100-msec dwell/acquisition time at each of the spray positions). For preparative LC/MS analysis, peak widths of 5–10 sec for analytical runs and 10–30 sec for preparative analyses are routinely observed. Thus, this duty cycle should have only limited impact on compound analysis at present. Recently, Cole [66] showed a variation on the multispray technique, incorporating a fast switching valve to "toggle" between chromatographic inlets. In this method, samples are introduced into a high-speed switching valve located outside of the ion source. Samples are sequentially transferred through the various valve positions, providing a temporally spaced flow streams into the ion source. Discussion of duty cycle and cross-contamination between channels were not made in this preliminary report.

In order to make parallel mass spectrometry more mainstream, more effort on the chromatographic inlet side will be required. Many groups have achieved flow splitting by incorporating a simple manifold splitting tee. This approach requires that each column be maintained at nearly identical back-pressures in order to maintain constant and identical flow through each of the columns in the array. To achieve constant and identical flow through the array of parallel columns requires not only excellent quality control over column selection, but careful plumbing as well. Furthermore, the variation in back-pressure becomes even more predominant in a gradient run.

Concomitant to the advances being made in mass spectrometric detection of parallel flow streams are the advances being made in parallel sampling and parallel chromatographic separations. Coffey et al. pioneered the Biotage Parallex™ HPLC system, a fully automated, high-throughput organic chemistry parallel purification system with parallel UV detection. In collaboration with scientists from GlaxoSmithKline, Coffey described an elegant parallel chromatography system couple with on-line mass spectrometric characterization to support combinatorial library purification [67]. The Biotage Parallex™ HPLC system catalyzed the development of a number of parallel chromatography workstations now available for both intermediate and final product purification. Parallel HPLC systems now available commercially consist of multiple HPLC pumps, multiprobe autosamplers, parallel UV detection, and parallel mass spectrometer ion source interfaces. To date, there are three commercially available systems to support parallel analysis and purification. They are instruments from Sepiatec, Inc. (Germany), Nanostream, Inc. (USA), and Eksigent (USA). The instrument vendors are continually being challenged to introduce more cost-effective and more compact systems. In particular, Gumm and co-workers introduced the concept of flow control for parallel column chromatography operation [68]. Rather than controlling flow to each column of the array by individual pumps, their technique involves precision control from a single-source HPLC binary pumping system to an array of columns, using a series of pressure feedback sensors, as shown in Figure 11-10.

The flow control unit works with both analytical and preparative columns and distributes flow evenly to the columns, independent of individual column

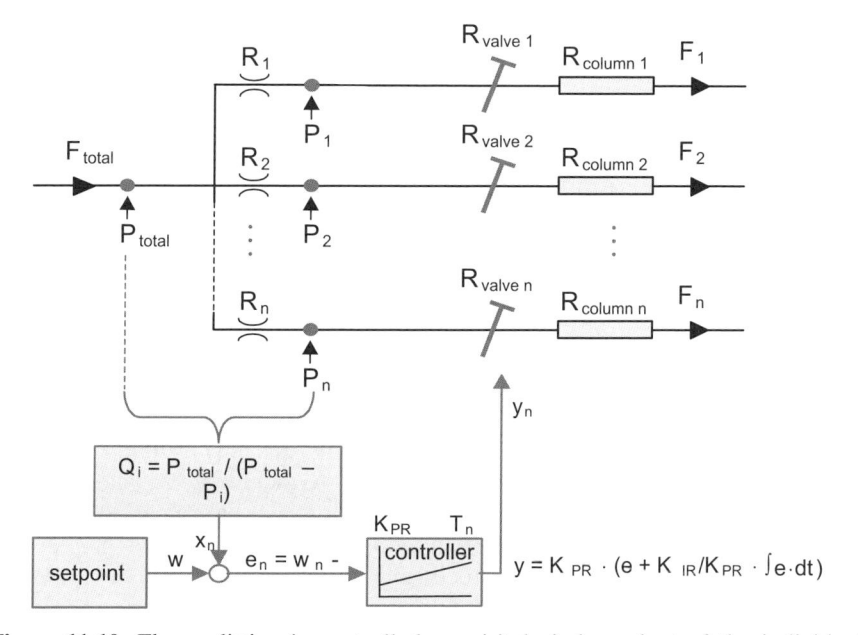

Figure 11-10. Flow splitting is controlled exquisitely independent of the individual column backpressures, as shown in the scheme.

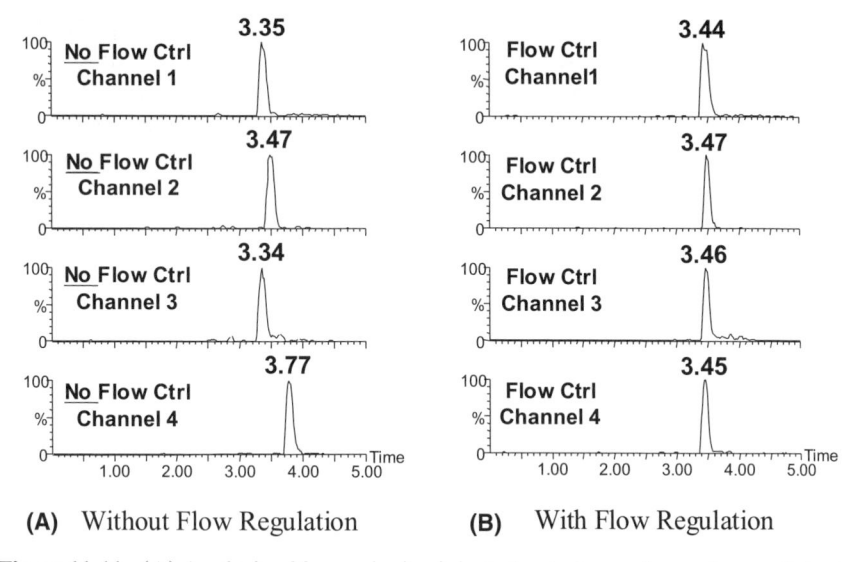

(A)   Without Flow Regulation   **(B)**   With Flow Regulation

Figure 11-11. (A) Analysis of fmoc-alanine injected onto four columns in parallel using a simple manifold tee to distribute the flow to the columns. (B) Repeat analysis of fmoc-alanine onto four columns using the flow controller to distribute the flow evenly to the array of columns.

backpressure. An example of the power of flow control is shown in Figure 11-11. In this example, the same compound was injected onto each of a parallel array of four analytical columns (all from the same lot of the same vendor). The retention time difference between columns was as great as 15%. Flow control, as described by Gumm and co-workers, showed that the retention time difference between columns could be reduced to as little as 2% without compromising peak shape or significantly affecting retention time.

11.8.3   Streamlining the Purification Process

The analytical community has focused primarily on developing higher-throughput analysis and purification tools that keep pace with parallel synthesis. The problem, however, is not just to provide fast methods for characterizing and purifying compound libraries. It is a process problem requiring integration from synthesis to compound screening, including automated analysis, post-data-acquisition processing, purification, post-purification analysis, and reformatting for biological screening and compound archiving. A seamless integration of these steps is necessary to achieve an efficient discovery process.

True high-throughput purification is only possible if the entire purification process is managed effectively. Between compound synthesis and compound registration are several important steps in addition to purification, including automated analysis, post-data-acquisition processing, purification, post-purification analysis, and reformatting for biological screening and compound archiving. A seamless integration of these steps is necessary to achieve an efficient discovery process. Following synthesis, a decision needs to be made as to what compounds should be (a) submitted directly for registration, (b) purified prior to submission, or (c) discarded outright. This is accomplished by performing a high-throughput analysis of the reaction mixture, most typically by LC/MS with UV and/or evaporative light scattering detection, as a method for assessing compound purity. Automated data processing facilitates the selection of compounds requiring purification prior to compound submission. Following library purification, a decision needs to be made as to what fractions should be submitted for post-purification LC/MS analysis to verify purity and assess sample quantity. Again, sophisticated sample tracking and automated sample list generation greatly facilitates the next step in the process. Ultimately, the post-purification LC/MS data are collated (electronically), and those compounds satisfying both the purity and quantity criteria set by the project team (or organization) are registered and plated for biological screening. Isbell et al. [69] reported on the time required for the entire purification process, beginning with the receipt of crude samples for purification, and then followed by purification, post-purification analysis, reformatting, dissolution, and registration of the "acceptably pure" fractions. Two significant benefits were realized from this technology: (1) the ability to collect only fractions likely to contain the product of interest and (2) the ability to more easily track fractions due to the optional plate mapping feature of the system.

In their work, a custom parallel LC/MS purification system was used for all analytical and preparative analyses. The robustness of the parallel LC/MS system was assessed purifying over 7800 reaction mixtures from several libraries over a period of seven months. When using technologies designed to facilitate fraction tracking and automated liquid handling, the time required to completely process samples, from receipt of sample plates for purification to submitting properly formatted plates of pure compounds for registration, required significantly more time than the LC/MS purification step itself. Overlooking these other important steps in the overall purification process may seriously underestimate the amount of time to purify a library and may have serious implications on plans focused on a large-scale purification effort.

11.9 ADME APPLICATIONS

The continuing quest for novel and safer drugs has led to the introduction of myriad new technologies within the pharmaceutical industry. Notably, advances in genomics, high-throughput screening, combinatorial chemistry, parallel synthesis, automation, and miniaturization have enabled large numbers of potent (active) and selective compounds to be identified at early stages of drug discovery. However, the fact that a compound is active and selective does not necessarily make it an attractive drug development candidate. To convert these "actives" into qualified clinical candidates has proved challenging. It has been reported that a significant number of compounds nominated for clinical development fail due to poor pharmacokinetics and toxicological properties (63% of all preclinical compounds) [70].

In order to identify chemotypes and lead compounds that have good pharmacokinetic and safety profiles (e.g., no hERG liability), it has been recognized that studies that assess absorption, distribution, metabolism, and elimination (ADME) should be initiated as early as possible in the discovery process [71–73]. The shift from late-stage optimization of ADME properties to a strategy of identifying potential liabilities early in the discovery process has taken hold within the pharmaceutical community, adding the dimension of structure–ADME relationships in parallel to structure–activity relationships as an integral part of the iterative drug discovery process.

Because of the large number of hits that are now routinely identified from screening compound collections and gene family compound libraries, the industry has recognized the need for high-throughput ADME assays. Fortunately, many ADME assays can be run in a high-throughput fashion, due principally to the widespread incorporation of liquid chromatography/mass spectrometry (LC-MS) and liquid chromatography/tandem mass spectrometry (LC-MS/MS) [74, 75]. LC/MS and LC/MS/MS have become the preferred techniques for ADME analyses due principally to enhanced sensitivity, selectivity, and ease of automation relative to traditional analytical methods. The selectivity advantages of LC/MS have made possible the ability to analyze endogenous and nonfluorescent probe substrates in cytochrome inhibition assays [76], enabled rapid permeability assessment (e.g., Caco-2 assay) [77], provided faster methods for assessing lipophilicity and solubility of drug leads, and provided much more facile assessment of liver metabolism [78, 79] for which many examples are highlighted below.

To achieve high throughput, it is important that these assays be brought forward into the discovery process as early as possible. One approach is to initiate ADME assays at the time of biological screening. Once compounds are registered, they are generally plated and arrayed in 96-well microtiter plates at a concentration of 10 mM in DMSO. Many *in vitro* ADME screens may be performed in micro titer plate format, and it is at the time of biological screening that a number of daughter plates may also be generated for high-throughput ADME, as shown in Figure 11-12. In addition to the metabolic stability assays, plasma protein binding can be performed in microtiter plate format using both the ultrafiltration method [80] and equilibrium dialysis method [81]. In addition, both solubility and log P screens have been performed in microtiter plate format [82, 83].

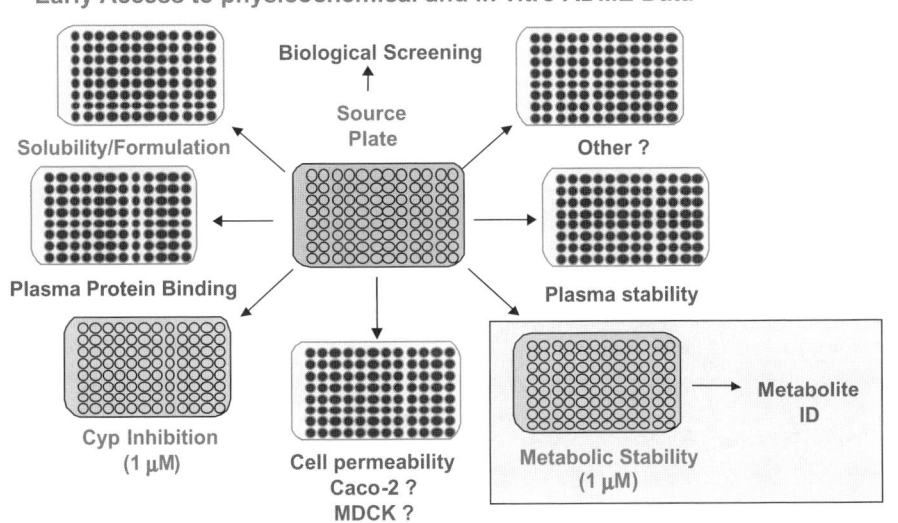

Figure 11-12. By coordinating plating for ADME analyses at the same time as biological screening, ADME analyses are streamlined.

11.10 FAST SERIAL ADME ANALYSES INCORPORATING LC-MS AND LC-MS/MS

Recently, analysis throughput has been improved significantly by shortening the HPLC run time. Samples can be analyzed one at a time or as mixtures in as little as 30 per sample by applying fast gradients compatible with mass spectrometric detection to assess ADME properties [84, 85]. These fast analyses are achieved using short chromatographic columns (e.g., 2.1 × 20 mm; 3- or 5-µm particle size). Generic gradients are typically employed [e.g., 5–95% organic in 1–2 min; mobile phase containing trifluoroacetic acid (0.035–0.05%), formic acid (0.1%), or ammonium acetate (10 mM)]. Flow rates are typically in the range of 1–5 mL/min.

11.10.1 Automated Data Processing Is Instrumental to Achieving High-Throughput ADME

Although great strides have been made in reducing chromatographic analysis times by the introduction of short, ballistic columns, a bottleneck to providing rapid turnaround of *in vitro* ADME information to project teams is data processing and reporting. The generation of analytical data using automated instrumentation has produced a bottleneck since data can be generated faster than it can be analyzed. The automated post-data acquisition analysis strategy

Liver Microsomal Stability Results

D0000138 Set1 RLM Assay 2002/09/26

Protocol: Microsomal Assay_020923

Note:

Color Code: >80% 40-80% <40%

Compound	%Remaining				Deviation				T1/2	Clint'
	0	5	15	30	0	5	15	30		
A	100	102.1	98.9	93.0	3.1	4.9	8.2	5.3	577	2.16
B	100	81.9	57.8	35.5	1.8	3.2	0.8	1.2	47	26.77
C	100	98.7	95.4	87.2	4.6	4.2	4.2	6.5	346	3.60
D	100	84.8	55.2	40.0	3.6	3.9	3.4	2.2	51	24.23
E	100	101.4	87.8	71.5	6.1	7.2	2.7	4.8	133	9.37
F	100	97.1	81.0	63.7	4.5	5.7	1.7	4.9	102	12.23
Buspirone	100	58.9	16.5	3.2	3.5	2.3	0.4	0.2	14	90.35
Proprano	100	18.2	0.3	0.1	2.9	1.8	0.1	0.1	7	177.97

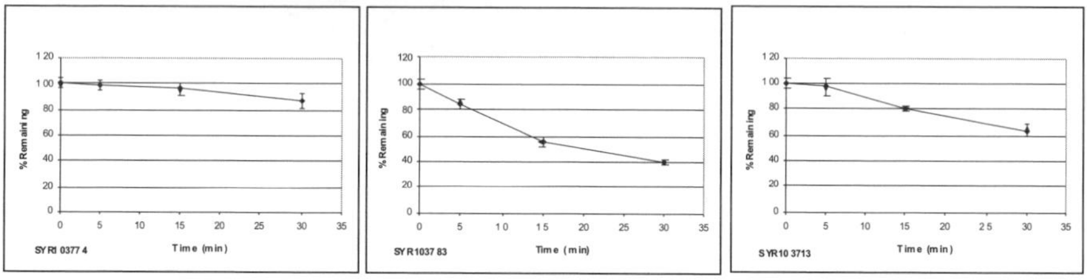

Figure 11-13. Time-course human liver microsomal incubation profiles for a number of positive and negative controls, as well as project compounds. Metabolic stability profiles are represented both in tabular format and graphically above.

is to extract the most appropriate information required for decision-making in as streamlined a manner as possible.

Time-course assessment of metabolic stability (e.g., four time-points and analyses in triplicate) generates 1152 samples for every plate of compounds submitted. Manual processing of so many samples would clearly render the data processing and data reporting rate-limiting. To address this, numerous groups have combined the power of vendor software programs that automate peak area determinations with visual basic programming to provide methods for data processing and data reporting [86–88]. Shown in Figure 11-13 is a partial summary report of a plate of eight reference compounds and 88 test compounds received from a drug discovery project. The percent remaining values of the parent compounds are color-coded for easy visualization: green, >80%; orange, 80–40%; red, <40%. The project chemist receives both (a) the summary report which helps "bin" the compounds into distinct classes of microsomal stability and (b) the time-course stability plots for all compounds submitted for HT microsomal stability analysis. This information helps the chemists prioritize compounds for further consideration as potential drug candidates.

11.10.2 Enhancing Throughput by Incorporating Pooling Strategies

Another approach is to profile multiple compounds simultaneously, known as cassette (or N-in-1) dosing. In essence, cassette dosing is a compound "pooling" strategy whereby compounds are profiled as mixtures so as to increase throughput, reduce the total number of samples to be analyzed, and hence reduce overall analysis times. Cassette dosing strategies have been used principally for rapid pharmacokinetic profiling of drug leads. Halm, Berman, and co-workers were the first to describe the application of cassette ("N-in-one") dosing to facilitate rapid pharmacokinetic screening [89]. Olah et al. [90] pushed the limits of the method to an $N = 22$ in a dog PK study, and Stevenson et al. [91] demonstrated the power of this approach for the *in vitro* cell permeability screening of compound libraries.

The one principal drawback with the cassette dosing strategy is that the risk for drug–drug interactions is exacerbated, which can lead to both false positives and false negatives [92]. Korfmacher effectively addressed this issue by implementing a variant of the cassette dosing technique, in essence a "cassette planning" strategy, and coined the cassette accelerated rapid rat screen (CARRS) technique as a means for increasing *in vivo* pharmacokinetic throughput [93]. In brief, this approach can be described as one in which drug candidates are dosed individually ($n = 2$ rats per compound) in batches of six compounds per set, and then samples are pooled across the two rats to provide a smaller number (six per compound) of test samples for analysis.

11.11 PARALLEL APPROACHES TO SPEEDING ADME ANALYSES

11.11.1 Nonindexed Parallel Mass Spectrometry

Parallel LC-MS methods involve injecting individual compounds onto multiple columns and detecting them simultaneously in a single mass spectrometer ion source. Xu et al. [94] presented a parallel *analytical* LC/MS for *in vitro* ADME analysis using a simple Valco manifold to split the flow from a binary HPLC system evenly between eight analytical columns. The generic high-throughput parallel LC-MS system, as shown in Figure 11-14, consists of a high-pressure binary solvent delivery pumping system, a multiple probe autosampler (generically, either a 8-channel Gilson or 4-channel Leap), a switching valve, and a single quadrupole mass spectrometer equipped with electrospray ionization (with or without MUX).

The system developed by Xu et al. was used for assessing the time-course metabolic stability (four time-points in triplicate) for hits identified from screening of lead generation libraries. For each compound, a total of 12 samples are generated (four time-points in triplicate) and a single plate of compounds therefore yields a total of 1152 samples requiring analysis. Single-column systems operated in the sequential sampling mode are capable of analyzing roughly one plate in a single day (assuming a 1-minute cycle time).

High Throughput Parallel LC/MS System

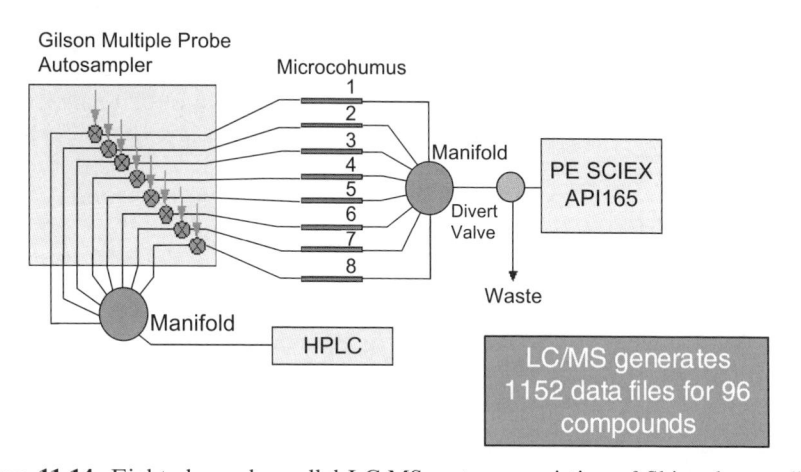

Figure 11-14. Eight-channel parallel LC-MS system consisting of Shimadzu gradient HPLC pumps, a Gilson 215 multiple probe autosampler, an AB/MDS-SCIEX API165 mass spectrometer equipped with a TurboIonSpray ion source, eight Peeke Scientific microbore columns (10-mm × 1-mm-i.d., 3-μm HQ-C18), and a Valco switching valve. The volume of the sample loops is 20 μL, and full loop injections were used for all experiments. Total mobile-phase flow rate is 2.0 mL/min (0.25 mL/min for each column). The eight-channel LC-MS system allows up to four plates of compounds (or 4 × 1152 samples) to be run in a single day.

On the other hand, a parallel array of eight columns enables up to eight plates to be analyzed in a single day on a single instrument (theoretical maximum throughput). In practice, the authors reported that their maximum throughput was four plates of compounds in a single day, approaching 5000 samples analyzed for metabolic stability in a single day, the highest sample throughput yet to be reported for the metabolic stability assay to date.

11.11.2   Indexed ("MUX") Parallel Mass Spectrometry

The commercially available multiplexed (MUX) electrospray interface, which introduces multiple LC flows directly into an "indexed" electrospray ion source, has also been applied successfully for high throughput ADME applications [95–97]. Yang et al. [98] identified the two main advantages of parallel LC-MS/MS using Micromass Ultima with MUX interface to be (a) parallel analysis and (b) four times the throughput relative to single-column systems. However, disadvantages were reported as (a) cross-talk between the sprayers (negligible at concentrations <100 ng/mL but as high as 0.08% at 1000 ng/mL), (b) sensitivity less than that of a single sprayer interface (about 3 × lower than the single sprayer interface), and (c) total cycle time longer than that of a single sprayer interface (hence not compatible with ultrafast chromatography). The MUX technique was validated for rabbit, rat, mouse, and dog plasma, and the authors concluded that the technique is well-suited for simultaneous method validations and early discovery studies.

11.12   AUTOMATED "INTELLIGENT" METABOLIC STABILITY AND METABOLITE ID

Until very recently, metabolic stability screening and metabolite identification (metabolite ID) have been sequential processes; that is, the metabolic stability assays are typically performed first to identify rapidly metabolized compounds and a follow-up metabolite ID study is performed next, typically at a 10-fold higher substrate concentration to ensure generation of sufficiently high-quality MS/MS information to support structure elucidation.

To perform detailed metabolism studies on candidate compounds has been both laborious and time-consuming. Tuning, method setup, assessing rates of metabolism, and metabolite ID typically have been distinct manual and independent processes requiring significant time by the investigator. In an effort to automate this process, prototype software was recently conceived and evaluated to automate metabolic stability assays by both Xu et al. [99] and Detlefsen et al. [100]. These software tools have been developed to provide full automation of the following: (a) on-line quantitation to determine the rate of parent loss as a function of incubation time, (b) "intelligent" selection (i.e., qualitative trigger) of compounds for detailed metabolite ID (based on percent loss of parent at a fixed time point), (c) selection of a suitable product ion for metabolite determination using precursor ion scanning, (d) creation of a custom optimized MS1 and precursor IDA, (e) analysis of both sample and control, and (f) metabolite data analysis by metabolite ID software.

A flow-scheme for intelligent metabolite ID is shown in Figure 11-15. Two sets of samples are generated. The first set contains stock solutions of the test compounds. The stock solutions are used to automatically optimize the MS and MS/MS parameters (i.e., determine the best precursor-product transition and the best precursor and neutral loss masses for metabolite ID analysis). The second set contains the incubated samples. Compounds are incubated with liver microsomes or liver hepatocytes at a substrate concentration of $1 \mu M$ over a limited time course (30 minutes). Following data acquisition, the software automatically calculates the peak areas at each time point and identifies compounds that dropped below the present parent stability threshold (user-definable). Immediately following the microsomal stability assessment, flagged samples are reinjected onto a high resolution chromatographic column and MS/MS scans are initiated, including multiple reaction monitoring (MRM), precursor ion scans, and neutral loss scans.

Shown in Figure 11-16 are the time-course stability plots for a number of standards and test compounds. In the example shown, all four of the standard

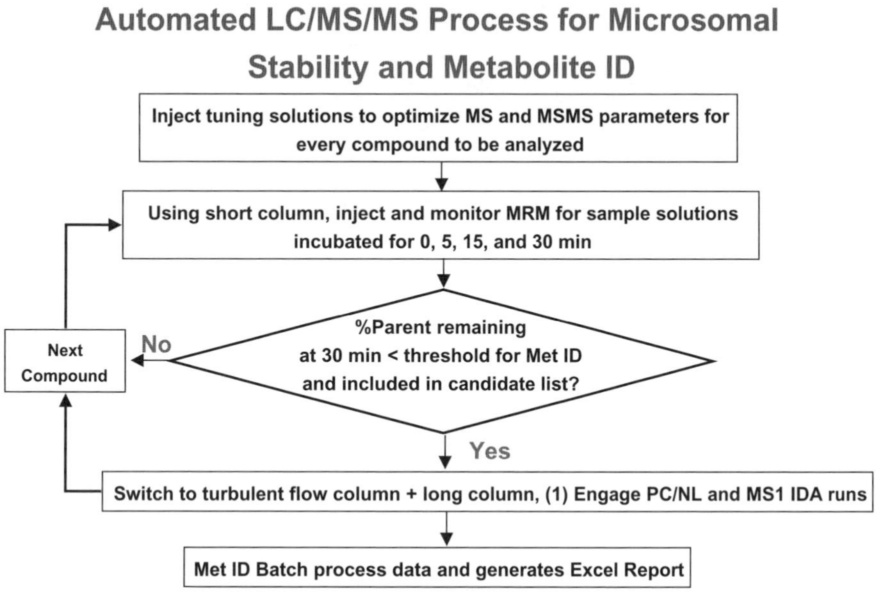

Automated LC/MS/MS Process for Microsomal Stability and Metabolite ID

Inject tuning solutions to optimize MS and MSMS parameters for every compound to be analyzed

↓

Using short column, inject and monitor MRM for sample solutions incubated for 0, 5, 15, and 30 min

↓

%Parent remaining at 30 min < threshold for Met ID and included in candidate list?

No → Next Compound

Yes ↓

Switch to turbulent flow column + long column, (1) Engage PC/NL and MS1 IDA runs

↓

Met ID Batch process data and generates Excel Report

Figure 11-15. Flow diagram for performing intelligent metabolite ID studies. Compounds are incubated with liver microsomes or hepatocytes at a test concentration of 1 μM over a limited time course. The percent of substrate remaining after 30-min incubation is determined on-the-fly. Compounds exhibiting poor metabolic stability are automatically queued for detailed metabolite ID studies. This approach enables metabolic stability and metabolite ID data to be generated in a completely automated, independent manner.

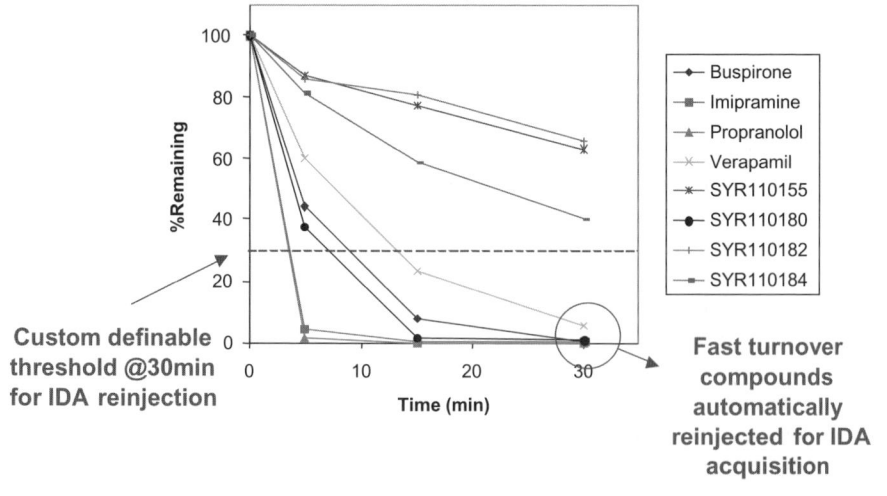

Custom definable threshold @30min for IDA reinjection

Fast turnover compounds automatically reinjected for IDA acquisition

Legend:
- Buspirone
- Imipramine
- Propranolol
- Verapamil
- SYR110155
- SYR110180
- SYR110182
- SYR110184

%Remaining — Time (min)

Figure 11-16. A user-defined threshold for percent parent stability is pre-set in the custom software. Compounds dropping below this pre-defined threshold are automatically selected for detailed metabolite identification by LC-MS/MS incorporating data-dependent acquisition, parent, and precursor ion scans.

42 MRM survey IDA MS² for Buspirone

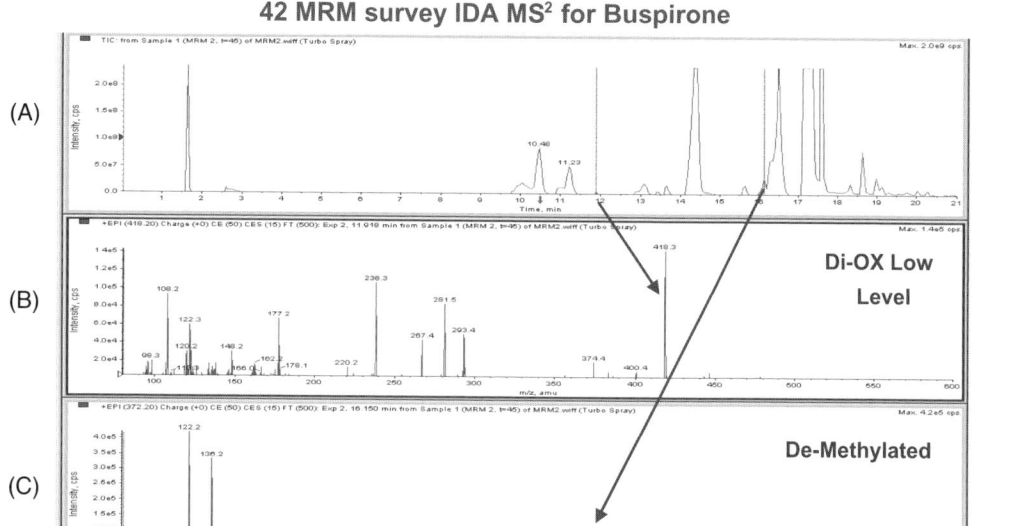

Figure 11-17. The software intelligently "flags" compounds falling below the pre-set stability threshold, automatically re-injects the compounds, and analyzes them in the MS/MS mode. (A) Base peak chromatograms from the MRM scans show a number of metabolites detected for buspirone. (B) MS/MS spectrum of one of the low-level hydroxylated metabolites. (C) MS/MS spectrum of one of the low-level N-demethylated metabolites.

compounds (buspirone, propanolol, imipramine, and verapimil) were found to be rapidly metabolized by liver microsomes, and the amount of substrate remaining after a 30-minute incubation was below the pre-set stability threshold, thereby triggering automated, detailed metabolite ID studies. A larger volume of the microsomal incubate was then loaded onto a trapping cartridge, so as to concentrate the sample, and eluted onto a high resolution chromatographic column to support detailed metabolite ID studies, including MRM, precursor, and neutral loss scans.

Figure 11-17 shows the results of automated, unattended MRM and subsequent MS/MS analysis of buspirone, one of the four substrates to be rapidly metabolized by liver microsomes. A total of 42 MRMs were monitored in this particular MS/MS analysis. The MRMs pre-selected represented the common phase I oxidative metabolites, including mono- and d-hydroxylation, N-oxide formation, and N-dealkylation, to name a few. For 21 of the MRM transitions, the mass corresponding to the expected oxidative metabolite was added to the mass of the precursor ion (e.g., buspirone MW = 386, mono-hydroxylated buspirone MW = 402), keeping the mass of the product ion of the MRM pair unchanged. For the remaining 21 MRM transitions, the mass of the expected metabolite was added both to the precursor ion and to the product ion. A number of metabolites of buspirone were identified by MRM scans, as shown in the top panel of Figure 11-17. Due to the high-sensitivity nature of MRM scans, it was possible to identify both major and minor metabolites; and due to the high sensitivity of the mass spectrometric detector, it was equally possible to generate high-quality MS/MS spectra of the metabolites, as well. The MS/MS spectra of two of the lower level metabolites are shown in the bottom two panels of Figure 11-17. The high quality of the MS/MS spectra of each of these candidate metabolites proved very useful information for pinpointing sites of metabolism. As this software technology enters the mainstream, it should indeed be possible to achieve significantly enhanced metabolite ID throughput.

11.13 CONCLUSIONS

Undeniably, one of the greatest challenges of analytical chemistry in an era of high-throughput drug discovery has been to balance the need for high throughput while maintaining an analytical standard of high quality. Significant advances in sample analysis and purification throughput have been achieved by incorporating fast chromatography coupled with mass spectrometry and by performing these analyses both in series and in parallel. Throughput has also been impacted dramatically by the ability to seamlessly link all of the processes downstream of library synthesis, including automating the uploading of sample lists for automated data acquisition, automating the assessment of compound purity, and automating post-purification analysis and associated sample handling. Fast emerging as the next bottleneck in drug discovery is to design and synthesize more "drug-like" molecules so as to increase the likelihood of their developmental success. Innovations in high throughput *in vitro* and *in vivo* ADME analysis are beginning to be realized, too, with the introduction of fast serial and fast parallel chromatography coupled with mass spectrometry. As the amount of analytical data increases for each and every molecule synthesized, the key will be how to glean from these data the information content that ultimately accelerates the drug discovery process.

Early ADME assessment of compounds is occurring at nearly every stage of the discovery process, from lead generation through late-stage lead optimization. This has occurred for two principal reasons, the first being an enlightened view of medicinal chemists as to the importance of optimizing on drug-like properties in addition to potency and selectivity early in the drug discovery process. Secondly, and perhaps more importantly, early ADME assessment is occurring due to innovations in analytical chemistry and the widespread proliferation of LC/MS technology. Automated sample preparation, data acquisition, and data processing have enabled ADME profiling studies to move into the high-throughput realm. It was only a few years ago that it was suggested that high-throughput ADME (unlike high-throughput analysis and purification to support combinatorial chemistry and parallel synthesis) would be difficult to achieve because, according to the anonymous author, "for most *in vitro* systems, the analytical requirements are rate-limiting, relying heavily on liquid chromatography coupled with mass spectrometry." Continued innovations in fast chromatography, parallel analysis, and automated, "intelligent" data processing and reporting have readily challenged this view.

REFERENCES

1. L. J. Deterding, M. A. Moseley, K. B. Tomer, and J. W. Jorgenson, Nanoscale separations combined with tandem mass spectrometry, *J. Chromatogr.* **554** (1–2), (1991), 73–82.

2. W. J. Henzel, J. H. Bourell, and J. T. Stults, *Anal. Biochem.* **187** (1990), 228.

3. P. Palma and A. Cappiello, Micro-HPLC split flow gradient elution in the analysis of peptides, *Annali di Chinica*, **82** (1992), 371.

4. D. B. Kassel, B. D. Musselman, and J. A. Smith, Primary structure dtermination of peptides and enzymatically digested proteins using capillary liquid chroamtography/mass spectrometry and rapid linked-scan techniques, *Anal. Chem.* **63** (1991), 1091–1097.

5. M. R. Wilkins, J. C. Sanchez, A. A. Gooley, R. D. Appel, I. Humphrey-Smith, and D. F. Hochstrasser, Progress with proteome projects: Why all proteins expressed by a genome should be identified and how to do it, *Biotechnol. Gen. Eng. Rev.*, **13** (1996), 19–50.

6. A. J. Link, J. Eng, D. M. Schieltz, E. Carmack, G. J. Mize, D. R. Morris, B. M. Garvik, and J. R, Yates, Iii, Direct analysis of protein complexes using mass spectrometry, *Nature Biotechnol.* **17** (1999), 676–682.

7. D. K. Han, J. Eng, H. Zhou, and R. Aebersold, Quantitative profiling of differentiation-induced microsomal proteins using isotope-coded affinity tags and mass spectrometry, *Nature Biotechnol.* **19** (2001), 946–951.

8. T. Stasyk and L. A. Huber, Zooming in: Fractionation strategies in proteomics, *Proteomics* **4** (2004), 3704–3716.

9. M. Austen, and C. Dohrmann, Phenotype-first screening for the identification of novel drug targets, *Drug Disc. Technol.* **10** (2005), 275–282.

10. S. M. Elbashir et al., Duplexes of 21-nucleotide RNAs mediate RNA interference in cultured mammalian cells, *Nature.* **411** (2001), 494–498.

11. P. J. Alaimo, M. A. Shogren-Knaak, and K. M. Shokat, Chemical genetic approaches for the elucidation of signaling pathways, *Curr. Opin. Chem. Biol.* **5** (2001), 360–367.

12. W. A. Colburn, Biomarkers in Drug Discovery and Development: From Target Identification Through Drug Marketing, *J. Clin. Pharmacol.* **43** (2003), 329–341.

13. D. B. Kassel, C. van Pelt, and K. Lim, Novel Analytical Strategies in Support of High Throughput Structural Chemistry, 51st ASMS Conference on Mass Spectrometry and Allied Topics, Book of Abstracts.

14. C. D. Wagner, J. T. Hall, K. A. Hoffman, W. L. White, and J. D. Williams, Open-access mass spectrometry utilizing advanced protein search processing integrated with a multi-site protein analysis LIMS, in 52nd ASMS Conference on Mass Spectrometry and Allied Topics, Nashville, TN (Book of Abstracts), 2004.

15. S. L. Evarts, T. C. Webb, and M. T. Cancilla, A dual stream LC ESI quadrupole-time-of-flight mass spectrometer to double the throughput of intact protein and peptide digest analyses, in 52nd ASMS Conference on Mass Spectrometry and Allied Topics, Nashville, TN (Book of Abstracts), 2004.

16. D. B. Kassel, T. G. Consler, M. Shallaby, P. Sekhri, N. Gordon, and T. Nadler, Direct coupling of an automated 2-dimensional microcolumn affinity chromatography–capillary HPLC system with mass spectrometry for biomolecule analysis, *Techniques in Protein Chemistry* VI, Academic Press, New York, 1995, pp. 39–46.

17. S. Kaur, L. McGuire, D. Tang, G. Dollinger, and V. Huebner, *J. Protein Chem.* **16** (1997), 505.

18. J. Berman, M. Green, E. Sugg, R. Anderegg, D. S. Millington, D. L. Norwood, J. McGeehan, and J. Wiseman, *J. Biol. Chem.* **267** (1993), 1434–1437.

19. M. Lambert, R. K. Blackburn, T. D. Seaton, D. B. Kassel, D. S. Kinder, M. A. Leesnitzer, D. M. Bickett, J. R. Warner, M. W. Andersen, J. G. Badiang, D. J. Cowan, M. D. Gaul, K. C. Glennon, M. H. Rabinowitz, R. W. Wiethe, J. D. Becherer, D. L. McDougald, D. L. Musso, R. C. Andrews, and M. L. Moss, Substrate specificity and novel selective inhibitors of TNF-α converting enzyme (TACE) from two-dimensional substrate mapping, *Comb. Chem. High Throughput Screen.* **8** (2005), 327–339.

20. R. Gerber, J. D. Ryan, and D. S. Clark, Cell-based screen of HMG-CoA reductase inhibitors and expression regulators using LC-MS, *Anal. Biochem.* **329** (2004), 28–34.

21. S. J. Thibodeaux, K. D. W. Roth, W.-K. Yeh, D. A. Briere, M. D. Michael, and J. Dage, Determination of K_i and *in vivo* Efficacy for 11β-hydroxysteroid dehydrogenases using LC/MS/MS, in Proceedings of the 52nd ASMS Conference on Mass Spectrometry and Allied Topics, Nashville, TN (Book of Abstracts), 2004.

22. R. Xu, B.-C. Sang, M. Navre, and D. B. Kassel, "Cell-Based Assay Screening 11β-Hydroxysteroid Dehydrogenase Inhibitors Using Liquid Chromatography/Tandem Mass Spectrometry Detection," *Rapid Commun. Mass. Spectrom.* **20** (2006), 1–5.

23. K. C. Nicolaou, J. A. Pfefferkorn, H. J. Mitchell, A. J. Roecker, S. Barluenga, G.-Q. Cao, R. L. Allfeck, and J. E. Lillig, Natural product-like combinatorial libraries based on privileged structures. 2. Construction of a 10,000-membered benzopyran library by directed split-and-pool chemistry using NanoKans and optical encoding, *J. Am. Chem. Soc.* **122** (2000), 9954–9967.

24. B. A. Kozikowski, T. M. Burt, D. A. Tirey, L. E. Williams, B. R. Kuzmak, D. T. Stanton, K. L. Morand, and S. L. Nelson, The effect of freeze/thaw cycles on the stability of compounds in DMSO, *J. Biomol. Screen.* **8** (2003), 210–215.

25. K. L. Morand, T. M. Burt, B. T. Regg, and T. L. Chester, Techniques for increasing the throughput of flow injection mass spectrometry, *Anal. Chem.* **73** (2001), 247–252.

26. J. Greaves, Operation of an academic open access mass spectrometry facility with particular reference to the analysis of synthetic compounds, *J. Mass Spectrom.* **37** (2002), 777–785.

27. J. Hogan, Directed combinatorial chemistry, *Nature* **384** (Suppl. 6604) (1996), 17–19.

28. J. J. Parlow and J. E. Normansell, Discovery of an herbicidal lead using polymer-bound activated esters in generating a combinatorial library of amides and esters, *Mol. Divers.* **1** (1996), 266–269.

29. M. G. Siegel, A. J. Shuker, C. A. Droste, P. J. Hahn, D. C. Jesudsan, J. H. I. McDonald, D. P. Matthews, C. J. Rito, and A. J. Thorpe, The use of high through-put synthesis and purification in the preparation of a directed library of adrenergic agents, *Mol. Divers.* **3** (1997), 113–116.

30. B. Yan, H.-U. Gemlich, S. Moss, G. M. Coppla, Q. Sun, and L. Liu, *J. Comb. Chem.* **1** (1999), 46.

31. C. M. Snively, G. Oskarsdottir, and J. Lauterbach, **2** (2000), 243.

32. J. C. Lindon, J. K. Nicholson, and I. D. Wilson, Directly coupled HPLC-NMR and HPLC-NMR-MS in pharmaceutical research and development, *J. Chromatogr. B* **748** (2000), 233–258.

33. J. A. Syage, B. J. Nies, M. D. Evans, and K. A. Hanold, Field-portable, high-speed GC/TOFMS, *J. Am. Soc. Mass Spectrom.* **12** (2001), 648–655.

34. C. Enjabal, J. Martinez, and J. L. Aubagnac, *Mass Spectrometry Reviews* **19** (2000), 139–161.

35. I. Hughes and D. Hunter, Techniques for analysis and purification in high-throughput chemistry, *Curr. Opin. Chem. Biol.* **5** (2001), 243–247.

36. M. Grieg, Use of automated HPLC-MS analysis for monitoring and improving the purity of combinatorial libraries, *Am. Lab.* **31** (1999) 28–32.

37. J. N. Kyranos, H. Cai, D. Wei, and W. K. Goetzinger, High-throughput high performance liquid chromatography/mass spectrometry for modern drug discovery, *Anal. Biotechnol.* (2001), 105–111.

38. J. N. Kyranos, H. Lee, W. K. Goetzinger and L. Y. Li, One-minute full-gradient HPLC/UV/ELSD/MS analysis to support high-throughput parallel synthesis, *J. Comb Chem.* **6** (2004), 796–804.

39. E. W. Taylor, W. Jia, M. Bush, and G. D. Dollinger, Accelerating the drug optimization process: Identification, structure elucidation, and quantification of in vivo metabolites using stable isotopes with LC/MS^n and the chemiluminescent nitrogen detector, *Anal Chem.* **74** (2002), 3232–3238.

40. C. E. Kibbey, Quantitation of combinatorial libraries of small organic molecules by normal-phase HPLC with evaporative light-scattering detection, *Mol. Divers.* **4** (1996), 247–258.

41. B. E. Hsu, E. Orton, S. Y. Tang, and R. A. Carlton, Application of evaporative light scattering detection to the characterization of combinatorial and parallel synthesis libraries for pharmaceutical drug discovery, *J. Chromatogr. B Biomed. Sci. Appl.* **725** (1999), 103–112.

42. W. K. Goetzinger, and J. N. Kyranos, Fast gradient RP-HPLC for high throughput quality control analysis of spatially addressable combinatorial libraries, *Am. Lab.* **30** (1998), 27–37.

43. L. A. Romanyshyn and P. R. Tiller, Ultra-short columns and balistic gradients: considerations for ultra-fast chromatographic liquid chromatographic-tandem mass spectrometric analysis, *J. Chromatogr. A* **928** (2001), 41–51.

44. J. Zhao, L. Zhang, and B. Yan, Strategies and methods for purifying organic compounds and combinatorial libraries, in B. Yan (ed.), *Analysis and Purification Methods in Combinatorial Chemistry*, John Wiley & Sons, 2004, pp. 255–280.

45. H. N. Weller, M. G. Young, S. J. Michalczyk, G. H. Reitnauer, R. S. Cooley, P. C. Rahn, D. J. Loyd, D. Fiore, and S. J. Fischman, High throughput analysis and purification in support of automated parallel synthesis, *Mol. Divers.* **3** (1997), 61–70.

46. L. Schultz, C. D. Garr, L. M. Cameron, and J. Bukowski, High throughput purification of combinatorial libraries, *Bioorg. Med. Chem. Lett.* **8** (1998), 2409.

47. C. Kibbey, An automated system for the purification of combinatorial libraries by preparative LC/MS, *Lab. Rob. Autom.* **9** (1998), 309–321.

48. J. Hocklowski, High throughput purification: Triage and optimization, in B. Yan (ed.), *Analysis and Purification Methods in Combinatorial Chemistry*, John Wiley & Sons, Hoboken, NJ, 2004, pp. 281–306.

49. B. Yan, N. Collins, J. Wheatley, M. Irving, K. Leopold, C. Chan, A. Shornikov, L. Fang, A. Lee, M. Stock, and J. Zhao, High throughput purification of combinatorial libraries I: A High throughput purification system using an accelerated retention window approach, *J. Comb. Chem.* **6** (2004), 255–261.

50. T. Karancsi, L. Godorhazt, D. Szalay, and F. Darvas, UV-triggered maincomponent fraction collection method and its application for high throughput chromatographic purification of combinatorial libraries, *J. Comb. Chem.* **7** (2005), 58–62.

51. L. Zeng, L. Burton, K. Yung, B. Shushan, and D. B. Kassel, Automated analytical/preparative high performance liquid chromatography-mass spectrometry for the rapid characterization and purification of compound libraries, *J. Chromatogr. A* **794** (1998), 3–13.

52. L. Zeng, X. Wang, T. Wang, and D. B. Kassel, New developments in automated PrepLCMS extends the robustness and utility of the method for compound library analysis and purification, **1** (1998), 101–111.

53. J. P. Kiplinger, R. O. Cole, S. Robinson, E. J. Roskanp, H. J. O'Connell, R. S. Ware, A. Brailsford, and J. Batt, *Rapid Commun. Mass Spectrom.* **12** (1998), 658–664.

54. G. Nemeth and D. Kassel, Existing and emerging strategies for the analytical characterization and profiling of compound libraries, in G. Trainor (ed.), *Annual Reports in Medicinal Chemistry*, Academic Press, 2001, pp. 277–292.

55. I. G. Popa-Burke, O. Issakova, J. D. Arroway, P. Bernasconi, M. Chen, L. Coudurier, S. Galasinski, A. P. Jadhav, W. P. Janzen, D. Lagasca, D. Liu, R. S. Lewis, R. P. Mohney, N. Sepetov, D. A. Sparkman, and C. N. Hodge, Streamlined system for purifying and quantifying a diverse library of compounds and the effect of compound concentration measurements on the accurate interpretation of biological assay results, *Anal. Chem.* **76** (2004), 7278–7287.

56. Z. Luo, Q. Zhang, Y. Oderaotoshi, and D. P. Curran, Fluorous mixture synthesis: A fluorous-tagging strategy for the synthesis and separation of mixtures of organic compounds, *Science*, **291** (2001), 1766–1769.

57. W. Zhang, Y. Lu, C. H.-T. Chen, D. P. Curvan, and S. Geib, Fluorous Synthesis of Hydrantoin-, Piperazinedione-, and Benzodiazepinedione-Fused Tricyclic and Tetracyclic Ring Systems, *Eur. J. Org. Chem.* (2006), 2055–2059.

58. W. Zhang, Y. Lu, C. H.-T. Chen, L. Zeng, and D. B. Kassel, Fluorous Mixture Synthesis of Two Libraries with Hydrantoin-, and Benzodiazepinedione-Fused Heterocyclic Scaffolds, *J. Comb. Chem.* **8** (2006), 687–695.

59. V. D. Di Biasi, N. Haskins, A. Organ, R. Bateman, K. Giles, and S. Jarvis, High throughput liquid chromatography/mass spectrometric analyses using a novel multiplexed electrospray interface, *Rapid Commun. Mass Spectrom.* **13** (1999), 1165–1168.

60. T. Wang, L. Zeng, J. Cohen, and D. B. Kassel, A multiple electrospray interface for parallel mass spectrometric analyses of compound libraries, *Comb. Chem. High Throughput Screening* **2** (1999), 327–334.

61. D. Kassel, Combinatorial chemistry and mass spectrometry in the 21st century drug discovery laboratory, *Chem. Rev.* **101** (2001), 255–268.

62. L.-F. Jiang and M. Moini, Development of multi-ESI-sprayer, multi-atmospheric pressure inlet mass spectrometry and its application to accurate mass measurement using time-of-flight mass spectrometry, *Anal. Chem.* **72** (2000), 20–24.

63. M. K. Bayliss, D. Little, D. N. Mallett, and R. S. Plumb, Parallel ultra-high flow rate liquid chromatography with mass spectrometric detection, using a multiplex electrospray source for direct, sensitive determination of pharmaceuticals in plasma at extremely high throughput, *Rapid Commun. Mass Spectrom.* **14** (2000), 2039–2045.

64. L. Zeng, and D. B. Kassel, Developments of a fully automated parallel HPLC/mass spectrometry system for the analytical characterization and preparative purification of combinatorial libraries, *Anal. Chem.* **70** (1998), 4380–4388.

65. R. Xu, C. Liu, C. Nemes, K. M. Jenkins, R. A. Rourick, and D. B. Kassel, Application of parallel liquid chromatography/mass spectrometry for high throughput microsomal stability screening of compound libraries, *J. Am Soc Mass Spectrom.* **13** (2002), 155–165.

66. R. Cole, 46th ASMS Conference on Mass Spectrometry and Allied Topics, Long Beach, CA (June 11–15th, 2000, LC/MS Workshop).

67. C. Edwards, J. Liu, T. J. Smith, D. Brooke, D. J. Hunter, A. Organ, and P. Coffey, Parallel preparative high-performance liquid chromatography with on-line molecular mass characterization, *Rapid Commun. Mass Spectrom.* **17** (2003), 2027–2033.

68. R. God and H. Gumm, Parallel HPLC in high-throughput analysis and purification, in B. Yan (ed.), *Analysis and Purification Methods in Combinatorial Chemistry*, John Wiley & Sons, Hoboken, NJ, 2004, pp. 307–320.

69. J. Isbell, R. Xu, Z. Cai, and D. B. Kassel, Realities of high-throughput liquid chromatography/mass spectrometry purification of large combinatorial libraries: A report on overall sample throughput using parallel purification, **4** (2002), 600–611.

70. R. A. Lipper, How can we optimize selection of drug development candidates from many compounds at the discovery stage? *Mod. Drug Discovery* **2** (1999), 55.

71. T. N. Thompson, Optimization of metabolic stability as a goal of modern drug design, *Med. Res. Rev.* **21** (2001), 412.

72. D. A. Smith and H. van de Waterbeemd, Pharmacokinetics and metabolism in early drug discovery, *Curr. Opin. Chem. Biol.* **3** (1999), 373.

73. E. H. Kerns, High throughput physicochemical profiling for drug discovery, *J. Pharm. Sci.* **90** (2001), 1838.

74. D. T. Rossi and M. Sinz, *Mass Spectrometry in Drug Discovery*. Marcel Dekker, New York, 2002.

75. B. L. Ackermann, M. J. Berna, and A. T. Murphy, Recent advances in use of LC/MS/MS for quantitative high-throughput bioanalytical support of drug discovery, *Curr. Top. Med. Chem.* **2** (2002), 53.

76. A. D. Rodrigues and J. H. Lin, Screening of drug candidates for their drug–drug interaction potential, *Curr. Opin. Chem. Biol.* **5** (2001), 396.

77. Y. Li, Y. G. Shin, C. Yu, J. W. Kosmeder, W. H. Hirschelman, J. M. Pezzuto, and R. B. van Breemen, Increasing the throughput and productivity of Caco-2 cell permeability assays using liquid chromatography-mass spectrometry: Application to resveratrol absorption and metabolism, *Comb Chem. High Throughput Screening* **6** (2003), 757.

78. M. S. Bryant, W. A. Korfmacher, S. Wang, C. Nardo, A. A. Nomeir, and C. C. Lin, Pharmacokinetic screening for the selection of new drug discovery candidates is greatly enhanced through the use of liquid chromatography-atmospheric pressure ionization tandem mass spectrometry, *J. Chromatogr. A* **777** (1997), 61–66.

79. P. J. Eddershaw and M. Dickins, Advances in drug metabolism screening, *Pharm. Sci. Technol. Today* **2** (1999), 13.

80. A. J. Weiss, Performing ADME earlier—a way to gain speed and productivity, Business Briefing: Future Drug Discovery, 2003.

81. I. Kariv, H. Cao, and K. R. Oldenburg, Development of a high throughput equilibrium dialysis method, *J. Pharm. Sci.* **90** (2001), 580.

82. M. J. Qian, A fast screening method to measure equilibrium solubility in early drug discovery process, in AAPS Annual Meeting and Exposition, Denver, CO, 2001.

83. D. M. Wilson, X. Wang, E. Walsh, and R. A. Rourick, High throughput log D determination using liquid chromatography-mass spectrometry, *Comb. Chem. High Throughput Screening* **4** (2001), 511–519.

84. J. Ayrton, G. J. Dear, W. J. Leavens, D. N. Mallett, and R. S. Plumb, Optimization and routine use of generic ultra-high flow-rate liquid chromatography with mass spectrometric detection for the direct on-line analysis of pharmaceuticals in plasma, *J. Chromatogr. A* **828** (1998), 199–207.

85. R. J. Scott, J. Palmer, I. A. Lewis, and S. Pleasance, Determination of a "GW cocktail" of cytochrome P450 probe substrates and their metabolites in plasma and urine using automated solid phase extraction and fast gradient liquid chromatography tandem mass spectrometry, *Rapid Commun. Mass Spectrom.* **13** (1999), 2305–2319.

86. J. L. Whitney, M. E. Hail, and D. J. Detlefsen, Automated metabolite confirmation using data-dependent LC/MS and intelligent chemometrics, in *50th ASMS Conference on Mass Spectrometry and Allied Topics, Orlando, FL*, ASMS, 2002.

87. A. Williams, Applications of computer software for the interpretation and management of mass spectrometry data in pharmaceutical science, *Curr. Top. Med. Chem.* **2** (2002), 99.

88. W. A. Korfmacher, C. A. Palmer, C. Nardo, K. Dunn-Meynell, D. Grotz, K. Cox, C. C. Lin, C. Elicone, C. Liu, and E. Duchoslav, Development of an automated mass spectrometry system for the quantitative analysis of liver microsomal incubation samples: A tool for rapid screening of new compounds for metabolic stability, *Rapid Commun. Mass Spectrom.* **13** (1999), 901–907.

89. J. E. Shaffer, K. K. Adkison, K. Halm, K. Hedeen, and J. Berman. Use of "*N*-in-one" dosing to create an *in vivo* pharmacokinetics database for use in developing structure-pharmacokinetic relationships, *J. Pharm. Sci.* **88** (1999), 313–318.

90. T. V. Olah, D. A. McLoughlin, and J. D. Gilbert, The simultaneous determination of mixtures of drug candidates by liquid chromatography/atmospheric pressure chemical ionization mass spectrometry as an *in vivo* drug screening procedure, *Rapid Commun. Mass Spectrom.* **11** (1997), 17.

91. C. L. Stevenson, P. F. Augustijns, and R. W. Hendren, Use of Caco-2 cells and LC/MS/MS to screen a peptide combinatorial library for permeable structures, *Int. J. Pharm.* **177** (1999), 103.

92. R. E. White and P. Manitpisitkul, Pharmacokinetic theory of cassette dosing in drug discovery screening, *Drug Metab. Dispos.* **29** (2001), 957.

93. W. A. Korfmacher, K. A. Cox, K. J. Ng, J. Veals, Y. Hsieh, S. Wainhaus, L. Broske, D. Prelusky, A. Nomeir, and R. E. White, Cassette-accelerated rapid rat screen: A systematic procedure for the dosing and liquid chromatography/atmospheric pressure ionization tandem mass spectrometric analysis of new chemical entities as part of new drug discovery, *Rapid Commun. Mass Spectrom.* **15** (2001), 335.

94. R. Xu, C. Liu, C. Nemes, K. M. Jenkins, R. A. Rourick, and D. B. Kassel, Application of parallel liquid chromatography/mass spectrometry for high throughput microsomal stability screening of compound libraries, *J. Am. Soc. Mass Spectrom.* **13** (2002), 155–165.

95. W. A. Korfmacher, C. A. Palmer, C. Nardo, K. Dunn-Meynell, D. Grotz, K. Cox, C. C. Lin, C. Elicone, C. Liu, and E. Duchoslav, Demonstration of the capabilities of a parallel high performance liquid chromatography tandem mass spectrometry system for use in the analysis of drug discovery plasma samples, *Rapid Commun. Mass Spectrom.* **13** (1999), 1991.

96. D. L. Hiller, et al., High throughput quantitation using indexed multiprobe electrospray technology in support of drug discovery, in *48th ASMS Conference on Mass Spectrometry and Allied Topics, Long Beach, CA*, ASMS, 2000.

97. M. K. Bayliss, D. Little, D. N. Mallett, and R. S. Plumb, Parallel ultra-high flow rate liquid chromatography with mass spectrometric detection using a multiplex electrospray source for direct, sensitive determination of pharmaceuticals in plasma at extremely high throughput, *Rapid Commun. Mass Spectrom.* **14** (2000), 2039.

98. L. Yang, T. D. Mann, D. Little, N. Wu, R. P. Clement, and P. J. Rudewicz, Evaluation of a four-channel multiplexed electrospray triple quadrupole mass spectrometer for the simultaneous validation of LC/MS/MS methods in four different preclinical matrixes, *Anal Chem.* **73** (2001), 1740–1747.

99. R. Xu, E. Duchoslav, A. Aparicio, E. B. Jones, and D. B. Kassel, Streamlined approaches to metabolic stability assessment and metabolite profiling in drug discovery, in *51st ASMS Conference on Mass Spectrometry and Allied Topics*, Montreal, Canada, Book of Abstracts, ASUS, 2003.

100. D. J. Detlefsen, J. L. Whitney, M. E. Hail, J. L. Josephs, S. Sanders, and K. D. Nugent, A total analysis solution for metabolic stability and detailed metabolite profiling, in *51st ASMS Conference on Mass Spectrometry and Allied Topics*, Montreal, Canada, Book of Abstracts, ASUS, 2003.

12

ROLE OF HPLC IN PREFORMULATION

Irina Kazakevich

12.1 INTRODUCTION

Preformulation is a bridge between discovery and development where development scientists participate in selection and optimization of lead compounds. It is very critical at this stage to evaluate the developability of potential drug candidates in order to select new chemical entities and decrease the number of failures during future drug development.

On average, only one out of ten new chemical entities (NCE) entering first-in-human testing reaches registration, approval, and marketing stage. The reasons for failures of development compounds include problems with biopharmaceutical properties, clinical safety, toxicology, efficacy, cost of goods, and marketing (see Figure 12-1) [1, 2]. The biopharmaceutical properties such as gastrointestinal and plasma solubility, lipophilicity (LogD), permeability, first-pass metabolism, systemic metabolism, protein binding, and *in vivo* bioavailability are related to the solubility, chemical stability, and permeability of drug candidates and have to be considered at discovery lead selection before recommendation to the development stage.

A major challenge in any drug discovery program is achieving reasonable bioavailability upon oral administration; therefore, any information that highlights potential problems with cell permeability and absorption is valuable when reviewing structural families as leads for drug discovery. Lipinski et al. [3] have reviewed 2245 compounds selected from the United States Adopted

HPLC for Pharmaceutical Scientists, Edited by Yuri Kazakevich and Rosario LoBrutto
Copyright © 2007 by John Wiley & Sons, Inc.

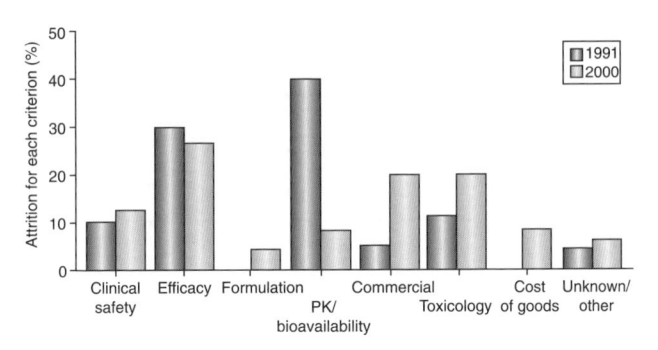

Figure 12-1. Reasons for attrition from 1991 to 2000. (Reprinted with permission from reference 1.)

Name (USAN), International Nonproprietary Name (INN), and World Drug Index (WDI), comparing calculated physical properties and clinical exposure. Four parameters were chosen that were associated with solubility and permeability, namely, molecular weight, octanol/water partition coefficient, the number of hydrogen bond donors, and the number of hydrogen bond acceptors. It was concluded that compounds are most likely to have poor absorption when molecular weight is >500, the calculated LogP is >5, the number of hydrogen bond donors is >5, and the number of hydrogen bond acceptors is >10. Lipinski has referred to this analysis as "rule of five" because the cutoffs for each of the four parameters were all close to five or a multiple of five. The rule of five can serve as qualitative absorption/permeability predictor.

The absorption of drug molecules in the gastrointestinal tract is dependent upon the pK_a of the compound and the pH of the gastrointestinal region (Figure 12-2). Almost 63% of all drugs are ionized in aqueous solution and can exist in a neutral or a charged state, depending on the pH of the local environment [4].

Based on the major goal of preformulation—identification of possible failure in future development—numerous studies are performed to fully characterize prospective drug candidates. The major analytical technique in each preformulation group is liquid chromatography. Ninety percent of all analytical equipment in preformulation groups are HPLC systems equipped with UV and MS detection systems. HPLC is a fast and reliable method for concentration and identity determination by UV and/or MS detection, respectively. The type of HPLC methods differ based on the specific preformulation tests that will be described below.

In the early stage of preformulation, characterization of the drug molecule involves ionization constants and partition coefficient determinations, aqueous and nonaqueous kinetic and equilibrium solubility determination, pH solubility profile, chemical stability assessment, and salt and polymorph screening. Assessment of biopharmaceutics and toxicological screening are also essential

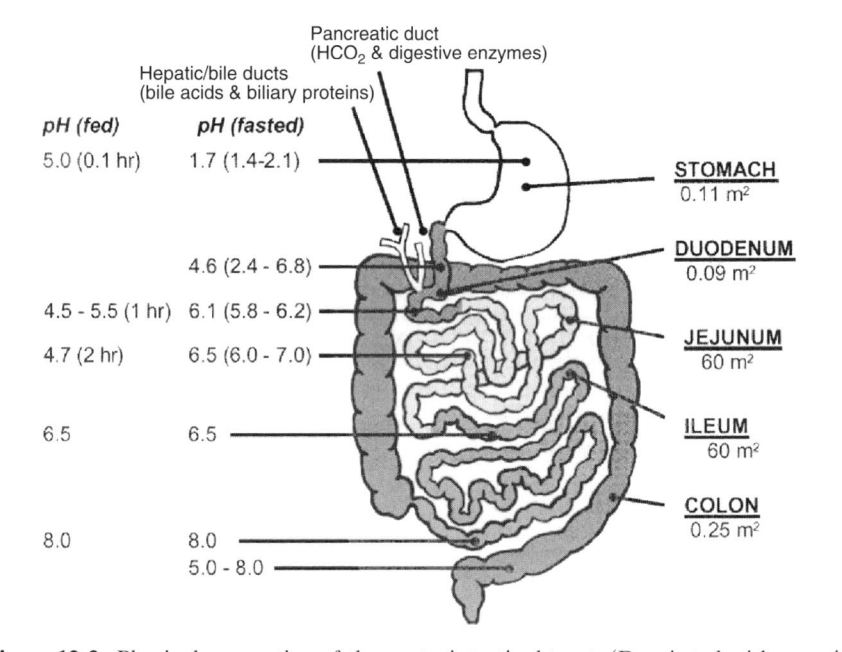

Figure 12-2. Physical properties of the gastrointestinal tract. (Reprinted with permission from reference 5.)

at this stage. At the later stage of preformulation, after recommendation of NCE to development, the development support from preformulation group involves a more detailed solid-state characterization program, elaborating on moisture sorption, compressibility, melting point, particle size, shape, and surface area assessments, as well as excipient compatibility and prototype formulation stability evaluation.

Further information on the role of preformulation in drug development process can be found in several excellent monographs [6–8] with the focus on pharmaceutical aspects of process development.

12.2   INITIAL PHYSICOCHEMICAL CHARACTERIZATION (DISCOVERY SUPPORT)

During the early discovery stage the medicinal chemists use *in vitro* activities and fast *in vivo* small animal studies to discover the best compound to develop. The support from development scientist consists of providing information about LogP, pK_a, and LogD for ionizable drugs and aqueous solubility. These physical characteristics can affect the absorption of drug candidate and, therefore, drug bioaivalability. The requirements for HPLC analysis at this stage are speed and efficiency of the separation. It is critical to mention that at the early stage of discovery, very little information is available about the properties of molecule and only a few milligrams of compound is available for characterization. Therefore, it is important to choose the most efficient column and the simplest mobile phase. Also, recommended is the use of more contemporary HPLC systems as UPLC from Waters employing columns with dimensions of $50 \times 2.1\,mm$, 1.8-μm particle size and the Fast 1200 system from Agilent with column dimensions of $50 \times 4.6\,mm$, 1.8-μm particle size, respectively, to enhance the turnaround time for sample analysis. Other platforms would include using Chromolith Speedrod® monolithic columns at high flow rates. Also, taking into consideration the short column length, gradient elution should be recommended for all HPLC methods at this stage of drug candidate characterization. The post-run equilibration time is not significant in the case where short columns are used, and dwell volume is improved significantly for a new generation of HPLC systems.

Many types of modeling techniques are available in the discovery phase of drug development, from structure activity relationships (SAR) to physiology based pharmacokinetics (PBPK) and pharmacokinetics-/pharmacodynamics (PK/PD) to help choosing some of the lead compounds. Some tests that are carried out by discovery include techniques related to structure determination, metabolism, and permeability: NMR, MS/MS, elemental analysis, PAMPA, CACO-2, and *in vitro* metabolic stability. Although they are important as a part of physicochemical molecular characterization under the biopharmaceutics umbrella, they will not be discussed here. The reader can find relevant information in numerous monographs [9, 10].

12.2.1 Ionization Constant, pK_a

Most potential drug candidates are weak bases or acids. Solubility and many other properties of the drug molecule is dependent on its ionization state. Acids are usually considered to be proton donors and bases are proton acceptors. Any drug molecule with basic functionality in aqueous media holds the following equilibrium:

$$BH^+ \leftrightarrow B + H^+ \tag{12-1}$$

where the ionization equilibrium constant could be expressed as

$$K_a = \frac{[B] \cdot [H^+]}{[BH^+]} \tag{12-2}$$

It is obvious from the above equilibrium that the ratio of ionic to nonionic form of the drug in the solution is controlled by the proton concentration, which is commonly represented by pH values (negative logarithm of proton concentration). Taking the negative logarithm of expression (12-2), the well-known Henderson–Hasselbalch equation could be obtained:

$$pK_a = pH + \log\frac{[BH^+]}{[B]} \tag{12-3}$$

This allows for the estimation of the prevailing drug form at a particular pH. Ionic form of any organic molecule is usually more soluble in aqueous media, while the neutral form is usually more hydrophobic and thus shows an increased affinity for lipids.

Variation of the ionization state of the molecule at different pH has typical sigmoidal shape (as shown in Figure 12-3). Corresponding expression for this dependence could be derived from equation (12-2) and the mass balance of the ionic and nonionic form of the drug:

$$q = [B] + [BH^+] \tag{12-4}$$

If one assumes quantity q equal to 100, then concentration of B or BH^+ forms will numerically be equal to the percentage of corresponding form in the solution and solving equation (12-3) with expression (12-4) one will get the expression for BH^+ concentration expressed as a percent of ionized form

$$[BH^+] = \frac{100 \cdot 10^{(pK_a - pH)}}{1 + 10^{(pK_a - pH)}} \tag{12-5}$$

The inflection point of this curve corresponds to the point where $pH = pK_a$, and it is a common way for the determination of the drug pK_a values.

Several different techniques are usually employed for pK_a determination. They were described in detail by Comer [11].

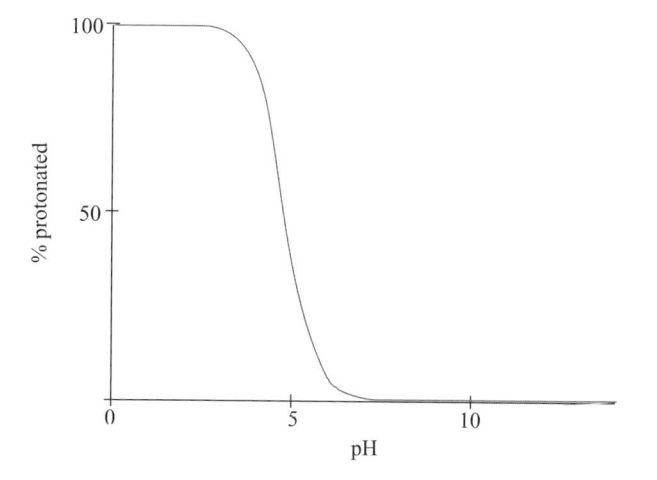

Figure 12-3. Dependence of the relative amount (in the form of a percent) of protonated form on the pH of aqueous media.

In practice the most common technique to determine pK_a value is by employing potentiometric titration based on the detection of the variations of either the conductivity or current at fixed applied potential at various pH values. The automated potentiometric titration system well known as a $GLpK_a$ or PCA200 from Sirius Analytical [12] is considered to be a good approach for pK_a determination with water-soluble drugs at pH 2–8 for the new drug candidates when the amount of drug substance is limited. For poorly water-soluble compounds it is advised to use $GlpK_a$ with D-Pass or Sirius Profiler SGA as a pH/UV method for determination of compounds that have inherently lower concentration in the solution media.

HPLC is another convenient method for measurement of the NCE pK_a values. As was shown by Melander and Horvath [13], the retention of any ionizable analyte closely resembles the curve shown in Figure 12-3. Chromatographic determination of the pK_a could be accurately performed with very limited amount of sample. Fast HPLC method with optimum analyte retention is suitable for this purpose, but the influence of the organic mobile-phase modifier on the mobile phase pH and analyte pK_a should be accounted for in order to provide the accurate calculation of the respective pK_a value. Detailed discussion of the HPLC-based methods for the pK_a determination is given in Chapter 4.

In the case of sufficient drug supply the old-fashioned solubility method can be used for pK_a determination based on the different equilibrium solubility at different pH values. This method is very precise, but time- and drug-consuming, and is described in detail in reference 6.

Drug substance often contains several ionizable groups that may significantly complicate experimental measurement of the pK_a. All different types of pK_a determination methods are essentially based on the measurement of the titration curve. If the pK_a values of several ionizable groups in the molecule are within 2 pH units from each other, experimental measurement become very tedious. Recent advancements in the molecular computational methods and developments of physicochemical databases for a large number of known compounds allow computer-based prediction of the pK_a values on the basis of known physicochemical correlations and fast computer screening of known values for related or structurally similar compounds from the database. Detailed discussion of these programs is given in Chapter 10.

12.2.2 Partition and Distribution Coefficients

One of the most important physicochemical parameters associated with oral absorption, central nervous system (CNS) penetration, and other pharmacokinetic parameters is lipophilicity of organic compounds, which determines distribution of a molecule between the aqueous and the lipid environments. The lipophilicity in the form of LogP was included in Lipinski's rule of five as one of the major characteristics of drug-like organic molecules. It was stated that LogP should be not more than five for drug candidates to have a good

TABLE 12-1. Preferable Dosing Form for Different LogP Regions

LogP		Dosing Form
Low	<0	Injectable
Medium	0–3	Oral
High	3–4	Transdermal
Very high	4–7	Toxic buildup in fatty acids

oral absorption property. In Table 12-1, some LogP values for various types of dosage forms are given.

The partition coefficient itself is a constant and is defined as the ratio of concentration of compound in aqueous phase to the concentration in an immiscible solvent, *as the neutral molecule.* In practical terms the neutral molecule exists for bases >2 pH units above pK_a and for acids >2 pH units below pK_a. In practice, log P will vary according to the conditions under which it is measured and the choice of partitioning solvent. LogP is the logarithm of distribution coefficient at a pH where analyte is in its neutral state. This is not a constant and will vary according to the protogenic nature of the molecule.

The choice of partition solvent has been a subject of debate. Different type of solvents have been used for the determination of partitioning coefficient [14], but the majority of the data are generated using water–*n*-octanol partitioning. Octanol was chosen as a simple model of a phospholipid membrane. However, it has shown serious shortcomings in predicting blood–brain barrier or skin penetration. Other solvents such as chloroform, cyclohexane, and propylene glycol dipelargonate (PGDP) have been used for modeling biological membranes.

Octanol is a hydrogen-bonding solvent, and thus it shows certain specificity in its ability to dissolve some components. For example, K_w^0 for phenol in hexane is only 0.11 while in octanol it is equal to 29.5. There were several attempts to rationalize solvent effects using solubility parameters [15], dielectric constant [16], and others, but none appear to be consistent. *n*-Octanol gives the most consistent results with other physicochemical properties and drug absorption in gastrointestinal tract.

The classical measurement of LogP is the shake flask method [17]. A known amount of drug is dissolved in a flask containing both octanol phase and aqueous buffer at controlled pH to ensure the existence of only nonionic form (at least two units from the drug pK_a). The flask is shaken to equilibrate the sample between two phases. There must be no undissolved substance present in both phases. After the system reaches its equilibrium, which is time- and temperature-dependent, the concentration of drug is analyzed by HPLC in both phases. Partitioning coefficient is calculated as

$$K_w^0 = \frac{c_0}{c_w} \tag{12-6}$$

This method allows for the accurate determination of K_w^0 only within the -1000 to $+1000$ region or approximately within six orders of magnitude span. These experiments could be complicated by solubility and equilibration kinetics and the properties of a substance. For example, if a studied compound has a property of nonionic surfactant, it will be mainly accumulated at the water–organic interface, and shaking of this two-phase system will create a stable emulsion difficult for analytical sampling. The ultracentrifugation at speed of 14,000 rpm for 15–20 min can be enough in most cases to separate two phases. Actual equilibration of the system is tested by several measurements of the equilibrium concentration at different time intervals.

Because of the wide range of partitioning coefficient values, in most cases the decimal logarithm of K_w^0 is used, and it is denoted as LogP:

$$LogP = log(K_w^0) \tag{12-7}$$

The biggest challenge for the use of HPLC in the LogP measurement is the determination of the drug concentration in the octanol phase. If the octanol solution is being injected onto the reversed-phase column, it can modify the stationary phase, shift the analyte retention, and lead to an incorrect measurement due to the retention shift. To avoid this problem the dilution in the corresponding mobile phase is recommended. Also, when LogP is more than four, the concentration of drug in water phase is very small, causing a detection problem with UV detection. This becomes even more troublesome if the compound of interest has a weak UV chromophore. The use of MS detection and proper ionization mode is recommended to increase the sensitivity.

Direct HPLC experiment can be used for estimation of LogP, but this technique is valid only for neutral molecules or for ionized molecules analyzed in their neutral state [18]. The following is a brief description of this method.

Compounds with known LogP is injected onto C18 hydrophobic column, and the respective retention factors are used to create a calibration curve. The estimation of LogP for unknown compounds can be made on the basis of this calibration curve. This method is straightforward, but requires the previous knowledge of pK_a values for ionizable compounds to avoid the possible ionization that will lead to incorrect determination of values of LogP. Recently, an automated isocratic liquid chromatography system, dedicated to the measurement of LogP, Profiler LDA, was introduced into the market by Sirius-Analytical, Ltd. There were numerous attempts to use the retention time of compound in correlation with its distribution properties in RP HPLC [19, 20]. The retention factor was used to calculate a distribution coefficient between stationary phase and mobile phase. In case of Sirius Profiler LDA automated system, a set of molecules with known LogP values was used to calibrate the system and convert the chromatographic retention time into octanol/water partition coefficients. The system could cover the LogP range from -1 to 5.5 by choosing between three different methods and different column lengths ranging from 1 to 25 cm, but was recently removed from the market. The well-known automated pH titrator from Sirius, GlpK$_a$, can be used as well to determine the octanol/water partition coefficient. The measurement is based on a two-phase acid/base titration in a mixture of water/octanol [21].

Partition coefficient discussed above represents oil/water equilibrium distribution of only neutral forms of a substance. The distribution at different pH is described by LogD, which is the logarithm of the ratio of the concentrations of all forms of analyte in oil and water phases at particular pH. Logarithm of distribution coefficient at pH 7.4 is often used to estimate the lipophilicity of a drug at the pH of blood plasma.

As follows from the definition, the distribution coefficient is dependent on the pH. It is usually assumed that in the oil-phase drug molecule could exist in only nonionic form; thus the distribution coefficient, D_w^0, for basic drug B could be written as

$$D_w^0 = \frac{[B]_{oil}}{[B]_{water} + [BH^+]_{water}} \tag{12-8}$$

If LogP and pK$_a$ for a studied drug is known, then it is possible to express D_w^0 as a function of pH of aqueous phase through these values using equations (12-3) and (12-6)–(12-8). Resulting expression is

$$Log(D_w^0(pH)) = LogP - Log[1 + 10^{pK_a - pH}] \tag{12-9}$$

Figure 12-4 represents the comparison of the pH dependencies of ionic form of a basic drug with LogD.

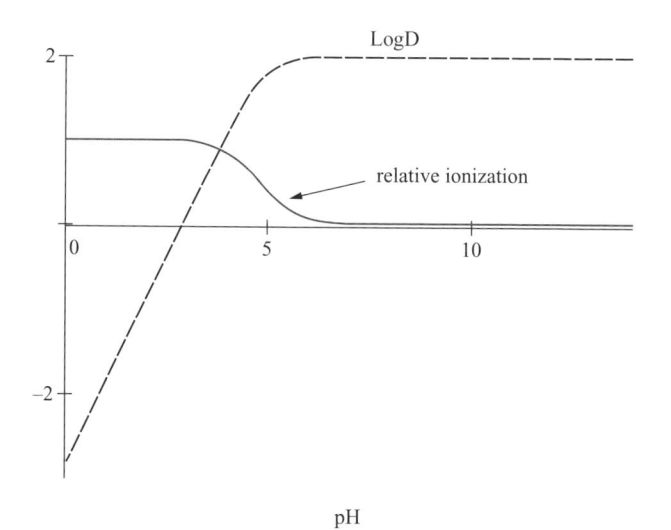

Figure 12-4. Normalized dependence of the protonated form of the base (solid) and its LogD dependence on the aqueous pH (dashed).

At high pH, the neutral form of a drug (basic compound) has a distribution coefficient equal to its partitioning coefficient. With the decrease of the pH of the aqueous phase, the degree of drug ionization increases, thus increasing its total concentration in the aqueous phase. As the pH decreases, the ionic equilibrium is shifted toward the protonated form of a drug, which continually increases its concentration in the aqueous phase and decreases its content in oil phase. There is no plateau region in the LogD curve at low pH for basic compounds (Figure 12-4). On the other hand, for acidic compounds, there is a plateau region in the LogD curve at low pH (pHs below the pK_a); and then as the pH increases, the more ionic equilibrium is shifted toward the ionized form of the acid, which continually increases its concentration in the aqueous phase and decreases its content in the oil phase. This results in the absence of plateau in the LogD curve at high pH (pH > pK_a) for acidic compounds.

These are only the theoretical dependencies; real behavior of actual molecule usually is significantly altered due to different types of intermolecular interactions. Molecular solvation, association, hydrogen bonding, and counterions all have a significant effect on drug ionization constant and partitioning and distribution coefficients. Detailed and comprehensive discussion of these effects could be found in the book by Avdeef [22].

12.2.3 Solubility and Solubilization

Aqueous solubility is one of the most important physicochemical properties of a new drug candidate because it affects both drug absorption and dosage form development. Only a drug in solution can be absorbed by the gastrointestinal track. The rate of dissolution and the intestinal permeability of the drug molecules are dependent on the aqueous solubility—that is, the higher the solubility, the faster the rate of dissolution. An excellent monograph describing the theory of solubility and solubility behavior of organic compounds was written by Grant and Higuchi [23]. For additional information on solubility, the reader can be referred to references 24–27.

Solubility is expressed as the concentration of a substance in a saturated solution at a defined temperature. The US Pharmacopeia (USP) gives the solubility definitions shown in Table 12-2.

Solubility measurements are generally carried out in the early stages of drug development because it affects drug bioavailability evaluation; in many cases, solubility-limited absorption has been reported. Only a compound that is in solution is available to cross the gastrointestinal membrane. The solubility measurements in aqueous buffered systems at different pHs are used to mimic gastrointestinal human or animal fluids. Solubility determination in DMSO is very important at the early stages of lead candidate selection because of the increasing use of 10 mM DMSO solution as a stock solution for biological testing for very slightly soluble lead candidates [29]. In general,

TABLE 12-2. Solubility Definitions by US Pharmacopeia [28]

Descriptive Term	Parts of Solvent Required for One Part of Solute
Very soluble	Less than 1
Freely soluble	From 1 to 10
Soluble	From 10 to 30
Sparingly soluble	From 30 to 100
Slightly soluble	From 100 to 1000
Very slightly soluble	From 1000 to 10,000
Insoluble	10,000 and over

aqueous solubility is measured in simple buffered aqueous media. In practice, the aqueous medium of the gastrointestinal track is a mixture of salts and sur-factants, and the recipes to mimic the fasted (fasted state simulated intestinal fluid, FaSSIF) [30] and fed state (fed state simulated intestinal fluid, FeSSIF) [31] may be used when the influence of gastrointestinal fluid on oral absorption of NCE is studied especially for *in vivo/in vitro* correlation experiments [32]. It was reported that for some compounds the solubility in FaSSIF and FeSSIF will be higher than the solubility in aqueous buffers at the same pH [33].

At the early stage of candidate selection the different experimental methods based on high-throughput solubility measurements are used to determine the apparent solubility of potential lead candidates as well as in silico predictions [34] to quickly assess aqueous solubility. These methods are described in details in references 5 and 35. In the later stages of preformulation when the drug candidate is in a well-characterized crystalline solid state, more precise determination of the equilibrium aqueous solubility is necessary for designing appropriate formulations. The old-fashioned shake flask method is recommended to measure equilibrium aqueous solubility [36] at this stage. The procedure is very simple. The compound in solid state is added to buffered solution in excess (saturated solution), and the suspension is shaken on a mechanical shaker until the system reaches the equilibrium between two phases, solid and liquid. Sometimes the equilibration time is very long and can vary from 2 hours to a few days or weeks, which is dependent upon the numerous factors that affect solubility. Solution stability may also be a concern, as an additional precaution the solutions should be protected from light when possible if they may be prone to photodegradation. To check the equilibrium condition, several HPLC measurements should be determined at several time points. The system is considered to be in equilibrium when the solubility measurements between several time points remain constant.

However, the equilibrium solubility values are very difficult to obtain, because they are affected by many factors such as crystalline form of a substance, particle size distribution, temperature, composition of aqueous phase,

TABLE 12-3. Variation of Aqueous Solubility in the Literature [37]

Compound	Solubility Range (g/mL)
Estradiol	0.16–5.00
Indomethacin	4.00–14.0
Griseofulvin	8.00–13.0
Progesterone	7.90–200
Digoxin	28.0–97.9
Riboflavine	66.0–99.9
Dexamethasone	89.1–121
Hydrocortisone	280–359

TABLE 12-4. Biopharmaceutical Classification of Drug Substances

Class	Solubility	Permeability
Class 1	High solubility	High permeability
Class 2	Low solubility	High permeability
Class 3	High solubility	Low permeability
Class 4	Low solubility	Low permeability

and even the amount of excess solids [37]. Table 12-3 shows some examples of reported aqueous solubility range for commercial drugs.

Aqueous solubility of ionizable molecules at different pH values is an important characteristic because it indicates the potential substance behavior in the stomach and intestinal tract and its potential impact on bioavailability. Moreover, it also provides important information for formulation scientists to define the class of a drug substance in the Biopharmaceutics Classification System (BCS), a regulatory guidance for bioequivalence studies. The BCS is a scientific framework proposed by the FDA to classify drug substances based on their aqueous solubility and intestinal permeability and defines important parameters in the selection of drug candidates into development. According to the BCS, drug substances are classified as shown in Table 12-4.

An objective of preformulation scientist is to determine the equilibrium solubility of a drug substance under physiological pH to identify the BCS class of drug candidate for further development. For BCS classification the test conditions are strictly defined by the FDA. The pH solubility profile of the test drug substance should be determined at 37°C in aqueous media with a pH in the range of 1–7.5. Standard buffer solutions described in the USP are considered to be appropriate for use in these studies. A number of pH conditions are used bracketing the pK_a value for the respective test substance. For example, for a drug with a pK_a of 5, solubility should be determined at pH = pK_a, pH = pK_a + 1, pH = pK_a − 1, pH = 1, and pH = 7.4 Concentration of the drug substance should be determined using a stability-indicating assay that can distinguish drug substance from its degradation products if observed. In order to be classified as highly soluble, the FDA BCS requires that the highest human dose be soluble in 250 mL of aqueous medium over a pH range 1–7.5 [38]. The identification of specific class for the drug candidate is critical for future development of dosage forms.

Different platforms are used for solubility measurements: UV; HPLC with UV detection; or HPLC with MS detection. UV spectrophotometry is the simplest and fastest method, unfortunately with limited applicability. In most cases the drug substance available for the study in the preformulation stage is not pure enough to provide an adequate absorbance–concentration relationship of drug substance itself. In this case, HPLC with UV detection is the most applicable technique to use. Fast gradient methods on short columns could be successfully used in most cases as described in Chapter 17. Some software programs such as ACD/LogD Sol Suite [39] can be used to estimate the solubility as a function of pH and can be used as a starting point to estimate the appropriate dilution of the different solutions prepared at the different pH values.

In some cases, drug substance does not have chromophores with a molar absorbtivity sufficient for accurate quantitation using UV detection. If HPLC with UV detection is used as a basic quantitation technique, then MS detection as a complementary technique is desirable in most cases. LC-MS is essentially preferable in most preformulation assays. High selectivity of the MS detector allows the use of fast gradient HPLC separation methods, which does not require significant development time. Practically in all assays used in preformulation, the quantitation of only drug substance is required and MS detection provides an accurate quantitation.

Identification of pharmaceutically acceptable vehicles that afford sufficient solubilization while maximizing physiological compatibility for preclinical pharmacokinetic evaluation is critical. The most frequently used solubilization techniques include pH manipulation for ionizable compounds; use of co-solvents such as PEG 400, ethanol, DMSO, and propylene glycol; micellar solubilization with surfactants such as Tween 80 or SLS; complexation with cylodextrins [40]. By using the solubilization techniques, the enhancement in solubility of poor water-soluble compounds can be significant compared to aqueous solubility and can facilitate the absorption of drug molecules in the gastrointestinal tract when delivered in solution form.

The requirements for HPLC methods include careful selection of the mobile phases to avoid sample precipitation or emulsification. At the same time, chromatographic conditions should provide positive retention of the drug substance so it won't elute with the void volume.

The solubility measurement at several time points can be used for preliminary solution stability evaluation of new drug candidates. If degradation is observed during the solubility evaluation, further HPLC method development should be oriented not only to determine drug substance concentration, but also on the separation of degradation products from the active.

12.3 CHEMICAL STABILITY

HPLC is a major tool in preformulation stability testing of potential drug candidate. The design of stability testing in the early stage of drug development is not strictly defined by FDA guidance, and different approaches are taken by different pharmaceutical companies. However, there are several major components to a comprehensive stability testing with a goal to achieve maximum information within the shortest period of time:

- Development of a sensitive and reliable HPLC method of separation
- Solution-state stability as a function of pH, temperature, and light
- Chemical solid-state stability evaluation as a function of temperature and humidity
- Identification of degradation products followed by structure elucidation and possible description of degradation mechanism

To achieve this goal, the best approach is to perform forced degradation studies at the preformulation stage of drug development with most viable candidates, which may include the free base or acid and several corresponding salt forms. The FDA and ICH guidance provides very little information about strategies and principles for conducting forced degradation studies, including problems with poorly soluble drugs and exceptionally stable compounds. The stressing condition should be regulated based on the requirements to produce enough degradation products to evaluate the possible routes of degradation, but not to unduly overstress the drug and obtain aberrant results. Sufficient exposure is achieved when a drug substance has degraded >10% from its original amount or after an exposure in excess of the energy provided by an accelerated storage condition. The goal is to mimic what would be observed in formal stability studies under ICH conditions [41]. Another major concern is related to the use of a co-solvent to dissolve the sufficient amount of drug for determination and detection of degradation products. In general, acetonitrile or methanol is used as common co-solvent for forced degradation studies. It was shown that when acetonitrile was used as a co-solvent compared to no co-solvent system, the number of degradation products increased and led to a consequent change in the degradation pathway [42]. The recommendation in this case is to prepare the samples in several co-solvents and compare the behavior of methanol versus acetonitrile for a specific drug candidate. Forced degradation studies based on FDA guidelines are carried out in solution. This involves conditions that are more severe than in accelerated solid-state stability testing. For example, these include temperatures in excess of 40°C, extreme high and low pH values, oxidation by 3% hydrogen peroxide, and light conditions exceeding ICH guideline [43].

As a part of discovery support, these forced degradation studies are performed on discovery batch material to identify future problems with drug candidates and to eliminate the recommendation of unstable molecules to develop or to help define proper storage conditions for early-phase material— that is, store at low temperature, protect from light, and ensure tight packaging. As a part of preformulation studies, this forced degradation testing is not a part of formal stability program for clinical batches, but sheds light in regard to possible thermolitic, hydrolitic, oxidative, and photolitic degradation mechanisms for the prospective drug candidate. At this stage it is critical to develop a suitable HPLC separation method, not only based on UV detection and peak purity check, but also one that is compatible with MS detection. Preferably, columns with 3-μm particles and not more than 15 cm in length (i.d. could be 3.0 or 4.6 mm) should be used, and mobile phases compatible with MS detection are recommended. As a starting point, a C8 column that is stable from 2 to 11 or a phenyl hexyl column that is stable from 2 to 10 could be selected and a gradient could be employed from 5% acetonitrile to 95% acetonitrile. 0.05 v/v% TFA could be used in both acetonitrile and water mobile phases. Development of stability-indicating methods are discussed in the method development chapter (Chapter 8). Despite the usual situation in the preformulation research environment when all tests should have been done yesterday, an analyst should carefully develop a stability-indicating HPLC method because in most cases the conditions of this method will be used as a starting point for most, if not all, further HPLC methods during the development process of a particular drug in the downstream formulation development. Unfortunately, the isolated drug substance and drug-product-related degradation products are not available at this early stage, and the peak purity analysis using UV diode array detection along with mass spectrometric detection should be performed.

Once the initial stability-indicating method is developed, the forced degradation studies are carried out and the pathways for degradation may be elucidated. Four major degradation processes are usually distinguished: oxidation, hydrolysis (H^+ or OH^-), photolysis (light), and catalysis (effect of trace metal ions, Fe^{2+}, Fe^{3+}, Cu^{2+}, Co^{2+}, etc.). Temperature is an integral part of all these processes. According to the Arrhenius equation, the reaction constant is related to the temperature as follows:

$$K = A \exp\left(-\frac{E_a}{RT}\right) \tag{12-10}$$

where E_a is activation energy, R is the gas constant, and T is temperature in degrees kelvin. The higher the temperature the higher the reaction constant, and this leads to the increase of the degradation rate.

Degradation is a chemical transformation of the drug substance and can be expressed as a chemical reaction with the specific kinetics. These reactions can have different orders, which are characterized by the different rate of parent compound decomposition. The most common are zero, first and second order reactions. It is not a subject of this chapter to discuss reaction kinetics in details; however, specific preformulation-related discussions can be found in reference 6, and a general approach with examples is very well described by Martin [44].

Zero-order reactions are usually of self-disintegration type, where decomposition is independent of the concentration of reactants (including drug substance). For this reaction the decrease of the drug substance amount has a linear dependence versus time.

In the first-order reaction, the decomposition is dependent on the concentration of one reactant (drug substance) and the decrease of the substance concentration is exponential. In the second-order reaction, the decomposition is dependent on the concentration of two reactants (e.g., drug substance and water in a hydrolytic degradation). The rate of the decrease of the substance amount is reciprocal to the drug concentration.

Usually the determination of the amount of drug substance at four or more different time points of the degradation experiment is necessary for the determination of the reaction order and construction of the degradation curve, which can then be used to determine the rate constant at a particular temperature.

If the reaction order is known, then rate constant could be calculated from just two points. For example, for the first-order reaction the rate constant is expressed as

$$\ln\left(\frac{[C]}{[C_0]}\right) = -Kt \qquad (12\text{-}11)$$

where $[C]$ is a drug concentration at time t, $[C_0]$ is the original drug concentration (at time $t_0 = 0$), and K is the reaction constant. Subtracting the same equations for time moments t_1 and t_2 from each other, it is possible to calculate the rate constant:

$$K = \frac{\ln\left(\frac{[C_1]}{[C_2]}\right)}{t_2 - t_1} \qquad (12\text{-}12)$$

Note. Only the ratio between initial concentration C_0 of parent compound to the concentration C_t at defined time point should be used in any kinetic calculations as described in detail by Martin [44]. It is not mathematically accurate to select the starting concentration as a base value and calculate all concentrational variations relative to the starting concentration.

TABLE 12-5. Rate Constant and Half-Life Equations

Order	Integrated Rate Equation	Half-Life Equation
0	$x = kt$	$t_{\frac{1}{2}} = \dfrac{a}{2k}$
1	$\log \dfrac{a}{a-x} = \dfrac{kt}{2.303}$	$t_{\frac{1}{2}} = \dfrac{0.693}{k}$
2	$\dfrac{x}{a(a-x)} = kt$	$t_{\frac{1}{2}} = \dfrac{1}{ak}$

Source: Reprinted with permission from reference 45, p. 289.

Based on the known rate constant, the half-life (the period of time required for a drug to decompose to one half the original concentration), can be determined as shown in Table 12-5.

Measurements of the rate constants for at least three different temperatures allows for the calculation of the activation energy and prediction of the temperature dependencies of the drug degradation based on the Arrhenius equation. The relationship between the rate constant and the temperature is given by the Arrhenius equation:

$$k = Ae^{-E_0/RT} \tag{12-13}$$

or

$$\log k = \log A - \frac{E_a}{2.303}\frac{1}{RT} \tag{12-14}$$

where k is the rate constant, R is the gas constant, A is an Arrhenius factor (constant), T is the temperature (in Kelvin), and E_a is activation energy. A plot of the logarithm of rate constant versus the reciprocal of the absolute temperature defines a straight line of slope $-E_a/R$ and intercept $\log A$ [45]. The activation energy can be determined at the different forced degradation conditions (heat, light, peroxide).

In all types of degradation assays the use of LC-MS detection is desirable since it allows for selective detection and quantitation and sometimes allows for structural elucidation of the degradation products. In some cases, tautomerization or intramolecular rearrangements could lead to the formation of degradation products with the same molecular weight. These molecules are usually indistinguishable from the parent compound using MS with molecular ion detection. The employment of LC-NMR technique may be needed to further elucidate the structures.

TABLE 12-6. Ionization Constants and Relative Usage Rate for the Most Common Counterions

Basic Drugs			Acidic Drugs		
Anion	pK_a	%	Cation	pK_a	%
Hydrochloride	−6.1	43	Potassium	16.0	10.8
Sulphate	−3.0	7.5	Sodium	14.8	62
Mesylate	−1.2	2.0	Calcium	12.9	10.5
Maleate	1.9	3.0	Magnesium	11.42	1.3
Phosphate	2.2	3.2	Diethanolamine	9.7	1.0
Salycilate	3.0	0.9	Zinc	9.0	3.0
Tartrate	3.0	3.5	Choline	8.9	0.3
Lactate	3.1	0.8	Aluminium	5.0	0.7
Citrate	3.1	3.0	Alternatives		8.8
Benzoate	4.2	0.5			
Succinate	4.2	0.4			
Acetate	4.8	1.4			
Alternatives		30.2			

12.4 SALT SELECTION

For ionic drugs the salt form can be considered as an alternative to increase the solubility. Drug substance usually is more soluble in aqueous media in its ionic form. Low solubility of the neutral form of the drug substance suggests the necessity to formulate it in the form of salt. The reader is referred to reference 46 for more information about the properties, selection, and use of salt forms for future drug development. Examples of commonly used salt counterions are shown in Table 12-6.

Salt form selection is mainly covered by solid-state charactezation methods, and HPLC is only used to determine the solubility and solid/solution stability of different salt forms. The requirements for HPLC method development is the same as for solubility/stability determination described previously, and the same HPLC method may be applied.

12.5 POLYMORPHISM

Polymorphism is an ability of the drug substance to form crystals with different molecular arrangements giving distinct crystal species with different physical properties such as solubility, hygroscopicity, compressibility, and others. This phenomenon is well known within pharmaceutical companies. The reader can find additional information in references 47 and 48. The determination of possible polymorphic transition and existence of thermodynamically unstable forms during preformulation stage of drug development is important. Typical methods used for solid-state characterization of polymorphism are DSC,

FT/IR, microscopy, and X-ray powder diffraction [49, 50]. HPLC is used to evaluate chemical stability of different polymorphic forms as well as for solubility determination, and this parameter is very critical for drug development, because the difference in solubility can lead to different bioavailability of solid dosage form, especially if the bioavailability is dissolution-limited. An example of how polymorphism can affect final product solubility can be shown on Abbott Laboratories products and on Norvir oral liquid and Norvir semisolid capsules, with Ritonavir as an active ingredient. Ritonavir was not bioavailable in the solid state, and both formulations contained ritonavir in ethanol/water solutions. At the time there was no crystal form control required from FDA for semisolid formulation, and only one form was identified at the development stage. After many successful lots of semisolid capsules, suddenly one lot did not pass the dissolution testing and when the content of the capsule was analyzed by microscopy and X-ray, the different polymorphic form of Ritonavir was identified with significantly low solubility compared to original crystal form [51]. The product was recalled from market and was reformulated. It was a rare example of a dramatic effect of the existence of multiple crystal forms of a commercial pharmaceutical and showed the importance of polymorphic screening for all type of pharmaceutical dosage forms. When the existence of polymorphism for new chemical entity is identified, the property of practical interest is the relative thermodynamic stability of the identified polymorphs; that is, are they monotrops (one is more stable than the other at any temperature) or enantiotrops (a transition temperature T_t exists below and above which the stability order is reversed)? Temperature dependence of the solubility for different polymorphic forms allows easy analysis of the existence of monotrops and enantiotrops and determination of transition temperature from the solubility ratio of the polymorphs [52]. As can be seen from Figure 12-5, intersecting solubility curves (dependence of the logarithm of the

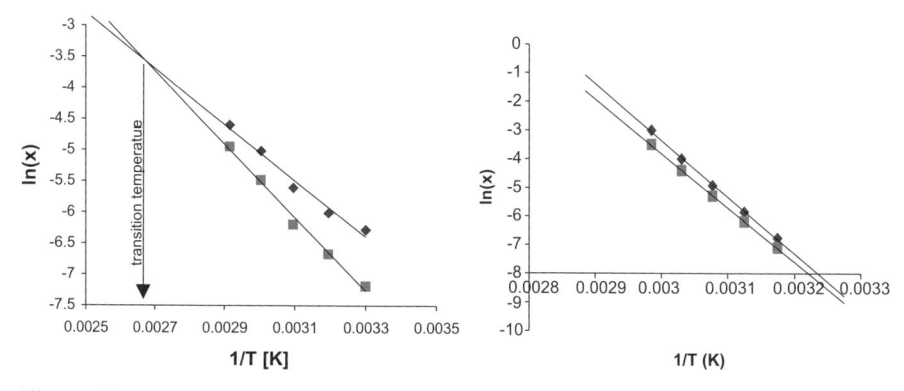

Figure 12-5. Intersecting solubility curves (dependence of the logarithm of the saturation concentration on the inverse temperature) indicate an enantiotropic nature of the polymorps, while parallel curves are indicative for monotropic polymorphs. The intercept for enantiotrops corresponds to the transition temperature.

saturation concentration on the inverse temperature) indicate an enantiotropic nature of the polymorps, while parallel curves are indicative for monotropic polymorphs. The intercept for enantiotrops corresponds to the transition temperature, T_t, which can be easily determined from the graph. In general, the most thermodynamically stable form that has a lower solubility and better stability is accepted for development. It was reported previously that the more thermodynamically stable polymorph is more chemically stable than a metastable polymorph due to different factors such as higher density, optimized orientation of molecules, and hydrogen bonding in the crystal lattice [48, 53].

The HPLC method development requirements using short columns and fast HPLC to determine the assay concentration for each polymorph at the different temperatures are the same as for solubility determination. However, for stability evaluation of the different polymorphs a stability-indicating HPLC method should be used.

12.6 PREFORMULATION LATE STAGE (DEVELOPMENT SUPPORT)

After a new chemical entity has been selected to move forward to development, the preformulation scientist supports the studies related to formulation, toxicology, and pharmacology.

Based on the previous knowledge about the properties of novel chemical entity obtained during the late discovery and nomination, the stability studies of API are performed based on ICH guidance [43].

Typically, by using a GMP batch with selected solid-state form, the solution-state stability and solid-state stability studies are performed at various conditions. In general, the three conditions used for solid-state stability evaluation—25/60, 40°C/75% RH, and 50°C dry conditions—at several time points up to 3 months (initial, 2 weeks, 6 weeks, 3 months) are reasonable to evaluate storage conditions of API and the impact of heat and humidity. For solution-state stability, it is important to evaluate (a) the stability at pH 1, 2, 4, 7, and 10, at ambient and elevated temperature, (b) the influence of ICH light, and (c) oxidation by peroxide. To support toxicology studies, the stability of API suspension at different strengths in aqueous solutions with corresponding excipients are evaluated after 1 day, 2 days, 3 days, and 8 days on potency and stability. HPLC techniques are used for all these types of stability testing, and GLP requirements are applied to HPLC methods. The system suitability for the method needs to be defined and the figures of merit such as linearity, LOD, LOQ, and solution stability in the diluent need to be performed to qualify this as a stability-indicating HPLC method at this stage.

Since the degradation products are not yet identified in this stage, it is advisable to use detection systems, which have universal response and also provide high sensitivity. MS is probably the most sensitive detector that also can provide relatively universal response for most ionizable compounds, although the degree of ionization may vary with the type of interface used. In some cases the use of evaporative light scattering detector is also advisable. This detector has a universal response for practically any molecule with molecular weight above 300 Da.

The key to a good stability-indicating assay is to select concentrations for analysis that allow the detection of degradation product peaks that are at least 0.1% of the parent peak, which is consistent with ICH impurities guideline [54].

Identification of dosage form composition during the design of Phase I clinical formulations is a key step in accelerating drug development, and this is performed through drug–excipient compatibility testing. This test is the most time- and labor-consuming. A proper design of experiments must be conducted. The amount of samples sometimes reaches 100 or more (conditions: 50°C, 40°C/75% RH, or 40°C plus water, for placebo, API, and binary mixtures of more then 30 excipients) for one time point, depending on the accepted scheme of compatibility testing. Despite the importance of drug–excipient compatibility testing, there is no universal protocol for this study. In general, the amount of the API in the binary mixture is determined on the basis of the expected drug-to-excipient ratio in the final formulation. To eliminate a time-consuming step of analyzing all binary mixtures (API + excipient), additional samples can be prepared such that different groups of samples containing four or five excipients together at the same ratio as in final form are analyzed first. If there is no change observed, the binary mixtures are not analyzed; however, if a change is observed, the respective binary samples (i.e., API + excipient 1, API + excipient 2) containing the excipients from the mixture (i.e., API + 5 excipients) should be analyzed to determine which excipient(s) had led to the degradation. Sometimes a combination of excipients leads to the degradation, and this should not be ruled out.

For many years, differential scanning calorimetry (DSC) was a standard method in preformulation to characterize drug–excipient compatibility based on the change of thermal curves [55, 56]. Despite the simplicity of DSC experiment and small quantity of API needed, an evaluation of thermograms can be difficult, and conclusions based only on DSC results can be misleading [56]. Recently [57], the use of stability-indicating HPLC methods was emphasized to be used for these excipient compatability studies in order to better characterize the API–excipient interactions by providing not only qualitative but also quantitative results for test substance and its related degradation products. A comparison between two methods, HPLC and DSC, was reported by Ceschel et al. [58]. Good correlation between DSC and HPLC results was demonstrated in the case of acetyl salicylic acid with a number of commonly used excipients. The incompatibility of magnesium stearate with acetyl salicylic acid was shown by DSC, and this was confirmed by HPLC. In the case of using HPLC as an analytical method for excipient compatibility studies, not only can the results be reported as potency or concentration of parent peak, but also

**TABLE 12-7. Example of Drug–Excipient Compatibility Testing Design.
Reprinted with Permission from reference 59.**

	1	2	3	4	5	6
Drug substance	200	25	25	25	25	25
Lactose		175				170
Mannitol			175			
Microcrystalline cellulose				175		
Dibasic calcium phosphate dihydrate					175	
Magnesium stearate						5
Sodium stearyl fumarate						
Stearic acid						
Potency remaining (% initial)	96.4	95.7	95.8	93.9	85.0	64.3
Hydrolysis product formed[a]	3.3	4.1	4.0	5.8	16.7	37.0

[a]Expressed as a percentage of the parent drug.

all the amounts (mass or area%) of unknown peaks can be determined. The excipients themselves should also be analyzed, and the chromatographic peaks of excipient should be recorded but not integrated against the API. Also, the excipient peaks should be resolved from both the API synthetic by-products and the degradation products formed due to the incompatibility of API with excipients. The placebos should also be put on stability. This study provides very critical information for formulators to guide their future development of novel dosage forms.

A very good example of a drug–excipient compatibility screening model was described by Serajuddin et al. [59]. They showed the importance of this test in the early stage of formulation development prior to Phase I and developed a protocol for this study (see Table 12-7). Table 12-7 shows compositions of the 17 drug–excipient blends stored at 50°C in closed vials with 20% added water (weights of all ingredients are in milligrams). The assay% and area% (using area normalization) were determined for each of the mixtures. It can be seen that some excipients caused significant degradation, and the major degradation pathway is hydrolysis.

Based on the reported results, it was advised to use the described model to perform drug–excipient compatibility testing prior to Phase I to eliminate potential future issues related to drug instability in final formulation. Because the drug–excipient compatibility testing is conducted at an early drug development stage when a fully validated HPLC method is not available, the same GLP HPLC method as for forced degradation studies can be used for this test as well.

Moreover, if adequate forced degradation studies (i.e., acid/base hydrolysis) are performed in the early preformulation stage, the identification of a potential degradation product that might arise during excipient compatibility

	Experiment									
7	8	9	10	11	12	13	14	15	16	17
25	25	25	25	25	25	25	25	25	25	25
			170				170			
170				170				170		
	170				170				170	
		170				170				170
5	5	5								
			5	5	5	5				
							5	5	5	5
65.4	65.3	38.1	77.9	81.9	77.6	81.8	90.0	92.9	88.1	78.3
36.7	36.3	33.7	21.8	15.4	20.1	15.3	9.7	6.9	11.7	21.6

studies might have been already determined (i.e., acid/base hydrolysis product). Note that an adequate HPLC method must be developed to properly retain any compounds that may be formed as a result of hydrolysis and are more polar and less hydrophobic than the API; also, the same HPLC method must be able to elute additional hydrophobic species—for example, dimers, which are formed during the excipient compatibility studies.

12.7 CONCLUSIONS

Discovery formulation support and early preclinical development support can be provided by the preformulation group as a part of the drug development process. The functions of this unit is to help discovery in physicochemical characterization of new drug molecules by providing information on solubility, stability, pK_a, and LogP/LogD as well as formulation support for PK animal studies to recommend a final candidate for selection to development. After candidate selection, the early preformulation unit provides a major source of information to formulation and analytical scientists regarding the properties of the recommended drug molecules. The development preformulation support provides the additional testing of prototype formulation and excipient compatibility samples as well as guidance for salt form selection and polymorphs screening. The major role of this unit is to bridge discovery and development stages. HPLC coupled with not only UV, but also other alternative detectors, is the predominate tool for analyzing drug substances with high speed and efficiency, which is required in the preformulation stage of drug development.

REFERENCES

1. I. Kola and J. Landis, Can the pharmaceutical industry reduce attrition rates?, *Nature Rev. Drug Discovery* **3** (2004), 711–715.

2. R. A. Lipper, How can we optimize selection of drug candidates from many compounds at the discovery stage, *Mod. Drug Discovery* **2** (1999), 55–60.

3. C. A. Lipinski, R. Lambardo, B. W. Dominy, and P. J. Feeney, *Adv. Drug Delivery Rev.* **23** (1997), 3–25.

4. H. van de Waterbeemd, H. Lennernas, and P. Artursson, *Drug Bioavailability*, Willey-VCH, New York, 2002, p. 22.

5. A. Avdeef, *Absorption and Drug Development*, Wiley Interscience, New York, 2003.

6. J. I. Wells, *Pharmaceutical Preformulation*, Ellis Horwood Chichester, UK, 1988.

7. J. T. Carstensen, *Pharmaceutical Preformulation*, Technomic Publishing Company, Lancaster, PA, 1998.

8. M. Gibson (ed.), *Pharmaceutical Preformulation and Formulation*, Interpharm/ CRC Press, Boca Raton, Florida, 2004, pp. 585.

9. H. van de Waterbeemd, H. Lennernäs, and P. Artursson, Pharmacokinetic/pharmacodynamic modeling in drug development, drug bioavailability, *Annu. Rev. Pharmacol. Toxicol.* **40** (2000), 67–95.

10. A. Avdeef, Absorption and drug development, in R. T. Borchart and C. R. Middaugh (eds.), *Pharmaceutical Profiling in Drug Discovery for Lead Selection*, AAPS Press, Tucson, AZ 2004, Wiley, Hoboken, NJ, 2003.

11. J. E. A. Comer, High-throughput measurement of log D and pKa in H. van de Waterbeemd, H. Lennernas, P. Artursson (eds.), *Drug Bioavalability*, Wiley, Hoboken, NJ, 2004, pp. 33–39.

12. *Applications and Theory Guide to pH—Metric pK_a and log P Determination*, Sirius Analytical Instruments Ltd., Forest Row, UK, 1992.

13. W. R. Melander and C. Horvath, Reversed-phase chromatography, in C. Horvath (ed.), *HPLC, Advances and Perspectives*, Vol. 2, Academic Press, New York, 1980, pp. 114–303.

14. A. J. Leo, P. Y. C. Jow, C. Silipo, and C. Hansch, *J. Med. Chem.* **18** (1975), 865.

15. S. Siekerski and R. Olszer, *J. Inorg. Nucl. Chem.* **25** (1963), 1351.

16. H. A. Mottola and H. Friezer, Distribution of certain 8-quinolinols and their copper(II) chelates in a series of organic solvent—aqueous pairs, *Talanta* **13** (1966), 55–65.

17. K. Valko, Measurements of physical properties for drug design in industry, in *Separation Methods in Drug Synthesis and Purification*, Vol 1, Elsevier, Amsterdam, 2000, pp. 539–542.

18. R. Lombardo, M. Y. Shalaeva, K. A. Tupper, and M. H. Arapham, ElogP(oct): A tool for lipophilicity determination in drug discovery, *J. Med. Chem.* **43** (2000), 2922–2928.

19. K. Valko, General approach for the estimation of octanol/water partition coefficient by RP HPLC, *J. Liq. Chromatogr.* **7** (1984), 1405–1424.

20. K. Valko, C. M. Du, C. Bevan, D. Reynolds, and M. H. Abraham, Rapid method for the estimation of octanol/water partition coefficient (log P_{oct}) from gradient RP-HPLC retention and a hydrogen bond acidity term, *Curr. Med. Chem.* **8** (2001), 1137–1146.

21. A. Avdeef and K. J. Box, *Sirius Technical Application Notes (STAN)*, Vol. 2, Sirius Analytical Instruments, Ltd., Forest Row, UK, 1995.

22. A. Avdeef, *Absorption and Drug Development: Solubility, Permeability, and Charge State*, Wiley, Hoboken, NJ, 2003.

23. D. J. W. Grant and T. Higuchi, *Solubility Behavior of Organic Compounds*, Wiley, New York, 1990.

24. S. H. Yalkowsky, *Solubility and Solubilization in Aqeous Media*, Oxford University Press, New York, 1999.

25. S. H. Yalkowsky and Y. He. *Handbook of Aqueous Solubility Data*, CRC Press, Boca Raton, FL, 2003.

26. D. J. W. Grant and H. G. Brittain. Solubility of pharmaceutical solids, in H. G. Brittain (ed.), *Physical Characterization of Pharmaceutical Solids*, Marcel Dekker, New York, 1995, pp. 321–386.

27. K. C. James, *Solubility and Related Properties*, Marcel Dekker, New York, 1986.

28. The USP 29-NF24, General Notices, Test and Assays.

29. C. A. Lipinski, *Solubility in Water and DMSO: Issues and Potential Solutions in Pharmaceutical Profiling in Drug Discovery for Lead Selection*, AAPS Press, Tucson, AZ, 2004, pp. 93–125.

30. E. S. Kostewicz, U. Brauns, R. Becker, and J. B. Dressman, Forecasting the oral absorption behavior of poorly soluble weak bases using solubility and dissolution studies in biorelevant media, *Pharm. Res.* **19** (2002), 345–349.

31. E. Galia, E. Nicolaides, D. Horter, R. Lobenberg, C. Reppas, and J. B. Dressman, Evaluation of various dissolution media for predicting *in vivo* performance of class I and II drugs, *Pharm. Res.* **15** (1998), 698–705.

32. S. Li, H. He, L. J. Parthiban, H. Yin, and A. T. M. Serajuddin, IV–IVC considerations in the development of immediate-release oral dosage form, *J. Pharm. Sci.* **94** (2005), 1396–1417.

33. B. L. Pedersen, A. Mullertz, H. Bronsted, and H. G. Kristensen, A comparison of the solubility of danazol in human and simulated gastrointestinal fluids, *Pharm. Res.* **17** (2000), 891–895.

34. W. L. Jorgensen and E. M. Duffy, Prediction of drug solubility from structure, *Adv. Drug Deliv. Rev.* **54** (2002), 355–366.

35. L. Pan, Q. Ho, K. Tsutsui, and L. Takahashi, Comparison of chromatographic and spectroscopic methods used to rank compounds for aqueous solubility, *J. Pharm. Sci.* **90** (2001), 521–529.

36. S. H. Yalkowsky and S. Banerjce, *Aqueous Solubility: Methods of Estimation for Organic Compounds*, Wiley, New York, 1992.

37. K. Kawakami, K. Miyoshi, and Y. Ida, Impact of the amount of excess solids on apparent solubility, *Pharm. Res.* **22** (2005), 1537–1543.

38. FDA 2000 Guidance for industry: Waiver of *in vivo* bioavailability and bioequivalence studies for immediate release solid oral dosage forms based on Biopharmaceutics Classifiaction System, Rockville, MD.

39. ACD/Log *D* Sol Suite, version 7.0, Advanced Chemistry Development, Inc., Toronto ON, Canada, www.acdlabs.com.

40. R. G. Strickley, Solubilizing excipients in oral and injectable formulations, *Pharm. Res.* **21** (2004), 201–230.

41. D. W. Reynolds, K. L. Facchine, J. F. Mullaney, K. M. Alsante, T. D. Hatajik, and M. G. Motto, Available quidances and best practices for conducting forced degradation studies, *Pharm. Technol.* **2** (2002), 48–58.

42. G. Laus, Kinetics of acetonitrile-assited oxidation of tertiary amines by hydrogen peroxide, *J. Chem. Soc. Perkin Trans.* **2** (2001), 864–868.

43. FDA, International Conference on Harmonization: Stability Testing of New Drug Substances and Products, Federal Register 59 (183) 48753–48759, September 22, 1994.

44. A. Martin, *Physical Pharmacy*, 4th edition, Lippincott Williams & Wilkins, Baltimore, 1993, pp. 284–323.

45. P. W. Atkins, *Physical Chemistry*, W. H. Freeman, San Franscisco, 1978.

46. P. H. Stahl and C. G. Wermuth (eds.), *Handbook of Pharmaceutical Salts*, Wiley-VCH, New York, 2002.

47. D. J. W. Grant, Theory and origin of polymorphism, in H. G. Brittain (ed.), *Polymorphism in Pharmaceutical Solids*, Drugs and the Pharmaceutical Sciences Series, Marcel Dekker, New York, Vol. 95, (1999), pp. 1–33.

48. S. R. Byrn, R. Pfeiffer, and J. Stowell, *Solid-State Chemistry of Drugs*, 2nd edition, SSCI, Inc., West Lafayette, IN, 1999.

49. H. G. Brittain, Methods for the characterization of polymorphs and solvates, in H. Brittain: *Polymorphism in Pharmaceutcal Sciences*, Marcel Dekker, New York, (1999), 227–278.

50. X. Chen, M. Carillo, R. C. Haltiwanger, and P. Bradley, Solid state characterization of mometazone furoate anhydrous and monohydrate forms, *J. Pharm. Sci.* **94** (2005), 2496–2509.

51. J. Bauer, S. Spanton, R. Henry, J. Quick, W. Dziki, W. Porter, and J. Morris, Ritonavir: An extraordinary example of conformational polymorphism, *Pharm. Res.* **18** (2001), 859–866.

52. C. H. Gu and D. J. Grant, Estimating the relative stability of polymorphs and hydrates from heats of solution and solubility data, *J. Pharm. Sci.* **90** (2001), 1277–1287.

53. S. R. Byrn, W. Xu, and A. W. Newman, Chemical reactivity in solid-state pharmaceuticals: Formulation implications, *Adv. Drug Deliv. Rev.* **48** (2001), 115–136.

54. FDA, ICH: Guideline on impurities in new drug substances, *Fed. Register* **61** (1996), 371–376.

55. J. T. Carstensen and C. T. Rhodes (eds.), *Drug Stability, Principles and Practices*, Marcel Dekker, New York, 2000.

56. M. Tomassetti, A. Catalani, V. Rossi, and S. Vecchio, Thermal analysis study of the interactions between acetaminophen and excipients in solid dosage forms in some binary mixtures, *J. Pharm. Biochem. Anal.* **37** (2005), 949–955.

57. *The USP*, 24th Revision. United States Pharmacopeia Convention, Inc., Rockville, MD, 2000.

58. G. C. Ceschel, R. Badiello, C. Ronchi, and P. Maffei, Degradation of components in drug formulations: A comparison between HPLC and DSC methods, *J. Pharm. Biochem. Anal.* **32** (2003), 1067–1072.

59. A. T. M. Serajuddin, A. B. Thakur, R. N. Ghoshal, M. G. Fakes, S. A. Ranadive, K. R. Morris, and S. A. Varia, Selection of solid dosage form composition through drug–excipient compatibility testing, *J. Pharm. Sci.* **88** (1999), 659–704.

13

THE ROLE OF LIQUID CHROMATOGRAPHY–MASS SPECTROMETRY IN PHARMACOKINETICS AND DRUG METABOLISM

RAY BAKHTIAR, TAPAN K. MAJUMDAR, AND FRANCIS L. S. TSE

13.1 INTRODUCTION

Recent advances in mass spectrometry have rendered it an attractive and versatile tool in industrial and academic research laboratories. As a part of this rapid growth, a considerable body of literature has been devoted to the application of mass spectrometry in clinical studies. In concert with separation techniques such as liquid chromatography, mass spectrometry allows the rapid characterization and quantitative determination of a large array of molecules in complex mixtures. Herein, we present an overview of the above techniques accompanied with several examples of the use of liquid chromatography–tandem mass spectrometry in pharmacokinetics/drug metabolism assessment during drug development.

Since the evolution of pharmaceutical research [1, 2], the stages of drug discovery and development have followed three predominant patterns: (i) the systematic and methodical approach by chemists to rationally design and synthesize a molecule to target a specific molecular system (e.g., ion channels, receptors, enzymes, DNA); (ii) the isolation and purification of the active ingredients of medicinal plants or microorganisms to screen their spectrum of

HPLC for Pharmaceutical Scientists, Edited by Yuri Kazakevich and Rosario LoBrutto
Copyright © 2007 by John Wiley & Sons, Inc.

activity using *in vitro* models; or (iii) the serendipitous discovery of a compound with a novel pharmacological action (e.g., the accidental discovery of antidepressants). Today, one of the increasingly popular and complementary approaches for drug discovery in the pharmaceutical industry is to perform massive parallel synthesis in solution or on a solid support. In addition, with the advent of functional genomics and proteomics, cell-based assays, and molecular biology, a multitude of therapeutic targets have been validated [3].

With an increasing number of potential molecular targets identified through the science of functional proteomics and genomics, diverse libraries of new chemical entities (NCEs) have to be generated and evaluated. Consequently, the rapid growth of combinatorial libraries has posed a need for faster, accurate, and sensitive analytical techniques capable of large-scale high-throughput screening (HTS). Although *in vitro* assays do not necessarily reflect the complexity of the *in vivo* interactions, the speed and simplicity of the former have rendered them an integral part of the screening process.

In recent years, the *in silico* and experimental modeling of pharmacokinetic/pharmacodynamic (PK/PD) relationship have become increasingly popular [4, 5]. The integration of PK (i.e., drug dose and biological fluid concentration) and PD (i.e., pharmacologic effect) provides a key determinant in understanding the dosing regimen and therapeutic effect of a potential drug compound. To this end, analytical assays also play a pivotal role in defining the PK/PD relation of NCEs. In many cases, both the drug concentration and PD biomarkers (*vide infra*) can be directly measured in peripheral fluids using specific analytical techniques.

Furthermore, samples generated from large-scale clinical trials along with the ambitious development timelines to get safe and efficacious drugs to market warrant the use of HT bioanalysis. Numerous improvements in speed, sensitivity, and accuracy, augmented with innovations in automation in conjunction with mass spectrometry (MS) detection, have allowed for versatile and multifaceted platforms [6–8].

13.2   IONIZATION PROCESSES

Mass spectrometry (MS) is playing an increasingly visible role in the molecular characterization of combinatorial libraries, natural products, drug metabolism and pharmacokinetics, toxicology and forensic investigations, and proteomics. Toward this end, electrospray ionization (ESI), atmospheric pressure chemical ionization (APCI), and atmospheric pressure photo-ionization (APPI) have proven valuable for both qualitative and quantitative screening of small molecules (e.g., pharmaceutical products) [9–14].

The utility of ESI (Figure 13-1) lies in its ability to generate ions directly from the solution phase into the gas phase. The ions are produced by application of a strong electric field to a very fine spray of the solution containing the analyte. The electric field creates highly charged droplets whose subse-

Liqud Phase Ionization

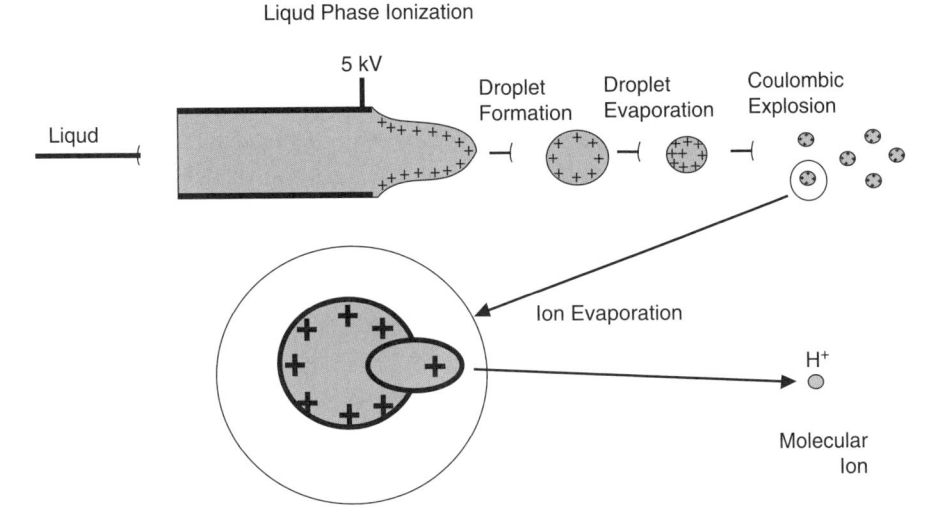

Figure 13-1. A simplified schematic of the ESI process. (Courtesy of Dr. P. Tiller.)

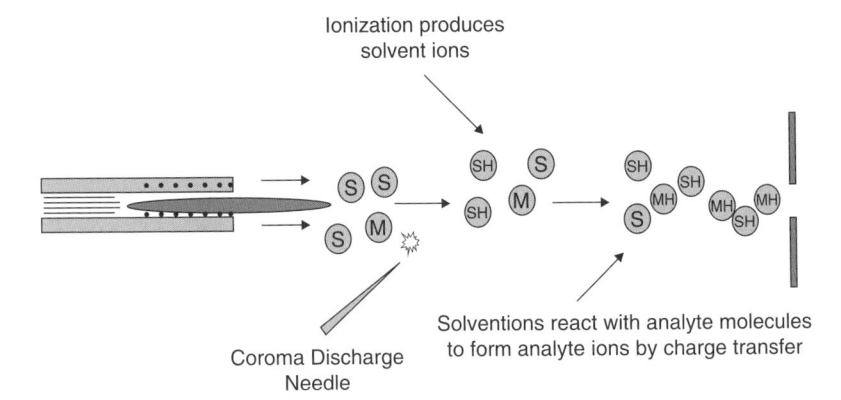

Figure 13-2. A simplified schematic of the APCI process. (Courtesy of Dr. P. Tiller.)

quent vaporization (or desolvation) results in the production of gaseous ions. The fact that ions are formed from solution has established the technique as a convenient mass detector for liquid chromatography (LC/MS) and for automated sample analysis. In addition, ESI-MS offers many tangible benefits over other mass spectrometric methods including the ability to qualitatively analyze low-molecular-weight compounds, inherent soft-ionization, excellent quantitation and reproducibility, high sensitivity, and its amenability to automation.

Analogous to the ESI interface, APCI (Figure 13-2), also referred to as the heated nebulizer (HN), induces little or no fragmentation to the analyte.

Therefore, the APCI spectrum also tends to be simpler in interpretation than the traditional electron ionization (EI), which results in extensive fragmentation of the precursor ion. As a result, APCI and ESI are referred to as "soft-ionizations," while EI is considered a "hard-ionization" technique. Generally, volatile and thermally stable compounds can be subjected to LC/APCI/MS analysis. In quantitative analysis, APCI provides a greater (i.e., in terms of linearity) dynamic range than ESI and it is considered rugged, easy to operate, and relatively tolerant of higher buffer concentrations (i.e., fewer matrix effects). In ESI, at about 10^{-5} M and higher, the ion signal becomes fixed and independent of sample concentration (plateauing effect) and may exhibit non-linearity at higher concentrations. In contrast, APCI can offer a wider linear dynamic range. For example, in our laboratory (data not shown) we have routinely developed reversed-phase LC/APCI/MS/MS assays ranging from 1.0 ng/mL to 10,000 ng/mL with a correlation coefficient of >0.996. Furthermore, APCI can accommodate flow rates of up to 2.0 mL/min and is effective in the analysis of medium- and low-polarity compounds [12]. In qualitative drug metabolism studies, a combination of APCI and ESI experiments can prove valuable in distinguishing certain oxidative biotransformations (e.g., *N*-oxidation *versus* hydroxylation) [15, 16]. In contrast to ESI, APCI is not suited for the analysis of biopolymers, proteins, peptides, and thermally labile species.

In the APCI process, electrons originating from a corona discharge needle ionize the analyte via a series of gas-phase ion-molecule reactions. For example, in the positive-ion mode, the energetic electrons start a sequence of reactions with the nebulizing gas (typically nitrogen), giving rise to nitrogen molecular ions. Using APCI, depending on the composition of the HPLC mobile phase, ions such as $[H_2O + H]^+$, $[CH_3OH + H]^+$, $[NH_3 + H]^+$, and/or $[CH_3CN + H]^+$ are formed via series of ion-molecule reactions with the nitrogen molecular ions. Subsequently, additional ionization is initiated by exothermic proton transfers from the protonated solvent ions to the neutral analyte molecules yielding [analyte + H]$^+$, [analyte + CH_3OH + H]$^+$, [analyte + NH_3 + H]$^+$ ions, and so on. In general, metal adduct ions are observed less commonly in APCI as opposed to ESI, where they are more prevalent. Greater sensitivity is attained if the solvent is polar and contains ions through the addition of an electrolyte. The desolvation process is then further enhanced by the heating element within the APCI assembly, which is maintained at 300–550°C. One of the drawbacks of APCI is its lack of compatibility with low eluent flow rates. The stability of the ionization response may be poor at low rates (i.e., less than 50 μL/min).

In contrast, ESI is compatible with miniaturized columns and amenable to sample-limited scenarios such as biochemical and biotechnological applications. ESI can be considered a flow-sensitive technique. The dimension of the primary droplets is dependent on the flow rate. Therefore, by using columns with a smaller internal diameter (i.d.) and consequently lower flow rates, the concentration of the analytes in the spray solution can vary and it can be considered a concentration-dependent ionization process. It is concentration-dependent in the sense that the surface charge density of the droplets in the gas phase is higher due to more effective desolvation of the droplets since lower flow rates are used. The use of solvent-buffer post-column addition also allows optimization for improved analyte ion current response. Increasing the flow rate increases droplet size, which decreases the yield of gas-phase ions from the charged droplets.

Recently, atmospheric pressure photo-ionization (APPI) [17–19] was introduced as a complementary ionization technique to ESI and APCI. APPI (Figure 13-3) is now commercially available by several MS vendors such as Agilent Technologies, Applied Biosystems (Sciex), Waters (Micromass), and Thermo Electron (Finnigan) Corporations. This technique can be used to ionize an analyte that otherwise is not easily ionizable using either APCI or ESI. In APPI, to increase ionization efficiency, a high-intensity UV radiation source is used (i.e., a 10-eV krypton discharge lamp) in a direct or an indirect mode. In the direct mode, often a molecular ion is generated by irradiation; while in the indirect mode, a dopant is used in conjunction with the analysis. A photoionizable dopant such as acetone or toluene is employed to mediate (as dopant photo-ions) the production of ions by proton or electron transfer. The dopant is introduced to the APPI ionization chamber by a separate pump at an optimized steady flow rate during analysis (e.g., 10–15% of the mobile phase flow rate, post-column). A number of excellent articles have recently been published on the applicability of APPI for the analysis of small molecules [17–19].

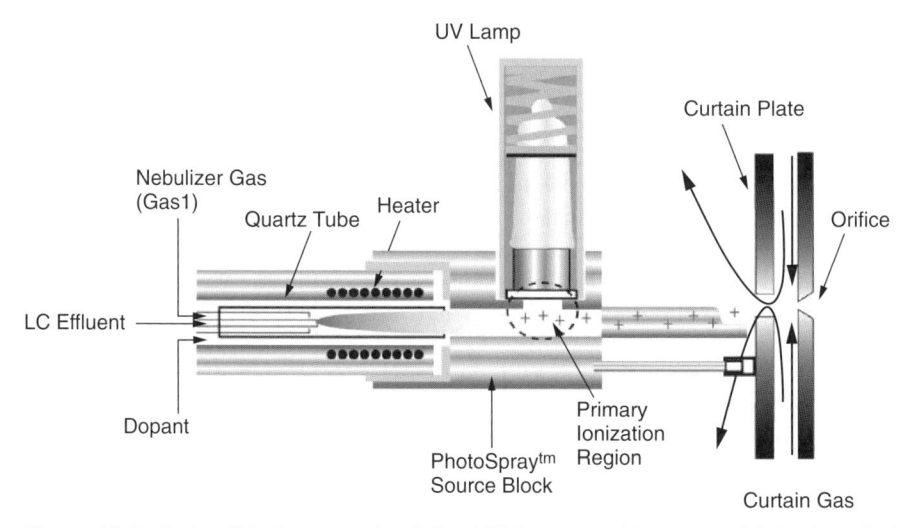

Figure 13-3. A simplified schematic of the APPI process. (Courtesy of Sciex/Applied Biosystems Corporation.)

13.3 TANDEM-MASS SPECTROMETRY (MS/MS)

For purposes of quantitative analysis, selected ion monitoring (SIM) and selected reaction monitoring (SRM) are two commonly utilized approaches. The latter is also referred to as multiple reaction monitoring (MRM). In both modes, considerable structural information is lost; nonetheless, these techniques are extremely powerful for target compound quantification in biological matrices, if the compound of interest is known.

In the SIM mode, the MS is tuned to a particular m/z window (preferably at unit resolution), which corresponds to the ion of interest (i.e., $[M + H]^+$, or a stable adduct such as $[M + X]^+$, where X = Na, K, NH_4, etc.). SIM may require a more elaborate chromatographic separation in order to minimize interference from endogenous species. However, in the SRM approach, higher selectivity and sensitivity are realized. Thus, shorter chromatographic runs (faster injection cycles) and limited sample pretreatment could be tolerated without significant loss in sensitivity.

In addition, due to lack of MS/MS capability, SIM has been more commonly performed on single quadrupole MS, while SRM has been broadly adapted on triple quadrupole (Figure 13-4) and ion-trap mass spectrometers. The increase in sensitivity and selectivity of SRM stem from the ion-chromatogram (i.e., LC-MS/MS) obtained by specific precursor-to-product ion transition for an analyte of interest (Figure 13-4). Conversely, in an SIM mode, the relative background noise due to the presence of other isobaric species (i.e., ions with a same m/z as the analyte of interest) can result in a lower signal-to-noise ratio for the analyte. Due to the widespread acceptance of SRM in quantitative analysis, the remaining part of this section focuses on a description of tandem-mass spectrometry (MS/MS), which is utilized in SRM (or MRM) experiments.

Tandem-mass spectrometry or collision-induced dissociation (CID) is one of the most widely used techniques for probing the structure of ions in the gas phase [20]. To this end, ease of application to various instrumental types, along with its experimental simplicity, account for the wide popularity of CID. In a typical CID experiment, a beam of ions with a specific m/z (denoted as the precursor or parent ion) is selected and collided with a neutral and nonreactive gas-phase target (e.g., argon, xenon, helium, nitrogen). These collisions result in subsequent fragmentation and product ions that are a direct consequence of dissociation of the precursor ion. Generally, the resulting fragmentation pattern is unique to a particular ion structure. The various CID techniques can be subdivided into categories based on the translational or collision energy of the precursor ion prior to collision with the target gas. The two main categories include low-energy CID, in the range of 1–300 eV (i.e., used in triple quadrupole and ion-trap instruments), and high-energy CID at approximately 1–25 KeV (i.e., used in guided-ion beam or sector instruments). Currently, one of the most common approaches is to perform MS/MS experiments on a triple quadrupole instrument. Tandem-MS experiments have been particularly popular for the qualitative and quantitative analysis of small mol-

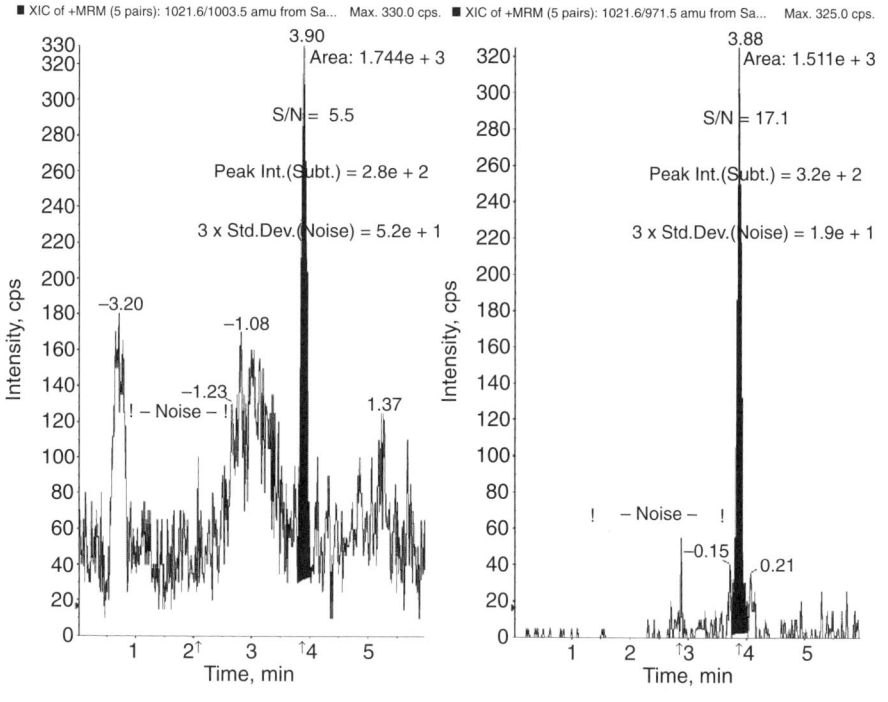

Figure 13-4. Representative MRM scans (plasma extract of a proprietary compound) using an API 5000 triple quadrupole unit (Sciex). Each panel contains a distinct MRM transition for the same compound: m/z 1021.6 → 1003.5 (left panel) and m/z 1021.6 → 971.5 (right panel). Signal-to-noise ratio is designated as S/N. Experimental conditions: ESI, positive ion mode, protein precipitation was used for sample preparation, injection volume was 10 μL, the column was a C_{18}, and dimension was 20 × 2.1 mm, using a linear gradient elution: 0 min (20% B)–6 min (90% B)–8 min (90% B), where B was 0.2% formic acid in acetonitrile and A was 0.2% formic acid in water; separation was performed at room temperature.

ecules such as pharmaceutical products in biological fluids [21–23]. In recent years the sensitivity and selectivity of MS/MS analysis of xenobiotics have been put to use in toxicokinetics, pharmacokinetics, metabolic, formulation, and early drug discovery studies.

13.4   SAMPLE PREPARATION USING AN OFF-LINE APPROACH

One of the critical steps in qualitative and quantitative analysis is the sample preparation procedure. Sample preparation step can affect specificity, sensitivity, accuracy, precision, and throughput of a bioanalytical procedure. In addition to development and optimization of the chemistry involved in sample processing, the use of semiautomated or fully automated protocols has been

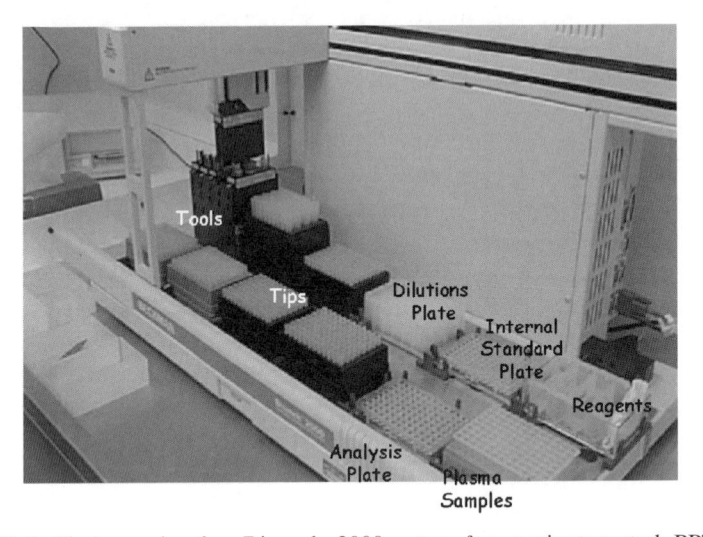

Figure 13-5. Photograph of a Biomek 2000 setup for semiautomated PPT, LLE, or SPE process in the authors' laboratory (also see www.beckmancoulter.com and reference 99).

implemented in recent years [24, 25]. The popularity of off-line sample processing in batch-mode has dramatically improved the throughput of this rate-limiting step.

Generally, there are three commonly used approaches for off-line sample processing: SPE (solid-phase extraction), LLE (liquid–liquid extraction), and protein precipitation (PPT). These three methods have been successfully used in conjunction with robotics for achieving an increase in sample preparation throughput. For example, Figure 13-5 is the photograph of a Beckman's Biomek 2000 (other models such as Biomek 3000 and Biomek FX are also applicable) for semi-automated sample preparation that can accommodate SPE, LLE, and PPT procedures. This scheme has been established for use with SPE, LLE, or PPT in a 96-well plate format to analyze pharmaceutical products in biological matrices (e.g., whole blood, plasma, serum, and cerebral spinal fluid (CSF)) in our laboratories (unpublished data).

13.4.1 SPE

In the 96-well SPE format, similar to the traditional manual procedure, issues such as the nature of the bonded-phase (e.g., ion exchange, C_2, C_8, C_{18}, cyano, phenyl, polymeric, strong or weak cation exchange, strong or weak anion exchange, mixed phases, etc.), solvent strength (for conditioning/washing of the phases, target analyte elution), and chemical characteristics (e.g., solubility, presence of the key functional groups) of the analyte(s) need to be addressed. A general scheme for initial development of an SPE method is outlined below. Depending on the structure of the compound (hydrophobicity and ionizable functionalities), specific steps to optimize sample recovery are needed.

- Condition sample for optimum retention
- Condition SPE bed with methanol
- Equilibrate SPE bed with water
- Load sample onto SPE bed
    (a) Cation exchange: wash with 2% formic acid [low pH (3)]
    (b) Anion exchange: wash with 50 mM NaOAc buffer [high pH (8–10)]
- Wash bed with 5% methanol
- Elute retained materials with an organic solvent (i.e., CH_3OH, CH_3CN, isopropyl alcohol, or a combination thereof)
    (a) Cation exchange: add 5% NH_4OH to eluent
    (b) Anion exchange: add 2% formic acid to eluent

Some of the most commonly utilized robotic modules for the 96-well SPE procedure are Tomtec Quadra (Tomtec, Hamden, CT, USA), Packard Multi-Probe (Packard Instruments, Meriden, CT, USA), Biomek (Beckman–Coulter, Fullerton, CA), and Tecan (Durham, NC, USA) units.

For example, we have successfully and routinely adopted the Tomtec Quadra technology in the development and validation of several off-line SPE assays in whole blood, plasma, and urine followed by MS detection. The Packard Multi-Probe liquid handling workstation (Figure 13-6) has also shown promise for off-line SPE procedures involving plasma and serum [26–28]. In addition, this unit as well as the Tecan and Biomek systems can be programmed for the initial sample (e.g., plasma) transfer step from vials to the 96-well blocks, buffer addition (if applicable), and to aliquot internal standard. The advantage of the above capabilities is a significant reduction in time and labor for the entire sample processing procedure. Possible technical problems such as carry-over by fixed-tip pipettes used to aliquot the biological fluid can be alleviated by incorporation of several wash cycles or their replacement with disposable pipette tips. In addition, possible inaccurate transfer of samples from the collection tubes to the 96-well blocks due to pipette tip clogging by endogenous protein clots or lipid layers should also be considered. Specific steps such as storage of the plasma samples at −80°C and/or centrifugation at 14,000 rpm prior to sample transfer can be considered for precluding fibrinogen clot formation.

13.4.2 PPT

Due to its ease of use and speed, PPT is one of the most common approaches in sample preparation in early drug discovery [25]. While PPT is fast, easy-to-apply, and applicable to a broad class of small molecules, it also suffers from

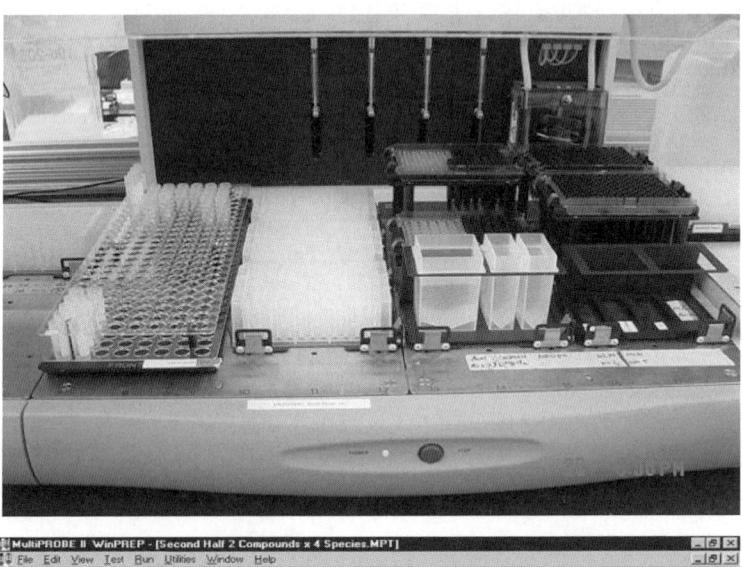

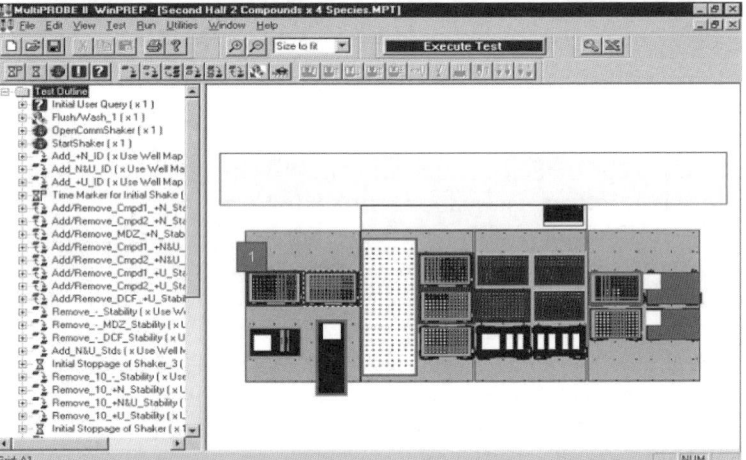

Figure 13-6. Photograph (top panel) of a Packard Multi-Probe (www.perkinelmer. com) platform for semiautomated PPT, LLE, or SPE process in the authors' laboratory. The bottom panel shows a typical layout using the corresponding operating software package.

several disadvantages. Briefly, in a PPT procedure, often an equal or higher volume (e.g., 1:3) of acetonitrile (or sometimes methanol) is added to a sample of plasma, which contains the test article as well as an internal standard. The sample is mixed and centrifuged, resulting in the formation of a protein pellet and its corresponding supernatant. The supernatant is transferred, dried, reconstituted, or directly injected onto a LC column. Clearly, this procedure is easily amenable to automation and is applicable to a host of structurally diverse group of small molecules. However, PPT lacks specificity and selectivity that SPE or LLE can offer. Consequently, significant matrix effect and ion suppression can be observed due to the presence of other endogenous molecules that compete with the analyte(s) during ionization [29–38]. In addition, compounds that are highly bound to the protein can yield low sample recovery in the PPT procedure. PPT procedure is also more demanding on the MS interface, which requires more frequent cleanup, due to the endogenous interference and contaminants. An example of PPT in drug development will be presented in the latter part of this chapter (STI571; *vide infra*).

13.4.3   LLE

Liquid–liquid extraction is another well-established and attractive approach, which has been useful for the analysis of xenobiotics in biological fluids. LLE can be designed to be highly selective yielding cleaner sample extracts. Briefly, LLE is a mass transfer procedure where an aqueous sample (e.g., analyte containing biological fluid) is in contact with an immiscible solvent that exhibits preferential selectivity toward one or more of the components in the aqueous sample (e.g., plasma or whole blood). In an SPE procedure (*vide supra*), a solid sorbent material such as an alkyl-bonded silica is packed into a cartridge, into a disk, or in a 96-well plate format, and it performs essentially the same function as the organic solvent in LLE. This is particularly critical in minimizing ion suppression by co-eluting matrix components, when an ESI interface is used for the LC/MS analysis. Due to a different mechanism of operation, ion suppression is not a major determinant for signal loss in APCI [37–40]). The ion suppression is exacerbated in some cases, where the chromatography results in low peak capacity factors [39]. This could be attributed to co-elution with polar species that had also partitioned into the immiscible solvent and were consequently injected onto the HPLC column. Based on a series of experiments reported by King and co-workers [35], the order of ESI response suppression is PPT > SPE > LLE, where liquid–liquid extraction yields the least amount of analyte ion loss.

13.5   AUTOMATED SAMPLE TRANSFER

Lastly, one of the labor-intensive steps in bioanalytical sample processing is the accurate initial transfer of plasma or whole blood from cryogenic vials (e.g., polypropylene tubes) to 96-well plates. This step is particularly laborious and time-consuming when a large number of samples (e.g., in 1000s) are subjected to analysis in late-stage clinical trials. The main bottleneck involves the "manual" uncapping and re-capping steps for each individual vial. In this regard, the Tomtec Corporation is in the process of final testing and commercialization of the "Formatter" (Figure 13-7), which is designed to alleviate the above bottleneck. According to the vendor (and tests during a demo by one

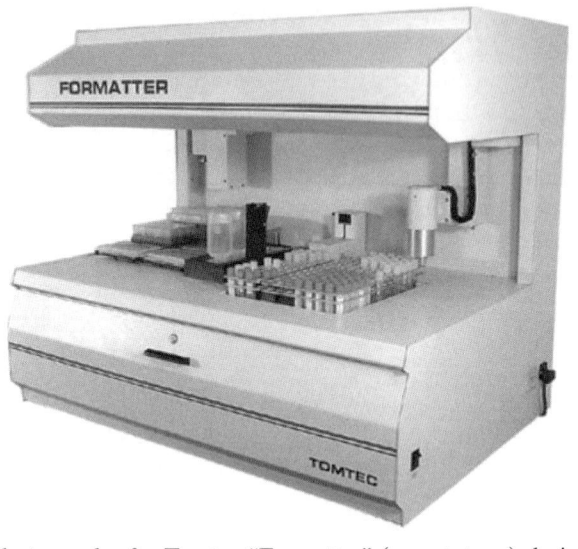

Figure 13-7. Photograph of a Tomtec "Formatter" (a prototype) designed to de-cap, aliquot, and re-cap sample vials containing biological fluids (e.g., plasma, serum, whole blood). For more details see www.tomtec.com.

of the authors), this unit will be able to accurately de-cap, aliquot, transfer, and re-cap automatically. In addition, the unit can pipette 10–450 μL accurately, has sample bar code tracking capability, and can interface with Watson-LIMS (www.tomtec.com). In a typical experiment, plasma samples are transferred from individual vials to a 96-well plate and subsequently processed by an appropriate extraction method on a Tomtec unit (PPT, SPE, LLE). Of course, steps such as vortexing and centrifugation still require a manual intervention; hence, sample extraction methods using a Tomtec are often referred to as "semiautomated."

13.6 SAMPLE PROCESSING USING AN ON-LINE APPROACH

In recent years, high-throughput and automated on-line sample extraction procedures have offered viable alternatives to improve efficiency for sample processing [41–46]. One such approach has been turbulent flow LC. In turbulent flow LC, single- or dual-column configurations have commonly been reported. Recently, a four-channel staggered injection system (e.g., Cohesive's multiplexing Aria LX-4 system) has been reported for decreasing the MS idle time and improving productivity (cycle time) [45]. A typical dimension of the extraction column can be 50 × 1.0 mm (i.d.), although smaller lengths can also be used. In the single-column configuration, a sample containing the analyte and internal standard is loaded on the extraction column at a

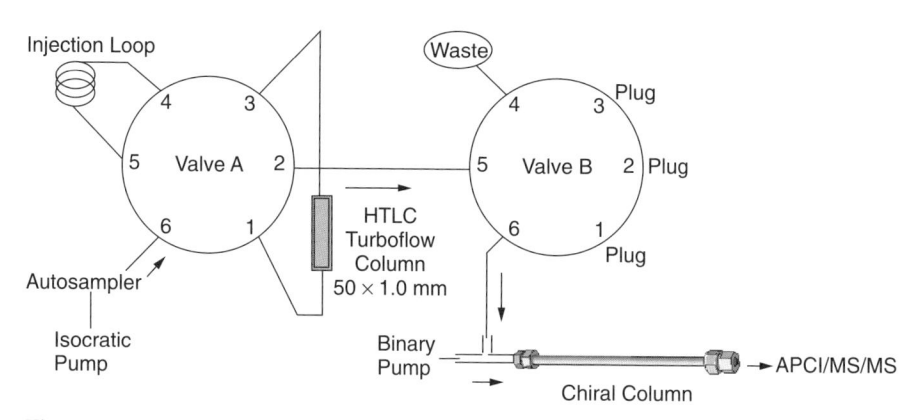

Figure 13-8. Simplified schematic of two-valve/two-column system (built in-house) used in the turbulent flow LC experiment [47]. The first column (e.g., Cyclone P, 50 × 0.5 mm) is used for the extraction step. In our laboratory, this design has been used in conjunction with chiral, standard narrow bore (e.g., 50 × 2 mm), and monolithic analytical columns.

high linear velocity (e.g., 5.0 mL/min). The analyte is retained via rapid diffusion into the packing, while other matrix components are washed into waste using an aqueous mobile phase. Subsequently, the analyte is eluted by a step or linear gradient and is detected by the mass spectrometer. In the dual-column configuration, a standard analytical column (e.g., C_{18} or C_8; 50- × 4.6-mm i.d.) is placed after the extraction column to improve chromatographic separation and sample cleanup. In our laboratory, we have successfully validated and applied the dual-column (Figures 13-8 and 13-9) configuration to perform racemic reversed-phase as well as chiral LC-MS/MS analysis. In the latter assay, we replaced the second column by one containing a chiral stationary phase. A full account of the assay optimization and validation of an on-line achiral–chiral column configuration has been reported elsewhere [47].

Direct sample injection has also been accomplished by using on-line C_{18} (4-mm i.d.) guard cartridges for cytochrome inhibition studies, capillary SPE, C_{18}-alkyl-diol-silica restricted access phase, and PROSPEKT™ SPE modules. A more recent instrument by Spark–Holland (Figure 13-10, the Symbiosis unit) consists of Shimadzu HPLC pumps, "conditioned stacker," an autosampler housing, and a "high-pressure dispenser" SPE station [48, 49]. This unit can be used as a stand-alone autosampler (LC mode) and/or an on-line sample-processing module (XLC mode) with a mass spectrometer. The companion software is embedded in the Sciex's Analyst® 1.4.1 and can be easily used within the sample batch design and as part of its acquisition method. The vendor also provides 96-well plate-screening cartridges containing *HySphere* SPE resins such as C_{18}, C_8, C_2, CN, ion exchange, and so on, in a 12 × 8 format (dimension: 2 × 10 mm).

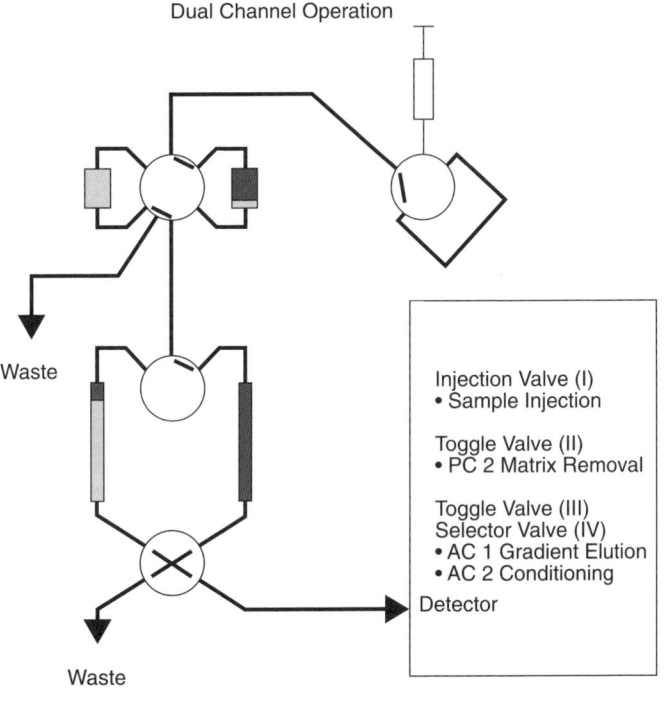

Figure 13-9. Configuration of a dual channel on-line sample extraction, the CTC Trio Valve system (LEAP Technologies, www.leaptec.com; also see reference 100 for technical details), which is commercially available. AC and PC signify analytical and processing columns, respectively.

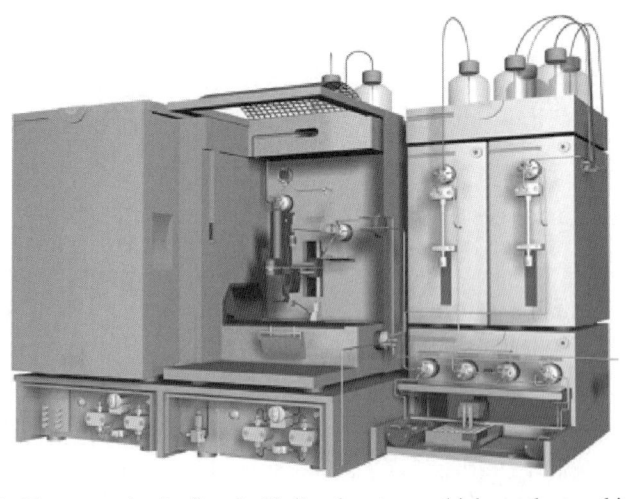

Figure 13-10. Photograph of a Spark–Holland system, which can be used in the off-line (LC) and on-line SPE (XLC) modes. This unit contains an autosampler, two Shimadzu pumps, a degasser, and SPE capability (for more details see www.sparkholland.com).

A caveat for all direct sample injection assays is an understanding of the analyte chemical stability in the biological fluid during the analysis period. Nonetheless, an increasingly growing body of literature is suggestive that direct injection of post-dose biological fluids for quantification purposes has become a routine and efficient procedure.

13.7 MATRIX EFFECT AND ION SUPPRESSION

Ion suppression or so-called matrix effect is a common problem in atmospheric pressure ionization (API) mass spectrometry [29–40]. There are differences in opinion as to the amount of ion suppression that is acceptable for an analytical method.

In some laboratories, ion suppression in an analytical method is not acceptable, but in other laboratories it is acceptable if there is no significant effect on the validity of the analytical data with appropriate quality controls. The phenomenon of ion suppression results in reduction of signal intensity. Consequently, the lower limit of quantification (LLOQ) for highly sensitive bioanalytical methods could be difficult to achieve. The problem can be further complicated by (a) differential ion suppression due to intersubject variability and (b) the use of blank bio-matrix with varying lot numbers (i.e., control blood obtained from different patients/subjects) in preparing the calibrators. Differences in ion suppression between the analytes and structurally different internal standards may also be problematic. This issue can be mitigated by the use of stable-isotope-labeled (SIL) analogues as internal standards. The extent of ion suppression is dependent on the methods for sample preparation and chromatographic separation. The supernatant produced by the PPT method is most likely to cause an ion suppression in ESI due to the lack of selectivity of PPT procedure (co-eluting endogenous compounds such as lipids, phospholipids, fatty acids, etc., that affect the ESI droplet desolvation process; for more details see reference 38 and technical notes on www.tandemlabs.com). The extracts obtained from solid-phase extraction (SPE) and liquid–liquid extraction (LLE) are relatively cleaner. Further studies on the molecular identities of co-eluting endogenous compounds, leading to ion suppression, are required to clearly delineate their contribution.

One of the widely used methods to qualitatively assess the matrix effect consists of post-column addition of analytes to the LC-eluent flowing from the column to the ESI interface of the mass spectrometer (Figure 13-11) [35]. Briefly, an analyte and the internal standard (IS) dissolved in the same LC eluent are infused (e.g., flow rate 10 µL/min) using a syringe pump, through a "tee-mixer," located between the column eluent (e.g., flow rate 200 µL/min) and the ESI interface of the mass spectrometer. An extract (using LLE or SPE) or supernatant (if PPT is used) from an analyte-free matrix, such as blank or control plasma, is injected via the autosampler, while the test article and the internal standard are introduced, post-column, to the MS ionization

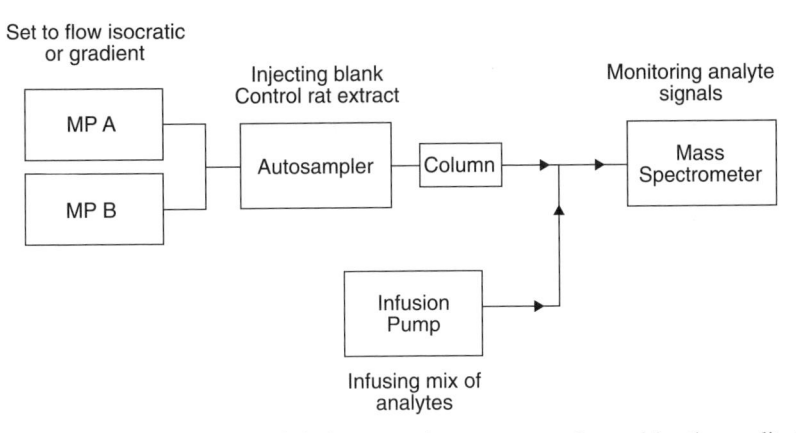

Figure 13-11. The post-column infusion experiment commonly used for the qualitative assessment of ion suppression and originally reported by King and co-workers [35] (MP signifies mobile phase).

source at a stable and continuous flow using an infusion pump. Since the test article and the internal standard are introduced to the MS at a constant flow, a steady ion response is obtained as a function of time. If there is an ion suppression, a drop in the MS ion signal is observed upon the injection of the extract obtained from a control plasma (or any other biological matrix). The infusion LC-MS/MS (MRM mode) extracted ion chromatograms of both the internal standard and the analyte are shown in Figure 13-12 for an off-line PPT plasma supernatant injection. In this example there is no significant ion suppression from 3–10 min. If there is a significant ion suppression, modification to a more selective (cleaner) sample preparation, adoption of a more elaborate chromatographic condition to separate ion suppressing agents from the analyte(s) of interest, and/or use of a stable label internal standard are recommended [29–40]. Figure 13-13 illustrates the significance of the choice of sample preparation on signal-to-noise ratio of an investigational compound, where an OASIS®-HLB sorbent (hydrophilic–lipophilic balanced co-polymer) (denoted as HLB in Figure 13-13) SPE and a strong anion-exchange SPE (MAX) (denoted as anion ex in Figure 13-13) yielded higher ion intensity than PPT. The count per second (CPS) signifies the detector (an electron multiplier) response. More efficient extraction was obtained with the anion-exchange SPE than with C_{18} SPE or using the PPT mode.

13.8   REGULATORY REQUIREMENTS FOR LC/MS METHOD VALIDATION

Validation of quantitative LC/MS methods used in the determination of small-molecule drugs and/or their metabolites in biological fluids is of paramount

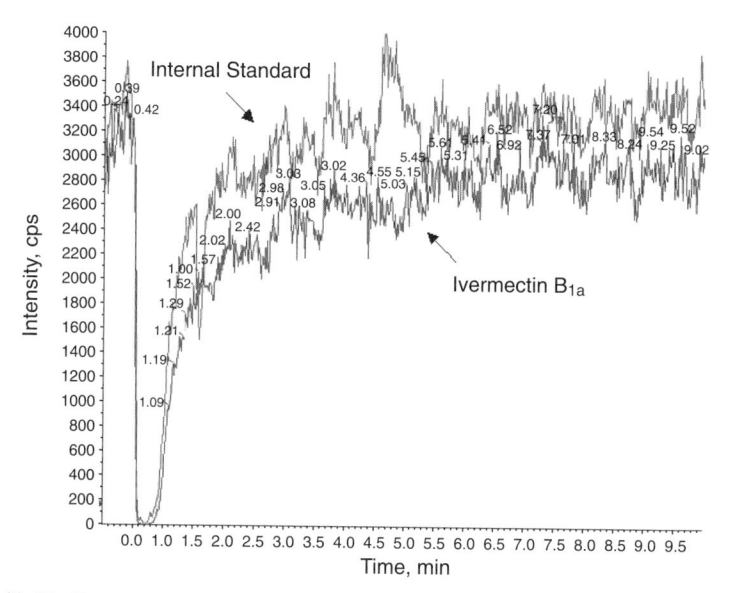

Figure 13-12. Representative MRM scans obtained using the ion-suppression infusion experiment, developed by King and co-workers [35]. A significant ion matrix effect is observed between 0.5 and 1 min using control rat plasma. The sample preparation was PPT. Note that this experiment needs to be performed "prior" to the method development and validation, so necessary changes to the sample preparation protocol and chromatographic method are made. Reprinted with permission from [101].

importance and a key determinant in obtaining reliable pharmacokinetics (PK) information. A properly developed and validated LC/MS method may often be used throughout the process of a drug's evaluation lifecycle. These could include early discovery PK studies, preclinical toxicology studies (e.g., dose-proportionality studies), salt/formulation selection (pharmaceutical research and development), clinical PK studies, and post-marketing surveillance (Figure 13-14) [50–53]. Hence, intra- and interlaboratory specificity, accuracy, precision, and ruggedness have to be established [54].

In order to bridge some of the regulatory filings (e.g., within the United States, Japan, and Europe), the US Food and Drug Administration (FDA) Center for Drug Evaluation and Research (CDER), the International Conference on Harmonization (ICH), Japan's regulatory agency, and the European Community have devised common as well as distinct requirements for bioanalytical method validation [55]. To this end, it is imperative to strictly follow these requirements during preclinical toxicology (e.g., toxicokinetics) and Phases I, II (a and b), and III clinical trials as well as during all the post-marketing PK studies (Figure 13-15).

Moreover, in introduction of generic or new formulations and/or to establish bioequivalence (BE) between two products, certain guidelines are followed to compare the systematic exposure of the test article to that of a

LC Conditions

Column: FluoroSep RP Phenyl, 2.1 × 50
      mm, 5 μm (ES Industries)
Mobile Phase: A = 5 mM aq. ammonium
      formate
      B = Methanol
Flow Rate: 400 μL/min.
Gradient: 95% A to 5% A in 2.5 mins

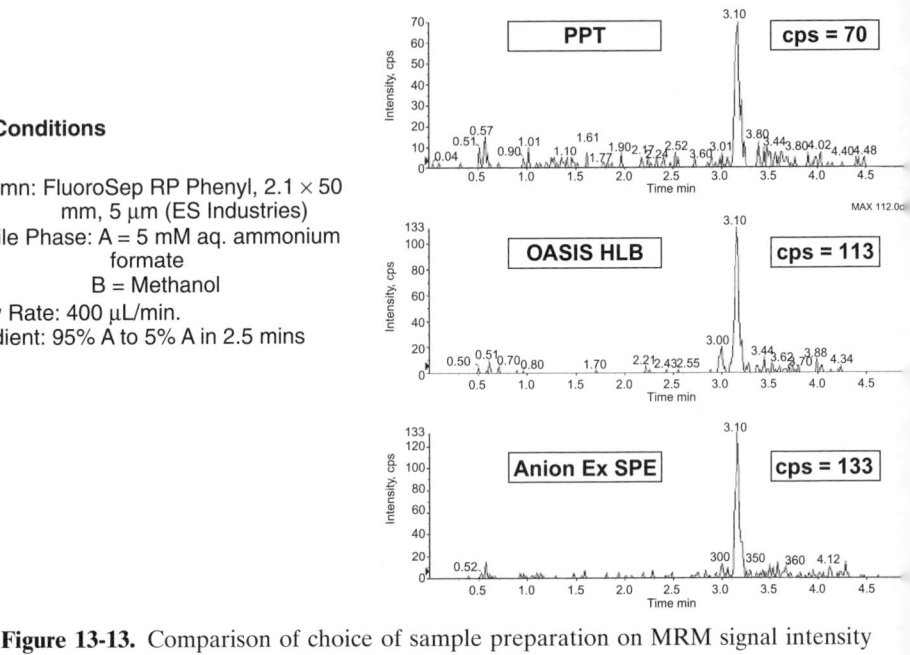

Figure 13-13. Comparison of choice of sample preparation on MRM signal intensity of an investigation compound. The injection volume was 40 μL. The count per second (CPS) signifies the detector (an electron multiplier) response. Protein precipitation (PPT), hydrophilic–lipophilic balanced co-polymer-based SPE (Oasis HLB- co-polymer of styrene, divinylbenzene and *n*-vinylpirrolidone monomers; the hydrophilic refers to the NVP monomer, and the lipophilic refers to the SDVB monomers), and strong anion exchange SPE (Max) (all in 96-well plate format) were used in control rat plasma (unpublished data).

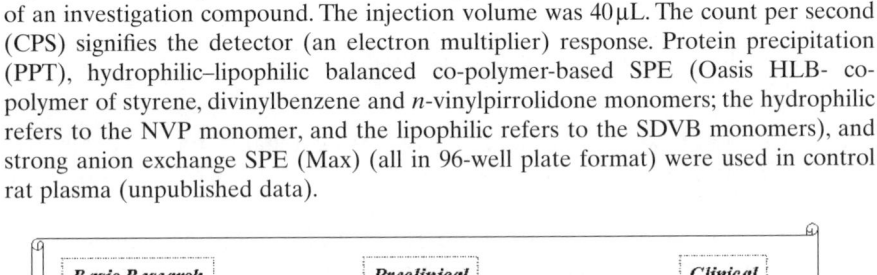
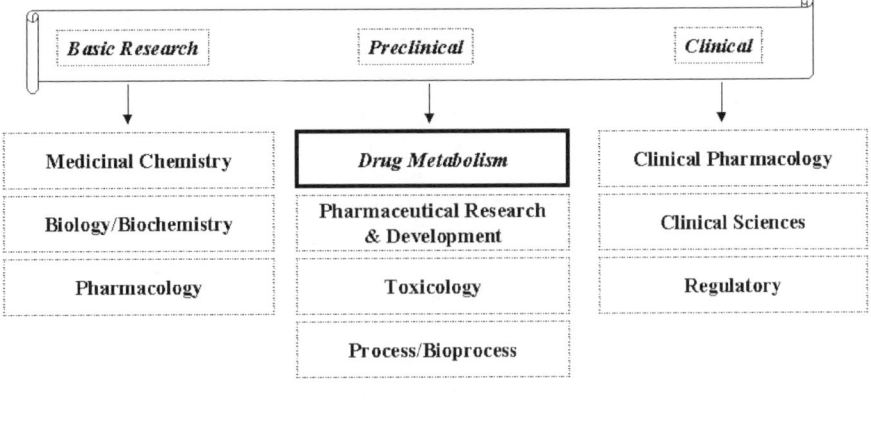

Figure 13-14. A generic flow chart for the process of drug discovery and development [102].

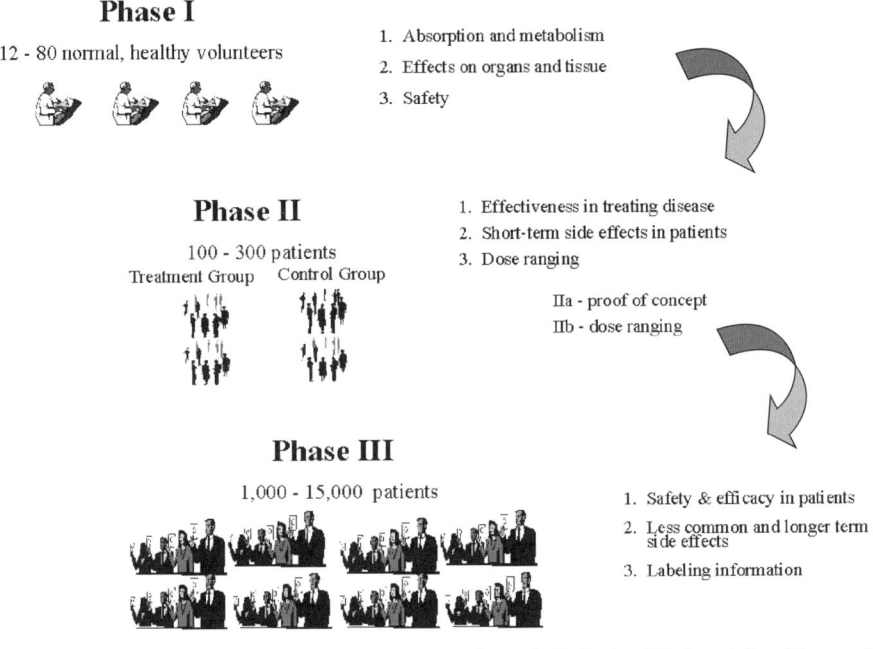

Phase I

12 - 80 normal, healthy volunteers

1. Absorption and metabolism
2. Effects on organs and tissue
3. Safety

Phase II

100 - 300 patients

Treatment Group   Control Group

1. Effectiveness in treating disease
2. Short-term side effects in patients
3. Dose ranging

IIa - proof of concept
IIb - dose ranging

Phase III

1,000 - 15,000 patients

1. Safety & efficacy in patients
2. Less common and longer term side effects
3. Labeling information

Figure 13-15. A simplified/generic representation of clinical trials involving Phases I (safety and tolerability in healthy participants and/or patients), II (to evaluate effectiveness of the drug in patients), and III (to perform expanded controlled and uncontrolled trials and gather benefit–risk data). About 70–90% of drugs entering Phase III studies successfully complete this phase of evaluation. For more details see the glossary at www.clinicaltrials.gov.

reference drug [56, 57]. The latter is a key part of the abbreviated new drug application (ANDA). New formulations could include sustained-release products that exhibit a lag time in quantifiable plasma concentration. An extensive discussion of the above topics is beyond the scope of this chapter; nonetheless, we will summarize some of the key points in the remaining segments of this section.

Some of the key validation characteristics include accuracy, precision, specificity, detection limit, quantification limit, dynamic range, linearity, matrix effect, sample, recovery, sample stability, and overall ruggedness of the assay. An evaluation of these requirements is an obligatory step in conducing reliable PK studies under the good laboratory practice guidelines (GLP, 21 CFR Part 58 issued by FDA). In addition, software validation pertaining to a regulated analytical laboratory involves installation qualification (IQ), operational qualification (OQ), and performance qualification (PQ), which are used to define and demonstrate process consistency and validity. The first step or IQ ensures that the system and all its components (including calibration) are installed correctly. The OQ involves testing critical parameters, the system's baseline setup, and variables. Lastly, the PQ is the final phase of validation where it examines the system to perform over long periods of time within the predetermined accepted tolerance. In the PQ phase, often the individual components of the system are not tested; instead, the system is treated as a whole (for more details see www.fda.gov; *General Principles of Software Validation; Final Guidance for Industry and FDA Staff*, January 11, 2002). The PQ is usually performed prior to the analysis of the samples.

To establish uniformity, each GLP lab has up-to-date standard operating procedures (SOPs), which all the involved bio-analytical scientists strictly follow, to ensure quality control (QC) and quality assurance (QA) compliance. Calibration curves consisting of single blank (drug free + internal standard), double blank (drug and internal standard free), and a minimum of six to eight nonzero calibrators covering the expected dynamic range are often used. QCs obtained from multiple lots of the bio-fluid (e.g., serum, whole blood, plasma, urine, saliva, etc.) are prepared for short- and long-term storage. In addition, stability tests relating to 3-cycles of freeze–thaw, autosampler, bench top, freezer storage, and stock solution (neat) are established.

In summary, proper validation and its documentation in a GLP, GCP (good clinical practice), and GMP (good manufacturing practice) setting, instrument qualification records, submission of audited reports, and archiving of electronic records (21 CFR Part 11, issued by FDA) are strictly followed in order to achieve successful regulatory filing. The following two case studies, Ritalin® [58] and Gleevec™ [59, 60], from our laboratory are representative applications of GLP compliance validated methods that were submitted for worldwide regulatory filing.

13.9 RITALIN®: AN APPLICATION OF ENANTIOSELECTIVE LC-MS/MS

Currently there is a trend toward the synthesis and large-scale production of a single active enantiomer in the pharmaceutical industry [61–63]. In addition, in some cases a racemic drug formulation may contain an enantiomer that will be more potent (pharmacologically active) than the other enantiomer(s). For example, carvedilol, a drug that interacts with adrenoceptors, has one chiral center yielding two enantiomers. The (–)-enantiomer is a potent beta-receptor blocker while the (+)-enantiomer is about 100-fold weaker at the beta-receptor. Ketamine is an intravenous anesthetic where the (+)-enantiomer is more potent and less toxic than the (–)-enantiomer. Furthermore, the possibility of *in vivo* chiral inversion—that is, prochiral → chiral, chiral → nonchiral, chiral → diastereoisomer, and chiral → chiral transformations—could create critical issues in the interpretation of the metabolism and pharmacokinetics of the drug. Therefore, selective analytical methods for separations of enantionmers and diastereomers, where applicable, are inherently important.

The first step in designing an enantioselective assay is to examine the chemical structure of the analyte and identify characteristics such as the role of π–π interactions, analyte pK_a, solubility, functional groups (e.g., polar versus hydrophobic moieties), inclusion–complexation propensity, steric interference, and hydrogen bonding capacity. Therefore, the process of enantioselective chromatography method development tends to be time-consuming and requires planning and careful experimentation. A good/methodical design of experiments needs to be implemented to understand the role of the organic modifier, pH of the mobile phase, and the choice of the bonded phase on the enantiomeric separation.

When developing a chiral LC/MS or LC-MS/MS method, several factors need to be taken into consideration. First, optimum separation efficiency (baseline resolution) is desired in order to facilitate the quantitative analysis of trace levels of enantiomers with high accuracy and precision. Second, the CSP (chiral stationary phase) should be rugged and exhibit durability when subjected to frequent analysis (>200 injections of extracts from biological fluids) at the analytical conditions (temperature, pH, type of mobile phase). Third, inorganic buffers (e.g., phosphate buffer) and/or highly aqueous mobile phases (particularly in the case of protein-derived CSPs), which can potentially lead to significant ion suppression during LC/MS experiments, are not preferred. Fourth, for quantitative analysis of a large number of samples, such as the ones encountered during a clinical trial, a relatively short chromatographic run is desired. The high-throughput analysis should be achieved without compromising sensitivity and chromatographic resolution of the critical pairs. Fifth, a drawback of a number of chiral chromatographic assays is that they may require derivatization (e.g., using Marfey's reagent, dinitrobenzoyl, dinitrophenyl, (–)-[1-(9-fluorenyl)-ethyl]chloroformate) of the analyte, which can be tedious and time-consuming (derivatization conditions, yield of derivatization, etc., need to be investigated). In accord with today's standards for high-throughput analysis, it is preferred to avoid a derivatization step where possible. Sixth, use of flammable solvents such as hexane in normal-phase chiral LC requires safety measures when utilized in conjunction with ESI or APCI interfaces. In one approach, in order to minimize the possibility of an ignition during the use of hexane within the heated high-voltage corona discharge environment, nitrogen can be used as the auxiliary gas, resulting in the displacement of the oxygen in the API housing.

An example from our own laboratory is the methylphenidate bioanalytical chiral LC-MS/MS assay in support of toxicokinetics (TK) and PK studies (*vide infra*). Attention-deficit hyperactivity disorder (ADHD) is a recognized medical problem characterized by symptoms of inattention, hyperactivity, and impulsivity. Methylphenidate (MPH; Ritalin®: methyl-alpha-phenyl-2-piperidinacetate hydrochloride) is prescribed for the treatment of ADHD. MPH has two chiral centers yielding enantiomers with distinct pharmacokinetic and pharmacodynamic properties in humans. It is known that the *d*-threo [2R,2'R]-MPH (i.e., (+)-threo) exhibits greater pharmacological activity than the

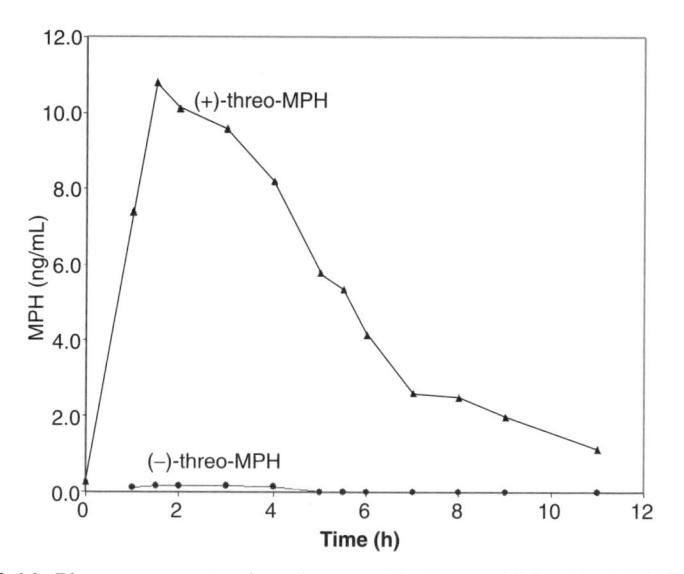

Figure 13-16. Plasma concentration–time profile for a child with ADHD after an oral administration of 17.5 mg of racemic MPH. The plasma concentrations were determined using a validated method described elsewhere. Reprinted with permission from [58].

l-enantiomer (*l*-threo [2S,2′S]-MPH) or (−)-threo). Although the precise mechanism of its action has not been fully defined, it is often postulated that MPH inhibits dopamine uptake via binding to the dopamine transporter (DAT) with high affinity. Hence, an increase in synaptic dopamine is linked to the reinforcing effects of MPH. Biotransformation of MPH is rapid and extensive, leading to mainly the de-esterified metabolite, alpha-phenyl-2-piperidine acetic acid (ritalinic acid), and this product needs to be quantified and separated from the analyte of interest in the analytical assay. Recently, using a vancomycin column (antibiotic chiral stationary phase), a relatively rapid and sensitive reversed-phase chiral LC/MS/MS assay was developed and validated for the determination of MPH enantiomers in rat, rabbit, dog, and human plasma [58]. For example, a validated LC-MS/MS method with a lower limit of quantification of 87 pg/mL in human plasma was reported [58]. Figure 13-16 depicts a representative plasma concentration–time profile, obtained using LC-MS/MS, for a child with ADHD subsequent to an oral administration of 17.5 mg of racemic form of MPH. The plasma levels for the pharmacologically inactive enantiomer was substantially lower than those of the active enantiomer, consistent with previous findings [58 and references cited therein].

13.10 GLEEVEC™ (STI571)

Enzymes are powerful biological catalysts that are essential for the proper maintenance and propagation of any organism. These properties make them excellent candidates as therapeutic targets to combat diseases of either genetic or pathogenic origin.

Protein-tyrosine kinases (PTKs) are enzymes that have the ability to transfer the terminal phosphate of an adenosine triphosphate (ATP) molecule to a protein substrate. PTKs are critical modulators of cellular signal-transduction pathways, which mediate cell proliferation, differentiation, and communication. If for any reason these signaling proteins are subjected to oncogenic mutation(s), a cellular deregulation may occur, yielding an imbalance between cell division, cell growth, and cell death (apoptosis). Hence, PTKs have emerged as important therapeutic targets for intervention in cancer.

The Philadelphia (Ph) chromosome is the consequence of a reciprocal translocation between chromosomes 9 and 22 yielding a fusion oncoprotein referred to as Bcr-Abl (~210 kDa). This molecular consequence leads to an elevated catalytic activity of Bcr-Abl, resulting in a resistance to apoptosis, cell transformation, and malignancy. A cytogenetic hallmark of chronic myeloid leukemia (CML), a clonal hematopoetic stem cell disorder, is the Ph chromosome and high activity of Bcr-Abl tyrosine kinase. The clinical chemistry manifestation of CML is elevated levels of white blood cells (e.g., $>20 \times 109/L$), and in some patients an increase in platelet counts (e.g., $>450 \times 109/L$) is observed. Therapeutic options for CML include allogeneic stem cell transplantation, interferon-alpha treatment, and chemotherapy with hydroxyurea or busulfan. Allogeneic stem cell transplantation requires the availability of a suitable donor and presents a risk of mortality in older patients. Chemotherapy-based methods often do not provide a cure, present potential toxic side effects, and lead to intolerability and/or resistance to the treatment. In addition, none of the agents used for CML is known to target the underlying cause of the disease [64].

Gleevec™ or imatinib mesylate, designated chemically as 4-[(4-methyl-1-piperazinyl)methyl]-N-[4-methyl-3-[[4-(3-pyridinyl)-2-pyrimidinyl]amino]-phenyl] benzamide methanesulfonate, is a PTK inhibitor that potently inhibits the Abelson (Abl) tyrosine kinase, and this was demonstrated in *in vitro* and *in vivo* studies [65]. Recently, the signal transduction inhibitor 571 (STI571 or Gleevec™) was approved by FDA in a record time for the treatment of patients at any of the three stages of CML: myeloid blast crisis, accelerated phase, and chronic phase after failure of interferon-alpha therapy. Gleevec™ has been referred to as a milestone for the drug development in cancer and an ideal targeted drug at the molecular level. With high specificity, it competitively inhibits the binding of ATP to the kinase activation domain of Bcr-Abl (Figure 13-17) [66].

Due to the fast-track status of STID571 (Gleevec™) by the US FDA, a significant improvement in throughput of bioanalytical methods was warranted. The duration from first time in man (FIM) to completion of the NDA filing was approximately 2.6 years. In order to complete the PK studies with sufficient speed to meet various target dates, a semiautomated procedure using protein precipitation was developed and validated.

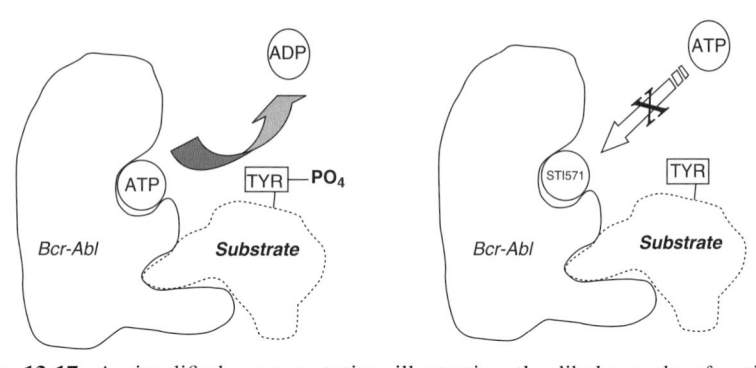

Figure 13-17. A simplified representation illustrating the likely mode of action of STI571 or Gleevec™. STI571 inhibits the binding of adenosine triphosphate (ATP) to the kinase activation domain of Bcr-Abl oncoprotein [64–66].

As was mentioned earlier, PPT using acetonitrile is a commonly used procedure for the treatment of plasma samples in drug analysis. Given the current paradigm shift toward automated and robotics systems, the utility of PPT in 96-well format was explored for the quantitative analysis of STI571 and CGP 74588 (the des-methyl or N-de methylated metabolite). The current method required a LC-MS/MS run time of about 2.5 min (injection-to-injection cycle). The typical batch sizes were two to four plates per day. The sample preparation step was approximately 2 hr per plate. Extensive sample clean-up was not needed to ensure column longevity due to the reduced column sample loading. Figures 13-18 and 13-19 show the corresponding structures/MRM transitions and a typical extracted ion chromatogram, respectively.

Figure 13-20 depicts the mean plasma concentration–time profile, obtained using LC-MS/MS, for patients subsequent to a daily oral administration of STI571 (free base) for 28 consecutive days (steady state). The sampling period was up to 48 hr post-dose. The dosing regimen was escalated from 25 to 600 mg/day until a favorable hematologic response was observed. PK/PD studies of STI571 in a Phase I trial in patients showed a dose-proportional exposure subsequent to daily oral administration of 25 to 1000 mg with a 1.5- to 3-fold drug accumulation at steady state [67].

In this regard, our laboratory demonstrated a method for the quantification of STI571 and its main metabolite, CGP 74588, in human plasma using a semi-automated PPT method and a relatively rapid LC/APCI/MS/MS analysis. The assay exhibited an excellent linearity from 4.00 to 10,000 ng/mL in human plasma. The method was utilized for the analysis of thousands of clinical samples. Furthermore, the method was routinely amenable to analysis of STI571 and CGP 74588 in cerebrospinal fluids (CSF), gastrointestinal stromal tumor (GIST) biopsy specimens, and toxicokinetic studies (data not shown).

In addition, a semiautomated procedure using solid-phase extraction (SPE) was also developed and validated. A Packard Multi-Probe and a SPE step in

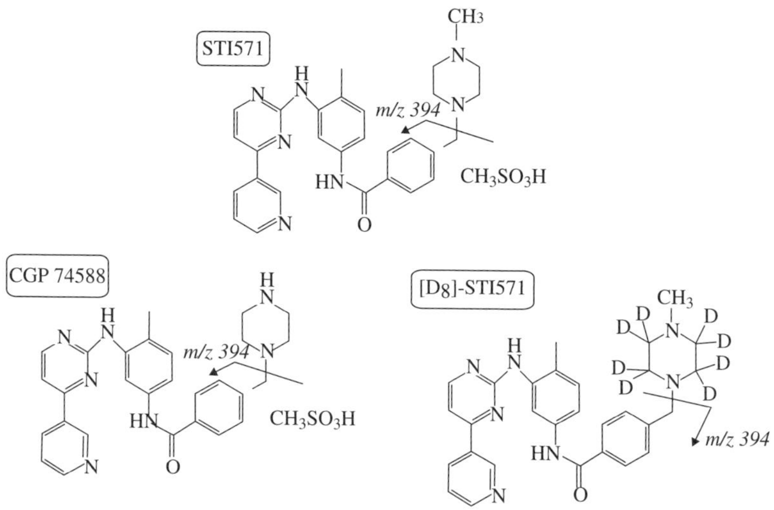

Figure 13-18. Structure of STI571, CGP 74588, and the internal standard. Arrows indicate that the product-ion that was selected for the multiple-reaction monitoring (MRM) experiment [59, 60]. For complete metabolic profile and disposition of STI571 (imatinib, Gleevec™) in humans, see reference 103.

a 96-well plate format were utilized. A 3M Empore octyl (C_8)-standard density 96-well plate was used for plasma sample extraction. A Sciex API 3000 triple quadrupole mass spectrometer with an APCI interface operated in positive ion mode was used for detection. Lower limits of quantification of 1.00 ng/mL and 2.00 ng/mL were attained for STI571 and its metabolite, CGP 74588, respectively. Automation of sample processing for Gleevec™ was extremely critical due to its fast-track status.

13.11 BIOMARKERS

According to a recent report [68], the cost of drug development from discovery to regulatory approval can exceed US $800 million. Furthermore, the time between the initial preclinical studies to marketing can range from 3.2 to as high long as 20 years (average of 8.5 years). In 2001, the R & D budget of the pharmaceutical industry was about US $30.3 billion while the National Institute of Health (NIH) budget for all its research areas was about US $20.3 billion. Clearly, the task of discovery and development of novel therapeutic agents has become increasingly costly, complex, inefficient, and competitive. To this end, validated biomarkers and their reproducible measurement have the potential to shorten the drug discovery process (*proof-of-concept*), determine toxicity in real time (e.g., to determine safety margins in preclinical

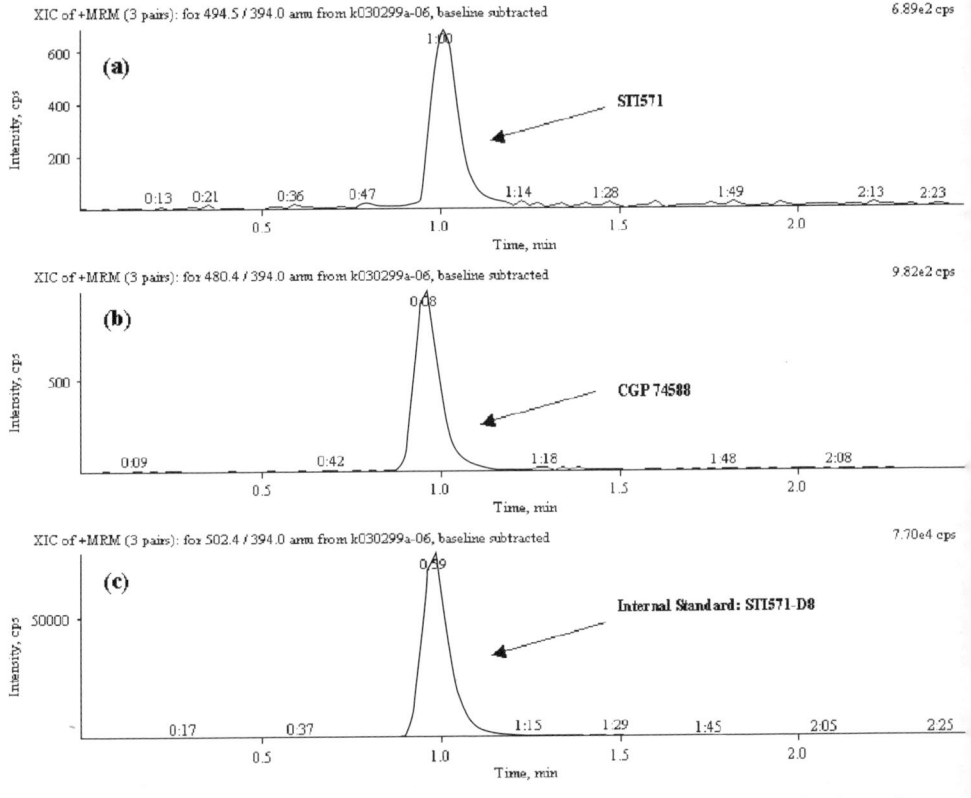

Figure 13-19. Representative LC-APCI/MS/MS ion-chromatograms resulting from the analysis of 1.00 ng/mL (LLOQ) of STI571 and 2.00 ng/mL (LLOQ) of CGP 74588 spiked with the internal standard (25 ng/sample). The injection volume was 10 μL. Excellent sensitivity was obtained for ~10 and 20 pg/mL of on-column injection for STI571 and CGP 74588, respectively. Panels a, b, and c correspond to the STI571, CGP 74588, and the internal standard, respectively [59, 60]. Reprinted with permission from [59].

species), and predict a patient's response to an investigational compound (e.g., efficacy biomarkers).

Recent advancements in genomics and proteomics have generated considerable interest in the discovery and validation of biomarkers in mechanism-based drug development [69–71]. These advances have been welcomed to reduce the cost, increase success rates, and accelerate timelines in the drug discovery and development process. Herein, a brief overview of the application of biological markers in early discovery, development, toxicological assessments, and efficacy studies in humans is presented.

A biomarker or biological marker is generally defined as an objectively determined characteristic, which is utilized in the assessment of normal biological processes, pathogenic events, and/or pharmacologic responses to a

day = 28

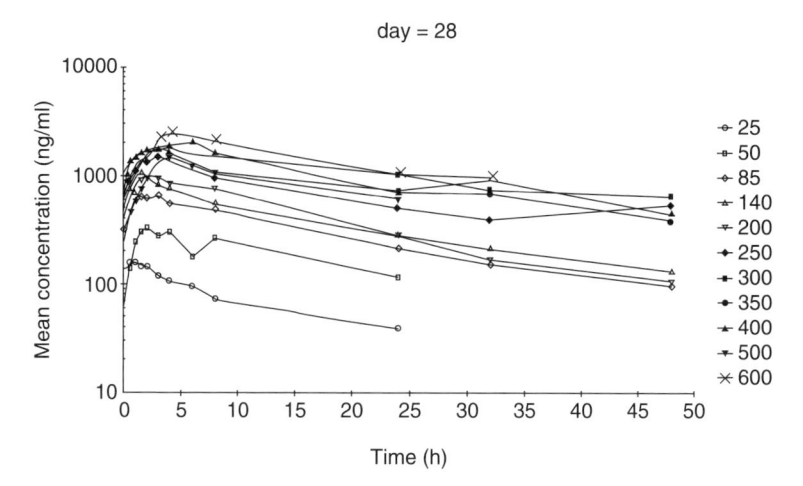

Figure 13-20. Mean plasma concentration–time profile for patients subsequent to a daily oral administration of Gleevec™ or STI571 (free base) for 28 consecutive days (steady-state). The sampling period was up to 48 hr post-dose. The dose regimen was escalated from 25 to 600 mg/day until a favorable hematologic response was observed. Reprinted with permission from [60].

therapeutic intervention [72]. Although the concept of biological markers has long been established in the clinical arena, recent technological advancements in genomics, molecular profiling, imaging, and bioinformatics have brought a new importance to the identification of disease biomarkers [72]. Biomarker research and validation is becoming an integral part of early discovery, safety assessment studies, preventive medicine, and clinical studies for compound prioritization, *proof-of-concept*, as prognostic indicators, and predictors of responses to drugs. Consequently, many ever-proliferating "-omics" disciplines have emerged including proteomics [73], genomics [74], metabolomics [75], metabonomics [76], pharmacogenomics [77], toxicogenomics [78], lipidomics [79], and glycomics [80].

In March 2004, the US FDA issued a white paper entitled Innovation or Stagnation? Challenge and Opportunity on the Critical Path to New Medical Products [81]. One of the central themes of this report has been to incorporate quantitative predictive tools for the assessment of safety and efficacy of a NCE along the *"critical path research"* (i.e., from preclinical development through FDA approval). An integral part of the proposed expeditious and early efficacy and toxicology assessments is the use of biomarkers. Furthermore, the application of a validated biomarker for dose–response readouts is critical in Phase II of a clinical trial. According to a recent report [82], Phase II has the highest attrition frequency, with an average industry success rate of 50% or lower from 1994 to 1998.

In the early stages of drug discovery (e.g., screening for lead compounds), animal (e.g., KO mice) or *in vitro* models are used to demonstrate *proof-of-*

concept, PK/PD evaluation/correlation, and/or prioritize the ranking of NCEs. As the lead optimization continues, the need for the type of biomarker may change, including preclinical efficacy biomarker, toxicity biomarker, and clinical biomarker. In clinical PK studies, it is possible to use certain biological markers for determination of the therapeutic index of an investigational drug. The therapeutic index relates to the dose of a drug needed to provide efficacy versus the dose leading to undesired side effects. To this end, reliable LC/MS methods in conjunction with biologically validated biomarkers can aid in estimation of dose–effect curves for such a determination.

One particular issue that complicates the validation of a biomarker LC/MS assay is their endogenous nature and interference with standard curve and quality control samples. Furthermore, disease state, interspecies differences, possible lack of reference standards, the correct sampling location (an invasive procedure versus obtaining peripheral fluids), and presence of variable forms of a biomarker can lead to additional challenges. Consequently, the standards routinely applied in a GLP LC/MS laboratories may not be easily applicable to analytical biomarker assay activities. For example, many biomarkers are endogenous and, in the case of macromolecules, heterogeneous (e.g., different glycoforms), which can lead to interference (i.e., even present at substantial levels in the control biological matrix) during an assay. Hence, obtaining analyte-free matrices to prepare standard curves can be challenging. In some cases, reference standards and appropriate internal standard can be difficult to obtain because of complex and costly synthetic routes. Lastly, a biomarker concentration in peripheral fluids can be much lower than that at the site of production/action (e.g., a receptor site), leading to analytical challenges [83, 84].

For example, the study of interconversion of cortisone to cortisol, catalyzed by 11-beta-hydroxysteriod dehydrogenase enzymes, is important in the pathogenesis of several common conditions such as Cushing's syndrome, diabetes, osteoporosis, and obesity [85]. Since healthy human subjects have circulating levels of cortisone and cortisol in their blood, an analyst encounters high background and interference from these endogenous glucocorticoids in the standard curve and quality control samples. Therefore, surrogate reference standards, namely synthetic stable label isotopes of cortisone and cortisol, have been utilized to develop a validated SPE and LC-MS/MS assay in human plasma [86]. In this approach, $M + 4$ stable-isotope-labeled cortisone and cortisol were used for preparation of calibrators and quality control samples. In addition, $M + 9$ stable-isotope-labeled cortisone and cortisol were used as internal standards. This method demonstrated the absence of interference, by virtue of mass resolution, from endogenous baseline cortisone and cortisol in control plasma samples [86].

Another example is the accurate determination of small biomarker molecules in blood. Detection of low-abundance–low-molecular-weight (MW) biomarkers (small molecules or peptides) in blood, using mass spectrometry, can often be challenging. These low-MW molecules tend to undergo noncovalent complexation with the abundant plasma proteins such as albumin and alpha-acid glycoprotein [87, 88]. Hence, during sample preparation/fractionation, which often includes albumin depletion (i.e., to decrease interfering background), these valuable small molecules are discarded along with the carrier proteins. Currently, there are ongoing proposals on techniques in harvesting the low-MW segment of the whole blood such as surface-modified nanoparticles that can selectively capture the biomarkers of interest followed by MS analysis [87, 88].

13.12 CONCLUSIONS

A myriad of published reports has now proven the broad and multifaceted applications of modern MS-based techniques for the analysis of small molecules [89–103]. Higher-throughput screening has been in demand and will continue to be one of the main objectives of industrial laboratories. However, bioanalytical scientists should bear in mind that the quality of science cannot be compromised at the expense of speed. To this end, poorly developed LC/MS-based methods that lack specificity, sensitivity, and/or ruggedness can lead to erroneous or misleading PK readouts.

ACKNOWLEDGMENTS

We are grateful to P. Tiller (Merck and Co.), Y.-Q. Xia (Bristol-Myers Squibb), A. Watt (Merck and Co.), T. Pereira (Merck and Co.), Waters, Finnigan, Sciex/Applied Biosystems, Beckman-Coulter, Spark–Holland, Tomtec, and LEAP Technologies for providing some of the figures and helpful discussions.

REFERENCES

1. J. Drews, Drug discovery: A historical perspective, *Science* **287** (2000), 1960–1964.
2. L. J. Gershell and J. H. Atkins, A brief history of novel drug discovery technologies, *Nature Rev. Drug Discovery* **2** (2003), 321–327.
3. M. Entzeroth, Emerging trends in high-throughput screening, *Curr. Opin. Pharmacol.* **3** (2003), 522–529.
4. J. Y. Chien, S. Friedrich, M. A. Heathman, D. P. de Alwis, and V. Sinha, Pharmacokinetics/pharmacodynamics and the stages of drug development: Role of modeling and simulation, *AAPS J.* **7** (2005), E544–E559.
5. D. K. Walker, The use of pharmacokinetic and pharmacodynamic data in the assessment of drug safety in early drug development, *Br. J. Clin. Pharmacol.* **58** (2004), 601–608.
6. G. L. Glish and R. W. Vachet, The basics of mass spectrometry in the twenty-first century, *Nature Rev. Drug Discovery* **2** (2003), 140–150.

7. W. J. Griffiths, A. P. Jonsson, S. Liu, D. K. Rai, and Y. Wang, Electrospray and tandem mass spectrometry in biochemistry, *Biochem. J.* **355** (2001), 545–561.

8. D. I. Papac and Z. Shahrokh, Mass spectrometry innovations in drug discovery and development, *Pharm. Res.* **18** (2001), 131–145.

9. Y. Hsieh, E. Fukuda, J. Wingate, and W. A. Korfmacher, Fast mass spectrometry-based methodologies for pharmaceutical analyses, *Comb. Chem. High Throughput Screening* **9** (2006), 3–8.

10. H. Keski-Hynnila, M. Kurkela, E. Elovaara, L. Antonio, J. Magdalou, L. Luukkanen, J. Taskinen, and R. Kostiainen, Comparison of electrospray, atmospheric pressure chemical ionization, and atmospheric pressure photoionization in the identification of apomorphine, dobutamine, and entacapone phase II metabolites in biological samples, *Anal. Chem.* **74** (2002), 3449–3457.

11. G. Wang, Y. Hsieh, and W. A. Korfmacher, Comparison of atmospheric pressure chemical ionization, electrospray ionization, and atmospheric pressure photoionization for the determination of cyclosporine A in rat plasma, *Anal. Chem.* **77** (2005), 541–548.

12. G. J. Van Berkel, An overview of some recent developments in ionization methods for mass spectrometry, *Eur. J. Mass Spectrom.* **9** (2003), 539–562.

13. J. Abian, A. J. Oosterkamp, and E. Gelpi, Comparison of conventional, narrow-bore and capillary liquid chromatography/mass spectrometry for electrospray ionization mass spectrometry: Practical consideration, *J. Mass Spectrom.* **34** (1999), 244–254.

14. R. Kostiainen, T. Kotiaho, T. Kuuranne, and S. Auriola, Liquid chromatography/atmospheric pressure ionization–mass spectrometry in drug metabolism studies. *J. Mass Spectrom.* **38** (2003), 357–372.

15. S. Ma, S. K. Chowdhury, and K. B. Alton, Thermally induced N-to-O rearrangement of *tert*-N-oxides in atmospheric pressure chemical ionization and atmospheric pressure photoionization mass spectrometry: Differentiation of N-oxidation from hydroxylation and potential determination of N-oxidation site, *Anal. Chem.* **77** (2005), 3676–3682.

16. W. Tong, S. K. Chowdhury, J. C. Chen, R. Zhong, K. B. Alton, and J. E. Patrick, Fragmentation of N-oxides (deoxygenation) in atmospheric pressure ionization: Investigation of the activation process, *Rapid Commun. Mass Spectrom.* **15** (2001), 2085–2090.

17. D. B. Robb, T. R. Covey, and A. P. Bruins, Atmospheric pressure photoionization: An ionization method for liquid chromatography–mass spectrometry, *Anal. Chem.* **72** (2000), 3653–3659.

18. K. A. Hanold, S. M. Fischer, P. H. Cormia, C. E. Miller, and J. A. Syage, Atmospheric pressure photoionization. 1. General properties for LC/MS, *Anal. Chem.* **76** (2004), 2842–2851.

19. A. Raffaelli and A. Saba, Atmospheric pressure photoionization mass spectrometry, *Mass Spectrom Rev.* **22** (2003), 318–331.

20. L. Sleno and D. A. Volmer, Ion activation methods for tandem mass spectrometry, *J. Mass Spectrom.* **39** (2004), 1091–1112.

21. A. E. Nassar, R. E. Talaat, and A. M. Kamel, The impact of recent innovations in the use of liquid chromatography–mass spectrometry in support of drug metab-

olism studies: Are we all the way there yet? *Curr. Opin. Drug Discovery Develop* **9** (2006), 61–74.

22. A. P. Watt, D. Morrison, and D. C. Evans, Approaches to higher-throughput pharmacokinetics (HTPK) in drug discovery, *Drug Discovery Today* **5** (2000), 17–24.

23. D. B. Kassel, Applications of high-throughput ADME in drug discovery, *Curr. Opin. Chem. Biol.* **8** (2004), 339–345.

24. M. Jemal, High-throughput quantitative bioanalysis by LC/MS/MS, *Biomed. Chromatogr.* **14** (2000), 422–429.

25. X. Xu, J. Lan, and W. A. Korfmacher, Rapid LC/MS/MS method development for drug discovery, *Anal. Chem.* **77** (2005), 389A–394A.

26. Y. Xu, L. Du, E. D. Soli, M. P. Braun, D. C. Dean, and D. G. Musson, Simultaneous determination of a novel KDR kinase inhibitor and its N-oxide metabolite in human plasma using 96-well solid-phase extraction and liquid chromatography/ tandem mass spectrometry, *J. Chromatogr. B* **817** (2005), 287–296.

27. C. J. Kitchen, A. Q. Wang, D. G. Musson, A. Y. Yang, and A. L. Fisher, A semiautomated 96-well protein precipitation method for the determination of montelukast in human plasma using high performance liquid chromatography/ fluorescence detection, *J. Pharm. Biomed. Anal.* **31** (2003), 647–654.

28. H. Song, K. Yan, X. Xu, and M. W. Lo, Quantitative determination of MK-0767, a dual alpha/gamma peroxisome proliferators-activated receptor (PPAR) agonist, in human plasma by liquid chromatography–tandem mass spectrometry, *J. Chromatogr. B* **810** (2004), 7–13.

29. B. K. Matuszewski, Standard line slopes as a measure of a matrix effect in quantitative HPLC-MS bioanalysis, *J. Chromatogr. B* **830** (2006), 293–300.

30. F. Beaudry and P. Vachon, Electrospray ionization suppression, a physical or a chemical phenomenon, *Biomed. Chromatogr.* **20** (2006), 200–205.

31. X, Xu, H. Mei, S. Wang, Q. Zhou, G. Wang, L. Broske, A. Pena, and W. A. Korfmacher, A study of common discovery dosing formulation components and their potential for causing time-dependent matrix effects in high-performance liquid chromatography tandem mass spectrometry assays, *Rapid Commun. Mass Spectrom.* **19** (2005), 2643–2650.

32. X. S. Tong, J. Wang, S. Zheng, J. V. Pivnichny, P. R. Griffin, X. Shen, M. Donnelly, K. Vakerich, C. Nunes, and J. Fenyk-Melody, Effect of signal interference from dosing excipients on pharmacokinetic screening of drug candidates by liquid chromatography/mass spectrometry, *Anal. Chem.* **74** (2002), 6305–6313.

33. H. Mei, Y. Hsieh, C. Nardo, X. Xu, S. Wang, K. Ng, and W. A. Korfmacher, Investigation of matrix effects in bioanalytical high-performance liquid chromatography/tandem mass spectrometric assays: Application to drug discovery, *Rapid Commun. Mass Spectrom.* **17** (2003), 97–103.

34. W. Z. Shou and W. Naidong, Post-column infusion study of the "dosing vehicle effect" in the liquid chromatography/tandem mass spectrometric analysis of discovery pharmacokinetic samples, *Rapid Commun. Mass Spectrom.* **17** (2003), 589–597.

35. R. Bonfiglio, R. C. King, T. V. Olah, and K. Merkle, The effects of sample preparation methods on the variability of the electrospray ionization response for model drug compounds, *Rapid Commun. Mass Spectrom.* **13** (1999), 1175–1185.

36. N. P. Sadagopan, W. Li, J. A. Cook, B. Galvan, D. L. Weller, S. T. Fountain, and L. H. Cohen, Investigation of EDTA anticoagulant in plasma to improve the throughput of liquid chromatography/tandem mass spectrometric assays, *Rapid Commun. Mass Spectrom.* **17** (2003), 1065–1070.

37. S. Souverain, S. Rudaz, and J. L. Veuthey, Matrix effect in LC-ESI-MS and LC-APCI-MS with off-line and on-line extraction procedures, *J. Chromatogr. A* **1058** (2004), 61–66.

38. R. King, R. Bonfiglio, C. Fernandez-Metzler, C. Miller-Stein, and T. Olah, Mechanistic investigation of ionization suppression in electrospray ionization, *J. Am. Soc. Mass Spectrom.* **11** (2000), 942–950.

39. T. M. Annesley, Ion suppression in mass spectrometry, *Clin. Chem.* **49** (2003), 1041–1044.

40. J. X. Shen, R. J. Motyka, J. P. Roach, and R. N. Hayes, Minimization of ion suppression in LC-MS/MS analysis through the application of strong cation exchange solid-phase extraction (SCX-SPE), *J. Pharm. Biomed. Anal.* **37** (2005), 359–367.

41. J. L. Herman, Generic method for on-line extraction of drug substances in the presence of biological matrices using turbulent flow chromatography, *Rapid Commun. Mass Spectrom.* **16** (2002), 421–426.

42. H. Zeng, J.-T. Wu, and S. E. Unger, The investigation and the use of high flow column-switching LC/MS/MS as a high-throughput approach for direct plasma sample analysis of single and multiple components in pharmacokinetic studies, *J. Pharm. Biomed. Anal.* **27** (2002), 967–982.

43. J. L. Herman, The use of turbulent flow chromatography and the isocratic focusing effect to achieve on-line cleanup and concentration of neat biological samples for low-level metabolite analysis, *Rapid Commun. Mass Spectrom.* **19** (2005), 696–700.

44. H. Zeng, Y. Deng, and J.-T. Wu, Fast analysis using monolithic columns coupled with high-flow on-line extraction and electrospray mass spectrometric detection for the direct and simultaneous quantitation of multiple components in plasma, *J. Chromatogr. B* **788** (2003), 331–337.

45. R. C. King, C. Miller-Stein, D. J. Magiera, and J. Brann, Description and validation of a staggered parallel high performance liquid chromatography system for good laboratory practice level quantitative analysis by liquid chromatography/tandem mass spectrometry, *Rapid Commun. Mass Spectrom.* **16** (2002), 43–52.

46. F. L. Sauvage, J. M. Gaulier, G. Lachatre, and P. Marquet, A fully automated turbulent-flow liquid chromatography-tandem mass spectrometry technique for monitoring antidepressant in human serum, *Ther. Drug Monit.* **28** (2006), 123–130.

47. Y. Q. Xia, R. Bakhtiar, and R. B. Franklin, Automated online-dual-column extraction coupled with teicoplanin stationary phase for simultaneous determination of (*R*)- and (*S*)-propranolol in rat plasma using liquid-chromatography/tandem mass spectrometry, *J. Chromatogr. B* **788** (2003), 317–329.

48. Y. Alnouti, K. Srinivasan, D. Waddell, H. Bi, O. Kavetskaia, and A. I. Gusev, Development and application of a new on-line SPE system combined with LC-MS/MS detection for high throughput direct analysis of pharmaceutical compounds in plasma, *J. Chromatogr. A* **1080** (2005), 99–106.

49. Y. Alnouti, M. Li, O. Kavetskaia, H. Bi, C. E. C. A. Hop, and A. I. Gusev, Method for internal standard introduction for quantitative analysis using on-line solid-phase extraction LC-MS/MS, *Anal. Chem.* **78** (2006), 1331–1336.

50. G. K. Webster, L. Kott, and T. D. Maloney, Considerations when implementing automated methods into GxP laboratories, *J. Assoc Lab Autom (JALA)* **10** (2005), 182–191.

51. V. P. Shah, K. K. Midha, J. W. A. Findlay, H. M. Hill, J. D. Hulse, I. L. McGilveray, G. McKay, K. J. Miller, R. N. Patnaik, M. L. Powell, A. Tonelli, C. T. Viswanathan, and A. Yacobi, Bioanalytical method validation—A revisit with a decade of progress, *Pharm. Res.* **17** (2000), 1551–1557.

52. R. Brown, M. Caphart, P. Faustino, R. Frankewich, J. Gibbs, E. Leutzinger, G. Lunn, L. Ng, R. Rajagopalan, Y. Chiu, and E. Sheinin, Analytical procedures and method validation: Highlights of the FDA's draft guidance, *LCGC* **19** (2001), 74–79.

53. S. Zhou, Q. Song, Y. Tang, and W. Naidong, Critical review of development, validation, and transfer for high throughput bioanalytical LC-MS/MS methods, *Curr. Pharm. Anal.* **1** (2005), 3–14.

54. D. H. Wilson, D. Sepe, and G. Barnes, Inter-laboratory differences in sirolimus results from six sirolimus testing centers using HPLC tandem mass spectrometry (LC/MS/MS), *Clin. Chim. Acta* **355** (2005), 211–213.

55. For more information, please see WWW.FDA.GOV and WWW.ICH.ORG.

56. M. L. Chen, V. Shah, R. Patnaik, W. Adams, A. Hussain, D. Conner, M. Mehta, H. Malinowski, J. Lazor, S. M. Huang, D. Hare, L. Lesko, D. Sporn, and R. Williams, Bioavailability and bioequivalence: An FDA regulatory overview, *Pharm. Res.* **18** (2001), 1645–1650.

57. K. K. Midha, M. J. Rawson, and J. W. Hubbard, The role of metabolites in bioequivalence, *Pharm. Res.* **21** (2004), 1331–1344.

58. R. Bakhtiar, L. Ramos, and F. L. S. Tse, Use of atmospheric pressure ionization mass spectrometry in enantioselective liquid chromatography, *Chirality* **13** (2001), 63–74 and references cited therein.

59. R. Bakhtiar, L. Khemani, M. Hayes, T. Bedman, and F. Tse, Quantification of the anti-leukemia drug STI571 (Gleevec) and its metabolite (CGP 74588) in monkey plasma using a semi-automated solid phase extraction procedure and liquid chromatography–tandem mass spectrometry, *J. Pharm. Biomed. Anal.* **28** (2002), 1183–1194.

60. R. Bakhtiar, J. Lohne, L. Ramos, L. Khemani, M. Hayes, and F. Tse, High-throughput quantification of the anti-leukemia drug STI571 (Gleevec) and its main metabolite (CGP 74588) in human plasma using liquid chromatography–tandem mass spectrometry, *J. Chromatogr. B* **768** (2002), 325–340.

61. T. Andersson, Single-isomer drugs: True therapeutic advances, *Clin. Pharmacokinet.* **43** (2004), 279–285.

62. I. Agranat, H. Caner, and J. Caldwell, Putting chirality to work: The strategy of chiral switches, *Nature Rev. Drug Discovery* **1** (2002), 753–768.

63. N. R. Srinivas, Simultaneous chiral analyses of multiple analytes: Case studies, implications and method development considerations, *Biomed. Chromatogr.* **18** (2004), 759–784.

64. M. Deininger, E. Buchdunger, and B. J. Druker, The development of imatinib as a therapeutic agent for chronic myeloid leukemia, *Blood* **105** (2005), 2640–2653.

65. B. J. Druker, Imatinib as a paradigm of targeted therapies, *Adv. Cancer Res.* **91** (2004), 1–30.

66. B. J. Druker, Imatinib mesylate in the treatment of chronic myeloid leukemia, *Expert Opin. Pharmacother* **4** (2003), 963–971.

67. B. Peng, M. Hayes, D. Resta, A. Racine-Poon, B. J. Druker, M. Talpaz, C. L. Sawyers, M. Rosamilia, J. Ford, P. Lloyd, and R. Capdeville, Pharmacokinetics and pharmacodynamics of imatinib in a Phase I trial with chronic myeloid leukemia patients, *J. Clin. Oncol.* **22** (2004), 935–942.

68. M. Dickson and J. P. Gagnon, Key factors in the rising cost of new drug discovery and development, *Nature Rev. Drug Discovery* **3** (2004), 417–429.

69. R. Kramer and D. Cohen, Functional genomics to new drug targets, *Nature Rev. Drug Discovery* **3** (2004), 965–972.

70. D. A. Lewin and M. P. Weiner, Molecular biomarkers in drug development, *Drug Discovery Today* **9** (2004), 976–983.

71. G. A. FitzGerald, Anticipating change in drug development: The emerging era of translational medicine and therapeutics, *Nature Rev. Drug Discovery* **4** (2005), 815–818.

72. R. Frank and R. Hargreaves, Clinical biomarkers in drug discovery and development, *Nature Rev. Drug Discovery* **2** (2003), 566–580.

73. J. T. Stults and D. Arnott, Proteomics, *Methods Enzymol.* **402** (2005), 245–289.

74. J. A. Bilello, The agony and ecstasy of "OMIC" technology in drug development, *Curr. Mol. Med.* **5** (2005), 39–52.

75. W. Weckwerth and K. Morgenthal, Metabolomics: From pattern recognition to biological interpretation, *Drug Discovery Today* **10** (2005), 1551–1558.

76. C. E. Thomas and G. Ganji, Integration of genomic and metabonomic data in systems biology: Are we "there" yet? *Curr. Opin. Drug Discovery Dev.* **9** (2006), 92–100.

77. W. Sadee and Z. Dai, Pharmacogenetics/genomics and personalized medicine, *Hum. Mol. Genet.* **14** (2005), R207–R214.

78. A. Luhe, L. Suter, S. Ruepp, T. Singer, T. Weiser, and S. Albertini, Toxicogenomics in the pharmaceutical industry: Hollow promises or real benefit, *Mutat. Res.* **575** (2005), 102–115.

79. M. R. Wenk, The emerging field of lipidomics, *Nature Rev. Drug Discovery* **4** (2005), 594–610.

80. Z. Shriver, S. Raguram, and R. Sasisekharan, Glycomics: A pathway to a class of new and improved therapeutics, *Nature Rev. Drug Discovery* **3** (2004), 863–873.

81. L. J. Lesko and J. Woodcock, Translation of pharmacogenomics and pharmacogenetics: A regulatory perspective, *Nature Rev. Drug Discovery* **3** (2004), 763–769.

82. J. Mervis, Productivity counts—but the definition is key, *Science* **309** (2005), 726–727.

83. J. W. Lee, V. Devanarayan, Y. C. Barrett, R. Weiner, J. Allinson, S. Fountain, S. Keller, I. Weinryb, M. Green, L. Duan, J. A. Rogers, R. Millham, P. J. O'Brien, J. Sailstad, M. Khan, C. Ray, and J. A. Wagner, Fit-for-purpose method development and validation for successful biomarker measurement, *Pharm. Res.* **23** (2006), 312–328.

84. W. A. Colburn and J. W. Lee, Biomarkers, validation and pharmacokinetic–pharmacodynamic modeling, *Clin. Pharmacokinet.* **42** (2003), 997–1022.

85. J. R. Seckl, 11Beta-hydroxysteroid dehydrogenase: Changing glucocorticoid action, *Curr. Opin. Pharmacol.* **4** (2004), 597–602.

86. A. Y. Yang, L. Sun, D. G. Musson, and J. J. Zhao, Determination of $M + 4$ stable isotope labeled cortisone and cortisol in human plasma by μElution solid-phase extraction and liquid chromatography/tandem mass spectrometry, *Rapid Commun. Mass Spectrom.* **20** (2006), 233–240.

87. L. A. Liotta, M. Ferrari, and E. Petricoin, Written in blood, *Nature* **425** (2003), 905.

88. D. H. Geho, C. D. Jones, E. F. Petricoin, and L. A. Liotta, Nanoparticles: Potential biomarker harvesters, *Curr. Opin. Chem. Biol.* **10** (2006), 56–61.

89. C. L. Fernandez-Metzler and R. C. King, The emergence and application of technological advances in biotransformation studies, *Curr. Top. Med. Chem.* **2** (2002), 67–76 and references cited therein.

90. W. A. Korfmacher, Principles and applications of LC-MS in new drug discovery, *Drug Discovery Today* **10** (2005), 1357–1367.

91. D. W. Johnson, Contemporary clinical usage of LC/MS: Analysis of biologically important carboxylic acids, *Clin. Biochem.* **38** (2005), 351–361.

92. W. A. Korfmacher (ed.), *Using Mass Spectrometry for Drug Metabolism Studies*, CRC Press, Boca Raton, FL, 2004.

93. R. C. Garner, Accelerated mass spectrometry in pharmaceutical research and development: A new ultrasensitive analytical method for isotope measurement, *Curr. Drug Metab.* **1** (2004), 205–213.

94. G. Deng and G. Sanyal, Applications of mass spectrometry in early stages of target drug discovery, *J. Pharm. Biomed. Anal.* **40** (2006), 528–538.

95. S. A. C. Wren and P. Tchelitcheff, UPLC/MS for the identification of β-blockers, *J. Pharm. Biomed. Anal.* **40** (2006), 571–580.

96. S. K. Balani, N. V. Nagaraja, M. G. Qian, A. O. Costa, J. S. Daniels, H. Yang, P. R. Shimoga, J. T. Wu, L. S. Gan, F. W. Lee, and G. T. Miwa, Evaluation of microdosing to assess pharmacokinetic linearity in rats using liquid chromatography–tandem mass spectrometry, *Drug Metab. Dispos.* **34** (2006), 384–388.

97. G. L. Erny and A. Cifuentes, Liquid separation techniques coupled with mass spectrometry for chiral analysis of pharmaceuticals compounds and their metabolites in biological fluids, *J. Pharm. Biomed. Anal.* **40** (2006), 509–515.

98. C. Pan, F. Liu, Q. Ji, W. Wang, D. Drinkwater, and R. Vivilecchia, The use of LC/MS, GC/MS, and LC/NMR hyphenated techniques to identify a drug degradation product in pharmaceutical development, *J. Pharm. Biomed. Anal.* **40** (2006), 581–590.

99. A. P. Watt, D. Morrison, K. L. Locker, and D. C. Evans, Higher throughput bioanalysis by automation of a protein precipitation assay using a 96-well format with detection by LC-MS/MS, *Anal. Chem.* **72** (2000), 979–984.

100. H. Wang and Z. Shen, Enantiomeric separation and quantification of pindolol in human plasma by chiral liquid chromatography/tandem mass spectrometry using staggered injection with a CTC Trio valve system, *Rapid Commun. Mass Spectrom.* **20** (2006), 291–297.

101. T. Pereira and S. W. Chang, Semi-automated quantification of ivermectin in rat and human plasma using protein precipitation and filtration with liquid chromatography/tandem mass spectrometry, *Rapid Commun. Mass Spectrom.* **18** (2004), 1265–1276.

102. J. F. Pritchard, M. Jurima-Romet, M. L. J. Reimer, E. Mortimer, B. Rolfe, and M. N. Cayen, Making better drugs: Decision gates in non-clinical drug development, *Nature Rev. Drug Discovery* **2** (2003), 542–553.

103. H. P. Gschwind, U. Pfaar, F. Waldmeier, M. Zollinger, C. Sayer, P. Zbinden, M. Hayes, R. Pokorny, M. Seiberling, M. Ben-Am, B. Peng, and G. Gross, Metabolism and disposition of imatinib mesylate in healthy volunteers, *Drug Metab. Dispos.* **33** (2005), 1503–1512.

14

ROLE OF HPLC IN PROCESS DEVELOPMENT

Richard Thompson and Rosario LoBrutto

14.1 RESPONSIBILITIES OF THE ANALYTICAL CHEMIST DURING PROCESS DEVELOPMENT

In the drug discovery area, a compound with desired therapeutic properties is identified, and its structure may be modified by synthetic alterations to enhance potency and specificity or to decrease toxicity and undesired side effects. The lead drug candidate is then transitioned into the drug development area. Only small amounts of drug (typically less than a gram) are required to support the required studies in the Drug Discovery area. However larger amounts are required to support the studies conducted in the Drug Development area. The amount required in the preclinical stage typically ranges from 20 to 2000 g. This material is required to support studies including subchronic toxicity, genotoxicity, ancillary pharmacology, early animal pharmacokinetics (PK), salt/form selection, and formulation development. As the drug candidate progresses through the various clinical stages, the drug requirements typically range from 1 kg to 200 kg. This material supports the various clinical studies as well as chronic toxicity, carcinogenicity, development and reproductive toxicity, and formulation development. Finally, tons of drug may be required upon successful approval and commercialization (Figure 14-1).

The synthetic pathway to the drug substance is likely to evolve during the various stages of development. It is highly unlikely that the synthetic process

HPLC for Pharmaceutical Scientists, Edited by Yuri Kazakevich and Rosario LoBrutto
Copyright © 2007 by John Wiley & Sons, Inc.

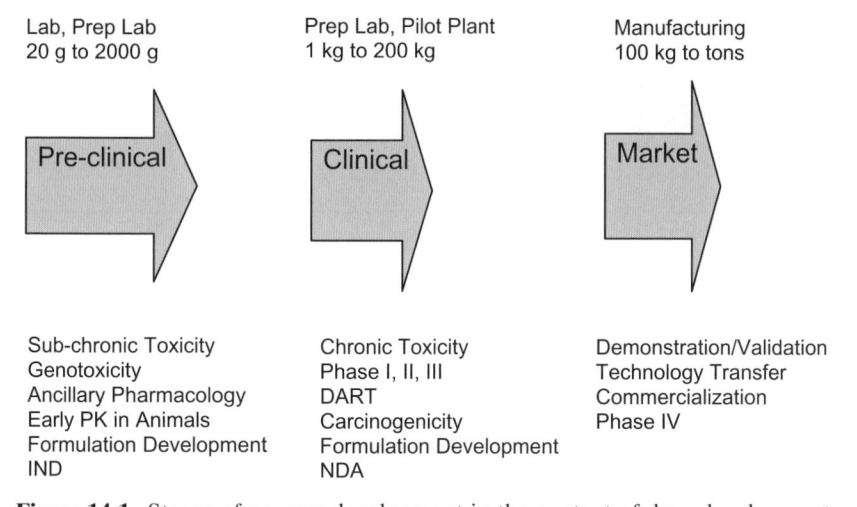

Lab, Prep Lab	Prep Lab, Pilot Plant	Manufacturing
20 g to 2000 g	1 kg to 200 kg	100 kg to tons

Pre-clinical | Clinical | Market

Sub-chronic Toxicity	Chronic Toxicity	Demonstration/Validation
Genotoxicity	Phase I, II, III	Technology Transfer
Ancillary Pharmacology	DART	Commercialization
Early PK in Animals	Carcinogenicity	Phase IV
Formulation Development	Formulation Development	
IND	NDA	

Figure 14-1. Stages of process development in the context of drug development.

utilized in drug discovery will be the same as that used to provide commercial scale quantities. The discovery chemist may utilize a large number of synthetic steps, use a number of reagents that are expensive or not practical at scale-up, use a number of chromatographic steps for purification, and experience very low yields. For scale up, the process development chemist must factor in safety, economical, and ecological considerations while producing a robust and reproducible synthesis. He must consider operating limitations such as heat and mass transfer. Economic factors will dictate minimization of the synthetic steps, maximization of yield, and choice of raw materials. In addition, the process must meet environmental, occupational health, and safety requirements.

Furthermore, the process development chemist must follow guidelines from the Food and Drug Administration (FDA) in relation to the control and identification of impurities in drugs that will be used in humans. Regulatory bodies require that the maximum possible human exposure to an impurity in a drug substance be supported by toxicological studies in animals that indicate no significant adverse effects. Consequently, impurities that exceed a 0.1% tolerance limit in clinical material must first be qualified in animal toxicological studies. Scale up of a synthesis, however, may generate a different impurity profile than observed for the smaller quantities prepared to support the toxicological studies. Kinetic factors, changes in raw materials, or changes in reaction conditions may result in the introduction of new or elevated impurities. These new impurities may be qualified in additional chronic toxicity and genotoxicity studies, but this strategy is often not economically feasible and is undertaken more as a last resort. A better strategy is to identify and then control impurities that are generated during the continuously evolving stages of process development.

As a consequence of process evolution and regulatory requirements, the analytical chemist supporting process development is faced with a number of challenges. He must evaluate the purity and stability of raw materials, intermediates, and drug substance. He must evaluate yield and impurity generation across the various synthetic steps. Impurities in drug substance, intermediates, and raw materials may require identification. Analytical methods may have to be adapted to accommodate process changes. Finally he must set specifications, validate analytical methods, provide regulatory documentation, and perform a technology transfer prior to drug approval and commercialization. To this end, HPLC is a critical tool to perform many of the above tasks. Most pharmaceutical compounds are amenable to analysis by HPLC. HPLC is a powerful technology that is capable of separating complex mixtures into individual components that can then be quantified. A well-developed HPLC method resolves and quantifies impurities from an analyte of interest in a reproducible, rugged, precise, and accurate fashion.

14.2 HPLC SEPARATION MODES

The multitude of available separation modes, mobile phases, and columns provide a plethora of parameters that can be manipulated to meet the criteria for a well developed HPLC method. Conversely, it also creates a dilemma in choosing the optimal parameters from the myriad of possibilities. The commonly utilized modes of HPLC in pharmaceutical development are reversed phase (RP) and normal phase (NP) for small organic molecules (<1000 Da). Sub-/supercritical chromatography (SFC) is utilized as an alternative to normal phase chromatography. Ion chromatography (IC) is utilized for ionic species. Other modes utilized include hydrophilic interaction and chiral separations. In-depth discussion of the theory and method development of RPLC, ion exchange, and NPLC are highlighted in the first section of the book. How each of these chromatographic modes could be applied for the analysis of compounds that the analytical chemist encounters within the framework of process development will be discussed in more detail.

14.2.1 Reversed-Phase Liquid Chromatography

Reversed-phase chromatography is the preferred HPLC mode in the pharmaceutical industry. Its popularity is in part derived from its mobile-phase compatibility with the typical polar drug substance, the higher efficiencies associated with this mode, shorter re-equilibration times, and the ability to run gradient methods covering a large range in polarity. This allows for its use in reaction monitoring, qualifying synthetic intermediates, and for stability and release testing of the drug substance. Several case studies of reversed-phase method development of drug substances and drug substance intermediates are given in Chapter 8 sections 8.2 and 8.5.

14.2.2 Normal-Phase Chromatography

Normal-phase or adsorption HPLC utilizes a polar stationary phase and a less polar mobile phase. Retention occurs through polar interactions, such as hydrogen bonding and dipole interactions, between the solute and the stationary phase. Retention is more predictable than for RP chromatography. Carboxylic acids tend to show the strongest retention followed by amines, ketones, ethers, aromatic hydrocarbons, and saturated hydrocarbons in decreasing order of retention. Good selectivity is often observed for positional and stereoisomers. The mobile phase is usually a mixture of a nonpolar solvent such as heptane or hexane(s) and a more polar solvent. The polar organic solvent can be chosen based on it physicochemical properties (hydrogen bonding capabilities, lipophilicity, and polarizability). There are a large number of options for eluent components, and extensive selectivity changes are observed with the use of various mobile-phase components. In addition, small changes of the polar organic solvent can cause large changes in retention, and this should be investigated during method development. Common solvents include ethanol, isopropanol, tetrahydrofuran, ethyl acetate, and dichloromethane. The level of water in the solvents needs to be controlled as well, since differences in retention may be observed. Note that heptane and methanol have limited miscibility and only a maximum of 5% methanol should be used. Mobile-phase miscibility should be checked prior to pumping a particular composition on the HPLC. A simple mixture of the solvents in the beaker should allow the chromatographer to discern if the two components are miscible. Additives such as triethylamine and trifluoroacetic acid are recommended to reduce retention and improve peak shape for the analysis of bases and acids, respectively, by reducing interactions of the solute with the highly active sites of silica. Commonly utilized stationary phases include cyano, diol, amino, and silica. However, unmodified silica possesses greater surface heterogeneity and is more retentive than the other three phases. Very little selectivity differences are observed as a function of the type of stationary phase.

When using NP the chromatographer must remember to convert their system to NP mode if RP mode was used previously. Any aqueous/buffer left in the system could precipitate out when the normal-phase solvents are pumped than the system. Water contamination in the mobile phase lines can also lead to water absorption on the column and change the chromatography significantly. It is generally recommended, that if the system was previously in RP mode, to flush the system with pure water for about 15 minutes at 2 mL/min. Then use IPA to flush the system for an additional 10 minutes at 2 mL/min. The system should then be flushed with the desired NP mobile phase for 5 minute at 2 mL/min. Then the NP column can be installed and equilibrated with the NP mobile phase.

Despite the popularity of RP chromatography, NP has its usefulness in the analysis of compounds during drug development. It can be used for polar solutes that are poorly retained in RP, nonpolar solutes that are strongly retained in RP, positional and stereoisomers, or solutes that are labile or possess poor solubility in RP mobile phases. An example of an RP-incompatible method involves reaction monitoring of a mesylation step:

$$RCOOH + MeSO_2Cl \rightarrow RCOOSO_2Me + RCOOOCR + HCl$$

The anhydride was formed as a side product in this reaction impacting yield. However, in an RP mobile phase, both the mesylate and anhydride would revert back to the carboxylic acid. Derivatization would produce the same product for both the mesylate and the anhydride. The reaction components were separated and quantified under NP conditions using a diol column with a 0.1 v/v% TFA in heptane/THF mobile phase (Figure 14-2). This method was used monitor the reaction such that the level of the carboxylic acid intermediate was less than 0.5% in the reaction mixture.

14.2.3 Sub-/Supercritical Chromatography

Sub-/supercritical fluid chromatography is essentially NP chromatography with the added advantage that the lower viscosity and higher diffusivity of the mobile phase results in higher column efficiencies allowing for rapid resolutions. The columns employed are the same as those utilized in conventional NP chromatography. Carbon dioxide is the most commonly used nonpolar eluent but requires a more polar modifier such as an alcohol for the elution of polar solutes. The modifier increases the polarity of the mobile phase and

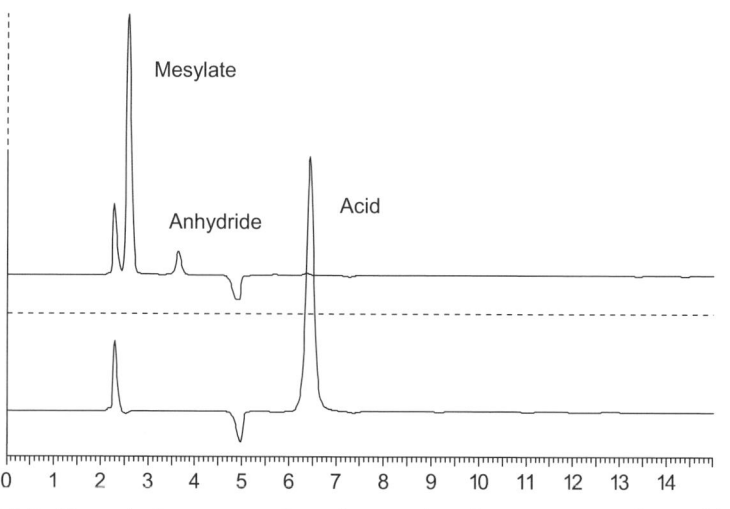

Figure 14-2. Normal-phase separation of a mesylate from corresponding acid. Chromatographic conditions: YMC Pack Diol 150 × 4.6 mm, 90% 0.1% TFA in heptane/10% 0.1% TFA in THF.

occupies active sites on the stationary phase, leading to reduced retention of solutes. As with conventional NP chromatography, the use of triethylamine and trifluoroacetic acid as additives is recommended for the analysis of amines and acids, respectively. The polar nature of most drug substance requires the use of high levels of organic modifier, and thus the mobile phase is most often in the subcritical state. Retention characteristics are the same as in conventional NP chromatography.

Subcritical fluid chromatography was applied for the resolution of a bromosulfone drug intermediate from various process-related compounds (Figure 14-3). Initial steps toward method development were performed in RP mode. However, significant fronting of the bromosulfone peak was observed, indicating on-column degradation that was later determined to occur through (a) nucleophilic substitution of the bromo group with a hydroxyl group to form

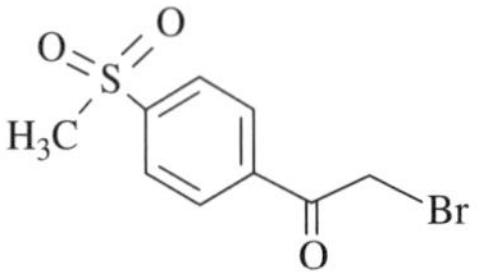

(1)
Bromosulfone

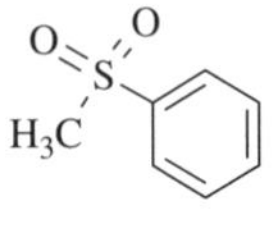

(2)
Phenylmethylsulfone

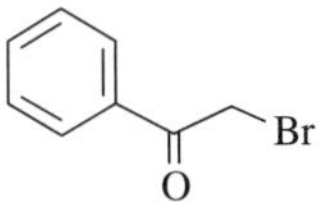

(3)
2-Bromoacetophenone

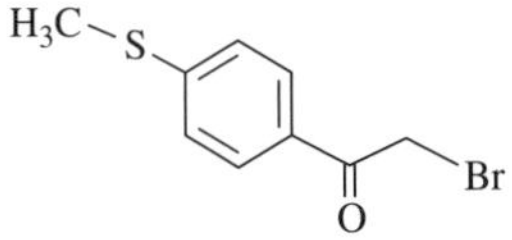

(4)
Bromosulfide

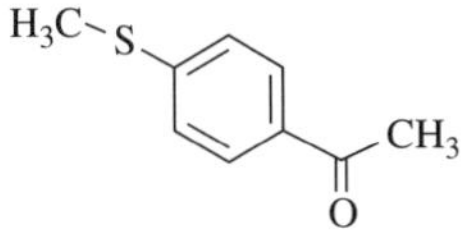

(5)
Ketosulfide

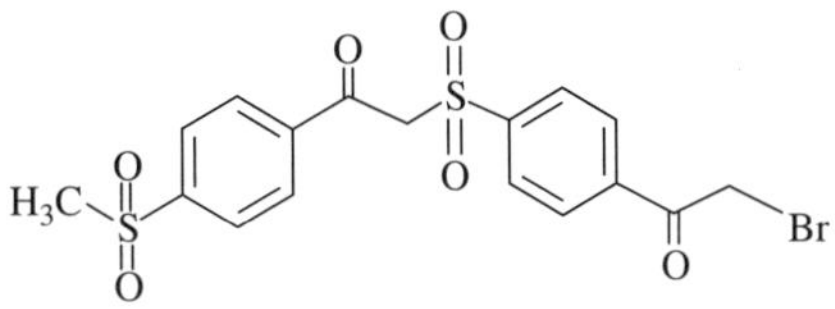

(6)
Bromosulfone Dimer

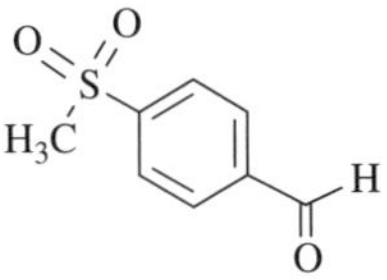

(7)
4-Methylsulfonylbenzaldehyde

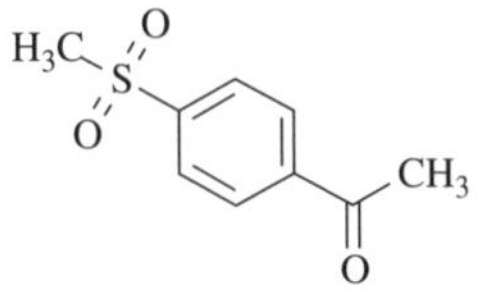

(8)
Ketosulfone

Figure 14-3. Structures of bromosulfone and process-related impurities. (Reprinted from reference 1, with permission.)

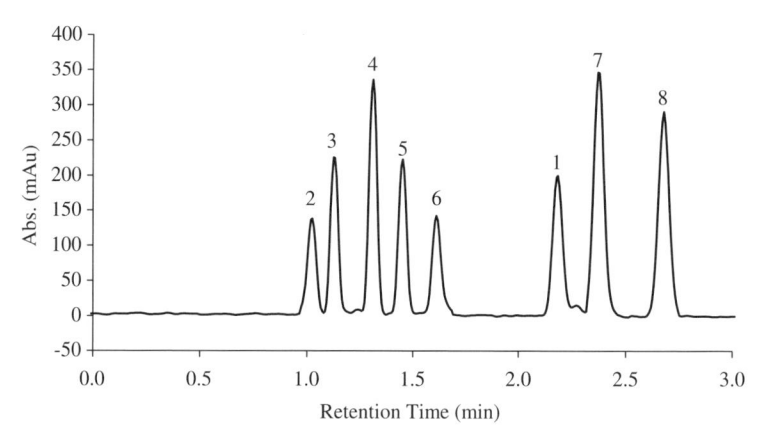

Figure 14-4. Separation of bromosulfone from process-related impurities by SFC. Chromatographic conditions: 50% Carbon dioxide/50% (80/20 methylene chloride/acetonitrile), Zorbax silica 250 × 4.6 mm. (Reprinted from reference 1, with permission.)

an alcohol and (b) addition to the ketone group to form a gem diol. Separation of bromosulfone from seven process-related compounds was achieved under subcritical conditions using a silica column and a mobile phase of carbon dioxide and 50% modifier (80/20 methylene chloride/acetonitrile) within three minutes (Figure 14-4) [1].

14.2.4 Hydrophilic Interaction Chromatography

Another option to conventional NP chromatography is hydrophilic interaction chromatography (HILIC). This mode utilizes a polar stationary phase with aqueous/organic modifier but with very high percentages of organic modifier. A simple acetonitrile/aqueous buffer mobile phase is commonly utilized in conjunction with a silica or amino stationary phase. Ammonium acetate is often used as a buffer salt in the mobile phase because it possesses good solubility at high organic content. At lower pHs, phosphoric acid can be utilized. An adsorbed water layer on the silica substrate is formed under these chromatographic conditions. Polar solutes partition from the highly organic bulk mobile phase into the adsorbed water layer where they can undergo polar interactions. In addition, positively charged solutes, such as amines, can undergo ionic interactions with charged silanol groups. As a consequence, retention of solutes increases with their increasing polarity. This mode is particularly useful for the separation of very polar solutes drug substance intermediates and/or raw materials that show minimal or no retention under RP conditions and are very strongly retained under NP conditions. Figures 14-5 and 14-6 depict the separation of nine very polar pyridine derivatives [2]. In process research environment, for example, one of these pyridine derivatives

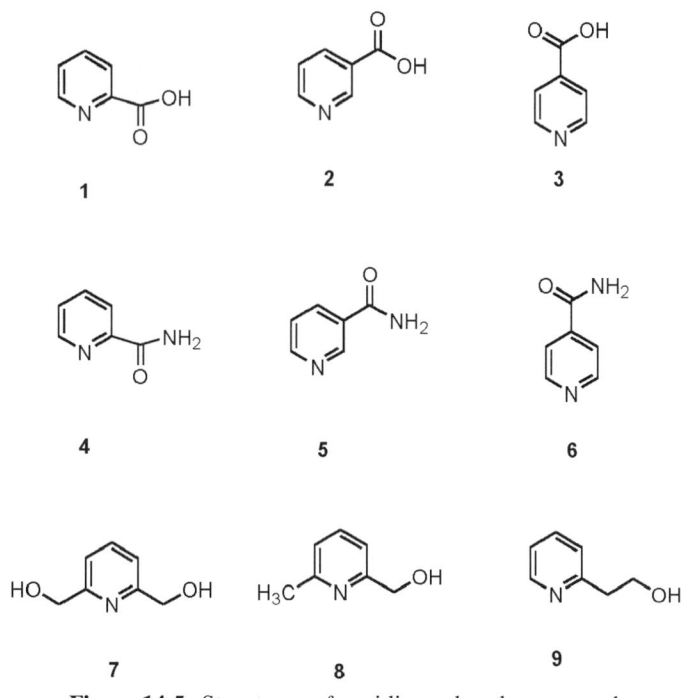

Figure 14-5. Structures of pyridine-related compounds.

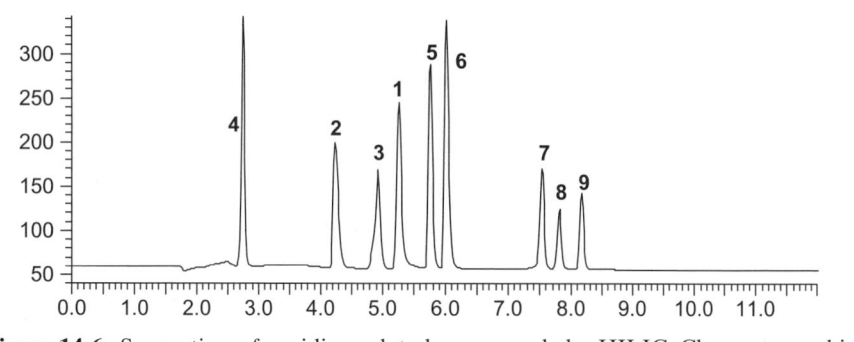

Figure 14-6. Separation of pyridine-related compounds by HILIC. Chromatographic conditions: Atlantis HILIC silica 3 μm, 150 × 4.6 mm. Mobile phase A: 0.1% phosphoric acid in D.I water. Mobile phase B: Acetonitrile. Gradient at 95% B to 60% B in 7 min and then hold 8 min.

could be a key raw material in a synthetic process. The possible isomeric forms of the key raw material should be well-resolved from the key raw material and needs to be controlled (sometimes a certain set of acceptance criteria are set for both the overall purity of the key raw material and maximum amount of undesired impurity) to avoid undesired reactions in the downstream processing.

14.2.5 Ion-Exchange Chromatography

Ion-exchange chromatography is useful for the separation of ionic or ionizable solutes and resolves solutes based on the strength of their ionic interactions with ionic functional groups on the stationary phase. The mobile phase is aqueous. The solute and the functional group on the stationary phase possess opposite charges, and the mobile phase contains a counterion with the same type of charge as the solute and thus effectively competes with the solute ion for ion pair interactions with the stationary phase. The retention of the solute is dependent upon the ionic size, charge magnitude, and polarizability of the solute and mobile-phase counterion as well as the ionic strength of the mobile-phase counterion. Gradients of counterion concentration can be employed. Retention is also dependent upon the mobile-phase pH and the dissociation constants of protolytic solute and mobile-phase species.

The stationary phase can be categorized as strong or weak ion exchangers. The capacity of strong ion exchangers is independent of pH, while the capacity of weak ion exchangers varies as a function of their protonated state. Strong ion exchangers include sulfonate functionalities for the analysis of cationic species and quaternary ammonium functionalities for the analysis of anionic species. Weak ion exchangers include carboxylate functionalities for the analysis of cationic species and amines for the analysis of anionic species. The functionalities are commonly attached to a polymeric matrix such as poly(styrene-divinylbenzene), polyacrylate, or polymethylacrylate.

Ion chromatography can be applied for the quantitation of inorganic impurities, drug substance counterions, and ionic synthetic impurities and degradation products. The most common forms of detection are by conductivity detection and indirect photometric detection (IPD), which allows for the use of conventional UV detectors. With IPD the mobile-phase anion possesses a significant chromophore. When a solute molecule, with a weaker chromophore, is eluted and passes through the detector cell, it is manifested as a negative peak. This form of detection can be used for analysis of ionic impurities in API [3–5]. Alendronate is a highly ionic bisphosphonate species that also possesses a primary amine functionality that can be derivatized with 9-fluorenylmethyl chloroformate (FMOC) and analyzed by conventional RPLC. However, alendronate does not possess a significant chromophore, and process-related impurities may also have low chromophores and may also not have an amine functionality that can be derivatized by FMOC. Such impurities would not be detected in the conventional RPLC method. To address this issue, an ion exchange method was developed to separate alendronate from similar bisphosphonates, synthetic impurities, and inorganic impurities (Figure 14-7) [4].

The addition of a compatible organic solvent may also influence selectivity, particularly when the stationary phase has a polymeric substrate. With these types of phases, the solute can undergo both hydrophobic and ion-exchange interactions. The addition of an organic solvent will result in increased

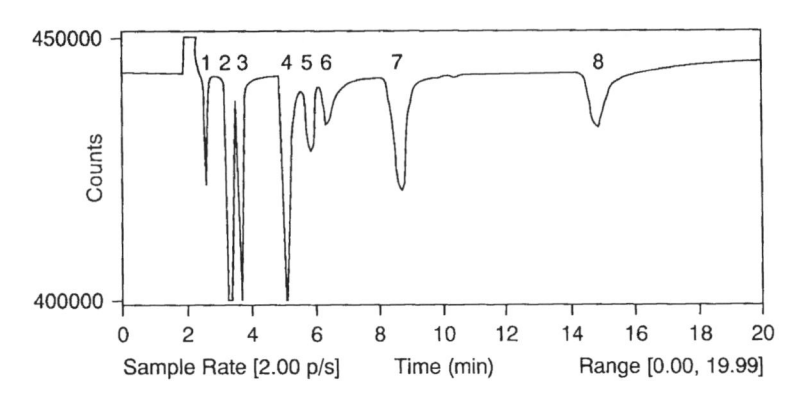

Figure 14-7. Separation of organophosphonates and process-related impurities. Chromatographic conditions: Hamilton PRP-X100, 250 × 4.6 mm, 1 mM trimesic acid (pH 5.5). 1, Phosphonopyrrolidine; 2, alendronate; 3, phosphite; 4, chloride; 5, methanesulfonate; 6, alendronate dimer; 7, etidronate; 8, clodronate. (Reprinted from reference 4, with permission.)

retention for solutes such as inorganic ions that only undergo ion interactions. Ions such as acetate and alendronate, which can undergo both types of interactions, may be more or less strongly retained depending on the ratio of hydrophobic to ion exchange interactions [4, 5].

14.2.6 Chiral Chromatography

Chiral separations can be considered as a special subset of HPLC. The FDA suggests that for drugs developed as a single enantiomer, the stereoisomeric composition should be evaluated in terms of identity and purity [6]. The undesired enantiomer should be treated as a structurally related impurity, and its level should be assessed by an enantioselective means. The interpretation is that methods should be in place that resolve the drug substance from its enantiomer and should have the ability to quantitate the enantiomer at the 0.1% level. Chiral separations can be performed in reversed phase, normal phase, and polar organic phase modes. Chiral stationary phases (CSP) range from small bonded synthetic selectors to large biopolymers. The classes of CSP that are most commonly utilized in the pharmaceutical industry include Pirkle type, crown ether, protein, polysaccharide, and antibiotic phases [7].

Pirkle-type phases are amino acid derivatives possessing an aromatic entity which can undergo π–π interactions with the solute. The aromatic entity can be either a π donor or π acceptor. The CSP and the solute form a π donor/acceptor pair. This complex is then stabilized by additional interactions such as hydrogen bonding, dipole interactions, or steric repulsion [8]. The Pirkle-type phases are most commonly used in normal-phase mode in order to enhance the π–π and hydrogen bond interactions. Hexane with an alcoholic modifier, such as isopropanol, is the mobile phase of choice. These phases have also been utilized in the reversed-phase mode but with poorer enantioselectivity and in some cases different elution orders indicating a change in the chiral recognition mechanism. These phases can also be utilized in super-/subcritical mode.

Crown ethers are heteroatomic macrocycles possessing a hydrophobic exterior and a hydrophilic cavity. Crown ethers show a strong affinity for primary amines through a hydrogen bonding interaction. The introduction of bulky groups, such as binaphthyl or carboxylate groups, onto the exterior of the crown ethers provides steric barriers and induces enantioselective interactions with solute molecules. Separations are performed in reversed-phase mode. Retention and selectivity is controlled by the concentration and type of counteranion in the mobile phase and the percent of organic modifier. One commercially available stationary phase contains a crown ether phase, with binaphthyl appendages, that is dynamically coated onto a silica substrate. An aqueous mobile phase is recommended when using this column. Retention increases with the chaotropicity and concentration of the counteranion [9]. A second commercially available phase utilizes a crown ether with carboxylate appendages and is covalently bonded to a silica substrate. Organic solvents can be used in the eluent. In an in-house study for a series of amines (drug substance intermediates), retention increased with organic content opposite for what is expected from a reversed-phase system. This behavior can be explained due to the fact that the primary interactions are hydrogen bonding and ion pairing, both of which would increase in strength with decreasing polarity of the mobile phase. Retention also increases with increasing deprotonation of the CSP's carboxylate groups as a consequence of increased sites for ion pair interactions [10].

The antibiotic glycopeptides—vancomycin, teicoplanin, and ristocetin A—have been extensively utilized as chiral selectors [11]. These macrocyclic antibiotics possess several characteristics that enable them to stereoselectively interact with solutes. They contain an aglycon bucket consisting of three or four macrocyclic rings. They also possess multiple stereogenic centers and a number of functional groups including sugars, aromatic rings, phenol groups, amide linkages, amine, moieties, and acid/esters moieties. As a consequence, they can interact with a solute through hydrogen bonding, dipole interactions, π–π interactions, hydrophobic interactions, electrostatic interactions, and steric hindrance. The phases can be used in normal-phase, reversed-phase, polar organic, and sub-/supercritical modes. These columns show very good selectivity to amino acids and other carboxylic acids but also resolve many neutral and basic solutes.

A number of proteins are commercially available as CSPs including α-acid glycoproteins (AGP, the major plasma binding protein for basic drugs), human serum albumin (HSA, the major plasma binding protein for weakly acidic drugs), bovine serum albumin (BSA), ovomucoid (OVM), and cellobiohydrolase (CBH) [12]. The proteins are bonded to silica and utilized in reversed-phase mode with an aqueous buffer/organic modifier eluent. Mobile-phase optimization is performed through variation of the pH, ionic strength, temperature, and organic modifier [13]. It is believed that chiral recognition occurs predominantly through hydrophobic interactions in an apolar calyx that is buried in the interior of the structure. In the calyx, additional interactions such as electrostatic interactions, hydrogen bonding, dipole interactions, and steric hindrance occur. The protein CSPs are very broad-based in the types of drugs that they can enantioseparate.

Several variations of the triphenylesters and triphenylcarbamates of amylose and cellulose are commercially available from Diacel. These polysaccharide phases show the broadest applicability of all of the commercially available CSP and are capable of resolving a large and diverse selection of chiral solutes [14, 15]. The more popular phases are the 3,5-dimethylphenyl-carbamates of amylose and cellulose (Chiracel OD and Chiralpak AD, respectively). For most of these phases, the polysaccharide is dynamically coated onto a silica substrate. A 3,5-dimethylphenylcarbamate derivative of amylose that is covalently bonded to silica was recently introduced (Chiralpak IA). The polysaccharide phases are very flexible in that they can be used in normal-phase, reversed-phase, polar organic, and sub-/supercritical mode. Chiral recognition on polysaccharide phases are attributed to shape-selective inclusion into the chiral grooves enhanced by additional interactions such as hydrogen bonding, dipole interactions, π interactions, and van der Waal forces, depending upon the chromatographic mode [16, 17]. Enantioselectivity can vary as a function of amylose versus cellulose, ester derivative versus carbamate derivative, mobile-phase components, temperature, and chromatographic mode.

A more detailed discussion of the stationary phase types and mechanism of interaction and separation theory in relation to chiral compounds is given in Chapter 22. A large number of chiral stationary phases are currently available to meet the needs of the pharmaceutical industry for determination of the enantiomeric purity of active pharmaceutical ingredients, raw materials, and metabolites. As a consequence, there are a multitude of options in terms of columns, separation mode, and separation conditions to explore in achieving an enantioseparation.

For chiral liquid chromatography method development, the first choice to be made is the separation mode. The popular options are reversed-phase and normal/subcritical mode. The reversed-phase mode generally offers the advantage of sensitivity. Peak efficiency tends to be greater in reversed-phase mode relative to the normal-phase mode because of faster mass transfer. Combined with the ability to use low-UV-cutoff mobile-phase solvents, one can generally detect 0.1% of the enantiomeric impurity. Moreover, premixed solvents may be used to increase the detection limits as this will lead to a flatter baseline (no pulsation due to the pump mixing will be observed). Subcritical mode also offers the same level of sensitivity but is hampered somewhat by instrumental limitations with respect to ruggedness and robustness. Normal-phase and the subcritical modes allow the analysts to take advantage of interactions such as hydrogen bonding and dipole interactions that are strongest in apolar media.

The polysaccharide phases are known to separate a large range of pharmaceutical compounds. Chiral screening should include at least the Chiralpak OD and AD columns. Other popular columns that can be utilized include protein and antibiotic columns in reversed-phase mode and crown ether stationary phases for the separation of primary amines. Chirbase (http://chirbase.u-3mrs.fr/chirbase), a database specializing in chiral chromatographic separations, offers comprehensive structural, experimental, and bibliographic information on both successful and unsuccessful separations. It lists over 100,000 separations. This database indicates that polysaccharide-based stationary phases are the most frequently utilized phases accounting for ~40% of the separations. This database can be utilized as a starting point for method development.

14.3 SAMPLE PREPARATION

Sample preparation is required for the removal of potential interferents, to increase or decrease the concentration of an analyte and to convert the analyte into a suitable form for separation and detection. Sample preparation can be performed manually or through automation. In the process support area, sample preparation is seldom more complex than a simple "dilute and shoot." In some cases it may be required to dilute the sample in a solution that quenches an ongoing reaction for in-process samples. In other cases, solid-phase extraction (SPE) may be required for analysis of certain species in the presence of an interfering component. The SPE sorbent is chosen either to retain the analytes of interest while the interfering component is unretained, or the interfering component is retained while the analytes of interest are unretained. As an example, a method for determination of azide in the presence of a triazole derivative utilized a cation-exchange SPE step prior to analysis on an anion-exchange column [18]. The triazole derivative was strongly retained on the cation-exchange cartridge. The sorbent for SPE can be normal-phase, reversed-phase, or ion-exchange packings. SPE can also be used for enrichment of low-level analytes.

Derivatization is another form of sample preparation. It is utilized for the analysis of labile analytes or to enhance retention or detection with a preferred type of detector. Derivatization can be performed to enhance detection by UV/Vis, fluorescence, or electrochemical detection. Consideration must be given to the stability of the derivatize to solvolysis and thermal degradation. In our labs alendronate, a bisphosphonate with a primary amine functionality, was derivatized with FMOC to enhance detection by UV/Vis as well as to increase retention in RPLC mode [19]. An acylchloride was derivatized with aniline to form a stable anilide derivative prior to analysis in the RPLC mode to quantitate the content of the corresponding carboxylic acid and other impurities [20]. The triflation of a drug intermediate alcohol formed an active trifluoromethanesulfonyl ester. This active ester was derivatized with tetrabutylammonium bromide to form the bromo analog prior to analysis by reversed-phase LC [21].

14.4   HPLC DETECTORS

The detectors utilized for HPLC are designed to respond to the solute being eluted. HPLC detectors can be classified into two broad categories: universal and selective. Selective detectors respond to some physicochemical property of the solute, while universal detectors respond to all solutes independent of their physicochemical properties. The ideal detector would be highly universal and highly sensitive, have a wide linear range, and not be affected by change in temperature or mobile phase composition. Commercially available detectors possess some of these characteristics but not all.

The most commonly utilized detectors used in process development are the UV/Vis detectors that can be fixed-wavelength, variable-wavelength, or diode array. These detectors are sensitive, have a wide linear range, and are relatively unaffected by temperature or mobile-phase composition. They respond to solutes containing double bonds, and compounds with unpaired electrons such as bromine, iodine, and sulfur. Their response, however, is not equivalent. A variable-wavelength detector uses a deuterium or xenon lamp source, and the desired wavelength is isolated by a monochromator. A diode array detector performs a simultaneous measurement of absorption as a function of analysis time and over a chosen wavelength range. Thus a UV spectrum is obtained for each eluted peak. The main advantage of a diode array detector is for method development where wavelength maxima of the drug substance and its impurities may be unknown or where the UV spectra can be used to track peaks as operating conditions are changed. If the solute of interest does not possess a significant chromophore, then indirect photometric detection can be utilized. In this mode it is the mobile phase that possesses a chromophore and absorbs light. The detector still measures the difference in absorption between the mobile phase and the solute. When an analyte without a significant chromophore passes through the detector cell, the absorption of the mobile phase is decreased and is recorded as a negative peak. An ion-exchange method for the resolution of alendronate from other bisphosphonates, ionic synthetic impurities, and inorganic impurities utilized indirect photometric detection (Figure 14-7) [4].

Fluorescence occurs when a compound absorbs radiation then emits it at a longer wavelength. It is highly selective. Fluorescence is exhibited by rigid molecules possessing a large number of delocalized π electrons. Electron-donating groups enhance while electron-withdrawing groups decrease fluorescence. Few drugs possess natural fluorescence but for those that do, fluorescence detection is an option that offers increased specificity and sensitivity over UV/Vis detectors. Fluorescence is more sensitive than UV/Vis detection, particularly for laser-induced fluorescence. Care must be taken in choosing a compatible mobile phase because fluorescence can be quenched by highly polar solvents or halide ions. Fluorescence efficiency is also dependent upon pH of the mobile phase.

Some solutes may not have a significant chromophore, and alternate detectors must be utilized. These detectors include refractive index, evaporative light-scattering, element-specific, electrochemical, and mass spectrometric detectors. Refractive index (RI) detectors monitor changes in the refractive index of the mobile phase that occur due to the presence of solute molecules. Detection is universal but less sensitive than UV detectors. It is suitable for solutes without significant chromophores. The refractive index of the mobile phase must be constant, and thus this mode of detection is not amenable to gradient elution. Slight variations in temperature will also change the refractive index of the mobile phase. Therefore, very good temperature control is required.

Evaporative light-scattering detectors (ELSD) require nebulization of the eluent after which the aerosol is transported through a heated tube allowing the mobile phase to be evaporated. The residual particles pass through a light beam, and scattered light is then detected at a fixed angle from the incident light. Volatile mobile-phase components such as trifluoroacetic acid, formic acid, acetic acid, and ammonium hydroxide must be used. The ELSD is a universal detector as long as the solute is less volatile than the mobile phase. The linear range is not wide. It is intermediate between UV and RI detectors in terms of sensitivity and can be utilized with gradient elution. ELSD is useful for detection of solutes that do not possess a significant chromophore [22] but should not be used for thermolabile solutes. A recently commercialized alternative to ELSD is a corona discharge detector. The HPLC effluent is similarly converted to an aerosol, the aerosol particles are then charged by a positive corona discharge, and the current from the charged particle flux is then measured. This detector is generally regarded as more sensitive than ELSD [23].

Chemiluminescent detectors (CLND) are very selective and sensitive. If a solute contains at least one nitrogen atom, it can be detected. The effluent is nebulized, and then it is oxidized by combustion in a high-temperature furnace. Nitrogen-containing solutes are converted into nitric oxide, which is then passed into a chamber where it reacts with ozone to produce excited-state nitrogen dioxide that emits a photon upon relaxation. The photon flux is then measured by a photomultiplier tube [24]. The signal generated is proportional to the number of nitrogen atoms in the solute molecule. This detection mode requires volatile mobile phases that are free of nitrogen-containing molecules (no acetonitrile). CLND have been determined to have a wider linear range and greater sensitivity than ELSD [24].

Electrochemical detection can be utilized for compounds that are ionic or readily oxidizable or reducible. Thus, this form of detection can be used for the analysis of inorganic ions, protolytic organic compounds such as amines and carboxylic acids, and other compounds such as phenols, thiols, and alcohols. Conductivity detectors measure differences in the equivalent conductance of the solute and ions in the mobile phase. The conductivity response is maximized through the use of ion suppressors that effectively eliminate the conductivity of the mobile-phase ions through chemical removal or electronic subtraction. The linearity range is wide, and detection is highly sensitive. In our labs, conductivity with ion suppression was utilized to detect residual levels (~0.1%) of choline (quaternary saturated amine) in drug substance [25]. Amperometric detection is less commonly utilized and is suitable for compounds that can be electrolytically oxidized such as phenols. This mode is not generally applied in the reductive mode due to interference from dissolved oxygen in the mobile phase. Amperometric detection is highly sensitive and selective.

Mass spectrometric detection is close to being a universal detector. Ionization techniques such as atmospheric pressure chemical ionization (APCI) and electrospray ionization (ESI) are routinely employed. These techniques allow the transfer of the LC effluent into the gas phase. With APCI, the eluent is converted to an aerosol by a sheath gas. The aerosol is then subjected to a chemical ionization plasma created by a corona discharge, leading to formation of solute ions. These ions are then transferred into the mass spectrometer. With ESI, the eluent is converted to charged droplets. ESI is preferred for compounds that are ionized in solution. APCI is better for compounds of medium polarity. Both techniques can be used in positive or negative ion mode. Positive ion mode is commonly used. Negative ion detection is useful for negatively charged ions such as acids. Nonpolar compounds are difficult to analyze with these atmospheric ionization techniques due to their soft ionization mechanisms. Atmospheric pressure photoionization is an emerging technology for the analysis of these nonpolar compounds. This technique is similar to APCI; however, a gas discharge lamp that emits photons in the vacuum UV region is utilized. Sensitivity can be increased by the use of dopants such as toluene or acetone added post-column to the eluent. The dopant is first ionized and then ionizes the analytes through further reactions [26].

Mass spectrometric detection in the process development area is generally performed with a single-quadropole, triple-quadrupole, or ion trap mass spectrometer. Other options include sector and time of flight spectrometers. A single quadrupole provides information pertaining to the mass to charge ratio (m/z) of the solute. Ion traps and triple quadrupoles provide additional information through tandem MS, allowing for a more definitive structural elucidation of the solute. Volatile buffers such as ammonium acetate or ammonium formate and low-pH mobile phases such as 0.1% formic or acetic acid are recommended to prevent blockages of sample cones or capillaries. The relative sensitivity of MS versus UV/Vis detection may differ by many orders of magnitude in either direction, depending upon the chromophoric properties and the ionizability of the analyte. It can be very selective when used in selected ion monitoring mode where it is detecting one specific mass/charge ratio. The use of MS is extremely valuable in identifying by-products of reactions, impurities in intermediates that may react further in downstream processing, and impurities that are formed during stability testing.

14.5   METHOD DEVELOPMENT

The approach to method development is dependent upon the physicochemical properties of the solute and any known potential impurities and the purpose of the method. The method may be required for an impurity profile, assay, in-process monitoring, or chiral/isomeric evaluation. Method development is usually dynamic. As more knowledge about the properties of the solute and potential impurities is gained, the method can be further optimized. Analytical laboratories supporting process development should be stocked with a variety of columns for RPLC, NPLC, ion-exchange, and chiral separations. Column switching capability is also an asset for method development. Column switching allows for analysis of the same sample with as many as six different columns in an overnight run to help speed method development. Initial development can be performed empirically, based on the chromatographer's experience, or through the use of simulations with one of the commercially available method development software packages. The parameters to explore for method development include separation mode, column selection, mobile-phase optimization, temperature, detection wavelength, sample diluent and concentration, injection volume, and sample preparation procedure. The use of an orthogonal chromatographic method, with the developed method as a check, is recommended. Having an orthogonal method minimizes the possibility of peak co-elution, particularly in cases where there is limited information available regarding the nature of impurities. An orthogonal method may be employed once the final synthesis is set during the development of a drug. The final synthesis is usually set for preparation of the clinical material used for Phase II clinical studies. When a method has been developed that is deemed appropriate for the purpose, system suitability parameters should be implemented and some degree of validation should be performed to ensure that the method meets the needs of the chromatographer.

For developing an impurity profile for raw materials, intermediates, or drug substance, communication with the process chemists regarding potential reaction by-products is always the best start. This information plus any garnered knowledge of the physicochemical properties of the solute and potential impurities such as pK_a, $\log P$ (octanol/water partition coefficients), solubility, and UV spectrum will determine the selection of the appropriate mode, column, mobile-phase, and other separation parameters. Given the potential for generation of impurities that are unanticipated by the process chemists, it is recommended that for early development a gradient method be employed. A gradient method will allow for coverage of a wide range of polarity and thus be able to capture early and late eluting impurities in the same run. For this reason, a reversed-phase method is the first choice. Most components can be eluted in a 10–90% gradient of organic modifier as long as there are no miscibility issues with the aqueous mobile phase. An isocratic hold at 90% organic should be performed especially in early development to detect the presence of any extremely hydrophobic impurities. Ideally the peaks of interest should be eluted with a capacity factor between 1 and 10.

The choice of a column is dictated in large part by the hydrophobicity (evaluated as $\log P$ when available) of the solute. C18 or C8 columns are commonly utilized in reversed-phase mode. Retention and selectivity for these phases can vary, depending upon whether they are conventional, polar end-capped, polar embedded, or hybrid silica. A high-carbon-load C18 or a graphitic column can be used to increase retention. A low-carbon-load C8 or a phenyl or cyano column can be used to decrease retention. Alkyl, phenyl, and cyano phases may offer different selectivities. Selectivity may also vary as a function of the substrate: silica versus polymer versus zirconia. The sheer volume of commercially available reversed-phase columns makes selection of the best column, for a particular separation, anything but a simple task. Much research has been performed toward the classification of reversed-phase columns. Approaches include regression of $\log k$ versus $\log P$, thermodynamic measurements of retention, and quantitative structure–retention relationships (QSRR) using experimentally determined or calculated molecular descriptors [27–35]. For example, classifications in terms of efficiency, hydrophobicity, silanol activity, and steric selectivity were used in the evaluation by principal component analysis of 69 columns differing in type of silica, pore size, end-capped/not end-capped, base deactivated/not base deactivated, and polar embedded [31]. Based on this classification, one can select four columns, which fall into separate categories ensuring selectivity differences, for initial method development. Similarly, classification of 28 columns in terms of selectivity based on hydrophobicity, steric selectivity, efficiency, and silanol activity using chemometric approaches led to the selection of eight columns of low, intermediate, and high hydrophobicity that were highly efficient and showed good steric selectivity [35]. One could then choose one column from each hydrophobic class for method development. One should also ensure that the selected columns are stable within the intended pH and temperature regions that they will be employed. A good understanding of the chemical stability of the stationary phases is essential.

An additional variable for varying selectivity is column temperature. Significant changes in selectivity may be observed when comparing separations at 10°C and 50°C. This depends on the nature of the analyte and its interaction with the stationary-phase and mobile-phase components. Elevated temperatures, however, may lead to unwanted compound degradation and should be avoided for labile components.

Mobile-phase composition is another major parameter for affecting selectivity in a separation. Points to consider include choice of organic solvent, mobile-phase pH, and use of additives. The three most commonly utilized organic solvents are acetonitrile, methanol, and THF. Acetonitrile is usually a good starting organic solvent as a consequence of its lower viscosity and UV cutoff. Method development can be performed with each of these three organic solvents or with mixtures of them. The pH of the mobile phase is also critical. Low pH protonates acids and bases, resulting in neutral acids and charged bases. Conversely, high pH deprotonates acids and bases, resulting in charged acids and neutral bases. In general, retention decreases with increasing charge on the solute. Buffers are recommended. Phosphate is a commonly utilized buffer with pK_a values of 2.1 and 7. Buffers such as acetate and formate are useful for detection modes requiring volatilization of the mobile phase. Care should be taken to avoid working within ±1.5 units of the pK_a of the solute, because this may result in poor retention precision. Many pharmaceutical compounds are acidic or basic, and a good starting point for method development is low pH. A low pH suppresses the ionization of acid solutes and the silanol sites of the stationary phase. High pH can be used to increase retention of bases (neutral form) or to take advantage of ion-exchange interactions (with bases in ionized form) to improve selectivity (however, bad peak shapes sometimes are the result due to strong silanophilic interactions). Retention can also be enhanced by the use of additives such as chaotropic anions (perchlorate, hexafluorophosphate) or by ion-pairing agents (hexanesulfonate) [36].

Knowledge of the UV spectra of the solute of interest can be applied to choice of wavelength for UV/Vis is detection where amenable. One can choose a wavelength near or at the UV maxima for detection. This choice suffers the disadvantage that unknown impurities in the intermediates and/or drug substance may not exhibit strong extinction coefficients at the chosen wavelength and may go undetected. An alternative is to work at a low wavelength such as 210 or 220 nm where most solutes possessing a chromophore will have significant absorption (π–π* bands for double bonds and n–σ* bands for amines and halogens). The choice of wavelength is also dictated by the UV cutoff of the mobile-phase components. Knowledge of the solubility of the solute as well as its compatibility directs the choice of diluent. A combination of adequate solubility and injection volume should be chosen such that ideally 0.05% of the solute can be detected with a signal-to-noise ratio of greater than 10 to 1.

For development of a weight percent assay, a short isocratic method can be implemented based on observations from the gradient method used for the impurity profile. One can use a shorter column such as a 5-cm column and keep retention of the solute of interest to around a capacity factor of 3 as long as it is still resolved from impurities observed in the impurity profile. Additionally, the elution of more hydrophobic species should not co-elute with the drug substance in later injections. During the method development of an isocratic method, the compound should be injected and then a suitable number of blanks injected to ensure that more hydrophobic impurities do not elute at the same time as the analyte peak in later injections.

A similar approach using an isocratic method can be applied to in-process monitoring, where the goal is to monitor the disappearance of the starting material and appearance of the product. In-process methods will be discussed in greater detail in Section 14.6.

The use of computer simulations is an alternative approach to method development. Computer-based expert systems are designed to mimic the thought processes of an experienced chromatographer. These systems contain a database that can used to evaluate chromatographic data and provide optimized conditions. Variables such as solute structure, column type, mobile-phase components, pH, and temperature can be inputted, and proposed optimized chromatographic conditions are outputted. This approach is generally faster and cheaper than performing all of the experiments necessary for method development. Systems with artificial intelligence can plan experiments, collect and evaluate data, and adjust chromatographic conditions in real time according to predefined decision schemes until a satisfactory separation is achieved. Further discussion of the different automated method development software available is given in Chapter 10.

14.6   IN-PROCESS MONITORING

In-process monitoring is implemented to maximize yield and minimize impurity generation during the various synthetic steps. An ideal in-process method should quickly evaluate a specific sample and provide results in a timely fashion such that changes may be triggered to maintain the reaction conditions at the optimal level required to secure production with high purity and maximum yield. Process analytics using on-line spectroscopic analysis can provide instantaneous feedback; however, the reaction mixture is often too complex to provide accurate results. Oftentimes, separation is required to evaluate levels of several components. Chromatography can provide the necessary separation, but the time lag of the analysis must be short enough to monitor the actual state of the reaction. An emphasis should be placed on providing near-real-time feedback by using methods with short run times. Ideally, this would be accomplished by reducing the run time without a concomitant loss in column efficiency or resolution.

One approach to achieving near-real-time feedback with chromatography is through the use of short columns with smaller particles. Small particles result in higher column efficiency, but with increased backpressure limiting the workable column length. Short columns (10 cm or less) with smaller particle sizes (1.5 to 3.5 μm) can result in comparable separations to longer columns (25 cm, 5-μm particles) but with one-half to one-fifth the run time. The efficiency of these shorter columns is equivalent to, and often superior to, the longer conventional columns. Shorter columns, however, are susceptible to instrumental band-broadening effects, and care must be taken to minimize the extra-column volume.

A nonporous silica C18 column was utilized in conjunction with an on-line HPLC to provide rapid feedback for a deprotecting step for a drug substance [37]. The conventional method had a run time of 35 minutes, not including sample preparation time and the time needed to sample the batch and transport the sample from the pilot plant to the analytical lab. The on-line method had a run time of 10 minutes, and the sample preparation for the subsequent sample was ongoing during each analysis point. As a consequence, the batch could be evaluated every 10 minutes as compared to every 60–90 minutes by the conventional method. On-line sampling was feasible only because the reaction mixture was a homogeneous solution (Figures 14-8 to 14-10).

Another approach is the use of monolithic columns consisting of silica based rods of bimodal pore structure. They contain macropores (~1–2 μm) and smaller mesopores (~10–20 nm) [38]. The macropores allow for low backpressure at high flow rates. The mesopores provide the needed surface area for interactions between the solute and stationary phase. The macropores result in higher total porosity as compared to porous silica particles. Flow rates of 5 mL/min can be tolerated on a 10-cm column without an appreciable loss in

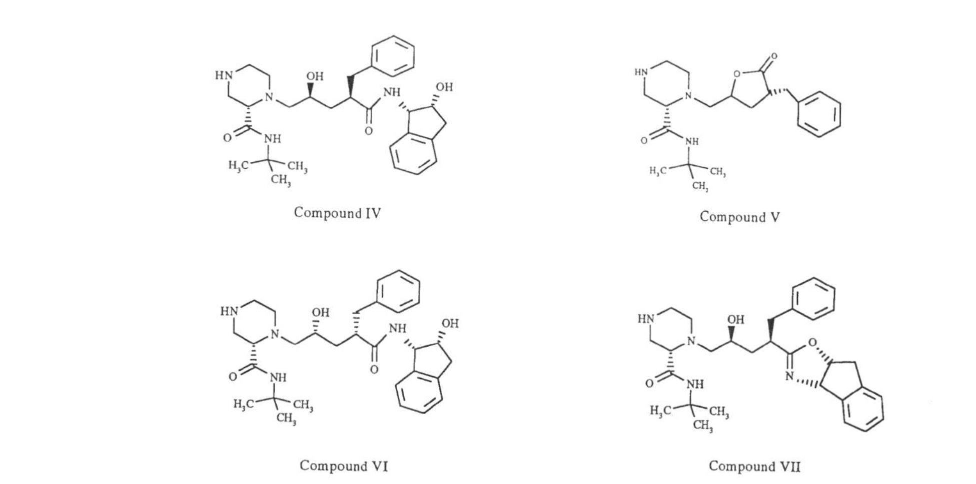

Compound I

Compound II

Compound III

Compound IV

Compound V

Compound VI

Compound VII

Figure 14-8. Structures of components in a deprotecting process. (Reprinted from reference 37, with permission.)

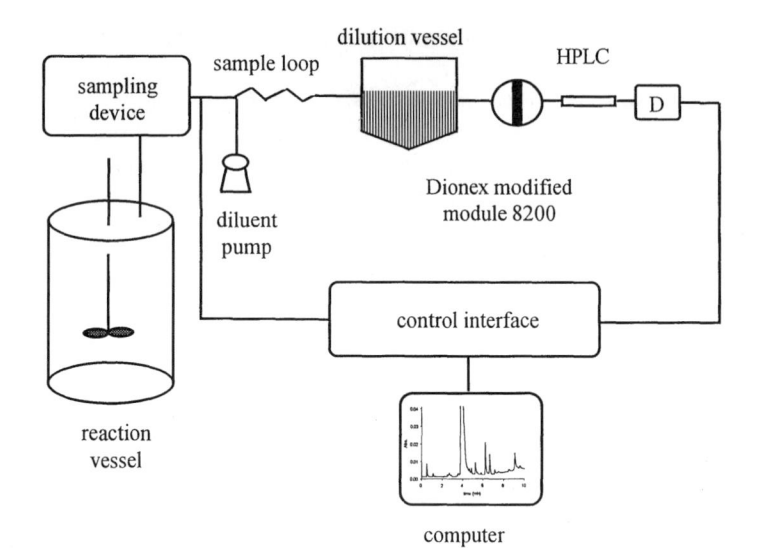

Figure 14-9. Schematics of an on-line HPLC system. (Reprinted from reference 37, with permission.)

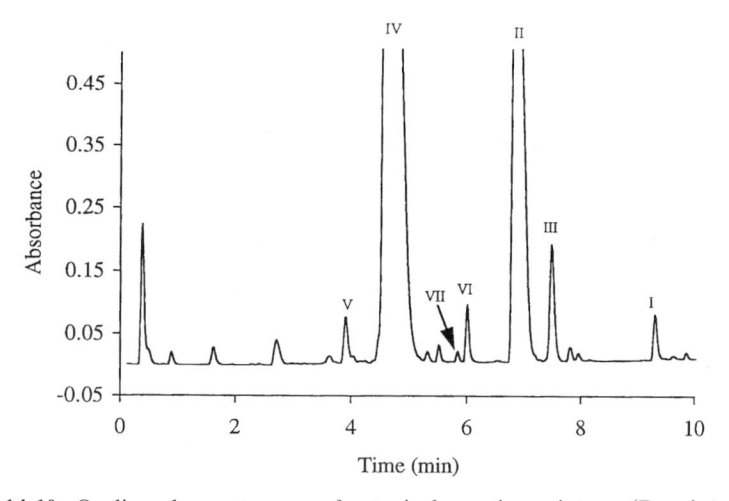

Figure 14-10. On-line chromatogram of a typical reaction mixture. (Reprinted from reference 37, with permission.)

column efficiency. Monolithic columns show similar selectivity to spherical particle columns but with shorter retention times due to the ability to use higher flow rates without compromising efficiency. Methods have been developed, with less than 10-minute run times, using 15-cm Chromolith columns for a number of in-process samples in our labs. One application was for monitoring a coupling reaction between a biarylpiperazine and an epoxide during the synthesis of an HIV drug candidate [39]. A second application was for catalyst screening [39].

14.7 IMPURITY IDENTIFICATION

A critical aspect of drug development is the control and identification of impurities in the active pharmaceutical ingredient. Regulatory requirements dictate the control of impurities and the identification, where possible, of impurities that exceed 0.1% [40–42]. The identification of these impurities provides information toward their potential toxicity. Impurities in the drug substance can originate from incomplete reactions, over-reactions, and side reactions and can also originate from impurities in starting materials, reaction solvents, catalysts, or residual solvents. Impurities may also occur as a result of degradation. Common degradation pathways are hydrolysis, dehydration, oxidation, dimerization, or a combination thereof. Delineation between process related and degradation impurities is not always possible. Impurities may be organic or inorganic in nature. Inorganic impurities may originate from the counterion if the drug substance is in salt form, from reaction vessels, filters, adsorbents, and tubing, and also from catalysts and inorganic reagents used in the synthesis. Control of impurities that appear in the drug substance is not only undertaken at the final isolation step but also through control of impurities in raw materials, intermediates, and solvents and at critical synthetic steps. Whenever possible, the level of impurities originating from the starting materials should be limited by in-process controls rather than at the final step. Chromatography as a consequence of its selectivity and precision is an excellent tool for the monitoring of impurities.

The identification of process-related impurities can ultimately lead to elucidation of their mechanistic pathways. The first step in this process entails determination of structures for the impurities using tools such as NMR or LC-MS. With an elucidated structure, one can then attempt to propose a mechanism for its formation and to identify the synthetic steps that lead to its formation. This information can then be utilized to manipulate synthetic conditions so as to minimize its formation or develop efficient recrystallization procedures to reduce impurity level while still maintaining acceptable yield.

The retention times of impurities found in a drug substance impurity profile should be first compared to those of known process-related impurities, intermediates, and starting materials which could have been carried forward to the final product. Where facile, as a first step, the process chemist can synthesize some of the potential impurities on a small scale. Potential impurities include the penultimate intermediates, products of over-, under-, and side reactions, and degradation products. Analogs may have also been synthesized during the drug discovery stage for screening that can also be potential impurities. These isolated impurities can then be correlated to impurities that appear in the chromatogram. However, a simple matching of retention times is not satisfactory confirmation of impurity identification. The use of an orthogonal chromatographic method is recommended. For example, one can match the prepared impurity versus an impurity that appears in the reference RP chromatographic method by using an orthogonal TLC method or a NPLC method. If the impurities match up in both chromatographic systems, then there is a level of confidence that the impurity observed in the reference chromatogram is the same as the synthesized potential impurity. Further confirmation can be provided by matching UV spectra using diode array analysis. The definitive confirmation, which can occur at a later stage in development, is through matching of MS or NMR spectra.

A logical second option is to evaluate samples where the relative concentration of impurities is higher. These samples include crude product, mother liquors, and stressed drug substance. Isolation of the impurity and sensitivity issues for identification can then be overcome. Care must be taken, however, that the impurity found in these samples are exactly the same as those found in the drug substance and not just co-eluters. Consequently, the use of orthogonal chromatographic methods and UV spectra matching should be strongly considered.

As a third option, impurities appearing in the reference chromatographic method may be isolated using analytical (under overloaded conditions) or semi-prep chromatography and then identified by MS and/or NMR. Mobile-phase components should be removed prior to analysis. Removal of mobile-phase components may entail evaporation, liquid–liquid extraction, freeze-drying, or solid-phase extraction. Identification by MS should be attempted first because its high sensitivity requires only small sample amounts. The less sensitive NMR requires the isolation of greater amounts of the impurity. An alternative to this approach is the use of LC-MS or LC-NMR.

LC-MS coupled with UV analysis is now routinely used in the process development area for the identification of impurities. The UV detection allows for cross-referencing of impurities because response factors may differ and changes in retention time may occur as a consequence of differences in instrumentation setup or changes in the chromatographic method to make it MS-compatible (e.g., switch from phosphate to formate buffer). LC-UV-MS using soft ionization techniques will simply provide a molecular mass of the impurity. The use of collision-induced dissociation can provide some fragmentation information. LC-UV-MS-MS will provide even more structural information. For example, a triple quadrupole instrument can isolate one ion, which is then fragmented by gas-phase collisions (argon or xenon) in the second quadrupole. The produced fragments are then analyzed by the third quadrupole. The increasing ease of use and the proliferation of LC-MS instrumentation in the analytical labs have led to its utilization as the first option for identification of impurities, particularly when small-scale synthesis of potential impurities is not trivial. For in-depth discussion of LC-MS, the reader is referred to Chapter 7. However, if the MS data do not provide sufficient structural information, then it may be necessary to isolate and identify by NMR or LC-NMR.

The introduction of NMR probes that can be interfaced with HPLC has resulted in the emergence of another powerful tool for the structural elucidation of impurities [43]. The intense signals from the protons of the solvents can overwhelm the weak signals of the impurities, rendering solvent signal suppression necessary. However, solvent signal suppression may suppress impurity signals lying under the solvent signal, and thus some information may be lost. To minimize this effect, simple binary mobile phases such as methanol/water or acetonitrile/water are used. Another alternative is to use only deuterated solvents, but this is a rather expensive choice given the price of these solvents. Additionally, there will still be some strong signals from the solvent as a consequence of contamination.

LC-NMR analysis can be performed in continuous-flow or stop-flow mode [43]. In continuous-flow mode a series of NMR spectra is collected rapidly as the HPLC eluent flows through the NMR probe. In stopped-flow mode, the flow is stopped for short intervals as the peak of interest passes through the probe, allowing for the collection of sequential spectra through the peak. This mode can lead to band broadening and affect the resolution of later eluting peaks. It is best employed for analysis of just one impurity in a chromatogram. An alternative is to store the peaks of interest in capillary loops. Stopped-flow allows for 2D NMR experiments such as correlation spectroscopy (COSY) to be performed. The main disadvantage of HPLC-NMR is sensitivity. Analysis of peaks at 0.1% to 1% is a challenge. This can be overcome to some extent by trapping chromatographic peaks on solid-phase extraction cartridges, which can be subsequently eluted with deuterated solvents [44]. This mode allows for greater flexibility in mobile-phase choice since the mobile-phase additives can be separated from the impurity peak on the SPE cartridge. For a detailed discussion of LC-NMR, the reader is referred to chapter 20.

Finally, impurity analysis can also be utilized to demonstrate illegal use of patented reaction routes. The impurity profile of a drug substance is influenced by the synthetic route and the source and quality of the starting materials. Identification of impurities in drug prepared by two different manufacturers may provide valuable insight into the manufacturing route and determine if patent infringement has occurred, because certain impurities may be indicators of a specific synthetic route.

14.8 ESTABLISHMENT OF HPLC SELECTIVITY BY STRESS STUDIES

According to the ICH guideline on stability testing, the purpose of stress testing is twofold [45]. First, it can be used to predict the stability of the molecule and from the degradation products establish degradation pathways. Second, it can validate the stability-indicating capability of the analytical method. As a consequence, stress studies are conducted at the very early stages of method validation, particularly because it can drive method optimization. Stress studies are performed under appropriate conditions for a specific drug substance and may include acid/base hydrolysis, thermal degradation, oxidation, and photo-degradation. The goal is to achieve 10–20% degradation of the drug substance. The initial stress conditions may not result in degradation. Options such as extended stress periods, increased acid/base/oxidant concentration, or increased temperature may be applied to induce degradation. However, excessive stress is counterproductive because it does not reflect the stability properties of the drug substance, and considerable time may be spent in optimizing the analytical method to resolve degradation products that will never be observed during storage of the drug. For stress studies a control sample should ideally be maintained and stored at 5°C or less.

Acid/base stress studies are generally performed in our labs with 0.1 N HCl and NaOH. Typically, the stress study is conducted at ~1 mg/mL of drug substance. A typical experiment would be to place 25 mg of drug substance in a 25-mL volumetric flask which is then diluted to volume with 0.1 N HCl or NaOH. An organic solvent such as acetonitrile or methanol may be required to dissolve the drug substance, and the volume is adjusted accordingly to maintain an ~1 mg/mL in 0.1 N acid or base. At the 2-hour interval, 1 mL of sample is removed and diluted to the normal HPLC injection volume with the appropriate HPLC diluent and analyzed. The degraded sample is run against a control sample, a diluent blank, and a blank containing the degrading agent. If inadequate degradation is observed, samples are periodically taken and analyzed over a 24-hour period. If too much degradation is observed, the experiment is repeated at a lower concentration of acid/base or within a shorter time interval. If >10% degradation is observed in less than an hour, the solutions may be put in the refrigerator to slow down the kinetics of the degradation. A similar procedure is employed for oxidative stress using 3% hydrogen peroxide as the starting point. Thermal stability is demonstrated by holding solid sample at ~10°C below the melting point of the compound, in a GC oven, until significant degradation is observed by the HPLC method. If thermogravimetric analysis or differential scanning calorimetry indicates that significant degradation occurs below the melting point, a lower temperature may be chosen to conduct the study. Photodegradation is generally performed with the drug substance in solution kept in a clear glass container on a benchtop exposed to the normal light source in the lab or placed in a qualified light chamber (i.e., cool white fluorescent and near-ultraviolet lamp).

14.8.1 Stability in Solution and Forced Degradation Studies (Process Intermediate Compound A)

In the following case study, during the course of the initial method development for an intermediate, some late eluting species (18–20 minutes) were presumed to be forming in the sample solutions (prepared in acetonitrile) in

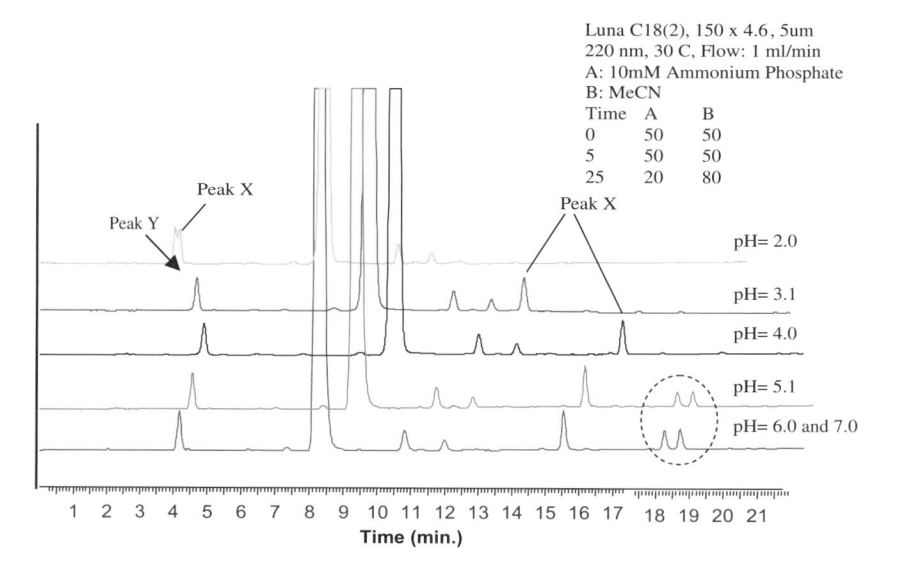

Figure 14-11. $^{w}_{w}$pH study on Luna C18 (2) for intermediate compound A. Method conditions are indicated in the figure.

ambient conditions under normal room light. In Figure 14-11 ($^{w}_{w}$pH 7, column temperature 30°C, for sample stored for 36 hours at room temperature) these potential degradation products formed in solution are circled. Further studies were performed in order to determine whether these degradation products originate from interactions with the diluent and/or are unstable in the diluent under normal light conditions at ambient conditions in the laboratory.

Two samples of the intermediate were prepared using acetonitrile and methanol as the diluents. Each sample preparation was immediately analyzed using an autosampler-tray-cooler-equipped HPLC for the initial control time-point, and portions were also stored under ambient conditions (clear volumetric flask and ambient room lighting) and under refrigerated conditions for 50 hours.

The samples prepared in acetonitrile and stored at ambient temperature showed increased levels of three major impurities (RRT = 1.76–1.78, RRT = 2.06, RRT = 2.10–2.11 Figure 14-12). The two impurities at RRT 2.06 and RRT 2.10–2.11 were not present in the refrigerated solution. API-ESI LC-MS analysis of these degradation products suggested that the RRT = 1.76–1.78 was a des-N-oxide impurity [M – 16 + H]. The RRT = 2.06, and RRT = 2.10–2.11 impurities had the same mass [M + 18 + H] and were presumed to be N-hydroxy adducts (OH group added across one of the double bonds on the ring) of the intermediate. These findings were further confirmed by using deuterated mobile phases which indicated the presence of an additional labile hydrogen atom.

With methanol as the diluent, the peaks eluting at RRT = 2.06 and RRT = 2.10–2.11 were not observed. However, other degradation products were formed (Figure 14-13). The formation of these additional degradation

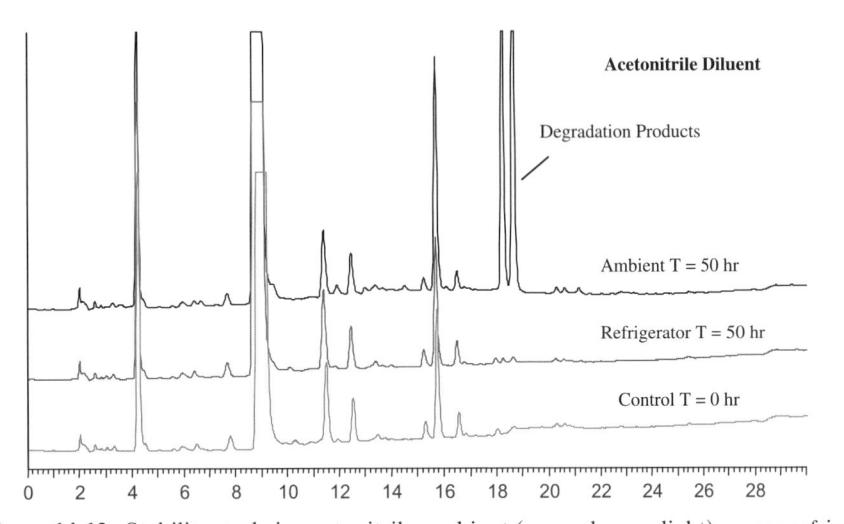

Figure 14-12. Stability study in acetonitrile, ambient (normal room light) versus refrigerated conditions.

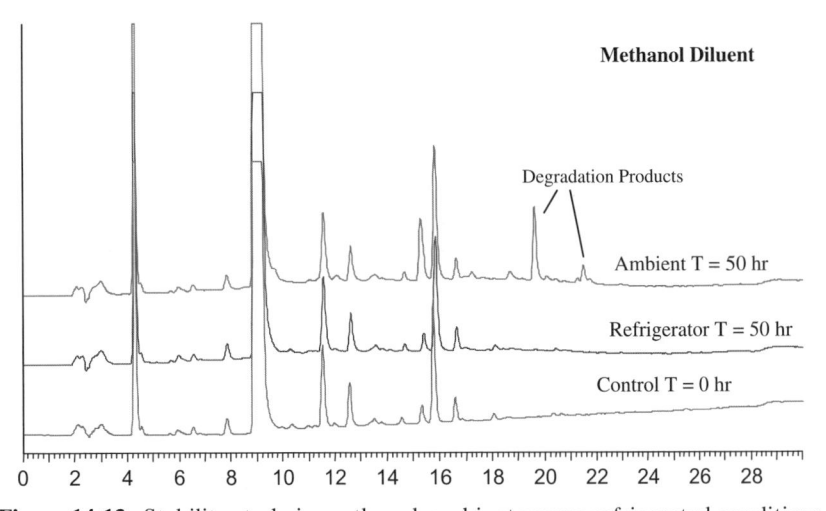

Figure 14-13. Stability study in methanol, ambient versus refrigerated conditions.

products in the methanol diluent was also inhibited at 4°C and was at lower levels than the degradation products formed in acetonitrile diluent. Two major impurities found when using the methanol diluent, RRT = 2.38 and RRT = 2.18, were identified. API-ESI LC-MS analysis of these degradation products suggested the RRT = 2.38 and the RRT = 2.18 are [M + 32 + H] species and are presumed to be *N*-methoxy adducts of the intermediate. These degradation products are at much lower levels than the degradation products formed in acetonitrile. These studies showed that this intermediate is more

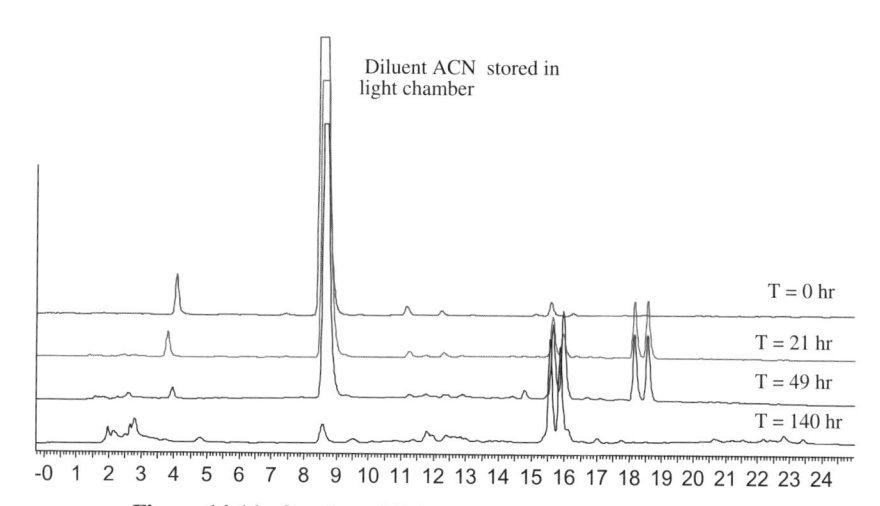

Figure 14-14. Overlay of light-stressed sample in acetonitrile.

stable in methanol than in acetonitrile; however, in both solvents, the sample solutions are stable in the refrigerator (4°C) for up to 50 hours. Solution light stability studies (dark and light chamber) in both solvents at ambient temperature were also performed in order to further determine whether the formation of the degradation products is photocatalyzed or generated at ambient temperature (~25°C).

Control samples were placed in an amber flask at ambient conditions, and the stress samples were placed in a light chamber employing white-light intensity and ultraviolet-light intensity at room temperature. The solutions were then analyzed after approximately 0, 21, 49, and 140 hours (Figures 14-14 to 14-16). The samples exposed to light in both diluents showed considerable degradation over the course of the study, while the samples protected from light at room temperature remained stable for up to 140 hours. As a result of this study, the method was modified to prescribe that all samples be prepared in amber glassware using methanol diluent.

Having established that the sample preparations are light-sensitive, it was necessary to also establish if the solid was light sensitive. A light stress study with a "pure" sample (recrystallized intermediate) was carried out over 141 hours where a sample of API solid was placed in a quartz glass dish and left in the light chamber at ambient conditions. A sample was analyzed at the following time points: 0, 24.5, 67.5, 113, and 141 hours. A control sample was placed in a similar dish and protected from light with aluminum foil. This sample was placed in a dark cabinet and analyzed after 143 hours. The solid under the light stress conditions showed significant degradation over the course of the study (total impurities grew from 1.3% to 2.2%), while the control sample stored in the dark for at least 143 hours demonstrated no significant growth in impurities. The intermediate should have a protect-from-

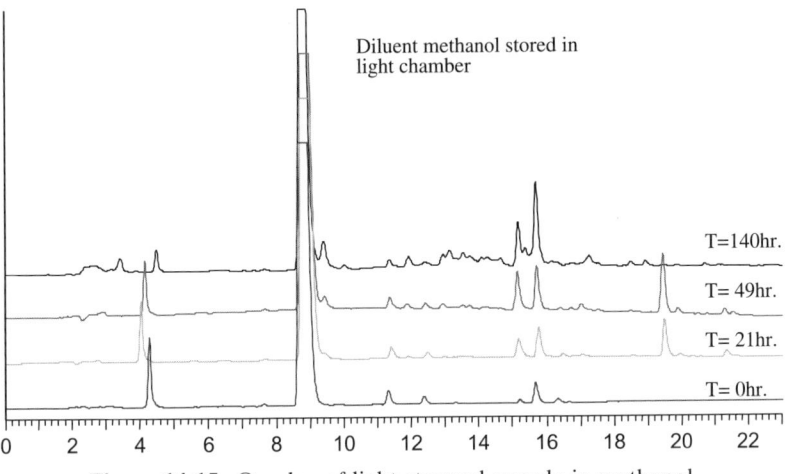

Figure 14-15. Overlay of light-stressed sample in methanol.

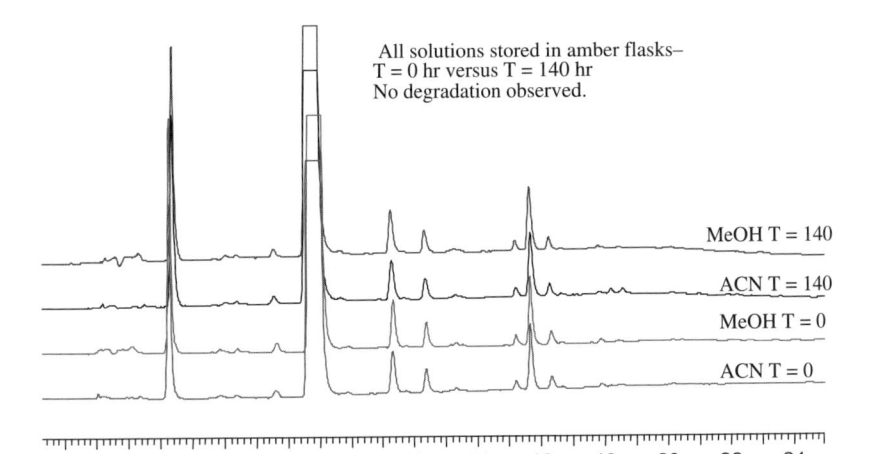

Figure 14-16. Overlay of dark control samples (stored in amber flasks) in methanol and acetonitrile: $T = 0$ versus $T = 140$ hr.

light requirement, and sample preparation should be performed in amber (Actinic) glassware with methanol diluent in order to prevent potential light-induced degradation products.

14.9   HPLC METHOD VALIDATION

Following establishment of selectivity, the HPLC method should be validated. The validation of an analytical method is meant to demonstrate the suitability of the method to perform its intended purpose. Validation is an important portion of the filed regulatory documentation to support an IND or an NDA [46–48]. ICH guideline Q2A defines the validation terms and Q2B describes methods for the determination of various validation parameters including specificity, linearity, accuracy, precision, detection and quantitation limits, robustness, and system suitability testing [47, 48]. The level of validation is stage-dependent, and the validation criteria increase as the projects evolves from preclinical to clinical to commercialization. While these requirements pertain particularly to the regulatory HPLC impurity profile, any HPLC method should be validated to some level to ensure acceptable accuracy and precision.

14.9.1 Prevalidation and System Suitability

Prior to performing a formal validation, the analytical chemist should have performed some prevalidation during method development. The expectation is that a well-developed HPLC method should subsequently be validated with no major surprises or failures. Prior to validation, specificity and some degree of robustness should be demonstrated. In addition, some form of system suitability criteria will have been established. System suitability evaluates the capability of an HPLC system to perform a specific procedure on a given day. It is a quality check to ensure that the system functions as expected and that the generated data will be reliable. Only if the system passes this test should the analyst proceed to perform the specific analysis. System suitability can be based on resolution of two specified components, relative standard deviation, tailing factor, limit of quantitation or detection, expected retention times, number of theoretical plates, or a reference check.

System resolution depends upon resolution achieved between the peak for the drug substance and the peaks directly preceding and succeeding it. The lowest resolution between any pair of peaks is also critical. A minimum resolution can be set for two peaks to establish system suitability. For assays, a maximum relative standard deviation for multiple injections of a standard can be used as a system suitability criterion. Minimum theoretical plates, a maximum tailing factor, a retention time range for a specific peak, and the ability to visually observe a peak for a 1000-fold dilution of the normal injection concentration are all appropriate system suitability criteria. Alternatively, a reference standard sample containing several low-level impurities can be injected, and various parameters such a resolution, retention time, and the ability to detect certain low level impurities can be used to establish system suitability. Also, if a sample is not available that contains an impurity that is known to form during stability, this impurity can be generated under stress conditions and used as a system suitability solution. For example, if a drug substance is known to oxidize readily at a particular moiety and the potential impurity elutes close to the active ingredient, the impurity can be generated with a suitable stress solution for a defined period of time and a critical resolution criterion can be set.

14.9.2 Validation

Once the method is determined to be optimum and system suitability criteria has been set, validation can be initiated. There may be different stages of validation, and this can vary from one pharmaceutical company to another. For example, some companies may have a three-tier validation strategy (validation testing to be available prior to the first drug substance delivery, validation to be available for regulatory methods when IND is filed, validation to be available for regulatory methods to be transferred to the manufacturing facility and to support the NDA filing) or a two-tier validation strategy (early phase before final synthesis is set and late phase after final synthesis is set).

Specificity is demonstrated through the resolution of the drug substance from impurities. Selectivity from starting materials, intermediates, by-products, and degradation products is a good starting point to demonstrate specificity. Stress studies and peak purity experiments using diode array analysis or MS can suffice for selectivity at the early stage of development. At later stages of development, isolated impurities will become available. They can then be utilized to establish selectivity and low-level linearity, and to determine their relative response factors. Peak purity can be assessed by visual examination of the peaks for Gaussian symmetry, by diode array or MS, or by use of an orthogonal LC method.

The stability of the sample solution should also be evaluated as a function of time and storage temperature. A suitable criterion is that sample solutions should be stable for at least 24 hours under defined storage conditions (e.g., amber glassware, low temperature). If greater solution stability is desired, the solution stability at refrigerated conditions should be compared to the solutions stored at ambient conditions. Linearity (typically correlation greater than or equal to 0.999) can be established from the limit of quantitation (LOQ) to 120% or 130% of injection concentration (minimum five serial dilutions) or segmented into low (e.g., LOQ to 1%, minimum three concentrations) and high concentration (e.g., 50–130%) linearity. Establishment of LOQ and limit of detection (LOD) is also critical for early stage validation. There is flexibility in setting LOQ and LOD. LOQ and LOD is generally established at a minimum signal-to-noise ratio (SNR) of 10 and 3, respectively. The target LOQ is generally 0.05% (one-half of the identification threshold defined in the ICH 3A guidelines). Other criteria that can be set are (a) a maximum deviation of 10–20% of the response factor of the LOQ solution compared to a 5X LOQ solution and (b) a maximum 10–15% RSD for area counts for a minimum of three injections for the LOQ solution.

The accuracy of the method can be determined by performing recovery experiments or by comparison to another analytical method (such as titration, DSC, and PSA). Spiking experiments, where increasing amounts of an impurity are introduced into the sample and the accuracy of the result versus theoretical is evaluated, are usually conducted at later stages of development where authentic isolated impurities become available. When the assay is being used for a minor component, recovery should be performed for at least three concentrations ranging from 120% of the impurity specification to the LOQ. The relative response factors of the impurities compared to the active drug substance should also be determined at this stage as well.

Validation of the precision of an HPLC method occurs at three stages. The first stage is injection precision based on multiple injections of a single preparation of a sample on a particular sample on a given day. The second stage is repeatability where multiple preparations of a sample are analyzed with multiple injections by the same chemist on the same day. The third stage requires analysis of multiple preparations by more than one analyst, on different instruments on different days. Robustness is also demonstrated at later stages of development and reflects the ability of the method to remain unaffected by minor variations in operating conditions such as injection amount, flow rate, column temperature, mobile-phase composition, and for different lots of columns.

14.10   TECHNOLOGY TRANSFER

The final step in process development for the analytical chemists is the transfer of his methods to the manufacturing division. This transfer usually occurs late in clinical phase III and prior to filing the NDA. The transfer is conducted to ensure that the method can be implemented and used correctly in the new lab. The method transfer serves as training for the receiving lab and requires that the receiving lab demonstrate the capability to perform the method. At this stage, method validation would be completed, although some companies may involve the receiving lab in the validation process as part of the transfer. Validation performed by the receiving lab qualifies it to use that method for its intended purpose.

The method transfer is executed through the authoring of a method documentation package that should include a detailed description of the method, some insight into the method development process, a validation summary, and a transfer protocol. The transfer protocol should clearly define the role and responsibilities of the transferring and receiving groups. It should list all of the instrumentation, columns or equivalent columns, chemicals, and samples required to run the method, outline the experiments that need to be performed, and the acceptance criteria for successful transfer of the method. The transfer documentation must be reviewed and agreed upon by both groups. Upon completion of the method transfer, data are compiled and analyzed and a final report is written. This report should indicate whether the method was successfully transferred and should list any deviations that were made to the original protocol.

14.11 CONCLUDING REMARKS

HPLC plays a significant role in the analytical aspect of process development. It is the most commonly used tool to determine the purity of the active pharmaceutical ingredient and to track impurity generation and yield during the process. There are a plethora of options available in terms of separation mode, stationary phase, and mobile phase to cover most of the wide range of diverse physiochemical properties associated with the active pharmaceutical ingredients, raw materials, intermediates, and impurities. This chapter has presented some of these options and how they can be applied. Procedures to develop HPLC methods and to ensure that the methods are precise and accurate have also been presented. Finally, how these HPLC methods fulfill regulatory requirements and how they are successfully transferred to manufacturing sites have been outlined.

REFERENCES

1. J. Xu, R. Thompson, B. Li, and Z. Ge, Application of packed column supercritical fluid chromatography for separation of bromosulfone from process related impurities, *J. Liq. Chrom. Relat. Technol.* **25** (2002), 1007–1018.

2. M. Yang and R. Thompson, A retention and selectivity model for hydrophilic interaction chromatography HILIC, submitted to J. Chrom. A (Aug 2006).

3. Q. Tu, T. B. Wang, X. Jia, and X. Bu, Speciation analysis of halogenides and oxyhalogens by ion chromatography with inductively coupled plasma mass spectrometer as element-specific detector, Merck Research Laboratories internal communication.

4. R. Thompson, N. Grinberg, H. Perpall, G. Bicker, and P. Tway, Separation of organophosphonates by ion chromatography with indirect photometric detection, *J. Liq. Chrom.* **17** (1994), 2511–2531.

5. M. See, R. Thompson, N. Grinberg, H. Perpall, G. Bicker, and P. Tway, Chromatographic analysis of residual acetate in bulk drugs, *J. Liq. Chromatogr.* **18** (1995), 137–154.

6. FDA's policy statement for the development of new stereoisomeric drugs, *Chirality* **4** (1992), 338.

7. P. Borman, B. Houghtflower, K. Cattanach, K. Crane, K. Freebairn, G. Jonas, I. Mutton, A. Patel, M. Sanders, and D. Thompson, Comparative performances of selected chiral HPLC, SFC, and CE systems with a chemically diverse sample set, *Chirality* **15** (2003), S1–S12.

8. C. Welch, Evolution of chiral stationary phase design in the Pirkle laboratories, *J. Chromatogr. A* **666** (1994), 3–26.

9. R. Thompson, Z. Ge, N. Grinberg, D. Ellison, and P. Tway, Mechanistic aspects of the stereospecific interaction for aminoindanol with a crown ether column, *Anal. Chem.* **67** (1995), 1580–1587.

10. R. Thompson, unpublished data.

11. T. Ward and A. Farris III, Chiral separations using the macrocyclic antibiotics: A review, *J. Chromatogr. A* **906** (2001), 73–89.

12. J. Haginaka, Protein-based chiral stationary phases for high-performance liquid chromatography enantioseparations, *J. Chromatogr. A* **906** (2001), 253–273.

13. R. Thompson, V. Prasad, N. Grinberg, D. Ellison, and J. Wyvratt, Mechanistic aspects of the stereospecific interactions of immobilized a1-acid glycoprotein, *J. Liq. Chrom. Relat. Technol.* **24** (2001), 813–826.

14. E. Yashima, Polysaccharide-based chiral stationary phases for high-performance liquid chromatographic enantioseparation, *J. Chromatogr. A* **906** (2001), 105–125.

15. K. Tachibana and A. Ohnishi, Reversed-phase liquid chromatographic separation of enantiomers on polysaccharide type chiral stationary phase, *J. Chromatogr. A* **906** (2001), 127–154.

16. T. O'Brien, L. Crocker, R. Thompson, K. Thompson, P. Toma, D. Conlon, B. Feibush, C. Moeder, G. Bicker, and N. Grinberg, Mechanistic aspects of chiral discrimination on modified cellulose, *Anal. Chem.* **69** (1997), 1999–2007.

17. H. Ding, N. Grinberg, R. Thompson, and D. Ellison, Enantiorecognition mechanisms for derivatized cellulose under reversed phase conditions, *J. Liq. Chrom. Relat. Technol.* **23** (2000), 2641–2651.

18. Y. Bereznitski, R. LoBrutto, and N. Grinberg, Trace analysis of sodium azide in an organic matrix, *J. Liq. Chrom. Relat. Technol.* **24** (2001), 2111–2120.

19. R. Thompson, unpublished data.

20. C. Machado, S. Thomas, D. Hegarty, R. Thompson, D. Ellison, and J. Wyvrat, Development of an indirect reversed phase method for the quality assessment of an acyl halide, *J. Liq. Chromatogr. Relat. Technol.* **21** (1998), 575–589.

21. V. Antonucci and L. Wright, Development of practical chromatographic methods for the analysis of active esters, *J. Liq. Chrom. Relat. Technol.* **24** (2001), 2145–2159.

22. S. Chen, H. Yuan, N. Grinberg, A. Dovletoglou, and G. Bicker, Enantiomeric separation of *trans*-2-aminocyclohexanol on a crown ether stationary phase using evaporative light scattering detection, *J. Liq. Chrom. Relat. Technol.* **26** (2003), 425–442.

23. R. Dixon and D. Peterson, Development and testing of a detection method for liquid chromatography based on aerosol charging, *Anal. Chem.* **74** (2002), 2930–2937.

24. M. Allgeier, M. Nussbaum, and D. Risley, Comparison of an evaporative light-scattering detector and a chemiluminescent nitrogen detector for analyzing compounds lacking a sufficient UV chromophore, *LC-GC North Am.* **21** (2003), 376–381.

25. R. Thompson, unpublished data.

26. D. Robb, T. Covey, and A. Bruins, Atmospheric pressure photoionization: An ionization method for liquid chromatography-mass spectrometry, *Anal. Chem.* **72** (2000), 3653–3659.

27. M. Reta, P. Carr, P. Sadek, and S. Rutan, Comparative study of hydrocarbon, fluorocarbon, and aromatic bonded RP-HPLC stationary phases by linear solvation energy relationships, *Anal. Chem.* **71** (1999), 3484–3496.

28. R. Kaliszan, M. van Straten, M. Markuszewski, C. Cramers, and H. Claessens, Molecular mechanism of retention in reversed-phase high-performance liquid

chromatography and classification of modern stationary phases by using quantitative structure–retention relationships, *J. Chromatogr. A* **855** (1999), 455–486.

29. R. Vervoort, A. Debets, H. Claessens, C. Cramers, and G. de Jong, Optimisation and characterization of silica-based reversed-phase liquid chromatographic systems for the analysis of basic pharmaceuticals, *J. Chromatogr. A* **897** (2000), 1–22.

30. R. Vervoort, E. Ruyter, A. Debets, H. Claessens, C. Cramers, and G. De Jong, Characterization of reversed-phase stationary phases for the liquid chromatographic analysis of basic pharmaceuticals by thermodynamic data, *J. Chromatogr. A* **964** (2002), 67–76.

31. D. Visky, Y. Vander Heyden, T. Ivanyi, P. Baten, J. De Beer, Z. Kovacs, B. Noszal, P. Dehouck, E. Roets, D. Massart, and J. Hoogmartens, Characterisation of reversed-phase liquid chromatographic columns by chromatographic tests. Rational column classification by a minimal number of column test parameters, *J. Chromatogr. A* **1012** (2003), 11–29.

32. J. Gilroy, J. Dolan, and L. Snyder, Column selectivity in reversed-phase liquid chromatography IV. Type-B alkyl silica columns, *J. Chromatogr. A* **1000** (2003), 757–778.

33. P. Dehouck, D. Visky, Y. Vander Heyden, E. Adams, Z. Kovacs, B. Noszal, D. Massart, and J. Hoogmartens, Characterisation of reversed-phase liquid chromatographic columns by chromatographic tests. Comparing column classification based on chromatographic parameters and column performance for the separation of acetylsalicylic acid and related compounds, *J. Chromatogr. A* **1025** (2004), 189–200.

34. N. Wilson, J. Gilroy, J. Dolan, and L. Snyder, Column selectivity in reversed-phase liquid chromatography VI. Columns with embedded or end-capping polar groups, *J. Chromatogr. A* **1026** (2004), 91–100.

35. E. Van Gyseghem, M. Jimidar, R. Sneyers, D. Redlich, E. Verhoeven, D. Massart, and Y. Vander Heyden, Selection of reversed-phase liquid chromatographic columns with diverse selectivity towards the potential separation of impurities in drugs, *J. Chromatogr. A* **1042** (2004), 69–80.

36. L. Pan, R. LoBrutto, Y. Kazakevich, and R. Thompson, Influence of inorganic mobile phase additives on the retention, efficiency, and peak symmetry of protonated basic compounds in reversed-phase liquid chromatography, *J. Chromatogr. A* **1049** (2004), 63–73.

37. Z. Ge, R. Thompson, D. DeTora, T. Maher, P. McKenzie, and D. Ellison, On-line HPLC monitoring of a deprotecting process, *J. Process Anal. Chem.* **3** (1997), 1–6.

38. H. Minakuchi, K. Nakanishi, N. Soga, N. Ishizuka, and N. Tanaka, Effect of skeleton size on the performance of octadecylsilylated continuous porous silica columns in reversed-phase liquid chromatography, *J. Chromatogr. A* **762** (1997), 135–146.

39. N. Wu, J. Dempsey, P. Yehl, A. Dovletoglou, D. Ellison, and J. Wyvratt, Practical aspects of fast HPLC separations for pharmaceutical process development using monolithic columns, *Anal. Chim. Acta* **523** (2004), 49–156.

40. ICH, Q3A: Impurities in new drug substances, 1996.

41. ICH Guideline: Impurities in new drug substances, *Federal Register* **61** (1996), 371ff.

42. International Conference on Harmonisation Steering Committee. Revised Guidance on "Impurities in new drug substances," *Federal Register* **68** (2003), 6924–6925.

43. K. Albert (ed.), *On-Line LC-NMR and Related Techniques*, John Wiley & Sons, Chichester, Sussex, UK, 2002.

44. N. Nyberg, H. Baumann, and L. Kenne, Application of solid-phase extraction coupled to an NMR flow-probe in the analysis of HPLC fractions, *Magn. Reson. Chem.* **39** (2001), 236–240.

45. International Conference on Harmonisation Steering Committee, Stability testing of new drug substances and products, 1999.

46. CDER Guideline on Validation of Chromatographic Methods, Reviewer guidance of chromatographic methods, US Food and Drug Administration, Center for Drugs and Biologics, Department of Health and Human Services, 1994.

47. ICH, Q2A: Validation of analytical methods: Definitions and terminology, October 1994.

48. ICH, Q2B: Analytical validation—methodology, November 1996.

15

ROLE OF HPLC DURING FORMULATION DEVELOPMENT

Tarun S. Patel and Rosario LoBrutto

15.1 INTRODUCTION

What is the definition of a formulation? Why is it needed? What is the importance of a formulation? These are some important questions that need to be addressed during the development of a potential drug product. The strict definition of the word is to specify a formula or to express a formula in systematic terms or concepts. The formula, in the current case, is a pharmaceutical dosage form. A formulation is needed to deliver the drug or the active pharmaceutical ingredient (API) to its targeted site. In order to overcome some of the physiochemical limitations of an API, it must be combined with inert ingredients (excipients) to safely and reproducibly deliver the API by the selected route of administration. Whether the delivery of a drug is oral (capsules, tablets, suspensions), by injection (intravenous and intramuscular injections), transdermal (patches), or by inhalation (metered dose inhaler, dry powder inhalation, nebulizer), the drug (active pharmaceutical ingredient) cannot be delivered by itself.

Formulation development is a complex process involving the physiochemical characterization of the API, identifying compatible excipients, developing a reliable manufacturing process, and thorough analytical characterization of the dosage form. The entire process starts early in the preformulation stage (covered in Chapter 12) and continues until the final market image is developed and launched (after approval from health authorities). The goal of any

HPLC for Pharmaceutical Scientists, Edited by Yuri Kazakevich and Rosario LoBrutto
Copyright © 2007 by John Wiley & Sons, Inc.

formulation development is to ensure that each batch manufactured meets the specifications for identity, strength, quality, and purity.

Formulation development has four generic phases that are somewhat aligned to the clinical phases: preclinical, clinical, registration, and commercial. During the preclinical stage, the first dosage form is developed to evaluate the safety and pharmacokinetic profile in humans. Generally due to limitations in the availability of an API and to minimize development time, this formulation is not optimized from a manufacturing perspective. For oral delivery, formulations such as powder in a bottle or a capsule are often selected. The information derived from preformulation studies is used to select the excipients for this first formulation. Depending upon the drug, its target, and its effect, a clinical trial may last anywhere from days to months. During this time, clinical programs are being updated and the product moves from one clinical phase to another (Preclinical, Phase I, Phase IIA, Phase IIB, Phase III, and Phase IV).

Preclinical studies are initial studies that are conducted to determine potential new treatments for specific disease indications. Animal models and *in vitro* assessments are used to help determine a treatment's safety prior to introducing it to humans. Phase I clinical trials are conducted to evaluate the safety of a drug, study absorption, and metabolism and how the drug should be administered (e.g., injection, oral), and different dose levels may also be evaluated. Phase 1 trials typically involve a small group ranging from 5 to 80 patients which depends on the disease indication. Phase II clinical trials are conducted to further evaluate the safety of a drug and to evaluate its efficacy, and further studies are performed to determine optimal dose levels. Phase II trials typically involve approximately 100 to 300 patients, depending on the disease indication. They usually contain a control and treatment group. Phase II is usually broken up into two subphases, Phase IIa (Proof of concept) and Phase IIb (dose ranging). Phase III clinical trials are conducted to confirm the safety and efficacy of a new drug and determine the labeling information. Trial participants are usually included in one of two study groups such that one will receive the new drug being evaluated, and the other group receives the already approved, current standard of treatment. Phase III trials can enroll anywhere from 50 to 15,000 patients.

In the early clinical phases, the safe and efficacious dose range is identified. Often based upon this new clinical information and marketing input, a new formulation must be developed, generally during Phase II. Before the pivotal Phase III clinical trials, the final market image should be available. This dosage form must be optimized, scaled-up, and validated.

During initial clinical trials, only a small number of units of dosage forms are manufactured and put on stability. When additional and longer clinical trials are taking place, the scale-up process of making the dosage form is further investigated. The scale-up procedure is not a single-fold scale-up from few units to commercial scale (millions of units) but is rather compromised of several iterative steps. The scale-up is generally performed in stages: few units (bench scale) to thousands (lab scale) to hundred thousands (pilot scale) to millions (validation and commercial).

During the scale-up, many attributes of a dosage form are defined. In addition, excipients may need to be changed since the original formulation may not be scalable. Ratios and physical attributes (appearance, color, embossing, size, etc.) of all material within the dosage form are defined at this point. Each time the process is scaled-up, the dosage form must be carefully characterized to ensure that the physiochemical properties and quality attributes have not changed.

During formulation development, these changes in the quality attributes are monitored employing several analytical methods (identification, assay, dissolution rate, related substances, content uniformity, etc.). HPLC plays a vital role in these analyses. Similar to formulation development, the health authorities recognize that the methods may change as one learns more about the drug product or changes occur in the formulation during the development of a given formulation.

15.2 PREREQUISITE FOR ANALYTICAL CHEMISTS DURING FORMULATION DEVELOPMENT

One of the strengths that an analytical chemist in the pharmaceutical industry has is the ability to propose degradation pathways without doing an experiment. Based on a given API structure and the environment that the API is going to reside in (solid state and/or solution state over time), degradation pathways are postulated based on known organic chemistry. An exact structure of degradation product is not necessary during the paper exercise, but impact of chemical stability must be addressed. In other words, which degradation pathway is most to least likely to occur, at certain conditions, must be discussed in order to direct formulation development from its inception.

An addendum at the end of this chapter highlights some common degradation pathways, along with reactions of some common functional groups [1] present in the literature. For an analytical chemist who works in a pharmaceutical research and development group, it is necessary to postulate degradation products and is essential that stability-indicating methods (SIMs) for new molecular entities (NMEs) be developed that can separate these potential degradation products from the active pharmaceutical ingredient.

15.2.1 Major Degradation Pathways in Pharmaceuticals

There are four major degradation pathways that an analytical chemists must focus on for pharmaceutical analysis: thermolytic (heat), hydrolytic (water), oxidative (oxygen, light, peroxide), and photolytic (UV and Vis light).

Many typical organic reactions that are potential degradation pathways are discussed in the addendum to this chapter as they pertain to an analytical chemist in drug substance and drug product sectors of pharmaceutical development. References 2–7 provide a more in-depth analysis of the major degradation pathways for pharmaceuticals.

All of the background information discussed in the addendum (Common Functional Groups) will help an analytical chemist with development of a stability-indicating method (SIM) that will assist in the development of an optimized formulation that will become a drug product in the marketplace. *Remember, an analytical chemist must be able to provide knowledge, not just data.*

15.3 PROPERTIES OF DRUG SUBSTANCE

By the time an active pharmaceutical ingredient (API) is made available to an analytical chemist in the formulation development group, most or all of the physical characteristics of an API has already been studied and the information should be available in some sort of a report from the drug substance group or preformulation group. Some of the key parameters that an analytical chemist in formulation development requires from such a report are the solubility and solution stability.

15.3.1 Solubility of Drug Substance in Presence of Formulation

Many times, depending upon the active-to-excipient ratio, the solubility of active in presence of excipients in certain solvents will be different. When an API is mixed with excipients, its solubility is usually lower when compared to its solubility in the same solvent by itself.

If the HPLC mode for the analysis of a given drug product is reversed-phase HPLC (RP-HPLC), then the solubility of an ionizable API in aqueous solutions (pH from 1.0 to 10), methanol, acetonitrile and in aqueous/organic mixtures should be determined by an analytical chemist during formulation development. On the other hand, if alternative modes of HPLC analysis for a given drug product such as normal phase is employed, then the solubility of API in IPA, and hexane, is also very important. The main reason that an analytical chemist supporting formulation development must know the solubility before the commencement of method development is that this information will provide an initial assessment of the type of mobile phase (separation or a potency method) and/or sample preparation solvent (sample preparation procedure) that will be used for drug product method development.

However, in both of the cases, the solubility listed on a drug substance report will be higher, when compared to the API with excipient diluted in the same solvent. This may not be true in all cases since it will mainly depend on the type and the amount of excipients present in the formulation. Some excipients may change the final pH of the solution after addition and/or some excipients will act as solubilizing agents for a given API. In both of these cases, an analytical chemist must know the properties of excipients in the formulation.

For analytical sample preparation, measurement of the final pH of the sample solution (excipients, API(s), and sample solvent mixed together) will be helpful in the development of any analytical procedure. If an API is known to be stable in acidic pH (pH 1–2), then an analytical chemist will try to utilize a certain sample solvent that has a pH in the required range. However, when a dosage form is dissolved in a sample solvent, the excipients present in the formulation (and even the API) will change the pH of the solution. The final pH of the solution must be measured in order to determine the optimal pH of sample solution to achieve longest solution stability. This is particularly important for a long sequence of injections on autosamplers for analysis, so solutions do not need to be made daily.

15.3.2 Solution Stability

The objective is to determine the degradation products in the solution, not in the dosage form. It is very important to determine how the active will behave in solution. Is an API going to degrade in the solution, and to what extent? These questions can be answered with solution stability studies.

The goal of solution stability for different tests can vary. For assay and degradation product test, no new degradation products (over time and above LOQ) should be formed in solution. Certain acceptance criteria (max % increase) may be set for impurities at particular levels during the solution stability study. For assay determination (assay, CU, and dissolution), the content of assay over time should not change more than 2.0%. Also, refer to Chapter 9, which discusses method validation.

15.4 PROPERTIES OF EXCIPIENTS

Excipients are the so-called inert ingredients contained in a pharmaceutical dosage form. Excipients are chosen based on their compatibility with the API and the function of a given excipient. Analytical chemist must spend as much time studying the properties of API as to study the properties of excipients. Even though the excipients are said to be inert, they are chemicals having functional groups (refer to Table 15-1). Some of the functional groups may react with your API to form degradation products that could not have been foreseen by just focusing on the structure of the API.

A special consideration must be given to the effect of water on a given drug product (or a formulation). Water presents a real problem for a formulation when the drug substance or the excipients are sensitive or susceptible to hydrolysis. Whether the excipients will dissolve in certain solvents or not, their interactions with the active (in solution and solid state) are very important.

TABLE 15-1. Common Excipients and Their Functional Groups [8]

Excipient	Functional Group(s)	Functional Category[a]
Talc	—OH	Anticaking agent, glidant, tablet and capsule diluent, and lubricant
Mg- or Ca-stearate	—COO—	Tablet and capsule lubricant
Lactose, cellulose	—O—, —OH	Diluent for dry-powder inhalers, tablets, and capsules
PEG	—O—, —OH	Plasticizer, solvent, suppository base, tablet and capsule lubricant
Polymethacrylates	—CO—O—	Film former, tablet binder and diluent, sustained release polymer in right combinations
Triethyl citrate	—CO—O—, —OH	Plasticizer

[a]R. C. Rowe, P. J. Sheskey, and P. J. Weller (eds.), *Handbook of Pharmaceutical Excipients*, 4th edition, 2003 [8].

15.5 IMPACT OF EXCIPIENTS ON DEGRADATION OF API(s)

Excipients are chemicals and they have functional groups. The functional group on the excipients may react with the API under certain conditions. These conditions could be under normal storage conditions or accelerated conditions. For these reasons, excipients must not be considered inert.

For RP-HPLC, if an API is interacting with any of the excipients, then the resulting degradation product could be of a higher molecular weight than the API. However, this does not necessarily mean that it will elute after API, since the retention of components in RPLC is dependent on their relative hydrophobicity and interactions with the stationary phase. If a more hydrophobic degradation product is formed, then such a degradation product will elute after API peak (usually true for RP-HPLC when neutral species are involved, but may not be true when ionizable species are involved).

In the example, shown in Figure 15-1, a development compound A has a carboxylic acid functional group (pK_a ~4.5). Three known synthetic by-products of the carboxylic acid active pharmaceutical ingredient were the alcohol, ethyl ester, and an aldehyde, which are all neutral. Three degradation products were observed in excipient compatibility stability and/or when different formulations were put on accelerated stability: neutral degradation product and two acidic degradation products (A and B). Note that the drug substance stored under similar conditions did not produce these three degradation products, indicating that some excipient(s) may be inducing degradation on the API. Using RP-HPLC and performing a pH study, by varying the mobile-phase pH, the ionogenic nature of these degradation products was elucidated.

The retention of degradation products A and B (acidic compounds) was dependent on the mobile-phase pH, while the retention time of the neutrals

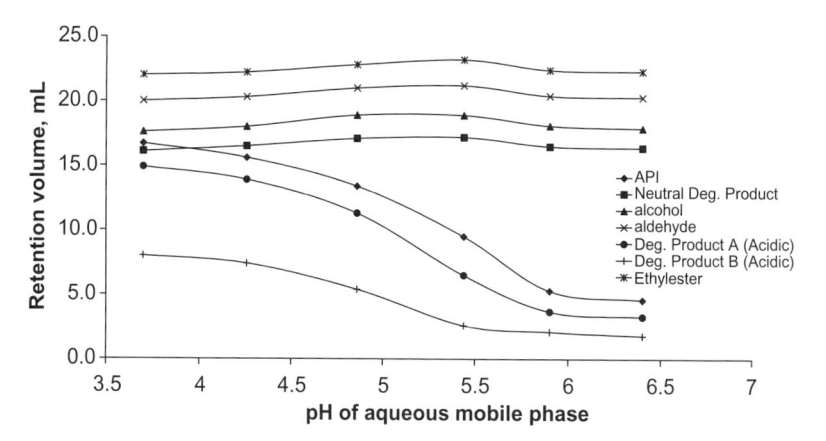

Figure 15-1. Dependence of retention volume on pH of an aqueous mobile phase in RP-HPLC for drug product A on a 150 × 3.0-mm column using a 0.7-mL/min flow rate.

(neutral degradation product) did not change (refer to Figure 15-1). The optimal pH of the aqueous phase was chosen to be pH 6.5 to avoid any significant changes in retention with minor variations in the pH of the aqueous portion of the mobile phase. Using this pH, the API elutes prior to all the neutral species (ethyl ester, alcohol, aldehyde, and the neutral degradation product). The neutral degradation product eluted prior to the aldehyde peak at all pH conditions, which indicated that this peak is less hydrophobic than the aldehyde and is a neutral compound (varying mobile-phase pH did not change its retention time; refer to Figure 15-1). This degradation product was later identified by LC-MS as a cyclic product. The two known neutral degradation products (alcohol and aldehyde) have lower molecular weight than the active, but both of them still elute after the active peak. The main reason is the retention in HPLC is dependent on the hydrophobic nature of the compound and not the molecular weight. However, all three neutral species elute in the order of their hydrophobicity (alcohol elutes first, then the aldehyde and then the ethyl ester, Figure 15-1). The acidic degradation products, A and B, show the greatest retention at low pH (analyzed in their predominately neutral form) and the least retention at high pH (analyzed in their predominately ionized form), which is typical retention behavior for acidic compounds in RP-HPLC.

Another example in which the excipient reacts with the API is the reaction between lactose and fluoxetine hydrochloride (refer to Figure A15-6 in the Addendum of this chapter). This is a typical example of the Maillard reaction [9]. Another example of the Maillard reaction is presented where Prozac (fluoxetine hydrochloride) and two generic drug products of Prozac were compared at accelerated conditions (40°C/75% RH), and the amount of degradation products (analyzed by HPLC) were found to be very different between the studied formulations (refer to Table 15-2) [10]. The authors

TABLE 15-2. Stability of Fluoxetine HCl Products at 40°C/75% RH [10]

		% Total Degradation Products by HPLC				
Product	Major Excipient	Initial	1 month	3 months	6 months	9 months
Prozac	Starch	0.17	0.19	0.21	0.23	0.23
Generic A	Lactose	0.43	0.47	0.60	0.90	1.10
Generic Z	Lactose	0.30	0.35	0.45	0.63	0.74

indicated that the difference in amounts of degradation products is not due to different dosage form (capsule or the tablet), but rather due to the type and the amount of excipients used in the formulation (in particular lactose). It is also interesting to note that the formulation containing starch instead of lactose gave the lowest amount of degradation products. One of the lessons learned here must be that if an API contains an amine functional group (primary or a secondary), avoid lactose (or a similar carbohydrate) in the formulation.

Also, if an API is prone to hydrolytic degradation, then the formulations in both capsules and tablets should be investigated. Capsules are made of gelatin, sugar, and water and contain about 10–15% moisture, and gelatin can absorb additional moisture. If gelatin capsules are placed in areas of high humidity, they may become malformed as they absorb moisture; and if capsules are placed in low humidity, they may become dry and brittle and may crack. The amount of degradation products formed could be influenced by the type of dosage form (tablets versus capsules) and the respective storage conditions for each of the dosage forms.

15.6   TEST METHODS FOR MOST COMMON DOSAGE FORMS IN WHICH HPLC IS THE PRIMARY TECHNIQUE

The list below is a summary of test methods that are most likely to be required during the testing of most common dosage forms (tablets, capsules, and solutions). The list is by no means an exhaustive (since test methods to determine moisture content, pH, sterility, particulate matter, and microbial testing are not listed) and the selection of test methods depends on an evaluation of a dosage form and also on the phase of drug development. For each type of test, most common techniques utilized are listed in square brackets. As can be seen, HPLC can be employed for all of the following test methods.

- Batch release and stability testing
  Identification [HPLC, UV, IR, NIR] (for release only)
  Assay [HPLC, NIR, CE, GC]
  Degradation products [HPLC, CE]

Content uniformity (not needed for parenterals and on stability) [UV, NIR, HPLC, CE]

Dissolution (not needed for parenterals) [UV, HPLC]

- Cleaning verification for active [HPLC, UV, ion mobility spectrometry, MS [11–13]]

Very sensitive and specific methods are needed

- Process validation

Analysis of blends [HPLC, NIR]

Cores and coated tablets [HPLC, NIR]

- Extractables/Leachables (for parenterals) [HPLC]
- Structure determination, trouble shooting [hyphenated HPLC techniques]

The most common HPLC test methods from the above list are selected and thoroughly described below to ensure that the dosage form for which these test methods are selected will meet the criteria for identity, quality, potency, and purity for a given dosage form.

15.6.1 Assay and Related Substances

The two most fundamental issues of importance in drug therapy are safety and efficacy. The impurities found in the bulk and dosage form may cause adverse effects by their pharmacological–toxicological profile, which in turn determines the safety of a drug. The impurities found in a drug product may possess unwanted pharmacological and/or toxicological effects [14]. Both the quality and safety of a drug are said to be assured when the related substances found in the drug product are controlled and monitored effectively. Therefore, the most important topic (from the view of an analytical chemist) during formulation development is the activities around a test method for related substances [15–17].

The related substances found in an API could have originated during the synthetic steps, from the original starting materials/intermediates or from impurities from the starting materials that reacted in the downstream chemistry (all of these are known as synthetic by-products). When a given API is utilized to manufacture a drug product, the degradation products found in the drug product must be identified, characterized, and/or qualified based on ICH guidelines [18]. The most important reason for this is to have a quality product, which is the basis of having a safe and an efficacious drug to begin with. The relationship between synthetic by-products, degradation products, and related substances is that related substances contain the sum of synthetic by-products (originating from chemical synthesis which do not change with time and conditions) and degradation products (increases with time and varies under different storage conditions). However, sometimes the synthetic by-products of the API can also be degradation products of the drug product.

A test method for related substances must be able to separate all known and unidentified components from a given drug product. The phrase that defines this process is called "stability-indicating."

15.6.2   Stability-Indicating Method (SIM)

What is a "stability-indicating method" (SIM)? The answer to this question is found in the FDA guidance document [19] as follows: "Validated quantitative analytical methods that can detect the changes with time in the chemical, physical, or microbiological properties of the drug substance and drug product, and that are specific so that the contents of active ingredient, degradation products, and other components of interest can be accurately measured without interference."

The above statement has lot of details in reference to "what is a SIM?" The statement starts with method validation (refer to Chapter 9). Next, most methods need to be specific (specificity, resolution of active from related substances, peak purity), reproducible (precision), quantitative (recovery, linearity, LOD, LOQ), and able to monitor a change in chemical, physical, and/or microbiological properties of drug products over time (refer to sections on stability testing and mass balance).

All assay and purity methods during formulation development should start with an identical method that has been developed and validated for the particular drug substance utilized to manufacture that drug product. The drug substance method should have demonstrated that the API is resolved from all potential degradation products, synthetic intermediates, and degradation products. When possible, the same drug substance method should be used for the drug product. This is especially helpful if the impurities (synthetic by-product) have been qualified (based on toxicological data and appropriate safety factor) in the drug substance. Therefore if a synthetic by-product is also a degradation product, the proposed specification limit of that degradation product in the drug product can be assigned. However, sometimes the same HPLC method cannot be used for the drug product due to potential interference of excipients and solubility of the excipients at a particular mobile-phase pH. Then a cross-correlation of the relative retention time (RRT) of the impurities in the drug substance which were qualified and the RRT of the impurities in the drug product should be determined. The use of LC-MS and/or LC-NMR is strongly encouraged. Also, if authentic synthetic by-products of the API are available, then elution can be confirmed using both the API and drug product methods.

Even if the same drug substance HPLC method is used for the drug product, forced decomposition studies must be performed again for the drug product to confirm the resolution of potential degradation products from the API. In addition, forced decomposition studies must also be performed for different dosage forms (capsule, tablet, suspension, injectable, etc.) of the same drug substance.

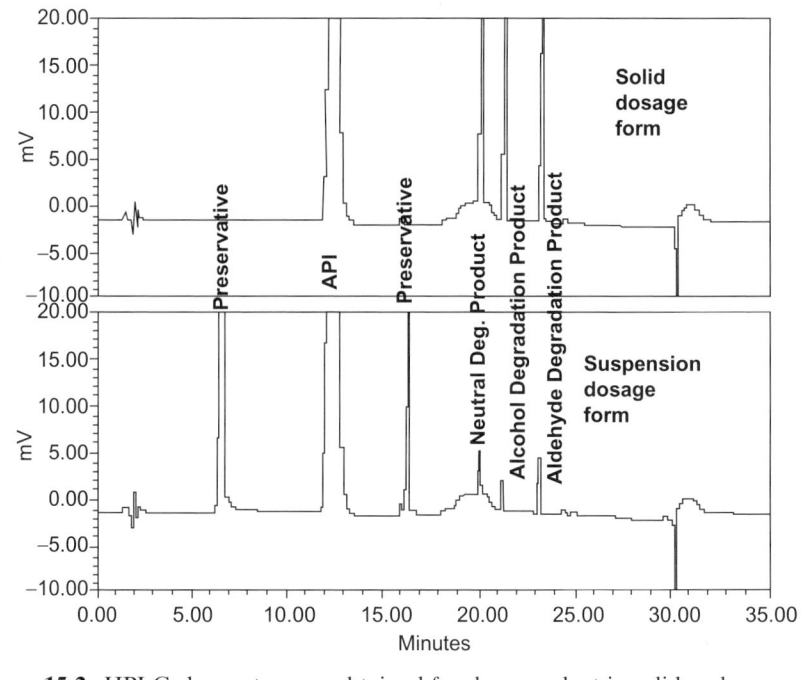

Figure 15-2. HPLC chromatogram obtained for drug product in solid and suspension dosage forms.

Different dosage forms of development compound A (capsule, tablet, suspension, injectable, and a combination product with aspirin) previously described in Section 15.5 were explored. The HPLC parameters/conditions remained very similar (only a change in the gradient composition was needed). The chromatogram in Figure 15-2 shows the observed peaks for solid and suspension dosage forms. The two extra peaks in the suspension drug product are preservatives that have UV chromophores, but are neutral. The suspension drug product also behaved very similar to the solid dosage form (capsule and tablet) in terms of which degradation products were formed. However, the amounts of each of these degradation products were found to be slightly higher than solid dosage form at identical storage conditions. The injectable drug product had a co-solvent, a surfactant, an anti-oxidant, and a buffer system where the API had its highest solubility. In this liquid environment, the API degraded in a different manner than the solid dosage forms and the suspension. One of the major degradation products in the injectable eluted before the active peak (labeled as degradation product B in Figure 15-1). Based on its retention behavior when pH was varied (refer to Figure 15-1), it indicated that it is also an acidic in nature.

Further in development, a combination product (capsule dosage form) of the development compound A with aspirin was proposed. The preformulation

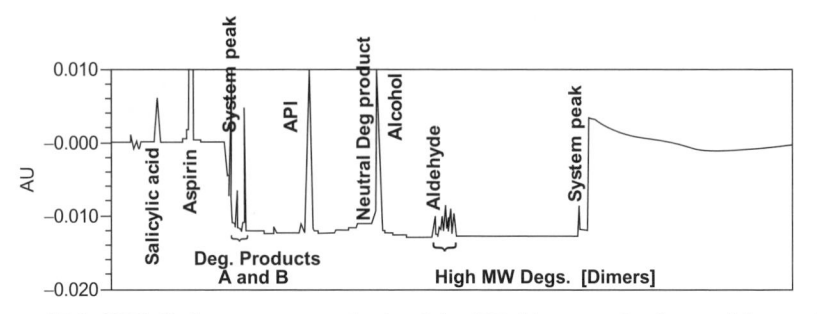

Figure 15-3. HPLC chromatogram obtained for DP (drug product) + aspirin combination drug product. (API = Development Compound A)

TABLE 15-3. Dual pH Method

Time (minutes)	Mobile Phase A	Mobile Phase B	Mobile Phase C	Gradient
Initial	100	0	0	Isocratic
5.0	0	70	30	Step
20.0	0	70	30	linear
35.0	0	30	70	linear
40.0	0	30	70	linear
40.1	100	0	0	linear
45.0	100	0	0	linear

Mobile phase A = 30% pH 1.2 solution + 70% acetonitrile.
Mobile phase B = 90% pH 6.5 + 10% acetonitrile.
Mobile phase C = 10% pH 6.5 + 90% acetonitrile.

studies were conducted and the challenge was the separation method for both components (aspirin and the development compound A). Aspirin is known to hydrolyze in the presence of water. In addition, aspirin and its potential degradation products are very hydrophilic, and they eluted very close to the void volume at mobile-phase pH values above their pK_a. Therefore, a dual pH mobile-phase system was employed to retain and separate the aspirin and the development compound A (refer to Figure 15-3). A pH 1.2 mobile phase was first utilized at an isocratic condition to elute the aspirin and its degradation product (salicylic acid). Once both of these peaks were eluted off the column (after the void volume), the pH of the mobile phase was increased to pH 6.5 (identical to pH stated in the last section for development product A) to obtain a similar known degradation profile as earlier in development for development product A. Once the gradient for development compound A was completed, the column was re-equilibrated with initial mobile phase (pH 1.2). This dual pH method (refer to Table 15-3) has a total run time of 45 minutes, which for preformulation studies was acceptable. This was the preliminary HPLC method and was further optimized during development for the combination drug product.

15.7   FORCED DECOMPOSITION

When developing stability-indicating methods, forced degradation testing is performed to demonstrate specificity of any separation method as well as to gain some insight into the degradation pathways. Forced degradation studies are more severe than stress studies, which are defined in health authority guidelines. For this particular reason, it is possible that the information gained during forced degradation may help facilitate formulation development, manufacturing processes, and packaging components in which the knowledge gained may improve overall performance of a drug product (longer shelf life, fewer degradation products, more robust manufacturing process, etc.), which in turn could improve its attributes (safety and/or efficacy).

According to the available guidance [20] from the FDA, forced decomposition studies are carried out for the following reasons [21]. (1) development and validation of stability-indicating methodology, (2) determination of degradation pathways of drug substances and drug products, (3) discernment of degradation products in formulations that are related to drug substances versus those that are related to non-drug substances (e.g., excipients), (4) structure elucidation of degradation products, (5) determination of the intrinsic stability of a drug substance molecule, (6) stability in solution and solid state, (7) involve conditions that are more severe than the accelerated conditions defined in guidance documents, (8) are not part of formal stability program, (9) include conditions that analyze thermolytic, hydrolytic, oxidative, and photolytic degradation mechanism in the drug substance and drug product.

The guidance on forced degradation is available, as stated above; however, the details of the investigations are left up to the pharmaceutical researcher. Forced decomposition testing within the pharmaceutical industry varies tremendously; this was demonstrated by Baertschi [22], who surveyed 20 pharmaceutical companies on the practices of forced decomposition studies.

During formulation development, forced decomposition would be the first study that must be repeated (for the reasons stated above) in the presence of API + excipients of the potential formulation. A drug product without the API is utilized as a control (placebo). The potential loss of API for each stress condition should be targeted at about 5–10%. If API is degraded by more than 10%, then it is possible that the degradation products may be created from the initial degradation products. Typical forced decomposition studies include heat (40°C, 50°C, and/or 60°C—solution and solid state), light (1200 KLux hours—solution and solid state), moisture (75% RH, open), acid (0.1 N HCl), base (0.1 N NaOH), and peroxide (3% H_2O_2). The conditions listed above in parentheses are not exhaustive, but rather the most common practiced conditions in the pharmaceutical industry [22]. Once these studies are completed, an analyst must be able to determine the major degradation products and their relative retention time (compared to API) for a given HPLC method. The HPLC method should be able to resolve all possible major degradation

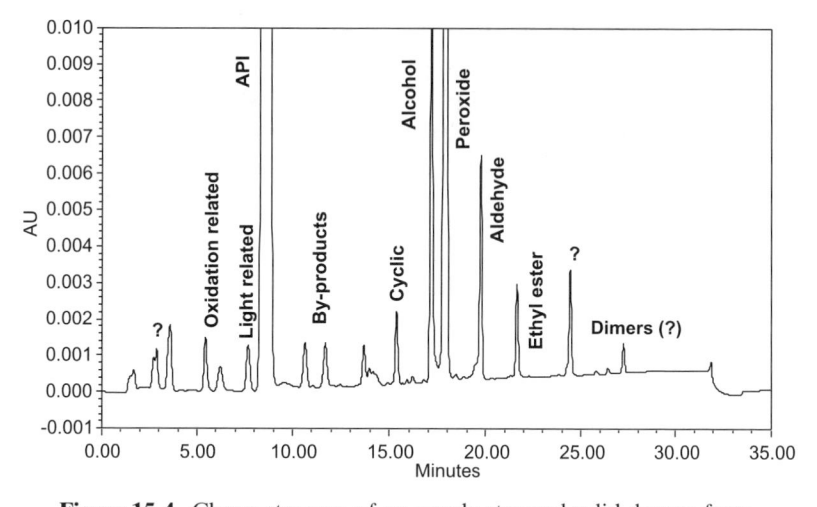

Figure 15-4. Chromatogram of an overly stressed solid dosage form.

products generated from the forced degradation studies (heat-, light-, water-, acid-, base-, and peroxide-related) from the API.

Initial solution forced decomposition experiments should be focused on determining the time required to generate a loss of about 5–10% of API at each forced degradation condition, and these forced degradation samples can be used during early-phase method development. Once the duration for each forced degradation condition is estimated, then all solution forced degradation experiments are repeated (for each condition) when the final method is developed (specificity and selectivity), and LC-MS is usually used to propose the structures/masses of major degradation products [23]. Based on these solution forced degradation studies, particular peaks from solid-state stability studies can be properly assigned and attributed to a particular forced degradation pathway. Figure 15-4 shows the separation of the degradation products from the API in development product A solid dosage form. This sample was a tablet dosage form that was stressed for a few months at 50°C/75% RH under normal light conditions in an open Petri dish. The labeled peaks were later identified by LC-MS as listed on the chromatogram. The peaks marked with question marks were never observed in real-time stability studies under intended and accelerated conditions.

Some approaches/examples for conducting forced degradation studies are given below: For a forced degradation acid study for a particular API the API is exposed to acidic conditions. The API (at a known concentration) is usually prepared in the sample preparation solvent, which gives 0.1 M HCl concentration in the final solution. Once this solution is prepared, it is injected every half hour or hour to determine the loss of API over time. If the API is susceptible to degradation under acidic conditions, then peak(s) of degradation products would increase over time and the API should decrease over time (once these peaks are separated in the chromatogram, peak purity is performed using PDA and/or mass spectrometry to determine if major degradation product(s) are co-eluting and to determine spectral homogeneity of the main component, API). An experiment for base hydrolysis is conducted in a fashion similar to that of the acid hydrolysis experiment. In both conditions, if no degradation is observed, the solutions could be exposed to accelerated temperatures (40°C or 50°C) for a certain time period (ie one day) to accelerate degradation. If no degradation is observed, then it can be assumed that this compound is not prone to acid/base hydrolysis.

To study oxidative degradation, an experimental design can be set up as follows: A pre-weighed excipient mixture, API and API plus excipients, is placed each into separate crimped headspace vials. This process is repeated two more times. This will result in three sets of vials (excipients, API, and API plus excipients) to be stored at three stress conditions (headspace of air, nitrogen, and oxygen). To generate the nitrogen (control) and oxygen stress conditions, two needles would be placed in the rubber septa: one to allow flow of gas in and one to allow flow of gas out to fill the headspace with the desired gas. For set one, transfer nitrogen into vials and for set two, transfer pure oxygen into vials for a few minutes and then remove the inlet and outlet needles. If molecular oxygen is playing a role to generate degradation products, then the highest oxidative degradation would be observed in the set in which the headspace vials have pure oxygen. The next highest amount would be in set number three, where the headspace has air (note that air contains approximately 78% nitrogen and 21% oxygen). For set number one in which the headspace vials have nitrogen, oxidative degradation is expected to be minimal. If for all three sets of conditions, the three types of samples have very similar degradation profile and amount of major degradation peaks, then it would be safe to assess that the molecular oxygen is not causing any oxidative degradation of excipient, API, and API + excipients. The above experiment can be performed for solid dosage form as well as liquid dosage form, where the gas passed through the needles must be bubbled through the liquid.

The above example would show oxidative degradation through molecular oxygen; however, oxidative degradation can occur also through trace metal impurities or radical initiation (peroxides residues from excipients in the presence of light) [24]. All possible experiments must be performed in order to determine how and what is the source causing the degradation. All possible experiments should be performed earlier in development. The other unwanted alternative would be to discover compatibility problems later in development; at that stage, time is money. Especially if the degradation product(s) must be identified and/or qualified this could cause delays in the clinical program.

During the development of a liquid dosage form of development compound C, it was observed that the solution changed color from slight yellow (initial color due to components in the formulation) to brown. What caused the change in color was the obvious question: From one or more of the excipients which have slight yellow color or is it from the degradation product(s) of the

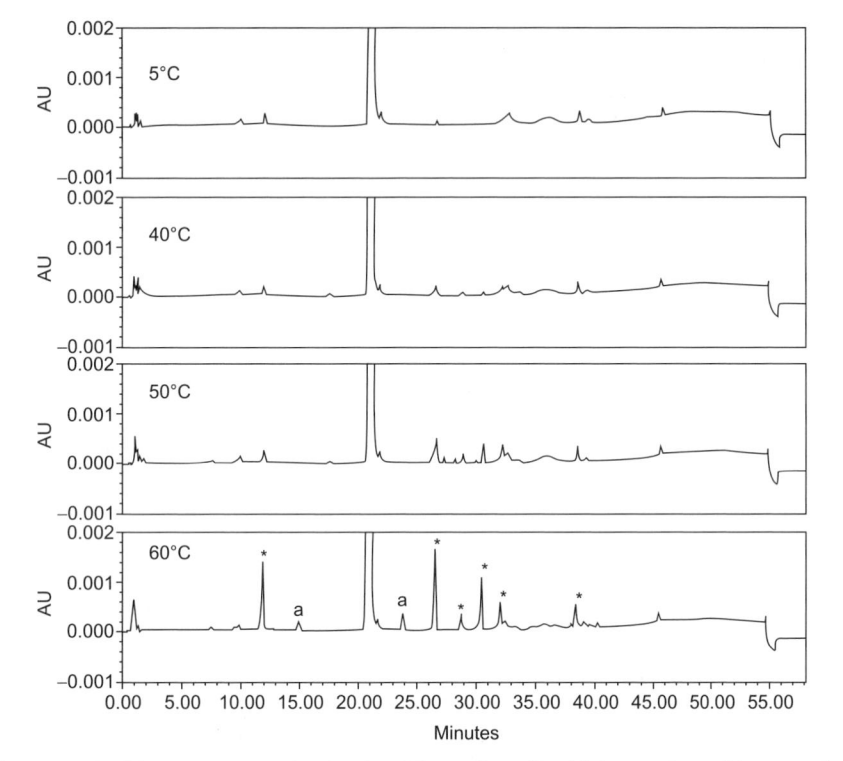

Figure 15-5. Chromatogram obtained at 254 nm for a liquid dosage form (drug product C) stored at different temperatures.

API (the impurities formed could be more conjugated in nature leading to colored impurity). A very small impurity, even at level of 0.05%, can result in colored material. A simple experiment was carried out in the lab. The solution was placed at different temperatures (5°C, 40°C, 50°C, and 60°C) in a controlled heated oven in sealed ampoules. Since the dosage form is stored in ampoules, the humidity is not a variable. If the nominal analytical concentration of the major peak in solution is injected (~0.2 mg/mL), then the response of the degradation products is small (less than 2 mV) at 254 nm UV (refer to Figure 15-5). The solution is observed to be slight yellow at 5°C, and for the solutions stored at higher temperatures the solutions became darker in color as the temperature approached 60°C. Other additional information is obtained from the stacked chromatograms shown in Figure 15-5 such that some degradation products increased as the temperature was raised from 5°C to 60°C (indicated by an asterisk). Also, additional degradation products were observed at 60°C (indicated by "a"). The peaks labeled "a" were only observed at 60°C and were not observed at any other temperature, which indicates that these peaks had different rate constants from the other peaks observed from 5°C to 50°C. Later in development, the highest temperature utilized for any temperature study was 50°C.

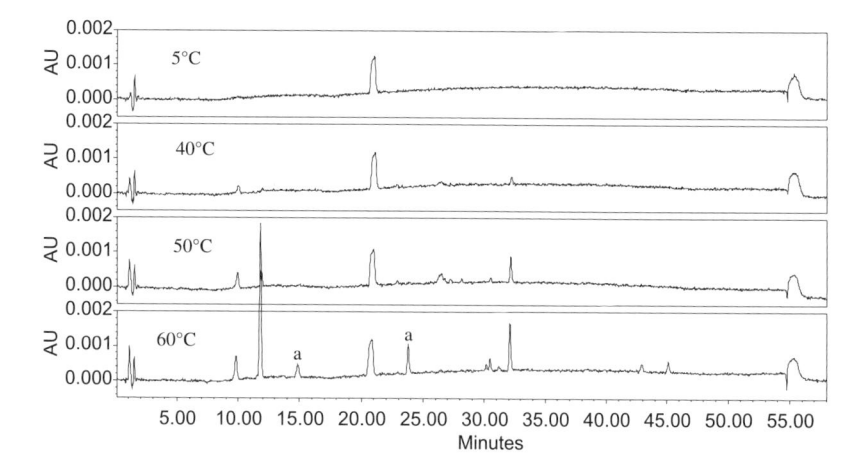

Figure 15-6. Chromatograms obtained at UV 410 nm for a liquid dosage form (drug product C) shored at different temperatures.

Since the response of the peaks was low, a greater concentration (more than 10 mg/mL) of the solution was injected into the same HPLC system to achieve greater mass on column. This is not advisable for routine analysis because the API may overload the column and carry-over may become an issue that is dependendent upon the API, its ionization state, and the mobile-phase conditions.

PDA analysis was employed for the above experiment, and a UV profile at 410 nm was extracted (refer to Figure 15-6). The major peak eluted at 21 minutes. The reason for choosing UV 410 nm is because yellow color absorbs in this region of the UV light. The color of the injected solutions was as follows: 5°C was faint yellow, 40°C was slight yellow, 50°C was dark yellow, and 60°C was brown. The color of the injected solution can be correlated to the peaks observed at UV 410 nm in Figure 15-6. The change of the color of the solution can be attributed to one or more the peaks observed at 50°C and 60°C. The response factors of these impurities should be taken into consideration, especially if using low wavelength detection. They may not have a significant response factor at low wavelength and may be underestimated. For this type of formulation, it would be recommended to store at 5°C or 25°C to prevent degradation of the drug product. The accelerated conditions in this case would then be 25°C or 40°C, respectively.

15.8 COMPATIBILITY OF EXCIPIENTS WITH API(s) (TYPE AND RATIO)

When an NME is discovered in research, it is an API that would be developed further into a drug product. However, an API alone does not go into clinical trials, but a drug product does. This means that a formulation must be

TABLE 15-4. Example of Compatibility Experimental Design

	1	2	3	4	5	6
Drug substance (I)	200	25	25	25	25	25
Lactose		175				170
Mannitol			175			
Microcrystalline cellulose				175		
Dibasic calcium phosphate dihydrate					175	
Magnesium stearate						5
Sodium stearyl fumarate						
Stearic acid						
Potency remaining[a] (% initial)	96.4	95.7	95.8	93.9	85.0	64.3
Hydrolysis product formed	3.3	4.1	4.0	5.8	16.7	37.0

[a]Compositions of drug–excipient blends used for I and assay results after 3 weeks of storage at 50°C in closed vials with 20% added water; weights of all ingredients are in milligrams.
Source: Reprinted from reference 25, with permission.

developed to effectively deliver that particular API. Compatibility testing of an API and excipients(s) is a key to accelerate drug development timelines because an unexpected stability issue later in development may increase development time and costs [25–27].

The drug–excipient compatibility screening model developed by Serajuddin et al. [25] could be used to determine potential stability problems due to inter-actions of API with excipients in solid dosage forms. Table 15-4 shows an example of binary and ternary mixtures employed by Serajuddin et al.

The design of the experiment should be able to determine the chemical nature of an excipient with the API, chemical stability based on the API–excip-ient ratios, and effect of temperature, water, and light. Not all of the stated variables are studied for a given API, since some information on the API would be known beforehand. For example, if the molecule is light-sensitive, all compatibility experiments would be performed in the absence of light and then the final solid dosage form would be film-coated to avoid light degrada-tion, or for an uncoated drug product the packaging material can be chosen to avoid penetration of light, or the coated material inside a bottle (if a bottle is chosen as a packaging material) could be covered by black color because black color absorbs light.

In the following excipient compatibility example, drug substance was put on accelerated stability with different binary ratios of varied excipients and stored at 2 weeks and 4 weeks at dry and humid conditions, 50°C (dry) and 50°C/75% relative humidity, respectively (Figure 15-7). With most of the excip-ients under dry conditions, it was observed that the level of increase of degra-dation products in the binary mixtures was less than 0.2%. However, under 50°C/75%RH conditions, most of the excipients showed increasing degrada-

	Experiment										
	7	8	9	10	11	12	13	14	15	16	17
	25	25	25	25	25	25	25	25	25	25	25
				170				170			
	170				170				170		
		170				170				170	
			170				170				170
	5	5	5								
				5	5	5	5				
								5	5	5	5
	65.4	65.3	68.1	77.9	81.9	77.6	81.8	90.0	92.9	88.1	78.3
	36.7	36.3	33.7	21.8	15.4	20.1	15.3	9.7	6.9	11.7	21.6

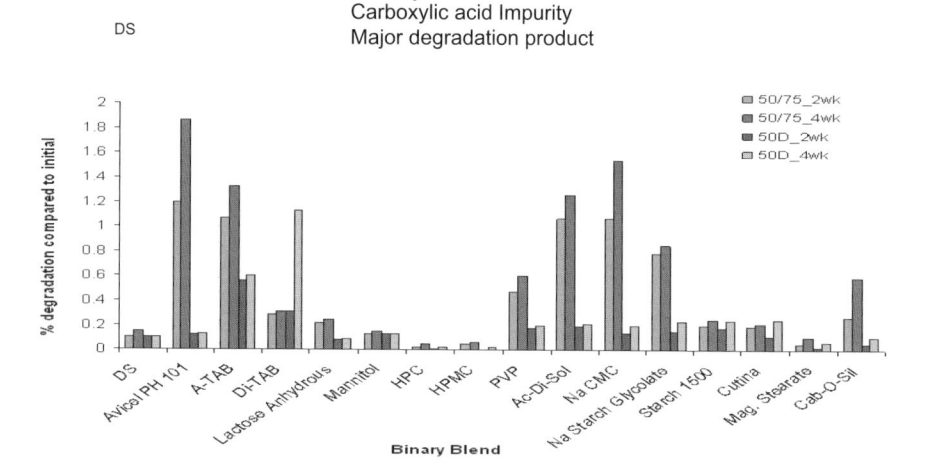

Figure 15-7. Excipient compatibility screening performed in early formulation development.

tion at 2 weeks and further degradation at 4 weeks, indicating that if these excipients are to be used for further formulations, the formulation should be protected from moisture. The major degradation product was identified by HPLC-MS as the carboxylic acid impurity.

This drug substance contains an amide bond; this can be amenable to acid/base hydrolysis, consequently leading to a carboxylic acid degradation product. This potential degradation product was predicted from looking at the structure prior to method development. Consequently, an acid stressed sample was used in the initial method development work to generate this potential degradation product, and measures were taken to adequately retain this potential degradation product using the proper pH of the mobile phase and eluent conditions. Looking at the structure of the API and predicting degradation products should be a common practice for chromatographers working in a drug product environment.

15.9   MASS BALANCE

The process of adding the assay value to the degradation products to achieve a total close to 100% of the initial value has been discussed in an ICH guideline [20]. An evaluation should be considered for both the assay and the degradation products, to review the adequacy of the mass balance. Any mass balance issue encountered during formulation development can be addressed by questioning either the recovery step of the sample preparation (assay and degradation products), adsorption on the column, or the detection technique (relative response factors of the degradation products) employed.

Recovery of an active or degradation product during the sample preparation is most likely due to adsorption on the undissolved excipients or capsule shells. The following case study [28] illustrates this point very well.

15.9.1   Case Study 1

A combination dosage form was under development where Starlix® and Metformin HCl were combined together. During spiked recovery experiments, it was discovered that the recovery of Metformin HCl in the presence of excipients and Starlix® was low at high excipients-to-Metformin ratios (in solution) as shown in Figure 15-8. Initially, it was thought that the sample extraction solvent is not dissolving/extracting the Metformin HCl in solution, so the extraction was performed with different sample solvents. Refer to Table 15-5 for results. None of the selected sample extraction solvents improved recovery of Metformin HCl. What if Metformin HCl was interacting with one of the excipients in solution? To answer this question, Metformin HCl recovery was performed with dry-mix placebo blend that contained all excipients and by removing one excipient at a time from the blend. Refer to Figure 15-9 for results. Figure 15-9 shows that when croscarmellose was removed from the dry-mix placebo blend, the recovery of Metformin HCl was increased and complete recovery was achieved. It was concluded that Metformin HCl is

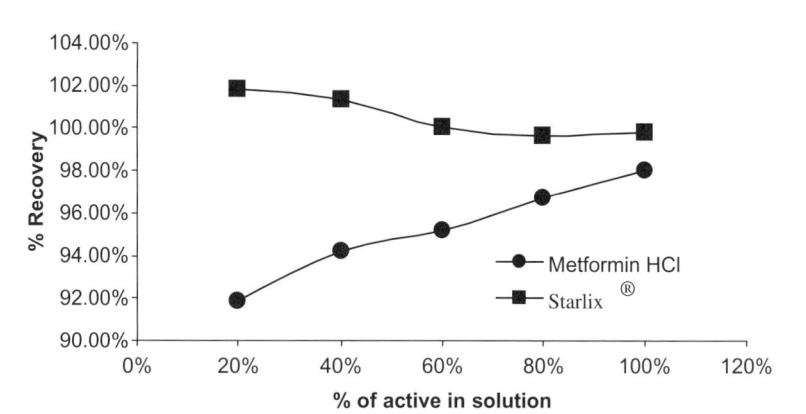

Figure 15-8. Percent recovery of Metformin HCl and Starlix® in different concentration solutions spiked with placebo.

TABLE 15-5. Recovery of Metformin HCl with Different Sample Extraction Solvents

Metformin in Solution (%)	Sample Extraction Solvent	Recovery (%)
20	Water/acetonitrile (50:50)	91.9
20	Water/methanol (50:50)	93.9
20	Water/ACN/10 mM OSAS,[a] pH 2.5	92.1
20	Water/ACN/10 mM SDS[b]	91.9
20	Water/ACN/10 mM OSAS/phosphate, pH 6.8	92.4
20	Water/ACN/20 mM OSAS/phosphate, pH 6.8	93.7
20	Water/ACN/triethylamine	93.7

[a]OSAS, octane sulfonic acid sodium salt monohydrate.
[b]SDS, sodium dodecyl sulfate.

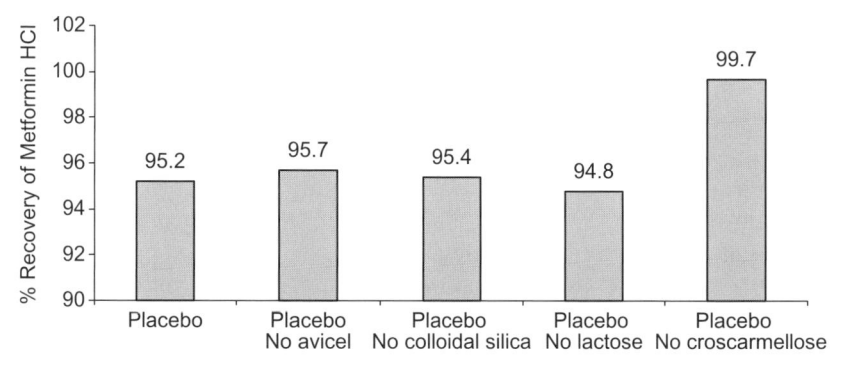

Figure 15-9. Percent recovery of Metformin HCl in the presence of different placebo blends.

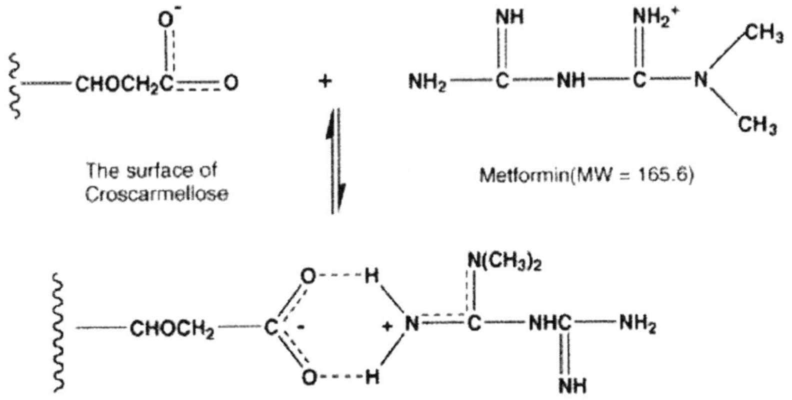

Figure 15-10. Functional group interaction scheme between croscarmellose and Metformin HCl.

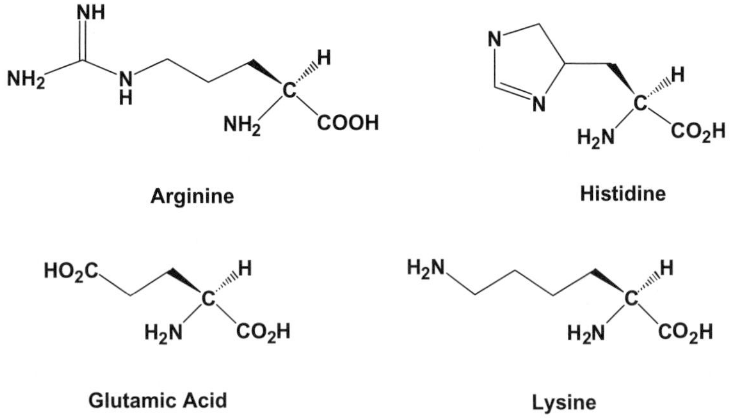

Arginine

Histidine

Glutamic Acid

Lysine

Figure 15-11. Structures of four (arginine, histidine, glutamic acid, and lysine) amino acids.

interacting with croscarmellose in solution. Figure 15-10 shows the proposed interaction between the excipient and Metformin HCl. Since croscarmellose in the dosage form cannot be avoided, the sample preparation procedure must be modified to increase the recovery of Metformin HCl in solution in the presence of croscarmellose. What if another compound with an amino functionality was added in the solution which can interact with croscarmellose rather than Metformin HCl? Few amino acids given in Figure 15-11 were evaluated

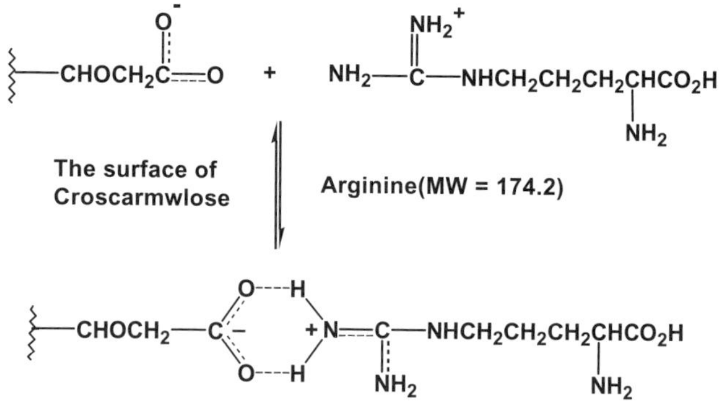

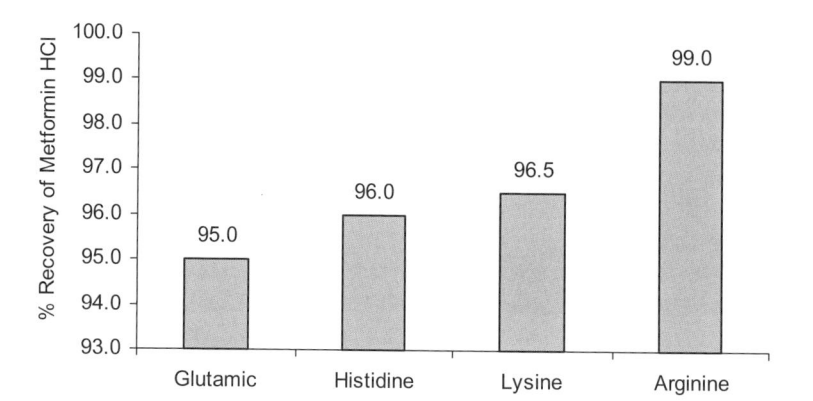

Figure 15-12. Functional group interaction scheme between croscarmellose and arginine.

Figure 15-13. Percent recovery of Metformin HCl in the presence of different amino acids.

for this purpose. Based on the observed data, as well as the structure, it is very obvious that arginine would have a greater probability of interaction with croscarmellose (refer to Figure 15-12). Thus, higher recoveries of Metformin HCl are observed (refer to Figure 15-13). Finally, the recovery experiment with and without arginine in solution was repeated at 20% and 100% Metformin HCl in solution to confirm that arginine inhibits the interaction of Metformin HCl with croscarmellose because arginine is believed to be competing for the

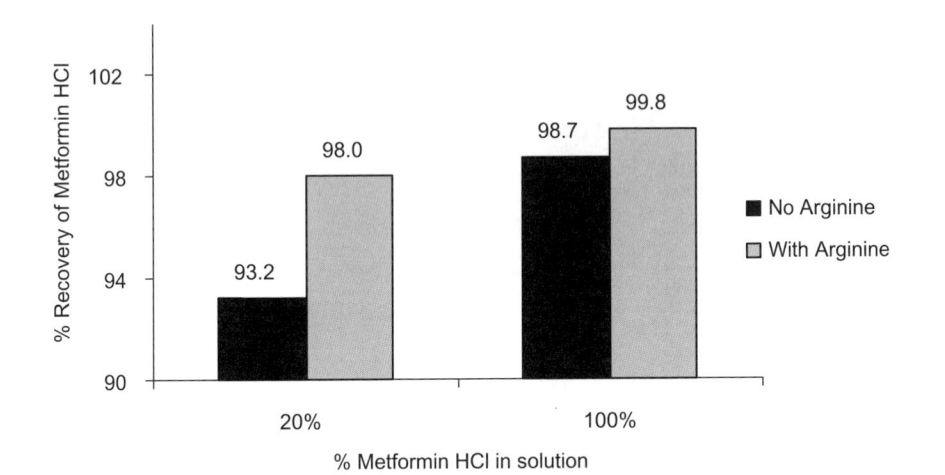

Figure 15-14. Percent recovery of Metformin HCl in the presence of arginine and without arginine at different concentrations of Metformin HCl in solution.

same functional group on the croscarmellose. Refer to Figure 15-14, which shows that arginine does play an important role in inhibiting the interaction of croscarmellose with Metformin HCl. This case study shows that the excipients not only play a role in solid-state chemistry but also may play a major role in solution chemistry and can aid in sample preparation of solid dosage forms.

15.9.2 Case Study 2

A second case study [29] highlighting the mass balance issue during formulation development is summarized here. Background information gained during the development is as follows. (1) Preformulation studies were conducted to identify excipients that are compatible with the drug. These studies included binary and ternary mixtures. (2) Capsule and tablet formulations were selected based on the compatibility study results, utilizing commonly used excipients— that is, binders, fillers, disintegrants, and lubricants. (3) A wet granulation process was utilized for both capsule and tablet formulations. However, in the case of tablet formulations, a direct compression process was also utilized for some selected formulations. (4) An RP-HPLC method was developed for analysis of drug product. The extraction medium consisted of acetonitrile and water (40/60, v/v) and was selected to ensure high solubility, disintegration, and extractability. (5) The analytical method was demonstrated to be accurate,

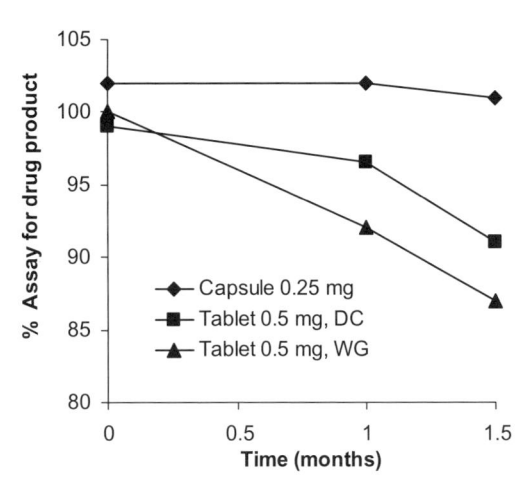

Figure 15-15. Percent assay of drug product (different dosage strengths) over 1.5 months of storage at 40°C/75% RH condition. (DC = Direct Compression, WG = wet granulation).

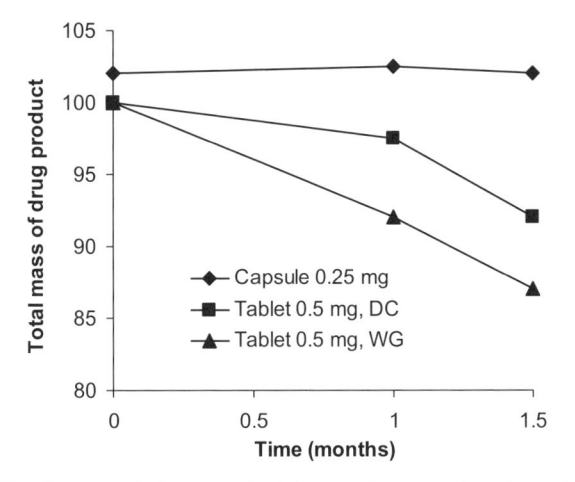

Figure 15-16. Total mass of drug product (percent assay plus degradation products) over 1.5 months of storage at 40°C/75% RH condition.

selective, specific, linear, and quantitative (excellent recovery of active from LOQ level to 130% of nominal value).

The API under development went through preformulation and initial clinical service dosage form as a capsule without significant loss of assay value (refer to Figure 15-15) and showed good mass balance (refer to Figure 15-16) during stability studies at the 40°C/75% RH condition. The only packaging material studied for all dosage forms was HDPE bottles with aluminum induction seal and child-resistance cap. Based on the data presented in Figures

15-15 and 15-16, it can be observed that the capsule dosage form has no assay loss and excellent mass balance while the tablet dosage form (direct compression or wet granulation) has significant assay loss and poor mass balance. The important question here is, Why is a capsule more stable than the tablet dosage form under identical storage condition and packaging material?

Investigation to answer this very important question started with the following review or evaluation: (1) drug substance solution and solid-state stability data, (2) analytical method for assay and degradation products, (3) the effect of temperature and humidity, (4) the effect of dosage form (capsule versus tablet), and (5) the effect of packaging material. The drug substance solution stability data reveals that the active in solution is most stable at acidic pH and becomes unstable as the pH increases. The stability of solid state drug substance (maleate salt) showed no loss of active at 40°C and 50°C for 4 weeks under dry and 75% RH condition ("solid as a rock"). The analytical extraction procedure was challenged with different solvents (different solvent strength, acidic, basic, etc.) and an increase in extraction time (from 30 minutes to 2 hours), and the solution was passed through different membrane filters and checked with centrifuged solution to confirm if any absorption occurred and no absorption was determined. Next, the detection method was challenged where PDA was utilized to scan from 200 to 400 nm (no other peaks were detected at any wavelength), along with utilization of LC/MS (ESI or APCI) (no additional impurities were observed by MS). Another question that arose was whether if all compounds were eluting off the column. Higher strength of organic solvent with isocratic hold for two hours as well as pure organic solvent for one hour in the mobile phase were investigated using PDA detection, and no additional peaks eluted off the column. In addition, radiolabeled study was performed to confirm that the loss is not due to the adsorption on to the column. During all of these studies, the original stability study (40C/75% RH and 50C/75% RH in HDPE bottles) with capsules and tablet dosage form was repeated. The repeated study confirmed the original findings of no mass balance issue for capsule dosage form while the tablets showed significant mass balance issues. Again, where is the API going? Why is capsule more stable than tablet, and why there is no missing mass for capsule?

The next study was conducted in a fashion identical to the initial study in HDPE bottles, except the tablet was placed inside the capsule (over-encapsulated tablet) with the tablets as a control experiment. Sure enough, the tablets that were not over-capsulated showed a mass balance issue, while the same batch of tablets inside a capsule showed no mass balance issue. This is very interesting, since the capsule is preventing the loss of assay value. The next obvious experiment was to study the packaging material. The experiment was set up such that tablets were stored in an amber glass bottle and HDPE induction seal bottles for 6 weeks at 40°C/75% RH. In addition, over-encapsulated tablets were stored in HDPE induction seal bottles for 6 weeks at 40°C/75% RH. Results are shown in Figure 15-17. It can be concluded from the data presented in Figure 15-17 that the active was possibly adsorbed to the

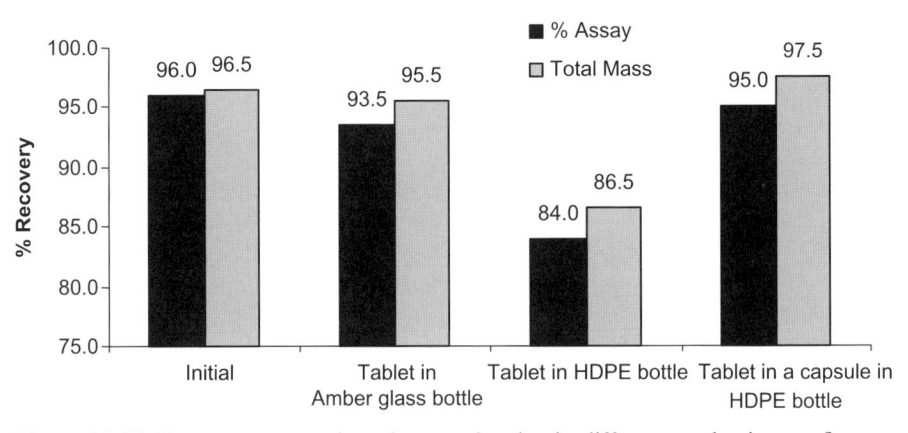

Figure 15-17. Percent assay and total mass of active in different packaging configuration 1.5 months when stored at 40°C/75% RH condition.

HDPE bottle. To confirm this finding, a solution of active was placed in two amber induction scaled glass bottles (one of the bottles contained cut pieces from HDPE bottle) and stored at 60°C for 4 weeks. The assay amount for the active in glass bottle (without cut HDPE pieces) was found to be excellent (no loss of assay), while in the glass bottle (with cut HDPE pieces) the assay value was found to be only about 40% (60% loss of assay). However, the degradation products were found to be only about 2%. In summary, the active is presumed to be interacting with HDPE material. The questions that arise are how does an active travel form a tablet dosage form to the HDPE bottle or is it being lost in headspace (volatility)? To show that the active is being lost due to volatility, an experiment was carried out where the blend was stored in an open glass volumetric flask and in an open glass Petri dish, and both of these blends were analyzed after few weeks when stored at 50°C/75% RH. The results showed little loss of assay value from a volumetric flask, while the open Petri dish showed significant loss of assay value. The interpretation of these results was that the volatile compound is heavier than air. This would prevent the volatile material to escape from the volumetric flask with narrow neck, but would easily escape from open Petri dish.

To further prove the hypothesis that the active is volatile a purge and trap setup was put together as illustrated in Figure 15-18. The results are tabulated in Table 15-6. The results confirmed that the active is volatile and is migrating to the plastic container to be held up in the container during stability studies. It was also confirmed that the volatile substance was trapped in Tenax™ (shown in Figure 15-18). Additional studies were conducted, and it was determined that the salt was being converted to free base and the free base is volatile! The solution to the problem was to formulate the active in a microenvironment, where it would remain as a salt in the solid state (at a certain pH);

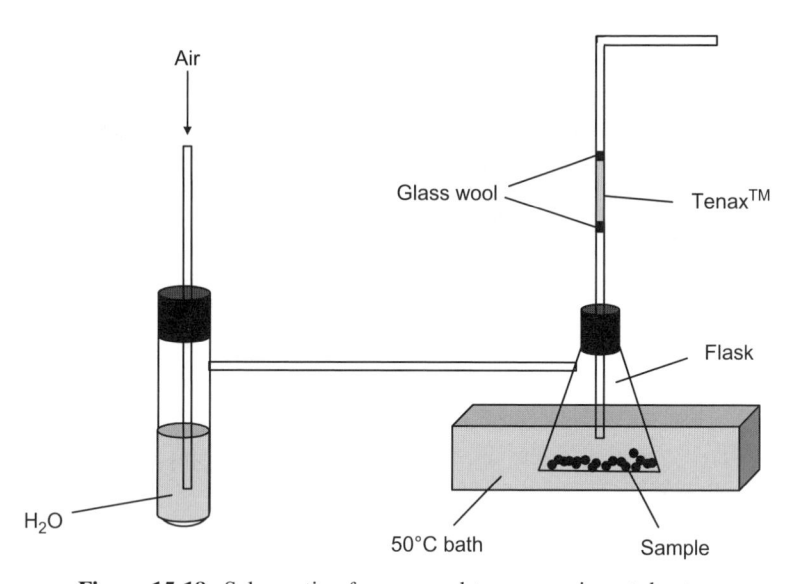

Figure 15-18. Schematic of purge and trap experimental setup.

TABLE 15-6. Purge and Trap Results

	Initial	Flask		Trap		%Total
Sample	Assay	%Assay	%Impurity	%Assay	%Impurity	Mass
Active (Salt)	100	99.8	0.1	0.03	0	99.9
Placebo	None detected					
Formulation Blend with Active (Salt)	97.9	94.9	1	0.6	0.03	96.5

thus the salt will not be converted to free base, which avoids the loss of assay in the packaging, leading to a very stable drug product.

15.9.3   Detection Considerations

Detection could be an issue if degradation products are formed, but are not detected by the detection technique. An example could be that an API having a strong UV chromophore could form a degradation product that has a weak UV chromophore at the specified wavelength. If the degradation product is present at very low levels it may not be detected if UV detection at that wavelength is employed. If a degradation product is detected, then the relative response factor could be very different (by a factor of 2 or more) and the degradation product could be underestimated. A correction factor should then be applied to normalize the level of the degradation product versus the API.

Additional detection techniques that can be employed to help solve mass balance issues with RP-HPLC are MS [30], chemiluminescent nitrogen-specific detector [31], evaporative light-scattering detector, ELSD [32], and corona charged aerosol detection [CAD] [33].

15.9.4 Mass Balance Concluding Remarks

Some degradation products are either very polar or very nonpolar in nature; this may present an issue in chromatography, where they may not be retained or strongly adsorbed on the column, respectively. One may also want to double check the reference standard for its purity, moisture content, and/or salt/acid–base ratio for calculation. An analytical chemist must remember to explore all possibilities if mass balance issue is observed (either during method development, validation and/or stability testing).

15.10 SUMMARY OF ASSAY AND RELATED SUBSTANCES

The following four steps have been proposed by Bakshi and Singh for a development of a SIM [34]: (1) critical study of an API to assess the likely decomposition route(s), (2) collection of information on physicochemical properties, (3) stress (forced decomposition) studies, and (4) preliminary separation studies on stressed samples.

However, a SIM during formulation development involves more than the above four steps, since the formulation development is a dynamic process where the formulation is optimized as the clinical program moves further along the development process. To continue the list from above, the following three steps must be added to the list: (5) compatibility studies with excipients, (6) stability trending (variables would be temperature, humidity, and packaging material), and (7) mass balance for assay. An analytical chemist must revisit the separation of all components in the related substances method after steps 3, 5, and 6 to ensure that the test method is truly a SIM.

In addition, HPLC can be employed as a reference method for unknown degradation products where an unknown degradation product can be identified based on its the retention behavior in RP-HPLC and the corresponding relative retention time is determined (retention of degradation product/retention of API).

15.11 UNIFORMITY OF DOSAGE UNITS

The term uniformity of dosage units (UDUs) is defined as the degree of uniformity in the amount of the active substance among dosage units [35]. The UDUs can be demonstrated by either of two methods: content uniformity (CU) or mass variation (MV). Mass variation can be utilized for drug

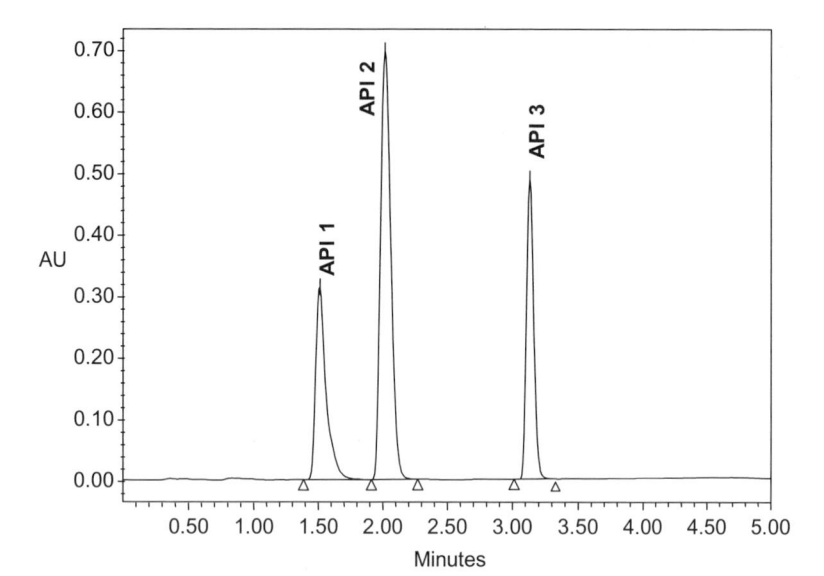

Figure 15-19. Chromatogram showing a separation of three APIs for a CU method by RP-HPLC for a combination product.

products containing 25% (w/w) active in a dosage form and has a label claim of 25 mg or higher.

The HPLC method (if a separation of two components is required) should be very simple because it is for the measurement of the API only. However, during the HPLC method there should be no interference at the retention time of the API from any placebo, capsule, or a sample solvent solution.

Since the HPLC method for CU must be fast, use of DryLab® or other commercially available software is recommended to develop a short run-time method. An example chromatogram in Figure 15-19 illustrates a fast LC run for a combination drug product containing three APIs. In addition, the use of HPLCs that can operate at higher pressures (up to 9000–15,000 psi) could also be used. More details on fast HPLC can be found in Chapter 17.

15.12 BLEND UNIFORMITY (BU)

The objective of blend analysis is to determine whether the blend is uniformly mixed or not (very similar to CU, except BU is for powders and CU is for dosage units). What one needs is quick answers during blend analysis, so fast and simple methods are preferable. There is no need to develop a new HPLC method for blend analysis, since the CU method can easily be employed here.

The process that explains how blend uniformity is performed is not the scope of this chapter. In summary, there are two ways to perform blend uniformity. Most common practice involves a thief that is utilized to sample aliquots of blend powder from top, middle, and bottom of a blender. The size of an opening in a thief is adjusted based on powder density to ensure that about one to three dosage units of powder is sampled. Readers interested in this topic are referred to references on the guidance document from the FDA [36] and the challenges of sampling and measuring blending uniformity samples [37, 38]. The second way to measure blend uniformity is to perform on-line measurement with NIR, which is part of the Process Analytical Technology (PAT) paradigm.

15.13 CLEANING VERIFICATION

The manufacturer must demonstrate that the cleaning procedure used will consistently minimize the probability of product adulteration which ensures the quality, purity, and safety of a given drug product. A number of possible adulteration sources are previously manufactured products, cleaning agents, solvents, and the water utilized during the cleaning cycle of the equipment. As defined by Harder [39], a validated cleaning process is

> A procedure whose effectiveness has been proven by a documented program providing a high degree of assurance that a specific cleaning procedure, when performed appropriately, will consistently clean a particular piece of equipment to a predetermined level of cleanliness.

There are two types of analysis needed for cleaning verification: (1) active (2) soap. Since residual amounts of active and soap are to be determined, the methods need to be very sensitive. For measuring the active, the assay or the content uniformity method can be employed (if the sensitivity of the existing method is acceptable). If the sensitivity of the current assay or content uniformity method is not acceptable, then modifications can be made to an existing method or a more sensitive test method is developed.

Inject a higher amount of API on the column.

Choose a wavelength near 200 nm to increase the signal.

Utilize micro-HPLC to obtain a sharp peak (increase S/N).

Utilize nonporous silica.

If the sensitivity is still not sufficient, then a different detection techniques or analytical technique may need to be employed (for example, LC-MS and fluorescence detection, capillary electrophoresis, ion mobility spectrometry, etc.). These alternate methods and/or detection techniques may have higher sensitivity than HPLC with UV detection. Since the new method will be a simple method (peaks from API(s) and/or functional excipients), development of these methods should be simple and fast. MS is generally more sensitive than UV when employed to monitor single mass unit (single ion monitoring). Details of cleaning verification are found in references 11–13 and 40–42.

15.14 EXTRACTABLES/LEACHABLES

What is the difference between Extractables and Leachables? The following definitions are taken directly from PQRI [43].

> Leachables in orally inhaled and nasal drug products (OINDP) are compounds which are present in the drug product due to leaching from container closure system components. Extractables are compounds that can be extracted from OINDP device components, or surfaces of the OINDP container closure system when in presence of an appropriate solvent(s) and/or condition(s). Leachables are often a subset of, or are derived directly or indirectly from, extractables. Extractables may, therefore, be considered as potential leachables in OINDPs. Some leachables may affect product quality and/or present potential safety risks, therefore regulatory guidance has provided some recommendations regarding the analysis and toxicological safety assessment (i.e., qualification) of such compounds.

Depending upon the dosage form, one or both (leachables/extractables) need to be evaluated. HPLC method development for these is quite challenging. There are many different extractables and/or leachables that are known from packaging components. Each company has its own way of dealing with this topic. In order to be more efficient, it is recommended that a company, in general, decide on what packaging components are going to go to be used for clinical trials and/or marketplace for all of their drug products. Once this is determined, it is possible to list all possible extractables/leachables from the known packaging components and try to separate only these compounds. Once the known extractables/leachables are separated, the final method would be a template method for these compounds for products in development and all future products as well. However, this is only the beginning because for every API, the method will need to be evaluated to confirm that the API and its known degradation products are resolved from all extractables/leachables eluting in the template method. If not, then the method conditions need to be adjusted to obtain adequate resolution of API and its synthetic by-products and the extractable/leachable peaks. Once this is performed, the method would need to be validated for the API. This topic is a major concern for a liquid product (suspensions and solutions) where the concentration of active in solution is very high (for low volume/high concentration injections or drinking solution) in which the liquid "as is" is injected into the HPLC system to determine the extractables and/or leachables. An example is given below for a liquid drug product.

An injectable drug product under development was promoted to begin Phase I clinical trials in which a set of catheters, needles, tubing, and a solution for dilution were identified by the clinical centers. The information on all of these components were gathered by the packaging experts, and few sets of these were shipped to analytical laboratories for a compatibility testing of the liquid drug product to determine leachables and to determine if their was any absorption of the API on the different infusion sets. Two major questions needed to be answered: (1) Which leachables are leached out of this particular set? and (2) How much of each leachable is being leached out? There was an HPLC method already developed which can separate the following nine leachables/extractables: (1) formylpiperdine, (2) diethylphthalate, (3) dibutylphthalate, (4) bis(3-*tert*-butyl-6-hydroxy-methylphenyl)-methane, (5) dioctylphthalate, (6) pentaerytritol-tetrakis, (7) trimethyl-2,4,6-tris(3,5-di-*tert*-butyl-4-hydroxybenzyl)-benzene(1,3,5-), (8) octadecyl-3-(3,5-di-*tert*-butyl-4-hydroxyphenyl)-propionate, (9) tris(2,4-*tert*-butyl-phenyl)-phosphite. The method consists of a C18 column with premixed organic–aqueous mobile phase (mobile phase A consists of low organic and high aqueous, and mobile phase B consists of high organic and low aqueous). The aqueous portion is adjusted to acidic pH before mixing it with organic portion. It is a linear gradient method, and detection is kept at 254 nm. The chromatogram in Figure 15-20 shows the separation of all nine components (peaks 1 to 9 in the same order as the listed components above) in 55 minutes. It also shows a blank injection of sample solvent to ensure that there is no interference from the

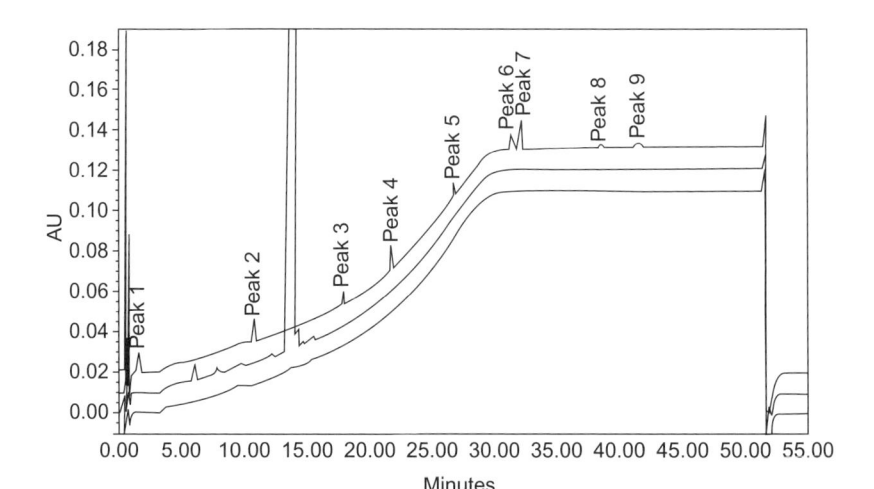

Figure 15-20. An example of chromatogram showing blank, reference solution containing nine known leachables and a sample solution containing the API. Peak numbers 1 to 9 are listed in the text.

sample solvent at the retention time of any of the nine known peaks. The existing method is checked by injecting a neat API solution that has not passed through the set of tubing and cathedar. A liquid drug product was injected to confirm that the peaks from the neat solution (as a control) are not interfering at the retention times of nine known leachables peaks. If there are any peaks from the drug product solution that interfere at the retention time of known leachables, then the method will need to be modified until there is a baseline separation of all peaks in question. In Figure 15-20, it is clear that all peaks from the drug product are resolved from the known nine leachables peaks. Once it is confirmed that the separation of known leachables and the drug product is satisfactory, then the method will need to validated for recovery, linearity, precision, LOD, and LOQ. The specificity was addressed already because there was no interference from the peaks of drug products at the retention time of known leachables.

The method validation in this case is straightforward where known amounts of leachables are spiked (LOQ to acceptable or specification level) into the drug product. The acceptable or specification level is dependent on the type of leachable and the total daily intake defined by the product dosing. Once they are mixed (hand shaking, mechanical shaking, and/or sonication), injection of the solution is made to calculate the amounts based on external reference standard. The recovery experiment must be performed for all known leachables in the presence each other and in the presence of all excipients from the drug product. The recovery, precision, and linearity experiments can be combined into one experiment. The spiked recovery at different (minimum five) concentrations can be used to validate the method in terms of recovery (spiked versus found), precision (%RSD of found values), and linearity (plot of found versus spiked and perform regression analysis). The LOD and LOQ is addressed in the same manner as the degradation product method by HPLC, except it is based on only one leachable. For analysis of leachables from a liquid drug product that has passed through the set of tubing, catheter, and so on, a reference standard is prepared made up of only one leachable (identified as the major leachable) and all other leachables are calculated (one of many ways to calculate them) based on this major leachable (external reference standard method). The analysis must be performed for a control drug product that has not passed through the set of tubing and catheter as well as at timed intervals where a drug product is collected after it has passed through the system for time t. In addition, the same experiment must be performed for a placebo to confirm whether a leachable is due to the placebo solution or due to the active present in the solution. The entire procedure could be explained in the Investigator's Brochure (IB) to confirm compliance from all clinical centers as well as to identify a common procedure for administrating the drug to the patients. For an analytical chemist, the above example is one of many examples where the final outcome of an experiment is seen in action in the real world!

15.15 DISSOLUTION

HPLC methods are preferred if excipients would interfere, if nonspecific detection techniques (mainly in UV) would be used, or when multiple APIs (combination product) are present in a drug product. Since dissolution sample set analysis can be very long due to six samples per bath as well as multiple time points for profile testing, fast run times are preferred to quickly determine the results. If a fast HPLC method for CU is available, then the identical HPLC method can be utilized for dissolution analysis.

15.16 METHOD DEVELOPMENT

HPLC method development has already been covered in Chapter 8. The focus of this chapter is to utilize HPLC and the data generated by this technique to help in developing a robust formulation for a drug product. However, the current section will discuss sample preparation solvent since it becomes an integral part of HPLC when we are discussing HPLC methods for a particular formulation. Any sample preparation solvent that is chosen for any HPLC method must be compatible with the HPLC solvents utilized for that particular test method. The current section will assume that a new molecular entity (NME) is utilized for a particular drug product.

The pros and cons of utilizing the information in hand for method development during formulation development are as follows for any HPLC method. During the formulation development stage, much of the physical characterization of an API has already been performed and is known. Many of the HPLC methods (identification, related substances, etc.) are also known for an API. The caveat of utilizing the identical test methods from a drug substance for a drug product is that the drug product has excipients that are not present when an API is analyzed. The groundwork has already been performed, but it is not as easy as it sounds to develop SIM methods for a drug product. Some examples are stated below to highlight this point: (1) Drug product has more than one API. (2) Drug product has one or more APIs and/or excipients that interfere in some of the detection methods (for example, more than one component absorbs in the same region of UV). (3) Combination product has an optimal wavelength of detection that is different for both compounds. (4) API forms degradation products when it interacts with excipients. In addition, it is interesting to note that for multiple APIs, how does one quantitate the unidentified peaks in the chromatogram, since it is not known from which API does it originates? The best possible answer would be to overestimate the unidentified peak by calculating it against the API (using 1% standard and assuming relative response factor of 1 for the unidentified peak), which gives the lowest response factor at that wavelength. Further solid-state stability studies could be performed by putting placebo, placebo with API 1, placebo with API 2, API 1, API 2, and combined APIs (no placebo) all alongside the formulation to discern which degradation products are generated from which API. Sometimes one API induces the degradation of the other API, and other times the two APIs may react in the solid state and the degradation products may contain structural features of both. Also, mass spectrometry may be used to discern which impurity is related to a particular API. For example, if API 1 has a chlorine group (isotope of Cl^{35} and Cl^{37}, 3:1 ratio) and API 2 does not and the degradation product does contain a chlorine, then the degradation product could potentially be related to API 1.

15.16.1 Sample Preparation Solvent

As much as HPLC method is important, sample preparation is as important during formulation development. Why is sample preparation solvent so important during formulation development? The main reason is that the chosen sample preparation solvent must be able to extract and dissolve the active and all degradation products from the sample matrix. The sample matrix is the actual dosage form (capsule, tablet, suspension, liquid solution, patch, etc.). Do remember that the analysis at this stage (during formulation development) is performed on the entire dosage form and not on an API alone. In addition, the sample preparation solvent must be suitable for chosen HPLC analysis. For RP, the sample solvent should be aqueous based with methanol and/or acetonitrile. For normal phase, the sample solvent should contain a nonpolar-solvent and some amount of a polar modifier (water must be avoided). To determine which sample preparation solvent and method of extraction will be utilized for a particular formulation depends on properties of drug substance and excipients as well as known chemistry of these ingredients in the solution.

15.17 METHOD VALIDATION

During formulation development, all experiments are considered GMP experiments. Therefore, all test methods that are utilized to determine any results must be developed for the intended purpose and they must be validated according to regulatory requirements. Method validation is covered in Chapter 9, so it will not be repeated here except for completeness of extraction that is specific to drug products.

15.17.1 Completeness of Extraction

For drug products containing constituents that are insoluble in the extraction medium used in the analytical procedure, it may be deemed adequate to perform a separate test for completeness of extraction (in addition to recovery experiments as described above). The completeness of extraction can be evaluated two ways: kinetically (over some elapsed time t) and thermodynamically (change in volume).

For kinetic experiments, when the regular recovery experiment is completed (when using a real drug product, not from spiked experiments), the solution that is left over is set aside for time t (usually 24 hours). After time t, the solution is re-shaken by hand and then re-injected to determine assay value.

For a thermodynamic experiment, it is necessary to perform the regular recovery experiment using a real drug product with more volume of sample solvent. For example, if the procedure is stated to extract the content of a drug product with 50 mL of solvent, then use 75 mL and/or 100 mL for this experiment.

The results from these two experiments (kinetic and thermodynamic) will show whether the regular extraction procedure is complete or not. Most likely, for modified-release drug products, time is essential (higher recovery over time, but watch out for solution stability!). The change in volume will have an impact if the solubility of an API is on the border of the solubility limit in that particular sample preparation solvent (in the presence of excipients). If the latter is the case, then the procedure should be modified to extract with higher volume of sample preparation solvent and/or change the pH or composition of the solvent.

Lastly, to really prove that the sample preparation procedure or the recovery procedure is completely extracting your API and degradation products, utilize homogenization (refer to section on automation) instead of mechanical shaking and/or sonication. This should be done only for small molecules. For proteins and peptides, this may lead to the generation of aggregates.

15.18   TESTING OF SAMPLES

Once test methods are developed and validated, test methods are utilized to test samples (clinical release and stability).

15.18.1   Clinical Release

During clinical phases of a development project, there will be few batches that will need a clinical release status, so they can be utilized in clinical trials in humans. For a clinical release, all test methods need to be written by an analytical chemist and approved by a quality assurance group. All test methods that are utilized to test a clinical batch needs method validation as described in Chapter 9. Everything at this stage is performed by following cGMPs.

15.18.2   Stability

There are three types of stability testing (technical, clinical, and registration) performed during formulation development. Requirements of a stability program increases going from technical to clinical to registration. The major requirements for each of these are summarized in Table 15-7. In addition, the

TABLE 15-7. Requirements of Stability Programs

Technical	Clinical	Registration
Accelerated (3 months)	Accelerated (6 months)	Accelerated (6 months) Japan specifics (if filing globally) (1 month open and closed) at 50°C (dry and 75% RH).
Long term (12 months)	Multiple long term depending upon climate zones (at least 24 months) 5°C and/or −20°C (6 months)	Multiple long term depending upon climate zones (at least 24 months) 5°C and/or −20°C (6 months)
One packaging (HDPE bottle with induction seal)	Multiple packaging	Multiple packaging (multiple count bottles, and multiple material blisters) Consumption test (for bottles) Light test (1200 K lux hours) Micro-testing at annual time points (to test the entire packaging operation and to track the changes over time)
Defines storage period and conditions for the first clinical batch after a major change in formulation or for a development of a new dosage form using the same API.	Confirms storage period and conditions for all clinical batches.	Storage period, packaging material, and storage conditions for a marketed product are defined for the final label.

reader is referred to available guidance documents from the FDA and ICH [44–48] for further reading of this topic.

In summary, data generated during the length of stability studies during formulation development should be able to (1) establish appropriate drug product storage and transport conditions, (2) determine clinical re-test periods, (3) assign shelf lives for marketed drug products, and (4) select an optimal package configuration for each of the worldwide climatic zones [47].

15.18.2.1 Stability Data Trending. Stability testing provides information about the quality of a drug product under the influence of a variety of environmental factors such as temperature, humidity, and light varying with time and packaging material. All analytical chemists must be able to evaluate trends in the data that they generate. One of the major parts of an IND and/or NDA filing is defining packaging and assigning expiration or re-test period for a given product. All of this information is coming directly from analytical chemists.

15.18.2.2 Stability Case Study. The HPLC stability-indicating method (assay and related substances) is used to determine the stability of the product; if all acceptance criteria are met for particular storage condition and time point, then the data can be used to justify the shelf life. Note that the drug product is stored at the intended storage condition and an accelerated storage condition at a minimum. The product must be stable at accelerated condition (+15°C of intended storage condition) in order to get four times the shelf life at proposed storage condition. Usually, accelerated stability samples are generated up to and including 6 months; and if all acceptance criteria are met in the specifications, then the product would be deemed stable for at least 2 years at the intended storage condition. The 2-year time point at the recommended storage condition would also have to be tested to confirm this. The drug product X was stored at 5°C (Figure 15-21), 25°C/60% RH (Figure 15-22), and 40°C/75% RH (Figure 15-23). If 5°C would be the intended storage condition, then 25°C/60% RH would be the accelerated condition. On the other hand, if 25°C/60% RH was the intended storage condition, then 40°C/75% RH would be the accelerated condition. It was observed that after 3 months at each storage condition, the synthetic by-products eluting at RRT 2.1–2.2 started to decrease and new degradation products were increasing (RRT 0.94, RRT 1.59), with the greatest degradation at 40°C/75% RH. This degradation process was accelerated at 40°C/75% RH. LC-MS analysis confirmed that the synthetic by-products at RRT 2.1–2.2 were homo- and heterodimers of the API. The degradation products at RRT 0.94 and 1.59 were subunits of these dimers and

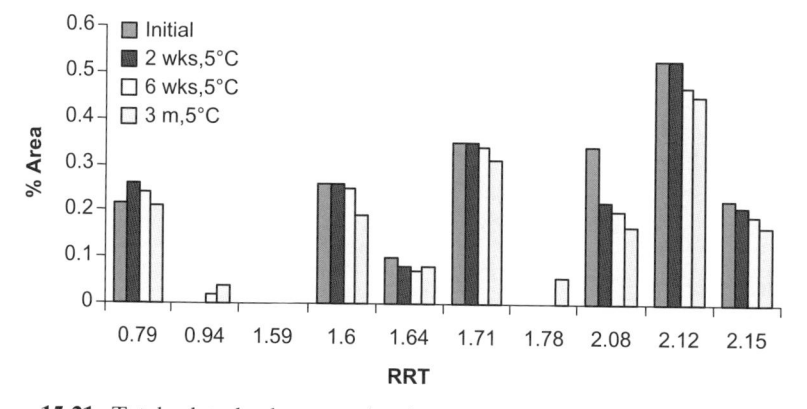

Figure 15-21. Total related substances for drug product X stored for 2 weeks (wks) at 5°C, 6 weeks (wks) at 5°C, and 3 months (m) at 5°C.

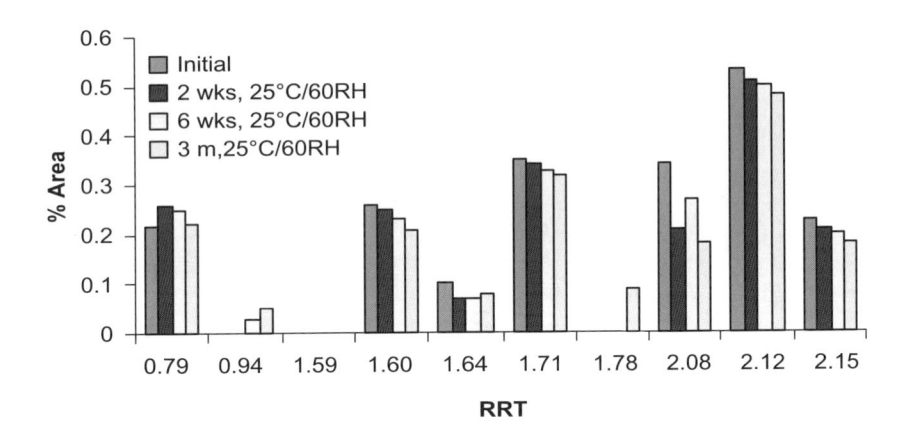

Figure 15-22. Total related substances for drug product X stored at 2 weeks (wks) 25°C/60% RH, 6 weeks (wks) 25°C/60% RH, and 3 months (m) 25°C/60% RH.

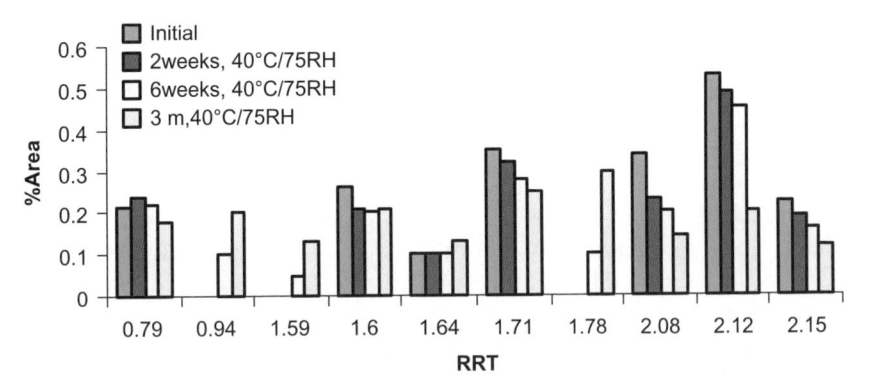

Figure 15-23. Total related substances for drug product X stored at 2 weeks (wks) 45°C/75% RH, 6 weeks (wks) 45°C/75% RH, and 3 months (m) 45°C/75% RH.

were confirmed by MS analysis. The API was deemed not to be stable at 40°C/75% RH for 3 months. Therefore, at this elevated temperature the combination of the excipients in the formulation had led to catalyzation of the cleavage of the bonds in the dimers to generate the more early eluting (less hydrophobic) species (subunits of the dimers).

15.19 AUTOMATION OPPORTUNITIES

Laboratory automation is widely used in the pharmaceutical industry to support drug discovery, nonclinical and clinical drug development, and quality control. Analytical applications for which automation is commonly utilized during formulation development include HPLC, dissolution, content uniformity, and tablet/capsule assay testing for release, technical, clinical, and registration stability analysis. Laboratory automation is used for preparation, introduction, analysis of samples, sample and reporting of results.

HPLC itself is now considered to be a fully automated system. Once the solutions are placed in vials and sample sequence is programmed into the software, the auto-injector injects as stated in the sample sequence and it runs as specified in the instrument method. Many types of software, once programmed, also allow integration, processing, and reporting of an entire HPLC run to be performed automatically once the run is completed.

The focus of automation is not solely on HPLC anymore (since it is automated), but also includes the sample preparation procedure for drug products since it involves many labor intensive steps (weighing, addition of sample solvent, extraction procedure, mixing, filtration, additional dilutions, second mixing, etc.). So, if the sample preparation procedures are automated and combined with HPLC, then the entire analysis is said to be fully automated whether it is for assay, CU, blend analysis, or even for dissolution testing. Such automated workstations are currently available from number of vendors (Caliper Life Science, SOTAX Inc.), but their ability to integrate with HPLC or UV spectrophotometer is left to the end users. This open-ended option depends on the end user requirements.

15.20 IMPLEMENTATION OF ALTERNATIVE TECHNOLOGIES

Hyphenated techniques utilizing HPLC plays very important role in drug development process, especially LC-MS. The power of LC combined with an universal detector is an optimal combination to determine related substances and peak homogeneity. Selected articles are referenced for the reader [30, 49, 50].

HPLC is currently playing a major role in the implementation of Process Analytical Technology (PAT) in the pharmaceutical industry. During the development of PAT models, the results obtained through HPLC analysis can be utilized as reference values. Fast and accurate HPLC methods must be developed. Several examples where HPLC is used as a reference method for NIR and Raman analysis are cited [51–53]. HPLC has been historically been an important off-line analytical tool in the pharmaceutical industry laboratory for measuring product quality; however, it can also play a valuable role in on-line process monitoring as well. On-line liquid chromatography can be used for process monitoring, automation, and control and can be considered a PAT application. On-line HPLC has been used for both small- and large-molecule processes for process monitoring to increase process knowledge and as an enabling technology for process automation and control to increase process efficiency and reduce process variability. One particular example was the design and implementation of on-line HPLC analyzers for closed-loop control of chromatographic purification processes in the production of Humulin, a biosynthetic insulin [54].

15.21 CHALLENGES AND FUTURE TRENDS

Challenges for HPLC method development are for combination products where more than one active is present in the same formulation. Since degradation products are of big concern from a safety perspective, HPLC will continue to be utilized as a major separation technique during formulation development to develop the most safe and efficacious formulations to be used for human use. Many different types of bonded phases are currently available for routine HPLC analysis; in addition, very selective and sensitive detection techniques can be integrated with HPLC to help an analytical chemist control the final quality of the drug product. Furthermore, the availability of this technique makes it the first and sometimes the only choice for the analysis of degradation products.

REFERENCES

1. F. A. Carey, *Organic Chemistry*, McGraw-Hill, New York, 1987.
2. K. C. Waterman, R. C. Adami, K. M. Alsante, A. S. Antipas, D. R. Arenson, R. Carrier, J. Hong, M. S. Landis, F. Lombardo, J. C. Shah, E. Shalaev, S. W. Smith, and H. Wang, Hydrolysis in pharmaceutical formulations, *Pharm. Dev. Technol.* **7** (2002), 113–146.
3. H. H. Tonnesen, Formulation and stability testing of photolabile drugs, *Int. J. Pharm.* **225** (2001), 1–14.
4. S. R. Thatcher, R. K. Mansfield, R. B. Miller, C. W. Davis, and S. W. Baertschi, Pharmaceutical photostability, *Pharm. Technol.*, Mar. 2001 (Part I) and Apr. 2001 (Part II).
5. S. W. Hovorka and C. Schoneich, Oxidative degradation of pharmaceuticals: Theory, mechanisms and inhibition, *J. Pharm. Sci.* **90** (2001), 253–269.
6. K. C. Waterman, R. C. Adami, K. M. Alsante, J. Hong, M. S. Landis, F. Lombardo, and C. J. Roberts, Stabilization of pharmaceuticals to oxidative degradation, *Pharm. Dev. Tech.* **7** (2002), 1–32.
7. S. R. Byrn, R. R. Pfeiffer, and J. G. Stowell, *Solid State Chemistry of Drugs*, 2nd edition, SSCI Inc., West Lafayette, IN, 1999.
8. R. C. Rowe, P. J. Sheskey, and P. J. Weller (eds.), *Handbook of Pharmaceutical Excipients*, 4th edition, 2003, published by Pharmaceutical Press (London, UK), and The American Pharmaceutical Association (Washington, DC, USA).
9. G. P. Ellis, The Maillard reaction, *Adv. Carbohydr. Chem.* **14** (1959), 63–134.
10. D. D. Wirth, S. W. Baertschi, R. A. Johnson, S. R. Maple, M. S. Miller, and K. Diana, Maillard reaction of lactose and Fluoxetine Hydrochloride, a secondary amine, *J. Pharm. Sci.* **87** (1998), 31–39.

11. C. Sottani, R. Turci, G. Micoli, M. L. Fiorentino, and C. Minoia, Rapid and sensitive determination of paclitaxel (Taxol®) in environmental samples by high-performance liquid chromatography tandem mass spectrometry, *Rapid Commun. Mass Spectrom.* **14** (2000) 930.

12. P. Sajonz, T. K. Natishan, Y. Wu, N. T. McGachy, and D. DeTora, Development and validation of a sensitive and robust wipe-test method for the detection and quantification of the antibiotic Ertapenem and its primary degradates in a pharmaceutical manufacturing environment, *J. Liq. Chromatogr. Relat. Technol.* **28** (2005), 713–725.

13. R. Munden, R. Everitt, R. Sandor, J. Carroll, and R. DeBono, IMS limit test improves cleaning verification and method development, Pharmaceutical Technology Europe, October 2002. http://www.ptemag.com/pharmtecheurope/

14. J. G. Hardman, L. E. Limbrid, P. B. Molinoff, R. W. Ruddon, A. G. Gilman (eds.), *The Pharmacological Basis of Therapeutics*, 9th edition, McGraw-Hill, New York, 1996.

15. S. Gorog (ed.), *Determination of Impurities in Drugs*, Elsevier Sciences, Amsterdam, 1999.

16. S. Ahuja, *Impurities Evaluation of Pharmaceuticals*, Marcel Dekker, New York, 1998.

17. R. N. Rao and V. Nagaraju, An overview of the recent trends in development of HPLC methods for determination of impurities in drugs, *J. Pharm. Biomed. Anal.* **33** (2003), 335–377.

18. ICH guidelines Q3B(R), Impurities in new drug products (revised), Feb. 2003.

19. FDA Guidance for Industry: Stability testing of drug substances and drug products (draft guidance), Food and Drug Administration, Rockville, MD, June 1998.

20. ICH guidelines QIA(R2) Stability testing of new drug substances and products (second revision), Feb. 2003.

21. D. W. Reynolds, K. L. Facchine, J. F. Mullaney, K. M. Alsante, T. D. Hatajik, and M. G. Motto, Available guidance and best practices for conducting forced degradation studies, *Pharm. Technol.* **Feb.** (2002), 48–56.

22. S. W. Baertschi, A stress testing benchmarking study *Pharm. Technol.* **Feb.** (2003), 60–72.

23. M. Bouma, B. Nuijen, M. T. Jansen, G. Sava, F. Picotti, A. Flaibani, A. Bult, and J. H. Beijnen, Development of a LC method for pharmaceutical quality control of the antimetastatic ruthenium complex NAMI-A, *J. Pharm. Biomed. Anal.* **31** (2003), 215–228.

24. T. Huang, M. E. Garceau, and P. Gao, Liquid chromatographic determination of residual hydrogen peroxide in pharmaceutical excipients using platinum and wired enzyme electrodes, *J. Pharm. Biomed. Anal.* **31** (2003), 1203–1210.

25. A. T. M. Serajuddin, A. B. Thakur, R. N. Ghoshal, M. G. Fakes, S. A. Ranadive, K. R. Morris, S. A. Varia, Selection of solid dosage form composition through drug–excipient compatibility testing, *J. Pharm. Sci.* **88** (1999), 696–704.

26. S. R. Byrn, W. Xu, and A. W. Newman, Chemical reactivity in solid-state pharmaceuticals: Formulation implications, *Adv. Drug Delivery Rev.* **48** (2001), 115–136.

27. M. J. Akers, Excipient—drug interactions in parenteral formulations, *J. Pharm. Sci.* **91** (2002), 2283–2300.

28. W. X. Huang, M. Desai, Q. Tang, R. Yang, R. V. Vivilecchia, and Y. Joshi, Elimination of Metformin–Croscramellose sodium interaction by competition, *Int. J. Pharm.* **311** (2006), 33–39.

29. E. Zannou, Q. Ji, A. Serajuddin, and Y. Joshi, Stabilization of maleate salt of a basic drug in solid dosage form by microenviromental pH adjustment, *Intl. J. Pharmaceutics* (Oct 2006), submitted.

30. Y. Wu, The use of liquid chromatography–mass spectrometry for the identification of drug degradation products in pharmaceutical formulations, *Biomed. Chromatogr.* **14** (2000), 384–396.

31. M. A. Nassbaum, S. W. Baertschi, and P. J. Jansen, Determination of relative UV response factors for HPLC by use of a chemiluminescent nitrogen-specific detector, *J. Pharm. Biomed. Anal.* **27** (2002), 983–993.

32. E. G. Galanakis, N. C. Megoulas, P. Solich, and M. A. Koupparis, Development and validation of a novel LC non-derivatization method for the determination of amikacin in pharmaceuticals based on evaporative light scattering detection, *J. Pharm. Biomed. Anal.* **40** (2006), 1114–1120.

33. P. H. Gamache, R. S. McCarthy, S. M. Freeto, D. J. Asa, M. J. Woodcock, K. Laws, and R. O. Cole, HPLC analysis of nonvolatile analytes using charged aerosol detection, *LC-GC North Am.* Feb. 1, 2005.

34. M. Bakshi and S. Singh, Development of validated stability-indicating assay methods—critical review, *J. Pharm. Biomed. Anal.* **28** (2002), 1011–1040.

35. European Pharmacopeia, Uniformity of dosage units, Chapter 2.9.40, 2005.

36. FDA, Guidance for Industry—Powder blends and finished dosage units stratified, in *Process Dosage Unit Sampling and Assessment*, Center for Drug Evaluation and Research, Rockville, MD, 2003.

37. D. Brone and F. Muzzio, Enhanced mixing in double-cone blenders, *Powder Technol.* **110** (2000), 179–189.

38. R. Hwang and S. Wu, Challenges of blend uniformity testing for tablet formulation, *Am. Pharm. Rev.* **7** (2004), 101–103.

39. S. W. Harder, The validation of cleaning procedures, *Pharm. Technol.* **8** (1984), 29.

40. M. J. Nozal, J. L. Bernal, J. J. Jiménez, M. T. Martín, and F. J. Diez, Development and validation of a liquid chromatographic method for determination of lacidipine residues on surfaces in the manufacture of pharmaceuticals, *J. Chromatogr. A* **1024** (2004), 115–122.

41. R. Klinkenberg, B. Streel, and A. Ceccato, Development and validation of a liquid chromatographic method for the determination of amlodipine residues on manufacturing equipment surfaces, *J. Pharm. Biomed. Anal.* **32** (2003), 345–352.

42. P. Yang, K. Burson, D. Feder, and F. Macdonald, Method development of swab sampling for cleaning validation of a residual active pharmaceutical ingredient, *Pharm. Technol.* **Jan.** (2005), 84–94.

43. Product Quality Research Institute (PQRI), Leachables and extractables working group, Spring 2002.

44. FDA guidance for industry, Stability testing of new drug substances and products, June 1998.

45. ICH Guidelines Q1C, Stability testing for new dosage products, Nov. 1996.

46. ICH Guidelines Q1D, Bracketing and matrixing designs for stability testing of new drug substances and products, Feb. 2002.

47. ICH Guidelines Q1E, Evaluation of stability data, Feb. 2003.

48. USP 29, General Chapter <1150>, Pharmaceutical stability, 2006.

49. C. Atsriku, D. G. Watson, J. N. A. Tettey, M. H. Grant, and G. G. Skellern, Determination of diminazene aceturate in pharmaceutical formulations by HPLC and identification of related substances by LC/MS, *J. Pharm. Biomed. Anal.* **30** (2002), 979–986.

50. A. Abdoh, M. M. Al-Omari, A. A. Badwan, A. M. Y. Jaber, Amlodipine besylate—excipients interactions in solid dosage form, *Pharm. Dev. Technol.* **9** (2004), 15–24.

51. Y.Wang, L. Parthiban, R. LoBrutto, R. Vivilecchia, M. Zheng, B. Reddy, and M. Dryfoos, NIR: Practical implementation in tablet content uniformity measurement, *PAT*, October 2005.

52. M. R. Smith, R. D. Jee, A. C. Moffat, D. R. Rees, and N. W. Broad, A procedure for calibration transfer between near-infrared instruments—a worked example using a transmittance single tablet assay for piroxicam in intact tablets, *Analyst* **129** (2004), 806–816.

53. D. S. Hausman, R. Thomas Cambron, and A. Sakr, Application of Raman spectroscopy for on-line monitoring of low dose blend uniformity, *Int. J. Pharm.* **298** (2005), 80–90.

54. http://www.pharmacychoice.com/News/article.cfm?Article_ID=6781, accessed May 16, 2006.

A15.1   ADDENDUM (COMMON FUNCTIONAL GROUPS)

A functional group is an atom (e.g., F) or group of atoms (e.g., OH) in a molecule that experiences chemical change (e.g., —OH to =O) under a prescribed set of reaction conditions (e.g., oxidation) [1]. The translation for an analytical chemist is that a molecule will be API(s), excipients(s), reagent(s), and solvent(s) and a set of conditions would be heat, light, moisture, and some external factor. An analytical chemist must be able to perform the following paper exercise: (1) Recognize general functional groups (alcohol, ether, nitro, etc.) for all molecules in the formulation (API(s) and excipient(s)). (2) Recognize major subunits (hydrophobic, hydrophilic, electrophile, nucleophile) that are present in pharmaceutical compounds (API(s) and excipient(s)). (3) Recognize acid/base properties, if any, and determine pK_a for each functional group(s). If multiple functional groups are present, then determine or estimate individual pK_a values of each ionizable group.

Search the literature (your company's databases, organic chemistry textbooks, articles on degradation, forced degradation, identification of degradation products, etc.) on how these functional group react (under what conditions and what could be the resulting degradation product). Some of the very important functional groups that may undergo reaction to form unwanted degradation products relevant to pharmaceutical chemists are highlighted below.

A15.1.1 Carbonyls

A15.1.1.1 Carboxylic Acids. Carboxylic acids (R—CO—OH) are an acidic class of compounds that contain only carbon, hydrogen, and oxygen having pK_a values in the range of 2 to 5.5. They are much stronger acids than water and alcohols. Carboxylic acids are weak acids. They tend to act as nucleophiles in the anionic form (R—COO⁻) and more electrophilic when protonated (R—CO—OH). They are not prone to oxidative degradation. Carboxylic acids can be "reduced" to a primary alcohol (R—CO$_2$—H → RCH$_2$OH).

Avoid alcohols (methanol, ethanol, etc.) in the sample preparation solvent due to the following reaction (especially under acidic conditions). This may lead to the formation of an ester degradation product in the diluent.

$$R—CO—OH + R'—OH \leftrightarrow R—CO—OR' + H_2O$$

A15.1.1.2 Lactones. Hydroxy acids (HO—(CH$_2$)$_3$—CO—OH), are compounds that contain both a hydroxyl and a carboxylic acid function within the same molecule, and have the capacity to form cyclic esters called lactones (refer to Figures A15-1 and A15-2). Lactones are subject to general acid- or base-

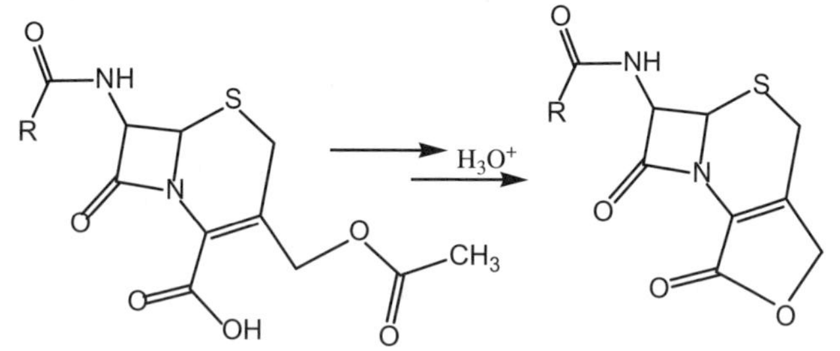

Figure A15-1. Lactone formation for cefuroxime sodium or cephalothin [2].

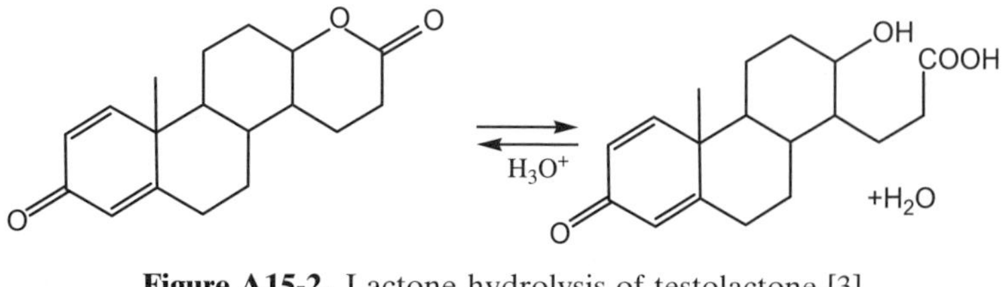

Figure A15-2. Lactone hydrolysis of testolactone [3].

catalyzed hydrolysis (reversible reaction in acid). They are, however, not prone to oxidative degradation. Lactones can be hydrolyzed (with water and acid or base) back to carboxylic acid and alcohol on the same molecule, since it is a reversible reaction. Depending on the conditions of the mobile phase, an on-column degradation may occur, leading to skewed peaks. Careful control of the temperature and the pH mobile phase must be investigated.

A15.1.1.3   Ester Hydrolysis.   Esters can be hydrolyzed (with water and acid or base) back to carboxylic acid and alcohol because it is a reversible reaction. Classic example in the pharmaceutical industry is aspirin hydrolysis to acetic acid and salicylic acid [4]. Molecules having ester functional groups are susceptible to hydrolysis by water, so avoid water in the formulation and processing. The excipients utilized for oral dosage forms are known to contain a certain amount of water, and some of them are known to be hygroscopic. Avoid these excipients in the formulation if ester functionality is present in API(s). If one or more of these excipients cannot be avoided, then control the reaction through the utilization of suitable packaging material (HDPE bottles with an aluminum induction seal containing desiccant would be the best, while triplex blisters would be the worse) and storage conditions (the rate of solid-state reaction could be suppressed at lower temperatures).

For the analysis of esters by RP-HPLC, it is very hard to avoid water because it will be present in the sample extraction solvent and even in the mobile phase. However, the kinetics of ester hydrolysis in solution can be studied by stopped flow injection experiments to determine compatibility with mobile phase. The same experiment must be repeated with different pH (from pH 1.2 to pH 6.8), since it is known that ester hydrolysis is pH-dependent. An example of pH-dependent ester hydrolysis is methylphenidate HCl. Methylphenidate HCl has an ester functional group, and its hydrolysis is pH-dependent. Refer to Table A15-1 for stability of methylphenidate HCl at different pHs [5]. The hydrolysis product of methylphenidate HCl is the free acid (refer to Figure A15-3) [5].

A15.1.1.4   Decarboxylation.   The loss of a molecule of carbon dioxide from a carboxylic acid is known as a decarboxylation reaction ($R—CO_2—H \rightarrow RH + CO_2$). The reaction

TABLE A15-1. Stability of Methylphenidate HCl at Different pHs [5]

pH of Solution	Temperature (°C)	Time (hours)	Methylphenidate Remaining (%)
1.7	100	20	100
3.7	100	20	100
4.9	100	20	84
5.7	100	20	49
8.9	100	20	0

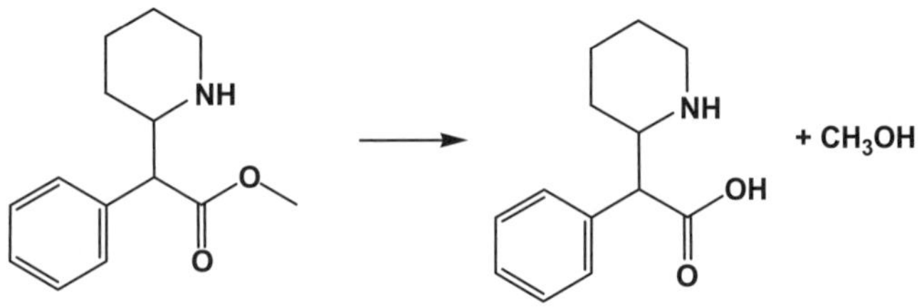

Figure A15-3. Ester hydrolysis of methylphenidate HCl [5].

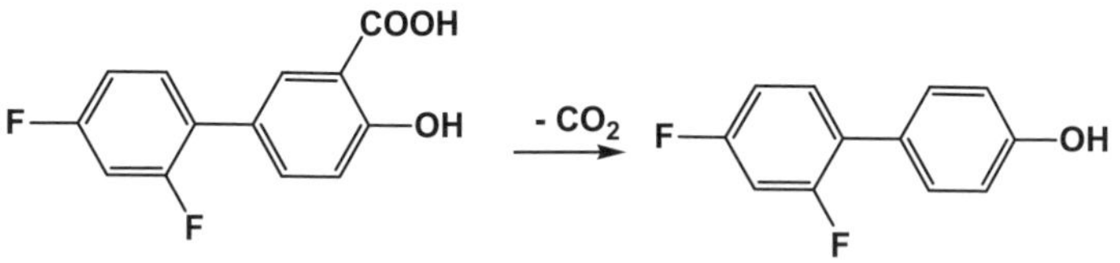

Figure A15-4. Decarboxylation of diflunisal [6].

$$R—CO—CH_2—CO_2H \rightarrow R—CO—CH_3 + CO_2$$

is either acid-catalyzed or thermally-driven. Therefore the use of acidic diluents must be evaluated to see if it has an impact on the solution stability. Also, the presence of this potential degradation product could be present at accelerated stability conditions. An example of decarboxylation in pharmaceuticals is for diflunisal [6]. Refer to Figure A15-4.

A15.1.1.5   Ketone and Aldehyde. Aldehydes and ketones are characterized by the presence of an acyl group (RCO^-) bonded either to hydrogen or to another carbon, respectively. In both cases, carbonyl carbon is electrophilic. A carbon that is beta to carbonyl can be nucleophilic (when deprotonated). Aldehydes tend to be more electrophilic than ketones, in general.

A primary alcohol functional group on a molecule can lead to an aldehyde, if it is oxidized $(RCH_2OH \rightarrow R—CO—H)$. The aldehyde can further be oxidized to the corresponding carboxylic acid. However, a secondary alcohol will form a ketone when oxidized.

Carbonyl often exists as gem-diol (hydrate) in a aqueous solution $(R—CO—R' + H_2O \leftrightarrow .R—C(OH)_2—R')$. The nucleophile (hydroxide ion, $HO—$) in a base-catalyzed reaction is much more powerful than a water molecule, the nucleophile in neutral media. Aldehydes react faster than ketones because equilibrium constants for hydration are more favorable for aldehydes than they are for ketones. Aldehydes $[CO(R')(H)]$ also react readily with amines (primary amines form an imine or Schiff bases $R—N=C(R')(H) + H_2O$ while secondary amines form an enamine (must have an aldehyde with hydrogen on alpha carbon).

If an aldehyde or a ketone is present in an API, then the use of an alcohol must be avoided in an acidic conditions to avoid the formation of an acetal $(R—CO—R' + 2R''—OH \leftrightarrow R—C(OR'')_2—R' + H_2O)$.

In summary, if an API has an aldehyde, ketone, alcohol, and/or carboxylic acid, be careful in developing a formulation (excipients and processing) and developing analytical procedures (sample preparation, mobile phase, and type of packaging chosen for stability studies) based on the specific examples presented in the above sections (aldehyde, ketone, alcohol, and carboxylic acid).

A15.1.1.6 Aldol Condensation. An aldehyde is partially converted to its enolate anion by bases such as hydroxide ion $(2(R—CH_2—CO—H) + OH^- + H_2O \leftrightarrow RCH_2—CH(OH)—CH(R)—COH + OH^-)$. The product ($\beta$-hydroxy aldehyde) formed in the reaction above undergoes dehydration upon heating, to yield an α,β-unsaturated aldehyde $(R—CH_2—CH=C(R)—COH)$ plus water. The process by which two molecules of an aldehyde combine to form an α,β-unsaturated aldehyde and a molecule of water (under heat and basic conditions) is called an aldol condensation.

A15.1.1.7 Amides and Lactams. Amides $(R—CO—NR_2')$ are fairly stable in water, but the amide bond is cleaved on heating in the presence of strong acid or bases. Nominally, this cleavage produces an amine and a carboxylic acid $(R—CO—NR_2' + H_2O + \text{strong acid or base} \rightarrow R—CO—OH + HN—R_2')$. An example here is acetaminophen, where it is hydrolyzed to form acetic acid plus *p*-aminophenol [7]. Amide hydrolysis is irreversible in acid because of protonation of the amine product.

Esters and amines react in 1:1 molar ratio to give amides $(R—CO—O—R' + R_2''NH \rightarrow R—CO—NR_2'' + R'—OH)$. This reaction under heat and presence of water can also occur intermolecularly (within the same molecule). A pharmaceutical example is Baclofen [8].

Lactams are cyclic amides and are analogous to lactones, which are cyclic esters. In contrast to their lactone counterparts, β-lactams (i.e., four-member cyclic amides) are relatively stable. The antibiotics (penicillin G and cephalexin) are examples of β-lactams. β-Lactam hydrolysis (in presence of water or a nucleophile) is known for antibiotics [9]. β-Lactams are known to form polymers under the "right" conditions. (An example is ampicillin, in which the primary amine (acting as a nucleophile) attacks the lactam of another ampicillin molecule to form a dimer. A third molecule of ampicillin may attack the lactam of the dimer to form a trimer, and so on).

Both amides and lactams are not particularly susceptible to oxidation.

A15.1.1.8 Carbamates. Carbamic esters $(R—NR'—CO—O—R''')$ hydrolyze to the corresponding carbamic acid $(R—NR'—CO—OH)$, followed by carbamic acid decarboxylation (much less likely to hydrolyze than esters or amides) to form CO_2 and an amine (RNR'). If this reaction is taking place in a given system (drug product plus packaging material), then the measurement of CO_2 by headspace GC may be an avenue to confirm this reaction.

A15.1.1.9 Imides. Compounds that have two acyl groups bonded to a single nitrogen are known as imides (R—CO—NH—CO—R′). The most common imides are cyclic ones (maleimide). Maleimide will convert to maleic acid under water and acid/base. Another example of imide hydrolysis is phenobarbital in which phenobarbital (a cyclic imide) is hydrolyzed to form urea and α-ethylbenzeneacetic acid.

A15.1.2 Nitrogen Functional Groups

Refer to previous sections for amides, lactams, carbamates, and imides.

A15.1.2.1 Amines. Most primary, secondary, and tertiary amines have pK_a values in the range of 7.5–11.5. Most aryl amines have pK_a values in the range of 4–6. The protonation state of amines is critical to an understanding of the degradation chemistry. Primary, secondary, and tertiary amines are nucleophilic and will react with electrophiles. Unprotonated amines are more easily oxidized and more volatile.

Meropenem, a secondary amine, reacts with bicarbonate to reversibly form a CO_2 adduct that exists in both the solid powder drug product and reconstituted solution [10].

Formaldehyde (HCOH) is a common impurity found from packaging material, resulting from reaction between carbohydrates and excipients. Since formaldehyde is a carbonyl, an API containing an amine is prone to reaction with this functionality (refer to Section A15.1.2.2).

Amines are known to react with molecular oxygen or peroxides. One of the important notes here is that certain excipients (e.g., povidone) have peroxides present in them [11]. One of the examples of secondary amine reacting with peroxides is Sertraline HCl (refer to Figure A15-5).

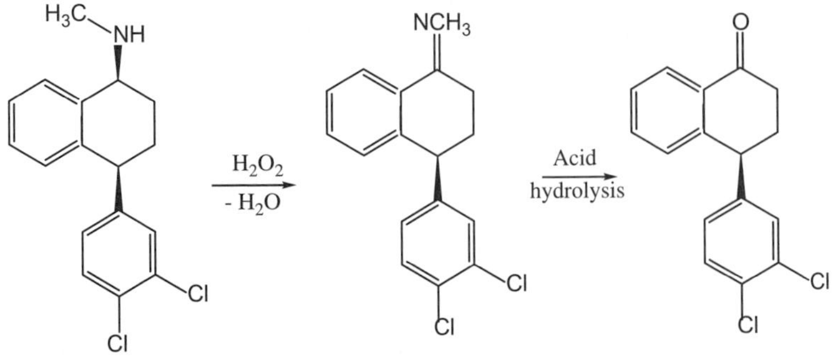

Figure A15-5. Sertraline hydrochloride oxidation, imine formation, and hydrolysis [12].

One of the most known formulation reaction in the pharmaceutical industry is the reaction between an amine (primary or secondary) and a carbohydrate (e.g., sugar) is known as the Maillard reaction (the browning reaction due to mixing of a reduced sugar and an amine) [13]. One of the examples of the Maillard reaction is fluoxetine HCl [14] Figure A15-6. The details of how to avoid this reaction in the formulation is outlined in Section 15.5 [Impact of excipients on Degradation products of API(s)].

A.15.1.2.2 Imines and Enamines. Primary amines react with aldehydes and ketones to form the corresponding *N*-alkyl- or *N*-aryl-substituted imines (R—CO—R′ + R″—NH$_2$ → R—C=(NR″)—R′ + H$_2$O). Imine formation is a

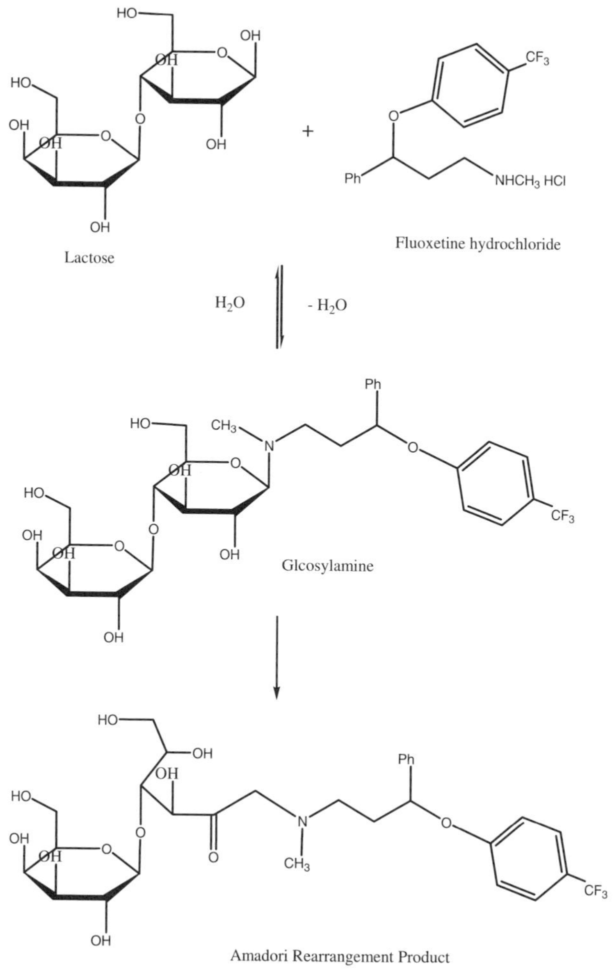

Figure A15-6. Maillard reaction between fluoxetine HCl and lactose [14].

reversible reaction. Imines are known to readily hydrolyze under acidic (and basic) conditions (imine + acid + water → aldehyde + primary amine).

Secondary amines react with aldehydes and ketones to form enamines. Enamines are also susceptible to acid-catalyzed hydrolysis.

A15.1.2.3 Nitriles. Nitriles are classified as carboxylic acid derivatives because they are converted to carboxylic acids on hydrolysis. Like the hydrolysis of amides, nitrile hydrolysis is irreversible in the presence of acids or bases. Acid hydrolysis yields an ammonium ion and a carboxylic acid, e.g., cimetidine [15]. Nitriles are also susceptible to oxidation by peroxides under mildly basic conditions (e.g., pH 7.5 to 8).

A15.1.3   Ethers, Thioethers

Both ethers and thioethers are hydrolyzed via acid-catalysis to the alcohol or thiol, but are reasonably stable at neutral and basic conditions. Hydrolysis of ethers and thioethers is usually only observed if attached to an aryl group or other cationic stabilizing groups (e.g., API cefamandole nafate thiother hydrolysis [16]).

Secondary thiol (R—S—R′) oxidizes to R—SO—R′, which further oxidizes to sulfone (R—SO_2—R′) (e.g., fluphenazine enanthate thioether [17]).

A15.1.3.1 Sulfonamides, Sulfonylureas, Epoxides, and Aziridines. Sulfonamides ($RSO_2NR''_2$) are susceptible to acid hydrolysis ($RSO_2NR''_2$ + water + acid → RSO_2OH), but not basic hydrolysis. Primary alcohols react rapidly only with N,N′-disubstituted sulfonamides to yield sulfonic esters ($RSO_2NR''_2$ + R′OH + acid → $RSO_2OR′$ + NHR''_2). Sulfonamides are not susceptible to oxidation because the sulfur is already fully oxidized.

Sulfonylureas (R—SO_2—NH—CO—NHR′) undergo hydrolysis (acid plus water) to form RSO_2NH_2 + CO_2 + H^+. An example of sulfonylurea hydrolysis is Glibenclamide [18].

Epoxides undergo nucleophilic attack by water (hydrolysis to form diol) or other nuclophiles. The oxirane ring contains significant ring-strain and ring opening relieves this strain.

Hydrolysis to the diol is catalyzed by both acid (preferably) and base. The diol formed from hydrolysis may react further by dehydration and tautomerization to form a ketone.

Mitomycin C contains an aziridine ring [19]. In aqueous solutions, aziridine ring opening occurs with retention of configuration at C2 and water attack at C1 from both faces yielding *cis*- and *trans*-mitosene.

A15.1.4   Alkyl/Aryl Halides

Alkyl/aryl halides are susceptible to hydrolysis (with heat), which occurs via S_N1 (RX + H_2O → ROH_2^+ → ROH + H^+) or S_N2 (RX + OH^- → ROH).

Hydrolysis of alkyl halides can be dramatically facilitated by the presence of a nitrogen or sulfur attached to the carbon alpha to the halide (neighboring group participation) effect (e.g., melphalan hydrolysis) [20].

A15.1.5 Hydroxyls

Hydroxyls (—OH) act as nuclophiles (although less nuclophilic than amines or thiols). Hydroxyls can be eliminated in a dehydration reaction and are not readily ionizable under normal pH conditions (e.g., pH 1–13).

Hydroxyls have often been observed to participate in intramolecular cyclization reactions to form lactones from carboxylic acids, esters, and thioesters, especially if lactone is a five- or six-membered ring [21].

Under oxidative stress testing, hydroxyls are oxidized to the corresponding ketone derivative [22]. Hydroxyl groups can also undergo elimination reactions to form the corresponding olefin.

A15.1.6 Thiols

Thiols (R—SH) can hydrolyze to the corresponding hydroxyl (R—OH) via acid or base catalysis, releasing hydrogen sulfide.

Thiols are susceptible to oxidation by peroxides, molecular oxygen, and other oxidizing processes (e.g., radical-catalyzed oxidation). Thiol oxidation commonly leads to disulfides (R—S—S—R), although further autoxidation to the sulfinic (R—SO$_2$H) and sulfonic acid (RSO$_3$H) may occur under basic conditions.

Thioesters (R—CO—SR′) undergo acyl transfer (nucleophilic acyl substitution similar to esters) to give a thiol and an ester (R—CO—S—(CH$_2$)$_x$—O—R′ + R″—OH → R—CO—OR″ + HS—(CH$_2$)$_x$—O—R′) in the presence of a strong acid (e.g., HCl).

A15.1.7 Phenols

The hydroxyl on the phenyl ring is strongly electron-donating into the phenyl ring, and is the key to the ring oxidizability. Abstraction of the proton (with a nucleophile such as a strong base) provides a resonance stabilized radical that can lead to reaction with molecular oxygen.

Epinephrine is an *o*-diphenol containing a hydroxyl group in the α-position. Oxidation is proposed to occur through the transient formation of epinephrine quinine with subsequent formation of adrenochrome [23].

A15.1.8 Olefins

When subjected to oxidation conditions, olefins (R—CR″ = CR‴—R‴′) can undergo expoxidation followed by S$_N$2 reaction, resulting in anti-hydroxylation by treatment with hydrogen peroxide and formic acid, common excipients impurities in drug product formulations.

Olefins are also susceptible to photochemical cycloaddition reactions. APIs containing olefins can dimerize with another molecule of API to form a 2 + 2 cycloaddition product [24].

Under light conditions, olefins undergo *cis–trans* isomerization reactions. Allylic centers are also quite susceptible to autoxidation chemistry. The alkylic proton has a weaker C–H bond dissociation energy due to the resonance stabilization energy of the resulting alkylic radical. Dihydroergotoxine methanesulfonate undergoes autoxidation of the olefin moiety to the corresponding autoxidation products [25].

A15.1.9 Dimerization

Many compounds [olefins, alcohols, and carboxylic acids (or other carbonyl chemistry)] will undergo dimerization reactions. Figure A15-7 shows how carboxylic acids can react with an alcohol to form a dimer [6] (note that it should be loss of water and not carbon dioxide). In RP-HPLC under basic conditions, the elution order would be diflunisal in its ionized form < descarboxydiflunisal < the dimer. Indoles have been shown to dimerize under acidic conditions, and phenols have shown to dimerize under free radical initiated oxidative conditions, usually ortho-phenols [1]. Due to the low bond dissociation energy of the benzylic C–H bond and ease of radical formation, dimerization can occur at the benzylic center. Nalidixic acid undergoes dimerization under thermolysis conditions to produce a dimeric structure [26].

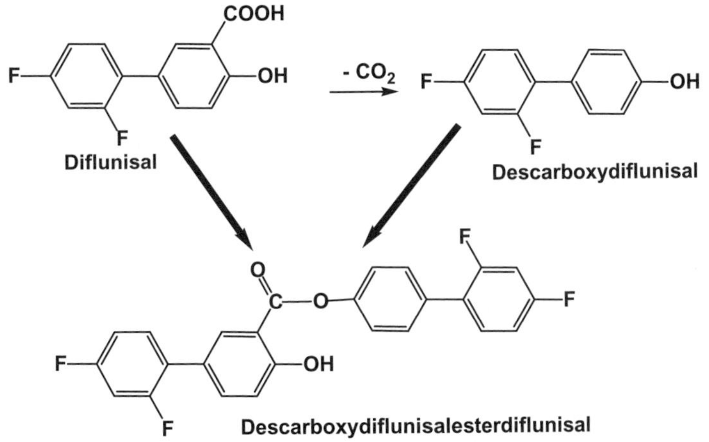

Figure A15-7. Dimer formation for diflunisal [6].

A15.1.10 Ring Transformations

Iorazepam can lose a molecule of water and rearrange to form an aromatic ring structure. Norfloxacin contains a piperazine ring that degrades under light conditions in the solution and solid state to form the ring-opened ethylene diamine derivative.

A15.1.10.1 Epimerization and Isomerization Reactions. Two stereoisomers that have multiple chiral centers but differ in configuration at only one of them are referred to as epimers. D-Glucose and D-mannose are epimeric at the carbon-2 position [1]. Under basic conditions, the degradation of an API etoposide occurs through epimerization of the *trans*-lactone ring to the *cis*-lactone ring, occurring through a planar enol intermediate [27].

Isomerization reaction involves a change in molecule's functional group without changing the chemical formula. An example here is the isomerization of D-glucose to D-fructose by way of enediol intermediate during glycolysis [1]. Cephalosporin antibiotic APIs will undergo isomerization of the olefin from the $\Delta 3$ position to the $\Delta 2$ position [28].

ADDENDUM REFERENCES

1. F. A. Carey, *Organic Chemistry*, McGraw-Hill, New York, 1987.

2. R. J. Simmons, Sodium cephalothin, in K. Florey (ed.), *Analytical Profiles of Drug Substances*, Vol. 1, Academic Press, New York, 1972, p. 329.

3. K. Florey, Testolactone, in K. Florey (ed.), *Analytical Profiles of Drug Substances*, Vol. 5, Academic Press, New York, 1976, p. 545.

4. K. Florey, Aspirin, in K. Florey (ed.), *Analytical Profiles of Drug Substances*, Vol. 8, Academic Press, New York, 1979, p. 31.

5. G. R. Padmanabhan, Methylphenidate HCl, in K. Florey (ed.), *Analytical Profiles of Drug Substances*, Vol. 10, Academic Press, New York, 1981, p. 485.

6. M. L. Cotton and R. A. Hux, Diflunisal, in K. Florey (ed.), *Analytical Profiles of Drug Substances*, Vol. 14, Academic Press, New York, 1985, p. 512.

7. J. E. Fairbrother, Acetaminophen, in K. Florey (ed.), *Analytical Profiles of Drug Substances*, Vol. 3, Academic Press, New York, 1974, p. 39.

8. S. Ahuja, Baclofen, in K. Florey (ed.), *Analytical Profiles of Drug Substances*, Vol. 14, Academic Press, New York, 1985, p. 539.

9. M. Bakshi and S. Singh, Development of validated stability-indicating assay methods—critical review, *J. Pharm. Biomed. Anal.* **28** (2002), 1011–1040.

10. O. Almarsson, M. J. Kaufman, J. D. Stong, Y. Wu, S. M. Mayr, M. A. Petrich, and J. M. Williams, Meropenem exists in equilibrium with a carbon dioxide adduct in bicarbonate solution, *J. Pharm. Sci.* **87** (1998), 663–666.

11. K. Hartauer, G. N. Arbuthnot, and S. W. Baertschi, Influence of peroxide impurities in povidone and crospovidone on the stability of raloxifene hydrochloride in tablets: Identification and control of an oxidative degradation product, *Pharm. Dev. Technol.* **5** (2000), 303–319.

12. K. M. Alsante, Pharmaceutical photostability conference, Research Triangle Park, NC, July 19, 2001.

13. G. P. Ellis, The Malliard reaction, *Adv. Carbohydr. Chem.* **14** (1959), 63–134.

14. D. D. Wirth, S. W. Baertschi, R. A. Johnson, S. R. Maple, M. S. Miller, and K. Diana, Maillard reaction of lactose and fluoxetine hydrochloride, a secondary amine, *J. Pharm. Sci.* **87** (1998), 31–39.

15. P. M. G. Bavin et al., Cimetidine, in K. Florey (ed.), *Analytical Profiles of Drug Substances*, Vol. 13, Academic Press, New York, 1984, pp. 164–165.

16. R. H. Bishra and E. C. Rickard, Cefamandole nafate, in K. Florey (ed.), *Analytical Profiles of Drug Substances*, Vol. 9, Academic Press, New York, 1980, pp. 138–141.

17. K. Florey, Fluphenazine enanthate, in K. Florey (ed.), *Analytical Profiles of Drug Substances*, Vol. 2, Academic Press, New York, 1973, p. 256.

18. P. Girgis, Takla glibenclamide, in K. Florey (ed.), *Analytical Profiles of Drug Substances*, Vol. 10, Academic Press, New York, 1981, pp. 344–346.

19. J. H. Beijnen et al., Mitomycin C, in K. Florey (ed.), *Analytical Profiles of Drug Substances*, Vol. 16, Academic Press, New York, 1987, p. 383.

20. L. Valentin, Feyns melphalan, in K. Florey (ed.), *Analytical Profiles of Drug Substances*, Vol. 13, Academic Press, New York, 1984, pp. 289–290.

21. Bontchev, Inhalt, *Pharm. Zeit* **33** (2004), 346–348.

22. G. S. Brenner, D. K. Ellison, and M. J. Kaufman, Lovastatin, in H. G. Brittain (ed.), *Analytical Profiles of Drug Substances and Excipients*, Vol. 21, Academic Press, New York, 1992, p. 299.

23. D. H. Szulczewski and W. Hong, Epinephrine, in K. Florey (ed.), *Analytical Profiles of Drug Substances*, Vol. 7, Academic Press, New York, 1978, p. 215.

24. T. H. Lowry and K. S. Richardson, *Mechanism and Theory in Organic Chemistry*, 3rd edition, Harper Collins, New York, 1987, p. 915.

25. W. Dieter Schoenleber et al. Dihyroergotoxine methanesulfonate, in K. Florey (ed.), *Analytical Profiles of Drug Substances*, Vol. 7, Academic Press, New York, 1978, p. 125.

26. P. E. Grubb, Nalidixic acid, in K. Florey (ed.), *Analytical Profiles of Drug Substances*, Vol. 8, Academic Press, New York, 1979, p. 382.

27. J. M. Holthuis et al, Etoposide. In K. Florey (ed.), *Analytical Profiles of Drug Substances*, Vol. 18, Academic Press, New York, 1989, pp. 140–141.

28. E. Popa, M. J. Huang, M. E. Brewstera, and N. Bodora, On the mechanism of cephalosporin isomerization, *J. Mol. Structure: Theochem.* **315** (1994), 1–7.

16

THE ROLE OF HPLC IN TECHNICAL TRANSFER AND MANUFACTURING

Joseph Etse

16.1 INTRODUCTION

Analytical technology transfer and manufacturing is the mechanism by which knowledge acquired about a process for making a pharmaceutical active ingredient or dosage form during the clinical development phase is transferred from research and development to commercial scale-up operation or shared between internal groups or with third parties. Analytical technology transfer guarantees that laboratories can routinely execute tests, obtain acceptable results, and be able to accurately and independently judge the quality of commercial batches. One of the most important analytical technology transfers is high-performance liquid chromatography (HPLC) methods. The success or failure of analytical technology transfers are judged on the merits of data generated using HPLC. Consequently, a major focus of regulatory authorities [1–3] is on methods transfer as a critical link in the drug development continuum. Depending on the structure of the pharmaceutical organization, transfer of analytical technology and manufacturing may occur at the end of the phase II clinical studies or during the transition from phase II to phase III. However, for a successful transfer of analytical technology to occur, the existence of HPLC methods that have been fully validated in accordance with the ICH guidelines on validation will be required [4–7]. A full description of method validation is provided in Chapter 9.

HPLC for Pharmaceutical Scientists, Edited by Yuri Kazakevich and Rosario LoBrutto
Copyright © 2007 by John Wiley & Sons, Inc.

16.2 PREREQUISITES FOR TRANSFER OF HPLC METHODS

16.2.1 Availability of Either Fully or Partially Validated Methods

A prerequisite for the transfer of analytical technology is the establishment of fully validated methods in accordance with the International Council on Harmonization (ICH) [4, 5], United States Pharmacopeia (USP) [6] and European Pharmacopeia (EP) [7] guidelines for method validation, the existence of a final synthetic process for the active pharmaceutical ingredients (APIs), and final market image of the pharmaceutical dosage form. Method development and validation usually parallels the API and pharmaceutical dosage form development. It progresses from very rudimentary Tier 1 methods with limited validation as shown in Table 16-1 through to Tier 2 methods and culminating in Tier 3/registration-type methods [8–10].

Differences between methods from Tier 1 through Tier 3 are due to the extent of validation of the analytical figures of merit that is performed [3]. During early development of the active pharmaceutical ingredient and early dosage form development, emphasis is placed on speed and quantitation of the API. At this stage, methods rely on the use of short columns, fast flows, and very minimum validation to quickly identify the most desirable synthetic route for the API that will produce an adequate impurity profile (overall yield may not be optimized at this stage) and most desirable prototype formulations and excipients that will ultimately lead to the selection of the final formula-

TABLE 16-1. Progressive Validation of Analytical Figures of Merit

	Progressive Method Development			
Analytical Figures of Merit	Discovery/ Phase I Tier 1	Phase II Tier 2	Phase III Tier 3	Phase IV Registration
1. Linearity	√	√	√	√
2. Range	√	√	√	√
3. Accuracy	—	√	√	√
4. Specificity/stress studies	—	√	√	√
5. Precision				
• Repeatability (injection)	√	√	√	√
• Intermediate precision (API)	—	√	√	√
• Intermediate precision (RS)	—	—	√	√
6. Robustness	—	—	√	√
7. Solution stability	√	√	√	√
8. Limit of detection (LOD)	—	—	√	√
9. Limit of quantitation (LOQ)	—	√	√	√

√, Validated; —, not validated; R&D, discovery research, API, active pharmaceutical ingredient; RS, related substance.

Source: Reprinted from *Am. Pharm. Rev.* Vol. **8**(1), (2005), 76, with permission.

tion. Typically, Tier 1 and Tier 2 methods include validation of some, but not all, of the analytical figures of merit as shown in Table 16-1. Tier 1 methods are the simplest methods in the sense that only linearity and precision may have been validated. As the synthesis scheme for the API becomes optimized with respect to improving the overall API yield and as the dosage form development evolves from prototype formulations to the more robust final market image formulations, the analytical methods employed also evolve and become increasingly robust and optimized for the quantitation of the API as well as degradation products and related substances. Once the final synthesis is set and final formulations are selected, more robust and fully validated Tier 3 methods [3] are established to ensure the successful transfer of analytical technology from research and development to commercial operation. Supplementary to this prerequisite is the identification of the commercial production site or launch site where the pharmaceutical dosage form will be manufactured. This is usually the stage in which all the drug development activities come together in a New Drug Application (NDA) for regulatory approval (Figures 16-1 and 16-2). Phase IV methods are usually slight variations of Tier 3 methods which include but are not limited to calculation formulas, the number of sample preparations for API, and the number of dosage units.

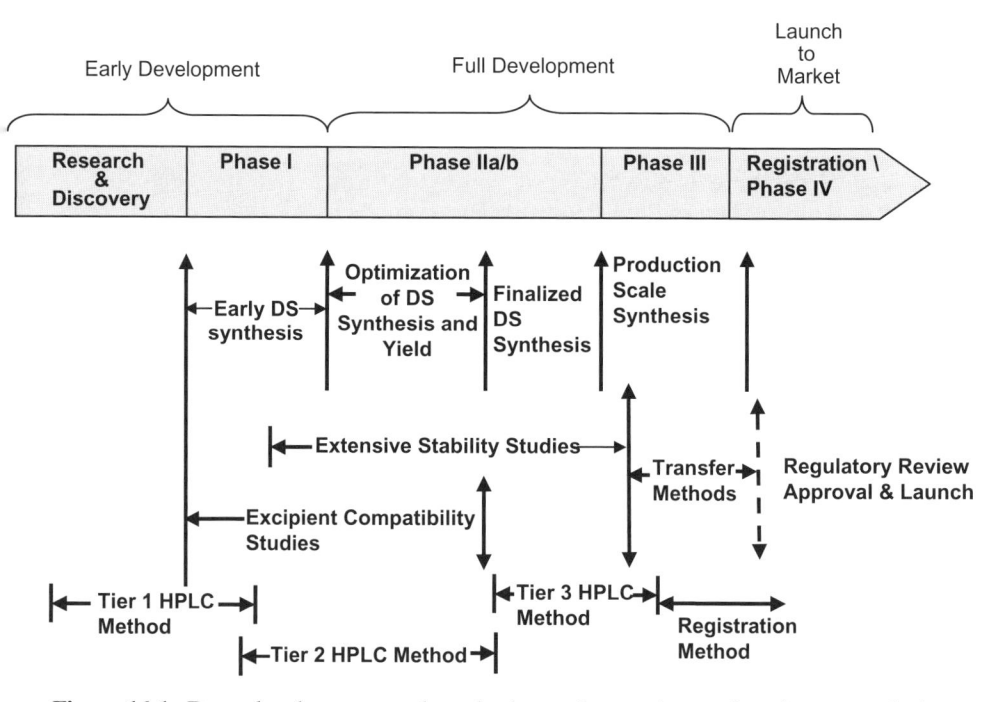

Figure 16-1. Drug development and method transfer continuum for pharmaceutical active ingredients (APIs).

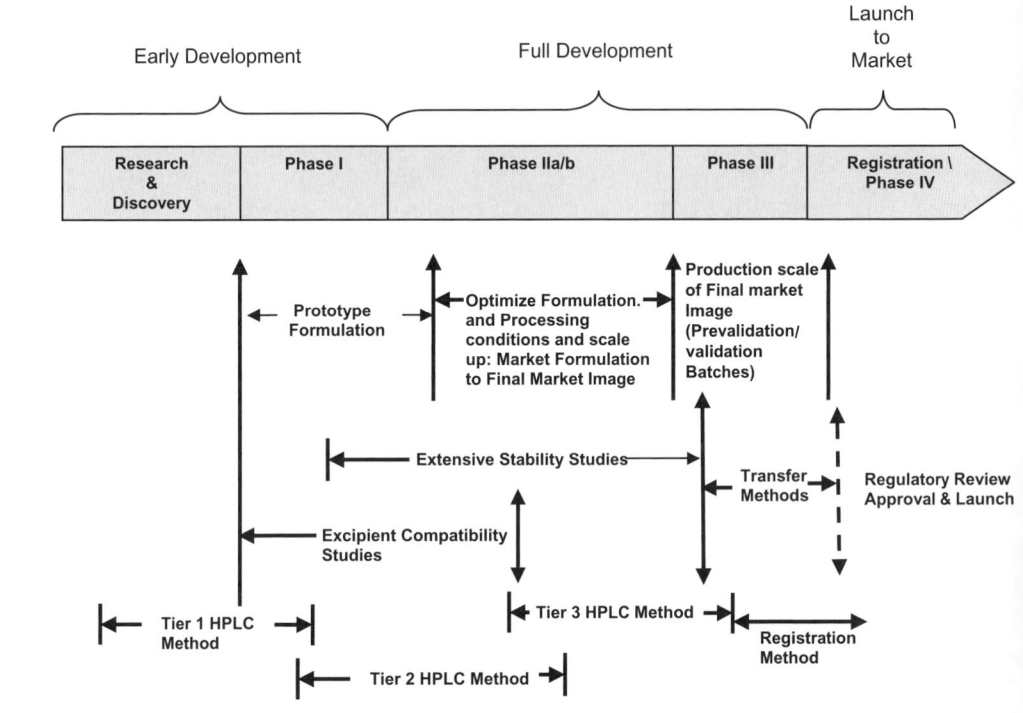

Figure 16-2. Drug development and method transfer continuum for pharmaceutical dosage forms.

16.2.2 Availability of the Finalized Pharmaceutical Active Ingredient (API), Known Degradation Products, By-products and Reference Standards

Besides the existence of validated HPLC methods, the availability of a finalized API synthetic scheme and optimized formulations is another prerequisite for ensuring a successful transfer of analytical technology from research and development to commercial operation (Figure 16-1). Evaluation of data generated from the HPLC analysis of the API provides the means by which determination is made about whether a validated API synthetic scheme exists and if the API can be made reproducibly during commercial operation.

Confirmation of the existence of a validated API synthetic process is based on the interpretation of acquired nuclear magnetic resonance (NMR) spectrum of the API in conjunction with definitive molecular formula for the API and its components based on mass spectroscopy (MS). Concluding that a reproducible API manufacturing process exists is based on whether purity profiles of subsequent drug substance batches retain the same profile of the API and its related substance peaks as was in the reference material. In that sense, data generated using HPLC and hyphenated HPLC techniques such as HPLC/MS, HPLC-MS/MS, and HPLC/NMR serve as the foundation for declaring the existence of reproducible API manufacturing process [11–13].

16.2.3 Availability of Drug Products Made by the Definitive Manufacturing Process

Once the API has been selected for further development, a clear definition and demonstration of the validated status of the manufacturing process for the pharmaceutical dosage form is required in order to initiate transfer of analytical technology and manufacturing process. Development of the pharmaceutical dosage form consists of a series of experimental activities that ultimately result in the transformation of the API into a dosage form (i.e., tablet, capsule, suspension, injectable, patch, creams, inhalation product) suitable for human use [14]. This is achieved through the manufacture of batches at the chosen commercial site for scale-up production (Figure 16-2). As the formulation development process evolves into the final market image dosage form, more robust HPLC methods as shown in Table 16-1 are established [3]. Assessment of whether or not the transfer of manufacturing process has been successful is usually done by sampling and analyzing batches for blend uniformity [15, 16]. Sampling of the batches can be done in a number of ways. Sampling by variables or the Bergum approach [17] are examples of sampling techniques that can be used to measure the quality characteristics of the batch on a continuous statistical scale. A simple method of measurement used to judge the quality of the batch is the *sample mean value* (SMV) and its corresponding *percent relative standard deviation* (%RSD). A sample mean value that is close to the expected target value of 100% with a corresponding low %RSD serves to prove the attainment of a homogeneous product that is uniform and not variable in the content of the API. Assessment of product quality using the Bergum approach relies on whether selected samples of a batch has an SMV and %RSD that match the Bergum acceptance criteria. The process is rejected if the SMV and %RSD lie outside the allowable Bergum acceptance criteria [17–19]. An example of how the Bergum acceptance criteria are applied for evaluating content uniformity data generated using HPLC is shown in Figure 16-3. A sample with an SMV of 95% will pass the Bergum criteria, provided that its corresponding %RSD is less than or equal to 3.5. Similarly, a sample with a %RSD of 2.0 must have a corresponding SMV of either 90% or 109% to pass the Bergum acceptance criteria. The general trend as seen with Bergum acceptance criteria is that an increase in the magnitude of the %RSD and its corresponding *sample mean value* (SMV) follow a bell-shaped distribution in such way that %RSD decreases to zero as the SMV becomes greater than or less than 100%.

16.2.4 Availability of Suitable Instruments and Personnel

A critical activity that precedes the start of analytical technology transfer is to assess whether suitable instrumentation and qualified personnel are available at the receiving laboratory. The receiving laboratory is the laboratory to which the analytical methods are transferred to. This assessment is accomplished through the organization of an analytical challenge meeting. The purpose of

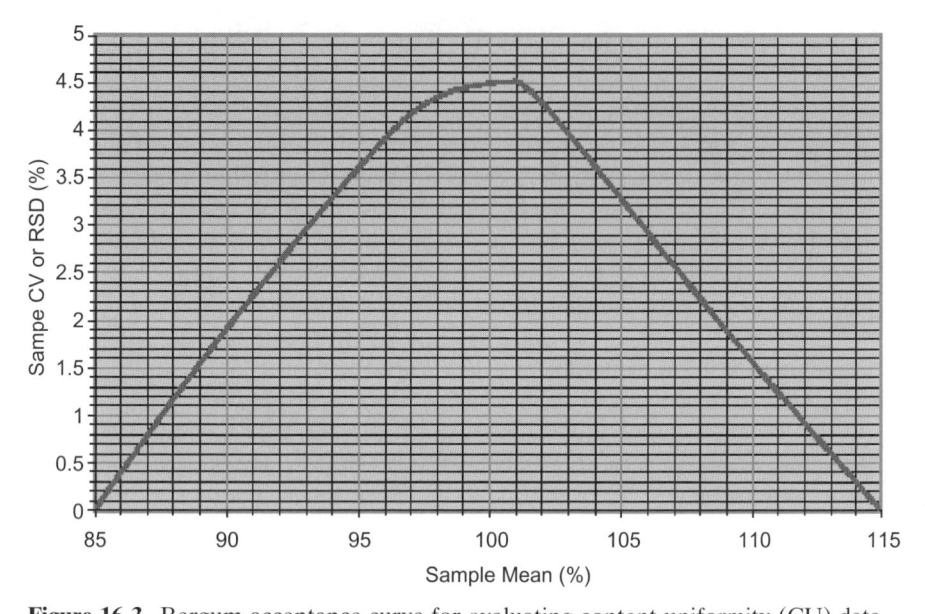

Figure 16-3. Bergum acceptance curve for evaluating content uniformity (CU) data. *Note:* The curve shows that for a *relative standard deviation* (%RSD) of 4.5%, a *sample mean value* (SMV) of 100% must be achieved for the manufacturing process to be judged as validated.

this meeting is usually to open up channels of communication between the participating laboratories to discuss the methods that are going to be transferred, to share and exchange knowledge about the idiosyncrasies of the methods, and to agree on the types of tests to perform and samples that will be used for the cross-over testing. Because of the potential impact of sample-to-sample variability on the agreement between data generated, participating labs should agree to use identical batches and also decide on the number and types of batches that should be tested [17–20]. Additionally, identification of key contact persons and assignment of responsibilities are other items that can be agreed upon during the analytical challenge meeting.

16.2.5 Availability of a Protocol Containing Predetermined Acceptance Criteria

The methods transfer protocol is the main driver that governs the conduct of the experiments and ensures that assessment of results generated is not unduly influenced by biases due to either (a) the analytical method or (b) inherent batch-to-batch variability of the active pharmaceutical ingredient or pharmaceutical dosage form. The methods transfer protocol establishes the predetermined acceptance criteria by which results will be judged to have either passed or failed the methods transfer. The criteria for assessment of success or failure contained in the methods transfer protocol is achieved through an iterative process of exchange of ideas and comments between the originating lab and the participating or receiving labs.

Since the aim of the protocol is to ensure the mitigation of problems, the essential elements of the protocol consists of sections that include (a) an Introduction, (b) treatment and disposition of data, (c) types of methods being transferred, (d) materials, reference standards, and reagents being used, (e) recommended type of equipment, (f) sample handling, (g) predetermined acceptance criteria, and (h) an Acknowledgment section. An example of a typical table of contents (TOC) of an analytical methods transfer protocol is discussed in Table 16-2.

TABLE 16-2. Table of Contents of an Example of Analytical Transfer Protocol

Section	Description
1. Introduction	• Aim and scope of the protocol • Brief description of dosage forms • Arguments for waivers, bracketing, and matrixing (if applicable)
2. Treatment and disposition of data	• Format for reporting data (i.e., number of decimal places • Archival of data • Handling and resolution of OOS or OOT results
3. Materials, standards, and reagents	• Batches to be tested including batch-specific data (storage conditions, shelf life, etc.) • Reference substances (including storage conditions and expiry dates) • Special handling instructions or precautions (if applicable) • Reagents and source of supplier(s)
4. Test methods and specifications	• Lists all applicable test methods and specifications • List methods that will not be transferred by cross-over testing
5. Acceptance criteria	• Lists pass or fail requirements • Enumerates the release acceptance requirements • Stipulates the number of required replicate determinations • Establishes the statistical assessment requirements
6. Method acknowledgment	• Feedback from participating labs • Signature of acceptance/approval (receiving and transfer laboratory)

OOS, out-of-specification results; OOT, out-of-trend results.
Source: Adapted from *Am. Pharm. Rev.* 8(1) (2005), 76, with permission.

16.2.5.1   Introduction. The introduction section lays down the purpose of the transfer and clearly identifies the originator lab where the method was developed as the reference lab (also designated as center A or originator lab) and identifies all the other labs participating in the transfer as participating or receiving labs. A brief description of the pharmaceutical dosage forms being transferred as well as any arguments for bracketing and matrixing of the testing plan is included in this section. Because a minimum of three batches of each dosage strength of a pharmaceutical dosage form is normally required to be to tested during the transfer activities [17–19], any argument that serves as justification for omitting the testing of certain dosage strengths through the application of bracketing and matrixing strategies becomes an important consideration that is addressed in this section.

Bracketing and matrix testing is the approach in which only the highest and lowest dosage strengths are tested in the cross-over experiments and carry the benefits of reducing the amount of testing required. When this approach is applied to a pharmaceutical dosage form consisting of 50-, 75-, 100-, 125-, and 150-mg dosage strengths, cross-over testing is performed using only the 50-, 100-, and 150-mg dose strengths. In that sense, passing results generated for the 50-, 100-, and 150-mg strengths automatically becomes surrogate and proof of the transfer for the 75- and 125-mg dose strengths. Usually, arguments for bracketing and matrixing are easier to justify if the analytical sample preparation steps for all dosage strengths are similar and if the dosage strengths are made from similar or identical granulation and the excipient-to-drug content ratio is dose and/or weight proportional.

16.2.5.2   Treatment and Disposition of Data. This section discusses treatment and disposition of data and establishes the mechanism by which data will be assessed as having passed or failed the predetermined acceptance criteria. It is important that this section also address (a) the mechanism by which *out-of-specification* (OOS) or *out-of-trend* (OOT) results will be handled and (b) procedures for reporting and archiving of data. Stipulation of how assessment of reported results and the archival and disposition of data will be handled should also be discussed under this section. Clear rules governing how assessment of data is handled takes on an ever-increasing significance because of the intense scrutiny regulatory agencies apply when reviewing any documented report that claims equivalency based on data comparison between different laboratories [20–22].

16.2.5.3   List of Materials, Standards, and Reagents. The impact that batch-to-batch variability of a pharmaceutical dosage form can have on the interpretation of results can be very challenging [17, 18]. The same challenges are also present for the API because different vendors may be used for the raw materials that may lead to different impurity profiles of the API. Hence, the purpose of this section of the protocol is to describe in unequivocal terms the analytical methods, test samples, standards, and reagents that should be used. An approach that greatly helps to alleviate potential differences that lead to non-agreement of data between the participating and originator labs due to the potential impact of batch-to-batch variability is to use either freshly manufactured or well-characterized batches. This can be accomplished by providing all the participating labs with identical samples derived from previously prepared composite samples or aliquots from the same batches of drug product (DP) and drug substance (DS). The prepared composite samples for the DP or DS are subdivided evenly into smaller lots that are distributed to all participating laboratories. Because the ultimate goal of the technical transfer is the transfer of analytical test methods and not the identification of inherent batch-to-batch variability that may exist among batches, it is essential to stipulate the lot number, date of manufacture and the required storage condition of the batches selected. For example, a storage condition of "5°C" or "Do not store above 30°C," together with any special handling instructions, and the appropriate "Re-test" or "Expiry date (i.e., 2 years, etc.)," should be provided [23, 24]. Additionally, it may also be necessary to stipulate and restrict the number of vendors/suppliers of reagents and reference standards as a means by which potential biases from such sources on the results generated can be eliminated or minimized. For example, if *in-house* reference standards are used, it may be necessary to quarantine that particular lot of reference standard material to ensure that adequate supplies are available at all times for the duration of the analytical methods transfer. Material Safety and Data Sheet (MSDS) classification of the reagents, standards, and the API should be provided to assure the types of adequate precaution that should be taken to avoid unintended exposure of analysts to potentially dangerous materials [23, 24].

16.2.5.4 Test Methods and Specifications.

This section tabulates all the test methods and specifications that are being transferred via an *interlaboratory* cross-over testing plan. In addition, a discussion of the rationale for transferring the methods should be provided in this section. When transfer of related substances methods is required, a number of steps should be considered to ensure that data generated will be reasonable and meaningful for the purposes of comparing data between the different laboratories. Transfer of related substances methods often presents a challenging situation because related substances tend to occur in low levels, especially in recently manufactured batches, or may be present at levels that are close to the quantitation limits of the method. To circumvent this, a very well-defined strategy can adopted by the participating labs to mitigate difficulties associated with the transfer of related substances. For example, clear instructions regarding whether a spike experiment involving the use of pre-prepared samples containing a known amount of the degradation product or the use of samples containing well-characterized amount of the degradation product of interest can be stipulated. Also, clear definition of the sample preparation in regard to sonication (power output, water level in batch), shaking, filtration steps, and so on, must be communicated [25–29].

Another aspect of this section of the protocol is to discuss the rationale for not transferring certain methods. In general, test methods such as microbial limit tests (MLT) and other types of test methods based on universally accepted pharmacopeia procedures are not typically subjected to *interlaboratory* cross-over testing. Instead, such methods are validated locally by demonstrating suitability for their intended purpose. Similarly, test methods that rely on the use of well-established techniques that are considered routine such as "appearance by visual examination" and "uniformity of dosage units by weights" are not normally subjected to *interlaboratory* cross-over testing. However, if some of the more common tests such as "Appearance of Solution" tests require special pharmacopeia procedures that are not routine, it may be necessary to consider the inclusion of such methods in the *interlaboratory* cross-over testing scheme. Any decision that is contemplated to either include or exclude a test method or methods from conventional cross-over testing should clearly be justified on the basis of sound scientific argument(s).

16.2.5.5 Acceptance Criteria. A most important section of the protocol is the section that outlines all the acceptance criteria by which results generated for each method being transferred will be judged as having fulfilled the requirements of the transfer. Since the interpretation of acceptance criteria is often based on the application of some type of statistical value, an important aspect of this section is to include clear instructions regarding the number of batches and the number of replicate determinations that has to be performed [18, 19]. Table 16-3 shows examples of acceptance criteria that may be applied

TABLE 16-3. Example of Acceptance Criteria for Assay, Content Uniformity, and Dissolution

Test	Replicates (n)	Acceptance Criteria
Assay	3	• Mean difference $\leq 2.0\%$ [%RSD, $n = 6$) $\leq 3.0\%$]
Content uniformity (CU)	10	• Results meet current USP <905> and/or harmonized USP, JP, and EP 2.6.1 requirements. • If %RSD at either site is >4.0%, then [STDev (Lab B)/STDev (Lab A)] $\leq 2.0\%$ • For each site, SMV = $\pm 3\%$ of the mean assay within each site
Dissolution	12	• Mean difference $\leq 7.5\%$

SMV, sample mean value; RSD, relative standard deviation; n, number of replicates; USP, United State Pharmacopeia; JP, Japanese Pharmacopeia; EP, European Pharmacopeia.

Source: Adapted from *Am. Pharm. Rev.* **8**(1) (2005), 77, with permission.

for comparing assay, content uniformity, and dissolution data between participating labs.

For the purposes of comparing assay, content uniformity, and dissolution data, simple statistics such as *sample mean value* (SMV) and *relative standard deviation* (%RSD) derived from experience of performing the tests over long periods of time can be used as acceptance criteria. Alternatively, more sophisticated statistics such as the *z-test, F-test* or *t-test* as shown in Table 16-4 can be applied [17–19, 25]. In the case of evaluating CU data, it can be concluded that results from two labs are equivalent based on applying the simple statistics of the difference between the SMV from Lab A and Lab B (Table 16-4) not to be more than 2.0%. In the other examples where the more sophisticated statistics such as *z-test, F-test,* or *t-test* are applied (Table 16-4), results from two labs are considered to be equivalent because the calculated statistics in each case (*z-calculated* values of 0.32/0.64, *F-calculated* value of 0.14, or *T-calculated* values of 0.30/0.60) are less than the predicted statistics (*z-critical* value of 1.64, *F-critical* values of 3.18, or *T-critical* value of 1.73) [19, 25].

16.2.5.6 Method Acknowledgment. Since the goal of the transfer is to achieve an efficient and an issue-free transfer, a meeting to discuss the HPLC methods with the participating labs prior to the start of the transfer activities is highly recommended. As previously discussed, the meeting provides opportunities for the participating labs (or receiving labs) to be made aware of any special features or idiosyncrasies of the methods. Hence, the primary focus of this section is to capture feedback and suggestions from discussions with the participating labs prior to the start of the transfer activities. Sometimes based on feedback from the discussions, methods can be further optimized to address special concerns or to accommodate well-established procedural practices at the receiving labs. For example, the test method for assay can be changed to permit the use of 20 instead of 10 sample composite for assay in order to reconcile practices in AR&D with quality control (QC). In other cases, a stipulated column oven temperature of 30°C can be changed to 35°C, and the initial isocratic hold time in a gradient method can also be modified in order to accommodate differences in instrument capabilities (dwell volumes) between QC and AR&D. Additionally, this section stipulates who needs to approve the protocol from both the participating and originator labs.

16.3 TYPES OF TECHNICAL TRANSFER

16.3.1 From Analytical Research and Development (AR&D) to Quality Control (QC) Lab of the Commercial Organization

Technical transfer from AR&D to QC constitutes the majority of technology transfers that are performed. It is the process by which a laboratory is

TABLE 16-4. Role of HPLC in the Comparison of CU Data Between Two Labs

Test Sample #	Lab A	Lab B
1	99.4	99.0
2	99.8	99.6
3	100.0	101.0
4	101.0	100.0
5	97.9	98.8
6	99.9	99.3
7	99.5	99.6
8	100.0	99.9
9	101.2	100.0
10	102.0	101.2
Average	**100.1**	**99.8**
Standard deviation	**1.1**	**0.8**
Mean difference	**0.30**	

z-Test: Two Sample for Means

	Lab A	Lab B
Minimum value	*99.4*	*99.0*
Mean	100.1	99.9
Known variance	1.27	0.6
Observations	9	9
Hypothesized mean difference	0	
z	0.46	
$P(Z \leq z)$ one-tail	**0.32**	
z Critical one-tail	1.64	
$P(Z \leq z)$ two-tail	**0.64**	
z Critical two-tail	1.96	

F-Test Two-Sample for Variances

Mean	100.1	99.8
Variance	1.27	0.60
Observations	10	10
df	9	9
F	2.11	
$P(F \leq f)$ one-tail	**0.14**	
F Critical one-tail	3.18	

t-Test: Two-Sample Assuming Equal Variances

	Lab A	Lab B
Mean	100.1	99.8
Variance	1.27	0.60
Observations	10	10
Pooled variance	0.94	
Hypothesized mean difference	0	
df	18	
t Stat	0.53	
$P(T \leq t)$ one-tail	**0.30**	
t Critical one-tail	1.73	
$P(T \leq t)$ two-tail	**0.60**	
t Critical two-tail	2.10	

qualified to utilize analytical methods for the routine release of products to the marketplace [20]. This type of transfer guarantees that commercial organization can routinely perform the tests and obtained acceptable results. From that perspective, the timely transfer of methods from AR&D to QC is regarded as providing a competitive advantage in accelerating the commercialization of the drug products. Transfer from AR&D to QC guarantees the proper training of QC chemists in the use of the analytical methodologies and avoids unnecessary delays due to analytical issues in the timely release of products to the marketplace [20]. For this type of transfer to occur, it is usually the originator of the method (AR&D) who undertakes the full validation of the methods and also initiates the transfer of the methods. Though timing of the transfer generally occurs prior to the regulatory submission of a New Drug Application (NDA), the transfer can also take place earlier in the drug development continuum around the time of filing an Investigational New Drug Application (IND). Transfer during the IND filing phase can occur if a project is transferred from one development center to another. A distinguishing feature of this type of transfer is that fully validated Tier 2 or Tier 3/ registration method (Table 16-1) must be available before the transfer can begin. The assessment of success or failure of the transfer under this type of transfer is usually based on extensive *interlaboratory* cross-over testing and comparison of results against predetermined acceptance criteria [26, 29].

16.3.2 Transfer from AR&D to Another AR&D Organization

Transfer from one AR&D organization to another AR&D unit occurs when projects are transferred mid-stream during the drug development continuum. There may be a number of reasons that could lead to this type of transfer. Notable among these are;

(a) Realignment of project portfolios due to significant reorganization of a company.

(b) Merger/acquisition situation requiring the divestiture of projects to avoid creation of monopolies.

(c) Change in a company's therapeutic area of focus or interest leading to either out-licensing or in-licensing of new development activities to bolster pipeline depreciation issues.

Unlike transfer from AR&D to QC, methods available at this time may not have been fully validated and may either be Tier 1 or Tier 2 methods (Table 16-1). Depending on the status of the project at the time of the transfer, the receiving lab may have to redevelop and revalidate the methods. Transfer under this mode can be prompted by legally mandated timeline, especially in a merger situation, to ensure the timely and complete transfer of technology to the receiving lab. Assessment of success or failure of transfer may not be based on extensive *interlaboratory* cross-over testing. Instead, the timely transfer of all available relevant documentation and the comprehension of the documentation by the receiving laboratory may be sufficient proof of the successful transfer of the analytical methods.

16.3.3   Transfer from AR&D to Contract Research Organization (CRO)

Transfer from AR&D to CRO is becoming an increasing means by which companies faced with severe capacity constraints find external partners to off-load some of their routine activities to free-up capacity from performing routine activities such as testing of stability samples and manufacture of comparator/positive control batches. Additionally, companies may engage in this type of transfer when they place greater emphasis on the performance of only certain activities such as early proof concept screening of compound and discovery support activities. Transfer under this category can be similar to the transfer from AR&D to QC or the transfer from one AR&D unit to another. When transfer is similar to the transfer from AR&D to QC, the originator of the method (AR&D) undertakes the full validation of the methods and initiates the transfer. A facet of this transfer is that methods at this stage can span the whole gamut from Tier 1 methods to the fully validated Tier 3 methods. When Tier 2 methods are involved, the CRO undertakes the task of completing the validation of the method with either full or partial participation in the validation efforts by the originator lab. Successful transfer in this case is based on the CRO completing the validation exercises in accordance with criteria defined in a protocol. With regard to transferring fully validated Tier 3-type methods, assessment of success or the transfer is similar to the transfer from AR&D to QC in the sense that success or failure of the transfer is predicated on the generation of results from extensive *interlaboratory* cross-over testing and the agreement of the results with a set of predetermined acceptance criteria.

16.4   DIFFERENT APPROACHES FOR TECHNICAL TRANSFER AND MANUFACTURING

16.4.1   Comparative Testing

Comparative testing is the most common approach employed for ensuring the transfer of analytical methods. Because this approach usually requires the availability of fully validated Tier 3/registration-type methods, a prerequisite for this approach is that the originator lab and participating laboratories agree to use the same fully validated methods and preselected and mutually agreed-upon products sourced from identical batches of the material. Considerations that are paramount to the success or failure of this approach are the precautionary measures that must be taken to either eliminate or at least minimize the potential impact of any inherent batch-to-batch variability on the results. To achieve this goal, material used for the *interlaboratory* cross-over testing is sourced from the same batch/lot of drug product or drug substance. One of the most popular means employed to ensure the elimination of potential influence of batch-to-batch variability on data interpretation is that the originator lab prepares a composite sample of the material to be tested. The composite sample is split into equal portions that are then supplied to the participating labs. Results generated by the participating labs are compared to the originator lab's results as the reference or gold standard lab [18–20]. Assessment of agreement of results is done by using a variety of statistical approaches [20, 25]. The simplest statistical approach that is often used is to compare results based on predetermined *relative standard deviation* (%RSD) and difference between *sample mean value* (SMV). Selection of the %RSD and SMV criteria can be based on accumulated historical data generated from analyzing several batches at the originator lab. Table 16-5 shows data generated for the transfer of an HPLC assay method from Lab A (the originator lab) to Lab B (the receiving lab). In this case, a predetermined acceptance criteria of difference between the SMV as ≤2.0% for assay was established prior to the start of the transfer activities. In addition, a %RSD criteria of ≤2.0% was also established. For the three dosage strengths (250 mg, 500 mg, and 750 mg tablet) tested in the *interlaboratory* cross-over testing, data generated by both labs agreed with the predetermined SMV and %RSD acceptance criteria, and so the transfer is judged to have been successful.

In other cases of determining whether a transfer has been successful involves the use of sophisticated statistical means (Table 16-4) such as the Student *t*-test, testing of equality of means from two groups (two sample

TABLE 16-5. Comparison of Assay Results for Three Replicates of Three Dosage Strengths

Drug Product Batch: Dosage strength:	A 250 mg		B 500 mg		C 750 mg	
Participating labs:	**Lab A**	**Lab B**	**Lab A**	**Lab B**	**Lab A**	**Lab B**
First value:	101.1	99.5	101.4	101.2	101.3	99.6
Second value:	100.4	98.9	101.3	100.7	100.6	100.5
Third value:	99.8	98.5	101.1	101.1	100.9	100.2
Mean value:	100.5	99.0	101.3	101.0	100.9	100.2
RSD:	0.7	0.5	0.1	0.3	0.4	0.7
Mean difference:	1.5		0.3		0.7	

Note: Lab A is the reference or originator lab. Acceptance criteria based on three replicate determination are as follows:
Sample mean value (SMV) = 95–105%; Difference between the SMV (i.e., mean difference) is ≤2%; RSD (n = 3) ≤2.0%. In this example, the transfer was successfully completed because all predetermined acceptance criteria were met.

t-test), one-way analysis of variance (*ANOVA*), *z-test, F-test*, and use of control charts [17, 25]. As previously discussed in Table 16-4, the sample mean values obtained at the originator and participating labs are considered to be equivalent because the calculated *P-value* for each lab is less than the *z-critical* or *t-critical* value. This situation is interpreted to mean that the transfer has been successful. Under this type of transfer, participating labs are certified as qualified to perform the tests only after the results they generate unequivocally agree with predetermined acceptance criteria and with the originator lab's data. Comparative testing is the conventional approach often used for the transfer of analytical methods from AR&D to either QC or contract research organizations (CRO).

16.4.2   Co-validation of Methods

Unlike in the case of comparative testing, which requires the existence of fully validated methods, there is no requirement for the existence of a validated method for transfer based on co-validation. In other words, availability of a fully validated method is not a prerequisite for this mode of transfer. This type of transfer is usually employed for transferring methods from one AR&D organization to another ARD organization or from an AR&D unit to a QC lab. Because the receiving labs in this type of transfer participate in all aspects of the final validation of the methods, completion of the validation is generally considered proof of the successful transfer of the methods. In that regard the receiving labs become immediately certified to perform the test and do not require any formal certification. Although this mode of transfer is not often used, it is gaining in popularity because of the benefits it offers in terms of reducing the amount of time required to complete a transfer exercise. The other benefits of this transfer is that the use of pre-selected batches is not required because validation experiments are shared between the labs. However, for this approach to work well, prior agreement is reached on the analytical figures of merit (i.e., linearity, accuracy, intermediate precision, etc.) that each lab will have to include or exclude from the validation (see Chapter 9 for more details on method validation). In some cases, acceptance criteria may be based on the originator's Method validation SOP for demonstrating accuracy, linearity, precision, repeatability, range, LOD and LOQ (Table 16-6). In the example shown in Table 16-6, independent achievement of the acceptance criteria by the participating labs is considered sufficient proof for declaring that the transfer has been successful.

16.4.3   Revalidation of Methods

Revalidation of methods is the approach by which methods are be transferred based on a complete revalidation of already existing fully validated methods. With this mode of transfer no prior agreement is required in deciding what analytical figures of merit require revalidation. Because use of preselected

TABLE 16-6. Validation of Analytical Figures of Merit; Example of Acceptance Criteria

Analytical Figures of Merit	Acceptance Criteria
Assay	
Recovery (3×3 levels of concentration)	95–105%
%RSD	$\leq 2.0\%$
Completeness of extraction	$\leq 1\%$
Linearity	$(n \geq 6)$
Correlation coefficient (r)	≥ 0.999
Y-intercept	$\leq 2.0\%$
Residual standard deviation	$\leq 2.0\%$
Range	70–130% of label claim
Precision/repeatability	$n \geq 6$, %RSD $\leq 2.0\%$
Intermediate	$n \geq 6$, %RSD $\leq 2.0\%$
LOD	$S/N \geq 3:1$
LOQ	$S/N \geq 10:1$ and
	%RSD$(n \geq 5) = 10$–20%

well-characterized batches is also not required, the decision to revalidate is based solely on the requirements of prevailing SOPs at the receiving lab which do not require any tacit approval by the originator lab.

The other aspect of this mode of transfer is that it does not normally require comparison of data generated at the receiving lab with data previously reported by the originator lab. The receiving labs are certified to perform the tests once they complete revalidation of the methods.

This mode of transfer is sometimes employed for transferring methods from one AR&D organization to another AR&D organization or from an AR&D unit to a QC lab or CRO. Though this type of transfer can be the simplest mode of transferring methods, it may also be the most time-consuming approach because all aspects of an already existing method will have to revalidated. On the contrary, if timing of the revalidation is coordinated in a manner that coincides with the validation of the methods by the originator lab, considerable time savings can be achieved because the receiving or participating labs will not have to wait for the originator lab to complete validation of the method before they start their own revalidation exercises. Both activities can occur in parallel and not in the usual sequential fashion that tends to characterize all the other modes of transfer. However, a potential drawback for not waiting before embarking on revalidation is that any time saving that may have been gained by the receiving lab would be erased when the originator lab decides to change the method in course of their validation exercise because of last-minute discovery of unexpected analytical issues. Under those circumstances, both labs may now have to divert resources to perform additional experiments in order to resolve the unexpected analytical issue. Though not a necessary requirement for this mode of transfer, it may still be prudent for the participating labs to show that by using their versions of the methods, comparable data with the originator is obtained for the assay of similar or identical batches of the same pharmaceutical dosage form. The importance of showing comparability is borne out by the fact that subtle differences in the acceptance criterion applied for accuracy—for example, 95–105% versus 95.0–105.0% stipulated in the different local SOPs—could potentially contribute to instances of nonagreement of results between the originator and receiving labs. This fact is illustrated by the example of an assay criterion set to 95.0%, in which case a value of 95.1% would fail this criterion but would pass a criterion of 95% simply because values obtained must first be rounded to the same number of decimal places as in the acceptance criteria before comparison against the acceptance criterion can be performed.

16.4.4   Waiver of Transfer

Waiver is the mode of transfer in which no *interlaboratory* cross-over testing is performed. This is the least used mode of transfer, and it is evoked only under certain special circumstances. An example of such a circumstance is when the method to be transferred can universally be acknowledged and accepted as relying on technologies that are considered standard and routine analytical techniques that have been in use at the receiving lab for a considerable period of time. Identification by visual examination and uniformity of dosage unit by a weighing operation are two examples that merit consideration for waiver transfer. Another circumstance under which a waiver may be justified is when it can be proven that an earlier version of a method has routinely been used to test products at the receiving lab facility. A strong case for waiver consideration is when it can be demonstrated that differences between the earlier and newer version of the method are attributable mainly to minor editorial revisions to the method and changes involving a switch from a shorter to a longer column of the same column chemistry or a change in the sample preparation procedure by changing the sequence of addition of extraction solvent or an increase in shake or sonication time to account for a more robust extraction of analytes.

A poignant argument that can be made for waiver of transfer is when personnel from the lab where the methods were originally developed and validated are transferred to a brand new lab in a completely different facility. In this case the granting of waiver is justified because the expertise required for performing the methods already exists among the personnel at this new site assuming that any new personnel performing testing in the different facility of the same company have been adequately trained using the intended analytical modalities.

In spite of such arguments, it must be noted that even when a waiver transfer is judged to be justified and acceptable, it may still be prudent to expect a minimum interlaboratory cross-over testing to provide continued assurance that the receiving lab can reliably and competently perform the tests.

16.5 POTENTIAL PITFALLS DURING TECHNICAL TRANSFER AND MANUFACTURING

Because the success or failure of any technical transfer is judged primarily on the merits of the agreement data between the originator and participating labs generated using HPLC, a number of factors must be taken into consideration to ensure the mitigation of differences in results.

16.5.1 Sample Handling

The impact of sample handling as innocuous as it may appear can have a significant impact on the interpretation of results. Figure 16-4 shows the presence of an unknown peak (i.e., leachable peak B) that appeared in the chromatogram of a sample that was prepared by a chemist in a receiving lab but was absent in the chromatogram from the originator lab. Follow-up investigation confirmed that the source of the unknown peak initially reported by

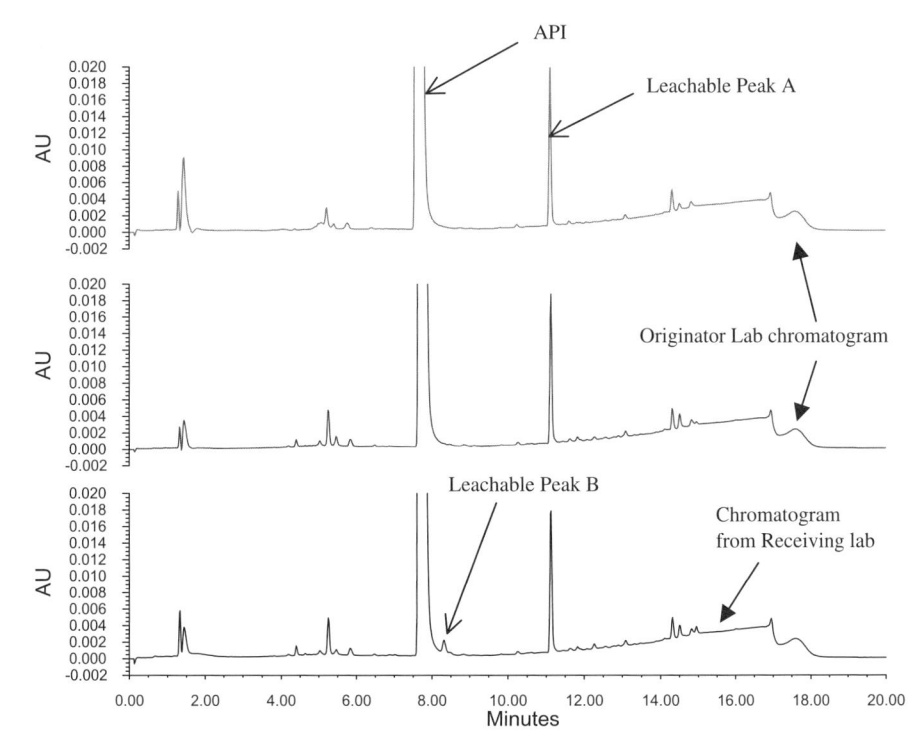

Figure 16-4. Leachable peaks from nitrile gloves found in the sample. HPLC conditions: Column temperature 35°C, column: YMC ODS-AQ, S-3, 120A, 3.0 mm × 150 mm, wavelength, 267 nm. Gradient Mobile phase A: Water/acetonitrile/TFA (950/50/1, v/v/v). Mobile phase B: Water/acetonitrile/TFA (50/950/1, v/v/v). Injection volume, 15 mL; flow rate, 0.8 mL/min.

the receiving lab as degradation peaks was indeed a leachable peak from the nitrile gloves that the chemist wore during the preparation of the samples. Incidents such as this goes to highlight the need to exercise caution in the choice of gloves that can be worn during sample preparation procedures [26, 29]. During the execution of sample preparation procedures, care should be taken to avoid spillage of solvents onto gloves, and steps should be taken to replace soiled gloves with new ones.

Sonication is a popular sample preparation procedure that is widely used for sample preparation of solid dosage forms such as tablets and capsules. However, due to the intense sound waves that sonicators produce, different models of sonicators occasionally produce different degrees of local hot spots [26, 27]. Also, sonication may also be dependent on where the solution is placed in the bath, and the water level in the bath. The chromatograms in Figure 16-5 show increased formation of two degradation product peaks A and B that occurred in a sample as the sonication time was extended from 20 to 30 minutes. Due to the potential formation of degradation peaks that could contribute to nonagreement between results from different labs, explicit instructions about the duration of the sonication time that should be applied to the samples, power output of the sonicator, level of water in the sonicator, and where samples should be placed within the sonicator should always be provided [26–29].

Another source that contributes to frequent differences in results during technical transfers is inefficient extraction. This is often attributed to the

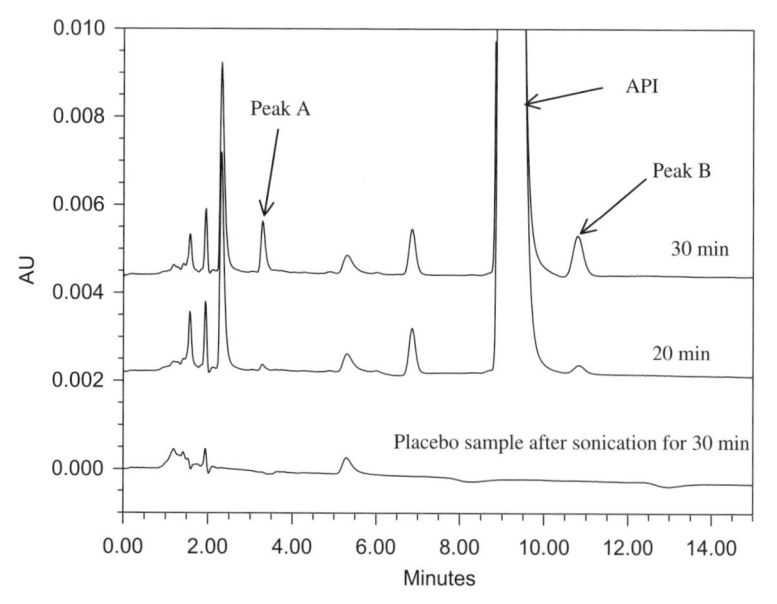

Figure 16-5. Degradation product peaks A and B formed from prolonged sonication of samples.

procedures employed in the preparation of the sample. For example, in a procedure that requires an initial sonication step followed by shaking, it was observed that different results were obtained when samples were sonicated for about 30 minutes with an accompanying occasional shaking of the flasks versus a combination of sonication followed by vigorous swirling of the flasks. As much as 3% difference between results was obtained from various studies in the literature [26, 28–30]. The type of extraction solvent and the manner or order in which the solvent is added to the sample has also been implicated in differences observed between results from different laboratories. For example, in the extraction of β-estradiol from tablets, results obtained varied based on how the extraction solvent was added. When the tablets were pre-wetted by the addition of water followed by the addition of the methanol, higher results were obtained than when a premixed extraction solvent of water/methanol was used [29]. Adsorption of analytes on the surfaces extraction vessel (i.e., glass or plastic) is another classical source that contributes to differences in results.

The order of addition of solvents in the preparation of mobile phase has been reported as causing significant differences in retention factors (k) of analytes, resulting in critical degradation peaks being missed and incorrect assignment of peaks [29, 30]. In the following example, it was noted that the preparation of the mobile phase required addition of 600 mL of methanol to a graduated flask, and then adding water to a target volume of 1 L gave a k value of 2.32 for a pivotal degradation peak. However, a k value of 3.05 for the same peak was obtained when the order of addition was reversed by making the mobile phase a 1-L volume by the addition of 400 mL of water first to the graduated flask and then adding methanol to the 1-L mark [10, 29, 30]. In this example, differences in the ensuring exothermic reaction due to the sequence of mixing water and methanol that caused volume expansion was attributed to be the cause for the observed k value differences. In another reported example, it was noted that chloroform containing no preservative versus chloroform containing 0.5% ethanol as preservative was responsible for the same types of compounds to elute differently [29–31] using normal-phase chromatography. Additionally, slight changes in pH of the mobile phase is also a well-known source of difference that results in resolution difference of a critical pair of components that arise from lab to lab. In some instances, slight changes in mobile-phase pH have been reported to even cause peak retention reversals [10, 26, 29]. Therefore, precise instructions for the preparation of mobile phases including pH adjustment must be provided to ensure that uniform procedures for mobile-phase preparation are followed.

16.5.2 Sample Type and Number of Replicate Determination

One of the challenges faced during the transfer of HPLC methods is to ensure that samples used in the transfer are from identical batches of materials. Associated with this challenge is the determination of the appropriate number of replicate tests that would be required to demonstrate comparability between results from different labs. This issue becomes even more important especially in situations when either out-of-specification (OOS) or out-of-trend (OOT) results are obtained. An OOT or OOS result triggers a thorough investigation of the reliability of the analytical methods as well as the manufacturing process in an attempt to identify an assignable cause [22]. In some cases an assignable cause is found, whereas in many other cases the investigation may not yield any assignable cause and thus leads to the performance of additional testing. An approach that has been proposed to aid in the selection of the appropriate number of tests to be performed is the determination of a K criterion for successive repeat testing [18]. Although the primary focus of the original application of the K criterion discusses the number of retests that will have to be performed in order to nullify an initial suspect result, the concept can be adopted in deciding the number of replicate tests that will have to be performed to assure batch-to-batch variability in the dosage form does not unduly influence interpretation of results during the transfer of methods. In the original concept, of the K criterion, it was used primarily to assign the number of additional tests that will have to be performed to assure a 95% chance of achieving agreement between results from different analysts. As shown in Table 16-7, the probability of performing additional tests to achieve agreement between results decreases as the number of repeat testing is increased. In other words, the likelihood of meeting the acceptance criterion becomes greater as more tests are performed. For this very reason, indeterminate testing till compliance is achieved is precisely the rationale behind predetermining the appropriate number of replicates tests to be performed becomes paramount in avoiding situations in which testing continues until agreement is achieved [22].

16.5.3 Time of Testing

As trivial as the time of testing may be, the time interval between when testing is performed has been identified as a potential source of difference between

TABLE 16-7. Number of Retests Required to Achieve Agreement Between Results

Number of Retest, K	True % of Suspect Results with 95% Chance of Meeting Acceptance Criteria
3	1.70
4	1.27
5	1.02
6	0.85
7	0.73
8	0.64

Source: Adapted from J. D. Hofer, *Pharm. Technol.* 27(11) (2003), 62.

results obtained for testing of identical materials [19]. To minimize the contribution of testing time differences, it is recommended that all participating labs perform testing at about the same time and usually a 1 to 2-week time difference may be considered acceptable. But when the time interval that had elapsed between when one lab completed testing versus other participating labs is more than 2–3 months apart, then it is advisable for all labs to repeat the test at about the same time.

16.5.4   Instrumental Issues

Tubing and fitting differences between HPLC instruments may contribute to instances of nonconvergence of results between labs. Depending on the instrumental setup and the length and diameter of the tubings used, performance of the HPLC system can be adversely impacted because of the creation of additional extra-column volumes [10, 31, 32]. Extra-column volumes dilute the chromatographic zone, leading to differences in retention behavior and peak efficiency and hence may lead to the potential for wrong peak assignments and subpar resolution between critical pairs. An additional factor to consider is fitting connections, since poor connection can led to a mixing chamber from between the end fitting and the column and lead to poorer chromatographic performance. This is especially true if methods are transferred between a Waters® instrument and a non-Waters instrument because Waters® end fittings tend to have a much deeper bore depth [10, 32].

Also, sometimes, interlaboratory differences in results have been traced to dwell volume differences between the different models of HPLC equipment [26, 30, 32]. This is particularly true of gradient HPLC methods in which the failure to make the appropriate adjustments to the gradient program due to the dwell volume differences results in differences in the retention behavior of the same analytes between labs [10]. In gradient methods, dwell volume can have a deleterious effect on the gradient program if care is not taken to ensure that appropriate adjustments are made to the gradient program to account for differences in dwell volume of instruments.

16.5.5   Column and Instrumental Issues

16.5.5.1   Column Oven Temperature. Though HPLC is regarded as mature and an established technique, problems are frequently encountered when methods are transferred between laboratories. One major uncontrollable variable that has been identified is the temperature of the chromatographic method [33]. Smith et al. reported that different methods of heating the column, the design of the column oven, and the lengths of connecting tubes and the nature of the tubing connecting the injector to the column had a marked effect on the effective internal temperature in the column and on both the retention and efficiency measurements. They found that the worst case of nonagreement of data between labs came from metal block ovens without a fan drive circulating air supply. The most efficient ovens are the ones that use circulating air ovens probably because these do not create local hot spots by being able to more efficiently dissipate the heat. Temperature-induced retention shifts have been reported to have caused serious issues leading to the wrong identification of critical peaks [34]. In *LC-GC Application Notebook* (February 2004, p. 60), shifts in retention times were observed as temperature was increased from ambient to 40°C. In that article, it was reported that the retention time of pyrene decreased from 13.1 to 10.0 min and that similar but much less pronounced shifts were noted for fluorene (7.40 min < 6.5 min < 6.1 min) and naphthalene (4.7 min < 4.2 min < 4.0 min). In order to circumvent minor shifts in retention with respect to temperature, Smith et al. recommended (a) specifying the length of tubing connecting the injector to the column and (b) the use of a calibrated thermometer to adjust column oven temperatures as an approach to minimize temperature-induced retention time shifts.

16.5.5.2 Columns Robustness. One of the significant hurdles to be addressed is specification of the column type to use. Currently, the USP classifies all C18 columns as type L1. Unfortunately, this classification does not provide any rational system to differentiate among the myriad differences that exist among C18 columns with respect to selectivity and retention characteristics [35–37]. Variability in selectivity and retention may exist even in the same silica-based columns sold by the same vendor or packed by different vendors [36]. The difference could potentially be attributed to variations in the base silica and/or the different packing procedures (i.e., solvents used) by the different vendors. The absence of a standardized column characterization system that can help to determine that a specified C18 column is equivalent to another coupled with the reluctance of column vendors to reveal the functionality of their stationary phases poses the greatest challenge in ensuring that equivalent results are obtained during transfer of HPLC methods if different L1-type columns are used. In addition, peak reversals from one C18-type column to another as well as the change in resolution as columns age are also issues that can lead to different labs obtaining different or similar results [26, 30, 36].

Column age can have multiple meanings:

1. Columns are brand new but have not been used (sitting in a box for multiple years).
2. Columns have been exposed to multiple volumes of mobile phase.

For point 1, the vendor may have changed the manufacturing or packing procedure for a particular bonded phase or column, respectively. The column vendor may have performed internal tests according to approved change control and other compliance procedures and have ensured with the probes they are using that the columns manufactured with the new process have not changed their inherent nature with respect to selectivity and retention of the probes compared to columns produced with prior manufacturing procedure. However, these manufacturing changes can inadvertently affect the selectivity for the separation performed in the pharmaceutical company, since the manufacturing test probes are inept in discriminating between the selectivity differences. It is recommended to discuss with the vendor if any manufacturing changes have been implemented for a particular stationary phase. If so, columns should be obtained from lots after the manufacturing change was implemented.

For point 2, column performance can be adversely affected by pumping multiple volumes of mobile phase through the column due to (a) changes in the bonded-phase stability at various temperatures, pH values, and pressures and (b) accumulation of retained analyte components on the stationary phase. The maximum pH, temperature, and backpressure for which the column can operate versus the maximum number of column volumes should be known in order to avoid discrepancy in the data generated between laboratories (new versus aged column). For method transfer it is suggested that brand new unused columns (manufactured less than one year) should be used.

Therefore, it is imperative to make sure column lot-to-lot reproducibility is demonstrated in the validation phase and are packed by the same vendor, and the vendor should be asked if any manufacturing changes have occurred for a particular column in order to minimize column robustness issues.

16.5.5.3 Column Frits.

Buildup of particulate matter on the column inlet frit leads to plugging that causes peak splitting and hence difficulty in the correct assignment of peaks [32, 36, 37]. Plugging due to particulate matter could be due to poorly filtered samples or the natural wear and tear of pump seals due to age-releasing materials into the flow path which eventually collects on column inlet frit. Similarly, particulate matter emanating from the normal wear and tear of injector values into the flow path accumulates on the column inlet frit. To reduce column plugging, solvent reservoirs should be kept clean and covered; preferably, samples should be filtered before injection and pump seals and autoinjector seals should routinely be replaced [29, 32, 37].

16.5.6 Differences in Chromatographic Data Acquisition Systems

Modern HPLC systems equipped with analog-to-digital converters (ADCs) and initial data acquisition and signal analysis are performed by either built-in microprocessor or attached computer [38, 39]. Depending on the data acquisition rate and initial analysis criterion (peak height, peak area, or sensitivity thresholds, etc.), the integration limits or area calculation could be erroneous, and in some cases the whole peaks could be missed.

Though most modern data acquisition systems can either sense or adjust integration parameters to accommodate baseline drift or permit manual setting of parameters to overcome drift-induced problems, it is still easy to mistake a system peak arising from the mobile phase as an impurity if the appropriate integration parameters are not chosen [37–39]. Thus, to minimize errors due to missing peaks or mistaken peaks, components of the data-acquisition system should be tested regularly to ensure that the chromatographic data system is operating properly [37–40]. Moreover, differences in pre- and post-data processing can cause variations in peak profiles and chromatographic performance (efficiency, peak width, S/N). For example, different values for digital filtering in pre-data processing and differences in values for data smoothing (Savitzy–Golay) in post-data processing can lead to differences in peak profiles. It is recommended to do post-data processing so original data is kept and no analytical information is lost.

16.6 CONCLUSION

Technical transfer and manufacturing is accomplished through sampling and analyzing of batches and statistical evaluation of data generated using HPLC. Statistical evaluation is done by comparing data generated against predetermined acceptance criteria. From the very conception of the drug development continuum, reliance on data generated using HPLC is ubiquitous. Progression from one phase of the drug development continuum to the next is fueled among other things by the assessment of data generated using either HPLC alone or hyphenated HPLC techniques (i.e., LC/MS, LC/NMR, etc.).

This heavy reliance on data generated using HPLC in the go/no-go decision gates along the drug development continuum highway requires that HPLC methods are properly validated and transferred in an efficient fashion. Efficient and seamless transfer of HPLC methods can avoid any delays in the regulatory submission process and the ultimate introduction of new pharmaceutical dosage form to the public.

Hence, it is proper to conclude that the role of HPLC in technical transfer and manufacturing is at the center of the universe of the drug development continuum aiding in the progression from discovery through development to lab, to pilot scale, and finally to commercial scale-up manufacture and production.

REFERENCES

1. Method concerns addressed in New Industry Tech Transfer Guide: Gold Sheet, *Pharm. Biotechnol. Control* **37**(7) (2003), 1–9.
2. C. H. Dubin, Scale-up & technology transfer, from the laboratory to commercial production, *Pharm. Formulation Quality* **Oct/Nov** (2003), 19–21.
3. J. Etse, utility of hplc as a tool in the transfer of analytical technology, *Am. Pharm. Rev.* **January/February 8**(1) (2005), 74–77.
4. ICH-Q2A Guideline for Industry, *Text on Validation of Analytical Procedures*, 1995.

5. ICH-Q2B Guidance for Industry, *Validation of Analytical Procedures*, Methodology, November 1996.

6. United States Pharmacopeia USP 28-NF 23, US Pharmacopeial Convention, Inc., 2005.

7. European Pharmacopeia, 5th edition, Council of Europe, 67075 Strasboury Cedex, France, 2005.

8. ICH-Q3A Guidance for Industry, *Impurities in New Drug Substances,* November 2003.

9. ICH-Q3B(R) Guidance for Industry, *Impurities in New Drug Products,* November 2003.

10. L. R. Sunder, J. J. Kirkland, and J. L. Glajch, *Practical HPLC Method Development,* 2nd edition, John Wiley & Sons, New York, 1997.

11. J. F. Pritchard, M. Juirma-Romet, M. L. J. Reimer, E. Mortimer, B. Rolfe, and M. N. Cayen, Making better drugs: decision gates in non-clinical drug development, *Nature Rev. Drug Discovery* **2**(7) (2003), 542–553.

12. T. Kenakin, A guide to drug discovery, Predicting therapeutic value in the lead optimization phase of drug discovery, *Nature Rev. Drug Discovery* **2** (2003), 429–438.

13. M. Freemantle, Scale-up synthesis of discodermolide, *Chem. Eng. News* **March 1** (2004), 33–35.

14. J. T. Carstensen, *Pharmaceutical Preformulation*, Technomic Publishing Company, Lancaster, PA, 1998.

15. PQRI Stratified Sampling Approach to Drug Uniformity, *The Pink Sheet* **May 13** (2002), 25.

16. US Food and Drug Administration Code of Federal Regulations (2004), *Part 211. Sampling and testing of in-process materials and drug products,* Title 21, Vol. 4, Section 211.110.

17. J. S. Bergum, Constructing acceptance limits for multiple stage test, *Drug Discovery Indust. Pharm.* **16**(14) (1990), 2153–2166.

18. J. D. Hofer, Considerations when determining routine sample size for a retest procedure, *Pharm. Technol.* **27**(11) (2003), 60–68.

19. S. Burke, Analysis of variance: Statistics and data analysis, *LC-GC Europe Online Supplement* **January 2** (2001), 9–12.

20. S. Scypinski, D. Roberts, M. Oates, and J. Etse, Pharmaceutical Research and Manufacturing Association acceptable analytical practice for analytical method transfer, *Pharm. Technol.* **March** (2002), 84–88.

21. US Food and Drug Administration Code of Federal Regulations (2004), *Part 211, Current good manufacturing practice for finished pharmaceuticals,* Title 21, vol. 4, subparts E, F, G and I.

22. Barr Court Decision, *United States of America, Plaintiff, v. Barr Laboratories, Inc.,* et al. Defendants, Civil Action No. 92–1744, 1992.

23. S. K. Ritter, Material safety data sheets eyed, *Chem. Eng. News* **83**(6) (2005), 24–26.

24. US Department of Labor, Occupational Safety & Health Administration: Regulations (Standards 29 CFR), Part 1910.1200 App A, Health Hazard Definitions.

25. S. Bolton and C. Bon, *Pharmaceutical Statistics; Practical and Clinical Applications,* 4th edition, Marcel Dekker, New York, 1997.

26. J. J. Kirschbaum and R. E. Majors, Trends in sample preparation, *LC-GC* **20**(12) (2003), 1098–1113.

27. L. A. Crum, Bubbles hotter than the sun, *New Scientist* **146** (1995), 36–40.

28. K. S. Suslick, ed., *Ultrasound: Its Chemical, Physical and Biological Effects*, VCH Publishers, New York, 1988.

29. J. J. Kirschbaum, Inter-laboratory transfer of HPLC methods: Problems and solutions, *J. Pharm. Biomed. Anal.* **7**(7) (1989), 813–833.

30. J. W. Dolan, *LC-GC* **2** (1984), 582–587.

31. J. L. Glajch, J. J. Kirkland, and J. J. Kohler, *Chromatogr.* **384** (1986), 81–90.

32. Technical Report, *A Troubleshooting Guide to Plumbing Problems in HPLC*, MAC-MOD Analytical Inc., Chadds Ford, PA.

33. R. M. Smith, P. V. Subba Rao, S. Dube, and H. Shah, Problems of the interlaboratory transferability of the measurement of the properties of a reversed-phase HPLC column, *Chromatogr. Suppl.* **57** (2003), S27–S37.

34. D. A. Isom, Effective temperature method development, *The Applications Notebook*, *LC-GC* **February** (2004), 60.

35. P. Petersson, RPLC Column classification and the development of a column selection tool, *ACD/Labs European Users' Meeting*, September 23–24, 2003, Obernai, France.

36. *Comparison Guide to C18 Reversed Phase HPLC Columns* (April 2003), 2nd edition, MAC-MOD Analytical Inc., Chadds Ford, PA.

37. J. W. Dolan, Understanding split peaks, The application notebook, *LC-GC* **February** (2004), 82–84.

38. E. Grushka and S. Levin, *Analytical Chemistry* **57** (1985), 1830–1835.

39. G. I. Ouchi, Chromatographic data acquisition: Analog to digital conversion and raw data storage, *LC-GC* **9**(7) (1991), 474–477.

40. G. I. Ouchi, Peak detection and integration, *LC-GC* **9**(9) (1991), 628–633.

PART III

HYPHENATED TECHNIQUES AND SPECIALIZED HPLC SEPARATIONS

17

DEVELOPMENT OF FAST HPLC METHODS

Anton D. Jerkovich and Richard V. Vivilecchia

17.1 INTRODUCTION

Developing fast high-performance liquid chromatography (HPLC) methods can improve work efficiency during research, development, or production of a drug substance or a drug product. HPLC is a key technique in all of these areas. Until recently, analysis times of greater than 30 minutes were common. Modern pharmaceutical R&D, with its high-throughput screening, demands high-throughput methods to deal with the large number of samples. To reduce production cycle time, fast HPLC methods are essential for on-line or at-line process control and for rapid release testing. Consider a GMP laboratory responsible for releasing a single batch of drug substance. Assuming a run time of 30 minutes and a total of 12 injections, a run time of 6 hours would be required to cover system suitability, calibration, and sample analysis. If the run time were 5 minutes, only 1 hour would be required for the analysis. With the advent of commercial chromatographic porous media of less than 5 μm and more recently in the 1- to 2-μm range, analyses times of less than 1–2 minutes have been demonstrated. Hundreds of samples which required days can now be analyzed in less than a day. This chapter will focus on how to optimize isocratic and gradient methods for speed without sacrificing resolution. In addition, the implication on selection of column dimensions and media particle size on the speed of methods development will also be discussed.

Reducing chromatographic media particle size allows the number of theoretical plates per second to be increased. However, due to the resolution

HPLC for Pharmaceutical Scientists, Edited by Yuri Kazakevich and Rosario LoBrutto

dependence on $N^{1/2}$, doubling of N will only increase resolution by $2^{1/2}$. As discussed below, a reduction in particle size can lead to a pressure limitation due to the inverse dependence of pressure drop to the square of the particle diameter and the maximum operating pressure of the chromatograph. The key to optimizing speed is to maximize selectivity, α. Maximizing selectivity for the critical separation pairs will allow the shortest column lengths and highest mobile-phase linear velocity. Short columns, 3–10 cm packed with particles in the 1- to 3-μm range, provide high-speed analyses while maintaining reasonable pressure drop. Due to the fast analysis time of these short columns, method development time can also be shortened. Multiple columns can be rapidly screened for optimizing selectivity. Short columns are especially useful when the components to be separated are known. However, when dealing with complex samples with unknown components such as forced decomposition or biological samples, using longer columns may be more judicious to achieve optimum separation of critical components. After selectivity optimization, the method can be optimized for speed by reducing column length. The discussion in this chapter will focus on optimizing speed of analysis and not on selectivity. The reader is referred to Chapters 4 and 8 on how to optimize selectivity.

17.2 BASIC THEORY

To understand how to optimize a separation for speed, it is worth revisiting some of the theoretical concepts developed earlier in this text. The analysis time, t_a, is the time it takes for all sample components to elute off a column at a certain flow rate and is given by

$$t_a = \frac{L}{u}(1+k) \qquad (17\text{-}1)$$

where L is the column length, u is the linear flow velocity of the mobile phase, and k is the retention factor of the latest-eluting peak. Notice here some obvious ways to increase the speed of analysis: The length of the column can be shortened, mobile phase can be pumped at a faster flow velocity, and one can ensure that the retention of sample components is not prohibitively long. Once any of these approaches are attempted, however, it is quickly seen that other important parameters of the separation are affected, principally the resolution and the column backpressure. These parameters must be considered when enhancing the speed of analysis. Ideally, the analyst would like to maximize both resolution and speed of analysis, while remaining within the pressure capabilities of the instrument. What is discovered, though, is the inevitable existence of a trade-off between resolution, analysis time, and backpressure. Resolution can be enhanced if more time is allowed; conversely, analysis time can be shortened, but at the expense of resolution. In addition, both resolution and speed are limited by the constraints of the instrumentation. The interrelationship between these factors will be considered, starting with the most important parameter describing the quality of our separation—resolution.

17.2.1 Resolution and Analysis Time

The practical goal of most separations is not to achieve the greatest resolution possible, but rather to obtain sufficient resolution to separate all components in the shortest amount of time. To optimize for speed, the starting condition is that there is a minimum resolution requirement for the separation. Resolution is a function of three parameters: column efficiency, or theoretical plates (N), selectivity (α), and the retention factor (k):

$$R_s = \left(\frac{\sqrt{N}}{4}\right)\left(\frac{\alpha-1}{\alpha}\right)\left(\frac{k_2}{1+k_2}\right) \qquad (17\text{-}2)$$

Selectivity and retention are influenced by the choice of column chemistry and the mobile phase and gradient conditions. Due to the trade-off between resolution and analysis time, any "excess" resolution that can be generated beyond the minimum requirement can theoretically be traded for shorter analysis times. In this regard, the power of selectivity cannot be underestimated, especially when α is close to 1. For example, Karger et al. [1] have shown that an increase in α from 1.05 to 1.10 can result in more than a three-fold reduction in analysis time. High selectivity also lessens the required theoretical plate count necessary to resolve all components, which allows use of a shorter column to speed up the analysis. Consequently, choosing a column or using mobile-phase conditions that produce a high relative selectivity between critical peak pairs can be very advantageous for achieving fast methods. In addition, resolution as well as analysis time depends on the retention factor. For isocratic conditions, the optimum k for resolution and speed occurs in the range of 1–10 [1]. For samples containing many components or with analytes of wide-ranging polarity, gradient elution must then be used to achieve reasonable analysis times. Optimizing selectivity and retention so as to maximize resolution and minimize analysis time in gradient separations is discussed further in Section 17.6.

Beyond these two parameters, the minimum resolution that must be achieved will require a certain number of theoretical plates, which can be expressed in terms of the column length and plate height, H, as

$$N = \frac{L}{H} \qquad (17\text{-}3)$$

From this equation, column efficiency scales directly with column length and is inversely proportional to the plate height. Solving this equation for L and substituting into equation (17-1) results in a useful expression that more clearly relates analysis time to the quality of the separation:

$$t_a = \frac{NH}{u}(1+k) \tag{17-4}$$

Note that if the plate height (H) remains constant, an increase in the required plate number (N) will require a proportional increase in the analysis time. This is because for a fixed plate height, an increase in plate number must be obtained by an increase in the column length. Here one encounters the trade-off between resolution and speed. While it is desirable to use a short column to limit analysis time, it is also seen that a longer column provides a higher plate count and resolution. However, resolution increases not with N, but with $\sqrt{N}$, meaning the gain in resolution from lengthening the column will always be proportionally less than the price paid in time. Consequently, for fast analyses, columns no longer than that which gives the minimum theoretical plates to adequately resolve all peaks should be used.

Note also that t_a varies with the ratio H/u. Equation (17-3) shows that reducing the plate height is one way to obtain higher theoretical plates without increasing the column length. Now it is seen that for a fixed plate number (the plates needed to achieve the resolution requirement), decreasing the plate height will shorten analysis times by allowing use of a shorter column. As discussed in the next section, though, plate height is dependent on the linear velocity. Thus, when optimizing for speed, the two must be considered together. The goal, then, is not just to reduce H, but to minimize H/u. This will favor both high resolution and short analysis times. Minimizing H/u, then, encompasses the heart of what is desired in a fast HPLC method—greatest *resolution per unit of time.*

Exploring this concept a little further, knowing that $H = L/N$ and $u = L/t_0$, substituting in these relationships results in

$$\frac{H}{u} = \frac{t_0}{N} \tag{17-5}$$

This is known as the "plate time" and has units of seconds. It is equivalent to the amount of time it takes to generate one theoretical plate. Its inverse would be "plates per second," N/t_0. Plates per second may also be expressed more generally as N/t for elution times other than the void time [2, 3]. These terms more effectively describe the criteria of resolution per unit time that are desired to be maximized (actually, N/t is proportional to resolution squared per time); unfortunately, they are not widely used in the literature, and for the sake of continuity will not be used in this discussion. The following sections will look at what influences plate height and velocity and how best to minimize H/u.

17.2.2 Plate Height and Band-Broadening

Plate height is a measure of peak-broadening and column performance: Reducing or eliminating sources of band-broadening should be a main goal when choosing columns and instrumentation, and otherwise developing methods. Plate height can also be described in terms of its dependence on the linear flow velocity, u, by the van Deemter equation [4]:

$$H = A + \frac{B}{u} + Cu \qquad (17\text{-}6)$$

where A, B, and C are the coefficients for "eddy" diffusion, longitudinal diffusion, and resistance to mass transfer, respectively. A plot of H versus u is often referred to as a van Deemter plot and is shown in Figure 17-1 along with plots of the individual terms that comprise it. While other, more complex and theoretically correct equations have been derived [5–8], the simplicity of the van Deemter equation makes it useful in understanding sources of band-spreading and how to minimize them. Each of the three terms in the equation represents a contribution to the broadening of a peak and will be examined in more detail.

The A term of the van Deemter equation is independent of the mobile-phase linear velocity and describes the broadening that occurs due to the multiple flow paths present within the column. Since these paths are of different lengths, molecules will travel different distances depending on what flow paths they experience. For a column bed of randomly packed particles, the A term is proportional to the particle diameter, d_p, and to a factor λ related to the packing structure:

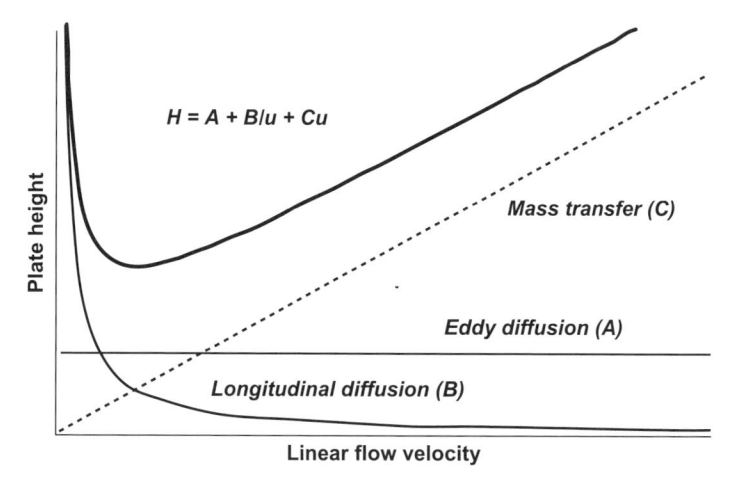

Figure 17-1. van Deemter plot showing contribution of individual terms.

$$A = \lambda d_p \tag{17-7}$$

The B term describes broadening due to axial molecular diffusion and is inversely proportional to the linear velocity. In other words, the faster an analyte zone migrates through the column, the less broadening due to diffusion it will experience. The B term coefficient is given by

$$B = 2\gamma D_M \tag{17-8}$$

where D_M is the diffusion coefficient of the analyte in the mobile phase, and γ is the tortuosity or obstruction factor, accounting for the obstruction to diffusion presented by the packing material.

The C term, or resistance to mass transfer term, is a complex agglomeration of all broadening that becomes worse with increasing flow velocity. Multiple contributions to the C term have been described; however, for the purposes of this discussion the focus will only be on the relationships relevant to improving resolution per unit time. In general, C is related to the diffusion coefficient D of the analyte in the medium through which mass transfer is taking place, and it is also related to the square of the distance d over which it occurs. Fast diffusion and short diffusional distances aid mass transfer and reduce band-spreading; hence, the C term takes the form

$$C \propto \frac{d^2}{D} \tag{17-9}$$

For example, for the mass transfer in the bulk mobile phase between stationary-phase particles, D becomes the diffusion of the analyte in the bulk mobile phase, D_M, and d becomes the distance between particles, which is roughly proportional to the particle diameter, d_p. The mobile-phase C term expression C_M can therefore be approximated as

$$C_M \propto \frac{d_p^2}{D_M} \tag{17-10}$$

When looking at the individual plate height equations, some important relationships are noticed. The B term worsens at slower flow velocities and with faster molecular diffusion. In contrast, C-term broadening worsens at faster velocities, but improves with faster molecular diffusion. These opposing phenomena are what cause the van Deemter curve to possess a minimum plate height at some intermediate velocity (the optimum velocity, u_{opt}). It can also be seen from Figure 17-1 that the increase in plate height is more abrupt at the low velocity end of the curve (where broadening is dominated by the B term) than it is at the high velocity side (where the C term is dominant). Since conditions that favor speed are desired, operating at velocities greater than the optimum velocity without significantly sacrificing efficiency is advantageous.

Although the B and C terms exhibit opposite relationships with analyte diffusion, the C-term relationship is mainly of interest because resistance to mass transfer is the dominant form of band-spreading at the faster velocities that are desired. Equations (17-9) and (17-10) imply that speeding up diffusion will increase mass transfer and help decrease plate height. The Wilke–Chang equation [9] shows that diffusivity is directly proportional to temperature and inversely proportional to viscosity:

$$D = 7.4 \times 10^{-8} \frac{\sqrt{\Psi_2 M_2} T}{\eta V_1^{0.6}} \qquad (17\text{-}11)$$

where T is temperature, η is the solvent viscosity, V_1 is the molar volume of the solute, M_2 is the molecular weight of the solvent, and Ψ_2 is a solvent association factor. Since mobile-phase composition largely dictates the selectivity of our separation, varying the viscosity of the mobile phase directly by the selection of solvents may not be an option. Raising the temperature of the mobile phase, then, is the most effective way to speed up diffusion. It also has the added benefit of lowering the mobile-phase viscosity, thereby increasing diffusion indirectly. This all serves to reduce the plate height at faster velocities. As shall be seen in the next section, raising the temperature also speeds up the analysis by lowering the pressure drop across the column.

The plate height relationships also show that the A term is dependent on the particle diameter, and the mobile-phase C-term is dependent on the particle diameter squared. Reducing the diameter of the packing material is therefore a powerful approach for reducing plate height. The minimum attainable plate height for a column, H_{min}—that is, the plate height occurring at the optimum velocity u_{opt}—will be proportional to d_p. When operating at velocities greater than u_{opt}, the quadratic dependence of C on d_p means that the reduction in plate height is especially significant. This makes sense, since mass transfer will improve as the distances molecules must travel become smaller. That is, smaller particles result in smaller interparticle spaces and thus shorter diffusional distances. By using a smaller particle size, the slope of the C-term side of the van Deemter curve will decrease dramatically, allowing operation at higher velocities without having to sacrifice as much in resolution compared to larger particles. This is illustrated in Figure 17-2, which shows the performance of columns packed with 1.7-, 3-, and 5-μm particles. Smaller plate heights *and* higher velocities are made possible, thus considerably reducing H/u. As a result, one should aim to keep the particle diameter as small as possible.

Since the goal is to reduce analysis time by minimizing H/u while holding N constant (at the minimum required plate count), the approximation can be made that $H \propto d_p$, and therefore $N \propto L/d_p$. This means as the particle diameter is reduced, the column length must also be reduced proportionally.

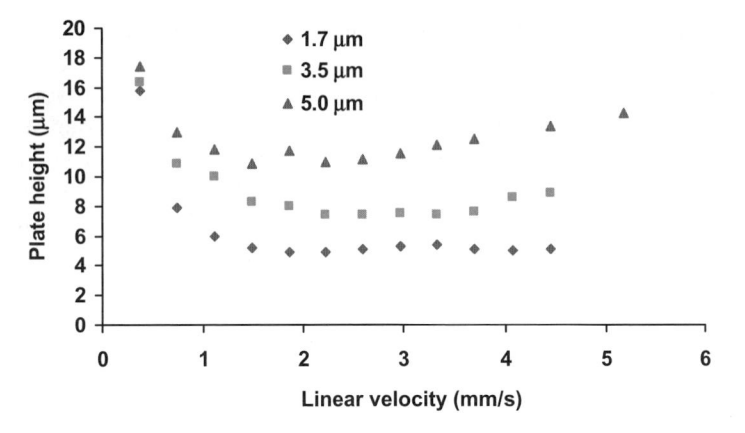

Figure 17-2. Performance of 2.1- × 100-mm columns packed with 1.7-, 3.5-, and 5-μm particles. Stationary phase was bridged-ethyl hybrid C18 prototype material in each case.

Holding L/d_p constant while both length and particle size are decreased is therefore one of the most effective means of achieving fast separations. This is the motivation seen in the evolution of chromatography columns over the last few decades. Where once the 25-cm column packed with 5-μm particles was the standard workhorse analytical column, now 10- and 15-cm columns packed with 3-μm particles are used. As column technology continues to improve, even shorter columns packed with particles <3 μm are being introduced.

To illustrate more clearly the effect of these variables on analysis time, reduced parameters can be used for the plate height and velocity. Reduced parameters effectively normalize the plate height and velocity for the particle diameter and the diffusion coefficient to produce dimensionless parameters that allow comparison of different columns and separation conditions. The reduced plate height and reduced velocity are expressed, respectively, as

$$h = \frac{H}{d_p} \tag{17-12}$$

$$v = \frac{ud_p}{D_M} \tag{17-13}$$

Solving these equations for H and u, respectively, and substituting into equation (17-4) yields

$$t_a = \frac{h}{v}\frac{d_p^2}{D_M}N(1+k) \tag{17-14}$$

The benefit of reducing the particle diameter on separation time is most evident here. It is also seen that increasing diffusion will speed up the analysis.

Now that the factors affecting plate height have been examined, it is time to turn to the effect of linear velocity and the limitation of pressure on the resolution per unit time.

17.2.3 Flow Velocity and Column Backpressure

It is known that increasing the linear flow velocity of the mobile phase will lead to faster separations. But since H is dependent on u, what velocity is needed to maximize the resolution per unit time (minimize H/u)? Using the van Deemter equation, H/u may be expressed as

$$\frac{H}{u} = \frac{A}{u} + \frac{B}{u^2} + C \tag{17-15}$$

From this equation, H/u approaches its minimum value of C as u becomes large. In other words, the separation should be performed at the fastest velocity possible. (Note also that this represents mathematically what was presented in the previous section; that is, in the case of optimizing for speed, the separation is dominated by the C-term.) This doesn't mean that the resolution itself will improve—on the contrary, since H generally increases with velocity when $u > u_{\text{opt}}$, resolution will worsen—but that the resolution per unit time is improving. Again, since the quality of the separation must not be sacrificed, the speed of analysis can be improved only to the point where resolution can no longer be sacrificed.

Of course, the ability to increase u depends on the pressure capabilities of the instrument, since pressure is directly proportional to velocity:

$$\Delta P = \frac{uL\eta\phi}{d_p^2} \tag{17-16}$$

where ΔP is the pressure drop across the column, η is viscosity, and ϕ is the flow resistance factor. Thus the speed of analysis is limited by the maximum pressure capability of the instrument. As a result, the most should be made of the pressure available by reducing the pressure drop across the column as much as possible.

Decreasing the column length lowers the pressure requirement proportionally, allowing use of the available pressure to gain an advantage in speed. Column efficiency, however, drops with use of a shorter column and at faster velocities. Care must therefore be taken to ensure that resolution between peaks is not lost when decreasing analysis time in this manner.

Lowering the viscosity of the mobile phase is another way to lessen the required pressure. This may be accomplished by raising the column temperature. Increasing temperature has the double advantage of allowing use of a higher flow velocity and speeding up diffusion, both of which appear in the denominator of equation (17-14). This is a strong motivator for the use of temperature above ambient conditions in order to speed up the separation. Of course, sample degradation, the boiling point of our mobile phase, stability of the stationary phase, and the capability of the column heater limit the maximum temperature that can be used. Temperatures up to about 70°C are considered routine; beyond that, columns and heaters specifically designed for high-temperature chromatography are needed. Much research has been done in the area of elevated-temperature chromatography, where interesting possibilities arise, such as the use of temperature gradients and purely aqueous mobile phases [10]. Chapter 18 elaborates on the use of temperature in chromatography for pharmaceutical applications.

The velocity we can obtain at a given pressure will also be limited by the resistance to flow presented by the column, known as the specific column permeability. In equation (17-16) the permeability is broken up into its two main components: the flow resistance parameter, ϕ, and the particle diameter squared, d_p^2, and can be expressed as

$$B_0 = \frac{d_p^2 \varepsilon}{\phi} \qquad (17\text{-}17)$$

where ε is the interstitial porosity of the column (i.e., the fraction of the total column volume occupied by the interparticle space), usually about 0.4. The flow resistance parameter is given by

$$\phi = \frac{185(1-\varepsilon)^2}{\varepsilon^2} \qquad (17\text{-}18)$$

and is purely a function of the porosity of the column—that is, the packing density. Its value is essentially fixed for a given column and out of the analyst's control. The quantity ϕ/ε, represented by the symbol Φ, has a value around 1000 for well-packed columns [11].

Reducing the particle diameter can be a powerful way to gain speed in separations. On the other hand, equation (17-16) shows an inverse quadratic relationship of pressure to the particle diameter. This strong dependence means that an enormous price in pressure is paid for reducing the particle diameter. However, it was stated previously that when reducing the particle diameter the column length can be reduced as well to keep L/d_p constant. Since pressure scales with column length, this eases the pressure requirement. But even keeping L/d_p constant, the pressure will still go up with $1/d_p$. Eventually, the upper pressure limit of the pump will be reached and it won't be possible to further reduce d_p without either a proportionally greater reduction in L, which reduces the efficiency, or a relatively smaller linear velocity, which cuts back on speed. Because u_{opt} increases in proportion to $1/d_p$, the maximum pressure of the system may not be able to reach a velocity beyond the optimum, and the plate height may suffer. Finally, as the column length becomes ever shorter, the column volume becomes smaller relative to the volume of the tubing, injector, and detector. In this circumstance, extra-column band-broadening becomes a significant issue (see Section 17.7 of this chapter). With standard commercial pumps having an upper pressure limit of ~400 bar (~6000 psi) and columns now being produced with particles <2 μm in size, that practical limit has now been reached. This demonstrates that the upper pressure limit of the instrument becomes the limiting factor for the resolution per unit time and speed of analysis that may be obtained.

There are a few approaches to the problem of the pressure limitation that have gained in popularity in recent years. Monolithic columns replace the bed of packed particles entirely and instead use a "monolithic" polymerized porous structure synthesized *in situ*. These columns are able to produce equivalent efficiencies of packed particle columns but with much lower flow resistances, enabling much higher velocities. Another approach is to develop instrumentation capable of much higher pressures than is possible with conventional systems. This technique is termed ultra-high-pressure liquid chromatography (UHPLC). With the pressure maximum now much higher, the limitations on particle size and column length can consequently be extended even further, allowing even greater resolution per unit time and speed of analysis. The subjects of monolithic columns and UHPLC are covered in the next two sections of this chapter.

17.3 MONOLITHIC COLUMNS

Both the speed and efficiency attainable by an HPLC column are ultimately limited by the maximum pressure capabilities of the instrument. In the case of particle-packed columns, decreasing the particle diameter leads to improved efficiency and speed; however, because ΔP is proportional to $1/d_p^2$, the price paid in pressure will always be proportionally greater than the gain in column performance. Monolithic columns are a viable alternative to particle-packed columns as a means to achieving efficient separations while overcoming the pressure limitation due to their comparably higher permeabilities (lower flow resistances) [12]. High efficiencies together with lower pressure drop make monolithic columns attractive options for fast HPLC.

17.3.1 Physical Properties and Preparation of Monolithic Columns

A monolithic column is a single interconnected skeletal stationary-phase support structure consisting of large-flow through-pores (1–3 μm). The key to high permeability is that this network support structure can be made in such a way that the ratio of the through-pore size to the skeleton size is much greater than can ever be obtained by a column packed with individual spherical particles. While the interstitial porosity of spherical particle-packed columns is typically ~0.4, monolithic columns exhibit external porosities of ~0.6–0.7. When the intraparticle pores in spherical particles and mesopores in the monolithic skeleton are included, the total porosities are on the order of ~0.65–0.75 for particulate columns and on the order ~0.80–0.90 for monolithic columns. The presence of the mesopores (10–25 nm) supplies sufficient surface area for retention, around 300 m²/g, which is comparable to most porous silica particles [13].

Monolithic columns are generally prepared by the *in situ* polymerization of either organic or inorganic monomers to form the skeletal support. Organic polymer monoliths are produced by nucleation and growth, followed by aggregation to form the network structure. Control of the polymerization kinetics determine the size of the macro- and mesopores. A main drawback of polymer monoliths, however, is that polymers tend to swell or shrink in the presence of organic solvent, which leads to poor chromatographic performance and a lack of mechanical stability under pressure-driven flow. Monolithic silica columns are prepared using a sol–gel method by hydrolytic polymerization of alkoxysilanes to form the skeleton. Physical features such as through-pore size and skeletal size can be more precisely controlled in the preparation of silica monoliths. In addition, the chemical and mechanical stability of silica monoliths is better than polymeric columns. However, due to shrinkage during solidification, silica monoliths cannot be prepared *in situ*, but must first be prepared in a mold, and then removed and encased in PEEK tubing before bonding of stationary phase takes place [13, 14].

17.3.2 Chromatographic Properties and Applications of Monolithic Columns

In addition to higher permeabilities, another advantage of monoliths is improved (that is, decreased) mass transfer broadening. In packed columns, flow occurs through the interstitial spaces between particles, while mobile-phase transport inside the pores occurs predominantly by diffusion. By contrast, due to the high porosity of monolithic columns, a much greater percentage of mobile phase transport is accomplished by flow. Where diffusion does occur, in the mesopores, the shorter diffusion path lengths afforded by the small skeleton sizes aids in mass transfer. This is especially true for large molecules such as proteins that have small diffusion coefficients. Accordingly, a silica-based monolith (Chromolith, Merck, Darmstadt, Germany) demonstrates efficiencies comparable to a column of identical dimensions packed with 3-μm particles, while exhibiting backpressures comparable to that of 11-μm particles [15]. Wu et al. [16] performed van Deemter analysis on columns packed with 3- and 5-μm particles and on a commercially available monolithic column (Figure 17-3). The minimum plate height of the monolithic column was similar to the 3-μm particle column; however, the slope of the high-velocity, *C*-term side of the plot was lower, enabling faster velocities.

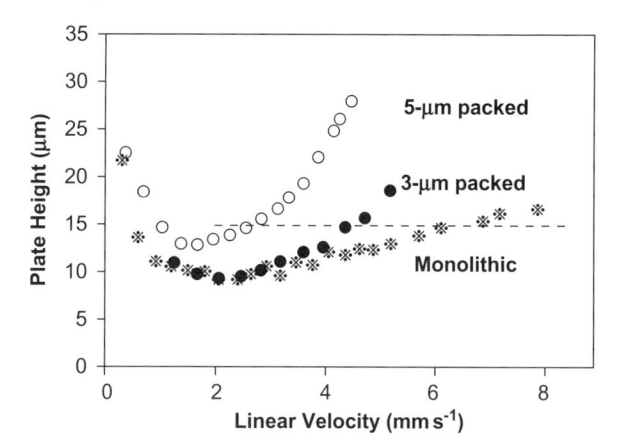

Figure 17-3. van Deemter curves for packed (YMC C_{18}) and monolithic (Chromolith) 4.6- × 100-mm columns. (Reprinted from reference 16, with permission from Elsevier.)

Silica monoliths have been applied to peptide separations [17, 18] as well as to small-molecule pharmaceutical development samples [16, 19].

Monolithic columns do have disadvantages. Although very high flow rates are used to speed up separations, this generates a considerable amount of solvent waste for ≥4.6-mm-bore columns. The number of phases and column sizes is very limited at present, as is the number of commercial manufacturers. Also, the technology of particle-packed columns is not static, but continues to improve as well. Monolithic columns have not yet demonstrated the performance capabilities exhibited by sub-2-μm particles and UHPLC. However, advances in monolithic column technology in the years to come promise to bridge that gap.

17.4 ULTRA-HIGH-PRESSURE LIQUID CHROMATOGRAPHY

The increase in resolution and speed of analysis afforded by reducing the particle diameter has resulted in a trend of using smaller particles in shorter columns. Columns packed with particles less than 2μm in size, however, challenge the pressure capabilities of conventional HPLC instrument technology, which operate up to ~400 bar. Since chromatographers generally should operate at or above the optimum flow velocity for a given column, even extremely short columns with these particles reach the system pressure limits before their full benefits can be realized. A straightforward approach to take full advantage of sub-2-μm particles is to develop instrumentation capable of the requisite pressures.

In 1997, the laboratory of James Jorgenson at the University of North Carolina [20, 21] was the first to demonstrate this approach by introducing

UHPLC. They utilized long (>50 cm) fused silica capillary columns packed with 1.5-μm nonporous silica reversed-phase particles and pressures as high as 4100 bar to achieve greater than 200,000 theoretical plates. The tremendous improvement in performance that was demonstrated over conventional columns (i.e., stainless steel tubes 3–4.6 mm in diameter packed with particles 3–5 μm in size) generated significant interest in this technique. A number of academic research labs have subsequently conducted research using UHPLC with nonporous particles, notably the laboratories of Milton Lee (Brigham Young University) [22–29] and Luis Colón (SUNY—Buffalo). [30, 31] The practical challenges and limitations of the technique have largely limited its use to research environments. However, the recent development of porous stationary-phase material in the sub-2-μm range [32, 33] and the introduction in 2004 of the first commercial instrumentation are steps toward making UHPLC a viable tool for pharmaceutical analysis.

Two pressure regimes have been described: very-high pressure LC (VHPLC), for the pressure range of about 400–1500 bar, and ultra-high pressure LC (UHPLC), for pressures >1500 bar. [20, 21, 34] This naming convention is not strictly adhered to, however, and it is often common to refer to anything above the conventional HPLC pressure limit of 400 bar as UHPLC. Ultra-high-pressure LC will find its greatest utility for complex samples containing dozens or even hundreds of components (e.g., samples of biological nature) where extremely high resolving power is needed. For such applications, long, highly efficient columns packed with micron-sized particles run at ultra-high pressures are desirable. Very-high-pressure LC is well-suited for applications where it is not so much high resolution that is needed, but fast analysis times. Samples containing less than 15–20 peaks, such as those encountered in pharmaceutical development (e.g., small-molecule pharmaceutical compounds and related impurities and degradation products), can be separated in a matter of seconds to minutes using short columns with particles 1–2 μm in size at pressures moderately higher than conventional HPLC. Although these columns offer only a marginal improvement in efficiency over conventional HPLC columns, their advantage lies in speed of analysis due to smaller particles, shorter column lengths, and higher pumping pressures (see Section 17.2 of this chapter). Of course, VHPLC used with longer columns can provide an improvement in efficiency for the separation of complex samples as well. Due to the challenges of constructing ultra-high-pressure instrumentation and manufacturing porous micron-sized stationary-phase materials, the logical first step is a chromatographic system in the very-high-pressure realm using 1.5- to 2-μm particles. This will allow a significant gain in speed and a reasonable improvement in separation power of analytical HPLC methods. Commercial products are now available that meet these goals. With continuing research and advances in instrument and column technology, it is hoped the goal of a commercial ultra-high-pressure LC system can be realized in the near future.

17.4.1 Instrument Considerations when Using Ultra-High Pressures

A number of concerns arise when using ultra-high pressures in chromatography. The most obvious is the engineering challenge associated with operating at such pressures. The pumps, pump seals, tubing, connections, valves, columns, and other hardware must be able to reliably withstand and operate at the pressures required. Careful consideration must be given to the pressure limitations of instrument components and the design of the system. This necessitates at a minimum a comprehensive improvement in existing HPLC technology and may require altogether new designs of instrument components, such as pumps, injectors and autosamplers, columns, and detectors.

For this reason, initially all UHPLC systems were custom-made instruments. Jorgenson described a constant pressure isocratic system consisting of a Haskel® air-driven pneumatic pump and high-pressure tubing and fittings capable of pressures up to 7000 bar [20]. A stainless steel static-split injection valve and column fittings were designed and constructed in-house. Slurry-packed fused silica capillary columns were prepared in-house as well. Similar systems with lower pressure capabilities were constructed in other labs [22, 30]. A largely custom-made constant flow gradient system with a pressure limit of 5000 bar has also been described [21]. While these instruments have been successful in an academic research environment, they lack the ruggedness, reliability, and ease of use required in an industrial setting. Toward that aim, Tolley et al. [34] modified a commercially available pump to achieve pressures in excess of 1200 bar for use with capillary columns 22 cm long packed with 1.5-μm nonporous particles. Finally, commercially available systems with upper pressure limits of 1000 bar have been introduced. Although this represents a moderate increase over the conventional HPLC pressure limit when compared to the systems just described, it allows use of sub-2-μm particles in a system capable of meeting the rigorous requirements for use in the pharmaceutical industry.

Special consideration must be given to the injection valve. The moving parts and sample handling requirements make sample introduction at ultra-high pressures a challenge. A number of parameters must be considered: pressure limitation, injection volume range, injection accuracy (i.e., delivery of mass on column), precision (i.e., peak area reproducibility), linearity of response versus injection volume, injection cycle time, and finally the amount of broadening to the sample plug caused by the injector. The first static-split UHPLC injectors accomplished an injection by applying pressure over several seconds to introduce a small plug of sample onto the head of the column. The actual injection volume is difficult to determine and reproducibility is poor, precluding use of this injector for quantitative analysis. A number of commercially available injectors capable of high pressures have been evaluated with UHPLC. One such system was a novel pressure-balanced rotary injection valve from Valco Instruments that was employed by Wu et al. [24]. It operated at pressures up to 1200 bar and demonstrated superior peak area reproducibility (<2% RSD, $n = 5$) compared to the custom-built static-split injectors. Recently, an air-actuated needle valve injection system, also from Valco, rated to withstand 2700 bar and capable of injection cycle times of <2 seconds was evaluated independently by both Anspach et al. [35] and Xiang et al. [36]. Injection reproducibility of ≤1.5% RSD ($n = 5$) was obtained at 2000 bar for injection volumes of 1–2.5 μL [35]. Given the challenges of performing injections under high-pressure conditions, it is strongly recommended that the injection performance of a high-pressure instrument is evaluated before developing methods for highly accurate quantitative analyses. Doing this during performance qualification (PQ) of the instrument or an initial vendor evaluation before purchasing is best. Precision (peak area, peak height, and retention time), accuracy (recovery of mass injected on column), and linearity are parameters that should be investigated. In order to routinely meet system suitability criteria, it is recommended that an instrument be able to achieve ≤1% RSD ($n = 6$) for peak area and peak height precision.

Due to the high efficiency afforded by sub-2-μm particles, it is crucial that extra-column broadening be kept to a minimum [37]. With fused silica capillary columns it is possible to introduce sample directly onto the head of the column and to perform on-column detection, thereby essentially eliminating extra-column sources of broadening. This is usually not the case with more conventional column dimensions and instrumentation. The tubing, connections, the injector, and the detector flow cell all add extra-column volume to the system which will contribute to analyte band-spreading. Extra-column broadening and instrument requirements for highly efficient columns is discussed further in Section 17.7.3.

Extremely narrow peak widths will present challenges for detection. The data acquisition rate must be sufficiently fast to sample enough data points to accurately define the peak. A good rule of thumb is to acquire at least 15 points per peak. Peaks on the order of 1 second wide will also challenge the cycle times of scanning mass spectrometers. For fast, highly efficient chromatography, a time-of-flight (TOF) mass analyzer may be the best option for very fast detection. Detector requirements for fast LC is covered in more detail in Section 17.7.2.

Currently the availability of columns packed with sub-2-μm stationary-phase material is rather limited. This situation will likely change as the use of UHPLC proliferates and the demand for such columns is more widespread. Until recently, the only stationary phase materials available in the sub-2-μm regime were nonporous silica particles. They are much easier to synthesize in this size range than are porous materials, and their mechanical strength allows them to be used at pressures up to 7000 bar [38]. They also have the advantage of greater efficiency than porous particles of equivalent size due to the lack of additional band-broadening contributions presented by the pores. All the initial work demonstrating the principles of UHPLC was consequently performed on nonporous materials. Their obvious drawback, however, is their poor loading capacity—up to 100 times less sample may be loaded onto non-porous particles compared to porous particles. This drastically limits the sensitivity obtainable using such columns and may require alternate detection schemes or derivitization of the sample. Also, very small injection volumes that challenge the capabilities of the injector may be required. As a result, the use of conventional columns packed with nonporous particles has been limited to specific applications in pharmaceutical analysis such as protein and peptide separations, where sample volumes are often very small and the slower molecular diffusivities make the absence of pores especially beneficial due to the decreased C-term band-broadening. The development of high-quality porous particles 1–2 μm in size for use with elevated pressures has therefore been a necessary and critical advancement for UHPLC [32, 33]. The superior loading capacity of these materials makes them practical for most pharmaceutical analyses. Such columns are still susceptible to the crushing of particles at high pressures, which will manifest itself as rising backpressure due to plugging of the column. Another problem that may arise is the compression of the packed bed inside the column, leaving a void at the column head which will result in distorted peak shapes. Care should be taken not to operate a column at a pressure higher than that at which the column was packed. To avoid this, columns should be packed at pressures several hundred bar greater than the maximum pressure at which it is to be used.

Finally, the safety of UHPLC must also be considered. Rupture or failure of seals, tubing, and fittings can present a potential danger to the user. With proper instrument design and normal safety precautions, however, UHPLC can be safe to use [38]. This is especially true of commercial instruments, which are no more dangerous than any other HPLC.

17.4.2 Chromatographic Effects of Ultra-High Pressures

Another concern is the potential for frictional heating inside the column. Forcing mobile phase through the column at such pressures will generate heat that may cause a significant rise in the temperature of the mobile phase [40, 41]. As heat is dissipated from the column walls, axial and radial temperature gradients will form within the column. The resulting differences in analyte diffusivity and retention within the column will lead to additional band-broadening. The power (heat) generated by flow through a packed bed is equal to the product of the flow rate (F) and the pressure drop across the column (ΔP):

$$P = F\Delta P \qquad (17\text{-}19)$$

Table 17-1 shows the power generated by pumping mobile phase through 100-mm long columns of various diameters packed with 1.5-μm particles at a linear velocity of 3 mm/s and a pressure of ~900 bar. One can see that at this pressure a standard-bore 4.6-mm-i.d. column generates 3.0 W of heat. For

TABLE 17-1. Power Generated Due to Frictional Heating of the Mobile Phase at a Linear Velocity of ~3 mm/sec in Columns of Varying Diameter Packed with 1.5-μm Particles[a]

Column Dimensions	Flow Rate	Power Generated
4.6 × 100 mm	2.0 mL/min	3.0 W
3.0 × 100 mm	0.85 mL/min	1.3 W
2.1 × 100 mm	0.41 mL/min	0.60 W
1.0 × 100 mm	92 μL/min	130 mW
0.30 × 100 mm	8.5 μL/min	13 mW

[a]In all cases, column backpressure is ~900 bar. A solvent viscosity of 1.0 cP was used for all calculations.

comparison, consider conditions typically encountered in conventional HPLC. A 4.6- × 100-mm column packed with 3-μm particles operating at 1 mL/min (corresponding to 1.5 mm/sec) will require 170 bar and generates only 0.19 W of heat. Columns larger than 2.1 mm in diameter would therefore be undesirable for pressures and conditions outlined in Table 17-1. As the particle size is reduced even further—to 1.0 μm, for example—or as the column is lengthened, the operating pressures become correspondingly greater and even more heat is generated. This pushes the largest usable column diameter to under 1.0 mm. Thus, the frictional heating effect serves as a strong motivator for the use of capillary columns in UHPLC. Patel et al. [42] investigated the effect of flow-induced heating in capillary columns up to 150 μm in diameter packed with 1.0-μm particles and found negligible effects on column efficiencies. Indeed, the vast majority of academic research in UHPLC has been performed in fused silica capillaries less than 100 μm in diameter.

Thought must also be given to the possible chromatographic effects arising from changes in the retention factor of analytes and the compressibility of the mobile phase as a function of pressure. It has been shown that retention factors for small molecules increase linearly with pressure [42]. This increase is moderate, however, and does not significantly affect analysis time. Because of the compressibility of the mobile phase at ultra-high pressures, a situation that is familiar in gas chromatography results: The volumetric flow rate will be greater at the outlet of the column than at the inlet as the compressed mobile phase expands at lower pressures. The magnitude of this difference will vary depending on the solvent composition and pressures used. In practice, this means the measured flow rate at the outlet of the column will be greater than the flow rate to which the pump piston is set. Other than some theoretical considerations, the changes of retention factor and flow rate with pressure will have no adverse effects on a chromatographic run. One consequence has been noted, however, for UHPLC systems that perform injections while the column is off-line of the pump or otherwise at atmospheric pressure [43]. In such systems, sample is introduced onto the head of the column and then pressure is subsequently applied to the column to start the run. When this occurs, the mobile phase inside the column becomes rapidly compressed in volume. This compression causes a surge in velocity at the head of the column, which contributes significantly to broadening of the sample band. An injector that performs injections while the column is pressurized must be used in order to circumvent this problem.

The use of ultra-high pressure in LC was found to have beneficial effects on protein recovery [44]. By using pressures >1600 bar, protein recovery was enhanced and carry-over from run to run was reduced and in some cases eliminated. While the mechanism of recovery is not known, it was postulated that ultra-high pressures improve desorption from the stationary phase surface by causing partial unfolding or deaggregation of the proteins.

17.4.3 UHPLC Applications

Isocratic separation of test compounds is a useful way to demonstrate the performance of a system. Basic chromatographic characteristics, such as theoretical plates, are easily measured and can be compared to what is expected from theory and to performance of other chromatographic systems. Figure 17-4 is a UHPLC chromatogram obtained under isocratic conditions on a 43-cm-long capillary column packed with 1.0-μm nonporous C18 particles (Eichrom

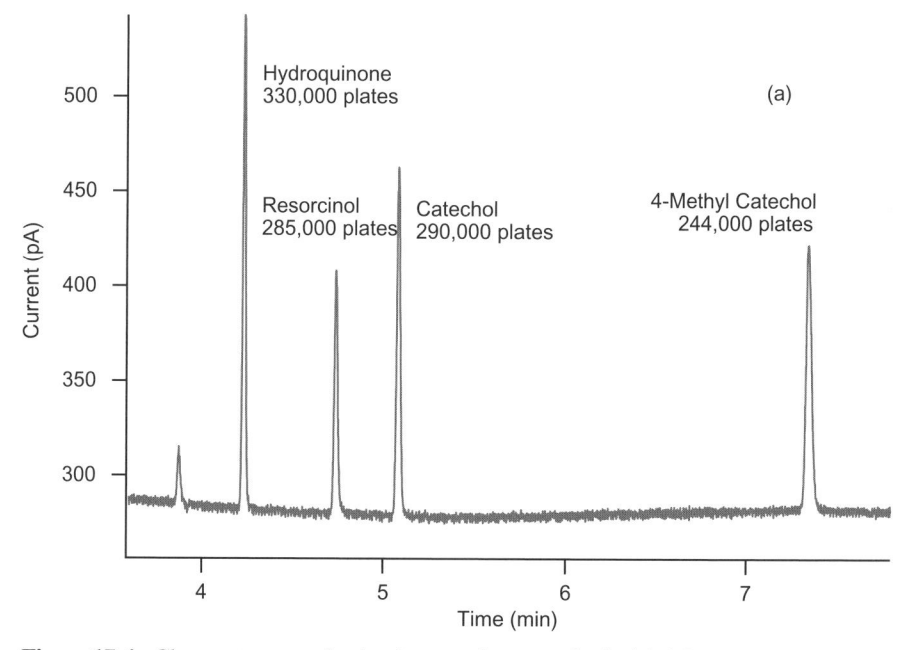

Figure 17-4. Chromatogram obtained on a column packed with 1.0-μm nonporous particles at a run pressure of 3000 bar. (Reprinted from reference 37, with permission.)

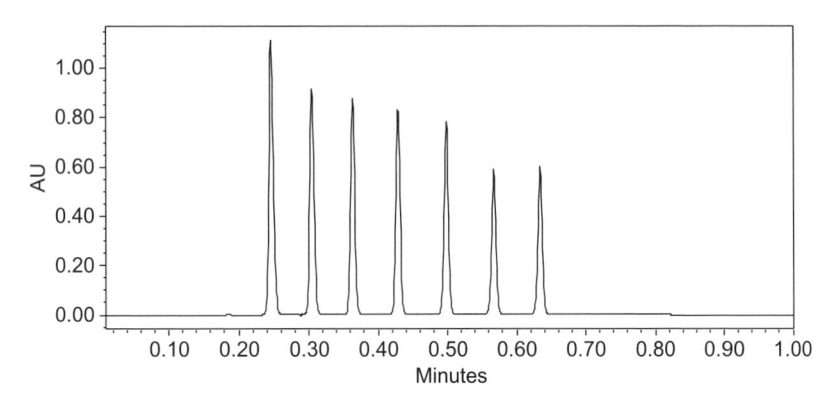

Figure 17-5. Gradient separation of, in order of elution, acetophenone, propiophenone, *n*-butyrophenone, valerophenone, hexanophenone, heptanophenone, and octanophenone, performed on a Waters Acquity UPLC™ instrument. Column: 2.1 × 100 mm, 1.7-μm ACQUITY BEH C18. Gradient: 50–90% acetonitrile in 1.0 minutes. Column temperature 35°C.

Scientific, Darien, IL). Five compounds—ascorbic acid (dead time marker), hydroquinone, resorcinol, catechol, and 4-methyl catechol—were eluted with a 10/90 (v/v) acetonitrile/water mobile phase containing 0.1% TFA and were detected with amperometric detection (+1.0 V versus Ag/AgCl). The chromatogram was obtained near the optimum linear velocity at a run pressure of 3000 bar. All compounds eluted in less than 8 minutes, with efficiencies ranging from a low of 244,000 plates for 4-methyl catechol to as high as 330,000 plates for hydroquinone. These correspond to about 570,000 and 770,000 plates/m, respectively—much higher than the 150,000 plates/m typically seen with conventional columns.

The potential for fast gradient separations is shown in Figure 17-5. A series of phenones was separated by an extremely fast gradient in the very-high pressure regime, at about 750 bar (11,000 psi). This separation was accomplished in less than 1 minute and was performed on a commercially available high pressure instrument and column packed with 1.7-μm porous bridged-ethyl hybrid C18 particles. All peaks are less than 2 seconds wide and are baseline resolved. The data acquisition rate was set at 20 pts/sec.

A more complex gradient UHPLC separation of a tryptic digest of the protein bovine serum albumin is shown in Figure 17-6. This sample contains hundreds of peptide fragments and requires a separation method with large peak capacity. The sample was run with gradient elution using constant-flow pumps at 3600 bar on a 38-cm-long capillary packed with 1.0-μm nonporous C18 particles. The peptides from the digest were tagged with the fluorophore tetramethylrhodamine isothiocyanate (TRITC) and detected by laser-induced fluorescence. Since it is not valid to calculate theoretical plates under mobile-phase gradient conditions, peak capacity is used as an alternative measure of the separating power of a system. Peak capacity is defined as the total number

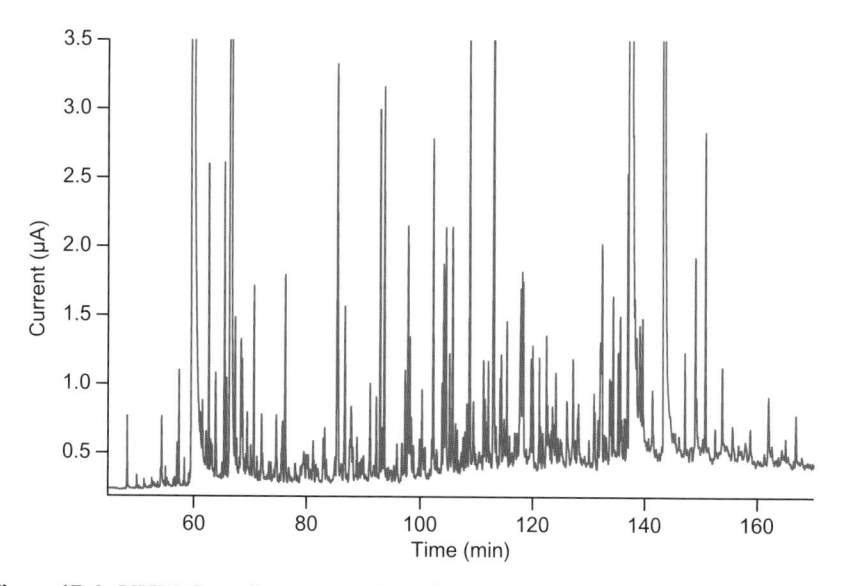

Figure 17-6. UHPLC gradient separation of a tryptic digest of bovine serum albumin. A peak capacity of 500 was obtained between 48 and 168 minutes. (Reprinted from reference 37, with permission.)

of peaks separable with unit resolution in a given separation space. In the chromatogram shown, the peak capacity between 48 and 168 minutes is approximately 500, with an average peak width of 14.5 seconds. This is significantly higher than the peak capacities of conventional HPLC columns packed with 5-μm particles, which tend to be below 200 for similar samples.

Low flow rates and narrow peak widths make capillary UHPLC particularly suitable for coupling with mass spectrometry via nanoelectrospray ionization [23, 45]. Tolley et al. [34] have used very-high pressures of around 1000 bar (15,000 psi) to separate bovine serum albumin digests on columns packed with 1.5-μm nonporous reversed-phase particles by gradient elution, with quadrupole/time-of-flight (Q-TOF) tandem mass spectrometry for detection. Figure 17-7 is a base peak index (BPI) chromatogram of 12.5 fmol of a BSA digest on a 150-μm × 22-cm column packed with 1.5-μm nonporous C18 particles. A 30-minute gradient from 1 to 45% acetonitrile in water with 0.5% formic acid was used. A 20-fold enhancement in sensitivity over nanoelectrospray MS/MS was observed.

17.4.4 Method Transfer Considerations

The significant gains in speed and resolution are strong motivators to transfer existing HPLC methods to a commercial ultra-high-pressure instrument. This can be accomplished fairly easily with a few simple steps. First, a UHPLC column that is appropriate for the separation must be selected. One with selectivity similar to that of the original column is preferred. Choice of column

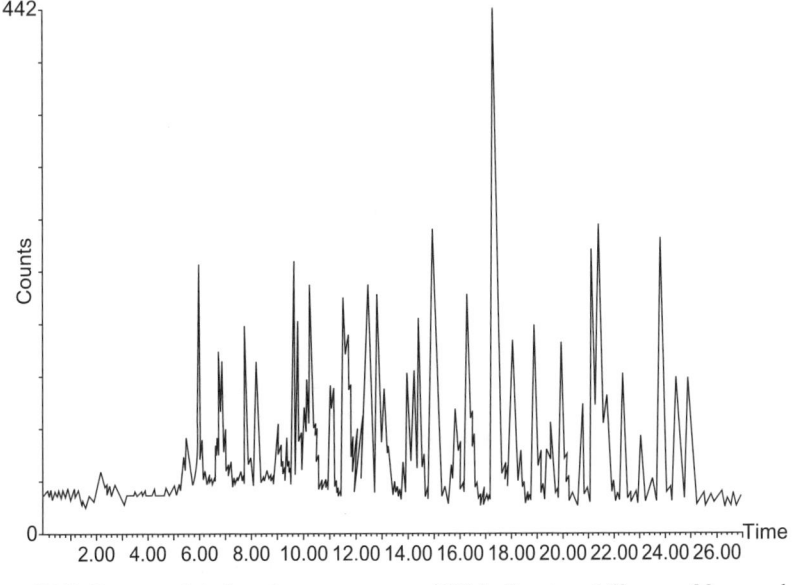

Figure 17-7. Base peak index chromatogram of BSA digest on 150-μm × 22-cm column packed with 1.5-μm nonporous C18 silica particles. (Reprinted from reference 34, with permission from American Chemical Society.)

diameter, length, and particle size should be made in accordance with the principles discussed in this chapter and with the overall goal of the method transfer (i.e., to speed up the existing separation or to obtain greater resolution of components).

Second, the method should be scaled geometrically to account for differences in column dimensions. The scaling equations can be found in Section 17.7.4. Once these parameters are scaled, the flow rate can then be increased to speed up the separation. With an increase in flow rate the gradient times must again be adjusted proportionally (e.g., doubling the flow rate requires gradient times to be halved). The optimum flow velocity for the separation must be kept in mind, however. A column with smaller stationary-phase particles will have a higher optimum velocity, so the new scaled method may not be at optimum conditions. The molecular weight of the analyte also plays a role: Large molecules will have a lower optimum velocity than small molecules due to slower diffusion in the mobile phase. At this point the method may be further optimized using standard method development strategies.

17.5 SEPARATIONS ON CHIPS

The miniaturization of chemical analysis systems has grown considerably in recent years due to the promise of faster analyses, the ability to analyze very small sample volumes, less reagent consumption, and exciting possibilities such as multiplexing and interfacing of multiple analytical techniques. These miniaturized systems are described by a number of terms such as "lab-on-a-chip," "micro-total analysis system (μTAS)," or "microfluidic" devices. Devices have been developed for numerous and varied applications incorporating many aspects of sample analysis: sample preparation, fluidic handling and manipulation, reactions or derivitization, separation, and detection. The focus here is on those systems that incorporate a separations step on-chip.

Lab-on-a-chip systems typically consist of microfabricated fluid channels photolithographically patterned onto a silicon or glass substrate. More recently, polymeric substrates such as poly(dimethylsiloxane) (PDMS) or poly(methyl methacrylate) (PMMA) have been used [46, 47]. The channel dimensions range from tens of micrometers in width and depth to less than a micrometer. Transport of fluid is accomplished most commonly by electroosmosis, and less so by pressure-induced pumping, which usually requires interfacing the chip to a benchtop pump. The difficulties in construction of valves on-chip is another limitation for pressure-based systems. Consequently, capillary electrophoresis is the easiest and most common separation principle employed on a chip. Separation channels packed with stationary-phase particles as well as porous monoliths have been prepared to perform chromatography and electrochromatography. While benchtop separation instruments typically analyze samples one at a time, multiple parallel separation channels may be patterned onto a chip to simultaneously analyze dozens of samples for increased sample throughput and lab productivity. Detection may be performed on-chip via fluorescence or electrochemical detection, for example, or the chip may be interfaced to a mass spectrometer, which is particularly amenable to the low flow rates of chip-based separations [48].

While a wide range of opportunities exist, such as environmental, clinical, and trace analysis, the principal application for labs-on-a-chip is in the analysis of biological samples. The miniaturized dimensions allow extremely small sample volumes to be analyzed, and a microchip format can allow chemical reaction, mixing, sample manipulation, and multiplexing to be performed. Single-cell analysis, immunoassays, protein and peptide separations, DNA analysis and sequencing, and polymerase chain reactions have all been performed on microchip devices [48].

There are still significant technical hurdles that must be overcome for microchips to develop into an accepted and widespread technique. The unreliability of electroosmotic pumping and other microfluidic processes contributes to the lack of robustness of these systems. The miniaturization and integration of other components, such as pumps, valves, and detection schemes, onto the chip is another essential step for future development. The reader is referred to the reviews cited in the references for research performed with microchips [46–48].

17.6   OPTIMIZING GRADIENT SEPARATIONS FOR SPEED

17.6.1   Advantages of Gradient Chromatography

Gradient chromatography is a very powerful method to control retention and resolution over a very broad analyte polarity range. Unlike isocratic methods where each component has a different retention factor and peak width, linear gradients elute peaks at constant retention factor and peak width. Consequently, sensitivity is uniform throughout the chromatogram. In addition, due to sample concentration at the head of the column during injection, larger injection volumes are possible, resulting in a further increase in sensitivity in gradient chromatography in comparison to isocratic methods. This advantage is particularly important as column volume is reduced by decreasing column length and/or diameter. Some commonly perceived disadvantages of gradient chromatography include poor robustness compared to isocratic methods and long equilibration times. Table 17-2 outlines these advantages and disadvantages. However, modern HPLC instruments and improved stability of chromatographic columns have largely overcome the perceived disadvantages of gradient methods.

17.6.2   Optimizing Instrumental Factors

Modern instruments have optimized system volumes to take advantage of short high-efficiency columns and are capable of generating fast reproducible gradient profiles. Instruments have also been optimized to reduce cycle time. The total cycle time in a gradient method is the sum of the gradient delay time (t_D), the gradient time (t_g), hold time (t_h), re-equilibration time (t_e), and injector cycle time (t_i). Figure 17-8 shows a schematic of the steps in a gradient method. The dwell volume, equilibration time, and the injector cycle time are the instrument-related factors that affect the overall gradient cycle time.

Gradient delay time, t_D, is calculated as

$$t_D = \frac{V_D}{F} \qquad (17\text{-}20)$$

where V_D is the dwell volume of the instrument and F is the flow rate. The dwell volume represents the total volume between the pump and the head of the column; it is composed of the volumes of the gradient mixer, the injector, and all the tubing connectors. The larger the dwell volume, the longer is the time

TABLE 17-2. Advantages and Perceived Disadvantages of Gradient Elution Chromatography

Advantages	Perceived Disadvantages
Wide polarity retention control	Methods not robust
Uniform sensitivity independent of retention	Complex method development
Flexible control of R_S and t_R	Long cycle time
All components eluted	Shorter column lifetime
Larger injection volumes	Method transfer problems

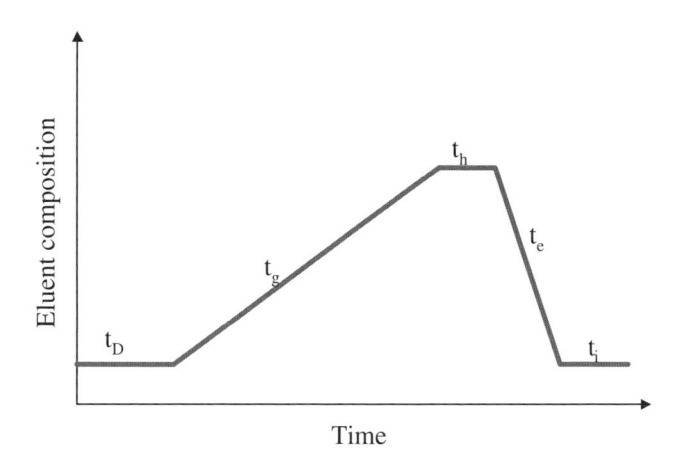

Figure 17-8. Individual contributions to a typical HPLC gradient cycle time.

for solvent delivered by the pump to reach the head of the column. This results in a longer gradient delay time as well as a longer column equilibration time. Most modern HPLC systems have dwell volumes from 0.7 to 1.1 mL.

Gradient methods have the reputation of being slow because of the need for column equilibration between injections. However, by optimizing column configuration, instrumentation, and method parameters, run times below 10 minutes are readily achievable. Column equilibration requires that a sufficient volume of mobile phase has been flushed through the system to return the column to the initial mobile phase conditions at the start of the gradient. Both the system dwell volume and the column volume must be taken into account. Equilibration time, t_e, is described by the following equation:

$$t_e = \frac{3V_D + 5V_M}{F} \qquad (17\text{-}21)$$

where V_M is the column volume (total volume of mobile phase in the column, approximately equal to $0.7\pi r^2 L$). The equilibration time can constitute a significant proportion of the total run time. It can be minimized by reducing dwell volume, reducing column volume by decreasing column length L, and increasing flow rate. Table 17-3 illustrates how flow rate column length, column diameter, and dwell volume affect equilibration time in comparison to a 4.6- × 150-mm column run at 1.0 mL/min. As shown, increasing the flow rate (either for the entire run or solely during the equilibration step at the end of the gradient run) directly decreases the equilibration time. Reducing column volume by reducing the column length threefold, for example, results in a 2.2-fold reduction in equilibration time. Reducing the column volume by decreasing the radius, however, does not have the same effect, since the flow rate must also be proportionally reduced in order to maintain equivalent linear velocity. Compare the 4.6- × 30-mm and 2.1- × 150-mm columns, which have about the same column volume, to the 4.6- × 150-mm column. Despite both columns having about five times smaller volume than the 4.6- × 150-mm dimension, in the first case the equilibration time is reduced and in the second case it actually

TABLE 17-3. The Effect of Column Dimensions, Flow Rate, and Dwell Volume on Equilibration Time

	Column Dimensions (mm)	Column Volume (mL)	Flow rate (mL/min)	Equilibration Time (min)
Comparison column	4.6 × 150	1.74	1.0	11.0
Effect of increasing flow	4.6 × 150	1.74	1.5	7.4
rate[a]	4.6 × 150	1.74	2.0	5.5
Effect of decreasing	4.6 × 50	0.58	1.0	5.0
column length[a]	4.6 × 30	0.35	1.0	3.9
Effect of decreasing	3.0 × 150	0.74	0.5	11.6
column diameter[a]	2.1 × 150	0.36	0.25	15.6
Effect of decreasing	3.0 × 150	0.74	0.5	8.0
column diameter with	2.1 × 150	0.36	0.25	8.4
optimized dwell volume[b]				

[a]The system dwell volume is assumed to be 0.7 mL.
[b]The system dwell volume is optimized to 0.1 mL.

increases. This emphasizes the importance of the gradient dwell volume, V_D, which has its biggest effect on reduction of equilibration time as column diameter is reduced. Since lower flow rates are used with smaller diameter columns, the contribution of dwell volume becomes significantly greater. Thus, system volumes must be optimized to take advantage of short-length and narrower-diameter columns. If the dwell volume is reduced to 0.1 mL (as it is in some modern fast HPLC instruments), equilibration time for the narrow diameter columns is decreased to within the range of the larger bore columns. See Section 17.7 for discussion of optimization of instrumental parameters for fast methods.

Column equilibration time must be determined as part of the method development. Equation (17-21) may not apply to all methods. For example, more than five column volumes may be necessary to re-equilibrate when using separation modes such as ion-pairing, or when switching to different pH mobile phases. Starting with 100% aqueous phase will also extend re-equilibration time and should be avoided.

17.6.3 Basic Parameters Controlling Speed and Resolution

17.6.3.1 Retention factor, k*. In isocratic chromatography, the retention factor of a solute remains constant as the solute migrates through the column. Under gradient conditions, however, the composition of the mobile phase is changing with time, and so is the retention factor. For gradient elution, then, the effective retention factor k^* is defined as the k value of the sample band when it has migrated halfway through the column. The retention factor k^* for a linear gradient is given by

$$k^* = 100 \frac{t_g}{\Delta\%} \frac{F}{V_M} \frac{1}{S} \qquad (17\text{-}22)$$

where $\Delta\%$ is the change is $\%B$ from gradient start to end (the gradient range), and S is a constant depending upon the molecular structure of the sample [49]. For molecular weights below 500, $S \cong 5$ and the equation reduces to

$$k^* = 20\frac{t_g}{\Delta\%}\frac{F}{V_M} \qquad (17\text{-}23)$$

The retention factor k^* is a constant for all solutes eluting in a linear gradient. This simple equation provides the basic insight for a starting point for developing or optimizing a method. Adjusting the parameters in the above equation such that k^* is about 5 for the target analyte provides a good starting point for method development. As in isocratic chromatography, the higher the retention factor, the longer the retention. In addition, largest changes in resolution occur at $k^* = 2$–10. Once the method has been optimized for selectivity, equation (17-23) provides insight on how to optimize for speed. If any single parameter is changed independently, k^* will change. Furthermore, a change in k^* may lead to a change in selectivity. Thus, when optimizing for speed, more than one parameter must be adjusted proportionally to hold k^* constant. In any method development, after the retention factor (values 2–8) and resolution have been optimized, the best approaches to optimize the speed of analysis according to equation (17-23) are as follows:

- Increase F with proportional decrease in t_g.
- Decrease V_M by decreasing column length, L, with a proportional decrease in t_g. If column diameter is also changed, the ratio of gradient volume to column volume ($t_g F/V_M$) must remain constant. Thus, F must be adjusted with change in column diameter.
- Decrease $\Delta\%$ with proportional decrease in t_g.

Other factors not covered by equation (17-23) are packing particle size and column temperature. These factors are also discussed below. These changes are discussed in more detail with examples in the resolution section. Keep in mind that these optimization approaches can be simulated in DryLab™ or similar software packages and should be used during the method optimization process.

17.6.3.2 Resolution. Maximizing selectivity, as part of the early stages of methods development, will result in the fastest methods since as flow rate is increased or column length is decreased, resolution will decrease. The resolution equation describes the key parameters. In equation (17-24), k^* is substituted for the isocratic retention factor, k, to give

$$R_S \cong \left(\frac{\sqrt{N}}{4}\right)(\alpha - 1)\left(\frac{1}{1 + (V_M\Delta\%/20t_gF)}\right) \qquad (17\text{-}24)$$

TABLE 17-4. Influence of Gradient Parameters on Resolution, Retention, and Run Time

Gradient Parameter	R_S	$k*$	Run Time
Decrease t_g	↓[a]	↓	↓
Increase t_g	↑[a]	↑	↑
Increase F	~[b]	↑	↓
Decrease tg, Increase F	↓[b]	**NC**[c]	↓↓
Decrease V_M (L)	↓[a]	↑	↓
Decrease V_M (L), Decrease t_g	↓	**NC**[c]	↓↓
Increase Δ%	↓[a]	↓	NC
Increase Δ%, Increase t_g	NC[c]	NC[c]	↑
Decrease Δ%	↑[a]	↑	NC
Decrease Δ%, Decrease t_g	**NC**[c]	**NC**[c]	↓

NC, no change.
[a] Assuming no change in selectivity. Changing only one parameter at a time will change gradient slope ($\Delta\% V_M / F\, t_g$) resulting in possible change in selectivity. As a result, resolution can increase, decrease, or remain unchanged for some peak pairs.
[b] Influence depends on the starting point on the H–u curve.
[c] Assuming the two parameters are changed in proportion to one another (gradient slope remains constant).

Table 17-4 summarizes the qualitative influence of changing the gradient parameters on resolution, retention factor, and run time. The bold rows of Table 17-4 offer the best approach to reducing analysis time and will be discussed in more detail. The nonbold rows represent parameters that should be optimized during the initial stages of method development to optimize resolution.

17.6.3.3 Decrease Gradient Time (t_g), Increase Flow Rate (F) Proportionally.

Total run time can be decreased through the proportional change of gradient time and flow rate. To decrease analysis time by a factor of two while keeping $k*$ constant, the flow rate should be doubled while the gradient time is reduced twofold. The proportional change in flow rate and gradient time is illustrated in Figure 17-9. In this example, run time was decreased from 24 to 6 minutes by a fourfold change in F and t_g. In general, increasing flow rate is limited by the pressure limitation of the HPLC equipment. Although Table 17-4 indicates resolution will be reduced, Figure 17-10 demonstrates that this decrease is minor and does not affect separation of critical peak pairs.

17.6.3.4 Decrease Column Volume (V_M) by Decreasing Column Length (L), and Decrease Gradient Time (t_g).

Changing the column volume by reducing column length while proportionally decreasing t_g is also very effective in reducing analysis time and is illustrated in Figure 17-11 (relatively difficult separation of diastereomers, $\alpha = 1.02$). Although resolution decreases somewhat, a threefold reduction in analysis time is achieved. Equations (17-23) and (17-24) can be used to calculate the minimum column length to

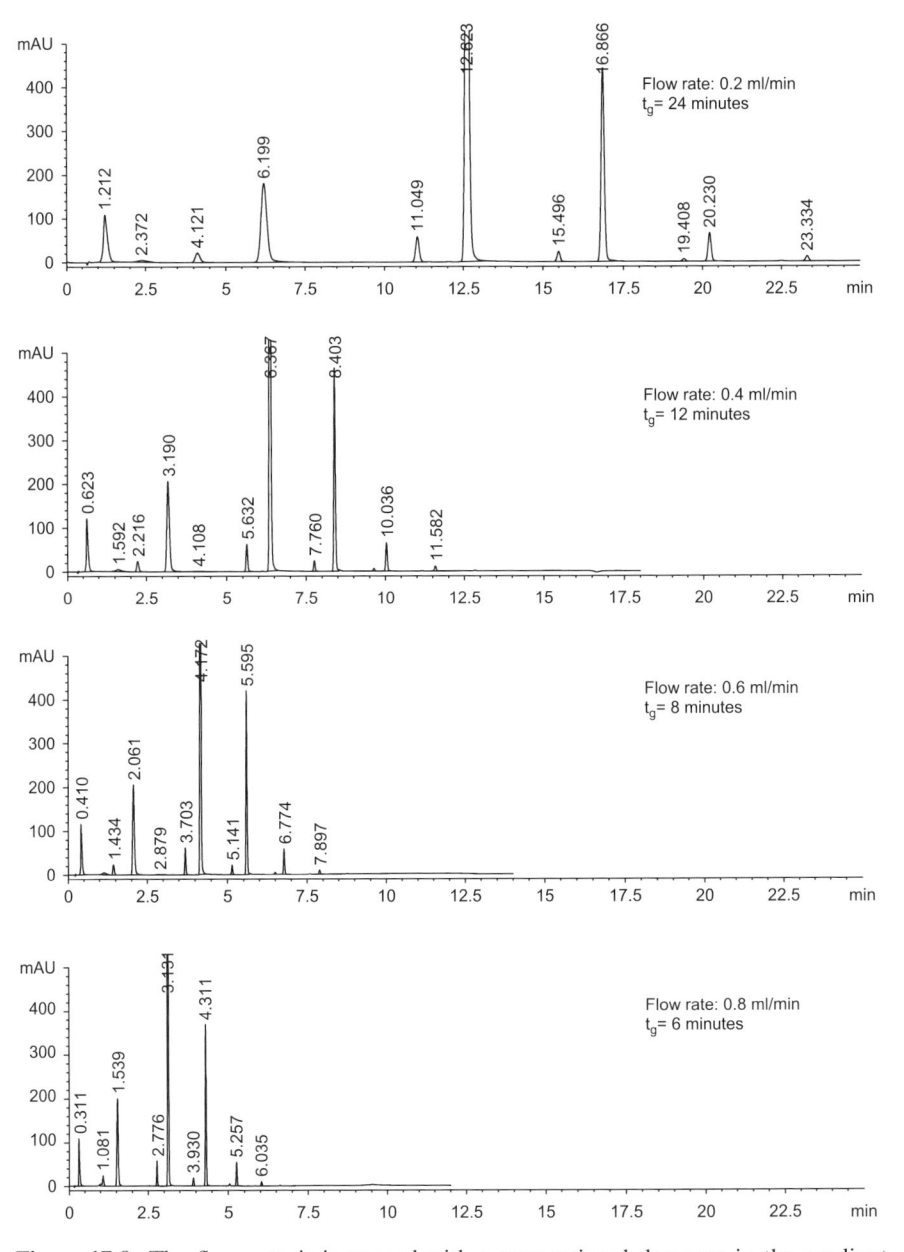

Figure 17-9. The flow rate is increased with a proportional decrease in the gradient time. All separations are performed on the XTerra MS-C18, 50 × 3.0 mm (3.5-μm particle); mobile phase A is water, mobile phase B is CH_3CN; the gradient is 30–90% B @ the flow rate and time intervals described. Note that the peak area changes by a factor of 4 while the peak height changes by a factor of only 1.2.

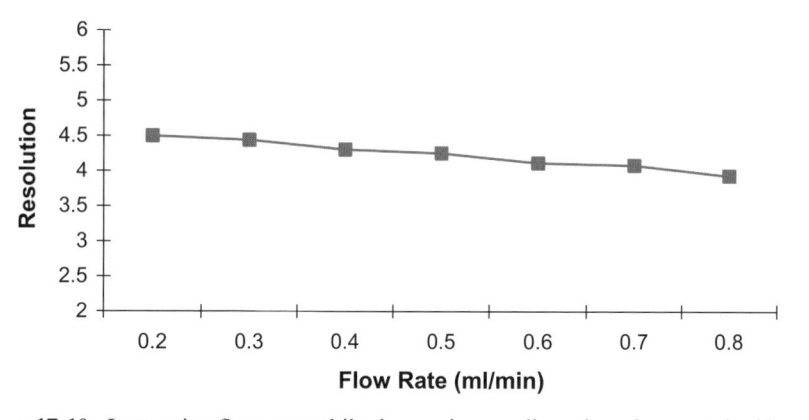

Figure 17-10. Increasing flow rate while decreasing gradient time does *not* significantly reduce resolution of the critical peak pair.

achieve a given resolution. For example, for a 3.0- × 150-mm column packed with 3.5-μm particles operated at 0.8 mL/min with a gradient of 10–60% B in 5 min, k^* can be calculated as

$$k^* = 20 \left(\frac{5\,\text{min}}{50\%} \right) \cdot \left(\frac{800\,\mu\text{L}/\text{min}}{742\,\mu\text{L}} \right) = 2.16$$

If the selectivity (α) of the critical peak pair is 1.2, the values for k^* and α can be substituted in equation (17-24) to calculate the minimum number of theoretical plates, N, needed to achieve a resolution of 1.5. Assuming that HETP is $\sim 2d_p$ and independent of L, the minimum L is calculated from $L = HN$. The minimum column length for this example would be 63 mm. In this case, the next longest commercially available column should be used. However, as theory predicts, reducing column length also reduces total plate count and hence decreases resolution. This is demonstrated in Figures 17-11 and 17-12. For a robust method, choose a column length longer than the minimum column length.

17.6.3.5   Decrease Gradient Range (Δ%), Decrease Gradient Time (t_g). The parameter Δ% can have an impact on the total run time if not optimized properly. For example, if in the beginning and at the end of the chromatographic run no peaks are eluting, run time is wasted. The starting and ending compositions of %B should be adjusted as part of the optimization of selectivity and resolution.

According to equation (17-23), no change in k^* results if the gradient range (Δ%) and gradient time (t_g) are changed proportionally. This suggests that resolution is independent of gradient starting or endpoint. To be true, all components must elute within the gradient. Figure 17-13 illustrates these

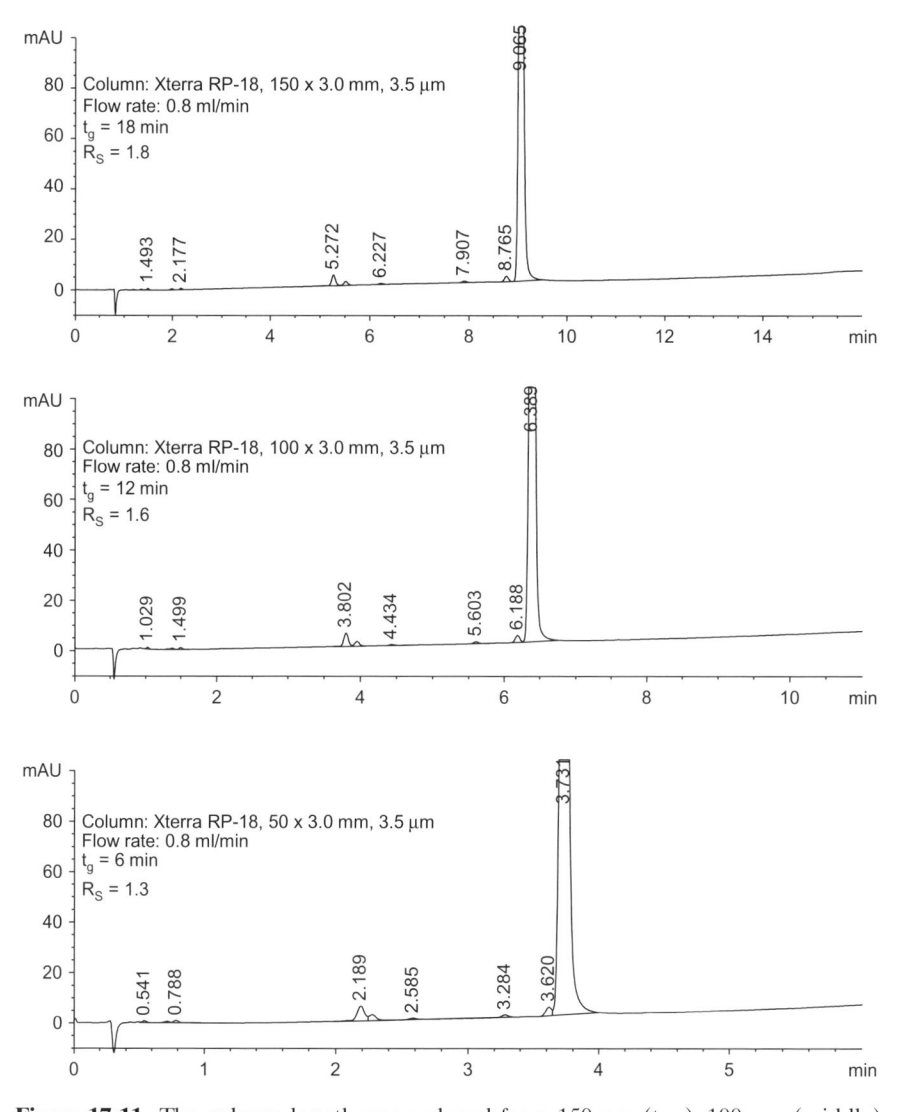

Figure 17-11. The column length was reduced from 150 mm (top), 100 mm (middle), 50 mm (bottom); the gradient time t_g was reduced proportionally.

points. Reducing the change in % B ($\Delta\%$) and a proportional decrease in gradient time (t_g) produced no change in solute retention time (t_R) or resolution (R_S). This is because all the components elute within the gradient and the gradient slope, ($\Delta\%/F \; t_g$), is constant in all three chromatograms. However, because the gradient time is shorter, the run time can be reduced. Essentially, the time spent running at >80% B can be eliminated as no peaks elute beyond

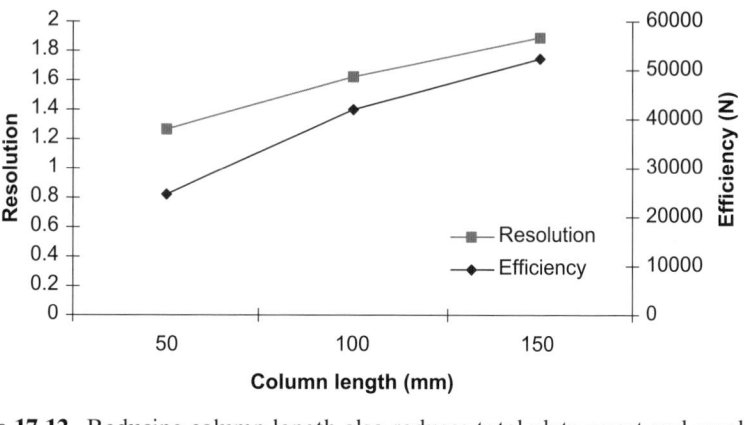

Figure 17-12. Reducing column length also reduces total plate count and resolution.

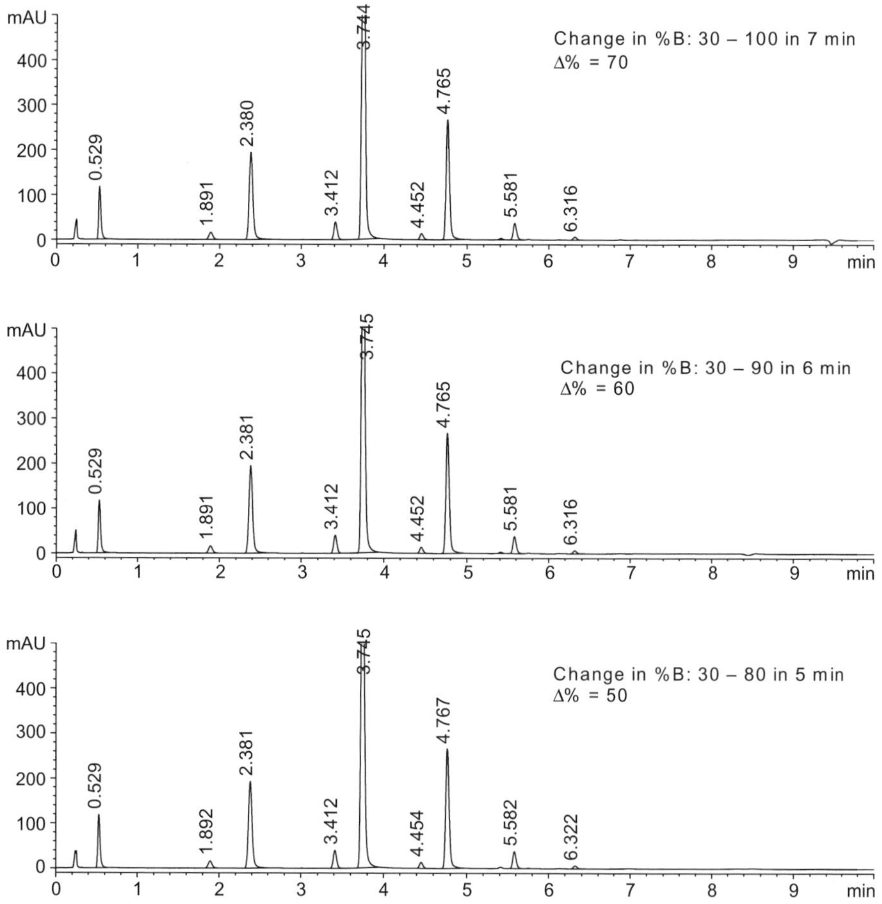

mAU

Change in %B: 30 – 100 in 7 min
Δ% = 70

400

300

200

100

0

0.529

1.891

2.380

3.412

3.744

4.452

4.765

5.581

6.316

0   1   2   3   4   5   6   7   8   9   min

mAU

Change in %B: 30 – 90 in 6 min
Δ% = 60

400

300

200

100

0

0.529

1.891

2.381

3.412

3.745

4.452

4.765

5.581

6.316

0   1   2   3   4   5   6   7   8   9   min

mAU

Change in %B: 30 – 80 in 5 min
Δ% = 50

400

300

200

100

0

0.529

1.892

2.381

3.412

3.745

4.454

4.767

5.582

6.322

0   1   2   3   4   5   6   7   8   9   min

Figure 17-13. Effect of changing gradient endpoint and time proportionally.

this point. The gradient range should be reduced only when no late or unretained peaks will elute outside of the gradient.

17.6.3.6 Decrease Column Length (L) and Particle Size (d$_p$). Another approach to speed up the separation for both gradient and isocratic chromatography is to reduce the packing particle size and decrease column length (see Section 17.2.2 for a more detailed discussion of this topic). In general, the length and particle diameter should be reduced together in proportion to one another so efficiency remains the same. Thus, a 100-mm column with 5-μm particle size can be reduced to a 50-mm column with 2.5-μm particles (see Figure 17-14). In addition, the loss of efficiency with increasing linear velocity is less significant with a well-packed column containing smaller particles, as predicted by the van Deemter equation. Thus, for high-speed analyses, it is desirable to operate at high linear velocity with short columns packed with small (≤3 μm) particles. This optimization is eventually limited by the pressure capabilities of the instrument and the commercial availability of short, small-particle columns.

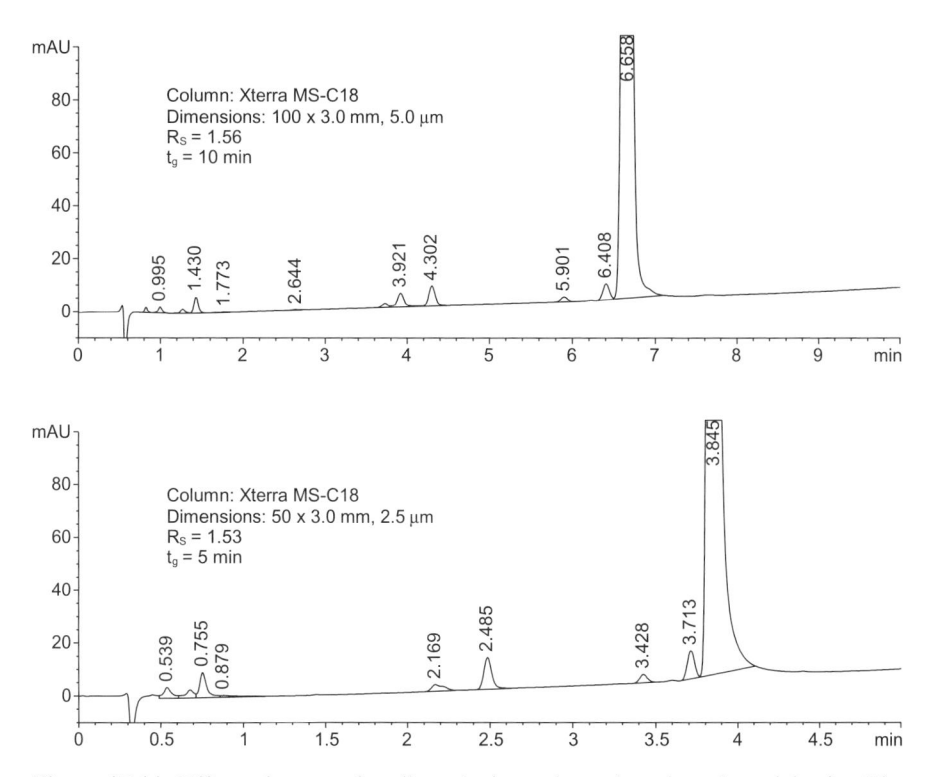

Figure 17-14. Effect of proportionally reducing column length and particle size. The separation is performed using 0.05% TFA in water (mobile phase A) and 0.05% TFA in acetonitrile (mobile phase B); the Δ% B is 30% over 6 minutes at a flow rate of 0.8 mL/min.

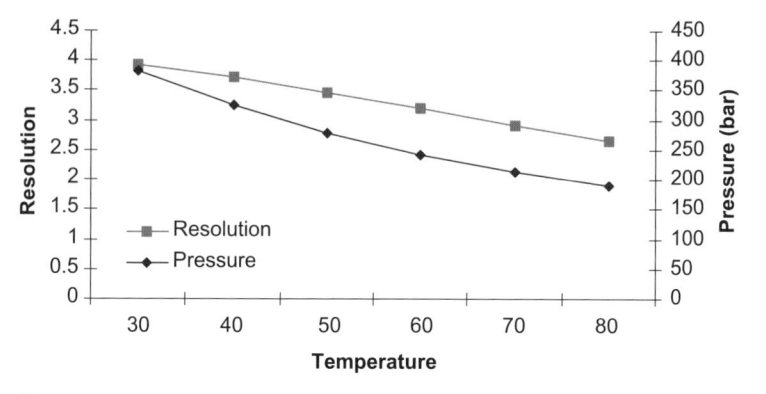

Figure 17-15. Increasing temperature results in a reduction in pressure with some loss of resolution.

17.6.3.7 Temperature. Column backpressure is often a limitation to speed in HPLC methods. In gradients the viscosity of the mobile phase can change significantly, leading to higher-pressure drops that may reach the instrument pressure limit. Also, as particle size is reduced, the column backpressure increases inversely with the square of particle size. Reducing column length only partially compensates for this pressure increase. Decreasing the mobile-phase viscosity by increasing the column temperature can reduce the pressure drop, allowing for a higher flow rate to speed up the separation (see Section 17.2.3). Figure 17-15 shows a graph of pressure drop and resolution as function of temperature. A temperature increase of 50°C reduces ΔP by 50%, but only reduces R_S by 30%. The application and separation conditions are described in Figure 17-9. However, increased temperature reduces column lifetime. In general, temperature above 60°C should not be applied.

Initial methods development should include an investigation of resolution as a function of temperature. The recommended minimum temperature is 40°C unless a lower temperature is required to meet the method objectives. DryLab™ can be used to model effect of temperature upon resolution. Four gradients should be run: two gradients differing in gradient slope (shallow and steep gradient) and each at two different temperatures (e.g., 35°C and 50°C).

17.7 INSTRUMENTAL REQUIREMENTS FOR OPERATING HIGH-EFFICIENCY COLUMNS

17.7.1 Extra-column Band-Broadening

Achieving the theoretically expected performance of high-efficiency columns requires proper instrument design to ensure that band broadening outside of the column is negligible. Sources of extra-column broadening can be classified into two categories: volumetric effects and electronic effects. Those associated with extra-column system volumes are treated in this section, effects due to electronics are covered under detector requirements in the following section, and broadening originating in the injector is singled out for separate discussion in Section 17.7.3.

The excessive volumes of the injector, connection unions, capillary tubing, and detector flow cell can degrade the efficiency of the column. The instrument band-spreading can adversely affect column performance, especially as the dimensions of the column and support particle size are reduced. Since peak volumes are much smaller in high-speed columns, extra-column contributions to band-broadening are more significant and thus need to be reduced as much as possible. The total peak width in volume units, W_V (defined here as $4\sigma_V$), is a result of accumulated band-broadening from both column and extra-column sources. Since mathematically it is the variances that are additive, not the broadening itself, the total peak width variance is expressed as the sum of the column (W_C^2) and extra-column (W_{EC}^2) peak width variances:

$$W_V^2 = W_C^2 + W_{EC}^2 \qquad (17\text{-}25)$$

where the total extra-column band spreading is given by the sum of the variances due to the injector (W_I^2), connecting tubing (W_{CT}^2) and the detector cell (W_D^2):

$$W_{EC}^2 = W_I^2 + W_{CT}^2 + W_D^2 \qquad (17\text{-}26)$$

and the band-spreading of the column in isocratic conditions is expressed by

$$W_C = \frac{4}{\sqrt{N}} V_M (1+k) \qquad (17\text{-}27)$$

Thus, if the column is highly efficient (large N), if column void volume (V_M) is small due to decreased dimensions (either by reducing length or diameter), or if the analyte is lightly retained (small k), the magnitude of the extra-column contribution to peak width will be large relative to the contribution of the column. A good rule of thumb is that, in general, extra-column band-spreading should contribute no more than 10% loss in efficiency of the column, which corresponds to ~5% drop in resolution. This means that W_{EC} should be no greater than about $\frac{1}{3}$ of peak width volume, W_V. The derivation of this rule is shown below:

$$N_{\text{system}} = \frac{9}{10} N_{\text{column}} = 16 \frac{V_R^2}{\left(\frac{10}{9} W_C^2 \right)}$$

$$W_V^2 = \frac{10}{9} W_C^2 = W_C^2 + \frac{W_V^2}{10}$$

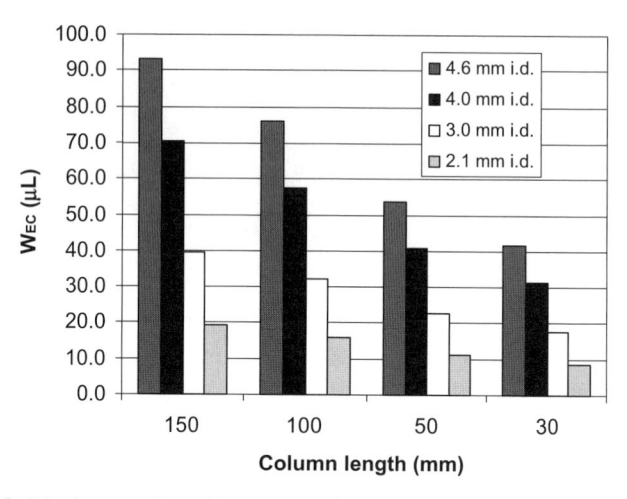

Figure 17-16. Maximum allowable extra-column broadening for ≤10% loss in efficiency for various column dimensions packed with 3.0-μm particles. Calculated with equations (17-25) and (17-27), assuming optimum column efficiency ($H = 2d_p$) and $k = 5$.

$$W_{EC}^2 = \frac{W_V^2}{10} \quad \text{and} \quad W_{EC} \approx \frac{1}{3} W_V$$

The minimum column dimensions that can be used without adversely affecting resolution can be determined from equations (17-25) and (17-27) and knowledge of the instruments band spreading. Figure 17-16 displays the maximum allowable instrument band spreading for ≤10% loss in efficiency for columns packed with 3.0-μm particles as a function of column dimensions. This plot emphasizes that extra-column band broadening must be optimized to achieve expected performance of columns, especially as column dimensions and particle size are reduced. It is calculated for isocratic elution using a k value of 5, which is approximately in the middle of the desirable range for retention factor (2–10). Peaks eluting earlier than $k = 5$ will require even smaller allowable extra-column broadening. For 3-μm particles, columns of 3.0- to 4.6-mm i.d. are recommended for instruments having W_{EC} of about 30 μL. Smaller particles (<2 μm) are not only more efficient, but must be used in smaller-diameter columns, and thus demand exceptionally low extra-column broadening. Recently, a few commercially available systems have been designed to operate such columns. In addition to low extra-column volumes, systems must operate in excess of 500 bar to achieve mobile-phase linear velocity in the optimum performance region.

The instrument band-spreading (W_{EC}) can be measured by replacing the column with a zero dead-volume restrictor, injecting a very small volume of sample, and measuring the resulting peak width. Table 17-5 provides band-

TABLE 17-5. Extra-column Volumes and Band-Spreading of Different Commercial Systems

System	Detector Cell Volume, Path Length	V_{EC} [µL] (Extra-column Volume)	W_{EC} [µL] (Extra-column Band Spreading)
Agilent 1100 DAD	13 µL, 10 mm	40	38
TSP VWD	15 µL, 10 mm	40	*
Shimadzu 2010	8 µL, 10 mm	90	*
Waters Acquity UPLC	500 nL, 10 mm	10	12
Waters Alliance 2695 DAD	8 µL, 10 mm	60	39
Waters Alliance 2690 UV 2487	10 µL, 10 mm	90	99

*Not measured.

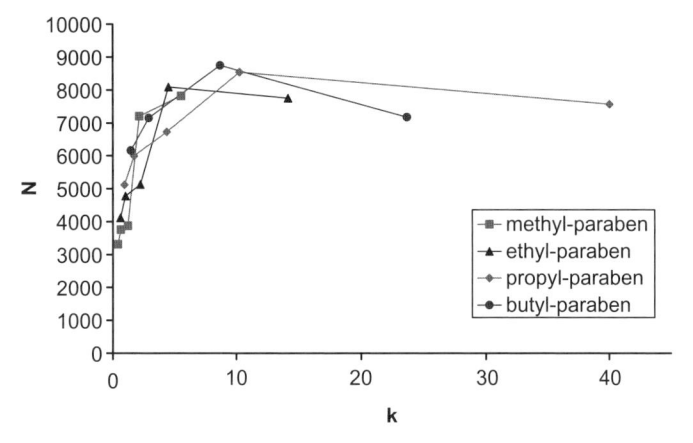

Figure 17-17. Theoretical plates versus retention factor for series of parabens separated by isocratic chromatography. Extra-column effects result in decreased efficiency for early-eluting components (low values of k).

spreading for some commercial instruments. While instrument band-spreading is related to the extra-column volume, it does not correlate exactly due to the complexity of flow through tubing, connections, injection systems, and detector flow cells.

The phenomenon of pre-concentration of analytes onto the head of the column in gradient chromatography makes the portion of W_{EC} before the column negligible, providing all of the components elute within the gradient. In isocratic methods, however, extra-column band-spreading is more critical, especially for the early eluting peaks. Figure 17-17 shows how efficiency decreases as retention factor becomes small due to extra-column effects.

Instrument dispersion can be reduced by optimizing injector and detector systems and reducing diameter of connection capillaries. The individual sources of volumetric extra-column broadening specified in equation (17-26) can now be examined in more detail. The dispersion in connection capillaries is given by the Taylor–Aris equation [50]:

$$W_{CT}^2 = \frac{\pi d_{cap}^4 F L_{cap}}{24 D_M} \qquad (17\text{-}28)$$

where d_{cap} is the diameter of the connection capillary, F is the volumetric flow rate, L is the tubing length, and D_M is the diffusion coefficient of the solute. Reducing the diameter of the capillary from 0.23 mm to 0.13 mm will reduce dispersion by factor of 96, assuming constant L and F. Furthermore, as the column diameter is reduced, lower volumetric flow rates are required and hence further reduction in capillary dispersion can be achieved by reducing capillary diameter. In addition, since unions are also used, ferrules must be properly set and capillary ends must be squarely cut to eliminate unnecessary gaps in the connection.

The injection volume contribution to total peak width is

$$W_I^2 = \frac{4}{3} V_I^2 \qquad (17\text{-}29)$$

where V_I is the injection volume [50]. The equation for the detector volume contribution takes the same form:

$$W_D^2 = \frac{4}{3} V_{FC}^2 \qquad (17\text{-}30)$$

where V_{FC} is the volume of the detector flow cell. These two sources are directly related to their respective volumes, and it is therefore important to keep them as small as possible. Again, the rule that broadening should not be greater than 1/3 total peak width volume can be applied to each individual contribution. [Remember, since it is variances that are additive, 1/3 of total peak width volume for each of the three individual contributions still results in a total extra-column contribution of 1/3 total peak width volume, according to equation (17-26).] Solving equations (17-29) and (17-30) for V shows that the volume of the injection and the volume of the detector flow cell should each not be greater than $W_V/\sqrt{12}$, which for convenience can be approximated as 1/3 of total peak width volume. Thus, a handy "1/3 rule" can be used for determining acceptable extra-column broadening: All individual contributions to broadening, as well as the injection and detector flow cell volumes, should be no more than 1/3 of total peak width volume to give ≤10% loss in column efficiency.

17.7.2 Detector Requirements

In order to accurately describe narrow peaks and not cause peak distortion in fast HPLC, detector parameters such as sampling rate and digital filtering must be properly set [51]. Sampling rate is defined as the number of data points per second that is collected to accurately describe the chromatographic features during data acquisition. To properly integrate a peak, a minimum of 15 data points across a Gaussian peak or 21 points for a non-Gaussian peak is required. The optimum sampling rate for the narrowest peak of interest can be calculated by the following formula:

$$\text{Sample rate} = \frac{n}{W} \tag{17-31}$$

where n is the minimum number of points/peak for accurate sampling rate and W is the narrowest peak width in seconds. For example, if a peak width is 1 second, the sampling rate must be equal to or larger than 20 Hz (1 Hz = 1 point per second) as illustrated in Figure 17-18. In most HPLC systems with UV detectors, the default value for sampling rate is 1 Hz.

The detector time constant (or digital filter for modern instruments) is used to remove high-frequency noise. If the detector time constant is too slow, the observed peaks will be broadened. The "1/3 rule" may be applied here as well: As a rule of thumb, the maximum detector time constant (seconds) tolerable is about 1/3 the standard deviation of the peak in seconds. Peaks 1–2 seconds in width require a time constant no larger than 0.1 second. Figure 17-19 illustrates these points. Although noise-free chromatograms are desirable, resolution and sensitivity can be adversely affected by excessively large time constants due to peak distortion.

Chromatographic noise can also be reduced through Savitzky–Golay smoothing of the unfiltered raw signal (time constant = 0). This method essentially performs a polynomial regression (of degree k) on a distribution of at least $k + 1$ neighboring data points to determine the smoothed value of each

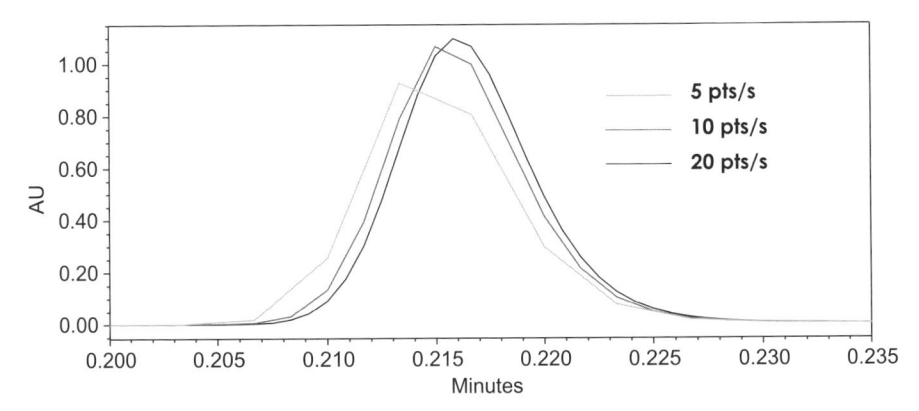

Figure 17-18. Effect of sampling rate on peak shape for a 1-second-wide peak with retention factor $k = 1$.

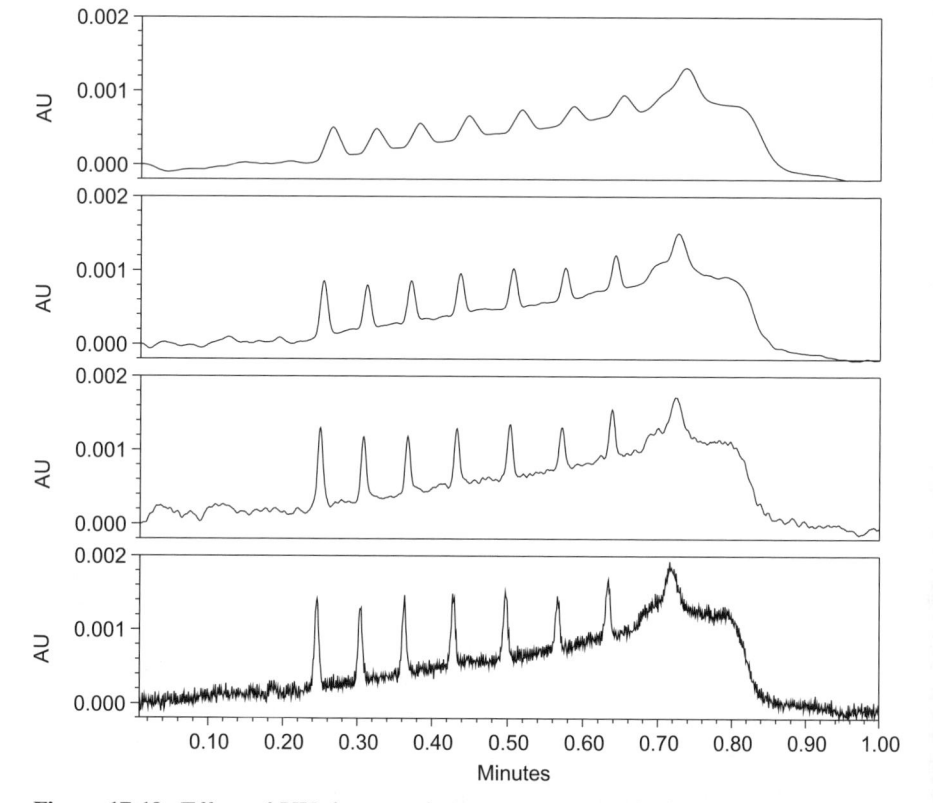

Figure 17-19. Effect of UV detector time constant on peak shape and noise for fast LC separation of phenones (0.05 μg/mL). Time constants, from top to bottom, are 0.5 sec, 0.2 sec, 0.1 sec, and 0 sec (no time constant). Peak widths are approximately 1 sec; sampling rate is 20 Hz; column is 2.1 mm × 50 mm.

data point [52]. In Figure 17-20, a 15-neighboring-point smooth removes noise and preserves the innate features of each peak.

17.7.3 Injection Considerations

The injection is a critical factor in fast LC methods and must be considered to maintain column efficiency. Injection volumes that are too large can cause volume overload of the column, which results in broad, flat-top shaped peaks with low plate counts that are more pronounced for earlier eluting components. As injection volume is increased, peak height should increase; however, peak width should remain the same. If peak width increases as well, this is indicative of volume overload. As column dimensions are reduced, the maximum injection volume must be reduced by the ratio of the column volumes [see equation (17-33) in Section 17.7.4]. For example, reducing

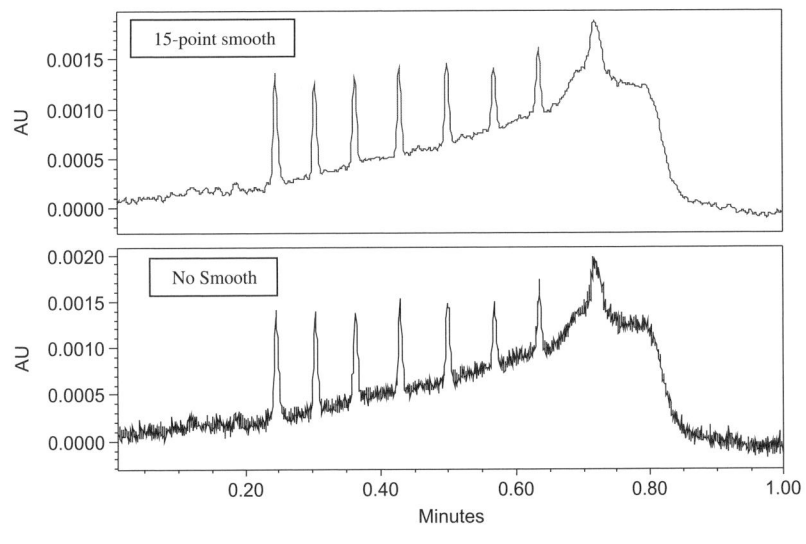

Figure 17-20. Savitzky–Golay smoothing.

column diameter from 4.6-mm to 2.1-mm i.d. requires an injection volume reduction by a factor of 4.8. As stated previously in Section 17.7.1, in isocratic mode the maximum injection volume should be ≤1/3 the peak width volume [53]. Considering a 2.1- × 50-mm column packed with 1.8-μm particles, the maximum injection volume should be ~4 μL (peak width volume = 13 μL at $k = 2$). Higher injection volumes are possible in the gradient mode due the concentration of the sample band at the head of the column. When we decrease injection volume to avoid volume overload, the sample concentration should be increased proportionally so that the same mass of analyte is being injected. This will avoid any loss in sensitivity.

Mass overload occurs when the stationary phase does not have the capacity to retain the amount of sample injected. This can occur even for small injection volumes if the concentration of sample is high enough. This results in a characteristic "shark-fin" peak shape, where peak tailing starts from the peak's apex. For example, in order to obtain sufficient sensitivity, analytes with weak UV molar absorptivity may require a large enough amount of sample to be injected that the stationary phase becomes overloaded. Injecting less amount of sample, either by a smaller injection volume or by diluting the sample, can solve the problem of mass overload. However, sensitivity will decrease in this case.

Another injection-related effect that can diminish the separation performance is the "diluent effect," also known as solvent mismatch. This occurs when the elution strength of the sample solvent is greater than the starting mobile-phase strength. The retention of the analyte on the stationary phase is less in the small plug of sample solvent than it is in the surrounding bulk mobile phase. Consequently, some analyte molecules will be "carried" ahead of the analyte zone by the plug of sample solvent as it migrates down the column. It usually results in peak distortion (either fronting or splitting) that is more prevalent for early-eluting components. Large injection volumes exacerbate this effect. Dissolving the sample in solvent weaker than the starting mobile phase or in the starting mobile phase itself is preferred, although it is not always possible due to the solubility of the sample. When the sample diluent must be stronger than the starting mobile phase, reducing the injection volume can lessen the effect, because the smaller plug of sample solvent is more rapidly diluted by the mobile phase.

17.7.4 Geometric Scaling Relationships

When changing the dimensions of the column, the method should be scaled geometrically to maintain equivalency. The flow rate is scaled to maintain the same linear velocity as the original method. This is done based on the ratio of the column diameters squared:

$$F_{\text{column}\,2} = F_{\text{column}\,1} \times \frac{d^2_{\text{column}2}}{d^2_{\text{column}1}} \tag{17-32}$$

Next, scale the injection volume according to the ratio of the two column volumes (V_M):

$$\text{inj. vol.}_{\text{column}\,2} = \text{inj. vol.}_{\text{column}\,1} \times \frac{V_{M\,\text{column}\,2}}{V_{M\,\text{column}\,1}} \tag{17-33}$$

Here column volume is defined as the volume of mobile phase in the column and may be calculated as $V_M \approx 0.7\pi r^2 L$, where r and L are the column radius and length, respectively, and 0.7 is the approximate fraction of the empty tube occupied by mobile phase for a column packed with porous particles.

Finally, the gradient times and total run time must be scaled. An equivalent gradient profile will have the same gradient slope; that is, it will deliver the same number of column volumes of mobile phase per gradient step. The number of column volumes per step can be calculated by multiplying the flow rate by the step duration, and dividing by the column volume for the particular column in question. The new gradient time t_g for a given step is therefore

$$t_{g\,\text{column}\,2} = t_{g\,\text{column}\,1} \times \frac{F_{\text{column}\,1}}{V_{M\,\text{column}\,1}} \times \frac{V_{M\,\text{column}\,2}}{F_{\text{column}\,2}} \tag{17-34}$$

If the flow rate has been scaled according to equation (17-32) so that linear velocity remains constant, then equation (17-34) simplifies and gradient time can be scaled in proportion to the column lengths:

$$t_{g\,\text{column}\,2} = t_{g\,\text{column}\,1} \times \frac{L_{\text{column}\,2}}{L_{\text{column}\,1}} \qquad (17\text{-}35)$$

The other parameters of the separation, such as the mobile-phase composition, gradient range ($\Delta\%B$), and column temperature, should be kept the same. A difference in particle size between columns does not affect the geometric scaling relationships; however, the location on the H versus u curve (i.e., column efficiency) and the column backpressure may change, which may require an increase or decrease in flow rate or column length.

17.8 CONCLUSIONS

The pharmaceutical analytical laboratory is increasingly under pressure to improve work throughput and shorten timelines, and HPLC methods can often be a bottleneck or otherwise be a limit to lab productivity. The benefits of fast HPLC methods are therefore obvious. A prerequisite to fast separations, however, is that adequate selectivity between components is achieved. Once that is realized, much can be done to speed up the analysis, as outlined in this chapter. Understanding the parameters governing isocratic and gradient separations can lead to significant improvements in the speed of current methods. Additionally, new exciting technologies entering the marketplace promise even greater speed of analysis than is currently possible with conventional columns and equipment. It is anticipated that run times of <5 minutes will become commonplace in the near future.

REFERENCES

1. B. L. Karger, L. R. Snyder, and C. Horvath, *An Introduction to Separation Science*, John Wiley & Sons, New York, 1973, p. 153.
2. J. C. Giddings, *Unified Separation Science*, John Wiley & Sons, New York, 1991, pp. 196–197.
3. H. Poppe, Some reflections on speed and efficiency of modern chromatographic methods, *J. Chromatogr. A* **778** (1997), 3–21.
4. J. J. van Deemter, F. J. Zuiderweg, and A. Klinkenberg, Longitudinal diffusion and resistance to mass transfer as causes of nonideality in chromatography, *Chem. Eng. Sci.* **5** (1956), 271.
5. J. C. Giddings, *Unified Separation Science*, John Wiley & Sons, New York, 1991, p. 266.
6. G. H. Kennedy and J. H. Knox, *J. Chromatogr. Sci.* **10** (1972), 549–556.
7. C. Horvath and H. J. Lin, Movement and band spreading of unsorbed solutes in liquid chromatography, *J. Chromatogr.* **126** (1976), 401–420.
8. A. L. Berdichevsky and U. D. Neue, Nature of the eddy dispersion in packed beds, *J. Chromatogr.* **535** (1990), 189–198.

9. U. D. Neue, *HPLC Columns: Theory, Technology, and Practice*, John Wiley & Sons, New York, 1997, p. 33.

10. T. Greibrokk and T. Anderson, High-temperature liquid chromatography, *J. Chromatogr. A* **1000** (2003), 743–755.

11. U. D. Neue, *HPLC Columns: Theory, Technology, and Practice*, John Wiley & Sons, New York, 1997, p. 32.

12. D. V. McCalley, Comparison of conventional microparticulate and a monolithic reversed-phase column for high-efficiency fast liquid chromatography for basic compounds, *J. Chromatogr.* **965** (2002), 51–64.

13. N. Tanaka, H. Kobayashi, K. Nakanishi, H. Minakuchi, and N. Ishizuka, Monolithic LC columns, *Anal. Chem.* **73** (2001), 420A–429A.

14. N. Tanaka, H. Kobayashi, N. Ishizuka, H. Minakuchi, K. Nakanishi, K. Hosoya, and T. Ikegami, Monolithic silica columns for high-efficiency chromatographic separations, *J. Chromatogr. A* **965** (2002), 35–49.

15. F. Svec and C. G. Huber, Monolithic materials: promises, challenges, achievements, *Anal. Chem.* **78** (2006), 2101–2107.

16. N. Wu, J. Dempsey, P. M. Yehl, A. Dovletoglou, D. Ellison, and J. Wyvratt, Practical aspects of fast HPLC separations for pharmaceutical process development using monolithic columns, *Anal. Chim. Acta* **523** (2004), 149–156.

17. L. Xiong, R. Zhang, and F. E. Regnier, Potential of silica monolithic columns in peptide separations, *J. Chromatogr. A* **1030** (2004), 187–194.

18. H. Minakuchi, N. Ishizuka, K. Nakanishi, N. Soga, and N. Tanaka, Performance of octadecylsilylated continuous porous silica column in polypeptide separations, *J. Chromatogr. A* **828** (1998), 83–90.

19. F. Gerber, M. Krummen, H. Potgeter, A. Roth, C. Siffrin, and C. Spoendlin, Practical aspects of fast reversed-phase high-performance liquid chromatography using 3 µm particle packed columns and monolithic columns in pharmaceutical development and production working under current good manufacturing practice, *J. Chromatogr. A* **1036** (2004), 127–133.

20. J. E. MacNair, K. C. Lewis, and J. W. Jorgenson, Ultrahigh-pressure reversed-phase liquid chromatography in packed capillary columns, *Anal. Chem.* **69** (1997), 983–989.

21. J. E. MacNair, K. D. Patel, and J. W. Jorgenson, Ultrahigh-pressure reversed-phase capillary liquid chromatography: Isocratic and gradient elution using columns packed with 1.0-µm particles, *Anal. Chem.* **71** (1999), 700–708.

22. J. A. Lippert, B. Xin, N. Wu, and M. L. Lee, Fast ultrahigh-pressure liquid chromatography: On-column UV and time-of-flight mass spectrometric detection, *J. Microcolumn Sep.* **11** (1999), 631–643.

23. N. Wu, D. C. Collins, J. A. Lippert, Y. Xiang, and M. L. Lee, Ultrahigh pressure liquid chromatography/time-of-flight mass spectrometry for fast separations, *J. Microcolumn Sep.* **12** (2000), 462–469.

24. N. Wu, J. A. Lippert, and M. L. Lee, Practical aspects of ultrahigh pressure capillary liquid chromatography, *J. Chromatogr. A* **911** (2001), 1–12.

25. D. C. Collins, Y. Xiang, and M. L. Lee, Comprehensive ultra-high pressure capillary liquid chromatography/ion mobility spectrometry, *Chromatographia* **55** (2002), 123–128.

26. Y. Xiang, N. Wu, J. A. Lippert, and M. L. Lee, Separation of chiral pharmaceuticals using ultrahigh pressure liquid chromatography, *Chromatographia* **55** (2002), 399–403.

27. Y. Xiang, B. Yan, C. V. McNeff, P. W. Carr, and M. L. Lee, Synthesis of micron diameter polybutadiene-encapsulated non-porous zirconia particles for ultrahigh pressure liquid chromatography, *J. Chromatogr. A* **1002** (2003), 71–78.

28. Y. Xiang, B. Yan, B. Yue, C. V. McNeff, P. W. Carr, and M. L. Lee, Elevated-temperature ultrahigh-pressure liquid chromatography using very small polybutadiene-coated nonporous zirconia particles, *J. Chromatogr. A* **983** (2003), 83–89.

29. Y. Gong, Y. Xiang, B. Yue, G. Xue, J. S. Bradshaw, H. K. Lee, and M. L. Lee, Application of diaza-18-crown-6-capped β-cyclodextrin bonded silica particles as chiral stationary phases for ultrahigh pressure capillary liquid chromatography, *J. Chromatogr. A* **1002** (2003), 63–70.

30. J. M. Cintrón and L. A. Colón, Organo-silica nano-particles used in ultrahigh-pressure liquid chromatography, *Analyst* **127** (2002), 701–704.

31. L. A. Colón, J. M. Cintrón, J. A. Anspach, A. M. Fermier, and K. A. Swinney, Very high pressure HPLC with 1 mm id columns, *Analyst* **129** (2004), 503–504.

32. J. S. Mellors and J. W. Jorgenson, Use of 1.5-μm porous ethyl-bridged hybrid particles as a stationary-phase support for reversed-phase ultrahigh-pressure liquid chromatography, *Anal. Chem.* **76** (2004), 5441–5450.

33. M. E. Swartz, UPLC™: An introduction and review, *J. Liq. Chromatogr. Relat. Technol.* **28** (2005), 1253–1263.

34. L. Tolley, J. W. Jorgenson, and M. A. Moseley, Very high pressure gradient LC/MS/MS, *Anal. Chem.* **73** (2001), 2985–2991.

35. J. A. Anspach, T. D. Maloney, R. W. Brice, and L. A. Colón, Injection valve for ultra-high-pressure liquid chromatography, *Anal. Chem.* **77** (2005), 7489–7494.

36. Y. Xiang, Y. Liu, S. D. Stearns, A. Plistil, M. P. Brisbin, and M. L. Lee, Pseudolinear gradient ultrahigh-pressure liquid chromatography using an injection valve assembly, *Anal. Chem.* **78** (2006), 858–864.

37. A. D. Jerkovich, R. LoBrutto, and R. V. Vivilecchia, The use of Acquity UPLC™ in pharmaceutical development, *LC-GC, UPLC Suppl.* **May** (2005), 15–21.

38. A. D. Jerkovich, J. S. Mellors, and J. W. Jorgenson, The use of micrometer-sized particles in ultrahigh pressure liquid chromatography, *LCGC North Am.* **21** (2003), 600–610.

39. Y. Xiang, D. R. Maynes, and M. L. Lee, Safety concerns in ultrahigh pressure capillary liquid chromatography using air-driven pumps, *J. Chromatogr. A* **991** (2003), 189–196.

40. I. Halász, R. Endele, and J. Asshauer, Ultimate limits in high-pressure liquid chromatography, *J. Chromatogr.* **112** (1975), 37–60.

41. M. Martin and G. Guiochon, Study of the pertinency of pressure in liquid chromatography III. A practical method for choosing the experimental conditions in liquid chromatography, *J. Chromatogr.* **110** (1975), 213–232.

42. K. D. Patel, A. D. Jerkovich, J. C. Link, and J. W. Jorgenson, In-depth characterization of slurry packed capillary columns with 1.0-μm nonporous particles using reversed-phase isocratic ultrahigh-pressure liquid chromatography, *Anal. Chem.* **76** (2004), 5777–5786.

43. A. D. Jerkovich, J. S. Mellors, J. W. Thompson, and J. W. Jorgenson, Linear velocity surge caused by mobile-phase compression as a source of band broadening in isocratic ultrahigh-pressure liquid chromatography, *Anal. Chem.* **77** (2005), 6292–6299.

44. J. W. Eschelbach and J. W. Jorgenson, Improved protein recovery in reversed-phase liquid chromatography by the use of ultrahigh pressures, *Anal. Chem.* **78** (2006), 1697–1706.

45. Y. Shen, N. Tolić, R. Zhao, L. Paša-Tolić, L. Li, S. J. Berger, R. Harkewicz, G. A. Anderson, M. E. Belov, and R. D. Smith, High-throughput proteomics using high-efficiency multiple-capillary liquid chromatography with on-line high-performance ESI FTICR mass spectrometry, *Anal. Chem.* **73** (2001), 3011–3021.

46. N. A. Polson and M. A. Hayes, Microfluidics: Controlling fluids in small places, *Anal. Chem.* **73** (2001), 312A–319A.

47. D. R. Reyes, D. Iossifidis, P. A. Auroux, and A. Manz, Micro total analysis systems. 1. introduction, theory, and technology, *Anal. Chem.* **74** (2002), 2623–2636.

48. P. A. Auroux, D. Iossifidis, D. R. Reyes, and A. Manz, Micro total analysis systems. 2. analytical standard operations and applications, *Anal. Chem.* **74** (2002), 2637–2652.

49. L. R. Snyder, J. J. Kirkland, and J. L. Glajch, *Practical HPLC Method Development*, 2nd edition, John Wiley & Sons, New York, 1997, p. 366.

50. U. D. Neue, *HPLC Columns: Theory, Technology, and Practice,* John Wiley & Sons, New York, 1997, p. 57.

51. J. C. Sternberg, *Advances in Chromatography*, Vol. 2, J. C. Giddings and R. A. Keller (eds.), Marcel Dekker, New York, 1966, p. 205.

52. A. Savitzky and M. J. E. Golay, Smoothing and differentiation of data by simplified least square procedures, *Anal. Chem.* **36** (1964), 1627–1639.

53. L. R. Snyder, J. J. Kirkland, and J. L. Glajch, *Practical HPLC Method Development*, 2nd edition, John Wiley & Sons, New York, 1997, pp. 51–52.

18

TEMPERATURE AS A VARIABLE IN PHARMACEUTICAL APPLICATIONS

Roger M. Smith

18.1 THE INFLUENCE OF TEMPERATURE ON CHROMATOGRAPHY

For much of the early development of liquid chromatography, separations were carried out at ambient temperature and many laboratories did not attempt to regulate or control the temperature of the column. Frequently, the column would be mounted on the side of the pump or detector and thus would be subjected to changes in the room temperature or changes due to external factors, such as sunlight. However, the influence of temperature on the retention times of analytes was well known and had been studied by a number of groups—in particular, Melander *et al.* [1]. They demonstrated that for most analytes there was a linear relationship between the retention factor of an analyte and the inverse of the absolute column temperature (see Chapter 1). However, for a few samples there has been an increase in retention with increasing temperature usually attributed to entropy effects. In the case of polyethylene glycol oligomers, the optimum separation was achieved with a negative temperature gradient [2]. The retention of leucine-phenylalanine at low pH and high % acetonitrile also increased with increasing temperature [3].

As a result, temperature can play an important role in pharmaceutical analysis. The precise and accurate control of temperature can improve reproducibility and method transferability (Section 18.2). In recent years, the use of

HPLC for Pharmaceutical Scientists, Edited by Yuri Kazakevich and Rosario LoBrutto
Copyright © 2007 by John Wiley & Sons, Inc.

elevated or unconventional temperatures have been examined as methods to alter selectivity and column efficiency, either with conventional mobile phases (Section 18.3) or with solvent free systems, such as superheated water (Section 18.4). Although normally the interest has been in elevated temperatures, sub-ambient chromatography has provided a number of interesting separations (Section 18.5).

18.2   EFFECTS ON METHOD TRANSFERABILITY AND REPRODUCIBILITY

As pharmaceutical analysis developed and the need for long-term repro-ducibility became more important, instrument manufacturers recognized the need for temperature stability and by the early 1990s started to include column ovens as an integral part of their instruments. In most cases the temperatures were controlled near or just above ambient because the aim was to ensure a reproducible result rather than to employ temperature as a method variable. However, even now, many chromatographers carry out separations at ambient temperature, partly on the assumption that the conditions in a heated or air-conditioned building are constant. The reality is often different and the tem-perature around a column can alter quite markedly. Sunlight can shine on a column, draughts can blow on the column, or the air-conditioning can be pro-grammed to lower the laboratory temperature overnight or at weekends as a cost-saving exercise. The result is that the retention of analyte compounds can move outside predefined retention windows and the system can show daily or long-term variations and poor reproducibility.

Of particular concern is that methods that have been developed and tested in one laboratory are often transferred to another laboratory in the same or a different company as the drug product moves from discovery through toxi-cology, stability studies, formulation, scale-up, and eventually to manufactur-ing quality control. Frequently, it is found that at each transfer, a new method optimization and a revalidation are required, each taking time and money. Sur-prisingly, relatively little research had addressed this problem. There are only few reports of interlaboratory collaborative studies where the target has been to assess the transferability of retention or resolution. In contrast, the trans-fer of quantitation has been repeatedly examined, but this is based on relative peaks areas to an internal or external standard measured under the same con-ditions. This usually compensates for differences in retention time. Typically, interlaboratory studies produce retention time reproducibility, which is much worse than intralaboratory measurements. A comparison of the analysis of forensic drugs in different UK labs [4] and in an international study [5] showed wide variation in relative and absolute retention times even through the mobile phases were closely specified and all the columns were from a single batch of packing material.

Within a single laboratory, the situation could be improved markedly by employing temperature control of the column, with an oven or water bath [6].

The values were then quite usable for quality control and identification especially when the system was calibrated with standards at frequent intervals. The main residual source of variation was then batch-to-batch differences between columns, although these differences have been reduced in recent years, and uncertainties in the preparation of the mobile phase, which can be reduced by close control of the protocols used.

As part of a project to develop a certified reference material for high-performance liquid chromatography [7], it was necessary to demonstrate that the proposed method would yield identical results in different laboratories and on different equipment. However, initial results using a specified temperature and columns from a single batch of packing material gave poor interlaboratory results, and temperature variations were suspected as a cause [8]. It was found that although the ovens in each laboratory were set to the same nominal temperature, different oven types, air, fan air, convection/conduction, and water bath gave significantly different results, the worst results coming from heaters where the column rested against a heated block. The effective temperature could be up to 6°C lower than the set value, and this could be monitored by using the changes in shape selectivity and hydrophobicity of a test mixture [9]. Similar observations of oven variability were made by Paesen and Hoogmartens [10]. The protocol for the CRM was then tightened so all the laboratories used a water bath or circulating water jacket and specified lengths of eluent preheating tubing. This gave interlaboratory and intralaboratory variations that were comparable and within the acceptable range [11]. Thus to achieve good transferability of a method, not only the obvious factors, such as column make and mobile phase, need to be defined, but also the method of maintaining a constant temperature needs to be specified.

Part of the cause of the problem is attributable to differences in the dissipation of the fictional heating generated by the movement of the mobile phase through the stationary phase. In a liquid bath, this heat is readily lost to the bath as the external temperature of the column is constant along its length, whereas in a noncontrolled or static air system the mobile phase elutes from the column at a higher temperature (2–3°C) than the inlet [12]. There is also an axial temperature gradient in each case. The effect of different temperature control was also examined by Welsch et al. [13], who found differences between air oven and water baths on normal-phase separations and also studied the effect of inlet temperatures. These effects were later studied for reversed-phase separations by Wolcott et al. [14], who suggested a number of temperature-related reasons for poor method transferability and suggested how different effects changed the temperature profile within the column.

18.3 ELEVATED TEMPERATURE AND PHARMACEUTICAL SEPARATIONS

Although temperature has been used for many years to alter separation properties, especially selectivity and efficiency, the operating range has usually been modest, typically up to 40–60°C. The principal aim has been to establish a controlled system sufficiently above ambient temperature so that day-to-day changes in the laboratory conditions have no effect on the separation. Although the use of non-ambient temperatures might offer advantages in bioanalytical methods, it has been noted that the selection often seems arbitrary and without specific justification [15]. However, Brinkman et al. [16] commented that temperature was a variable that should be considered in method development. This comment was echoed in a recent review of the use of moderate temperature changes for drug assays by Zhu et al. [17], who noted that "temperature should be considered as a useful variable to control resolution only when components in a mixture are of different types."

For many years, analysts have been deterred from applying significantly elevated temperatures because of concern about the volatility of mobile phases and the stability of stationary phases and analytes. More recently as a spin-off from work with supercritical fluid chromatography, many laboratories learned how to handle separations in pressurized columns (up to 300 bar), and hardware with pressure-resistant detector flow cells became available. As a consequence, the expertise and equipment were commercially available, which could control mobile phases above their boiling point. This has enabled the examination of separations under conditions up to 250°C based on either conventional mobile phases or less common solvents, such as superheated water [18]. Temperatures above 80°C, where pressure has to be applied to prevent the mobile phase from boiling, are usually termed either pressurized, superheated, or subcritical conditions, the latter two terms being more frequently applied to separations with just water as the mobile phase. Either the separations can be isothermal or a temperature gradient can be employed, which generates an effect similar to gradient elution, speeding up the later components [19, 20]. However, concern is often expressed that the mass of a packed HPLC column might cause the internal temperature to lag behind the oven setting but as long as the internal temperature is reproducible, a valid method can be developed. A number of early studies employed packed capillary columns with a low thermal mass [19].

Three main aims have driven these studies: the use of temperature as a variable to optimize separations, an interest in improved efficiency, and the potential for "green" separations methods, such as superheated water chromatography, which can eliminate the organic solvent from the mobile phase.

18.3.1 Effect of Temperature on Selectivity

Although temperature has been proposed as a variable in altering selectivity, it has not been widely used, because the majority of analytes show very similar changes on changing temperature (especially over the limited conventional temperature range). Significant differences may be observed if temperature can cause ionization changes or if analytes with very different functional groups are present. However, care must be taken in these situations because relative retention changes with column temperature could result in a lack of method robustness, especially if caused by ionization changes.

A few application and studies have examined temperature effects, such as the selectivity dependence of the carotenoids [21] on different columns from 25–45°C. Studies of the prediction of the influence of temperature and solvent strength on the separation of 47 basic acidic and neutral drugs compounds were reported by Zhu et al. [22] in an interlaboratory collaborative study. More recently the influence on temperature on selectivity has been reviewed by Dolan [23, 24].

The changes in retention and selectivity can also be exploited in the thermally tuned tandem column concept by Mao and Carr [25], in which the temperatures of two sequentially linked columns containing different stationary phases can be altered to provide the optimum separation. The technique was applied to the separation of barbiturates, phenylthiohydantoin amino acids [26], and selected basic pharmaceuticals, such as antihistamines (Figure 18-1) [27].

Berthod et al. [28] examined the effect of temperature on chiral separations between 5°C and 45°C using four macrocyclic glycopeptides phases; and although the efficiencies increased with temperature, in 83% of cases the chiral selectivity decreased.

18.3.2   Effect of Temperature on Separation Efficiency

One of the reported advantages of raising the temperature of a chromatographic separation is an increase in peak efficiency. This is usually attributed

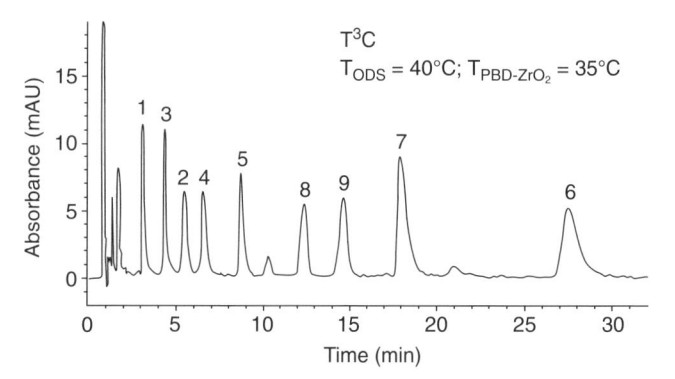

Figure 18-1. Separation of antihistamines on linked columns with different temperatures. ODS at 40°C and PBD-ZrO$_2$ at 35°C. Mobile phase 40:60 methanol/pH 7 buffer. Solutes: 1, pheniramine; 2, chlorpheniramine; 3, thenyldiamine; 4, bromopheniramine; 5, cyclizane; 6, pyrrobutamine; 7, chlorcyclizane; 8, thonylamine; 9, meclizane. (Reprinted from reference 28, with permission. Copyright 2001, American Chemical Society.)

to a reduction in the viscosity of the eluent and an increase in the diffusion rate of the analyte as the temperature is increased. A higher diffusion rate should reduces the mass transfer term effect (the C term) in the van Deemter equation but can also worsen the influence of longitudinal molecular diffusion in the column (the B term) [29]. The improvement in efficiency is normally regarded as most significant for larger analytes, such as biological and synthetic macromolecules, whose size reduces their mobility [30]. For smaller molecules the effects are relatively small, and often an increase in efficiency can be attributed to a reduction in the retention factor on raising the temperature.

A second factor, which influences peak shape and apparent efficiency, is the temperature of the incoming mobile phase relative to the column temperature. The presence of this underestimated factor may have obscured or confused previous studies of efficiency. Frequently, it has been claimed that the mobile phase in a high-temperature separation must be heated to the same temperature as the column; otherwise, peaks distortion and broadening are observed [18, 31]. This was demonstrated by Thompson and coworkers [32, 33], who reported the band-broadening effect of a thermal mismatch and advocated the use of narrow bore columns to reduce the effects. Guillarme et al. [34] also demonstrated the need for some preheating of the mobile phase and considered the length of tubing required for effective preheating (within 5°C of the oven) particularly with high flow rates.

However, even early studies including those by Cooke et al. [35] and by Poppe and Kraak [36] demonstrated that using a mobile phase slightly cooler than the column temperature can improve column efficiency. Usually, differences of 10–20°C gave the highest efficiency. For example, Mayr and Welsch [12], who found the highest efficiency for the separation of five hormone steroids was obtained with the incoming mobile phase at 10°C and a column at 30°C (Figure 18-2). Spearman et al. [9] reported that in one case reducing the inlet to 37°C below the column temperature optimized the results.

The effect is thought to have two origins. First, a cooler mobile phase and by implication cooler sample causes an initial sample focusing at the head of the column. Second, the cooler eluent flow reduces the analyte mobility at the center of the column, thus balancing the enhanced temperature and hence mobility in the center of the column caused by friction heating (Figure 18-3) [14]. At its most serious, not only efficiency but also peak distortion has been observed caused apparently by a temperature in-balance. The selection of the optimum column inlet temperature is not totally clear, and this is an area of ongoing research.

In such a situation the internal diameter of the column might also effect the equilibration process but Molander et al. [37] found that even using a temperature gradient, the differences were minimal for columns narrower than 4.6mm internal diameter. A recent study has found that elevated temperatures, up to 70°C, markedly improved the efficiency and peak shapes of bases with intermediate pH eluents [38].

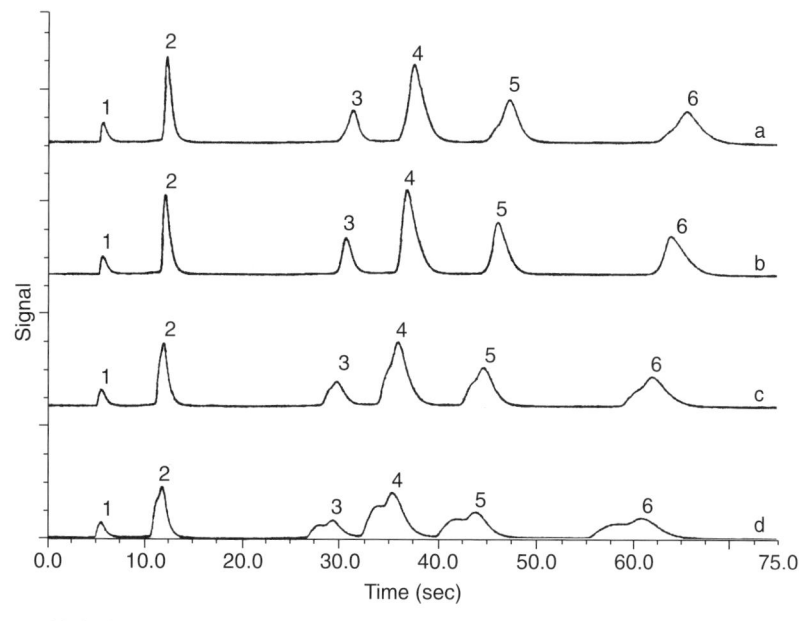

Figure 18-2. Separation of hormones on ChromSpher UOP column at 30°C at different eluent inlet temperatures: a, 5°C; b, 10°C; c, 22°C; d, 30°C. Compounds: 1, thiourea; 2, hydrocortisone; 3, nortestosterone; 4, dehydro-17a-methyltestosterone; 5, testosterone; 6, 17a-methyltestosterone. (Reprinted from reference 12, with permission from Elsevier.)

18.3.3 Other Temperature Effects

A side effect of a lack of temperature control is that changes can alter the refractive index of the mobile phase, causing baseline disturbances and reducing sensitivity The problem is principally with refractive index detection [39], but it can also influence spectroscopic detectors and their light path can be distorted. Temperature has also been reported to alter the nature of some stationary phases. For example, it caused a change in the chiral selectivity of the resolution of dihydropyrimidone acid and its methyl ester on amylose and cellulose stationary phases [40].

18.3.4 Applications of Elevated Temperatures

Almost all the high-temperature work on pharmaceutical compounds has employed reversed-phase separations. In a series of studies since the 1990s, Greibrokk and co-workers [41, 42] have examined the role of elevated temperature and temperature gradients. Many were devoted to polymer separations where the use of a temperature gradient speeded up the larger oligomers and provided clear advantages because of the complexity of the sample. As part of the optimization of the conditions for a separation on a packed

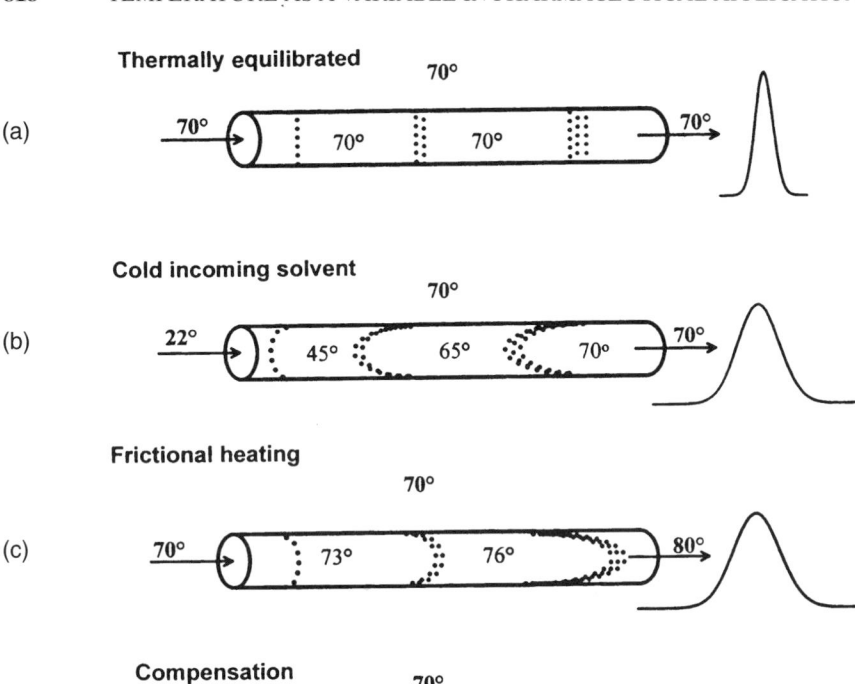

Figure 18-3. Band-broadening due to thermal effects. (a) Ideal case, no thermal effects; (b) effect of incoming mobile phase that is at a lower temperature than the column; (c) effect of frictional heating; (d) combined effect of cold incoming mobile phase and frictional heating. An oven temperature of 70°C is assumed. (Reproduced from reference 14, by permission from Elsevier.)

capillary, Tran et al. [3] included the effect of temperature on a range of compounds, including naphthalene, acenaphthene, ibuprofen, butylparaben, diethyl phthalate, monoethyl phthalate, amitriptyline, propranolol, amphetamine, all-*trans*-retinol, 13-*cis*-retinol, and DL-leucine-DL-phenylalanine.

A few applications have employed conventional packed columns, although recent developments in new thermally stable stationary-phase materials have generated a renewed interest and the temperature stability of the different stationary-phase materials has been reviewed by Claessens and van Straten [43]. The new materials have included stable metal oxide materials, based on zirconia (Figures 18-4 and 18-5) and titania [44, 45] and hybrid phases combining silica and methylene or ethyl bridges [46]. These have been applied in a number of applications to pharmaceutical compounds (Table 18-1).

One of the most interesting thermally stable groups of stationary phase materials has been the polybutadiene, carbon and phenyl-coated zirconia

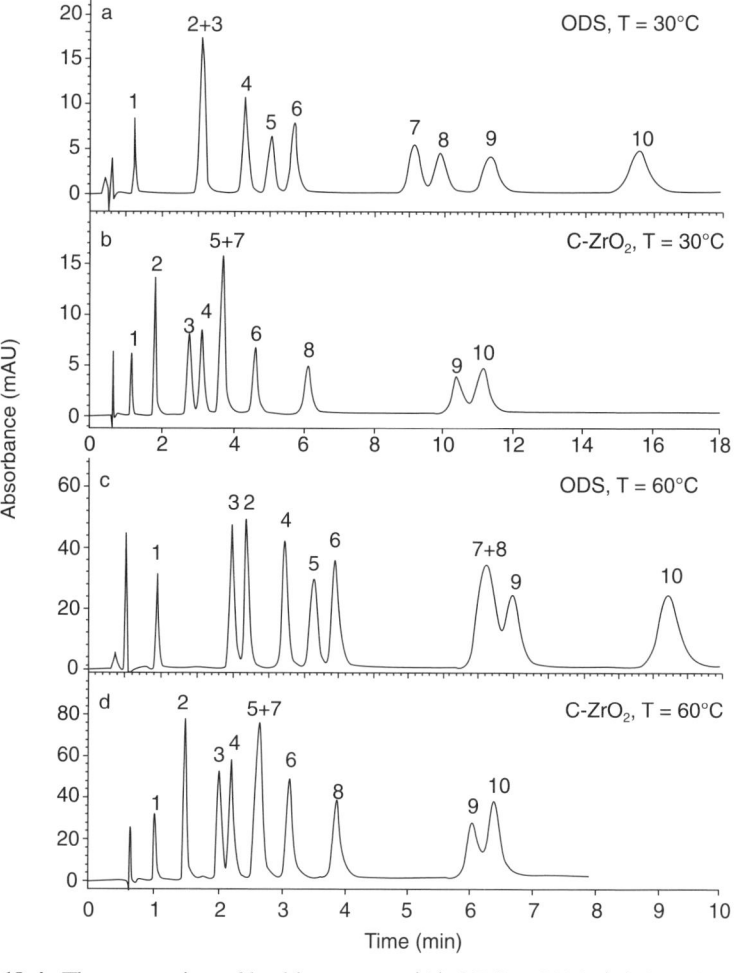

Figure 18-4. The separation of barbiturates on (A) ODS at 30°C, (B) C-ZrO$_2$ at 30°C, (C) ODS at 60°C, (D) C-ZrO$_2$ at 60°C. Mobile-phase 20/80 acetonitrile. (Reproduced from reference 27, with permission. Copyright 2001, American Chemical Society.)

phases developed by Carr and colleagues [48, 49]. They reported that at high temperatures, the zirconia material offered a much higher stability than silica-based columns [50]. Under these conditions the reduced solvent viscosity gave advantages as flow rates as high at 5 ml/min were feasible [51]. The PBD zirconia column has been used for the separation of tricyclic antidepressants [50] and lidocaine, quinidine, norephrine, tryptamine, amitriptyline, and nortriptyline. Some selectivity changes with temperature were noted. The low viscosity at 100°C also enabled very small 1-μm particles to be used for the separation of benzdiazepines (Figure 18-6) [52]. Guillarme applied these techniques to the separation of a series of caffeine derivatives, including

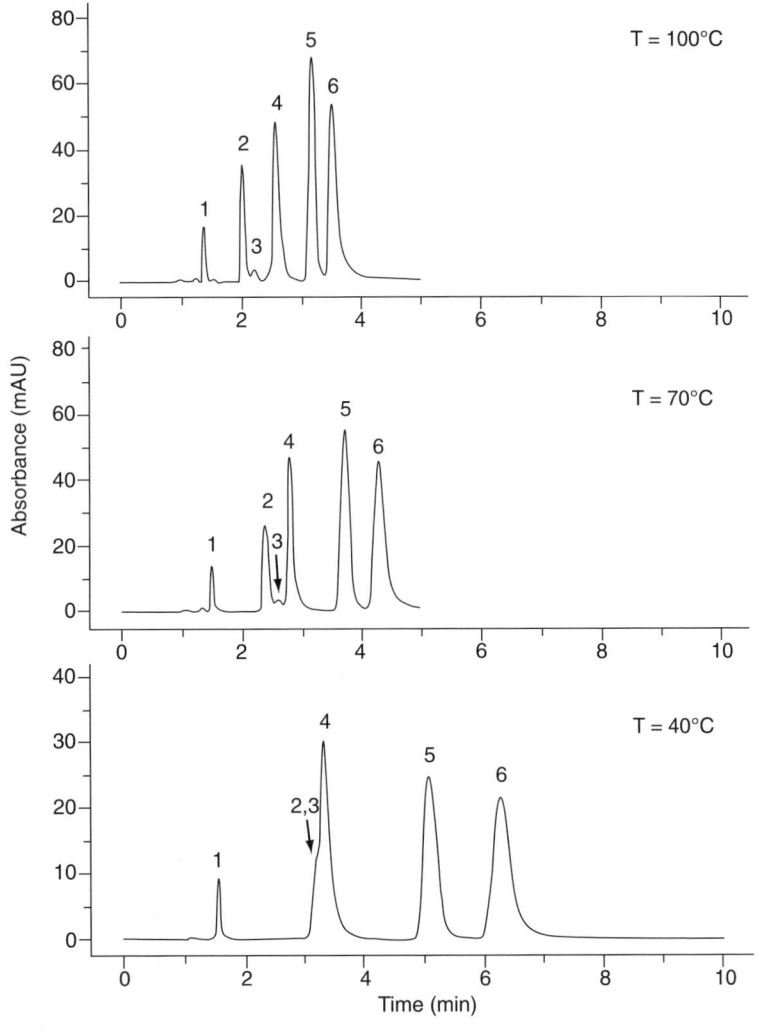

Figure 18-5. The separation of therapeutic tricyclic antidepressants on PBD-coated zirconia at different temperatures. Solutes: 1, lidnocaine; 2, quinidine; 3, norephedrine; 4, tryptamine; 5, amitriptyline; 6, nortriptyline. (Reproduced from reference 52, with permission. Copyright 1997, American Chemical Society.)

theobromine, theophylline, and caffeine when separation on a PBD zirconia column at 150°C took place in less than 1 minute compared to 7 minutes at 40°C on a Hypercarb column (Figure 18-7) [35].

In a recent study, Marin et al. [53] used a set of test compounds including amitriptyline, salicylic acid, and ibuprofen to compare the temperature stability of six stationary phases at temperatures up to 150°C.

TABLE 18-1. Pharmaceuticals Separated at Elevated Temperature Using Conventional Mobile Phases

Analyte	Mobile Phases	Stationary Phase	Temperature	Reference
Tricyclic antidepressants	—	PBD zirconia	40–100°C	49
Benzodiazepines	Acetonitrile–water	PBD zirconia (nonporous)	100°C	53
Barbiturates	Acetonitrile–water	ODS + silica + PDB zirconia	30–80°C	27
Basic pharmaceuticals				28
Vitamin A and retinoids	Acetonitrile–water	Suplex pKb-100	25–60°C	47

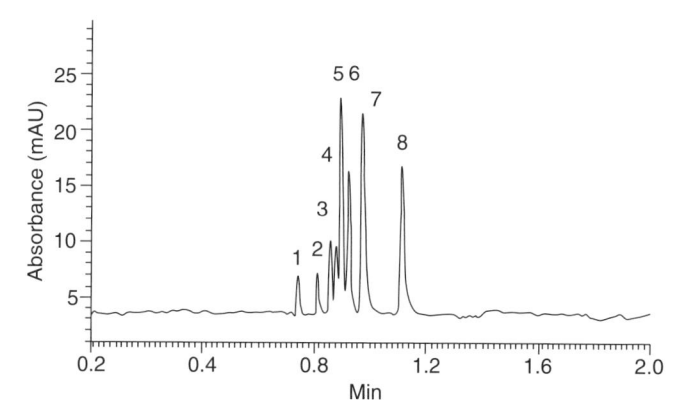

Figure 18-6. UPLC chromatography of benzodiazepines on a 14.5-cm × 50-μm column packed with 1-μm polybutadiene-encapsulated non-porous zirconia particles. Eluent pH 7 buffer–acetonitrile 68:22 at 100°C. Peaks: 1, uracil; 2, clorazepate; 3, flunitrozepam; 4, clonazepam; 5, chlordiazepoxide; 6, oxazepam; 7, clorazepate; 8, diazepam. (Reproduced from reference 53, with permission from Elsevier.)

18.4 SUPERHEATED WATER CHROMATOGRAPHY

With an increased interest and awareness of the impact of society and industry on the environment, there has been a significant attempt in recent years to reduce or replace the usage of organic solvents. Much early work in this area concentrated on the application of supercritical and subcritical carbon dioxide, but in recent years superheated (or subcritical/pressurized hot) water (SHW) has become of interest for both chromatography and extraction [43, 54]. The earliest work was reported by Guillemin et al. [55], who used the term thermal aqueous liquid chromatography. As well as using SHW for the separation of

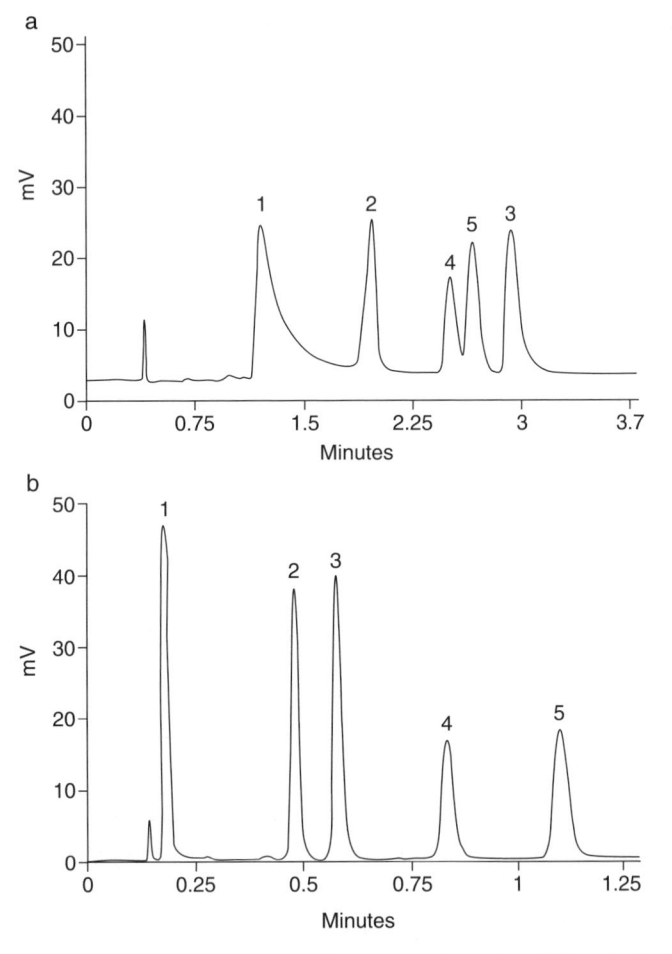

Figure 18-7. Effect of temperature on the separation of caffeine derivatives on a Hypercarb column (1 mm × 100 mm). (a) Column at 100°C, mobile phase: acetonitrile; (b) Column at 180°C, mobile phase: water/acetonitrile 70/30. Samples: 1, hypoxantine; 2, theobromine; 3. theophylline; 4, caffeine 5, β-hydroxyethyltheophylline. (Reproduced from reference 35, with permission.)

alcohols, carbohydrates, and phenol, they also looked at iprodine and used an on-line FID detector for analysis.

Water has interesting and unusual thermal properties, which have only recently been significantly exploited by chromatographers [56–58]. As the temperature is increased, thermal motion weakens the hydrogen-bonding so that the polarity of water is reduced (Figure 18-8). At 200°C, water has a polarity similar to that of methanol; in addition, the viscosity also drops markedly with temperature and the diffusion rate increases. However, the vapor pressure remains low and by 250°C has only reached 30 bar, well within the normal

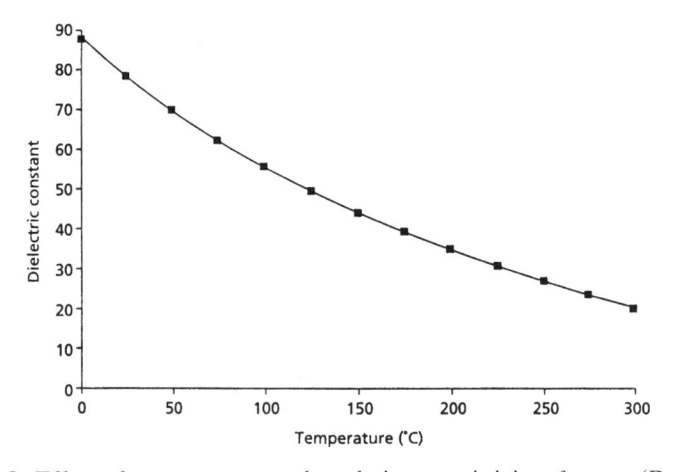

Figure 18-8. Effect of temperature on the relative permittivity of water. (Reproduced from reference 59, with permission.)

capabilities of HPLC systems and markedly below the 300–350 bar usually needed in SFC. However, the density is largely unaltered so that the water in effectively incompressible. Hence the pressure applied to the system has a minimal effect, as long as it is sufficiently high to prevent gasification, it does not influence separations.

The principal advantages in the use of superheated water are that it is relatively easy to attain and the back-pressures required on the column are small. Thus even a modest length of narrow bore tubing can be employed to provide sufficient resistance to prevent boiling in the column and at these pressures many conventional spectroscopic flow cells can be used. Because of the high temperatures, there have been concerns about the thermal stability of the analytes, but of the numerous examples, there have been few reports of instability or a tendency for accelerated hydrolysis or oxidation, of the reported examples, only aspirin has hydrolyzed. Compounds which might be expected to be labile to oxidation or hydrolysis, such as the paraben antioxidants, have chromatographed without problems even up to 200°C [59].

Because of its solvent properties, SHW up to 250°C has also been used for extractions mainly of environmental samples [59]. At higher temperatures >350°C, the critical point of water can be achieved, but by that point the conditions are severe and will probably cause analyte degradation.

18.4.1 Columns for Superheated Water Chromatography

The principal limitation of the use of superheated water has been the thermal instability of conventional ODS-silica-based stationary phases, which are unstable above 70°C or 80°C. Early work concentrated on PS-DVB columns, which were stable up to 220°C. Then zirconia-based PBD and ODS bonded phases became available with stabilities up to 140°C or 180°C, respectively (Section 18.3.4). In addition, PGC columns can be used up to 200°C, and many of these materials have been compared to conventional column materials [60, 61]. Hybrid phases, such as ODS-X-Terra, can also be employed up to 150°C.

18.4.2 Detectors in Superheated Water Chromatography

Superheated water also offers some novel advantages in detection because the absence of an organic solvent reduces low-wavelength spectroscopic absorption, eliminates the solvent peaks from NMR spectra, and eliminates the solvent signal from flame detectors. This has enabled a wide range of unique HPLC detection methods to be employed.

UV and fluorescent spectroscopy can be employed down to 190 nm because there is no solvent interference. Mass spectrometry is easy because the water provides good ionization. Flame ionization detection (FID) is of particular interest because potentially it offers a sensitive and universal detector. A number of different interfaces have been used, including heated capillaries, which have been examined by Miller and Hawthorne [62], Ingelse et al. [63], and others [64, 65], who separated a range of analytes including alcohols, amino acids, and phenols. An alternative method employing a cold nebulization of the eluent has been introduced by Bone et al. [66]. They were able to detect both aliphatic and aromatic alcohols, polymers, carbohydrates, parabens, and steroids.

By using heavy water (deuterium oxide) as the eluent, on-line NMR spectroscopic detection is simplified as negligible solvent signals are detected to interfere with the sample signals. This method can be used for drug analysis (Figures 18-9 and 18-10) [67]. By stopping the mobile-phase flow, the peak can be held in the NMR spectrometer cell, thereby increasing sensitivity or enabling more complex data analysis, such as COSY. This method was also combined with on-line mass spectroscopy for a number of model drugs [68] and was used to understand the mechanism of an unexpected selective deuterium exchange that occurred during the separation of some sulfonamides [69]. The combination of detectors using SHW as the eluent has been extended, and a train of four on-line detectors (UV spectroscopy, FT-IR, 1H-NMR, and MS) were applied to model pharmaceuticals [70] and ecdysteroids in plant extracts [71].

18.4.3. Pharmaceutical Applications of Superheated Water Chromatography

One attraction of SHW is that it can be used for reversed-phase separations and is therefore readily applicable to a wide range of pharmaceutical compounds including barbiturates, sulfonamides, analgesics and steroids (Table 18-2), and anticancer drugs, including 5-fluorouracil, methotrexate, and

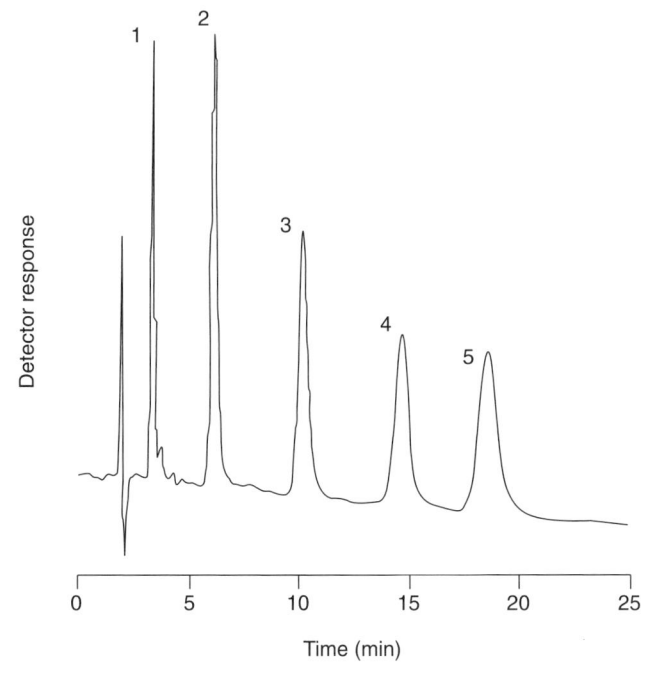

Figure 18-9. Separation of barbiturates on PS-DVB column at 200°C with water as the eluent. Samples; 1, barbitone; 2, phenobarbitone; 3, talbarbitone; 4, amylobarbitone; 5, heptabarbitone. (Reproduced from reference 59, with permission.)

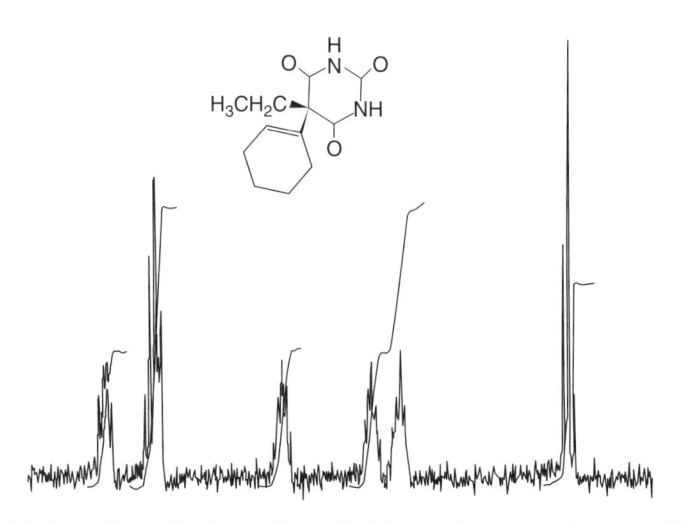

Figure 18-10. Stop flow LC-NMR of heptabarbitone after separation as in Figure 18-9 with D_2O as the eluent at 200°C. (Reproduced from reference 59, with permission.)

TABLE 18-2. Pharmaceutical Compounds Separated Using Hot and Superheated Water

Analyte	Mobile Phase	Column	Temperature	Reference
Barbiturates	Deuterium oxide	PS-DVB	200°C	69
Sulfonamide	Buffered water pH 3–12	PS-DVB	70–190°C	72
Sulfonamide	Deuterium oxide	PS-DVB	160–200°C	71
Steroids	Water	Zirconia PDB	170–200°C	62
Analgesics paracetamol, caffeine, and phenacetin	Deuterium oxide	Novapak C18	80–130°C	70
Anticancer drugs	Water pH 11.5 and 3.6	PS-DVB	Up to 160°C	73
Caffeine, paracetamol, amitriptyline, and phenacetin	Deuterium oxide	Oasis HLB Xterra	185°C 85°C	72
Paracetamol, antipyrine, and caffeine	Water	Hypercarb, PS-DVB and zirconia PBD	Up to 225°C	61
Paracetamol, salicylamide, methyl paraben, phenacetinethyl paraben.	Water	PS-DVB	75–185°C	74

etoposide (Figure 18-11) [75]. The method could be applied to even relatively nonpolar pharmaceutical compounds, such as the steroids [62]. In a related study, Tajuddin and Smith demonstrated the on-line coupling of SHW extraction with SHW chromatography for the separation of a series of pharmaceuticals [76]. The drugs could be sequentially released from the extraction by stepwise temperature increases (Figure 18-12).

SHW has also been applied to the separation of nutraceuticals, natural products, and biochemicals, including the water-soluble vitamins, thiamine, riboflavin, and pyidoxine (Table 18-3) without significant thermal degradation.

18.6   SUBAMBIENT SEPARATIONS

As well as selectivity changes at low temperatures, such as those reported by Sander and Wise [78, 79] for the separation of PAHs (Figure 18-13), subambient column temperatures can also alter chromatographic separations, by reducing the rate of the racemization of enantiomers and structural isomeri-

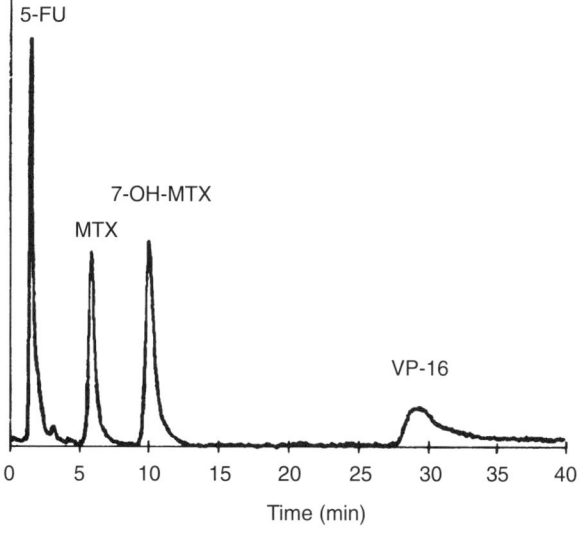

Figure 18-11. Separation of the anticancer drugs: 5-fluorouracil (5-FU), methotrexate (MTX), 7-hydroxymethotrexate (7-OH-MTX), and etopoxide (VP-16) using super-heated water and a PS-DVB column at 80°C. (Reproduced from reference 75, with permission. Copyright 2001, American Chemical Society.)

sation, such as *cis–trans* interconversions. However, as backpressure rises, eluent viscosity increases and diffusion decreases.

A classic example is the anomerization of the carbohydrates: Sucrose will give two broad peaks at room temperature but a single sharp peak by 50°C. At 10°C it will give two resolved peaks. Lower temperature separation will also result in the separation of the xylose-mannose and rhammnose-arabinose pairs on anion exchange chromatography [80]. Early work on the Pirkle chiral column with propanolol gave improved peak shapes and chiral resolution as the temperature was reduced from 21°C to –24°C [81]. Improved low-temperature chiral selectivity was also reported by Kersten [82] in the separation of beta-amino-3-pyridylpropanoic acid at subambient temperatures. Reducing the temperature to close to the freezing point of the eluent enabled isomeric dipeptides containing proline at the C-terminus to be resolved [83, 84]. Potential anti-arthritic protein kinase inhibitors also showed enhanced chiral resolution at subambient temperatures in a study on Chiralcel OD by Whatley [85].

Normal-phase separations at –30°C enabled the determination of the impurity profile of a mesylated ester, which underwent in-column cyclization at room temperature, to be determined [86]. Subambient temperatures down to –10°C were used by LoBrutto et al. [87] to generate a single product from the on-column derivatization of an acetylenic aldehyde with diethylamine.

"Supercritical" fluid chromatography using carbon dioxide as the eluent is often carried out subcritically at 20°C or 25°C, because the more dense eluent

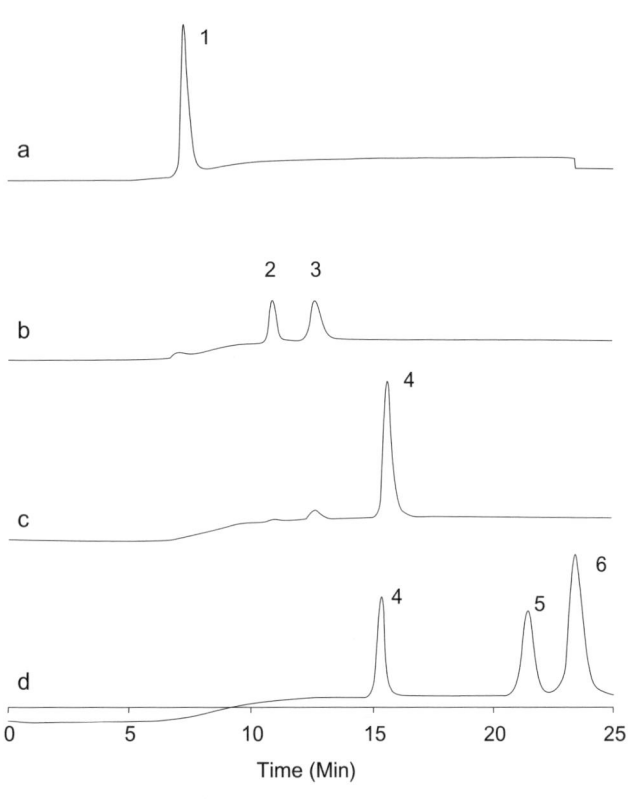

Figure 18-12. Separation of test mixture and fractions after extraction and trapping and sequential elution at increasing temperatures. Separation on PS-DVB column at 75–185°C at 15°C/min. Analytes: 1, paracetamol; 2, salicylamide; 3, caffeine; 4, methyl paraben; 5, phenacetin; 6, ethyl paraben. Separations: a, direct injection of original mixture of 1–6 without trapping; b, fraction untrapped at ambient temperature; c, fraction released from trap at 70°C; d, released at 90°C; e, released at 110°C. (Reproduced from reference 76, with permission from Royal Society of Chemistry.)

TABLE 18-3. Superheated Water Chromatography of Nutraceuticals and Natural Products

Analyte	Mobile Phase	Stationary Phases	Temperature	Reference
Ginger	Deuterium oxide	Xterra RP 18	50–130°C	75
Ecdysteroids	Deuterium oxide	C8- XTerra or C18 X-Terra	160°C	73
Water-soluble vitamins	Deuterium oxide	PS-DVB	Up to 200°C	76
Kava	Deuterium oxide	zirconia PBD	80–160°C	77

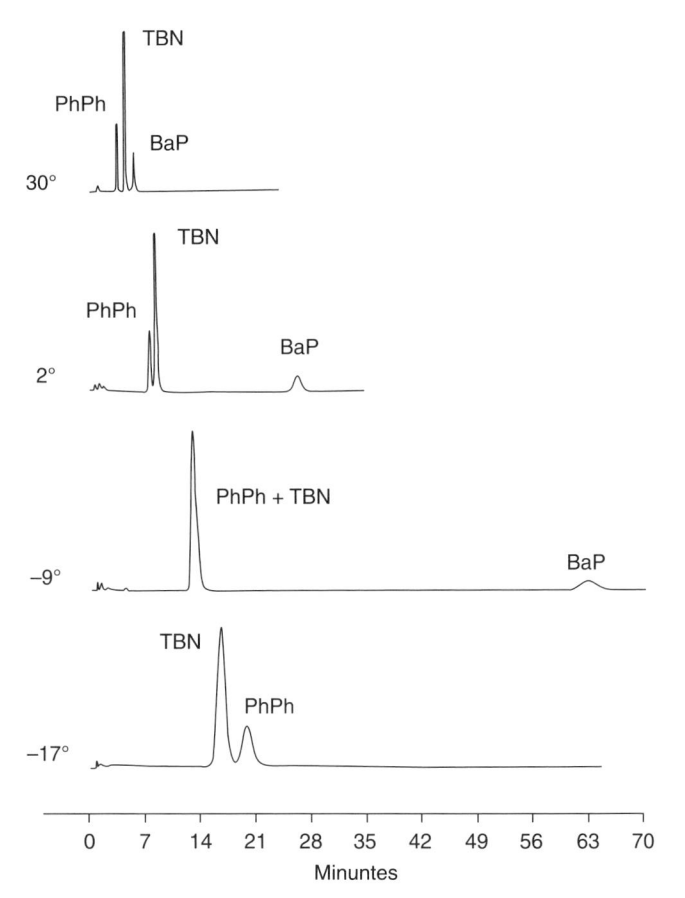

Figure 18-13. Separation of phase selectivity test mixture of phenanthro[3,4-*c*]phenan-threne (PhPh), 1,2:3,4:5,5:7,8-tetrabenzonaphthalene (TBN), and benzo[*a*]pyrene (BaP) on polymeric C18 phase (Vydac) at subambient temperatures. (Reproduced from reference 81, with permission. Copyright 1989, American Chemical Society.)

enables a greater sample loading and more column interactions, especially useful in chiral separations [88]. These conditions can also change the selectivity in "entropically driven" chiral separations, resulting in a reversal of the elution order of some pharmaceuticals [89].

The use of carbon dioxide as the mobile phase also means that it is possible to carry out assays considerably below subambient temperatures. At −50°C, Gasparrini et al. used a DACH-DNB column to resolve the enantiomers of the thermally enantiolabile 2-methyl-1-(2,2-dimethylpropanoyl) naphthalene, which can undergo rotation around the CO–CAr bond [54]. Reducing the temperature resulted in negligible degradation in column performance.

18.7 CONCLUSION

Temperature is an important and often ignored parameter in method optimization. A lack of temperature control can result in poor inter- and intralaboratory reproducibility. Increased temperatures can speed up and alter separations and may improve efficiency and throughput, especially of macromolecules. High-temperature work using superheated water can eliminate organic solvents from the mobile phase, simplifying detection and solvent interferences in detection. At lower temperature the reduction in molecular motion can resolve interconverting chiral and structural analytes.

REFERENCES

1. W. R. Melander, B. K. Chen, and C. Horváth, Mobile phase effects in reversed-phase chromatography. VII. Dependence of retention on mobile phase composition and column temperature, *J. Chromatogr.* **318** (1985), 1–10.

2. T. Andersen, P. Molander, R. Trones, D. R. Hegna, and T. Greibrokk, Separation of polyethylene glycol oligomers using inverse temperature programming in packed capillary liquid chromatography, *J. Chromatogr. A* **918** (2001), 221–226.

3. J. V. Tran, P. Molander, Y. Greibrokk, and E. Lundanes, Temperature effects on retention in reversed phase liquid chromatography, *J. Sep. Sci.* **24** (2001), 930–940.

4. R. Gill, A. C. Moffat, R. M. Smith, and T. G. Hurdley, A collaborative study to investigate the retention reproducibility of barbiturates in HPLC with a view to establishing databases for drug identification, *J. Chromatogr. Sci.* **24** (1986), 153–159.

5. R. Gill, D. M. Osselton, R. M. Smith, and T. G. Hurdley, Retention reproducibility of basic drugs in high performance liquid chromatography on a silica column with a methanol–ammonium nitrate eluent. Interlaboratory collaborative study, *J. Chromatogr.* **386** (1987), 65–77.

6. R. M. Smith, T. G. Hurdley, J. P. Westlake, R. Gill, and M. D. Osselton, Retention reproducibility of basic drugs in high performance liquid chromatography on a silica column with a methanol-ammonium nitrate buffer. Batch-to-batch reproducibility of the stationary phase, *J. Chromatogr.* **455** (1988), 77–93.

7. K. K. Unger, C.du Fresne von Hohenesche, H. Engelhardt, F. Steiner, R. M. Smith, C. A. Cramers, H. A. Claessens, J. Jiskra, R. Arras, K. Bischoff, S. Lamotte, D. Sanchez, M. Sieber, U. Berger, S. Bowadt, and A. Boenke, The method dependent certification of an high performance liquid chromatography (HPLC) column for its shape selectivity, hydrophobicity and ion exchange activity, Certification Report: CRM-722, Bureau of Community Reference, Institute of Reference Methods and Measurements, IRMM, Geel, Belgium, 2003.

8. R. M. Smith, P. V. Subba Rao, S. Dube, and H. Shah, Problems of the interlaboratory transferability of the measurement of the properties of a reversed-phase HPLC column, *Chromatographia* **57** (Suppl) (2003), S-27–S-37.

9. L. Spearman, R. M. Smith, and S. Dube, Monitoring effective column temperature by using shape selectivity and hydrophobicity and the effects of the mobile phase temperature, *J. Chromatogr. A* **1060** (2004), 147–151.

10. J. Paesen and J. Hoogmartens, Column heating and resolution—a case study, *LC-GC* **10** (1992), 364–365 and *LC-GC, Internat.* **5** (1992), 18–20.

11. R. M. Smith and S. Dube, Certified Reference material for HPLC, *Chromatographia* **61** (2005), 325–332.

12. G. Mayr and T. Welsch, Influence of viscous heat dissipation on efficiency in high-speed high-performance liquid chromatography, *J. Chromatogr. A* **845** (1999), 155–163.

13. T. Welsch, M. Schmid, J. Kutter, and A. Kalman, Temperature of the eluent: A neglected tool in high performance liquid chromatography, *J. Chromatogr. A* **728** (1996), 299–306.

14. R. G. Wolcott, J. W. Dolan, L. R. Snyder, S. R. Bakalyar, M. A. Arnold, and J. A. Nichols, Control of column temperature in reversed-phase liquid chromatography, *J. Chromatogr. A* **869** (2000), 211.

15. E. Reid, Use of non-ambient temperatures in separatory runs, *Chromatographia* **52** (Suppl) (2000), S55.

16. U. A. T. Brinkman, H. M. Hill, K. Borner, and E. Reid, Unorthodox temperature conditions: Opinions and supplementary observation from Bioanalytical Forum participants, *Chromatographia* **52** (Suppl) (2000), S57–S59.

17. C. Zhu, D. M. Goodall, and S. A. C. Wren, Elevated temperature HPLC: Principles and applications to small molecules and biomolecules, *LC-GC Eur.* **17** (2004), 530–540.

18. B. Ooms, Temperature control in high performance liquid chromatography, *LC-GC Internat.* **9** (1996), 574–585.

19. N. M. Djordjevic, P. W. J. Fowler, and F. Houdiere, High temperature and temperature programming in high-performance liquid chromatography: instrumental considerations, *J. Microcol. Sep.* **11** (1999), 403–413.

20. B. A. Jones, Temperature programmed liquid chromatography, *J. Liq. Chromatogr.* **27** (2004), 1331–1352.

21. C. M. Bell, L. C. Sander, and S. A. Wise, Temperature dependence of carotenoids on C_{18}, C_{30} and C_{34} bonded stationary phases, *J. Chromatogr. A* **757** (1997), 29–39.

22. P. L. Zhu, L. R. Snyder, J. W. Dolan, N. M. Djordjevic, D. W. Hill, L. C. Sander, and T. J. Waeghe, Combined use of temperature and solvent strength in reversed-phase gradient elution. I. Predicting separation as a function of temperature and gradient conditions, *J. Chromatogr. A* **756** (1996), 21–39.

23. J. W. Dolan, The Importance of temperature, *LC-GC Eur.* **15** (2002), 2–5.

24. J. W. Dolan, Temperature selectivity in reversed-phase high performance liquid chromatography, *J. Chromatogr. A* **965** (2002), 195–205.

25. Y. Mao and P. W. Carr, Adjusting selectivity in liquid chromatography by the use of the thermally tuned tandem column concept, *Anal. Chem.* **72** (2000), 110–118.

26. Y. Mao and P. W. Carr, Separation of barbiturates and phenylthiohydantoin amino acids using the thermally tuned tandem column concept, *Anal. Chem.* **73** (2001), 1821–1830.

27. Y. Mao and P. W. Carr, Separation of selected basic pharmaceuticals by reversed-phase and ion-exchange chromatography using thermally tuned tandem columns, *Anal. Chem.* **73** (2001), 4478–4485.

28. A. Berthod, B. L. He, and T. E. Beesley, Temperature and enantioseparation by macrocyclic glycopeptide chiral stationary phases, *J. Chromatogr. A* **1060** (2004), 205–214.

29. F. V. Warren and B. A. Bidlingmeyer, Influence of temperature on column efficiency in reversed-phase liquid chromatography, *Anal. Chem.* **60** (1988), 2821–2824.

30. F. D. Antia and C. Horávth, High-performance liquid chromatography at elevated temperature: examination of conditions for the rapid separation of large molecules. *J. Chromatogr.* **435** (1988), 1–15.

31. M. S. McCown, D. Southern, B. E. Morrision, and D. Garteiz, Solvent properties and their effect on gradient elution high-performance liquid chromatography. II. Temperature gradients in high performance liquid chromatography columns, *J. Chromatogr.* **352** (1986), 483–492.

32. J. D. Thompson, J. S. Brown, and P. W. Carr, Dependence of thermal mismatch broadening on column diameter in high-speed liquid chromatography at elevated temperatures, *Anal. Chem.* **73** (2001), 3340–3347.

33. J. D. Thompson and P. W. Carr, A study of the critical criteria for analyte stability in high-temperature liquid chromatography, *Anal. Chem.* **74** (2002), 1017–1023.

34. D. Guillarme, S. Heinisch, and J. L. Rocca, Effect of temperature in reversed phase liquid chromatography, *J. Chromatogr. A* **1052** (2004), 39–51.

35. N. H. C. Cooke, B. G. Archer, K. Olsen, and A. Berick, Comparison of 3 and 5 micrometer column packings for reversed phase liquid chromatography, *Anal. Chem.* **54** (1982), 2277–2283.

36. H. Poppe and J. C. Kraak, Influence of thermal conditions on the efficiency of high-performance liquid chromatographic columns, *J. Chromatogr.* **282** (1983), 399–412.

37. P. Molander, R. Olsen, E. Lundanes, and T. Greibrokk, The impact of column inner diameter on chromatographic performance in temperature gradient liquid chromatography, *Analyst* **128** (2003), 1341–1345.

38. S. M. C. Buckenmaier, D. V. McCalley, and M. R. Euerby, Rationalisation of unusual changes in efficiency and retention with temperature shown for bases in reversed-phase high-performance liquid chromatography at intermediate pH, *J. Chromatogr. A* **1060** (2004), 117–126.

39. K. Choiket and G, Rozing, The physicochemical causes of baseline disturbance in HPLC-Part II: Column temperature and refractive index detection, *LC-GC Eur.* **16** (2003), 811–818.

40. F. Wang, T. O'Brien, T. Dowling, G. Bicker, and J. Wyvratt, Unusual effect of column temperature on chromatographic enantioseparation of dihydropyrimidone acid and methyl ester in amylose chiral stationary phase, *J. Chromatogr. A* **958** (2002), 69–77.

41. T. Greibrokk, Heating or cooling LC columns, *Anal. Chem.* **74** (2002), 374A–378A.

42. T. Greibrokk and T. Andersen, High temperature liquid chromatography, *J. Chromatogr. A* **1000** (2003), 743–755.

43. H. A. Claessens and M. A. van Straten, Review on the chemical and thermal stability of stationary phases for reversed-phase liquid chromatography, *J. Chromatogr. A* **1060** (2004), 23–41.

44. J. Nawrocki, C. Dunlap, A. McCormick, and P. W. Carr, Part I. Chromatography using ultra-stable metal oxide-based stationary phases for HPLC, *J. Chromatogr. A* **1028** (2004), 1–30.

45. J. Nawrocki, C. Dunlap, J. Li, J. Zhao, C. V. McNeff, A. McCormick, and P. W. Carr, Part II. Chromatography using ultra-stable metal oxide-based stationary phases for HPLC, *J. Chromatogr. A* **1028** (2004), 31–62.

46. K. D. Wyndham, J. E. O'Gara, T. H. Walter, K. H. Glose, N. L. Lawrence, B. A. Alden, G. S. Izzo, C. J. Hudalla, and P. C. Iraneta, Characterisation and evaluation of C18 HPLC stationary phases based on ethyl-bridged hybrid organic/inorganic particles, *Anal. Chem.* **75** (2003), 6781–6788.

47. P. Molander, T. E. Gundersen, C. Haas, T. Greibrokk, R. Blomhoff, and E. Lundanes, Determination of retenoids by packed-capillary chromatography with large-volume on-column focusing and temperature optimization, *J. Chromatogr. A* **847** (1999), 59–68.

48. C. J. Dunlap, P. W. Carr, C. V. McNeff, and D. Stoll, Zirconia stationary phases for extreme separations, *Anal. Chem.* **73** (2001), 598A–607A.

49. J. Li and P. W. Carr, Effect of temperature on the thermodynamic properties, kinetic performance, and stability of polybutadiene-coated zirconia, *Anal. Chem.* **69** (1997), 837–843.

50. J. Li and P. W. Carr, Evaluation of temperature effects on selectivity in RPLC separations using polybutadiene-coated zirconia, *Anal. Chem.* **69** (1997), 2202–2206.

51. J. Li, Y. Hue, and P. W. Carr, Fast separations at elevated temperatures on polybutadiene-coated zirconia reversed-phase materials, *Anal. Chem.* **69** (1997), 3884–3888.

52. Y. Xiang, B. Yan, C. V. McNeff, P. W. Carr, and M. L. Lee, Synthesis of micron diameter polybutadiene-encapsulated non-porous zirconia particles for ultrahigh pressure liquid chromatography, *J. Chromatogr. A* **1002** (2003), 71–78.

53. S. J. Marin, B. A. Jones, W. D. Felix, and J. Clark, Effect of high-temperature on high-temperature liquid chromatography column stability and performance under temperature-programmed conditions, *J. Chromatogr. A* **1030** (2004), 255–262.

54. J. W. Coym and J. G. Dorsey, Superheated water chromatography: A brief review of an emerging technique, *Anal. Lett.* **37** (2004), 1013–1023.

55. C. L. Guillemin, J. L. Miller, and J. Dubois, Thermal Aqueous liquid chromatography—The TALC Technique, *JHRC and CC* **4** (1981), 280–286.

56. R. M. Smith and R. J. Burgess, Superheated water—A clean eluent for reversed-phase high performance liquid chromatography, *Anal. Commun.* **33** (1996), 327–329.

57. R. M. Smith and R. J. Burgess, Superheated water as an eluent for reversed-phase high performance liquid chromatography, *J. Chromatogr. A* **785** (1997), 49–55.

58. R. M. Smith, R. J. Burgess, O. Chienthavorn, and J. R. Stuttard, Superheated water: a new look at chromatographic eluents for reversed-phase liquid chromatography *LC-GC Internat.* **12** (1999), 30–36.

59. R. M. Smith, Extractions with superheated water, *J. Chromatogr. A* **975** (2002), 31–46.

60. I. D. Wilson, Investigation of a range of stationary phases for the separation of model drugs by HPLC using superheated water as the mobile phase, *Chromatographia* **52** (Suppl) (2000), S28–S34.

61. S. F. Fields, C. Q. Ye, D. D. Zhang, B. R. Branch, X. J. Zhang, and N. Okafo, Superheated water as eluent in high-temperature high-performance liquid chromatographic separation of steroids on a polymer-coated zirconia column, *J. Chromatogr. A* **913** (2001), 197–204.

62. D. J. Miller and S. B. Hawthorne, Subcritical water chromatography with flame ionization detection, *Anal. Chem.* **69** (1997), 623–627.

63. B. A. Ingelse, H. G. Janssen, and C. A. Cramers, HPLC-FID with superheated water as the effluent: Improved methods and instrumentation, *J. High Resolut. Chromatogr.* **21** (1998), 613–616.

64. Y. Yang, A. D. Jones, J. A. Mathis, and M. A. Francis, Flame ionization detection after splitting the water effluent in subcritical water chromatography, *J. Chromatogr. A* **942** (2001), 231–236.

65. R. Nakajima, T. Yarita, and M. Shibukawa, Analysis of alcohols by superheated water chromatography with flame ionization detection, *Bunseki Kagaku* **52** (2003), 305–309.

66. J. R. Bone, R. M. Smith, and B. L. Sharp, Analyte detection system, GB Patent GB20000021567 20000902.

67. R. M. Smith, O. Chienthavorn, I. D. Wilson, and B. Wright, Superheated deuterium oxide reversed-phase chromatography coupled to proton nuclear magnetic resonance spectroscopy, *Anal. Commun.* **35** (1998), 261–263.

68. R. M. Smith, O. Chienthavorn, I. D. Wilson, B. Wright, and S. D. Taylor, Superheated heavy water as the eluent for HPLC-NMR and HPLC-NMR-MS of model drugs, *Anal. Chem.* **71** (1999), 4493–4497.

69. R. M. Smith, O. Chienthavorn, S. Saha, I. D. Wilson, B. Wright, and S. D. Taylor, Selective deuterium exchange during superheated heavy water chromatography-nuclear magnetic resonance spectroscopy–mass spectrometry of sulfonamides, *J. Chromatogr. A* **886** (2000), 289–295.

70. D. Louden, A. Handley, S. Taylor, I. Sinclair, E. Lenz, and I. D. Wilson, High temperature reversed-phase HPLC using deuterium oxide as a mobile phase for the separation of model pharmaceuticals with multiple on-line spectroscopic analysis (UV, IR, 1H-NMR and MS), *Analyst* **126** (2001), 1625–1629.

71. D. Louden, A. Handley, R. Lafont, S. Taylor, I. Sinclair, E. Lenz, T. Orton, and I. D. Wilson, HPLC analysis of ecdysteroids in plant extracts using superheated deuterium oxide with multiple on-line spectroscopic analysis (UV, IR, 1H NMR, and MS), *Anal. Chem.* **74** (2002), 288–294.

72. O. Chienthavorn and R. M. Smith, Buffered superheated water as an eluent for reversed-phase high performance liquid chromatography, *Chromatographia* **50** (1999), 485–489.

73. T. Teutenberg, O. Lerch. H-J. Götze, and P. Zinn, Separation of selected anticancer drugs using superheated water as the mobile phase, *Anal. Chem.* **73** (2001), 3896–3899.

74. R. Tajuddin and R. M. Smith, On-line coupled superheated water extraction (SWE) and superheated water chromatography (SWC), *Analyst* **17** (2002), 883–885.

75. S. Saha, R. M. Smith, E. Lenz, and I. D. Wilson, Analysis of a ginger extract by high-performance liquid chromatography coupled to nuclear magnetic resonance spectroscopy using superheated deuterium oxide as the mobile phase, *J. Chromatogr. A* **991** (2003), 143–150.

76. O. Chienthavorn, R. M. Smith, S. Saha, I. D. Wilson, B. Wright, S. D. Taylor, and E. M. Lenz, Superheated water chromatography-nuclear magnetic resonance spectroscopy and mass spectrometry of vitamins, *J. Pharm. Biomed. Anal.* **36** (2004), 477–482.

77. R. M. Smith, O. Chienthavorn, I. D. Wilson, B. Wright, and E. M. Lenz, Superheated water chromatography–nuclear magnetic resonance spectroscopy of kava lactones, *Phytochem. Anal.* **16** (2005), 217–221.

78. L. C. Sander and S. A. Wise, The influence of column temperature on selectivity in reversed-phase liquid chromatography for shape-constrained solutes, *J. Sep. Sci.* **24** (2001), 910–920.

79. L. C. Sander and S. A. Wise, Subambient temperature modification of selectivity in reversed-phase liquid chromatography, *Anal. Chem.* **61** (1989), 1749–1754.

80. C. Panagiotopoulos, R. Sempere, R. Lafont, and P. Kerherve, Sub-ambient temperature effects on the separation of monosaccharides by high-performance anion-exchange chromatography with pulse amperometric detection-application to marine chemistry, *J. Chromatogr. A* **920** (2001), 13–22.

81. W. H. Pirkle and J. A. Burke, Enantioseparations by subcritical fluid chromatography at cryogenic temperatures *J. Chromatogr.* **557** (1991), 173–185.

82. B. S. Kersten, HPLC Chiral optimization of a unique beta-amino acid and its ester, *J. Liq. Chromatogr.* **17** (1994), 33–48.

83. D. E. Henderson and C. Horvath, Low temperature high-performance liquid chromatography of cis-trans proline dipeptides, *J. Chromatogr. A* **368** (1986), 203–213.

84. D. E. Henderson and J. A. Mello, Physicochemical studies of biologically active peptides by low temperature reversed-phase high-performance liquid chromatography, *J. Chromatogr. A* **499** (1990), 79–88.

85. J. A. Whatley, Chiral resolution of protein-kinase inhibitors by reversed-phase high-performance liquid chromatography on cellulose tris-3,5-dimethylphenylcarbamte, *J. Chromatogr. A* **697** (1995), 263–269.

86. J. O. Egekeze, M. C. Danielski, N. Grinberg, G. B. Smith, D. R. Sidler, H. J. Perall, G. R. Bicker, and P. C. Tway, Kinetic analysis and subambient temperature chromatography of an active ester, *Anal. Chem.* **67** (1995), 2292–2295.

87. R. LoBrutto, Y. Bereznitski, T. J. Novak, L. Dimichele, L. Pan, M. Journet, J. Kowal, and N. Grinberg, Kinetic analysis and subambient temperature on-line on-column derivatization of an active aldehyde, *J. Chromatogr. A* **995** (2003), 67–78.

88. R. M. Smith, Chiral chromatography using sub- and supercritical fluids, in C. L. Berger and K. Anton (eds.), *Supercritical fluid chromatography with packed columns. Techniques and Applications*, Marcel Dekker, New York, 1998, Chapter, 8, pp. 223–249.

89. R. W. Stringham and J. A. Blackwell, Entropically driven chiral separations in supercritical fluid chromatography. Confirmation of isoelution temperature and reversal of elution order, *Anal. Chem.* **68** (1996), 2179–2185.

19

LC/MS ANALYSIS OF PROTEINS AND PEPTIDES IN DRUG DISCOVERY

Guodong Chen, Yan-Hui Liu, and Birendra N. Pramanik

19.1 INTRODUCTION

The modern drug discovery process, in general, involves the identification of a biochemical target (usually protein target), screening of synthetic compounds or compound libraries from combinatorial chemistry/natural sources for a lead compound, and optimization of the lead compound (activity, selectivity, pharmacokinetics, etc.) for recommending a potential clinical candidate. The ultimate goal is to develop highly potent compounds (small molecules) that bind noncovalently with target proteins and produce the desired therapeutic response with minimal side effects [1].

In addition, the discovery of DNA structures by Francis Crick and James Watson laid a foundation for the $30 billion-a-year biotechnology industry that has produced some 160 drugs and vaccines, treating everything from breast cancer to diabetes. Recent advances in recombinant DNA technology have provided means to produce and develop protein products as novel drugs, vaccines, and diagnostic agents. For example, INTRON A (interferon α-2b) is one of the first recombinant protein drugs introduced on the market. This synthetic *E. coli* recombinant DNA-derived protein functions as a natural interferon produced by the human body as part of the immune system in response to the presence of enemy cells. It not only interferes with foreign invaders that may cause infections, but also prevents the growth and spread of other diseased

HPLC for Pharmaceutical Scientists, Edited by Yuri Kazakevich and Rosario LoBrutto
Copyright © 2007 by John Wiley & Sons, Inc.

cells in the body. This protein drug is effective in treating hepatitis C virus and a variety of tumors. ENBREL (etanercept) is another protein drug used for treatment of rheumatoid arthritis. It is produced from a Chinese hamster ovary mammalian cell expression system. This protein drug is a dimeric fusion protein consisting of the extracellular ligand-binding portion of the human 75-kilodalton (kDa) tumor necrosis factor receptor (TNF). TNF is one of the chemical messengers that are involved in the inflammatory process. Too much TNF produced in the human body overwhelms the human immune system's ability to control inflammation in the joints. ENBREL binds to and inactivates some TNF molecules before they can trigger inflammation, thus reducing inflammatory symptoms [2, 3].

One of difficulties encountered in producing large quantities of biologically active proteins is the elimination of microheterogeneity related to these proteins. The therapeutic proteins and the drug target proteins are usually associated with post-translational modifications, such as phosphorylation [4], glycosylation [5], aggregation, and disulfide bond formation [6], with all contributing to the heterogeneity of the proteins. These post-translational modifications control many biological activities/processes. Therefore, characterization of proteins with respect to assessment of purity and structure is an integral part of the overall efforts toward drug development, including submission of the analytical data to the regulatory agencies. Furthermore, progress in genomics and proteomics research has generated new proteins that require rapid characterization by analytical methods [7].

19.2   GENERAL STRATEGIES FOR ANALYSIS OF PROTEINS/PEPTIDES

The analytical strategies for protein characterization rely heavily on high-performance liquid chromatography (HPLC) and/or electrophoretic separation of proteins/peptides, followed by other detection methods [e.g., mass spectrometry (MS)].

19.2.1   HPLC Methods in Proteins/Peptides

Achieving good separation of proteins/peptides is always one of many challenges in chromatographic separations. Proteins are highly complex molecules with enormous amount of structural diversity, including hydrophobic/hydrophilic and anionic/cationic interactions. The differences in physical, chemical, and functional properties of proteins/peptides provide the molecular basis for their separations. There are five basic chromatographic separation methods, including size-exclusion chromatography, ion-exchange chromatography, reversed-phase chromatography, hydrophobic interaction chromatography (HIC), and affinity chromatography (detailed discussions on the first three techniques are provided in Part I of this book) [8, 9].

Size-exclusion chromatography (often referred to as gel filtration or gel permeation chromatography) is a chromatographic process involving separation of proteins on the basis of their differential apparent molecular sizes [10]. The column packing materials usually consist of particles with well-controlled pore size. When mobile-phase liquid flows through these particles, the proteins (solutes) with different size can get into and out of the pores with different accessibility. For a specific size-exclusion column with a specific pore size, proteins with molecular weights above the exclusion limit (in daltons) of the column are too large to enter the pores and are excluded from the column. Proteins with molecular weights less than the exclusion limit can have different access to pores of particles and elute after the void volume, depending on their size and shape. In theory, there is a linear relationship between the logarithm of protein molecular size (molecular weight) and the elution volume of the protein. A calibration curve based on this linear relationship can be used to determine the molecular weight of proteins, assuming that the protein is globular and symmetrical in shape, and there is no other interaction between the protein and column. In practice, denaturants (e.g., 0.1% SDS) are sometimes used in the mobile phase to disrupt possible formation of undesired protein aggregates in solution and promote uniformity in conformations of proteins. Thus, the separation can be performed in near-ideal situations to obtain more accurate molecular weight determination of proteins using this approach.

Several parameters should be given special consideration in method development of size-exclusion chromatography. Although its nature of separation requires no interactions between the proteins and stationary phase, the column packing material often exhibits anionic and hydrophobic characters. The addition of salts to the mobile phase can suppress these column effects. However, a higher concentration of salts (>0.5 M) might promote hydrophobic interactions between proteins and the column. Amount of salts added to the mobile phase should be carefully adjusted. Another factor is pH value. The formation of silanolate anions from column can be minimized by carrying out experiments at pH values less than 7. Typical experimental conditions include mobile phases with low ionic strength buffers (<0.1 M) in near-physiological pH ranges—that is, 50 mM phosphate buffer with 100 mM KCl (pH 6.8). Flow rates can vary from 0.5 mL/min to 1.0 mL/min, although a better resolution can be achieved with slower flow rates. The sample injection volume and analyte concentration is also critical for optimum performance. The loading capacity is very low for size-exclusion chromatography. Generally, the sample injection volume should not exceed 5% of the column bed volume in order to maintain good resolution. Protein samples should be concentrated without causing precipitation prior to analysis. Once an appropriate method is developed, size-exclusion chromatography can be an excellent method for separation of protein complexes. It is also suitable for buffer exchange as a desalting procedure in protein purifications (salts can be easily separated from proteins by size-exclusion chromatography) and estimation of the molecular weight of proteins. A key advantage of this technique is that the biological activity of proteins is maintained during the separation.

Ion-exchange chromatography relies on reversible, electrostatic (or ionic) interactions between charged proteins/peptides in the mobile phase and charged ion-exchange group on the stationary phase [11]. Proteins/peptides normally possess either net positive or negative charges depending on pH. They are positively charged at pH values below their pI (isoelectric point) and negatively charged at pH values above their pI. For acidic proteins and peptides (pI < 6), they are normally separated using anion-exchange columns because they are negatively charged. Basic proteins and peptides (pI > 8) are usually chromatographed on cation-exchange column because they are positively charged. The choice of pH is important for optimum separation results. The pH of the mobile phase is typically set at least one pH unit away from the pK_a of its ion-exchange resin in order to keep 90% of the full charge on the column. For anion-exchange column, the pH is chosen to be lower than the pK_a. For cation exchangers, the pH is set to be higher than the pK_a. Other key parameters include the ionic strength of the mobile phase. The salts used in the buffer solution are the counterions that might bind to the ion-exchange column in competition with proteins/peptides. Thus, if a protein/peptide is strongly bound to the ion-exchange column, a stronger counterion can be used to improve the elution. Some common counterions with their relative strength include $Cs^+ > K^+ > NH_4^+ > Na^+$ and $PO_4^{3-} > CN^- > HCOO^- > CH_3COO^-$. The unique feature of ion-exchange chromatography is that the biological activity of proteins is almost always preserved, and this separation method can also be used to concentrate dilute protein samples.

More recently, another related technique — chromatofocusing — has emerged as a chromatographic technique complementary to electrophoretic methods for pI determination. Chromatofocusing is an ion-exchange technique in which a pH gradient is established across the column, allowing for the eventual separation of amphoteric substances (i.e., proteins) based on their pI. The main advantages of chromatofocusing are high loadability of the column, high resolution power allowing separation of two proteins (i.e., protein and a degradation product variant) differing less than 0.05 pI units, and the high efficiency due to both gradient elution mode and special focusing effect of the polyampholytes. Furthermore, peptides and proteins are less likely to precipitate in chromatofocusing than in isoelectrical focusing.

Reversed-phase (RP) chromatography is a hydrophobic separation technique based on the interaction between the nonpolar regions of proteins/peptides and the stationary phase [12]. It typically utilizes volatile organic solvents (acetonitrile, etc.) as mobile phases under acidic pH conditions. It provides high speed and high efficiency and is compatible with MS detection. This technique is the most widely used HPLC method in the separation of peptides and proteins.

There are a number of factors to be considered in method development of RPLC for separation of proteins and peptides. Appropriate pore size is one of primary considerations in selecting a column. For proteins greater than 10 kDa, large pore size (300 Å) is necessary to reduce restriction of the protein into the stationary phase and avoid poor recoveries and decreased efficiencies. Polypeptides (<10 kDa) can be effectively separated using a column with a small pore size (<150 Å). The hydrophobicity of the protein is also important when choosing a column. In general, C18 column is used for hydrophilic proteins/small peptides, and C4 or C5 bonded phase is used for hydrophobic proteins/large polypeptides. The use of C4/C5 column for hydrophobic proteins may reduce undesired protein absorption on the column because more retentive C18 column for hydrophobic proteins can lead to irreversible binding of the protein to the column. The most commonly used mobile phase in RPLC involves acetonitrile solution with 0.1% trifluoroacetic acid (TFA). In addition, alcohols such as isopropanol are sometimes used for large and more hydrophobic proteins to enhance the elution and improve recovery. Note that all mobile phase reagents should be of the highest quality to avoid the appearance of ghost peaks from solvent impurities. Some ion-pairing reagents are often used to optimize resolution and retention. For example, hydrophobic, anionic ion-pairing reagents (i.e., TFA and pentafluoropropionic acid) can complex with positively charged basic residues and influence the chromatography. On the other hand, hydrophobic, cationic ion-pairing reagents (i.e., triethylamine acetate) interact with negatively charged groups (i.e., carboxylic acid, free carboxyl terminus at pH > pK_a) and effect their retention. Thus, manipulation of ion-pairing reagent and pH value provides alternative approaches in optimizing RPLC. Variation of flow rate and gradient rate can have an impact on the chromatography as well. An increase in flow rate or a decrease in gradient rate improves resolution, although it may result in a loss of sensitivity. Typically, a shallower gradient is employed to maintain good resolution—that is, 0.25% to 4% per minute. Column temperature also affects the separation. Higher column temperature usually improves column efficiency, peak shape, and resolution. However, it may lead to the loss of biological activity of the protein.

Hydrophobic interaction chromatography involves weak interactions of hydrophobic patches on the surface of the intact protein and nonpolar groups on the stationary phase [13]. This technique uses aqueous mobile phases of high ionic strength and neutral pH. It does not denature or unfold proteins and can be used to detect protein conformational changes. Key factors affecting protein separations include column, salt, mobile-phase pH, and temperature. Most columns used in HIC are made of silica-based stationary phases with modified aryl groups, diol derivatives, and short alkyl chains. The overall hydrophobicity of the stationary phase is determined by both the nonpolar character of the bonded ligands and their density. Strong column-solute interactions should be avoided to reduce denaturation. The type and concentration of salt are critical in HIC. One of considerations in choosing a salt is its surface tension. Salts with higher surface tension values may lead to the increase in solute retention. The amount of proteins bound to the column also increases with increasing of salt concentration. More hydrophobic proteins should be separated using salts with higher surface tensions. Commonly used salts with relative surface tension include KCl < NaCl < Na_2HPO_4 < $(NH_4)_2SO_4$ < Na_3PO_4, with typical concentrations ranging from 1 M to 3 M in order to maximize selectivity or column capacity. The pH value in HIC is usually maintained in the neutral range (pH 5–8). Appropriate pH for the optimization of resolution/selectivity in HIC can only be made empirically since proteins differ significantly in their susceptibility to denaturation with changing of pH. Another important parameter in developing HIC method is temperature. In general, proteins tend to be more stable at lower temperatures. To maintain the conformations of proteins, the lowest temperature sufficient for separation should be used in the HIC technique.

As an illustration of HIC technique, the recombinant human growth hormone (hGH) and methionyl hGH (met-hGH) were well-separated by the HIC technique [14]. The optimized conditions were found to be 1 M ammonium phosphate dibasic, pH 8.0/propanol (99.5:0.5) and 0.1 M sodium phosphate dibasic, pH 8.0/propanol (97.5:2.5) for mobile phase A and B, respectively, with a descending gradient from 100% A to 100% B in 30 minutes at a column (TSK-phenyl 5PW, 75 × 7.5 mm) temperature of 30°C. Note that the addition of a small amount of propanol as organic modifiers significantly decreases elution time while maintaining resolution and efficiency. This HIC method allowed separation of several hGH variants from the main hGH peak while retaining their native structures.

Affinity chromatography is based on reversible, specific binding of one biomolecule to another [15]. The analyte to be purified is specifically and reversibly adsorbed to a ligand (binding substance) that is immobilized by a covalent bond to a chromatographic bed material (matrix). The choice of ligand is a critical factor in affinity chromatography, because it determines the interaction mode between the solute and the ligand. There are two types of ligands: specific ones and multifunctional ones. Specific ligands include potent binders of single classes of peptides or proteins, such as enzyme substrates/ inhibitors and antigens/antibodies. Examples of multifunctional ligands include (a) concanavalin A that binds to some specific carbohydrate residues and (b) nucleotides that bind to enzymes. The chromatography steps involve sample loading in which samples are applied under favorable conditions for their specific binding to the ligand. Analytes of interest are consequently bound to the ligand while unbound substances are washed away. Recovery of molecules of interest can be achieved by changing experimental conditions to favor desorption (elution). Various elution techniques used include changes in mobile-phase composition (e.g., ionic strength, pH) and disruption of ligand/solute complex using competitive ligands in the mobile phase. The separation of analytes depends on their native conformations (for proteins) and relative binding affinities for the immobilized ligand on the column. The affinity interactions can be extremely specific, an antibody binding to its antigen, and so on. This technique is a powerful tool in investigating protein–protein,

protein–peptide, and drug–protein interactions. Its applications in inhibitor screening using affinity chromatography–MS methods in drug discovery will be discussed later in this chapter.

19.2.2   MS Methods for Protein Characterization

MS is another powerful analytical technique for protein characterization. This technique measures mass-to-charge ratios of ions in the gas phase, providing both molecular weight (MW) information and structural information [16]. The introduction of electrospray ionization (ESI) [17, 18] and matrix-assisted laser desorption/ionization (MALDI) [19] or soft ionization [20] has revolutionized applications of MS in protein characterization, making it quite straightforward to analyze proteins with molecular weight of over 1 million daltons (Da). ESI forms multiple-charged ions for proteins/peptides by spraying the sample solution through a nozzle under a strong electrical field. The molecular weight of a protein can be calculated from a group of $[M + nH]^{n+}$ ions in the ESI spectrum with a better precision. Also, multiple-charge ions appear at m/z values which are only fractions of the actual molecular weight of the analyte. This allows one to observe high-molecular-weight proteins beyond the normal mass range of a mass spectrometer. In addition, ESI operates at atmospheric pressure, which allows the direct on-line analysis by interfacing HPLC with MS. The MALDI technique has high ionization efficiencies for proteins and can achieve a mass range of over 500 kDa when coupled with a time-of-flight (TOF) mass analyzer. In this technique, proteins are mixed with an IR or UV absorbing matrix in large excess and the mixed sample is deposited on a sample target, dried, and inserted into the mass spectrometer for laser irradiation. In contrast to multiple-charge ions in ESI, the singly charged ions are the most abundant species in the MALDI-MS spectrum. Higher sensitivity (lower femtomole) can be achieved with MALDI-MS analysis.

The very first step in protein characterization is the molecular weight determination. With multiple-charge ions formed in ESI, a deconvoluted mass spectrum can be generated to give an average molecular weight of the protein by calculating from successive multiple-charged ions. For example, Figure 19-1 shows an ESI mass spectrum of a recombinant interferon α-2b (antiviral protein drug) with a charge distribution of +9 to +13. The deconvoluted spectrum (Figure 19-1, insert) gives a molecular weight of 19,266.3 Da for this protein. The mass measurement precision and accuracy are enhanced by the use of all the observed multiple-charged ions (typically better than 0.01% for masses up to 100 kDa) [21]. The MALDI-MS technique can also be employed to analyze intact proteins with high tolerance of impurities (salts, etc.). Figure 19-2 illustrates a MALDI-TOF mass spectrum of 1 pmol of anti-IL-5 MAB protein with an average molecular weight of 146.5 kDa [1]. The singly charged molecular ion $[M + H]^+$ is observed at m/z 146,485, along with a doubly charged molecular ion.

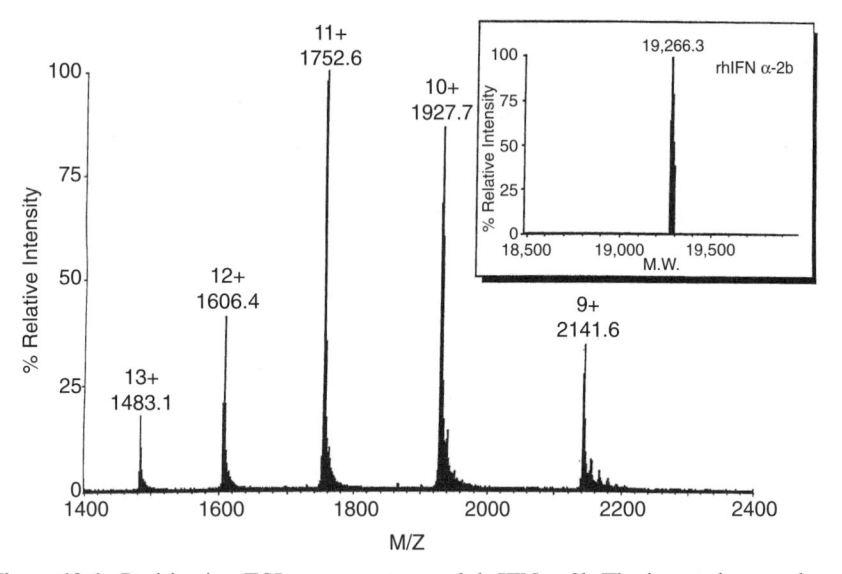

Figure 19-1. Positive ion ESI mass spectrum of rh-IFN-α-2b. The insert shows a deconvoluted spectrum.

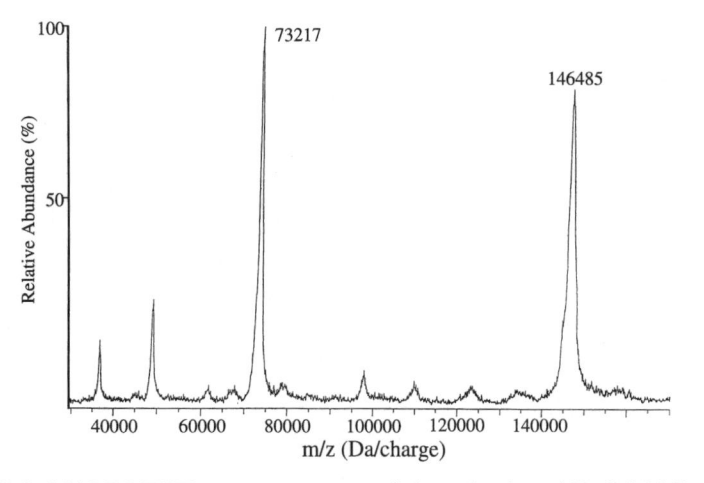

Figure 19-2. MALDI-TOF mass spectrum of 1 pmol of anti-IL-5 MAB protein. (Reprinted from reference 1, with permission of the Thomson Corporation.)

The protein identification or sequence determination of a protein can be achieved using two different approaches: "top-down" [22, 23] and "bottom-up" [24]. A top-down experiment involves high-resolution measurement of an intact molecular weight and direct fragmentation of protein ions by tandem mass spectrometry (MS/MS) [25]. This approach surveys an entire protein sequence with 100% coverage. Post-translational modifications such as glycosylation and phosphorylation tend to remain intact during MS/MS fragmentation at the protein level. The fragment ions obtained allow the protein identification by database retrieval, quick positioning of the N- and C-termini, confirmation of large sections of sequences, and partial or exact localization of modifications. This is a preferred method for protein identifications. However, there are some obstacles that need to be overcome before this approach can be widely accepted as a standard in protein identifications. These challenges include accessibility of expensive MS instrumentation for accurate mass measurements of large proteins, development of suitable MS instrumentation for efficient MS/MS data acquisition in automatic fashion, and appropriate database search algorithm. In contrast to the top-down methodology, the bottom-up experiment refers to the process in which proteins are digested into smaller peptides under enzymatic cleavages without measuring the accurate mass value of the intact protein. These enzymatic digested peptides (tryptic peptides, etc.) often can be unique in terms of their mass, amino acid composition/sequence, and separation characteristics. They can be separated/detected and either (a) directly searched against a genome or protein database for protein identification (peptide mass mapping) or (b) further dissociated in a tandem mass spectrometric experiment to generate fragment ions for database search (sequence tagging) [26, 27]. The principal fragment ions in polypeptide ions are b ions (N-terminus) and y ions (C-terminus) resulted from cleavages of amide bonds under collision-induced dissociations [28]. These are amino acid-specific fragment ions and can be used to derive sequences of polypeptides. Further database search based on the MS/MS information can lead to identification of proteins. The general sequence coverage from this approach (5–70%) is far less than 100% from top-down approach. Post-translational modifications are likely to be lost during MS/MS fragmentation at the peptide level. In spite of these limitations, the bottom-up approach has become a current standard method in protein identifications because of its high-throughput format and well-refined methodology—for example, mature instrumentation and excellent software development [29]. Some specific examples using this approach will be described in the following sections.

19.3   APPLICATIONS FOR BIOTECHNOLOGY PRODUCTS AND DRUG TARGETS

19.3.1   Biotechnology Products Development

The production of biologically important proteins by recombinant DNA techniques and development of modified counterparts is a very challenging field. Certain criteria of safety, quality, and efficacy are required for the development and approval of these protein products as therapeutic agents. The presence of structural variations during the different steps in the protein production process could affect the protein's biological properties and alter the safety, potency, and stability of the protein product. The development of sensitive analytical techniques for the analysis of therapeutic proteins is essential for the quality control and structural characterization of recombinant protein products. Two examples are illustrated below, including recombinant human granulocyte-macrophage colony stimulating factor (rh-GM-CSF) and interferon alpha-2b (rh-IFN-α-2b).

19.3.1.1 rh-GM-CSF. GM-CSF belongs to a group of interacting glycoproteins that regulate the differentiation, activation, and proliferation of multiple blood-cell types from progenitor stem cells. This particular glycoprotein is essential for the proliferation and differentiation of progenitor cells into mature granulocytes and macrophages [30]. It enhances the production and function of white blood cells with its potential clinical applications for follow-up treatment for patients who have gone through chemo or radiation therapy for tumors, as well as bone marrow transplantation. GM-CSF has been cloned and expressed in various cell lines that include yeast, Chinese hamster ovary, and *E. coli*. The *E. coli* derived GM-CSF used in this study contains 127 amino acid and has a molecular weight of ~14,477.6 Da.

One of the first measurements performed to characterize a protein is determination of the molecular weight. It is an important physical parameter that can be used to confirm primary structure and identity of the protein, characterize post-translational modifications, and determine batch-to-batch reproducibility in the production of recombinant proteins. The mature protein sequence for human GM-CSF with four cysteine residues is shown in Table 19-1 [31]. Figure 19-3A displays the ESI-MS spectrum of rh-GM-CSF, containing a series of multiply-charged ions ranging from the 7+ to the 16+ charge state that correspond to molecular ions of the protein. The measured average molecular weight (14,472 Da, as shown in the insert) suggests the presence of two disulfide bonds in the rh-GM-CSF because the calculated averaged molecular weight of rh-GM-CSF derived from the sequence is 14,477.6 Da

TABLE 19-1. Amino Acid Sequence of rh-GM-CSF from *E. Coli*

APARSPSPSTQPWEHVNAIQEARRLLNLSRDTAAEMNETVEVI
-T_1-→------T_2--------------------------→---T_3-----→------------------------------
-------V_1--------------→----V_2----→---------V_3------------→-V_4->-V_5->------------
SEMFDLQEPTC^{54}LQTRLELYKQGLRGSLTKLKGPLTMMASHYK
-----------T_4---------------→--T_5---→--T_6-→--T_7--→T_8>--------T_9--------→--------
V_6>--V_7----→------V_8------→--------------------V_9------------------------------
QHC^{88}PPTPETSC^{96}ATQIITFESFKENLKDFLLVIPFDC^{121}WEPVQE
---------------T_{10}---------------------→-T_{11}->----------T_{12}----------------→--------
---------------→------V_{10}--------→-V_{11}->-----------V_{12}------------→-V_{13}-→--------

*The T_n and V_n indicate expected tryptic and *S. aureus* V8 protease peptides, respectively.

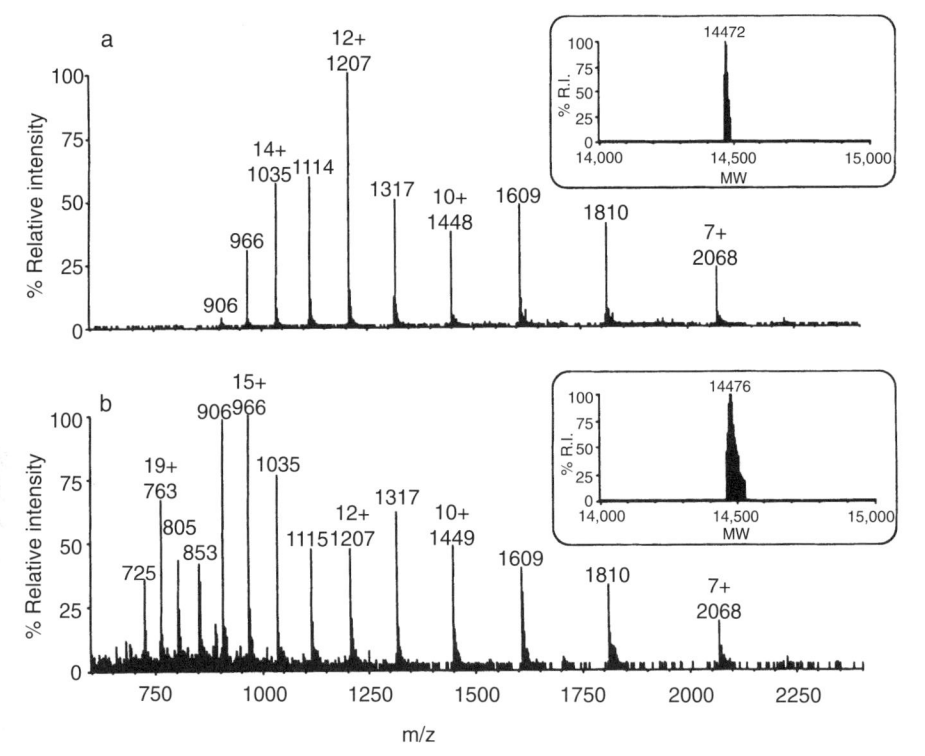

Figure 19-3. Positive ion ESI mass spectra of rh-GM-CSF. (A) In 1% HCOOH and (B) after treatment with β-mercaptoethanol. The deconvoluted spectra are shown in the inserts. (Reprinted from reference 31, with permission of the Protein Society.)

(without accounting for existing disulfide bonds). This was further supported by ESI-MS analysis of rh-GM-CSF after reduction with β-mercaptoethanol, as shown in Figure 19-3B. The 4-Da mass shift of the measured molecular weight of reduced rh-GM-CSF (14,476 Da) from nonreduced rh-GM-CSF confirms the presence of two disulfide bonds in the protein molecule. In addition, the charge state distribution is also shifted to higher charge states (17+, 18+, 19+, 20+) for the reduced form, indicating a more open form of protein structure for protonations upon disulfide-bonds reduction. Furthermore, the molecular weight information obtained from ESI-MS spectrum has a higher accuracy of mass measurement (generally better than 0.01%).

The primary structural information of the protein can be obtained by enzymatic cleavage of the protein into smaller peptide fragments, followed by MS determination of the molecular weights of the resulting mixture peptides (peptide mass mapping). In this case, peptide mass mapping involved enzymatic digestion of the rh-GM-CSF with either trypsin or *Staphylococcus aureus* V8 protease, followed by MS analysis of digestion mixtures. Trypsin selectively cleaves rh-GM-CSF at the C-terminal side of argine (R) and lysine (K), while V8 protease specifically cleaves the peptide bond on the C-terminal side of glutamic acid (E) residues. It is important to note that an enzymatic digest of a large protein can yield fragments of incomplete digestion. For example, trypsin does not cleave at a lysine-proline (K-P) bond, and R-P bonds are marginally more susceptible. Also, peptide fragments that contained two contiguous basic sites (K-K, K-R, R-R, etc.) are observed with R or K on the N-terminal. This results from the poor exoprotease activity of typsin. Similarly, V8 protease can produce incomplete digestion products; Asp (D) is occasionally cleaved. The expected peptide fragments from enzymatic cleavages of rh-GM-CSF with trypsin or V8 are shown in Table 19-1. For tryptic digest of unmodified rh-GM-CSF (V0), the mass values of the majority of the observed signals could be matched with the molecular ions of the tryptic peptides predicted from amino acid sequence (Table 19-2), with the exception of the cysteine-containing fragments T_4 (DTAAEMNETEVISEMFDLQEPTC[54] LQTR), T_{10} (QHC[88]PPTPETSC[96]ATQIITFESFK), and T_{12} (DFLLVIPFDC[121] WEPVQE). These peptide fragments (T_4, T_{10}, T_{12}) are interconnected by disulfide bonds with an isotopically averaged mass of 7614.6 Da, as illustrated in Figure 19-4. This disulfide-linked core peptide was detected at m/z 7613.3 by Cs^+ liquid secondary-ion MS, indicating the presence of this core peptide and two disulfide bonds in rh-GM-CSF. Furthermore, these peptide fragments were released after treatment of the tryptic digests with dithiothreitol (reducing reagent), and subsequent MS analysis of the mixture yielded signals at m/z 3202.3, 2466.8, and 1951.8 corresponding to their free sulfhydryl forms as T_4, T_{10}, and T_{12}, respectively, thus confirming the presence of two disulfide bonds in rh-GM-CSF. The assignment of the cysteine-containing peptides was also confirmed by MS analysis of a tryptic digest of rh-GM-CSF in which the cystine residue were S-alkylated with 4-vinylpyridine in the presence of tri-N-

TABLE 19-2. Tryptic Digest of rh-GM-CSF (V0) and Its Variants (V1 and V2)

Code	Sequence	Expected Mass Value	Ions Observed (V0)	Ions Observed (V1)	Ions Observed (V2)
T_1	APAR	413	+	+	+
T_2	SPSPSTQPWEHVNAIQEAR	2134	+	+	+
T_3	(R)LLNLSR	715	+[a]	+	+
T_4	DTAAEMNETEVISEM[46]FDLQEPTC[54]LQTR	3202	+	+	3218
T_5	LELYK	665	+	+	+
T_6	QGLR	473	+	+	+
T_7	GSLTK	505	+	+	+
T_8	LK	259	[b]		
T_9	GPLTM[79]M[80]ASHYK	1236	+	1252	1252
T_{10}	QHC[88]PPTPETSC[96]ATQIITFESFK	2466		+	+
T_{11}	ENLK	502	+	+	+
T_{12}	DFLLVIPFDC[121]WEPVQE	1950		+	+

[a]Also as RLLNLSR.
[b]Observed as T_{8-9} at m/z 1477.

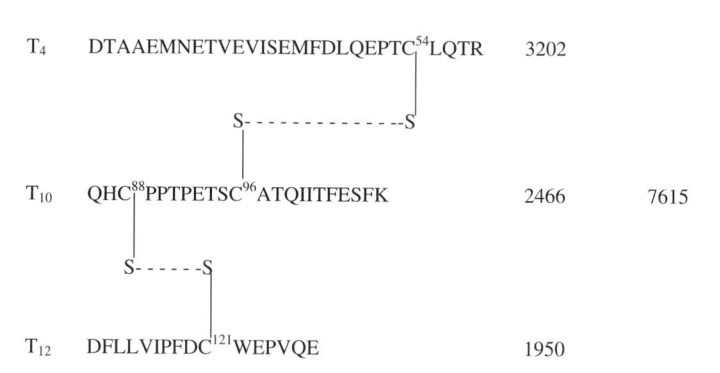

Figure 19-4. Amino acid sequence and calculated average mass values of the tryptic peptides comprising the disulfide-linked core peptide in rh-GM-CSF.

butylphosphine [32]. The resulting pyridylethyl cysteine tryptic peptides were observed as strong ions with masses 106 Da higher than the unmodified peptides (data not shown).

Although tryptic peptide mass mapping of rh-GM-CSF demonstrated the presence of two disulfide bonds and suggested two possible combinations of disulfide pairing (i.e., exact modification site) as C54-C88/C96-C121 or C54-C96/C88-C121, the assignment of the disulfide pairing was not possible due to absence of a tryptic site between C88 and C96 residues of T_{10}. Thus, V8 protease was employed to digest rh-GM-CSF and cleave the protein between each half-cystine residue at the C-terminal side of glutamic acid. The MS analysis of the V8 protease digest of rh-GM-CSF confirmed the presence of most of the predicted peptides (Table 19-3). The ions at m/z 2272 and 3036 corresponded to the disulfide-linked peptides V_8-SS-V_{10} (PTC^{54}LQTRLE-SS-TSC^{96}ATQIITFE) and $V_{7,8}$-SS-V_{10} (MFDLQE PTC^{54}LQTRLE-SS-TSC^{96}ATQIITFE), arising from incomplete cleavage at Glu(51). These MS signals disappeared upon dithiothreitol (DTT) reduction reaction, thus suggesting a Cys(54)–Cys(96) disulfide bond. The absence of digested peptides V_1 and V_7 was likely due to the incomplete cleavages, as indicated by the presence of V_{1-2} and V_{7-8} peptides. Interestingly, V_9 and V_{12} peptides were not observed in the spectra despite their hydrophobic character based on primary structures. This signal suppression may arise from contributions of peptide's secondary or tertiary structure affecting its hydrophobic character [31]. To overcome the difficulty in detecting absent peptides, the mixture of digested V8 peptides was separated by HPLC and isolated fractions were analyzed by MS. All 13 V8 peptide fragments were revealed. V_1 peptide was observed as V_{1-2} at m/z 2302, while V_7 peptide was shown as part of V_{7-8} at m/z 1824 due to incomplete cleavages. V_9 peptide was not only seen at m/z 3712 as expected, but was identified as V_9-SS-V_{12-13} (LYKQGLRGSLTKLKGPLTMMASHYK QHC^{88}PPTPE-SS- NLKDFLLVIPFDC^{121}WEPVQE, m/z 6017.6) and V_9-SS-V_{11-13} (LYKQGLRGSLTKLKGPLTMMASHYKQHC^{88}PPTPE-SS-

TABLE 19-3. V8 Protease Digest of rh-GM-CSF (V0) and Its Variant (V2)

Code	Sequence	Expected Mass Value	Ions (V0)	Ions (V2)
V_1	APARSPSPSTQPWE	1511		
V_2	HVNAIQE	810	+	+
V_3	ARRLLNLSRDTAAE	1586	+	+
V_4	MNE	393	+	+
V_5	TVE	347	+	+
V_6	VISE	447	+	+
V_7	MFDLQE	782		
V_8	PTC^{54}LQTRLE	1060	+	+
V_9	LYKQGLRGSLTKLKGPLTMMASHYKQHC^{88}PPTPE	3713		
V_{10}	TSC^{96}ATQIITFE	1214	+	+
V_{11}	SFKE	510	+	+
V_{12}	NLKDFLLVIPFDC^{121}WE	1852		
V_{13}	PVQE	472	+	+
V_{7-8}	M^{46}FDLQEPTC^{54}LQTRLE	1824	+	1840
LQE-V_8	LQEPTC^{54}LQTRLE	1431	+	+
V_{7-8}-SS-V_{10}	M^{46}FDLQEPTC^{54}LQTRLE-SS-TSC^{96}ATQIITFE	3036	+	3052
LQE-V_8-SS-V_{10}	LQEPTC^{54}LQTRLE-SS-TSC^{96}ATQIITFE	2641	+	
V_8-SS-V_{10}	PTC^{54}LQTRLE-SS-TSC^{96}ATQIITFE	2272	+	
V_{12}	NLKD	488	+	

SFKENLKDFLLVIPFDC^{121}WEPVQE, m/z 6508.7) [31]. These data clearly established another pairing of disulfide bond between Cys(88) and Cys(121).

For a recombinant protein, post-translational modifications such as phosphorylation, oxidation, deamidation, and sulfation are known to occur. The GM-CSF variants were first observed after SDS polyacrylamide gel electrophoresis (SDS-PAGE) of an *E. coli* derived GM-CSF preparation as a hazy band located slightly above the band corresponding to unmodified GM-CSF (V0). The haze was further separated and purified by preparative reversed-phase HPLC. Typically, a Rainin Dynamax C4 column (300 Å, 4.1 × 250 mm) was run at a flow rate of 30 mL/min on a Rainin autoprep preparative HPLC system. Samples were eluted using a linear gradient of 27% to 72% acetonitrile in 0.1% trifluoroacetic acid (TFA) over a 30-min period. A Knauer variable wavelength detector set at 280-nm absorbance was used to monitor peaks. Fractions were taken manually based on UV absorption and retention time. Isolated fractions containing two GM-CSF variants V1 and V2 were diluted threefold and re-chromatographed separately on a Rainin Dynamax C4 column (300 Å, 2.1 × 250 mm) at a flow rate of 10 mL/min on a Rainin autoprep HPLC system using a linear gradient of 27% to 72% acetonitrile in 0.1% TFA. These two variants, V1 and V2, were found to have comparable biological activity to the parent GM-CSF (V0). Further structural identification work was carried out on isolated fractions using MS methods.

The peptide mass mapping strategy using trypsin and V8 protease was applied to solve structural identification problems of the variants. The comparison of the trypsin and V8 protease digest of the native GM-CSF (V0) and its variant V1 and V2 demonstrated that one or two methionine residues in V0 have been converted to methionine sulfoxides (Tables 19-2 and 19-3). In the case of V1, tryptic peptide T_9 had a mass increase of 16 Da (m/z 1252, Table 19-2), suggesting oxidation of Met(79) or Met(80). In the case of V2, however, both the tryptic peptide T_4 (m/z 3218) and T_9 (m/z 1252) had a mass shift of 16 Da with respect to T_4 and T_9 in V0 (Table 19-2). Therefore, V2 contains two methionine sulfoxides: one at Met(46), the other at Met(79) or Met(80). The assignment of Met(46) oxidation was further confirmed by a mass increase of 16 Da for V8 protease peptides V_{7-8} and V_{7-8}-SS-V_{10}. No tandem MS experiments were attempted to differentiate oxidation sites between Met(79) and Met(80) at that time because of instrumentation limitations, although these experiments would have provided detailed information on the exact modification sites. An example on this approach using modern instrumentation is illustrated in the case of rh-IFN-α-2b. The structural assignments of V1 and V2 were further supported by MS studies of chemically modified proteins VS-1, VS-2, VS-3, and VS-4 that have different degrees of oxidation of the four methionine residues in rh-GM-CSF amino acid sequence (data not shown). In these experiments, GM-CSF was treated with H_2O_2 under optimized conditions to produce oxidized proteins. The preferential oxidation of Met(79) was observed in the mapping experiments of permethylated GM-CSF, where an unusual cleavage at Met(79)-Met(80) yielded a signal at m/z 1306 and a weak signal 16 Da higher.

It is evident from the discussions above that mass spectrometric method in combination with enzymatic digestion offers a convenient approach to the characterization of GM-CSF and its variants. ESI-MS method demonstrated a mass accuracy of better than 0.01% for a recombinant protein. The mass spectral data of the enzymatic digest of GM-CSF and its variants allow the precise determination of the molecular weights of the peptides, leading to the identification of sites of covalent modifications, the disulfide bonding pattern, and confirmation of the cDNA-derived sequence of the protein.

19.3.1.2 rh-IFN-α-2b. Interferon α-2b (IFN-α-2b) is an *E. coli* recombinant DNA-derived therapeutic protein that is used as an anticancer agent and in the treatment of chronic hepatitis B and C [33]. It is a 165-amino acid protein, containing four cysteines at positions 1, 29, 98, and 138. These four cysteines form two disulfide bonds. Cysteine 1, the N-terminal amino acid, is linked to cysteine 98; cysteine 29 is linked to cysteine 138 (Figure 19-5). The molecular weight of IFN-α-2b is calculated to be 19,265 Da from its cDNA amino acid sequence [34]. The sequence and disulfide mapping of IFN-α-2b has been successfully carried out using the same peptide mass mapping method as described in the case of rh-GM-CSF—for example, enzymatic digestion with trypsin on purified protein and mass analysis of digested peptide mixtures [35].

It is not unusual that the *E. coli* expression of IFN-α-2b produces several isoforms in addition to the target protein, as shown in its reversed-phase

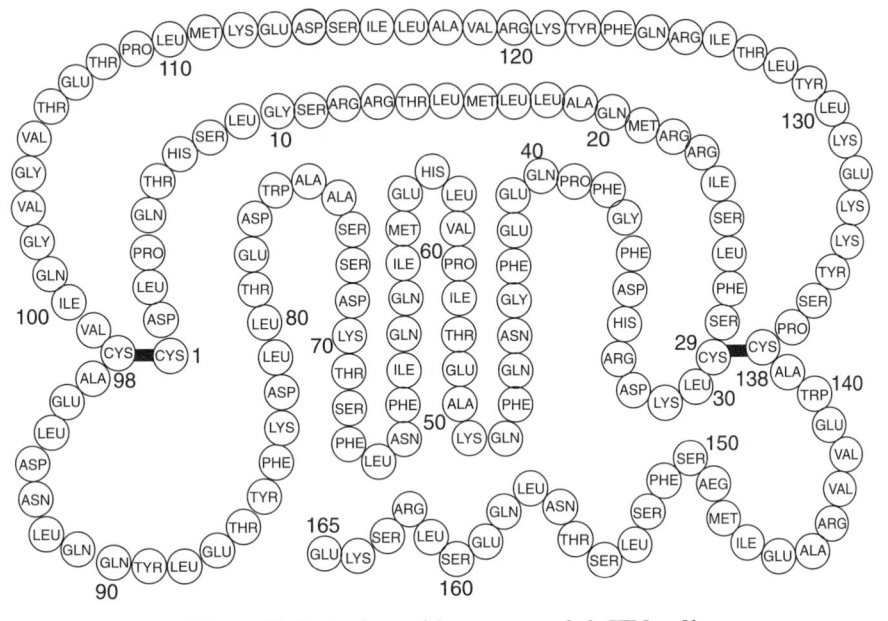

Figure 19-5. Amino acid sequence of rh-IFN-α-2b.

HPLC chromatogram (Figure 19-6). Two of the three isoforms, Iso-2 and Iso-3, were predicted to be incorrectly folded forms of the target protein with scrambled disulfides. The third isoform, Iso-4, was thought to be reduced IFN-α-2b containing four free cysteine sulfhydryls (SH). The level of Iso-4 was observed to decrease during the purification process, suggesting that Iso-4 may refold back to IFN-α-2b. Earlier RP-HPLC data provided experimental evidence that IFN-α-2b could be reduced with DTT to Iso-4, and Iso-4 might be re-oxidized to IFN-α-2b. In addition to these isoforms, a fourth component, a variant of IFN-α-2b, was detected either co-eluting with or as a small shoulder eluting in front of the target protein peak (peak 1). The separation of this shoulder peak from IFN-α-2b depended on the HPLC column load; for example, better separation was obtained with lower column loads as illustrated in Figure 19-7. The exact structures of these isoforms and the variant of IFN-α-2b can only be obtained using mass spectrometry in conjunction with RP-HPLC.

The initial studies was carried out using on-line RP-HPLC coupled with a single quadrupole ESI-MS to measure the molecular weights of IFN-α-2b components. The mass spectrum showed that other than IFN-α-2b, peak 1 in Figure 19-7c contained a protein with a MW of 19,281 Da that was 16 Da higher than the predicted MW of 19,265 Da for IFN-α-2b. This higher mass component corresponds to oxidation of one of the five methionine amino acids present in IFN-α-2b. The oxidation of a methionine is also indicated by the fact that this component elutes earlier than the parent protein. It is well known

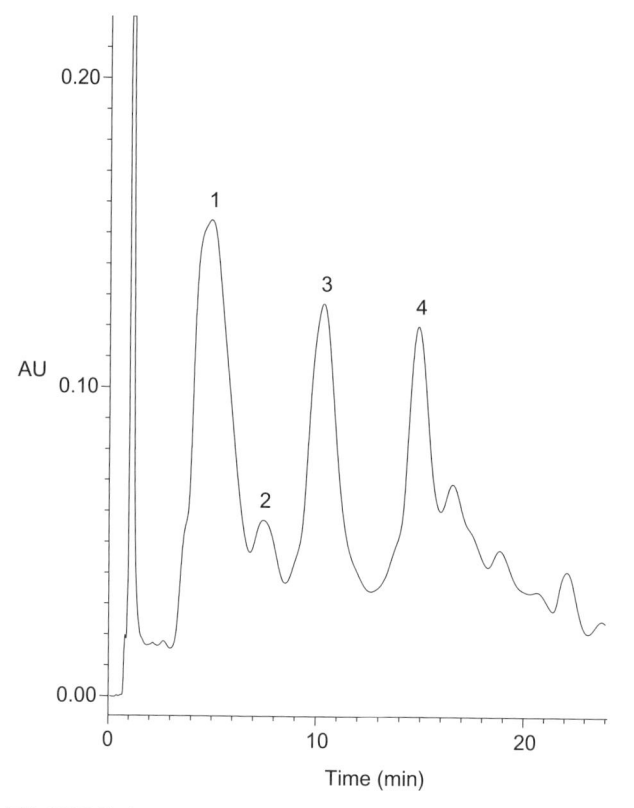

Figure 19-6. RP-HPLC chromatographic profile of an "in-process" sample from *E. coli* recombinant DNA derived IFN-α-2b. Peak 1 is IFN-α-2b. Isoform peak 2 and 3 are putative scrambled disulfides. Isoform peak 4 is a putative open disulfide. The HPLC was run under a linear gradient of 49–65% B (10:90 H_2O:CH_3CN/0.1% TFA) over 24 minutes with the UV set at 214 nm. The mobile phase A was water with 0.1% TFA and the flow rate was set at 0.2 mL/min. The column used was Vydac C8 column at 30°C (2.1 mm × 50 mm, 5 μm, 300 Å).

that proteins containing an oxidized methionine are more hydrophilic and they tend to elute earlier on RP-HPLC than the parent protein [36, 37]. This oxidized variant is present at approximately <2% by HPLC peak area normalization. The dynamic range of the mass spectrometer was large enough to detect the presence of this variant as well as the more abundant IFN-α-2b, even at more diluted column loads as shown in Figure 19-7.

HPLC peaks 2 and 3 in Figure 19-6 corresponded to the predicted scrambled disulfides of IFN-α-2b, Iso-2 and Iso-3. They were expected to have the same MW of 19,265 Da as that of IFN-α-2b (peak 1). However, the measured MWs were found to be different from those predicted for an incorrectly folded form of IFN-α-2b. The determined MW of Iso-2 (M_r = 19,310 Da) was 45 Da

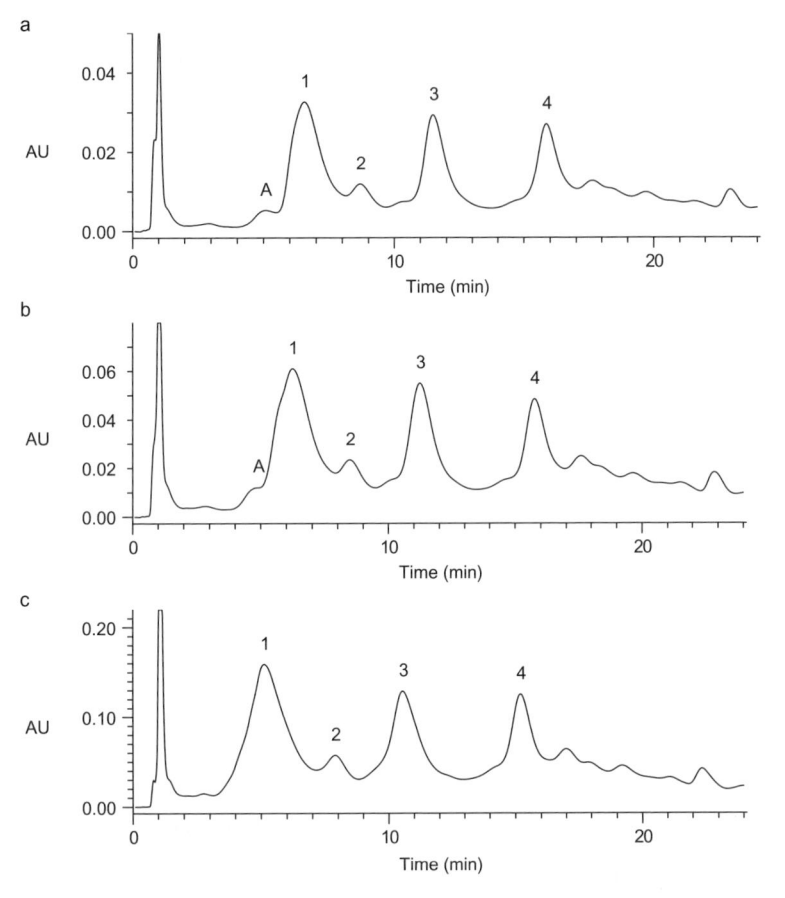

Figure 19-7. RP-HPLC chromatograms showing dependence of the early eluting variant, peak A, on column load. (a) Peak A and peak 1 resolved with a column load of 3 μg of proteins. (b) Peak A and peak 1 partially resolved with a column load of ~6 μg of proteins. (c) Peak A and peak 1 co-eluting with a column load of ~15 μg of protein.

higher than that of IFN-α-2b. This increased mass suggests the possibility of acetylation of the N-terminus of the reduced target protein since the acetyl group, $CH_3CO—$, corresponds to a mass addition of 42 Da. The MW of Iso-3 ($M_r = 19{,}643$) was 378 Da higher than that of IFN-α-2b. The protein MW information obtained from MS studies indicated that neither peak 2 nor peak 3 corresponded to the postulated scrambled disulfides of IFN-α-2b. They are most likely to be post-translationally modified IFN-α-2b.

HPLC peak 4, Iso-4, in Figure 19-6 corresponded to the putative reduced IFN-α-2b containing four free cysteine sulfhydryls ($M_r = 19{,}269$ Da). It was expected to have an MW that was 4 Da higher than that of the target protein. The mass spectrum of peak 4 revealed that this symmetrical HPLC peak actually consisted of two co-eluting components. The MW of one of the components, at 19,269 Da, corresponded to the reduced IFN-α-2b, that is, the predicted Iso-4. However, the MW of the second component, at 19,336 Da, is 71 Da higher than that of the target protein. No obvious post-translational modification could be proposed.

The above approach using RP-HPLC/ESI-MS to determine the MW of the isoforms is a powerful tool in monitoring the production process of IFN-α-2b. It provided insight into the potential structures of two of the four isoforms and the variant that were present at various stages in the production of the target protein. However, the structure and the identification of the post-translational modifications in Iso-2, Iso-3, and Iso-4 could not be determined solely based on this approach. To fully characterize the post-translational modifications, individual isoforms were isolated from an early step in the purification of IFN-α-2b, followed by extensive MS characterization. This was demonstrated in the case of Iso-4.

The first step was to verify the MW of the isolated protein Iso-4 using triple quadrupole ESI-MS. The MW of isolated Iso-4 was found to be 72 Da higher than that expected for IFN-α-2b. The next step involved RP-HPLC/ESI-MS analysis of tryptic digests of the control IFN α-2b and IFN Iso-4 in order to identify the nature of the modification. The peptide mass mapping results are displayed in Figure 19-8 and Table 19-4. Comparison of the ESI-MS peptide maps of the two proteins shows differences in the N-terminal peptide fragments. The N-terminal peptide fragment of IFN-α-2b, T_1 ([1]CDLPQTH SLGSR[12]), is linked with peptide T_{10} (or $T_{9,10}$ and $T_{9,10,11}$) through the disulfide bond formed between Cys-1 and Cys-98. These disulfide-linked peptide fragments—for example, T_1-ss-T_{10} (m/z 4617)—were largely absent in the Iso-4 digest shown in Figure 19-8b. Instead, the Iso-4 tryptic peptide map revealed two new peptide fragments at m/z 1314 and 1384, respectively. These peptide fragments corresponded to the N-terminal peptide fragment T_1 and T_1 + 70 Da. The mass difference of 70 Da in these peptide fragments is in agreement with the mass difference (70 Da) between Iso-4 and IFN-α-2b when the mass increase of 2 Da resulted from reduction of the disulfide bond is considered.

The amino acid sequence of the modified peptide and the site of the modification in Iso-4 was further determined by RP-HPLC/ESI-MS/MS studies of the doubly charged molecular ions of the T_1 (m/z 658) and the T_1 + 70 Da (m/z 693) peptides (Figure 19-9). Tandem MS data of the doubly charged ion for T_1 + 70 demonstrated that the peptide fragment was indeed the N-terminal tryptic peptide fragment, T_1, of IFN-α-2b with a 70-Da modification group residing on the N-terminal cysteine. The observation of the more prominent N-terminal fragment ions of the modified T_1 peptide, which were shifted by 26 Da compared with those of the T_1 peptide of IFN-α-2b, implied a rapid loss of 44 Da (CO_2). This suggested that a labile carboxyl group could be a part of the 70-Da modification moiety. This assumption was further confirmed by observation of the loss of 44 Da from T_1 + 70 using a higher orifice potential (80 V) for peptide mass mapping of Iso 4 using MS. No such loss was detected for T_1 peptide under the same orifice condition. Product ion spectrum of the

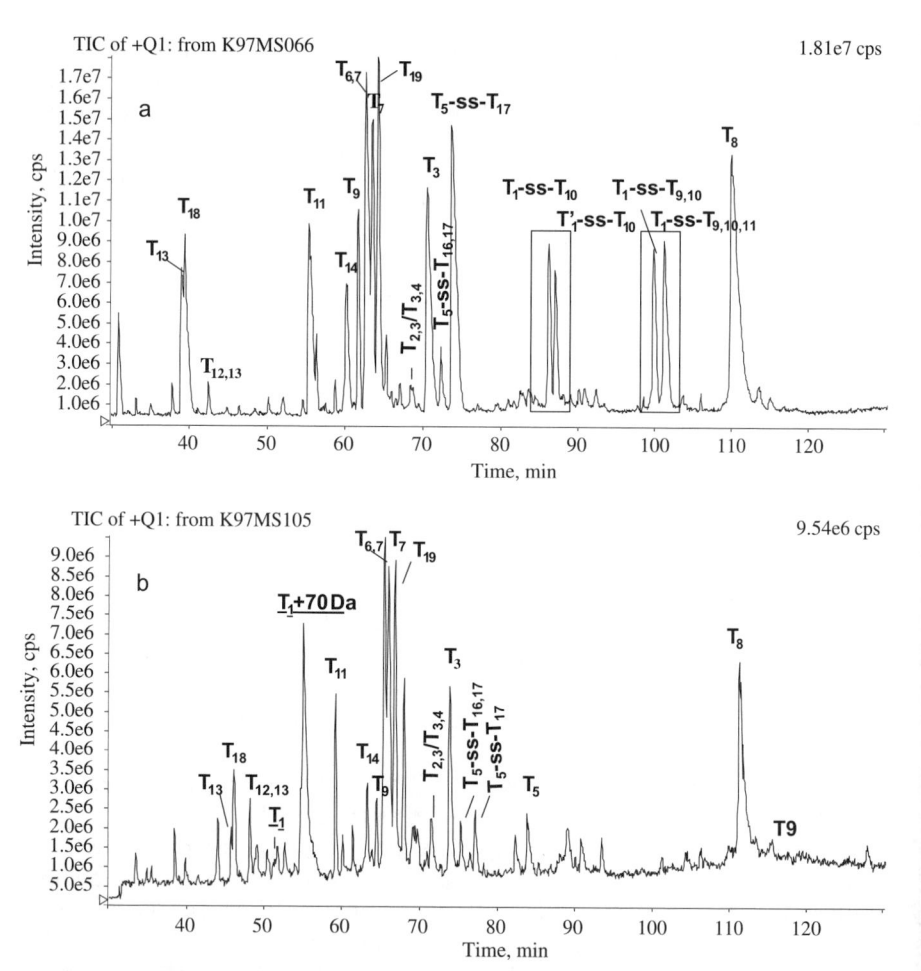

Figure 19-8. Peptide mass mapping by RP-HPLC/ESI-MS. (a) Total ion chromatogram (TIC) of the trypsin digested IFN-α-2b showing the intact N-terminal peptide disulfide fragments, T_1-ss-T_{10} and T_1-ss-$T_{9,10}$. (b) TIC of the trypsin digested Iso-4 displaying the absence of the intact N-terminal peptide disulfide fragments, T_1-ss-T_{10} and T_1-ss-$T_{9,10}$, and the appearance of a T_1 + 70 Da peptide fragment. The tryptic peptides was first desalted with 5% mobile phase B (CH$_3$CN/0.08% TFA), followed by a gradient run on a Supelcosil LC-18-DB column (1 mm × 300 mm, 100 Å) with a 5–95% B in 150 minutes (40 μL/min with a mobile phase A: water with 0.1% TFA).

doubly charged ion of T_1 + 26, generated from the high orifice ESI-MS experiment, exhibited the N-terminal fragment ions of b_2 + 26, b_3 + 26, and a_2 + 26. As expected, the second series of fragment ions—that is, b_2 + 70, b_3 + 70, and a_2 + 70—were absent.

The elemental composition of the 70-Da post-translational modification group was determined by accurate mass measurement using high-resolution

TABLE 19-4. Tryptic Peptide Fragments of IFN-α-2b

Cys-Asp-Leu-Pro-Gln-Thr-His-Ser-Leu-Gly-Ser-Arg-Arg-
-----------------------------T_1----------------------------------→-T_2-→
-Thr-Leu-Met-Leu-Leu-Ala-Gln-Met-Arg-Arg-
----------------------------T_3---------------→-T_4-→
-Ile-Ser-Leu-Phe-Ser-Cys-Leu-Lys-Asp-Arg-
--------------------T_5--------------------→---T_6--→
-His-Asp-Phe-Gly-Phe-Pro-Gln-Glu-Glu-Phe-Gly-Asn-Gln-Phe-Gln-Lys-
--T_7--→---
Ala-Glu-Thr-Ile-Pro-Val-Leu-His-Glu-Met-Ile-Gln-Gln-Ile-Phe-Asn-Leu-Phe-Ser-
Thr-Lys-
--T_8--
-Asp-Ser-Ser-Ala-Ala-Trp-Asp-Glu-Thr-Leu-Leu-Asp-Lys-
----------------------------T_9---
-Phe-Tyr-Thr-Glu-Leu-Tyr-Gln-Gln-Leu-Asn-Asp-Leu-Glu-Ala-Cys-Val-Ile-Gln-Gly-
Val-Gly-Val-Thr-Glu-Thr-Pro-Leu-Met-Lys-
---T_{10}--
-Glu-Asp-Ser-Ile-Leu-Ala-Val-Arg-Lys- -Tyr-Phe-Gln-Arg- -Ile-Thr-Leu-Tyr-Leu-
Lys-
------------------------T_{11}-----------→T_{12}-- --------T_{13}--------- --------------T_{14}------------------
-Glu-Lys-Lys- -Tyr-Ser-Pro-Cys-Ala-Trp-Glu-Val-Val-Arg- -Ala-Glu-Ile-Met-Arg-Ser
---T_{15}-->-T_{16}- ------------------------T_{17}------------------------ ----------- -T_{18}-------→ -------------
Phe-Ser-Leu-Ser-Thr-Asn-Leu-Gln-Glu-Ser-Leu-Arg- -Ser-Lys- -Glu
------------------------T_{19}------------------------------------- -- T_{20}--→--T_{21}---

MALDI-TOF-MS. Clearly, the 70-Da modification group was a pyruvate ($C_3H_2O_2$). Pyruvic acid ($CH_3COCOOH$), like acetic acid (CH_3COOH) and other common acids, forms a strong amide bond through the carboxyl group (C-1) with the N-terminal amine group in proteins [38, 39]. This amide bond is generally stable to mild acidic and base conditions. However, the pyruvate bond in Iso-4 appeared to be labile under mild acidic conditions. In addition, the modification of the protein through C-1 of pyruvic acid is not likely to generate a labile carboxyl group in the modification moiety as observed in the MS/MS studies. This information led to the hypothesis that the puruvation of IFN Iso-4 involved a unique chemistry in which a ketimine link was likely formed between C-2 of pyruvic acid and the N-terminal cysteine amino group. This ketimine bond is reversible under mild acidic conditions as illustrated in Figure 19-10. The absence of the disulfide bond between Cys-1 and Cys-98 in Iso-4 favors formation of the cyclic pyruvate intermediate (B) rather than formation of the ketimine (imine) intermediate (A). This hypothesis was confirmed by comparing the product ion spectrum of the T_1 peptide fragment of the Iso-4 with that of a synthetically prepared T_1 peptide fragment that was derivatized with pyruvic acid. The MS/MS analysis of this pyruvated synthetic peptide generated the same fragmentation pattern as that of the N-terminal tryptic peptide of Iso-4. The N-terminal fragment ion of $b_2 + 26$ (m/z 245),

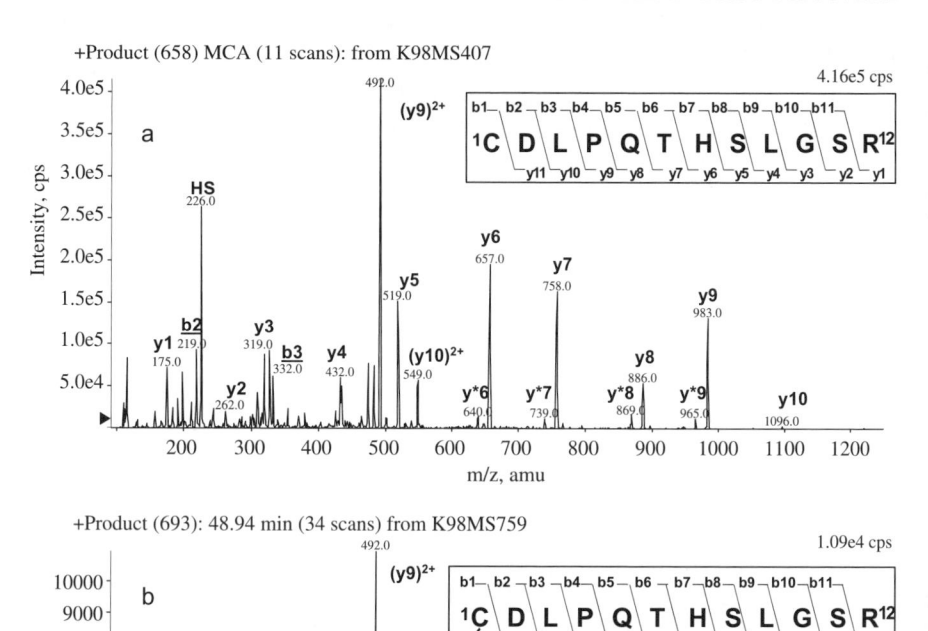

Figure 19-9. LC/ESI-MS/MS product ion mass spectra of the doubly charged ions of the N-terminal tryptic peptide T_1 of (a) IFN-α-2b (m/z 658) and (b) Iso-4 (m/z 693). N-terminal Cys-Asp was identified as the modified fragment.

generated by MS/MS of T_1 + 70 from Iso-4 and the synthetic peptide, was further dissociated in the ion trap mass spectrometer, producing fragment ions at m/z 102 and 130 resulted from cleavage of CO—CH and the amide bond of Cys-1, respectively. This multiple-stage MS analysis (MS^n) further supported the original hypothesis.

In addition, the DNPH (2,4-dinitrophenylhydrazine) [38] and NADH (dihydronicotinamide adenine dinucleotide) [36] studies with purified Iso-4 provided the evidence that the 70 Da moiety was a pyruvate derivative ($C_3H_2O_2$). In the DNPH study, treatment of Iso-4 with acid and 2,4-dinitrophenylhydrazine produced the 2,4-dinitrophenylhydrazone of the pyruvic acid liberated from Iso-4. In the NADH study, the amount of NAD^+

Figure 19-10. The pyruvate formation with the N-terminal cysteine. The C-2 carbonyl in pyruvic acid initially forms a ketimine intermediate (A). The sulfhydryl (SH) group of Cys-1 generated from the reduced cysteine 1–98 disulfide bond in Iso-4 tends to favor the formation of the more thermodynamically stable cyclic thiazolidine pyruvate intermediate (B).

produced is proportional to the amount of pyruvic acid liberated from the mild acid hydrolysis of Iso-4. These procedures were also applied to purified IFN-α-2b as a control. The control experiments demonstrated that the pyruvate derivative was the active component measured in the Iso-4 experiments.

To verify that Iso-4 was interconvertible with IFN-α-2b, a sample of purified Iso-4 was treated under mild acidic conditions in an attempt to convert it to IFN-α-2b. The MW measurement of the converted protein by RP-HPLC/ESI-MS confirmed that Iso-4 could be converted to IFN-α-2b under mild acidic conditions. Furthermore, the IFN-α-2b obtained from the conversion of Iso-4 was enzymatically digested with trypsin and studied by RP-HPLC/ESI-MS to assess the status of disulfide bonds. The presence of the two disulfide-bonded peptide fragments, T_1-ss-T_{10} and T_5-ss-T_{17}, revealed the correctly folded IFN-α-2b.

The other isoforms, Iso-2 and Iso-3, expressed in the *E. coli* fermentation of IFN-α-2b, were characterized using a similar approach. The isolated Iso-2 and Iso-3 were enzymatically digested with trypsin, and the resulted peptide mixtures were mass-mapped using RP-HPLC/ESI-MS. The results indicated that Iso-2 was a correctly folded IFN-α-2b acetylated on the amino group of the N-terminal cysteine. Iso-3 was similarly determined to be a glutathionated form (Cys-98) of the partially reduced IFN-α-2b that was pyruvated on the N-terminal cysteine. The complete structures for IFN-α-2b, Iso-2, Iso-3, and Iso-4 are shown in Figure 19-11.

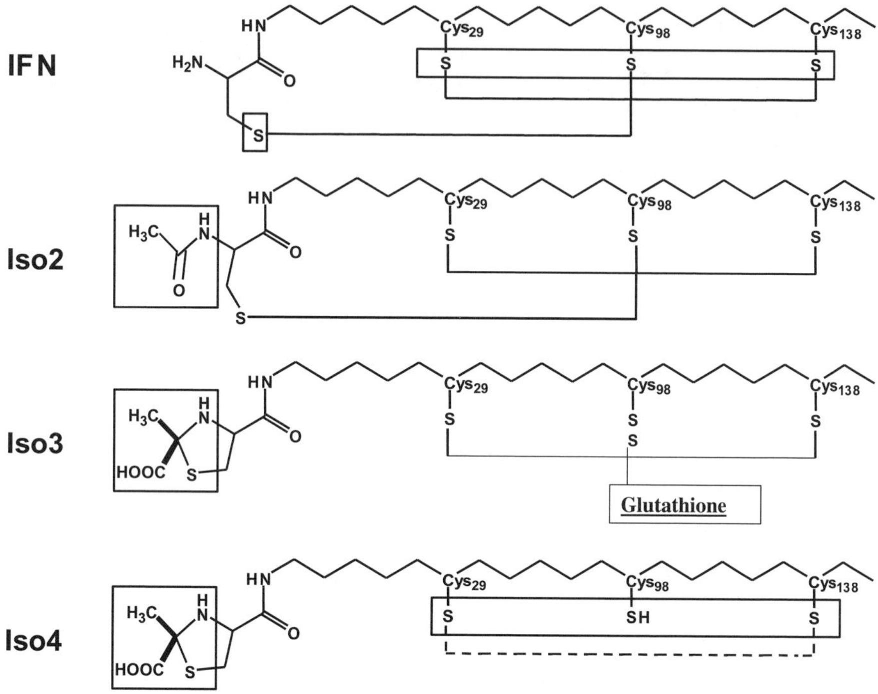

Figure 19-11. The structures of IFN-α-2b and its three isoforms, Iso-2, Iso-3, and Iso-4. The solid line indicates the disulfide bond formation, while the dashed line indicates the reduced disulfide bond or partial disulfide bond formation.

The pyruvic modification of the N-terminal cysteine of *E. coli* derived recombinant IFN-α-2b via a ketimine linkage has not been reported previously. There were only two cases in the literature that involved the ketimine formation of the pyruvic acid C-2 carbonyl group and the amino group of the N-terminal cysteine amino acid, including the post-translational modification of the Ner protein of the bacteriophage Mu [38] and the β-chain of hemoglobin A_{1b} [39] with pyruvic acid. The chemistry of pyruvic acid attachment to Iso-4 from this study has a significant impact on the production of IFN-α-2b. It led to the development of a reproducible conversion procedure from Iso-4 to IFN-α-2b in the production process, resulting in a five- to sevenfold increase in the production yield.

19.3.2   Protein Glycosylation and Phosphorylation

19.3.2.1   Glycosylation. Carbohydrate modifications of proteins (glycosylation) are key factors in modulating protein structures and functions within cells. Glycosylation affects probably more than half of all proteins in a eukaryotic cell [40]. In the extracellular environment, the oligosaccharide moieties of glycoproteins are implicated in a wide range of cell–cell and cell–matrix recognition events which exert effects on cellular recognition in infection, cancer, and immune response. There are many instances where glycan structures have been shown to have significant importance in the biological function of a protein. For example, glycosylation of Asn-319 on rabies virus glycoprotein is essential for the secretion of soluble rabies virus glycoprotein [41]. Changes in levels and types of glycosylation can be associated with disease. It has been illustrated that detecting changes in glycan structure may be used as a diagnostic for aggressive breast cancer [42]. Glycan profiling of normal and diseased forms of a glycoprotein has provided new insights for future research in rheumatoid arthritis, prion disease, and congenital disorders of glycosylation [43–47]. In all these diseases, differences in glycosylation indicate that there are cellular or genetic changes that affect the activity of specific glycotransferases. Glycosylation also represents the most common modification for recombinant protein products expressed in mammalian and insect cell lines. Carbohydrate modifications of recombinant proteins have significant impacts on their solubility, immunogenicity, resistance to proteolysis, circulatory half-life, and thermal stability, all of which will affect the use of the recombinant proteins as therapeutic entities or as drug targets. The important roles that glycoproteins play in biology and medicine have stimulated a rapid expansion of the field of glycobiology and brought up the need to develop rapid and accurate analytical methods to characterize the glycoproteins.

Glycosylation occurs in the endoplasmic reticulum (ER) and Golgi compartments of the cell and the reactions are catalyzed by membrane-bound glycotransferases and glycosidases [48, 49]. All mammalian N-linked oligosaccharides share a common trimannosyl core $Man_3GlcNAc_2$ derived from a biosynthetic precursor $Glc_3Man_9GlcNAc_2$ that is added cotranslationally to polypeptides in the ER. There are three types of N-linked oligosaccharides: high mannose-type, complex-type, and hybrid-type. For N-linked glycoprotein, the attachment of glycan structures to proteins usually occurs at an Asn-Xaa-Ser/Thr consensus. Xaa may be any amino acid except proline. O-oligosaccharide biosynthesis is initiated in the Golgi by the addition of a single sugar to serine or threonine. There are at least seven O-linked oligosaccharide core structures, four of which are particularly widespread in mammalian glycoproteins [49].

Carbohydrates are polymers with a wide diversity of glycan structures which comes from the variation in the type, number, and position of individual sugar residues, the degree of branching, and the level of acetylation, methylation, sialylation, phosphorylation, and sulfation. The populations of sugars attached to an individual protein will depend on the cell type in which the glycoprotein is expressed and on the physiological status of the cell, and they may be developmentally and disease-regulated. A glycoprotein usually exists as complex mixtures of glycosylated variants (glycoforms) due to (a) the diversity of oligosaccharides attached to the glycoprotein and (b) the occupancy of each glycosylation site. Complete structural characterization of a glycoprotein requires the determination of the peptide primary sequence and the glycosylation sites, as well as the definition of the attached oligosaccharides in terms of their linear sequencing, branching, linkage, configurations, and the positional isomers.

In general, glycoprotein is enzymatically digested such that each glycosylation site is located within a separate peptide [50]. HPLC separation of the peptides coupled with MS precursor ion scans of sugar specific oxonium ions, such as m/z 163 (protonated Hex), m/z 204 (protonated HexNAc), or m/z 366 (protonated Hex-HexNAc), allows the glycopeptides to be identified from the mixture of peptides for further studies [51–54]. In some cases, where tandem MS is not available, fragmentation induced by internal energy transfer to the ion during ionization process (in source fragmentation during electrospray or post-source decay in MALDI) can also generate these marker ions in the low m/z range of the mass spectra for glycopeptides identification [55, 56]. Dissociation of a glycopeptide in a CID experiment will provide the information on primary sequence of the peptide, the type of sugar attached, and the amino acid residue that was modified by the glycosyl group [57–59]. However, identification of glycopeptides from a peptide mixture by the above MS approaches can sometime be problematic, primarily due to the poor ionization efficiency of glycopeptides compared to their unmodified forms and the extensive gas-phase deglycosylation to locate the site of sugar attachment [52, 57, 60]. To overcome these problems, several strategies have been developed:

1. Removal of glycans through β-elimination for O-linked glycans or enzymatic digestion using *N*-glycosidase for N-linked glycans. O-linked glycan elimination converts Ser to Ala and Thr to aminobutytic acid, where 16-Da mass losses will be observed and can be served as a marker for the site of sugar attachment. Deglycosylation by *N*-glycosidase converts Asn to Asp with a mass increase of 1 Da. *N*-glycosidase F with simultaneous partial (50%) or full ^{18}O-labeling of glycosylated asparagine residues has been used to magnify the mass difference for glycosylation site identification [61–64]. The glycan elimination with subsequent changes in peptide MW simplifies the downstream determination of the glycosylation site by MS/MS. Furthermore, the ^{18}O/^{16}O labeling method can be used to determine the degree of occupancy of each *N*-glycosylation site.

2. Affinity capture of the N-linked glycopeptides or glycoproteins via lectin column-mediated affinity purification followed by MS analysis [63–65]. Kaji et al. [64] developed a strategy termed isotope-coded glycosylation-site-specific tagging (IGOT), which combined the lectin affinity purification and *N*-glycosidase mediated [18]O labeling method. Applying IGOT, they characterized the N-linked high-mannose and/or hybrid-type glycoproteins from an extract of *C. elegans* proteins and were able to identify 250 glycoproteins with the simultaneous determination of 400 unique *N*-glycosylation sites by using multidimensional LC-MS.

3. "Top-down" sequence analysis of whole glycoprotein ions using CID and ion/ion proton transfer in a quadrupole ion trap MS [66]. This approach eliminated gas-phase deglycosylation of N-linked oligosaccharide in ribonuclease B, and the glycosylation site was identified to be Asn-Leu-Thr at residues 34–36 [66].

As an example of general structural characterization of glycoproteins, MS analysis of CHO cell-derived interleukin-4 (IL-4) was illustrated. IL-4 is T-cell-derived lymphokine that mediates the growth, proliferation, and differentiation of B- and T-lymphocytes and myeloid cells [67]. CHO cell-derived IL-4 is a 129-amino acid glycoprotein that contains two potential N-glycosylation sites at Asn-38 and Asn-105 [68]. Compositional analysis of the oligosaccharide moieties of CHO IL-4, carried out by high-performance anion-exchange chromatography coupled with pulsed amperometric detection, resulted in the following molar concentrations: Man 1.9, GlcN 3.9, Gal 2.1, sialic acid 1.9, Fuc 0.7, and GalN < 0.04.

Analysis of the intact CHO IL-4 by ESI-MS was first attempted to provide preliminary information on the glycan components [69]. The ESI mass spectrum contained three envelopes of multiply charged ions ranging from the 8+ to the 10+ charge state, each comprising eight peaks corresponding to the individual glycoforms of the protein (Figure 19-12A). The deconvoluted mass spectrum revealed several components, with signals at 17,019 Da and 17,309 Da corresponding to sialylated glycoforms (Figure 19-12B), since their mass separation concurred with the incremental mass of the sialic acid unit (NeuAc; 291 Da). The presence of these sialylated components, combined with the carbohydrate compositional results, indicated the likely identity of these major glycans as the fucosylated biantennary oligosaccharide in the mono- and di-sialylated forms (theoretical MW of IL-4: 14,963 Da, not accounting for the existing disulfide bonds). The signal at 16,727 Da corresponded to the asialo biantennary oligosaccharide, whereas a glycoform lacking a unit of galactose and fucose yielded the signal at 16,417 Da (Figure 19-12B). Other, higher mass signals in the deconvoluted ESI mass spectrum indicated the presence of tri- and tetraantennary glycans containing up to three additional lactosamine units (Hex-HexNAc, in-chain mass of 365 Da). Overall, the ESI-MS analysis provided the correct assignment of the two major glycoforms in CHO IL-4 and also allowed the detection of signals arising from more complex glycans.

In order to assess the size of the carbohydrate component, deglycosylated CHO IL-4 by N-glycanase was analyzed by ESI-MS. The deconvoluted spectrum displayed an MW of 14,955 Da, which conformed well to the theoretical MW of the protein (MW 14,963 Da) when considering the presence of three disulfide bonds in the protein.

Mapping of the primary structure of CHO IL-4 was carried out by tryptic hydrolysis followed by measurement of the resulting peptide fragments by on-line HPLC/ESI-MS. Since the primary sequence of IL-4 was known, this MS mapping approach could confirm the cDNA-derived protein sequence,

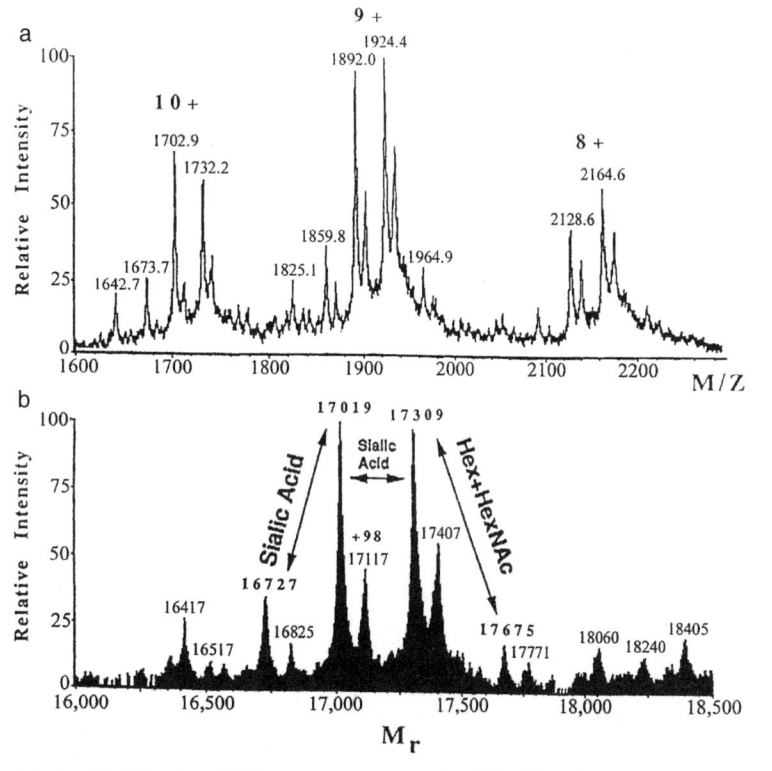

Figure 19-12. Positive-ion ESI mass spectrum of CHO IL-4. (A) Raw spectrum and (B) deconvoluted spectrum. (Reprinted from reference 69, with permission of John Wiley & Sons, Ltd.)

and also allow identification of any posttranslational modification(s), by comparing the ESI-derived mass values with the calculated MW of the predicted tryptic peptides. Figure 19-13 exhibits the amino acid sequence of IL-4 with all the peptide fragments from trypsin digestion. The major difference between the LC/MS profiles of CHO IL-4 tryptic digest and deglycosylated CHO IL-4 tryptic digest was the observed pattern and molecular mass values for HPLC peaks 11 and 14 (Table 19-5), thus making them good candidates for glycopeptide-containing fractions. The monosaccharide composition results and the presence of sialic acid residues, which was shown at the molecular mass level (Figure 19-12B), indicated the presence of a complex-type N-linked oligosaccharide. Since the types of complex N-linked carbohydrate structures typically present in mammalian proteins contain a defined number of sequence and branching variation, the ESI-derived glycopeptide masses should reveal the type of the attached carbohydrate. For HPLC peak 11, the resulting glycopeptide molecular masses determined from charge states were 5286, 5577, and 5868 Da. These masses were in agreement with the assignment

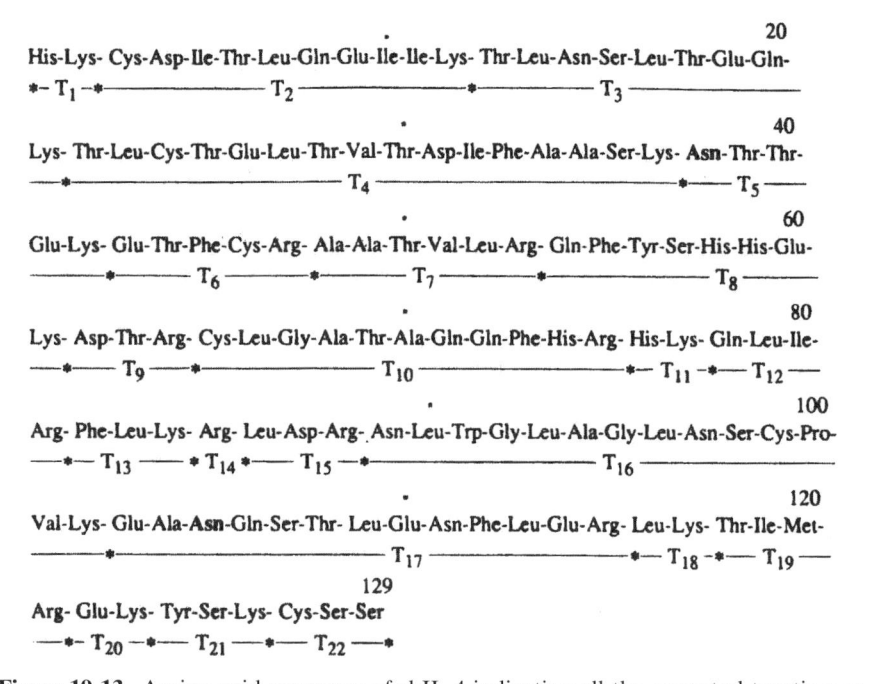

Figure 19-13. Amino acid sequence of rhIL-4 indicating all the expected tryptic peptides Tn. The potential glycosylation sites at Asn[38] and Asn[105] are indicated in bold. (Reprinted from reference 69, with permission of John Wiley & Sons, Ltd.)

TABLE 19-5. ESI-MS Analysis of the Tryptic Digest of CHO IL-4 for HPLC Peak 11 & 14

HPLC Peak	Sequence Position	Expected M_r	Observed Mr for CHO IL-4	Observed Mr for Deglyc. IL-4	Tryptic Peptide
11	38–47 and 89–102	2698.2	—	2698.5	$T_{5,6}$-S-S-T_{16}
	22–42 and 65–75	3516.0	5286, 5577, and 5868	—	$T_{4,5^*}$-S-S-T_{10}
14	38–47 and 89–115	4230.8	7682 and 7976	4231.6	$T_{5,6}$-S-S-$T_{16,17}$

Source: Reprinted from reference 69, with permission of John Wiley & Sons, Ltd.

of the $T_{4,5}$–T_{10} disulfide-linked peptide containing an asialo biantennary oligosaccharide (calculated MW 5286.6 Da), accompanied by two glycoforms with one or two NeuAc units 291 Da apart. Several weak glycopeptide signals corresponding to variations in the Hex-HexNAc content of the carbohydrate were also observed in the ESI mass spectrum, indicating either additional branching of the asialo biantennary structure or arm extension of the asialo biantennary glycan prior to capping with NeuAc groups.

Identification of the CHO IL-4 glycosylation site was provided by HPLC/ESI-MS analysis of the V8 protease digest, where $V_{4,5}$ peptide (residues 27–43) containing a sialylated biantennary N-linked oligosaccharide was observed, with ESI-derived MW of 3929 and 4220 Da (data not shown). These glycopeptides were absent in the analysis of the N-glycanase-treated IL-4, where multiply charged ions corresponding to the $V_{4,5}$ peptide were detected with an ESI-derived MW of 1868 Da. These data are consistent with the presence of a biantennary complex-type carbohydrate at the Asn-38 residue of peptide fragment V_4, thus confirming that N-glycosylation occurs at Asn-38 rather than the other potential site of Asn-105. Thus, it seems that the presence of the Asn-38-containing tryptic peptide T_5 resulted in partial inaccessibility of the adjacent Lys-37 tryptic site and shifted the disulfide-indicative ESI signals to higher m/z values owing to incorporation of the T_5-CHO glycopeptide (T_{5*}) into the adjacent disulfide-linked peptide T_4.

Detection of low-mass sugar-specific oxonium ions in the ESI mass spectrum can assist in the identification of glycopeptide-containing fractions among the HPLC peaks of a glycoprotein proteolytic digest. Production of these ions can be induced by increasing the orifice potential, which controls the extent of fragmentation in the de-clustering region of a mass spectrometer. For example, at a higher orifice potential of 110-V, low-mass sugar oxonium ions at m/z 204 (HexNAc$^+$), m/z 274 (NeuAc$^+$), m/z 366 (Hex-HexNAc$^+$) and m/z 657 (NeuAc-Hex-HexNAc$^+$) were observed for CHO IL-4 tryptic peak 11 (data not shown), indicating that this HPLC peak contained sialylated oligosaccharides of the complex type. Similar monitoring of these sugar-diagnostic ions revealed the other fractions (peak 14, e.g.) containing glycopeptide fragments from CHO IL-4. Thus, all glycopeptide-containing fractions during the HPLC/ESI-MS analysis of CHO IL-4 were rapidly identified by carrying out the ESI-MS experiments at an elevated orifice potential, without having to search the ESI spectra of each individual peak for signal patterns characteristic of glycopeptides.

This HPLC/ESI-MS approach proved to be useful in detecting several glycoforms in CHO IL-4. Not only were the main asialo and sialylated biantennary glycoforms detected, but also additional signals indicative of higher branching were well-separated. This rapid assessment of glycosylation at the molecular level is invaluable for an initial batch-to-batch evaluation of mammalian cell-derived proteins. It provides real-time monitoring of the existing glycoform distribution and also allows for the detection of any changes that may arise from varying the conditions of the production process.

19.3.2.2 *Phosphorylation.*

Phosphorylation. Phosphorylation is a posttranslational modification that is reflected in close to 30% of eukaryotic gene products and almost 2% of the human genome-encoded protein kinases. Protein phosphorylation plays an essential role in intercellular communication during development, in physiological responses and homeostasis, and in the functioning of the nervous and immune systems. Reversible phosphorylation regulates many diverse cellular processes such as growth, metabolism, proliferation, motility, and differentiation [70–72]. Mutation and deregulation of the proteins, such as protein kinases, play causal roles in human diseases. The features of all cancers, deregulated cell growth and apoptosis, is a result of defective signaling pathways [73]; protein kinases are essential elements in the signaling pathways that mediate cell growth and programmed cell death. Therefore, a complete catalog and characterization of phosphorylated proteins will afford the possibility of developing agonists and antagonists of these enzymes for use in disease therapy [74, 75].

In eukaryotic cells, protein phosphorylation happens mostly on serine, threonine, and tyrosine residues (it could also happen on histidine, arginine, lysine, cysteine, glutamic acid, and aspartic acid to a much lesser extent). A comprehensive analysis of protein phosphorylation involves the identification of the phosphoproteins, the localization of the residues that are phosphorylated, and the quantitation of phosphorylation. MS-based approach for characterization of phosphorylated proteins is based on the lability of the phosphor moiety of the phosphorylated peptides upon low-energy collisional activation in a tandem mass spectrometric experiment. The detection of a characteristic loss allows the identification of phosphorylated peptides from an unseparated peptide mixture or during on-line HPLC experiments. In the positive ion mode, a neutral loss of 98 Da (H_3PO_4 or HPO_3 and H_2O) from the phosphopeptide can be used to confirm the existence of a phosphopeptide [76, 77]. The fragment ion at m/z −79 (PO_3^-), however, is more frequently used in the negative ion mode for phosphorylation specific precursor ion scanning [78–81]. The advantage of precursor ion scanning of m/z −79 includes its applicability to all phosphopeptides with phosphorylation occurring on serine, threonine, or tyrosine. Precursor ion of m/z 216 has also been used for detection of tyrosine specific phosphorylation. Mann and co-workers [82] monitored precursors of m/z 216.043 (immonium ion of phosphotyrosine) by using quadrupole TOF mass spectrometer in the positive ion mode and showed that the quadrupole TOF was ~fivefold more sensitive than the triple quadrupole instrument in monitoring precursors of this ion. FTMS coupled with electron capture dissociation (ECD) has been applied to characterize the phosphopeptides or intact phosphoproteins in recent studies [83]. In this technique, dissociation is induced by electron recombination with the protons of the multiply charge peptide or protein, where labile modifications such as phosphorylation of serine and threonine remain intact. This process allows unambiguous assignment of the modification site in the peptide/protein.

Although powerful, there are challenges of mapping phosphorylation sites solely relying on the use of tandem MS. For example, in the commonly used positive ion detection mode of peptide MS analysis, signal suppression of phosphate containing peptides is often evident; the inherent lability of the phosphate group undergoing neutral loss of HPO_3 (80 Da) upon CID can make the identification of phosphorylation site difficult; for long phosphopeptides,

phosphopeptides present in low abundance, and peptide phosphorylated at substoichiometric levels, it is difficult to achieve full sequence coverage of the peptide during CID experiment, making the phosphorylation site determination ambiguous. Hence, there is a need to enrich/purify the phosphorylated peptides before MS studies.

One approach is to use immobilized metal ion affinity chromatography (IMAC) for selective enrichment of phosphopeptides [84, 85]. To prevent binding of nonphosphorylated peptides to the IMAC column, Ficarro et al. [86] converted the peptides to methyl esters before IMAC step. In their studies, a total of 216 peptide sequences defining 383 phosphorylation sites from a whole-cell lysate of *Saccharomyces cerevisiae* were determined. Affinity-based enrichment of tyrosine-phosphorylated proteins has also been explored using immunoprecipitation with anti-phosphotyrosine antibody [87]. Even though it has been difficult to identify proteins that are phosphorylated on serine/threonine due to the lack of the antibodies, a recent study by Gronborg et al. [88] reported the identification of phosphoserine/threonine proteins using phosphor-specific antibodies. As an alternative to IMAC and the immunoprecipitation, several groups have applied β-elimination—Michael addition reaction or by phosphoamidate chemistry to chemically convert the phosphorylated residues to moieties with an "affinity handle" for subsequent affinity or covalent chemistry-based purification [89–91]. Carbodiimide-catalyzed condensation of cysteamine with a phosphate group formed phosphoramidate with a free sulfhydryl group [89]. β-elimination of phosphoserine or phosphothreonine followed by Michael addition also generates a free sulhydryl group [90]. The free sulfhydryl moiety formed bases for affinity enrichments and enabled the incorporation of isotope-coded affinity tags to allow quantitative comparison of phosphopeptide levels between different cellular states [92, 93]. A recent approach that also relies on the well-established β-elimination is designed by Shokat and co-workers [94] to create specific proteolysis sites that could recognize serine and threonine phosphorylation. In their work, phosphoserine and phosphothreonine residues are chemically transformed into lysine analogs, aminoethylcysteine, and β-methylaminoethylcysteine, respectively, which can be cleaved by lysine-specific protease to map sites of phosphorylation.

Identification of phosphorylation sites is a critical step toward understanding the function and regulation of many protein kinases and kinase substrates. Protein kinases mediate most of the signal transduction in eukaryotic cells. By modification of substrate activity, protein kinases also control many other cellular processes, including metabolism, transcription, cell cycle progression, apoptosis, and differentiation [73, 95, 96]. In our study of the mitogen activated kinase, MEK1 (MAP/ERK kinase), "off-pathway" phosphorylation occurs during insect cell expression of the protein. Identification of these spurious phosphorylation sites resulted from protein expression is important for developing effective strategies in protein purification, mutagenesis, and dephosphorylation [97].

MEK1 was overexpressed in insect cells and purified by metal-chelating chromatography and gel filtration chromatography. The protein is composed of 429 amino acids and the theoretical MW of MEK1 is 47,534 Da. The initial MS analysis on the intact protein gave measured MWs of 47,550 Da and 47,630 Da, indicating the presence of one major phosphate group. To further characterize and identify the phosphorylation site, the protein sample was digested with Glu-C or a combination of enzymes, such as Glu-C/Trypsin, Glu-C/Chymotrypsin, and Trypsin/Asp-N with the substrate:enzyme ratio at 20:1. The reaction was carried out overnight at 37°C. Protein digestion by two enzymes was achieved by adding the second enzyme after the first digestion was complete. The mixture was incubated for another 4–12 hours. The resulting peptides were examined in both positive and negative ion modes. Peptide mass mapping, precursor ion scan of m/z −79 (PO_3^-), and MS/MS analysis of the enzyme cleavage products were performed using a triple quadrupole mass spectrometer.

Table 19-6 shows the results of precursor ion scans on MEK1 treated with a combination of enzymes. Precursor ion scan detected a phosphorylated Glu-C fragment (m/z 3458). However, MS/MS sequencing was not successful due to the size of this peptide. The approach of using two different enzymes was subsequently employed. Initial Glu-C cleavage followed by trypsin digestion successfully cleaved the peptides into smaller fragments. The precursor ion scan resulted in a phosphorylated peptide (m/z 2309) from the same region of MEK1. However, sequencing of this peptide (m/z 2309) did not give satisfactory results. Glu-C cleavage followed by Chymotrypsin digestion produced a peptide (m/z 1057) that was captured by precursor ion scan (Figure 19-14). The peptide MW (MW 1056 Da) matched the same sequence region as the one obtained by the above two approaches. And also, the mass of the peptide was adequate for sequencing. MS/MS data of the doubly charged ion at m/z 529.50 identified this peptide as phosporylated MEK1 (328–336). Ser-334 was identified as the phosphorylation site, which was evident by the addition of 80 Da (HPO_3) to y_3, y_4, y_5, and y_7 as well as to b_6 and b_8 fragments (Figure 19-15). To reconfirm the phosphorylation site, another enzyme combination was chosen; trypsin digestion followed by Asp-N led to the detection of a phosphorylated peptide at m/z 1246 through precursor ion scan. The MS/MS data of m/z 1246 further established Ser-334 as the phosphorylation site. Using a

TABLE 19-6. Precursor Ion Scan Results of Enzymatic Cleavages of MEK1

Enzyme	Precursor Ion Scan Result (Da)	Matching Sequence
Glu-C	3458	[317]GDAAETPPRPRTPGRPLSSYGMDSRPPMAIFE[348]
Glu-C/trypsin	2309	[328]TPGRPLSSYGMDSRPPMAIFE[348]
Glu-C/chymotrypsin	1057	[328]TPGRPLSSY[336]
Trypsin/Asp-N	1246	[328]TPGRPLSSYGM[338]

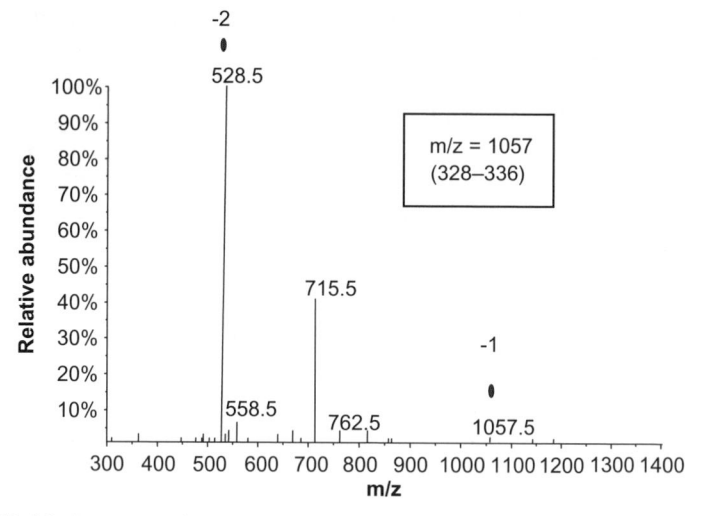

Figure 19-14. Precursor ion scan mass spectrum of Glu-C/chymotrypsin-digested MEK1.

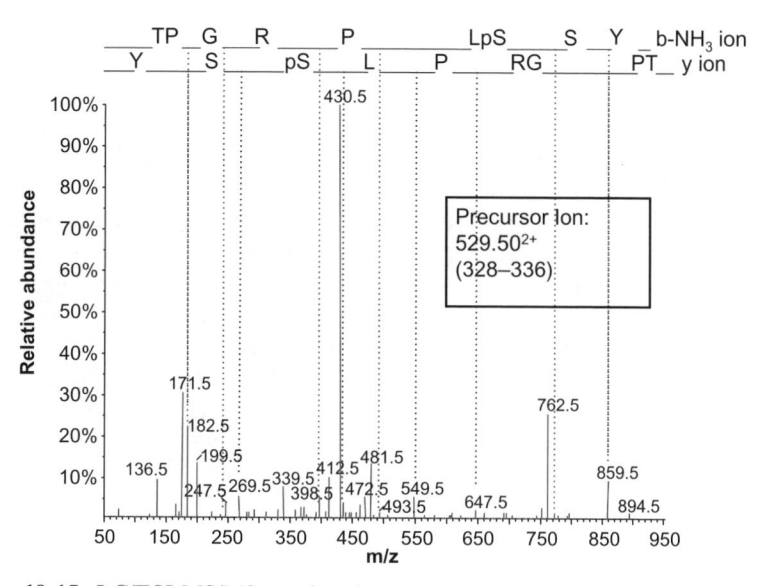

Figure 19-15. LC/ESI-MS/MS product ion mass spectrum of phosphopeptides from Glu-C/chymotrypsin-digested MEK1.

combination of enzymes and MS/MS analysis of phosphorylated peptides, all the experimental data were cross-checked with each other and resulted in the positive identification of Ser-334 as the phosphorylation site. This residue resides in a proline-rich region of MEK1 that may have regulatory importance.

19.3.3 Microwave-Assisted Methods for Proteins/Peptides

A microwave is a form of electromagnetic energy, consisting of an electrical field and a magnetic field, although only the electrical field transfers energy to heat a substance. The microwave couples directly with the molecules that are heating at a high rate (about 10^{-9} sec), leading to a rapid rise in temperature. The result is an instantaneous heating of a substance either by dipole rotation or ionic conduction. A typical chemical reaction can be speeded up by as much as 1000-fold under microwave irradiation. Microwave-assisted chemistry is a relatively new field. It has been shown to accelerate synthetic organic reactions [98, 99]. Application of microwave technology to faster analysis of proteins/peptides is an emerging field that is gaining more attention.

One of initial studies was microwave-assisted Akabori reaction for peptide analysis [100]. The classical Akabori reaction [101], devised in 1952 for the identification of C-terminus amino acids, involved the heating of a linear peptide in the presence of anhydrous hydrazine in a sealed tube for several hours. The C-terminus group is liberated as free amino acid and can be distinguished from the remaining amino acid residues that have been converted to hydrazides. The reaction mechanism is illustrated in Figure 19-16.

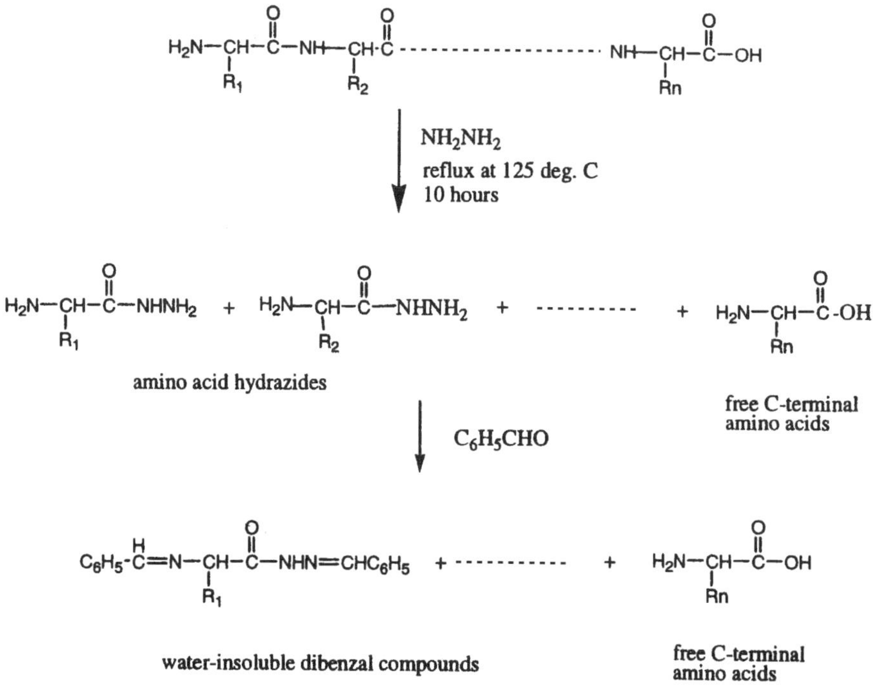

Figure 19-16. Akabori hydrazinolysis of oligopeptides. (Reprinted from reference 100, with permission of Elsevier Science, Inc.)

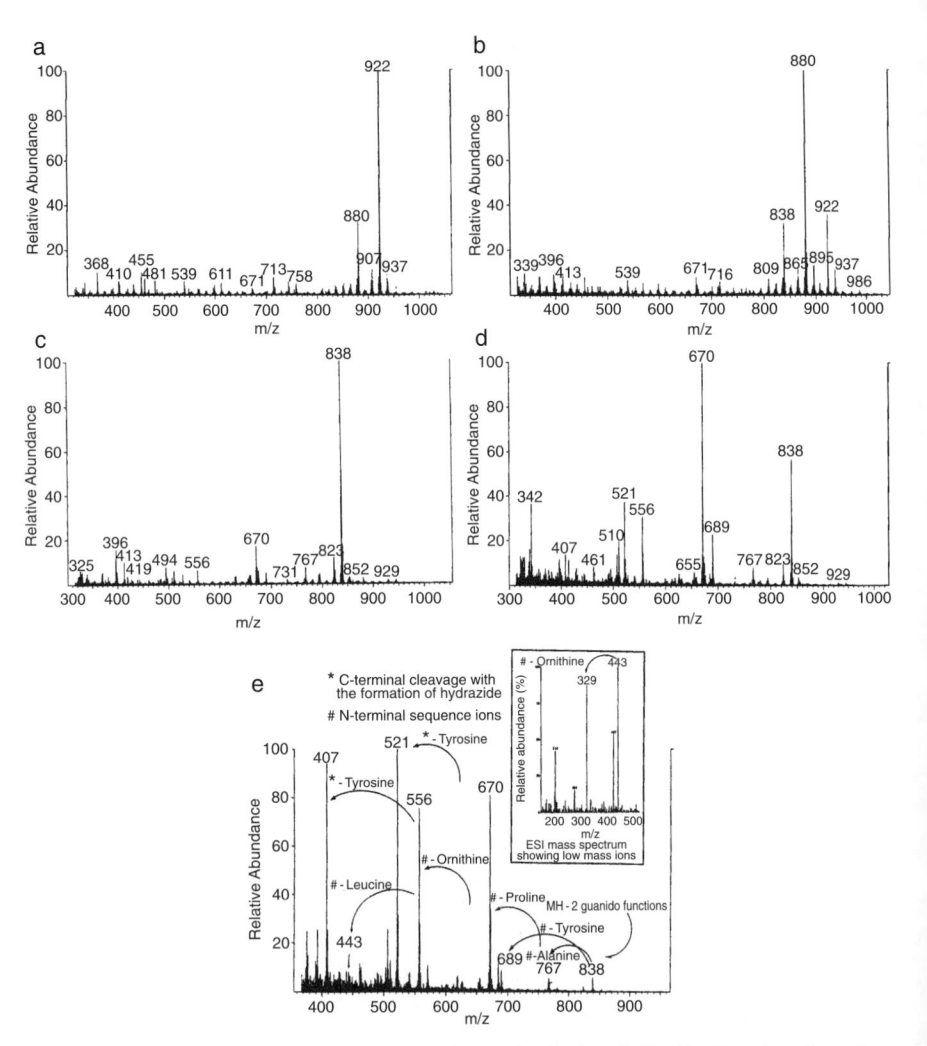

Figure 19-17. FAB-MS spectra of the hydrazinolysis of H-Ala-Pro-Arg-Leu-Arg-Phe-Tyr-OH at (a) $t = 2\,\text{min}$, (b) $t = 8\,\text{min}$, (c) $t = 20\,\text{min}$, (d) $t = 30\,\text{min}$, (e) $t = 50\,\text{min}$. (Reprinted from reference 100, with permission of Elsevier Science, Inc.)

In the microwave-assisted methodology, the linear polypeptide and hydrazine solution, contained in a loosely covered conical flask, was exposed to a few minutes of irradiation using an unmodified domestic microwave oven. To monitor the hydrazinolysis progress, dimethyl sulfoxide was added to dilute the reaction mixture that allowed aliquots of the reaction mixture to be drawn every few minutes over a period of about an hour. Then, the aliquots were analyzed by mass spectrometry. As an illustration, Figure 19-17 displayed fast-atom bombardment (FAB) mass spectra of microwave-assisted Akabori

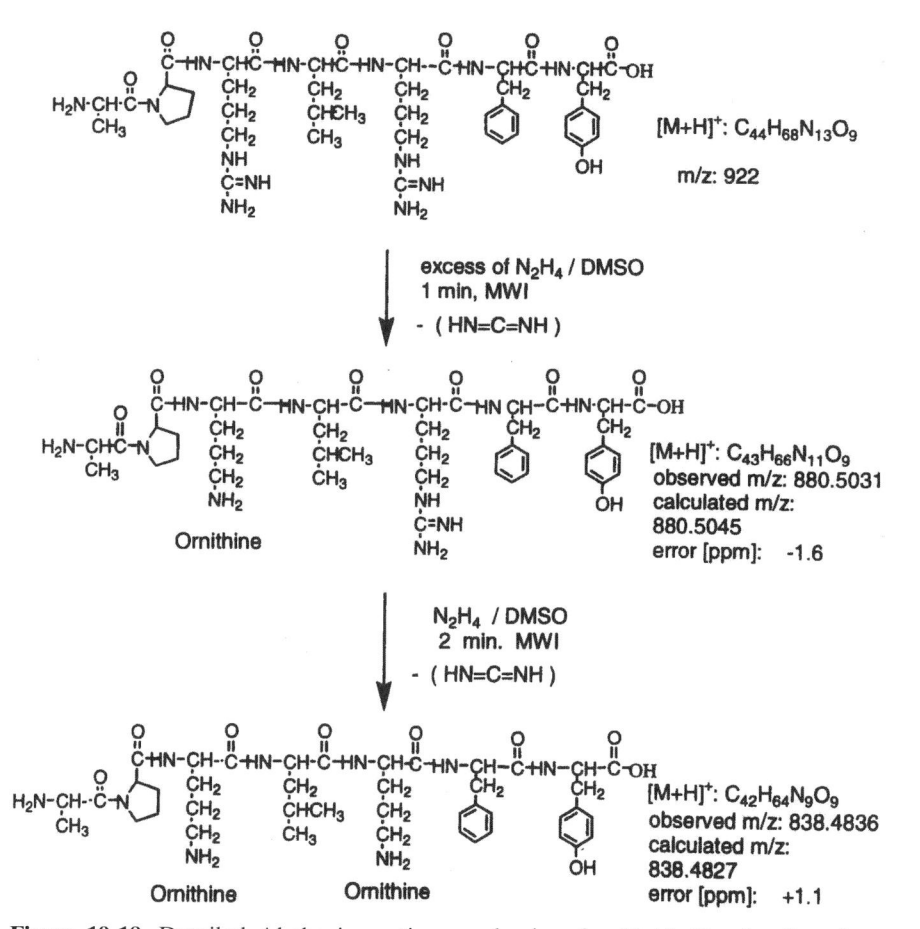

Figure 19-18. Detailed Akabori reaction mechanism for H-Ala-Pro-Arg-Leu-Arg-Phe-Tyr-OH. (Reprinted from reference 100, with permission of Elsevier Science, Inc.)

reaction products for the hydrazinolysis of a heptapeptide, H-Ala-Pro-Arg-Leu-Arg-Phe-Tyr-OH, sampled at 2, 8, 10, 30, and 50 minutes. The detailed reaction mechanism was summarized in Figure 19-18. The initial Akabori cleavage, involving the loss of C-terminal tyrosine from the modified heptapeptide (m/z 838), led to the formation of the hexapeptide hydrazide at m/z 689. Accurate mass measurement confirmed the expected elemental composition. Two additional ions were also generated by first-order Akabori cleavage: the tetrapeptide at m/z 521 and the tripeptide at m/z 407 at 30-minute intervals (Figure 19-17d). In all three fragment ions, the same C-terminal amino acid (tyrosine) was lost with different N-terminal sequence. In addition to C-terminus Akabori cleavage, microwave-assisted hydrazinolysis generated sequential cleavage from the N-terminus of the modified heptapeptide (m/z 838), yielding a series of ions at m/z 767, 670, 556, 443, and 329 (Figure 19-17e).

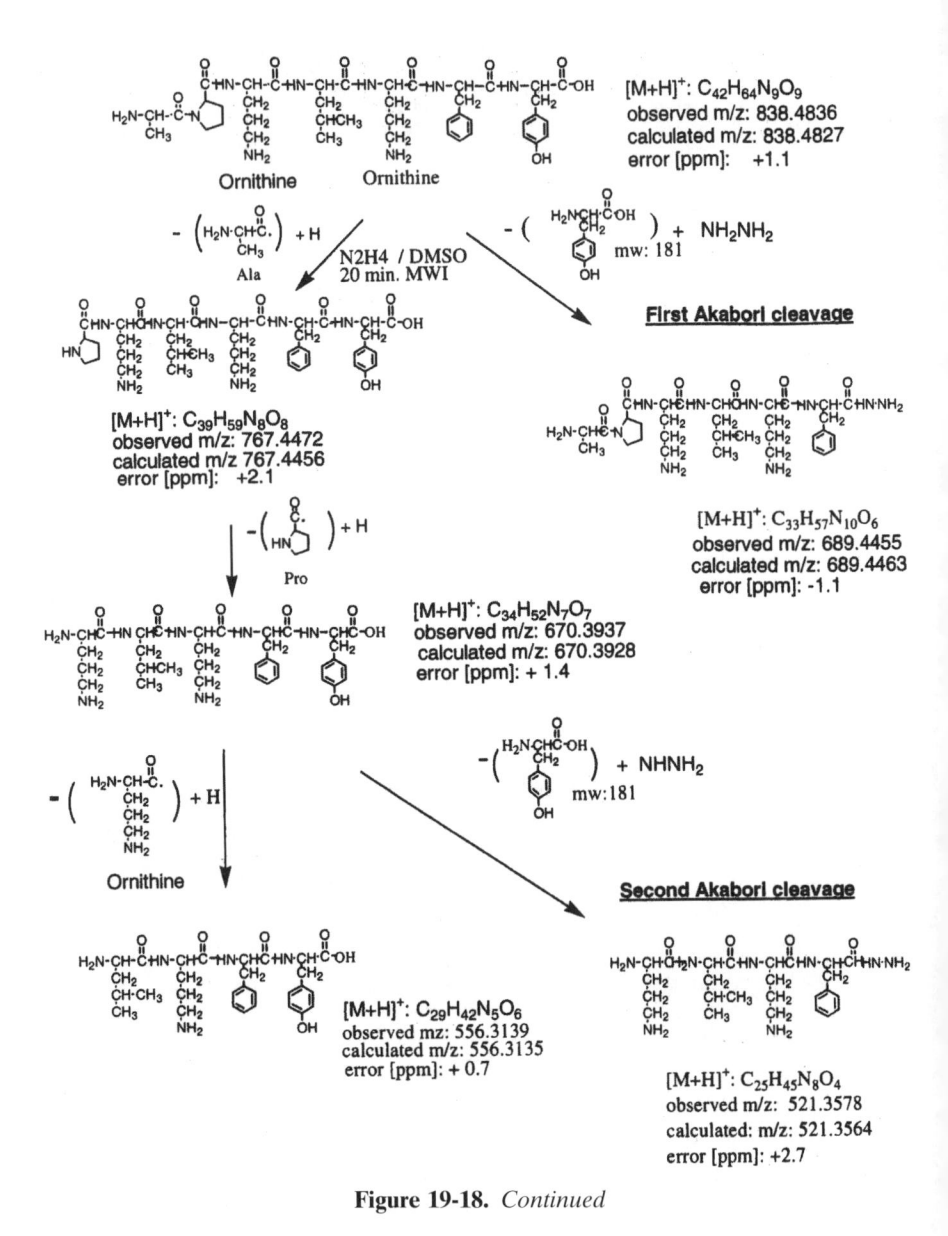

Figure 19-18. *Continued*

Their proposed fragment ion structures were supported by accurate mass measurements, as shown in Figure 19-18. Clearly, microwave-assisted Akabori reaction can lead to rapid identification of C-terminus amino acid in a polypeptide including its amino acid sequence information at both C-terminus and the N-terminus. It was also found that microwave-assisted hydrazinolysis of N-terminal substituted polypeptides followed the same pattern as the unsubstituted peptides, while traditional Edman degradation approach would be unsuccessful. Furthermore, the presence of arginine and amino acids containing β-SH, COOH, and $CONH_2$ groups in their side chains can be rapidly confirmed because they are susceptible to modifications by hydrazine. For example, the carboxy (glutamic acid and aspartic acid) and the carbox-amido groups (asparagine and glutamine) can be converted to the carboxy-hydrazide group ($—CONHNH_2$), resulting in peaks with increased m/z values at +14 and +15, respectively. The replacement of the $—SH$ group in cysteine-containing peptides by $—NHNH_2$ generates a peak with decreased m/z value (–2). The loss of 42 Da ($NHC{=}NH$) often indicates the presence of arginine in the peptides—for example, conversion to ornithine residue.

Microwave-assisted Akabori reaction was extended to cyclic oligopeptides for the rapid determination of the sequence of amino acids [102]. The traditional Edman degradation is not particular attractive to sequencing of cyclic peptides since selective hydrolysis of peptide bonds by this method is not easy to achieve because of the lack of free N-terminus. The other approach using tandem mass spectrometry also proves to be difficult because the indiscriminate ring-opening pathways often gives a set of acylium ions of the same mass-to-charge ratio [103, 104]. In the case of glycine-containing cyclic peptides, microwave-assisted hydrazinolysis led to selective ring-opening at glycine residue to generate the corresponding open chain hydrazide(s) in a few minutes. The reaction mixtures were analyzed by RP-HPLC/ESI-MS and MS/MS for sequence determination. For example, a nonapeptide, cyclo (-Phe-His-Trp-Ala-Val-Gly-His-Leu-Leu-), treated with 98% hydrazine under microwave irradiation for a few minutes, generated a linear oligopeptide hydrazide (m/z 1093). Since there was no arginine, aspargine, cysteine, and other amino acids that react with hydrazine under microwave irradiation, no additional peaks resulted from modifications appeared in the mass spectrum. The RP-HPLC/ESI-MS/MS product ion spectrum of this component (m/z 1093) gave characteristic b ions and y ions, revealing the sequence of amino acids in the cyclonoapeptide (Figure 19-19). The cleavage site was determined to be at the amide bond of glycine. This is in contrast to direct MS/MS analysis of the cyclonoapeptide ions which fragment randomly and may not be useful for sequence determination. A number of other non-glycine-containing cyclic peptides were also investigated under microwave irradiation [105]. Those cyclopeptides were all opened to form linear hydrazides in about 10–20 minutes. The ring opening occurred substantially at a single amide bond. Subsequent RP-HPLC/ESI-MS/MS studies all yielded correct amino acid sequences.

Another application area in microwave technology is the use of microwave irradiation for the enzymatic digestion of proteins [106]. As discussed in the early part of this chapter, enzymatic cleavage to produce smaller peptide fragments of protein samples is an important step in structural characterization of proteins [24]. Traditional enzymatic digestion method usually takes several hours, whereas microwave-assisted digestion occurs in minutes. The initial

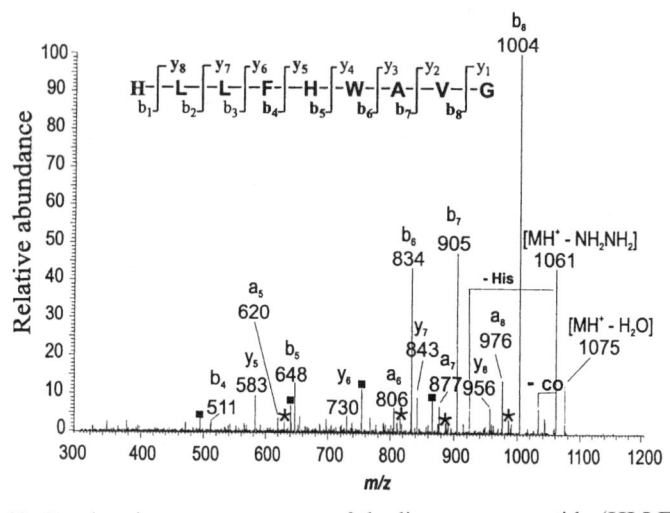

Figure 19-19. Product ion mass spectrum of the linear nonapeptide (HLLFHWAVG-NHNH$_2$). The peaks marked with an asterisk represent bn-H$_2$O ions, whereas the peaks marked with a black square represent the internal fragment ions. (Reprinted from reference 102, with permission of Elsevier Science, Inc.)

studies were carried out on bovine cytochrome c, a global protein relatively resistant to enzymatic cleavage under non-denaturing conditions. The protein was treated with trypsin at a 1:25 protease-to-protein ratio and the solution was subjected to microwave irradiation for about 10 minutes. The products were analyzed by MALDI-MS, as shown in Figure 19-20a. Most of the expected tryptic peptides were observed in the spectrum. The extensive sequence coverage was displayed in Figure 19-20b. The result was similar to what was obtained using traditional digestion approach that took about 6 hours. Several other proteins, including bovine ubiquitin, lysozyme, myoglobin and IFN-α-2b, were shown to exhibit the same accelerated proteolytic cleavages under microwave irradiation. The action mechanism involved in the observed rate acceleration of the enzymatic cleavage of proteins under microwave irradiation was studied at different microwave temperatures at different irradiation time intervals. The results suggest that the rapid increase in the reaction temperature is at least partially responsible for the large acceleration of digestion observed under microwave conditions [106]. This strategy using microwave irradiation for protein digestion has significant implications in proteomics research in which protein digestion is often a tedious step in protein identification. For those tightly folded proteins, they are known to require long hours for adequate proteolysis by enzymes under conventional conditions. Clearly, this approach can dramatically enhance proteolysis rates under microwave irradiation, improve the efficiency of protein digestion, thus protein identification.

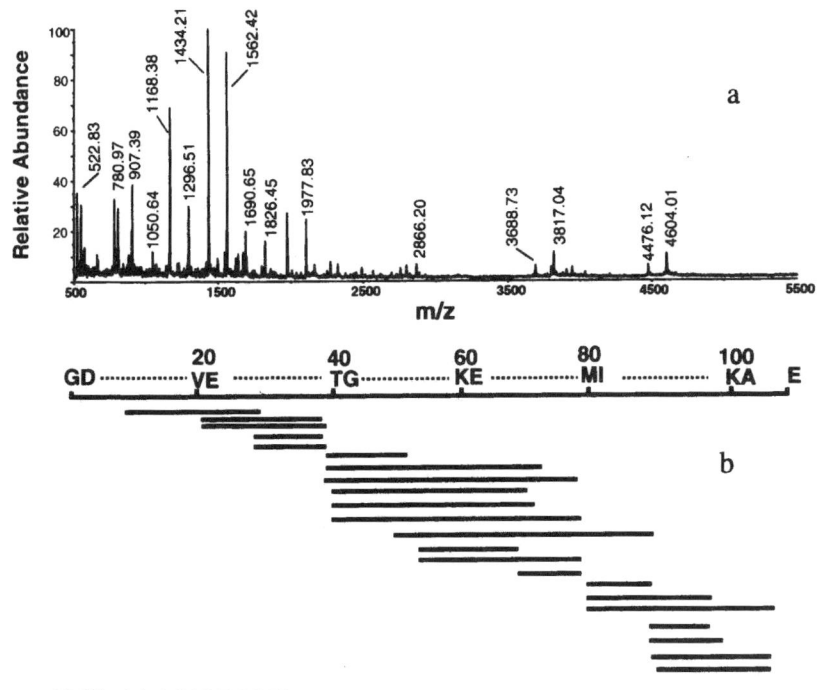

Figure 19-20. (a) MALDI-MS spectrum of tryptic fragments of cytochrome c after 10 min of microwave irradiation at 1 : 25 protease-to-protein ratio by weight. (b) Tryptic fragments in (a) are represented in the form of horizontal lines, showing the total sequence coverage of cytochrome c. (Reprinted from reference 106, with permission of the Protein Society.)

Overall, microwave-assisted methods are effective approaches in analysis of proteins/peptides, especially when used with other analytical techniques, such as HPLC and MS methods.

19.3.4 Drug–Protein Interaction by Affinity-Based HPLC/MS

Many important biological functions are mediated through noncovalent interactions between bio-macromolecules and other components in the cell. These noncovalent interactions may involve enzyme and substrate, protein and ligand, protein and protein, and antigen and antibody [107, 108]. In a drug discovery process, it is desirable to have a good understanding of potential drug candidates that have affinities toward a particular target protein. The investigation of drug–protein interactions can be achieved by combining affinity-based HPLC and MS, providing rapid screening techniques for lead generation and optimization.

One of techniques is frontal affinity chromatography (FAC) coupled to MS [109]. In this method, a column is prepared with an immobilized biological target protein. A sample containing potential drug ligands is continuously infused through the column. No change in mobile phase is required for separation and elution. The order of elution of ligands from the column parallels their affinities for the target protein, with the tightest binding ligands eluting last. The eluted ligands are monitored by ESI-MS, allowing analysis of mixtures in a single run. The unknown binding components to a target can be identified by MS. In addition, dynamic equilibrium processes underlying FAC/MS allow the ready determination of affinity dissociation constants (Kd's) of individual active ligands in the mixture, based on the relationship between the elution volume and the amount of ligand absorbed to the target [110].

Kelly et al. [111] carried out FAC/MS experiments using the polyol pathway target sorbitol dehydrogenase (SDH) to evaluate the applicability of FAC/MS for lead optimization and structure–activity profiling. The SDH was chemically biotinylated by modification of the primary amine groups with Sulfo-NHS-LC-Biotin. This reagent contains a sulfo-N-hydroxy-succinimide that reacts with amines to form a stable amide bond and is soluble in aqueous solvents. The NHS was coupled to biotin through a spacer arm that limited steric hindrance of the protein once it was immobilized. Approximately 15 pmol of protein was immobilized onto a 0.5-μL column packed with 200-μm streptavidin beads. K_d values were measured for selected ligands, ranging from 2.2 μM to 1.3 nM. They correlated well with existing values. Rank ordering of multiple binders to SDH was shown for three different subsets of ligands that were in agreement with their K_d values (Figure 19-21). Compound class-related nonbinders were shown not to bind with the SDH target. Isomers of different affinities were resolved in the correct rank order by employing MS/MS detection [111]. This technique offers resolution of multiple binders and the rank order for a wide range of binding affinities and class-specific structural trends.

Another affinity-based HPLC/MS approach involves size-exclusion chromatography and MS [112]. In this approach, drug ligands and target protein are incubated under native conditions. Active compounds will form tightly bound noncovalent complexes with the target protein. The complexes will then be separated from the unbound (inactive) molecules by size-exclusion chromatography. Either complexes or unbound molecules can be analyzed using MS. Wabnitz and Loo [113] demonstrated the effectiveness of the strategy in determining both the strong binding affinity and the relative affinity rank ordering of ligands toward a specific target protein, Co^{2+}-substituted peptide deformylase (Co-PDF). They used a micro-size-exclusion column (MicroSpin G-25) to retain low-MW ligands and exclude high-MW compounds. The column had a cutoff of MW 3000 Da. The separation was carried out using a bench-top micro-centrifuge. The column was equilibrated with 100-μL 25 mM ammonium bicarbonate (pH 8) for 1 minute at 3000 rpm. About 100-μL aliquot of the incubation mixture was loaded onto the column and centrifuged. The protein, whether bound or unbound to the low-MW ligands, would pass the column. The unbound ligands would be retained on the column and

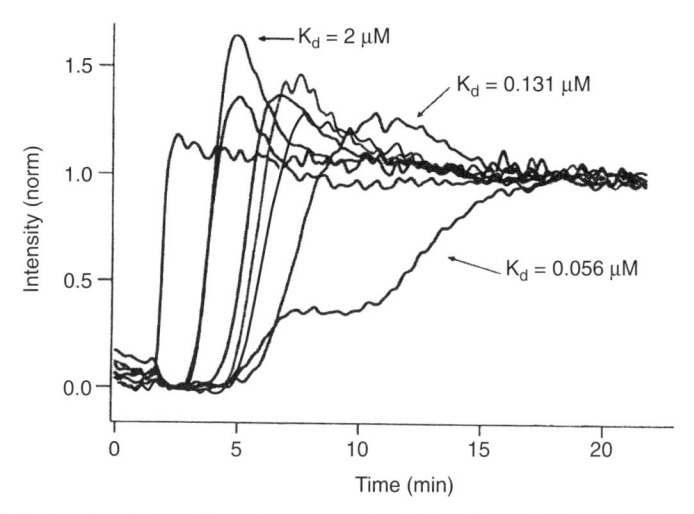

Figure 19-21. FAC-MS profiles of compounds eluting from the SDH column as measured using single ion monitoring of molecular ions. A nonbinding compound elutes first as the void marker because it shows no affinity to the target. The elution order of eight components analyzed as a mixture reflects their relative binding strengths, as confirmed by IC_{50} and K_d values. (Reprinted from reference 111, with permission of the American Chemical Society.)

removed by eluting with 100-μL CH_3CN/H_2O (4:1) solution at 14,000 rpm for 1 minute. The resulting fractions were dried and analyzed by capillary HPLC/MS. The schematic flow chart is shown in Figure 19-22. The relative rank order of binding for different ligands was also determined, as illustrated in Figure 19-23. The results showed an absence of ligand C when compared with the control (b), indicating ligand C as the stronger binder to Co-PDF. This technique is relatively simple, as compared to immobilized protein affinity column, and it provides greater indication of binding under native conditions with considerable sensitivity (picomole range). In addition, the analysis of small molecule ligands is more straightforward than analyzing the protein/drug ligand complex. This methodology can potentially lead to high-throughput format that may be complementary to other screening techniques used in drug discovery.

19.3.5 Multidimensional HPLC in Proteomics

Proteomics is an emerging field of intensive research in the post-genomic era that involves the global analysis of gene expression, including identification, quantification, and characterization of proteins [114, 115]. Although proteins are a translated version of genes, the complexity of proteins is enormous. As many as 1 million proteins can exist in the proteome. An estimated 20,000 proteins are expressed in a particular type of cells at any time. All proteins do not

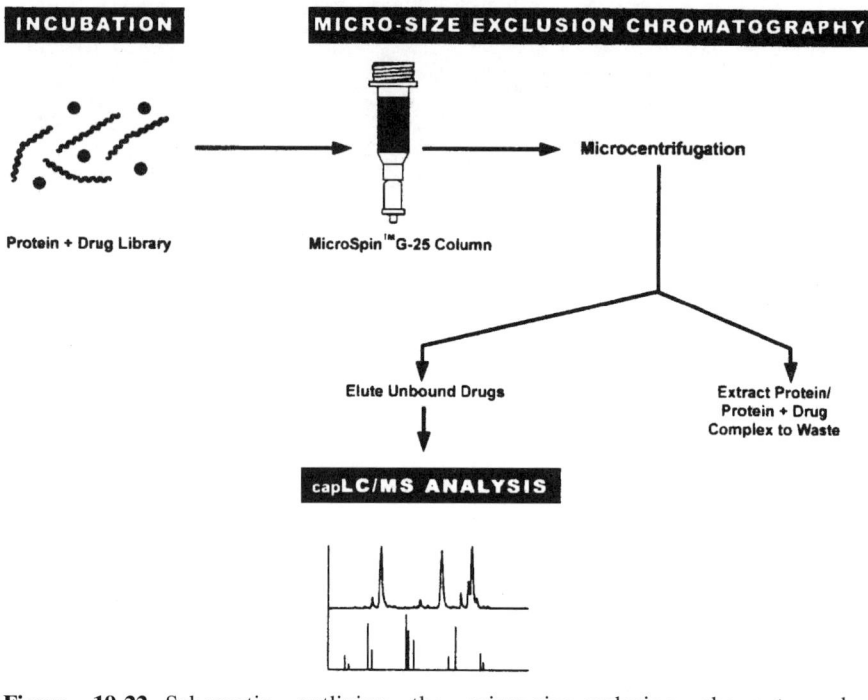

Figure 19-22. Schematic outlining the micro-size-exclusion chromatography-capLC/MS approach for ligand drug screening for a specific target macromolecule. (Reprinted from reference 113, with permission of John Wiley & Sons, Ltd.)

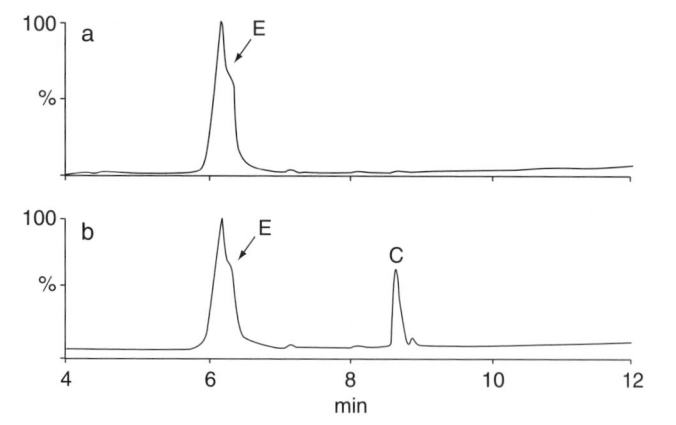

Figure 19-23. (a) Chromatogram of the micro-size-exclusion results of ligand mixture E/C together with Co-PDF, indicating ligand C as the stronger binder to Co-PDF. (b) Chromatogram of the control experiment of ligand mixture E/C with the exclusion Co-PDF. (Reprinted from reference 113, with permission of John Wiley & Sons, Ltd.)

need to be expressed in every single cell type in a particular organism at all times. The proteome level varies with time, depending on the genome, the environment of the cell, and the cell history. Many post-translational modifications of proteins take place, leading to different cell functions. Analysis of such complicated systems is a challenge in proteomics research.

To reduce the complexity of proteins expressed in the cell and overcome the differences in expression rates of proteins, prefractionation techniques have to be deployed before further protein characterization [116]. Two current approaches include electrophoretic and chromatographic methods.

Notably, a multicompartment electrolyzer device based on permeability through a barrier via surface-charge modulation has been developed in preparative electrokinetic fractionation processes [117]. The device consists of a stack of chambers between an anodic and a cathodic reservoir, accommodating up to eight flow chambers (six for sample collection and two as electrodic reservoirs). The apparatus has two orthogonal flows: (1) a hydraulic flow that transports all samples into and out of the electrical field until steady-state conditions are established for all components and (2) an electrical flow that allows each protein to reach the chamber under isoelectric conditions. Two isoelectric membranes facing each flow chamber function by continuously titrating the protein of interest to its pI, thus retaining them with pIs between their limiting values. This approach allows much higher sample loads without interference from the most abundant species.

On the chromatographic methods for prefractionation, Fountoulakis' group initiated several approaches, including ion-exchange, HIC, and affinity chromatography, to enrich low-copy-number gene products [118–120]. The affinity chromatography on heparin gels was used to enrich certain protein fractions in the bacterium *Haemophilus influenzae* [120]. Heparin is a highly sulfated glucosaminoglycan with affinity for a broad range of proteins, including nuclear acid-binding proteins and protein synthesis factors. The existence of sulfate groups in heparin allows it to function as a cation exchanger as well. In their studies, about 160 cytosolic proteins bound with different affinities to the heparin matrix were highly enriched prior to separation/identification. Consequently, more than 110 new protein spots detected in the heparin fraction were identified.

The prefractionation technique provides improvements in separation resolution, increased sensitivity, and enhanced ability of loading a much higher amount of sample in any narrow pH interval in gel electrophoresis. Once the complexity of proteins is reduced by prefractionation, individual fractions can be subjected to either 2D gel electrophoresis or multidimensional HPLC for further separation, followed by MS analysis.

Since the beginning of 2D gel electrophoresis about 30 years ago [121], it has become a method of choice for high-resolution protein separation [122]. On a 2D gel, up to thousands of proteins expressed by an organism or cell can be separated followed by identification with MS [123]. Traditionally, a 2D gel-matching procedure (gel imaging) can be used to compare two sets of protein mixtures run under standardized conditions from different cell states. The difference spots are analyzed for further investigation. However, this approach can introduce ambiguities simply because not all protein spots on two given gels can be correctly matched, even under the most standardized conditions. Misleading protein identifications can also occur because proteins are not identified on the basis of their origin using this matching procedure. A modified approach is to pick all the spots and determine the difference after MS analysis. In general, all gel spots are cut and digested with enzymes. The identities of proteins are obtained from either peptide mass mapping or sequence tagging on the resulted polypeptides in combination with database search. The sequence tagging is usually carried out by capillary RP-HPLC/MS and MS/MS methods.

In spite of the success and wide use, 2D gel is a relatively slow, labor-intensive and cumbersome technique, even with the availability of prefabricated strips for the first dimension (isoelectric focusing) and the pre-cast gels used in the second dimension. Another limitation is its inability to resolve lower abundance (less stable) proteins, membrane proteins, highly acidic or basic proteins, very large or small proteins, and hydrophobic proteins like G-coupled protein receptors with several transmembrane regions.

To address these issues in 2D gels and study those "difficult" proteins separated by 2D gels, alternative separation methods such as multidimensional HPLC have been introduced and implemented. Opiteck et al. [124] studied native and non-native proteins expressed by E. coli using 2D HPLC coupled with MS. The first-dimensional HPLC was a size-exclusion column, followed by RP-HPLC. The size-exclusion chromatography was performed under either denaturing or non-denaturing conditions. Peaks eluted from the first dimension were automatically injected onto RP-HPLC column for separation of similarly sized proteins on the basis of their hydrophobicities. RP-HPLC also served to desalt the samples for further analysis. Following the chromatographic separation, the analytes were deposited into 96-well plate via a fraction collector. The entire 2D HPLC was performed in a few hours and interfaced directly to MS for measurements of intact proteins. The identification was achieved using off-line bottom-up digestion on isolated fractions. The advantage of this 2D HPLC experiment includes that separations in the first dimension can be performed with buffers incompatible with MS analysis, since RP-HPLC is the second-dimensional HPLC. The overall process is highly automatic and can be implemented in a high-throughput format.

Another approach is the "shotgun" proteomics in which a complex mixture of proteins is enzymatically digested in solution to generate a complex mixture of peptides and the mixture of peptides is separated on-line either by 1D or 2D HPLC/MS [125, 126]. The MS/MS spectra for individual eluting peptides are obtained and searched against database for protein identification (similar to bottom-up experiment). The advantages of this approach include that peptides are easier to separate than proteins, native structures of the protein do not have to be maintained during the analysis, the slow digestion step is performed only once, and almost all proteins (very large, very hydrophobic, very basic) will give rise to peptides with sufficient size for identification. However, the complexity of the sample is increased tremendously because a number of peptides can be produced from each protein. The information about the integrity of the original protein or its post-traditional modifications may be lost to some extent. In an effort to reduce the number of peptides in tryptic digests, Reginer and co-workers [127] used affinity chromatography selection of peptides on the basis of low-abundance amino acid residues. They focused on the ability of concanavalin A lectin column to select glycopeptides from tryptic digests of N-type glycoproteins. Concanavalin A has a high affinity for N-type hybrid and high-mannose oligosaccharides, slightly lower affinity for complex di-antenary oligosaccharides, and virtually no affinity for complex N-type tri- and tetra-antenary oligosaccharides. The RP-HPLC analysis of gly-copeptides captured by the affinity column has shown a great simplicity. Applying affinity selection strategies for some specific peptides would allow a rapid reduction in sample complexity in a single step and still get enough peptides for protein identifications.

In addition to identification of proteins, quantitation of changes in protein expression between different states is another frontier in proteomics. It is not uncommon to have a 10-fold change in protein expression between normal state and disease state. In other cases, a subtle change of less than 2-fold in protein expression may be seen. Some approaches based on isotope tagging of peptides have been developed to measure the changes in expression levels between two proteomes in a single experiment. Particularly, various methods are available for relative quantitation of peptides from a proteolytic digestion of complex proteins, including isotope-coded affinity tag (ICAT) [128], N-terminal labeling of peptides [129], and using $^{18}O/^{16}O$-labeled water [130]. The ICAT method is a popular approach to quantify the relative changes of proteins. The ICAT reagent consists of a biotin group followed by a linker and is terminated with a cysteine-reactive group. The method involves *in vitro* derivatization with biotinylated tags of proteins from different samples, followed by mixing tagged proteins (light-form and heavy-form stable isotopes, e.g., deuterium), trypsin digestion, and purification on a monomeric avidin column (binds to the biotin in the tag) by affinity chromatography. The cysteine-containing tagged peptides containing light and heavy stable isotopes (e.g., deuterium) have different molecular mass and appear at different m/z values in the mass spectrum. Quantitation is achieved by measuring the relative amount of the light-isotope- and heavy-isotope-tagged peptides by MS. A common issue in current chemical tagging approaches is that the dynamic range of the method may not coincide with the expression level that may spread over a few orders of magnitude. In addition, the labeling reaction kinetics may not favor the labeling of peptides if there is a very small amount of proteins. It is likely that the current ICAT and other tagging approaches are to be pursued further to address these issues in terms of increasing the sensitivity and the scope of techniques.

19.3.6 Characterization of Adenovirus Structural Proteins for Gene Therapy

Recent advances in gene therapy technology have provided new tools for treatment of cancer and other serious diseases. There are several approaches in gene therapy, including the restoration of normal gene function of nonfunctional suppressor genes in damaged cells by gene substitution, the enhancement of antitumor activity of natural defense mechanisms of the tumor patient, and the selective eradication of tumor cells by inducing apoptosis using additive gene insertion. All these approaches involve nucleic acids as the active pharmaceutical reagent encoding proteins with therapeutic potential or proteins that interfere with the expression of damaged target genes. These therapeutic genes can be transferred into the damaged cells by viral or nonviral vector systems.

The adenovirus is an icosahedral, nonenveloped, double-stranded DNA virus that infects a broad range of mammalian cell types without undesired integration of the therapeutic gene into the host cell genome. The use of recombinant adenovirus (rAd) vectors to deliver potentially therapeutical genes to target cells has become a preferred method in gene therapy [131]. The commonly used recombinant adenovirus is derived from an adenovirus serotype 5 (Ad 5) virus that has had the E1 coding sequence replaced with a 1.4-kb full-length human p53 cDNA. This 200×10^6 Da virus has at least 11 structural proteins with a wide mass range, from less than 10,000 Da to more than 100,000 Da. Each of these proteins comprises 1% to 45% of the total viral protein content. The infectivity of the virus depends on the assembly of these structural proteins in forming a complete virion. In the previrion stage, empty capsid is essentially noninfectious, mainly consisting of hexon (II), penton-base (III), and fiber (IV) proteins in the absence of internal protein V and VII. Once the internal precursor protein pVII is added and a 23-kDa viral proteinase is activated, it will lead to the formation of virion. The viral proteinase is critical for virion maturation/infectivity since it cleaves precursor proteins (pVI, pVII, pVIII, pIIIa, and pX) to form their mature forms. Clearly, an effective analytical method is needed to monitor the viral production/purification and assess the maturation stage in gene therapy.

Traditionally, density gradient centrifugation is a primary method for viral purification [132]. However, it is not suitable for large-scale viral production. Recent new methodologies developed for production and purification of viruses mostly involve chromatographic methods, including size-exclusion chromatography [133], affinity chromatography [134], ion-exchange chromatography [135], and reversed-phase high-performance liquid chromatography [136]. All these methods lack accurate identification of viral proteins and their possible post-translational modifications at the proteome level. The sodium dodecylsulfate polyacrylamide gel electrophoresis (SDS-PAGE) with N-terminal amino acid sequencing has also been used to identify viral proteins [137]. However, most of the rAd proteins were found to be N-terminally blocked, and the mobility of some viral proteins in SDS-PAGE could be shifted due to certain post-translational modifications, leading to inaccurate results.

In our laboratory, we have developed a novel MALDI-MS based assay in combination with separation techniques to rapidly identify and confirm the presence of viral proteins (immature or mature) and impurities in the rAd vector [138]. The approach combines powerful multidimensional analytical techniques to fully characterize viral proteins, including RP-HPLC, SDS-PAGE, MALDI-MS, MS/MS, and database searching methods (Figure 19-24). This assay involves dissociation/separation of intact viruses by RP-HPLC, separation of the SDS-dissociated viruses by SDS-PAGE, and enzymatic digestion of the dissociated viral proteins from the RP-HPLC fractions and the gel bands, followed by MALDI-MS, MALDI-post source decay (PSD) studies, and database search.

Replication-deficient recombinant adenoviruses derived from type 5 expressing human p53 transgene were produced in HEK293 cells growing in serum-containing medium. The concentrated virus lysates were thawed and filtered, and the salt concentration was adjusted to 260 mM NaCl. It was then adsorbed to a DEAE-Fractogel anion-exchange column equilibrated with 50 mM sodium phosphate (pH 7.5), 260 mM NaCl, 2 mM $MgCl_2$ and 2% sucrose, and eluted with a 280–600 mM NaCl gradient. The fractions were pooled and separated on a Superdex-200 size exclusion column.

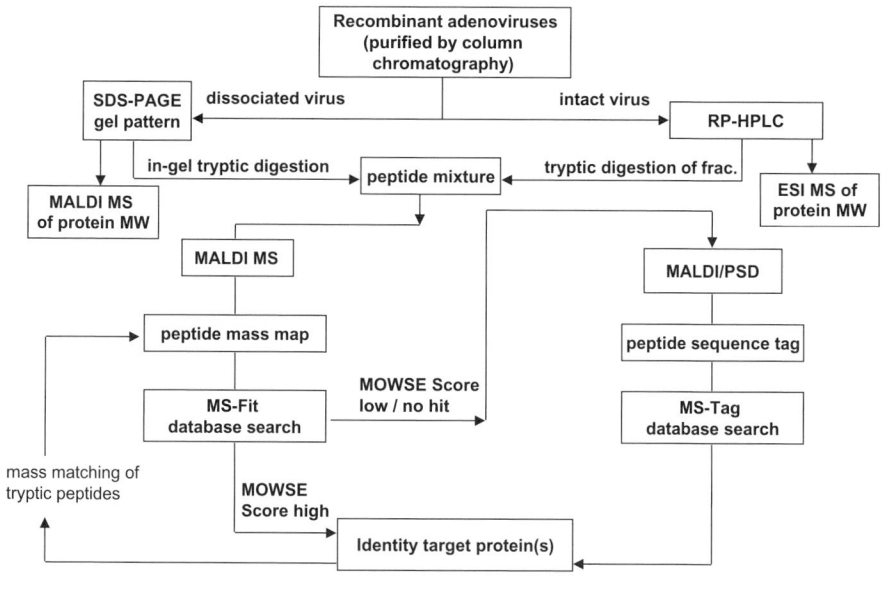

Figure 19-24. Schematic diagram of multidimensional analysis of recombinant adenoviral proteins. (Reprinted from reference 138, with permission of Elsevier Science B.V.)

For SDS-PAGE, adenovirus structural proteins dissociated from 1.0×10^{12} particles/mL of virus were loaded onto each well of Novex precast gradient 4% to 20% acrylamide gels and stained with Coomassie Blue.

Preparative RP-HPLC separation of viral proteins was performed on a 4.6- × 250-mm Jupiter C4 column at 50°C. Approximately 6 mL of the column-purified virus preparation (~1 × 10^{12} particles/mL) was loaded onto the column. Each chromatographic peak was collected manually for further MS characterization.

19.3.6.1   Identification of SDS-PAGE Separated Adenoviral Proteins. As previously reported, the recombinant adenoviral proteins were initially extracted from gel bands of SDS-PAGE for the determination of their MWs by MALDI-MS [139]. This step provided mass measurements of viral proteins with far better accuracy than those obtained from SDS-PAGE (a mass accuracy of 0.1% was normally obtained for MALDI-MS). However, depending on the stage of maturation of the adenovirus, there are possibilities of the presence of precursor proteins, propeptides, and post-translational modified proteins in the viral particle. To gain detailed structural information of viral proteins, tryptic digestion of each gel band followed by peptide mass mapping and protein identification through database search was carried out to provide additional information on the primary structures of these proteins.

The proteins separated by SDS-PAGE with MWs ranging from 10,000 Da to 100,000 Da were identified as adenoviral proteins from the SwissProt database when searched against all taxonomy. The results were summarized in Table 19-7.

It was noticed that a gel band located in between band V and band VI was observed and labeled as the 31-kDa band. This band is one of the three major bands (II, IIIa, and 31 kDa) observed only in the empty capsids, which was not detected in complete viral particles [136]. MALDI-MS analysis of trypsin digested 31-kDa band followed by database search using MS-Fit (peptide mass

TABLE 19-7. Adenoviral Protein Identification from SDS-PAGE by MALDI-MS

Gel Band	Calculated MW (Da)	Measured MW (Da)	MS-Fit Search Sequence Coverage (%)
II	107,876	108,074	45
III	63,292	63,338	38
IIIa	63,501	63,526	50
V	41,446	41,518	43
pVIII (31 kDa)	24,687	24,620	43
VI	22,100	22,123	32
VII	19,412	19,482	52
Propeptide of pVIII (15 kDa)	12,114	12,058	52

Source: Reprinted from reference 138, with permission of Elsevier Science B.V.

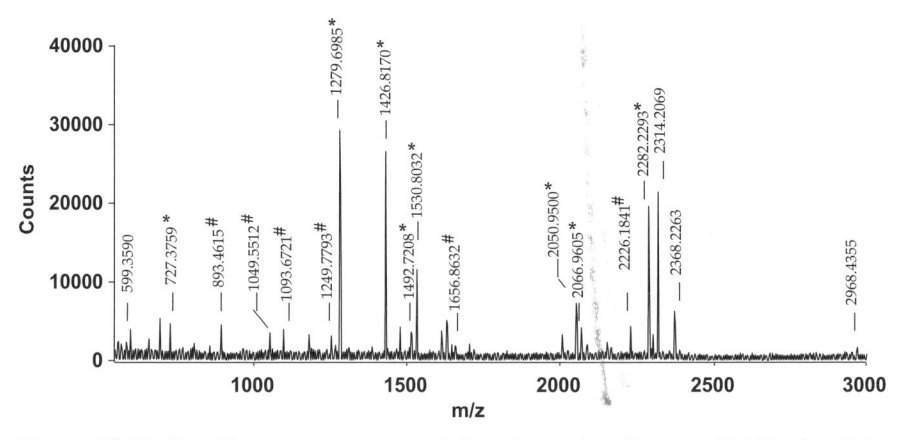

Figure 19-25. Peptide mass mapping of in-gel trypsin digested 31-kDa band by MALDI-MS. (Reprinted from reference 138, with permission of Elsevier Science B.V.)

MSKEIPTPYM	WSYQPQMGLA	AGAAQDYSTR	INYMSAGPHM	ISRVNGIRAH	RNRILLEQAA

OX
Ad protease

ITTTPRNNLN	PRSWPAALVY	QESPAPTTVV	LPRDAQAEVQ	MTNSGAQLAG G|FRHRVRSPG	

QGITHLTIRG	RGIQLNDESV	SSSLGLRPDG	TFQIGGAGRP	SFTPRQAILT	LQTSSSEPRS

GGIGTLQFIE	EFVPSVYFNP	FSGPPGHYPD	QFIPNFDAVK	DSADGYD

Figure 19-26. Sequence and sequence coverage of pVIII (matched sequences are highlighted in boldface). Oxidation of Met[101] was detected by peptide mapping. (Reprinted from reference 138, with permission of Elsevier Science B.V.)

mapping algorithm) showed that the tryptic peptides generated from the 31-kDa band matched the sequence of pVIII with 43% sequence coverage (Figures 19-25 and 19-26). In Figure 19-25, the signal with asterisk (*) corresponded to the tryptic peptides from pVIII. The remaining signals represented peptides from trypsin autolysis and background ions. The measured MW of this protein at 24,620 Da is consistent with the identity of this band as pVIII.

The presence of pVIII in empty capsid confirms that the viral particle is at an early stage of virus assembly. Based on this finding and the uniqueness of pVIII to the empty capsid of the adenovirus, an assay was developed to quantify the empty capsid contaminants by measuring the amount of pVIII detected in SDS-PAGE during the recombinant adenovirus preparation [136].

19.3.6.2 MALDI-MS Analysis of Recombinant Adenoviral Proteins Isolated from RP-HPLC. Another approach in structural analysis of adenoviral proteins is to inject intact viruses onto RP-HPLC and collect fractions for further mass analysis. Compared with SDS-PAGE assay, RP-HPLC is

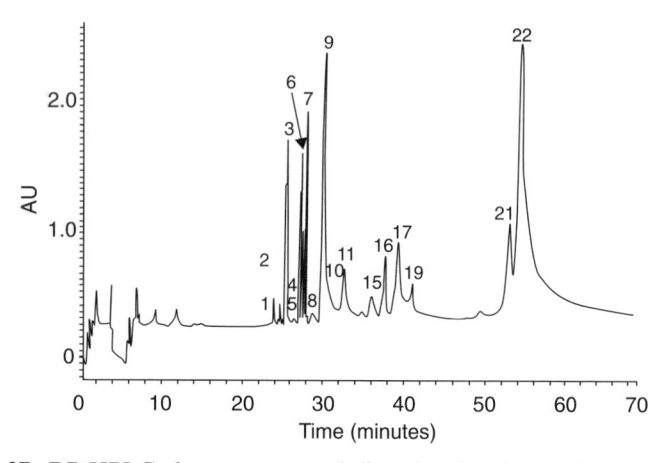

Figure 19-27. RP-HPLC chromatogram of dissociated polypeptides of recombinant adenoviruses. (Reprinted from reference 138, with permission of Elsevier Science B.V.)

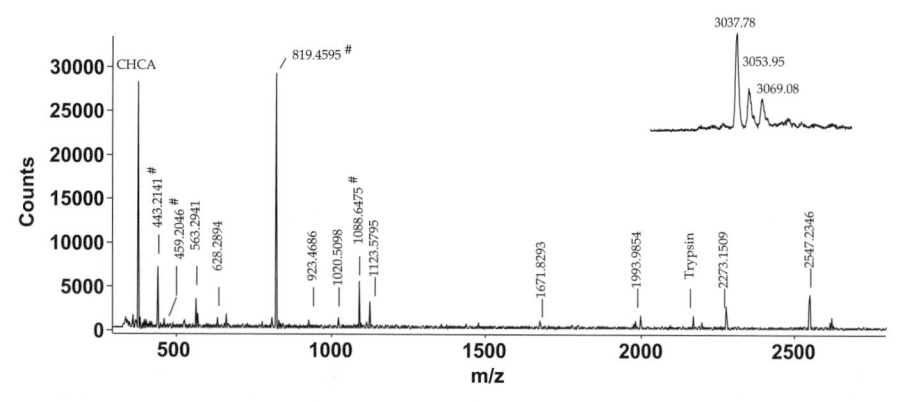

Figure 19-28. Peptide mass mapping of trypsin digested RP-HPLC fraction 2 by MALDI-MS. (Reprinted from reference 138, with permission of Elsevier Science B.V.)

highly sensitive in providing faster and reproducible results. Furthermore, it can capture the information of smaller adenoviral polypeptides generated during the maturation process of the virus. Figure 19-27 exhibits a typical RP-HPLC chromatogram from adenovirus along with 17 identified adenoviral proteins using this approach. As an example, identification of fractions 2 and 6 will be illustrated.

The isolated fraction 2 has a measured molecular weight of 3037 Da. This fraction was trypsin digested and resulted peptides were analyzed by MALDI-MS (Figure 19-28). The MS-Fit search resulted in four protein identification with low and undistinguishable MOWSE scores, including a very low MOWSE score (48.7) on an adenovirus related core protein (precursor protein X, pX). The matched peptides coverage was 15%, which is far less than the commonly

------------------------------propeptide--Late L2 MU Core Protein--
MALTCRLRFP VPGFRGRMHR   RRG<u>MAGH</u>| GLT GG| MRRAHHRR RRASHRR<u>MRG</u>
                | Ad protease
                ox

------------------------------propeptide-----------------------------
G |ILPLLIPLI AAAIGAVPGI   ASVALQAQRH

* Substrate specificity of the enzyme: (M,I,L)XGG| X or (M,I,L)XGX| G

Figure 19-29. Adenoviral proteinase cleavage sites of pX and the sequence coverage of pX by MALDI-MS and MALDI-PSD-MS studies. (Reprinted from reference 138, with permission of Elsevier Science B.V.)

acceptable coverage (about 30%). It was well-documented that for those MS-Fit database search with low MOWSE score or no hit, peptide sequence tags from MALDI-PSD studies could be used to search database with MS-Tag (sequence tag searching algorithm) for protein identification [27]. In this case, a MALDI-PSD experiment (MS/MS) was performed on the base peak of tryptic peptide ion at m/z 819.5. The obtained data were used to search against SwissProt database using MS-Tag program. Adenoviral protein pX was unambiguously identified. With additional sequence information, a new MS-Fit search was carried out and amino acid coverage of 88% was achieved, consistent with the sequence tag result.

In addition, the sequence information obtained from MALDI-PSD studies of fraction 2 indicated that the adenoviral polypeptide covers the N-terminal portion of the authentic propeptide region of pX (1–32), from amino acid residues 2 to 27 (Figure 19-29). The cleavage at positions 27/28 (MAGH/G, "/" designates cleavage site) conforms to one of the viral proteinase cleavage consensus sequences, which is (M, I, L)XGX/G (X represents any amino acid). This novel cleavage site of pX by adenoviral proteinase was not reported until very recently [140]. The detection of propeptide of pX confirms the adenoviral proteinase activity. The pX has an important role in facilitating DNA packaging into the virion. The information obtained from the studies is quite valuable for understanding the virion packaging and adenoviral proteinase activity. Methionine oxidation was also detected on the propeptide of pX (-RM(-oxidation)HR-, m/z 443.2141 versus m/z 459.2046) (Figure 19-28).

This methodology was also applied to the identification of fraction 6. The data from MALDI-MS spectrum of the tryptic digested fraction 6 were used for database search using MS-Fit, and no hit was found to be related to adenoviral proteins. The base peak of digested peptide ion at m/z 1405.6859 was selected for MALDI-PSD studies. The database search against SwissProt using MS-Tag indicates that fraction 6 is a propeptide of minor capsid protein VI precursor (pVI). The calculated mass of parent ion is 1363.6753 Da. The mass difference (42 Da) suggests an N-terminal acetylation of the propeptide of pVI, Ac-MEDINFASLAPR. Again, the presence of propeptide of pVI (1–33) is another indication of the adenoviral proteinase activity.

The study of the proteome of the recombinant adenovirus type 5 vectors demonstrated an important application of separation techniques in combination with MS methods in the drug discovery process. With completely sequenced adenovirus genome available, this approach provides a chemically well-defined method of characterization of structural proteins of recombinant adenoviral vectors. The information of protein MWs, tryptic peptide mass mapping, and sequence tags of tryptic peptides derived from HPLC/MS resulted in the identification of 17 adenoviral proteins/polypeptides in the purified virion. The rapid and accurate identification of viral proteins from recombinant adenoviruses in this study is significant since it provides direct evidence of the maturation stage of adenoviruses, which is closely related to viral infectivity and efficacy in gene therapy.

19.4 CONCLUSIONS

LC/MS analysis of proteins and peptides is an important part of drug discovery process, as illustrated in this chapter. The combination of various HPLC techniques and advanced MS methods provides unique analytical capabilities of structural identifications for therapeutic proteins and target proteins. The continuous evolution of proteomics research provides both an opportunity and a challenge for further developments in separation techniques and MS characterization methods. It is expected that these analytical techniques will continue to play important roles in drug discovery in the future.

ACKNOWLEDGMENT

The authors would like to thank Dr. John J. Piwinski for his support on the projects.

REFERENCES

1. B. N. Pramanik, P. L. Bartner, and G. Chen, The role of mass spectrometry in the drug discovery process, *Curr. Opin. Drug Discovery Dev.* **2** (1999), 401–417.
2. S. Nanda and J. M. Bathon, Etanercept: A clinical review of current and emerging indications. *Expert Opin. Pharmacother.* **5** (2004), 1175–1186.
3. N. J. Olsen and C. M. Stein, New drugs for rheumatoid arthritis, *N. Engl. J. Med.* **350** (2004), 2167–2179.
4. E. H. Fischer, Cellular regulation by protein phosphorylation, *Bull. Inst. Pasteur* **81** (1983), 7–31.

5. J. Axford, The impact of glycobiology on medicine, *Trends Immunol.* **22** (2001), 237–239.

6. W. J. Wedemeyer, E. Welker, M. Narayan, and H. A. Scheraga, Disulfide bonds and protein folding, *Biochemistry* **39** (2000), 4207–4216.

7. J. Godovac-Zimmermann and L. R. Brown, Perspectives for mass spectrometry and functional proteomics, *Mass Spectrom. Rev.* **20** (2001), 1–57.

8. R. W. A. Oliver, *HPLC of Macromolecules: A Practical Approach*, IRL Press at Oxford University Press, Oxford, 1989.

9. K. Gooding and F. Regnier, *HPLC of Biological Macromolecules*, Marcel Dekker, New York, 1990.

10. G. B. Irvine, High-performance size-exclusion chromatography of peptides, *J. Biochem. Biophys. Methods* **56** (2003), 233–242.

11. T. Kawai, K. Saito, and W. Lee, Protein binding to polymer brush, based on ion-exchange, hydrophobic, and affinity interactions, *J. Chromatogr. B Analyst. Technol. Biomed. Life Sci.* **790** (2003), 131–142.

12. P. D. McDonald, Improving our understanding of reversed-phase separations for the 21st century, *Adv. Chromatogr.* **42** (2003), 323–375.

13. J. A. Queiroz, C. T. Tomaz, and J. M. Cabral, Hydrophobic interaction chromatography of proteins. *J. Biotechnol.* **87** (2001), 143–159.

14. M. Girard and N. Mousseau, Analysis of human growth hormone by hydrophobic interaction HPLC. Method development, validation, and comparative selectivity to reversed phase HPLC, *J. Liq. Chrom. Relat. Technol.* **22** (1999), 1103–1116.

15. N. E. Labrou, Design and selection of ligands for affinity chromatography, *J. Chromatogr. B Analyt. Technol. Biomed. Life Sci.* **790** (2003), 67–78.

16. R. G. Cooks, G. Chen, and P. Wong, Mass spectrometers, in G. L. Trigg (ed.), *Encyclopedia of Applied Physics*, VCH Publishers, New York, 1997, pp. 289–330.

17. J. B. Fenn, M. Mann, C. K. Meng, S. F. Wong, and C. M. Whitehouse, Electrospray ionization for mass spectrometry of large biomolecules, *Science* **246** (1989), 64–71.

18. B. N. Pramanik, A. K. Ganguly, and M. L. Gross, *Applied Electrospray Mass Spectrometry*, Marcel Dekker, New York, 2002.

19. F. Hillenkamp, M. Karas, R. C. Beavis, and B. T. Chait, Matrix-assisted laser desorption/ionization mass spectrometry of biopolymers, *Anal. Chem.* **63** (1991), 1193A–1203A.

20. K. Tanaka, H. Waki, Y. Ido, S. Akita, Y. Yoshida, and T. Yoshida, Protein and polymer analyses up to m/z 100,000 by laser ionization time-of-flight mass spectrometry, *Rapid Commun. Mass Spectrom.* **2** (1988), 151–153.

21. R. D. Smith, J. A. Loo, C. G. Edmonds, C. J. Barinaga, and H. R. Udseth, New developments in biochemical mass spectrometry: Electrospray ionization, *Anal. Chem.* **62** (1990), 882–899.

22. F. W. McLafferty, Tandem mass spectrometry, *Science* **214** (1981), 280–287.

23. K. Hirayama, R. Takahashi, S. Akashi, K. Fukuhara, N. Oouchi, A. Murai, M. Arai, S. Murao, K. Tanaka, and I. Nojima, Primary structure of paim I, an alpha-amylase inhibitor from *Streptomyces corchorushii*, determined by the combination of Edman degradation and fast atom bombardment mass spectrometry, *Biochemistry* **26** (1987), 6483–6488.

24. W. J. Henzel, T. M. Billeci, J. T. Stults, S. C. Wong, C. Grimley, and C. Watanable, Identifying proteins from two-dimensional gels by molecular mass searching of peptide fragments in protein sequence databases, *Proc. Natl. Acad. Sci. USA* **90** (1993), 5011–5015.

25. N. L. Kelleher, Top-down proteomics, *Anal. Chem.* **76** (2004), 197A–203A.

26. D. F. Hunt, J. R. Yates III, J. Shabanowitz, S. Winston, and C. R. Hauer, Protein sequencing by tandem mass spectrometry, *Proc. Natl. Acad. Sci. USA* **83** (1986), 6233–6237.

27. J. R. Yates III, Mass spectrometry and the age of the proteome, *J. Mass Spectrom.* **33** (1998), 1–19.

28. P. Roepstorff and J. Fohlman, Proposal for a common nomenclature for sequence ions in mass spectra of peptides, *Biomedical Mass Spectrom.* **11** (1984), 601.

29. R. W. Nelson, The use of bioreactive probes in protein characterization, *Mass Spectrom. Rev.* **16** (1997), 353–376.

30. D. Metcalf, G. R. Johnson, and A. W. Burgess, Direct stimulation by purified GM-CSF of the proliferation of multipotential and erythroid precursor cells, *Blood* **55** (1980), 138–147.

31. A. Tsarbopoulos, B. N. Pramanik, J. Labdon, P. Reichert, G. Gitlin, S. Patel, V. Sardana, T. L. Nagabhushan, and P. P. Trotta, Isolation and characterization of a resistant core peptide of recombinant human granulocyte-macrophage colony-stimulating factor (GM-CSF); confirmation of the GM-CSF amino acid sequence by mass spectrometry, *Protein Sci.* **2** (1993), 1948–1958.

32. P. C. Andrews and J. E. Dixon, A procedure for *in situ* alkylation of cysteine residue on glass fiber prior to protein microsequence analysis, *Anal. Biochem.* **161** (1987), 524–528.

33. R. Bordens, S. E. Grossberg, P. P. Trotta, and T. L. Nagabhushan, Molecular and biologic characterization of recombinant interferon-α-2b, *Semin. Oncol.* **24** (1997), S9–S51.

34. M. Streuli, S. Nagata, and C. Weissmann, At least three human type alpha interferons: Structure of alpha 2, *Science* **209** (1980), 1343–1347.

35. B. N. Pramanik, A. Tsarbopoulos, J. E. Labdon, P. P. Trotta, and T. L. Nagabhushan, Structural analysis of biologically active peptides and recombinant proteins and their modified counterparts by mass spectrometry, *J. Chromatogr.* **562** (1991), 377–389.

36. E. P. Marbach and M. H. Weil, Rapid enzymatic measurement of blood lactate and pyruvate: Use and significance of metaphosphoric acid as a common precipitant, *Clin. Chem.* **13** (1967), 314–325.

37. G. W. Becker, P. M. Tackett, W. W. Bromer, D. S. Lefeber, and R. M. Riggin, Isolation and characterization of a sulfoxide and a desamido derivative of biosynthetic human growth hormone, *Biotechnol. Appl. Biochem.* **10** (1988), 326–337.

38. K. Rose, M. G. Simona, L. A. Savoy, P. O. Regamey, B. N. Green, G. M. Clore, A. M. Gronenborn, and P. T. Wingfield, Pyruvic acid is attached through its central carbon atom to the amino terminus of the recombinant DNA-derived DNA-binding protein Ner of Mu, *J. Biol. Chem.* **267** (1992), 19101–19106.

39. D. Prome, Y. Blouquit, C. Ponthus, J. C. Prome, and J. Rosa, Structure of the human adult hemoglobin minor fraction A1b by electrospray and secondary ion mass

spectrometry. Pyruvic acid as amino-terminal blocking group, *J. Biol. chem.* **266** (1991), 13050–13054.

40. R. Apweiler, H. Hermjakob, and N. Sharon, On the frequency of protein glycosylation, as deduced from analysis of the SWISS-PROT database, *Biochim. Biophys. Acta* **1473** (1999), 4–8.

41. B. S. Wojczyk, M. Stwora-Wojczyk, S. Shakin-Eshelman, W. H. Wunner, and S. L. Spitalnik, The role of site-specific *N*-glycosylation in secretion of soluble forms of rabies virus glycoprotein, *Glycobiology* **8** (1998), 121–130.

42. M. V. Dwek, H. A. Ross, and A. J. Leathem, Proteome and glycosylation mapping identifies post-translational modifications associated with aggressive breast cancer, *Proteomics* **1** (2001), 756–762.

43. P. M. Rudd, T. Elliott, P. Cresswell, I. A. Wilson, and R. A. Dwek, Glycosylation and the immune system, *Science* **291** (2001), 2370–2376.

44. R. Peracaula, G. Tabares, L. Royle, D. J. Harvey, R. A. Dwek, P. M. Rudd, and R. de Llorens, Altered glycosylation pattern allows the distinction between prostate-specific antigen (PSA) from normal and tumor origins, *Glycobiology* **13** (2003), 457–470.

45. R. Peracaula, L. Royle, G. Tabares, G. Mallorqui-Fernandez, S. Barrabes, D. J. Harvey, R. Dwek, P. M. Rudd, and R. de Llorens, Glycosylation of human pancreatic ribonuclease: Differences between normal and tumor states, *Glycobiology* **13** (2002), 227–244.

46. P. M. Rudd, A. H. Merry, M. R. Wormald, and A. R. Dwek, Glycosylation and prion protein, *Curr. Opin. Struct. Biol.* **12** (2002), 578–586.

47. M. Butler, D. Quelhas, A. J. Critchley, H. Carchon, H. F. Hebestreit, R. G. Hibbert, L. Vilarinto, E. Teles, G. Matthijs, E. Scholler, P. Argibay, D. J. Harvey, R. A. Dwek, J. Jaeken, and P. M. Rudd, Detailed glycan analysis of serum glycoproteins of patients with congenital disorders of glycosylation indicates the specific defective glycan processing step and provides an insight into pathogenesis, *Glycobiology* **13** (2003), 601–622.

48. R. Kornfeld and S. Kornfeld, Assembly of asparagines-linked oligosaccharides, *Annu. Rev. Biochem.* **54** (1985), 631–664.

49. H. Schachter, Glycoproteins, in J. Montreuil, J. F. G. Vliegenthart, and H. Schachter (eds.), *New Comprehensive Biochemistry*, Vol. 29a, Elsevier, Amsterdam, 1995, pp. 123–454.

50. H. J. An, T. R. Peavy, J. L. Hedrick, and C. B. Lebrilla, Determination of *N*-glycosylation sites and site heterogeneity in glycoproteins, *Anal. Chem.* **75** (2003), 5628–5637.

51. M. J. Huddleston, M. F. Bean, and S. A. Carr, Collisional fragmentation of glycopeptides by electrospray ionization LC/MS and LC/MS/MS: Methods for selective detection of glycopeptides in protein digests, *Anal. Chem.* **65** (1993), 877–884.

52. S. A. Carr, M. J. Huddleston, and M. F. Bean, Selective identification and differentiation of N- and O-linked oligosaccharides in glycoproteins by liquid chromatography–mass spectrometry, *Protein Sci.* **2** (1993), 183–196.

53. P. T. Jedrezejewski and W. D. Lehmann, Detection of modified peptides in enzyme digests by capillary liquid chromatography/electrospray mass spectrometry and a programmable skimmer CID acquisition routine, *Anal. Chem.* **69** (1997), 294–301.

54. J. Colangelo, V. Licon, J. Benen, J. Visser, C. Bergmann, and R. Orlando, Characterization of the N-linked glycosylation site of recombinant pectate lyase, *Rapid Commun. Mass Spectrom.* **13** (1999), 2382–2387.

55. A. W. Guzzetta, L. J. Basa, W. S. Hancock, B. A. Keyt, and W. F. Bennett, Identification of carbohydrate structures in glycoprotein peptide maps by the use of LC/MS with selected ion extraction with special reference to tissue plasminogen activator and a glycosylation variant produced by site directed mutagenesis, *Anal. Chem.* **65** (1993), 2953–2962.

56. I. Mazsaroff, W. Yu, B. D. Kelley, and J. E. Vath, Quantitative comparison of global carbohydrate structures of glycoproteins using LC-MS and in-source fragmentation, *Anal. Chem.* **69** (1997), 2517–2524.

57. E. K. Fridriksson, A. Beavil, D. Holowka, H. J. Gould, B. Baird, and F. W. McLafferty, Heterogeneous glycosylation of immunoglobulin E constructs characterization by top-down high-resolution 2-D mass spectrometry, *Biochemistry* **39** (2000), 3369–3376.

58. P. Teng-umnuay, H. R. Morris, A. Dell, M. Panico, T. Paxton, and C. M. West, The cytoplasmic F-box binding protein SKP1 contains a novel pentasaccharide linked to hydroxyproline in *Dictyostelium, J. Biol. Chem.* **273** (1998), 18242–18249.

59. T. Kurahashi, A. Miyazaki, Y. Murakami, S. Suwan, T. Franz, M. Isobe, N. Tani, and H. Kai, Determination of a sugar chain and its linkage site on a glycoprotein TIME-EA4 from silkworm diapause eggs by means of LC-ESI-Q-TOF-MA and MS-MS, *Bioorg. Med. Chem.* **10** (2002), 1703–1710.

60. J. F. Nemeth, G. P. Hochensang, L. J. Marnett, and R. M. Caprioli, Characterization of the glycosylation sites in cyclooxygenase-2 using mass spectrometry, *Biochemistry* **40** (2001), 3109–3116.

61. J. Gonzalez, T. Takao, H. Hori, V. Besada, R. Rodriguez, G. Padron, and Y. Shimonishi, A method for determination of N-glycosylation sites in glycoproteins by collision-induced dissociation analysis in fast atom bombardment mass spectrometry: Identification of the positions of carbohydrate-linked asparagine in recombinant α-amylase by treatment with peptide-N-glycosidase F in ^{18}O-labeled water, *Anal. Biochem.* **205** (1992), 151–158.

62. B. Kuster and M. Mann, ^{18}O-Labeling of N-glycosylation sites to improve the identification of gel-separated glycoproteins using peptide mass mapping and database searching, *Anal. Chem.* **71** (1999), 1431–1440.

63. L. Xiong and F. Regnier, Use of a lectin affinity selector in the search for unusual glycosylation in proteomics, *J. Chromatgr. B* **782** (2002), 405–418.

64. H. Kaji, H. Saito, Y. Yamauchi, T. Shinkawa, M. Taoka, J. Hirabayashi, K. Kasai, N. Takahashi, and T. Isobe, Lectin affinity capture, isotope-coded tagging and mass spectrometry to identify N-linked glycoproteins, *Nature Biotech.* **21** (2003), 667–672.

65. J. Bunkenborg, B. J. Pilch, A. V. Podteleinikov, and J. R. Wisniewski, Screening for N-glycosylated proteins by liquid chromatography mass spectrometry, *Proteomics* **4** (2004), 454–465.

66. G. Reid, J. L. Stephenson, Jr., and S. A. McLuckey, Tandem mass spectrometry of ribonuclease A and B: N-linked glycosylation site analysis of whole protein ions, *Anal. Chem.* **74** (2002), 577–583.

67. A. Zlotnik, J. Ransom, G. Frank, M. Fischer, and M. Howard, Interleukin 4 is a growth factor for activated thymocytes: Possible role in T-cell ontogeny, *Proc. Natl. Acad. Sci. USA* **84** (1987), 3856–3860.

68. H. V. Le, L. Ramanathan, J. E. Labdon, C. Mays-Ichinco, R. Syto, N. Arai, P. Hoy, Y. Takebe, T. L. Nagabhushan, and P. P. Trotta, Isolation and characterization of multiple variants of recombinant human interleukin 4 expressed in mammalian cells, *J. Biol. Chem.* **263** (1988), 10817–10823.

69. A. Tsarbopoulos, B. N. Pramanik, and T. L. Nagabhushan, Structural analysis of the CHO-derived interleukin-4 by liquid-chromatography/electrospray ionization mass spectrometry, *J. Mass Spectrom.* **30** (1995), 1752–1763.

70. P. Cohen, The regulation of protein function by multisite phosphorylation—a 25 year update, *Trends Biochem. Sci.* **25** (2000), 596–601.

71. J. D. Graves and E. D. Krebs, Protein phosphorylation and signal transduction, *Pharmacol. Ther.* **82** (1999), 111–121.

72. T. Hunter, Signaling—2000 and beyond, *Cell* **100** (2000), 113–127.

73. P. Blume-Jensen and T. Hunter, Oncogenic kinase signaling, *Nature* **411** (2001), 355–365.

74. P. Cohen, Timeline: Protein kinases—The major drug targets of the twenty-first centry? *Nature Rev. Drug Discovery* **1** (2002), 309–315.

75. C. C. Kumar and V. Madison, Drugs targeted against protein kinases, *Expert Opin. Emerging Drugs* **6** (2001), 303–315.

76. T. Covey, B. Shushan, R. Bonner, W. Schroder, and F. Hucho, in H. Jornvall, J.-O. Hoog, and A.-M. Gustavsson (eds.), *Methods in Protein Sequence Analysis*, Birkhauser Verlag, Basel, 1991, p. 249.

77. E. J. Chang, V. Archambault, D. T. McLachlin, A. N. Krutchinsky, and B. T. Chait, Analysis of protein phosphorylation by hypothesis-driven multi-stage mass spectrometry, *Anal. Chem.* **76** (2004), 4472–4483.

78. M. J. Huddleston, R. S. Annan, M. F. Bean, and S. A. Carr, Selective detection of phosphopeptides in complex mixtures by electrospray liquid chromatography/ mass spectrometry, *J. Am. Soc. Mass Spectrom.* **4** (1993), 710–717.

79. R. S. Annan, M. J. Huddleston, R. Verma, R. J. Deshaies, and S. A. Carr, A multidimensional electrospray MS-based approach to phosphopeptide mapping, *Anal. Chem.* **73** (2001), 393–404.

80. G. Neubauer and M. Mann, Mapping of phosphorylation sites of gel-isolated proteins by nanoelectrospray tandem mass spectrometry: Potentials and limitations, *Anal. Chem.* **71** (1999), 235–242.

81. J. Schlossmann, A. Ammendola, K. Ashman, X. Zong, A. Huber, G. Neubauer, G. X. Wang, H. D. Allescher, M. Korth, M. Wilm, F. Hofmann, and P. Ruth, Regulation of intracellular calcium by a signalling complex of IRAG, IP_3 receptor and cGMP kinase I^β, *Nature* **404** (2000), 197–201.

82. H. Steen, B. Kuster, and M. Mann, Quadrupole time-of-flight versus triplequadrupole mass spectrometry for the determination of phosphopeptides by precursor ion scanning, *J. Mass Spectrom.* **36** (2001), 782–790.

83. S. D. Shi, M. E. Hemling, S. A. Carr, D. M. Horn, I. Lindh, and F. W. McLafferty, Phosphopeptide/phosphoprotein mapping by electron-capture dissociation mass spectrometry, *Anal. Chem.* **73** (2001), 19–22.

84. L. Andersson and J. Porath, Isolation of phosphoproteins by immobilized metal (Fe^{3+}) affinity chromatography, *Anal. Biochem.* **154** (1986), 250–254.

85. M. C. Posewitz and P. Tempst, Immobilized gallium (III) affinity chromatography of phosphopeptides, *Anal. Chem.* **71** (1999), 2883–2892.

86. S. Ficarro, M. L. McCleland, P. T. Stukenberg, D. J. Burke, M. M. Ross, J. Shabanowitz, D. F. Hunt, and F. M. White, Phosphoproteome analysis by mass spectrometry and its application to *Saccharomyces cerevisiae*, *Nat. Biotechnol.* **20** (2002), 301–305.

87. A. Pandey, A. V. Podtelejnikov, B. Blagoev, X. R. Bustelo, M. Mann, and H. F. Lodish, Analysis of receptor signaling pathways by mass spectrometry: Identification of Vav-2 as a substrate of the epidermal and platelet-derived growth factor receptors, *Proc. Natl. Acad. Sci. USA* **97** (2000), 179–184.

88. M. Gronborg, T. Z. Kristiansen, A. Stensballe, J. S. Andersen, O. Ohara, M. Mann, O. N. Jensen, and A. Pandey, A mass spectrometry-based proteomic approach for identification of serine/threonine-phosphorylated proteins by enrichment with phosphor-specific antibodies: Identification of a novel protein, Frigg, as a protein kinase A substrate, *Mol. Cell. Proteomics* **1** (2002), 517–527.

89. H. Zhou, J. D. Watts, and R. A. Aebersold, A systematic approach to the analysis of protein phosphorylation, *Nat. Biotechnol.* **19** (2001), 375–378.

90. Y. Oda, T. Nagasu, and B. T. Chait, Enrichment analysis of phosphorylated proteins as a tool for probing the phosphoproteome, *Nat. Biotechnol.* **19** (2001), 379–382.

91. W. J. Qian, M. B. Goshe, D. G. Camp II, L.-R. Yu, K. Tang, and R. D. Smith, Phosphoprotein isotope-coded solid-phase tag approach for enrichment and quantitative analysis of phosphopeptides from complex mixtures, *Anal. Chem.* **75** (2003), 5441–5450.

92. D. Bonenfant, T. Schmelze, E. Jacinto, J. L. Crespo, T. Mini, M. N. Hall, and P. Jenoe, Quantitation of changes in protein phosphorylation: a simple method based on stable isotope labeling and mass spectrometry, *Proc. Natl. Acad. Sci. USA* **100** (2003), 880–885.

93. D. T. Mclachlin and B. T. Chait, Improved β-elimination-based affinity purification strategy for enrichment of phosphopeptides, *Anal. Chem.* **75** (2003), 6826–6836.

94. Z. A. Knight, B. Schilling, R. H. Row, D. M. Kenski, B. W. Gibson, and K. M. Shokat, Phosphospecific proteolysis for mapping sites of protein phosphorylation, *Nat. Biotechnol.* **21** (2003), 1047–1396.

95. E. S. Lander and R. A. Weinberg, Genomics: Journey to the center of biology, *Science* **287** (2000), 1777–1782.

96. G. Manning, D. B. Whyte, R. Martinez, T. Hunter, and S. Sudarsanam, The protein kinase complement of the human genome, *Science* **298** (2002), 1912–1934.

97. S. Wang, Y. H. Liu, C. K. Smith, and B. N. Pramanik, Mass spectrometric characterization of MEK1 phosphorylation during expression in insect cells, presented at the 52nd ASMS Conference on Mass Spectrometry, May 23–27, 2004, Nashville, Tennessee.

98. R. C. Richter, D. Link, and H. M. S. Kingston, Microwave-enhanced chemistry, *Anal. Chem.* **73** (2001), 31A–37A.

99. P. Lidstrom, J. Tierney, B. Wathey, and J. Westman, Microwave-assisted organic synthesis, *Tetrahedron Lett.* **57** (2001), 9225–9283.

100. A. K. Bose, Y. H. Ing, N. Lavlinskaia, C. Sareen, B. N. Pramanik, P. L. Bartner, Y. H. Liu, and L. Heimark, Microwave enhanced Akabori reaction for peptide analysis, *J. Am. Soc. Mass Spectrom.* **13** (2002), 839–850.

101. S. Akabori, K. Ohno, and K. Narita, On the hydrazinolysis of proteins and peptides: A method for the characterization of carboxy-terminal amino acids in proteins, *Bull. Chem. Soc. Japan* **25** (1952), 214–218.

102. B. N. Pramanik, Y. H. Ing, A. K. Bose, L. K. Zhang, Y. H. Liu, S. N. Ganguly, and P. L. Bartner, Rapid cyclopeptide analysis by microwave enhanced Akabori reaction, *Tetrahedron Lett.* **44** (2003), 2565–2568.

103. L. Ngoka and M. L. Gross, Multistep tandem mass spectrometry for sequencing cyclic peptides in an ion-trap mass spectrometer, *J. Am. Soc. Mass Spectrom.* **10** (1999), 732–746.

104. J. Kuroda, T. Fukai, and T. Nomura, Collision-induced dissociation of ring-opened cyclic depsipeptides with a guanidino group by electrospray ionization/ion trap mass spectrometry, *J. Mass Spectrom.* **36** (2001), 30–37.

105. Y. H. Ing, L. K. Zhang, P. L. Bartner, A. K. Bose, and B. N. Pramanik, Rapid cyclopeptide analysis by microwave enhanced Akabori reaction. presented at the 51st ASMS conference on mass spectrometry and allied topics, Montreal, Canada, 2003.

106. B. N. Pramanik, U. A. Mirza, Y. H. Ing, Y. H. Liu, P. L. Bartner, P. C. Weber, and A. K. Bose, Microwave-enhanced enzyme reaction for protein mapping by mass spectrometry: A new approach to protein digestion in minutes, *Protein Science* **11** (2002), 2676–2687.

107. B. N. Pramanik, P. L. Bartner, U. A. Mirza, Y. H. Liu, and A. K. Ganguly, Electrospray ionization mass spectrometry for the study of non-covalent complexes: An emerging technology, *J. Mass Spectrom.* **33** (1998), 911–920.

108. A. K. Ganguly, B. N. Pramanik, G. Chen, and A. Tsarbopoulos, Detection of noncovalent complexes by electrospray ionization mass spectrometry, in B. N. Pramanik, A. K. Ganguly, and M. L. Gross (eds.), *Applied Electrospray Mass Spectrometry*, Marcel Dekker, New York, 2002, pp. 361–387.

109. D. C. Schriemer, D. R. Bundle, L. Li, and O. Hindsgaul, Micro-scale frontal affinity chromatography with mass spectrometric detection: A new method for the screening of compound libraries, *Angew. Chem. Int. Ed.* **37** (1998), 3383–3387.

110. K. Kasai, Y. Oda, M. Nishikata, and S. Ishii, Frontal affinity chromatography: Theory for its application to studies on specific interactions of biomolecules, *J. Chromatogr. Biomed. Appl.* **376** (1986), 33–47.

111. M. A. Kelly, T. J. McLellan, and P. J. Rosner, Strategic use of affinity-based mass spectrometry techniques in the drug discovery process, *Anal. Chem.* **74** (2002), 1–9.

112. Y. M. Dunayevskiy, J. J. Lai, C. Quinn, F. Talley, and P. Vouros, Mass Spectromic identification of ligands selected from combinatorial libraries using gel filtration, *Rapid Commun. Mass Spectrom.* **11** (1997), 1178–1184.

113. P. A. Wabnitz and J. A. Loo, Drug screening of pharmaceutical discovery compounds by micro-size exclusion chromatography/mass spectrometry, *Rapid Commun. Mass Spectrom.* **16** (2002), 85–91.

114. M. R. Wilkins, J. C. Sanchez, A. A. Gooley, R. D. Appel, I. Humphery-Smith, D. F. Hochstrasser, and K. L. Williams, Progress with proteome projects: Why all

proteins expressed by a genome should be identified and how to do it, *Biotechnol. Genet. Eng. Rev.* **13** (1996), 19–50.

115. M. Tyers and M. Mann, From genomics to proteomics, *Nature* **422** (2003), 193–197.

116. P. G. Righetti, A. Castagna, and B. Herbert, Prefractionation techniques in proteome analysis, *Anal. Chem.* **73** (2001), 320A–326A.

117. P. G. Righetti, E. Wenisch, A. Jungbauer, H. Katinger, and M. Faupel, Preparative purification of human monoclonal antibody isoforms in a multi-compartment electrolyser with immobiline membranes, *J. Chromatogr.* **500** (1990), 681–696.

118. M. Fountoulakis, H. Langen, C. Gray, and B. Takacs, Enrichment and purification of proteins of *Haemophilus influenzae* by chromatofocusing, *J. Chromatogr. A.* **806** (1998), 279–291.

119. M. Fountoulakis, M. F. Takacs, and B. Takacs, Enrichment of low-copy-number gene products by hydrophobic interaction chromatography, *J. Chromatogr. A.* **833** (1999), 157–168.

120. M. Fountoulakis and B. Takacs, Design of protein purification pathways: Application to the proteome of *Haemophilus influenzae* using heparin chromatography, *Protein Expr. Purif.* **14** (1998), 113–119.

121. P. H. O'Farrell, High resolution two-dimensional electrophoresis of proteins, *J. Biol. Chem.* **250** (1975), 4007–4021.

122. M. Hamdan and P. G. Righetti, Assessment of protein expression by means of 2-D gel electrophoresis with and without mass spectrometry, *Mass Spectrom. Rev.* **22** (2003), 272–284.

123. J. Klose and U. Kobalz, Two-dimensional electrophoresis of proteins: An updated protocol and implications for a functional analysis of the genome, *Electrophoresis* **16** (1995), 1034–1059.

124. G. J. Opiteck, S. R. Ramirez, J. W. Jorgensen, and M. A. Moseley, Comprehensive two-dimensional high-performance liquid chromatography for the isolation of overexpressed proteins and proteome mapping, *Anal. Biochem.* **258** (1998), 349–361.

125. W. H. McDonald and J. R. Yates III, Shotgun proteomics and biomarker discovery, *Dis. Markers* **18** (2002), 99–105.

126. W. S. Hancock, S. L. Wu, and P. Shieh, The challenges of developing a sound proteomics strategy, *Proteomics* **2** (2002), 352–359.

127. M. Geng, J. Ji, and F. E. Reginer, Signature-peptide approach to detecting proteins in complex mixtures, *J. Chromatogr. A.* **870** (2000), 295–313.

128. S. P. Gygi, B. Rist, S. A. Gerber, F. Turecek, M. H. Gelb, and R. Aebersold, Quantitative analysis of complex protein mixtures using isotope-coded affinity tags, *Nat. Biotechnol.* **17** (1999), 994–999.

129. M. Munchbach, M. Quadroni, G. Miotto, and P. James, Quantitation and facilitated *de novo* sequencing of proteins by isotopic N-terminal labeling of peptides with a fragmentation-directing moiety, *Anal. Chem.* **72** (2000), 4047–4057.

130. I. I. Stewart, T. Thomson, and D. Figeys, ^{18}O labeling: A tool for proteomics, *Rapid Commun. Mass Spectrom.* **15** (2001), 2456–2465.

131. L. J. Henry, D. Xia, M. E. Wilke, J. Deisenhofer, and R. D. Gerard, Characterization of the knob domain of the adenovirus type 5 fiber protein expressed in *Escherichia coli, J. Virol.* **68** (1994), 5239–5246.

132. M. Green and M. Pina, Biochemical studies on adenovirus multiplication, VI. properties of highly purified tumorigenic human adenoviruses and their DNA, *Proc. Natl. Acad. Sci. USA* **51** (1964), 1251–1259.

133. M. Albrechtsen and M. Heide, Purification of plant viruses and virus coat proteins by high performance liquid chromatography, *J. Virol. Methods* **28** (1990), 245–256.

134. M. Njayou and G. Quash, Purification of measles virus by affinity chromatography and by ultracentrifugation: A comparative study, *J. Virol. Methods* **32** (1991), 67–77.

135. P. W. Shabram, D. D. Giroux, A. M. Goudreau, R. J. Gregory, M. T. Horn, B. G. Huyghe, X. Liu, M. H. Nunnally, B. J. Sugarman, and S. Sutjipto, Analytical anion-exchange HPLC of recombinant type-5 adenoviral particles, *Hum. Gene Ther.* **8** (1997), 453–465.

136. G. Vellekamp, F. W. Porter, S. Sutjipto, C. Cutler, L. Bondoc, Y. H. Liu, D. Wylie, S. Cannon-Carlson, J. T. Tang, A. Frei, M. Voloch, and S. Zhuang, Empty capsids in column-purified recombinant adenovirus preparations, *Hum. Gene Ther.* **12** (2001), 1923–1936.

137. B. G. Huyghe, X. Liu, S. Sutjipto, B. J. Sugarman, M. T. Horn, H. M. Shepard, C. J. Scandella, and P. Shabram, Purification of a type 5 recombinant adenovirus encoding human p53 by column chromatography, *Hum. Gene Ther.* **6** (1995), 1403–1416.

138. Y. H. Liu, G. Vellekamp, G. Chen, U. A. Mirza, D. Wylie, B. Twarowska, J. T. Tang, F. W. Porter, S. Wang, T. L. Nagabhushan, and B. N. Pramanik, Proteomic study of recombinant adenovirus 5 encoding human p53 by matrix-assisted laser desorption/ionization mass spectrometry in combination with database search, *Int. J. Mass Spectrom.* **226** (2003), 55–69.

139. U. A. Mirza, Y. H. Liu, J. T. Tang, F. Porter, L. Bondoc, G. Chen, and B. N. Pramanik, Nagabhushan, Extraction and characterization of adenovirus proteins from sodium dodecylsulfide polyacrylamide gel electrophoresis by matrix-assisted laser desorption/ionization mass spectrometry, *J. Am. Soc. Mass Spectrom.* **11** (2000), 356–361.

140. F. Blanche, B. Monegier, D. Faucher, M. Duchesne, F. Audhuy, A. Barbot, S. Bouvier, G. Daude, H. Dubois, T. Guillemin, and L. Maton, Polypeptide composition of an adenovirus type 5 used in cancer gene therapy, *J. Chromatogr. A.* **921** (2001), 39–48.

20

LC-NMR OVERVIEW AND PHARMACEUTICAL APPLICATIONS*

Maria Victoria Silva Elipe

20.1 INTRODUCTION

The most widely used analytical separation technique for the qualitative and quantitative determination of chemical mixtures in solution in the pharmaceutical industry is high-performance liquid chromatography (HPLC). However, conventional detectors used to monitor the separation, such as UV, refractive index, fluorescence, and radioactive detectors, provide limited information on the molecular structure of the components of the mixture. Mass spectrometry (MS) and nuclear magnetic resonance (NMR) are the primary analytical techniques that provide structural information on the analytes. NMR is widely recognized as one of the most important methods of structural elucidation, but it becomes cumbersome for the analysis of complex mixtures that require time-consuming sample purification before the NMR analysis. During the last two decades, hyphenated analytical techniques have grown rapidly and have been applied successfully to many complex analytical problems in the pharmaceutical industry. The combination of separation technologies with spectroscopic techniques is extremely powerful in carrying out qualitative and quantitative analysis of unknown compounds in complex matrices in all the stages of drug discovery, development, production, and manufacturing in the pharmaceutical industry. The HPLC (or LC) and MS (LC-MS) or NMR (LC-NMR) interface increases the capability of solving

*This chapter is an update reprinted from the reference 40, reprinted with permission from Elsevier, copyright 2003.

HPLC for Pharmaceutical Scientists, Edited by Yuri Kazakevich and Rosario LoBrutto
Copyright © 2007 by John Wiley & Sons, Inc.

structural problems of mixtures of unknown compounds. LC-MS has been one of the most extensively applied hyphenated techniques for complex mixtures because MS is more compatible with HPLC and has higher sensitivity than NMR [1–3]. Recent advances in NMR technology have made NMR more compatible with HPLC and MS and have enabled LC-NMR and even LC-MS-NMR (or LC-NMR-MS or LC-NMR/MS) to become routine analytical tools in many laboratories in the pharmaceutical environment. The present chapter provides an overview of the LC-NMR and LC-MS-NMR hyphenated analytical techniques with (a) a description of their limitations together with examples of LC-NMR and LC-MS-NMR to illustrate the data generated by these hyphenated techniques and (b) extensive references toward the application in the pharmaceutical industry (drug discovery, drug metabolism, drug impurities, degradation products, natural products, food analysis, and pharmaceutical research). This chapter is not meant to imply that LC-MS-NMR will replace LC-MS, LC-NMR, or NMR techniques for structural elucidation of compounds. LC-MS-NMR together with LC-MS, LC-NMR, and NMR are techniques that should be available and applied in appropriate cases based on their advantages and limitations.

20.2 HISTORICAL BACKGROUND OF NMR

The first part of this section (Section 20.2.1) will provide the reader with historical overview of NMR and with a brief description of the most typical experiments used in NMR for the structural elucidation of organic compounds. The second part of this section (Section 20.2.2) will focus mainly on the improvements carried out in the NMR as a hyphenated analytical technique for the elucidation of organic compounds and an understanding of the need to develop LC-NMR for the analysis of complex mixtures.

20.2.1 Historical Development of NMR

In 1945 NMR signals in condensed phases were detected by the physicists Bloch [4] at Stanford and Purcell [5] at Harvard, who received the first Nobel Prize in NMR. Work on solids dominated the early years of NMR because of the limitations of the instruments and the incomplete development of theory. Work in liquids was confined to relaxation studies. A later development was the discovery of the chemical shift and the spin–spin coupling constant. In 1951 the proton spectrum of ethanol with three distinct resonances showed the potential of NMR for structure elucidation of organic compounds [6]. Scalar coupling provides information on spins that are connected by bonds. Spin decoupling or double resonance, which removes the spin–spin splitting by a second radiofrequency field, was developed to obtain information about the scalar couplings in molecules by simplifying the NMR spectrum [7]. Initial manipulation of the nuclear spin carried out by Hahn [8] was essential for further development of experiments such as insensitive nuclei enhanced by polarization transfer (INEPT) [9], which is the basis of many modern pulse sequence experiments. During the 1960s and 1970s the development of super-conducting magnets and computers improved the sensitivity and broadened the applications of the NMR spectrometers. The Fourier transform (FT) tech-nique was implemented in the instruments by Anderson and Ernst [10] in the 1960s, but it took time to become the standard method of acquiring spectra. Another milestone which increased the signal-to-noise (S/N) ratio was the dis-covery of the nuclear Overhauser effect (NOE) by Overhauser [11], which improves the S/N in less sensitive nuclei by polarization transfer. The three-fold enhancement generally observed for the weak carbon-13 (^{13}C) signals was a major factor in stimulating research on this important nuclide. Several years later, the proton–proton Overhauser effect was applied to identify protons that are within 5 Å of each other. In the 1970s Ernst [12] implemented the idea of acquiring a two-dimensional (2D) spectrum by applying two separate radiofrequency pulses with different increments between the pulses, and after two Fourier transformations the 2D spectrum was created. Two-dimensional experiments opened up a new direction for the development of NMR, and Ernst obtained the second Nobel Prize in NMR in 1991. 2D correlation exper-iments are of special value because they connect signals through bonds. Exam-ples of these correlation experiments are correlation spectroscopy (COSY) [12], total correlation spectroscopy (TOCSY) [13], heteronuclear correlation spectroscopy (HETCOR) [14], and variations. Other 2D experiments such as nuclear Overhauser effect spectroscopy (NOESY) [15] and rotating frame Overhauser effect spectroscopy (ROESY) [16] provide information on protons that are connected through space to establish molecular conforma-tions. In 1979 Müller [17] developed a novel 2D experiment that correlates the chemical shift of two spins, one with a strong and the other with weak mag-netic moment. Initially the experiment was applied to detect the weak ^{15}N nuclei in proteins, but was later modified to detect the chemical shift of ^{13}C nuclei through the detection of the protons attached directly to the carbons [18]. The heteronuclear multiple quantum correlation (HMQC) experiment gives the same data as the HETCOR, but with greater sensitivity. Heteronu-clear single quantum correlation (HSQC) [19] is another widely used experi-ment that provides the same information as the HMQC and uses two successive INEPT sequences to transfer the polarization from protons to ^{13}C or ^{15}N. Heteronuclear multiple bond correlation (HMBC) [20] experiment gives correlations through long-range couplings, which allows two and three ^{1}H–^{13}C connectivities to be observed for organic compounds. In 1981 a 2D incredible natural abundance double quantum transfer experiment (INADE-QUATE) [21] was developed and defines all the carbon–carbon bonds, thus establishing the complete carbon skeleton in a single experiment. However, due to the low natural abundance of adjacent ^{13}C nuclei, this experiment is not very practical. All of these experiments became available with the develop-ment of computers in the 1980s. With the accelerated improvements in elec-tronics, computers, and software in the 1990s, the use of the pulsed field gradients as part of the pulse sequences was developed [22] and applied to improve solvent suppression and to decrease the time required to acquire 2D experimental data.

This brief historical introduction is intended to give a simplified overview of some of the critical milestones of NMR mainly in chemical applications, excluding the innovations in the field of proteins, solid state, and magnetic resonance in clinical medicine. To find out more details, see the articles written by Emsley and Feeney [23], Shoolery [24], and Freeman [25], and their included references.

20.2.2 Historical Development of LC-NMR

As mentioned at the end of the historical development of NMR section, the development of the pulse field gradients extended the applications of NMR. One of the areas not mentioned is the hyphenated techniques. NMR is one of the most powerful techniques for elucidating the structure of organic compounds. Before undertaking NMR analysis of a complex mixture, separation of the individual components by chromatography is required. LC-MS is routinely used to analyze mixtures without prior isolation of its components. In many cases, however, NMR is needed for an unambiguous identification. Even though hyphenated LC-NMR has been known since the late 1970s [26–33], it has not been widely implemented until the last decade [34–40].

The first paper on LC-NMR was published in 1978 [26] using stop-flow to analyze a mixture of two or three known compounds. At that time, the limitations in the NMR side—for example, sensitivity, available NMR solvents, software and hardware, and resolution achieved only with sample-spinning—made direct coupling to the HPLC difficult. Watanabe and Niki [26] modified the NMR probe to make it more sensitive, introducing a thin-wall teflon tube of 1.4 mm (inner diameter) and thereby transforming it into a flow-through structure. The effective length and volume of this probe were about 1 cm and 15 μL, respectively. Two three-way valves connected this probe to the HPLC detector. This connection needed to be short to minimize broadening of the chromatographic peaks. During the stop-flow mode, the time to acquire an NMR spectrum on each peak was limited to two hours to avoid excess broadening of the remaining chromatographic peaks. The authors also mentioned that use of tetrachloroethylene or carbon tetrachloride as solvents, along with ETH-silica as a normal-phase column, limited the applications for this technique. Because solvent suppression techniques were not available at that time, the authors [26] recognized that more development was required in the software and hardware of the NMR side to include the use of reverse-phased columns and their solvents, which in turn would broaden the range of applications. A year later, Bayer et al. [27] carried out on-flow and stop-flow experiments with a different flow-probe design on standard compounds. They used normal phase columns and carbon tetrachloride as solvent. One of their observations was that the resolution of the NMR spectra in the LC-NMR system was poorer than for the uncoupled NMR system, which made the measurement of small coupling constants difficult. The first application of on-flow LC-NMR was carried out in 1980 to analyze mixtures of several jet fuel samples [28]. Deuterated chloroform and Freon-113 and normal-phase columns were the common conditions used for LC-NMR [29–33], limiting the application of this technique.

The use of reversed-phase columns in LC-NMR complicates the NMR analysis because of (1) the use of more than one protonated solvent, which will very likely interfere with the sample, (2) the change in solvent resonances during the course of the chromatographic run when using solvent gradients, and (3) small analyte signals relative to those of the solvent. In 1995 Smallcombe et al. [41] overcame these problems by developing the solvent-suppression technique, which greatly improved the quality of the spectra obtained by on-flow or stop-flow experiments. The optimization of the WET (water suppression enhanced through T1 effects) solvent suppression technique generates high-quality spectra and effectively obtains 1D on-flow and stop-flow spectra and 2D spectra for the stop-flow mode, such as WET-TOCSY, WET-COSY, WET-NOESY, and others [41].

During the last few years, more progress has been achieved by hyphenating LC-NMR to MS. The LC-NMR-MS or LC-NMR/MS (referred to as LC-MS-NMR in this chapter) has expanded the structure-solving capabilities by obtaining simultaneously MS and NMR data from the same chromatographic peak. There are some compromises that have to be taken into account because of the differences between MS and NMR, such as sensitivity, solvent compatibility, and destructive versus nondestructive technique, discussed below. LC-MS has been used for many years as a preferred analytical technique; however, with the development of electrospray ionization techniques, LC-MS has been routinely used for the analysis of complex mixtures in the pharmaceutical industry. LC-MS-NMR is a combination of LC-MS with electrospray and LC-NMR presented below.

20.3 LC-NMR

20.3.1 Introduction

The decision to use either NMR or LC-NMR for the analysis of mixtures in the pharmaceutical industry depends on factors related to their chromatographic separation and the ability of NMR to elucidate the structure of organic compounds whether hyphenated or not. The major technical considerations of LC-NMR, discussed below, are NMR sensitivity, NMR and chromatographically compatible solvents, solvent suppression, NMR flow-probe design, and LC-NMR sensitivity or compatibility of the volume of the chromatographic peak with the volume of the NMR flow cell for better detection. Figure 20-1 shows the schematic setup of the LC-NMR connected to other devices, such as radioactivity detector and MS (see Section 20.4).

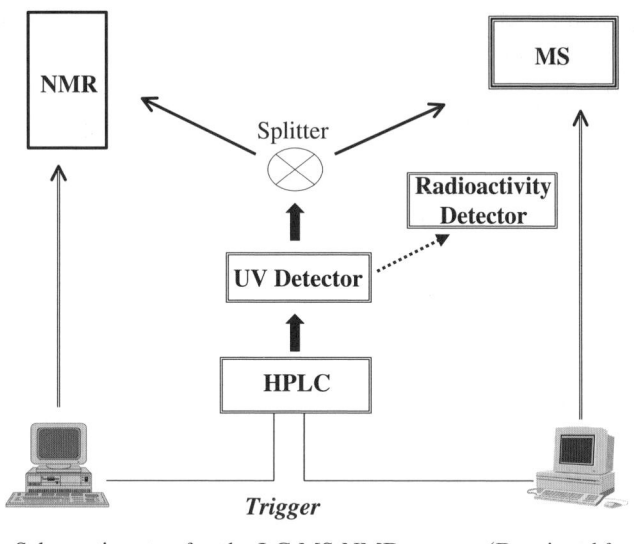

Figure 20-1. Schematic setup for the LC-MS-NMR system. (Reprinted from reference 40, copyright 2003, with permission from Elsevier.)

20.3.1.1 NMR Sensitivity. NMR is a less sensitive technique compared to MS and hence requires much larger samples for structural analysis. MS analysis is routinely carried out in the picogram range. Modern high-field NMR spectrometers (400 MHz and higher) can detect proton signals from pure demonstration samples well into the nanogram range (MW 300 Da). With the cryoprobes (for Bruker NMR instruments) or cold probes (for Varian NMR instruments), depending on the NMR vendor currently available, the sensitivity of NMR markedly improves. The samples in the low nanogram range can be detected. In the high nanogram range, structural analysis can be carried out. For real-world samples, however, purity problems become more intrusive with diminishing sample size and can be overwhelming in the submicrogram domain, even by the interference of the impurities from the deuterated solvent used for the NMR studies. This places a current practical lower limit for most structural elucidation by NMR, which is estimated by the writer to be close to 500 nanograms (MW 300 Da).

Although several other important nuclides can be detected by NMR, proton (^{1}H) remains the most widely used because of its high sensitivity, high isotopic natural abundance (99.985%), and ubiquitous presence in organic compounds. Of comparable importance is carbon (^{13}C), 1.108% abundance, which, because of substantial improvements in instrument sensitivity, is now utilized as routinely as proton. Fluorine (^{19}F), 100% abundance, is less used since it is present in only about 10% of pharmaceutical compounds. Another consequence of the intrinsic low sensitivity of NMR is that virtually all samples require signal averaging to reach an acceptable signal-to-noise level. Depending on sample size and amount of sample for the structural analysis, signal averaging may range anywhere from several minutes to several days. For metabolites in the 1- to 10-µg range, for example, overnight experiments are generally necessary.

20.3.1.2 NMR and Chromatographically Compatible Solvents. Liquid NMR requires the use of deuterated solvents. Conventionally the sample is analyzed as a solution using a 5- or 3-mm NMR tube depending on the NMR probe, which requires *ca.* 500 or 150 µL respectively of deuterated solvents. The increased solvent requirements for LC-NMR make this technique highly expensive. Deuterium oxide (D_2O) is the most readily available, reasonably priced solvent (over \$300/L). The cost of deuterated acetonitrile (CD_3CN) is decreasing and varies depending on the percentage of included D_2O, but is still over \$1000/L. Deuterated methanol (CD_3OD) is even more expensive. Deuterated solvents for normal-phase columns are not readily available, but those that are readily available have even more prohibitive prices. This necessitates the use of reversed-phase columns. Another factor to be concerned with is compatibility of the HPLC gradient-solvent system with the NMR operations. An HPLC gradient-solvent system greater than 2–3%/min causes problems in optimizing the magnetic field homogeneity (shimming) due to solvent mixing in the flow cell. A gradient-solvent system greater than 3%/min may take days for the mixture to equilibrate in the flow cell before NMR experiments can be carried out. Recently, with the new technology developments in solid-phase extraction (SPE) as SPE-NMR and capillary-based HPLC as capLC-NMR or microflow NMR (see Section 20.3.3), the amount of deuterated solvents needed is much less and is in the microliter to milliter range to pump the analyte of interest to the flow cell for the NMR analysis. These developments make the hyphenated NMR techniques economically more accessible.

20.3.1.3 Solvent Suppression. During the LC-NMR run, the solvent signal in the chromatographic peak is much larger than those of the sample and needs to be suppressed. This applies even with deuterated solvents. In the case of acetonitrile, the two ^{13}C satellite peaks of either the protonated or residual protonated methyl group for CH_3CN or CD_3CN also require suppression because they are typically much larger than signals from the sample. With the optimization of the WET solvent suppression technique by Smallcombe et al. [41] in 1995, the quality of spectra generated during LC-NMR has been greatly improved and is routine. The WET solvent suppression technique is the standard technique for LC-NMR because it has the capability of suppressing several solvent lines without minimum baseline distortions, compared with others such as presaturation or watergate. One disadvantage of suppressing the solvent lines is that any nearby analyte signal will also be suppressed, resulting in loss of structural information. With the development of SPE-NMR and capLC-NMR or microflow NMR (see Section 20.3.3), the solvent suppression is not as dramatic as for conventional LC-NMR improving the quality of the NMR data.

20.3.1.4   NMR Flow-Probe Design. Conventional NMR flow cells have an active volume of 60 μL (i.e., corresponds to the length of the receiver coil around the flow cell) and a total volume of 120 μL. This means that NMR will only "see" 60 μL of the chromatographic peak. If the flow rate in the HPLC is 1 mL/min, when 4.6-mm columns are used, only 3.6 sec of the chromatographic peak will be "seen" by NMR. Chromatographic peaks are generally much wider than 4 sec, indicating that less than half of the chromatographic peak will be detected. This is one of the disadvantages of LC-NMR compared with conventional 3-mm NMR probes where the amount of sample "seen" by the NMR receiver coil is independent of the width of the chromatographic peak. Recently, NMR flow cells with an active volume of 10, 30, 60, and 120 μL are commercially available. Applications using solid-phase extraction (SPE) as SPE-NMR will be more appropriate for 10- or 30-μL flow cells (see Section 20.3.3). Microcoil NMR flow cells for capLC-NMR or microflow NMR have an active volument of 1.5 μL for applications of samples in low concentration (see Section 20.3.3).

20.3.1.5   LC-NMR Sensitivity. Because NMR is a low-sensitivity technique, which requires samples in the order of several micrograms, analytical HPLC columns have to be saturated when injecting samples in that range. This will affect the chromatographic resolution and separation since resolution often degrades when sample injection is scaled-up to that level. Another factor that can affect chromatographic performance is the use of deuterated solvents. In many cases, analytes show broad chromatographic peaks and occasionally different retention times when using deuterated solvents due to different polarity and hydrogen bonding of deuterated versus nondeuterated solvents. When this occurs, more chromatographic development is required in order to obtain reasonable resolution. One way to increase the LC-NMR sensitivity is by decreasing the flow rate to less than 1 mL/min. At flow rate lower than 1 mL/min, a greater portion of the chromatographic peak will be "seen" by NMR. However, this is only possible if the pump of the LC system is accurate at rates lower than 1 mL/min. In the case of the SPE-NMR, the LC-NMR sensitivity can be improved by concentrating the chromatographic peak into the SPE cartridge by injecting the sample several times (see Section 20.3.3). For capLC-NMR or microflow NMR, the LC-NMR sensitivity can improved if the sample is concentrated in a volume of 5 μL.

20.3.2   Modes of Operation for LC-NMR

The HPLC is connected by red polyether ether ketone (PEEK) tubing to the NMR flow cell which is inside the magnet. With shielded cryomagnets or ultrashielded magnets the HPLC can be as close as 30–50 cm to the magnet versus

1.5–2 m for conventional magnets. Normally a UV detector is used in the HPLC system to monitor the chromatographic run. Radioactivity or fluorescent detectors can also be used to trigger the chromatographic peak(s) of interest.

There are four general modes of operation for LC-NMR: on-flow, stop-flow, time-sliced, and loop collection. These modes described below are automated by software that controls the valves of the HPLC to stop the flow when needed, depending on the mode of operation selected for LC-NMR.

20.3.2.1 On-Flow. On the on-flow or continuous-flow mode, the chromatographic run continues without stopping at any point of the run. The chromatographic peaks are flowing through the NMR flow cell while NMR spectra are being acquired. In this mode, the NMR experiments require more amount of sample to analyze "on the fly" because the resident time in the NMR flow cell is very short (3.6 sec at 1 mL/min) during the chromatographic run, which limits this approach to 1D NMR spectra acquisition only. This mode can be used to analyze the major components of the mixture and, in many cases, to rapidly identify the major known compounds of the mixture.

20.3.2.2 Stop-Flow. On the stop-flow mode, the chromatographic peak is analyzed under static conditions. The chromatographic peak of interest is submitted directly from the HPLC to the NMR flow cell. Stop-flow requires the calibration of the delay time, which is the time required for the sample to travel from the UV detector of the HPLC to the NMR flow cell, which depends in turn on the flow rate and the length of the tubing connecting the HPLC with the NMR. Because the chromatographic run is automatically stopped when the chromatographic peak of interest is in the flow cell, the amount of sample required for the analysis can be reduced compared to the on-flow mode and 2D NMR experiments, such as WET-COSY, WET-TOCSY, and others [41], can be obtained since the sample can remain inside the flow cell for days. It is possible to obtain NMR data on a number of chromatographic peaks in a series of stops during the chromatographic run without on-column diffusion that causes loss of resolution, but only if the NMR data for each chromatographic peak can be acquired in a short time (30 min or less if more than four peaks have to be analyzed, and less than two hours for the analysis of no more than three peaks). The use of commercially available cryoprobes or cold probes improves the sensitivity of the stop-flow mode (see Section 20.3.1.1).

For instance, stop-flow is the preferred mode for the analysis of metabolites when the chromatography is reasonable or the metabolite is unstable. One example is the analysis of the major metabolites of compound I (Figure 20-2), a *ras* farnesyl transferase inhibitor in rats and dogs [42]. Preliminary studies by LC-NMR using a linear solvent gradient [5–75% B 0–25 min, 75–95% B 25–35 min, A: D_2O, B: ACN (acetonitrile), 1 mL/min, 235 nm, BDS Hypersil C18 column 15 cm × 4.6 cm, 5 μm] indicated that even with the use of

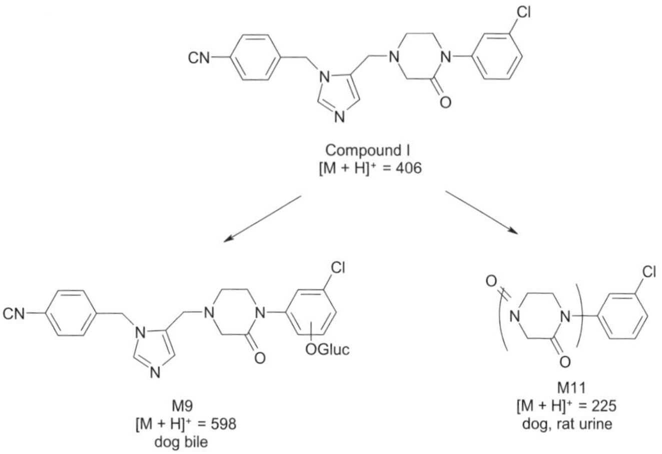

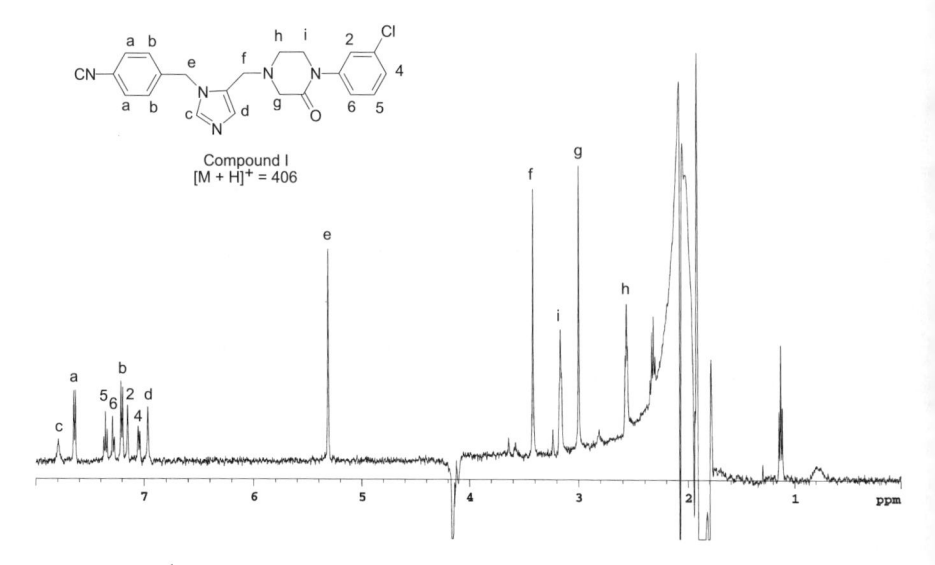

Figure 20-2. Structure of compound I, a *ras* farnesyl transferase inhibitor in rats and dogs, and proposed structures by MS of its major metabolites in dog bile (M9) and dog and rat urine (M11). (Reprinted from reference 40, copyright 2003, with permission from Elsevier.)

Figure 20-3. ¹H NMR spectrum of compound I in stop-flow. (Reprinted from reference 40, copyright 2003, with permission from Elsevier.)

protonated acetonitrile in the solvent mixture, all the resonances were visible (Figure 20-3). Figures 20-4A and 20-4B are the UV chromatograms from a small injection of dog bile and dog urine for metabolites M9 (retention time 10 min) and M11 (retention time 21 min), respectively. These small injections

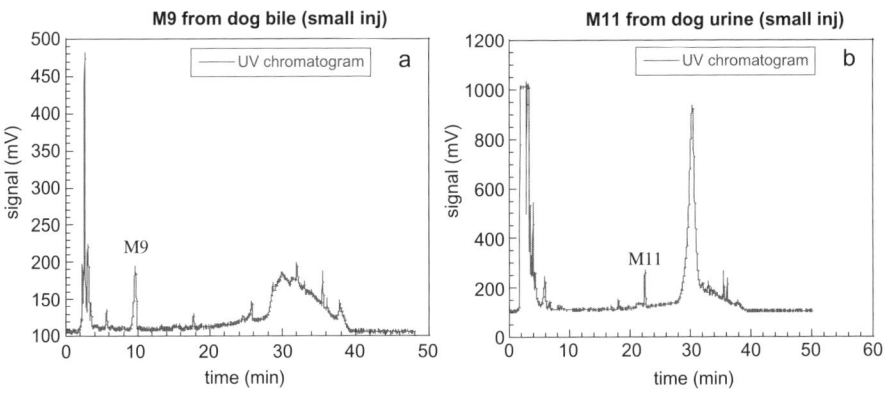

Figure 20-4. UV chromatograms from small injections of the dog bile containing metabolite M9 (A) and dog urine containing metabolite M11 (B). (Reprinted from reference 40, copyright 2003, with permission from Elsevier.)

were carried out to identify the UV chromatographic peaks of the analytes of interest to determine if there were other chromatographic peaks that could interfere the NMR studies by stop-flow. Metabolite M11 was also found in rat urine. To analyze the structures of M9 and M11 by NMR, larger injections of dog bile, dog urine, and rat urine were carried out for the stop-flow experiments. The ^{1}H NMR spectrum on the LC-NMR system (Varian Inova 500 MHz equipped with an ^{1}H–^{13}C pulse field gradient indirect detection microflow NMR probe with a 60-µL flow cell, Palo Alto, CA) of M9 (Figure 20-5) revealed the presence of a 1,2,4-trisubstituted aromatic ring in the 3-chlorophenyl ring and the glucuronide moiety. Neither of the two possibilities for the position of the glucuronide moiety ring, positions 4 or 6, could be distinguished. NOE experiments on the LC-NMR were not successful because of problems with the solvent suppression. The sample was collected and the NOE was performed (Varian Unity 400 MHz, equipped with a 3-mm ^{1}H–^{13}C pulse field gradient indirect detection Nalorac probe, Palo Alto, CA) over a weekend (Figure 20-6). Even though the collected sample contained more impurities, the NOE experiment showed that the glucuronide moiety was attached at C-4 by irradiating the methylene at i which elicited NOE signals from H-2 and H-6, thus eliminating the C-6 possibility (Figure 20-6). LC-MS on M11 indicated it to be only the 1-(3-chlorophenyl)piperazinone moiety with an additional oxidation on the piperazinone ring. The ^{1}H NMR spectrum on the LC-NMR system of M11 lacked the isolated methylene signal on the piperazine ring (Figure 20-7), indicating it to be the (1-(3-chlorophenyl)piperazine-2,3-dione).

Recently, a radioactive volatile metabolite M3 with a small molecular weight was studied using LC-NMR [43]. Conventional NMR was not possible because the radioactivity of the sample was lost when the fraction containing the metabolite was evaporated to dryness prior to the NMR studies. In this

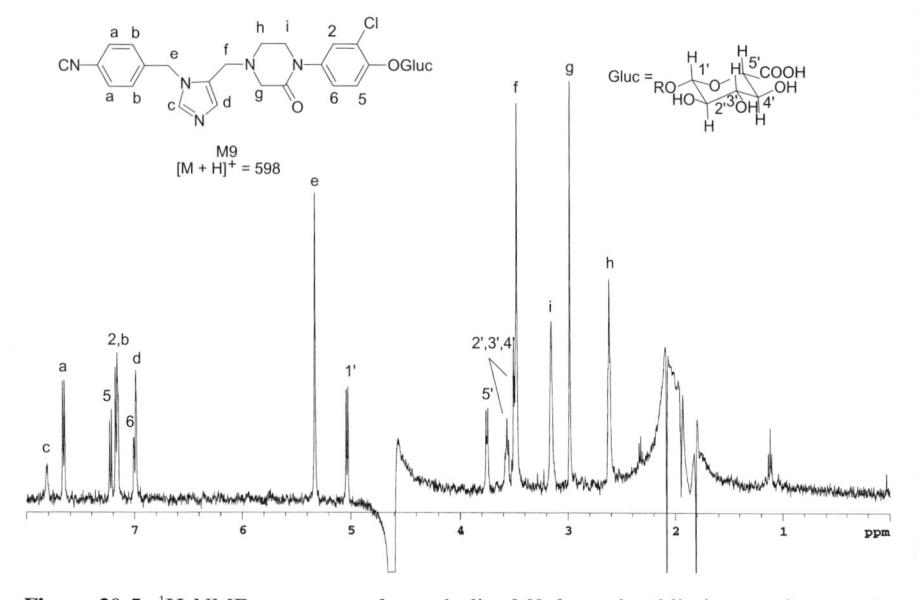

Figure 20-5. ^{1}H NMR spectrum of metabolite M9 from dog bile in stop-flow mode. (Reprinted from reference 40, copyright 2003, with permission from Elsevier.)

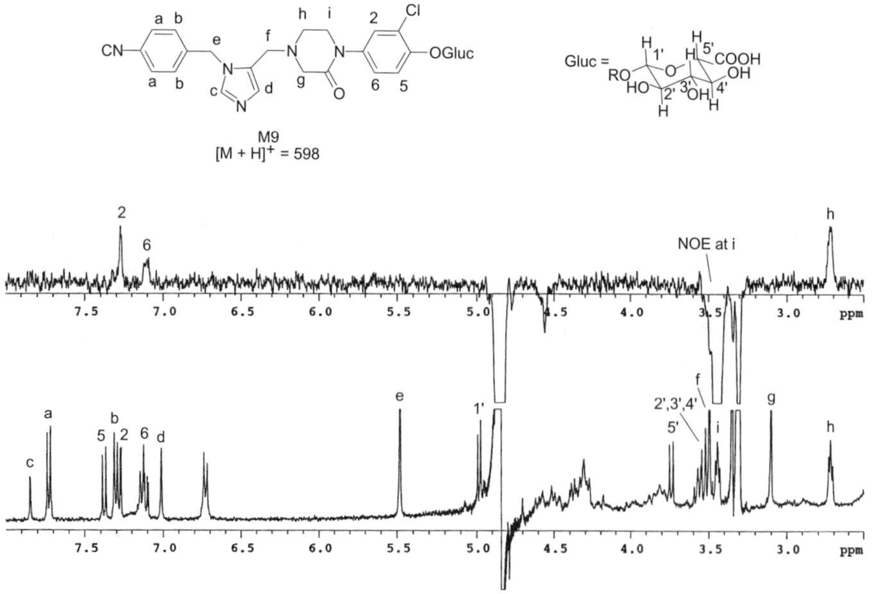

Figure 20-6. ^{1}H NMR (bottom) and 1D NOE spectra at i (top) of M9 from dog bile recovered from LC-NMR. (Reprinted from reference 40, copyright 2003, with permission from Elsevier.)

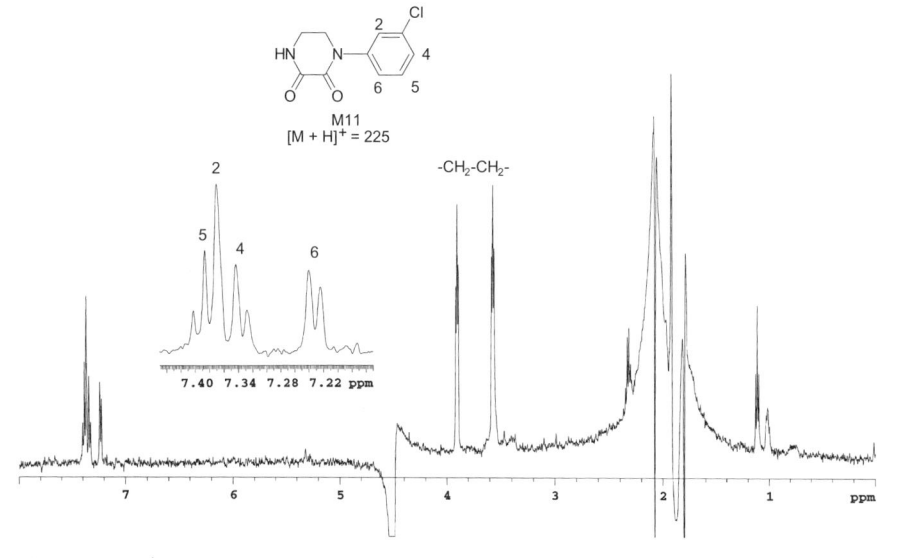

Figure 20-7. ^{1}H NMR spectrum of metabolite M11 from dog urine in stop-flow mode. (Reprinted from reference 40, copyright 2003, with permission from Elsevier.)

example, the LC-MS was not informative, suggesting a molecular weight less than 200 Da. LC-NMR was one of the alternatives used to solve this structural problem. To be able to identify the UV chromatographic peak corresponding to the radioactive metabolite, a radioactivity detector equipped with a liquid cell (Radiomatic C150TR, Packard) was connected on-line to the LC-UV system of the LC-NMR. Figure 20-1 shows the schematic diagram for this setup. Small injections were carried out initially to identify the metabolite UV chromatographic peak with the radioactive peak prior to the stop-flow experiments (Figure 20-8). Stop-flow experiments were triggered by UV because the transfer delay from the UV to the NMR was shorter than from the radioactive detector to the NMR, due to the thicker tubing used in the liquid cell of the radioactivity detector. ^{1}H NMR spectrum revealed the presence of the *p*-fluorophenyl ring with the characteristic splitting pattern, indicating that the compound was drug-related. The downfield shift of the *ortho* protons at 7.91 ppm suggested the presence of a carbonyl substituent (Figure 20-8). The presence of a singlet at 4.85 ppm, integrating for approximately two protons, was consistent with a methylene that was flanked by the carbonyl and a hydroxyl group (Figure 20-8). These features thus led to proposing the structure for M3 as the *p*-fluoro-α-hydroxyacetophenone (Figure 20-8).

20.3.2.3 Time-Sliced Mode.

The time-sliced mode involves a series of stops during the elution of the chromatographic peak of interest. A time-sliced mode is used when two analytes elute together or with close retention times, or when the separation is poor. Depending on the NMR vendor, the software can be

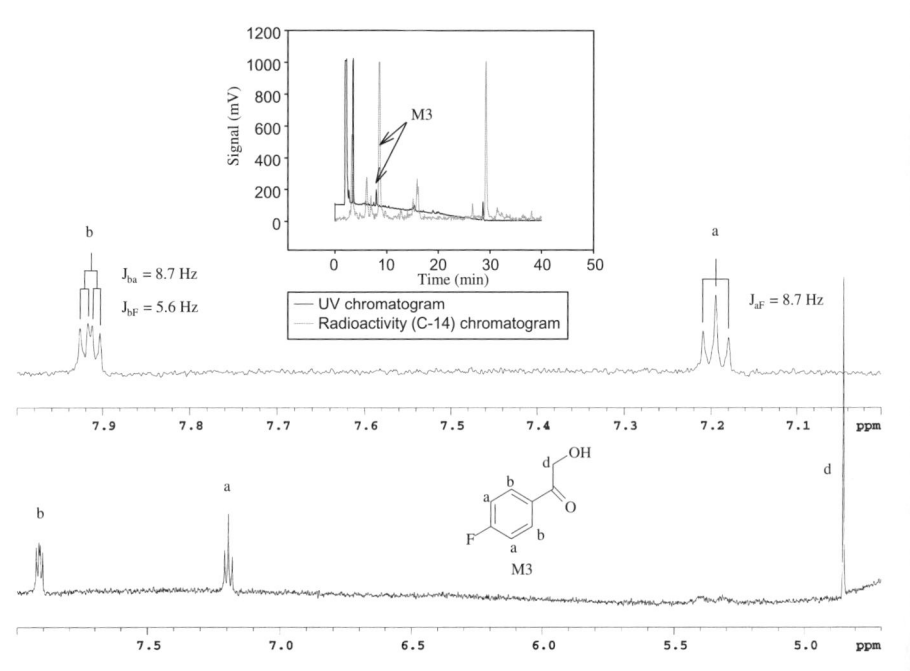

Figure 20-8. UV-radioactive (C-14) chromatograms of the fraction containing metabolite M3 (top) and expanded sections of the ^{1}H NMR spectrum of metabolite M3 acquired for one day (bottom).

designed to automate this mode, but sometimes the analyst may prefer to do it manually.

20.3.2.4 Loop Collection. On the loop collection mode, the chromatographic peaks of interest are automatically stored in loops controlled by the software for later off-line NMR study. Then the stored chromatographic peaks are transferred to the NMR flow cell individually for NMR studies. The software is designed to send the stored chromatographic peaks to the NMR flow cell in the same or different order as they were stored from the chromatographic run. Loop collection can be used when there is more than one chromatographic peak of interest in the same run. In this case the analytes must be stable inside the loops during the extended period of analysis. Capillary tubing should be used to avoid peak broadening with concomitant loss of analyte "seen" by the NMR spectrometer. Loop collection can be used in connection with SPE for SPE-NMR analysis (see Section 20.3.3).

20.3.3 Other Analytical Separation Techniques Hyphenated with NMR

Recently, other chromatographic techniques have been coupled on-line to NMR for additional applications in the pharmaceutical environment, such as size-exclusion chromatography (SEC) as SEC-NMR for the characterization of polymer additives [44], capillary electrophoresis (CE) as CE-NMR for small volume samples [45–47], capillary electrochromatography (CEC) as CEC-NMR, capillary zone electrophoresis (CZE) as CZE-NMR for on-flow identification of metabolites with small volume samples [46, 48–52], and gel-permeation chromatography (GPC) as GPC-NMR and supercritical fluid chromatography (SFC) as SCF-NMR for polymer separation and identification [53] as examples. CE-NMR and CEC-NMR are techniques that work with very small-volume NMR probes with capillary separations. Solid-phase extraction (SPE) as SPE-NMR is becoming a popular technique for trace analysis. In SPE-NMR, the chromatographic peaks are trapped into trap cartridges using multiple injections to increase the concentration of the chromatographic peaks, and then the cartridges are dried with nitrogen to remove all residual solvents. With this technique, deuterated solvents are only used to flush each peak from the cartridge to the NMR flow cell, creating a sharp eluting band (25- to 30-μL eluting volume) that requires the use of small NMR flow cells, such as 10- or 30-μL flow cells. SPE-NMR allows increasing the sensitivity compared with regular LC-NMR. The recent use of cryogenic flow probe with the SPE-NMR application improves tremendously the sensitivity of NMR [54]. SPE-NMR has been applied for trace analysis [55], microbial metabolites [56], and natural products [54, 57, 58]. Lately, more developments have been carried out to hyphenate capillary-based HPLC (capLC) with NMR as capLC-NMR or microflow NMR and the use of commercial microcoil NMR probes [46, 59–61]. With microcoil NMR probes, the range of sample used in capLC-NMR could reach the nanogram level (low nanogram level only for detection limit but not for structural analysis) [46, 59–61]. With this technique, the volume of the chromatographic peak is comparable to the volume of the microcoil NMR flow cell. The volume observed for a commercial microcoil NMR flow cell is approximately 1.5 μL, and there is a wider range of solvent gradient variation than in the standard LC-NMR. CapLC-NMR can be used without a column for analysis of low concentrated pure compounds, such as 1 μg, or with the column to study mixtures of compounds. One of the requirements for capLC-NMR is that the sample has to be soluble in a volume of approximately 5 μL or less, which is not always possible. The delay time between the UV detector of the cap-LC and the NMR flow cell has to be calibrated for all chromatographic conditions due to the changes of viscosity of the different solvent compositions, which has an effect on the pump of the cap-LC. More recently, the development of multiple coils connected in parallel may be applicable to acquire NMR data of several samples at the same time [39, 62–64]. So far, four samples can be run at the same time, but recent developments are going toward analysis of 96-well plates emulating techniques such as LC-MS [39]. CapLC-NMR with single or multiple solenoidal microcoils can also be used with other capillary techniques such as capillary electrophoresis (CE) [63, 64], capillary isotachophoresis [63, 65, 66], and others [63].

20.3.4   Applications of LC-NMR

There are many examples in the literature for applications of LC-NMR in the pharmaceutical industry. In the area of natural products, LC-NMR has been applied to screen plant constituents from crude extracts [54, 57, 67, 68] and to analyze plant and marine alkaloids [69–72], flavonoids [73], sesquiterpene lactones [74, 75], saponins [58, 76], vitamin E homologues [77], and antifungal and bacterial constituents [56, 78, 79] as examples. In the field of drug metabolism, LC-NMR has been extensively applied for the identification of metabolites [42, 80–88] and even polar [89] or unstable metabolites [43]. And finally, LC-NMR has been used for areas such degradation products [90–93], drug impurities [94–102], drug discovery [103, 104], and food analysis [105–107].

20.4   LC-MS-NMR (OR LC-NMR-MS OR LC-NMR/MS)

20.4.1   Introduction

The capability of analyzing a complex mixture in a chromatographic run by the hyphenation of several techniques, such as NMR and MS, to HPLC is becoming more popular in the pharmaceutical industry. NMR and MS data on the same analyte are crucial for structural elucidation. When different isolates such as metabolites are analyzed by NMR and MS, one cannot always be certain that the NMR and the MS data apply to the same analyte, especially when the analytes have been isolated using analytical columns and prep columns for the MS and NMR analysis, respectively. HPLC conditions are not always reproducible when analytical and prep-HPLC columns are used to isolate different amounts of the analytes of interest. To avoid this ambiguity, LC-MS and LC-NMR are combined. MS data should be obtained initially because with NMR, data collection in the stop-flow mode can take hours or days, depending on the complexity of the structure and the amount of sample. This is why it is preferable to designate this operation as LC-MS-NMR rather than LC-NMR-MS or LC-NMR/MS.

Since MS is considerably more sensitive than NMR, a splitter is incorporated after the HPLC to direct the sample to the MS and NMR units separately. In the example below, the MS used in these studies is a classic LCQ instrument (ThermoFinnigan, CA). A custom-made splitter was used with a splitting ratio of 1/100 (Acurate™, LC Packings, CA). It was designed to deliver 1% of the sample initially to the MS and the balance 20 seconds later to the NMR. With a flow rate of 1 mL/min, the final flow rate going to the NMR will be 0.990 mL/min, and the final flow rate going to the MS will be 0.010 mL/min. Electrospray is the only source of ionization that will work with such low flow rate (10 µL/min) in LCQ. Figure 20-1 depicts the scheme of the LC-MS-NMR system used in the example for this chapter. The technical considerations of LC-MS-NMR are the same as LC-NMR (see Section 20.3) plus the effect of using deuterated solvents for the MS of the LC-MS-NMR.

For the last 4–5 years, the LC-NMR-MS system has been commercially available only for the Bruker NMR instruments. For the Varian NMR instruments, the system has recently become available. The work presented here has been carried out by the author using a custom design of the LC-MS-NMR system on a Varian NMR instrument as explained above.

20.4.1.1 The Use of Deuterated Solvents. Another consideration for the LC-MS-NMR is the use of deuterated solvents needed for NMR. Analytes with exchangeable or "active" hydrogens can exchange (i.e., equilibrate) with deuterium (^2H) at different rates. The analyst should be alert to this possibility because it could result in the appearance of several closely spaced molecular ions with pseudo-molecular ions increased, depending on the number of exchangeable hydrogens being deuterated. If the compound of interest exchanges all the active hydrogens for deuteriums, the pseudo-molecular molecular ion will be $[M + {}^2H]^+$ or $[M - {}^2H]^-$ in positive or negative mode, respectively, where M is the molecular weight with all the exchangeable hydrogens deuterated. When buffers or other compatible solvents for MS are needed, it is recommendable to use deuterated buffers to avoid the suppression of additional solvent lines in the NMR spectra (see Section 20.3.1.3).

20.4.2 Modes of Operation for LC-MS-NMR

As mentioned in the section of modes of operation for LC-NMR (Section 20.3.2), with the use of shielded cryomagnets, the location of the MS instrument will follow the same rule as for the HPLC. The most common modes of operation for LC-MS-NMR are on-flow and stop-flow. With stop-flow, the MS instrument can also be used to stop the flow on the chromatographic peak of interest that is to be analyzed by NMR. These two modes are presented here with an example. In the loop collection mode, the MS of the LC-MS-NMR system may also monitor the trapping of the chromatographic peak inside the loop.

In the last few years, there have been relatively few examples in the literature dealing with the application of LC-MS-NMR in the pharmaceutical industry. The author of this chapter has been interested in evaluating this technology to determine the *pros* and *cons* and to decide which cases are suitable for this application. To illustrate these modes of operation, a group of flavonoids was chosen. Eight flavonoids were selected to mimic a real complex mixture of compounds of similar structure that may present some ambiguity in their analysis that can be resolved by this hyphenated technique versus the individual nonhyphenated techniques. Figure 20-9 shows the eight flavonoids (Aldrich) chosen for this example. These compounds have simple structures composed primarily of aromatic protons; some have low-field aliphatic protons which would not be hidden under the NMR solvent peaks. Phenolic protons exchange rapidly with D_2O, so that each compound will only show one pseudo-molecular ion. Flavonoids are natural products with important

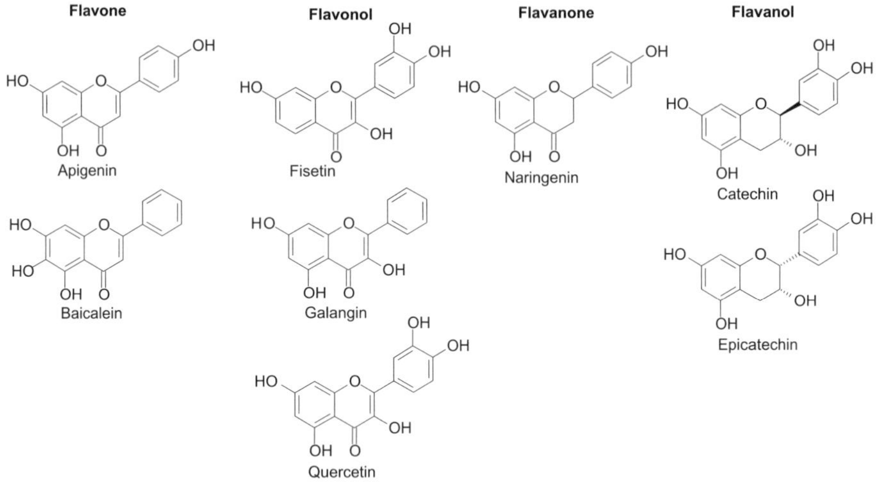

Figure 20-9. Structures of eight flavonoids used for the LC-MS-NMR technology development studies. (Reprinted from reference 40, copyright 2003, with permission from Elsevier.)

biological functions acting as antioxidants, free radical scavengers, and metal chelators and are important to the food industry.

The chromatographic conditions are as follows: 35–50% B 0–10 min, 50–80% B 10–15 min; A, D_2O; B, ACN; 1 mL/min, 287 nm, Discovery C18 column 15 cm × 4.6 cm, 5 μm. Stock solutions of each compound were prepared at 1 μg/μL in ACN : MeOH 1 : 1.

A Varian Unity Inova 600-MHz NMR instrument (Palo Alto, CA) equipped with a $^1H\{^{13}C/^{15}N\}$ pulse field gradient triple resonance microflow NMR probe (flow cell 60 μL; 3 mm O.D.) was used. Reversed-phase HPLC of the samples was carried out on a Varian modular HPLC system (a 9012 pump and a 9065 photodiode array UV detector). The Varian HPLC software was also equipped with the capability for programmable stop-flow experiments based on UV peak detection. An LCQ classic MS instrument, mentioned in the previous section, was connected on-line to the HPLC-UV system of the LC-NMR by contact closure. The ^2H resonance of the D_2O was used for field-frequency lock, and the spectra were centered on the ACN methyl resonance. Suppression of resonances from HOD and methyl of ACN and its two ^{13}C satellites was accomplished using a train of four selective WET pulses, each followed by a B_o gradient pulse and a composite 90-degree read pulse [41].

20.4.2.1 On-Flow. The on-flow experiment was carried out on a mixture of eight flavonoids (Figure 20-9) (20 μg each). MS and NMR data were obtained during this on-flow experiment. The UV chromatogram is depicted in Figure 20-10. Table 20-1 and Figures 20-12A–D show the pseudo-molecular ion information [M – $^2H]^-$, where M is the molecular weight with all the hydroxyl

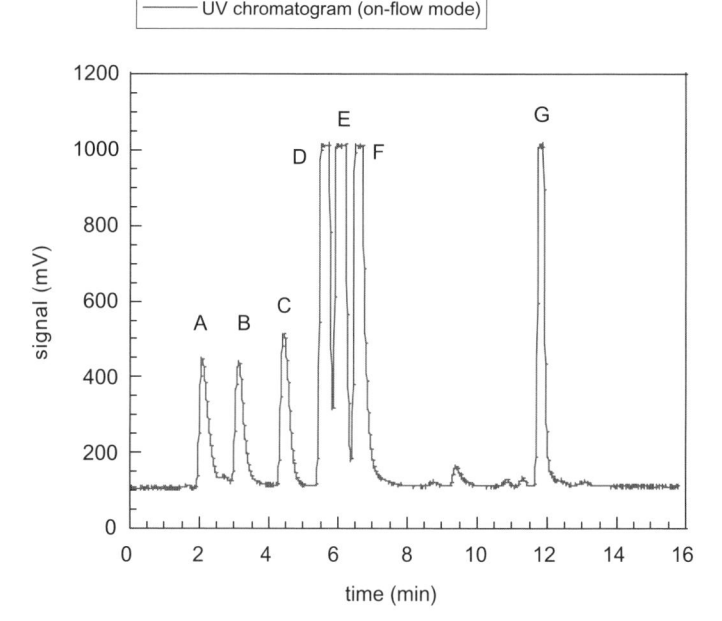

Figure 20-10. UV chromatogram of the on-flow experiment injecting a mixture of eight flavonoids (A: catechin + epicatechin; B: fisetin; C: quercetin; D: apigenin; E: naringenin; F: baicalein; G: galangin). (Reprinted from reference 40, copyright 2003, with permission from Elsevier.)

TABLE 20-1. MS Data of Flavonoids in Negative Mode from the On-Flow Run in the LC-MS-NMR

Peak	Compound	MW[a]	M[b]	m/z, [M-^{2}H]$^-$
A	Catechin + Epicatechin	290	295	293
B	Fisetin	286	290	288
C	Quercetin	302	307	305
D	Apigenin	270	273	271
E	Naringenin	272	275	273
F	Baicalein	270	273	271
G	Galangin	270	273	271

[a]Molecular weight.
[b]Molecular weight with all the hydroxyl protons deuterated.
Source: Reprinted from reference 40, copyright 2003, with permission from Elsevier.

protons deuterated, in negative mode for the eight flavonoids obtained in this on-flow experiment. Figure 20-11 is the 2D data set (time versus chemical shift) where each ^{1}H NMR spectrum was acquired for 16 scans and decreasing the delays (total time per spectrum of 20 sec) to obtain more spectra during

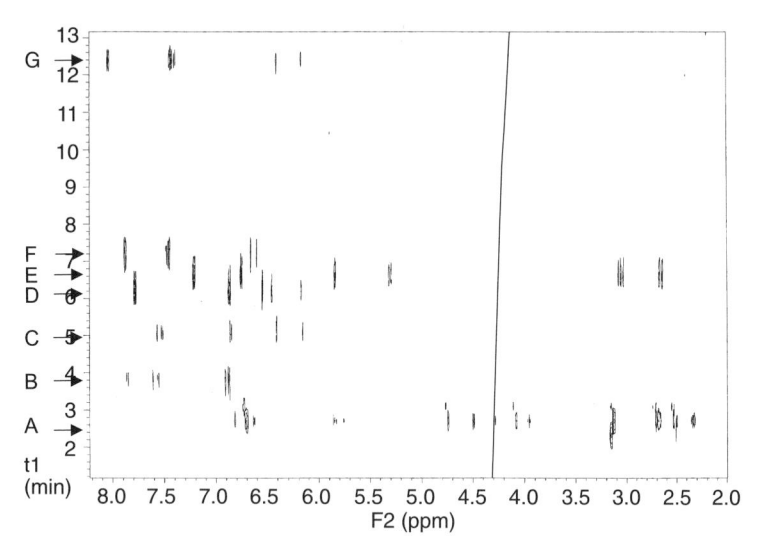

Figure 20-11. 2D data set (time/min versus chemical shift/ppm) for the on-flow experiment injecting a mixture of eight flavonoids (A: catechin + epichatechin; B: fisetin; C: quercetin; D: apigenin; E: naringenin; F: baicalein; G: galangin). (Reprinted from reference 40, copyright 2003, with permission from Elsevier.)

the chromatographic run and have more data points for the ^{1}H NMR spectra of the different components of the chromatographic run. Figures 20-12A–D depict the ^{1}H NMR traces of each flavonoid extracted from the 2D data set. Notice that catechin and epicatechin co-elute under these conditions (peak A of the UV chromatogram of Figure 20-10). Distinguishing these diastereomers by MS alone is not feasible (Table 20-1 and Figure 20-12A) because both have the same pseudo-molecular ion information. Differences in the NMR spectra would be expected and are, in fact, observed (Figure 20-12A). The ability of LC-MS-NMR to distinguish signals from the individual diastereomers is illustrated in Figures 20-11 and 20-12A. The protons H-2 and H-3 in catechin and H-2a and H-3a in epichatechin show different chemical shifts because of the slightly different local chemical environment around the chiral centers C-2 and C-3 for catechin and C-2a and C-3a for epicatechin as diasteromers. Those differences are enough for NMR to be able to distinguish well the diasteromers of organic molecules. The ^{1}H NMR spectrum of naringenin in Figure 20-12C shows the ability of NMR to analyze a mixture of two components in different ratio (X indicates the signals coming from apigenin as the minor component of this chromatographic peak). In this particular case, NMR shows clearly the presence of the two components of the mixture and MS only shows the major component. Assignments can be easily carried out based on the different ratios of the NMR signals for both compounds. This is another advantage of NMR versus MS.

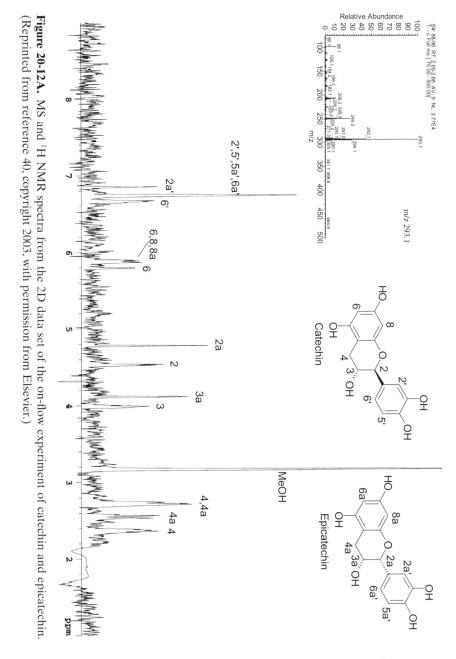

Figure 20-12A. MS and ¹H NMR spectra from the 2D data set of the on-flow experiment of catechin and epicatechin. (Reprinted from reference 40, copyright 2003, with permission from Elsevier.)

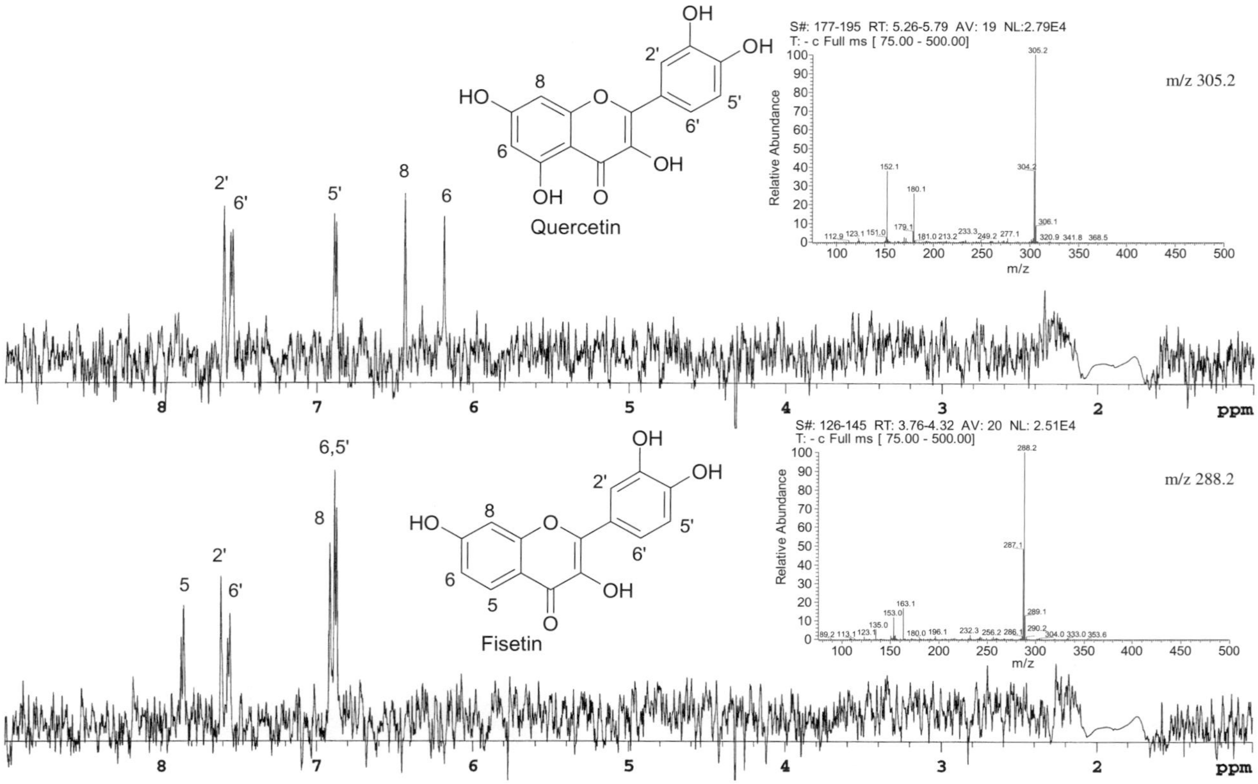

Figure 20-12B. MS and ^{1}H NMR spectra from the 2D data set of the on-flow experiment of fisetin (bottom) and quercetin (top). (Reprinted from reference 40, copyright 2003, with permission from Elsevier.)

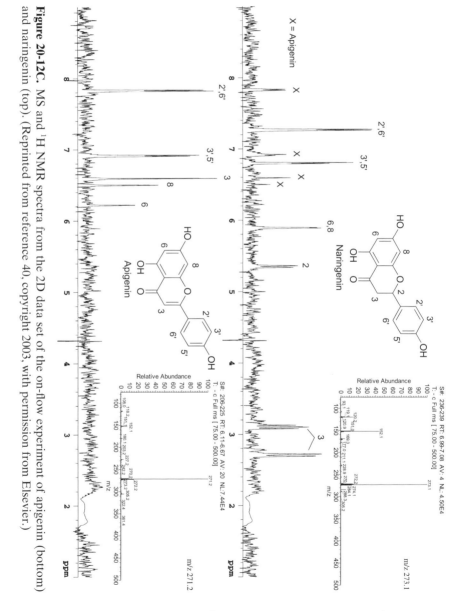

Figure 20-12C. MS and ^{1}H NMR spectra from the 2D data set of the on-flow experiment of apigenin (bottom) and naringenin (top). (Reprinted from reference 40, copyright 2003, with permission from Elsevier.)

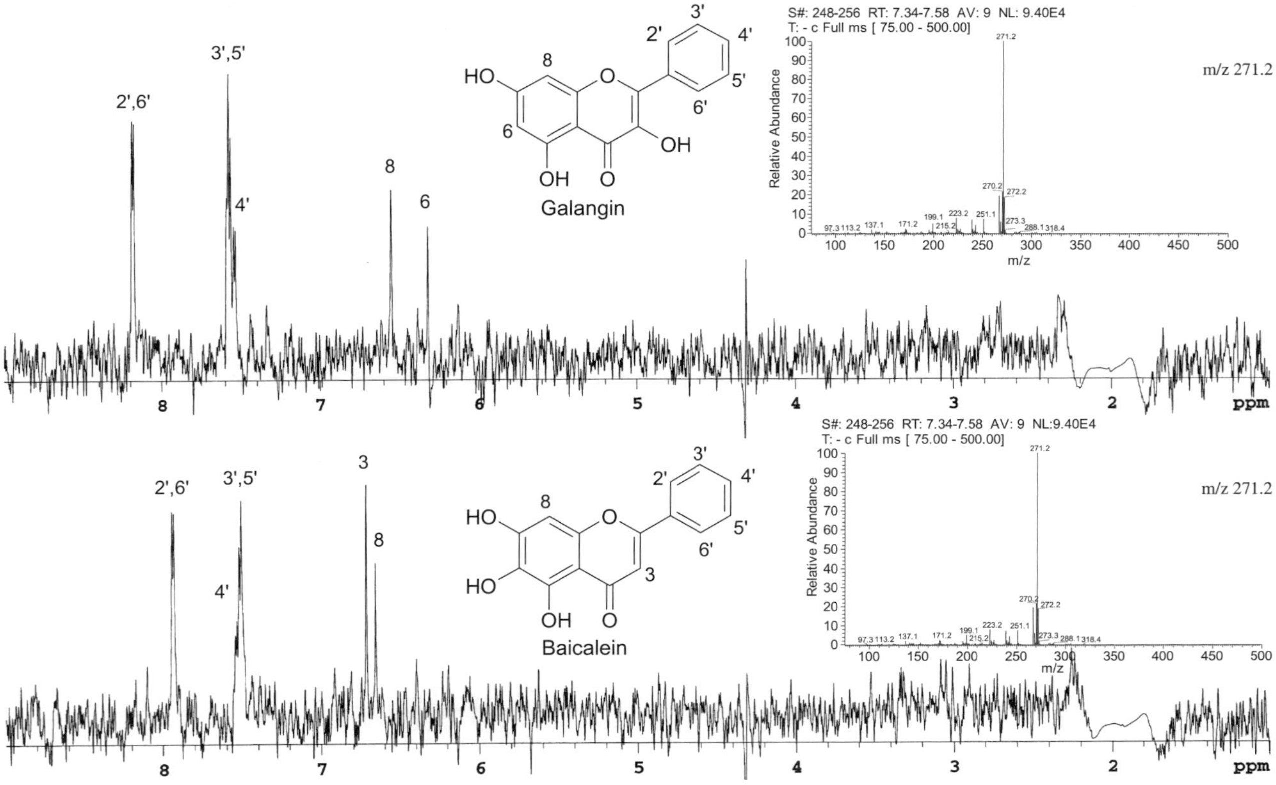

Figure 20-12D. MS and ¹H NMR spectra from the 2D data set of the on-flow experiment of baicalein (bottom) and galangin (top). (Reprinted from reference 40, copyright 2003, with permission from Elsevier.)

20.4.2.2 Stop-Flow. Two stop-flow experiments were carried out on apigenin (10 μg) (Figure 20-9) using, independently, the UV peak maximum or
the pseudo-molecular ion chromatographic peak seen in the total ion chromatogram (TIC) for the MS instrument to trigger the stop-flow. Since the
Varian software automatically triggers the stop-flow with the UV peak, this
mode was used as a reference point. When the MS was used to trigger the
stop-flow, it was carried out manually with a chronometer while monitoring
the molecular ion of apigenin in negative mode (m/z 275). After peak detection in the UV or MS and a time delay of about 52 sec or 20 sec, respectively,
the HPLC pump was stopped, trapping the peak of interest in the LC-NMR
microprobe. ^{1}H NMR stop-flow spectra were acquired using an acquisition
time of 1.5 sec, a delay between the successive pulses of 0.5 sec, a spectral width
of 9000 Hz, and 32 K time-domain data points. The methyl resonance of ACN
was referenced to 1.94 ppm. These two experiments were carried out injecting
10 μg of apigenin and acquiring ^{1}H NMR spectra for ~4.5 min (128 scans),
giving rise to the same quality of ^{1}H NMR spectra of apigenin (Figure
20-13).

These experiments indicated that for sample mixtures, the on-flow mode of
LC-MS-NMR is useful for obtaining structural information on the major components. If more detailed analysis is required, or the amount of sample is small
and the compound(s) cannot be isolated because of instability or volatility,
stop-flow is the mode of choice. LC-MS and LC-NMR chromatographic

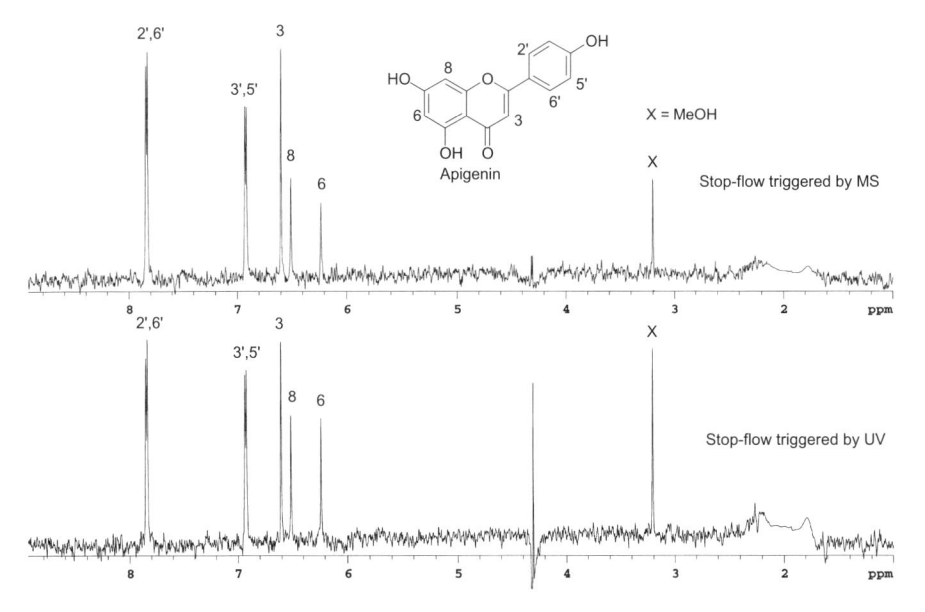

Figure 20-13. ^{1}H NMR spectra of apigenin triggering the stop-flow by UV (bottom)
and by MS (top). (Reprinted from reference 40, copyright 2003, with permission from
Elsevier.)

conditions must be compatible; in addition, prior evaluation of the LC conditions on the LC-MS-NMR system is required to ensure consistency with the chromatographic resolution needed in the LC-NMR part of the system. The sample must ionize well by electrospray to obtain MS data. When stop-flow mode is triggered by MS, prior MS information of the chromatographic peak(s) of interest is needed in deuterated solvent(s) to evaluate the suitability of the system to provide structural information.

20.4.3 Applications of LC-MS-NMR

There are examples in the literature for the application of LC-MS-NMR in the pharmaceutical industry. In the area of natural products, this technique has been applied as a rapid screening method of searching unknown marine natural products in chromatographic fractions [108] and for the separation and characterization of natural products from plant origin [109, 110]. Another application is in the area of combinatorial chemistry [111]. In the field of drug metabolism, LC-MS-NMR has been extensively applied for the identification of metabolites [112–120]. And finally, LC-MS-NMR has been used for areas such pharmaceutical research [35, 121, 122], drug discovery [123], degradation products [101], and food analysis [124, 125].

20.5 CONCLUSIONS

The hyphenation of analytical and spectroscopic techniques has enhanced the ability to solve structural problems. LC-MS had been the only hyphenated technique for qualitative analysis of structures on mixtures until recent developments in NMR. Prior to the last decade, NMR could be applied only to reasonably pure compounds. LC-NMR has expanded the capability to solve structural problems in complex mixtures. LC-NMR, however, is not comparable to LC-MS because of its lower sensitivity, the need of expensive deuterated solvents, the need of solvent suppression of the residual protonated solvents, and the compatibility of the volume of the chromatographic peak with the volume of the NMR flow cell. To overcome some of these problems, more development has led to the hyphenation of solid-phase extraction (SPE) and capillary HPLC (capLC) with NMR as SPE-NMR and capLC-NMR or microflow NMR, respectively. In SPE-NMR, the chromatographic peaks are concentrated by multiple injections into trap cartridges and flush to the NMR flow cell using deuterated solvent with negligible or no solvent suppression required. In the capLC-NMR or microflow NMR, the amount of solvent used is minimal and the volume of the chromatographic peak is comparable to the volume of the NMR flow cell, but the suppression of the residual protonated solvents must still be carried out.

Within the last decade hyphenated LC-MS, LC-NMR and LC-MS-NMR have become available analytical techniques. Since MS is a destructive technique (in contrast to NMR) and requires far less sample than NMR, a splitter is incorporated on-line to direct the bulk of the sample to the less sensitive technique. In addition to the advantage of having MS and NMR information on the same chromatographic peak, the combination of these two techniques with different sensitivities must deal with other issues such as the effect of deuterated solvents on the MS, the limitation of source of ionization on the MS compatible with low flow rates, and the timing which depends on the slower NMR technique. There is still room for improvement for LC-MS-NMR, and the next decade will define the areas where this hyphenated technique is best suited.

LC-MS-NMR cannot replace LC-MS, LC-NMR or even NMR techniques for the structural elucidation of compounds. There will always be cases where purification of the analyte(s) is required, when the structural problem is too complex or the separation of the chromatographic peak is not suitable. LC-MS-NMR. LC-MS, LC-NMR, and NMR have to be available to the analyst to choose the appropriate technique for each structural problem. The success rate of problems will depend on choosing the right technique, depending on the difficulty and nature of the problem. Each technique has its own advantages and limitations, and it is in the hands of the analyst to choose the one(s) that will help to solve the structural problems.

ACKNOWLEDGMENTS

The author dedicates this chapter to her parents, Joaquin Silvia Garcia and Maria Elipe Ruiz, for their love and dedication to their children (the author and her siblings). The author is thankful to Dr. Ray Bakhtiar (Drug Metabolism of MRL at Rahway) for the preparation of Figure 20-1 and for his support and encouragement in writing this chapter; Dr. Byron H. Arison (Drug Metabolism of MRL at Rahway) for his interest, support, encouragement, and constructive discussions during the course of this work; and D. Knapp and U. Parikh (Medicinal Chemistry of MRL at Rahway) for technical help connecting the radioactivity and MS detectors on-line to the LC-NMR system.

REFERENCES

1. M. S. Lee and E. H. Kerns, LC/MS applications in drug development, *Mass Spectrom. Rev.* **18** (1999), 187–279.

2. Y. Wu, The use of liquid chromatography–mass spectrometry for the identification of drug degradation products in pharmaceutical formulations, *Biomed. Chromatogr.* **14** (2000), 384–396.

3. N. J. Clarke, D. Rindgen, W. A. Korfmacher, and K. A. Cox, Systematic LC/MS metabolite identification in drug discovery, *Anal. Chem.* **73** (2001), 430A–439A.

4. F. Bloch, W. W. Hansen, and M. E. Packard, Nuclear induction, *Phys. Rev.* **69** (1946), 127.

5. E. M. Purcell, H. C. Torrey, and R. V. Pound, Resonance absorption by nuclear magnetic moments in a solid, *Phys. Rev.* **69** (1946), 37–38.

6. J. T. Arnold, S. S. Dharmatti, and M. E. Packard, Chemical Effects on Nuclear Induction Signals from Organic Compounds, *J. Chem. Phys.* **19** (1951), 507.

7. W. A. Anderson, Nuclear magnetic resonance spectra of some hydrocarbons, *Phys. Rev.* **102** (1956), 151–167.

8. E. L. Hahn, Spin echoes, *Phys. Rev.* **80** (1950), 580–594.

9. G. A. Morris and R. Freeman, Enhancement of nuclear magnetic resonance signals by polarization transfer, *J. Am. Chem. Soc.* **101** (1979), 760–762.

10. R. R. Ernst and W. A. Anderson, Application of Fourier-transform spectroscopy to magnetic resonance, *Rev. Sci. Instrum.* **37** (1966), 93–102.

11. A. W. Overhauser, Polarization of nuclei in metals, *Phys. Rev.* **92** (1953), 411–415.

12. W. P. Aue, E. Bartholdi, and R. R. Ernst, Two-dimensional spectroscopy. Application to nuclear magnetic resonance, *J. Chem. Phys.* **64** (1976), 2229–2246.

13. L. Braunschweiler and R. R. Ernst, Coherence transfer by isotropic mixing: Application to proton correlation spectroscopy, *J. Magn. Reson.* **53** (1983), 521–528.

14. R. Freeman and G. A. Morris, Experimental chemical shift correlation maps in nuclear magnetic resonance, *J. Chem. Soc. Chem. Commun.* (1978), 684–686.

15. J. Jeener, B. H. Meier, P. Bachmann, and R. R. Ernst, Investigation of exchange processes by two-dimensional NMR spectroscopy, *J. Chem. Phys.* **71** (1979), 4546–4553.

16. A. Bax and D. G. Davis, Practical aspect of two-dimensional transverse NOE spectroscopy, *J. Magn. Res.* **63** (1985), 207–213.

17. L. Müller, Sensitivity enhanced detection of weak nuclei using heteronuclear multiple quantum coherence, *J. Am. Chem. Soc.* **101** (1979), 4481–4484.

18. A. Bax and S. Subramanian, Sensitivity-enhanced two-dimensional heteronuclear shift correlation NMR spectroscopy, *J. Magn. Reson.* **67** (1986), 565–569.

19. G. Bodenhausen and D. J. Ruben, Natural abundance nitrogen-15 NMR by enhanced heteronuclear spectroscopy, *Chem. Phys. Lett.* **69** (1980), 185–189.

20. A. Bax and M. F. Summer, ^1H and ^{13}C assignments from sensitivity-enhanced detection of heteronuclear multiple-bond connectivity by 2D multiple quantum NMR, *J. Am. Chem. Soc.* **108** (1986), 2093–2094.

21. A. Bax, R. Freeman, and T. A. Frienkiel, An NMR technique for tracing out the carbon skeleton of an organic molecule, *J. Am. Chem. Soc.* **103** (1981), 2102–2104.

22. R. E. Hurd, Gradient-enhanced spectroscopy, *J. Magn. Reson.* **87** (1990), 422–428.

23. J. W. Emsley and J. Feeny, Milestone in the first fifty years of NMR, *Prog. Nuclear Magn. Reson. Spect.* **28** (1995), 1–9.

24. J. N. Shoolery, The development of experimental and analytical high resolution NMR, *Prog. Nuclear Magn. Reson. Spect.* **28** (1995), 37–52.

25. R. Freeman, Pioneers of High-Resolution NMR, *Concepts Magn. Reson.* **11** (1999), 61–70.

26. N. Watanabe and E. Niki, Direct-coupling of FT-NMR to high performance liquid chromatography, *Proc. Jpn. Acad. Ser. B* **54** (1978), 194–199.

27. E. Bayer, K. Albert, M. Nieder, E. Grom, and T. Keller, On-line coupling of high-performance liquid chromatography and nuclear magnetic resonance, *J. Chromatogr.* **186** (1979), 497–507.

28. J. F. Haw, T. E. Glass, D. W. Hausler, E. Motell, and H. C. Dorn, Direct coupling of a liquid chromatograph to a continuous flow hydrogen nuclear magnetic resonance detector for analysis of petroleum and synthetic fuels, *Anal. Chem.* **52** (1980), 1135–1140.

29. J. F. Haw, T. E. Glass, and H. C. Dorn, Continuous flow high field nuclear magnetic resonance detector for liquid chromatography analysis of fuel samples, *Anal. Chem.* **53** (1981), 2327–2332.

30. J. F. Haw, T. E. Glass, and H. C. Dorn, Analysis of coal conversion recycle solvent by liquid chromatography with nuclear magnetic resonance detection, *Anal. Chem.* **53** (1981), 2332–2336.

31. J. F. Haw, T. E. Glass, and H. C. Dorn, Conditions for quantitative flow FT-^{1}H NMR measurements under repetitive pulse conditions, *J. Magn. Reson.* **49** (1982), 22–31.

32. E. Bater, K. Albert, M. Nieder, and E. Grom, On-line coupling of liquid chromatography and high-field nuclear magnetic resonance spectrometry, *Anal. Chem.* **54** (1982), 1747–1750.

33. J. F. Haw, T. E. Glass, and H. C. Dorn, Liquid chromatography/proton nuclear magnetic resonance spectrometry average composition analysis of fuels, *Anal. Chem.* **55** (1983), 22–29.

34. K. Albert, Liquid chromatography-nuclear magnetic resonance spectroscopy, *J. Chromatogr. A* **856** (1999), 199–211.

35. J. C. Lindon, J. K. Nicholson, and I. D. Wilson, Directly coupled HPLC-NMR and HPLC-NMR-MS in pharmaceutical research and development, *J. Chromatogr. B* **748** (2000), 233–258.

36. R. P. Hicks, Recent advances in NMR: Expanding its role in rational drug design, *Curr. Med. Chem.* **8** (2001), 627–650.

37. J. C. Lindon, J. K. Nicholson, and I. D. Wilson, Direct coupling of chromatography to NMR spectroscopy, *Prog. Nuclear Magn. Reson. Spectrosc.* **29** (1996), 1–49.

38. M. V. Silva Elipe, LC-NMR and LC-MS-NMR: Recent technological advancements, *Encyclopedia of Chromatography*, Marcel Dekker, New York, 2002, pp. 1–13.

39. K. Albert, *On-Line LC-NMR and Related Techniques*, John Wiley & Sons, Chichester, England, 2002.

40. M. V. Silva Elipe, Advantages and disadvantages of nuclear magnetic resonance spectroscopy as hyphenated technique, *Anal. Chim. Acta* **497** (2003), 1–25.

41. S. H. Smallcombe, S. L. Patt, and P. A. Keifer, WET solvent suppression and its applications to LC NMR and high-resolution NMR spectroscopy, *J. Magn. Res. Ser. A* **117** (1995), 295–303.

42. R. Singh, I.-W. Chen, L. Jin, M. V. Silva, B. H. Arison, J. H. Lin, and B. K. Wong, Pharmacokinetics and metabolism of *RAS* farnesyl transferase inhibitor in rats and dogs: *In vitro–In vivo* correlation, *Drug Metab. Dispos.* **29** (2001), 1578–1587.

43. M. V. Silva, S.-E. W. Huskey, and B. Zhu, Application of LC-NMR for the study of the volatile metabolite of MK-0869, a substance P receptor antagonist, *J. Pharm. Biomed. Anal.* **30** (2003), 1431–1440.

44. M. Ludlow, D. Louden, A. Handley, S. Taylor, B. Wright, and I. D. Wilson, J. Size-exclusion chromatography with on-line ultraviolet, proton nuclear magnetic

resonance and mass spectrometric detection and on-line collection for off-line fourier transform infrared spectroscopy, *Chromatography A* **857** (1999), 89–96.

45. N. Wu, T. L. Peck, A. G. Webb, R. L. Magin, and J. V. Sweedler, Nanoliter volume sample cells for ¹H NMR application to on-line detection in capillary electrophoresis, *J. Am. Chem. Soc.* **116** (1994), 7929–7930.

46. M. E. Lacey, R. Subramanian, D. L. Olson, A. G. Webb, and J. V. Sweedler, High-resolution NMR spectroscopy of sample volumes from 1 nL to 10 µL, *Chem. Rev.* **99** (1999), 3133–3152.

47. J. Schewitz, R. Pusecker, P. Gfrörer, U. Gotz, L.-H. Tseng, K. Albert, and E. Bayer, Direct coupling of capillary electrophoresis and nuclear magnetic resonance spectroscopy for the identification of a dinucleotide, *Chromatographia* **50** (1999), 333–337.

48. K. Pusecker, J. Schewitz, P. Gfrörer, L.-H. Tseng, K. Albert, E. Bayer, I. D. Wilson, N. J. Bailey, G. B. Scarfe, J. K. Nicholson, and L. C. Lindon, On-flow identification of metabolites of paracetamol from human urine using directly coupled CZE-NMR and CEC-NMR spectroscopy, *Anal. Commun.* **35** (1998), 213–215.

49. K. Pusecker, J. Schewitz, P. Gfrörer, L.-H. Tseng, K. Albert, and E. Bayer, On-line coupling of capillary electrochromatography, capillary electrophoresis, and capillary HPLC with nuclear magnetic resonance spectroscopy, *Anal. Chem.* **70** (1998), 3280–3285.

50. J. Schewitz, P. Gfrörer, K. Pusecker, L.-H. Tseng, K. Albert, E. Bayer, I. D. Wilson, N. J. Bailey, G. B. Scarfe, J. K. Nicholson, and J. C. Lindon, Directly coupled CZE-NMR and CEC-NMR spectroscopy of metabolite analysis: Paracetamol metabolites in human tissue, *The Analyst* **12** (1998), 2835–2837.

51. P. Gfrörer, J. Schewitz, K. Pusecker, L.-H. Tseng, K. Albert, and E. Bayer, Gradient elution capillary electrochromatogrpahy and hyphenation with nuclear magnetic resonance, *Electrophoresis* **20** (1999), 3–8.

52. P. Gfrörer, L.-H. Tseng, E. Rapp, K. Albert, and E. Bayer, Influence of pressure upon coupling pressurized capillary electrochromatography with nuclear magnetic resonance spectroscopy, *Anal. Chem.* **73** (2001), 3234–3239.

53. S. A. Korhammer and A. Bernreuther, Hyphenation of high-performance liquid chromatography (HPLC) and other chromatographic techniques (SFC, GPC, GC, CE) with nuclear magnetic resonance (NMR): A review, *Fresen. J. Anal. Chem.* **354** (1996), 131–135.

54. V. Exarchou, M. Godejohann, T. A. van Beek, I. P. Gerothanassis, and J. Vervoort, LC-UV-solid-phase extraction-NMR-MS combined with a cryogenic flow probe and its applications to the identification of compounds present in Greek oregano, *Anal. Chem.* **75** (2003), 6288–6294.

55. J. A. De Koning, A. C. Hogenboom, T. Lacker, S. Strhoschein, K. Albert, and U. A. Th. Brinkman, On-line trace enrichment in hyphenated liquid chromatography-nuclear magnetic resonance spectroscopy, *J. Chromatogr. A* **813** (1998), 55–61.

56. T. Onaka, M. Kobayashi, Y. Ishii, K. Okumura, and M. Suzuki, Applications of solid-phase extraction to the analysis of the isomers generated in biodesulfurization against methylated dibenzothiophenes, *J. Chromatogr. A* **904** (2000), 193–202.

57. N. T. Nyberg, H. Baumann, and L. Kenne, Application of solid-phase extraction coupled to an NMR flow-probe in the analysis of HPLC fractions, *Magn. Reson. Chem.* **39** (2001), 236–240.

58. N. T. Nyberg, H. Baumann, and L. Kenne, Solid-phase extraction NMR studies of chromatographic fractions of saponins from *Quillaja Saponaria, Anal. Chem.* **75** (2003), 268–274.

59. M. E. Lacey, Z. J. Tan, A. G. Webb, and J. V. Sweedler, Union of capillary high-performance liquid chromatography and microcoil nuclear magnetic resonance spectroscopy applied to the separation and identification of terpenoids, *J. Chromatogr. A* **922** (2001), 139–149.

60. R. Subramanian, W. P. Kelley, P. D. Floyd, Z. J. Tan, A. G. Webb, and J. V. Sweedler, A microcoil NMR probe for coupling microscale HPLC with on-line NMR spectroscopy, *Anal. Chem.* **71** (1999), 5335–5339.

61. M. E. Lacey, J. V. Sweedler, C. K. Larive, A. J. Pipe, and R. D. Farrant, ¹H NMR characterization of the product from single solid-phase resin beads using capillary NMR flow probes, *J. Magn. Reson.* **153** (2001), 215–222.

62. E. MacNamara, T. Hou, G. Fisher, S. Williams, and D. Raftery, Multiplex sample NMR: An approach to high-throughput NMR using a parallel coil probe, *Anal. Chim. Acta* **397** (1999), 9–16.

63. A. M. Wolters, D. A. Jayawickrama, and J. V. Sweedler, Microscale NMR, *Curr. Opin. Chem. Biol.* **6** (2002), 711–716.

64. A. M. Wolters, D. A. Jayawickrama, A. G. Webb, and J. V. Sweedler, NMR detection with multiple solenoidal microcoils for continuous-flow capillary electrophoresis, *Anal. Chem.* **74** (2002), 5550–5555.

65. A. M. Wolters, D. A. Jayawickrama, C. K. Larive, and J. V. Sweedler, Capillary isotachophoresis/NMR: Extension to trace impurity analysis and improve instrumental coupling, *Anal. Chem.* **74** (2002), 2306–2313.

66. A. M. Wolters, D. A. Jayawickrama, C. K. Larive, and J. V. Sweedler, Insights into the cITP process using on-line NMR spectroscopy, *Anal. Chem.* **74** (2002), 4191–4197.

67. J.-L. Wolfender, S. Rodriguez, and K. Hostettmann, Liquid chromatography coupled to mass spectrometry and nuclear magnetic resonance spectroscopy for the screening of plant constituents, *J. Chromatogr. A* **794** (1998), 299–316.

68. J.-L. Wolfender, K. Ndjoko, and K. Hostettmann, The potential of LC-NMR in phytochemical analysis, *Phytochem. Anal.* **12** (2001), 2–22.

69. S. C. Bobzin, S. Yang, and T. P. Kasten, Applications of Liquid chromatography-nuclear magnetic resonance spectroscopy to the identification of natural products, *J. Chromatogr. B* **748** (2000), 259–267.

70. S. C. Bobzin, S. Yang, and T. P. Kasten, LC-NMR: A New tool to expedite the dereplication and identification of natural products, *J. Ind. Microb. Biotech.* **25** (2000), 342–345.

71. G. Bringmann, M. Wohlfarth, H. Rischer, M. Heubes, W. Saeb, S. Diem, M. Herderich, and J. Schlauer, A photometric screening method for dimeric naphthylisoquinoline alkaloids and complete on-line structural elucidation of a dimer in crude plant extract, by the LC-MS/LC-NMR/LC-CD triad, *Anal. Chem.* **73** (2001), 2571–2577.

72. G. Bringmann, M. Wohlfarth, H. Rischer, J. Schlauer, and R. Brun, Extract screening by HPLC coupled to MS-MS, NMR, and CD: A dimeric and three monomeric naphthylisoquinoline alkaloids from *Ancistrocladus Griffithii, Phytochemistry* **61** (2002), 195–204.

73. F. D. P. Andrade, L. C. Santos, M. Datchler, K. Albert, and W. Vilegas, Use on on-line liquid chromatography-nuclear magnetic resonance spectroscopy for the rapid investigation of flavonoids from *Sorocea bomplandii*, *J. Chromatogr. A* **953** (2002), 287–291.

74. O. Spring, R. Zipper, I. Klaiber, S. Reeb, and B. Vogler, Sesquiterpene lactones in *Viguiera Eriophora* and *Viguiera Puruana* (Heliantheae; Asteraceae), *Phytochemistry* **55** (2000), 255–261.

75. O. Spring, R. Zipper, S. Reeb, B. Vogler, and F. B. Da Costa, Sesquiterpene lactones ans a myoinositol from glandular trichomes of *Viguiera Quinqueremis* (Heliantheae; Asteraceae), *Phytochemistry* **57** (2001), 267–272.

76. T. Renukappa, G. Roos, I. Klaiber, B. Vogler, and W. Kraus, Application of high-performance liquid chromatography coupled to nuclear magnetic resonance spectrometry, mass spectrometry and bioassay for the determination of active saponins from *Bacopa Monniera* Wettst., *J. Chromatogr. A* **847** (1999), 109–116.

77. S. Strohschein, C. Rentel, T. Lacker, E. Bayer, and K. Albert, Separation and identification of tocotrienol isomers by HPLC-MS and HPLC-NMR coupling, *Anal. Chem.* **71** (1999), 1780–1785.

78. E. F. Queiroz, J.-L. Wolfender, K. K. Atindehou, D. Traore, and K. Hostettmann, On-line identification of the antifungal constituents of *Erythrina vogelii* by liquid chromatography with tandem mass spectrometry, ultraviolet absorbance detection and nuclear magnetic resonance spectrometry combined with liquid chromatographic micro-fractionation, *J. Chromatogr. A* **974** (2002), 123–134.

79. P. Kleinwächter, K. Martin, I. Groth, and K. Dornberger, Use of coupled HPLC/^{1}H NMR and HPLC/ESI-MS for the detection and identification of (2*E*,4*Z*)-decadienoic acid from a new *Agromyces* species, *J. High Resol. Chromatogr.* **23** (2000), 609–612.

80. A. E. Mutlib, J. T. Strupczewski, and S. M. Chesson, Application of hyphenated LC/NMR and LC/MS techniques in rapid identification of *In Vitro* metabolites of iloperidone, *Drug Metab. Dispos.* **23** (1995), 951–964.

81. J. C. Lindon, J. K. Nicholson, U. G. Sidelmann, and I. D. Wilson, Directly coupled HPLC-NMR and its application to drug metabolism, *Drug Metab. Rev.* **29** (1997), 705–746.

82. J. P. Shockor, I. S. Silver, R. M. Wurm, P. N. Sanderson, R. D. Farrant, B. C. Sweatman, and J. C. Lindon, Characterization of *In Vivo* metabolites from human liver microsomes using directly coupled HPLC-NMR: Application to a phenoxathiin monoamine oxidase-A inhibitor, *Xenobiotica* **26** (1996), 41–48.

83. W. J. Ehlhardt, J. M. Woodland, T. M. Baughman, M. Vandenbranden, S. A. Wrighton, J. S. Kroin, B. H. Norman, and S. R. Maple, Liquid Chromatography/Nuclear Magnetic resonance spectroscopy and liquid chromatography/mass spectrometry identification of novel metabolites of the multidrug resistance modulator LY335979 in rat bile and human liver microsomal incubations, *Drug Metab. Dispos.* **26** (1998), 42–51.

84. A. E. Mutlib, H. Chen, G. A. Nemeth, J. A. Markwalder, S. P. Seitz, L. S. Gan, and D. D. Christ, Identification and characterization of Efavirenz metabolites by liquid chromatography/mass spectrometry and high field NMR: Species differences in the metabolism of Efavirenz, *Drug. Metab. Dispos.* **27** (1999), 1319–1333.

85. G. J. Dear, I. M. Ismail, P. J. Mutch, R. S. Plumb, L. H. Davies, and B. C. Sweatman, Urinary metabolites of a novel quinoxaline non-nucleoside reverse transcriptase inhibitor in rabbit, mouse and human: Identification of fluorine NIH shift metabolites using NMR and tandem MS, *Xenobiotica* **30** (2000), 407–426.

86. K. E. Zhang, B. Hee, C. A. Lee, B. Liang, and B. C. M. Potts, Liquid chromatography–mass spectrometry and liquid chromatography-NMR characterization of *in vitro* metabolites of a potent and irreversible peptidomimetic inhibitor of rhinovirus 3C protease, *Drug Metab. Dispos.* **29** (2001), 729–734.

87. M. B. Fisher, D. Jackson, A. Kaerner, S. A. Wrighton, and A. G. Borel, Characterization by liquid chromatography–nuclear magnetic resonance spectroscopy and liquid chromatography–mass spectrometry of two coupled oxidative-conjugative metabolic pathways for 7-ethoxycoumarin in human liver microsomes treated with alamethin, *Drug Metab. Dispos.* **30** (2002), 270–275.

88. T. Prueksaritanont, R. Subramanian, X. Fang, B. Ma, Y. Qiu, J. H. Lin, P. G. Pearson, and T. A. Baillie, Glucuronidation of statins in animals and humans: A novel mechanism of statin lactonization, *Drug Metab. Dispos.* **30** (2002), 505–512.

89. G. J. Dear, R. S. Plumb, B. C. Sweatman, P. S. Parry, A. D. Roberts, J. C. Lindon, J. K. Nicholson, and I. M. Ismail, Use of directly coupled ion-exchange liquid chromatography–mass spectrometry and liquid chromatography-nuclear magnetic resonance spectroscopy as a strategy for polar metabolite identification, *J. Chromatogr. B* **748** (2000), 295–309.

90. S. X. Peng, B. Borah, R. L. M. Dobson, Y. D. Liu, and S. Pikul, Application of LC-NMR and LC-MS to the identification of degradation products of a protease inhibitor in dosage formulations, *J. Pharm. Biomed. Anal.* **20** (1999), 75–89.

91. W. Feng, H. Liu, G. Chen, R. Malchow, F. Bennet, E. Lin, B. Pramanik, and T-M. Chan, Structural characterization of the oxidative degradation product of an antifungal agent SCH 56592 by LC-NMR and LC-MS, *J. Pharm. Biomed. Anal.* **25** (2001), 545–557.

92. N. Wu, W. Feng, E. Lin, G. Chen, J. Patel, T-M. Chan, and B. Pramanik, Quantitative and structural determination of pseudoephedrine sulfate and its related compounds in pharmaceutical preparations using high-performance liquid chromatography, *J. Pharm. Biomed. Anal.* **30** (2002), 1143–1155.

93. J.-L. Wolfender, L. Verotta, L. Belvisi, N. Fuzzati, and K. Hostettmann, Structural investigation of isomeric oxidized forms of hyperforin by HPLC-NMR and HPLC-MS^n, *Phytochem. Anal.* **14** (2003), 290–297.

94. J. K. Roberts and R. J. Smith, Use of liquid chromatography–Nuclear magnetic resonance spectroscopy for the identifiaction of impurities in drug substances, *J. Chromatogr. A* **677** (1994), 385–389.

95. E. A. Crowe, J. K. Roberts, and R. J. Smith, ^{1}H and ^{19}F LC/NMR: Applications to the identification of impurities in compounds of pharmaceutical interest, *Pharmaceutical Sci.* **1** (1995), 103–105.

96. N. Mistry, I. M. Ismail, M. S. Smith, J. K. Nicholson, and J. C. Lindon, Characterization of impurities in bulk drug batches of fluticasone propionate using directly coupled HPLC-NMR spectroscopy and HPLC-MS, *J. Pharm. Biomed. Anal.* **16** (1997), 697–705.

97. S. D. McCrossen, D. K. Bryant, B. R. Cook, and J. J. Richards, Comparison of LC detection methods in the investigation of non-UV detectable organic impurities in a drug substance, *J. Pharm. Biomed. Anal.* **17** (1998), 455–471.

98. B. C. M. Potts, K. F. Albizati, M. O'Neil Johnson, and J. P. James, Applications of LC-NMR to the identification of bulk drug impurities in GART inhibitor AG2034, *Magn. Reson. Chem.* **37** (1999), 397–400.

99. N. Mistry, I. M. Ismail, R. D. Smith, M. Liu, J. K. Nicholson, and J. C. Lindon, Impurity profiling in bulk pharmaceutical batches using 19F NMR spectroscopy and distinction between monomeric and dimeric impurities by NMR-based diffusion measurements, *J. Pharm. Biomed. Anal.* **19** (1999), 511–517.

100. K. W. Sigvardson, S. P. Adams, T. B. Barnes, K. F. Blom, J. M. Fortunak, M. J. Haas, K. L. Reilly, A. J. Repta, and G. A. Nemeth, The isolation and identification of a toxic impurity in XP315 drug substance, *J. Pharm. Biomed. Anal.* **27** (2002), 327–334.

101. I. D. Wilson, L. Griffiths, J. C. Lindon, and J. K. Nicholson, HPLC/NMR and related hyphenated NMR methods, *Prog. Pharm. Biomed. Anal.* **4** (2002), 299–322.

102. G. J. Sharma and I. C. Jones, Critical investigation of coupled liquid chromatography–NMR spectroscopy in pharmaceutical impurity identification, *Magn. Reson. Chem.* **41** (2003), 448–454.

103. S. E. Peng, Hyphenated HPLC-NMR and its applications in drug discovery, *Biomed. Chromatogr.* **14** (2000), 430–441.

104. G. R. Eldridge, H. C. Vervoort, C. M. Lee, P. A. Cremin, C. T. Williams, S. M. Hart, M. G. Goering, M. O'Neil-Johnson, and L. Zeng, High-throughput method for the production and analysis of large natural product library for drug discovery, *Anal. Chem.* **74** (2002), 3963–3971.

105. M. Careri and A. Mangia, Multidimensional detection methods for separation and their application in food analysis, *Trends Anal. Chem.* **15** (1996), 538–550.

106. K. Albert, G. Schlotterbeck, L.-H. Tseng, and U. Braumann, Application of on-line capillary high-performance liquid chromatography–nuclear magnetic resonance spectrometry coupling for the analysis of vitamin A derivatives, *J. Chromatogr. A* **750** (1996), 303–309.

107. S. Strohschein, G. Schlotterbeck, J. Richter, M. Pursch, L.-H. Tseng, H. Händel, and K. Albert, Comparison of the separation of *Cis/Trans* isomers of tretinoin with different stationary phases by liquid chromatography–nuclear magnetic resonance coupling, *J. Chromatogr. A* **765** (1997), 207–214.

108. M. Sandvoss, A. Weltring, A. Preiss, K. Levsen, and G. Wuensch, Combination of matrix solid-phase dispersion extraction and direct on-line liquid chromatography–nuclear magnetic resonance spectroscopy–tandem mass spectrometry as a new efficient approach for the rapid screening of natural products: Application to the total asterosaponin fraction of the starfish *Asterias rubens*, *J. Chromatogr. A* **917** (2001), 75–86.

109. J. Fritsche, R. Angoelal, and M. Dachtler, On-line liquid-chromatography–nuclear magnetic resonance spectroscopy–mass spectrometry coupling for the separation and characterization of secoisolariciresinol diglucoside isomers in flaxseed, *J. Chromatogr. A* **972** (2002), 195–203.

110. N. J. C. Bailey, P. D. Stanley, S. T. Hadfield, J. C. Lindon, and J. K. Nicholson, Mass spectrometrically detected directly coupled high performance liquid chromatography/nuclear magnetic resonance/mass spectrometry for the identification of

xenobiotic metabolites in maize plants, *Rapid Commun. Mass Spectrom.* **14** (2000), 679–684.

111. R. M. Holt, M. J. Newman, F. S. Pullen, D. S. Richards, and A. G. Swanson, High-performance chromatography/NMR spectrometry/mass spectrometry: Further advances in hyphenated technology, *J. Mass Spectrom.* **32** (1997), 64–70.

112. K. I. Burton, J. R. Everett, M. J. Newman, F. S. Pullen, D. S. Richards, and A. G. Swanson, On-line liquid chromatography coupled with high field NMR and mass spectrometry (LC-NMR-MS): A new technique for drug metabolite structure elucidation, *J. Pharma. Biomed. Anal.* **15** (1997), 1903–1912.

113. G. B. Scarfe, I. D. Wilson, M. Spraul, M. Hofmann, U. Braumann, J. C. Lindon, and J. K. Nicholson, Application of directly coupled high-performance-nuclear magnetic resoance-mass spectrometry to the detection and characterisation of the metabolites of 2-Bromo-4-Trifluoromethylaniline in rat urine, *Anal. Commun.* **34** (1997), 37–39.

114. G. B. Scarfe, B. Wright, E. Clayton, S. Taylor, I. D. Wilson, J. C. Lindon, and J. K. Nicholson, ¹⁹F-NMR and directly coupled HPLC-NMR-MS investigations into the metabolism of 2-Bromo-4-Trifluoromethylaniline in rat: A urinary excretion balance study without the use of radiolabelling, *Xenobiotica* **28** (1998), 373–388.

115. G. J. Dear, J. Ayrton, R. Plumb, B. C. Sweatman, I. M. Ismail, I. J. Fraser, and J. Mutch, A rapid and efficient approach to metabolite identification using nuclear magnetic resonance spectroscopy, liquid chromatography/mass spectrometry and liquid chromatography/nuclear magnetic resonance spectroscopy/sequential mass spectrometry, *Rapid Commun. Mass Spectrom.* **12** (1998), 2023–2030.

116. G. B. Scarfe, B. Wright, E. Clayton, S. Taylor, I. D. Wilson, J. C. Lindon, and J. K. Nicholson, Quantitative studies on the urinary metabolic fate of 2-Chloro-4-Trifluoromethylaniline in the rat using ¹⁹F-NMR spectroscopy and directly coupled HPLC-NMR-MS, *Xenobiotica* **29** (1999), 77–91.

117. J. P. Shockcor, S. E. Unger, P. Savina, J. K. Nicholson, and J. C. Lindon, Application of directly coupled LC-NMR-MS to the structural elucidation of metabolites of the HIV-1 reverse-transcriptase inhibitor BW935U83, *J. Chromatogr. B* **748** (2000), 269–279.

118. G. J. Dear, R. S. Plumb, B. C. Sweatman, J. Ayrton, J. C. Lindon, J. K. Nicholson, and I. M. Ismail, Mass directed peak selection, an efficient method of drug metabolite identification using directly coupled liquid chromatography-mass spectrometry-nuclear magnetic resonance spectroscopy, *J. Chromatogr. B* **748** (2000), 281–293.

119. G. B. Scarfe, J. K. Nicholson, J. C. Lindon, I. D. Wilson, S. Taylor, E. Clayton, and B. Wright, Identification of the urinary metabolites of 4-Bromoaniline and 4-Bromo-[*carbonyl*-¹³C]-acetaniline in rat, *Xenobiotica* **32** (2002), 325–337.

120. D. Bao, V. Thanabal, and W. F. Pool, Determination of tacrine metabolites in microsomal incubate by high-performance liquid chromatography-nuclear magnetic resonance/mass spectrometry with a column trapping system, *J. Pharma. Biomed. Anal.* **28** (2002), 23–30.

121. F. S. Pullen, A. G. Swanson, M. J. Newman, and D. S. Richards, On-line liquid chromatography/nuclear magnetic resonance mass spectrometry—A powerful

spectroscopic tool for the analysis of mixture of pharmaceutical interest, *Rapid Commun. Mass Spectrom.* **9** (1995), 1003–1006.

122. J. C. Lindon, J. K. Nicholson, and I. D. Wilson, Directly coupled HPLC-NMR and HPLC-NMR-MS in pharmaceutical research and development, *J. Chromatogr. B* **746** (2000), 233–258.

123. O. Corcoran and M. Spraul, LC-NMR'MS in drug discovery, *Drug Discov. Today* **8** (2003), 624–631.

124. I. F. Duarte, M. Spraul, M. Godejohann, U. Braumann, and A. M. Gil, Application NMR and hyphenated NMR spectroscopy for the study of beer components, *Magn. Reson. Food Sci.* **286** (2002), 151–157.

125. A. M. Gil, I. F. Duarte, M. Godejohann, U. Braumann, M. Maraschin, and M. Spraul, Characterization of the aromatic composition of some liquid foods by nuclear magnetic resonance and liquid chromatography with nuclear magnetic resonance and mass spectrosmetric detection, *Anal. Chim. Acta* **488** (2003), 35–51.

21

TRENDS IN PREPARATIVE HPLC

Ernst Kuesters

21.1 INTRODUCTION

Directly from its beginning—now 100 years ago, when Michail Tswett developed the principles [1, 2] with the isolation of chlorophyll—chromatography has always been a *preparative technology*, and its value in producing compounds of high purity cannot be overemphasized. It was Paul Karrer [3] who stated very early *". . . it would be a mistake to believe that a preparation purified by crystallization should be purer than one obtained from chromatographic analysis. In all recent investigations chromatographic purification widely surpassed that of crystallization."* and Leslie Ettre, although not distinguishing between analytical and preparative separations, denoted chromatography as *"the separation technique of the 20th century"* [4]. From a historical point of view, the beginnings of preparative isolation of natural compounds were cumbersome. For example, it is reported [5] that six years of work and processing of 30 tons of strawberries was needed to finally obtain 35 mL of an oil, the essence of the fruit. This situation changed dramatically in the 1960s with the theoretical understanding of the chromatographic process, the development of high-performance liquid chromatography, and the synthesis of highly selective stationary phases. As a result of these improvements, the isolation of natural compounds with preparative chromatography on production scale (e.g., drug substances from fermentation processes) is still state of the art, even after 100 years.

Today, preparative HPLC has also become a powerful technology in pharmaceutical development and production either for isolation of impurities, for

HPLC for Pharmaceutical Scientists, Edited by Yuri Kazakevich and Rosario LoBrutto
Copyright © 2007 by John Wiley & Sons, Inc.

TABLE 21-1. Order of Magnitude and Purpose of Purified Amounts Obtained from Preparative Chromatography

Column Type	I.D. (mm)	Purpose	Amount of Stationary Phase (g)	Amount of Product (g)
Analytical	1–5	Isolation of reference substances (MS or NMR)	0.2–3	0.0002–0.003
Analytical— semipreparative	5–10	Starting materials for toxicology	0.003–25	0.003–0.1
Semipreparative —preparative	10–40	Intermediates for lab synthesis	25–100	0.1–5
Pilot plant	100–300	Manufacturing of drug substances for pharmaceutical development	100–1000	20–5000
Production	300–1,500	Manufacturing of trade products	1,000–4,000,000	kg-tons

chromatographic purifications, or as part of a scale-up process and subsequently has been reviewed in a lot of monographs [6–10]. The term *preparative amount* thus covers the range from milligram quantities (amounts for structure elucidation, analytical characterization, toxicology, or reference material) to large-scale production of tons of intermediates and drug substances. The separations therefore can be performed on all types of columns, starting from analytical ones up to production scale columns with 1-m i.d and several meters in length. Typical applications are summarized in Table 21-1.

The success of preparative HPLC on a production scale has been made possible because of significant improvements made in several areas like (i) column technology (today, mainly compressed columns are used), (ii) packing materials (pressure stable spherical particles with high homogeneity, either nonchiral or chiral), and (iii) the understanding of the nonlinear process in preparative HPLC (overloaded conditions) which resulted in new methods to determine the adsorption isotherms and which consequently led to new concepts like displacement chromatography and simulated moving bed (SMB) chromatography, where the knowledge of such adsorption isotherms is a prerequisite for the design of the corresponding separation process.

The aim of this chapter is to highlight current developments in these various fields of preparative HPLC, with particular emphasis on applications that have been developed at Chemical & Analytical Development at Novartis Pharma AG. Drug substance purifications from biological and synthetic sources are presented, along with the separation of chiral and/or achiral molecules on chiral stationary phases and typical isolations of by-products. Special attention is given to the determination of adsorption isotherms and their interplay with respect to the layout of chromatographic processes as well as the choice of technology. The applications have been selected in such a way that a broad variety of technologies like multiple injection, recycling, displacement, and SMB chromatography is covered. On-line detection tools have to fulfill other demands in preparative chromatography than in analytical chromatography. A special section has been devoted to this aspect below, and an instrument that was developed in-house is presented.

21.2 METHOD DEVELOPMENT IN PREPARATIVE HPLC

Since chromatography scales up linearly and independently from the selected technology (rationales when making a choice will be given later on), the column containing the stationary phase is still the heart of the system. Method development will therefore always start with the selection of the best stationary and mobile-phase composition to achieve an optimum in productivity, which does not necessarily mean an optimum in selectivity. For example, a high selectivity of $\alpha > 10$ has been obtained for the enantiomeric separation of β-blocking agents like pindolol using amylose- or cellulose-derived stationary phases, but the poor solubility of the racemates in the mobile phase (hexane/2-propanol mixtures) will never result in an economic separation process. This situation can be significantly improved by (i) solvent switch and (ii) adding of bases or acids, which leads to higher solubility and productivity, although the selectivity decreases. Figure 21-1 shows the separation of the enantiomers of pindolol under different conditions [11, 12]. Even though the addition of TFA clearly results in very distorted isotherms, the situation from the point of view of the preparative separation is much improved, with the throughput increasing from 322 to 860 g of racemate per kilogram of chiral stationary phase per day. Nevertheless, as a rule of thumb, in most cases higher productivities have

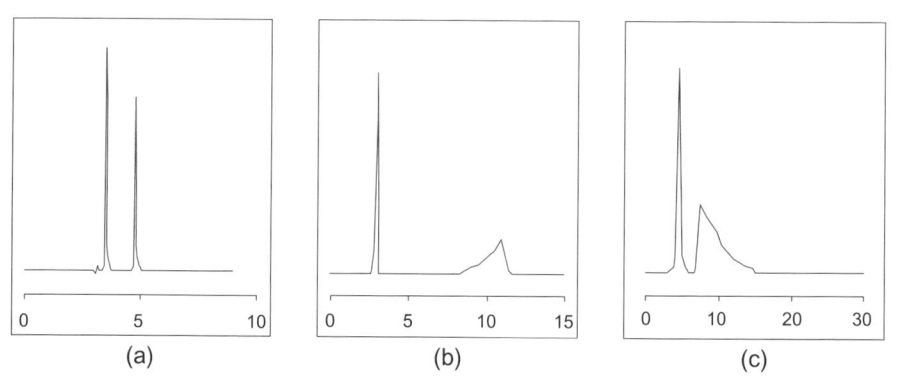

(a)   (b)   (c)

Figure 21-1. The effect of mobile-phase additives on pindolol on Chiralcel-OD (analytical column). Mobile phase: (a) Methanol/diethylamine = 99.9/0.1, 20°C. (b) Hexane/ethanol/trifluoroacetic acid = 60/40/0.5, 40°C. (c) Conditions as for (b), but 25-mg load. (Reprint from reference 12, with permission.)

been obtained under separation conditions where high selectivities have been identified. Therefore, in parallel, parameters like solubility of the sample in the mobile phase, capacity of the stationary phase, stability, and work-up of product containing fractions have to be determined. Once a robust system has been developed, the possibilities of scale-up (solubility of sample, stability of product in mobile phase, work-up, etc.) are investigated in the next step. And finally the adsorption isotherms are measured as a guide to the appropriate and economic technical realization on pilot plant or production scale.

21.2.1 Optimization of Selectivity

The first step, the search for an appropriate chromatographic system, can be explored with the aid of analytical columns or even more easily in the case of straight-phase chromatography with thin-layer chromatography (TLC). In the case of chiral separations with chiral stationary phases (CSP), a quick survey of separation strategies is provided by using electronic databases like Chirbase in advance. Since each type of column overloading will result in a loss of separation, the method development should start with the search for a sufficient peak resolution R_s. Under analytical conditions, the peak resolution R_s is the result of the interplay of selectivity or separation factor α, retention time, and column performance according to equation (21-1):

$$R_S = \frac{1}{4}(\alpha - 1)\frac{k}{(1 + k)}\sqrt{N} \qquad (21\text{-}1)$$

where α is the separation factor (selectivity) $= k_2/k_1$ for $k_2 > k_1$; k_1 and k_2 are the capacity factors of substance 1 and 2, respectively; and N is the plate number.

A rough estimation nicely highlights the contribution and importance of a well-developed separation factor. Whereas changes in k from 3 to 5 only improve the peak resolution by 10.7% and a doubling of N by 41.4%, the increase of selectivity from 1.2 to 2.2 will result in an improvement of 83.3%. Since in most cases the technical parameters like particle size and pressure are given and used under optimum conditions, the search for high selectivity cannot be overemphasized.

The main parameters to optimize the separation factor and peak resolution, respectively, are as follows:

- Appropriate stationary phase (which not only seeks for the appropriate polarity of the material; the "same stationary phase" from different supplier may have a significant influence on the selectivity because of differences in the manufacturing process).
- Appropriate mobile phase (which includes the choice and composition of solvents, additives, and pH value).

• Temperature. Especially the latter parameter should not be underestimated. Although, as a rule of thumb, achiral separations are often performed at elevated temperatures, it is generally believed that separations on chiral stationary phases should best be performed at lower temperatures. Nevertheless, sometimes it turns out that chiral separations are entropy controlled and better selectivities are obtained at higher temperatures [13–16].

Once the right set of parameters has been identified, computer-aided optimization using *modified sequential simplex* or *central composite design* methods can be applied to further fine-tune the separation under investigation, as has been published for the optimization of reverse-phase HPLC [17–20] and chiral separations [21–23].

21.2.2   Scale-Up of Analytical Methods

21.2.2.1   Overloading. The fundamental difference between preparative chromatography and analytical chromatography is the sample amount being injected. In analytical chromatography the sample amount is extremely small with regard to the amount of stationary phase (<1 : 10,000) and the chromatography is consequently performed in the linear range of the adsorption isotherms of the components being separated. A rough calculation at that point nicely demonstrates that a simple linear enlargement will never provide an economic process. Therefore the injection amount will successively be increased, which in the first instance will result in an adequate increase of peak heights and peak areas while leaving the retention times and separation factors unaffected. A further increase of the sample amount then will result in an overloading of the column and in deformed and moving peaks as a consequence of a shift in the nonlinear range of the adsorption isotherms. Concave isotherms will provide broader tailing peaks with shorter retention times, whereas convex isotherms will show broader fronting peaks with greater retention times. The separation of course will become poorer; nevertheless, as long as it is sufficient, the process will become more and more economic. The increase of the injected quantity until the two peaks touch is called touching-band optimization [24], and an example is given in Figure 21-2 for the separation of an artificial mixture of epothilone A and B.

This optimization approach has the advantage of being fast and simple, but it often overlooks specific effects that happen at larger loads. These effects concern the displacement of one product by another and have been described by Guiochon and co-workers [25–28] and Cox and co-workers [29–31]. The interplay of adsorption isotherm, peak form, and capacity factor k during overloading of a column is depicted in Figure 21-3 [32].

Sometimes, during the course of determining the capacity of the stationary phase and the adsorption isotherms, it turns out that significant preparative amounts of reference material can easily be obtained even with analytical

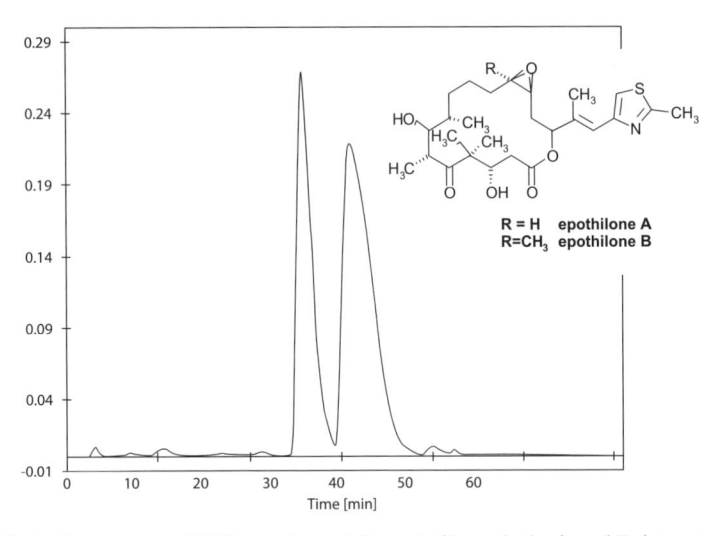

Figure 21-2. Separation of 247 mg of epothilone A (first eluting) and B (structure given below) on a semipreparative reversed-phase ODS column (25-cm × 2.0-cm i.d.). Particle size 11 μm, mobile phase acetonitrile/water = 4/6 (V/V), flow rate 15 mL/min, UV detection 250 nm.

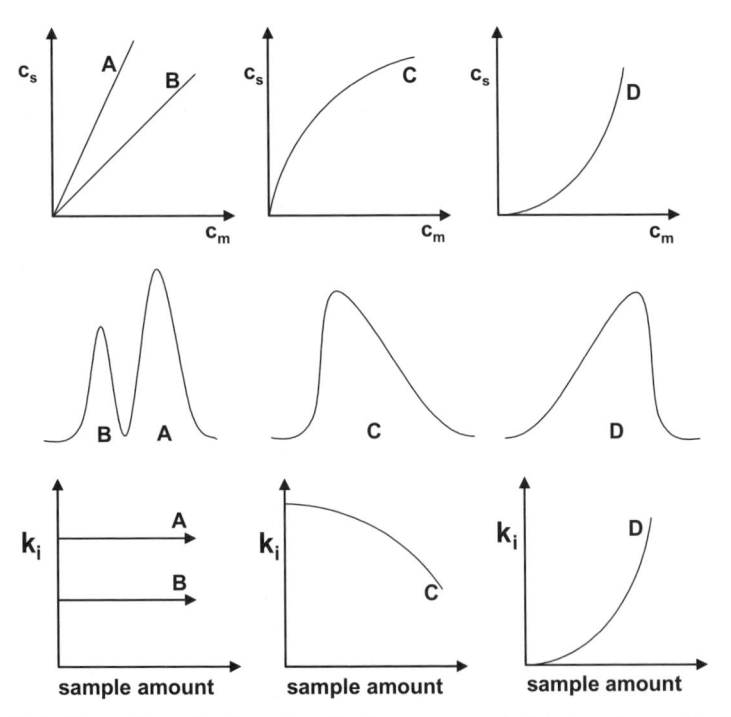

Figure 21-3. The effect of adsorption isotherm on peak form and capacity factor k during overloading of a column. c_s and c_m = concentration of substance in the stationary and mobile phase; A, B, C, D refer to substance A, B, C, D, respectively.

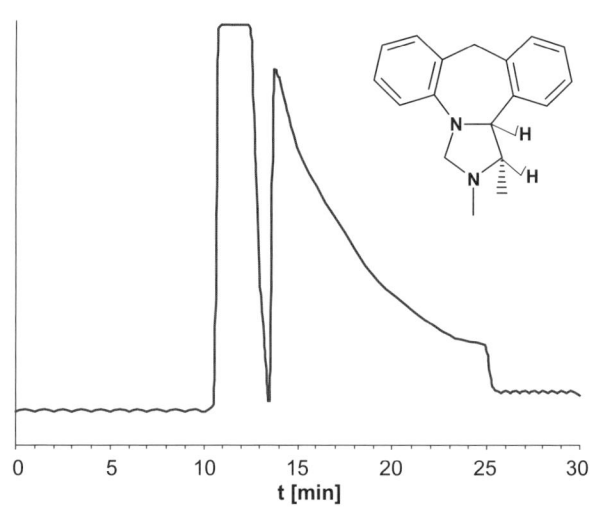

Figure 21-4. Preparative enantioseparation of a morphanthridine analogue on an analytical Chiralpak-AD column (250-cm × 4.6-mm i.d.). Mobile phase Hexane/2-propanol = 85/15 (V/V), 0.5 mL/min; temperature 40°C, UV detection 290 nm, injection amount 100 mg/250 μL hexane/2-propanol = 1/1 (V/V). (Reprint from reference 33, with permission.)

columns. Given the good solubility of a racemic morphanthridine in the mobile phase and the large separation factor, the author decided to estimate the capacity of the CSP for the given separation [33]. The injection amount systematically increased to estimate the final value for which a baseline separation could be observed. To obtain on-scale peaks, UV detection was carried out at 290 nm, and the automatic injection device was replaced by a manual loop with different volume sizes. After several runs the endpoint was the injection of 100 mg of racemate dissolved in 250 μL of hexane/2-propanol = 1/1 (V/V). The preparative chromatogram of this run is shown in Figure 21-4. It is obvious from the individual peak shapes that both enantiomers follow different adsorption isotherms. Whereas for the first eluting enantiomer, a linear adsorption isotherm is observed, the corresponding one for the second eluting enantiomer is much more complex. Nevertheless, both enantiomers are separated to baseline and completely eluted within 15 min. It is therefore obvious that even without further optimization, a daily yield of 9.6 g of resolved racemate can be achieved using an automatically injection device with repetitive injection. Based on this result, several interesting production scenarios can be derived. Just by increasing the inner diameter of the column, the production of ton amounts/year with a daily mobile phase consumption of less than 1 m^3 may be easily achieved. The results of the calculations are summarized in Table 21-2. As can be taken from Table 21-2, a respectable amount of 96 kg of racemate can be resolved per day on a column containing 30 kg of CSP. In a typical pilot plant environment, such a column belongs to the smaller ones and also

TABLE 21-2. Calculated Production Scenarios for a Preparative Enantioseparation of a Morphanthridine Analogue on Chiralpak-AD

Amount of CSP	Analytical Column	Pilot Plant Column	Production Column
	3 g	3 kg	30 kg
Batch Elution Mode			
Resolved racemate/day	9.6 g	9.6 kg	96.0 kg
Resolved racemate/year	3.5 kg	3.5 tons	35.0 tons
Solvent consumption/day	0.72 L	0.72 m^3	7.2 m^3
SMB Mode			
Resolved racemate/day	NA	19.2 kg	192.0 kg
Resolved racemate/year	NA	7.0 tons	70.0 tons
Solvent consumption/day	NA	72 L	0.72 m^3

NA, not applicable with respect to preparative method.

the daily mobile-phase consumption of 7.2 m^3 is not a technical hurdle. A fully automated chromatographic system would consequently provide a yearly production of 35 tons of resolved racemate. Later on (Section 21.4.4) it is shown that in most cases where conventional batch elution chromatography is compared with simulated moving bed (SMB) applications with the same amount of CSP, productivity can double and solvent savings up to 80–90% are achieved. Assuming such a production scenario for the above-mentioned morphanthridine analogue, a daily production of 192 kg (corresponding to 70 tons/year) reflects a feasible order of magnitude. In addition, a daily solvent consumption of 720 L is negligible from a production point of view.

21.2.2.2 Solubility and Self-Displacement. In the previous scenario, the feed concentration was gradually increased. This kind of overloading, called concentration overloading, comes to an end when the solubility product of the solute is achieved. A further increase of sample amount can then only be achieved with volume overloading, the injection of larger feed volumes into the column. Very often in practice the combination of both types of overloading comes into operation. In the case of an excellent selectivity in combination with a poor sample solubility, the addition of a more polar solvent to the feed solution may help to achieve a higher productivity. As a result of the slightly modified chromatographic system, a partial self-displacement is observed, visualized by a doubling of the eluting peaks. Since, in addition, the retention is shifted to shorter retention times, this improvement will also come to an end when the first compound leaves the column unretained with t_0. Therefore sometimes the reverse occurs—for example, when a good sample solubility meets excellent elution conditions. To avoid peak elution during the injection period, the polarity of the feed solution is changed by addition of a further solvent in such a way that the solubility of the feed solution decreases and takes significantly larger injection volumes into account. Injection times of 30 min. and longer are acceptable as long as the sample stays retained at the top of the column. After the injection is finished, the solutes are eluted with the mobile phase that has a better solubility. An example of this approach has recently been published for the purification of discodermolide [34] (Figure 21-4). A 38-g sample of crude product (82.4%) was dissolved in 11.2 L of 2-propanol and diluted with 78.4 L of water. After injection of this feed solution onto a column containing 15 kg of ODS-RP-18 reversed-phase phase silica gel, the drug substance was eluted with a mixture of acetonitrile/water = 25/75 (V/V) in an isocratic mode. It is noteworthy that in the large-scale synthesis of 60 g discodermolide, 39 steps (26 steps in the longest linear sequence) and several chromatographic purifications were involved. A chromatographic purification of such a "small" amount of a highly active drug substance which delivered sufficient material for early-stage human clinical trials is the method of choice, since extremely pure material is obtained on pilot plant equipment in a very short time. Figure 21-5 shows a semipreparative purification of discodermolide during method development on a lab-scale column and highlights the effectiveness of the purification step.

21.2.2.3 Purity of Solvents, Stability of Products and Work-up. The quality aspect of the solvents used as mobile phases should not be forgotten, since the evaporation residue from the mobile phase can be significant. Assuming an average product concentration of 1–2 g/L mobile phase, it becomes obvious that an evaporation residue of 10 mg/L solvent leads to 1 g of evaporation

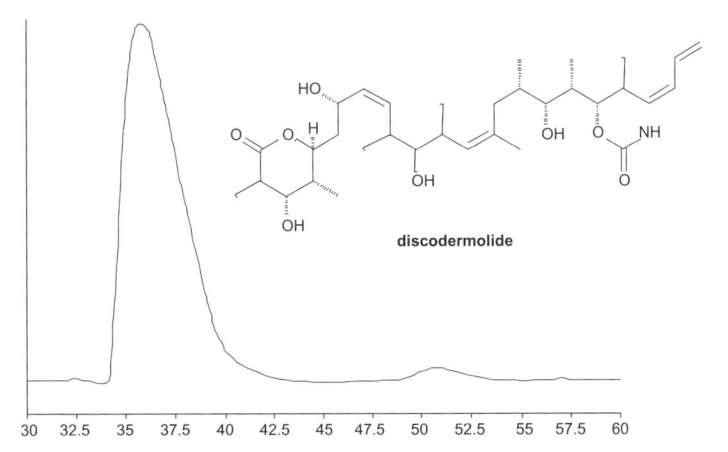

discodermolide

Figure 21-5. Purification of 101 mg of crude discodermolide on 46 g of YMC-OD-A 5–15 μm (column: 250-mm × 20-mm i.d.). The drug substance is dissolved in 31.4 mL of 2-propanol and 220.6 mL of water are added. The feed solution is pumped with a flow rate of 10 mL/min onto the column, and the compounds are eluted afterwards with a mixture of acetonitrile/water = 2/1 (V/V), flow rate 15 ml/min; UV detection 220 nm.

residue in 100 g of product. Solvents that are used in preparative chromatography should therefore have an evaporation residue of $<10^{-4}$ g/L. To ensure a good quality of the product, it is therefore sometimes necessary to purify the solvents in advance prior to their use as mobile phase. This not only will have an influence on the product quality, but also may, in addition, by removing heavy metals and/or stabilizers, have an impact on the resolution and therefore also affect the ruggedness of the chromatographic process. As has been shown by Dingenen [35], the switch from one supplier to another can lead to the complete loss of selectivity in a chromatographic step.

Once a chromatographic system has been identified for a preparative purpose, the stability and work-up procedure of the product-containing fractions should be investigated. Sometimes it turns out that the products cannot be isolated by simple removal of the solvents, because of thermal instability or too basic or acidic conditions in the mobile phase. In such a case an appropriate extraction procedure from the mobile phase may help to isolate the products.

21.2.3 Adsorption Isotherms and Their Determination

The most common technique used in preparative chromatography is still isocratic batch elution. However, more sophisticated technologies like recycling, gradient elution, displacement, or the simulated moving bed (SMB) process are being increasingly applied to enhance productivity and yields. A fair comparison between these rivaling technologies is only possible on the basis of real occurring concentration profiles that agree excellently with the theoretical predictions. The substantial progress that has been achieved in modeling preparative chromatography was reviewed recently [36–38].

The underlying equilibrium-dispersion model, for which the mass balance for solute i in a N component mixture and a volume element is given in equation (21-2), has been very often successfully applied to quantify chromatographic processes under overloaded conditions.

$$\frac{\partial c_i}{\partial t} + F\frac{\partial q_i(c)}{\partial t} + u\frac{\partial c_i}{\partial x} = D_{api}\frac{\partial^2 c_i}{\partial x^2}, \quad i = 1, \ldots, N \text{ with } c = (c_1, c_2, \ldots, c_N) \quad (21\text{-}2)$$

In this equation, c is the concentration in the fluid phase and q is the quantity in the solid phase. The column porosity ε (expressed as phase ratio $F = (1 - \varepsilon)/\varepsilon$) defines the fraction of the fluid phase in the column. Furthermore, u stands for the linear velocity and t and x are the time and space coordinates, respectively. All contributions leading to band-broadening are lumped in a simplifying manner into an apparent dispersion coefficient, D_{ap}. In equation (21-2), it is assumed that the two phases are constantly in equilibrium expressed by the adsorption isotherms. Due to the nonlinear character of the isotherm equations, the solution of equation (21-2) requires the use of numerical methods. The Godunov method is a good choice, because it exploits quantitatively the knowledge about numerical dispersion effects that are caused by usage of finite difference approximations. The method allows the application of rather coarse grids leading to fast calculations [39]. The adaption to simulate multicolumn countercurrent processes has been reported in detail [40]. The application of the model and these numerical solutions allows the simulation of elution chromatography, recycling chromatography, simulated moving bed chromatography, and annular chromatography on a personal computer within a few minutes. A systematic investigation (theoretical simulation on the basis of determined adsorption isotherms and experimental verification) to compare the different chromatographic modes has recently been published by Seidel-Morgenstern for the separation of a binary mixture consisting of two isomers of a steroid [41, 42].

The concentrations of component i in the liquid and in the solid phases, C_i and q_i, respectively, are related through the adsorption isotherms [equation (21-3)].

$$q_i = f(C_1, C_2, \ldots, C_N), \quad i = 1, \ldots, N \tag{21-3}$$

The knowledge of these adsorption isotherms is the main prerequisite for applying the mathematical models to simulate preparative HPLC, displacement or simulated moving bed chromatography. Several methods (e.g., frontal analysis, elution by characteristic point, minor disturbance method, adsorption–desorption, and chromatogram fitting) are available for the determination of the equilibrium data and have been reviewed by Nicoud and Seidel-Morgenstern [43] and very recently by Seidel-Morgenstern [44]. It is beyond the scope of this chapter to describe all methods with their benefits and drawbacks in detail, and the interested reader is referred to the literature [i.e., 39–44]. Nevertheless, three methods (given below) that we have used in our laboratories are briefly summarized to illustrated the underlying principles.

21.2.3.1 The Elution by Characteristic Point Method (ECP).

An easy and simple method to measure the adsorption isotherms for pure components is the ECP method suggested by Cremer and Huber [45]. This method evaluates chromatograms recorded after injecting samples of large size on a column. As a basic requirement for the applicability of the ECP method, the column has to be very efficient. Under these conditions, thermodynamics determine the shape of the chromatographic profiles and kinetic effects can be neglected. If a large sample size is injected on the column, usually the front of the obtained chromatogram is sharpened and the tail is dispersed. The concentration–time relation of the dispersed tail (Figure 21-6a) is completely defined by the course of the adsorption isotherm in equation (21-4), where t_R represents the retention time, t_0 the void volume, and F the phase ratio.

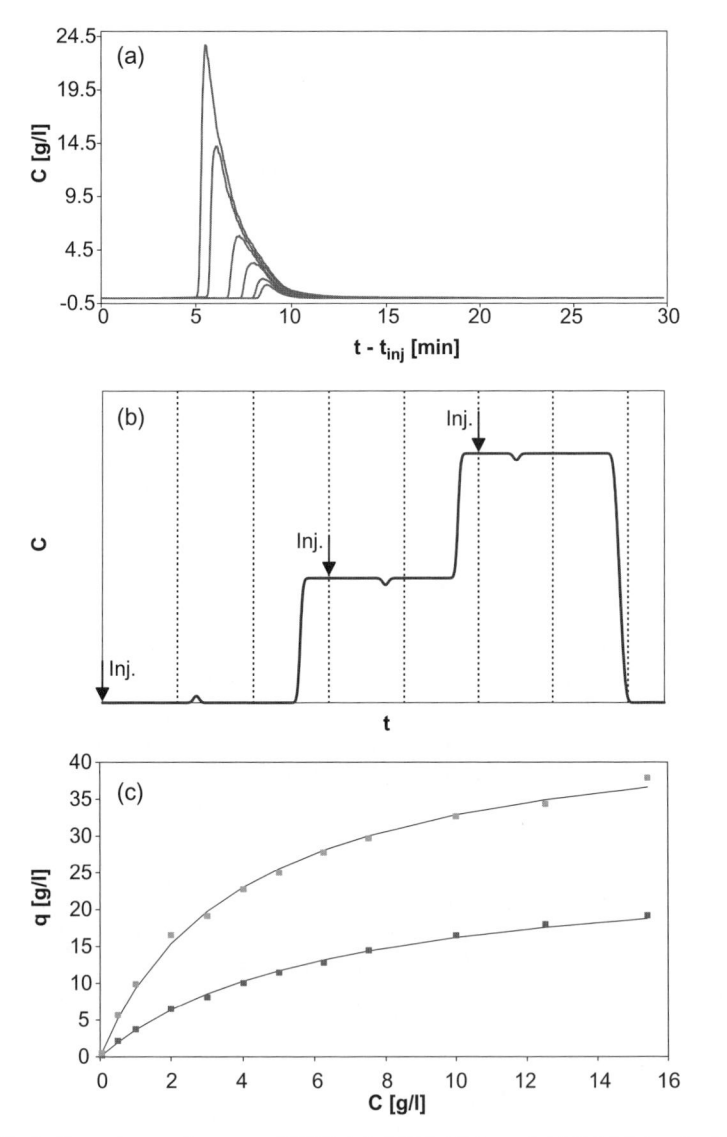

Figure 21-6. Experimental setup of ECP (a), MDM (b), and ADM (c) method for the determination of adsorption isotherms. The concentration–time relation of the dispersed tail in the ECP approach (a) is completely defined by the course of the adsorption isotherm, as can be visualized by the injection of increasing samples amounts. Solvent injections at defined concentrations will result in pulses in the MDM approach (b) which are linked to the adsorption isotherms. Although very precise during application of the ADM method, the data points of the adsorption isotherms (c) have to be measured individually.

$$\frac{dq_i}{dC_i}\bigg|_C = \frac{t_{R_i}(C_i) - t_{inj} - t_0}{t_0 F} \tag{21-4}$$

21.2.3.2 The Minor Disturbance Method (MDM). The principle of the MDM method is based on a stepwise saturation of the column with different known feed concentrations. After reaching equilibrium, small samples possessing a different concentration are injected and the corresponding retention times are measured. Figure 21-6b illustrates the principle of the perturbation method for a single component dissolved in a nonadsorbable eluent. At zero time a small (analytical) sample size is injected without preloading on the column. In the following steps the column is saturated at different concentrations and small amounts of pure eluent are injected at the times marked with arrows. Possible deviations of the retention times at higher concentrations are caused by the nonlinearity of the adsorption isotherm. Since the method depends only on the analysis of times, no detector calibration is necessary. To determine the competitive isotherms for a binary mixture, the same procedure can be applied, saturating the column with different solutions of known concentration of the two components. At each plateau a perturbation induces then two pulses. Using the column mass balance equation and the coherence condition introduced in the frame of the equilibrium theory [46], equation (21-5), being the derivative of the adsorption isotherms, can be derived. In other words, the principle of the MDM method is the determination of parameters of an isotherm model from measured retention times.

$$t_{R_{i,k}} = t_0\left(1 + F\frac{dq_i}{dC_i}\bigg|_k\right), \quad i, k = 1, 2 \tag{21-5}$$

21.2.3.3 The Adsorption—Desorption Method (ADM). Although time- and sample-consuming, the ADM method leads directly to the adsorption isotherms and has often proved to be the most precise method. After saturation of the column with defined increasing solute concentrations C_{E_i}, the corresponding amounts of solutes m_i in the column of volume V are obtained after desorption in each step with the same solvent mixture (Figure 21-6c). Equilibrium conditions assumed, the corresponding concentrations in the stationary phase q_{E_i} are obtained according to equation (21-6) (ε denotes the porosity and phase ratio, respectively):

$$q_{E_i}(C_{E_1}, C_{E_2}, \ldots, C_{E_N}) = \frac{m_i - \varepsilon V C_{E_i}}{(1-\varepsilon)V}, \quad i = 1, \ldots, N \tag{21-6}$$

The experimental setup of the above-mentioned approaches are summarized in Figure 21-6.

To model the adsorption equilibrium, a suitable isotherm equation has to be chosen. For mixtures, the model equations are usually coupled to take into

account the competition for available adsorption sites. The so-called multi-
Langmuir equation (21-7) was found to represent a lot of experimental data
satisfactorily.

$$q_i = \frac{a_i C_i}{1 + \sum_{j=1}^{N} b_j C_j}, \quad i = 1, \ldots, N \tag{21-7}$$

For enantiomeric separations, the modified competitive Langmuir equation
(21-8) was found to represent several sets of experimental data satisfactorily
[47]. This equation considers noncompetitive and competitive adsorption at
different types of adsorption sites. Other useful equations are described and
reported in the literature [48, 49].

$$q_i = \frac{a_i C_i}{1 + \sum_{j=1}^{N} b_j C_j} + \lambda_i C_i, \quad i = 1, \ldots, N \tag{21-8}$$

21.2.3.4 Curiosities. The following example may, in addition, illustrate the
importance of known adsorption isotherms. The enantiomeric separation of
3-benzyloxycarbonyl-2-*t*-butyloxazolidinone on the CSP Chiralcel-OD by
Francotte [50] revealed a concave adsorption isotherm for the first eluting
enantiomer and a convex one for the second eluting antipode (Figure 21-7).
With increasing sample amounts, the first enantiomer will therefore be shifted
to shorter retention times while the second enantiomer is shifted to longer
retention times. Good solubility of the racemate and a high capacity of the

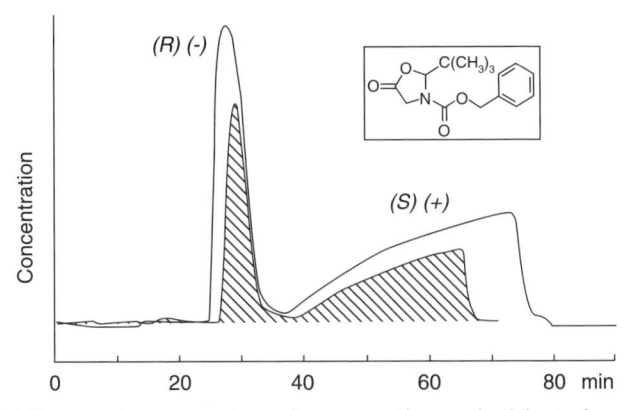

Figure 21-7. Preparative enantiomeric separation of 3-benzyloxycarbonyl-2-*t*-
butyloxazolidinone on Chiralcel-OD (50 cm × 5 cm); mobile phase hexane/2-propanol
= 8/2 (V/V), 50 mL/min; injection amounts 2 g (hatched area) and 3 g. (Reprint from
reference 50, with permission.)

stationary phase are fortuitous. In exceptional cases, where the concave adsorption isotherm crosses the convex one, even a reversal of the elution order is obtained and can be used to achieve a higher productivity as has been demonstrated by Roussel et al. [51] for the separation of the enantiomers of 3-(2-propylphenyl)-4-methyl-4-thiazolin-2-one on microcrystalline cellulose triacetate.

21.3 COLUMNS AND STATIONARY PHASES

In the past, preparative HPLC has been dominated by the use of irregular particles of large size, broad size distribution, and low mechanical stability. Since many improvements with respect to design and manufacturing of silica-based particles have been achieved, nowadays the field of preparative HPLC is dominated by the use of spherical particles with narrow distribution size, good mechanical stability, and high loadability. The loadability is determined by the following parameters: surface area, pore size, size distribution, and in special cases (e.g., enantiomeric separations with CSP) ligand density. These parameters are systematically optimized by the manufacturers [52] of stationary phases, and highly efficient columns are obtained and good packing of the column provided. An improvement in the methodology of column packing automatically results in reaching the required efficiency with shorter bed lengths and in a better productivity.

21.3.1 Stationary Phases

The most widely used packing materials in preparative HPLC are the silica-based particles. Although irregular particles are still available, for preparative columns most applications tend to use spherical packings, since better packings are obtained and for additional reasons mentioned below. Underivatized silica and C18 reversed-phase material (for most applications) are available in packed column as well as bulk quantities. Aside from silica, columns based upon other spherical packings are available, like organic polymers based upon poly (styrene-divinylbenzene) (PS-DVB). These materials have excellent separation properties in the field of peptide and protein purification. The columns can be used for or cleaned with caustic solutions, where silica-based material often has shortcomings. In addition, the manufacturing process has meanwhile been improved in such a way that mechanical stability is achieved comparable to that exhibited by silica-based stationary phases. It is out of the scope of this chapter to list all stationary phases with their advantages and limitations being used in preparative HPLC, and the interested reader is referred to the literature [53]. Nevertheless, two types of stationary phases have emerged during the last years which seem to be cornerstones of new innovations. Their importance is still increasing and they are therefore discussed in a little bit more detail:

- Chiral stationary phases for the separation of chiral and achiral compounds
- Preparative monoliths

21.3.1.1 Chiral Stationary Phases (CSP). The direct separation of enantiomers by preparative HPLC is now widely used, and a large number of CSP are commercially available. As a method to produce both enantiomers of a drug candidate directly at the beginning of the clinical development, it is becoming more and more attractive because it allows the rapid and easy supply of amounts for biological testing, for toxicological studies, and even, in a later stage, for clinical testing. In addition, data on the activity and toxicity profiles of the individual enantiomers are meanwhile systematically required by health authorities for new drugs submitted for registration. In addition, the concurrent development of simulated moving bed chromatography (a chromatographic system that ideally separates two component mixtures, see later) was fortunate for the boom in enantiomeric separations now reaching a production scale. Several reviews have been published [54–58] introducing CSP based on naturally occurring polymers (e.g., cellulose and amylose), synthetic chiral polymers (e.g., poly(meth)acrylamides), and chirally modified silica gels (e.g., "Pirkle phases," classifiable into π-acceptor and π-donor phases). While some 10–20 years ago it was generally believed that each chiral separation problem needed its own CSP for resolution, the applications of the last years have clearly revealed that up to 90% of all chiral separations can be performed with the aid of about 4 CSP. The "Daicel columns"—in particular, Chiralcel-OD and Chiralpak-AD (Figure 21-8)—have demonstrated their superior status in the field, and several applications are mentioned in Table 21-4.

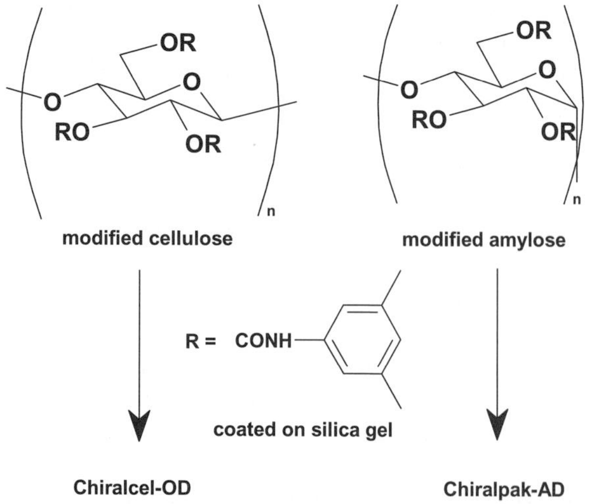

Figure 21-8. Structure of Chiralcel-OD and Chiralpak-AD.

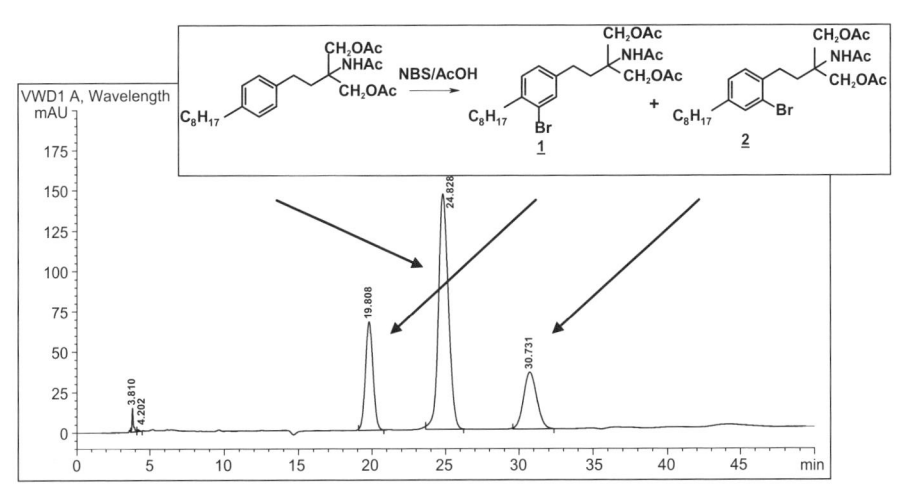

Figure 21-9. Separation of Br isomers of a drug intermediate on Chiralpak-AD (250 × 4.6 mm); mobile phase n-hexane/2-propanol = 100/5, 0.8 mL/min; temperature 30°C, UV detection 210 nm.

Whereas the separation of racemates on these two CSP are obvious, recent applications demonstrate that achiral isomers, especially aromatic compounds with substituents in different positions, are extraordinarily well separated on Chiralcel-OD and Chiralpak-AD as well (Figure 21-9). It is to be expected that further examples will follow and more and more achiral separation problems will be solved in the future on CSP.

21.3.1.2 Monoliths. Very recently, both silica-based and polymeric monolith preparative columns were introduced [59]. The positive feature of monoliths is their high permeability; thus, for preparative chromatography, they can be operated at high flow rates and still exhibit their good efficiency. Monolithic silica rods, offered by Merck (Darmstadt, Germany), are porous monoliths consisting of a skeleton with interconnecting macropores. Inside the silica skeleton a large number of mesopores is present. The mesopores determine the surface area of the sorbent, which is necessary for a high maximum loadability. The independent control of macro- and mesopores is a prerequisite for achieving a material useful for preparative chromatography. The monolithic silica rods are prepared via sol–gel process [60]. By varying the amount of polyethylene oxide in the starting sol mixture, the size of the macropores can be influenced (typically 3 mm). The controlled formation of the mesopores is achieved by immersion of the silica in an aqueous ammonium hydroxide solution. The duration and temperature of the process determine the mesopore size. Preparative applications have recently been published with the purification of 45 mg cyclosporine A from fermentation broth on a PrepROD column (100 × 25 mm i.d.) within a few minutes [61]. And by using eight columns simultaneously in a SMB unit, the separation of a 1.3-kg mixture consisting of χ- and δ-tocopherol from vegetable oil could be achieved in one day.

In the field of polymer-based monolith columns, BIA (Lubliana, Slovenia) has expanded its line of methacrylate copolymer convective interaction media (CIM) columns. The 800-mL column, based upon a poly(glycidylmethacrylate-co-ethyleneglycoldimethacrylate) polymer, was functionalized with a diethylamino group to be used for anion exchange separations. With a dynamic protein-binding capacity of 20 to 60-g protein/mL wet support, this column is focused on industrial scale biochromatography and is the first cGMP-compliant, industrial-scale monolith with a Drug Master File and other documentation for scale-up from research purification.

21.3.2 Particle Size, Shape, and Distribution

As has been outlined in the preceding section [equation (21-1)], the efficiency of the column is linked with the number of plates and with the particle size of the stationary phase, respectively. Theoretical work has shown [62] that there is an optimum particle size that depends on the conditions of the purification: the selectivity of the phase system, the isotherms, and so on. Accordingly, it is not possible to define an absolute optimum particle size. Nevertheless, most industrial applications are published with stationary phases using particles between 10 and 30 μm. From a practical point of view (pressure reasons), it is very unlikely that material with less than 5 μm will be used. The same is true for material with larger particles than 30 μm. A larger impact is noticed with respect to the particle size distribution. As has been demonstrated by Colin [63], a column with an artificial mixture of 3- and 8-μm particles exhibits a three times larger pressure drop than a comparable column with exactly 6-μm particles. From an economic point of view, it is necessary to run the equipment while using its full pressure capabilities, and high flow rates will then contribute to the productivity directly. In other words, a packing material with a large size distribution is not a good choice because the pressure capability of the equipment is used to overcome the flow resistance created by the small particles rather than speed up the separation. Whereas spherical particles are made directly at the right size with a very narrow size distribution, angular particles are obtained by crushing and sieving, which yields a broader particle size. Since the latter is not desirable, as mentioned before, spherical particles are very often advantageous. Nevertheless, angular material is sometimes used, especially when the efficiency is sufficient and the price of the material is more attractive.

21.3.3 Columns and Packing Procedures

For a given type of stationary phase, the efficiency of a column is mainly determined by the column length and the packing procedure. The quality of a packing technique can easily be derived from well-defined parameters: (i) the efficiency expressed in terms of reduced plate height, (ii) the reproducibility of the filling procedure (an important factor for the setup of SMB systems), and (iii) the long-term stability of the column to ensure continuous operation. It is meanwhile common practice to use the *dry filling* approach for materials with a particle size above 25 μm ± 5 μm and the *slurry method* for smaller particles. Both methods and their advantages have been described in detail by Dingenen [64]. Of the problems associated with increasing the column size, the redistribution of particles seems to be the major one. This is related to the loss of wall support, or, in other words, the existence of unstable regions formed in the bed during the packing process. They correspond to bridges of particles surrounding empty spaces. If these bridges collapse (because of shear forces, mechanical vibrations, etc.), redistribution takes place, resulting in reduced efficiency because diffusion takes place in these voids, resulting in band distortion and loss of separation power. The technology to fill large columns should avoid the formation of such voids. This hurdle can be overcome by using compression techniques. This does not mean that the redistribution will not happen, but the consequences are eliminated. Several compression methods have meanwhile been described in the literature [64] and are used for preparative HPLC. Nevertheless, it should be pointed out that most applications are performed with equipment using dynamic axial compression. With this approach, the column is packed and operated with a high piston pressure. The pressure is always maintained on the bed during column operation, and the piston always pushes on the bed. It is obvious that under these conditions the efficiency of the column can be held, since the formation of voids is permanently corrected. It has been demonstrated that columns operated under dynamic axial compression showed no loss in efficiency after days, whereas the efficiency dropped by 50–70% for columns with the same material which were operated without piston pressure after the packing. By means of the axial compression technique, it is also possible to reproducibly fill columns. Furthermore, the packed bed is stable and the bed length can easily be adjusted over a broad range just by choosing the desired amount of slurry. In addition, it is possible to remove the packing in a fast and clean way from the column, and finally the technology is easily scaled-up from semipreparative columns to large-diameter columns for industrial applications.

21.4 CHOICE OF PREPARATIVE LC TECHNOLOGY

From a process-engineering point of view, there is now a better understanding of the development of concentration profiles in chromatographic columns under overloaded conditions available. This includes in particular the quantitative description of displacement and tag-along effects caused by competitive adsorption. Since it is now possible (as mentioned before) to simulate concentration profiles on a personal computer, the choice of the appropriate mode of chromatography is easily achieved. Nevertheless, in pharmaceutical development, the equipment is very often given, and the chromatographic method will be adjusted accordingly. In addition, the amount to be purified has also a great influence on the chosen technology. The following section is intended to briefly introduce the different routes of preparative chromatography that are mainly used on pilot plant and production scale.

21.4.1 Classical Batch Elution

The most common approach used, especially in early development when small quantities (several kilogram) have to be purified, is classical batch elution. The lack of a need at that early stage to optimize the separation very often leads to suboptimized processes that seem to be disadvantageous in comparison with an excellent designed countercurrent process. Nevertheless, this comparison will become more favorable for batch elution when the full capacity of the column is being used. It should not be forgotten for preparative runs in isocratic mode that the process can be optimized in such a way that several separations can be performed successively on a column until the compounds of the first injection elute. The net elution time is then identical with the time interval between two injections. An appropriate application for the separation of a racemate of a drug substance intermediate on a CSP is shown in Figure 21-10.

21.4.2 Recycling Chromatography

In the case of low separation factors, recycling chromatography is often used to allow higher injection amounts. The technology nicely mimics longer columns without having the drawback of higher backpressure, and it can easily be adapted to conventional equipment. For the closed-loop recycling approach, a connection between detector outlet and pump inlet was first demonstrated by Porter and Johnson [65, 66] (a schematic diagram is given in Figure 21-11). In the period of recycling, the sample is reinjected in the column several times after passing the pump. By switching the four-port valve, the recycling procedure can be stopped and the samples will be eluted. In its peak shaving approach the switching process of the four-port valve can be arranged in such a way that pure side fractions are collected and the area of incomplete separation is again recycled. Both approaches therefore offer a solution to problems in preparative chromatography where under normal batch elution only partially resolved products are obtained. Since no fresh mobile phase is required during the recycling process, the solvent savings in recycling chromatography are considerable.

Theoretical treatments on recycling chromatography have been published by Chizhkov [67], Martin [68], and Coq [69]. Seidel-Morgenstern and Guiochon [70] developed a mathematical model to design recycling and

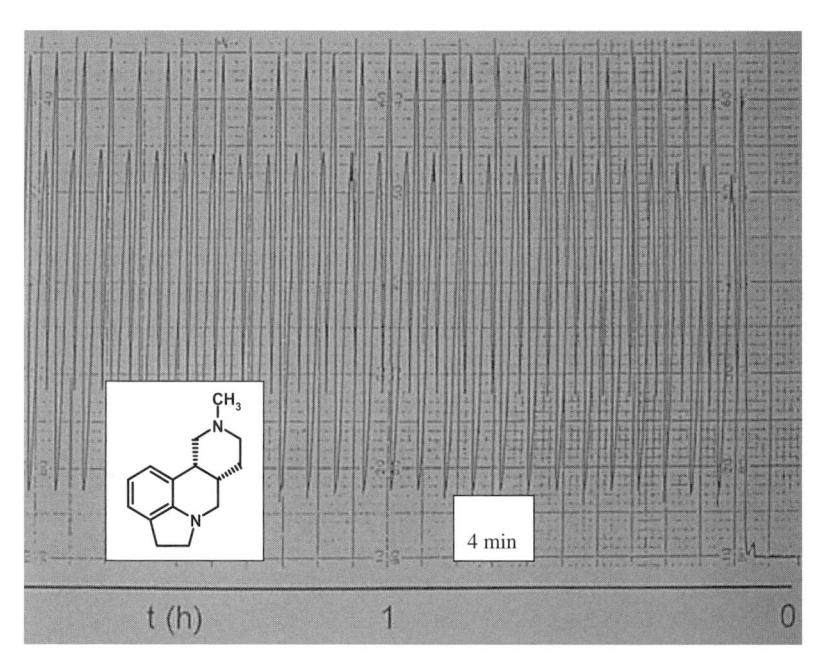

Figure 21-10. Enantiomeric separation of a drug substance intermediate on a chiral stationary phase [semipreparative column (34-cm × 10-cm i.d.) containing 1.8 kg of CSP]; 5 g of racemate is injected every 4 minutes onto the column; this is the identical time interval for the peak widths of both enantiomers after elution.

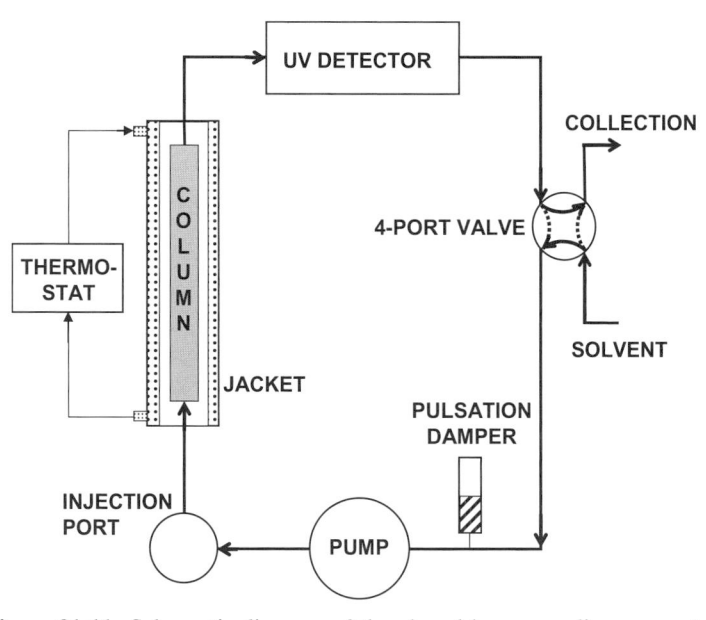

Figure 21-11. Schematic diagram of the closed-loop recycling apparatus.

peak-shaving chromatography under overloaded conditions. From a practical point of view, enantiomeric separations with CSP are very attractive, with chemical purity of the racemate provided. The first separation of enantiomers was performed by Schlögl. Others followed, recognizing the time and mobile-phase savings in the technology. An excellent review of preparative chromatographic resolution of racemates on CSP on a production scale by closed-loop recycling chromatography has recently been published by Dingenen and Kinkel [71]. The separation of kilogram quantities within one week is reported for hetrazepine, α-(2,4-dichlorophenyl)-1*H*-imidazole-1-ethanol, benztriazole derivatives, γ-aryl keto esters, and an alkylated 2-piperazinecarboxamide on various CSP. An outstanding preparative example has been described by Dingenen and Kinkel for the enantiomeric separation of 2,2'-dihydroxy-1, 1'-binaphthyl, whose enantiomers can be separated on a chiral polyacrylamide-diol-silica copolymer, prepared from *N*-acryloyl-(*S*)-phenylalanine diethylamide. By peak shaving of the first-eluted enantiomer starting in the first cycle, 1.35 g of racemate dissolved in 1.3 mL of eluent were separated within 12 min. The productivity and solvent consumption has been calculated for a semipreparative column (250-mm × 100-mm i.d.) filled with 1 kg of CSP. More than 18 kg of racemate can be separated per week with a total solvent consumption of 3760 L, an amount easily handled in a pilot plant.

Examples from our group have been worked out for the separation of 1-phenylethanol on Chiralcel-OD [72] and for the separation of a 2,6-dimethyl-8α-aminoergoline intermediate on Chiralpak-AD [73]. For the latter racemate, 0.5 g has been separated on 192 g Chiralpak-AD after three recyclings (Figure 21-12).

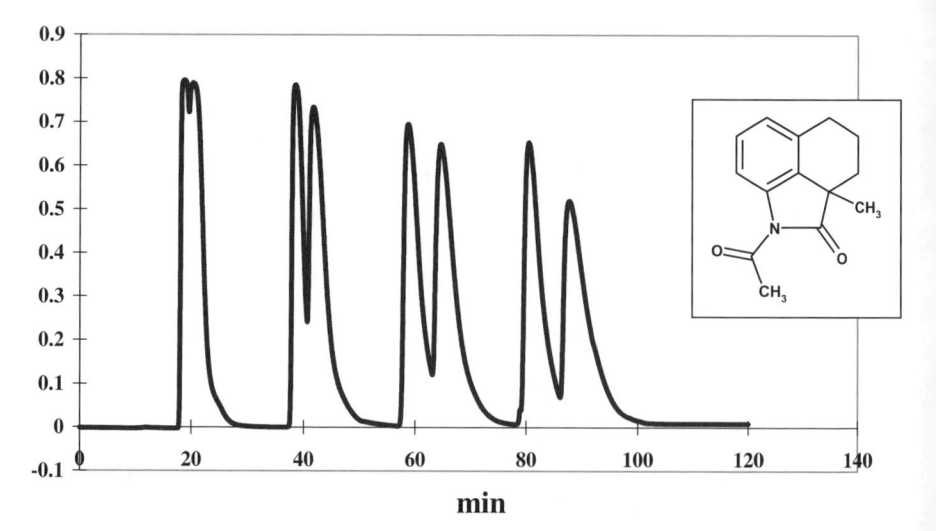

Figure 21-12. Separation of 0.5 g of racemic 2,6-dimethyl-8α-aminoergoline intermediate on 192 g of Chiralpak AD.

21.4.3 Displacement Chromatography

In displacement chromatography [74–76] the packed column is equilibrated with a mobile phase that has a very low affinity to the stationary phase. Then the feed solution containing the mixture dissolved in the mobile phase is injected in such a way that the components are adsorbed at the top of the column. In the next step the solution of a displacer substance that has stronger affinity to the stationary phase than any of the feed components is pumped into the column. The components of the feed arrange themselves upon the action of the displacer front moving down the column into a "displacement train" of adjoining square wave concentration pulses of the pure substances, all moving with the same velocity. After the product zones have passed, the column is regenerated by removing the displacer and re-equilibrating with the mobile phase. This additional operational step does not contribute directly to the separation, but is an undesirable feature of the technique. The relationship between the final pattern in displacement development and the isotherms of the displacer and the feed components is illustrated in Figure 21-13. As can be taken from the figure, the concentrations of fully developed zones of the components are determined by the intersections of the individual adsorption isotherms with the operating line. The requirement for complete displacement development to occur is that the isotherms should be convex and that the operating line drawn as the chord of the displacer isotherm intersects the isotherms of all feed components. Having in mind all additional prerequisites (like the search for a suitable displacer, determination of adsorption isotherms, etc.), it becomes obvious that the successful development of a chromatographic displacement process is more time-consuming than in the classical elution mode. On the other side, the increase in productivity (as a result of higher injection amounts) is very often many-fold in comparison to preparative chromatography under nonlinear conditions.

21.4.3.1 Examples. The majority of published displacement processes deal with the purification of peptides [77] and proteins [78–80] and the separation of racemates [81–83]. An example from our group may again highlight the potential of displacement chromatography for the purification of peptides. The peptide backbone of a calcitonin analogue is produced by solid-phase synthesis. Following cleavage from the resin and purification by conventional reversed-phase elution chromatography, the peptide is glycosylated in the presence of acetic acid and dimethylformamide. The crude product (Figure 21-14a) with a content of 75% contains the di-glucose form of the peptide and unreacted peptide (starting material) as the most crucial impurities which have to be removed.

Estimation of adsorption isotherms, the search for an appropriate displacer, scale-up, and optimization of the method with an analytical column will be described elsewhere [84]. The result of this development report can be summarized as follows. After equilibration of the reversed-phase column

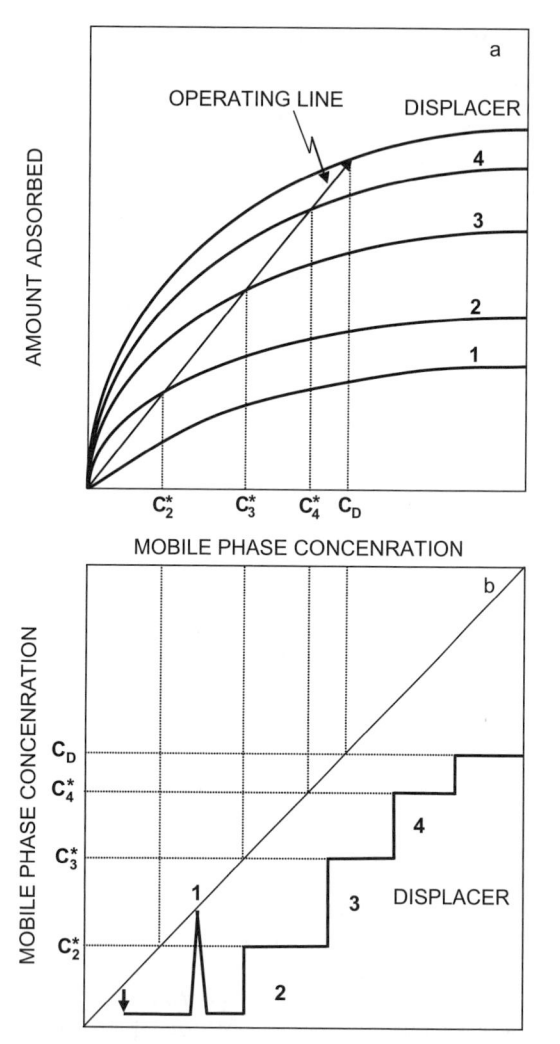

Figure 21-13. Relationship between the final pattern in displacement development and the isotherms of the displacer and feed components. The intersection of the operating line with the feed component isotherms determines their concentrations in the displacement train. The isotherm of component 1 lies beneath the operating line, hence it is eluted in the mobile phase. (Reprint from reference 74, with permission.)

(Hyperprep ODS 8 μm) with mobile phase (28% acetonitrile and 0.2% phosphoric acid), the feed was dissolved in the mobile phase at a concentration of 20 mg/mL, and the total feed mass was about 1/140th of that of the stationary phase in the column. The displacer solution contained benzyldimethylhexadecyl ammonium chloride dissolved at 5 mg/mL in a solution of 32% acetonitrile and 0.2% phosphoric acid. Upon breakthrough of the peptide

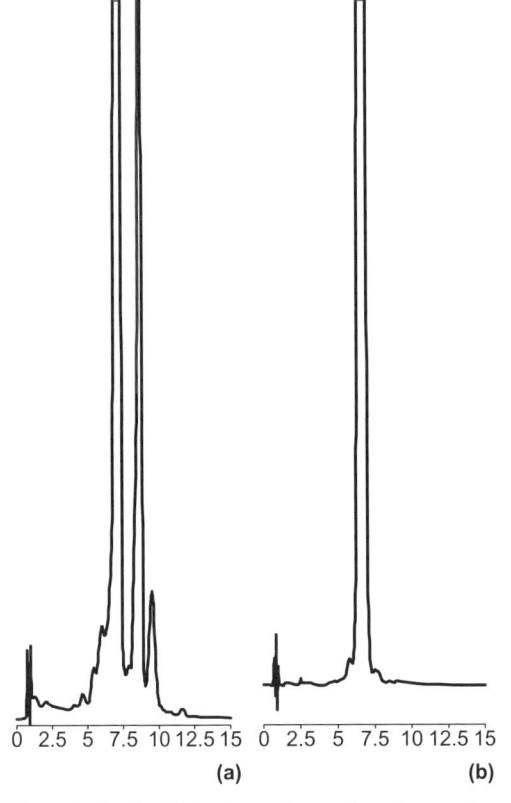

Figure 21-14. HPLC analysis of calcitonin analogue (crude reaction product) after gly-cosylation (a) and after displacement chromatography (b).

front—monitored by UV detection at 250 nm—fractions were collected and analyzed. Finally, the column was regenerated with 70% acetonitrile and 0.4% phosphoric acid at the beginning and ending with pure methanol. The purifi-cation effect is illustrated in Figure 21-14b, demonstrating that a final purity of 99.7% for the drug substance has been achieved. The process has been scaled-up, and several runs on pilot plant equipment have been performed. A careful comparison of drug substance chromatography in elution and dis-placement mode has been explored and production scenarios calculated. The results are summarized in Table 21-3.

The direct comparison has been made between a typical run in a 50-cm × 30-cm i.d. column running in the elution mode and the results of a displace-ment run on a 25-cm × 15-cm i.d. DAC column. In both cases, it is possible to carry out two runs per day (about eight per week is more realistic). Produc-tion rates with the two methods are quite similar. However, in displace-ment, eight times less sorbent is employed, and solvent consumption is much lower. Two advantages of the elution method are that it requires no

**TABLE 21-3. Comparison of an Elution Run on a
50-cm × 30-cm i.d. Column with a Displacement Run on a
25-cm × 15-cm i.d. Column for the Purification of a
calcitonin analogue**

	Elution	Displacement
Sorbent in column	30.0 kg	2.8 kg
Product per run	10–12 g	9.8 g
Runs per week	8	8
Total liquid pumped through column	450 L	54 L
Acetonitrile	160 L	13 L
Methanol	—	18 L
Displacer	—	105 g

preconcentration of the raw reaction product and that it operates at room temperature. However, preconcentration is a simple procedure and adds very little to the overall cost of the displacement method: enough material for about 14 displacement runs can be generated from one concentration run on a column of the size used for displacement. Moreover, preconcentration can be performed on a cheaper and coarser sorbent, if necessary.

21.4.4 Simulated Moving Bed Chromatography

Counter current processes have proven to be superior for the separation of binary mixtures in comparison to batch processes. Several disadvantages of batch elution chromatography (e.g., discontinuous injection of feed and operation, noneffective adsorbent utilization, and high product dilution) have been overcome by simulated moving bed (SMB) chromatography. This technique was originally developed by Broughton [85–87] to separate structurally related hydrocarbons. Whereas the technique has been used for many years in the petrochemical and sugar [88–90] industries, the first successful SMB enantioseparation was reported in 1992 by Negawa and Shoji (Table 21-4). It has since been used for the separation of several racemates (recently reviewed by Schulte and Strube [91]) and also for the purification of natural products obtained from fermentation processes [92]. A short summary of the principles of this new and promising technology is given below, followed by examples where special emphasis was laid on the topic of method development and design of an SMB process.

21.4.4.1 The True Moving Bed. The principle of a true moving bed is schematically illustrated in Figure 21-15 for the separation of a racemate on a chiral stationary phase, being the ideal problem for the separation of a binary mixture. There is countercurrent contact between the solid phase and the eluent which move in opposite directions. The racemate is injected in the middle of the column. Chiral discrimination provided for by the sorbent;

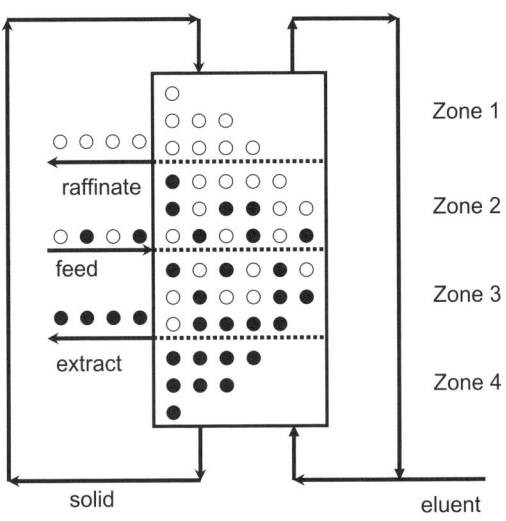

solid   eluent

Figure 21-15. Schematic of a true moving bed for the enantioseparation of a racemate on a CSP.

appropriate choice of the solid and liquid flow rates will therefore separate the racemate into two fractions. The less retained enantiomer (A) goes up in the direction of the mobile phase and is collected in the raffinate stream, whereas enantiomer B moves in the direction of the sorbent and is collected in the extract stream.

21.4.4.2 The Simulated Moving Bed. The technical realization of a SMB system is shown in Figure 21-16. The adsorbent is fixed in columns connected in series. Two incoming and two outgoing liquid streams divide the unit into four sections or zones. Each of the four zones (I, . . . , IV) consists of at least one column and has to fulfill distinct tasks [93]. At the feed point and at the eluent point the feed mixture to be processed and the fresh mobile phase are introduced into the unit, respectively. At the points of withdrawal the extract stream enriched with the more retained component and the raffinate stream enriched with the less retained component leave the unit. The desorbent flowing out of zone IV is recycled to zone I. The flow of adsorbent is simulated by shifting the inlet and outlet positions at a constant time interval in the direction of the liquid flow. The shifting of the incoming and outgoing ports mimics an apparent solid flow opposite to the direction of the liquid flow. The shifting period is linked to the solid flow rate of the equivalent true moving bed shown in Figure 21-15. In order to achieve separation of the feed components the internal flow rates of the liquid phase within the four zones and the shifting period corresponding to the apparent flow of the solid phase have to be specified. These operating conditions can be achieved by computer simulation of the concentration profiles (e.g., with the *help software* from

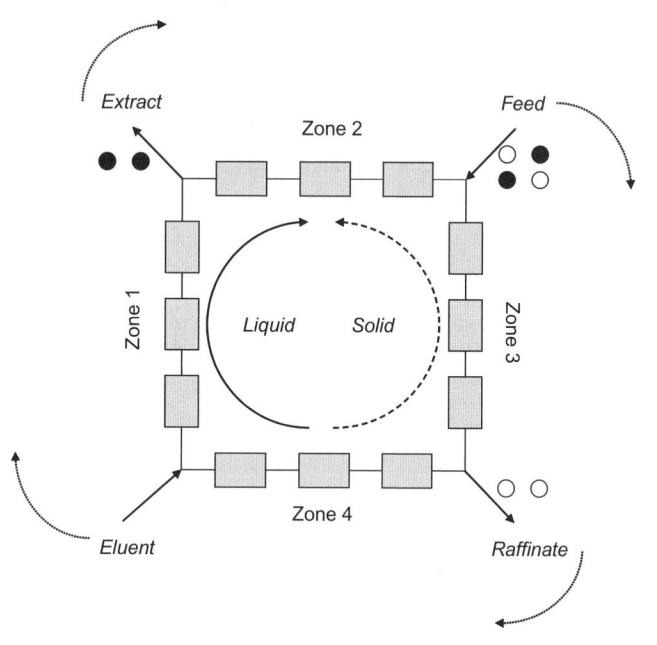

Figure 21-16. Schematic representation of a four-zone SMB chromatographic system.

Novasep, Champigneulles, etc., or with the *SMB guide software* from Knauer, Berlin, etc.) along the columns on the basis of known competitive adsorption isotherms or even easier empirically with the help of a chart that depicts visually the interrelationships between the system flow rates and the SMB design criteria as described by Antia [94]. The steady-state concentration profile in such a SMB system is schematically illustrated in Figure 21-17 for an artificial two-component mixture with linear conditions assumed.

21.4.4.3 Operating Conditions. A powerful tool to analyze the migration of concentration fronts in a fixed bed is the classical equilibrium theory. The basic equations of this theory are equation (21-2) and equation (21-3), assuming all D_{ap} to be zero. These equations have to be fulfilled for periodically changing boundary conditions in the multicolumn arrangement. Recently, several substantial contributions to analyze the SMB process exploiting the analogy to the more simple true moving bed process have been presented by Morbidelli and co-workers [95, 96]. The most interesting result for design purposes is the derivation of expressions for the linear velocities, u_j, or the corresponding volumetric flow-rates, $\dot{V}_j$, required in the four zones ($j =$ I, . . . , IV) to achieve the complete separation of a binary feed into two streams containing the pure components. The region of flow rates fulfilling this condition can be specified as a function of the four net ratios, m_j, related to the volumetric flow rates of the liquid and the solid. The latter flow rate depends for

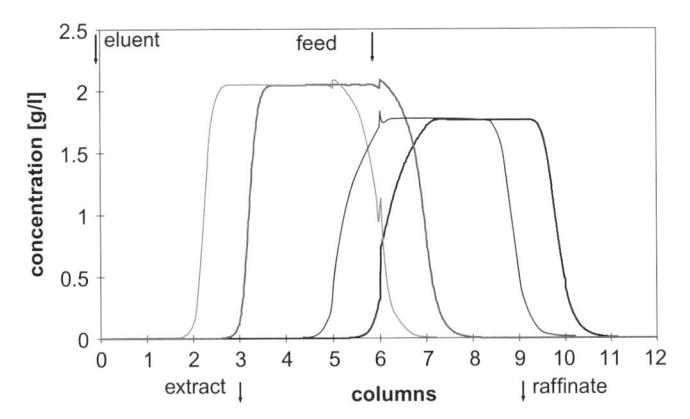

Figure 21-17. Concentration profile of an artificial two-component mixture (A and B) with linear conditions assumed. The SMB system consists of 12 columns (3 per zone). The profiles symbolize the movement of the peaks from the start to the end of a tact. More columns in combination with shorter shifting times will result in an optical "nonmoving steady state" concentration profile. It is obvious from the figure that an SMB system makes optimal use of the total stationary phase in the system and has to be much more economic in comparison to classical batch elution.

the SMB process on the shifting time t_{shift}. For the net flow ratios, equation (21-9) holds.

$$m_j = \frac{\dot{V}_j}{\dot{V}_{solid}} - \frac{\varepsilon}{1 - \varepsilon}, \quad j = I, \ldots, IV \qquad (21\text{-}9)$$

Most critical for a successful separation are the net flow ratios for the regions II and III upstream and downstream of the feed position, that is, m_{II} and m_{III}. Explicit equations are given to calculate regions in the m_{II}, m_{III} plane for the constant selectivity Langmuir model and modified variable selectivity Langmuir models [97]. Performing a superposition of two regions calculated with the Langmuir model approximate equations were derived for a Bi-Langmuir model [98]. The size and shape of the regions of complete separation depend on the isotherm model and the feed concentration. The more concentrated the feed is, the smaller the region of available operating parameters becomes. Usually the region has a triangular shape in the m_{II}, m_{III} plane.

21.4.4.4 Examples. The following examples from different areas will highlight the potential of SMB chromatography. The first one deals with the enantiomeric separation of a racemate, being an ideal two component separation problem. The second example shows the possibility of drug substance purification from a crude mixture, which could be achieved with two SMB process steps. The last application, an enantiomeric separation of a drug substance intermediate, again demonstrates that SMB chromatography on chiral stationary phases has meanwhile been successfully implemented on production scale. Finally, an overview of published applications on chiral stationary phases and drug substance purifications is briefly summarized in Table 21-4.

Example 1. Enantiomeric Separation of 1-Phenoxy-2-Propanol. The following example, the enantiomeric separation of 1-phenoxy-2-propanol, roughly illustrates the individual steps for the design of a suitable SMB process. Details, especially for the determination of the adsorption isotherms and the description of the SMB system, can be taken from the literature [124].

Step 1. Identification of a Suitable Chromatographic System. Chiralcel-OD has been recommended for the separation of the racemate using a mixture of n-hexane/2-propanol as mobile phase [125].

Step 2. Determination of Adsorption Isotherms. After injection of a small sample of hexane, the porosity of the stationary phase has been determined according to $\varepsilon = 0.68$. The MDM method was chosen for the determination of the adsorption isotherms, where solutions containing known concentrations of the racemate were pumped through the column. After reaching equilibrium indicated by constant plateau concentrations, the two retention times obtained by the injection of a small perturbation into the system were measured. The primary data were converted using equation (21-4) into two total derivatives, dq/dC, for each plateau. By fitting these differentials to the theoretical values based on the isotherm equation (21-8), the free parameters were determined. Figure 21-18 shows the total derivatives obtained from the retention times measured. The corresponding best fit of theoretical results for the chosen isotherm model is also shown in Figure 21-18, indicating a good agreement between theoretical and experimental data.

Step 3. Operating Conditions from the m_{II}/m_{III} Plane. To specify appropriate operating conditions for the separation of the racemic mixture, at first the region of complete separation in the (m_{II}/m_{III}) plane was calculated using equilibrium theory and the determined adsorption isotherms. Due to the strong nonlinearity of the isotherms, the region was found to be small for higher feed concentrations. The corresponding region of complete separation is shown in Figure 21-19. The consideration of the relative position of the operating point in the m_{II}/m_{III} plane allows an estimate of the purities of the two outlets. The operation conditions for pure enantiomers with the highest productivity are marked with a bullet.

Step 4. Experimental Verification. To demonstrate the usefulness of the described procedure for the specification of appropriate operating conditions and to verify the proposed mathematical model, experimental results were compared with numerical predictions. Figure 21-20 illustrates the internal concentration profiles of both enantiomers along the columns of the SMB obtained for the operating point marked in Figure 21-19 after reaching the cyclic steady state. The predicted profiles are shown just after and

TABLE 21-4. Summary of SMB Separations Reported in the Literature (Only Enantiomeric Separations on CSP and Purification of Intermediates and Drug Substance in Pharmaceutical Development)

Substance	Type of Separation e.s./p.p.[a]	Company	Stationary Phase	Productivity (g/Enantiomer or Product/Day kg Stationary Phase)	Solvent Consumption (l/g Enantiomer or Product)	Ref.
1-Phenyl-ethanol	e.s.	Daicel	Chiralcel-OD	70	3.33	99
	e.s.	1. Novasep/ Novartis	CTA	35	0.80	100–103
		2. Novartis	Chiralcel-OD	37	0.50	
		3. University of Porto	CTA	59	0.80	
"Sandoz-Epoxide"	e.s.	Novasep	Chirosol Proline	4	0.99	104
Threonine	e.s.	Novasep/ Boehringer Ingelheim	CTA	1453	0.18	63
WEB 2170	p.p.	Novasep	LiChrospher Si	549	0.08	105
	e.s.	University of Singapore	Cellulose triacetate	123	0.59	106
Praziquantel						

TABLE 21-4. *Continued*

Substance	Type of Separation e.s./p.p.[a]	Company	Stationary Phase	Productivity (g/Enantiomer or Product/Day kg Stationary Phase)	Solvent Consumption (l/g Enantiomer or Product)	Ref.
Aminoglutethimide	e.s.	Novasep/ Novartis	Chiralcel-OJ	160	1.55	107
Guaifenesin	e.s.	Novartis	Chiralcel-OD	77	0.67	108, 109
Formoterol	e.s.	Novartis	Chiralcel-OJ	11	10.25	67
EMD 53986	e.s.	1. Novasep/ Merck 2. Novasep/ Merck 3. Merck/ University of Dortmund	Polyacrylamide Chiralpak AD Chiralpak AD	319 432 375	2.54 2.60 1.48	68, 110, 111

Compound		Company	CSP			Ref.
EMD 77697	e.s.	Merck/University of Dortmund	Chiralcel OD	451	1.64	69
(quinuclidine oxime methyl ether)	e.s.	ETH Zürich/Merck	Chiraspher	103	1.82	112
(tert-butyl oxazolidinone, 2,2,2-trichloroethyl carbamate)	e.s.	SmithKline	Chiralpak AD	258	0.65	113
(2-[(dimethylamino)methyl]-1-(3-methoxyphenyl)cyclohexanol)	e.s.	UCB pharma	Chiralpak AD	600	0.29	114
Tramadol	e.s.	UOP	Chiralcel-OD	31	n.m.	115, 116
3-Chloro-1-phenyl-propanol	p.p.	AWD	LiChrospher Si	44	4.18	117
Cyclosporine A Paclitaxel	p.p.	Purdue University	Polystyrene-divinylbenzene	n.m.	7.4	118

TABLE 21-4. Continued

Substance	Type of Separation e.s./p.p.[a]	Company	Stationary Phase	Productivity (g/Enantiomer or Product/Day kg Stationary Phase)	Solvent Consumption (l/g Enantiomer or Product)	Ref.
[chemical structure]	e.s.	Daicel/Nissan	Chiralcel-OF	268	0.44	119, 120
3R,5S-enantiomer of DOLE						
Ascomycin derivative	p.p.	Novartis	Zorbax LP	50	2.33	121
	e.s.	Merck	Chiralpak AD	311	0.59	122
EMD 122347 [chemical structure]						
[chemical structure]	e.s.	UOP/Daicel	Chiralcel-OD	92	n.m.[b]	123
Propranolol						

[a] e.s., enantiomeric separation; p.p., product purification.
[b] n.m., not mentioned.

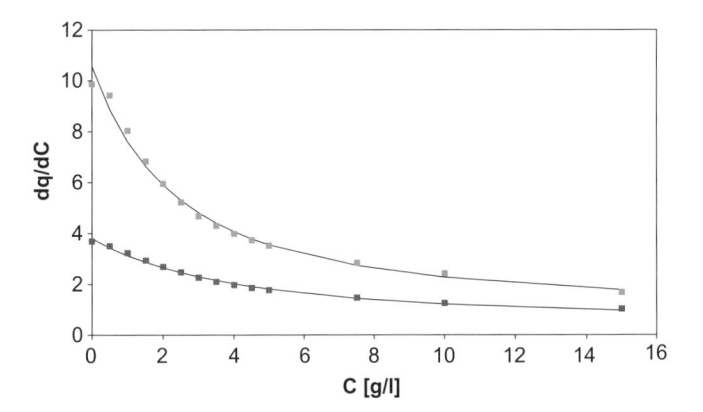

Figure 21-18. Total local derivatives of the isotherms belonging to different equilibrium states. Experimental data (squares) and theoretical results (solid lines) using the modified competitive Langmuir isotherm equation (21-8). (From reference 124, with permission.)

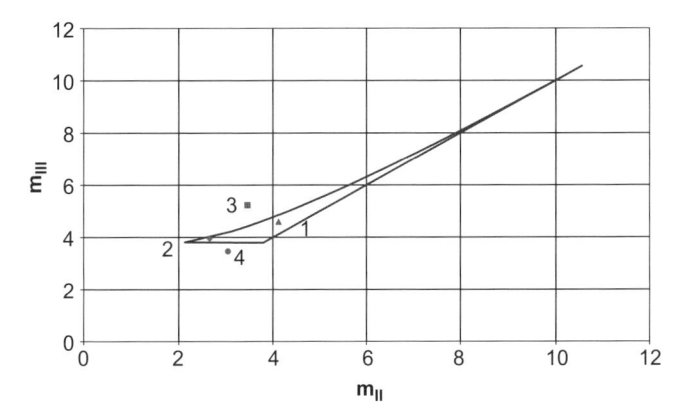

Figure 21-19. Region of complete separation (solid line) in the (m_{II}/m_{III}) plane for a feed concentration of 25 g racemate/L calculated with the parameter of the MDM adsorption isotherms and for the given lab-scale SMB unit (12 columns each 100-mm × 16-mm i.d., 3 in each zone). The four points mentioned illustrate the following scenarios: 1, robust conditions to obtain pure enantiomers; 2, less robust system with highest productivity; 3, area of pure extract and impure raffinate; 4, area of pure raffinate and impure extract. 1-△; 2-◇; 3-□; 4-○. (From reference 124, with permission.)

before a switching operation. The experimental data were measured at $t_{shift}/2$. Figure 21-20 nicely demonstrates that the experimental profiles are well represented by the model predictions.

Example 2. Purification of an Ascomycin Derivative. The second example describes the use of SMB technology for the purification of a semisynthetic

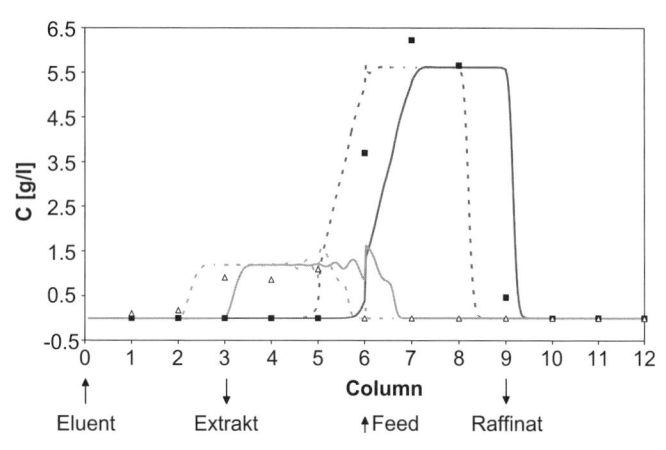

Figure 21-20. Experimental and predicted internal concentration profiles for the separation of 1-phenoxy-2-propanol using operating point 1 from Figure 21-19. Experimental data ((−)-enantiomer, ■; (+)-enantiomer, △) were taken at half-time period in the cyclic steady state. Theoretical data ((−)-enantiomer, thick lines; (+)-enantiomer, thin lines) illustrate the band profiles along the columns just after switching (dashed lines) and just before switching (solid lines). (From reference 124, with permission.)

crude ascomycin derivative from its by-products. The drug substance AD is manufactured by fermentation of ascomycin and chemically mediated cyclization [126]. The main goal of the purification process is the removal of the by-products 19-epi-AD, 9-epi-AD and ascomycin (the structures are given in Figure 21-21). Two SMB separations were performed on Zorbax LP, where in the first step the polar by-products and in the second one the apolar by-products were removed. Key elements of the whole process were the design of the first crystallization to obtain a product feasible for SMB chromatography and the specification of operating parameters for the two corresponding SMB steps. Starting from a crude product with an assay of only 45% (Figure 21-21), an overall yield for the whole process of 81% was achieved with a final purity of >98%. Figure 21-22 illustrates the internal concentration profiles for AD and 9-epi-AD along the columns of the SMB unit obtained for the chosen operating conditions. Technical details can be taken from the literature [127].

Example 3. Enantiomeric Separation of DOLE. The interest in SMB for performing large-scale separations of enantiopure drugs has been recognized, and several pharmaceutical and fine chemical companies have already developed SMB processes. Whereas most of them have described in the past only applications obtained on pilot plant scale with columns up to 5-cm i.d. (Table 21-5) [128], meanwhile the first applications on production scale have been announced [129]. DAICEL has set up an SMB production plant with columns

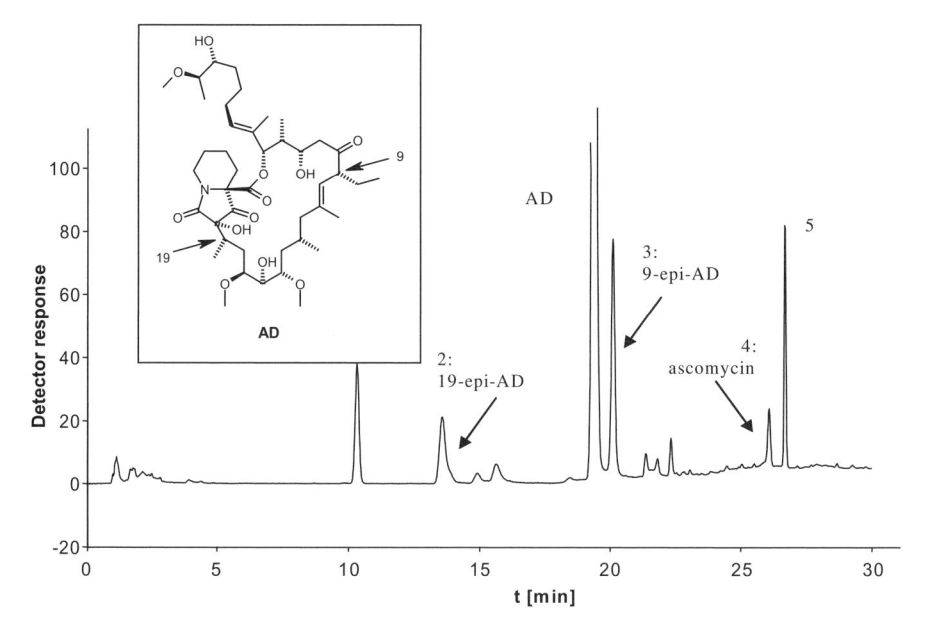

Figure 21-21. HPLC analysis of crude AD after chemical-mediated cyclization of ascomycin (in the case of 9-epi-AD and 19-epi-AD, the absolute configuration of C-9 and C-19, respectively, is opposite). The assay is only 45%.

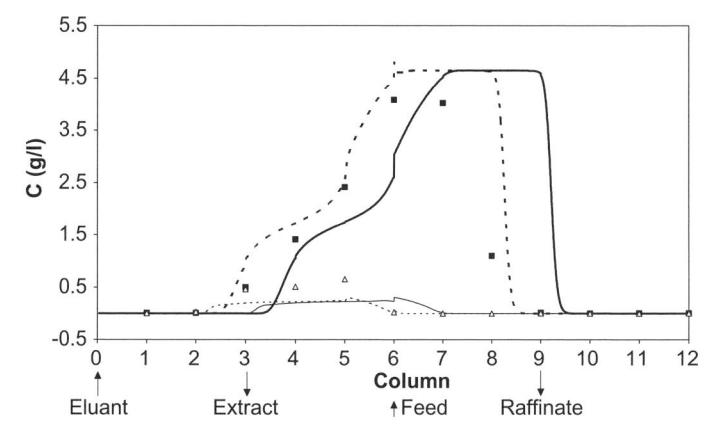

Figure 21-22. Experimental and predicted internal concentration profiles for the separation of AD and 9-epi-AD. Experimental data (AD, ■; 9-epi-AD, △) were taken at half-time period in the cyclic steady state. Theoretical data (AD, thick lines; 9-epi-AD, thin lines) illustrate the band profiles along the columns just after switching (dotted lines) and just before switching (solid lines). (From reference 127, with permission.)

TABLE 21-5. SMB Units Currently Used for Early Development and Phase I Manufacturing in the Pharmaceutical and Fine Chemical Industry

Column Size (I.D.)	Amounts to be Separated	Companies Engaged
5 cm	100 g–30 kg	Aerojet, AstraZeneca, Aventis, Bayer HealthCare, Carbogen, Chiral Technologies, Finorga, GSK, Merck KGaA, Novartis, Novasep, Numico Research, Pfizer, Schering AG, UCB, UPT
2.5 cm	10–100 g	Macfarlan Smith, Fluka

TABLE 21-6. Calculated Production Scenarios for Some Applications from Table 21-4, Assuming SMB Systems with Eight Columns and 20-cm Bed Height, an Average Density of the Stationary Phase of 0.6 kg/L, and 300 Days of Operation per Year

Desired Enantiomer	Ton/year on SMB System with Eight Columns		
	10-cm i.d.	45-cm i.d.	80-cm i.d.
Sandoz-Epoxide	0.1	2.7	8.5
Praziquantel	0.3	5.6	17.8
DOLE	0.6	12.3	38.8
EMD 53986	1.0	19.8	62.5
Tramadol	1.4	27.5	86.8
WEB 2170	3.3	66.5	210.2

of 10-cm i.d. for the production of hundreds of kilograms per year. UCB Pharma announced in 1997 its decision to install an SMB made of columns of 45-cm i.d. to replace a classical chemical process by SMB technology and Aerojet presented in 2000 an SMB unit of the same size [130].

DOLE, the ethyl ester of quinoline mevalonic acid (structure shown in Table 21-4), is a pharmaceutical intermediate of the cholesterol-reducing drug substance NK-104. The SMB process on Chiralcel-OF has been monitored at Daicel on pilot plant scale for more than 1 year while demonstrating the stability of the unit and of the chiral stationary phase (references mentioned in Table 21-4). Given a productivity of 0.268 kg enantiomer/day * kg CSP, interesting production scenarios can be calculated. Assuming an SMB system consisting of eight columns with each 20 cm of bed height, an average density of the stationary phase of 0.6 kg/L and 300 days of operation per year will yield a production of ~40 ton/year of the desired enantiomer. Further examples have been calculated on the basis of Table 21-4 and are summarized in Table 21-6 for SMB systems of different size. The results demonstrate that production on a multi-ton scale is feasible and surely implemented in some companies.

21.5 DETECTION TOOLS

Detection tools have to fulfill other demands in preparative chromatography than in analytical chromatography. For example, UV detection, the classical approach in analytical HPLC, is of minor value since preparative HPLC is normally performed under overloaded conditions, which often means that one large broad peak is monitored during elution. Since the economy of the process step is linked directly to a precise fractioning, suitable on-line technologies should be developed to simplify the fractioning process (e.g., while reducing the total number of fractions) and to improve the yield. Specific detectors being used in preparative HPLC are the polarimeter (for chiral separations), the MS spectrometer (where specific masses can be detected), or analytical tools that have been modified in a way that the key result (e.g., limit of a critical by-product) can be made available very fast (as with on-line HPLC).

21.5.1 On-Line HPLC Detection

Tailor-made fractioning has been achieved by on-line monitoring of the main stream, leaving the preparative column with the aid of a modified analytical HPLC system that analyzes the stream composition in a very short time. A small proportion of the main stream leaving the column flows through a by-pass from which the samples are taken automatically (Figure 21-23). To really obtain on-line analysis or real-time results, the analytical HPLC run has to be accelerated significantly. The most important modifications to implement are therefore as follows:

- Develop isocratic in-process control methods to avoid washing and re-equilibration after each analysis.
- Use short analytical columns with small particle size under maximum flow rates.
- The in-process control methods should be performed at high temperatures (up to 100°C) to avoid high back pressures and to speed up elution.
- The general recommendation to minimize dead volumes in a chromatographic system is of utmost importance in analytical on-line equipment. The whole system (injector, column, and detector) should be equipped with the shortest possible tubes.

With such an optimized system, analytical runs within 20–60 sec are possible. From the by-pass of the main stream, samples are automatically injected in time intervals of Δt (e.g., between 20 and 60 sec). Once the impurity (Imp.) to be removed has reached its critical value, fractioning into another vessel takes place. Ideally, the outlet tube between sampling point and fraction vessel is adjustable and designed in such a way that the analytical result is

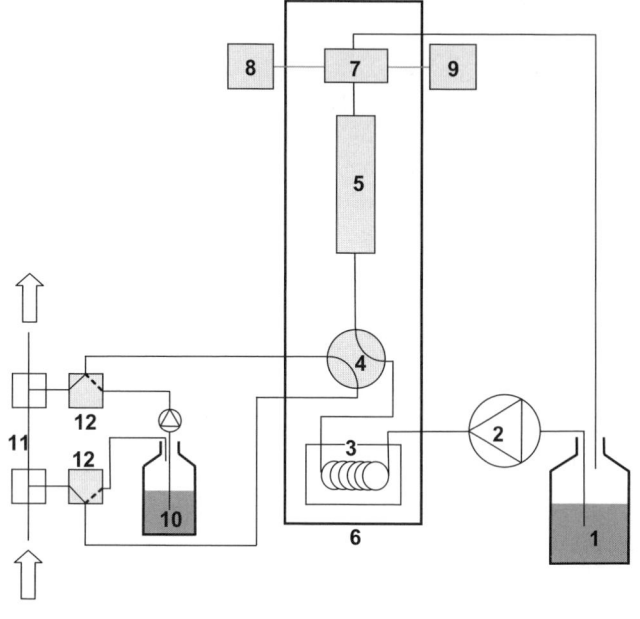

Figure 21-23. Flow scheme of an on-line HPLC system developed in-house (1, mobile phase; 2, HPLC pump; 3, heat exchanger; 4, injector with 2-μL loop; 5, HPLC-column; 6, heating oven; 7, UV cell; 8, UV lamp; 9, UV detector; 10, reference sample; 11, restrictor; 12, valves).

available prior to leaving of the solutes into the fraction vessels. The schematic in Figure 21-24 illustrates the procedure. For lucidity, the artifical preparative peak that elutes within 200 sec is analyzed in 20-sec intervals. The result of the on-line monitoring indicates that the prior eluting by-product (I) has reached its critical value after 40 sec. Depending on the requirement for the by-product in the final product (P), "fraction 3" may now be added to the product, re-chromatographed, or discarded. Real runs sometimes show peak widths up to one hour. Given a realistic sampling interval of 40 sec, this nicely illustrates that the preparative peak is then theoretically segmented in 90 portions, whereas only three to five real fraction vessels are needed in practice.

Further progress can be expected since new column materials (e.g., monolithic columns) and further improved HPLC systems (e.g., ultra performance systems) have been developed to speed up analytical runs significantly. Their use as routine on-line detection tools will improve the productivity of preparative HPLC not only in applications where broad overloaded peaks are observed, but also in cases where high concentrations of different solutes elute close to each other, an inherent aspect of displacement chromatography systems for example.

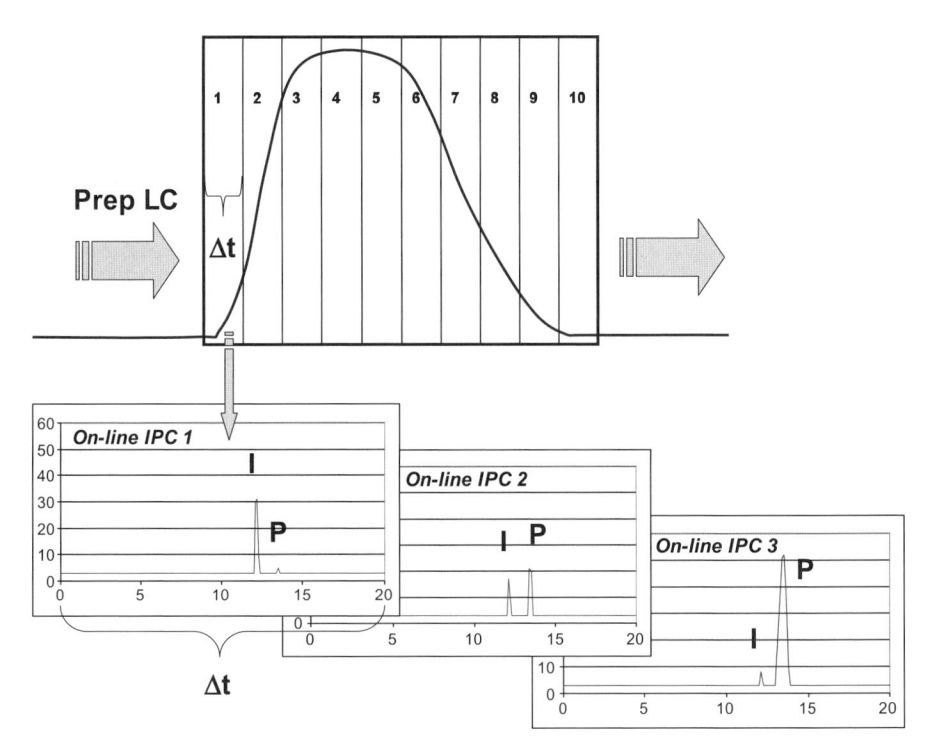

Figure 21-24. Schematic of on-line monitoring of a preparative LC run. From the by-pass of the main stream, samples are automatically injected in time intervals of Δt (every 20 sec). Once the impurity (I) to be removed has reached its critical value, fractioning into another vessel takes place.

21.5.2 Preparative HPLC-MS

Scientists who are involved in compound isolation (in particular by-products and degradation products) from complex matrices are more and more attracted by a new generation of autopurification HPLC-MS systems. These fully automated (injection, peak detection, and collecting) machines can simply shift between analytical and preparative column (the same particle size and column length will provide identical separations) and use MS detection. Post-column split flow technology maximizes the amount of sample being diverted to the fraction collector (99.9%). The remaining 0.1% is channeled to the mass detector and ancillary detectors. The software allows the mass detector to react to a mass chromatogram that meets the defined criteria (e.g., search for expected molecular masses of intermediates and products) and commands the fraction collector to deposit the analyte into collection vessels. The quick identification of reaction products using this technology is shown in Figure 21-25 for the bromination of a drug substance intermediate. Note how easily the isotope pattern of bromo derivatives from a quickly performed

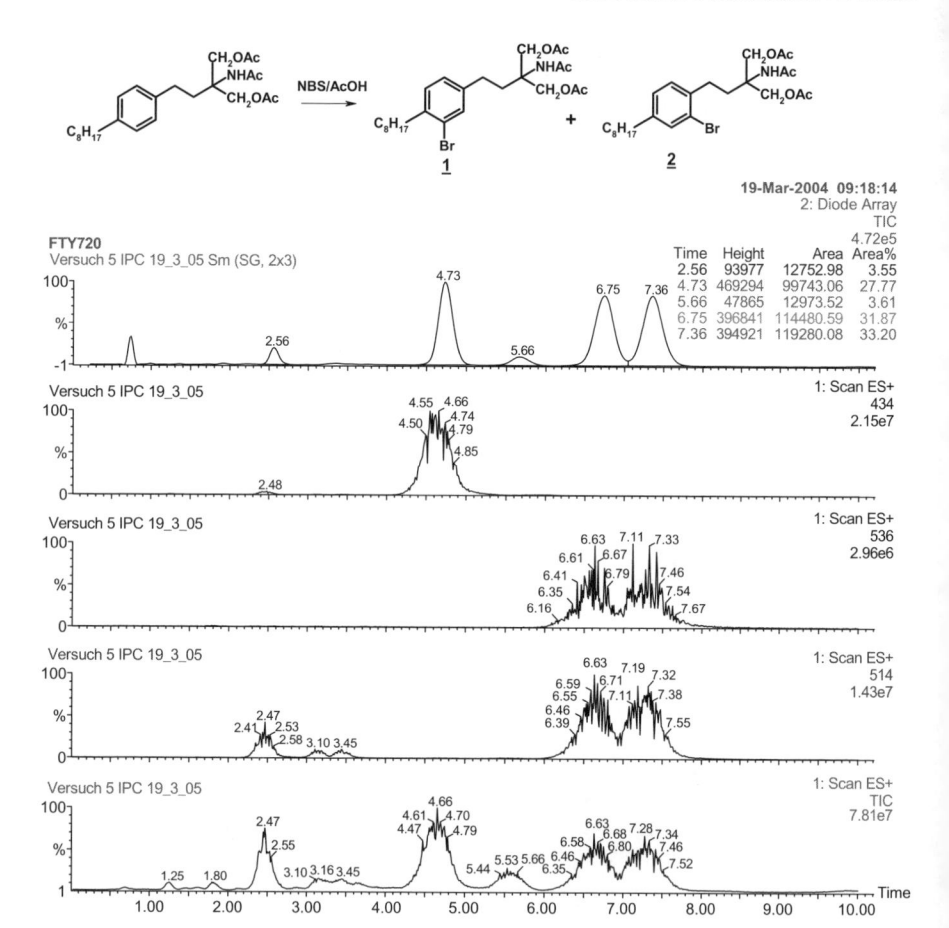

Figure 21-25. In-process control of a bromination reaction. The kinetics of the process is simply been monitored with UV detection (tray 1) after identification of the starting material ($m/s = 434$, tray 2) and the expected products ($m/s = 514$, tray 4 and $m/s = 536$ for the sodium adducts, tray 3). Another advantage is that sometimes compounds (especially those occurring in traces and being less UV-sensitive) can be made visible with the unselective mass scan (tray 5).

in-process control reveals several aliphatic and aromatic products that fulfilled the requirement of the expected molecular mass of $m/s = 514$. It is obvious that the identification from reaction mixtures containing several by-products in addition will become more cumbersome.

21.6 CONCLUSION

Preparative HPLC has become a well-known technique that is applied to the preparation of numerous purified compounds in the pharmaceutical industry.

The process is now well understood, as has been demonstrated by its successful modeling under a wide variety of experimental conditions and the excellent agreement observed between the model predictions and the experimental results [131, 132]. The theory of preparative chromatography is essentially complete; nevertheless, further improvements are to be expected. The important unsolved issues that remain so far have been summarized by Guiochon [133] as follows: *"They are: (i) in the thermodynamics of equilibrium:*

- *how can we rapidly and accurately measure equilibrium isotherms?*
- *how can we simply and accurately model these isotherms?*
- *how can we use molecular modeling to calculate these isotherms?*
- *more specifically, how can we predict the competitive isotherms of, for example, two enantiomers on a chiral stationary phase*

and (ii) in the mass transfer kinetics across the bed of a chromatographic column:

- *what are the key parameters controlling diffusion through a particle?*
- *how can we accelerate this rate?"*

Although the particle size seems to be well-optimized, the development of more efficient stationary phases has not come to an end and will provide new materials and concepts in the near future. The development of stationary phases, such as nonparticulate columns with bimodal or multimodal pore geometries (monoliths), indicates a new field of technical innovation. Further improvements in column filling are also to be expected, since preliminary results indicate that a significant reduction in the reduced HETP [2x reduced, abbr. not explained] of packed columns can be achieved [134] and can be made visible by accompanying NMR studies while detecting the packing contribution to the band-broadening [135, 136]. The existing technologies and operation modes, respectively, will further be fine-tuned, especially in the field of SMB applications. ModiCon [137, 138], the concept of modifying feed concentrations in SMB, has been developed as well as VariCol [139, 140], which makes use of an asynchronous column shifting. Especially the latter approach was described as a powerful technique to enhance the flexibility of the SMB processes. By analyzing the four zones for particular components, column segments can be removed periodically, allowing a real and significant reduction in mobile-phase usage. And last but not least, the development of a three-fraction separation SMB unit (through the addition of one pump) has been presented and will further widen the applicability of this type of continuous chromatography. Finally, combinatorial approaches are gaining more and more attraction. Increased productivities have been reported for hybrid approaches of preparative HPLC and crystallization [141], for the combination of displacement and simulated moving bed chromatography, and for integrated SMB-racemization operation. The comparison of production costs of the different implementations (e.g., overloaded elution, recycling, SMB, etc.) is so far difficult to achieve (because of missing carefully optimized conditions for each approach). In combination with further purification steps, this comparison will become more complex, highlighting the importance of a well-understood chromatographic process step.

ACKNOWLEDGMENTS

This chapter has only been made possible with the help and ideas of many colleagues at Chemical & Analytical Development of Novartis Pharma AG. In particular, the development of the displacement chromatographic process for the purification of a calcitonin analogue was performed by Dr. Firoz D. Antia [142]. I am grateful to Firoz for being allowed to present the key data prior to publication of the development strategy and the technical details. The setup of the on-line detection HPLC system has been realized and brought into operation by Dr. Heribert Hellstern. The systematic investigation into adsorption isotherms and their determination is based on the initial work of Dr. Christian Heuer. Last but not least, the technical assistance of Emil Schmid and Markus Blatter is appreciatively acknowledged.

REFERENCES

1. M. S. Tswett and T. Protok, *Varshav. Obshch. Estestvoistpyt., Otd. Biol.* **14** (1903, publ. 1905), 20. Reprinted and translated in G. Hesse, H. Weil, *Michael Tswett's erste chromatographische Schrift*, Woelm, Eschewegen, 1954.

2. M. Tswett, *Ber. Dtsch. Botan. Ges.* **24** (1906), 316–326 and 384–393.

3. P. Karrer, *Helv. Chim. Acta* **22** (1939), 1149.

4. L. S. Ettre, *Chromatographia* **51(1/2)** (2000), 7.

5. K. P. Dimick, *LC-GC Mag.* **8** (1990), 782.

6. B. A. Bidlingmeyer (ed.), *Preparative Liquid Chromatography*, Elsevier, Amsterdam, 1987.

7. G. Ganetsos and P. E. Barketr (ed.), *Preparative and Production Scale Chromatography*, Marcel Dekker, New York, 1993.

8. G. Guichon, S. Golshan-Shirazi, and A. M. Katti, *Fundamentals of Preparative Nonlinear Chromatography*, Academic Press, Boston, 1994.

9. *Preparative Chromatography* (coordinated by H. Colin), *Analusis Mag.* **26** (1998).

10. A. S. Rathore and A. Velayudhan (ed.), *Scale-Up and Optimization in Preparative Chromatography*, Chromatographic Science Series, Vol. 88, Marcel Dekker, New York, 2003.

11. F. Geiser, *HPLC 97* (1997), Birmingham.

12. G. B. Cox, *Analusis Mag.* **26/7** (1998), 70.

13. S. Jönsson, A. Schön, R. Isaksson, C. Pettersson, and G. Pettersson, *Chirality* **4** (1992), 505.

14. E. Küsters, V. Loux, E. Schmid, and P. Floersheim, *J. Chromatogr.* **666** (1994), 421.

15. K. Cabrera and D. Lubda, *J. Chromatogr.* **666** (1994), 433.

16. E. Küsters and C. Spöndlin, *J. Chromatogr. A* **737** (1996), 333.

17. J. C. Berridge, *Techniques for the Automated Optimization of HPLC Separations*, Wiley, New York, 1985.

18. P. J. Schoenmakers, *Optimization of Chromatographic Selectivity*, Elsevier, Amsterdam, 1987.

19. A. G. Wright, A. F. Fell, and J. C. Berridge, *Chromatographia* **24** (1987), 533.

20. A. G. Wright, A. F. Fell, and J. C. Berridge, *J. Chromatogr.* **458** (1988), 335.

21. A. F. Fell, T. A. G. Noctor, J. E. Mama, and B. J. Clark, *J. Chromatogr.* **434** (1988), 377.

22. K. M. Kirkland and D. A. McCombs, *J. Chromatogr.* **666** (1994), 211.

23. T. A. G. Noctor, B. J. Clarke, and A. F. Fell, *Anal. Proc.* **23** (1986), 441.

24. J. H. Knox and H. M. Pyper, *J. Chromatogr.* **363** (1986), 1.

25. S. Golshan-Shirazi and G. Guiochon, *Anal. Chem.* **60** (1988), 2364.

26. S. Golshan-Shirazi and G. Guiochon, *Anal. Chem.* **60** (1988), 2634.

27. G. Guiochon and S. Ghodbane, *J. Phys. Chem.* **92** (1988), 36.

28. S. Ghodbane-Shirazi and G. Guiochon, *J. Chromatogr.* **452** (1988), 209.

29. L. R. Snyder, J. W. Dolan, and G. B. Cox, *J. Chromatogr.* **484** (1989), 437.

30. J. E. Eble, R. L. Grob, and L. R. Snyder, *J. Chromatogr.* **405** (1987), 1.

31. L. R. Snyder, J. W. Dolan, D. E. Antle, and G. B. Cox, *Chromatographia* **24** (1987), 82.

32. F. Eisenbeiss et al., in K. K. Unger (ed.), *Handbuch der HPLC, Teil 2: Präparative Säulenflüssig-Chromatographie*, GIT Verlag, Darmstadt, Germany, 1994, p. 4.

33. E. Küsters and J. Nozulak, *Chromatographia* **47** (1998), 440.

34. S. J. Mickel, D. Niederer, R. Daeffler, A. Osmani, E. Küsters, E. Schmid, K. Schaer, R. Gamboni, W. Chen, E. Loeser, F. R. Kinder, K. Konigsberger, K. Prasad, T. M. Ramsey, O. Repic, R.-M. Wang, G. Florence, I. Lyothier, and I. Paterson, *Org. Process Res. Dev.* **8** (2004), 122.

35. J. Dingenen, in G. Subramanian (ed.), *A Practical Approach to Chiral Separations by Liquid Chromatography*, VCH, Weinheim, 1994, 124.

36. F. G. Helfferich and P. W. Carr, *J. Chromatogr.* **629** (1993), 97.

37. G. Guiochon, *J. Chromatogr. A* **965** (2002), 129.

38. G. Guiochon and B. Lin, *Modelling for Preparative Chromatography*, Academic Press, Amsterdam, 2003.

39. P. Rouchon, M. Schonauer, P. Valentin, and G. Guiochon, *Sep. Sci. Technol.* **22** (1987), 1793.

40. H. Kniep, Ph. D. thesis, University of Magdeburg, 1998.

41. A. Seidel-Morgenstern, *Analusis Mag.* **26/7** (1998), 46.

42. Heuer, H. Kniep, T. Falk, and A. Seidel-Morgenstern, *Chem. Ing. Tech.* **69** (1997), 1535.

43. R. M. Nicoud and A. Seidel-Morgenstern, *Isolat. Purif.* **2** (1996), 165.

44. A. Seidel-Morgenstern, *J. Chromatogr. A* **1037** (2004), 255.
45. E. Cremer and H. Huber, *Angew. Chem.* **73** (1961), 461.
46. F. G. Helfferich and G. Klein, *Multicomponent Chromatography*, Marcel Dekker, New York, 1970.
47. R. M. Nicoud and A. Seidel-Morgenstern, *Isolat. Purif.* **2** (1996), 165.
48. K. Mihlbachler, K. Kaczmarski, A. Seidel-Morgenstern, and G. Guiochon, *J. Chromatogr. A* **955** (2002), 35.
49. F. Gritti and G. Guiochon, *J. Colloid Interface Sci.* **264** (2003), 43.
50. E. Francotte, *J. Chromatogr.* **666** (1994), 565.
51. C. Roussel, J.-L. Stein, F. Beauvais, and A. Chemlal, *J. Chromatogr.* **462** (1989), 95.
52. D. Sanchez, *Analusis Mag.* **26/7** (1998), 33.
53. R. E. Majors, *LCGC North Am.* **22/7** (2004), 606.
54. E. Francotte and A. Junker-Buchheit, *J. Chromatogr.* **576** (1992), 1.
55. E. Francotte, *J. Chromatogr. A* **666** (1994), 565.
56. W. H. Pirkle, in S. Ahuja (ed.), *Chromatography and Separation Chemistry*, ACS, Washington, D.C., 1986, p. 101.
57. V. R. Meyer, *Chromatographia* **24** (1987), 639.
58. M. Zief, *Chromatogr. Sci.* **40** (1988), 337.
59. At Pittcon 2004.
60. K. Nakanishi, *J. Porous Mater.* **4** (1997), 67.
61. M. Schulte, D. Lubda, A. Delp, and J. Dingenen, *J. High Resolut Chromatogr.* **23/1** (2000), 100.
62. S. Golshan-Shirazi and G. Guiochon, *Anal. Chem.* **61** (1989), 1368.
63. H. Colin, in G. Ganetsos and P. E. Barketr (ed.), *Preparative and Production Scale Chromatography*, Marcel Dekker, New York, 1993, p. 11.
64. J. Dingenen, *Analusis Mag.* **26/7** (1998), 18.
65. R. S. Porter and J. F. Johnson, *Nature* **183** (1959), 391.
66. R. S. Porter and J. F. Johnson, *Nature* **184** (1959), 978.
67. V. P. Chizhkov, *Usp. Khim.* **40** (1971), 161.
68. M. Martin, F. Verillon, C. Eon, and G. Guiochon, *J. Chromatogr.* **125** (1976), 17.
69. B. Coq, J. L. Rocca, and J. Vialle, *J. Liq. Chromatogr.* **4** (1981), 237.
70. A. Seidel-Morgenstern and G. Guiochon, *Chem. Eng. Sci.* **48** (1993), 2787.
71. J. Dingenen and J. N. Kinkel, *J. Chromatogr.* **666** (1994), 627.
72. E. Küsters, *Chimica Oggi/Chem. Today* **7/8** (1996), 39.
73. E. Waldvogel, P. Engeli, and E. Küsters, *Helv. Chim. Acta.* **80** (1997), 2084.
74. C. Horváth, A. Nahum, and J. H. Frenz, *J. Chromatogr.* **218** (1981), 365.
75. F. D. Antia and C. Horváth, *J. Chromatogr.* **484** (1989), 961.
76. F. D. Antia and C. Horváth, *Ber. Bunsenges. Phys. Chem.* **93** (1989), 961.
77. F. D. Antia and C. Horváth, in C. T. Mant and R. S. Hodges (ed.), *High-Performance Liquid Chromatography of Peptides and Proteins*, CRC Press, Boca Raton, FL, 1991, p. 809.
78. A. A. Shukla, K. M. Sunasara, R. G. Rupp, and S. M. Cramer, *Biotechnol. Bioeng.* **68/6** (2000), 672.

79. K. M. Sunasara, F. Xia, R. S. Gronke, and S. M. Cramer, *Biotechnol. Bioeng.* **82/3** (2003), 330.

80. R. Freitag and S. Vogt, *Cytotechnology* **30** (1999), 159.

81. P. L. Camacho-Torralba, M. D. Beeson, G. Vigh, and D. H. Thompson, *J. Chromatogr.* **646** (1993), 259.

82. P. L. Camacho-Torralba and G. Vigh, *J. Chromatogr.* **691** (1995), 213.

83. G. Quintero, M. Vo, G. Farkas, and G. Vigh, *J. Chromatogr.* **693** (1995), 1.

84. F. D. Antia, manuscript in preparation.

85. D. B. Broughton and D. B. Carson, *Petroleum Refiner* **38** (1959), 130.

86. D. B. Broughton and C. G. Gerhold, U. S. Patent No. 2,985,589, May 23, 1961.

87. D. B. Broughton, *Chem. Eng. Prog.* **64** (1968), 60.

88. H. J. Bieser and A. J. de Rosset, *Die Stärke* **29** (1977), 392.

89. R. W. Neuzil and J. W. Priegnitz, U. S. Patent No. 4,024,331, May 17, 1977.

90. R. M. Nicoud, in G. Subramanian (ed.), *Bioseparation and Bioprocessing*, Vol. 1, VCH, Weinheim, 1998, p. 4.

91. M. Schulte and J. Strube, *J. Chromatogr. A* **906** (2001), 399.

92. U. Voigt, R. Hempel, J. N. Kinkel, and R. M. Nicoud, WO Patent 97/34018, 1997.

93. D. M. Ruthven and C. B. Ching, *Chem. Eng. Sci.* **44** (1989), 1011.

94. F. D. Antia, *Chromatogr. Sci. Ser.* **88** (2003), 173.

95. M. Mazzotti, G. Storti, and M. Morbidelli, *AIChE J.* **40** (1994), 1825.

96. M. Mazzotti, G. Storti, and M. Morbidelli, *AIChE J.* **42** (1996), 2784.

97. M. Mazzotti, G. Storti, and M. Morbidelli, *J. Chromatogr. A* **769** (1997), 3.

98. Gentillini, C. Migliorini, M. Mazotti, and M. Morbidelli, *J. Chromatogr. A* **805** (1998), 37.

99. M. Negawa and F. Shoji, *J. Chromatogr.* **590** (1992), 111.

100. R. M. Nicoud, G. Fuchs, P. Adam, M. Bailly, E. Küsters, F. Antia, R. Reuille, and E. Schmid, *Chiraliy* **5** (1993), 267.

101. E. Küsters, G. Gerber, and F. D. Antia, *Chromatographia* **40** (1995), 387.

102. T. Pröll and E. Küsters, *J. Chromatogr. A* **800** (1998), 135.

103. L. S. Pais, J. M. Loureiro, and A. E. Rodrigues, *J. Chromatogr. A* **827** (1998), 215.

104. R. M. Nicoud, M. Bailly, J. N. Kinkel, R. Devant, T. Hampe, and E. Küsters, in R. M. Nicoud (ed.), *Simulated Moving Beds—Basics and Applications*, INPL, Nancy, 1993, p. 65.

105. J. Blehaut, F. Charton, and R. M. Nicoud, *LC-GC Int.* **9** (1996), 228.

106. C. B. Ching, B. G. Lin, E. J. D. Lee, and S. C. Ng, *J. Chromatogr.* **634** (1993), 215.

107. J. N. Kinkel, M. Schulte, R. M. Nicoud, and F. Charlton, in *Chiral Europe 1995*, Symposium Proceedings, p. 121.

108. E. Francotte and P. Richert, *J. Chromatogr. A* **769** (1997), 101.

109. E. Francotte, P. Richert, M. Mazzotti, and M. Morbidelli, *J. Chromatogr. A* **796** (1998), 239.

110. M. Schulte, R. Ditz, R. M. Devant, J. N. Kinkel, and F. Charlton, *J. Chromatogr.* **769** (1997), 93.

111. J. Strube, A. Jupke, A. Epping, H. Schmidt-Traub, M. Schulte, and R. Devant, *Chirality* **11** (1999), 440.

112. D. Seebach, M. Hoffmann, A. Sting, J. N. Kinkel, M. Schulte, and E. Küsters, *J. Chromatogr.* **796** (1998), 299.

113. D. W. Guest, *J. Chromatogr. A* **760** (1997), 159.

114. E. Cavoy, M.-F. Deltent, S. Lehoucq, and D. Miggiano, *J. Chromatogr. A* **769** (1997), 49.

115. M. J. Gattuso, B. McCulloch, D. W. House, W. M. Baumann, and K. Gottschal, *Chim. Oggi* **50** (1996), 6.

116. M. Gattuso, B. McCulloch, D. W. House, W. M. Baumann, and K. Gottschall, *Pharm. Technol. Eur.* **6** (1996), 20.

117. U. Voigt, R. Hempel, J. N. Kinkel, and R. M. Nicoud, WO Patent 97/34018 (1997).

118. D. J. Wu, Z. Ma, and N. H. L. Wang, *J. Chromatogr. A* **855** (1999), 71.

119. S. Nagamatsu, K. Murazumi, H. Matsumoto, and S. Makino, in *Chiral Europe* 1996, Symposium Proceedings, p. 97.

120. S. Nagamatsu, K. Murazumi, and S. Makino, *J. Chromatogr. A* **832** (1999), 55.

121. E. Küsters, Ch. Heuer, and D. Wieckhusen, *J. Chromatogr. A* **874** (2000), 155.

122. M. Schulte and R. Devant, German Patent Application P 199 31 755.0, 1999.

123. M. J. Gattuso and S. Makino, presented at 6th International Symposium on Chiral Discrimination (1995), poster presentation.

124. C. Heuer, E. Küsters, T. Plattner, and A. Seidel-Morgenstern, *J. Chromatogr. A* **827** (1998), 175.

125. T. Shibata, K. Mon, and Y. Okamoto, in A. M. Krstulovic (ed.), *Chiral Separations by HPLC*, 1989.

126. G. Koch, R. Jeck, O. Hartmann, and E. Küsters, *Org. Res. Dev.* **5** (2001), 211.

127. E. Küsters, C. Heuer, and D. Wieckhusen, *J. Chromatogr. A* **874** (2000), 155.

128. M. Juza, *sp²*, August 2004, p. 16.

129. S. R. Perrin and R. M. Nicoud, in G. Subramanian (ed.), *Chiral Separation Techniques: A Practical Approach*, Wiley, 2001, p. 253.

130. M. McCoy, *Chem. & Eng. News*, June 19 (2000), p. 17.

131. G. Guiochon, S. G. Shirazi, and A. M. Katti, *Preparative and Nonlinear Chromatography*, Academic Press, Boston, 1994.

132. S. Khattabi, D. E. Cherrak, K. Mihlbachler, and G. Guiochon, *J. Chromatogr. A* **893** (2000), 307.

133. G. Guiochon, *J. Chromatogr. A* **965** (2002), 129.

134. D. E. Cherrak, J. Lee, and G. Guiochon, in preparation.

135. U. Tallarek, E. Baumeister, K. Albert, E. Bayer, and G. Guiochon, *J. Chromatogr. A* **696** (1995), 1.

136. U. Tallarek, K. Albert, E. Bayer, and G. Guiochon, *AIChE J.* **42** (1996), 3041.

137. Z. Zhang, M. Mazzotti, and M. Morbidelli, *Korean J. Chem. Eng.* **21/2** (2004), 454.

138. H. Schramm, M. Kaspereit, A. Kienle, and A. Seidel-Morgenstern, *J. Chromatogr. A* **1006/1–2** (2003), 77.

139. O. Ludemann-Homburger, R. M. Nicoud, and M. Bailly, *Sep. Sci. Technol.* **35/12** (2000), 1829.

140. O. Ludemann-Homburger, G. Pigrini, R. M. Nicoud, D. S. Ross, and O. Terfloth, *J. Chromatogr. A* **947** (2002), 59.

141. H. Lorenz, D. Sapoundjiev, and A. Seidel-Morgenstern, *Eng. Life Sci.* **3/3** (2003), 132.

142. Present address: Firoz D. Antia, Merck & Co., Inc, 126 East Lincoln Avenue, Rahway, NJ.

22

CHIRAL SEPARATION

Nelu Grinberg, Thomas Burakowski, and Apryll M. Stalcup

22.1 INTRODUCTION

Chirality plays a major role in biological processes, and the enantiomers of a bioactive molecule often possess different biological effects. For example, all pharmacological activity may reside in one enantiomer of a molecule, or enantiomers may have identical qualitative and quantitative pharmacological activity. In some cases, enantiomers may have qualitatively similar pharmacological activity, but different quantitative potencies. Since drugs that are produced by chemical synthesis are usually a mixture of enantiomers, there is a need to quantify the level of the isomeric impurity in the active pharmaceutical ingredient. Accurate assessment of the enantiomeric purity of substances is critical because isomeric impurities may have unwanted toxicological, pharmacological, or other effects. Such impurities may be carried through a synthesis and preferentially react at one or more steps and yield an undesirable level of another impurity. The determination of a trace enantiomeric impurity in a sample of a single enantiomer drug substance in the presence of a range of other structurally related impurities and a large excess of the major enantiomer remains challenging.

The history of enantiomeric separation starts with the work of Pasteur. In 1848 he discovered that the spontaneous resolution of racemic ammonium sodium tartrate yielded two enantiomorphic crystals. Individual solutions of these enantiomorphic crystals led to a *levo* and *dextro* rotation of the polarized light. Because the difference of the optical rotation was observed in solution, Pasteur suggested that like the two sets of crystals, the molecules are

HPLC for Pharmaceutical Scientists, Edited by Yuri Kazakevich and Rosario LoBrutto
Copyright © 2007 by John Wiley & Sons, Inc.

mirror images of each other and the phenomenon is due to the molecular asymmetry [1].

While Pasteur made the historical discovery, subsequent advances in the resolution of enantiomers by crystallization were based on empirical results. Several attempts to separate enantiomers using paper chromatography were met with unsystematic results. In 1952 Dalgliesh postulated that three points of simultaneous interaction between the enantiomeric analyte and the stationary phase are required for the separation of enantiomers [2].

Developments in the field of life sciences and in the pharmaceutical industry brought enantiomeric separation to a new level. In the late 1950s/early 1960s, many of the drugs were synthesized and used in a racemic form. An example with tragic consequences was the use of thalidomide, a sedative and a sleeping drug used in the early 1960s which produced severe malformations in newborn babies of women who took it in the early stage of pregnancy. Later it was demonstrated that only the (S)-enantiomer possesses teratogenic properties [3].

Introduction of gas chromatography gave a burst to the field of enantiomeric separation. In 1966 a group from the Weizmann Institute of Science in Israel reported the first successful separation of enantiomers using gas chromatography.

In a letter addressed to Emanuel Gil-Av after the publication of the first separation of enantiomers on a chiral separation of enantiomers on a chiral gas chromatography (GC) stationary phase [4], A. J. P. Martin wrote: "As you no doubt know, I had not expected such attempts to lead to much success, believing that the substrate-solvent association would normally be too loose to distinguish between the enantiomers." At the time there were just several reports on the separation of enantiomers using chromatographic methods. Later developments in HPLC gave an additional boost to the field. Today, there are over 60 types of rugged, well-characterized columns capable of separating enantiomers. Unfortunately, there is a great deal of trial and error in choosing a particular column for a chiral separation. Therefore this chapter will summarize a rationale for choosing a stationary phase that is based on the relationship that exists between the analytes and the chiral stationary phases.

22.1.1 Enantiomers, Diastereomers, Racemates

Chirality is due to the fact that the stereogenic center, also called the chiral center, has four different substitutions. These molecules are called asymmetrical and have a C_1 symmetry. When a chiral compound is synthesized in an achiral environment, the compound is generated as a 50:50 equimolar mixture of the two enantiomers and is called racemic mixture. This is because, in an achiral environment, enantiomers are energetically degenerate and interact in an identical way with the environment. In a similar way, enantiomers can be differentiated from each other only in a chiral environment provided under the conditions offered by a chiral stationary/mobile phase [5]. The separation of enantiomers using chiral stationary/mobile phases involves the formation of transient diastereomeric complexes between the enantiomeric analytes and the chiral moiety present in the chromatographic column. Thus, diastereomers are chiral molecules containing two or more chiral centers with the same chemical composition and connectivity. They differ in stereochemistry about one or more chiral centers. If two stereoisomers are not enantiomers of one another, they can in principle be separated in an achiral environment—that is, using a nonchiral stationary phase [5].

22.2 SEPARATION OF ENANTIOMERS THROUGH THE FORMATION OF DIASTEREOMERS

Formation of diastereomers for chromatographic purposes can be generated in two ways: transient diastereomers, which occur between the enantiomers and the chiral stationary phase (CSP) during the chromatographic process. Such a process is also called direct separation. The second way is to generate long-lived diastereomers that are formed by chemical reaction between the enantiomer and a chiral derivatizing reagent prior the chromatography. Such a process is called indirect separation. Indirect separation of enantiomers is usually a good technique when everything in direct separation fails. However, it requires suitable functionality in the enantiomers for reaction with a chiral derivatizing agent. The effectiveness of this approach may also depend on a variety of other conditions such as structural rigidity and the spatial relationship between the stereogenic centers of the enantiomers and the chiral center introduced through derivatization.

When two chiral compounds, racemic A and racemic B, react to form a covalent bond between them without affecting the asymmetric center, the stereochemical course of the reaction can be as follows [6]:

$$[(\pm) - A] + [(\pm) - B] \rightarrow [+A + B] + [+A - B] + [-A + B] + [-A - B]$$

where the first and the last products constitute an enantiomeric pair and the second and the third products constitute a second enantiomeric pair. In contrast, the first and the third products and the second and fourth products are diastereomeric pairs. In a chiral environment, one should be able to separate all of these four products. However, because diastereomers possess slightly different physicochemical properties, achiral chromatography of this mixture should lead to two peaks (corresponding to the two diastereomers).

Indirect approaches such as chiral derivatization with chiral derivatizing reagents (CDR) offers a variety of advantages. For instance, CDRs are cheaper than chiral columns. Separation of the product diastereomers is generally more flexible than the corresponding enantiomeric separation because achiral columns can be used in conjunction with various mobile-phase compositions. Depending on the functional groups on the enantiomers, there is a variety of CDRs on the market (chiral anhydrides, acid chlorides, chloroformates, isocianates, isothiocianates, etc.) which can be applied, which in turn can change the selectivity of a chromatographic system.

There are also disadvantages to the chiral derivatization approach including extra validation. For instance, the derivatizing reagent has to be optically pure, or the analysis can generate false-positive results. In addition, special care needs to be taken that the chiral center of the enantiomers or derivatizing agent is not racemized during the derivatization reaction. Furthermore, unequal detector response of the diastereomers must be corrected via standard procedures [7]. Often, the derivatization requires a long reaction time, which adds to the analysis time.

22.2.1 Mechanism of Separation

The separation of diastereomeric pair is due to the effect of their nonequivalent shape, size, polarity, and so on, on their relative solvation and sorption energies [8]. Their interaction with a particular stationary phase is dependent upon their molecular structure and availability of functional groups able to interact with the stationary phase. For instance, unsaturated bicyclic alcohols, which are capable of internal hydrogen bonding, show shorter retention than epimers or dihydro derivatives, which cannot undergo such types of interactions [9] (Figure 22-1). The compounds of Figure 22-1 were separated by gas chromatography on a 12-ft × $\frac{1}{4}$-in. column packed with 23% by weight of Ucon No. 50HB 2000 available from Union Carbide on Celite. As the number of double bonds increases in the molecules, the possibility of intramolecular hydrogen bonds between the hydroxyl groups and the double bond increases. Simultaneously, the potential for hydrogen bond formation between the compounds and the stationary phase decreases. As a consequence, the retention time of each isomer decreases as the number of double bonds in the molecules increases [10–13].

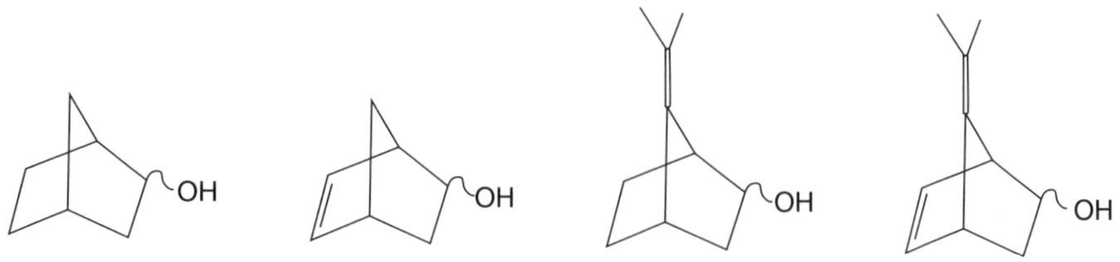

	I	II	III	IV
endo	104	79	38	30
exo	96	94	28	33

Figure 22-1. Retention time of bicyclic alcohols. The numbers under each structure represent the retention time in minutes. (Reprinted from reference 9, with permission.)

There are few differences between the separation in gas chromatography [14–16] and the separation in liquid chromatography (LC), because it is assumed that the differential solvation of the diastereomeric compounds during the LC separation does not play a very important role [17]. Helmchen et al. [18] explained the separation of diastereomeric amides using LC with a silica gel stationary phase under normal-phase conditions. In order to explain their separation, the authors made some assumptions:

1. Secondary amides adopt essentially the same conformation in polar solutions and in the adsorbed state (on silica gel).
2. In the adsorbed state, a parallel alignment of the planar amide group and the surface of silica gel is preferred.
3. Apolar groups (i.e., alkyl, aryl) outside the amide plane cause a disturbance of this preferred arrangement in proportion to their steric bulk in a direction perpendicular to the amide plane. Such groups are classified as large and small by indices L and S, respectively.
4. That member of a diastereomeric pair in which both faces of the amide plane are more shielded than the least shielded face in the other member is eluted first.
5. There is an attractive interaction between small polar groups and the silica gel, particularly if they are hydrogen bond donors not internally bonded to the amide group. Formally, such groups are assigned to the S (small) class.

The actual magnitude of the interaction of a given substituent with the adsorbent depends on the adsorbent, other substituents present, and the type and rigidity of the backbone of the diastereomeric analytes. Although no serious attempts at quantification have been made, repulsive interactions toward silica and alumina can be ranked roughly as H < methyl < phenyl = ethyl < tert-butyl < trifluoromethyl < α-naphthyl < 9-anthryl = pentafluoroethyl < heptafluoroethyl. Size and hydrophobicity are both relevant; incorporation of polar functionality (hydroxyl, carbalkoxy, cyano) leads to attractive rather than repulsive interactions with silica.

22.2.2 General Concepts for Derivatization of Functional Groups

As noted previously (Section 22.2), derivatization with a chiral derivatizing reagent (CDR) requires the presence of suitable functionality (e.g., —OH, Ar—OH, —SH, —COOH, —CO—, —NH$_2$, —NRH) within the chiral analyte to serve as a reactive site. Before addressing specific issues with regard to CDR and analyte classes, it may be helpful to review general considerations for achiral derivatization in chromatographic assays.

Desirable achiral derivatization reaction properties include fast, unidirectional reactions with no or minimal side reactions. In addition, both the reagent and the product should be stable. Most derivatization methods use an excess of reagent which can present as an interfering chromatographic peak. Of course, incorporating a derivatization step in an assay requires additional materials, time, and effort as well as additional method validation.

In the case of chiral derivatization, there are some unique considerations in addition to the ones noted above for achiral derivatization. Extra validation is required to establish the optical purity of the derivatizing agent. In addition, nonracemization of either the analyte or the derivatizing reagent during the derivatization must be confirmed. Excess reagent must be used to eliminate any potential chiral discrimination in the derivatization reaction. The presence of more than one type of reactive group (e.g., amine and alcohol) must be considered if the selected reagent has different reaction potentials for each moiety. In some cases, chiral derivatization may be coupled with achiral derivatization. If more than one reactive functional group is present in the analyte, usually the derivative in which the two stereogenic centers are in closest proximity yields the most favorable diastereomeric pair for separation by achiral chromatography. Also, derivatives that incorporate the most structural rigidity (e.g., amides versus esters) tend to be the most amenable to separations by achiral chromatography.

22.3 MOLECULAR INTERACTIONS

Generally speaking, there are three properties involved in an intermolecular interaction: the probability of the interaction occurring, the strength of the interaction, and the type of interaction. These properties will be discussed in the following sections.

22.3.1 The Probability of Molecular Interactions

Achieving enantiomeric discrimination requires understanding the interactions between the selector and the selectand. In his Ph.D. thesis [19], Feibush postulated that attaining an enantiomeric separation on a chromatographic chiral system required that certain conditions should exist:

> *A necessary condition for having a difference in the standard free energy of the two enantiomers in solution is that the solvent is chiral. The fact that the solvent is chiral is in itself not sufficient to sustain such difference. A certain solute–solvent correlation should exist to cause the difference in the behavior of the enantiomers. There should be strong (solute–solvent) interactions, such as π-complexation, coordinative bonds, [and] hydrogen bonds, to form associates between the asymmetric solvent/solute molecules. Such association can be regarded as short-living diastereomers. When the bonds that form these associates are in immediate proximity of their asymmetric carbons, a difference in the behavior of the enantiomers in the active phase is possible. We search for active phases and enantiomeric solutes that can form associates through (preferably) more than one hydrogen bond, and*

where these bonds are formed in the immediate proximity of the asymmetric carbons. In an associate formed through a single H-bond, free rotation of the bonded molecules still exists, on the other hand, more bonds prevent this possibility to a large extent, and a solute–solvent associate with a preferred conformation is formed. In addition, having more H-bonds between the asymmetric solute and solvent increases the interaction between these neighboring molecules and increases the population of (the selective) associates where asymmetric carbon are in close proximity. With the increase of the relative population of these particular associates from all the possible associates, an increase in the gap of the free solvation energy of the enantiomers is expected, which enables their GC separation.

This model can also be extended to enantiomeric separation using liquid chromatography.

Yet enantioerecognition is still a matter of debate [20–22]. More recently, Sundaresan and Abrol [23] proposed a novel stereocenter recognition (SR) model for describing the stereoselectivity of biological and other macromolecules toward substrates that have multiple stereocenters, based on the topology of substrate stereocenters. The SR model provides the minimum number of substrate locations interacting with receptor sites that need to be considered for understanding stereoselectivity characteristics. According to this model, the substrate locations and receptor sites can have binding, nonbinding, or repulsive interactions that may occur in a many-to-one or one-to-many fashion. The interactions between the two chiral entities must involve a minimum number of locations in the correct geometry. The model predicts that stereoselectivity toward a substrate with N stereocenters in a linear structure involves $N + 2$ substrate locations distributed over all stereocenters in the substrate, such that at least three locations per stereocenter effectively interact with one or more receptor sites.

In building models of possible enantioselective associates, conformational searching during docking of the selectands (enantiomeric solutes) with the selector (chiral solvent or ligand) is necessary. Usually it is not known which conformation of a ligand interacts more favorably with a particular receptor, and the flexibility of the ligand plays a major role in such computational approaches [24]. Associations where each of the pairing partners is not in its preferred conformation play only a minor role in the overall interaction between the selectand and the selector, and their contribution to the enantioselectivity is minimal.

In Figure 22-2, the diastereomeric associates between the selectand/selector are formed through one, two, or three substituents of the asymmetric carbon. The chirality of the selector or the selectand can arise from an asymmetric carbon, the molecular asymmetry, or the helicity of a polymer. Also, the bonds between substituents of the selectand and the selector can involve a single bond, but could also involve multiple bonds or surfaces. Such bonds represent the leading interactions between selectand and selector. Only when the leading interactions take place and the asymmetry of the two bodies are

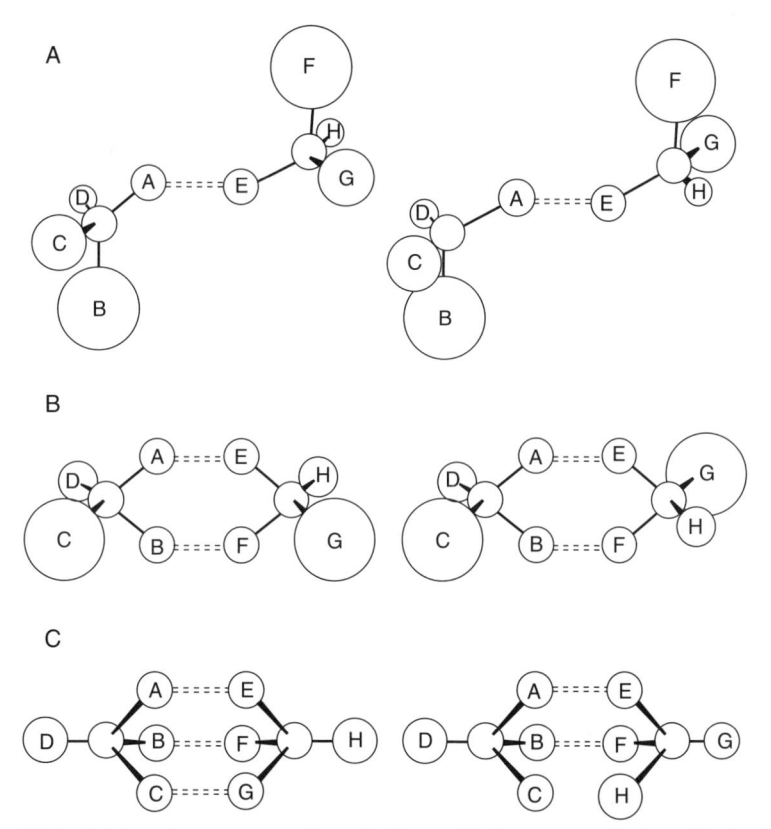

Figure 22-2. Schematic representation of selectand/selector associations. Dashed lines represent the leading interactions between the two chiral entities. (Reprinted from reference 25, with permission.)

brought in close proximity do the secondary interactions (e.g., van der Waals, steric hindrance, dipole–dipole) become effectively involved.

The secondary interactions can affect the conformation and the formation energy of the diastereomeric associates. If the interaction between the selectand and the selector takes place through one leading interaction (Figure 22-2A), then the enantioselectivity of the system is governed by the position of unbounded substituents B, C, and D of the selectand relative to the substituents F, G, and H of the selector. One particular enantiomer will interact more strongly with a particular selector if the contour and polarity of the two molecules are better complements of each other. When the interaction between the selectand and the selector occurs through two leading interactions (Figure 22-2B), the enantioselectivity of the system is determined by the effective size of the groups that do not participate in interactions. If, for example, G of the selector is an alkyl and H a hydrogen substituent, and C of the selectand is an alkyl group and D a hydrogen, then one enantiomer has the larger G and C groups in *syn* arrangement and the other in *anti* arrangement. In a variety of cases involving interactions through hydrogen bonding or ligand metal complexes, the enantiomer whose larger nonbonded groups are positioned *syn* to the corresponding larger group of the selector will elute last from a chromatographic column, as compared to the opposite isomer that forms the *anti* arrangement [25].

The solvation energy of one enantiomer in the active chiral phase can be described as the contribution of all possible forms of solvent/solute associates. These associates are in equilibrium with fast interconversion rates. Each form contributes to the total free energy according to its particular formation energy and its particular molar fraction [25, 26]. These complexes between the selector and selectand should also be as mutually exclusive as possible, to prevent a given interaction from occurring at multiple sites in the diastereomeric complexes [5].

22.3.2   The Types of Molecular Interactions

Chiral separations generally rely on the formation of transient diastereomeric complexes with differing stabilities. Complexes are defined as two or more compounds bound to one another in a definite structural relationship by forces such as hydrogen bonding, ion pairing, metal-ion-to-ligand attraction, π-acid/π-base interactions, van der Waals attractions, and entropic component desolvation. In the following sections, the most important types of molecular interactions in chiral separations are discussed.

22.3.3   Chiral Separation Through Hydrogen Bonding

Hydrogen bonding is a donor–acceptor interaction specifically involving hydrogen atoms [27]. When a covalently bonded hydrogen atom forms a second bond to another atom, the second bond is referred to as a hydrogen bond.

A hydrogen bond is formed by interaction between the partners R—X—H and :Y—R′ according to

$$R—X—H + :Y—R' \leftrightarrow R—X—H \cdots Y—R'$$

where R—X—H is the proton donor and :Y—R′ makes an electron pair available for the bridging bond. Hydrogen bonding can be regarded as a preliminary step in a Brønsted acid–base reaction, which would lead to a dipolar reaction product R—X$^-$$\cdots$H—Y$^+$—R′.

According to their bonding energy, hydrogen bonds can be subdivided into three categories: strong, moderate, and weak hydrogen bonds. Strong hydrogen bonds are formed by groups in which there is a deficiency of electron density in the donor group, (i.e., —O$^+$—H, >N$^+$—H) or an excess of electron density in the donor group (i.e., F⁻, O⁻—H, O⁻—C, O⁻—P, N⁻<). They are referred to as forced strong H-bonds [27].

Moderate hydrogen bonds are generally formed by neutral donor and acceptor groups, such as —O—H, =N—H, or —N(H)—H and O=, O=C, or N=, in which the donor X is electronegative relative to hydrogen and the Y atom (the acceptor) has a lone pair of unshared electrons. These are the most common hydrogen bonds and are essential contributors to the structure and function of biopolymers.

Weak hydrogen bonds are formed when the hydrogen atom is covalently bonded to a slightly more electrically neutral atom relative to hydrogen (e.g., C—H, Si—H) or when the acceptor group has no lone pair but has π electrons, (e.g., C=C or an aromatic ring). Although F is a very electronegative atom, F—C or F—S groups are only weak acceptors. These interactions have energies and geometries similar to those of van de Waals complexes, and they are distinguished from them by evidence of a directional involvement of the X—H bond.

The H-bond is generally assumed to be linear with θ between 175–180°. The geometrical requirement can, in certain cases, lead to arrangements in which a covalently bonded H-atom is located close to more than one potential acceptor atom, leading to a bifurcated hydrogen bond [28]. Such complexes have lower stability than those with a single hydrogen bond. An example of a bifurcated hydrogen bond between two drug enantiomers and amylose carbamate stationary phase is presented in Figure 22-3. The right-hand side enantiomer undergoes a bifurcated hydrogen bond with the amylose phase, forming a complex less stable than that from the left-hand side. As a consequence, the enantiomer forming the bifurcated hydrogen bond eluted earlier from the chromatographic column [29].

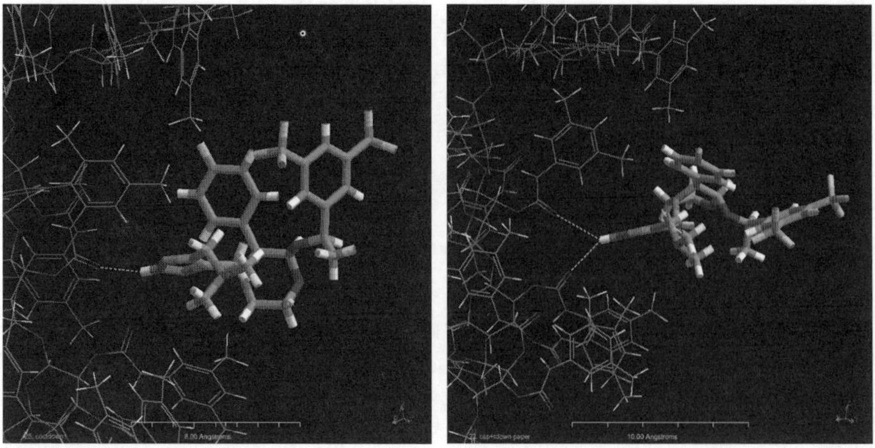

Figure 22-3. Interaction of two drug enantiomers with amylose carbamate stationary phase. (Reprinted from reference 29, with permission.)

The strength of hydrogen bonds depends on the solvent conditions in which the complex occurs. For instance, in the presence of an ionic medium (which generates an electric field), H-bonds of the solvate become polarized and, consequently, their symmetry can change from a symmetrical to an asymmetrical H-bond. The change in symmetry leads to weakening of the H-bonds between the solvate molecules. Furthermore, when the pK_a value of a dissolved molecule is larger than that of the protonated solvent, the addition of a strong acid leads the H^+ ions to become attached preferentially to the dissolved molecule $[(BH \cdots B)^+]$. When the pK_a of the dissolved molecules is smaller than that of the solvent, the addition of strong bases should favor H-bonds between the dissolved molecules $[(BH \cdots B)^-]$ [30].

The amide groups are one of the most important functional groups involved in designing chiral phases that involve hydrogen bonding. For this reason, a discussion of the amide structure is critical to understanding the interactions involved between the selectand and the selector. Furthermore, the amide group constitutes the backbone of linear peptide chains. The dimensions of a typical peptide group is given in Figure 22-4. The presence of an asymmetric center at the C^α carbon atom, along with the presence of only an L amino acid residue, results in an inherent asymmetry of the polypeptide chain [31].

Two configurations of the planar peptide bond are possible; the C^α can be in either *trans* or *cis* configuration, forms that are in equilibrium:

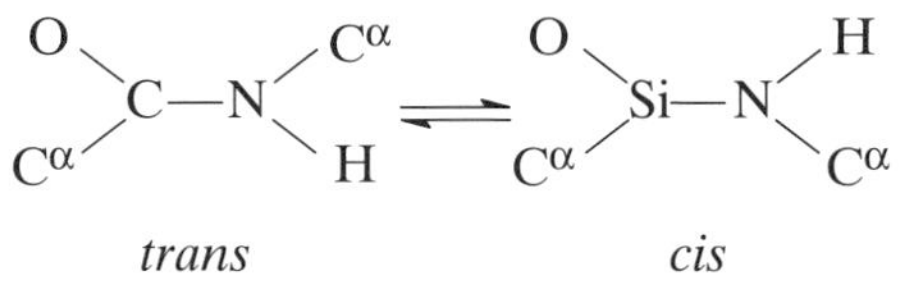

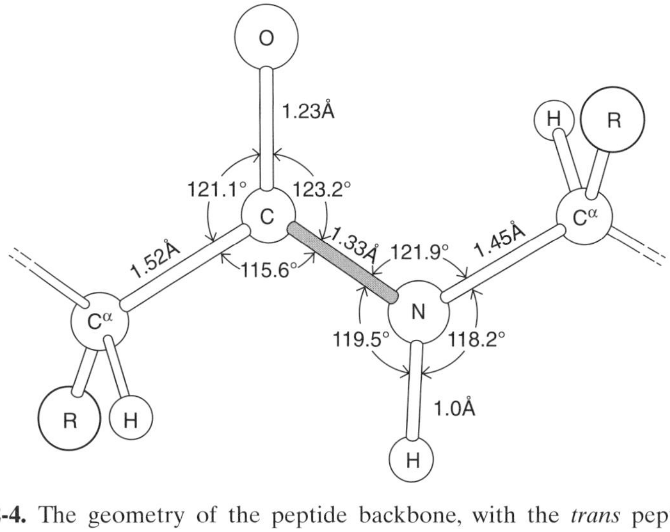

Figure 22-4. The geometry of the peptide backbone, with the *trans* peptide bond, showing all the atoms between two C^α atoms of adjacent residues. (Reprinted from reference 31, with permission.)

The *trans* form is energetically favored, due to less repulsion between non-bonded atoms [31]. For an amide group to hydrogen bond with another molecule able to undergo such interaction, the H···N distance should be ≤2.3 Å and the N···O distance should be ≤3.2 Å. An H-bond between N—H of an amino acid residue in the sequence m and C=O of a residue of the sequence number n is designated as $m \rightarrow n$ [32].

In the following section, we will present several chiral phases employed either in GC or in normal-phase HPLC for which the hydrogen-bonding interactions discussed above governs the interactions between the selectand and the selector. It should be noted that the interactions occurring in GC are similar to those occurring in normal-phase HPLC.

The first successful chiral phases used under GC conditions were N-trifluoro-acetyl (TFA)-L-α-amino acid esters. These phases separated racemates of the more volatile members of the same compounds [33]. Replacing the N-TFA moiety of the selector with trichloroacetyl reduced the enantioselectivity by half, while substituting with isobutyryl caused a total loss of the chiral separation.

The use of N-TFA ester derivatives of dipeptides as chiral phases significantly improved the enantioselectivity [34]. The chiral recognition was observed for a wider class of compounds, and substitution of TFA with acyl groups did not affect the selectivity.

The diamide stationary phase contained two hydrogen-bonding sites, a C5 and a C7 site, where hydrogen bonding selector/selectand associations could take place [25]:

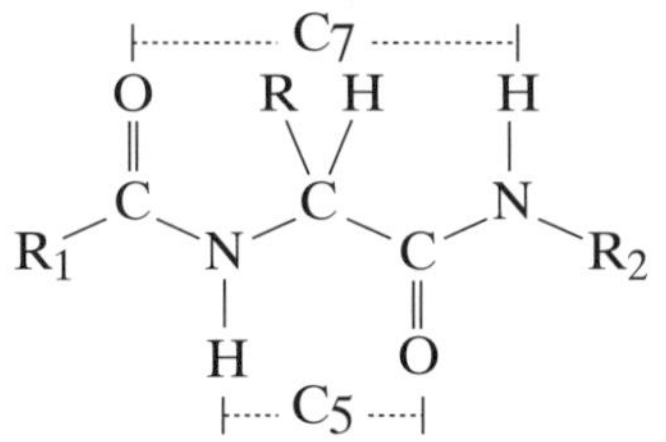

The structure of the diamide phase, derived from IR measurements of crystalline N-acetyl-L-leucylmethylamide (Figure 22-5) appeared to be similar to an anti-parallel β-sheet of poly-L-alanine. X-ray diffraction of the D,L-leucyl derivative showed the C5:C5 association, while the C7 site involved three molecules in the antiparallel arrangement. Figure 22-6 shows a C5:C5 associate of the L-diamide selector with L- and D-α-amino acid derivatives [35]. The back of the selector is flanked by a neighboring molecule through a C7:C7 associate as part of hydrogen bond network of the chiral stationary phase. The N-TFA-L-α-amino acid ester had the C5 site but was missing the C7 site; as a consequence, it formed a less organized hydrogen bond network [35].

A different association of the diamide-α-amino acid derivative is based on a C5:C7 parallel β-sheet arrangement, and it is shown in Figure 22-7. In this

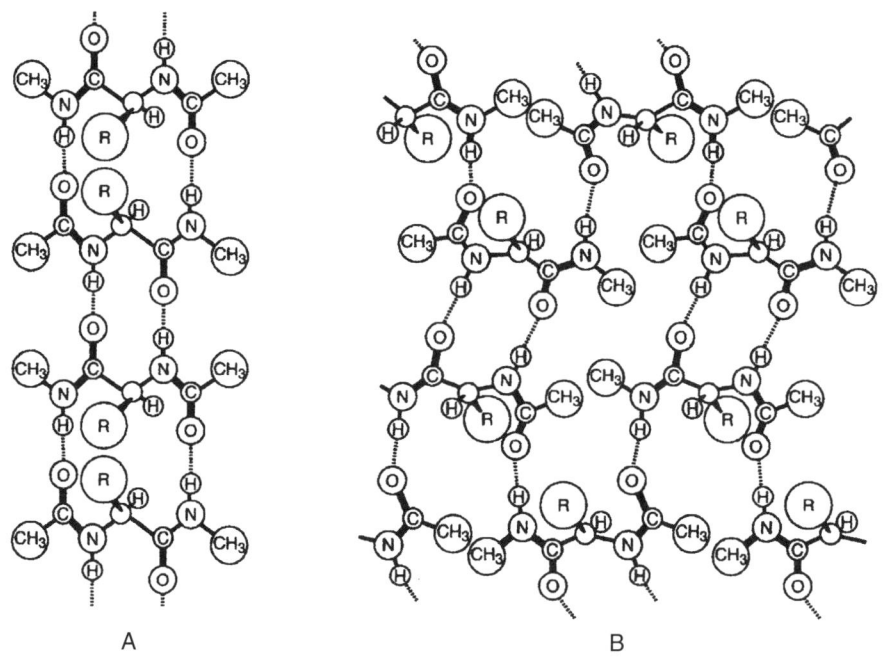

A   B

Figure 22-5. (A) The structure of *N*-acetyl-ʟ-leucylmethyl amide derived from IR spectra. (B) The structure of *N*-acetyl-ᴅ,ʟ-leucylmethylamide derived from X-ray diffraction (R = isobutyl). (Reprinted from reference 35, with permission.)

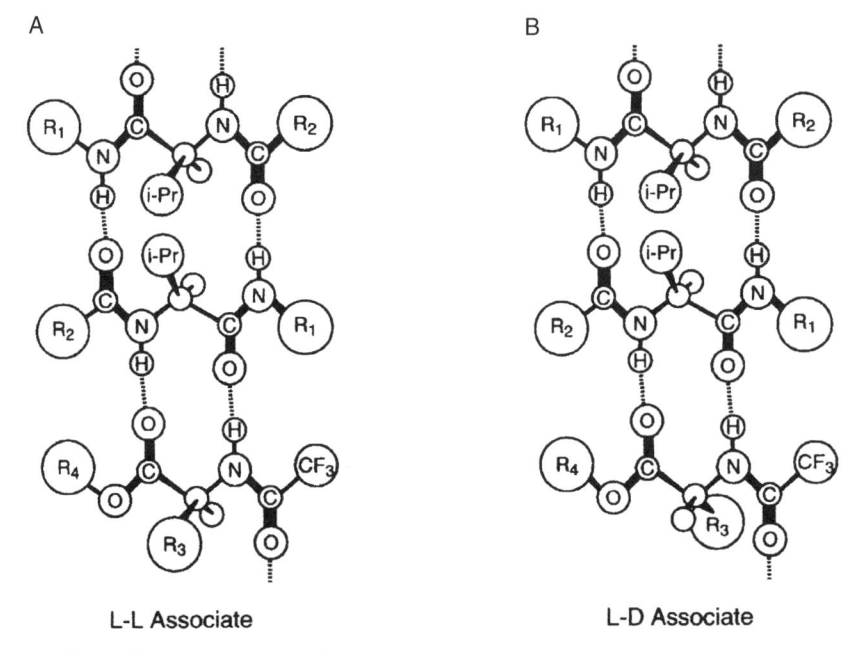

A   B

L-L Associate   L-D Associate

Figure 22-6. The hydrogen bond association of the ʟ-diamide phase in its antiparallel β-sheet conformation with (A) N-TFA-ʟ-α-amino acid alkyl ester and (B) *N*-TFA-ᴅ-α-amino acid alkyl ester. (Reprinted from reference 35, with permission.)

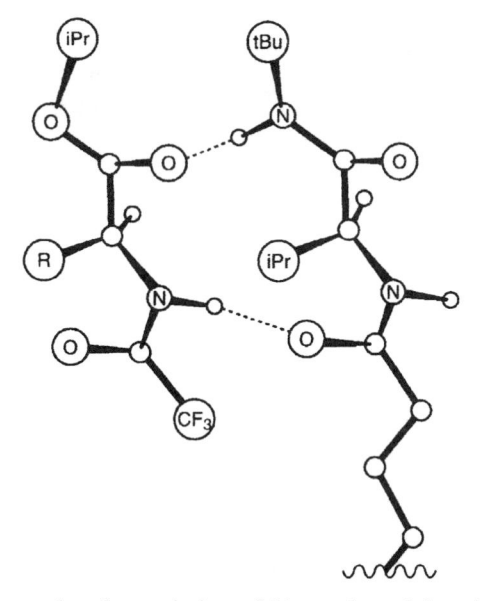

Figure 22-7. Hydrogen bond association of *N*-acetyl-L-valyl-tertbutylamide phase in its parallel β-sheet conformation with the *N*-TFA-α-amino acid isopropyl ester. (Reprinted from reference 36, with permission.)

arrangement, the alkyl substituents of the asymmetric carbons of the diamide phase and the α-amino acid solute are in close proximity (*syn* in the L:L associate), while the L:D association on opposite sides of the molecules is *anti* [36]. *N*-TFA-γ-amino acid esters have only a C7 hydrogen-bonding site and with a diamide phase can give C5:C7 and/or C7:C7 association with the C5 or C7 site of the phase.

The alkyl substituent of the asymmetric carbon of the D-enantiomer is *syn* to the R group of the L-diamide in either the C5:C7 or C7:C7 association. In general, all L-α-amino acid derivatives with an apolar R group, as well as D-γ-amino acid derivatives, interact more strongly with the L-diamide than their antipode; as a consequence, they elute last from the column. The main feature of these complexes is that the alkyl groups at the asymmetric carbons are in the *syn* position, yielding a more retained enantiomer than those in *anti* (Figure 22-7).

This principle also governs the separation on the commercially available Chirasil-Val® [37, 38]. In Chirasil-Val®, the chiral entity was incorporated in a polysiloxane backbone for higher thermal stability. Some of the compounds separated on Chirasil-Val® contained only groups, such as *N*-TFA-proline esters, that are able to accept hydrogen bonding. To undergo such an interaction, the diamide phase has to have a conformation where both NH groups point toward the selectand in a conformation similar to the α-helix structure of proteins [36].

The introduction of the diamide derivatives for enantiomeric separation was a step forward in designing selectors able to undergo hydrogen bonding interactions with a wide variety of selectands. The selector developed by Dobashi and Hara involved (R,R)-N,N'-diisopropyltartaramide (DIPTA). In the initial experiments, the selector was used as an additive in a nonaqueous mobile phase [39]. Enantiomers of α- and β-hydroxy carboxylic acid and α-amino acids were resolved with this chiral phase. Although addition of the selector to the mobile-phase complicates the interactions between the selectand and the selectors, through the introduction of secondary chemical equilibria, two conclusions could be drawn: (1) An increase in bulkiness of the N-alkyl-β-hydroxycarboxamides enhanced the separation. The bulkiness of the N- and O-alkyl groups of N-acyl-α-amino acid esters and amides had a similar effect. (2) An increase in bulkiness of the N-alkyl groups of N-alkyl-α-hydroxycarboxamides reduced the separation factors, and a similar effect was encountered for N-alkyl groups of N-dialkyl-β-hydroxycarboxamides. To improve the separation, aliphatic β-hydroxycarboxylic acids were derivatized to α-naphthylamides. Variation in the separation factor due to increased bulkiness of the alkyl substituents is likely related to preferential conformations of the derivatives. Specifically, the increased bulkiness of substituents causes the *threo* derivatives to adopt a *gauche* conformation (I) with regard to the two hydroxy groups, whereas the *erythro* derivatives adopt an *anti* conformation [39]:

(I)   (II)

The retention of the enantiomers in the column arises mainly from the equilibrium between the chiral selector:selectand. A large excess of chiral additive causes the equilibrium to shift to the association side. An increase in the polarity of the medium decreases the strength of the hydrogen bonding between the selectand and the selector and shifts the equilibrium towards the dissociation side. Subsequently, the same selector was bound to a silica support and packed into an HPLC column; it was also incorporated into a polysiloxane backbone and used as a chiral phase in gas chromatography in a similar manner previously used for Chirasil-Val® [40, 41].

A variation of these types of chiral stationary phases was reported by Anderson et al. [42], who synthesized a series of network polymeric stationary phase based on para-substituted N,N'-dialkyl-L-tartaramide dibenzoates.

These chiral phases also operate through hydrogen bonding between the analyte enantiomers and the chiral stationary phase, in a manner similar to the ones developed by Dobashi and Hara [39].

Another type of chiral phase based on hydrogen-bonding interactions is the polyacrylamide-type phases. Developed by Blaschke, the phase is comprised of a polyacrylamide that incorporates phenylalanine ethyl ester. The phase has a helical structure, and the interactions are based on hydrogen bonding between the polar groups of the enantiomer and the CO—NH groups of the polymer [43, 44].

In an effort to resolve a broad class of racemic heterocyclic drugs such as barbiturates, succinimides, glutaramides and hydantoins, a chiral stationary phase was developed that could undergo simultaneous triple hydrogen bonds with these analytes (Figure 22-8) [45]. The active part of the selector is a 2,6-pyridinediyl-bis(alkanamide), which is a complementary base that forms highly selective base pairs with these types of drugs. Chromatographic retention times (under normal-phase conditions) were directly linked to the formation of the base pairs. Compounds that can form the base pairs have substantial retention times, while closely related compounds that contain groups interfering with the base-pairing site elute in the void volume.

22.3.4 Chiral Separation Through Inclusion Compounds

Inclusion complexing partners are classified as hosts and guests [46]. There are two types of hosts that were successfully employed in the chromatographic separation of enantiomers: hosts that have a hydrophobic interior and hosts with a hydrophilic interior. The hydrophilic interior means that the cavity contains heteroatoms such as oxygen, where lone-pair electrons are able to participate in bonding to electron acceptors such as an organic cation (e.g., chiral crown ethers). In contrast, a host with a hydrophobic interior cavity is able to include hydrocarbon-rich parts of a molecule [47]. This type of host is found in the cyclodextrins.

22.3.4.1 Cyclodextrins. Cyclodextrins (CDs) were first isolated in 1891 as a degradation product of starch, and they were later characterized by Saenger as cyclic oligosaccharides [48]. If the amylose fraction of starch is degraded by glucosyltransferases, one or several turns of the amylose helix are hydrolyzed off and their ends are joined together, producing cyclic oligosaccharides called cyclodextrins. Because these enzymes are not specific, the hydrolysis produces a number of CDs with a variable number of sugar units. The most abundant are α-, β-, and γ-cyclodextrin (α-CD, β-CD and γ-CD, respectively) with six, seven, and eight glucose rings, respectively, also called cyclohexa-, cyclohepta-, and cyclooctaamylose (or CA6, CA7, CA8). Beyond these homologues, three more CDs have been characterized with 10, 14 (ε-CD and ι-CD, respectively), and 26 glucose rings. Larger homologues were synthetically produced [49]. The chemical structure of CA7 is depicted in Figure 22-9.

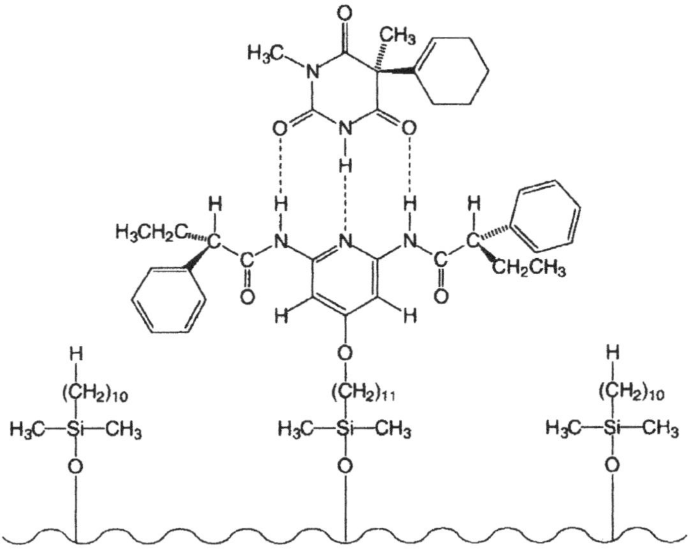

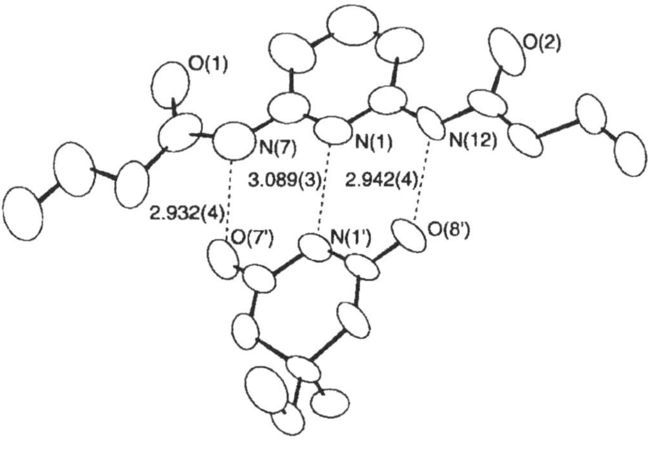

Figure 22-8. Structure of the complex between the stationary phase, a derivative of N,N'-2,6-pyridinediylbis[(S)-2-phenylbutanamide] boned to silicagel and (S)-hexobarbital (top). X-ray structure of the 1:1 complex of N,N'-2,6-pyridinediyl-bis(butanamide) and bemegride. (Reprinted from reference 45, with permission.)

Structures such as CA6, CA7, and CA8 have a doughnut shape and are able to host small molecules inside their cavity. Similar to amylose, the glucose units in the CAs are linked by $\alpha(1 \rightarrow 4)$ bonds that adopt a $^4C_1$ chair conformation. They may be considered as rigid building blocks giving fairly limited conformational freedom of the macrocycle in rotation of the C6–O6 groups and limited rotational movements about the glucosidic link C1(n)-O4(n − 1)-C4

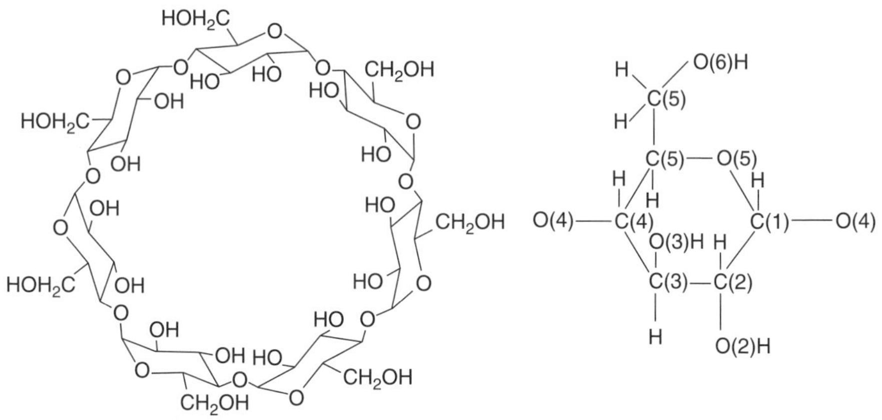

Figure 22-9. Chemical structure of CA7 (β-CD) where the numbering of glucose unit (1–7) is performed counterclockwise (left). Atom numbering scheme for a glucose unit (right). (Reprinted from reference 49, with permission.)

(n − 1). All glucose groups are aligned in *cis* configuration with the secondary O2 and O3 hydroxyls on one side, connected by O2(n)···O3(n − 1) hydrogen bonds, and the primary O6 hydroxyls on the other side. The smaller CA6 to CA8 have the overall shape of a hollow, truncated cone with the wide side occupied by O2 and O3 and the narrow side occupied by O6 [49].

There are a number of requirements for chiral discrimination using CDs. In cases where inclusion complexation is required, there must be a relatively tight fit between the complexed moiety and the CD. In addition, the chiral center or one substituent of the chiral center must be close to and interact with the 2- and 3-hydroxyl groups located at the rim of the CD cavity [50]. For example, the inclusion complexes of guests *d*- and *l*-propranolol with β-CD are placed identically within the CD cavity, and the structures are over-laid identically to the point of chiral carbon (Figure 22-10). The hydroxyl group attached to the chiral carbon is in the same position for the *d*- and *l*-enantiomer placed for optimal hydrogen bonding to a 3-hydroxyl group of the CD. Differences between the two complexes can be observed with respect to their secondary amine group. In the *d*-propranolol complex, the nitrogen is placed between the 2- and 3-hydroxyl groups at distances of 3.3 and 2.8 Å, respectively, which is well in the range of the length of a hydrogen bond. The amine in the *l*-propranolol complex is positioned less favorably for hydrogen bonding. The distances to the closest 2- and 3-hydroxyl group of CD are 3.8 and 4.5 Å, respectively. These findings suggest that the complex of *d*-propra-nolol with β-CD has higher stability than the complex with the *l*-propranolol. Thus, under chromatographic conditions with β-CD as chiral bonded phase, the *d*-enantiomer will be retained longer in the column.

Empirical rules for successful chiral recognition candidates using cyclodex-trins selectors have evolved based on extensive chromatographic data. For

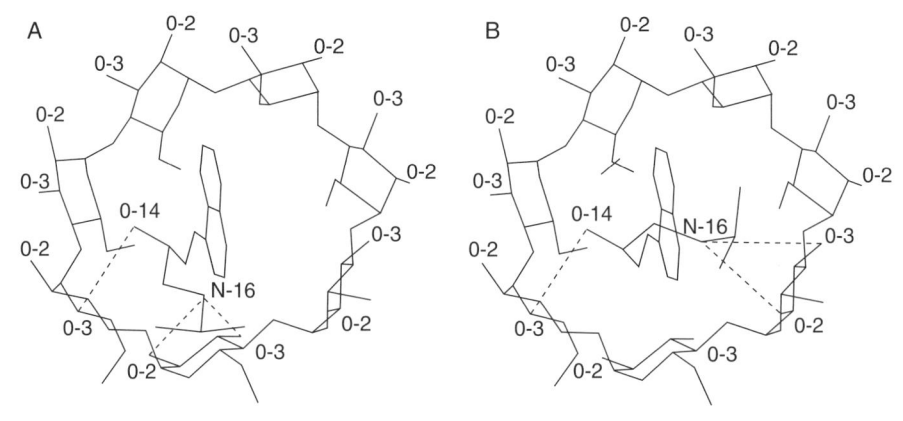

Figure 22-10. Computer projections of inclusion complexes of (A) *d*-propranolol and (B) *l*-propranolol in β-CD. Dashed lines represent potential hydrogen bonds (Reprinted from reference 50, with permission.)

instance, the presence in the guest molecule of at least one aromatic ring enhances chiral recognition with β-CD, although two appear to be more beneficial, particularly if the chiral center is positioned between the two rings or between a single aromatic ring and a carbonyl [50]. The enhanced chiral recognition was attributed to increased molecular rigidity [51]. Similar conclusions were reported by Armstrong et al. [52] for the separation of metallocene enantiomers where the chiral centers, upon inclusion, were located near or at the rim of β-CD. The metal ion was found to have no direct contact with the cyclodextrin; the interaction is called "second-sphere coordination" [53]. More linear metallocene enantiomers have to be complexed in a bent or skewed position to obtain optimum orientation. If the chiral center is buried between two bulky groups, however, the enantiomeric separation vanishes. Potential for hydrogen bonding between the enantiomers and the secondary hydroxyls of the CD should exist [54], although enantiodiscrimination, using mobile phases containing β-CD as additives, has been reported for terpene enantiomers, which lack hydrogen-bonding moieties [55–57]. The stoichiometry of complexation between the guest and the host CD in free solution can vary (e.g., from 1:1 to 1:2 guest:CD) [50, 56, 58]. For example, inclusion complexes between β-CD and (*S*)-(+)- and (*R*)-(−)-fenopren (Figure 22-11) [59] occur in the crystal structure through a 2:1 stoichiometry in which the (*S*)-(+) isomer is sandwiched in a dimer between two molecules of β-CD arranged head-to-tail, while the (*R*)-(−) isomer is sandwiched between two molecules of β-CD arranged in a head-to-head arrangement. The carboxylic group of the (*S*)-(+) isomer forms hydrogen bonding with the secondary hydroxyl groups of β-CD while (*R*)-(−) does not [59].

The chromatographic separation of enantiomers using CDs is usually performed using aqueous–organic mobile phases. The apparent pH of these

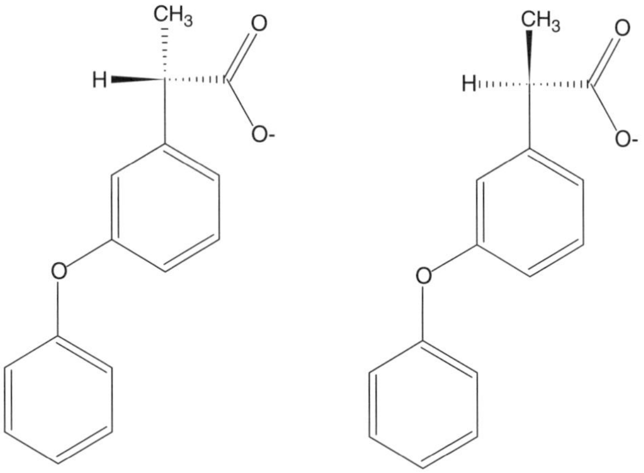

Figure 22-11. Chemical structure of (S)-(+)-(left) and (R)-(−)-fenopren (right). (Adapted from reference 59.)

mobile phases must be carefully controlled in order to handle the charge of the enantiomeric analytes. For example, separation of nicotine and nicotine analogues [54] could not be achieved at pH values lower than 5. This was a consequence of the protonation of nitrogens in the analyte molecules. At higher pH values, complete separation could be achieved, indicating that enantiomeric separation required the nitrogens to be partially deprotonated. Simultaneously, the hydrogen bonding between the β-CD and the analytes occurs through O—H to N.

The concentration of organic modifier in a hydroorganic mobile phase also influences retention. For instance, retention of analytes decreased as the amount of acetonitrile in the hydroorganic mobile phase increased up to a point, after which the retention started increasing again. Such behavior may indicate a change in retention interactions with the increase amount of acetonitrile in the mobile phase. No reversal of elution order was observed, indicating that no change in the enantioselective interactions occurred [54].

Polar organic mobile phases, such as mixtures of methanol and acetonitrile with small amounts of acetic acid and triethylamine, can also be effective for the separation of enantiomers mediated by the CDs. Under these conditions, the interior of the CD cavity is occupied by acetonitrile. The overwhelming concentration of acetonitrile renders its displacement by the enantiomeric analytes basically impossible. Acetonitrile is a polar aprotic solvent, with limited capacity for hydrogen bond formation. As a consequence, under these conditions, analytes are thought to undergo hydrogen bonding with the secondary hydroxyl groups located at the rim of the CDs. The addition of methanol and traces of acetic acid and triethylamine allows solute retention to be modulated through solvent mediation of the hydrogen bond strength [60].

Derivatized CDs have also been used successfully in HPLC. Armstrong and co-workers [61, 62] synthesized several derivatized β-CDs and used them as chiral stationary phases under normal-phase conditions. Under these conditions, inclusion is unlikely. A number of substituted derivatives were prepared including acetic anhydride, (R)- and (S)-1-(1-naphthyl)ethyl isocyanate, 2,6-dimethylphenyl isocyanate and p-toluoyl chloride. The presence of aromatic substitution provides possibilities for π–π interaction with the aromatic substituents of the enantiomeric analytes [61, 62]. For example, in (R)-(−)-, or (S)-(+)-1-(1-naphthyl)ethyl carbamate of β-CD, the naphthyl ethyl moiety has some π donor character. Incorporation of 3,5-dinitrophenyl substituents on chiral analytes promotes formation of a π–π complex. At the same time, the carbamate functionality that links the aromatic group to the β-CD produces sites that are able to undergo hydrogen bonding as well as dipole stacking with the enantiomeric analytes. An illustration of a possible association complex formed between a chiral analyte and the derivatized CD is shown in Figure 22-12 [61]. For clarity, only one naphthylethyl carbamate substituent is shown in Figure 22-12. The degree of substitution actually achieved is between three and eight substituents. Other orientations include positioning the phenyl ring of the solute over the cyclodextrin cavity, which results in a variety of interactions which can contribute to enantiomeric recognition.

22.3.4.2   Crown Ethers. Crown ethers can be described as heteroatomic macrocycles with repeating units of (—X—C_2H_4—) where the heteroatom X is usually oxygen, but may also be sulfur or nitrogen. They can also

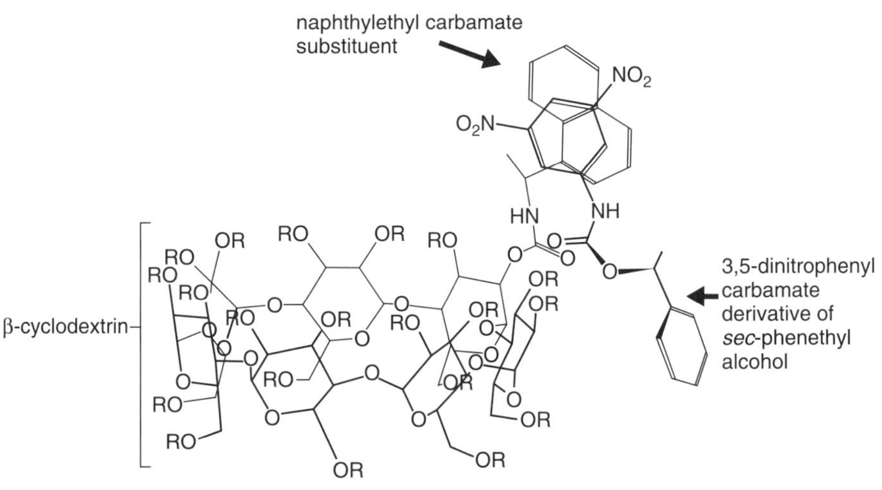

Figure 22-12. Schematic illustrating likely π–π and dipole stacking interactions between the 3,5-dinitrophenyl carbamate derivative or *sec*-phenyl alcohol and the naphthyl carbamoylated β-CD stationary phase. (R = H or naphthylethyl carbamate). (Reprinted from reference 61, with permission.)

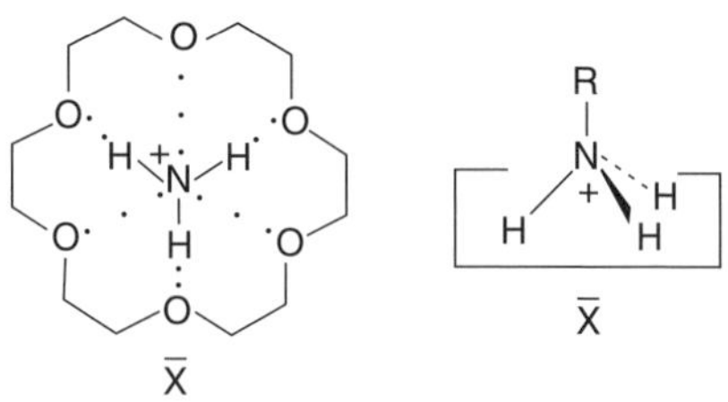

Figure 22-13. The structure of the complex between 18-crown-6 and alkyl ammonium salts. (The R group in the left structure is not included). (Reprinted from reference 65, with permission.)

incorporate aromatic moieties which enhance their lipophilicity. These types of compounds were first synthesized by Charles J. Pedersen, who named this class of compounds "crown ethers" [63]. The simple crown ether compound, 18-crown-6, was also found to complex with alkyl ammonium salts [64]. The structure of the complex of 18-crown-6 with ammonium and alkyl ammonium salts is presented in Figure 22-13.

Each oxygen atom possesses two unshared electron pairs. All six oxygens of the cyclic ether are turned inward to provide dipole-to-ion attractive interactions between host and guest. The main source of interaction is pole–dipole attraction between $^+NH\cdots O$ and $^+N\cdots O$. Three hydrogen bonds between the ammonium hydrogens and the crown ether oxygens can be formed. The ethylene units of the crown ether are turned outward and form a lipophilic barrier around the hydrogens of the hydrophilic ammonium ion. While the host molecule is roughly planar, and the nitrogen of the guest is situated slightly out of the plane at the apex of a shallow tripod, it may be argued that the association between the crown ether and the ammonium is not really an inclusion complex. The alkyl group attached to the nitrogen extends along the axis perpendicular to the plane of the cyclic ether (Figure 22-13). The ammonium ion can complex at either of the two faces of the cyclic polyether. The counterion, X^-, in a nonpolar environment, ion pairs with N^+ from the face opposite that occupied by the ammonium ion [65]. Such complexes constituted the start for Cram's complexes with chiral ammonium salts. To achieve enantiomeric separation, Cram introduced additional functional groups such as naphthalene rings into the crown ether structure, which provided additional interactions capable of discriminating between the enantiomers. The host that contained two chiral elements (Figure 22-14) provided the highest chiral selectivity [66].

In Figure 22-15, the four planes of the four naphthalene rings are perpendicular to the plane of the oxygen atoms, and form walls along the sides of the macrocycle. The space above, below, and along the side the macrocycle is divided by the four walls into four equivalent cavities, two above and two below the macrocycle [65]. The chiral cavities possess a pocket on one side (left side, Figure 22-15) and a barrier on the other (right side, Figure 22-15).

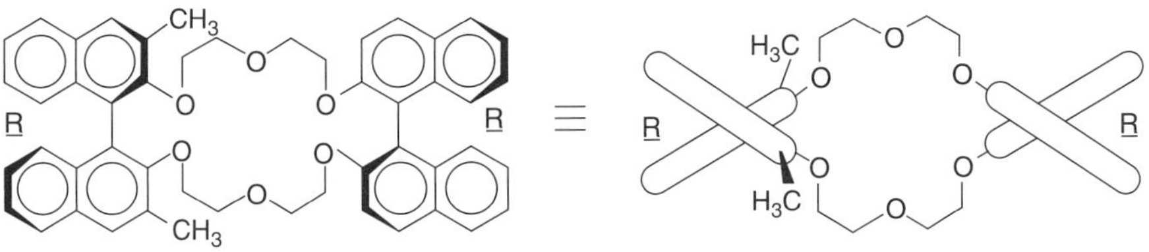

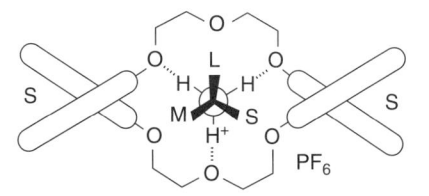

Figure 22-14. The structure of chiral crown ether. (Reprinted from reference 66, with permission.)

Figure 22-15. Interaction of a chiral 18-crown-6 with a chiral alkyl ammonium salt. (Reprinted from reference 67, with permission.)

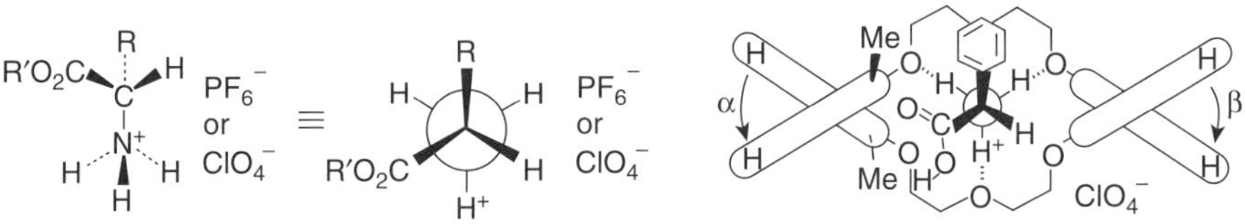

Figure 22-16. Structure of (L)-phenyl glycine (left) and the complex with chiral crown ether (right). (Reprinted from reference 68, with permission.)

Thus, when crown ether from Figure 22-14 complexes an optically active ammonium salt whose asymmetric center is adjacent to the primary ammonium groups, the same complex is formed whether the ammonium group complexes from the top or from the bottom of the host. In the complex, the large (L), medium (M), and small (S) groups attached to the chiral center must distribute themselves into the two equivalent cavities. In Figure 22-15, L is distributed in one cavity and M and S are distributed in the second. In the more sterically stable diastereomeric complex, molecular models predicted that M would reside in the pocket and S would reside against the barrier. This model is referred to as the three-point binding model and was confirmed by paramagnetic resonance (pmr) spectra [67].

Another example of such complexes is the interactions between the chiral 18-crown-6 and phenyl glycine, which is presented in Figure 22-26 [68]. Later, the chiral 18-crown-6 was immobilized on silica gel and polystyrene resins and used as a stationary phase in liquid chromatography for the separation of amino ester salts. Despite the fact that baseline separation was obtained

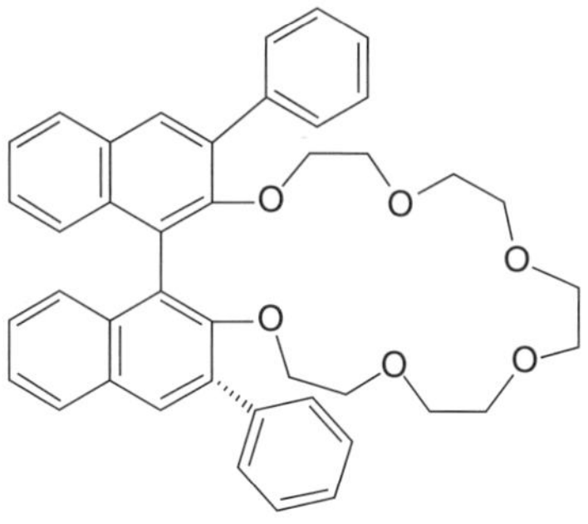

Figure 22-17. Structure of Crownpak CR (–). (Reprinted from reference 74, with permission.)

between the enantiomeric pairs, the chiral recognition on the bonded phase was smaller than in solution [69, 70].

Shimbo et al. used a chiral 18-crown-6 (Figure 22-17) dynamically coated on a reversed-phase stationary phase [71, 72]. This crown ether is commercially available under the trade name Crownpak CR.

This crown ether is able to resolve a large number of enantiomeric amines, amino alcohols, and amino acids using reversed phase conditions. It was found that additives such as perchlorate play an important role in chiral separation. This observation is compatible with the theory of chaotropicity. An anion with high chaotropicity is characterized by high polarizability. Such anions are able to break the structure of water, making it more lipophilic. In the hydration shell of such anions, the water's protons are directed in toward the anion [73]. In a series of anions such as ClO_4^-, CF_3COO^-, NO_3^-, and $H_2PO_4^-$ the retention factor of amino alcohols such as *cis* and *trans* amino indanol (at a constant pH of 2) increases in the following order: $ClO_4^- > CF_3COO^- > NO_3^- > H_2PO_4^-$. The selectivity factor, however, was not influenced by the nature of the chaotropic agent [74].

For more hydrophobic analytes, the retention can be modulated by the addition of organic modifiers such as methanol in the mobile phase. However, there is not a linear relationship between the amount of the organic modifier in the mobile phase and the retention factor of the enantiomeric analytes, indicating multiple types of retention interactions [75].

A different type of crown ether used to separate enantiomers is the one derived from 18-crown-6 tetracarboxylic acid, covalently immobilized on silica gel via reaction between 18-crown-6 tetracarboxylic acid and amino propyl silica gel [76]. The structure of 18-crown-6 tetra carboxylic is presented in Figure 22-18 [77]. The enantioselectivity on this chiral phase is improved by the addition of triethylamine into the mobile phase and operating at high methanol concentrations; however, enhanced selectivity may come at the expense of greatly increased retention times [78].

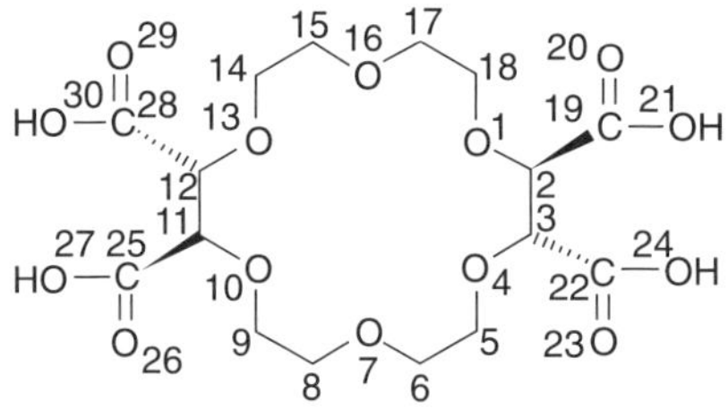

Figure 22-18. Structure of 18-crown-6-tetracarboxylic. (Reprinted from reference 77, with permission.)

22.3.5 Charge Transfer

Charge transfer complexes are an electron donor/electron acceptor association for which an intermolecular electronic charge transfer is observed [79–82]. Aromatic interactions have been suggested to consist of van der Waals, hydrophobic, and electrostatic forces [80]. The electrostatic component has been suggested to arise from interactions of the quadrupole moments of the aromatic rings. The edge–face geometry can be considered as a CH–π interaction found in benzene in the solid and liquid state and is commonly observed between aromatic residues in proteins [80]. Aromatic rings can also act as hydrogen bond acceptors. Energy calculations show that there is a significant interaction between a hydrogen bond donor (such as >NH group of an amine or amide) and the center of a benzene ring, which acts as a hydrogen bond acceptor. This aromatic hydrogen bond arises from small partial charges centered on the ring carbon and the hydrogen atom.

The formation of a donor–acceptor complex is described as an equilibrium process characterized by equilibrium constant. The presence of a solvent affects the complexation constant describing the equilibrium between the individual components of the complex. This is due to a competition of the solvent molecules toward each component of the complex. The solvent does not have to be a charge-transfer competitor. Competitive interactions such as hydrogen bonds can also affect the equilibrium. When the equilibrium constant of the complexation is quite low, the influence of the solvent is very significant, due to its overwhelming concentration compared to the concentration of the complex. For example, dioxane or ether are known to be effective n-donors; chloroform and methylene chloride have proved to participate in hydrogen bonds with π-donor molecules; and carbon tetrachloride behaves as an electron acceptor [83].

22.3.5.1 Chiral Separation Through Charge Transfer.
The first chiral separation to employ solely π–π charge transfer interaction was reported by Newman and co-workers [84, 85]. The authors used R-(−)-2-(2,4,5,7-tetranitro-9-fluorenylidene-iminoxy)-propionic acid (TAPA) to resolve racemic mixtures of 1-naphthyl-*sec*-butyl ether, as well as hexahelicenes, by crystallization.

H
‖
O ► C ◄ CO₂H
N ‖
X

NO₂ 7 8 ‖9 1 2 NO₂

6 3

5 4
NO₂ NO₂

X = methyl, R(–)-TAPA
X = ethyl, R(–)-TABA
X = isopropyl, R(–)-TAIVA
X = butyl, R(–)-TAHA

Figure 22-19. The structure of TAPA and its homologous. (Reprinted from reference 88, with permission.)

Later, Klemm and co-workers [86, 87] achieved partial resolution of aromatic compounds by low-pressure chromatography on silica gel impregnated with TAPA. The separation was attributed to π–π complexation between TAPA and the enantiomers. Mikes et al. [88] used a column packed with an (R)-(–)-TAPA aminopropyl-bonded silica support to accomplish the full resolution of helicenes. The authors extended their study to other homologues of TAPA (Figure 22-19). These compounds were coated on silica gel or ion-paired to an aminopropyl-bonded phase, and they were used in the HPLC separation of helicenes. To describe the selective interactions that occur between the stationary phase and the helicenes, the authors assumed that the 2,4,5,7-tetranitro-9-fluorenylidene moieties of the selector are laying down on the silica surface, while the X groups point away from the surface and above the plane of the fluorenyl ring.

When the (R)-(–)TAPA/P(+)-helicene complex is formed, the semicavity of P-(+) helicene can enclose the hydrogen and the methyl substituents of the asymmetric TAPA, while these substituents tend to lift the M(–)selectand off the selector (Figure 22-20). In this conformation, where the (R)-(–)-TAPA and M-(–)helicene molecules are parallel to each other, the substituents of the asymmetric carbon sterically hinder the π–π overlap and impair the inter-actions of the M-(–)-isomer. In other complex conformations where the selectand–selector molecules are antiparallel to each other, both M-(–)- and P-(+)-helicenes can form readily π–π overlapping complexes, but these do not involve the asymmetric carbon and are not enantioselective. This implies that the P-(+)-helicene/TAPA complexes can be formed in a wider range of ori-entations (with respect to the π–π axis) than can be formed with the M-(–)-isomer and thus have larger complexation constants and elute last from the column. A gradual increase of the size of the alkyl group X, on the asymmet-ric carbon, beyond the size of the semirigid cavities of the P-(+)-[6]-[14]-helicenes impairs the particular π–π selector–selectand interaction and consequently gradually diminishes enantioselectivity (Figure 22-20). An

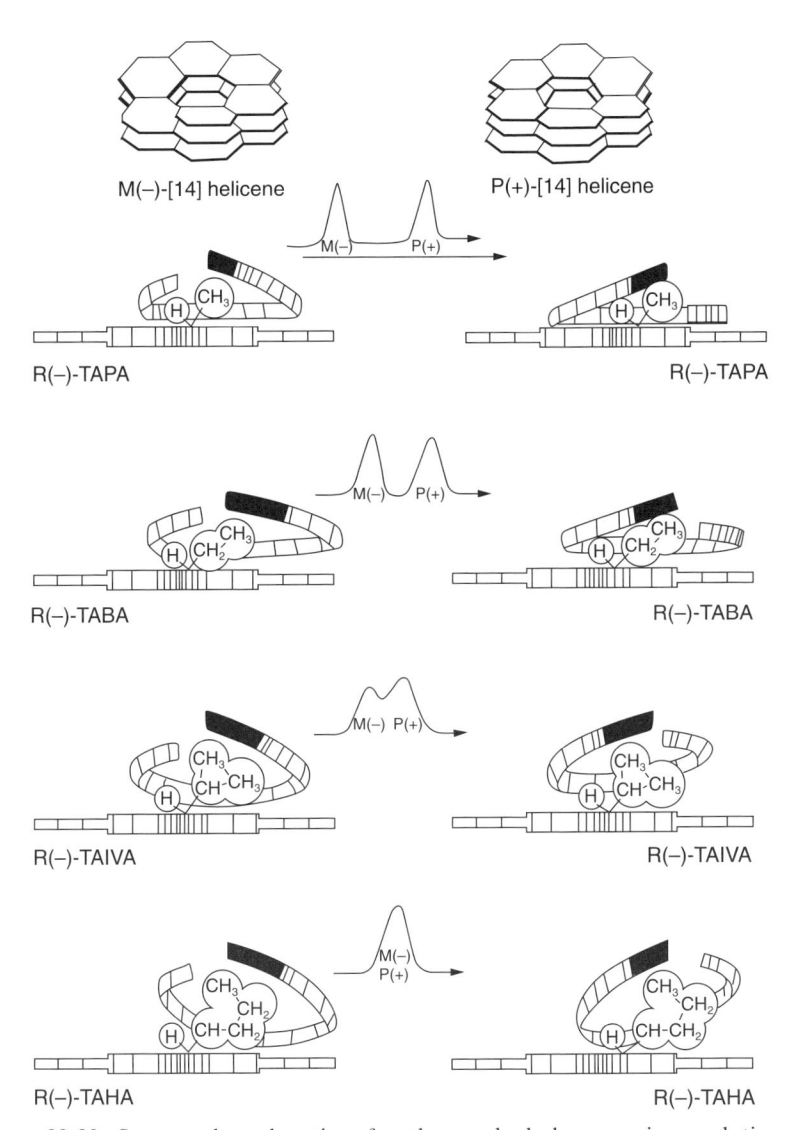

Figure 22-20. Suggested explanation for the gradual decrease in resolution with increase in size of the ligands at the asymmetric carbon of TAPA. (Reprinted from reference 88, with permission.)

increase in the polarity of the mobile phase similarly affects the retention of both enantiomers, resulting in no change in the selectivity factor α [89].

22.3.5.2 Chiral Separation Through Combination of Charge Transfer and Hydrogen Bonding and Electrostatic Interactions.

The separations described above involve solely charge transfer complexes between a chiral π-acceptor and π-donor. In the following section, we will present separations occurring through a combination of charge transfer and hydrogen bonds or electrostatic interactions.

In 1965, Raban and Mislow [90] postulated that nuclei placed in an asymmetric magnetic field should show NMR nonequivalence. In 1966, Pirkle [91] first reported the validity of the prediction when it was shown that (S)-1-phenylmethylamine caused ^{19}F-NMR nonequivalency of 2,2,2-trifluoro-1-phenylethanol in a carbon tetrachloride solution. In later studies, 2,2,2-trifluoro-1-(9-anthryl)ethanol, an NMR shift reagent, was used as a mobile-phase additive to separate 2,4-dinitrophenyl methyl sulfoxide on a silica gel column [92]. Later, one enantiomer of this fluoroalcohol was covalently attached to silica gel and used for resolution of a large number of solutes including sulfoxides, lactones, derivatives of alcohols, amines, amino acids, hydroxy acids, and mercaptans [93].

The model used to describe complex formation between the selectand and the selector consists of three simultaneous points of interaction first described by Dalgliesh [2] and illustrated in Figure 22-21 [94]. The hydroxyl of the fluoroalcohol hydrogen bonds to either the carbonyl oxygen or the dinitrobenzoyl (DNB) group of the alternate basic site, B, depending on which of these two sites is the most basic. The carbonyl hydrogen of the CSP interacts at the remaining basic site. The final interaction is π–π bonding between the anthryl and the DNB groups. Elution orders of configurationally known solutes support this model. Controlling the conformational mobility of the chiral selector on the CSP can enhance chiral recognition. For instance, chiral phases incorporating L-proline were designed to separate the enantiomers of N-(3,5-dinitrobenzoyl)amino acid esters and related analytes. Separation factors as high as eight were obtained for N-(3,5-dinitrobenzoyl)leucine amides [95]. The structure of the analyte and proline chiral phase is presented in Figure 22-22. The interaction between the proline CSP and the leucine derivative is presented in Figure 22-23 [96]. The trimethylacetyl group of the chiral selector

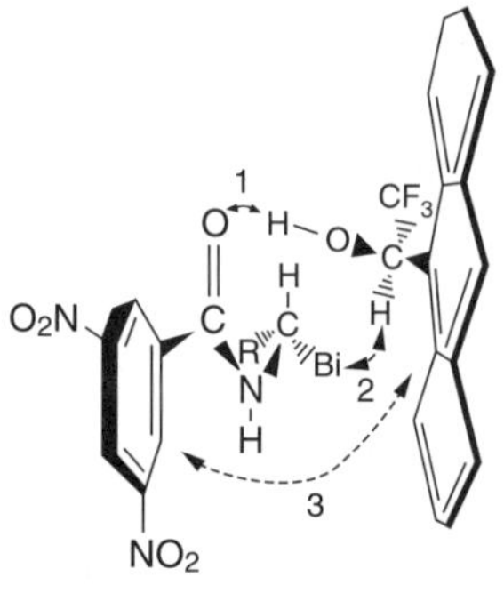

Figure 22-21. Three-point interaction of the most stable complex between 2,2,2-trifluoro-1-(9-anthryl)ethanol stationary phase and DNB derivative. (Reprinted from reference 94, with permission.)

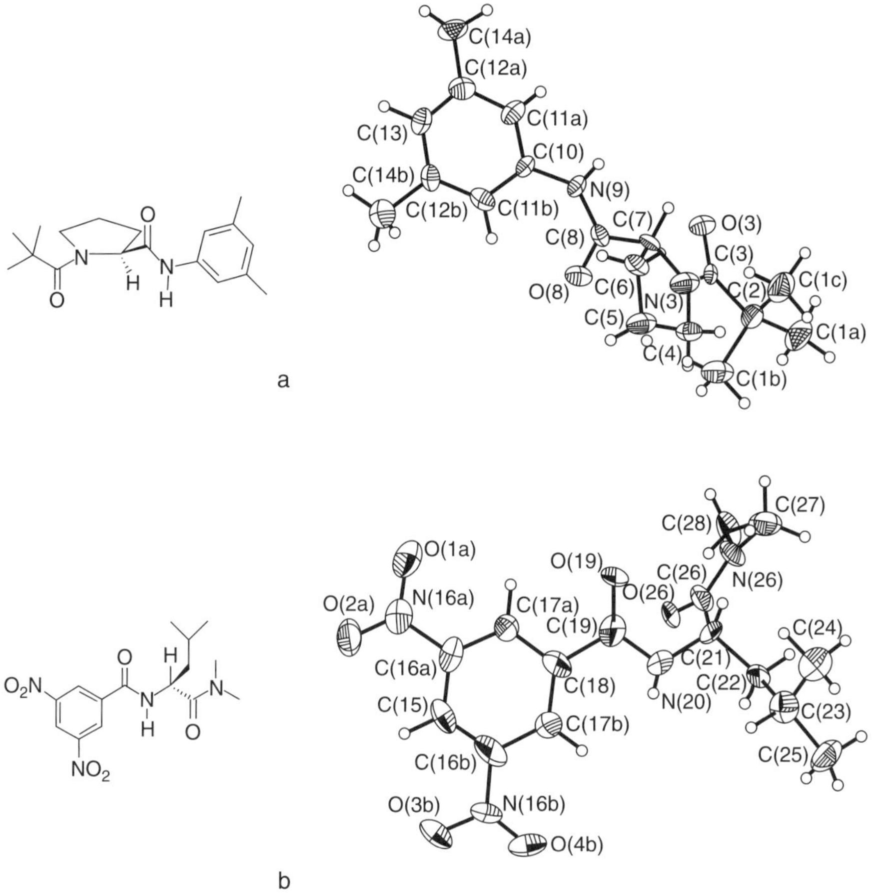

Figure 22-22. Structure of (a) L-proline stationary phase and (b) *N*-(3,5-dinitrobone-zoyl)leucine amides. The right ORTEP diagram illustrates showing the numbering system depicting the conformations present in the solid state complex. (Reprinted from reference 96, with permission.)

exists in the *trans*-rotamer. The 3,5-dimethylanilide group is planar and predominates as the Z-rotamer, which places H9 syn- to H7. The 3,5-dimethylanilide group is perpendicular to the plane of the five-membered proline ring. The two amide carbonyl oxygens are *anti* to one another.

The notion of reciprocity in chiral recognition has played an important role in the design of chiral selectors. In principle, if a single molecule of a chiral selector has different affinities for the enantiomers of another substance, then a single enantiomer of the latter will have different affinities for the enantiomers of the initial selector. In an effort to design a chiral stationary phase capable of separating naproxen, Pirkle et al. [97] first designed two stationary phases in which the carboxyl function of naproxen was linked to a silica matrix

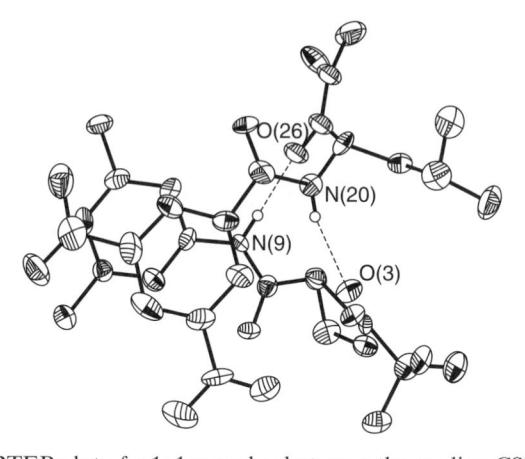

Figure 22-23. ORTEP plot of a 1:1 complex between the proline CSP and the leucine derivative. (Reprinted from reference 96, with permission.)

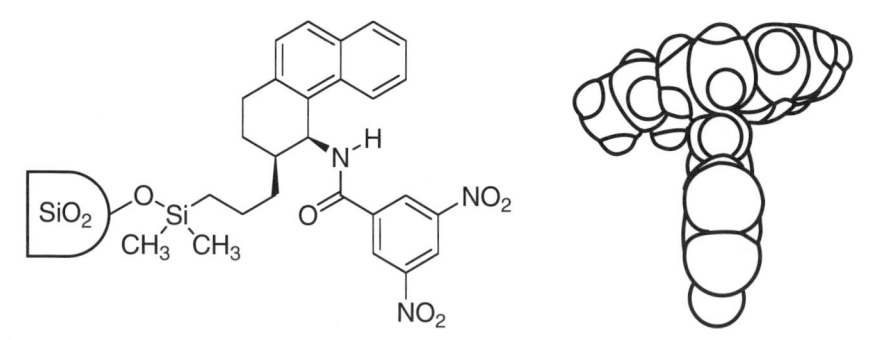

Figure 22-24. CSP for enantiomeric separation of Naproxen. Chemical structure (left) and a CPK model (right). (Adapted from reference 98, with permission.)

through either an ether link or an amide link. It was found that the enantiomers of 3,5-dintrobenzamide of α-(1-naphthyl)ethylamine could be separated on both stationary phases. Thus, a stationary phase was generated which incorporated the structure from Figure 22-24.

In Figure 22-24, the CSP has a cyclohexyl ring that contains the stereogenic center bearing the dinitrobenzamide group, controls orientation of the naphthyl moiety, and confers a high degree of conformational rigidity. The chiral selector can be viewed as a semirigid framework holding a π-acceptor 3,5-dinitrobenmzamide group perpendicular to a π-acceptor polynuclear aromatic group. The amide N—H serves as the hydrogen-bond donor and is situated in the cleft formed by the two aromatic systems. This selector is capable of simultaneous face-to-face and face-to-edge π–π interactions with an aromatic group present in the analyte. The face-to-face interaction presented to the analyte's aromatic substituents enhances its ability to simultaneously participate in the

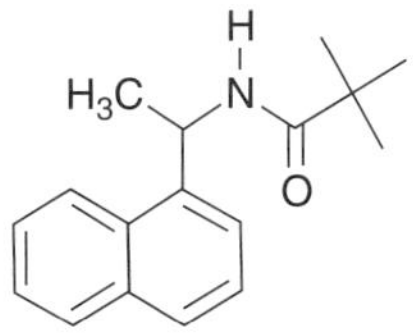

Figure 22-25. Structure of pivalamide derivative of 1-(1-naphthyl)ethylamine. (Reprinted from reference 99, with permission.)

face-to-edge interaction [98]. Further studies showed that this stationary phase separates not only enantiomers of underivatized naproxen, but also a large variety of other compounds [99]. The enantiomeric separation of the pivalamide derivative of 1-(1-naphthyl)ethylamine (Figure 22-25) on this stationary phase, along with an NMR spectroscopy study, elucidated the interaction between these enantiomeric analytes and the chiral moiety of the stationary phase [99].

The electron-rich naphthyl ring of the analyte is expected to participate in a face-to-face π–π interaction with the electron-deficient 3,5-dinitrobenzamide (DNB) ring of the selector. The carboxamide oxygen in the analyte is expected to participate in a hydrogen bond to the acidic 3,5-DNB amide proton of stationary phase. An edge-to-face π–π interaction between the naphthyl ring of the analyte and the π-cloud of the naphthyl group of the stationary phase is proposed as the third of the binding interactions responsible for the observed enantioselectivities. The (S)-enantiomer of the pivalamide derivative of 1-(1-naphthyl)ethylamine is believed to undergo these interactions simultaneously with the (S) stationary phase from a low-energy conformation, whereas the (R) enantiomer cannot. The homochiral [i.e., (S,S) or (R,R)] complex was found to be more stable than the heterochiral complex, since (S,S) and (as a consequence) the CSP preferentially retain the pivalamide derivative of (S)-α-(1-naphthyl)ethylamine. Figure 22-26 depicts the proposed most stable (S,S) complex between the pivalamide of (S)-α-(1-naphthyl)ethylamine and the chiral selector of Figure 22-24.

The selectivity of these phases can be changed by changing the solvent polarity. Such a change in polarity of the mobile phase can lead to a change in elution order of the two enantiomers [100].

Another type of CSP able to undergo charge transfer interaction is the one developed by Lindner's group [101]. In order to determine the interactions between the quinine CSP and the enantiomeric analytes, a detailed computational study was undertaken of the interaction of this stationary phase with 3,5-dinitrobenzoyl derivatives of leucine (Figure 22-27) [102].

The basis of the interactions in the complex consists of electrostatic interactions between the quinuclidin's ammonium ion and the selectand's carboxylate, the π–π interactions between the quinoline ring of the selector and the dinitrobenzoyl of the selectand, and the steric repulsion between the leucine side chain and the carbamate moiety [102].

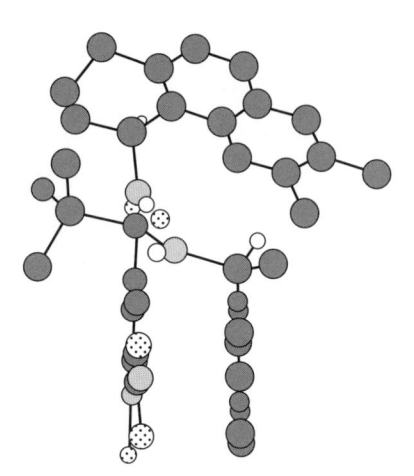

Figure 22-26. Proposed chiral recognition model for the more stable (*S*,*S*)-complex between (*S*)-pivalamide of 1-(1-naphthyl)ethylamine and the chiral phase of Figure 22.31. (Reprinted from reference 99, with permission.)

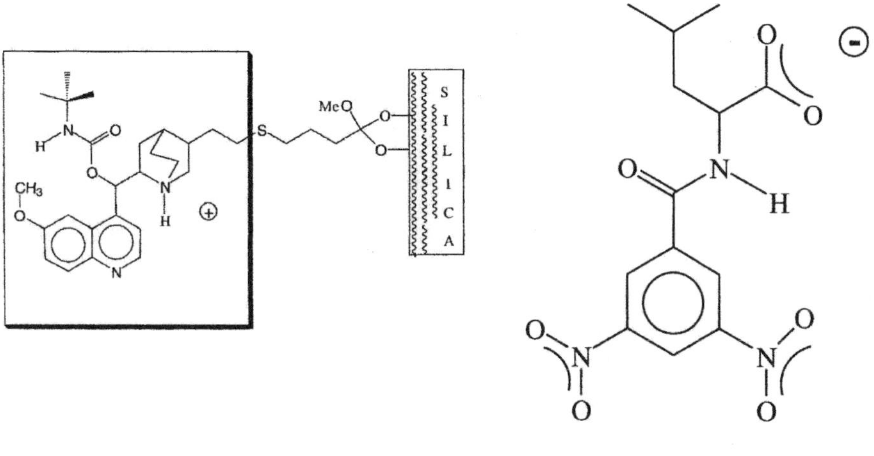

a

b

Figure 22-27. The structure of the quinine stationary (a) phase and 3,5-dinitrobeonzoyl leucine (b). (Reprinted from reference 102, with permission.)

22.4 MIXED TYPES OF INTERACTION

In Section 22.3 the main types of interactions occurring between the enantiomeric analytes and the stationary phase (hydrogen bonding, charge transfer, and inclusion complexes) was described. In the following section, enantiomeric separation based of combinations of these interactions will be presented. These types of stationary phases include polysaccharides, antibiotic phases (macrocyclic antibiotics), and protein phases.

22.4.1 Polysaccharide Phases

Many polysaccharides serve as extracellular elements in the cell walls of unicellular microorganisms and higher plants and in the outer surfaces of animal cells. There are many different structural polysaccharides. Cellulose is the most abundant extracellular structural polysaccharide of all biomolecules, plant or animal. Structurally, cellulose is a polysaccharide composed of several thousand molecules of D-glucose joined by β(1→4) glycosidic linkages. Complete hydrolysis of all glycosidic bonds of cellulose yields D-glucose.

Starch is a mixture of a water-dispersible fraction called amylose and a second component, amylopectin. Amylose is a polysaccharide made up of about 100 to several thousand D-glucose units joined together by α(1→4) glycosidic bond [103]. This apparent difference in the structure of cellulose and amylose results in polymeric structures with very different properties. Because of their β linkages, the D-glucose chains in cellulose assume an extended conformation and undergo side-by-side aggregation due to hydrogen bonding between the hydroxyl groups of adjacent cellulose chains. Due to the geometry of their α(1→4) linkage, the main chains of D-glucose units in amylose tend to assume a coiled, helical conformation [104]. The chiral discrimination ability of cellulose was recognized in the early 1950s, when Dalgliesh tried to separate aromatic amino acids using paper chromatography and postulated the three-point interaction principle as a principle for chiral discrimination [2]. Later on, cellulose was used in thin-layer chromatography to separate enantiomers of tryptophan and tryptophan derivatives using aqueous mobile phases [105–107]. Potato starch was also used to separate biphenyl atropisomers and other optical isomers [108]. The application of these polysaccharides to the separation of enantiomers is, however, rather limited due to their poor enantioselectivity.

Modification of hydroxy groups through esters or carbamate formation greatly enhanced the enantioselective properties of these polysaccharides. In 1973, Hesse and Hagel reported [109] for the first time the complete separation of Tröger's base on a column filled with microcrystalline cellulose triacetate. The triacetate cellulose is believed to preserve a structure closely related to native cellulose (form I) (CTA-1). The sorption of Tröger's base on the cellulose triacetate is not achieved by simply adhering to one glucose moiety but rather by insertion between two such moieties. The authors called this phenomena inclusion chromatography and stated that it was due to the tight packing of the crystalline form of cellulose triacetate. This stationary phase has been used in conjunction with mobile phases containing a mixture of water/ethanol for the separation of various racemates such as nonpolar and aromatic pharmaceuticals. Cellulose triacetate has two types of adsorption sites which differ in the adsorption/desorption rate: a "slow" site and a "quick" site. The slow type binds via an inclusion process and is of critical importance to chiral recognition [110]. It was suggested that for enantiomeric separation on cellulose triacetate the enantiomeric analytes should possess a phenyl or a cycloalkyl group close to the chiral center [43]. When CTA-1 was coated on silica gel from a solution of CTA-1 for HPLC use, a new stationary phase was obtained with different properties and chiral recognition ability. The new chiral phase has greater advantages than CTA-1 in column efficiency and durability.

The chiral phases derived from cellulose triacetate afforded the separation of a small number of classes of enantiomers. Derivatization of hydroxyl groups with aromatic moieties such as benzyl esters improved the chiral recognition capabilities. Such types of stationary phases were coated on silica gel for improved efficiency. The columns are commercially available under the trade name Chiracel OB and OJ. They are cellulose tribenzoate and tris(4-methylbenzoate), respectively. The effect of the substituents on the aromatic moiety was systematically studied, and it was observed that the position and the inductive effect of the substituent affect the enantioselectivity. 2-Substituted derivatives showed a low degree of resolution. Among four substituted tribenzoates (alkyl, halogen, trifluoromethyl, and methoxy groups), the derivatives having electron-donating substituents showed better chiral recognition ability than those having electron-withdrawing substituents. Highly electron-donating groups such as methoxy-substituted benzoate showed low chiral recognition due to the high polarity of the substituent [111]. Based on X-ray structural analysis, it was shown that cellulose tribenzoate has a 3/2 helical structure regardless of the preparation conditions [112]. Thermodynamic studies of the separation of a chiral diol using cellulose tris(4-methylbenzoate) (Chiracel OJ) stationary phase revealed that under HPLC conditions and the use of hexane/alcohols as mobile phases, the interaction is entropy-driven at low temperatures and is dominated by inclusion; at higher temperatures it is enthalph-driven. Differential scanning calorimetry and infrared spectroscopy experiments revealed that the transition between the enthalpic and entropic regions of van't Hoff plots are a result of a change in the conformation of stationary phase [113].

Substitution of the hydroxyl groups of cellulose or amylose with aromatic groups through a carbamate moiety brought a new dimension to the separation of enantiomers, giving more possibilities of interactions between the enantiomeric analytes and the chiral stationary phase. Most of the cellulose tris(phenylcarbamate) forms a lyotropic liquid crystal in high concentration solutions and shows high crystallinity under a polarizing microscope when they are cast from a solution. This indicates that coating phenylcarbamates on the silica surface are arranged in a regular fashion. Such an ordered structure seems to be very important for chiral recognition [114].

The mechanism of interaction between the enantiomeric analytes and the phenylcarbamates has been proposed based on chromatographic, computational, and spectroscopic studies. For example, the retention time of acetone on a phenyl carbamate tends to increase as the electron-withdrawing power of the substituent on the phenyl ring increases. At the same time, the retention of the first isomer of 2,2,2-trifluoroantryl ethanol decreases as the electron-withdrawing group on the phenyl ring of the phenylcarbamate increases. Such behavior indicates that the main chiral adsorption sites are probably the polar carbamate groups, which are able to interact with a racemate through hydrogen bonding to the NH and C=O groups. The nature of the substituent on the phenyl groups influences the polarity of the carbamate residues, which must change the resolving ability. That is why acetone is more strongly adsorbed on CSP with more acidic NH protons through hydrogen bonding. When the substituent at the phenyl moiety of the carbamate is an electron-donating group such as methyl, the electron density at the carbonyl oxygen atom of the carbamate increases. Therefore, the alcohols are more strongly adsorbed on the chiral stationary phase via hydrogen-bonding interactions. When the substituent on the phenyl group of the phenyl carbamate CSP is a polar group, such as nitro or methoxy, the racemates may interact with the substituent, which will result in a decrease in chiral recognition [115].

Amylose tris-(3,5-dimethylphenyl)carbamate has been reported to be a left-handed fourfold (4/1–4/3) helix. The optical resolving abilities of the amylose carbamate are different from those of the corresponding cellulose carbamate, being complementary to each other. The optimized structure of amylose tris-3,5-dimethylphenyl carbamate (ADMPC) has a left-handed 4/3 helix similar to that of amylose triisobutyrate, and the glucose residues are regularly arranged along the helical axis. A chiral helical groove with polar carbamate groups exists along the main chain. The polar carbamate groups are preferably located inside the polymer chain, and the nonpolar aromatic groups are placed outside the polymer chain, so that the polar enantiomer may interact predominantly with the carbamate residues in the groove through hydrogen bond formation. Differential scanning calorimetry indicated that the behavior can be attributed to the breakage of hydrogen bonds between the ADMPC chains, triggering a conformational change. Molecular modeling suggested that the most retained enantiomer interacts with the stationary phase through a hydrogen bond between the triazole proton and the C=O groups of the stationary phase, as well as through the inclusion of the compound in the cleft of the stationary phase. The other enantiomer exhibits a bifurcated hydrogen bond between the triazolic proton and the C=O groups of the stationary phase, leading to a less stable complex [29].

In order to maximize such interactions, these stationary phases can be used in conjunction with mobile phases containing mixtures of hexane/alcohols. It was found that linear alcohols such as ethanol and n-propanol produced a longer retention time than the branched alcohols. The reason for such a phenomenon is attributed to the behavior of linear alcohols, which self-associates in the presence of a nonpolar solvent such as hexane. Such association is at the expense of the interaction of these alcohols with the stationary phase, leaving more sites of interactions for the two enantiomers. In contrast, branched alcohols such as isopropanol exist mainly as monomers, interacting with the stationary phase at the expense of the two enantiomers [29]. Sometimes, changes from linear to branched alcohols are accompanied by a reversal in the elution order of the two enantiomers, due to a modification of the steric environment caused by the branched alcohols [115].

The use of additives such as small amounts of acids of bases in the mobile phase is often used to achieve separations on these in conjunction with these stationary phases. The net result is an improved kinetics of mass transfer and improved peak shape [116].

The polysaccharide stationary phases derived from cellulose and amylose are today among the most popular CSPs due to their versatility.

22.4.2 Antibiotic Phases

The glycopeptide macrocycle antibiotics are classes of compounds with high enantioselective properties. The include vancomycin, teicoplanin, and ristocetin A [117]. Such high enantioselectivity properties are due to their amphoteric character, their molecular structure that in solution accentuates the enantioselective interactions, and their hydrophilic and hydrophobic functional groups, which make these groups soluble in aqueous and organic solvents [118]. Such properties make them amenable for use as chiral selectors in either capillary electrophoresis or HPLC. These macrocyclic antibiotics are all members of the glycopeptide family of antibiotics and possess a closely related chemical structure. All members of this group were obtained from various species of *actinomycetes*, typically isolated from soil samples of various origins, all being active against gram-positive bacteria [119]. A summary of the sources and uses of some of these antibiotics is given in Table 22-1.

There have been over a hundred different glycopeptide antibiotics identified in the fermentation broth of various bacteria. All have a heptapeptide

TABLE 22-1. Producing Organism and Uses of the Vancomycin-Group Antibiotics

Antibiotic	Producing Organism	Use
Vancomycin	*Streptomyces orientalis*	Clinical: against severe infections
Ristocetin	*Nocardia lurida*	Clinical: diagnosis of von
	Proactinomyces fructifery	Willebrand's disease; study of blood platelet aggregation
Avoparcin	*Streptomyces candidas*	Animal feed additive
Actinoidin	*Nicordia actrinoides*	Investigation of ristocetin-induced platelet aggregation
Actaplanin	*Actinoplanes missouriensis*	Animal feed additive
A35512	*Streptomices candidus*	Animal feed additive
Teichomycin A2	*Actinoplanes teichomyceticus*	Investigational

Source: Reprinted from reference 119, with permission.

core of seven amino acid residues with the side-chain residues 2 and 4, 4 and 6, and 5 and 7 covalently joined, along with sugar substituents at various positions. In a classification scheme based on the type of residues at positions 1 and 3, vancomycin and eremomycin are assigned to group I with other compounds bearing aliphatic residues at these positions (Figure 22-28). Group II compounds, such as avoparcin, have individual aromatic residues at positions 1 and 3, whereas group III compounds, such as ristocetin and teicoplanin, have aromatic residues at positions 1 and 3 that are covalently joined (Figure 22-29) [120].

Sheldrick et al. [121] and Brown et al. [122, 123] studied the association between vancomycin and acetyl-D-alanyl-D-alanine using nuclear magnetic resonance spectroscopy. An analysis of binding as a function of pH showed that while the most stable complex is formed between the peptide anion and the vancomycin cation, weaker complexes are formed between uncharged peptide and the vancomycin cation and between the peptide anion and the uncharged vancomycin. Nieto and Perkins [124] and Perkins [125] studied the extent to which vancomycin and ristocetin-peptide interactions depend on the length and composition of the peptide chain by measuring the association constant of a range of synthetic peptides. The authors found that the prerequisites for the synthetic peptides to form stable complexes are as follows:

- There should be three amide linkages.
- The terminal carboxyl group must be free.
- The two COOH-terminal peptides must either be glycine or have a D-configuration for favorable interaction with vancomycin; a D-Ala residue being especially favorable in both positions.
- The third residue from the COOH terminus should have an L-configuration for optimal binding.

Armstrong et al. [126] were the first to use these macrocyclic antibiotics as chiral stationary phases. The authors bound vancomycin, rifamycin B, and thiostrepton on a silica matrix. These stationary phases appeared to be multimodal, and they can be used in normal-phase as well as in reversed-phase chromatography. The diversity of functional groups present in these macrocyclic antibiotics allow for different types of interactions between the CSP and the enantiomeric analytes. Thus the enantioseparation can take place through π–π complexation, hydrogen bonding, inclusion in a hydrophobic pocket, dipole stacking, steric interactions, or a combination thereof. The advantage of these CSPs is that all these interactions are available, allowing for separation of large classes of racemates. Cavazzini et al. [127] studied the mechanism of chiral discrimination teicoplanin-based chiral stationary phases. The authors demonstrated that the existence of a free carboxyl moiety in the enantiomeric analyte is very important for the formation of the complex between enantiomers and the aglycone basket of teicoplanin. Their conclusions are similar to those

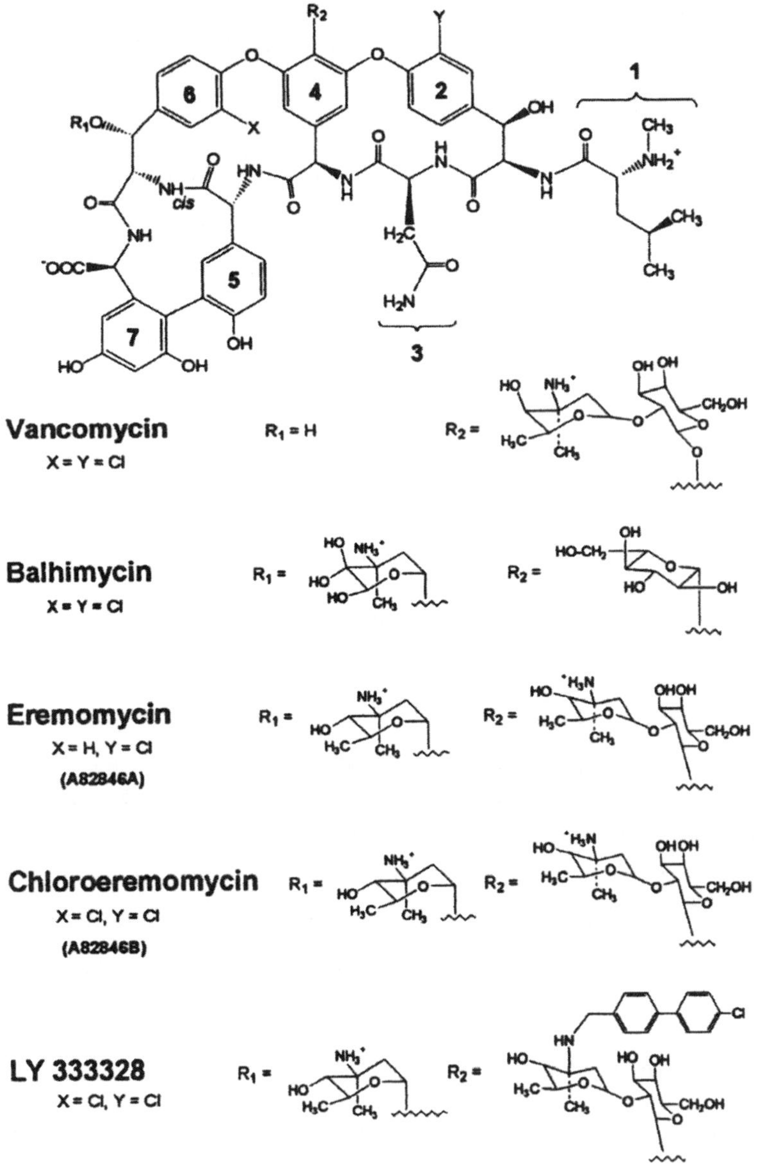

Figure 22-28. Group I antibiotics. (Reprinted from reference 120, with permission.)

obtained by Nieto and Perkins [124] and Perkins [125] for vancomycin and ristocetin. Additionally, the increased possibility of making a hydrogen bond between the amidic hydrogen of the acetylated compounds and the amidic groups of the stationary phase has been proposed to be of major importance for the stability of aglycone–D-enantiomer complex.

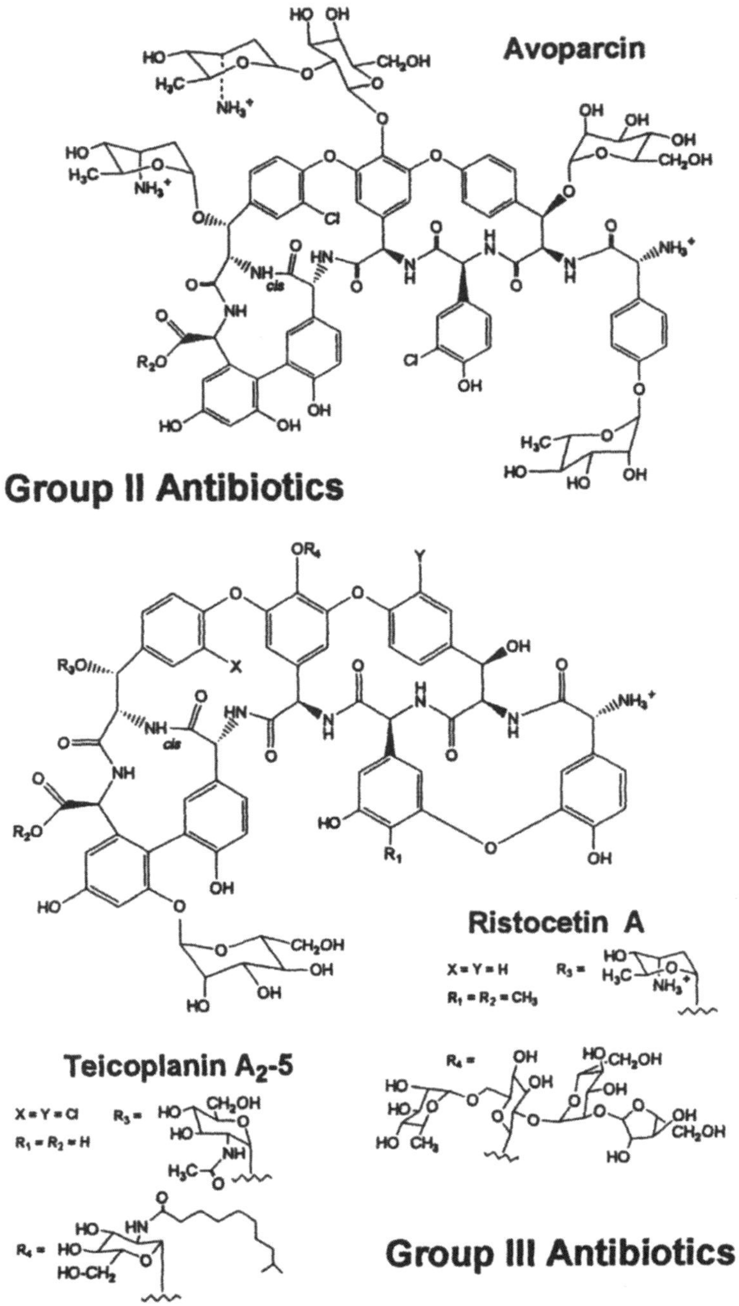

Figure 22-29. Group II and III antibiotics. (Reprinted from reference 120, with permission.)

The loading of the chiral phase on the silica matrix has an effect on retention, enantioselectivity, and efficiency. Thus, retention and enantioselectivity increased with increased CSP loading, but efficiency decreased. Similar results were obtained under both reversed-phase and normal-phase conditions [128].

Under reversed-phase elution conditions, the nature of organic modifier has a very important role on the separation. Thus, methanol proved to provide better enantioselectivity than acetonitrile for the separation of ergotamine on a vancomycin and teicoplanin stationary phase [129]. The buffer concentration also has an influence on the retention of the enantiomeric analytes. An increase in buffer concentration produces a decrease in the retention of the two enantiomers. It was found that in some cases the two stationary phases, teicoplanin and vancomycin, are complementary; for a poor separation on vancomycin, the stationary phase produced a baseline separation on teicoplanin and vice versa [129].

In conclusion, macrocyclic antibiotics have a wide application on the separation of wide classes of racemates, due to their diverse structure.

22.4.3 Protein Phases

Proteins are biopolymers consisting of amino acids linked together through peptide bonds arranged in a certain sequence. This sequence of amino acids constitutes the primary structure of protein molecules.

Many proteins contain only amino acids and no other chemical groups, and they are called simple proteins. However, other kinds of proteins contain other chemical components in addition to amino acids. These types of proteins are called conjugated proteins. The non-amino acid part of a conjugated protein is called the prosthetic group. Conjugated proteins are classified according to the nature of their prosthetic groups. For example, lipoproteins contain lipids, glycoproteins contain sugar groups, and metalloproteins contain metals such as iron, copper, or zinc [130].

The amide chains linking different amino acids are free to rotate about single bonds to the connecting α carbon. This twisting of amide planes about connecting α carbons gives the main chain backbone of a protein its right-handed helical conformation, also called an α helix. In the α helix, the carbonyl oxygens are in a favorable position to make hydrogen bonds with the hydrogen atom from a nitrogen located three residues away. Many α helices are amphipathic, in that they predominantly have side chains along one side of the helix and polar residues along the remainder of their surface. Such helices have substantial hydrophobic moment and often aggregate with each other or with other nonpolar surfaces [131].

When two or more almost fully extended polypeptide chains are brought together side by side, regular hydrogen bonds can form between the peptide backbone amide NH and the carbonyl oxygen of adjacent chains. Such an arrangement is called a β sheet. Since each backbone peptide group has its

NH and carbonyl groups in a *trans* orientation, it is possible to extend a β sheet into a multistranded structure by adding successive chains to the sheet. β sheets can occur in two different arrangements with the same N-to-C polypeptide sense to produce parallel β sheets. Alternatively, the chains can be aligned with opposite N-to-C senses to produce an antiparallel β sheet. The α helix and β sheet represent the secondary structure of a protein.

Association between elements of the secondary structure form structural domains with properties determined both by the chiral properties of the polypeptide chain and by the packing requirements which effectively minimize the molecule's hydrophobic surface area. Association of domains in proteins results in the formation of the protein's tertiary structure. Furthermore, protein subunits can pack together to form quaternary structure, which can either serve a structural role or provide a structural basis for modification of the protein's functional properties [132].

Molecular recognition in protein–ligand complexes is responsible for the selective binding of a low-molecular-weight ligand to a specific protein which can be characterized by a binding constant. All these interactions are usually noncovalent in nature. The experimentally determined binding constants, K_i, are typically in the range of 10^{-2} to 10^{-12} M^{-1}, which corresponds to a negative free energy of binding of $10–80 kJmol^{-1}$ [133].

Binding of small molecular ligands to proteins is an important biochemical and biological process that is used as a basis for drug design. The interactions between ligands and proteins involve a wide range of interactions including electrostatic, van der Waals, steric, hydrophobic and hydration forces [134] related to the active sites of the protein.

Such ligand–substrate binding can cause large-scale conformational changes in proteins, and such changes have been observed in many experimentally determined three-dimensional structures. Binding also induces more subtle changes in proteins, including changes in atomic mobility and low-frequency vibrational motion [135]. The binding of a small molecule ligand to a protein requires shape and property complementarity. In the course of binding, the ligands and protein have to adapt to each other to achieve a successful recognition process. The small molecule ligand is usually the more flexible partner, and thus can adopt a large variety of different low-energy conformations upon interaction with a protein [133].

Ligand-induced conformational changes in the protein can result in both cooperative and antagonistic binding effects on other ligands binding to the same protein. An example of a cooperative interaction is the enhancement of the interactions of progesterone and testosterone with human serum albumin (HSA) by the palmitate ion.

In contrast to cooperative binding, antagonistic binding has been proposed to explain why low concentrations of some analytes decreased the binding of others. An example is the inhibition of the interaction of chlorophenoxybutyrate with human serum albumin (HSA) caused by a low concentration of long-chain fatty acids [136].

Sundaresan and Abrol [137] proposed a general stereocenter recognition model for stereoselectivity of proteins toward substrates that have multiple stereocenters based on the topology of substrate stereocenters. The model provides the minimum number of substrate locations that need to enter into binding, nonbonding, or repulsive interactions with receptor sites for stereoselectivity to occur. According to this model, a substrate location may interact with multiple receptor sites, or multiple substrate locations may interact with a single receptor site, but a stereoselective receptor has to offer, in correct geometry, at least as many interactions as the required minimum number of substrate locations. An enantioselective recognition of molecules with one chiral center requires a protein to interact with a minimum of three substrate locations, while stereoselectivity toward a substrate with two or three stereocenters requires interactions with a minimum of four of five substrate locations, respectively. The authors demonstrate the general applicability of the model by applying it to experimental examples [137].

The inherent chirality of proteins makes them very good candidates for the separation of enantiomers. Proteins which can tolerate organic solvents, as well as high temperatures, and which can function over a wide range of pH are useful as chiral stationary phases. Table 22-2 shows some of the characteristics of these proteins [138].

Initial attempts to immobilize proteins involved physical adsorption of the protein to silica gel at a pH equal to the isoelectric point of the protein [139]. The limited elution conditions available for these types of enantiomeric separations makes such an approach unsatisfactory.

Stable stationary phases for enantiomeric separation are obtained by covalent binding of proteins to silica gel via the protein's free amino (belonging to a Lys or Arg, or terminal amino group) or carboxyl groups. The immobilization of proteins through amino groups involves the use of a porous modified silica gel with amino propyl or glycidoxypropyl. The support is further activated with either N,N'-disuccinimidylsuberate (DSS), tresyl chloride, or glutaraldehyde [140]. The glycidoxypropylsilica can be hydrolyzed and then oxidized to yield an aldehyde, which is further reacted with the protein in the presence of cyanoborohydride.

TABLE 22-2. Characteristics of Some Proteins Used as Chiral Stationary Phases

Protein	Molecular Weight	Isoelectric Point	S–S Bridges	% Carbohydrate	Sialic Acid Residues	Mechanism of Interaction
AGP	41,000	2.7	2	45	14	Cationic
BSA	66,000	4.7	17	—	—	Anionic–hydrophobic
ACHT	21,600	8.3, 8.7	—	—	—	Hydrogen bonding–hydrophobic
OVM	28,800	3.9–4.5	8	30	0.3	Hydrophobic

Source: Reprinted from reference 138, with permission.

The immobilization of a protein via its carboxylic group involves activation of the protein's carboxyl group with N-hydroxysulfosuccinimide, followed by the reaction of the activated protein with amino propyl silica gel. In this case an amide group is generated between the protein and the silica-based stationary phase [140]. The bonding of proteins to silica gel-based matrices, however, leads to a conformational change of the protein (which may impair the enantioerecognition [141]), as well as a blockage of the different functional groups involved in the recognition process. At the same time the enantioselectivity may change according to the method used to bound the protein to the matrix—for example, through the amino or carboxylic group [142].

The pore size of the silica gel matrix influences both the capacity factor and the enantioselectivity. Haginaka and Takehira [143] reported on the enantiomeric separation of benzoin, chlorpheniramine, and ibuprofen on an ovoglycoprotein stationary phase immobilized on silica gels with an average pore size of 12, 20, and 30 nm, respectively. The higher retention and enantioselectivity was obtained on the 12-nm pore size stationary phase. For the same pore size silica gel, a linear correlation was obtained between the capacity factor and the amount of protein loaded on the stationary phase. At the same time, the higher the protein loading, the higher the enantioselectivity.

One of the advantages of protein-based stationary phases is that the chromatography is performed under an aqueous organic mobile phase. Under such conditions, a variety of mobile-phase parameters, such as pH, ionic strength, organic modifier, and the type of organic modifier, can be varied for optimization of the chromatographic parameters (i.e., the retention). The influence of pH on the enantioseparation is related to the charge of the analyte as well as to the pI of the protein stationary phase. Acidic compounds show a decrease in k' upon increasing the pH. Such behavior can be explained by the decrease in charge interactions between the acidic analytes and the protein stationary phase. Similarly, amine compounds show a decrease in k' with a decrease in pH. For neutral compounds, such effects are rather small [141, 144].

Increasing ionic strength will modify the Debye–Hückel screening of the electrostatic interactions between the protein and the enantiomeric analytes, yielding a decrease in retention of the enantiomers [141]. At higher ionic strengths, hydrophobic interaction prevails, and the retention mechanism becomes more complex. The type of ions (sphere of their hydration) in the mobile phase influences the retention of the two enantiomers. Hedeland et al. [145] studied the separation of several β-blocking drugs on a cellobiohydrolase stationary phase. The authors found that an exchange of Na^+ with K^+ influences the retention of the two enantiomers, but has very little effect on enantioselectivity.

Protein phases have been shown to undergo a conformational change with increasing temperature. The impact of this conformational change on retention is generally more pronounced for the more retained enantiomer, because it has more interactions with the stationary phase than the early eluting enantiomer. In some examples, van't Hoff plots showed nonlinearity for the most retained enantiomer, while it was nearly linear for the least retained enantiomer [141, 146]. An inversion of the elution order with temperature was also observed in some instances, indicating a change from an entropically driven separation to an enthalpically driven separation. The entropic part of the separation was observed at lower temperatures, while the enthalpic part was noted at higher temperatures. The protein surface contains hydrophobic calyxes, where molecules of different polarity can be included. The difference in the entropy of the interaction of one enantiomer relative to the other becomes positive, along with the compensating enthalpy. Such phenomena are strongly influenced by the pH of the mobile phase [141, 146].

The organic solvents often used as modifiers in the aqueous mobile phase consist of *n*-propanol, isopropanol, methanol, and acetonitrile. They are efficient agents for modulating the hydrophobic interaction between the analytes and the protein stationary phase [144]. An increase in the organic modifier in the aqueous organic mobile phase will decrease the retention, but will have minimal effect on the overall enantioselectivity [141].

22.5 LIGAND EXCHANGE

Ligand exchange chromatography (LEC) is the typical example of complexation chromatography. Complexes formed during LEC encompass a metal cation associated with ligands (anions or neutral molecules) able to donate electron pairs to a vacant orbital of the metal [147]. The term ligand exchange was introduced Helfferich in 1961 when he described the substitution of organic diamine molecules with metal-ion-coordinated ammonia molecules in a polymeric phase [148]. The technique was further developed by Rogozhin and Davankov [149] for the separation of enantiomers. This technique is applicable for those enantiomers which are able to form metal complexes with the moiety bound onto the stationary phase. Enantiomeric analytes such as amino acids and hydroxy acids were successfully separated into enantiomers using LEC. In their article the authors described the model of interaction during the enantiomeric separation with LEC:

> *In such a process, one ligand (the optically active one) should be rigidly attached to a stationary phase, while the other (racemic) ligand should be able to move with the mobile phase. The metal atom forming the complex may be combined with either ligand. The important point is that the complex generated should be kinetically labile, i.e. readily decomposed and reformed.*

On this basis the first stationary phase using LEC consisted of a polymeric stationary phase obtained by copolymerizaiton between styrene-*p*-divinylbenzene and L-proline. The stationary phase was loaded with an aqueous solution of $CuSO_4$ dissolved in ammonium hydroxide. Using water as mobile phase, the chiral phase afforded the enantiomeric separation of D,L-

Figure 22-30. Typical model for sorption of proline enantiomers on L-proline or L-hydroxyproline incorporating polystyrene type sorbents. (Reprinted from reference 150, with permission.)

proline [149]. A typical model of the interaction between proline's enantiomers and a polystyrene base polymeric phase containing L-proline or L-hydroxy is presented in Figure 22-30. Retention of L-proline is diminished by the steric interaction with the water molecule coordinated in the axial position of the Cu(II) ion. Retention of D-proline is enhanced by the hydrophobic interaction with the nonpolar polystyrene chain [150]. These polymeric-based chiral stationary phases were characterized by poor efficiency, due to the poor mechanical properties of the polymeric phase.

Later developments in the field of silica based stationary phases led to more rugged and efficient LEC stationary phases. It was shown that the hydrophobic spacer through which the chiral ligand is bonded to the silicagel surface plays a major role in the enantioselectivity. Roumeliotis et al. [151] showed that when L-hydroxyproline was bonded to the silica through a spacer containing three carbons, the enantioselctivity for the α-amino acids was poorer than that obtained with a spacer with eight carbons. On increasing the *n*-alkyl chain length of the spacer the hydrophobic character of the surface increases. At the same the terminating chiral groups extend further into the pore space and become more flexible in the solvated state. The change in hydrophobicity and steric orientation of the bonded moiety, relative to other vicinal surface groups, may be expected to give rise to a unique enantioselectivity of the phase system. At a constant pH the retention increases with the increase of the length of the spacer. The pH of the mobile phase influences the stability constant of the complexes and consequently the retention as well as the enantioselectivity. The retention of hydrophobic analytes can be controlled by amounts of organic modifier such as acetonitrile or methanol. The flow rate of the mobile phase has major role on the resolution. The slower is the flow rate, the higher the resolution [152].

Ligand exchange chromatography is a very powerful method for separating enantiomers. However, it is limited to enantiomeric compounds that are able to undergo metal complexes with the chiral stationary phase such as amino acids, amino acid derivatives, and amino alcohols.

22.6   CHIRAL MOBILE PHASES

Chiral mobile phases were used extensively in chromatography in the 1960s, 1970s, and 1980s. Active compounds that were able to form ion-pair, metal, and inclusion complexes were added to the mobile phase to engender chiral selectivity in regular reversed-phase or normal-phase chromatography using achiral columns [153]. The preference for using these mobile phase additives under normal-phase or reversed-phase conditions was dictated by the number of chiral compounds with high optical purity which were commercially available or could be synthesized, purified, and characterized. At the time, commercially silica-based chiral phases commercially available were limited in number, and their preparation and lot-to-lot variability showed lack of robustness far from an acceptable level.

The enantiomeric separation with chiral mobile phases consists of the addition of an active compound in the mobile phase which is constantly pumped though the chromatographic system. The active ingredient contributes to a specific secondary chemical equilibrium, interacting with the enantiomers in the mobile phase as well as in the stationary phase, leading to the formation of diastereomeric complexes potentially in both phases. This affects the overall distribution of the analyte between the stationary phase and the mobile phase, affecting its retention and the overall enantiomeric separation. The rates of formation of the diastereomeric complexes should be similar to the diffusion rates to minimize excessive chemical contribution to the band-broadening.

The use of chiral mobile phases has both advantages and disadvantages. For example, the multiple equilibria occurring in the mobile phase and in the stationary phase complicates elucidation of the separation mechanism. The presence of the chiral mobile phase additive can also complicate detection. For instance, additives with relatively high UV absorbance decrease the detection limit of the separated enantiomers when using UV detection. Furthermore, resolved enantiomers enter in the detector cell in the form of complexes with the chiral resolving ligand. These complexes are diastereomers and therefore may differ in molecular absorptivity, as well as other properties. As a consequence, it is necessary to have a separate calibration curve for each enantiomer.

Conversely, the use of chiral mobile phases has advantages which make it very appealing. Many chiral additives are readily available or can be easily synthesized. Achiral stationary phases, which are significantly cheaper than chiral phases, can be used. The approach also offers more flexibility than direct separation with chiral stationary phases because the chiral mobile phase additives can often be easily washed out of the chromatographic system and replaced with another additive for subsequent separations.

22.6.1   Chiral Mobile-Phase Retention Mechanisms

To understand the mechanism of enantiomeric separation using chiral additives, let us assume that the compound L is a chiral additive that was added into the mobile phase, being constantly pumped through the chromatographic system. Under the chromatographic conditions, the compound (ligand), L, is distributed between the mobile phase and the stationary phase according to the equilibrium:

$$L_m \leftrightarrow L_s \qquad (22\text{-}1)$$

where the subscript m relates to the mobile phase and the subscript s relates to the stationary phase. When analyte A is injected into the chromatographic system, it will undergo the following equilibria:

$A_m + L_m \leftrightarrow AL_m$ Formation of diastereomers in the mobile phase   (22-2)

$AL_m \leftrightarrow AL_s$ Distribution of diastereomeric complex
between phases   (22-3)

$A_m \leftrightarrow A_s$ Distribution of analyte itself between phases   (22-4)

$A_s + L_s \leftrightarrow AL_s$ Formation of diastereomeric complex
in the stationary phase   (22-5)

Each of the equilibrium processes described by equations (22-1)–(22-5) can be characterized by the following equilibrium constants:

$$K_1 = \frac{[L_s]}{[L_m]} \qquad (22\text{-}6)$$

$$K_2 = \frac{[AL_m]}{\{[A_m] \cdot [L_m]\}} \qquad (22\text{-}7)$$

$$K_3 = \frac{[AL_s]}{[AL_m]} \qquad (22\text{-}8)$$

$$K_4 = \frac{[A_s]}{[A_m]} \qquad (22\text{-}9)$$

$$K_5 = \frac{[AL_s]}{\{[A_s] \cdot [L_s]\}} \qquad (22\text{-}10)$$

The overall distribution coefficient, D, of each of the species present in the chromatographic system is given by

$$D = \frac{\{[A_s] + [AL_s]\}}{\{[A_m] + [AL_m]\}}$$

which, by substitution from equations (22-6), (22-7), (22-9), and (22-10), leads to

$$D = \frac{\{K_4(1 + K_1 K_5[L_m])\}}{\{1 + K_2[L_m]\}} \tag{22-11}$$

The numerator from equation (22-11) represents complexation in the stationary phase, while the denominator represents the complexation in the mobile phase.

Using the rough approximation that the retention factor is proportional to the equilibrium constant of the dominating process in the column expression, we obtain

$$k = \Phi \frac{K_4\{(1 + K_1 K_5[L_m])\}}{1 + K_2[L_m]} \tag{22-12}$$

where Φ is the proportionality coefficient often regarded as the phase ratio.

Considering equation (22-12), several situations can be encountered:

1. If K_1 is very large, then the adsorption of the ligand onto the stationary phase is a dominating factor and the stationary phase can act as a chiral stationary phase. If the product $K_1 K_5[L_m]$ is large, then the interaction with the ligand will occur mainly in the stationary phase, leading to a stronger retention of the enantiomeric analytes and consequently a large value for k. Under these conditions, $K_2[L_m]$ will approach zero and the distribution coefficient becomes

$$D = K_4(1 + K_1 K_5[L_m]) \tag{22-13}$$

Under these circumstances, additional additives is no longer required in the mobile phase.

Such cases can be encountered in the case of thin-layer chromatographic separation of amino acids, using copper complexes of long chain amino acids as chiral additives via a ligand exchange approach. The copper complexes of alkyl amino acid chiral additives are so strongly adsorbed on the RP stationary phase that they act as a chiral stationary phase [154–156].

2. If $K_2[L_m]$ is large and K_1 approaches zero, then the interaction with the ligand will occur mainly in the mobile phase and the distribution coefficient becomes

$$D = \frac{K_4}{1 + K_2[L_m]} \tag{22-14}$$

causing smaller k values. Under these circumstances, the ligand has very little interaction with the stationary phase, and the complexation with the analytes occurs mainly in the mobile. This case is exemplified by enantiomeric separations using cyclodextrins in the mobile phase [157]. In principle, it is possible to choose an additive which has a certain K_1 to yield a certain distribution coefficient, D, between the stationary and mobile phases.

22.6.2   Selectivity with Chiral Mobile-Phase Additives

The selectivity factor, α, for a pair of enantiomers is

$$\alpha = \frac{k_R}{k_S} = \frac{D_R}{D_S} \qquad (22\text{-}15)$$

with $k_R > k_S$ and $D_R > D_S$, which leads to

$$\alpha = \frac{K_4(R)[1 + K_1 K_5(R)]}{K_4(S)[1 + K_1 K_5(S)]} \cdot \frac{1 + K_2(S)[L_m]}{1 + K_2(R)[L_m]} \qquad (22\text{-}16)$$

The first term of equation (22-16) represents the complexation of the enantiomers in the stationary phase, while the second term represents the complexation in the mobile phase. If the formation constants of the complexes are sufficiently high, the enantioselectivity of the chromatographic system is roughly given by the ratio of the complexation enantioselectivity in the stationary phase and mobile phase.

22.6.3   Chiral Additives with Chiral Stationary Phases

Assuming that analyte, A, is introduced into an HPLC system with a chiral stationary phase with an R configuration, the following equilibrium can be written:

$$A_m \leftrightarrow A_s$$

where the subscripts m and s stand for the species in the mobile phase and in the stationary phase, respectively. The above equilibrium is characterized by the equilibrium constant:

$$K_6 = \frac{[A_s]}{[A_m]} \qquad (22\text{-}17)$$

As the analyte enters in the stationary phase, it will interact with the immobilized chiral ligand according to

$$A_S + L_S \leftrightarrow A_s L_s \qquad (22\text{-}18)$$

with an equilibrium constant:

$$K_7 = \frac{[A_s L_s]}{[A_s][L_s]} \tag{22-19}$$

For a bonded chiral stationary phase, the retention factor can be written as

$$k = \Phi \cdot \frac{[A_s] + [A_s][L_s(R)]K_7}{A_m} = \Phi \cdot \frac{[A_s]}{[A_m]} \cdot (1 + K_7)[L_s(R)] \tag{22-20}$$

The selectivity factor for a pair of enantiomers introduced into a chiral column of R configuration, α_s, is

$$\alpha_s = \frac{k'(R)}{k'(S)} \quad \text{with } k'(R) > k'(S) \tag{22-21}$$

which can be written

$$\alpha_s = \frac{K_7(R)}{K_6(S)} \cdot \frac{1 + K_7(R)[L_s(R)]}{1 + K_7(S)[L_s(R)]} \tag{22-22}$$

The addition of an additive of the opposite configuration into the mobile phase will lead to a selectivity factor α:

$$\alpha = \frac{\alpha_s}{\alpha_m} = \frac{K_6(R)K_4(S)}{K_6(S)K_4(R)} \cdot \frac{1 + K_1 K_5(S)}{1 + K_1 K_5(R)} \cdot \frac{1 + K_2(R)[L_m(S)]}{1 + K_2(S)[L_m(S)]} \cdot \frac{1 + K_7(R)[L_s(R)]}{1 + K_7(S)[L_s(R)]} \tag{22-23}$$

The first term in equation (22-23) being a ratio of constants, can be considered as constant, C. If the product $K_1 K_5 \ll 1$, then equation (22-23) can be further simplified as

$$\alpha = C \cdot \frac{1 + K_2(R)[L_m(S)]}{1 + K_2(S)[L_m(S)]} \cdot \frac{1 + K_7(R)[L_s(R)]}{1 + K_7(S)[L_s(R)]} \tag{22-24}$$

To further simplify equation (22-24), we can denote the terms $K_2(R)[L_m(S)]$, $K_2(S)[L_m(S)]$, $K_7(R)[L_s(R)]$ and $K_7(S)[L_s(R)]$ as $K_2(R)_m$, $K_2(S)_m$, $K_7(R)_s$ and $K_7(S)_s$, respectively. The subscripts m and s correspond to species in the mobile phase and stationary phase, respectively.

Several situations can be encountered. If $K_2(R)_m > K_2(S)_m$ and $K_7(R)_s < K_7(S)_s$, then equation (22-24) becomes

$$\alpha = C \cdot \frac{1 + K_2(R)_m}{1 + K_2(S)_m} \cdot \frac{1 + K_7(R)_s}{1 + K_7(S)_s} \tag{22-25}$$

and, as a consequence, an inversion of elution will occur. If $K_7(R)_s \ll 1$ and $K_2(S)_m \ll 1$, then

$$\alpha = C \cdot \frac{1 + K_2(R)_m}{1 + K_7(S)_s} \qquad (22\text{-}26)$$

The numerator in equation (22-26) represents the processes occurring in the mobile phase, while the denominator represents the processes occurring in the stationary phase. Such a situation can be realized by combining a chiral stationary phase in a push–pull mode with a chiral mobile phase of opposite configuration, where two enantiomers of the chiral selector are involved, one for the chiral stationary phase and the other for the chiral mobile phase. The most selective chiral chromatographic system should be encountered when one enantiomer binds to the immobilized chiral selector in the stationary phase, whereas the other enantiomer predominantly associates with the chiral mobile-phase additive [158]. The above treatment is applicable to all applications regarding the use of chiral mobile phases.

22.6.4 Interactions with Chiral Mobile Phases

A variety of interaction chemistries have been exploited to affect chiral separations with chiral mobile-phase additives including ligand exchange, inclusion complexation, and ion pairing. Basically, the type of interactions between the enantiomeric analytes and the chiral additive is the same as in the case of chiral bonded phases when using the same selectand. The primary difference is where such interactions take place during the separation process. For instance, as noted above, in the case of inclusion complexation with cyclodextrins, the principal interactions occur in the mobile phase and account for why increasing the concentration of the chiral additive decreases the retention of the enantiomers [157]. In the case of ligand-exchange and ion-pair approaches, the main interactions occur in the stationary phase [159]. For instance, in a reversed-phase system where the chiral additive is a copper complex of L-proline [160], the ternary complex between copper, proline, and the L-enantiomer of the amino acid is additionally stabilized by interactions with the hydrocarbon stationary phase (Figure 22-31) [159].

In the case of ion-pair complexes between the chiral additive and the enantiomeric analytes, their interaction should be maximized by adjusting the mobile-phase polarity. Solvents of lower dielectric constant favor ion-pair formation.

Enantiomeric separation through mobile-phase additives is a very powerful method that unfortunately lost partially its use with the advancements in chiral stationary phase technology. With the development of capillary electrophoresis, however, this technique faces a revival showing its power to separate enantiomers.

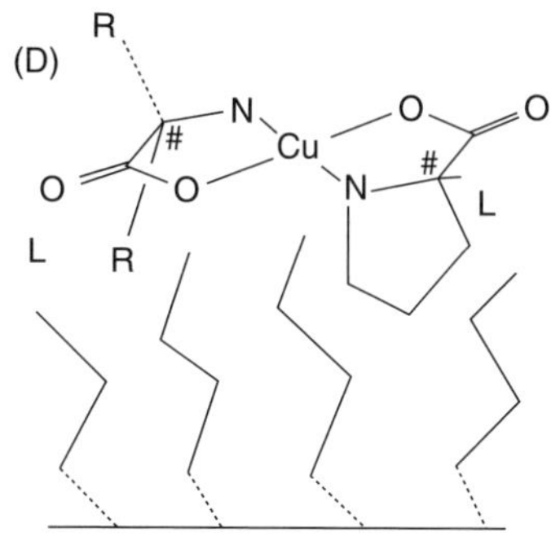

Figure 22-31. Ternary complex of copper, L-proline, and amino acids. (Reprinted from reference 159, with permission.)

22.7 METHOD DEVELOPMENT FOR CHIRAL SEPARATION

Method development for chiral separation is a multidisciplinary task. It requires knowledge of stereochemistry, organic chemistry, and separation techniques. Separation of enantiomers is not linked to a certain technique (i.e., GC, HPLC, etc.) but rather to an understanding of the specific interactions between the enantiomeric analytes and a certain chiral stationary phase. Knowing these types of relationships will enable one to easily understand the formation of transient diastereomeric complexes between enantiomers and a chiral stationary phase during a chromatographic separation as well as their stereochemical relationship within the complex. Once such dependencies are established, development of a method for the separation of enantiomers becomes an easy process. Based on such a relationship, chiral stationary phases can be divided in five categories [161]:

Type I: The solute chiral stationary-phase complexes are formed by attractive interactions such as hydrogen bonding, π–π interactions, dipole stacking, and so on, between the solute and the CSP.

Type II: The primary interaction for the formation of the transient diastereomers occurs through attractive interactions, while the secondary interaction occurs through inclusion complexes.

Type III: The primary interactions between the enantiomers and CSP occur through inclusion into chiral cavities.

Type IV: The diastereomeric complexes take place through metal complexes also known as chiral ligand exchange mechanism.

Type V: The interactions are based on a combination of polar and hydrophobic interactions.

Type I includes the CSP developed by Pirkle's group and is designed to operate through a combination of hydrogen-bonding, π–π interactions (see section on charge transfer, Section 22.3.5). To achieve such interactions, mobile phases such as a combination of hexane/alcohols are often used. The solute retention can be regulated by a careful combination of these solvents. Addition of small amounts of methylene chloride or acetonitrile can reduce the retention and improve efficiency of the chromatographic peaks [161].

Type II includes the polysaccharide, the acrylamide CSPs developed by Blaschke's group [44], and the quinine-based CSP [101, 102]. The interaction between the enantiomeric analytes and the stationary phases occurs through hydrogen-bonding, ion pairing as well as π–π, dipole–dipole, and inclusion interactions [101, 102, 113, 154, 162]. The mobile-phase conditions of the enantiomer's elution are similar to those used with Type I stationary phases. In the case of polysaccharide CSPs, the polar modifier should be carefully chosen because in some instances a change from linear to branched alcohols can yield an inversion of elution order of the two enantiomers [115]. This is due to a change in the configuration of the cavities in the stationary phase.

Type III includes chiral phases that interact with the enantiomers through inclusion interactions. This category includes cyclodextrins (α, β and γ cyclodextrins), as well as crown ethers. The leading interaction for the formation of the transient diastereomers is the insertion of the analytes in the cavity, and the stabilization of the complexes occurs through the secondary interaction, which can be either steric or hydrogen bonding. The driving force can be hydrophobic (cyclodextrins) or hydrogen bonding (crown ethers) [163]. The mobile phases used with such stationary phases can be mixtures of buffers in combinations with alcohols, acetonitrile, dimethylformamide, or dimethylsulfoxide [161]. Combinations of polar organic solvents such as acetonitrile methanol with small amounts of additives such as acetic acids can be used. In the case of crown ethers, the mobile phases consisted of a combination of water with small amounts of perchloric acid, or trichloroacetic acid.

Type IV includes chiral phases that usually interact with the enantiomeric analytes through the formation of metal complexes. There are usually used to separate amino acid enantiomers. These types of phases are also called ligand exchange phases. The transient diastereomeric complexes are ternary metal complexes between a transitional metal (usually Cu^{2+}), an amino acid enantiomeric analyte, and another compound immobilized on the CSP which is able to undergo complexation with the transitional metal (see also the ligand exchange section, Section 22.5). The two enantiomers are separated based on the difference in the stability constant of the two diastereomeric species. The mobile phases used to separate such enantiomeric analytes are usually aqueous solutions of copper (II) salts such as copper sulfate or copper acetate. To modulate the retention, several parameters—such as the pH of the mobile phase, the concentration of the copper ion, or the addition of an organic modifier such as acetonitrile or methanol in the mobile phase—can be varied.

Type V includes chiral stationary phases based on immobilized proteins as well as polysaccharide phases such as cellulose and amylose carbamate. They are used in conjunction with aqueous buffered mobile phases. The interaction between the stationary phase and the analytes is based on hydrophobic interaction as well as electrostatic interaction in the case of proteins. The retention of the analytes can be controlled by the addition of organic modifiers such methanol, ethanol, and 2-propanol.

The strategy for development of a method for the separation of enantiomers has to take into account the solubility of the analyte in different solvents. For instance, an enantiomeric mixture soluble in water should be amenable to separation on a chiral phase operating in reversed conditions. Conversely, enantiomeric analytes soluble in organic solvents should be good candidates for a separation on a chiral phase operating under normal-phase conditions, assuming that the analyte has the proper functional groups to interact with that particular chiral phase. For enantiomeric mixtures soluble in aqueous solutions, the first strategy is to analyze the structure, determining whether or not the enantiomers have the proper functional groups able to form metal complexes. Such compounds can be separated through a ligand exchange mechanism on a Type IV stationary phase. If such a possibility does not exist, the structure of the enantiomeric analytes should be examined for groups close to the stereogenic center which can be easily protonated (such as primary amines) or deprotonated (such as carboxyls) under the specific conditions of the mobile phase. In the case of primary amines, crown ether can be the CSP of choice, while for compounds bearing a carboxyl in the molecule, the Type V CSP can be a choice. For compounds that are neutral, cyclodextrins CSP can be a choice.

Compounds soluble in organic solvents can be amenable for separation on Type I and II CSPs. A scheme for the selection of the procedure for the development of chiral separation is given in Figure 22-32 [164].

22.8 CONCLUDING REMARKS

Separation of enantiomers is a technique driven mainly by the needs of pharmaceutical industry to produce drugs with controlled enantiomeric purity. Enantiomeric separation involves more than knowledge of chromatography; it requires an in-depth assessment of the stereochemistry of enantiomeric analytes and chiral stationary phase, as well as the interactions involved therein. In this situation, chromatography is just a tool that helps to separate enantiomers. That is why this chapter presents the main types of interactions occurring between the selectands and the selectors. Understanding these relationships, chiral separation becomes a logical process and trial and error is minimized.

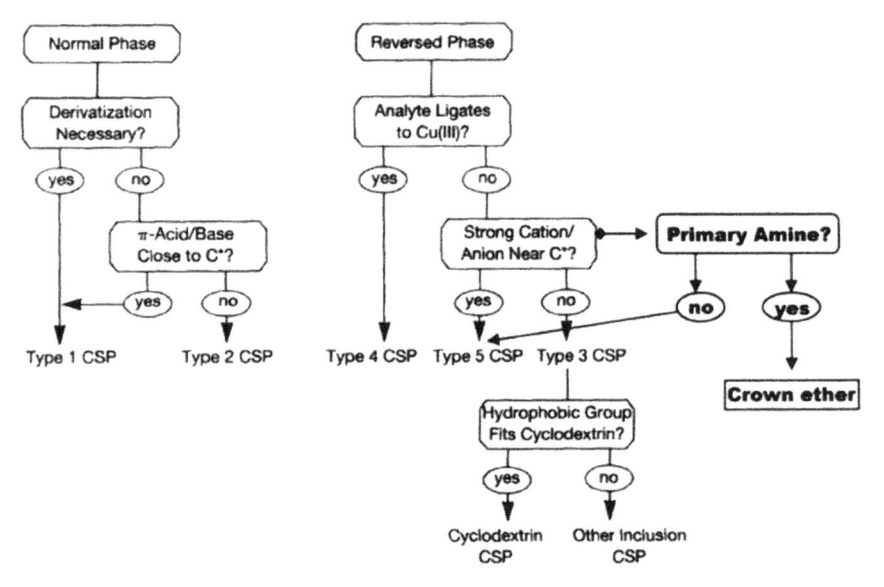

Figure 22-32. Strategy for the separation of enantiomers using type I, II, II, IV, and V stationary phases. (Adapted from reference 164.)

REFERENCES

1. B. Feibush and N. Grinberg, The history of enantiomeric resolution, in M. Zief and L. J. Crane (eds.), *Chromatographic Chiral Separations*, Marcel Dekker, New York, 1988, p. 1.

2. C. E. Dalgliesh, The optical resolution of aromatic amino acids on paper chromatograms. *J. Chem. Soc.* (1952), 3940.

3. S. G. Allenmark, *Chromatographic Enantioseparation. Methods and Applications*, John Wiley & Sons, New York, 1988, p. 13.

4. E. Gil-Av, B. Feibush, and R. Charles-Sigler, Separation of enantiomers by gas chromatography with an optically active stationary phase, *Tetrahedron Lett.* **7** (1966), 1009.

5. W. H. Pirkle and T. C. Pochapsky, Theory and design of chiral stationary phases for direct chromatographic separation of enantiomers, in K. K. Unger (ed.), *Packings and Stationary Phases in Chromatographic Techniques*, Marcel Dekker, New York, 1990, p. 783.

6. J. Gal, Indirect methods for the chromatographic resolution of drug enantiomers, in I. W. Wainer (ed.), *Drug Stereochemistry Analytical Methods and Pharmacology*, Marcel Dekker, New York, 1993, pp. 65–106.

7. W. Lindner, Indirect separation of enantiomers by liquid chromatography, in M. Zief and L. J. Crane (eds.), *Chromatographic Chiral Separation*, Marcel Dekker, New York, 1988, pp. 91–130.

8. B. Feibush, Chiral separation of enantiomers via selector/selectand hydrogen bondings, *Chirality* **10** (1998), 382–395.

9. C. H. DePuy and P. R. Story, Gas chromatographic evidence for intramolecular hydrogen bonding with double bonds, *Tetrahedron Lett.* **1** (1959), 20.

10. D. Nurok, G. L. Taylor, and A. M. Stephen, Separation of diastereomers by gas–liquid chromatography: Esters of butane-2,3-diol, *J. Chem. Soc. B* (1968), 291.

11. L. P. Kuhn, The Hydrogen bond. I. Intra- and intermolecular hydrogen bonds in alcohols. *J. Am. Chem. Soc.* **74** (1952), 2492.

12. Y. Gault and H. Felkin, Separation d'alcools acycliques diastereomeres par chromatographie gaseuse, *Bull. Soc. Chim. France* **3** (1965), 742.

13. E. Gil-Av and D. Nurok, Resolution of optical isomers by gas chromatography of diastereomers, in *Advances in Chromatography*, J. C. Giddings and R. A. Keller (eds.), Marcel Dekker, New York, 1974, pp. 99–172.

14. H. C. Rose, R. L. Stern, and B. L. Karger, Studies on the mechanism of separation of diastereomeric esters by gas–liquid chromatography; effect of bulk dissymmetry and distance between the optical centers, *Anal. Chem.* **38** (1966), 469.

15. B. L. Karger, R. L. Stern, and W. Keanr, Gas–liquid chromatographic separation of diastereomeric amides of racemic cyclic amines, *Anal. Chem.* **39** (1967), 228.

16. B. Feibush, Interpretation and correlation of bulkiness chirality and separation coefficients in the resolution of diastereomers by gas-chromatography, *Anal. Chem.* **43** (1971), 1098.

17. W. H. Pirkle and J. Finn, Separation of enantiomers by liquid chromatographic methods, in J. D. Morrison (ed.), *Asymmetric Synthesis*, Vol. 1, *Analytical Methods*, Academic Press: New York, 1983, pp. 87–124.

18. G. Helmchen, G. Nill, D. Flockerzi, W. Schuhle, and M. S. K. Yousef, Extreme liquid chromatographic separation effects in the case of diastereomeric amides containing polar substituents, *Angew. Chem. Int. Ed. Engl.* **18** (1979), 62.

19. B. Feibush, Separation of enantiomers by asymmetric phases in gas liquid chromatography, Ph.D thesis, 1967, The Weizmann Institute of Science, Rehovot, Israel.

20. V. A. Davankov, The nature of chiral recognition: Is it a three-point interaction? *Chirality* **9** (1997), 99.

21. W. H. Pirkle, On the minimum requirements of chiral recognition. *Chirality* **9** (1997), 103.

22. T. D. Booth, D. Wahnon, and I. W. Wainer, Is chiral recognition a three point process? *Chirality* **9** (1997), 96.

23. V. Sundaresan and R. Abrol, Biological chiral recognition: The substrate's perspective, *Chirality* **17** (2005), S30.

24. N. Brooijmans and I. D. Kuntz, Molecular recognition and docking algorithms, *Annu. Rev. Biophys. Biomol. Struct.* **32** (2003), 335.

25. B. Feibush, Chiral separation of enantiomers via selector/selectand hydrogen bondings, *Chirality* **10** (1998), 382.

26. S. Topiol, A general criterion for molecular recognition: Implications for chiral interactions, *Chirality* **1** (1989), 69.

27. G. J. Jeffrey, *An Introduction to Hydrogen Bonding*, Oxford University Press, New York, 1997, p. 11.

28. I. Olovsson and P.-G. Jonsson, X-ray and neutron diffraction studies of hydrogen bonded systems, in P. Schuster, G. Zundel, and C. Sandorfy (eds.), *The Hydrogen*

Bond. Recent Developments in Theory and Experiments. II. Structure and Spectroscopy, Vol. II, North Holland, Amsterdam, 1976, p. 393.

29. Y. Bereznitski, R. Lobrutto, N. Variancaval, R. Thompson, K. Thopmpson, P. Sajonz, L. S. Crocker, J. Kowal, D. Cai, M. Journet, T. Wang, J. Wyvratt, and N. Grinberg, Mechanistic aspects of chiral discrimination on an Amylose tris(3,5-dimethylphenyl)carbamate, *Enantiomer* **7** (2002), 305.

30. G. Zundel, Easily polarizable hydrogen bonds—their interactions with the environment—ir continuum and anomalous large proton conductivity, inP. Schuster, G. Zundel, and C. Sandorfy *The Hydrogen Bond. Recent Developments in Theory and Experiments. II. Structure and Spectroscopy*, Vol. II, North Holland, Amsterdam, 1976, p. 683.

31. T. E. Creighton, *Proteins Structure and Molecular Properties*, W. H. Freeman, New York, 1993, p. 5.

32. C. Toniolo, Intramolecular hidrogen-bonded peptide conformations, *CRC Crit. Rev. Biochem.* **9** (1980), 1.

33. E. Gil-Av, B. Feibush, and R. Cherles-Sigler, Separation of enantiomers by GLC with an optical active stationary phase, *Tetrahedron Lett.* **10** (1966), 1009.

34. B. Feibush and E. Gil-Av, Interaction between asymmetric solutes and solvents. Peptide derivatives as stationary phases in gas liquid partition chromatography, *Tetrahedron* **26** (1970), 1361.

35. U. Beitler and B. Feibush, Interaction between asymmetric solutes and solvents. Peptide derivatives as stationary phases in gas–liquid partition chromatography, *J. Chromatogr.* **123** (1976), 149.

36. B. Feibush, A. Balan, B. Altman, and E. Gil-Av, Chiral solute–solvent systems. Selective interaction between N-lauroyl-L-valine amides and N-trifluoroactyl esters of enantiomers of 2-amino-alkan-1-ols, α-, β-, and γ-amino acids, *J. Chem. Soc. Perkin Trans II* **9** (1979), 1230.

37. B. Koppenhoefer and E. Bayer, Chiral recognition in gas cgromatographic analysis of enantiomers on chiral polysiloxanes. *J. Chromatogr. Library* **32** (1985), 1.

38. B. Koppenhoefer and E. Bayer, Chiral recognition in the resolution of enantiomers by GLC, *Chromatographia* **19** (1984), 123.

39. Y. Dobashi and S. Hara, Extended scope of chiral recognition applying hydrogen bond association in nonaqueous media (R,R)-N,N'-disopropyltartaramide (DIPTA) as a widely applicable resolving agent, *J. Am. Chem. Soc.* **107** (1985), 3406.

40. Y. Dobashi and S. Hara, A chiral stationary phase derived from (R,R)-tartramide with broaden scope of application to the liquid chromatographic resolution of enantiomers, *J. Org. Chem.* **52** (1987), 2490.

41. Y. Dobashi, K. Nakamura, T. Saeki, M. Matsuo, S. Hara, and A. Dobashi, New chiral polysiloxane derived from (R,R)-tartramide for enantiomer resolution by capillary gas chromatography, *J. Org. Chem.* **56** (1991), 3299.

42. S. Anderson, S. Allenmark, P. Moller, B. Persson, and D. Sanchez, Chromatographic separation of enantiomers on N,N'-diallyl-L-tartardiamide-based network—Polymeric chiral stationary phases, *J. Chromatogr.* **741** (1996), 23.

43. G. Blaschke, Chromatographic resolution of chiral drugs on polyamides and cellulose triacetate, *J. Liq. Chromatogr.* **9** (1986), 341.

44. G. Blaschke, Substituted polyacrylamides as chiral phases for the resolution of drugs, in *Chromatographic Chiral Separation*, M. Zieff and L. J. Crane (eds.), Marcel Dekker, New York, 1988, p. 179.

45. B. Feibush, A. Figueroa, R. Charles, K. Onan, P. Feibush, and B. L. Karger, Chiral separation of heterocyclic drugs by HPLC: Solute-stationary phase base-pair interactions, *J. Am. Chem. Soc.* **108** (1986), 3310.

46. D. J. Cram and G. M. Lein, Host–guest complexation. 36. Spherand and litium and sodium complexation rates equilibria, *J. Am. Chem. Soc.* **107** (1985), 3657.

47. S. G. Allenmark, *Chromatographic Enantioseparation Methods and Applications*, Ellis Horwood, New York, 1988, p. 69.

48. W. Saenger, Cyclodextrin inclusion compounds in research and industry, *Angew. Chem. Int. Ed.* **19** (1980), 344.

49. W. Saenger, J. Jacob, K. Gessler, T. Steiner, D. Hoffman, H. Sanbe, K. Koizumi, S. M. Smith, and T. Takaha, Structure of the common cyclodextrins and their larger analogues—Beyond the doughnut, *Chem. Rev.* **98** (1998), 1787.

50. D. W. Armstrong, T. J. Ward, R. D. Armstrong, and T. E. Beesley, Separation of drug stereoisomers by formation of β-cyclodextrin inclusion complexes, *Science* **232** (1986), 1132.

51. S. M. Han, Y. I. Han, and D. W. Armstrong, Structural factors affecting the chiral recognition and separation on β-cyclodextrin bonded phase, *J. Chromatogr.* **441** (1988), 376.

52. D. W. Armstrong, W. DeMond, and B. P. Czech, Separation of metallocene enantiomers by liquid chromatography: Chiral recognition via cyclodextrin bonded phases, *Anal. Chem.* **57** (1985), 481.

53. K. Harata, Structural aspects of stereodiscrimination in the solid state, *Chem. Rev.* **98** (1998), 1803.

54. J. I. Seeman and H. V. Secor, Enantiomeric resolution and chiral recognition of racemic nicotine and nicotine analogues by β-cyclodextrins complexation. Structure–enantiomeric resolution relationship in host guest interaction, *Anal. Chem.* **60** (1988), 2120.

55. A. K. Chatjigakis, I. Clarot, P. J. P. Cardot, R. Nowakowski, and A. Coleman, Reversed phase chromatographic study of the inclusion selectivity of terpene derivatives with β-cyclodextrines in water/cosolvent mixtures, *J. Liq. Chrom Rel. Technol.* **22** (1999), 1267.

56. C. Moeder, T. O'Brien, R. Thompson, and G. Bicker, Determination of stochiometric coefficient and apparent formation constant for α- and β-CD complexes of terpenes using reversed phase liquid chromatography, *J. Chromatogr.* **736** (1996), 1.

57. I. Clarot, D. Cledat, S. Battu, and P. J. P. Cardot, Chromatographic study of terpene derivatives on porous graphitic carbon stationary phase with β-cyclodextrin as mobile phase modifier, *J. Chromatogr. A* **903** (2000), 67.

58. S. M. Z. Al-Kindy, F. E. O. Sulman, and A. A. Al-Hamadi, Fluorescence enhancement of coumarin-6-sulfonyl chloride amino acid derivatives in cyclodextrin media, *Anal. Sci.* **17** (2001), 539.

59. J. A. Hamilton and L. Chen, Crystal structures of inclusion complexes of b-cyclodextrin with (S)-(+)- and (R)-(−)-febpren, *J. Am. Chem. Soc.* **110** (1988), 4379.

60. A. Berthod, L. Berthod, and D. W. Armstrong, Selectivity of a native β-cyclodextrin column in the separation of catechins, *J. Liq. Chrom Rel. Technol.* **28** (2005), 1669.

61. A. M. Stalcup, S.-C. Chang, and D. W. Armstrong, Effect of the configuration of the substituents of derivatized β-cyclodextrin bonded phases on enantioselectivity in normal phase liquid chromatography, *J. Chromatogr. A.* **540** (1991), 113.

62. D. W. Armstrong, A. M. Stalcup, M. L. Hilton, J. D. Duncan, J. J. R. Faulkner, and S.-C. Chang, Derivatized cyclodextrins for normal phase liquid chromatographic separation of enantiomers, *Anal. Chem.* **62** (1990), 1610.

63. C. J. Pedersen, The discovery of crown ethers, Nobel Lecture, 1987.

64. C. J. Pedersen, Cyclic polyethers and their complexes with metal salts, *J. Am. Chem. Soc.* **89** (1967), 7017.

65. D. J. Cram and J. M. Cram, Host–guest chemistry, *Science* **183** (1974), 803.

66. D. S. Lingenfelter, R. G. Helgeson, and D. J. Cram, Host–guest complexation. 23. High chiral recognition of amino acid and ester guests by hosts containing one chiral element, *J. Org. Chem.* **46** (1981), 393.

67. D. J. Cram, R. C. Helgeson, L. R. Sousa, J. M. Timko, M. Newcomb, P. Moreau, F. deJong, G. W. Gokel, D. H. Hoffman, L. A. Domeieo, S. C. Peacock, K. Madan, and L. Kapplan, Chiral recognition in complexation of guests by designed host molecules, *Pure Appl. Chem.* **43** (1975), 327.

68. C. B. Knobler, F. C. A. Gaeta, and D. J. Cram, Source of chiral recognition in complexes with phenylglycine as guest, *J. Chem. Soc. Chem. Commun.* (1988), 330.

69. G. Dotsevi, Y. Sogah, and D. J. Cram, Chromatographic optical resolution through chiral complexation of amino ester salts by a host covalently bound to silica gel, *J. Am. Chem. Soc.* **97** (1975), 1259.

70. G. Dotsevi, Y. Sogah, and D. J. Cram, Host–guest complexation. 14. Host covalently bound to polystyrene resin for chromatographic resolution of enantiomers of amino acid and ester salts, *J. Am. Chem. Soc.* **101** (1979), 3035.

71. T. Shimbo, T. Yamaguchi, H. Yanagishita, D. Kitamoto, K. Sakaki, and M. Sugiura, Improved crown ether-based stationary phase, *J. Chromatogr.* **625** (1992), 101.

72. T. Shimbo, T. Yamaguchi, K. Nishimura, and M. Sugiura, Chromatographic separation of racemic amino acids by use of chiral crown ether-coated reversed-phase packings, *J. Chromatogr.* **405** (1987), 145.

73. Y. Hatefi and W. G. Hanstein, Solubilization of particulate proteins and nonelectrolytes by chaotropic agents, *Proc. Natl. Acad. Sci. USA* **62** (1969), 1129.

74. R. A. Thompson, Z. Gee, N. Grinberg, D. Ellison, and P. Tway, Mechanistic aspects of the stereospecific interaction for aminoindanol with crown ether column, *Anal. Chem.* **67** (1995), 1580.

75. S. Chen, H. Yuan, N. Grinberg, A. Dovletoglou, and G. Bicker, Enantiomeric separation of trans-2-aminocyclohexanol on a crown ether stationary phase using evaporative light scattering detection, *J. Liq. Chromatogr. Rel. Technol.* **25** (2003), 425.

76. M. H. Hyun, J. S. Jin, and W. Lee, Liquid chromatographic resolution of racemic amino acids and their derivatives on a new chiral stationary phase based on crown ether, *J. Chromatogr.* **822** (1998), 155.

77. E. Bang, J.-W. Jung, W. Lee, D. W. Lee, and W. Lee, Chiral recognition of (18-crown-6)tetracarboxylic acid as a chiral selector determined by NMR spectroscopy, *J. Chem. Soc. Perkin Trans II* (2001), 1685.

78. J. S. Jin, A. M. Stalcup, and M. H. Hyun, Impact of triethylamine as mobile phase additive on the resolution of racemic amino acids on an (+)-18-crown-6-tetracarboxyllic acid derived stationary phase, *J. Chromatogr. A* **933** (2001), 83.

79. C. J. Bender, Theoretical models of charge-transfer complexes, *Chem. Soc. Rev.* **15** (1986), 475.

80. M. L. Waters, Aromatic interactions in model systems, *Curr. Opin. Chem. Biol.* **6** (2002), 736.

81. W. B. Jennings, F. M. Farrell, and J. F. Malone, Attractive intermolecular edge-to-face aromatic interactions in flexible organic molecules, *Acc. Chem. Res.* **34** (2001), 885.

82. M. Levitt and M. F. Perutz, Aromatic rings act as hydrogen bond acceptor, *J. Mol. Biol.* **201** (1988), 751.

83. R. Foster, *Organic Charge-Transfer Complexes*, Academic Press, London, 1969, p. 182.

84. M. S. Newman, W. B. Lutz, and D. Lednicer, A new reagent for resolution by complex formation; the resolution of phenanthro-[3,4-*c*]phenanthrene, *J. Am. Chem. Soc.* **77** (1955), 3420.

85. M. S. Newman and D. Lednicer, The synthesis and resolution of hexahelicene, *J. Am. Chem. Soc.* **78** (1956), 4765.

86. L. H. Klemm, K. B. Desai, and J. J. R. Spooner, Optical resolution of 9-*sec*-butylphenantrene by molecular complexation chromatography, *J. Chromatogr.* **14** (1964), 297.

87. L. H. Klemm and D. Reed, Optical resolution by molecular complexation chromatography, *J. Chromatogr.* **3** (1959), 364.

88. F. Mikes, G. Boshart, and A. Gil-Av, Resolution of optical isomers by high performance liquid chromatography, using coated and bonded chiral charge transfer complexing agents as stationary phase, *J. Chromatogr.* **122** (1976), 205.

89. F. Mikes, The resolution of chiral compounds by modern liquid chromatography. Ph.D thesis, Organic Chemistry Department, 1975, The Weizmann Institute of Science, Rehovot, Israel.

90. M. Raban and K. Mislow, The determination of optical purity by nuclear magnetic resonance, *Tetrahedron Lett.* **48** (1965), 4249.

91. W. H. Pirkle, The nonequivalence of physical properties of enantiomers in Optically active solvents. Differences in nuclear magnetic resonance spectra. I, *J. Am. Chem. Soc.* **88** (1966), 1837.

92. W. H. Pirkle and D. L. Sikkenga, Resolution of optical isomers by liquid chromatography, *J. Chromatogr.* **123** (1976), 400.

93. W. H. Pirkle, D. W. House, and J. M. Finn, Broad spectrum resolution of optical isomers using chiral high-performance liquid chromatographic bonded phases, *J. Chromatogr.* **192** (1980), 143.

94. W. H. Pirkle and J. Finn, Separation of enantiomers by liquid chromatographic methods, in J. D. Norrison (ed.), *Asymmetric Synthesis*, Vol. 1. *Analytical Methods*, Academic Press, New York, 1983, p. 87.

95. W. H. Pirkle and P. G. Murray, Chiral stationary phase design. Use of intercalative effects to enhance enantioselectivity, *J. Chromatogr.* **641** (1993), 11.

96. W. H. Pirkle, P. G. Murray, and S. R. Wilson, X-ray crystallographic evidence support of a proposed chiral recognition mechanism, *J. Org. Chem.* **61** (1996), 4775.

97. W. H. Pirkle, C. J. Welch, and B. Lamm, Designed, synthesis and evaluation of an improved enantioselective naproxen selector, *J. Org. Chem.* **57** (1992), 3854.

98. W. H. Pirkle and C. J. Welch, Use of simultaneous face to face and face to edge π–π interactions to facilitate chiral recognition, *Tetrahedron Asymmetry* **5** (1994), 777.

99. W. H. Pirkle and S. R. Selness, Chiral recognition studies: Intra- and intermolecular 1H{1H}-nuclear Overhauser effects as effective tools in the study of bimolecular complexes, *J. Org. Chem.* **60** (1995), 3252.

100. M. Caude, A. Tambute, and L. Siret, Chiral stationary phase derived from tyrosine, *J. Chromatogr.* **550** (1991), 357.

101. M. Lammerhofer and W. Lindner, Quinince and quinidine derivatives as chiral selectors. Beush type chiral stationary phases for high performance liquid chromatography based on chincona carbamates and their applications as chiral anion exchanger, *J. Chromatogr.* **741** (1966), 33.

102. S. Schefzig, W. Lindner, K. B. Lipkowitz, and M. Jalaie, Enantiomeric discrimination by a quinine-based chiral stationary phase: A computational study, *Chirality* **12** (2000), 7.

103. F. A. Carey, *Organic Chemistry*, McGraw-Hill, New York, 1996, p. 1034.

104. A. L. Lehninger, *Principles of biochemistry*, Worth Publishers, New York, 1982, p. 277.

105. M. Lederer, Adsorption chromatography. VI. Further studies on the separation of D- and L-tryprophan on cellulose with aqueous solvents, *J. Chromatogr. A.* **510** (1990), 367.

106. M. Lederer, Adsorption chromatography. VII. Chiral separation on cellulose with aqueous solvents, *J. Chromatogr. A.* **604** (1992), 55.

107. H. T. K. Xuan and M. Lederer, Adsorption chromatography. IX. Chiral separation with aqueous solvents and liquid–liquid systems, *J. Chromatogr. A.* **635** (1993), 346.

108. T. Shibata, K. Mori, and Y. Okamoto, Polysaccharide phases, in A. M. Krstulovic (ed.), *Chiral Separation by HPLC. Applications to Pharmaceutical Compounds*, Ellis Horwood, New York, 1989, P. 336.

109. G. Hesse and R. Hagel, Eine volstandige Racemattrennung durch Elutions— Chromatographie an Cellulose-tri-acetat, *Chromatographia* **6** (1973), 277.

110. A. M. Rizzi, Band broadening in high-performance liquid chromatographic separations of enantiomers with swollen microcystalline cellulose triacetate packings. I. Influence of capacity factor, analyte structure, flow velocity and column loading, *J. Chromatogr. A.* **478** (1989), 71.

111. Y. Okamoto, R. Aburatini, and K. Hatada, Chromatographic chiral resolution. XIV. Cellulose tribenzoate derivatives as chiral stationary phases for high performance liquid chromatography, *J. Chromatogr. A* **389** (1987), 95.

112. H. Steinmeier, P. Zugenmaier, Homogeneous and heterogeneous cellulos triester and a cellulose triurethane: synthesis and structural investigation of the crystaline state. *Carbohydr. Res.* **164** (1987), 97.

113. T. O'Brien, L. Crocker, T. R. K. Thompson, P. H. Toma, D. A. Conlon, F. B. C. Moeder, G. Bicker, and N. Grinberg, Mechanistic aspects of chiral discrimination on modified cellulose, *Anal. Chem.* **69** (1997), 1999.

114. Y. Okamoto and E. Yashima, Polysaccharide derivatives for chromatographic separation of enantiomers, *Angew. Chem. Int. Ed.* **37** (1998), 1020.

115. T. Wang, Y. W. Chen, and A. Vailaya, Enantiomeric separation of some pharmaceutical intermediates and reversal of elution order by high performance liquid chromatography using cellulose and amylose tris-3,5-demethylphenylcarbamate) derivatives as stationary phase, *J. Chromatogr. A* **902** (2000), 345.

116. R. W. Stringham, The use of polysaccharide phases in the separation of enantiomers, in E. Grushka and N. Grinberg (eds.), *Advances in Chromatography*, Taylor & Francis, Boca Raton, FL, 2006, p. 257.

117. D. L. Boger, Vancomycin, teicoplanin and ramoplanin: Synthetic and mechanistic studies, *Med. Res. Rev.* **21** (2001), 356.

118. K. H. Ekborg-Ott, L. Youbang, and D. W. Armstrong, High enantioselective HPLC separations using the covalent bonded macrocyclic antibiotic, ristocetin A, chiral stationary phase. *Chirality*, **10** (1998), 434.

119. J. C. J. Barna and D. H. Williams, The structure and mode of action of glycopeptide antibiotics of the vancomycin group, *Annu. Rev. Microbiol.* **38** (1984), 339.

120. P. J. Loll and P. H. Axelsen, The structural biology of molecular recognition by vancomycin, *Annu. Rev. Biophys. Biomol. Struct.* **29** (2000), 265.

121. G. M. Sheldrick, P. G. Jones, O. Kennard, D. H. Williams, G. A. Smith, Structure of vancomycin and its complex with acetyl-D-alanyl-D-alanine, *Nature* **271** (1978), 223.

122. J. P. Brown, J. Feeney, and A. S. V. Burgen, A nuclear magnetic resonance study of the interaction between cancomycin and actyl-D-alanyl-D-alanine in aqueous solution, *Mol. Pharmacol.* **11** (1975), 119.

123. J. P. Brown, L. Terenius, J. Feeney, and A. S. V. Burgen, A structure–activity study by nuclear magnetic resonance of peptide interactions with vancomycin. *Mol. Pharmacol.* **11** (1975), 126.

124. M. Nieto and H. R. Perkins, Modifications of the acyl-D-alanyl-D-alanine terminus affecting complex formation with vancomycin, *Biochem. J.* **123** (1971), 789.

125. H. R. Perkins, Specificity of combination between mucopeptide precursors and vancomycin or ristocetin, *Biochem. J.* **111** (1969), 195.

126. D. W. Armstrong, Y. Tang, S. Chen, Y. Zhou, C. Bagwill, and J.-R. Chen, Macrocyclic antibiotics as a new class of chiral selectors for liquid chromatography, *Anal. Chem.* **66** (1994), 1473.

127. A. Cavazzini, G. Nadalini, F. Dondi, F. Gasparini, A. Ciogli, and C. Villani, Study of mechanism of chiral discrimination of amino acids and their derivatives on a teicoplanin-based chiral stationary phase, *J. Chromatogr. A* **1031** (2004), 143.

128. K. H. Ekborg-Ott, X. Wang, and D. W. Armstrong, Effect of selector coverage and mobile phase composition on enantiomeric separation with ristocetin A chiral stationary phase, *Microchem. J.* **62** (1999), 26.

129. E. Tesarova, K. Zaruba, and M. Flieger, Enantioseparation of semisynthetic ergot alkaloids on vancomycin and teicoplanin stationary phases, *J. Chromatogr. A* **844** (1999), 137.

130. A. L. Lehninger, *Principles of biochemistry*, Worth Publisher, New York, 1982, p. 121.

131. T. E. Creighton, *Proteins Structure and Molecular Properties*, W. H. Freeman, New York, 1993, p. 182.

132. G. Zubay, Biochemistry, Addison-Wesley, Reading, MA, 1983, p. 8.

133. H. J. Bohm and G. Klebe, What can we learn from molecular recognition in protein–ligand complexes for the design of new drugs? *Angew. Chem. Int. Ed.* **35** (1996), 2589.

134. A. W. Adamson and A. P. Gast, *Physical Chemistry of Surfaces*, John Wiley & Sons, New York, 1997, p. 246.

135. C. Y. Yang, R. Wang, and S. Wang, A systematic analysis of the effect of small-molecule binding on protein flexibility of the ligand binding sites, *J. Med. Chem.* **48** (2005), 5648.

136. W. A. Tao and R. K. Gilpin, Liquid chromatographic studies of the effect of phosphate on the binding properties of silica-immobilized bovine serum albumin, *J. Chromatogr. Sci.* **39** (2001), 205.

137. V. Sundaresan and R. Abrol, Towards a general model for protein–substrate stereoselectivity, *Protein Sci.* **11** (2002), 1330.

138. S. R. Narayanan, Immobilized proteins as chromatographic supports for chiral resolution, *J. Pharm. Biol. Anal.* **10** (1992), 251.

139. P. Erlandsson, L. Hansson, and R. Isaksson, Direct analytical and preparative resolution of enantiomers using albumin adsorbed to silica as a stationary phase, *J. Chromatogr.* **370** (1986), 475.

140. J. Haginaka, Protein-based chiral stationary phases for high performance liquid chromatography enantioseparations, *J. Chromatogr. A* **906** (2001), 253.

141. M. S. Waters, D. R. Sidler, A. J. Simon, C. R. Midaugh, R. Thompson, L. J. August, G. Bicker, H. J. Perpall, and N. Grinberg, Mechanistic aspects of chiral discrimination by surface immobilized α1-acid glycoprotein, *Chirality* **11** (1999), 224.

142. J. Haginaka, Y. Okazaki, and H. Matsunaga, Separation of enantiomers on a chiral stationary phase based on ovoglycoprotein. V. Influence of immobilization method on chiral separation, *J. Chromatogr.* **840** (1999), 171.

143. J. Haginaka and H. Takehira, Separation of enantiomers on a chiral stationary phase based on ovoglycoprotein. I. Influence of the pore size of base silica materials and bound protein amounts on chiral resolution, *J. Chromatogr.* **773** (1997), 85.

144. S. Allenmark, Separation of enantiomers by protein-based chiral phases, in G. Subramanian (ed.), *A Practical Approach to Chiral Separations by Liquid Chromatography*, VCH, Weinheim, 1994, p. 183.

145. M. Hedeland, S. Jonsson, R. Isaksson, and C. Petterson, Unexpected difference in enantioselectivie retention on cellulase (CBH I) silica stationary phase caused by exchange of potassium for sodium in the mobile phase, *Chirality* **10** (1998), 513.

146. R. K. Gilpin, S. E. Ethesham, and R. B. Gregory, Liquid chromatographic studies of the effect of temperature on the chiral recognition of tryptophan by silica-immobilized bovine albumine, *Anal. Chem.* **63** (1991), 2825.

147. V. A. Davankov, Ligand exchange chromatography of chiral compounds, in D. Cagniant (ed.), *Complexation Chromatography*, Marcel Dekker, New York, 1992, p. 197.

148. F. Helfferich, Ligand exchange: A novel separation technique, *Nature* **189** (1961), 1001.

149. S. V. Rogozhin and V. A. Davankov, Ligand chromatography on asymmetric complex-forming sorbent as a new method for resolution of racemates, *Chem. Commun.* (1971), 490.

150. V. A. Davankov, Chiral selectors with chelating properties in liquid chromatography: Fundamental reflections and selective review of recent development, *J. Chromatogr. A* **666** (1994), 55.

151. P. Roumeliotis, A. A. Kurganov, and V. A. Davankov, Effect of the hydrophobic spacer in bonded [Cu(L-hydroxyprolyl)alkyl]⁺ silicas on retention and enantioselectivity and enantioselectivity of α-amino acids in high performance liquid chromatography. *J. Chromatogr. A* **266** (1983), 439.

152. A. M. Rizzi, Efficiency in chiral high performance ligand exchange chromatography. Influence of complexation process, flow rate and capacity factor, *J. Chromatogr. A* **542** (1991), 221.

153. N. Nimura, Ternary complexation, in A. M. Krstulovic (ed.), *Chiral Separation by HPLC*, Ellis Horwood, New York, 1993, p. 107.

154. Y. Bereznitski, R. Thompson, E. O'Neill, and N. Grinberg, Thin-layer chromatography—A useful technique for the separation of enantiomers, *J. AOAC Int.* **84** (2001), 1242.

155. N. Grinberg, Chiral separation in thin-layer chromatography, in N. Grinberg (ed.), *Modern Thin-Layer Chromatography*, Marcel Dekker, New York, 1990, pp. 398–402.

156. K. Gunter, Thin layer chromatographic resolution via ligand exchange, *J. Chromatogr.* **448** (1988), 11.

157. C. Moeder, T. O'Brien, R. Thompson, and G. Bicker, Determination of stochiometric coefficient and apparent formation constant for α- and β-CD complexes of terpenes using reversed-phase liquid chromatography, *J. Chromatogr.* **1–2** (1996), 1.

158. K. J. Duff, H. L. Gray, R. J. Gray, and C. C. Bahler, Chiral stationary phases in concert with homologous chiral mobile phases additives: Push/pull model, *Chirality* **5** (1993), 201

159. V. A. Davankov, Separation of enantiomers, in V. A. Davankov, J. D. Navratil, and H. F. Walton, (eds.), *Ligand Exchange Chromatography*, CRC Press, Boca Raton, FL, 1988, p. 67.

160. E. Gil-Av, A. Tishbee, and P. E. Hare, Resolution of underivatized amino acids by reversed phase chromatography, *J. Am. Chem. Soc.* **102** (1980), 5115.

161. I. W. Wainer, Proposal for the classification of high performance liquid chromatographic stationary phases: How to choose the right column, *TRAC* **6** (1987), 125.

162. T. Loughlin, R. Thompson, G. Bicker, P. Tway, and N. Grinberg, Use of subcritical fluid chromatography for the separation of enantiomers using packed column cellulose based stationary phase, *Chirality* **8** (1996), 157.

163. I. W. Wainer, Drug stereochemistry, in I. W. Wainer (ed.), *Analytical Methods and Pharmacology*, 2nd edition, Marcel Dekker, New York, 1993, p. 139.

164. S. Ahuja, Chiral separation methods, in S. Ahuja (ed.), *Chiral Separation, Applications and Technologies*, American Chemical Society, Washington D.C., 1997, p. 138.

CHEMICAL AND DRUG COMPOUND INDEX

A

Alkylsulfates, adsorption isotherms of, 203

Adenosine triphosphate (ATP), 627

Acebutolol, 211

Acetic acid, 1006

Acetonitrile, 1006

N-Acetyl-L-leucylmethylamide, 998, 999

N-Acetyl-L-valyl-*tert*-butylamide, 1000

Actaplanin, 1022t

Actinoidin, 1022t

Alendronate, 649, 650

Acetyl-D-alanyl-D-alanine, vancomycin association with, 1023

Adrenaline retention factor, dependence on amphiphilic ion concentration, 204

Aldehydes, 726–727

Alcohols. *See also* Bicyclic alcohols branched, 1022
in reversed-phase liquid chromatography, 841

Aldol condensation, 726–727

Alkyl amides, 435

Alkyl amines, 435

Alkylbenzenes
retention of, 50, 51, 52, 54
selectivity between, 52t

Alkyl halides, 730–731

Alkylpyridines, retention of, 54

Alkyl-substituted polyaromatic hydrocarbons (PAHs), separation of positional isomers of, 253–254

N-Alkyl-β-hydroxycarboxamides, in diamide derivative enantioseparation, 1001

Amphipathic α helices, 1026

Amadori rearrangement, 729

1-Aminoindan, mass spectra of, 428

5-Aminoindazole
mass spectra of, 428
separation from 1-aminoindan, 426–427

Aminoglutethimide, enantiomeric SMB separation of, 968t

Amide groups, hydrogen bonding and, 997, 998–1000

Amides, 727. *See also* Alkyl amides

HPLC for Pharmaceutical Scientists, Edited by Yuri Kazakevich and Rosario LoBrutto
Copyright © 2007 by John Wiley & Sons, Inc.

SUBJECT INDEX

HPLC for Pharmaceutical Scientists, Edited by Yuri Kazakevich and Rosario LoBrutto
Copyright © 2007 by John Wiley & Sons, Inc.